FREE RADICALS

VOLUME I

REACTIVE INTERMEDIATES IN ORGANIC CHEMISTRY

Edited by GEORGE A. OLAH

Case Western Reserve University

A series of collective volumes and monographs on the chemistry of all the important species of organic reaction intermediates:

CARBONIUM IONS

Edited by George A. Olah of Case Western Reserve University and Paul v. R. Schleyer of Princeton University: Vol. I (1968), Vol. II (1970), Vol. III (1972), Vol. IV (1973), Vol. V (in preparation)

RADICAL IONS

Edited by E. T. Kaiser of the University of Chicago and L. Kevan of the University of Kansas (1968)

NITRENES

Edited by W. Lwowski of New Mexico State University (1970)

CARBENES

Edited by Maitland Jones, Jr., of Princeton University and Robert A. Moss of Rutgers University, the State University of New Jersey: Vol. I (1973), Vol. II (in preparation)

FREE RADICALS

Edited by J. K. Kochi of Indiana University: Vol. I (1973), Vol. II (1973)

Planned for the Series

CARBANIONS

By M. Szwarc of the State University of New York, College of Forestry, Syracuse

ARYNES

Edited by M. Stiles of the University of Michigan

20772
2/3/77

£8·75

FREE RADICALS

Edited by

JAY K. KOCHI

Department of Chemistry
Indiana University
Bloomington, Indiana

VOLUME I

A Wiley-Interscience Publication

JOHN WILEY & SONS, New York · London · Sydney · Toronto

Library of Congress Cataloging in Publication Data:

Kochi, Jay K.
Free Radicals.

(Reactive intermediates in organic chemistry)
Includes bibliographical references.
CONTENTS: v. 1., pt. 1. Dynamics of elementary processes.
1. Radicals (Chemistry) I. Title. II. Series.

QD471.K546 541′.224 72-6105
ISBN 0-471-49701-0

Printed in the United States of America

10 9 8 7 6 5 4 3 2 1

Authors of Volume I

Robert G. Bergman, *Division of Chemistry and Chemical Engineering, California Institute of Technology, Pasadena, California*

Alwyn G. Davies, *Department of Chemistry, University College London, London, England*

Hanspeter Fischer, *Ciba-Geigy Chemical Company, Basel, Switzerland*

John F. Garst, *Department of Chemistry, The University of Georgia, Athens, Georgia*

Keith U. Ingold, *Division of Chemistry, National Research Council of Canada, Ottawa, Canada*

J. Alistair Kerr, *Department of Chemistry, The University of Birmingham, Birmingham, England*

Jay K. Kochi, *Department of Chemistry, Indiana University, Bloomington, Indiana*

Thomas Koenig, *Department of Chemistry, University of Oregon, Eugene, Oregon*

Brian P. Roberts, *Department of Chemistry, University College London, London, England*

Glen A. Russell, *Department of Chemistry, Iowa State University, Ames, Iowa*

Harold R. Ward, *Metcalf Research Laboratories, Brown University, Providence, Rhode Island*

James W. Wilt, *Department of Chemistry, Loyola University, Chicago, Illinois*

Introduction to the Series

Reactive intermediates have always occupied a place of importance in the spectrum of organic chemistry. They were, however, long considered only as transient species of short life-time. With the increase in chemical sophistication many reactive intermediates have been directly observed, characterized, and even isolated. While the importance of reactive intermediates has never been disputed, they are usually considered from other points of view, primarily relative to possible reaction mechanism pathways based on kinetic, stereochemical, and synthetic chemical evidence. It was felt that it would be of value to initiate a series that would be primarily concerned with the reactive intermediates themselves and their impact and importance in organic chemistry. In each volume, critical, but not necessarily exhaustive coverage is anticipated. The reactive intermediates will be discussed from the points of view of: formation, isolation, physical characterization, and reactions.

The aim, therefore, is to create a forum wherein all the resources at the disposal of experts in the field could be brought together to enable the reader to become acquainted with the reactive intermediates in organic chemistry and their importance.

As the need arises, it is anticipated that supplementary volumes will be published to present new data in this rapidly developing field.

Preface

When asked to organize a monograph on free radicals, I agreed to do it because I believed that this type of book would fill a definite need. The object was to present an up-to-date and authoritative account of free radicals written by a group of experts active in the field. It soon became apparent, however, that the organization of the subject was not as trivial as I had imagined, and none of the usual classifications was adequate. I finally decided that a multifaceted approach was feasible in which five major areas would be covered:

Structure and Energetics
Free Radicals with Heteroatoms
Dynamics of Elementary Processes
Free Radical Chain Reactions
Synthesis

The first four categories are included in the present book. The last, which has, alas, fallen by the wayside, is perhaps the hardest to write about and was intended as a distillation of our knowledge about free radicals, to exploit them for chemical synthesis. Unfortunately, synthetic chemists have by and large eschewed this intermediate as being too reactive for their designs. Quite the contrary is true, and we wish to show in the following pages that free radicals as reactive intermediates do offer a number of unique advantages. One of our hopes for this book, therefore, is to stimulate chemists into expanding their repertoire of synthetic methods and their approaches to synthetic problems to include this versatile intermediate.

Overlapping coverage and duplication are inevitable in a book such as this, and no doubt detract from a unified presentation and lead to inefficiency. I believe, however, that each author has a message to convey and must feel free to develop it as he perceives it. No two authors approaching a problem from different points of view will interpret the data in the same way. Hopefully, such occurrences will be sufficiently minimal to augment rather than weaken the presentation. Most of the chapters were written as self-contained entities and are intended to be read as such. The treatment of free radicals in this book is not necessarily comprehensive, since the emphasis is primarily directed toward a critical analysis and unification of what is known and pointing out the direction for future developments.

I wish to express my thanks to all of the participants whose contributions made this endeavor possible.

JAY K. KOCHI

Bloomington, Indiana
April 1972

Contents

VOLUME I

Part I. Dynamics of Elementary Processes

VOLUME II

Part II. Free Radical Chain Reactions

Part III. Structure and Energetics

Part IV. Free Radicals with Heteroatoms

FREE RADICALS

VOLUME I

Part 1

Dynamics of Elementary Processes

CHAPTER 1

Rate Processes in the Gas Phase

J. ALISTAIR KERR
Department of Chemistry, The University of Birmingham, Birmingham, England

I. INTRODUCTION

This chapter summarizes the present state of knowledge of quantitative kinetics of free radical reactions in the gas phase. In essence it presents an up-to-date version of an earlier summary[1] entitled "Reactions of Alkyl Radicals" with the modification of including reactions of other radicals as well as alkyls.

Two significant changes have occurred in the decade following the previous article. The first is the enormous increase in the amount of quantitative kinetic data that has become available. Whereas ten years ago it was feasible to list all the rate constants of the reactions of alkyl radicals larger than methyl in a review of this type, to attempt the analogous task here would be formidable. The days have passed when textbooks and broadly based reviews can summarize all the measured rate constants. The classic monograph[2] of E. W. R. Steacie, *Atomic and Free Radical Reactions*, which dealt with the whole of quantitative gas kinetics at that time, would now run to several volumes. This large increase in the number of measurements has led to the development of compilations and evaluations of kinetic data. The former consist of logically and consistently arranged sets of tables of experimental kinetic data, sometimes with brief comments, while the latter attempt a critical assessment of data concerned with a limited number of reactions. So far three compilations dealing with radical reactions have appeared[3–5] and one evaluation of unimolecular reactions, a section of which covers decompositions of radicals.[6] It is clear that this type of effort will become increasingly important.* In this article selections of rate constants and Arrhenius parameters for the various types of radical reactions are given with a view to illustrating any general trends that exist. The reactions are limited to polyatomic monoradicals and concentrate on alkyl and substituted alkyl radicals but with some mention of heterocentered radicals. Lack of space precludes any discussion of biradicals or carbenes.

The second important change to occur in free radical kinetics during the past ten years is the development of quantitative approaches to the relations between kinetics and thermodynamics. While the existence of such connections has been recognized for some time, it is only by the recent work of Benson and co-workers[6–9] that they have been applied in a systematic way. This has come about by the development of straightforward and reliable methods of estimating the enthalpies, entropies, and heat capacities of molecules and free radicals. At the same time this group has fostered the application of transition state theory, again in a systematic way,[10] and although the main effort here concerns unimolecular reactions of molecules, some significant applications arise in free radical chemistry. These methods and general underlying philosophy are dealt with in Chapter 17, but their importance is worth emphasizing here and they do crop up at various points in this chapter.

II. MEASUREMENTS OF RATE CONSTANTS

The first point to note is that little data exist on absolute measurements of rate constants of free radical reactions in the gas phase. Most of the absolute data have been obtained for combination and disproportionation reactions by the rotating sector technique. This experimental approach is well documented and it suffices to say here that the method is not very accurate and not well suited to measuring temperature coefficients, particularly of low activation energy reactions. Errors in determining rate constants by rotating sector experiments in excess of $\pm 20\%$ are not uncommon.[11]

Absolute rate constants for the combination of methyl radicals have been measured by a variety of techniques[1,12] but seldom have they been exploited to study other reactions. The method most likely to be developed is that

* The International Council of Scientific Unions has set up a committee on data for science and technology (CODATA), and a task force in chemical kinetics under the chairmanship of S. W. Benson (Stanford Research Institute) has been created to coordinate data projects in chemical kinetics.

involving kinetic spectroscopy whereby the methyl combination reaction[13,14] and reaction with nitric oxide[14] have been studied by making direct measurements of methyl concentrations via their absorption at 216.4 nm in flash photolyses experiments. The absorption coefficients for methyl radicals at 216.4 nm have been measured and it is probable that absorptions of other radicals will be identified and measured in this way.

So far no gas-phase rate constants have been reported from ESR determinations of concentrations of polyatomic radicals, but considering the wealth of data derived by this method on atomic reactions, an eventual breakthrough seems likely.

The vast majority of rate constants discussed in this chapter have been measured relative to some other reaction of the radical, often the combination reaction. Where the rates of combination of the radicals have not been measured they are usually assigned a value based on the results of analogous reactions. The importance of the absolute measurements of the rates of radical combinations can hardly be overemphasized.

As regards the accuracy of rate constant measurements of free radical reactions in the gas phase, it has to be admitted that these are not among the most accurate measurements in physical chemistry. There are very few free radical reactions for which the activation energy is known to better than ± 0.5 kcal/mole with a corresponding error in the A factor [± 0.4 to ± 0.2 in log A (liters/mole-sec) for second-order reactions over the temperature range 300–500°K]. Most measurements of Arrhenius parameters of free radical reactions involve considerably larger errors than those quoted above. The precision of the data, as reflected in the standard deviations of least-mean-squares treatments, is usually considerably better than these errors; the differences arise from systematic errors which are particularly difficult to detect. Usually several independent determinations of the Arrhenius parameters for a radical reaction are necessary before reasonable estimated errors can be assigned with confidence.

Fortunately the position is much better for relative rate constants for a series of related reactions since many of the systematic errors are canceled. It is thus possible to make meaningful interpretations of the factors affecting the relative rates of related reactions.

III. SOURCES OF RADICALS

On the whole, ketones and aldehydes[1] have lost favor to azo compounds for the production of thermally equilibrated alkyl radicals.[15] A range of azo compounds is now commercially available and they are excellent clean sources of the corresponding alkyl radicals which can be generated in relatively high concentrations from high-intensity photolyses in the near ultraviolet region of the spectrum. The only minor disadvantage is that some of the radicals are lost to further reactions by addition to the double bond in the parent compound, but as the additions appear to have normal activation energies for double bond addition (5–8 kcal/mole), a suitable adjustment of the concentrations of reactants circumvents the difficulty.

Many valuable kinetic and thermochemical data on alkyl radicals have been gleaned by Rabinovitch and co-workers from detailed studies of vibrationally excited radicals produced by the additions of hydrogen atoms to olefins.[16] This method is particularly well suited to the formation of large alkyl radicals for which the ketone, aldehyde, or azo precursors are not readily available.

A large amount of information has been reported on the reactions of halogenoalkyl radicals, especially CF_3 and CCl_3 radicals.[17] In general, fluorinated or partially fluorinated alkyl radicals can be produced from the photolyses of the corresponding ketones, many of which are readily available. Other good sources of fluoroalkyl radicals are the photolyses of the iodides CF_3I and C_2F_5I, for example. The reactions of chloroalkyl radicals have been studied from the photolyses of the bromides or chlorides themselves—for instance, CCl_3 radicals from CCl_3Br or CCl_4 photolyses. Chloroketones do not photolyse to yield the corresponding chloroalkyl radicals but instead give chlorine atoms.[18] The addition of chlorine atoms to ethylene and chloroethylenes yields chloroethyl radicals, and some results on the reactions of the radicals have been derived from detailed analyses of these rather complex systems.[19]

The only oxygen-centered radicals which have reasonably convenient sources are alkoxy radicals which are formed either by the pyrolyses of peroxides and alkyl nitrites[20] or by the photolyses of esters.[21]

Many kinetic results on free radical reactions have been reported from systems in which the radicals of interest arise not from primary precursors but from reactions of sensitizer radicals with a substrate. Thus what little quantitative kinetic information there is on alkyl peroxy radicals is derived from photooxidation studies of alkyl radicals.[22] Also in this class of sources is the generation of $SiCl_3$ radicals in the chain sequence involving $SiHCl_3$ and an alkyl halide (RX):

$$R + SiHCl_3 = RH + SiCl_3$$

$$SiCl_3 + RX = SiCl_3X + R$$

This system, which reports the first gas-phase rate constants for silicon-centered radicals, depends on the fact that whereas in carbon compounds $D(C—H) > D(C—Cl)$ or $D(C—Br)$, the reverse is true for silicon compounds.[23]

IV. COMBINATION REACTIONS

Table 1 lists absolute rate constants for radical combination reactions and is more or less complete, with the omission of some earlier rate constants for methyl radical combination[1,12] as well as data on other combinations now known to be unreliable. Most of the results have been obtained over short ranges of temperature by the rotating sector technique. Within the limitations of the method the reactions appear to have zero or very low activation energies and high A factors, with associated rate constants close to the collision rates, based on simple collision theory assuming some "reasonable" value of the collision diameters for the radicals.

As mentioned in Section I, it is instructive to compare the kinetic data for the forward and reverse reactions with the thermodynamics.[6,7] Taking the example of the methyl combination reaction

$$CH_3 + CH_3 \underset{k_{\text{dissocn}}}{\overset{k_{\text{combn}}}{\rightleftharpoons}} C_2H_6$$

the comparison is made via the equilibrium constant

$$K = k_{\text{combn}}/k_{\text{dissocn}}$$

$$A_{\text{combn}}/A_{\text{dissocn}} = \exp(\Delta S^\circ_{\text{combn}}/R); \qquad \Delta H^\circ_{\text{combn}} = E_{\text{combn}} - E_{\text{dissocn}}$$

The entropies, heat capacities, and enthalpies of the species involved in these reactions are now readily available and it is a relatively simple matter to calculate $\Delta S^\circ_{\text{combn}}$ and $\Delta H^\circ_{\text{combn}}$ with reasonable accuracy.[7] Thus $\Delta H^\circ_{\text{combn}} = -88.2$ kcal/mole and $\Delta S^\circ_{\text{combn}} = -37.9$ cal/mole-deg at a temperature of 298°K and for a standard state* of 1 atm. A recent evaluation of the several reported rate constants for the dissociation of ethane[6] recommends the result $\log k_{\text{dissocn}}$ (sec^{-1}) $= 16.7 - 89.5/\theta$, where $\theta = 2.303RT$ in kilocalories per mole. Taking the mean temperature of the dissociation studies as 800°K and an average value of $\Delta C_{p(\text{combn})} = -2.8$ cal/mole-deg between 300 and 800°K, the entropy and enthalpy changes become $\Delta S^\circ_{\text{combn}} = -32.3$ cal/mole-deg and $\Delta H^\circ_{\text{combn}} = -88.0$ kcal/mole both at 800°K and for a standard state* of 1 mole/liter. These values in conjunction with the above k_{dissocn} lead to predicted Arrhenius parameters of $E_{\text{combn}} = 1.5$ kcal/mole and $\log A_{\text{combn}}$ (liters/mole-sec) $= 9.8$ again at 800°K, which are in reasonable agreement with the recombination rate data for methyl radicals measured by the sector and other techniques at lower temperatures.

When this type of comparison is made between the kinetic data for the combination of C_3 or C_4 radicals and the dissociation of the corresponding

* A full discussion of the corrections for different standard states in kinetics and thermodynamics is given in reference 7.

dimer molecules, serious discrepancies arise.[6,9,24] Let us consider the case of isopropyl radicals:

$$2i\text{-}C_3H_7 \underset{k_{\text{dissocn}}}{\overset{k_{\text{combn}}}{\rightleftharpoons}} (CH_3)_2CHCH(CH_3)_2$$

The kinetics of the dissociation reaction have been studied by Tsang[24c] using a competitive shock-tube technique from which he finds $\log k_{\text{dissocn}}$ ($\sec^{-1}$) $= 16.1 - 76.0/\theta$ at a mean temperature of 1000°K. The thermodynamics of the system can be estimated[7]; $\Delta H^{\circ}_{\text{combn}} = -77.7$ kcal/mole and $\Delta S^{\circ}_{\text{combn}} = -46.0$ cal/mole-deg at 300°K and $\Delta H^{\circ}_{\text{combn}} = -76.6$ kcal/mole and $\Delta S^{\circ}_{\text{combn}} = -35.8$ cal/mole-deg at 1000°K for a standard state of 1 atm, taking an average value of $\Delta C_{p(\text{combn})} = 1.5$ cal/mole-deg between 300 and 1000°K. Thus from the previous expression for the equilibrium constant, the above rate constant for the dissociation reaction, and the thermodynamic data at 1000°K, it can be calculated that at 1000°K $E_{\text{combn}} = 1.4$ kcal/mole and $\log A_{\text{combn}}$ (liters/mole-sec) $= 8.3$. While the estimated activation energy for the combination of isopropyl radicals is reasonably compatible with the rotating sector data measured at a lower temperature, the estimated A factor is much too low.

This can only mean that there is an error in the thermodynamic data of one or both sets of kinetic parameters. The least likely source of error would appear to be in the thermodynamics, particularly the enthalpy data. It has been pointed out that rotational barriers of CH_3 rotations in the radicals,

TABLE 1

Arrhenius Parameters for Radical Combination Reactions[a]

Ref.	Reaction	Log A (liter/mole-sec)	E (kcal/mole)	Temp (°K)
b	$2CH_3 \rightarrow C_2H_6$	10.34	0	398–448
c		10.42	0	298
d		10.39	0	293
e		[10.01]	0	800–1500
f	$C_2H_5 + H \rightarrow C_2H_6$	10.56	—	298
g	$2C_2H_5 \rightarrow C_4H_{10}$	[11.2]	[2]	323–423
h		10.4	0	400
i	$2CH_2{=}CHCH_2 \rightarrow (CH_2{=}CHCH_2)_2$	9.93	—	298
j	$2i\text{-}C_3H_7 \rightarrow C_6H_{14}$	[10.83]	0	354–442
h		9.9	0	400
k	$2t\text{-}C_4H_9 \rightarrow C_8H_{18}$	9.52	0	354–388
l	$2CF_3 \rightarrow C_2F_6$	10.36	0	400
m		[9.81]	0	298–333
n	$2CF_2Cl \rightarrow (CF_2Cl)_2$	10.09	0	486–541

TABLE 1—(*continued*)

Ref.	Reaction	Log *A* (liter/mole-sec)	*E* (kcal/mole)	Temp (°K)
o	$2CCl_3 \rightarrow C_2Cl_6$	9.66	0	303–426
p		9.59	0	385–456
w		[10.9]	0	350–446
q	$2CHCl_2CHCl \rightarrow (CHCl_2CHCl)_2$	10.20[v]	0	298–321
r	$2C_2Cl_5 \rightarrow C_4Cl_{10}$	8.61[v]	0	360–520
s	$2CF_3Cl_2 \rightarrow (CF_3CCl_2)_2$	9.82	0	405–423
t	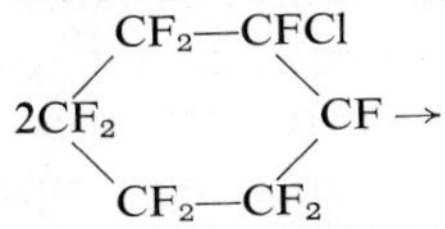 $\rightarrow (C_6F_{10}Cl)_2$	8.40[v]	0	314
u	$2(CH_3)_3Si \rightarrow ((CH_3)_3Si)_2$	[11.25]	0	317–399

[a] Measured by the rotating sector technique unless otherwise indicated; results in parentheses seem doubtful.

[b] A. Shepp, *J. Chem. Phys.*, **24**, 939 (1956).

[c] Reference 14, flash photolysis and kinetic spectroscopy involving CH_3 absorption at 216 nm.

[d] Reference 13, same method as reference 14.

[e] T. C. Clark, T. P. J. Izod, M. A. Di Valentin, and J. E. Dove, *J. Chem. Phys.*, **53**, 2982 (1970), shock-tube study with time-of-flight mass spectrometric monitoring of CH_3 concentration.

[f] M. J. Kurylo, N. C. Peterson, and W. Braun, *J. Chem. Phys.*, **53**, 2776 (1970), resonance fluorescence and absorption by H atom of α-Lyman radiation.

[g] A. Shepp and K. O. Kutschke, *J. Chem. Phys.*, **26**, 1020 (1957).

[h] Reference 7, estimated value.

[i] H. E. van den Bergh and A. B. Callear, *Trans. Faraday Soc.*, **66**, 2681 (1970), flash photolysis and kinetic spectroscopy involving absorption by allyl radical.

[j] E. L. Metcalfe and A. F. Trotman-Dickenson, *J. Chem. Soc.*, 4620 (1962).

[k] E. L. Metcalfe, *J. Chem. Soc.*, 3560 (1963).

[l] P. B. Ayscough, *J. Chem. Phys.*, **24**, 944 (1956).

[m] T. Ogawa, G. A. Carlson, and G. C. Pimentel, *J. Phys. Chem.*, **74**, 2090 (1970), flash photolysis followed by rapid scan infrared spectroscopy involving CF_3 absorption.

[n] J. R. Majer, C. Olavesen, and J. C. Robb, *Trans. Faraday Soc.*, **65**, 2988 (1969).

[o] G. R. De Maré and G. Huybrechts, *Trans. Faraday Soc.*, **64**, 1311 (1968).

[p] M. L. White and R. R. Kuntz, *Int. J. Chem. Kinet.*, **3**, 127 (1971).

[q] P. B. Ayscough, F. S. Dainton, and B. E. Fleischfresser, *Trans. Faraday Soc.*, **62**, 1838 (1966).

[r] S. Dusoleil, P. Goldfinger, A. M. Mahieu-Van der Auwera, G. Martens, and D. Van der Auwera, *Trans. Faraday Soc.*, **57**, 2197 (1961).

[s] Reference 11.

[t] Reference 29.

[u] P. Cadman, G. M. Tilsley, and A. F. Trotman-Dickenson, *J. Chem. Soc.*, D, 1721 (1970).

[v] Composite rate constant for combination and disproportionation.

[w] J. M. Tedder and J. C. Walton, *Trans. Faraday Soc.*, **63**, 2464 (1967).

isopropyl and *t*-butyl, of 8 kcal would significantly reduce the estimated entropies of these radicals and would reconcile the thermodynamic and kinetic data, but this suggestion was not considered likely.[9] There is a growing body of experimental evidence, from a variety of techniques including shock-tube measurements, very low pressure pyrolyses experiments involving a mass spectrometer, and classical kinetic measurements in static reaction vessels during the early stages of pyrolysis, to suggest that the dissociation kinetic data are reliable. The inference is that the experimental rate constants for the combination of the larger alkyl radicals could be wrong and these reactions may well involve significant activation energies. What is really needed is a reliable experimental method of measuring rate constants and temperature coefficients of radical combination reactions.

The high A factors for combinations and the reverse dimer dissociations are difficult to explain on the basis of transition state theory. They correspond to an extremely "loose" transition state in which the two alkyl fragments are rocking almost freely against each other.[25] Benson has postulated that the transition states in these reactions must involve considerable ionic character.[26] For methyl combination the following structures are envisaged:

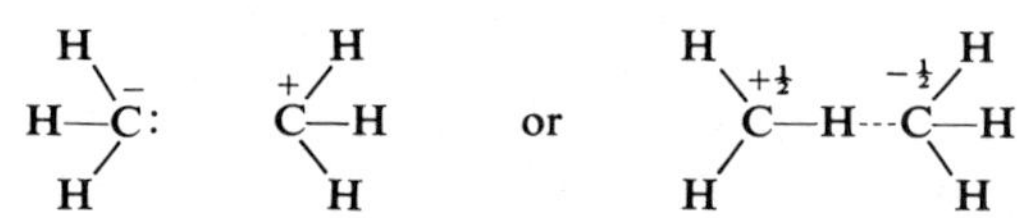

The participation of ionic configurations reduces the electron density between the associating atoms, thereby decreasing the frequencies of the corresponding rocking modes in the transition state to unusually low values. Hence the entropies of the transition state for combination are raised to the abnormally high values required by the observed high rate constants for the reactions.

There is now a vast body of experimental information on cross-combination ratios of rate constants for different radical pairs,[1,12,27] where three combination reactions are involved:

$$2R_a = (R_a)_2$$
$$2R_b = (R_b)_2$$
$$R_a + R_b = R_aR_b$$

The cross-combination ratio of rate constants is given by $\phi = k_{ab}/(k_{aa}k_{bb})^{1/2}$ and it invariably lies close to the value of 2, irrespective of the radicals or, apparently, the temperature. The relation also holds good when the ratio is expressed in terms of the sum of the rate constants for combination and disproportionation of the radical pairs.[28] A value of $\phi = 2 \pm 0.2$ would be expected for reactions of zero activation energy on the basis of the collision theory. In view of the above discussion of the combination rates of the larger alkyl radicals the consistency of the ϕ values, for, say, methyl and isopropyl radicals, with temperature is rather surprising. The data on the temperature variation of ϕ for such a pair of radicals are not extensive, however, and warrant further study.

It has recently been shown that the sum of the absolute rate constants for autocombination and autodisproportionation of the radicals listed in Table 1 decreases more or less linearly with the molecular weight of the radicals.[29] This interesting empirical relation in fact yields better estimates of the rate constants than any of the several attempts made to calculate them from transition state theory.[30]

The pressure dependence of the rate constants for methyl radical combination has been studied in a novel way by following the change of the cross-combination ratio ϕ(Me, *i*-Pr) with pressure.[31] It is safe to assume that combination of methyl with isopropyl and isopropyl with isopropyl will not be pressure dependent except at very low pressures, and consequently any variation in ϕ will reflect the pressure effect on the methyl–methyl combination reaction. All previous studies of this pressure effect have been made by relating the combination to hydrogen abstraction, a method which is much more susceptible to heterogeneous effects[32] than the cross-combination study. It appears that the falloff region for methyl radical combination is shifted toward much lower pressures than formerly believed. At 368°K the pressure ($P_{1/2}$) at which the rate constant falls to half of its limiting high-pressure value is about 0.05 torr. In addition $P_{1/2}$ shows a strong temperature dependence and this can only be explained by transition state theory if the model for the activated complex is modified such that the three-dimensional free rotation of the two methyl groups is replaced by a low-frequency bending.

V. DISPROPORTIONATION REACTIONS

A few absolute rate measurements have been made on disproportionation reactions of alkyl or halogenoalkyl radicals, but the vast majority of rate constants have been measured relative to the combination reaction.[3,5] Thus for ethyl radicals we have

$$2C_2H_5 = C_4H_{10} \tag{1}$$

$$2C_2H_5 = C_2H_6 + C_2H_4 \tag{2}$$

and the ratio of rate constants, k_2/k_1, can be expressed as Δ(Et, Et), the radical receiving the hydrogen atom being written first. The usefulness of this notation is seen by considering a mixed radical system, say ethyl and isopropyl, where, in addition to reactions 1 and 2, the following radical–radical reactions take place:

$$2i\text{-}C_3H_7 = C_6H_{14} \qquad (3)$$

$$2i\text{-}C_3H_7 = C_3H_8 + C_3H_6 \qquad (4)$$

$$C_2H_5 + i\text{-}C_3H_7 = C_5H_{12} \qquad (5)$$

$$C_2H_5 + i\text{-}C_3H_7 = C_2H_6 + C_3H_6 \qquad (6)$$

$$i\text{-}C_3H_7 + C_2H_5 = C_3H_8 + C_2H_4 \qquad (7)$$

Thus $\Delta(i\text{-}C_3H_7, i\text{-}C_3H_7) = k_4/k_3$, $\Delta(C_2H_5, i\text{-}C_3H_7) = k_6/k_5$, and $\Delta(i\text{-}C_3H_7, C_2H_5) = k_7/k_5$.

This system is used in Table 2 which lists a selection of Δ values for alkyl, halogenoalkyl, and a few heterocentered radicals.

The ratios are usually independent of temperature over the short ranges employed, although in a few cases Δ increases with temperature. Thus for $\Delta(C_2H_5, C_2H_5)$,[33] $\Delta(i\text{-}C_3H_7, i\text{-}C_3H_7)$,[34] and $\Delta(t\text{-}C_4H_9, t\text{-}C_4H_9)$,[35] the variation corresponds to a difference in activation energies $E_{\text{combn}} - E_{\text{dispn}} \simeq$ 0.3 kcal/mole. It has been argued that such temperature dependences probably arise from the T^n factors which appear in the separate rate constants, and by setting Δ proportional to $T^{-0.7}$ the activation energy differences disappear.

Disproportionation reactions of CH_3 radicals of the type

$$2CH_3 = CH_4 + CH_2$$

are not observed and this is usually ascribed to the unfavorable energetics in producing a carbene instead of an olefin. The reactions of CF_2H and CF_2Cl radicals are exceptions

$$2CF_2H = CF_2H_2 + CF_2$$

$$2CF_2Cl = CF_2Cl_2 + CF_2$$

and arise by virtue of the high thermodynamic stability of the CF_2 radical, making the disproportionations 30–40 kcal/mole exothermic.

Until recently there was considerable debate on the nature of the transition state in disproportionation reactions. One school of thought postulated that disproportionations and combinations proceed through the same transition state,[1,33,36] involving the excited dimer which could rearrange to give the disproportionation products or be quenched to give the combination product. Thus for two ethyl radicals we have

$$2C_2H_5 \longrightarrow C_4H_{10}^* + M \longrightarrow C_4H_{10}$$

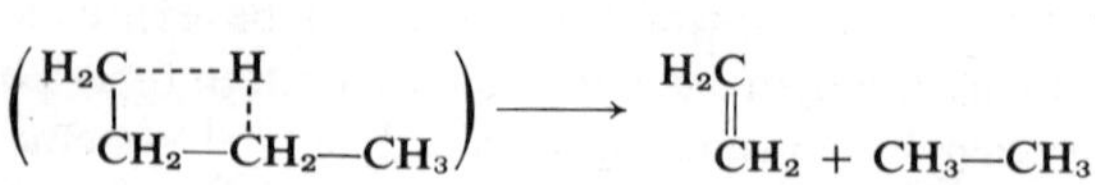

TABLE 2

Selected Values of Disproportionation–Combination Rate Constant Ratios[a]

Radicals			Radicals		
R′	R″	Δ(R′,R″)	R′	R″	Δ(R′, R″)
CH_3	C_2H_5	0.04	Vinyl	Vinyl	1.1
CH_3	n-C_3H_7	0.06	Allyl	Allyl	0.008
CH_3	i-C_3H_7	0.16	cyclo-C_6H_7[b]	cyclo-C_6H_7[b]	0.45
CH_3	cyclo-C_6H_7[b]	0.27	CH_3	CH_3O	1.9
C_2H_5	C_2H_5	0.14	CF_2H	CF_2H	0.15
C_2H_5	i-C_3H_7	0.20	CF_2Cl	CF_2Cl	0.15
C_2H_5	t-C_4H_9	0.54	CCl_3	C_2H_5	0.24
C_2H_5	cyclo-C_6H_7[b]	0.38	n-C_3F_7	C_2H_5	0.2
n-C_3H_7	n-C_3H_7	0.14	CH_3O	CH_3O	~67
i-C_3H_7	C_2H_5	0.12	C_2H_5O	C_2H_5	2.3
i-C_3H_7	i-C_3H_7	0.60	C_2H_5O	C_2H_5O	12
i-C_3H_7	t-C_4H_9	0.6	CH_3S	CH_3S	0.4
i-C_3H_7	cyclo-C_6H_7[b]	0.52	NH_2	C_3H_7	0.2
n-C_4H_9	n-C_4H_9	0.14	NO	C_2H_5	0.2
i-C_4H_9	i-C_4H_9	0.08	NO	i-C_3H_7	0.2
s-C_4H_9	s-C_4H_9	0.63			
t-C_4H_9	C_2H_5	0.3			
t-C_4H_9	t-C_4H_9	2.3			
t-C_4H_9	cyclo-C_6H_7[b]	1.3			

[a] For a full list of results see references 3 and 5. Where several values have been reported for the same pair of radicals the value selected above reflects the present author's preference. For reactions with more than one possible pathway the quoted ratios refer to the sum for the total pathways.

[b] Cyclohexadienyl.

The other postulate is that disproportionation occurs by the more straightforward head-to-tail association of the radicals:

$$2C_2H_5 \longrightarrow CH_3CH_2\text{---}H\text{---}CH_2\text{=\!=}CH_2 \longrightarrow CH_3CH_3 + CH_2{=}CH_2$$

and consequently the transition states for combination and disproportionation must be different.

The arguments put forward by Benson[26] in favor of different transition states are compelling and are now generally accepted. They are briefly as follows: (1) The dimer or head-to-head transition state for disproportionation is difficult to envisage for highly branched radicals, say *t*-butyl, and cannot be reconciled with Δ values in excess of unity. (2) If there is a common transition state involving the excited dimer molecule, Δ values should decrease with increasing total pressure and should be dependent on the nature of the quenching molecule. Results for $\Delta(C_2H_5, C_2H_5)$ indicate that the value is remarkably insensitive to large changes in pressure and even to changes of phase.[33] (3) There is direct evidence to refute the decomposition of n-C_4H_{10} to give C_2H_4 and C_2H_6 from a study of the isotopic compositions of the products of the pyrolysis of mixtures of n-C_4H_{10} and n-C_4D_{10} under inhibited and uninhibited conditions.

Indirect support for the head-to-tail mechanism comes from some calculations of the rate constants for alkyl radical disproportionations[37] based on the empirical bond-energy-bond order method developed by Johnston.[38] The method will be discussed more fully in relation to hydrogen transfer reactions of radicals with molecules; it suffices to say here that on the basis of the head-to-tail transition state

$$\mathrm{R + H{-}Ol \longrightarrow \underset{}{R}\text{-----}\underset{n_1}{H}\text{-----}\underset{n_2}{Ol} \longrightarrow R{-}H + Ol}$$

where n_1 and n_2 are the bond orders in the transition state and assuming conservation of total bond order in the reaction, order of magnitude agreement has been achieved between the calculated and experimental rate constants.

Bradley and Rabinovitch have suggested an alternative interpretation of the mechanisms of disproportionation and combination of alkyl radicals.[39] The transition states for combination and disproportionation consist of much more loosely bonded interacting radicals than in the transition state for metathesis. Similar activated complexes for combination and disproportionation of a given pair of radicals are proposed with the saddle points on the energy diagrams close together in energy and configuration. Although some reactions may proceed through both configurations, Bradley and Rabinovitch prefer to leave the configuration of the transition state unspecified.[39] They reject, however, the description of the complex for disproportionation as an excited dimer molecule and suggest that different structures may be involved, including couplings of the radicals through three-center hydrogen atom bonds already suggested for combination reactions.[40] The terms "head-to-head" and "head-to-tail" to describe the configuration of the activated complexes are discarded in the Bradley–Rabinovitch model.

The concept of a common activated complex in disproportionation and combination leads to the prediction of a linear relationship between log Δ and $(\Sigma S^\circ_{\text{dispn}} - S^\circ_{\text{combn}})$, the difference between the standard molar entropies of the products of disproportionation and combination. In highly exothermic reactions $(\Sigma S^\circ_{\text{dispn}} - S^\circ_{\text{combn}})$, the high temperature limit of $(\Sigma G^\circ_{\text{dispn}} - G^\circ_{\text{combn}})/T$ should characterize the activated complex. This semiempirical correlation was first suggested by Bradley[36] and other applications soon followed.[15b,41] Some authors have plotted log (Δ/H) where H is the number of hydrogen atoms available for disproportionation in the radical which donates the hydrogen atom, while others have simply plotted log Δ where Δ is the uncorrected experimental ratio. Approximate correlations are obtained in both cases and while little quantitative significance can be attached to this approach, it nevertheless is a useful guide for indicating grossly divergent experimental data.

The few Δ values reported for alkoxy radicals do not correlate with the entropy differences of the products in the same way as the alkyl radicals.[20] It has been suggested that the high disproportionation–combination ratios of alkoxy radicals are due to the polar nature of alkoxy radicals that favors disproportionation rather than combination.[42] Qualitative support for the importance of polarity in these reactions can be taken from the more normal ratio for CH_3S radicals [$\Delta(CH_3S, CH_3S) = 0.2$] which should be much less polar than CH_3O on the basis of the electronegativities of sulfur and oxygen atoms.

One of the difficulties in accepting the "head-to-tail" transition state for disproportionation is the high A factor (10^{10-11} liters/mole-sec) in relation to the normal range of A factors ($10^{8.5\pm1}$ liters/mole-sec) for hydrogen transfer reactions between molecules and radicals. The mechanism involving excited dimer formation was suggested to account for the very loose transition state which must be involved to explain the high A factor in disproportionations, but as we have seen, this model has to be rejected on other grounds. The idea that some type of ionic complexing occurs at close approach of the radicals, made to account for the loose complex in radical combinations, is again tenable and at present seems the best rationale for the high A factors in disproportionation reactions.[26] The participation of ionic configurations in both combinations and disproportionations is qualitatively consistent with the wide variation in the Δ values observed for the autodisproportionation of ethyl, isopropyl, and *t*-butyl radicals and for the cross-disproportionations of methyl, ethyl, isopropyl, and *t*-butyl with cyclohexadienyl radicals.[43] There is a parallel variation in the ionization potentials of the radicals, which are 10.0, 8.8, 7.9, and 7.4 eV, respectively, for methyl, ethyl, isopropyl, and *t*-butyl.

Disproportionation reactions of large alkyl radicals (C_6–C_8) have been studied in systems involving the generation of chemically activated radicals by hydrogen atom addition to olefins.[16] The radicals can be thermalized in the presence of high pressures of bath gases and the disproportionations studied from the yields of alkane and dimer. In general the measured Δ values for the large alkyl radicals are in accord with the correlations involving the entropy differences of the products.

An additional interesting feature of these studies concerns the ratios for the cross-disproportionations of *t*-butyl with 4,4-dimethylpentyl-2 (44DMP2·) and 3,4,4-trimethylpentyl-2 (344TMP2·) radicals, which have been found to be $\Delta(t\text{-}C_4H_9, 44DMP2\cdot) = 1.2$, $\Delta(44DMP2\cdot, t\text{-}C_4H_9) = 0.8$, $\Delta(t\text{-}C_4H_9, 344TMP2\cdot) = 1.4$, and $\Delta(344TMP2\cdot, t\text{-}C_4H_9) = 0.7$. In both cases *t*-butyl abstracts hydrogen atoms more readily from the larger radical than vice versa, despite there being more abstractable hydrogen atoms in *t*-butyl than in either of the larger radicals. The most likely explanation is that steric hindrance is playing an important part in such reactions.[43a]

VI. TRANSFER REACTIONS

A vast amount of kinetic data on hydrogen atom transfer reactions of radicals with molecules has been compiled,[3–5] and the general characteristics are well established and documented.[44–46] Here it is possible only to highlight some of the distinguishing features with a few selected examples of reactions and rate constants.

Trotman-Dickenson[45] has emphasized the importance of the tunnel effect[38] which leads to underestimation of the activation energies by some 10 to 20% in hydrogen transfer reactions. At the same time he has pointed out the unsatisfactory situation of recognizing the tunnel effect and yet not being able to make any quantitative allowance for it, in the interpretation of these reactions.

Table 3 summarizes the Arrhenius parameters and related data which are now available for a range of radicals reacting with C—H bonds in alkanes. Atomic reactions are included for comparison. The only absolute rate constants in Table 3 are those for OH radicals, from kinetic spectroscopy, where the concentrations of the radicals can be measured directly. For most of the other radicals the rates of hydrogen abstraction have been measured relative to the radical combination reaction. The most significant figures are the relative rate constants, as the errors in several of the Arrhenius parameters are still quite large. Table 3 is a useful guide to the relative reactivities of the attacking atoms and radicals; the most reactive radicals are the least selective, as shown by the rate constants. For all attacking species the reactivities of the alkanes run methane $<$ primary $<$ secondary $<$ tertiary in keeping with the sequence of C—H bond strengths.

Most of the systems in Table 3 fit the Polanyi relation,[47,48] although sometimes not very precisely. The activation energy is related to the enthalpy change by the equation

$$E = \alpha \Delta H^\circ + C$$

where α and C are constants which vary for each attacking species. For hydrogen-abstractions from alkanes, where the bond being formed is the

TABLE 3

Hydrogen Transfer Reactions of Radicals and Alkanes[a]: X + RH = XH + R

Ref.	Radical R		Me	Prim	Sec	Tert	$D(X—H)$[b]
	$D(R—H)$[b]		104	98	95	92	
	Radical X						
c	F	k (298°K)	0.7	1	1	1.3	135.8
		Log A	10.5	10.0	9.8	9.8	
		E	1.2	0.3	0	0	
c	Cl	k (523°K)	0.03	1	2	3	103.1
		Log A	9.8	10.2	10.4	10.3	
		E	3.8	0.9	0.6	0.1	
d	OH	k (500°K)	0.04	1	5	8	119
		Log A	8.9	8.8	9.2	9.1	
		E	5.0	1.6	0.9	0.2	
e	CF_3	k (455°K)	0.04	1	7	22	106
		Log A	8.4	8.1	8.0	8.1	
		E	11.0	7.5	5.5	4.7	
f	CH_3O	k (455°K)	0.04	1	2	5	104
		Log A	8.2	7.7	8.0	8.0	
		E	11.0	7.1	5.2	4.1	
e	CH_3	k (455°K)	0.03	1	10	64	104
		Log A	8.2	8.4	8.6	8.4	
		E	14.5	11.8	10.2	8.1	
g	CCl_3	k (455°K)	—	1	40	250	96
		Log A	—	7.9	7.9	7.6	
		E	—	13.4	10.3	7.9	
e	C_2H_5	k (455°K)	0.006	1	8	32	98
		Log A	8.1	7.8	7.7	7.6	
		E	17.7	12.6	10.3	8.9	
c	Br	k (300°K)	0.003	1	640	2.5×10^4	87.4
		Log A	10.5	10.1	10.5	10.3	
		E	18.4	13.4	10.0	7.5	
h	I	k (300°K)	4×10^{-4}	1	16	6200	71.4
		Log A	12.0	11.1	11.2	11.1	
		E	35.0	26.4	24.8	21.0	

[a] The values of log A (liters/mole-sec) and E (kcal/mole) are for a *single* H atom of the specified type and the values of k (liters/mole-sec) are for a *single* H atom of the specified type relative to k for a primary H atom.

[b] J. A. Kerr and A. F. Trotman-Dickenson, *Handbook of Chemistry and Physics*, 52nd ed., Chem. Rubber Publ. Co., Cleveland, Ohio, 1972.

[c] G. C. Fettis and J. H. Knox, *Prog. React. Kinet.*, **2**, 3 (1964).

[d] N. R. Greiner, *J. Chem. Phys.*, **53**, 1070 (1970); D. D. Drysdale and A. C. Lloyd, *Oxid. Combust. Rev.*, **4**, 157 (1970).

[e] References 3 and 5.

[f] Reference 20.

[g] Reference 11.

[h] D. M. Golden and S. W. Benson, *Chem. Rev.*, **69**, 125 (1969).

same for each set of reactions with a particular attacking atom or radical, the relation takes the form

$$E = \alpha[D(\text{C—H}) - C^1]$$

in which D(C—H) is the strength of the alkane bond being broken. As might be expected, the highly exothermic reactions of fluorine atoms with alkanes do not fit into the Polanyi relation pattern.

Unfortunately the application of the Polanyi and related empirical approaches to activation energies in hydrogen transfer reactions[38] does not extend beyond alkanes. Table 4 lists a selection of results for reactions involving hydrogen transfer from other types of molecules, from which it is apparent that the activation energies are not very sensitive to the enthalpy changes in exothermic reactions.

It has long been recognized that activation energies in hydrogen transfer reactions are influenced by polar effects as well as by the enthalpy change for the reaction. In the reactions of CF_3 and CH_3 radicals with the polar molecules HCl and $SiHCl_3$, the enthalpy difference, $\Delta H^\circ_{CF_3} - \Delta H^\circ_{CH_3} = -2$ kcal/mole, would imply that $E_{CF_3} < E_{CH_3}$. The observed opposite trends, $E_{CF_3} > E_{CH_3}$, have been interpreted as showing a polar repulsion between the electrophilic CF_3 radical and the substrate molecule which increases the activation energy. A second type of polar effect on activation energies is revealed in the reactions of the same pair of radicals with nonpolar molecules. In general $E_{CH_3} - E_{CF_3} = 2$–3 kcal/mole for nonpolar alkanes or silanes. In terms of the Polanyi relations for CH_3 and CF_3 radicals the difference in bond strengths $D(CF_3\text{—H}) - D(CH_3\text{—H}) = 2$ kcal/mole would account for only 1 kcal/mole difference in the activation energies. The additional 1–2 kcal/mole lowering in E_{CF_3} is undoubtedly real, and has been ascribed to a polar attraction between the electron density associated with the C—H bond in the substrate and the electrophilic CF_3 radical. Attempts have recently been made to partition activation energies into enthalpy and attractive or repulsive polar contributions in hydrogen transfer reactions, but the lack of accurate enthalpy data has hindered a proper quantitative test.[49]

The a priori calculation of activation energies for radical–molecule reactions is still beyond the present scope of quantum mechanics, despite the advent of large computers.[38] One of the more successful of several empirical or semiempirical attempts at calculating activation energies is the bond energy-bond order method of Johnston[38] and co-workers. The method involves no adjustable parameters but rather empirical relations of a non-kinetic type. For the class of reactions:

$$R^1 + HR \longrightarrow \underset{n_1\quad n_2}{R^1\text{---H---R}} \longrightarrow R^1H + R$$

TABLE 4

Arrhenius Parameters for Some Hydrogen Transfer Reactions of Radicals and Molecules[a]

Reaction	T_m (°K)	Log A (liters/mole-sec)	E (kcal/mole)	$\Delta H°$ (kcal/mole)
$CH_3 + HCl \rightarrow CH_4 + Cl$	334	8.6	2.3	−0.9
$CH_3 + HBr \rightarrow CH_4 + Br$	413	8.95	2.9	−16.6
$CH_3 + HI \rightarrow CH_4 + I$	561	9.50	2.3	−32.6
$CF_3 + HCl \rightarrow CF_3H + Cl$	394	8.44	5.5	−2.9
$CF_3 + HBr \rightarrow CF_3H + Br$	386	8.78	2.9	−18.6
$CF_3 + HI \rightarrow CF_3H + I$	454	8.73	0.5	−34.6
$CH_3 + C_2H_5OH \rightarrow CH_4 + C_2H_5O$	463	7.90	9.4	0
$CH_3 + C_2H_5OH \rightarrow CH_4 + C_2H_4OH$	463	8.60	9.7	−8
$CH_3 + C_2H_5CHO \rightarrow CH_4 + C_2H_5CO$	412	8.9	7.5	−18
$CF_3 + (CH_3)_2ND \rightarrow CF_3D + (CH_3)_2N$	367	7.88	4.7	−10
$CCl_3 + HCl \rightarrow CCl_3H + Cl$	365	8.65	11.3	7
$CH_3O + HCOOCH_3 \rightarrow CH_3OH + COOCH_3$	428	9.2	8.2	?
$i\text{-}C_3H_7 + 1,3\text{-cyclo-}C_6H_8 \rightarrow C_3H_8 + \text{cyclo-}C_6H_7$	361	8.9	6.4	−25
$i\text{-}C_3H_7 + (i\text{-}C_3H_7)_2N_2 \rightarrow C_3H_8 + C_3H_6N_2C_3H_7$	324	7.0	6.7	−2
$t\text{-}C_4H_9 + t\text{-}C_4H_9CHO \rightarrow i\text{-}C_4H_{10} + t\text{-}C_4H_9CO$	496	6.8	4.3	−6
$C_2H_5 + SiHCl_3 \rightarrow C_2H_6 + SiCl_3$[b]	364	8.54	5.3	−12
$CF_3 + CH_3SiCl_3 \rightarrow CF_3H + CH_2SiCl_3$[c]	468	8.31	9.4	?
$C_2F_5 + C_2H_6 \rightarrow C_2F_5H + C_2H_5$	424	9.18	8.7	−5
$CH_3S + CH_3CHO \rightarrow CH_3SH + CH_3CO$	693	9.0	4.8	−4
$CF_3 + CF_3CF_2H \rightarrow CF_3H + C_2F_5$	616	8.13	10.1	−3

[a] The A factors refer to *total* H abstraction at the site of the molecule indicated in the reaction; for sources of data see references 3 and 5 unless otherwise stated.

[b] J. A. Kerr, A. Stephens, and J. C. Young, *Int. J. Chem. Kinet.*, **1**, 371 (1969).

[c] T. N. Bell and U. F. Zucker, *Can. J. Chem.*, **48**, 1209 (1970).

the potential energy of activation can be calculated by resolving the energy of the complex into three parts, E_{RH}, E_{R^1H}, and E_{RR}, where E is the energy of the respective bond. The sum of the bond orders (n) of the complex is assumed to be unity and the bond orders are calculated from Pauling's empirical bond length-bond order relation. The bond energies are derived from single bond energies. The calculated potential energy of activation (V) is then related to the experimental activation energy (E) by the expression

$$E = V + \theta RT$$

where θRT is the zero point energy of the transition state minus the sum of the zero point energies of the reactants. These calculations yield activation energies which agree with the experimental energies within ± 2 kcal/mole.[38,50]

The A factors in hydrogen transfer in atom–molecule reactions usually fall within the range $10^{11.0 \pm 0.5}$ (liters/mole-sec) and those for radical–molecule reactions within the range $10^{8.5 \pm 0.5}$ (liters/mole-sec), both in agreement with transition state theory calculations. Benson has developed a useful method of estimating bimolecular A factors[7] which can be illustrated by considering the reaction

$$CH_3 + C_2H_6 = CH_4 + C_2H_5 \qquad (8)$$

The difficulty in carrying out a full transition state theory calculation rests in calculating the entropy of the complex CH_3---H---CH_2CH_3, but a lower limit can be set to this entropy by assuming it is a very tight complex with a structure resembling $CH_3CH_2CH_3$ as a model. In calculating $S^{\ddagger}(C_3H_9\cdot)$ from $S^{\circ}(C_3H_8)$ the translational and rotational entropy differences can be neglected, leaving only contributions from the electron spin and from changes in symmetry. The external symmetry numbers are $\sigma(C_3H_9\cdot) = 1$ and $\sigma(C_3H_8) = 2$, making a change in entropy of R ln 2 in going from C_3H_8 to the complex. The electron spin in the complex also contributes R ln 2 to the entropy and hence

$$S^{\ddagger}(C_3H_9\cdot) \geqslant S^{\circ}(C_3H_8) + R \ln 2 + R \ln 2$$

The entropy change in reaching the transition state is given by

$$\begin{aligned} \Delta S^{\ddagger} &\geqslant S^{\ddagger}(C_3H_9\cdot) - S^{\circ}(CH_3) - S^{\circ}(C_2H_6) \\ &\geqslant -25.4 \text{ cal/mole-deg} \end{aligned}$$

The A factor is then calculated from the transition state theory expression $A = (e^2kT/h)\exp(-\Delta S^{\ddagger}/R)$ and at 400°K this yields $A_8 \geqslant 10^8$ (liters/mole-sec), in reasonable agreement with the experimental value of $10^{8.3}$ (liters/mole-sec). The merit of these simple calculations is that they throw immediate suspicion on any experimental A factors which are much lower than the estimates.

Although there is a lot of kinetic information on carbon-centered radicals abstracting hydrogen atoms from a large range of molecules, there is comparatively little known of the analogous reactions of heterocentered radicals. One of the major difficulties is in finding convenient sources of such radicals. Relevant information can often be derived from studies of the reverse reactions and thermodynamic data on the equilibrium constant. This approach is illustrated by considering the reaction

$$CH_3 + H_2S \longrightarrow CH_4 + HS \tag{9}$$

for which the rate constant has been found to be[5]

$$\log k_9 \text{ (liters/mole-sec)} = 7.7 - 2.9/\theta$$

The entropy and enthalpy changes can be calculated[7] to be $\Delta S_9^\circ = -4.1$ cal/mole-deg and $\Delta H_9^\circ = -14$ kcal/mole, and from the previous equations relating ΔS° to the A factors and ΔH° to the activation energies it can be calculated that

$$\log k_{-9} \text{ (liters/mole-sec)} = 8.6 - 16.9/\theta$$

Arrhenius parameters for the transfer of halogen atoms in radical–molecule reactions are listed in Table 5. Most of the reactions are close to thermal neutrality with the exception of those involving halogen molecules which are 15–30 kcal/mole exothermic. Several of the experimental A factors are abnormally high and these are associated with high activation energies as well (compare CH_3 with CCl_4 and C_2Cl_6), which tends to suggest that internal compensation of the Arrhenius parameters has occurred. Secondary reactions may be the cause of systematic errors in these systems.

An extensive range of relative rate constants and relative Arrhenius parameters has been reported for the highly exothermic chlorine and bromine transfer reactions of $SiCl_3$ radicals with alkyl halides.[5]

$$CH_3Cl + SiCl_3 = CH_3 + SiCl_4 \tag{10}$$

$$RX + SiCl_3 = R + SiCl_3X \tag{11}$$

The temperature variations of k_{10}/k_{11} yield $E_{11} - E_{10}$ and A_{10}/A_{11}. About a dozen different alkyl halides (RX) have been studied. On the whole there is not a wide variation in the A factors, and the activation energy differences are in accord with the differences in C—X bond strengths. It will be necessary, however, to place the results on an absolute basis before they can be assessed properly.

Related to atom transfer reactions of radicals and molecules are group displacement reactions in which a radical attacks a molecule with the displacement of a second radical. The little quantitative data that exist on such radical–molecule reactions are given in Table 6. These reactions are difficult to study since they must compete with the faster hydrogen transfer reactions of the radicals. Benson has made the point that displacement reactions

TABLE 5

Arrhenius Parameters for the Transfer of Halogen Atoms in the Reactions of Radicals with Molecules[a]

Reaction	T_m (°K)	Log A (liters/mole-sec)	E (kcal/mole)
$CH_3 + CCl_4 \rightarrow CH_3Cl + CCl_3$	391	10.4	12.9
$CH_3 + CCl_3CN \rightarrow CH_3Cl + CCl_2CN$	391	9.9	10.4
$CH_3 + C_2Cl_6 \rightarrow CH_3Cl + C_2Cl_5$	391	8.8	10.1
$CH_3 + C_6H_5CCl_3 \rightarrow CH_3Cl + C_6H_5CCl_2$	391	7.3	7.6
$CH_3 + (CCl_3)_2CO \rightarrow CCl_4 + CCl_2COCCl_3$	391	9.6	9.7
$CH_3 + Br_2 \rightarrow CH_3Br + Br$	413	—	0.9
$CH_3 + CF_3Br \rightarrow CH_3Br + CF_3$	391	10.3	12.5
$CH_3 + CF_2Br_2 \rightarrow CH_3Br + CF_2Br$	391	8.0	6.4
$CH_3 + CCl_3Br \rightarrow CH_3Br + CCl_3$	391	10.2	7.1
$CH_3 + CCl_2Br_2 \rightarrow CH_3Br + CCl_2Br$	391	10.8	7.6
$CH_3 + CBr_4 \rightarrow CH_3Br + CBr_3$	391	11.2	7.9
$CH_3 + CF_3I \rightarrow CH_3I + CF_3$	391	10.8	7.5
$CF_3 + Cl_2 \rightarrow CF_3Cl + Cl$	453	9.89	3.6
$CF_3 + CH_2Cl_2 \rightarrow CF_3Cl + CH_2Cl$	651	8.49	11.8
$CF_3 + CHCl_3 \rightarrow CF_3Cl + CHCl_2$	441	8.33	11.0
$CF_3 + CCl_4 \rightarrow CF_3Cl + CCl_3$	478	9.08	12.0
$CF_3 + C_6H_5CCl_3 \rightarrow CF_3Cl + C_6H_5CCl_2$	411	9.49	9.7
$CF_3 + Br_2 \rightarrow CF_3Br + Br$	526	9.36	0.7
$CF_3 + CH_3Br \rightarrow CF_3Br + CH_3$	478	10.53	8.4
$CF_3 + C_6H_5Br \rightarrow CF_3Br + C_6H_5$	332	4.91	1.7
$CF_3 + C_6F_5Br \rightarrow CF_3Br + C_6F_5$	346	4.38	0.3
$CF_3 + I_2 \rightarrow CF_3I + I$	?	9.42	0
$CF_3 + CH_3I \rightarrow CF_3I + CH_3$	406	6.59	3.3
$CF_3 + C_6H_5I \rightarrow CF_3I + C_6H_5$	386	7.33	3.1
$CCl_3 + Cl_2 \rightarrow CCl_4 + Cl$	365	8.74	5.0
$CCl_3 + CCl_3Br \rightarrow CCl_4 + CCl_2Br$	523	10.2	18.6
$C_2H_5 + Cl_2 \rightarrow C_2H_5Cl + Cl$	?	10.1	1.0
$C_2H_5 + I_2 \rightarrow C_2H_5I + I$	556	9.50	0.2
$NF_2 + F_2 \rightarrow NF_3 + F$	319	7.26	9.7
$NF_2 + ClF_3 \rightarrow NF_3 + ClF_2$	468	9.17	23.5

[a] For sources of data see references 3 and 5.

involving attack of the radicals on saturated molecules at the elements carbon to fluorine, of the first row of the periodic table, would be expected to show appreciable activation energies since they will require expansion of the valence shell of the attacked atom, which is energetically unfavorable.[7] On the other hand, displacement reactions from boron compounds should

TABLE 6

Group Displacement Reactions of Radicals[a]

Reaction	T_m (°K)	Log A (liters/mole-sec)	E (kcal/mole)
$CH_3 + CF_3COCF_3 \rightarrow CH_3COCF_3 + CF_3$	437	8.7	7.0
$CH_3 + C_2F_5COC_2F_5 \rightarrow CH_3COC_2F_5 + C_2F_5$	485	—	7
$CD_3 + CH_3HgCH_3 \rightarrow CD_3HgCH_3 + CH_3$	453	5.95 (log k)	—
$CH_3 + CD_3SH \rightarrow CH_3SH + CD_3$	438	7.73	7.6
$CH_3 + CF_3COCH_3 \rightarrow C_2H_6 + CF_3CO$	523	—	14
$CH_3 + CH_3COCOCH_3 \rightarrow CH_3COCH_3 + CH_3CO$	423	7.7	5.6
$CH_3 + CH_3CH{=}CHCHO \rightarrow CH_3CH{=}CHCH_3 + CHO$	458	8.8	7.5
$CH_3 + CH_3HgCH_3 \rightarrow C_2H_6 + (Hg + CH_3)$	471	4	1
$CF_3 + (CH_3)_4Sn \rightarrow CF_3Sn(CH_3)_3 + CH_3$	423	4.0 (log k)[b]	—

[a] For sources of data see references 3 and 5.

[b] T. N. Bell and A. E. Platt, *J. Chem. Soc.*, D, 325 (1970).

be faster on the basis of this argument, which is supported by the recent finding that the reaction

$$CF_3 + B(CH_3)_3 \longrightarrow CF_3B(CH_3)_2 + CH_3$$

is relatively rapid.[50a]

Displacement reactions involving carbonyl compounds do not require expansion of the valence shell of the atom which is attacked, as the reaction can proceed by addition to the C═O bond:

$$CH_3 + CF_3COCF_3 \longrightarrow [CF_3\text{—}\underset{\underset{\dot{O}}{|}}{\overset{\overset{CH_3}{|}}{C}}\text{—}CF_3]^{\ddagger} \longrightarrow CF_3 + \overset{\overset{CH_3}{|}}{C}OCF_3$$

VII. ADDITION REACTIONS

Radical addition reactions to unsaturated molecules and the reverse radical decomposition reactions continue to be the most difficult classes of radical reactions to study.[1] No new general approach to these systems has been developed during the past ten years but a steady accumulation of data continues from established methods or their adaptations.

Cvetanović and Irwin have studied the addition of methyl radicals to a series of olefins by a variant of the mass-balance method.[51] Small amounts of biacetyl were photolyzed in the presence of the olefin and a reference hydrocarbon, isobutane. Under the experimental conditions the important reactions were:

$$CH_3COCOCH_3 + h\nu \longrightarrow 2CH_3CO \qquad (12)$$

$$CH_3CO \longrightarrow CH_3 + CO \qquad (13)$$

$$CH_3 + (CH_3)_3CH \longrightarrow CH_4 + (CH_3)_3C \qquad (14)$$

$$CH_3 + Ol \longrightarrow R' \qquad (15)$$

$$CH_3 + Ol \longrightarrow CH_4 + R'' \qquad (16)$$

Radical–radical reactions can be neglected provided the concentrations of olefin (Ol) and isobutane are far in excess of the methyl concentration. A steady-state treatment yields:

$$f = \left(\frac{R_{CO}}{R_{CH_4}} - 1\right)^{-1} = \frac{k_{16}}{k_{15}} + \frac{k_{14}}{k_{15}}\frac{[RH]}{[Ol]}$$

so that plots of f against [RH]/[Ol] yield k_{16}/k_{15} and k_{14}/k_{15}. In the limiting case with [Ol] = 0, the expression becomes

$$\frac{R_{CO}}{R_{CH_4}} = 1$$

and this was confirmed experimentally, although minor corrections were necessary for additional CH_3 reactions at low temperatures and low RH concentrations. The advantage in using biacetyl as a CH_3 source, in preference to acetone,[1] is that the ratio CO/CH_4 is unity with biacetyl even though all the CH_3CO radicals do not decompose by reaction 13. This method undoubtedly has yielded the most reliable set of relative rate constants for CH_3 additions to olefins. One minor disadvantage, which applies to all mass-balance methods, is that it gives no information on the site of radical addition to the olefin when two different adduct radicals can be formed.

The "aldehyde method" of studying alkyl radical additions to olefins makes use of a radical scavenger with which the adduct radicals readily react to form a stable product, thereby preventing further additions and other reactions of the radicals.[1] The radical scavenger in that case is the parent aldehyde from which the radicals were produced. Other examples of scavenger molecules for this purpose are acetaldehyde[52] and hydrogen sulfide.[53]

The Arrhenius parameters for CF_3 additions to a range of alkenes, alkynes, and aromatic compounds have been determined by Szwarc and co-workers from the photolysis of hexafluoroazomethane in the presence of a large excess of a reactive hydrocarbon (RH = 2,3-dimethybutane).[53a]. In the absence of an unsaturated substrate the only important reactions are

$$CF_3N{=}NCF_3 + h\nu \longrightarrow 2CF_3 + N_2$$

$$CF_3 + RH \longrightarrow CF_3H + R \qquad (17)$$

and the product ratio (CF_3H/N_2) is 2. When the reaction is carried out in the presence of the unsaturate (Ol), the ratio (CF_3H/N_2) is found to be less than 2 owing to the reaction

$$CF_3 + Ol = CF_3Ol \qquad (18)$$

and the rate constant ratio can be calculated from

$$k_{18}/k_{17} = ([RH]/[Ol])\{(CF_3H/N_2) - (CF_3H/N_2)_u\}/(CF_3H/N_2)_u$$

where $(CF_3H/N_2)_u$ and (CF_3H/N_2) are the ratios of products in experiments with and without unsaturate present, respectively. The experiments were carried out over the temperature range 338–453°K and Arrhenius parameters for reaction 18 were derived relative to 17. Unfortunately there is no reliable measurement of k_{17} and this leaves the absolute determinations of E_{18} and A_{18} in some doubt, although the relative rate constants for CF_3 additions by this method should be reliable to within a few percent.

Some interesting conclusions can be drawn from the results of these rate studies of CF_3 additions. The range of reactivities of alkenes is large, a factor of about 200 between vinyl fluoride, the slowest, and butadiene, the fastest, and this is due almost entirely to differences in activation energies, the A factors remaining essentially constant. The A factors for CF_3 addition to alkynes are about 3.5 times larger than those for the corresponding alkenes. Szwarc and co-workers have also studied CF_3 additions to unsaturated compounds in the liquid phase and in general have found results similar to those in the gas phase.[53a]

Tedder and Walton have studied the rate of addition of CCl_3 radicals to olefins using the photolysis of CCl_3Br as the radical source[17]

$$CCl_3Br + h\nu \longrightarrow CCl_3 + Br \tag{19}$$

$$CCl_3 + Ol \longrightarrow CCl_3Ol\cdot \tag{20}$$

$$CCl_3Ol\cdot + CCl_3Br \longrightarrow CCl_3OlBr + CCl_3 \tag{21}$$

$$Br + Ol \longrightarrow BrOl\cdot \tag{22}$$

$$BrOl\cdot + CCl_3Br \longrightarrow BrOlBr + CCl_3 \tag{23}$$

$$2CCl_3 \longrightarrow C_2Cl_6 \tag{24}$$

The above reaction scheme was proposed and tested under various conditions. Under conditions where $[CCl_3Br] > [Ol]$ it was established that reaction 24 was the main chain-terminating step, and the rate constants for addition were calculated from the steady-state expression:

$$[CCl_3OlBr]/[C_2Cl_6] = k_{20}[Ol]/(k_{24}\phi I_a)^{1/2} \tag{25}$$

where the rate of reaction 19 is defined as $R_{19} = \phi I_a = R_{C_2Cl_6}$. The products CCl_3OlBr and C_2Cl_6 were analyzed by gas chromatography and the Arrhenius parameters were determined from the temperature coefficient of (25), assuming ϕI_a to be temperature independent. The original results indicated that the *A* factors for many of the CCl_3 additions were abnormally low.

Subsequent studies of the photolysis of CCl_3Br alone revealed that ϕI_a was markedly temperature dependent.[54] The original data on CCl_3 additions to olefins have recently been reinterpreted[55] and the rate constants were calculated from the equation:

$$k_{20}/k_{24}^{1/2} = R_{CCl_3OlBr}/R_{C_2Cl_6}^{1/2}[Ol]$$

The recalculated data bring the results for CCl_3 additions into line with those for other radicals.[56]

Although the kinetic data on CCl_3 additions to olefins have required recalculation, the relative rate constants were substantially unaffected and more importantly the relative proportions of attack of CCl_3 radicals at the two sites in asymmetrical olefins, as determined by analyses of the products CCl_3OlBr, remained unaltered.

A selection of Arrhenius parameters for a variety of radicals adding to unsaturated molecules is given in Table 7. An evaluation of atom and radical additions is presently being carried out.[56] Here again it should be emphasized that these Arrhenius parameters involve relatively large errors; for instance, none of the activation energies is known to better than ± 1 kcal/mole. Notwithstanding these large experimental errors, it is fairly clear that the activation energies for alkyl additions fall within the narrow range of 6 ± 2 kcal/mole. The greater reactivity of CF_3 radicals is reflected in lower activation energies, the *A* factors being similar to those for alkyl additions. CCl_3

radical additions to hydrocarbon olefins follow much the same pattern of reactivity as for CH_3 radicals.

Table 8 summarizes relative rate constants for atoms, radicals, and carbenes adding to the series of olefins. The accuracy of the relative rate constants is much better than that of the absolute rate constants. The atoms Br, O, S, Se, and NF_2 radicals show typical electrophilic trends, the rates of addition increasing progressively with double bond substitution and a distinct selectivity being shown. At the other end of the scale CH_3 radicals display much less selectivity, although the reactions are very much slower, and here the governing factors appear to be the stability of the adduct radicals and steric effects. The tendency toward electrophilic property in the atoms and radicals of Table 8 decreases along the sequence Br to CH_3. The position of hydrogen atoms, just to the left of CF_3, is interesting and somewhat surprising. The greater reactivity of butadiene compared with the monoolefins is probably due to the formation of a resonance-stabilized allylic radical from the addition.[51]

Once again the principles of thermodynamic equilibria can be applied to test if the kinetic data for the addition reactions are in accord with those for the reverse radical decompositions and the enthalpy and entropy changes. Kinetic data on the radical decompositions (see next section) are usually only available for alkyl radicals and seldom for halogenoalkyls, but on the whole the agreement between the forward and reverse reactions is within the limits set by the thermodynamic data.[6,24a,57] These comparisons show that the A factors for the radical decomposition reactions indicate a relatively tight transition state. The rehybridizations from sp^2 to sp^3 bonding in the olefin and the radical in a radical addition to an olefin would be expected to occur at relatively close contact of the reactants, and this is supported by transition state theory calculations on the radical decomposition reactions.[7] The implication is that the transition state in the decomposition reaction resembles the reactant radical and consequently $\Delta S^{\ddagger}$ is quite small although always positive in the direction of decomposition. On this basis it is reasonable to set a lower limit on the A factor for the decomposition of (ekT/h) corresponding to $10^{13.35}$ sec^{-1} at 400°K. It is now possible to set a lower limit to the A factor for the addition reactions since the entropy changes can be estimated readily to within ± 2 cal/mole-deg. Most radical addition reactions have entropy changes within the range -35 to -40 cal/mole-deg[56] and consequently should have A factors of at least $10^{7.5}$ to $10^{6.5}$ liters/mole-sec. As can be seen from Table 7 and elsewhere[56] these estimates are reasonable for most radical addition reactions except for CF_2Br additions to olefins and for CH_3 addition to CO, where the A factors are abnormally low. This line of reasoning also led to the original data on CCl_3 additions being suspected and recalculated.[55]

The position of radical attack on addition to asymmetrical olefins is an interesting question which has been discussed at length.[17,58] Until recently there has been a distinct lack of quantitative data on this question for radical additions in the gas phase. In general it appears that most atoms and radicals add preferentially to the least substituted carbon atom of the double bond to produce the more thermodynamically stable radical. If it is purely an energy effect, then it would appear that in terms of a simple Hammond postulate, the more exothermic reaction has the lower activation energy. Thus the addition of hydrogen atoms to propene, at room temperature in the pressure range 40–250 torr, gives 94% isopropyl and 6% *n*-propyl radicals.[58]

It is now well established, however, mainly from the studies of Tedder and co-workers, that for radical additions to fluorinated olefins the predictions of the position of addition, based on the stabilities of the adduct radical, break down.[54–56]. The radicals CF_3, CCl_3, $n\text{-}C_3F_7$, CF_2Br, and SF_5 all add faster to the least substituted end of the fluoroethylenes, whereas the sequence of radical stabilities almost certainly runs $-\dot{C}H_2 > -\dot{C}HF > -\dot{C}F_2$. So far there is not even a qualitative explanation of these positions of addition.

The additions of alkyl radicals to allene occur exclusively at the terminal carbon atoms to give vinylic radicals and not at the central carbon atom to produce the more stable allylic radicals, also in contrast to the above predictions.[59,60] This may arise from the unique geometry of allene in which the π bonds are mutually perpendicular. In the transition state the incipient radical from center attack resembles a primary rather than an allylic radical since the orbital of the odd electron does not overlap with the π orbital of the remaining double bond.

Considerable attention has been given to finding an appropriate molecular property as an index for the reactivity of unsaturated molecules toward radical addition.[17,58] Correlations between rate constants and molecular properties

TABLE 7

Arrhenius Parameters for Some Radical Addition Reactions to Unsaturated Molecules[a]: R· + U → RU·

Ref.	Radical	Unsaturate	T_m (°K)	Log *A* (liters/mole-sec)	*E* (kcal/mole)
b	CH_3	$CH_2{=}CH_2$	403	8.94	7.9
c	C_2H_5	$CH_2{=}CH_2$	415	7.7	6.9
c	$i\text{-}C_3H_7$	$CH_2{=}CH_2$	415	7.8	6.9
c	$n\text{-}C_4H_9$	$CH_2{=}CH_2$	415	8.1	7.3
c	$t\text{-}C_4H_9$	$CH_2{=}CH_2$	415	7.5	7.1
d	$i\text{-}C_5H_{11}$	$CH_2{=}CH_2$	378	7.5	6.4
b	CH_3	$CH_3CH{=}CH_2$	403	8.52	7.4
b	CH_3	$(CH_3)_2C{=}CH_2$	403	8.46	6.9
e	CH_2F	$CH_2{=}CH_2$	372	7.6	4.3

TABLE 7—(*continued*)

Ref.	Radical	Unsaturate	T_m (°K)	Log A (liters/mole-sec)	E (kcal/mole)
f	CF_3	$CH_2{=}CH_2$	383	8.4	2.4
g	CF_3	$CH_3CH{=}CH_2$	395	8.78	1.9
h	CF_2Br	$CH_2{=}CH_2$	427	5.3	1.0
i	CCl_3	$CH_2{=}CH_2$	391	7.6	6.3
j	CCl_3	$CH_3CH{=}CH_2$	452	8.9	6.5
k	CH_3	$CF_2{=}CF_2$	403	8.95	5.7
l	CH_3	$CH_2{=}CHCl$	413	9.77	9.0
g	CF_3	$CH_2{=}CHF$	395	8.79	3.7
m	CCl_3	$CH_2{=}CHF$	390	5.9	4.4
m	CCl_3	$CF_2{=}CF_2$	398	7.5	6.1
c	CH_3	$CH{\equiv}CH$	415	8.4	7.8
n	CH_3	$CH_3C{\equiv}CH$	422	8.7	8.8
c	C_2H_5	$CH{\equiv}CH$	415	7.8	7.2
c	i-C_3H_7	$CH{\equiv}CH$	415	7.6	6.9
o	t-C_4H_9	$CH{\equiv}CH$	420	6.2	5.3
p	CF_3	$CH{\equiv}CH$	395	9.50	4.2
q	CH_3	$CH_3N{=}NCH_3$	360	7.59	5.9
r	CH_3	$C{\equiv}O$	304	5.58	3.9
p	CF_3		380	8.52	4.4

[a] For a comprehensive list see reference 56. The above Arrhenius parameters are based on measured or assumed rate constants for the radical combination reaction (Table 1). Where addition can give rise to two different radicals the parameters refer to the sum of both processes.

[b] Reference 51.

[c] Reference 1.

[d] K. W. Watkins and L. A. O'Deen, *J. Phys. Chem.*, **73**, 4094 (1969).

[e] J. M. Sangster and J. C. J. Thynne, *Trans. Faraday Soc.*, **65**, 2110 (1969).

[f] J. M. Sangster and J. C. J. Thynne, *J. Phys. Chem.*, **73**, 2746 (1969).

[g] J. M. Pearson and M. Szwarc, *Trans. Faraday Soc.*, **60**, 553 (1964).

[h] J. M. Tedder and J. C. Walton, *Trans. Faraday Soc.*, **66**, 1135 (1970).

[i] J. M. Tedder and J. C. Walton, *Trans. Faraday Soc.*, **60**, 1769 (1964), and reference 55.

[j] J. M. Tedder and J. C. Walton, *Trans. Faraday Soc.*, **63**, 2679 (1967), and reference 55.

[k] J. M. Sangster and J. C. J. Thynne, *Int. J. Chem. Kinet.*, **1**, 571 (1969).

[l] A. M. Hogg and P. Kebarle, *J. Am. Chem. Soc.*, **86**, 4558 (1964).

[m] J. M. Tedder and J. C. Walton, *Trans. Faraday Soc.*, **62**, 1859 (1966).

[n] R. R. Getty, J. A. Kerr, and A. F. Trotman-Dickenson, *J. Chem. Soc.*, A, 1360 (1967).

[o] J. A. Garcia Dominguez and A. F. Trotman-Dickenson, *J. Chem. Soc.*, 940 (1962).

[p] G. E. Owen, J. M. Pearson, and M. Szwarc, *Trans. Faraday Soc.*, **61**, 1722 (1965).

[q] D. G. L. James and R. D. Suart, *Trans. Faraday Soc.*, **65**, 175 (1969).

[r] J. A. Kerr and J. G. Calvert, *J. Phys. Chem.*, **69**, 1022 (1965).

TABLE 8

Relative Rate Constants for Addition of Atoms and Radicals to Olefins[a]

Olefin	$Br(^2P)$[b] (298°K)	$O(^3P)$ (298°K)	$S(^3P)$ (298°K)	$Se(^3P)$ (298°K)	$:CCO(^3\Sigma)$[d] (296°K)	NF_2[c] (373°K)	$:CH_2(^1A_1)$ (297°K)	H (298°K)	CF_3 (338°K)	$:CH_2(^3\Sigma g)$ (297°K)	CH_3 (453°K)
=	1	1	1	1	1	1	1	1	1	1	1
=–	18	5.8	6.8	3.5	5.7	4.4	1.3	1.8	1.2	1.0	0.7
=––	23	5.8	9.7	7.1	7.0	4.3	1.6	1.9	1.2	1.6	0.6
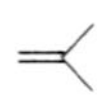	384	25	51	45	50	20	2.0	4.4	3.7	2.9	1.1
	95	24	16	24	9.1	10.3	1.4	0.8	0.9	0.94	0.23
	99	28	19	56	10.6	10.4	1.4	1.1	1.0	0.89	0.33
	—	79	92	—	100	34	2.1	—	—	1.8	0.32
	—	102	128	—	250	88	2.2	1.5	1.3	2.7	0.20
=–=	—	24	93	98	210	—	3.9	8.7	8.7	19.0	12.0
$E_{C_2H_4}$ (kcal/mole)	~2.9[e]	~2.6	—	2.8	—	15.5	—	3.2	2.4[f]	—	7.9

[a] Based on $k_{C_2H_4} = 1$ and taken from reference 51 unless otherwise stated.
[b] P. I. Abell, *Trans. Faraday Soc.*, **60**, 2214 (1964).
[c] A. J. Dijkstra, J. A. Kerr, and A. F. Trotman-Dickenson, *J. Chem. Soc.*, A, 105 (1967).
[d] D. G. Williamson and K. D. Bayes, *J. Am. Chem. Soc.*, **90**, 1957 (1968).
[e] Reference 6.
[f] Reference 53.

are less satisfactory than between activation energies and molecular properties since the former implies a uniformity of A factor which may vary somewhat with substitution at the double bond. One such correlation is that between the activation energies for ethyl radical addition and the carbon atom localization energies of the olefins calculated from a simple LCAO molecular orbital method.[61]

The reactions of radicals with molecular oxygen, which are addition reactions in the present context, fall within the scope of combustion kinetics and could well occupy a separate review. It is generally recognized that atoms and radicals react rapidly with oxygen but few rates have been measured and the chemistry and products of the reactions are still subjects of lively debate. McMillan and Calvert[22] have reviewed the available data on alkyl radical reactions and suggest a rate constant of 4.7×10^{10} liters^2/mole^2-sec (298–498°K) for the third-order reaction

$$CH_3 + O_2 + M \longrightarrow CH_3O_2^* + M$$

and a rate constant of 2.5×10^9 liters/mole-sec at 298°K for the reaction

$$C_2H_5 + O_2 \longrightarrow C_2H_5O_2$$

VIII. DECOMPOSITION REACTIONS

Decompositions of free radicals have recently been considered as part of a comprehensive evaluation of unimolecular reactions,[6] and unimolecular reactions of alkyl radicals have been reviewed.[24a] In addition a detailed account of radical sources, experimental methods, and procedures in studying the decomposition reactions of radicals has been presented.[57] Only the main features are outlined here.

The two major problems in studying these reactions are in finding suitable sources of the radicals and in monitoring the concentrations of the radicals. No new general approaches to these problems have been forthcoming, but careful exploitation of two established sources of radicals has been particularly fruitful. Reinvestigations of the mercury-photosensitized decompositions of alkanes have yielded valuable data on the decompositions of small alkyl radicals, especially the effect of pressure on the rate constants.[62] Rabinovitch and co-workers have made extensive studies of the decompositions of chemically activated alkyl radicals formed by the addition of hydrogen or deuterium atoms to olefins.[16] The method is illustrated by the addition of hydrogen to *cis*-3-methylhexene-2, which gives predominantly the 3-methyl hexyl-3 radical:

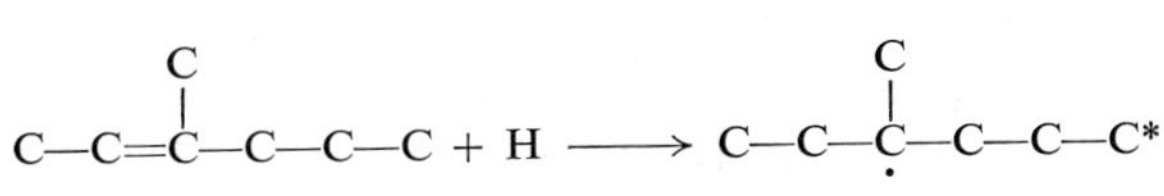

The radical decomposes by two routes:

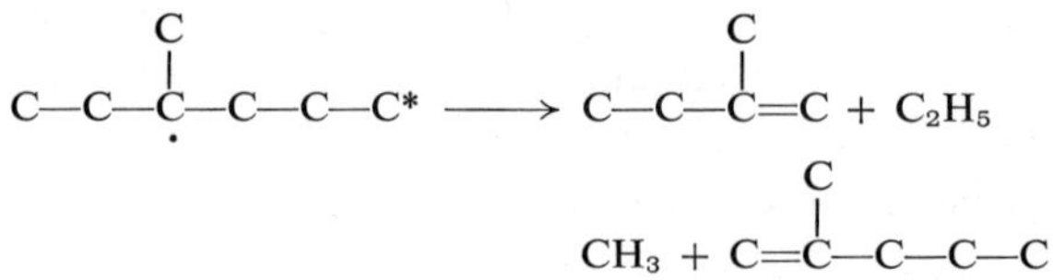

and relative rates for the decompositions can be obtained from analyses of the olefin products. A series of these competitive studies has been made on C_5 to C_8 alkyl radicals formed by the additions of hydrogen atoms to the appropriate olefin, and the rate constants have been interpreted by the RRKM theory of unimolecular reactions. This approach does not yield Arrhenius parameters for the thermal decomposition reactions of the radicals, but useful data on the thermochemistry of the alkyl radicals are derived and also differences in critical energies for the reverse CH_3 and C_2H_5 radical additions to the olefins.

Table 9 summarizes kinetic and thermodynamic data for a selection of decomposition reactions of radicals. The experimental data on the decomposition reactions are for the most recent study in cases where several sets of data exist. These parameters are compared with values calculated from the given data for the reverse addition reactions and the overall enthalpy and entropy changes by the procedures previously mentioned. On the whole the agreement between the calculated and experimental parameters for the radical decompositions is quite good. Where discrepancies arise in this type of comparison the most likely source of error is in the kinetic data for one or the other or both reactions.

For alkyl radicals the activation energies for the decomposition reactions lie within the ranges 42 $\pm$ 3 kcal/mole for C—H rupture and 32.5 $\pm$ 4 kcal/mole for C—C rupture, the difference being in line with the expected difference in C—H and C—C bond dissociation energies. The A factors lie within the rather narrow range $10^{14.1 \pm 0.5}$ sec^{-1} and, as previously pointed out, indicate a relatively "tight" transition state corresponding to entropies of activation of 3 $\pm$ 2 cal/deg-mole. Detailed arguments have been put forward to rationalize "tight" transition states in alkyl radical decompositions.[6,24a,25]

Reasonably extensive data exist on the kinetics of the decompositions of chloroalkyl and difluoroamino radicals and there is quite good agreement with the data for the reverse additions to the olefins and estimates of the thermodynamics of the systems.[57] The same may not be said for the decompositions of oxygen-containing radicals, particularly alkoxy radicals. An extreme example is the decomposition of the t-butoxy radical:

$$t\text{-}C_4H_9O \longrightarrow CH_3 + CH_3COCH_3$$

for which the activation energy has been variously reported to be from 9 to 23 kcal/mole. The higher values cannot be reconciled with the enthalpy change for the reaction, which is estimated to be 4 $\pm$ 3 kcal/mole.

TABLE 9

Selected Kinetic and Thermodynamic Data on Decomposition of Radicals[a]

Reaction	Experimental[b]			Reverse[c]					Calculated[b,d]	
	Log A	E	Ref.	Log A	E	Ref.	ΔH°_{298}[e]	ΔS°_{298}[e]	Log A	E
$C_2H_5 = C_2H_4 + H$	14.4	40.9	24a	11.16	3.3	56	38.6	20.0	14.10	41.7
n-$C_3H_7 = C_3H_6 + H$	14.1	37.0	24a	10.30	3.7	56	36.2	22.4	13.66	39.5
i-$C_3H_7 = C_3H_6 + H$	13.3	36.9	24a	10.54	2.2	56	39.7	24.0	14.24	41.4
t-$C_4H_9 = i$-$C_4H_8 + H$	15.5	43.6	24a	10.53	1.4	56	41.7	24.8	14.51	42.8
n-$C_3H_7 = C_2H_4 + CH_3$	13.5	31.4	24a	8.94	7.9	56	25.5	30.0	13.80	32.7
s-$C_4H_9 = CH_3 + C_3H_6$	14.6	32.6	24a	8.52	7.4	56	26.6	33.2	13.94	33.2
i-$C_4H_9 = CH_3 + C_3H_6$	12.4	31.0	24a	8.09	8.5	56	25.1	34.3	13.86	28.7
n-$C_4H_9 = C_2H_5 + C_2H_4$	13.6	28.7	24a	7.7	6.9	56	22.7	33.9	13.00	28.1
$C_2H_4Cl = Cl + C_2H_4$	13.0	23.9	57	10.3	1.5	56	20.2	23.7	13.66	21.1
$C_2H_4NF_2 = NF_2 + C_2H_4$	12.9	13.7	57	7.6	12.9	56	−0.4	—	—	11.5
$CH_3CO = CH_3 + CO$	10.3	15.0	57	5.6	3.9	56	13.0	29.2	10.51	16.6

[a] For a complete list see references 6, 24a, and 57.
[b] A in sec^{-1}, E in kcal/mole.
[c] A in liters/mole-sec, E in kcal/mole.
[d] $T_m = 500°K$.
[e] Standard state 1 atm, taken from references 6 and 56.

IX. ISOMERIZATION REACTIONS

It was originally postulated by Kossiakoff and Rice[63] that long-chain alkyl radicals could undergo isomerization by hydrogen atom transfer reactions involving cyclic transition states:

$$\dot{C}_1—C_2—C_3—C_4—C_5—C_6 \longrightarrow C_1—C_2—C_3—C_4—\dot{C}_5—C_6$$

Substantial qualitative evidence for 1,4 and 1,5 intramolecular hydrogen migration is available but conflicting results have been obtained for 1,2- and 1,3-hydrogen shifts in propyl and butyl radicals. McNesby and co-workers[64] have shown by isotopic labeling experiments that isomerizations in propyl and butyl radicals compete unfavorably with decomposition. On the other hand, there have been repeated claims that such reactions occur in hydrocarbon pyrolyses.[65]

Important results have recently been reported on isomerizations of vibrationally excited alkyl radicals from the additions of hydrogen atoms to olefins.[66] The systems involve a range of C_6 to C_9 alkyl radicals and several variations of the Kossiakoff–Rice reaction have been observed. A shorthand notation is employed, illustrated by the isomerization:

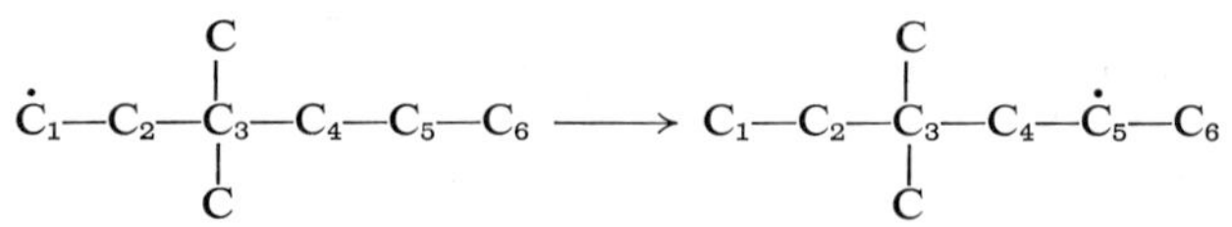

which is described as a 6*sp* process. The number refers to the ring size of the cyclic transition state, the first letter, *p*, *s*, or *t*, to the primary, secondary, or tertiary C—H bond broken, and the second letter to the C—H bond formed. The types of isomerization observed were 5*ss*, 6*ss*, 7*ss*, 5*ps*, 6*ps*, 7*ps*, 5*sp*, and 6*sp*. Processes involving ring sizes smaller than 5 were not important in these and other systems, confirming that isomerizations of propyl and butyl radicals are unlikely under normal conditions. Rate constants were obtained for the isomerization reactions as illustrated by the 3-methyl hexyl-2 radical formed by the reaction

The radical contains about 45 kcal/mole of excess internal energy from the exothermicity and activation energy of the addition reaction, and it can either isomerize or decompose:

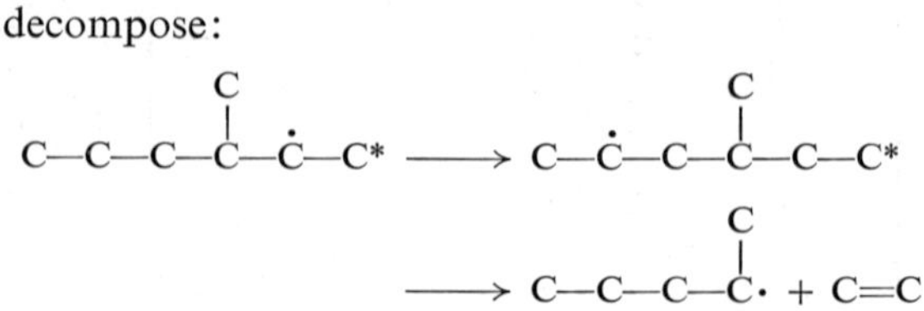

The original or isomerized radical can also be collisionally quenched to yield stabilized radicals and in the presence of "getter" radicals such as methyl, a proportion will form combination products:

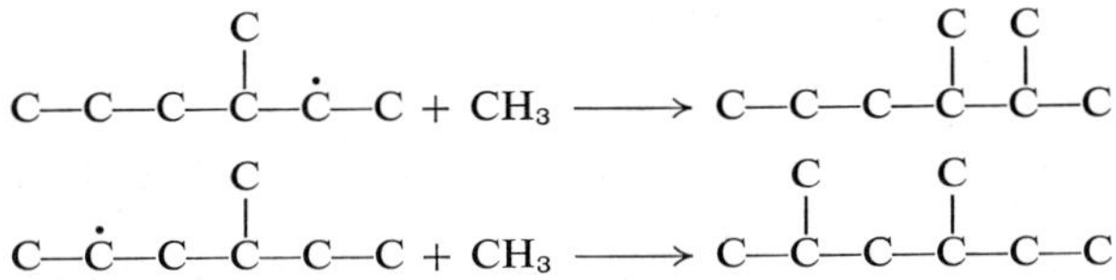

The steady-state concentration ratio of stabilized parent radicals relative to stabilized isomerized radicals, S/S_I, is given by the ratio of the two combination products which are analyzed. The high-pressure limiting rate constant for the isomerization is then obtained from the slope of a plot of $\log(S/S_I)$ against log(total pressure). It has also been shown that isomerization reactions are very much faster than the decomposition reactions for these large alkyl radicals.

So far only two quantitative kinetic studies of Kossiakoff–Rice isomerizations have been reported from thermal systems. For the 5*sp* isomerization of *n*-pentyl[67] the rate constant has been reported to be $\log k\ (\text{sec}^{-1}) = 7.15 - 10.8/\theta$ and for the 6*sp* isomerization of *n*-hexyl,[68] $\log k\ (\text{sec}^{-1}) = 7.30 - 8.3/\theta$. It is difficult to explain the abnormally low *A* factors which correspond to an unusually tight activated complex.

Isomerization of vinylic radicals, formed by the addition of alkyl radicals to acetylenes,[69] has been suggested to involve Kossiakoff–Rice-type isomerization followed by internal addition[12]:

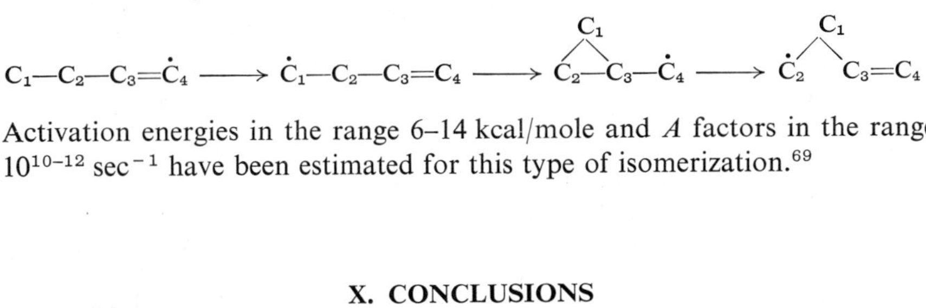

Activation energies in the range 6–14 kcal/mole and *A* factors in the range 10^{10-12} sec^{-1} have been estimated for this type of isomerization.[69]

X. CONCLUSIONS

The volume of kinetic data on the reactions of free radicals in the gas phase has grown to such an extent that data compilations and evaluations are essential. The vast majority of the results form a coherent pattern in terms of the thermodynamics of the systems and in terms of transition state theory. On the experimental side there is a great need for accurate measurements of temperature coefficients of radical–radical reactions while in theoretical kinetics much remains to be achieved by way of calculating activation energies for free radical reactions a priori from the properties of the reactants.

REFERENCES

1. J. A. Kerr and A. F. Trotman-Dickenson, *Prog. React. Kinet.*, **1**, 107 (1961).
2. E. W. R. Steacie, *Atomic and Free Radical Reactions, ACS Monograph No. 125*, 2nd ed., Reinhold, New York, 1954.
3. A. F. Trotman-Dickenson and G. S. Milne, *Tables of Bimolecular Gas Reactions*, NSRDS-NBS 9, 1967.
4. V. N. Kondratiev, *Handbook of Rate Constants of Gas Phase Reactions*, Nauka, Moscow, 1970.
5. E. Ratajczak and A. F. Trotman-Dickenson, *Supplementary Tables of Bimolecular Gas Reactions*, University of Wales Institute of Science and Technology, Cardiff, Wales, 1970.
6. S. W. Benson and H. E. O'Neal, *Kinetic Data on Gas Phase Unimolecular Reactions*, NSRDS-NBS 21, 1970.
7. S. W. Benson, *Thermochemical Kinetics*, Wiley, New York, 1968.
8. S. W. Benson, F. R. Cruickshank, D. M. Golden, G. R. Haugen, H. E. O'Neal, A. S. Rogers, R. Shaw, and R. Walsh, *Chem. Rev.*, **69**, 279 (1969).
9. S. W. Benson and H. E. O'Neal, *Int. J. Chem. Kinet.*, **1**, 221 (1969).
10. H. E. O'Neal and S. W. Benson, *J. Phys. Chem.*, **71**, 2903 (1967); **72**, 1866 (1968).
11. F. B. Wampler and R. R. Kuntz, *Int. J. Chem. Kinet.*, **3**, 483 (1971).
12. S. W. Benson and W. B. DeMore, *Ann. Rev. Phys. Chem.*, **16**, 397 (1965).
13. H. E. van den Berg, A. B. Callear, and R. J. Nostram, *Chem. Phys. Letters*, **4**, 101 (1969).
14. N. Basco, D. G. L. James, and R. D. Suart, *Int. J. Chem. Kinet.*, **2**, 215 (1970).
15. (a) W. E. Morganroth and J. G. Calvert, *J. Am. Chem. Soc.*, **88**, 5387 (1966); (b) J. O. Terry and J. H. Futrell, *Can. J. Chem.*, **46**, 664 (1968); (c) K. W. Watkins and L. A. O'Deen, *J. Phys. Chem.*, **73**, 4094 (1969).
16. (a) C. W. Larson and B. S. Rabinovitch, *J. Chem. Phys.*, **50**, 871 (1969); **52**, 5181 (1970). (b) E. A. Hardwidge, C. W. Larson, and B. S. Rabinovitch, *J. Am. Chem. Soc.*, **92**, 3278 (1970) and references cited therein.
17. J. M. Tedder and J. C. Walton, *Prog. React. Kinet.*, **4**, 37 (1967).
18. J. R. Majer and J. P. Simons, *Adv. Photochem.*, **2**, 137 (1964).
19. J. A. Franklin, G. Huybrechts, and C. Cillien, *Trans. Faraday Soc.*, **65**, 2094 (1969) and references cited therein.
20. P. Gray, R. Shaw, and J. C. J. Thynne, *Prog. React. Kinet.*, **4**, 63 (1967).
21. M. H. J. Wijnen, *J. Am. Chem. Soc.*, **82**, 3034 (1960).
22. G. R. McMillan and J. G. Calvert, *Oxid. Combust. Rev.*, **1**, 83 (1965).
23. J. A. Kerr, B. J. A. Smith, A. F. Trotman-Dickenson, and J. C. Young, *J. Chem. Soc. (A)*, 510 (1968).
24. (a) H. M. Frey and R. Walsh, *Chem. Rev.*, **69**, 103 (1969); (b) M. P. Halstead, R. S. Konar, D. A. Leathard, R. M. Marshall, and J. H. Purnell, *Proc. Roy. Soc.*, **A310**, 525 (1969); (c) W. Tsang, *J. Chem. Phys.*, **44**, 4283 (1966); *Int. J. Chem. Kinet.*, **2**, 311 (1970).
25. B. S. Rabinovitch and D. W. Setser, *Adv. Photochem.*, **3**, 1 (1964); F. H. Dorer and B. S. Rabinovitch, *J. Phys. Chem.*, **69**, 1952 (1965); J. W. Simons, B. S. Rabinovitch, and F. H. Dorer, *ibid.*, **70**, 1076 (1966).
26. S. W. Benson, *Adv. Photochem.*, **2**, 1 (1964).
27. J. A. Kerr, *Ann. Rep. Chem. Soc.*, **64**, 73 (1967); **65**, 189 (1968).
28. A. R. Blake, J. F. Henderson, and K. O. Kutschke, *Can. J. Chem.*, **39**, 1920 (1961); J. O. Terry and J. H. Futrell, *ibid.*, **45**, 2327 (1967).

29. L. Bretrand, G. R. De Maré, G. Huybrechts, J. Olbregts, and M. Toth, *Chem. Phys. Letters*, **5**, 183 (1970).
30. H. S. Johnston and P. Goldfinger, *J. Chem. Phys.*, **37**, 700 (1962); J. C. Walton, *J. Phys. Chem.*, **71**, 2763 (1963); A. D. Stepukhovich and V. A. Ultiskii, *Russ. Chem. Rev.*, **35**, 204 (1966).
31. J. Grotewold, E. A. Lissi, and M. G. Neumann, *J. Chem. Soc.* (*A*), 375 (1968); F. Casas, C. Previtali, J. Grotewold, and E. A. Lissi, *ibid.*, 1001 (1970).
32. J. Konstantatos and C. P. Quinn, *Trans. Faraday Soc.*, **65**, 2693 (1969).
33. M. Matsuoka, P. S. Dixon, A. P. Stefani, and M. Szwarc, *Proc. Chem. Soc.*, 304 (1962); P. S. Dixon, A. P. Stefani, and M. Szwarc, *J. Am. Chem. Soc.*, **85**, 2551 (1963).
34. R. Klein, M. D. Scheer, and R. Kelley, *J. Phys. Chem.*, **68**, 598 (1964).
35. D. G. L. James and R. D. Suart, *Trans. Faraday Soc.*, **65**, 175 (1969).
36. J. N. Bradley, *J. Chem. Phys.*, **35**, 748 (1961).
37. R. L. Thommarson, *J. Phys. Chem.*, **74**, 938 (1970).
38. H. S. Johnston, *Gas Phase Reaction Rate Theory*, Ronald Press, New York, 1966.
39. J. N. Bradley and B. S. Rabinovitch, *J. Chem. Phys.*, **36**, 3498 (1962).
40. O. K. Rice, *J. Phys. Chem.*, **65**, 1588 (1961).
41. R. A. Holroyd and G. W. Klein, *J. Phys. Chem.*, **67**, 2273 (1963); A. F. Trotman-Dickenson, *Proc. Chem. Soc.*, 249 (1964).
42. D. F. Dever and J. G. Calvert, *J. Am. Chem. Soc.*, **84**, 1362 (1962).
43. D. G. L. James and R. D. Suart, *Trans. Faraday Soc.*, **64**, 2735 (1968).
43a. R. S. Konar, *Int. J. Chem. Kinet.*, **2**, 419 (1970).
44. A. F. Trotman-Dickenson, *Gas Kinetics*, Butterworths, London, 1955.
45. A. F. Trotman-Dickenson, *Adv. Free Radical Chem.*, **1**, 1 (1965).
46. J. M. Tedder, *Q. Rev.*, **14**, 336 (1960).
47. A. F. Trotman-Dickenson, *Chem. Ind.*, 379 (1965).
48. J. A. Kerr, *Chem. Rev.*, **66**, 465 (1966).
49. J. A. Kerr and D. M. Timlin, *Int. J. Chem. Kinet.*, **3**, 69 (1971); T. N. Bell and A. E. Platt, *J. Phys. Chem.*, **75**, 603 (1971).
50. O. P. Strausz, E. Jakubowski, H. S. Sandhu, and H. E. Gunning, *J. Chem. Phys.*, **51**, 552 (1969); E. Jakubowski, H. S. Sandhu, H. E. Gunning, and O. P. Strausz, *ibid.*, **52**, 4242 (1970).
50a. T. N. Bell and A. E. Platt, *Chem. Commun.*, 325 (1970).
51. R. J. Cvetanović and R. S. Irwin, *J. Chem. Phys.*, **46**, 1694 (1967).
52. N. Yohoyama and R. K. Brinton, *Can. J. Chem.*, **47**, 2987 (1969).
53. J. M. Sangster and J. C. J. Thynne, *J. Phys. Chem.*, **73**, 2746 (1969).
53a. P. S. Dixon and M. Szwarc, *Trans. Faraday Soc.*, **59**, 112 (1963); J. M. Pearson and M. Szwarc, *ibid.*, **60**, 553 (1964); G. E. Owen, J. M. Pearson, and M. Szwarc, *ibid.*, **61**, 1722 (1965).
54. H. W. Sidebottom, J. M. Tedder, and J. C. Walton, *Trans. Faraday Soc.*, **65**, 755 (1969).
55. J. A. Kerr and M. J. Parsonage, *Int. J. Chem. Kinet.*, **4**, 243 (1972); H. W. Sidebottom, J. M. Tedder, and J. C. Walton, *ibid.*, **4**, 249 (1972).
56. J. A. Kerr and M. J. Parsonage, *Evaluated Kinetic Data on Gas Phase Addition Reactions: Reactions of Atoms and Radicals with Alkenes, Alkynes, and Aromatic Compounds*, Butterworths, London, 1972.
57. J. A. Kerr and A. C. Lloyd, *Q. Rev.*, **22**, 549 (1968).
58. R. J. Cvetanović, *Adv. Photochem.*, **1**, 115 (1963).
59. R. R. Getty, J. A. Kerr, and A. F. Trotman-Dickenson, *J. Chem. Soc.* (*A*), 979 (1967).
60. R. N. Haszeldine, K. Leedham, and R. B. Steele, *J. Chem. Soc.*, 2020 (1954).

61. J. E. Bloor, A. C. R. Brown, and D. G. L. James, *J. Phys. Chem.*, **70**, 2191 (1966).
62. L. F. Loucks and K. J. Laidler, *Can. J. Chem.*, **45**, 2767, 2795 (1967).
63. A. Kossiakoff and F. O. Rice, *J. Am. Chem. Soc.*, **65**, 590 (1943).
64. J. R. McNesby, C. M. Drew, and A. S. Gordon, *J. Chem. Phys.*, **24**, 1260 (1966); W. M. Jackson and J. R. McNesby, *J. Am. Chem. Soc.*, **83**, 4891 (1961); W. M. Jackson and J. R. McNesby, *J. Chem. Phys.*, **36**, 2272 (1962); W. M. Jackson, J. R. McNesby, and B. de B. Darwent, *ibid.*, **37**, 2256 (1962).
65. R. S. Konar, R. M. Marshall, and J. H. Purnell, *Trans. Faraday Soc.*, **64**, 405 (1968).
66. E. A. Hardwidge, C. W. Larson, and B. S. Rabinovitch, *J. Am. Chem. Soc.*, **92**, 3278 (1970) and references cited therein.
67. L. Endrenyi and D. J. LeRoy, *J. Phys. Chem.*, **70**, 4081 (1966).
68. K. W. Watkins and L. A. Ostreko, *J. Phys. Chem.*, **73**, 2080 (1969).
69. J. A. Garcia Dominguez and A. F. Trotman-Dickenson, *J. Chem. Soc.*, 940 (1962); R. R. Getty, J. A. Kerr, and A. F. Trotman-Dickenson, *ibid.*, A, 1360 (1967).

CHAPTER 2

Rate Constants for Free Radical Reactions in Solution

KEITH U. INGOLD
Division of Chemistry, National Research Council of Canada, Ottawa, Canada

I. INTRODUCTION

The rate constant, k_a, for the radical–molecule reaction

$$R\cdot + AB \xrightarrow{k_a} RA + B\cdot$$

is defined by the relation

$$\frac{-d[AB]}{dt} = \frac{d[RA]}{dt} = k_a[R\cdot][AB]$$

In order to determine k_a it is necessary to know the rate of reaction

$$(-d[AB]/dt = d[RA]/dt)$$

and the concentration of both reagents, R· and AB. Since most free radicals are highly reactive species they are generally present only in low concentrations, and this makes the measurement of rate constants for bimolecular reactions involving radicals a fairly difficult task. For unimolecular radical reactions,

$$R\cdot \xrightarrow{k_b} R'\cdot$$

the problems are less severe because the rate constant can be determined from measurements of the *relative* radical concentration at different times, that is,

$$\frac{-d[R\cdot]}{dt} = k_b[R\cdot]$$

which on integration from $t = 0$ gives

$$\log_e\left(\frac{[R\cdot]}{[R\cdot]_0}\right) = -k_b t$$

In contrast to the difficulty of measuring the absolute magnitude of bimolecular rate constants it is generally a fairly simple job to measure relative rate constants for radical–molecule reactions since the method of competing reactions eliminates the requirement for the radical concentration. This method can be illustrated by considering the generation of the radical in the presence of two reactants,

$$\mathrm{R\cdot + AB \xrightarrow{k_a} RA + B\cdot}$$

$$\mathrm{R\cdot + CD \xrightarrow{k_c} RC + D\cdot}$$

The relative rates of consumption of the two reagents (or of formation of products) and their relative concentrations serve to define k_a/k_c.

$$\frac{-d[\mathrm{AB}]/dt}{-d[\mathrm{CD}]/dt} = \frac{d[\mathrm{RA}]/dt}{d[\mathrm{RC}]/dt} = \frac{k_a[\mathrm{AB}]}{k_c[\mathrm{CD}]}$$

If R· is a highly reactive radical, then k_a and k_c will both be large and the radical will not differentiate as sharply between AB and CD as would a less reactive radical. That is, as a general rule the more reactive a radical, the less selective will it be in its reactions.

In recent years, the relative rate constants for the reactions of quite a wide variety of radicals with an even wider variety of substrates have been determined in the liquid phase. It is obvious that only one member of a set of relative rate constants need be measured in order to put the whole set on an absolute base. Such an absolute base is required if we wish to compare the reactivities of different radicals toward the same substrate rather than the reactivities of different substrates toward the same radical.

Before considering the different methods that have been used to measure rate constants for radical reactions it is appropriate to consider first just how fast a bimolecular reaction in solution can be.

II. DIFFUSION-CONTROLLED REACTIONS

When two species react with one another at every encounter[1] chemical change can take place only as fast as the reactants can diffuse together. Such a process is called a *diffusion-controlled* reaction. Second-order rate constants for such reactions can, in principle, be calculated from the laws of diffusion.[2] The observed bimolecular rate constant for a diffusion-controlled reaction is given by

$$k_{\mathrm{diff}} = (4\pi N/1000)(\sigma_{\mathrm{A}} + \sigma_{\mathrm{B}})(D_{\mathrm{A}} + D_{\mathrm{B}})$$

where D_{A} and D_{B} are the diffusion coefficients of the reactants A and B in the solvent medium, and σ_{A} and σ_{B} are molecular radii defined in terms of the distance across which reaction can take place between A and B. In the absence of diffusion coefficient data, it is common to use the Stokes–Einstein equation in the form

$$D = kT/6\pi\eta r$$

where η is the viscosity of the solution and r, the diffusion radii, is identified with the radii for reaction.[3] Combination of these two equations gives a simple relation between the rate constant and the solvent viscosity.

$$k_{\text{diff}} = 8RT/(3 \times 10^3 \eta)\ M^{-1}\ \text{sec}^{-1}$$

The solvents used in radical kinetic studies are normally of fairly low viscosity (e.g., cyclohexane, benzene, carbon tetrachloride, water, etc.) and the diffusion-controlled rate constants calculated from the abovementioned equation at ambient temperatures are generally approximately $8 \times 10^9\ M^{-1}\ \text{sec}^{-1}$.[4] Measured rate constants for bimolecular radical reactions may be as large as the diffusion-controlled limit but are more usually approximately $2 \times 10^9\ M^{-1}\ \text{sec}^{-1}$ or less. For radical–radical reactions an experimental rate constant of this magnitude raises the interesting question as to whether theory has overestimated k_{diff} or whether the theory is correct but the reaction only occurs in one-quarter of the encounters because only then will the total spin be in the singlet state—which is a necessity for reaction to take place. This raises the point as to whether spin inversion is fast or slow compared with the duration of a radical–radical encounter in solution. This is a subject of vigorous debate at the present time.[5]

III. RATE CONSTANTS FOR RADICAL–RADICAL REACTIONS

A. Experimental Methods

The normal procedure for determining an absolute rate constant for a radical–molecule reaction (k_a) or for a unimolecular radical reaction (k_b) is to measure the rate constant for the appropriate radical–radical reaction ($2k_t$)

$$\text{R}\cdot + \text{R}\cdot \xrightarrow{2k_t} \text{molecular products}$$

and then determine the rate constant ratio, $k_a/(2k_t)^{1/2}\ M^{-1/2}\ \text{sec}^{-1/2}$ or $k_b/(2k_t)^{1/2}\ M^{1/2}\ \text{sec}^{-1/2}$. Only occasionally can the rate constant for a radical–molecule reaction or for a unimolecular reaction be measured directly since it requires either that the radical be so stable (e.g., di-*t*-butyl nitroxide, 2,4,6-tri-*t*-butylphenoxy, diphenyl picryl hydrazyl, etc.) that its concentration can be determined readily, or special experimental techniques such as pulse radiolysis or flash photolysis.

Methods for measuring $2k_t$ can be divided simply according to whether a chain process with the appropriate termination step is, or is not, required. The principal methods requiring a chain are (1) the rotating sector (RS) technique and (2) the photochemical aftereffect (PA) technique. The principal methods of measuring $2k_t$ which may involve, but do not require, a chain are (3) steady-state electron spin resonance (SSESR) spectroscopy, (4) kinetic absorption spectroscopy (KAS), and (5) kinetic electron spin resonance (KESR) spectroscopy. These methods are described in this order below. One or two other methods have, on special occasions, been used (e.g., a flow technique for homogeneous polymerizations[6]) but they are of limited utility and are not included in the following discussion.

1. Rotating Sector

Classical methods (RS and PA) for measuring the rate constants of radical–radical reactions require a chain process that can be initiated photochemically or by ionizing radiation.[7] The rate of reaction should be readily measurable and it is preferable that it can be monitored continuously by some physical change in the system (e.g., temperature, volume, or pressure change). The RS method can best be illustrated by considering two chain reactions that have been used to measure rate constants for the self-reactions of alkyl radicals.

The overall reaction of thiols with triethyl phosphite can be represented by[8]

$$\mathrm{RSH} + \mathrm{P(OEt)_3} \longrightarrow \mathrm{RH} + \mathrm{SP(OEt)_3}$$

This reaction is a radical chain process[9] and Burkhart[10] has shown that, for the photoinitiated reaction, when the molar ratio of triethyl phosphite to thiol is sufficiently large, the chain can be represented by

Initiation:

$$\text{initiator} \xrightarrow{h\nu} \mathrm{R\cdot}$$

Propagation:

$$\mathrm{R\cdot} + \mathrm{RSH} \xrightarrow{(k_p)_{\mathrm{SH}}} \mathrm{RH} + \mathrm{RS\cdot}$$

$$\mathrm{RS\cdot} + \mathrm{P(OEt)_3} \xrightarrow{\text{fast}} \mathrm{R\cdot} + \mathrm{SP(OEt)_3}$$

Termination:

$$\mathrm{R\cdot} + \mathrm{R\cdot} \xrightarrow{2k_t} \text{molecular products}$$

The rate of thiol disappearance is given by

$$\frac{-d[\mathrm{RSH}]}{dt} = \frac{(k_p)_{\mathrm{SH}}[\mathrm{RSH}]R_i^{1/2}}{(2k_t)^{1/2}}$$

where R_i is the rate of chain initiation.

The reaction of alkyl halides with trialkyltin hydrides is also a radical chain process.[11]

$$\mathrm{RX} + \mathrm{R'_3SnH} \longrightarrow \mathrm{RH} + \mathrm{R'_3SnX}$$

Carlsson and Ingold[12] have shown that with alkyl bromides and iodides the chain can be represented by

Initiation:

$$\text{initiator} \xrightarrow{h\nu} \mathrm{R\cdot}$$

Propagation:

$$\mathrm{R\cdot + R'_3SnH \xrightarrow{(k_p)_{SnH}} RH + R'_3Sn\cdot}$$

$$\mathrm{R'_3Sn\cdot + RX \xrightarrow{fast} R'_3SnX + R\cdot}$$

Termination:

$$\mathrm{R\cdot + R\cdot} \xrightarrow{2k_t} \text{molecular products}$$

$$\frac{-d[\mathrm{RX}]}{dt} = \frac{(k_p)_{\mathrm{SnH}}[\mathrm{R'_3SnH}]R_i^{1/2}}{(2k_t)^{1/2}}$$

With alkyl chlorides the chain is normally terminated by the self-reaction of trialkyltin radicals.[12]

The rates of both the thiol–phosphite and halide–tin hydride reactions are proportional to the square root of the rates of chain initiation; that is, the overall rates are proportional to the square root of the light intensity. This makes these reactions suitable for the application of the rotating sector technique.

The theory behind the RS method for separating propagation and bimolecular termination rate constants has been clearly reviewed by Burnett and Melville[13] and the following is merely a brief resumé.

The reaction rate under steady photolysis is proportional to the square root of the light intensity, I.

$$R_{\mathrm{steady}} = KI^{1/2}$$

If the light is flashed on the reaction cell intermittently, the rate must decrease. Suppose the ratio of dark to light periods is 3:1 as would arise if a solid disk with a 90° sector cut out of it was rotated between the light and the reaction vessel. If this sectored disk is rotated very slowly, then the rate during the light periods will be R_{steady} and during the dark periods will be zero, that is,

$$R_{\mathrm{slow}} = (KI^{1/2})/4 = R_{\mathrm{steady}}/4$$

If the disk is rotated sufficiently rapidly, the effect of the flickering light must be the same as that due to a steady light with an intensity equal to $I/4$. That is,

$$R_{\mathrm{fast}} = K(I/4)^{1/2} = R_{\mathrm{steady}}/2$$

At some intermediate speeds of rotation the rate must vary with the rotation speed. The duration of the light flash, λ, in this intermediate region is related to the average lifetime, τ, of the chains, where τ is defined by the equation

$$\tau = (2k_tR_i)^{-1/2}$$

The intermediate rates, denoted by R_λ, have a complicated dependence on λ and τ. Values of $R_\lambda/R_{\mathrm{steady}}$ have been tabulated in terms of λ/τ by Burnett and

Melville.[13,14] Since λ is known from the rotation speed, τ is readily obtained and this gives $2k_t$ once R_i is known. Substitution of these quantities into the steady-state rate expression yields k_p.

The rate constants for the self-reactions of alkyl radicals (i.e., the termination rate constants) reported by Burkhart[10] and by Carlsson and Ingold[12] are in reasonable agreement. They are listed in Table 1 together with selected data for similar reactions that have been obtained by other workers using KAS and ESR methods principally. This table does not, by any means, list all the data available. However, it serves to illustrate the fact that in solvents of normal viscosity, $2k_t$ values for simple alkyl radicals lie in the range 10^8–10^{10} M^{-1} sec^{-1} and are commonly approximately 2×10^9 M^{-1} sec^{-1}. A relatively complete, but uncritical, list of reported rate constants for alkyl radical

TABLE 1

Some Selected Rate Constants for Alkyl Radical Self-Reactions at 25°

		$2k_t \times 10^{-9}$ (M^{-1} sec^{-1})		
Radical	Method	In cyclohexane	In benzene	Ref.
Methyl	RS	8.9	—	12
Methyl	KESR	11[a,b]	—	18
Ethyl	KESR	6.4[c]	—	19
n-Propyl	RS	3.4	0.42	10
n-Pentyl	RS	—	0.86	10
n-Hexyl	RS	2.2	—	12
neo-Pentyl	KESR	4.0[a]	—	18
Cyclopentyl	KESR	0.4[d]	—	20
Cyclohexyl	RS	1.2	0.72	10
Cyclohexyl	RS	2.7	—	12
Cyclohexyl	KESR	2.5[e]	—	20
Cyclohexyl	KAS[f]	2.5	—	21
t-Butyl	RS	2.1	—	12
t-Butyl	KESR	4.4	1.5	17
t-Butyl	KESR	8.1[a]	—	18
Benzyl	RS	4.0	3.6	10
Benzyl	KAS[f]	2.0	—	22
Cumyl	KESR	16	16	17

[a] In di-*t*-butylperoxide.

[b] At −17°.

[c] Extrapolated from low temperature (−140 to −177°) data in liquid ethane using the relation $2k_t = 2.6 \times 10^{10}\, e^{-830/RT}$ M^{-1} sec^{-1}.

[d] In cyclopentane at −50°.

[e] At 10°.

[f] Radicals generated by pulse radiolysis.

self-reactions covering the literature to mid-1968 has been given by Denisov.[16] However, with a few notable exceptions the early work has tended to underestimate $2k_t$ values quite considerably. For instance, early work[16] on the cumyl radical suggested that $2k_t$ was approximately $5 \times 10^6\ M^{-1}\ \text{sec}^{-1}$ whereas later work[17] has indicated a value of $1.6 \times 10^{10}\ M^{-1}\ \text{sec}^{-1}$ which is greater than for any other simple alkyl radical.[18–22]

2. *Photochemical Aftereffect Method*

Strictly speaking a photochemical aftereffect refers to some property of an illuminated system which can be observed when the illumination is interrupted. Provided the radicals are generated photochemically both the KAS and KESR methods are special cases of the PA method. However, it is more usual to restrict the term to experiments in which the overall rate of a photoinitiated chain reaction is followed when the illumination is cut off. The radicals take a finite time to decay and so the rate does not fall to zero (or to its dark value) instantly. The decay of the radicals is governed by the relation

$$\frac{-d[\text{R}\cdot]}{dt} = 2k_t[\text{R}\cdot]^2$$

If the light is cut off ($t = 0$) when the system is at a steady state, the radical concentration at time t, $[\text{R}\cdot]_t$, is given by

$$\frac{1}{[\text{R}\cdot]_t} - \frac{1}{[\text{R}\cdot]_{\text{steady}}} = -2k_t t$$

Since the average chain lifetime under steady-state conditions is given by

$$\tau = (2k_t R_i)^{-1/2} = (2k_t[\text{R}\cdot]_{\text{steady}})^{-1}$$

the expression for the radical concentration at time t can be written as

$$\frac{[\text{R}\cdot]_{\text{steady}}}{[\text{R}\cdot]_t} = 1 - \frac{t}{\tau}$$

that is,

$$\frac{R_{\text{steady}}}{R_t} = 1 - \frac{t}{\tau}$$

A plot of R_{steady}/R_t against time will yield a straight line of slope $1/\tau$.

Although the theory of the PA method for measuring $2k_t$ is simple, the experimental difficulties are considerable if the reaction has a termination rate constant which is much above $10^5\ M^{-1}\ \text{sec}^{-1}$. With small values of $2k_t$ the lifetime becomes sufficiently long for an accurate set of rate measurements to be made before reaction ceases. The PA method has proved particularly useful for studying autoxidations in which chains are terminated by the self-

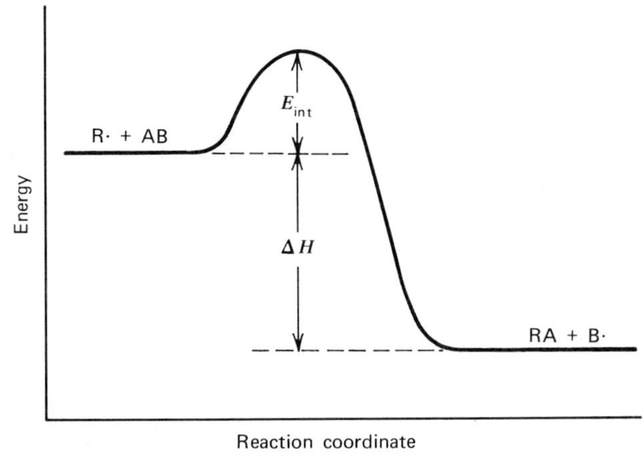

reaction of *t*-peroxy radicals for which $2k_t$ values are in the range 1×10^3 to 3×10^4 M^{-1} sec^{-1}.[23–25] The rate of the autoxidation is followed by the oxygen consumption.

A special case of the PA method which is particularly applicable to autoxidations is known as the chemiluminescence method. A small amount of light is emitted during the self-reaction of many peroxy radicals[26]

$$2ROO\cdot \longrightarrow \text{molecular products} + h\nu$$

Since the amount of light emitted is related to the peroxy radical concentration, it provides a ready method for monitoring the decay of the peroxy radicals when the initiating light is switched off. This method has been largely developed by Russian workers[26] and it appears to yield fairly accurate rate constants.[25] A number of variations of the basic technique have been developed,[26] including some that are not PA methods at all since they do not require that the oxidation be photoinitiated.

The PA method has been applied to other reactions also such as vinyl polymerizations and the addition of halocarbons to olefins,[27,28] for example,

$$\cdot CCl_3 + RCH{=}CH_2 \longrightarrow R\dot{C}HCH_2CCl_3$$

$$R\dot{C}HCH_2CCl_3 + CCl_4 \longrightarrow RCHClCH_2CCl_3 + \cdot CCl_3$$

Unfortunately, in most cases the reactions studied have had rather high termination rate constants and although many ingenious approaches have been developed to follow the reaction during the first few seconds immediately after the cessation of irradiation,[7] it would seem that the required data could generally have been obtained more easily, and more accurately, by the RS method.

3. *Steady-State Electron Spin Resonance*

In a radical reaction under steady-state conditions the concentration of the radicals which are undergoing the bimolecular self-reaction is given by

$$[\mathrm{R}\cdot]_{\mathrm{steady}} = (R_i/2k_t)^{1/2}$$

provided that this self-reaction is the only means by which radicals of all types are lost from the system. The minimum concentration of radicals that can be measured readily by ESR spectroscopy is about 10^{-7} *M*. For a diffusion-controlled termination, therefore, the steady-state concentration of radicals should be measurable if $R_i \geqslant \sim 10^{-5}$ *M* $\sec^{-1}$.

In spite of the apparent simplicity of this method for measuring $2k_t$ it has not often been applied, principally because of the difficulty of determining R_i with sufficient precision. The first successful application of this method to a diffusion-controlled termination was Fessenden and Schuler's[29] radiolysis of liquid ethane with 2.8 MeV electrons. Dose rate measurements and the absolute ethyl radical concentration gave $2k_t = 6 \times 10^8$ M^{-1} $\sec^{-1}$ at $-175°$. This value was later confirmed by Fessenden[19] using other ESR techniques. By way of contrast, the thermal decomposition of azobisisobutyronitrile (AIBN) at 120–130° gave an apparent $2k_t$ value for 2-methyl-2-cyanopropyl radicals of only 2×10^8 M^{-1} $\sec^{-1}$.[30]

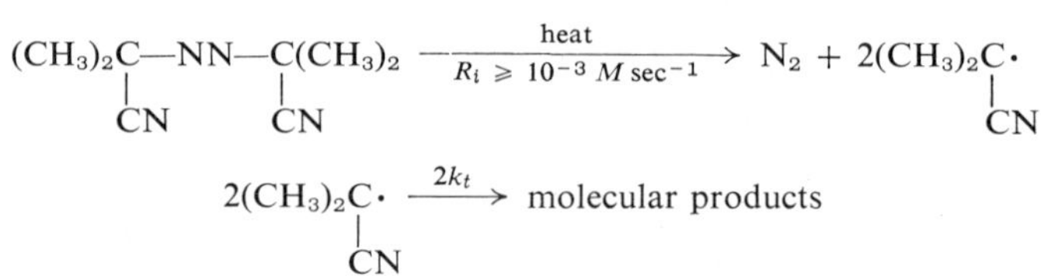

This value of $2k_t$ is an order of magnitude less than the currently accepted value of 2×10^9 M^{-1} $\sec^{-1}$ (in benzene) determined by Weiner and Hammond using the kinetic ESR method.[17,31] The low value can probably be ascribed to error in the estimated rate of radical formation from the azo compound.

The SSESR method has also been applied successfully to chain processes terminated by relatively slow radical–radical reactions. Thus, in 1961 Lebedev and co-workers[32] measured the concentration of the cumylperoxy radicals present during the azo-isobutyronitrile-initiated autoxidation of cumene. In the temperature range 70–90° the measured concentrations were in good agreement with those calculated from the known values of R_i and a value of $2k_t$ determined in 1954 by Melville and Richards[33] using the RS method. The agreement between theory and experiment has since been confirmed by Thomas[34] who reports that at 74° with 0.05 *M* [AIBN], the calculated peroxy radical concentration is 6.9×10^{-6} *M* and the measured concentration is 4.1×10^{-6} *M*. Later work has shown that this small discrepancy is due to an error in the value taken for $2k_t$ rather than to an error in the ESR work.

An entirely different SSESR method which yields relative rate constants for the self-reactions of two different radicals has been developed by Weiner.[35] This method depends for its operation on the generation of the two radicals in the same reaction step. It has proved particularly useful for measuring $2k_t$ values for ketyl radicals relative to the $2k_t$ values of radicals that can be formed by hydrogen loss to ketones in their triplet state, for example phenoxy radicals. The principle of the method is illustrated by the following series of equations:

$$>\!C{=}O \xrightarrow{h\nu} >\!C{=}O^{*1} \longrightarrow >\!C{=}O^{*3}$$

$$>\!C{=}O^{*3} + ArOH \longrightarrow >\!\dot{C}OH + ArO\cdot \quad (\text{rate} = R_i)$$

$$\left.\begin{array}{l} >\!\dot{C}OH + >\!\dot{C}OH \xrightarrow{(2k_t)_a} \\ >\!\dot{C}OH + ArO\cdot \xrightarrow{(2k_t)_b} \\ ArO\cdot + ArO\cdot \xrightarrow{(2k_t)_c} \end{array}\right\} \text{molecular products}$$

Under steady illumination the concentration of $>\!\dot{C}OH$ and ArO· will reach a steady state, that is,

$$\frac{d[>\!\dot{C}OH]}{dt} = 0 = R_i - (2k_t)_a[>\!\dot{C}OH]^2 - (2k_t)_b[>\!\dot{C}OH][ArO\cdot]$$

$$\frac{-d[ArO\cdot]}{dt} = 0 = R_i - (2k_t)_c[ArO\cdot]^2 - (2k_t)_b[>\!\dot{C}OH][ArO\cdot]$$

Hence,

$$\frac{[>\!\dot{C}OH]^2}{[ArO\cdot]^2} = \frac{(2k_t)_c}{(2k_t)_a}$$

This rate constant ratio is obtained simply by measuring the relative concentrations of ketyl and phenoxy radicals under steady illumination. Since $2k_t$ values for a number of ketyl radicals and semidione radicals (α-keto-ketyl radicals) have already been determined by KESR spectroscopy,[36–38] the

Weiner SSESR technique should prove of considerable value. It has already been applied successfully to a number of phenoxyl radicals.[35]

4. Kinetic Absorption Spectroscopy

Optical absorption spectroscopy can be used to follow the decay of radicals generated by a flash of light (flash photolysis) or a pulse of ionizing radiation (pulse radiolysis). The method was first used by Porter and Windsor[39] in 1957 to study short-lived alkyl radicals in solution. The flash photolysis of 10^{-2} *M* solutions of benzyl chloride in viscous paraffin resulted in the appearance of a transient absorption with maxima at 3175 and 3060 Å. An identical spectrum was obtained from benzyl formate, and the same bands, with additional absorption due to other species, appeared after photolysis of benzyl alcohol and benzyl methyl ether. The spectra were therefore assigned to the benzyl radical. The spectrum could be detected in hexane but the decay was too rapid for kinetic study with the techniques available at that time. In the paraffin with a viscosity of 163 cP, decay was second order with a rate constant of $7.8 \times 10^4 \ \varepsilon \ M^{-1} \ \text{sec}^{-1}$ where ε is the extinction coefficient at the 3175 Å maximum.

This report illustrates the two major problems associated with the KAS method. The first difficulty is to identify the transient intermediate, and this may require studies on a great many compounds of generally similar structure. The second difficulty is to determine the extinction coefficient of the radical, which is a necessity if true rate constants are to be obtained. In spite of these problems, the KAS method has been used by a number of workers with great success, particularly for radicals generated by pulse radiolysis.[40] Some data obtained by this method are listed in Table 1.

5. Kinetic Electron Spin Resonance

The problems associated with the KAS method can be overcome by using electron spin resonance spectroscopy to monitor the radicals since this allows not only for the unequivocal identification of the radicals but also gives a simple and straightforward measure of their concentration. The KESR method was used first by Piette and Landgraf[41] in a study of the relatively long-lived radicals present in UV-irradiated alkyl hydroperoxides. These are peroxy radicals rather than alkoxy radicals as was originally suggested. The first application of the KESR method to short-lived alkyl radicals was Fessenden's[19] study of ethyl radicals in liquid ethane. The output from a 2.8-MeV electron Van de Graaff generator was pulsed and the radical concentration was measured by a special sampling technique at various times after the end of a series of identical pulses. In this way a complete decay curve was constructed yielding $2k_t = 3.6 \times 10^8 \ M^{-1} \ \text{sec}^{-1}$ at $-177°$. A rather similar value ($3.1 \times 10^8 \ M^{-1} \ \text{sec}^{-1}$) was obtained by a variation of the rotating sector method in which the radical concentration was monitored over a range of pulse durations; from -140 to $-177°$, $2k_t = 2.6 \times 10^{10} e^{-830/RT}\ M^{-1}\ \text{sec}^{-1}$.

In spite of Fessenden's successful combination of pulse radiolysis with KESR spectroscopy his technique has rarely been used by other workers (see, however, reference 20) because of the complicated nature of the radical-generating equipment. Fortunately, even radicals that decay with a diffusion-controlled rate constant can be generated photolytically in sufficiently high concentrations for the KESR method to be used. This was first demonstrated by Weiner and Hammond[31] for 1-cyanoalkyl radicals produced by photolysis of azobisalkylnitriles. The experimental technique involved chopping the light beam that fell on a solution of the azo compound in the ESR spectrometer cavity. The light chopping was done with a rotating sector which also triggered a computer of average transients (CAT) to which was fed the voltage from the ESR spectrometer. At successive time intervals this voltage is stored in a set of memories. A large number of identical decays are sampled and the results for the various time intervals are added to the appropriate memories. The process is continued until sufficient data are accumulated to provide a satisfactory signal-to-noise ratio. This general technique for monitoring fairly rapid changes in radical concentration was first applied by Piette and co-workers.[42]

The usefulness of the photolytic KESR method depends on the ability to generate the desired radical in sufficient concentration by photolysis of readily available compounds. The principal direct sources of alkyl radicals are azo compounds.[17,31]

$$\text{RNNR} \xrightarrow{h\nu} 2\text{R}\cdot + \text{N}_2$$

and acyl peroxides[43,44]

$$\underset{\substack{\| \\ \text{O}}}{\text{RC}}\text{OO}\underset{\substack{\| \\ \text{O}}}{\text{CR}} \xrightarrow{h\nu} 2\text{R}\cdot + 2\text{CO}_2$$

The most useful indirect method for generating alkyl radicals is to photolyze di-*t*-butylperoxide[43,45–51] or hydrogen peroxide[52] in the presence of a suitable substrate, for example,

$$(\text{CH}_3)_3\text{CO}\cdot + (\text{CH}_3)_3\text{CH} \longrightarrow (\text{CH}_3)_3\text{COH} + (\text{CH}_3)_3\text{C}\cdot$$

$$(\text{CH}_3)_3\text{CO}\cdot + (\text{CH}_3\text{CH}_2\text{CH}_2\text{CH}_2)_3\text{B} \longrightarrow (\text{CH}_3)_3\text{COB}(\text{CH}_2\text{CH}_2\text{CH}_2\text{CH}_3)_2 + \text{CH}_3\text{CH}_2\text{CH}_2\text{CH}_2\cdot$$

Other photolytic methods suitable for radical generation for the KESR method also exist.

Another variation of the KESR technique is the flow method in which radicals are generated in a flowing stream of liquid just outside the spectrometer cavity. If the flow of the solution is stopped suddenly, the decay of the radicals can be followed. The radicals may be generated in a single stream, for example, phenoxy radicals from phenols in a solution passed over a plug of silver oxide, or lead dioxide,[53]

$$\text{ArOH} \xrightarrow[\text{PbO}_2]{\text{AgO or}} \text{ArO}\cdot$$

Alternatively, two streams of reagent may be mixed, for example, ceric ammonium sulfate (or nitrate) with phenols[54] or hydroperoxides[55–57]

$$\text{ROOH} \xrightarrow{\text{Ce}^{\text{IV}}} \text{ROO}\cdot$$

or titanium(III) chloride with hydrogen peroxide and an alcohol.[58,59]

$$\text{H}_2\text{O}_2 \xrightarrow{\text{Ti}^{\text{III}}} \text{HO}\cdot \xrightarrow{\text{CH}_3\text{OH}} \dot{\text{C}}\text{H}_2\text{OH}$$

For stop-flow kinetic studies it is necessary that the radicals be formed entirely outside the spectrometer. This puts a practical upper limit of about 10^6–10^7 M^{-1} sec^{-1} on the decay rate constants that can be measured by this method.

B. Rate Constants for Bimolecular Self-Reactions of Carbon-Centered Radicals

As we have already seen (Table 1), small alkyl radicals, including resonance-stabilized radicals such as benzyl, react with one another at ambient temperatures in solvents of normal viscosity with rate constants that are generally approximately 2×10^9 M^{-1} sec^{-1}. Small differences in apparent rate constants between alkyl radicals are probably not particularly significant since the accuracy of any current experimental method is not likely to be greater than about $\pm 50\%$.

The rate constants listed in Table 1 refer to the overall reaction

$$\text{R}^\cdot + \text{R}^\cdot \xrightarrow{2kt} \text{molecular products}$$

This process may be a radical combination, for example,

$$\text{CH}_3\cdot + \text{CH}_3\cdot \longrightarrow \text{C}_2\text{H}_6$$
$$\text{C}_6\text{H}_5\text{CH}_2\cdot + \text{C}_6\text{H}_5\text{CH}_2\cdot \longrightarrow \text{C}_6\text{H}_5\text{CH}_2\text{CH}_2\text{C}_6\text{H}_5$$

More usually, the reaction involves both combination and disproportionation, for example,

$$\text{C}_2\text{H}_5\cdot + \text{C}_2\text{H}_5\cdot \begin{cases} \xrightarrow{k_c} \text{C}_4\text{H}_{10} \\ \xrightarrow[k_d]{} \text{C}_2\text{H}_4 + \text{C}_2\text{H}_6 \end{cases}$$

There is relatively little change in k_d/k_c ratios with solvent or temperature.[60–65] For the ethyl radical at 0°, the k_d/k_c ratio is 0.130 in the gas phase, 0.177 in isooctane, 0.185 in toluene, 0.202 in isopropanol, and 0.26–0.27 in ethylene glycol.[60] This increase in k_d/k_c has been attributed to the increase in internal pressure, P_s, of the solvent, the volume of the transition state in the disproportionation being somewhat smaller than in combination, $V_c^{\ddagger} - V_d^{\ddagger} \sim 2.4 - 2.8$ cm^3/mole at 0°.[60] The relation between k_d/k_c and P_s is[66,66a]

$$\log(k_d/k_c) = -1.08 + (4.76 \times 10^{-3})P_s^{1/2}$$

For the ethyl radical the extent of disproportionation increases very slightly with decreasing temperature both in the gas phase and in solution. The temperature effect can be represented as[60] $k_d/k_c \propto T^{-0.7}$, or[62] $k_d/k_c = 0.087 \exp(371/RT)$, or[66a] $k_d/k_c = 5 \exp(290/RT)$.

The two competing processes *may* pass through a common set of intermediates of the type implied by the equilibrium of the two partially bonded subspecies shown below.[66]

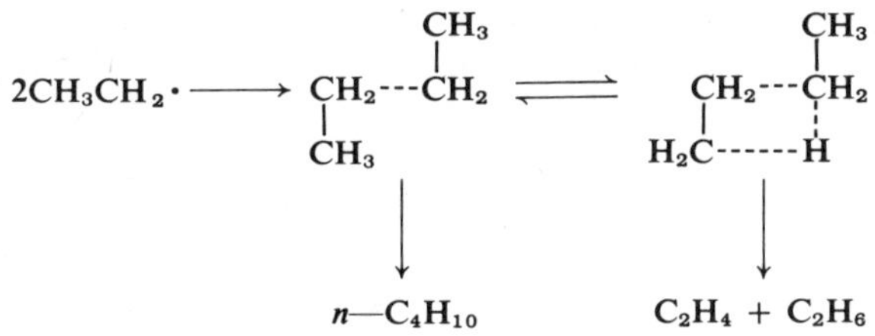

However, it seems more probable that combination and disproportionation proceed by way of different transition states.[66a]

The extent of disproportionation of simple alkyl radicals increases with an increase in the number of β-hydrogens available for transfer. Typical values of k_d/k_c are, 1.1 cyclohexyl,[67] 0.14 *n*-propyl,[65] 1.2 isopropyl,[65] and 7.2 *t*-butyl.[65] These and other values of k_d/k_c can be normalized by dividing by the number of β-hydrogen atoms available for disproportionation (i.e., 6 for two ethyl radicals, 18 for two *t*-butyl radicals). These normalized ratios (k_d/k_c per β-hydrogen) are 0.03, 0.1, and 0.4 for primary, secondary, and tertiary alkyl radicals,[65] which indicates that the factors affecting disproportionation are more than simple statistical ones.[68] This suggests that orientation of the radicals may influence combination, that is, for a large radical rotation may be slow relative to the duration of an encounter, so that with a radical such as *t*-butyl relatively few encounters are correctly oriented for combination (see also references 69 and 70). On the other hand, disproportionation is much less favored relative to combination when the radical is stabilized by delocalization of the unpaired electron. Thus, k_d/k_c per β-hydrogen is 0.023 for the secondary 1-phenylethyl radical,[71] and for tertiary radicals values of

0.005 for cumyl,[72] 0.005 for several (but not all) parasubstituted cumyl radicals,[73] and 0.008 for 2-cyano-2-propyl radicals[74] have been reported. It is clear that these k_d/k_c ratios cannot be explained simply on the basis of radical orientation during encounters. Additional work, including studies on dissimilar radical pairs,[75] will be required before we can properly understand k_d/k_c ratios.

Of course, not all carbon radicals have $2k_t$ values of approximately 10^8 to 10^{10} M^{-1} sec^{-1}. At the opposite extreme are stable[76] carbon radicals such as tri-*p*-nitrophenylmethyl [$(p\text{-}NO_2C_6H_4)_3C\cdot$] and tetraphenylcyclopentadienyl

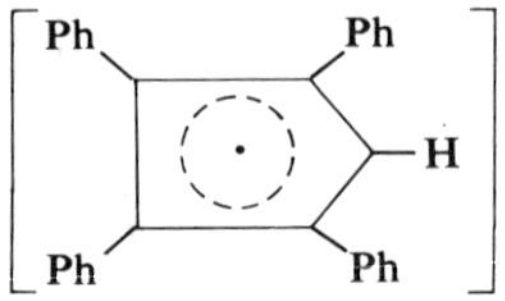

for which $2k_t$ is zero because they do not yield thermodynamically stable dimers or disproportionation products.[77,78] Other "stable" radicals exist in equilibrium with molecular products, the best known example being triphenylmethyl.[77–80]

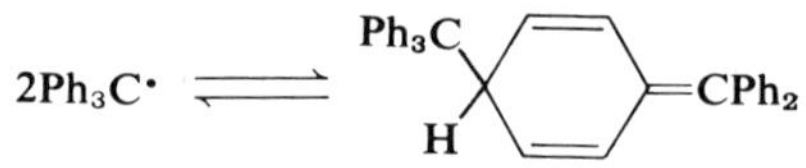

Ziegler and co-workers[81] studied the kinetics of the dissociation of "hexaphenylethane" by the rate of oxygen and nitric oxide absorption. The activation energy for decomposition was 19 ± 1 kcal/mole, which is considerably higher than its heat of dissociation of 11–12 kcal/mole. This means that the combination of triphenylmethyl radicals must proceed with an activation energy of about 7 kcal/mole. This conclusion has been verified by D'iachkovskii et al.[82] by electron spin resonance. These workers report that in toluene the equilibrium constant is given by

$$K = 8.8 \times 10^4 \exp(-11200/RT)\ M$$

and the rate constant for radical decay by

$$k = 3.9 \times 10^7 \exp(-6950/RT)\ M^{-1}\ \text{sec}^{-1}$$

Of equal interest are carbon radicals that dimerize to give a stable product rather slowly because of a high entropy of activation rather than because of a high energy of activation. That is, there are radicals for which the duration of an average encounter in a normal solvent is not sufficient to ensure that a configuration which would permit reaction is achieved. Thus, the radicals diffuse apart without reaction even though there is no real potential energy barrier to be crossed. A simple example of such a radical is tri-*t*-butylcycloheptatrienyl, the dimerization of which in pentane can be represented[83] over a temperature range from -112 to $+22°$ by

$$2k_t = 5.4 \times 10^6 \exp(-1700/RT)\ M^{-1}\ \text{sec}^{-1}$$
$$= 3 \times 10^5\ M^{-1}\ \text{sec}^{-1} \qquad \text{at } 22°$$

The small value of $2k_t$ is due principally to the small preexponential factor for the dimerization.

There are other relatively simple alkyl radicals that have $2k_t$ values significantly smaller than the diffusion limit, for example, trichloromethyl for which $2k_t$ is only approximately $10^8\ M^{-1}\ \text{sec}^{-1}$ [16,18,84] (Table 2). Since this radical is no more hindered than a *t*-butyl radical and since it is not resonance stabilized to any great extent, this low value must be ascribed to polar factors. To form C_2Cl_6 requires that the positive ends of two dipoles, $\overset{\delta-}{Cl_3}\overset{\delta+}{C}\cdot$, be forced together. This must introduce some barrier that hinders the combination.

Not all carbon radicals containing heteroatoms in the α position have small $2k_t$ values (see Table 2). For example, even in water the 1-hydroxyethyl radical has $2k_t \approx 2 \times 10^9\ M^{-1}\ \text{sec}^{-1}$.[86–88] Since this radical must interact strongly with the solvent it is clear that such an interaction does not necessarily reduce $2k_t$. This fact suggests that any report that a $2k_t$ value for a particular alkyl radical varies significantly from one solvent to another *of similar viscosity* should be treated with some caution. Unsubstituted alkyl radicals are not expected to interact very strongly with solvents (except perhaps with aromatic solvents) because they are relatively nonpolar species. Therefore $2k_t$ is unlikely to change significantly with solvent unless the viscosity also changes.[89] Thus, Sauer and Mani[93] have reported that $2k_t$ for cyclohexyl radicals is $2 \times 10^9\ M^{-1}\ \text{sec}^{-1}$ in cyclohexane and 1.4×10^9 in water, while for hexyl radicals $2k_t$ is 3.1×10^9 in *n*-hexane and 1.2×10^9 in water. This does not imply that solvent effects will also be negligible when the unpaired electron is itself centered on a heteroatom. Very large solvent effects have been observed in this case, for example, for the HOO· radical in hydrocarbon solvents,[94] $2k_t \sim 10^9\ M^{-1}\ \text{sec}^{-1}$ but in water $2k_t \sim 10^6\ M^{-1}\ \text{sec}^{-1}$.[95] Solvent effects on the reactivity of other peroxy radicals have also been reported.[25]

Other examples of relatively small alkyl radicals with low $2k_t$ values include certain α-hydroxyalkyl (ketyl) radicals[35,38,96,97] and certain semidione radicals[36,37] (see Table 2).[97a]

Since α substituents have relatively little effect on $2k_t$ we would not expect more remote substituents to have any effect whatever. However, fairly significant changes in $2k_t$ have been produced by the addition of parasubstituents

TABLE 2

Some Selected Rate Constants for the Self-Reactions of α-Substituted Carbon Radicals at ~25°

Radical	Method	Solvent	$2k_t \times 10^{-9}$ (M^{-1} sec^{-1})	Ref.
$Cl_3C\cdot$	Various	Various	0.1	16
$Cl_3C\cdot$	KESR	*t*-Butyl peroxide	0.1	18
$(CH_3)_2\dot{C}CN$	SSESR	Cyclohexane	8	17
$(CH_3)_2\dot{C}CN$	KESR	Benzene	1.9	17
$\overline{CH_2(CH_2)_4\dot{C}}CN$	KESR	Benzene	3.6	17
$CH_3\dot{C}HOH$	KAS[a]	Water	1.4	86, 87
$CH_3\dot{C}HOH$	RS	Water	2.0	88
$\overline{CH_2(CH_2)_4\dot{C}}OH$	KAS[a]	Cyclohexanol	0.34	96
$(C_6H_5)_2\dot{C}OH$	KAS[b]	Isopropanol	0.059	97
$(C_6H_5)_2\dot{C}OH$	KESR	Benzene	3.6	35, 38
$(C_6H_5)_2\dot{C}OH$	KESR	Isopropanol	0.22	35, 38
$CH_3CO\dot{C}(OH)CH_3$	KESR	Isopropanol	0.78	36
$C_6H_5CO\dot{C}(OH)C_6H_5$	KESR	Isopropanol	0.66	36
$O{=}\overline{CC(CH_3)_2CH_2C(CH_3)_2\dot{C}}OH$	KESR	Isopropanol	0.03	36
$CH_3CO\dot{C}(OH)OC_2H_5$	KESR	Isopropanol	0.60	37
$CH_3CO\dot{C}(OH)OH$	KESR	Isopropanol	0.38	37
$(CH_3)_3SiCH_2\cdot$	KESR	*t*-Butyl peroxide	4.6	18
$(CH_3)_3GeCH_2\cdot$	KESR	*t*-Butyl peroxide	2.5	18
$(CH_3)_3SnCH_2\cdot$	KESR	*t*-Butyl peroxide	1.9[c]	18

[a] Radicals generated by pulse radiolysis.
[b] Radicals generated by flash photolysis.
[c] At −40°.

to cumyl radicals[17] and to benzophenone ketyl radicals.[38] The reason for these substituent effects is currently unknown.

The largest group of carbon radicals for which $2k_t$ values are significantly less than 10^9 M^{-1} sec^{-1} is comprised of the macroradicals which are formed during the polymerization of vinyl monomers. The available rate constants for free radical polymerizations have been summarized several times.[7,16,98,99] The data reported by different workers often show a wide variation, which is perhaps not too surprising as the chain termination rate constants depend both on radical chain length and solution viscosity (see below). Some typical $2k_t$ values (in M^{-1} sec^{-1} units) for polymerizing vinyl monomers at ambient temperatures are as follows: styrene, 5×10^7; vinyl acetate, 3×10^7; methyl methacrylate, 2.5×10^7; *n*-butyl methacrylate (bulk), 6.5×10^5; methyl acrylate, 5×10^6; and acrylonitrile 1.8×10^6. However, because of the dependence of $2k_t$ on chain length and viscosity these rate constants have no intrinsic significance.

Although the chain termination rate constants for vinyl polymerization are all much less than 10^9 M^{-1} sec^{-1} it would appear that they are still "diffusion controlled."[6] That is, the rate of diffusion of a macroradical must be less than that of a small alkyl radical, and even when the two macroradicals are joined in an encounter, chemical reaction requires segmental diffusion of the active centers on the macromolecular chains to within the separation distance of a few angstroms required for chemical reaction. The chain-length dependence of the rate constant for such a reaction is quite large.[6] Thus, it has been predicted that a radical with a chain length of 100 will have a $2k_t$ value that is only one-fifth of the rate constant for the reaction of this radical with a radical of unit chain length. Such behavior has been observed in the low-temperature polymerization of methyl methacrylate.[100]

Evidence that the termination reactions in the solution polymerization of methyl methacrylate, other alkyl methacrylates, and vinyl acetate are diffusion controlled under all experimental conditions is provided by the fact that $2k_t$ values are inversely proportional to the viscosity down to the lowest attainable viscosities.[101,102] This important point was not realized prior to 1958, it being assumed that termination only became diffusion controlled when sufficient polymer had formed to raise the viscosity to a certain level. Activation energies for termination are generally approximately 3 kcal/mole or less, which is close to the activation energy for diffusion.[6,7] Furthermore, the calculated diffusion-controlled rate constants for macroradicals of the size present in typical polymerization experiments are found to be about 10^7 M^{-1} sec^{-1}.[6]

Several studies have shown that in polymerizations $2k_t$ decreases as the degree of conversion increases.[103] A detailed examination of methyl methacrylate polymerization[102] has shown that in the initial stages $2k_t$ increases slightly with increasing conversion and then undergoes a sharp decrease when the viscosity of the system starts to increase. This decrease in $2k_t$ produces a sharp increase in the overall rate of polymerization. This autoacceleration is a property common to many free radical polymerizations and is known as the "gel effect."[103]

The self-reaction of macroradicals may be strongly retarded below a certain critical temperature if reaction requires a segmental rotation of the polymer chain backbone. Below a certain temperature such rotation cannot occur and $2k_t$ is markedly reduced. This has been observed in the low-temperature polymerization of methyl methacrylate and has been interpreted as a "glass transition" in the isolated dissolved macroradical occurring at approximately $-50°$.[100]

C. Rate Constants for Bimolecular Self-Reactions of Heteroatom-Centered Radicals

1. Fast Reactions

There are no data available on the rate constants for the self-reactions of radicals with the unpaired electron centered on any element of Group I, II, or

TABLE 3

Some Selected Rate Constants for Fast Self-Reactions of Atoms and Heteroatom Radicals at $\sim 25°$

Radical	Method	Solvent	$2k_t \times 10^{-9}$ (M^{-1} sec^{-1})	Ref.
H·	—	Water	20[a]	16
Br·	KAS[b]	CCl_4	20	104
I·	—	CCl_4	19[a]	16
HO·	KAS[c]	Water	10	105
N_3·	KAS	Water	9[d]	106[b], 107[c]
$(CH_3)_3CO$·	KESR	*t*-Butyl peroxide	1.3	50
$(CH_3)_3CO$·	RS	CCl_4	0.2	108
C_6H_5O·	KAS[b]	Water	5.6	109
p-$CH_3C_6H_5O$·	KAS[b]	Water	6.5	109
C_6H_5O·	SSESR	Benzene	1.2	35, 110
C_6H_5O·	SSESR	Chlorobenzene	0.1	35, 110
p-$CH_3C_6H_5O$·	SSESR	Benzene	3.0	35, 110
C_6H_5NH·	KAS[b]	Water	1.5[e]	111
$(CH_3CH_2)_2N$·	KESR	Pentane	5	112
$(CH_3)_3Si$·	KESR	*t*-Butyl peroxide	5.5	18
$(CH_3)_3Ge$·	KESR	*t*-Butyl peroxide	3.6[f]	18
$(CH_3)_3Sn$·	KESR	*t*-Butyl peroxide	2.8[f]	18
$(CH_3)_3Sn$·	RS	Cyclohexane	3.1	12
(*n*-$C_4H_{10})_3Sn$·	RS	Cyclohexane	1.4	12
$(C_6H_5)_3Sn$·	RS	Cyclohexane	2.8	12

[a] Mean value from a number of different workers.
[b] Radicals generated by flash photolysis.
[c] Radicals generated by pulse radiolysis.
[d] Yields an N_6 species which decays by a first-order process to give nitrogen with a rate constant of 3.6×10^3 sec^{-1}.
[e] This rate constant refers to one or both of the following reactions: $2C_6H_5\dot{N}H \rightarrow$ products, or $C_6H_5\dot{N}H + C_6H_5\dot{N}H_2^+ \rightarrow$ products.
[f] At $-65°$.

III of the periodic table. *With a few notable exceptions*, radicals with the unpaired electron centered on an element from any of the other groups of the periodic table react at, or within an order of magnitude of, the diffusion-controlled limit. Data illustrative of this point are summarized in Table 3. The $2k_t$ values for the atoms H·,[16] Br·,[104] and I·,[16] and for the radicals HO·[105] and N_3·[106,107] are larger than for most simple alkyl radicals presumably because of the higher diffusivity of these small species. The $2k_t$ value for the *t*-butoxy radical is still rather uncertain.[49-51,108] Most unhindered phenoxy radicals,[35,109,110] anilino radicals,[111] amino radicals,[112] and radicals with the unpaired electron on silicon,[18,113] germanium,[18] and tin[12,18] react at close to the diffusion-controlled rate. It is worth noting that there is as yet little information on any radicals with the unpaired electron centered on sulfur[114] or phosphorus.

2. *Slow Reactions*

The majority of free radicals that undergo slow self-reactions have the unpaired electron centered on oxygen or nitrogen. This category includes certain hindered phenoxy radicals, all tertiary and many secondary peroxy radicals, and certain nitroxides, iminoxy radicals, and amino radicals. These species are each considered briefly below.

a. Phenoxy Radicals. The phenoxy radical itself and many simple substituted phenoxy radicals have $2k_t$ values in the range 10^8–10^9 M^{-1} sec^{-1} (see Table 3). However, the 4-phenylphenoxy and 4-methoxyphenoxy radicals have $2k_t$ values of 6.8×10^7 and 3.4×10^7, respectively, in benzene and 3.8×10^6 and 2.8×10^6, respectively, in chlorobenzene.[35,110] Such low values result from the fact that formation of the initial keto dimer is reversible and the rate-controlling step is enolization of this dimer.[110] The enolization is acid catalyzed and in the presence of acid the $2k_t$ values for such phenoxy radicals rise to the diffusion-controlled limit.[110]

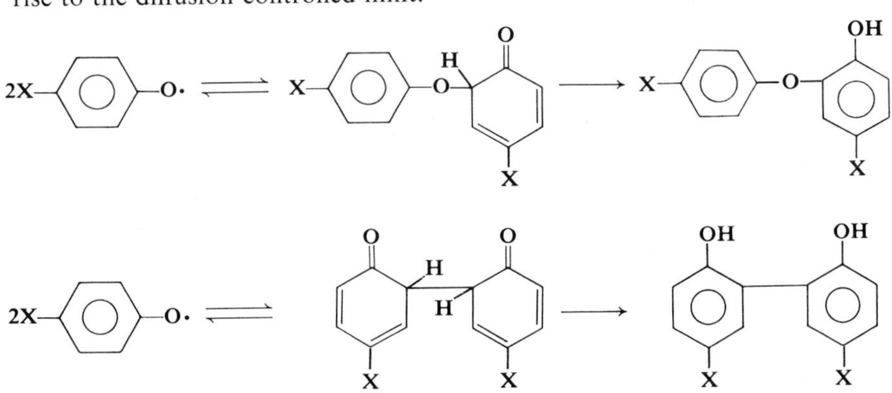

More dramatic reductions in the rate constant can be achieved by the addition of two *t*-butyl groups to the aromatic ring in the ortho positions:

Phenoxyl substituents	Solvent	$2k_t^{25°}$ (M^{-1} sec^{-1})	Ref.
2,6-Bu_2^t	Benzene	6×10^7	110
2,6-Bu_2^t	Chlorobenzene	1×10^8	112
2,6-Bu_2^t-4-Pr^i	Chlorobenzene	3.1	115
2,6-Bu_2^t-4-Pr^i	Benzene	2.2	115
2,6-Bu_2^t-4-Pr^i	Cyclohexane	4.2	116
2,6-Bu_2^t-4-Bu^s	Benzene	0.62	115
2,4,6-Bu_3^t		Stable	

As this list shows, the $2k_t$ values may range from 10^7 to 10^8 M^{-1} sec^{-1} for 2,6-di-*t*-butylphenoxy to zero for tri-*t*-butylphenoxy which is completely stable in solution in the absence of oxygen and other reactive species. The 2,6-di-*t*-butyl-4-isopropylphenoxy radical is of intermediate stability. Decay is second order and yields disproportionation products.

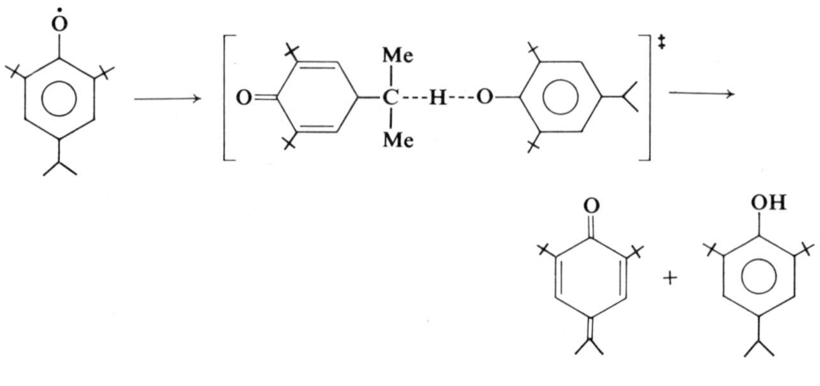

It is not surprising that such a hydrogen transfer is relatively slow. A similar mechanism applies to most other 2,6-di-*t*-butyl-4-alkylphenoxy radicals (e.g., 4-*s*-butyl,[115] 4-cyclohexyl,[115] 4-diphenylmethyl[116]).

The slow decay of 2,6-di-*t*-butyl-4-methylphenoxy was originally reported to be first order with a decay rate constant of 1.3×10^{-2} sec^{-1} in benzene at room temperature.[117] First-order kinetics have been supported by a group of Russian workers.[118] More recently, Stebbins and Sicilio[119] have reported second-order decay with a rate constant of 4.6×10^2 M^{-1} sec^{-1} in benzene at 24°. Our own results indicate a first-order decay with rate constants of 1.1×10^{-2} sec^{-1} in chlorobenzene, 0.9×10^{-2} in di-*t*-butyl peroxide, and 1.2×10^{-2} in cyclohexane at room temperature.[112] In addition to the slow decay of these hindered phenoxyls there is also a rapid, but reversible, dimerization to yield the head-to-tail dimer.[25] The overall mechanism can be represented as[110]

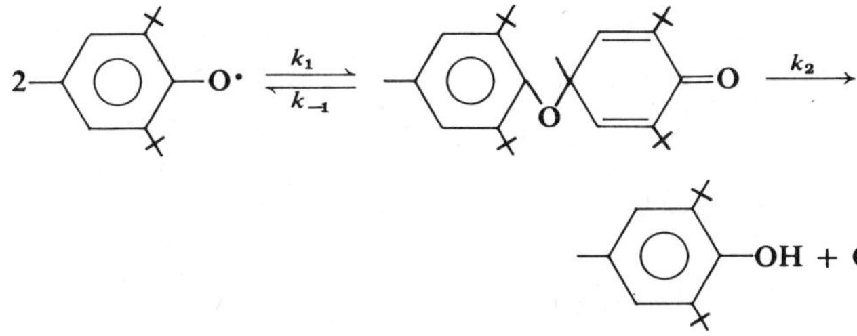

If k_1 is fast and $k_{-1} \gg k_2$, then the decay of the phenoxy radical follows first-order kinetics.[110] That is,

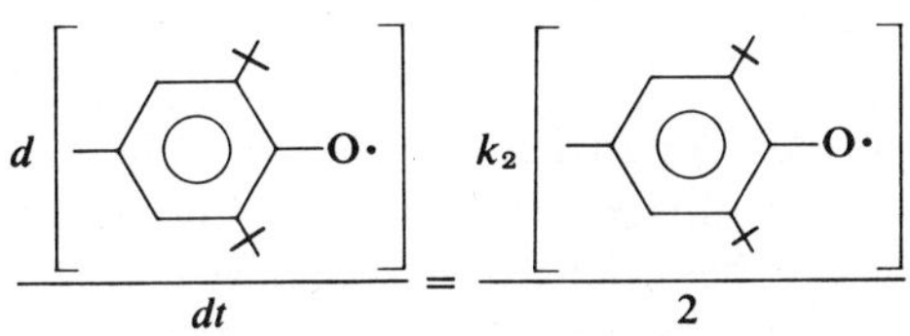

b. Peroxy Radicals. The self-reactions of peroxy radicals have been very thoroughly studied because this reaction is responsible for chain termination in hydrocarbon autoxidations.[25] Termination rate constants are highly dependent on the structure of the peroxy radical. For primary peroxy radicals, $RCH_2OO\cdot$, $2k_t$ values are about 2–4 $\times$ 10^8 M^{-1} sec^{-1}, for secondary peroxy radicals $2k_t$ is in the range 1×10^6 to 4×10^7 M^{-1} sec^{-1}, and for tertiary peroxy radicals $2k_t$ is approximately 1×10^3 to 5×10^4 M^{-1} sec^{-1}.[25,120] With the possible exception of the hydroperoxy radical, $HOO\cdot$, in non-aqueous media, no peroxy radicals terminate at rates even approaching the diffusion-controlled limit.

The slowness of the reaction between tertiary peroxy radicals can be accounted for on the basis of the following mechanism:

$$2ROO\cdot \underset{k_{-1}}{\overset{k_1}{\rightleftharpoons}} ROOOOR \xrightarrow{k_2} [RO\cdot + O_2 + \cdot OR]_{\text{cage}}$$

$$[RO\cdot + O_2 + \cdot OR]_{\text{cage}} \xrightarrow{\text{combination}} ROOR + O_2$$

$$[RO\cdot + O_2 + \cdot OR]_{\text{cage}} \xrightarrow{\text{diffusion}} 2RO\cdot + O_2$$

Formation of the tetroxide is exothermic by 9 kcal/mole and it has a negative entropy of formation of approximately 33 to 40 gibbs/mole.[25] The irreversible decomposition of the tetroxide, reaction 2, has an activation energy of 16–18 kcal/mole and a preexponential factor of 10^{15}–10^{17} sec^{-1} (which suggests that this step should perhaps be written as $RO_4R \rightarrow RO\cdot + RO_3\cdot \rightarrow$

2RO· + O_2).[25] Diffusion out of the solvent cage is four or five times as fast as combination so only about 20% of *t*-peroxy radical encounters that lead to reaction are actually chain terminating. The overall activation energy for chain termination is about 7–9 kcal/mole and so it is not surprising that $2k_t$ is small. Differences in the overall termination rate constants for different *t*-alkylperoxy radicals are apparently due to variations in k_2.

Essentially all encounters of secondary (and primary) alkylperoxy radicals that lead to reaction are chain-terminating events. In 1957 Russell[121] suggested that an intermediate tetroxide decomposed by way of a cyclic transition state in which one of the α-hydrogens is transferred to give the observed products, ketone, alcohol, and oxygen.

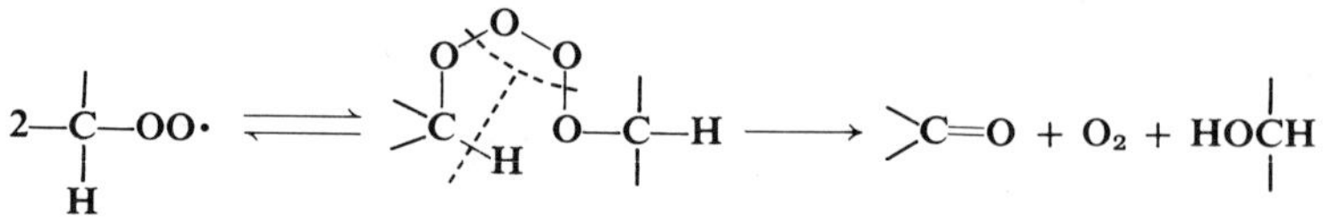

This mechanism provides an attractive explanation for the difference between secondary and tertiary alkylperoxy radicals. However, in spite of the considerable evidence that has accrued in recent years[25,120] in support of this mechanism it is by no means as firmly established as is the mechanism for the self-reaction of tertiary peroxy radicals.

c. Nitroxides. These radicals have an inherently stable electronic arrangement around the oxygen and nitrogen atoms[122] and most stable organic nitroxides show little tendency to dimerize at the O—N center at room temperature. Structural factors are important in determining the stabilities of organic nitroxides. Thus di-*t*-butyl nitroxide and 4,4′-dimethoxydiphenyl nitroxide are stable in the sense that they can be isolated and stored indefinitely. However, the vast majority of organic nitroxides are unstable. For example, diphenyl nitroxide in solution decomposes over the course of several days to diphenylamine and *N*-phenyl-*p*-benzoquinone-*N*-oxide.[123]

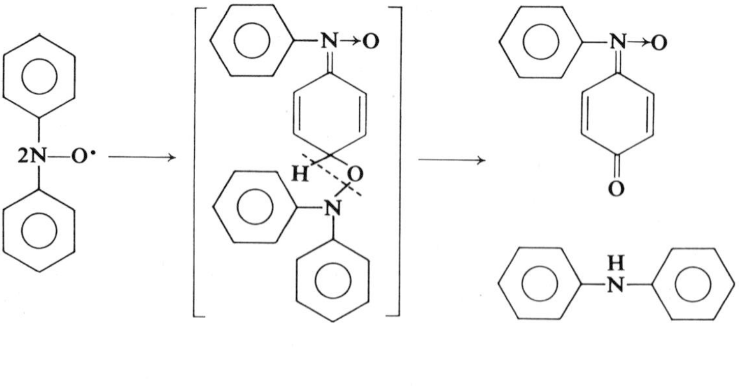

A similar slow reaction occurs with *t*-butyl phenyl nitroxide.[124]

Dialkyl nitroxides, other than di-*t*-alkyl nitroxides, are much less stable than diaryl nitroxides. Bimolecular rate constants and activation parameters have been measured for the self-reactions of several alkyl nitroxides in a number of solvents.[125] The results obtained with diethyl nitroxide, for which the overall reaction is

$$2(CH_3CH_2)_2NO\cdot \xrightarrow{2k_t} (CH_3CH_2)_2NOH + CH_3CH{=}\overset{\overset{\large O}{\uparrow}}{N}CH_2CH_3$$

are summarized in Table 4. It can be seen that the low rate constants arise

TABLE 4

Bimolecular Rate Constants and Activation Parameters for the Self-Reaction of Diethyl Nitroxide

Solvent	$2k_t^{25°} \times 10^{-3}$ (M^{-1} sec^{-1})	$\log_{10} A_t$ (M^{-1} sec^{-1})	E_t (kcal/mole)
Isopentane	33	3.8	−0.93
CF_2Cl_2	15.5	5.0	1.0
Benzene	13.6	5.2	1.4
n-Propanol	7.2	5.4	2.1
Methanol	1.5	4.6	1.9
Water	0.8	—	—

from the small preexponential factor in the Arrhenius equation ($A_t \sim 10^5$ M^{-1} sec^{-1}) rather than from a large activation energy for the overall reaction. The activation energy, E_t, is in fact negative in isopentane, that is, $2k_t$ increases with a decrease in temperature.

Diethyl nitroxide was shown to exist in equilibrium with a diamagnetic dimer at low temperatures in isopentane and in CF_2Cl_2.

$$2(CH_3CH_2)_2NO\cdot \underset{k_{-1}}{\overset{k_1}{\rightleftharpoons}} \{(CH_3CH_2)_2NO\}_2 \tag{1}$$

The evidence for dimer formation is that the nitroxide ESR signal could be reversibly increased or decreased by raising or lowering the temperature in the range −100 to −145°. The variation in the equilibrium constant with temperature could be represented by the van't Hoff relation,

$$K_1 = \frac{k_{-1}}{k_1} = \exp\left(\frac{\Delta S_1}{R}\right) \exp\left(\frac{-\Delta H_1}{RT}\right)$$

In isopentane (which is a poor solvent for polar nitroxide radicals so that solvent effects on the nitroxide self-reaction are minimized) $\Delta S_1 = 25$ gibbs/mole and $\Delta H_1 = 7.6$ kcal/mole. Assuming the dimer is an intermediate in the overall reaction,

$$\{(CH_3CH_2)_2NO\}_2 \xrightarrow{k_2} (CH_3CH_2)_2NOH + CH_3CH{=}N(O)CH_2CH_3 \qquad (2)$$

then $2k_t = k_2/K_1$ from which $k_2 = 10^{9.2}\, e^{-6700/RT}$ sec^{-1} in isopentane. The preexponential factor for reaction 2 is very low for a unimolecular reaction, which suggests that reaction occurs by a complex fission perhaps involving a cyclic transition state. If the dimer is formed by O—O bonding, the transition state for reaction 2 could be represented as

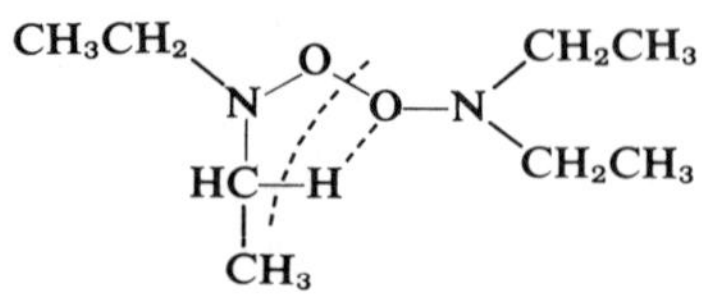

Analogous cyclic transition states can be written for the two other possible dimeric structures, namely, $\gt$NON$\lt$$\rightarrow$O and O$\leftarrow$N—N$\rightarrow$O. Extensive stretching of an α-C—H bond in the transition state is indicated by a large deuterium isotope effect when these hydrogens are replaced by deuterium, $(k_2)_H/(k_2)_D \sim 6$–14.

The preexponential factor for the overall reaction can be calculated from the measured value of ΔS_1 and the estimated entropy loss for the dimer which accompanies the formation of the cyclic transition state (-8.4 gibbs/mole). The calculated value of A_t is $10^{5.8}\ M^{-1}$ sec^{-1} in isopentane, which is in reasonable agreement with the measured value of $10^{3.8}\ M^{-1}$ sec^{-1} for diethyl nitroxide and with a mean value of $10^{4.5}\ M^{-1}$ sec^{-1} for other dialkyl nitroxides with the structure $(RCH_2)_2NO\cdot$, in this solvent. The agreement between the measured and calculated values of A_t is not so good in the more strongly solvating CF_2Cl_2. The proposed reaction mechanism[125] is very similar to the Russell mechanism for the self-reaction of the (isoelectronic) secondary peroxy radicals.[121]

Rate constants for the decay of some other dialkyl nitroxides are given in Table 5.[125] For two primary alkyl groups attached to nitrogen the values of $2k_t^{25°}$ are in the range 10^4–$10^5\ M^{-1}$ sec^{-1}, but for two secondary alkyl groups (diisopropyl nitroxide) $2k_t^{25°} = 4\ M^{-1}$ sec^{-1}. Di-*t*-butyl nitroxide is completely stable.

Monoalkyl nitroxides undergo much faster self-reactions than the dialkyl compounds (Table 5).

$$2RNHO\cdot \xrightarrow{2k_t} RNHOH + RNO$$

TABLE 5

Bimolecular Rate Constants for the Self-Reactions of Nitroxides in Benzene at 25°

Radical	$\log_{10} 2k_t$ (M^{-1} sec^{-1})	Radical	$\log_{10} 2k_t$ (M^{-1} sec^{-1})
$\overline{(CH_2)_5\text{—}NO}\cdot$	5.1	$CH_3NHO\cdot$	7.6
$\overline{(CH_2)_4\text{—}NO}\cdot$	4.6	$(CH_3)_2CHNHO\cdot$	7.1
$(CH_3)_2NO\cdot$	4.3	$(CH_3)_3CNHO\cdot$	6.8
$(CH_3CH_2)_2NO\cdot$	4.1	$C_6H_5NHO\cdot$	6.5
$(C_6H_5CH_2)_2NO\cdot$	4.0		
$[(CH_3)_2CH]_2NO\cdot$	0.6		
$[(CH_3)_3C]_2NO\cdot$	Completely stable		

For R = *t*-butyl this reaction is reversible and in benzene the change in the equilibrium constant with temperature gives $\Delta S = -8.2$ gibbs/mole and $\Delta H = 10.4$ kcal/mole. This measured exothermicity is in excellent agreement with a value of 10 kcal/mole calculated from thermochemical data for the analogous reaction with R = methyl.[125] The reaction is also reversible for R = phenyl with $\Delta S = 2.9$ gibbs/mole and $\Delta H = 7.2$ kcal/mole in benzene. In this case, but not in the *t*-butyl case, there is a slow bimolecular disappearance of the nitroxide ESR signal after the equilibrium concentration has been reached. The final product is azoxybenzene, but the mechanism of its formation is uncertain.

$$\text{PhNHOH} + \text{PhNO} \underset{2k_t}{\rightleftharpoons} 2\text{PhNHO}\cdot \xrightarrow{k_{azo}} \text{PhN}{=}\overset{\overset{\text{O}}{\uparrow}}{\text{N}}\text{Ph} + H_2O$$

The rate constant for this slow decay of phenyl nitroxide, k_{azo}, in benzene is given by

$$k_{azo} = 2.2 \times 10^6 \exp(-8300/RT)\ M^{-1}\ \text{sec}^{-1}$$
$$= 1.8\ M^{-1}\ \text{sec}^{-1} \quad \text{at } 25°$$ [125]

d. Iminoxy Radicals.[126] Diaryliminoxy radicals, arylalkyliminoxy radicals [e.g., $Ph(CH_3)C{=}NO\cdot$], and fairly hindered dialkyliminoxy radicals [e.g., $(\text{iso-Pr})_2C{=}NO\cdot$] decay rather slowly and principally by second-order processes. Values of $2k_t$ are generally in the range 10^2–10^5 M^{-1} sec^{-1} at ambient temperatures. There is a kinetically first-order component to the decay of some of these radicals at low concentrations and the majority of unhindered dialkyliminoxy radicals decay principally by first-order kinetics. The rate constants for first-order decay of a wide variety of unhindered dialkyliminoxy radicals are in the range 0.03–0.08 sec^{-1} at ambient temperatures. The similarity in the first-order constants rule out reaction via β-scission or intramolecular hydrogen abstraction. The reason for the first-order decay is apparently the same as in the case of 2,6-di-*t*-butyl-4-methylphenoxy, that is, the dialkyliminoxy radicals are in equilibrium with an unstable dimer,

$$2R_2C{=}NO\cdot \rightleftharpoons (R_2C{=}NO)_2 \xrightarrow{\text{slow}} \text{stable products}$$

Products indicative of N—N coupling [$R_2C{=}O + N_2$ and $R_2C{=}N(\rightarrow O)\text{-}N{=}CR_2$], of O—N coupling [$R_2C{=}NON(\rightarrow O){=}CR_2$], and of O—C coupling [$R_2C{=}NOC(R_2)N{=}O$] have been isolated.

Activation parameters for the decay of some iminoxy radicals in benzene are listed below:

Radical	E_t (kcal)	A_t (sec^{-1})	A_t (M^{-1} sec^{-1})
$(CH_3)_2C{=}NO\cdot$	17 ± 3	$10^{11\pm2}$	—
$CH_3(PhCH_2)C{=}NO\cdot$	20 ± 5	$10^{13\pm4}$	—
$(\text{iso-Pr})_2C{=}NO\cdot$	5 ± 3	—	$10^{5\pm2}$
$(Ph)_2C{=}NO\cdot$	2 ± 3	—	$10^{5\pm2}$
$Ph(CH_3)C{=}NO\cdot$	5 ± 3	—	$10^{9\pm2}$

The first-order process has a much higher activation energy than the second-order processes.

e. Amino Radicals. Steric effects are also important in the self-reactions of amino radicals.[112] Thus, $2k_t$ for diisopropylamino is approximately 5×10^6 M^{-1} sec^{-1} in the range -70 to $-10°$ in pentane. At higher temperatures the radical decays with first-order kinetics.

D. Rate Constants for Bimolecular Self-Reactions of Some Radical Ions

The self-reactions of radical ions are generally much slower than diffusion controlled. This is not unexpected since electrostatic repulsion should reduce the frequency of close encounters. There are several interesting examples of this effect. Thus, the pulse radiolysis of glycolic acid in aqueous solution at various pH's has yielded the following bimolecular rate constants for radical decay[127]:

$$\text{pH} = 1;\quad HO\dot{C}HCOOH,\quad 2k = 1.1 \times 10^9\ M^{-1}\ sec^{-1}$$

$$\text{pH} = 7.2;\quad HO\dot{C}HCOO^-,\quad 2k = 8.5 \times 10^8\ M^{-1}\ sec^{-1}$$

$$\text{pH} = 12;\quad {}^-O\dot{C}HCOO^-,\quad 2k = 1.5 \times 10^7\ M^{-1}\ sec^{-1}$$

Similar studies on aqueous hydrogen peroxide have yielded[95]:

$$HOO\cdot + HOO\cdot \longrightarrow H_2O_2 + O_2, \qquad 2k = 7.6 \times 10^5\ M^{-1}\,\mathrm{sec}^{-1}$$

$$HOO\cdot + O_2\cdot^- \xrightarrow{H_2O} H_2O_2 + O_2 + OH^-, \qquad 2k = 8.5 \times 10^7\ M^{-1}\,\mathrm{sec}^{-1}$$

$$O_2\cdot^- + O_2\cdot^- \xrightarrow{2H_2O} H_2O_2 + O_2 + 2OH^-, \qquad 2k < 100\ M^{-1}\,\mathrm{sec}^{-1}$$

Other examples of slow bimolecular self-reactions of radical ions include

$$2(Ph_2C{=}CH_2)^{\overset{\cdot}{-}} \longrightarrow (Ph_2C{=}CH_2)_2{}^{2-}$$

$2k \sim 10^5\ M^{-1}\,\mathrm{sec}^{-1}$ in hexamethylphosphoramide,[128]

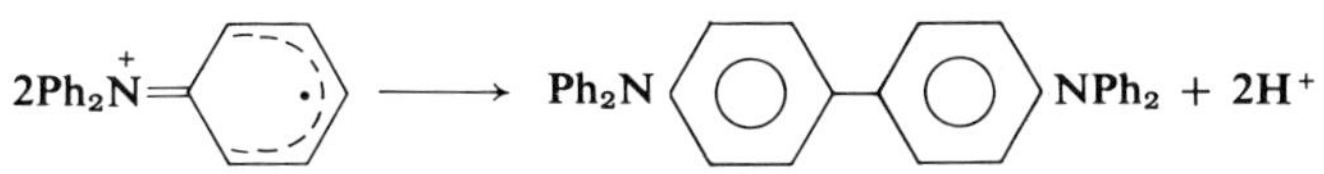

$2k \sim 10^3\ M^{-1}\,\mathrm{sec}^{-1}$ in acetonitrile,[129] and

$$2(CH_3)_2NH^{\overset{+}{\cdot}} \longrightarrow \text{molecular products}$$

$2k_t \sim 6 \times 10^6$ to $5 \times 10^7\ M^{-1}\,\mathrm{sec}^{-1}$ in 2–4 M sulfuric acid in acetic acid at 30°.[130]

Since protonation and deprotonation are extremely rapid processes the experimental data do not necessarily allow a distinction to be made between a reaction involving two radical ions and one involving a radical ion and the corresponding neutral species. The chemistry of radical ions and the kinetics of some of their reactions have been reviewed.[131]

E. Rate Constants for Reactions between Two Different Free Radicals

There have been relatively few rate constants measured for reactions between two different free radicals in solution[16] and even fewer that are likely to be accurate to better than an order of magnitude. If two free radicals R· and R′· have self-reaction rate constants that are diffusion controlled, then the cross reaction of R· + R′· will be equally fast. Even if one of the radicals is completely stable, it may react with a second radical with a rate constant that is close to the diffusion-controlled limit. This can be illustrated by the fact that galvinoxyl is an extremely efficient trap for alkyl,[132,133] alkoxy,[132] and alkylperoxy[134,135] radicals.

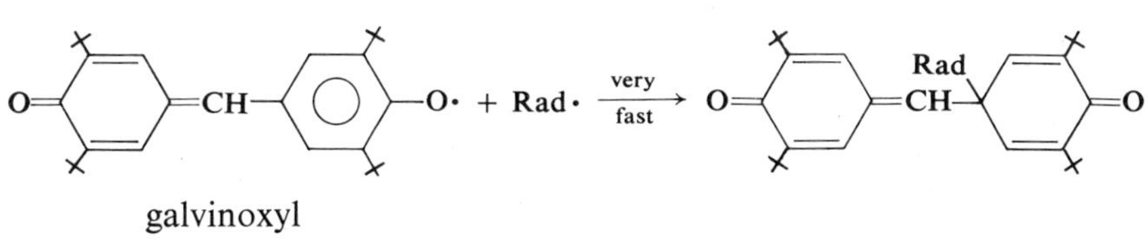

galvinoxyl

Rad· = alkyl, alkoxy, or alkylperoxy.

Galvinoxyl itself is quite stable in the absence of oxygen. Similarly, alkyl radicals are very rapidly trapped by stable dialkyl nitroxides,[136,137,137a] $k \sim 10^7\ M^{-1}\ sec^{-1}$

$$R\cdot + R_2'NO\cdot \longrightarrow R_2'NOR, \quad k \sim 10^7\ M^{-1}sec^{-1};$$

by certain metal ions, particularly copper(II),[138,139] for example,

$$CH_3CH_2\dot{C}HCH_3 + Cu^{II} \longrightarrow \text{Mixed Butenes} + H^+ + Cu^I, \quad k \sim 10^8\ M^{-1}sec^{-1};^{140}$$

by sodium naphthalenide,[143]

$$R\cdot + Naph\cdot^- \longrightarrow R\!:^- + Naph\ (+\ R\text{–}Naph\!:^-), \quad k \sim 10^{10}\ M^{-1}\ sec^{-1};$$

and by oxygen,[20,120,137a,144,145]

$$R\cdot + O_2 \longrightarrow ROO\cdot, \quad k \sim 10^8\text{–}10^{10}\ M^{-1}\ sec^{-1}.$$

Another example of rapid "cross" reactions is provided by peroxy radicals. As has already been mentioned, secondary peroxy radicals terminate autoxidations about 1000 times faster than tertiary peroxy radicals. However, the cross-reactions between secondary and tertiary peroxy radicals have rate constants that are about one-quarter as great as those for the self-reaction of the secondary peroxy radical.[25]

The available evidence indicates that the reaction between two dissimilar free radicals is likely to be fast irrespective of the values of the rate constants for the self-reactions of the two radicals.

F. Conclusion

The bimolecular self-reactions of radicals will, in general, be diffusion controlled if dimerization through the atom formally bearing the unpaired electron yields stable products, provided steric effects are absent and provided the radical is not a strong dipole.

$$2AB\cdot \longrightarrow ABBA$$

If the dimer is thermodynamically unstable, this reaction may be reversible, in which case an equilibrium concentration of AB· may be attained. A diffusion-controlled process is also likely if an exothermic disproportionation can occur with a hydrogen transfer from the atom adjacent to that bearing the unpaired electron.

$$2\underset{|}{\overset{}{A}}\text{—}B\cdot \longrightarrow A\text{—}B + A\text{=}B \quad (\text{2 H–A–B}\cdot \longrightarrow \text{H–A–B–H} + A\text{=}B)$$

Radical–radical reactions will be slower than diffusion controlled if there is steric hindrance to dimerization or disproportionation and also when these processes are only slightly exothermic. Reaction is also likely to be less than diffusion controlled when the radicals are highly polar and when extensive bond reorganization is required to yield stable products, for example,

$$\begin{array}{l}2\text{A—}\underset{\displaystyle\text{H}}{\underset{|}{\text{B}}}\text{—C}\cdot \longrightarrow \text{A—}\underset{\displaystyle\text{H}}{\underset{|}{\text{B}}}\text{—C—H} + \text{A=B}\rightarrow\text{C}\end{array}$$

$$2\text{A=B—C}\cdot \longrightarrow \text{A=B—C—A—B=C}$$

IV. RATE CONSTANTS FOR RADICAL-MOLECULE REACTIONS

A. Bimolecular Homolytic Substitutions (S_H2 Reactions)

There is a large class of radical–molecule reactions that result in the net replacement of one radical or atom by another radical or atom:

$$\text{R}\cdot + \text{AB} \longrightarrow \text{RA} + \text{B}\cdot$$

This process is referred to by the symbols S_H2 (substitution, homolytic, bimolecular[146]). The majority of such reactions involve the attack of R· on a univalent and, hence, peripheral or terminal atom such as hydrogen or halogen. Such processes are known also as *atom abstractions* or as *metathesis reactions*. Homolytic substitution can also occur at multivalent atoms. These processes need not involve high activation energies if A is coordinatively unsaturated as is the case if A forms part of a multiply bonded system or if A has low-energy vacant orbitals available for it to increase its coordination number (e.g., boron). Activation energies are normally high for S_H2 reactions at coordinatively saturated multivalent atoms (e.g., sp^3-hybridized, 4-coordinate carbon) and such reactions are therefore slow.

Bimolecular homolytic substitutions may be divided according to whether the adduct [RAB]· represents a transition state (S_H2, synchronous)

$$\text{R}^\cdot + \text{AB} \longrightarrow [\text{RAB}]^{\ddagger}_{\cdot} \longrightarrow \text{RA} + \text{B}^\cdot$$

or a discrete intermediate (S_H2, stepwise)

$$\begin{array}{rcl}\text{R}^\cdot + \text{AB} & \longrightarrow & [\text{RAB}]^\cdot \\ [\text{RAB}]^\cdot & \longrightarrow & \text{RA} + \text{B}^\cdot\end{array}$$

Atom abstractions are synchronous processes, but substitution at a coordinatively unsaturated atom is more likely to occur stepwise. Indeed, there are many reactions which yield such stable adduct radicals that substitution requires attack by a second radical, for example, many homolytic aromatic substitutions.[147]

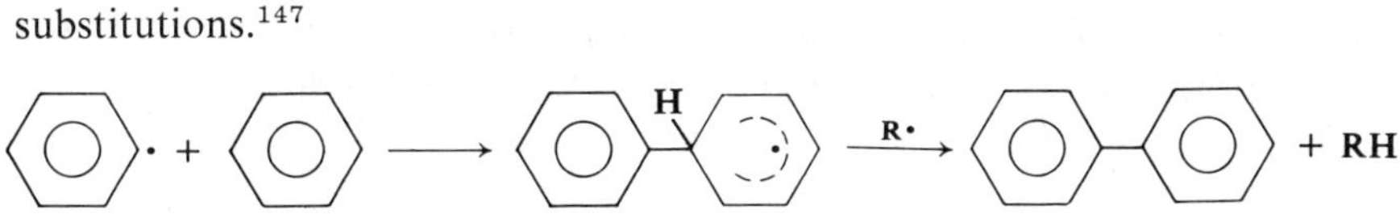

However, for our present purposes, it is more convenient to divide S_H2 reactions into processes occurring at univalent atoms and those occurring at multivalent atoms.

1. S_H2 Reactions at Univalent Atoms

a. Estimation of Entropy and Energy of Activation. According to transition state theory the preexponential factor (*A* factor) for a bimolecular reaction is given by $(ekT/h)\exp(\Delta S^{\ddagger}/R)$ where $\Delta S^{\ddagger}$ is the entropy change involved in forming a mole of transition state complex from the two reactants.[148] The calculation of $\Delta S^{\ddagger}$ (which is always negative) requires a knowledge of the structure of the $[RAB]\cdot^{\ddagger}$ complex. In the gas phase, a lower limit to $\Delta S^{\ddagger}$ can be calculated by comparison of $RAB^{\ddagger}$ with structurally related stable molecules. The problem is more complicated in solution because of unknown changes in the degree of solvation of reactants and transition state. However, in relatively nonpolar and nonpolarizable solvents the gas-phase calculations provide a reasonable approximation to the real situation.

In the gas phase, experiment and theory are generally in reasonably good agreement and it is possible to specify approximate preexponential factors for various types of reaction.[148] For an atom abstraction from a relatively simple molecule by an atom (e.g., $H\cdot + CH_4 \rightarrow H_2 + CH_3\cdot$) the preexponential factor is approximately $10^{11.0 \pm 0.5}\ M^{-1}\ sec^{-1}$, while for an atom abstraction by a radical (e.g., $CH_3\cdot + C_2H_6 \rightarrow CH_4 + C_2H_5\cdot$) the preexponential factor is approximately $10^{8.5 \pm 0.5}\ M^{-1}\ sec^{-1}$.[148] (A similar value applies for radical addition to an unconjugated double bond.) The preexponential factor will generally be lower in abstractions from more complex molecules. In particular, the preexponential factor will be low if abstraction yields a resonance-stabilized radical such as allyl or benzyl. This is because a relatively free rotation in the reactant must be "frozen out" in the transition state so that the unpaired electron in the developing radical can be delocalized over the entire π-electron system. Thus, in hydrogen abstraction from toluene the hydrogen must be removed in a direction perpendicular to the plane of the aromatic ring. The overall effect of structure on preexponential factors in abstractions can be illustrated by Golden and Benson's data[149] for hydrogen abstraction from hydrocarbons by iodine atoms in the gas phase.

$$I\cdot + RH \longrightarrow IH + R\cdot$$

Values of $\log_{10} A$ (per active hydrogen in $M^{-1}\ sec^{-1}$ units) for some typical substrates are as follows: 11.1, $C_2H_5CH_2$—H; 10.9, $(CH_3)_2CH$—H; 10.9, $(CH_3)_3C$—H; 9.7, CH_2=$CHCH_2$—H; 9.0, CH_2=$CHC(CH_3)H$—H; 9.2, $(CH_2$=$CH)_2CH$—H; and 7.9, $C_6H_5CH_2$—H. A detailed description of the methods that may be used to calculate approximately $\Delta S^{\ddagger}$ values for a wide variety of transition states has been given by Benson.[148]

The activation energy, E, for a radical–molecule reaction is more difficult to predict than the activation entropy. Of course, E cannot be less than the endothermicity of the reaction, ΔH. For a reversible process,

$$R\cdot + AB \rightleftharpoons RA + B\cdot$$

it is convenient to consider the activation energy in the exothermic direction as being the intrinsic activation energy for the process, E_{int}. The activation energy in the endothermic direction is then equal to $E_{int} + \Delta H$ (see figure on p. 45). If the strengths of the R—A and A—B bonds are known, then ΔH is known and all that remains is to determine E_{int}. Unfortunately, there is no simple correlation between E_{int} and ΔH that is applicable to all classes of S_H2 reactions. Quantum mechanical estimates of activation energies have been virtually confined to such simple systems as the $H\cdot + H_2$ reaction. Although recent semiempirical calculations give an activation energy for this reaction in reasonable agreement with the experimental value of 8 kcal/mole, it is not clear what significance this has in view of the approximations employed.

For limited classes of exothermic S_H2 reactions several empirical correlations relating E_{int} to ΔH have been proposed. Evans and Polanyi[150] originally suggested a relation which nowadays is usually put in the form

$$E_{int} = E_0 + \alpha(\Delta H)$$

and is referred to as the Polanyi equation. For a homologous series of reactions the adjustable parameters, E_0 and α, are constant. A greatly simplified version of the Polanyi relation has been proposed by Semenov[151] who pointed out that for many (though not for all) gas-phase S_H2 reactions $E_0 \approx 11.5$ kcal/mole and $\alpha \approx 0.25$. Other empirical equations have also been proposed.[152,153] For limited extrapolation or interpolation the Polanyi relation can be useful but it is obviously an oversimplification since it predicts a negative rather than a zero activation energy for highly exothermic reactions.[154]

A rather elaborate, empirical method of calculating E_{int} to within about ± 2 or 3 kcal which is based on the idea that the sum of the "bond orders" is conserved in a reaction has been developed by Johnston and Parr.[155] However, as Benson and DeMore[156] have pointed out, since about 90% of all reported gas-phase, exothermic, atom abstraction reactions have activation energies of 8 ± 3 kcal/mole, it is going to take a fairly sophisticated rule to have much predictive value. A more accurate method for calculating activation energies has recently been proposed by Zavitsas.[156a]

The failure of the Polanyi equation to predict activation energies for hydrogen abstractions from a nonhomogeneous series of compounds has been documented by Benson and DeMore[156] using the following approximately thermoneutral reactions as examples:

$$CH_3\cdot + H\overset{*}{-}CH_3 \longrightarrow CH_4 + {}^*CH_3\cdot, \qquad E = 14.5 \text{ kcal/mole}$$

$$CH_3\cdot + H-OCH_3 \longrightarrow CH_4 + CH_3O\cdot, \qquad E \sim 8 \text{ kcal/mole}$$

$$CH_3\cdot + H-Cl \longrightarrow CH_4 + Cl\cdot, \qquad E = 3.8 \text{ kcal/mole}$$

This reaction series also serves to illustrate the general observation that intrinsic activation energies tend to be lower for reactions of comparable exothermicity if the attacking radical is nucleophilic and the leaving radical is electrophilic (or vice versa) than when attacking and leaving radicals are both nucleophilic or electrophilic. This can be understood either in terms of decreased repulsion between attacking and leaving radicals[155] (i.e., a "tighter" transition state with consequent decrease of E_{int}) or in terms of a stabilization of the transition state by a strong dipolar contribution of the type[148]

$$\begin{bmatrix} (+\tfrac{1}{2}) & & (-\tfrac{1}{2}) \\ R & \cdot\, A\, \cdot & \dot{B} \end{bmatrix}^{\ddagger} \quad \text{or} \quad \begin{bmatrix} (-\tfrac{1}{2}) & & (+\tfrac{1}{2}) \\ \dot{R} & \cdot\, A\, \cdot & B \end{bmatrix}^{\ddagger}$$

To sum up, in the gas phase the A factor for an atom abstraction can generally be estimated with an accuracy of approximately $10^{\pm 0.5}\ M^{-1}\ \text{sec}^{-1}$ and the activation energy can be estimated to within about ± 2 kcal/mole. This means that even under favorable conditions the rate constant at ambient temperatures can only be estimated to within $10^{\pm 0.5} \times 10^{\pm 2000/4.6 \times 300} \approx 10^{\pm 2.0}\ M^{-1}\ \text{sec}^{-1}$, which is rather unsatisfactory. The estimation of rate constants for atom abstractions in solution is in an even less advanced state and about the only thing that can be done when such a quantity is required is to measure it experimentally. This applies with redoubled force to S_H2 reactions at multivalent atoms, there having been no theoretical calculations of rate constants for any such reactions even in the gas phase.

b. Rate Constants for Hydrogen Atom Abstractions. Very reactive hydrogen atom abstracting agents (e.g., F·, Cl·, ·OH) react with most organic substrates at rates which are close to the diffusion-controlled limit. Although relative reactivities of different molecules toward these atoms and radicals can be obtained readily by the method of competing reactions (which was outlined at the beginning of this chapter), it is extremely difficult to measure an absolute rate constant. The most successful method involves the rapid generation of a high concentration of the radical in the presence of a large excess of a reactive substrate. If the radical reacts principally with this substrate, it will decay according to the first-order kinetic law and its decay can be monitored by some of the methods used for following radical–radical reactions. The resulting pseudo-first-order rate constant (sec^{-1} units) is converted to the true bimolecular rate constant merely by dividing by the molar concentration of the reagent. This method has been most widely and successfully applied in pulse radiolysis studies.[40]

With less reactive hydrogen atom-abstracting agents [e.g., $(CH_3)_3CO\cdot$ and $(CH_3)_3COO\cdot$], it is generally possible to set up a chain reaction with a bimolecular termination step involving the radical of interest.[157] Measurement of the overall reaction rate, the rate of initiation, and the rate constant for chain termination by, for example, the rotating sector technique will yield the rate constant for chain propagation, that is, the rate constant for a radical–molecule reaction. The method of competing reactions can then be used to obtain the rate constants for reaction with other molecules. The usual procedure is to carry out a reaction in the presence of the two substrates and measure their relative rates of consumption or the relative rates of formation of products. The development of modern analytical methods, particularly gas chromatography, has made this procedure simple and versatile.

Two interesting variations of the analytical competitive procedure are in common use. One method relies on the change in overall reaction rate which is produced by the addition of the second substrate; the other relies on a change in the product. These two methods are illustrated by specific examples below.

The autoxidation of tetralin, RH, at temperatures in the range 20–100° and at oxygen pressures above approximately 150 mm can be represented by the following chain reaction:

Initiation:

$$\text{initiator} \longrightarrow \text{R}\cdot \qquad (\text{rate} = R_i)$$

Propagation:

$$\text{R}\cdot + \text{O}_2 \longrightarrow \text{ROO}\cdot$$

$$\text{ROO}\cdot + \text{RH} \xrightarrow{k_p} \text{ROOH} + \text{R}\cdot$$

Termination:

$$\text{ROO}\cdot + \text{ROO}\cdot \xrightarrow{2k_t} \text{molecular products}$$

the overall rate of oxidation is given by

$$\frac{-d[\text{O}_2]}{dt} = \frac{k_p[\text{RH}]R_i^{1/2}}{(2k_t)^{1/2}}$$

The rotating sector method has been used to determine $2k_t$ and k_p over a range of temperature.[158] In the presence of a small quantity (say $\sim 10^{-4}$ *M*) of 2,6-di-*t*-butyl-4-methylphenol the rate of oxidation is decreased and termination by the self-reaction of tetralylperoxy radicals is completely suppressed.[159]

The oxidation can then be represented by

Initiation:

$$\text{initiator:} \longrightarrow \text{R}\cdot \quad (\text{rate} = \text{R}_i)$$

Propagation:

$$\text{R}\cdot + \text{O}_2 \longrightarrow \text{ROO}\cdot$$

$$\text{ROO} + \text{RH}\cdot \xrightarrow{k_p} \text{ROOH} + \text{R}\cdot$$

Termination:

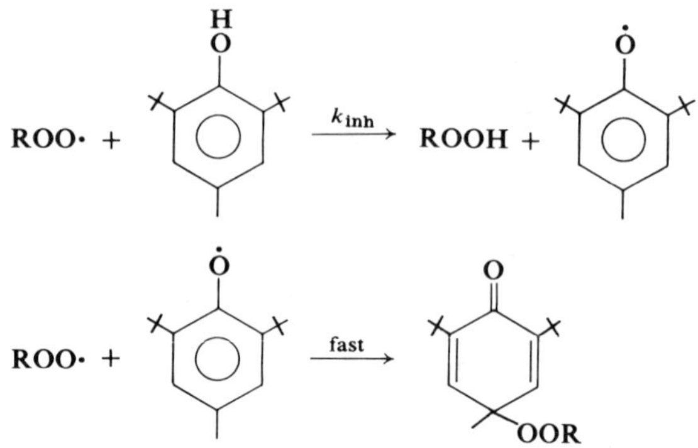

The inhibited (inh) oxidation rate is given by

$$\left(\frac{-d[\text{O}_2]}{dt}\right)_{\text{inh}} = \frac{k_p[\text{RH}]R_i}{2k_{\text{inh}}[\text{AH}]}$$

where AH represents the hindered phenol. Since R_i is known and k_p has been determined as 13.6 M^{-1} sec^{-1} at 65° the measurement of the rate of inhibited oxidation yields k_{inh}, the rate constant for tetralylperoxy radical attack on the phenol, as 2×10^4 M^{-1} sec^{-1}.[159]

The free radical polymerization of a vinyl monomer, M, such as styrene can be represented by

Initiation:

$$\text{initiator} \longrightarrow \text{M}\cdot \qquad (\text{rate} = R_i)$$

Propagation:

$$\text{M}_{n-1}\cdot + \text{M} \longrightarrow \text{M}_n\cdot$$

Termination:

$$\text{M}_n\cdot + \text{M}_m\cdot \xrightarrow{2k_t} \text{molecular products}$$

If a substrate, SH, which is capable of transferring hydrogen to the polymeric radical is added to the polymerizing monomer, a process known as chain transfer will occur[160]:

Transfer:

$$\text{M}_n\cdot + \text{SH} \xrightarrow{k_{\text{tr}}} \text{M}_n\text{H} + \text{S}\cdot$$

$$\text{S}\cdot + \text{M} \xrightarrow{\text{fast}} \text{SM}\cdot$$

In a polymerizing system, the average degree of polymerization, $\bar{P}$, is equal to the number of monomers incorporated into the average polymer molecule and this is equal to the ratio of the rate of propagation to the sum of the rates of all processes which prevent the further growth of a polymer molecule,[161] that is,

$$\bar{P} = k_p[M_n\cdot][M]/(2k_t[M_n\cdot]^2 + k_{tr}[M_n\cdot][SH])$$

This equation can be simplified to,

$$1/\bar{P} - (k_tR_i)^{1/2}/k_p[M] = C[SH]/[M]$$

by assuming a steady-state radical concentration. The transfer constant for SH, $C = k_{tr}/k_p$, can be obtained by measuring $\bar{P}$ at known rates of chain initiation and monomer concentration. Provided k_p has been measured, the rate constant for the attack of the polymeric radical on SH is readily obtained. Thus, in polymerizing styrene at 60° the transfer constant for toluene[162] is 1.25×10^{-5} which, when combined with $k_p = 145\ M^{-1}\ sec^{-1}$ at this temperature,[7] yields $k_{tr} = 1.8 \times 10^{-3}\ M^{-1}\ sec^{-1}$ for hydrogen abstraction from toluene by the polystyryl radical.

In both of the examples above, it should be clear that the usual analytical approach which relies on measuring the relative rates of consumption of the two reagents could not yield reliable rate constants. That is, the very large differences in rate constants between tetralin and the phenol and between styrene and toluene would make measurement of the extent of consumption of the less reactive component in each mixture extremely difficult.

The relative reactivities of many hydrocarbon types of C—H bonds toward some of the more common radicals and atoms are listed in Table 6. This table also includes estimates of the absolute rate constant for hydrogen abstraction from the side chain of toluene at 40° together with the preexponential factor and activation energy for this process. Almost all the data in the table can therefore be put on an absolute basis. However, the paucity of reliable determinations of S_H2 rate constants in solution must be pointed out. Thus, there do not appear to be any reliable measurements on chlorine or bromine atoms, or on methyl or phenyl radicals. Fortunately, for the first three of these radicals reliable rate constants are available from gas-phase studies. It seems reasonable to assume that the gas-phase data will be approximately applicable to solutions, *provided* the solvent is relatively nonpolar and nonpolarizable so that it does not complex the radical (or substrate) to any appreciable extent.[163] If complexing occurs, the S_H2 rate constants will probably be smaller in solution than in the gas phase. There have been no measurements on the rate of phenyl radical attack on hydrocarbons, but the high C—H bond strength in benzene (112 kcal/mole[165]) and the comparatively low selectivity of this radical suggest that hydrogen abstraction from

TABLE 6

Relative Reactivities of Various Types of Hydrocarbon C—H Bonds toward Common Radicals and Atoms in Solution and Absolute Rate Constants for the Reaction with Toluene

Radical *D*[R—H] kcal/mole Temperature	$C_6H_5\cdot$[a] 112 60°	$CH_3\cdot$[c] 104 65°	$(CH_3)_3CO\cdot$[h] 104 40°	$(C_6H_5)_2CO$[3k] 104 22°	$Cl\cdot$[l] 103 40°	$CCl_3\cdot$[p] 96 40°	$(CH_3)_3COO\cdot$[s] 90 30°	$Br\cdot$[u] 87 40°	Polystyryl[y] 82 60°
C—H Bond:									
Primary aliphatic	0.12	0.4[d]	0.1	<0.01	0.8[m]	—	~0.0006[t]	~0.0001[v]	—
Secondary aliphatic	1.0	4.0[d]	~1.2	0.5	~3[m]	~5[q]	0.025	~0.005	0.05
Tertiary aliphatic	4.8	9.4[d]	4.4	3.0	3.6[m]	160[q]	0.25	~0.2	—
Primary allylic	1.6	0.6[e]	~1.6	—	—	—	1.2	—	—
Secondary allylic	3.3	6.4[e]	~7	—	—	—	8.3	~40	—
Tertiary allylic	13	26[e]	18	—	—	—	10	—	—
Toluene (α)	(1)	(1)	(1)	(1)	(1)	(1)	(1)	(1)	(1)
Ethylbenzene (α)	4.6	4.1	3.2	4.6	2.5	50	8.3	17	8
Cumene (α)	9.7	13	6.9	9.9	5.5	260	13	37	20
Diphenylmethane (α)	7.7	—	4.7	—	2.0	50	22	10	25
Tetralin (α)	12	12	16[i]	—	4.1	—	20	36	—
Triphenylmethane (α)	39	—	9.6	—	7.2	160	23	18	84
Toluene:									
$\text{Log}(k^{40°}/\text{H})(M^{-1}\ \text{sec}^{-1})$	~6[b]	1.5[f,g]	3.8[j]	4.8[j]	9.1[n]	−0.3[r]	−1.6	5.0[w]–5.5[x]	−4.2[z]
$\text{Log}(A/\text{H})(M^{-1}\ \text{sec}^{-1})$	—	8.1[f,g]	7.7[j]	—	~10	7.8[r]	6.8	~10	~6.4[z]
E (kcal/mole)	—	9.5[f,g]	5.6[j]	—	1–2[o]	11.6[r]	12	7.2[x]	~15[z]

[a] Reference 166. [b] See text.

[c] Reference 167 unless otherwise noted. For additional data see W. A. Pryor, D. L. Fuller, and J. P. Stanley, *J. Am. Chem. Soc.*, **94**, 1632 (1972).

[d] Reaction in gas phase. Absolute rate constants were calculated from A and E values for $\cdot CH_3$ attack on alkanes quoted in reference 168. The absolute rate constant for $\cdot CH_3$ attack on the toluene side chain was calculated to be 106 M^{-1} sec^{-1} per H at 65° from gas-phase data in reference 169.

[e] Reference 170. Allylic reactivities were calculated relative to toluene from the relation $k_{toluene}/k_{isooctane} = 2.5$ (see reference 171).

[f] Reaction in gas phase, reference 169.

[g] The data in reference 172 give $\log k^{40°}/H = 1.8$, $\log A/H = 7.0$, and $E = 7.4$.

[h] References 173 and 174 unless otherwise noted.

[i] Reference 175. [j] Best estimates as given in reference 25.

[k] Triplet benzophenone. Reference 176. The acetophenone triplet seems to be more reactive; see, for example, H. Lutz and L. Lindquist, *Chem. Comm.*, 493 (1971).

[l] Reference 177. [m] Reference 164.

[n] Based on the assumption that cyclopentane has the same reactivity toward chlorine atoms in the gas phase as in noncomplexing solvents. The gas-phase value should probably be revised (see reference 178).

[o] Estimated to be in this range from reference 179. [p] Reference 180.

[q] Reaction in gas phase. The absolute rate constant for $\cdot CCl_3$ attack on cyclohexane is given by $\log k = 8.8 - (10700/2.3\ RT) = 23\ M^{-1}\ sec^{-1}$ (for 12 *sec*-alkyl hydrogens) at 40° (reference 181). The absolute rate constant for $\cdot CCl_3$ attack on cyclopentane is given by[181a] $\log k = 9.0 - (10700/2.3\ RT) = 37\ M^{-1}\ sec^{-1}$ (for ten *sec*-alkyl hydrogens) at 40°, and for attack on 2,3-dimethylbutane by $\log k = 7.9 - (8200/2.3\ RT) = 160\ M^{-1}\ sec^{-1}$ (for two *tert*-alkyl hydrogens) at 40°. These values have been compared with the absolute rate constant for attack on toluene in solution from reference 182. (Data in reference 183 treated similarly yield relative rate constants of 3.2 for primary aliphatic H and 1260 for secondary aliphatic H.) See also H. W. Sidebottom, J. M. Tedder, and J. C. Walton, *Int. J. Chem. Kin.*, **4**, 249 (1972); J. A. Kerr and M. J. Parsonage, *ibid.*, **4**, 243 (1972).

[r] Reference 182.

[s] Mostly unpublished data of S. Korcek, J. H. B. Chenier, J. A. Howard, and K. U. Ingold. Similar relative and absolute rate constants have been reported in reference 184 for cumylperoxy radical attack on the alkyl aromatic hydrocarbons. The relative rate constants for attack by primary and secondary alkylperoxy radicals are very similar, but the absolute rate constants are about ten times larger than those found with the tertiary alkylperoxy radicals. Thus, $\log k^{40°}/H$ for benzylperoxy radical abstraction from toluene is ~ -0.8.

[t] At 100°. Reference 185.

[u] From data by various authors collected in reference 166. For the important case of abstractions from bromoalkanes see reference 186.

[v] Extrapolated data for bromine atom abstraction from ethane in the gas phase (reference 187) yield $k/H = 10\ M^{-1}\ sec^{-1}$ at 40°.

[w] Reference 180. This rate constant is based on the assumption that the tertiary hydrogens of isobutane in the gas phase and 2,3-dimethylbutane in the liquid phase have an equal reactivity toward bromine atoms.

[x] Reference 188. Data in the gas phase that have been criticized.[189]

[y] Reference 162. Rather similar values have been reported in reference 190.

[z] See reference 175. For the analogous reaction of benzyl radicals with *n*-deuteriotoluene, $\log(k^{40°}/H) = -3.9$, $\log(A/H) = 10{,}0$, and $E = 19.9$ (reference 191). The pre-exponential factor and activation energy both seem rather large.

toluene occurs with a rate constant of approximately $10^6\ M^{-1}\ \text{sec}^{-1}$. The rate constant for the reaction,

$$C_6H_5\cdot + CH_3OH \longrightarrow C_6H_6 + \dot{C}H_2OH$$

has been estimated to be at least $1.4 \times 10^5\ M^{-1}\ \text{sec}^{-1}$ at room temperature.[165a]

In Table 6 the radicals have been arranged in order of R—H bond strength. The relative reactivities are on a "per hydrogen" basis. Thus, the standard "toluene (α)" refers to a single primary benzylic hydrogen in toluene. This nomenclature is used because a primary benzylic hydrogen in *p*-xylene is generally rather more reactive than one in toluene. Such minor structural effects on reactivity are generally less important in aliphatic hydrocarbons and have been explored less in olefins.

The least selective radical in Table 6 is chlorine,[192] D[Cl—H] = 103 kcal/mole, followed closely by the phenyl, methyl,[196] and *t*-butoxy radicals and the benzophenone triplet. Peroxy radicals, bromine atoms, and polystyryl radicals have somewhat similar selectivities tending to discriminate strongly against aliphatic hydrogens, abstraction of which would be endothermic or, at best, thermoneutral. The high selectivity for trichloromethyl radical attack on alkyl aromatic hydrocarbons is quite remarkable in view of the exothermicity of these reactions. Perhaps the most interesting feature of Table 6 is the rather monotonous pattern of C—H reactivities toward the different radicals, particularly in exothermic processes. The similarities in the reactivity patterns must mean that the relative heats of reaction for the various substrates are of more importance in determining their reactivities than steric or polar effects arising from changes in the attacking radical.

The presence of a heteroatom adjacent to the carbon from which a hydrogen atom is abstracted rarely has a dramatic effect on reactivity. (See also Chapter 7.)

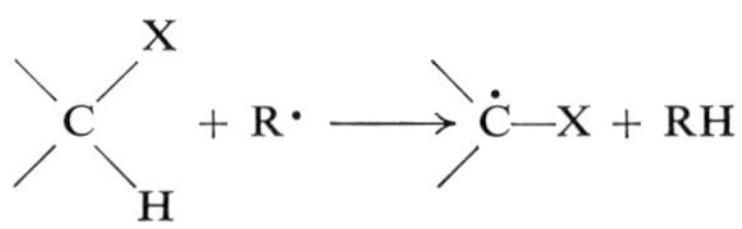

The maximum effect of an α-heteroatom (which is comparable to that produced by an aromatic ring in the α position) occurs when X = O (ethers and alcohols), S (sulfides), and N (amines). Thus, for hydrogen abstraction by the *t*-butylperoxy radical at 30° the rate constants per α-(secondary)hydrogen are[199–201]: ethylbenzene, 0.1 ($M^{-1}\ \text{sec}^{-1}$); di-*n*-butyl ether, 0.016; tetrahydrofuran, 0.085; tetrahydrothiophene, 0.025; while cyclohexane is only 0.0003. The rate constants per α-(tertiary)hydrogen for abstraction by *t*-butylperoxy are: cumene, 0.18; diisopropyl ether, 0.055; 2,5-dimethyltetrahydrofuran, 0.4; isopropanol, 0.009; and isobutane, 0.003. Similarly, the relative reactivities of some primary hydrogens toward the phenyl radical at 60° are[166]: toluene, 1.0; dimethyl ether, 0.52; methanol, 0.48; dimethyl sulfide, 1.8; and trimethylamine, 4.0. Other heteroatoms (e.g., Cl, Br, Si) and functional substituents [e.g., NO_2, C(=O)O, C≡N, C=O] generally activate the hydrogens on adjacent carbons but to only a relatively small extent compared with a phenyl group or double bond. Thus, for *t*-butylperoxy radical attack on the tertiary hydrogen of 2-substituted propanes at 30° the relative rate constants are[199–201]: hydrogen (i.e., no substituent), 1.0 (absolute value is $2.3 \times 10^{-4}\ M^{-1}\ \text{sec}^{-1}$); NO_2, 1.0; Cl, 2.0; Br, 2.0; CH_3, 10; $CH_3C(O)O$, 8; C≡N, 33; $CH_3OC(O)$, 40; and C=O, 87; while for the phenyl substituent the relative rate constant is 500–600. Similarly, the relative reactivities of the primary hydrogens in XCH_3 toward the phenyl radical at 60° are[166]: X = alkyl, 1.0; NO_2, 1.4; Cl, 2.0; $(CH_3)_3Si$, 2.1; C=O, 2.4; $CH_3OC(O)$, 2.5; and for a phenyl substituent, 7.7.

If the heteroatom or functional substituent is attached to a benzylic carbon, its activating effect is reduced. This is illustrated in Table 7 for the attack on α-substituted toluenes by *t*-butylperoxy radicals,[202] bromine atoms,[203] and phenyl radicals.[204] The first two radicals are strongly electrophilic and the relative rate constants correlate fairly well with one another.[202] The highly reactive phenyl radical is only very weakly electrophilic and it exhibits quite a different pattern of reactivity. The relative rate constants, k/k_0, for the reactions of bromine atoms with α-monosubstituted toluenes can be correlated by means of the Hammett equation,

$$\log k/k_0 = \rho\sigma$$

using Brown's electrophilic para-substituent constants, σ_p^+.[203] The ρ^+ value is -2.46 at 77°. The *t*-butylperoxy radical data correlate in the same way, yielding $\rho^+ = -1.2$. The smaller value of ρ^+ suggests that *t*-butylperoxy is less electrophilic than bromine atoms. The correlation with σ_p^+ suggests that both inductive and resonance effects influence the reactivity of the labile hydrogens.

Since α substituents on toluene do not have a dramatic effect on hydrogen atom reactivity, it is no surprise to find that more remote substituents on the aromatic ring have an even smaller influence. (See also Chapter 7.) Thus, the relative rates of attack on ring-substituted toluenes by bromine atoms at 80° can be correlated with the σ^+ substituent constants but the ρ^+ value is only -1.38.[205–207] Similarly, the ρ^+ value for *t*-butylperoxy radical attack on ring-substituted toluenes at 30° is -0.56,[208] only half the value for attack on α-substituted toluenes.

Substituent effects on hydrogen abstraction from ring-substituted toluenes and analogous compounds such as phenols, thiols, and anilines have been

TABLE 7

Rate Constants Relative to Toluene (per α-Hydrogen) for Reactions of α-Substituted Toluenes with *t*-Butylperoxy Radicals, Bromine Atoms, and Phenyl Radicals

Toluene derivative	$Bu^tOO\cdot$ at 30°	$Br\cdot$ at 77°	Phenyl at 60°
$PhCH_2OCH_2Ph$	33	159[a]	13
$(Ph)_3CH$	28	102	39
$(Ph)_2CHCH_3$	7.3	66	18
$PhCH(CH_3)_2$	18	57.5	9.7
$PhCH_2OPh$	11	46	11.4
$PhCH_2CH_3$	11	25.2	4.5
$PhCH_2Ph$	9.4	17.6	7.7
$PhCH_2OCOCH_3$	0.75	1.37	—
$PhCH_3$	1.0	1.0	1.0
$PhCH_2Cl$	0.9	0.46	6.2
$PhCH_2COOCH_3(Bu^t)$[b]	3.0	0.27	3.8
$PhCH_2CN$	1.1	0.22	24
$PhCH_2Br$	0.7	0.18	5.1
$PhCH_2OH$	6.7	—	14
$PhCH_2CH{=}CH_2$	20.5	—	7.2
$PhCH_2COCH_2Ph$	5	—	20
$PhCH_2OCOPh$	0.95	—	3.8
$PhCHO$	94.5	—	22

[a] Benzyl methyl ether.

[b] Methyl phenyl acetate was used for the reactions of bromine atoms and phenyl radicals while *t*-butyl phenyl acetate was used for *t*-butylperoxy radicals.

studied extensively.[209] The ρ values which are derived from such studies are useful in indicating the electrophilicity (or otherwise) of the attacking radical. Under standard conditions the magnitude of ρ is characteristic for each radical, which has proved extremely useful in identifying the active abstracting agent in certain cases. For example, ρ measurements on substituted toluenes have been used to show that bromine atoms are the effective species in *N*-bromosuccinimide bromination[205,206] and that chlorine atoms make an important contribution to *t*-butyl hypochlorite chlorinations.[210,211] However, it must be emphasized that since ρ^+ values in hydrogen abstraction are rarely larger (more negative) than -2, a ring substituent will rarely change the rate of abstraction, relative to the base hydrocarbon, by more than an order of magnitude. Much more dramatic effects are, of course, observed in heterolytic reactions.[212]

The halogenation of organic substrates with *N*-chloroamines and *N*-bromoamines in strong acids is a free radical chain process in which hydrogen is abstracted by a protonated amino radical, $\mathrm{H{-}\overset{\diagdown}{\underset{\diagup}{N}}\cdot^{+}}$ (an aminium radical[130,213,214]). Suitable substrates include saturated straight-chain carboxylic acids, esters, ethers, alcohols, amines, amides, and so on.[130,213,215] Halogenations of this type are of unusual synthetic importance for a number of reasons,[213,215] including, in particular, the high selectivity for the $\omega - 1(\mathrm{CH_2})$ position in 1-substituted straight-chain alkanes.

$$\mathrm{H{-}\overset{\diagdown}{\underset{\diagup}{N}}\cdot^{+} + CH_3(CH_2)_nX \longrightarrow H{-}\overset{\diagdown +}{\underset{\diagup}{N}}H + CH_3\dot{C}H(CH_2)_{n-1}X} \qquad \text{(often > 90\%)}$$

$$\mathrm{CH_3\dot{C}H(CH_2)_{n-1}X + H{-}\overset{\diagdown +}{\underset{\diagup}{N}}Cl \longrightarrow CH_3CHCl(CH_2)_{n-1}X + H{-}\overset{\diagdown}{\underset{\diagup}{N}}\cdot^{+}}$$

The rate constant for the hydrogen abstraction has been estimated to be in the range 7×10^2 to 1×10^4 M^{-1} sec^{-1} at 30° in solutions of sulfuric acid (2 M and 4 M) in acetic acid.[130] Although this propagation step (which is rate controlling) is not particularly fast the overall halogenation is quite rapid because of the relatively low rate constant for the chain termination step (6×10^6 to 5×10^7 M^{-1} sec^{-1}) which involves a bimolecular reaction of two aminium radicals.[130] The high selectivity for the $\omega - 1$ position is presumably due to the fact that X is protonated in the solvents employed, that is,

$$\mathrm{H{-}\overset{\diagdown}{\underset{\diagup}{N}}\cdot^{+} + CH_3(CH_2)_n\overset{+}{X}H \longrightarrow H{-}\overset{\diagdown +}{\underset{\diagup}{N}}H + CH_3\dot{C}H(CH_2)_{n-1}\overset{+}{X}H}$$

The attacking radical cation will be very sensitive to both the inductive and field effects due to $\mathrm{XH^+}$ and will tend to attack the molecule at a site remote from this group.

The abstraction of hydrogen attached to heteroatoms

$$\mathrm{{-}XH + R\cdot \longrightarrow {-}X\cdot + RH}$$

may be very much faster or very much slower than abstraction from carbon, principally because the X—H bond strength may be considerably different from the C—H bond strength in structurally related compounds. Hydrogen abstractions from heteroatoms in solution have not received much attention in general, though a few specific reactions (e.g., $\mathrm{ROO\cdot + ArOH \rightleftharpoons ROOH + ArO\cdot}$) have been explored in considerable detail. An appreciable fraction of the available data is summarized in Table 8 which lists most of the commonly used "good" hydrogen donors.

The three principal radicals produced during the irradiation of aqueous solutions with ionizing radiation are the hydroxyl radical, hydrogen atoms, and the solvated electron, e_{aq}^{-}. A comprehensive list of absolute rate constants for reactions involving these species has been compiled by Anbar and Neta.[238] A limited selection of their data is given in Table 9. These data (which could not conveniently be included in Table 6, 7, or 8) cover hydrogen abstractions, halogen abstractions, additions to multiple bonds, and electron transfer processes.

The solvated electron reacts rapidly with good electron acceptors but very slowly with poor acceptors. The hydrogen atom and hydroxyl radical may abstract, add, or undergo an electron transfer. Most interesting is the fact that the reactions of the hydroxyl radical with aromatic compounds normally involve an initial addition to the aromatic ring. This is true not only of

TABLE 8

Absolute Rate Constants (Log M^{-1} sec^{-1} Units) for Hydrogen Atom Abstraction from Some Heteroatoms by Some Common Radicals

Hydrogen donor	$CH_3\cdot$ 25°	$(CH_3)_3CO\cdot$ 30°	$ROO\cdot$[n] 65°	Polystyryl·[r] 60°
$C_6H_5O\underline{H}$	4.2[a]	7.3[k]	3.7	−0.7[s]
2,4,6-*t*-$Bu_3C_6H_2O\underline{H}$	4.0[a]	6.3[k]	4.2	−0.2[s,t]
$(CH_3)_3COO\underline{H}$	4.5[b]	7.0[l]	~3.3[o]	—
$R_2NO\underline{H}$	—	—	5.7[p]	—
$C_6H_5N\underline{H}_2$	—	—	~2.5	−1.5[u]
2,4,6-*t*-$Bu_3C_6H_2N\underline{H}_2$	—	—	3.4	—
$C_6H_5N\underline{H}CH_3$	—	—	3.6	—
$(C_6H_5)_2N\underline{H}$	—	—	4.3	—
$CH_3(CH_2)_3S\underline{H}$	6.1[c]	—[11]	—	3.5[v]
2,4,6-*t*-$Bu_3C_6H_2S\underline{H}$	—	—	~3.8[q]	—
n-$Bu_3Ge\underline{H}$	5.2[d]	—	—	—
n-$Bu_3Sn\underline{H}$	6.8[e,f]	~8[m]	—	—
$\underline{H}Cl$	7.1[g,h]	—	—	—[w]
$\underline{H}Br$	7.9[g,i]	—	—	—
$\underline{H}I$	8.9[g,j]	—	—	—

[a] At 60° in heptane. Estimated from data in reference 216 on the assumption that $k^{60°}$/H for $CH_3\cdot$ abstraction of a *sec*-aliphatic H is 120 M^{-1} sec^{-1} (Table 6). This reference includes measurements of tritium isotope effects. The activation energy differences and relative preexponential factors for abstraction of the phenolic H and of H from heptane are given. Abstraction of the hydroxylic H from simple alcohols proceeds at about 1% of the rate of abstraction of the phenolic H.

[b] At 40° in heptane. Calculated from the expression (reference 217) $k_{OH}/k_{CH} = 0.03 \exp(7000/RT) = 2350$ at 40° where k_{OH}/k_{CH} is the relative rate of H abstraction by $\cdot CH_3$ from OH and from a single CH in $(CH_3)_3COOH$. From data in Table 6, k_{CH} (per H) at $40° = 0.4$ antilog $1.5 = 12.6\ M^{-1}\ sec^{-1}$, hence $k_{OH} = 2350 \times 12.6 = 3 \times 10^4\ M^{-1}\ sec^{-1}$. In water the relevant expression is $k_{OH}/k_{CH} = 87 \exp(-1100/RT)$, yielding $k_{OH} = 190\ M^{-1}\ sec^{-1}$ at 40°. The H-bonded H is clearly much more difficult to abstract.

[c] Estimated rate constant for abstraction by the *sec*-alkyl radical which is formed in the addition of an *n*-butylthiyl radical to 1-pentene (reference 218).

[d] Reference 85.

[e] Reference 12.

[f] Values for other alkyl radicals include: 1×10^6, *n*-hexyl; 1.2×10^6, cyclohexyl; and 7.4×10^5, *t*-butyl.

[g] Gas phase.

[h] Calculated from the relation $\log k = 8.6 - (2100/2.3\ RT)$ (reference 219).

[i] Calculated from the relation $\log k = 8.9 - (1400/2.3\ RT)$ (reference 220, see also references 221–223).

[j] Calculated from the relation $\log k = 9.5 - (800/2.3\ RT)$ (reference 224).

[k] At 122° $k_{\text{abstraction}}/k_{\beta\text{-scission}} = 540$ for phenol and 56 for 2,4,6-tri-*t*-butylphenol in CCl_4 (reference 225), while for toluene at this temperature $k_{\text{abst}}/k_{\beta\text{-scission}} = 0.116$ per α-H (reference 226). The absolute rate constant for abstraction from toluene at 122° ≈ $4 \times 10^4\ M^{-1}\ sec^{-1}$ per α-H (Table 6).

[l] Attack of the cumyloxy radical on cumene hydroperoxide (reference 25).

[ll] Triplet acetophenone is quenched by butanethiol ($k = 1.4 \times 10^7\ M^{-1}\ sec^{-1}$), but attack is on the nonbonding electrons of the sulfur and not on the hydrogen. [R. G. Zepp and P. J. Wagner, *Chem. Comm.*, 167 (1972).]

[m] Triplet ketones, which are probably of similar reactivity to *t*-butoxy, are reduced by tributyltin hydride. For acetone the rate constant is $8 \times 10^8\ M^{-1}\ sec^{-1}$ at 25° (reference 227a) and for benzophenone $5 \times 10^7\ M^{-1}\ sec^{-1}$ (reference 227b).

[n] Poly(peroxystyryl)peroxy radicals. The rate constant data for H abstraction from phenol and from many substituted phenols which are not listed here are contained in references 25, 159, 228, and 229. Some of the early rate data were revised in later publications. The activation energy for H abstraction from phenols is small (reference 230) and the reverse reaction is fast (references 230, 231, and references cited therein). The data for H abstraction from aniline and from many substituted anilines not listed are contained in references 229 and 232.

[o] Reference 233.

[p] Reference 137. A dialkyl and a diaryl hydroxylamine both gave about the same rate constant.

[q] Reference 229.

[r] From measured transfer constants taking $k^{60°} = 145\ M^{-1}\ sec^{-1}$ for styrene polymerization (reference 7).

[s] Reference 234.

[t] 2,6-Di-*t*-butylphenol.

[u] Reference 235.

[v] Reference 236. The transfer constant for thiophenol has been reported to be much smaller than for alkyl mercaptans (reference 237). This seems unlikely to be correct and suggests rather that the phenylthiyl radical may be inefficient at reinitiating the polymerization (see reference 218).

[w] The transfer constant is reported to be zero at 100° (reference 237).

TABLE 9

Absolute Rate Constants for Reactions of HO·, H·, and e_{aq}^- in Aqueous Solution at Room Temperature[a]

Reactant	HO· $k \times 10^{-8}$ (M^{-1} sec^{-1})		H· k $\times 10^{-8}$ (M^{-1} sec^{-1})		e_{aq}^- $k \times 10^{-8}$ (M^{-1} sec^{-1})
CO_2	<0.01		<0.01		77
Fe^{2+}	3	(et)	0.2		2
H_2	0.5	(ab)	—		<0.1
H_2O_2	0.2	(ab)	0.8	(ab)	120
I^-	80	(et)	—		—
I_2	—		400	(ab)	510
O_2	—		200	(add)	190
Acetic acid	0.14	(ab)	0.0013[b]	(ab)	1.8
Acetone	0.6	(ab)	0.017[b]	(ab)	60
Aniline	50	(add)	18		0.2
Benzene	35	(add)	0.12[f]		0.1
Benzoate ion	60[c]	(add)	10[d]		30
2-Bromoethanol	4.6	(ab)	0.17 1.7	(H ab); (Br ab)	16
t-Butanol	3	(ab)	0.001	(ab)	—
Chloroform	0.08	(ab)	—		250
Dioxane	20	(ab)	0.05	(ab)	—
Ethanol	11	(ab)	0.25[b]	(ab)	<0.001
Ethylene	41[e]	(add)	32[e]	(add)	<0.025
Hydroquinone	120	(add)	—		<0.1
Methanol	5	(ab)	0.029[b]	(ab)	<0.0001
Phenol	~70	(add)	42		<0.04
Phenylacetate ion	26	(add)	8		0.3
2-Propanol	13	(ab)	0.78[b]	(ab)	—
Toluene	30	(add)	—		0.1

[a] Data from reference 238 unless otherwise noted: (ab) = abstraction; (add) = addition; (et) = electron transfer.
[b] Reference 239.
[c] Reference 240.
[d] Reference 241.
[e] Reference 145.
[f] Reference 241a. See also P. Neta and R. H. Schuler, *J. Am. Chem., Soc.* **94**, 1056 (1972).

benzene but also of substrates such as benzoic acid and the benzoate ion,[240] phenylacetic acid,[242] anisole,[243] toluene,[244] and even phenol and hydroquinone.[245–247] However, with aniline initial ·OH addition occurs 80% on the amino group and only 20% on the ring.[248] In both cases water is eliminated to produce the anilino radical. Similarly,[248] with alkylamines ·OH adds to the nitrogen and this is followed by the elimination of water to yield the amino radical. The cyclohexadienyl radicals produced by ·OH addition to aromatic rings may enter into radical–radical reactions or, if they have appropriate structural features, they may undergo an acid-catalyzed heterolytic bond rupture in which the hydroxide ion is eliminated from the ring and a proton (or a proton plus CO_2 or H_2CO) is lost from the side chain, for example,

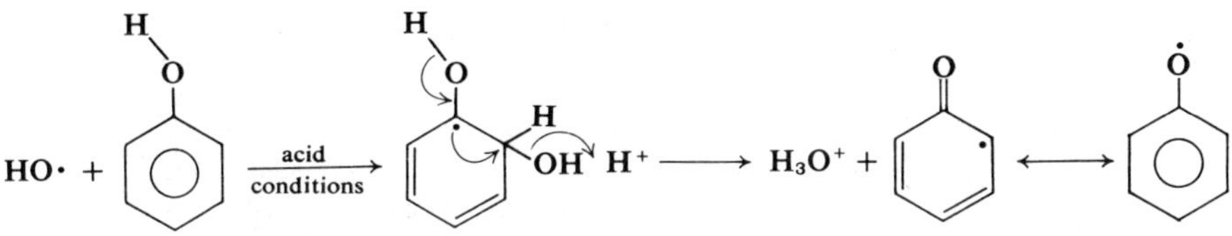

The mechanism of the attack of hydroxyl on aromatic compounds is analogous to electrophilic substitutions.

Pryor and co-workers[249,250] have generated hydrogen atoms by a variety of nonradiolytic methods. They obtained relative reactivities for hydrogen atom abstractions from hydrocarbons which agree with the results of radiolytic studies. However, their results for alcohols differ somewhat from those found by radiolysis. Thus, Pryor et al.'s most reliable data[250] for methanol, ethanol, and isopropanol yield relative reactivities of 1:2.8:4.7, respectively, whereas the radiolytic results quoted in Table 9 yield 1:9:27, and a direct ESR method for measuring the rates of reaction of hydrogen atoms with organic compounds in aqueous solution[241a] yields relative reactivities of 1:16:41, ($k_{CH_3OH} = 1.6 \times 10^6\ M^{-1}\ sec^{-1}$).[251] Although such discrepancies are disturbing, the possible changes that may be necessary to the absolute rate constants for these reactions are within the expected experimental error.

c. Rate Constants for Halogen Atom Abstractions. The abstraction of halogen atoms has received much less attention than the abstraction of hydrogen atoms. Alkyl radicals appear to abstract halogen from Cl_2, Br_2, and I_2 at essentially the diffusion-controlled limit, but they only abstract hydrogen from HCl, HBr, and HI. Perhalocarbons such as CCl_4 and CBr_4 are good halogen donors to alkyl radicals (see Table 10). Bromine is transferred much more readily than chlorine. The rate constant ratios for the abstraction of bromine from CCl_3Br and of chlorine from CCl_4 are about 50 for bridgehead

TABLE 10

Absolute Rate Constants for Halogen Abstractions from Perhalocarbons by Alkyl Radicals (M^{-1} sec^{-1} Units)[a]

Radical	CCl_4 60°	CBr_4 60°
Polystyryl	1.3	61500[c]
Poly(methyl methacrylate)	0.17	230
Poly(methyl acrylate)	0.26	850
Poly(vinyl acetate)	> 2300	> 90000
Methyl	1300[b]	—

[a] Calculated from the chain transfer constants and vinyl polymerization propagation constants given on pp. 157 and 95, respectively, of reference 7, unless otherwise noted.

[b] At 100°. Calculated from the ratio of the rate of H abstraction from toluene to Cl abstraction from CCl_4 given in reference 254. The rate constant for H abstraction from toluene at 100° was calculated to be 1×10^3 M^{-1} sec^{-1} from data in Table 6. A rate constant of $\sim 10^3$ M^{-1} sec^{-1} at 65° for Cl abstraction can also be calculated from data in reference 255.

[c] Redetermined by R. A. M. Thomson and I. R. Walters [*Trans. Faraday Soc.*, **67**, 3046 (1971)]. The original value (200 M^{-1} sec^{-1}) was apparently in error because the chain transfer constant was measured on a reaction carried to a fairly high conversion. The possibility that some of the other rate constants listed in this column are in error for the same reason cannot be overlooked.

alkyl radicals, about 150 for phenyl radicals, about 500 for *sec*-alkyl radicals, and about 1700 for benzyl radicals.[253]

The abstraction of iodine from alkyl iodides by carbon radicals is a much faster process than the abstraction of bromine and is very much faster than the abstraction of chlorine. Thus, the chain transfer constants in styrene at 60° are 4×10^{-6} for *n*-butyl chloride, 6×10^{-6} for *n*-butyl bromide (probably principally hydrogen abstraction in both cases), and 1.9×10^{-4} for *n*-butyl iodide.[235] Rapid iodine atom transfer has been shown to take place between aryl radicals and aryl iodides,[256–258] between aryl radicals and alkyl iodides,[259] and between alkyl radicals and alkyl iodides[255,260–262] (see Table 11).

The relative reactivities of a series of substituted phenyl iodides toward phenyl radicals can be correlated with the Hammet equation with $\rho = 0.57$.[258] This is the first reported example of an abstraction by a carbon radical giving a positive ρ value.[264] The transition state must involve the development of anionic character on the carbon from which the iodine is

TABLE 11

Absolute Rate Constants for Some Halogen Atom Abstractions[a] (M^{-1} sec^{-1} Units) $R\cdot + R'X \rightarrow RX + R'\cdot$

$R\cdot$	R'	X = I	X = Br	X = Cl
CH_3	CH_3	1.3×10^3	~1.5	—
CH_3	C_2H_5	5.4×10^4	—	—
CH_3	$CH_3CH_2CH_2$	4×10^5 [b]	—	—
CH_3	$(CH_3)_2CH$	2.5×10^5	—	—
CH_3	$(CH_3)_3C$	5.0×10^6	—	—
CH_3	$C_6H_5CH_2$	2.3×10^7	2.0×10^3	—
CH_3	CH_2Cl	1.9×10^7	4.2×10^2	—
CH_3	$CHCl_2$	—	3.9×10^4	—
CH_3	CCl_3	—	2.2×10^6	$4 \times 3.3 \times 10^2$
CH_3	CF_3	6.0×10^7	—	—
$CH_2{=}CHCH_2$	$CH_2{=}CHCH_2$	3.0×10^7 [b]	—	—
C_6H_5	C_6H_5	7×10^7 [c]	—	—

[a] Data at 65° from reference 254 unless otherwise noted. Calculated assuming that the rate constant for H abstraction from toluene is 3×10^2 M^{-1} sec^{-1} at 65° (see Table 6).
[b] At 100°, reference 262.
[c] At 25°, products unspecified, reference 263. At 60° this reaction is about 50 times as fast as the Cl abstraction from CCl_4.[258]

removed ($[\overset{\delta-}{R}\cdots I\cdots\overset{\delta+}{Rad}]^{\ddagger}$) whereas in hydrogen abstractions, which all have negative ρ values, cationic character is produced ($[\overset{\delta+}{R}\cdots H\cdots\overset{\delta-}{Rad}]^{\ddagger}$). Phenyl radical abstraction from alkyl iodides is much faster than from aryl iodides, being comparable to the rate of abstraction of bromine from CCl_3Br.[259]

tert-Butyl hypochlorite is an extremely efficient chlorine atom donor toward alkyl radicals. For example, a cis source of 9-decalyl radicals gives a short-lived radical which changes into the same radical as is obtained from a trans source but yields cis products on reaction with *t*-butyl hypochlorite[265] or oxygen.[266] The more stable radical obtained from a trans source yields both cis and trans products on reaction with these reagents. The high rate constant of the chlorine abstraction is indicated by the fact that the radical from a cis source could not be trapped by tin hydrides[267,267a] ($k_{R\cdot + R'_3SnH} \leqslant 3 \times 10^6$ M^{-1} sec^{-1}).[12] The lifetime of the short-lived radical from a cis source has been estimated[266] to be approximately 10^{-8}–10^{-9} sec.

Further evidence for a very fast abstraction of chlorine from *t*-butyl hypochlorite is provided by Zavitsas' experiments[268] with chlorine-36 labeled hypochlorite. In the reaction with toluene, 2,3-dimethylbutane, and neopentane in CCl_4 all the radioactivity appeared in the alkyl chloride product and none in the CCl_4. Walling and Kurkov's[269] kinetic study of chlorinations with *t*-butyl hypochlorite yielded rate constant ratios for the reactions

$$R\cdot + t\text{-BuOCl} \xrightarrow{k_p} RCl + t\text{-BuO}\cdot$$

$$R\cdot + R\cdot \xrightarrow{2k_t} \text{molecular products}$$

At 30° for R· = cyclohexyl, $2k_t/k_p^2 = 3.5 \times 10^{-2}\ M\ \text{sec}^{-1}$, which yields $k_p = 7.5 \times 10^5\ M^{-1}\ \text{sec}^{-1}$ if $2k_t$ is taken to be $2 \times 10^9\ M^{-1}\ \text{sec}^{-1}$. Similarly, for R· = benzyl, $2k_t/k_p^2 < 7.5 \times 10^{-4}$, giving $k_p > 1.6 \times 10^6\ M^{-1}\ \text{sec}^{-1}$. However, it seems rather unlikely that the rate constant for the benzyl radical would be larger than for the cyclohexyl radical and, moreover, rate constants in the range of about $10^6\ M^{-1}\ \text{sec}^{-1}$ would appear to be too low to account for the trapping of the *cis*-decalyl radical. However, the kinetics which yielded these rate constant ratios are by no means firmly established.[108]

Alkyl radicals abstract chlorine from *t*-butyl hypochlorite more readily than they abstract bromine from CCl_3Br. For example, Walling and Padwa[270] showed that the irradiation of an equimolar mixture of hypochlorite and CCl_3Br containing cyclohexane yields 97% cyclohexyl chloride and only 3% bromide. However, if a high $[CCl_3Br]/[\text{hypochlorite}]$ ratio is maintained by adding the hypochlorite slowly to an irradiated solution of cyclohexane in CCl_3Br, the yield of bromide can be increased to 94%. Toluene, on the other hand, yields only 3% benzyl bromide even under these latter conditions, possibly because attack of the benzyl radical on CCl_3Br is an endothermic process. Similar experiments with a more quantitative kinetic aim in view have been carried out by Zavitsas and Ehrenson.[271] Their results with nonresonance-stabilized alkyl radicals generally support Walling and Padwa's conclusions. At 0° the rate constant ratios for chlorine abstraction to bromine abstraction, k_{Cl}/k_{Br}, are 20 for the ethyl radical, 8 for isopropyl, 1 for cyclohexyl (which seems too low), and 4 for *t*-butyl. In these cases, the energy of activation is reported to favor bromine abstraction by 0.2–3.3 kcal/mole while the preexponential factor favors chlorine abstraction by a factor of 10 to 2×10^3. However, results reported for the cyclopentenyl radical ($k_{Cl}/k_{Br} = 0.1$ at 0°, $A_{Cl}/A_{Br} = 2.8 \times 10^{-7}$) are difficult to credit since there is no obvious reason for the preexponential factor to favor bromine abstraction by such a large amount. This result is also inconsistent with Walling and Padwa's toluene experiments.

The reduction of alkyl and aryl halides by trialkyltin hydrides has been more thoroughly studied than any other halogen abstraction reaction.[11,12,272] The rates of abstraction of the halide by a trialkyltin radical

$$R'_3Sn\cdot + RX \longrightarrow R'_3SnX + R\cdot$$

increase in the order RF < RCl < RBr < RI and, for any particular halogen, the reactivities increase in the order aryl < primary alkyl < secondary alkyl < tertiary alkyl < benzyl halide. Some rate constants for halogen abstraction by the tri-*n*-butyltin radical are given in Table 12.

TABLE 12

Absolute Rate Constants for Halogen Abstraction by the Tri-*n*-Butyltin Radical in Cyclohexane at 25° (M^{-1} sec^{-1} Units)[12]

$CH_3(CH_2)_3CH_2Cl$	8.5×10^2	$CH_3(CH_2)_4CH_2Br$	1.9×10^7
cyclo-$C_6H_{11}Cl$	2.0×10^3	cyclo-$C_6H_{11}Br$	2.2×10^7
$(CH_3)_3CCl$	1.6×10^4	$(CH_3)_3CBr$	8.5×10^7
$C_6H_5CH_2Cl$	6.4×10^5	CH_3I	2.5×10^9

Rate constants for halogen atom abstraction by pentacyanocobaltate(II) in aqueous methanol at 25° follow a pattern similar to that exhibited by the trialkyltin radical (see Table 13).[273]

TABLE 13

Absolute Rate Constants for Halogen Abstraction by Pentacyanocobaltate(II) in Aqueous Methanol at 25° (M^{-1} sec^{-1} Units)[273]

$C_6H_5CH_2Cl$	4.9×10^{-4}	CH_3I	9.5×10^{-3}
$C_6H_5CH_2Br$	2.3	CH_3CH_2I	5.9×10^{-2}
p-$NO_2C_6H_4CH_2Br$	1.05×10^2	$(CH_3)_2CHI$	1.2
$C_6H_5CH_2I$	3.8×10^3	$(CH_3)_3CI$	9.1

$$Co(CN)_5^{3-} + RX \longrightarrow Co(CN)_5X^{3-} + R\cdot$$

The same is true for halogen abstraction by ethylenediaminechromium(II) complexes in aqueous dimethylformamide at 25°.[274]

$$Cr^{II}(en)_2^{2+} + RX \longrightarrow Cr^{III}(en)_2^{2+}X^- + R\cdot$$

However, in this case the measured rate constants depend on the ratio of ethylenediamine to chromium(II), which was attributed to different formation constants for various $Cr^{II}(en)_n$ complexes. Thus, with isopropyl chloride and chromium(II) the rate constants varied from 2.1×10^{-3} M^{-1} sec^{-1} at an [en]/[Cr] ratio of 1, to 1.65×10^{-2} M^{-1} sec^{-1} at a ratio of 8. Some relative rate constants are: *n*-BuCl, 1.0; iso-PrCl, 3.6; *t*-BuCl, 29; *n*-BuBr, 140; iso-PrBr, 1100; *t*-BuBr, 5800; *n*-BuI, 10,000; iso-PrI, 63,000. In contrast to the above results, trialkylsilyl radicals abstract halogens in the order Br > Cl > F, I.[275] The low reactivity of the iodides is due to the weakness of the $R_3Si—I$ bond.

2. *S_H2 Reactions at Multivalent Atoms*

Steric effects are rarely very important in bimolecular homolytic substitutions at univalent, exterior atoms (H, Cl, etc.). However, for S_H2 reactions at multivalent, interior atoms steric effects can be of overriding importance. The present discussion has been restricted to three examples drawn from reference 146 that serve to illustrate this point. A more complete discussion is given by Davies and Roberts in the chapter on S_H2 reactions at metal centers (Chapter 10).

a. The Autoxidation of Organoboranes. Rate constants for the rate-controlling chain-propagation reaction in the autoxidation of organoboranes at 30° are given in Table 14.

TABLE 14

Absolute Rate Constants for the Autoxidation of Organoboranes at 30°[276,277]

$$ROO\cdot + \rangle BR \longrightarrow ROOB\langle + R\cdot$$

	Rate constants (M^{-1} sec^{-1})				
Organoborane	R = *n*-Bu	*s*-Bu	*t*-Bu	$PhCH_2$	PhCH(Me)
R_3B	2×10^6 [a]	8×10^4	—	5×10^6	—
R_2BOBR_2	3×10^5	7×10^4	—	—	—
R_2BOOR	3×10^4	1.5×10^4	—	—	—
R_2BOR'	5×10^3 [b]	2×10^3 [c]	—	—	—
$(RBO)_3$	1×10^3	5×10^4	3×10^4	1×10^6	4×10^6

[a] The rate constant for *t*-butoxy attack on this borane is $\sim 10^7$ M^{-1} sec^{-1}.[278]
[b] R′ = Et.
[c] R′ = *n*-Bu.

These reactions are extremely fast compared with rate constants for the analogous propagation reactions in hydrocarbon autoxidations. For a specific alkyl group the rate constants decrease with increasing oxygen substitution at the boron atom. This can be ascribed to the decreasing Lewis acidity of the boron atom caused by $p\pi$–$p\pi$ interaction with oxygen lone pairs. As a result, the activation energy for displacement, relative to the trialkyl borane, increases since the peroxy radical must disrupt the $\bar{B}{=}O^+$ π bond to some extent in forming the transition state for this reaction. Steric retardation of this S_H2 reaction by the branched alkyl groups becomes more apparent as the number of alkyl groups attached to boron increases. Thus, for a single alkyl group on boron, *sec*-butyl is 50 times as reactive as *n*-butyl, but for two alkyl groups on boron, *sec*-butyl is only $\frac{1}{2}$ to $\frac{1}{4}$ as reactive as *n*-butyl, and for three groups it is only $\frac{1}{25}$ as reactive.

b. The Tributyltin Radical-Induced Decomposition of Peroxides and Disulfides. Some rate constants for the attack of tributyltin radicals on peroxides

$$n\text{-Bu}_3\text{Sn}\cdot + \text{ROOR} \xrightarrow{2k_p} n\text{-Bu}_3\text{SnOR} + \text{RO}\cdot$$

peresters,

$$n\text{-Bu}_3\text{Sn}\cdot + \text{ROOR}' \xrightarrow{k_p} n\text{-Bu}_3\text{SnOR}' + \text{RO}\cdot$$

and disulfides,

$$n\text{-Bu}_3\text{Sn}\cdot + \text{RSSR} \xrightarrow{2k_s} n\text{-Bu}_3\text{SnSR} + \text{RS}\cdot$$

are listed in Table 15 together with the exothermicities of these elementary steps where they were measured.[279,280] Attack on the peroxide linkage of diethyl peroxide, diacetyl peroxide, dibenzoyl peroxide, and various *t*-butyl peresters occurs at a rather similar rate, whereas attack on the peroxide linkage of di-*t*-butyl peroxide is very much slower. Similarly, di-*t*-butyl disulfide is less reactive than the other disulfides. The low reactivity of the di-*t*-butyl compounds must be due primarily to steric shielding of the reaction center by the bulky *t*-butyl groups.

The disulfides are more reactive toward the tributyltin radical than the peroxides or peresters. This is so in spite of the fact that the heat liberated in breaking an S—S bond is much less than that liberated in breaking an O—O bond. It seems unlikely that the greater reactivity of the disulfides could be due solely to steric effects since it is difficult to believe that the peroxide link in diethyl peroxide or diacetyl peroxide is sterically hindered to any significant extent. The principal factor causing the disulfides to be more reactive than the peroxides is probably the ease with which an intermediate or transition state can be formed having nine electrons around sulfur. That is, the adduct radical [n-Bu$_3$Sn—$\dot{\text{S}}$(R)SR], or radical anion [RSSR]$^{\overset{\cdot}{-}}$, with nine electrons around sulfur is formed more easily than the adduct with nine electrons around oxygen because lower lying *d* orbitals are available to accept the extra electron.

TABLE 15

Rate Constants for Attack of Tributyltin Radicals on Peroxides,[a] Peresters,[a] and Disulfides[b]

Substrate			
RX—	—XR′	$k_p(k_s) \times 10^{-4}$ (M^{-1} sec^{-1})	ΔH (kcal/mole)
$(CH_3)_3CO$—	—$OC(CH_3)_3$	~0.005[c]	—
CH_3CH_2O—	—OCH_2CH_3	2.5	39
$C_6H_5C(O)O$—	—$O(O)CC_6H_5$	2	—
$CH_3C(O)O$—	—$O(O)CCH_3$	7	56
$(CH_3)_3CO$—	—$O(O)CCH_3$	4	57
$(CH_3)_3CO$—	—$O(O)CCH(CH_3)_2$	7	—
$(CH_3)_3CO$—	—$O(O)CC_6H_5$	14	48
$(CH_3)_3CS$—	—$SC(CH_3)_3$	4	9
$CH_3(CH_2)_3S$—	—$S(CH_2)_3CH_3$	55	15
$C_6H_5CH_2S$—	—$SCH_2C_6H_5$	35	—
C_6H_5S—	—SC_6H_5	450	—

[a] At 10° in benzene unless otherwise noted. Reference 279.
[b] At 30° in cyclohexane. Reference 280.
[c] At 40°.

c. The Alkyl Radical–Trifluoromethanesulfenyl Chloride Reaction. The free radical chain reaction of alkanes with CF_3SCl yields both alkyl chlorides and

$$R\cdot + CF_3SCl \longrightarrow RCl + CF_3S\cdot$$

$$R\cdot + CF_3SCl \longrightarrow RSCF_3 + Cl\cdot$$

alkyl sulfides.[281] Although absolute rate constants for these reactions are unknown the products provide an excellent illustration of the importance of steric effects. Thus, primary alkyl radicals from *n*-butane and isobutane give sulfides almost exclusively, only small amounts of chloride being formed. The secondary radical from *n*-butane gives sulfide and chloride with the former predominating, while the tertiary radical from isobutane gives both products but in this case the chloride predominates.

$$CF_3SCl + n\text{-}C_4H_{10} \xrightarrow{h\nu} CH_3(CH_2)_2CH_2SCF_3(13\%) + CH_3(CH_2)_2CH_2Cl(1\%) \quad [+\ CF_3SSCF_3(28\%)]$$

$$\phantom{CF_3SCl + n\text{-}C_4H_{10}} \xrightarrow{h\nu} CH_3CH_2CH(SCF_3)CH_3(46\%) + CH_3CH_2CHClCH_3(12\%)$$

$$CF_3SCl + (CH_3)_3CH \xrightarrow{h\nu} (CH_3)_2CHCH_2SCF_3(12\%) + (CH_3)_2CHCH_2Cl(1\%) \; [+\, CF_3SSCF_3(47\%)]$$

$$ \xrightarrow{h\nu} (CH_3)_3CSCF_3(24\%) + (CH_3)_3CCl(33\%)$$

These products suggest that alkyl radicals intrinsically prefer to attack the sulfur atom of CF_3SCl and the primary alkyl radicals do so almost exclusively. Secondary and tertiary alkyl radicals experience steric difficulty in attacking sulfur and they therefore tend to attack the more exposed chlorine atom.

B. Radical Addition to Multiple Bonds

The literature on relative and absolute rate constants for radical additions to multiple bonds is even more extensive than the literature on S_H2 reactions. This is because of the economic importance of processes such as the polymerization and copolymerization of vinyl monomers,

$$\sim CH_2\cdot + CH_2{=}CH_2 \longrightarrow \sim CH_2\cdot$$

and the addition of compounds such as halogens, thiols, silanes, polyhalomethanes, and so on, to double bonds,

$$ACH_2CH_2\cdot + AB \longrightarrow ACH_2CH_2B + A\cdot$$

$$A\cdot + CH_2{=}CH_2 \longrightarrow ACH_2CH_2\cdot$$

The subject of vinyl polymerization has been reviewed many times[6,7,98,103,161] and it is sufficient for the present purpose to list the propagation rate constants that have been reported for some common vinyl monomers (Table 16). With the exception of vinyl chloride (for which the data may be slightly in error) it is apparent that the activation energies for propagation are in the range 6–10 kcal/mole and the preexponential factors are about $10^{7.5 \pm 1.0}$ M^{-1} sec^{-1}. These values are typical of reactions involving radical additions to double bonds.

Since radical additions to double bonds are described by Abell in this book (Part II, Chapter 13), further discussion of the subject is restricted to a brief survey of the addition of methyl radicals to unsaturated systems. This reaction has been studied in solution in considerable detail by Szwarc and coworkers.[282] Acetyl peroxide served as the source of methyl radicals.

$$CH_3C(O)OOC(O)CH_3 \longrightarrow 2(CH_3C(O)O\cdot) \longrightarrow 2CH_3\cdot + 2CO_2$$

It was decomposed in isooctane with which the methyl radicals reacted to yield methane,

$$CH_3\cdot + C_8H_{18} \xrightarrow{k_{abst}} CH_4 + C_8H_{17}\cdot$$

TABLE 16

Propagation Rate Constants for the Polymerization of Some Vinyl Monomers[6,7,98]

Monomer	k (M^{-1} sec^{-1}) (60°)	log A (M^{-1} sec^{-1})	E (kcal/mole)
Acrylamide	18,000[a]	—	—
Vinyl chloride	12,300	6.5	3.7
N,N-Dimethylacrylamide	11,000[b]	—	—
Vinyl acetate	3,700	8.4	7.3
Methyl acrylate	2,090	8.0	7.1
Acrylonitrile	1,960	—	—
Methyl methacrylate	734	7.0	6.3
Styrene[c]	145	6.7	7.3
Butadiene	100	8.1	9.3
Isoprene	50	8.1	9.8

[a] At 25°.
[b] At 50°.
[c] Reference 6 gives $k^{60°} = 176$, log $A = 7.0$, and $E = 7.8$.

TABLE 17

Relative Rate Constants for Methyl Radical Additions at 65°[129]

Compound	$\frac{k_{add}}{k_{abst}}$	Compound	$\frac{k_{add}}{k_{abst}}$
Ethylene	34	Benzene	~0.4
Propylene	22	Diphenyl	2
Butene-1	22	Naphthalene	9
Isobutene	36	Phenanthrene	11
trans-Butene-2	6.9	Chrysene	23
cis-Butene-2	3.4	Pyrene	50
Trimethylethylene	5.6	Dibenzanthracene	150
Styrene	796	Benzanthracene	206
1,1-Diphenylethylene	1590	Benzopyrene	270
1,2-Diphenylethylene (trans)	104	Anthracene	330
1,2-Diphenylethylene (cis)	29	Naphthacene	3700
1,1,2-Triphenylethylene	48	Pyridine	1.2
Tetraphenylethylene	8	Quinoline	12
Butadiene	2090	Quinone	2000

In the presence of an aromatic or olefinic compound, A, the yield of CO_2 is unaffected but the yield of methane is decreased because of the reaction:

$$CH_3\cdot + A \xrightarrow{k_{add}} CH_3A\cdot$$

Measurement of the $[CH_4]/[CO_2]$ ratio yields the rate constant ratio k_{add}/k_{abst}. A few of the many relative rate constants obtained by Szwarc at 65° are given in Table 17. It can be seen that additions to terminal olefins are fastest. Increasing the degree of substitution around the double bond eventually causes a decrease in the rate of methyl radical addition in spite of the fact that the adduct radical, $CH_3A\cdot$, is becoming more stable. This steric effect in the phenyl-substituted ethylenes is particularly noteworthy. The activation energies and preexponential factors for the abstraction and additions are similar[170]:

Olefin	$E_{add} - E_{abst}$ (kcal/mole)	A_{add}/A_{abst}
Ethylene	−0.6	13.0
Propylene	−1.1	4.3
Isobutene	−0.5	16.2
trans-Butene-2	−1.9	0.4
cis-Butene-2	−1.9	0.2

The addition rate constants in Table 17 can be calculated by multiplying k_{add}/k_{abst} by 127 M^{-1} sec^{-1}, which is the value of k_{abst} calculated from the data in Table 6.

The relative rates of addition to the aromatic compounds can be correlated with the *localization energies* of the compounds. The localization energy is defined as the π-bonding energy necessary to isolate an electron from the resonating system and localize it on a particular atom. Such localization leaves a resonating system with one less electron and one less atom which is similar to the π-bond structure of the transition state for the methyl radical addition. The π-bond energies of the aromatic compound and localized system can be calculated by simple molecular orbital methods.[283] The difference between the two energies should be proportional to the energy required to get from the aromatic compound to the transition state for methylation and hence should be proportional to the activation energy for the reaction. The logarithms of the rate constants for methylation are therefore proportional to the localization energies.

V. RATE CONSTANTS FOR UNIMOLECULAR RADICAL REACTIONS

A. Intramolecular Rearrangements[284,285,285a]

Unimolecular radical reactions which do not involve fragmentation and for which there are at least some reliable absolute rate data are conveniently subdivided into four classes: (1) hydrogen atom shifts; (2) 1,2-phenyl group shifts; (3) ring-opening and ring-closing rearrangements; and (4) cis-trans isomerizations of allylic radicals. Some reactions illustrative of these processes are discussed below.

1. Hydrogen Atom Shifts

In the free radical decomposition of hypochlorites containing long side chains there is the possibility of intramolecular hydrogen abstraction by the alkoxy radical. The products are governed by a three-way competition, for example,

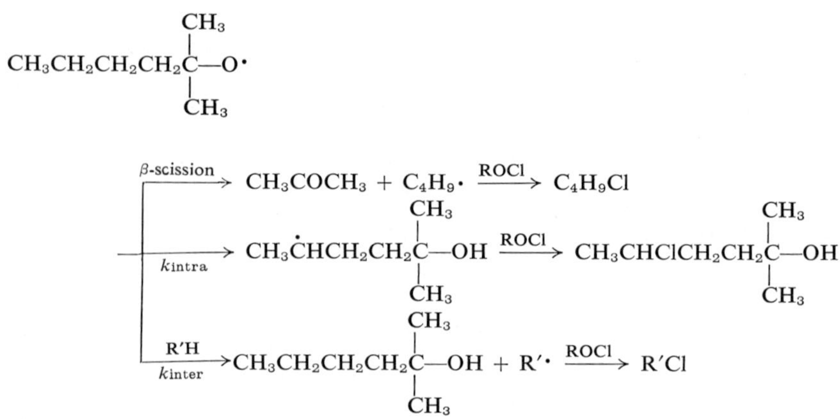

Experimentally it has been found that the intramolecular process is strongly favored.[286–288] For example,[288] the photodecomposition of 1.6 *M* butyldimethylcarbinyl hypochlorite at 0° in cyclohexane as solvent to give the above alkoxy radical yields 2% acetone and butyl chloride, 81% 2-methyl-5-chloro-2-hexanol, and 8% 2-methyl-2-hexanol and cyclohexyl chloride. The preference for intramolecular abstraction is striking. Taking the cyclohexane concentration to be about 8 *M* and the rate constant for hydrogen abstraction from cyclohexane by a *t*-alkoxy radical to be $5 \times 10^4\ M^{-1}\ \text{sec}^{-1}$ at 0° (cf. Table 6 and reference 25) yields the rate constant for the intramolecular abstraction, $k_{\text{intra}} \approx 5 \times 10^4 \times 8 \times 81/8 \approx 4 \times 10^6\ \text{sec}^{-1}$. Estimated A factors for these two reactions are[156] $A_{\text{inter}} = 10^{8.5 \pm 0.5}\ M^{-1}\ \text{sec}^{-1}$ and $A_{\text{intra}} = 10^{11.5 \pm 0.5}\ \text{sec}^{-1}$.[289] Therefore a rate constant ratio $k_{\text{inter}}/k_{\text{intra}}$ of about $10^{-3}\ M^{-1}$ implies that the two reactions have similar activation energies. Yields of intramolecular products from other long-chain hypochlorites indicate that attack on the δ-hydrogen through a quasi-six-membered cyclic transition state is strongly preferred, only small yields of ε-chloro products are formed, and there is no detectable attack at other positions.

Rather similar results have been obtained with other radicals that are active in hydrogen atom abstractions such as aminium radicals. For example,[130] for the di-*n*-butylaminium radical the ratio of intramolecular abstraction to intermolecular abstraction from *n*-butyl chloride yields $k_{inter}/k_{intra} < 10^{-3}\ M^{-1}$. The rate constant for the intermolecular reaction is in the range 7×10^2 to $1 \times 10^4\ M^{-1}\ sec^{-1}$ at 30° giving an intramolecular rate constant of at least $10^6\ sec^{-1}$.

In contrast, the much less reactive peroxy radical appears not to favor intramolecular abstraction to nearly the same extent as *t*-alkoxy or aminium radicals. Thus, among the branched-chain alkanes examined by Rust,[290] 2,4-dimethylpentane gave the highest yield of dihydroperoxide on autoxidation. At 100° the rate constant ratio $k_{inter}/k_{intra} = 1.3 \times 10^{-2}\ M^{-1}$ for this hydrocarbon[291] but the ratio is much larger (i.e., less intramolecular abstraction) with 2,3-dimethylbutane, 2,3-dimethylpentane, 2,5-dimethylhexane, and 2,6-dimethylheptane. In the autoxidation of isopropyl ether and benzyl ether at 30°, $k_{inter}/k_{intra} = 0.48$ and $0.35\ M^{-1}$, respectively.[199,292] Intramolecular hydrogen abstraction is also disfavored for the isomerization of the *n*-hexyl radical to the *sec*-hexyl radical in the gas phase.[289a] In this case it was shown that the activation energy was in the usual range for abstraction of a secondary hydrogen by an alkyl radical (8.3 kcal/mole). The preexponential factor was, however, only $2 \times 10^7\ sec^{-1}$, which is very much lower than would be predicted merely on the basis of the freezing out of internal rotations in the transition state. It would appear, therefore, that the relative importance of intramolecular and intermolecular abstractions is highly dependent on the structure and, perhaps, on the reactivity of the radical.

2. *1,2-Phenyl Group Shifts*

The 2,2,2-triphenylethyl radical has long been known to rearrange to the 1,1,2-triphenylethyl radical with great rapidity.

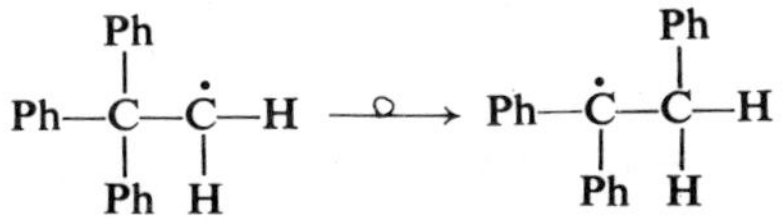

In 1966, Kaplan[293] showed that in the reaction of 2,2,2-triphenylethyl chloride with triphenyltin hydride at 70–130° the unrearranged radical could be trapped. At 100° the yield of Ph_3CCH_3 was 76% with 2.6 *M* [Ph_3SnH] and 40% with 0.75 *M* [Ph_3SnH], the other major product being Ph_2CHCH_2Ph.

Combining these product yields with an estimated rate constant of 5×10^7 M^{-1} sec^{-1} for hydrogen abstraction from the tin hydride by the $Ph_3CCH_2\cdot$ radical[12] yields a rate constant of 5×10^7 sec^{-1} for the isomerization at 100°. A similar rate constant has been calculated for the rearrangement at 70°.[294]

Many other examples of 1,2-aryl group shifts in carbon radicals are known[284,285] but usually the reported data do not permit the rate constant for the rearrangement to be calculated.[295]

Rearrangements involving 1,2-aryl group shifts are not restricted to carbon radicals.[296] For example triarylmethoxy radicals rearrange rapidly to carbon radicals,

$$Ph_3C\dot{O} \longrightarrow Ph_2\dot{C}OPh$$

Similarly, the 1,1-diphenylethoxy radical rearranges at 30° much more rapidly than it undergoes β-scission since none of the expected β-scission product, benzophenone, can be detected[297]:

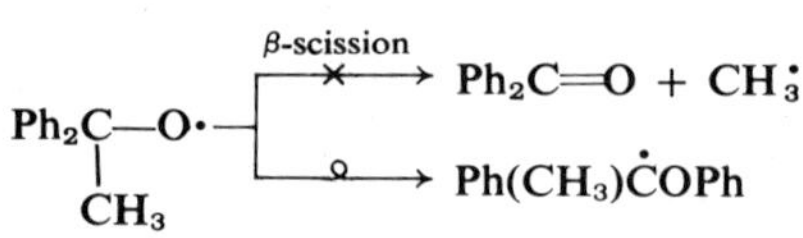

The β-scission of this radical should be at least as fast as that for cumyloxy [$Ph(CH_3)_2CO\cdot$, $k_\beta^{30°} \sim 3 \times 10^4$ sec^{-1}] so the rate constant for the 1,2-phenyl shift must be greater than 10^6 sec^{-1}.

3. Ring-Opening and Ring-Closing Rearrangements

The opening of the cyclopropyl ring in cyclopropylcarbinyl radicals to yield the allylcarbinyl radical is a very facile process.

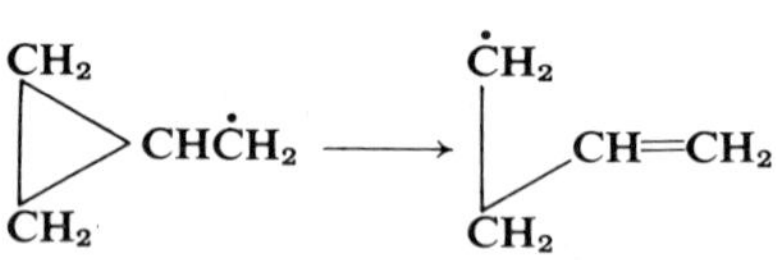

Kochi et al.[43] have shown by ESR studies of these radicals that this simple rearrangement is rapid at temperatures higher than −120° and that the reverse process does not occur at low temperatures, though it has been observed at higher temperatures.[298] Analogous ring openings of substituted cyclopropylcarbinyl radicals and ring closings of substituted allylcarbinyl radicals have been reported by several authors.[299–303]. The cyclopropylcarbinyl radical may also be an intermediate (or transition state) in a more complex rearrangement,[303,304] for example,

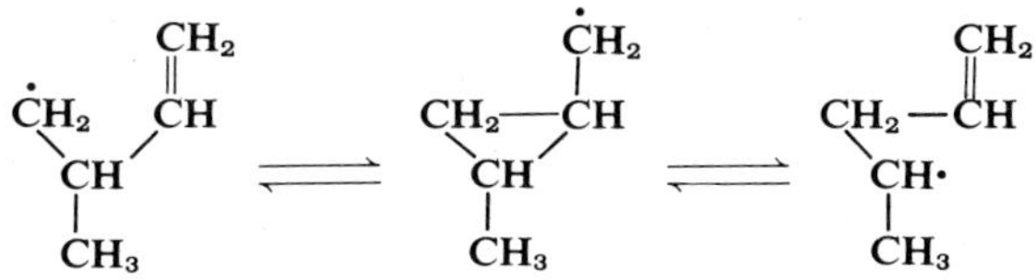

In a few cases the reported data permit the calculation of the rate constants for the rearrangement. Thus, the opening of the cyclopropyl ring as the cyclocholestanyl radical rearranges to the cholesteryl radical[301] has been estimated to have a rate constant of 10^8 sec^{-1} at 30°.[12]

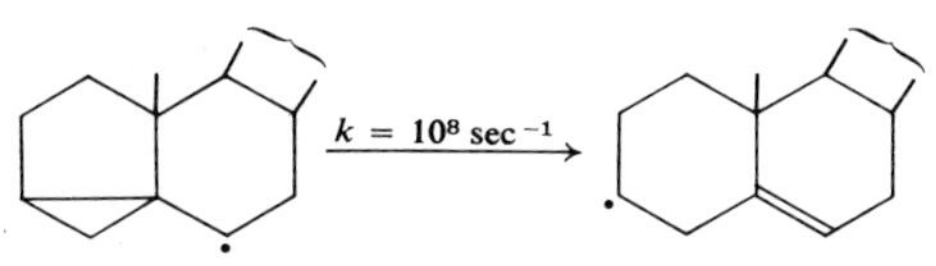

The opening of the cyclopropane ring during the rearrangement of the nortricyclyl radical to the norbornenyl radical[299] has been estimated[12] to have a similar rate constant. The reverse ring-closing reaction has a rate constant of about 4×10^7 sec^{-1}.[303]

A ring-closure reaction that has proved of great value in mechanistic studies aimed at identifying the occurrence of free radical processes[143,274,305,306] is the cyclization of the 5-hexenyl radical to yield the cyclopentylmethyl radical.

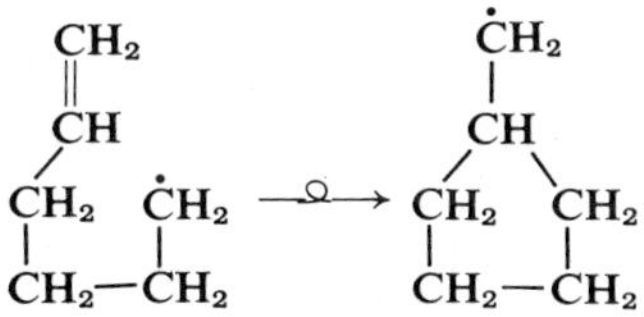

The rate constant for this rearrangement has been estimated[12] to be 10^5 sec^{-1} at 40° on the basis of the yields of 1-hexene and methylcyclopentane formed in the reduction of 5-hexenyl bromide with tributyltin hydride.[306] A similar value has been estimated from product yields in the reaction of 5-hexenyl fluoride with sodium naphthalenide,[143] it being assumed that the 5-hexenyl radical is reduced by sodium naphthalenide in a diffusion-controlled reaction. Electron spin resonance spectroscopy has been used to show that cyclization of the 5-hexenyl radical is rapid at temperatures above −55°.[44,65,307]

The analogous cyclization of the 4-(1-cyclohexenyl)butyl radical,[308]

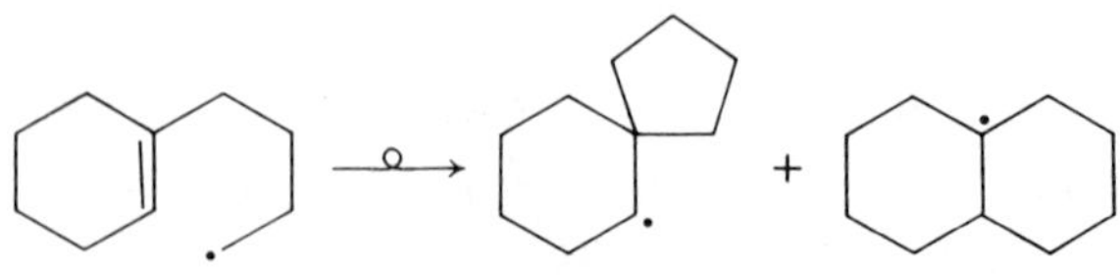

occurs with a rate constant of about 4×10^4 sec^{-1} at 65°.[12,309] A rate constant of the same order of magnitude can be calculated from Wilt et al.'s data[310] for the cyclization

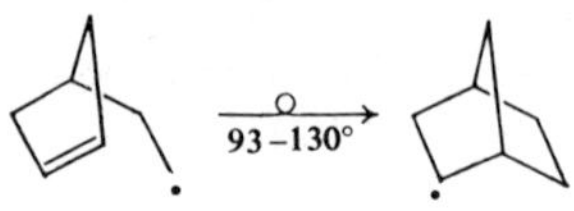

by assuming a rate constant of about 10^7 M^{-1} sec^{-1} for abstraction from the hydride.[12] This cyclization has an activation energy approximately 7 to 8 kcal/mole greater than that for hydrogen abstraction from tributyltin hydride by the primary alkyl radical.

The cyclization of the γ,γ-diphenylallylcarbinyl radical involves attack at the ortho position of one of the aromatic rings and yields, after loss of a hydrogen atom, 1-phenyl-3,4-dihydronaphthalene.[300]

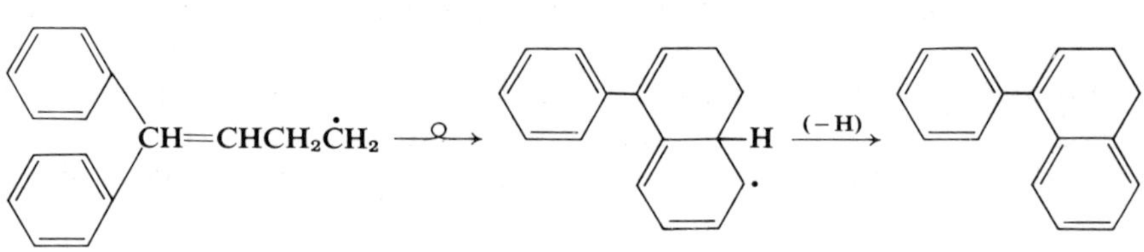

From product yields when the diphenylallylcarbinyl radical is generated at 125° in the presence of triethyltin hydride it is possible to estimate a rate constant of approximately 6×10^5 sec^{-1} for this cyclization. Other cyclizations involving intramolecular alkyl radical addition to aromatic rings are known.[311]

The three preceding rate constants for alkyl radical cyclizations may be compared with the rate constants for analogous intermolecular addition reactions. Thus, from data in Table 17 we calculate that at 65° the rate constant for methyl radical addition to butene-1 is $22 \times 127 = 2.8 \times 10^3$ M^{-1} sec^{-1} [312] while for addition to benzene the rate constant is approximately 50 M^{-1} sec^{-1}. The "effective" concentration of the double bond or aromatic ring for the intramolecular cyclizations is therefore extremely high. That is, intramolecular cyclizations as well as intramolecular hydrogen atom abstractions proceed much more readily than their intermolecular counterparts. It should perhaps be added that intramolecular cyclizations involving alkyl radical addition to triple bonds[313] and alkoxy radical addition to double bonds[314] have also been reported.

4. *Cis-trans Isomerizations of Allylic Radicals*

There is a dearth of experimental data that would allow the rate constant of this interesting radical isomerization to be calculated. *tert*-Butyl hypochlorite reacts with cis or trans olefins to give allylic chlorides with complete

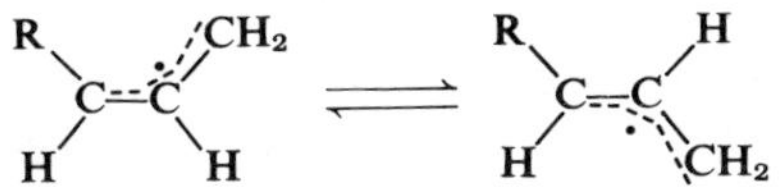

retention of configuration.[174] Allyl radicals also retain their configuration in the presence of oxygen and benzenethiol.[315] Tin hydrides, on the other hand, react with allylic chlorides to give olefins with only a partial retention of configuration,[316] and Golden[317] has shown that the experimental data are consistent with a rate constant which he calculated, on theoretical grounds, should be given by

$$k = 10^{13.5} \exp(-16700/RT) \text{ sec}^{-1}$$

B. Fragmentation Reactions[285a]

The unimolecular decomposition of a radical to yield a molecule and a smaller radical usually occurs by β-scission as in the well-known fragmentation of *t*-butoxy to acetone and a methyl radical.

$$(CH_3)_3CO\cdot \longrightarrow CH_3COCH_3 + CH_3\cdot$$

However, for a few radicals an α-scission is also a possibility. The α-scission process for a radical $R\dot{A}B$,

$$R\dot{A}B \longrightarrow RA + B\cdot$$

is clearly a stage in the homolytic substitution reaction

$$R\cdot + AB \longrightarrow R\dot{A}B \longrightarrow RA + B\cdot$$

The intermediate adduct radical is usually too short-lived to be identified and often, probably, is no more than a transition state in the overall reaction. However, it has been identified by ESR spectroscopy in one case; namely, Kochi and Krusic[48] have detected the *t*-butoxytrimethylphosphoranyl radical in the reaction of *t*-butoxy radicals with trimethylphosphine:

$$(CH_3)_3CO\cdot + (CH_3)_3P \longrightarrow (CH_3)_3CO\dot{P}(CH_3)_3$$

A signal due to the methyl radical was also identified, which implies that the phosphoranyl radical undergoes α-scission:

$$(CH_3)_3CO\dot{P}(CH_3)_3 \longrightarrow (CH_3)_3COP(CH_3)_2 + CH_3\cdot$$

The rate of this process was not determined.[318,318a]

The β-scission process represents the reverse of radical addition to a multiple bond,

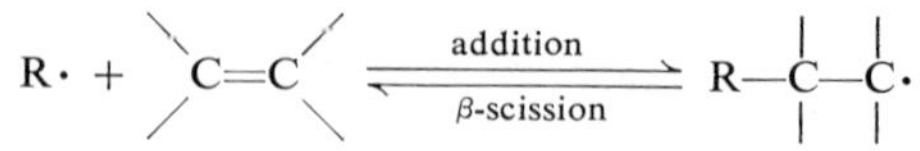

Absolute rate constants for β-scission can, in principle, be calculated from the equilibrium constant K $(= k_{add}/k_\beta)$ together with the rate constant for the addition reaction.[103] Because such equilibria are extremely difficult to study experimentally it is usual to measure β-scission rate constants by some competitive procedure. Walling and Padwa's[320] work on the β-scission of a series of t-alkoxy radicals represents an interesting example of this procedure. A series of t-alkyl hypochlorites (principally 1,1-dimethylalkyl hypochlorites) were photodecomposed in CCl_4 in the presence of cyclohexane. The relative yields of the products of β-scission and of attack on cyclohexane were determined.

$$R_1R_2R_3CO\cdot \xrightarrow[k_a]{C_6H_{12}} R_1R_2R_3COH + C_6H_{11}\cdot \xrightarrow{R_1R_2R_3COCl} C_6H_{11}Cl$$

$$R_1R_2R_3CO\cdot \xrightarrow[k_\beta]{} R_1R_2C{=}O + R_3\cdot \xrightarrow{R_1R_2R_3COCl} R_3Cl$$

The rate constant ratios k_β/k_a are given in Table 18. The results indicate increased β-scission in the order methyl < chloromethyl < ethyl, benzyl < isopropyl < t-butyl. With the exception of benzyl this order corresponds to increasing stabilization of the expelled alkyl radical. The anomalous position of benzyl was shown later by Walling and McGuinness[211] to be due to the incursion of a chlorine atom chain. That is, the cyclohexyl chloride was formed from a cyclohexyl radical produced by attack of a chlorine atom on cyclohexane rather than by the attack of the $PhCH_2(CH_3)_2CO\cdot$ radical. Since the chlorine atom chain probably interfered to varying extents with the other hypochlorites the k_β/k_a rate constant ratios are not as reliable as might be wished. Absolute values of k_β can be calculated if the reasonable assumption is made that k_a is the same for all these t-alkoxy radicals. For the t-butoxy radical $k_a^{40°} \sim 1 \times 10^5\ M^{-1}\ sec^{-1}$ (see Table 6[321]).

The variation with temperature of the rate constant for β-scission of t-butoxy radicals in CCl_4 can be represented by[25]

$$k_\beta = 3.4 \times 10^{12} \exp(-13900/RT)\ sec^{-1}$$

Relative rate constants for the different modes of β-scission of t-alkoxy

TABLE 18

Rate Constants for β-Scission of Alkoxy Radicals in CCl_4 at 40° [a]

Reaction	k_β/k_a (*M*)	
$(CH_3)_3CO\cdot \rightarrow CH_3COCH_3 + CH_3\cdot$	0.02	
$CH_2Cl(CH_3)_2CO\cdot \rightarrow CH_3COCH_3 + CH_2Cl\cdot$	0.12	
$C_6H_5(CH_3)_2CO\cdot \rightarrow C_6H_5COCH_3 + CH_3\cdot$	0.48	
$C_2H_5(CH_3)_2CO\cdot \rightarrow CH_3COCH_3 + C_2H_5\cdot$	2.1	(0.53)[b]
$i\text{-}C_3H_5(CH_3)_2CO\cdot \rightarrow CH_3COCH_3 + i\text{-}C_3H_5\cdot$	76	(68)[b]
$C_6H_5CH_2(CH_3)_2CO\cdot \rightarrow CH_3COCH_3 + C_6H_5CH_2\cdot$	(2.0) 250[c]	(> 70)[b]
$t\text{-}C_4H_9(CH_3)_2CO\cdot \rightarrow CH_3COCH_3 + t\text{-}C_4H_9\cdot$	> 300	
$\overline{\lfloor(CH_2)_5—C}(CH_3)O\cdot \rightarrow \overline{\lfloor(CH_2)_5—C}{=}O + CH_3\cdot$	10	
$\overline{\lfloor(CH_2)_4—C}(CH_3)O\cdot \rightarrow \overline{\lfloor(CH_2)_4—C}{=}O + CH_3\cdot$	120	

[a] Data from Walling and Padwa [320] unless otherwise noted.

[b] In the presence of olefin which will have partially or wholly suppressed the Cl atom chain.

[c] Data of Walling and McGuiness.[211]

radicals have been measured by several groups of workers.[286,287,322–324] For example,[287] the 2-ethyl-3-methyl-2-butoxy radical breaks down in three different ways:

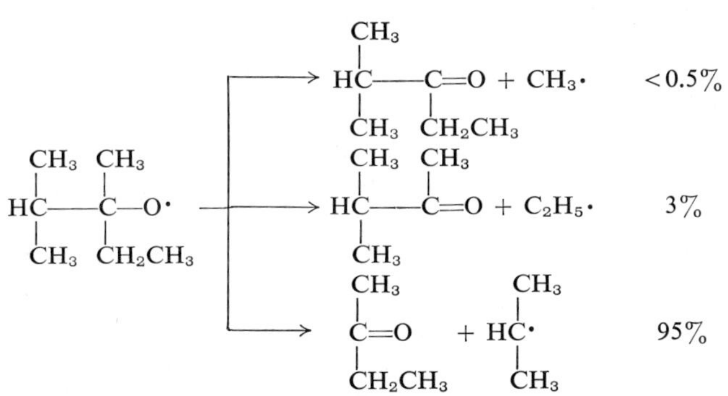

Some of these relative rate data could be put on an absolute basis by making various assumptions. Thus, it seems reasonable to suppose that the methyl group will cleave from *t*-butoxy and from other *t*-alkoxy radicals at the same rate provided the ketone which is formed is not conjugated with a double bond or aromatic ring.

As a final example of β-scission, the rate constant for decomposition of the acetoxy radical has been estimated to be $1.6 \times 10^9\ sec^{-1}$ at 60° with an activation energy of 6.6 kcal/mole.[325]

$$CH_3\dot{C}O_2 \longrightarrow \dot{C}H_3 + CO_2$$

This reaction is exothermic by about 10 kcal/mole. The decarboxylation of other acyloxy radicals is probably equally fast but the benzoyloxy radical decarboxylates at a much lower rate. Decarboxylation of the *t*-butoxycarbonyl radical,

$$t\text{-BuO}\dot{C}{=}O \rightarrow t\text{-Bu}\cdot + CO_2$$

is also a relatively slow process,

$$\log_{10} k = 10.8 - 7710/2.3\ RT\ sec^{-1}.$$ [325a]

The rates and activation parameters for the fragmentation and isomerization of many radicals in the gas phase have been summarized by Kerr and Lloyd.[326]

REFERENCES AND NOTES

1. In solution, in contrast to the vapor, there are many collisions per encounter.
2. A. M. North, *The Collision Theory of Chemical Reactions in Liquids*, Methuen, London, 1964; *Q. Rev.* (*London*), **20**, 421 (1966); *Prog. High Polym.*, **2**, 95 (1968).
3. For small molecules the Stokes–Einstein equation should probably be replaced by $D = kT/4\pi\eta r$ [E. McGlaughlin, *Trans. Faraday Soc.*, **55**, 28 (1959)].
4. For a listing of k_{diff} at 20° in a wide variety of solvents see J. G. Calvert and J. N. Pitts, Jr., *Photochemistry*, Wiley, New York, 1966, p. 627.
5. See, e.g., E. L. Allred and R. L. Smith, *J. Am. Chem. Soc.*, **91**, 6766 (1969); P. S. Engel and P. D. Bartlett, *ibid.*, **92**, 5883 (1970); S. A. Weiner, *ibid.*, **93**, 425 (1971).
6. A. M. North, *International Encyclopedia of Physical Chemistry and Chemical Physics*, Topic 17, *Macromolecules*, C. E. H. Bawn, Ed., Vol. 1, *Kinetics of Free Radical Polymerization*, Pergamon Press, Oxford, 1966.
7. C. Walling, *Free Radicals in Solution*, Wiley, New York, 1957, Chap. 3.
8. F. W. Hoffmann, R. J. Ess, T. C. Simmons, and R. S. Hanzel, *J. Am. Chem. Soc.*, **78**, 6414 (1956).
9. C. Walling and R. Rabinowitz, *J. Am. Chem. Soc.*, **79**, 5326 (1957).
10. R. D. Burkhart, *J. Phys. Chem.*, **70**, 605 (1966); *J. Am. Chem. Soc.*, **90**, 273 (1968); *J. Phys. Chem.*, **73**, 2703 (1969); R. D. Burkhart, R. F. Boynton, and J. C. Merrill, *J. Am. Chem. Soc.*, **93**, 5013 (1971).
11. H. G. Kuivila, L. W. Menapace, and C. R. Warner, *J. Am. Chem. Soc.*, **84**, 3584 (1962); L. W. Menapace and H. G. Kuivila, *ibid.*, **86**, 3047 (1964).
12. D. J. Carlsson and K. U. Ingold, *J. Am. Chem. Soc.*, **90**, 7047 (1968).
13. G. M. Burnett and H. W. Melville, "Investigation of Rates and Mechanisms of Reactions," in *Techniques of Organic Chemistry*, Vol. VIII, Part II, S. L. Friess, E. S. Lewis, and A. Weissberger, Eds., Interscience, New York, 1963, Chap. 20.

14. Values are also tabulated for other dark-to-light ratios and there are tables which make allowance both for any concurrent thermal initiation and for any first-order termination process. More complete tables are available.[15]
15. J. A. Howard and K. U. Ingold, *Can. J. Chem.*, **43**, 2729 (1965).
16. E. T. Denisov, *Russ. Chem. Rev.*, **39**, 31 (1970).
17. S. A. Weiner and G. S. Hammond, *J. Am. Chem. Soc.*, **91**, 986 (1969).
18. G. B. Watts and K. U. Ingold, *J. Am. Chem. Soc.*, **94**, 491 (1972).
19. R. W. Fessenden, *J. Phys. Chem.*, **68**, 1508 (1964).
20. B. Smaller, J. R. Remko, and E. C. Avery, *J. Chem. Phys.*, **48**, 5174 (1968).
21. M. Ebert, J. P. Keene, E. J. Lande, and A. J. Swallow, *Proc. Roy. Soc.* (*London*), **287A**, 1 (1965).
22. R. J. Hagemann and H. A. Schwartz, *J. Phys. Chem.*, **71**, 2694 (1967).
23. D. G. Hendry, *J. Am. Chem. Soc.*, **89**, 5433 (1967).
24. J. A. Howard, K. Adamic, and K. U. Ingold, *Can. J. Chem.*, **47**, 3793 (1969).
25. J. A. Howard, *Adv. Free Radical Chem.*, **4**, 49 (1971).
26. V. Ya. Shlyapintokh et al., *Chemiluminescence Techniques in Chemical Reactions*, Consultants Bureau, New York, 1968.
27. W. I. Bengough and R. A. M. Thomson, *Trans. Faraday Soc.*, **57**, 1928 (1961).
28. Kh'. S. Bagdasar'yan and A. F. Revzin, *Kinet. Katal.*, **4**, 844 (1963).
29. R. W. Fessenden and R. H. Schuler, *J. Chem. Phys.*, **39**, 2147 (1963).
30. A. A. Bichutinskii, A. I. Prokof'ev, and V. A. Shabalkin, *Russ. J. Phys. Chem.*, **38**, 534 (1964).
31. S. Weiner and G. S. Hammond, *J. Am. Chem. Soc.*, **90**, 1659 (1968).
32. Ya. S. Lebedev, V. F. Tsepalov, and V. Ya. Shlyapintokh, *Dokl. Akad. Nauk SSSR*, **139**, 1409 (1961).
33. H. W. Melville and S. Richards, *J. Chem. Soc.*, 944 (1954).
34. J. R. Thomas, *J. Am. Chem. Soc.*, **85**, 591 (1963).
35. S. A. Weiner, *J. Am. Chem. Soc.*, **94**, 581 (1972).
36. S. A. Weiner, E. J. Hamilton, Jr., and B. M. Monroe, *J. Am. Chem. Soc.*, **91**, 6350 (1969).
37. T. Fujisawa, B. M. Monroe, and G. S. Hammond, *J. Am. Chem. Soc.*, **92**, 542 (1970).
38. G. S. Hammond and S. A. Weiner, *Intra-Sci. Chem. Rep.*, **3**, 241 (1969).
39. G. Porter and M. W. Windsor, *Nature*, **180**, 187 (1957).
40. M. S. Matheson and L. M. Dorfman, *Pulse Radiolysis*, M.I.T. Press, Cambridge, Mass., 1969; B. H. J. Bielski and J. M. Gebicki, *Adv. Radiat. Chem.*, **2**, 177 (1970). Both of these references provide comprehensive lists of absolute rate constants for various classes of reactions involving species such as $H\cdot$, $HO\cdot$, $O\cdot^-$, $HOO\cdot$, $O_2\cdot^-$, e_{aq}^-, etc., which have been generated by pulse radiolysis.
41. L. H. Piette and W. C. Landgraf, *J. Chem. Phys.*, **32**, 1107 (1960).
42. L. H. Piette, G. Bulow, and K. Loeffler, *Div. Pet. Chem.*, *Preprints*, ACS Meeting, Philadelphia, April 1964, **9** (2), C-9 (1964).
43. J. K. Kochi, P. J. Krusic, and D. R. Eaton, *J. Am. Chem. Soc.*, **91**, 1877, 1879 (1969).
44. J. K. Kochi and P. J. Krusic, *J. Am. Chem. Soc.*, **91**, 3940 (1969).
45. P. J. Krusic and J. K. Kochi, *J. Am. Chem. Soc.*, **90**, 7155 (1968).
46. P. J. Krusic and J. K. Kochi, *J. Am. Chem. Soc.*, **91**, 3938 (1969).
47. P. J. Krusic and J. K. Kochi, *J. Am. Chem. Soc.*, **91**, 3942 (1969).
48. J. K. Kochi and P. J. Krusic, *J. Am. Chem. Soc.*, **91**, 3944 (1969).
49. The *t*-butoxy radicals are difficult[50] or perhaps impossible[51] to see by ESR spectroscopy.

50. S. Weiner and G. S. Hammond, *J. Am. Chem. Soc.*, **91**, 2182 (1969).
51. M. C. R. Symons, *J. Am. Chem. Soc.*, **91**, 5924 (1969); H. J. Hefter, T. A. Hecht, and G. S. Hammond, *ibid.*, **94**, 2793 (1972).
52. R. Livingston and H. Zeldes, *J. Chem. Phys.*, **44**, 1245 (1966); *J. Am. Chem. Soc.* **88**, 433 (1966).
53. W. G. B. Huysmans and W. A. Waters, *J. Chem. Soc.* (*B*), 1047 (1966); 1163 (1967).
54. T. J. Stone and W. A. Waters, *J. Chem. Soc.*, 213 (1964).
55. B. H. J. Bielski and E. Saito, *J. Phys. Chem.*, **66**, 2266 (1962).
56. J. R. Thomas, *J. Am. Chem. Soc.*, **87**, 3935 (1965).
57. J. R. Thomas and K. U. Ingold, *Adv. Chem. Ser.*, **75**, 258 (1968).
58. W. T. Dixon and R. O. C. Norman, *J. Chem. Soc.*, 3119 (1963).
59. R. O. C. Norman and P. R. West, *J. Chem. Soc.* (*B*), 389 (1969).
60. P. S. Dixon, A. P. Stefani, and M. Szwarc, *J. Am. Chem. Soc.*, **85**, 2551 (1963); C.-M. Backman, S. Claesson, and M. Szwarc, *Trans. Faraday Soc.*, **66**, 3061 (1970).
61. P. S. Dixon, A. P. Stefani, and M. Szwarc, *J. Am. Chem. Soc.*, **85**, 3344 (1963).
62. S. Kodama, S. Fujita, J. Takeishi, and O. Toyama, *Bull. Chem. Soc. Japan*, **39**, 1009 (1966).
63. S. Fujita, S. Kodama, and O. Toyama, *Bull. Chem. Soc. Japan*, **39**, 1323 (1966).
64. J. E. Bennett, B. Mile, A. Thomas, and B. Ward, *Adv. Phys. Org. Chem.*, **8**, 1 (1970).
65. R. A. Sheldon and J. K. Kochi, *J. Am. Chem. Soc.*, **92**, 4395 (1970).
66. A. P. Stefani, *J. Am. Chem. Soc.*, **90**, 1694 (1968).
66a. See, however, H. A. Gillis, *Can. J. Chem.*, **49**, 2861 (1971).
67. W. A. Cramer, *J. Phys. Chem.*, **71**, 1171 (1967).
68. See also R. S. Konar, *Int. J. Chem. Kinet.*, **2**, 419 (1970).
69. O. Dobis, J. M. Pearson, and M. Szwarc, *J. Am. Chem. Soc.*, **90**, 278 (1968).
70. K. Chakravorty, J. M. Pearson, and M. Szwarc, *J. Am. Chem. Soc.*, **90**, 283 (1968).
71. F. D. Greene, M. A. Berwick, and J. C. Stowell, *J. Am. Chem. Soc.*, **92**, 867 (1970).
72. S. F. Nelsen and P. D. Bartlett, *J. Am. Chem. Soc.*, **88**, 137 (1966).
73. J. R. Shelton, C. K. Liang, and P. Kovacic, *J. Am. Chem. Soc.*, **90**, 354 (1968).
74. A. F. Bickel and W. A. Waters, *Rec. Trav. Chim. Pays-Bas*, **69**, 1490 (1950).
75. See, e.g., R. C. Neumann, Jr., and E. S. Alhadeff, *J. Org. Chem.*, **35**, 3401 (1970); R. A. Sheldon and J. K. Kochi, *J. Am. Chem. Soc.*, **92**, 5175 (1970).
76. Stable is used in the sense that the reaction

$$R\cdot + R\cdot \longrightarrow \text{molecular products}$$

does not occur. It does not mean that R· will not react with a molecule or a different free radical.
77. Reference 7, Chap. 10.
78. A. R. Forrester, J. M. Hay, and R. H. Thomson, *Organic Chemistry of Stable Free Radicals*, Academic Press, London, 1968, Chap. 2.
79. H. Lankamp, W. Th. Nauta, and C. MacLean, *Tetrahedron Letters*, 249 (1968).
80. H. A. Staab, H. Brettschneider, and H. Brunner, *Chem. Ber.*, **103**, 1101 (1970).
81. K. Ziegler and L. Ewald, *Ann.*, **473**, 163 (1929); K. Ziegler, L. Ewald, and P. Orth, *ibid.*, **479**, 277 (1930); K. Ziegler, P. Orth, and K. Weber, *ibid.*, **504**, 131 (1933); K. Ziegler, L. Ewald, and A. Seib, *ibid*, **504**, 162, 182 (1933); K. Ziegler and A. Luttringhaus, *ibid.*, **504**, 189 (1933); K. Ziegler, A. Seib, K. Knoevenagel, P. Herte, and F. Andreas, *ibid.*, **551**, 150 (1942).
82. F. S. D'iachkovskii, N. N. Bubnov, and A. E. Shilov, *Dokl. Akad. Nauk SSSR*, **122**, 629 (1958).
83. M. L. Morrell and G. Vincow, *J. Am. Chem. Soc.*, **91**, 6389 (1969).

84. The value of $\sim 10^9\ M^{-1}\ sec^{-1}$ reported in reference 85 is certainly too large.
85. D. J. Carlsson, K. U. Ingold, and L. C. Bray, *Int. J. Chem. Kinet.*, **1**, 315 (1969).
86. L. M. Dorfman and I. A. Taub, *J. Am. Chem. Soc.*, **85**, 2370 (1963).
87. J. W. Fletcher, P. J. Richards, and W. A. Seddon, *Can. J. Chem.*, **48**, 3765 (1970).
88. W. A. Seddon and A. U. Allen, *J. Phys. Chem.*, **71**, 1914 (1967).
89. The viscosity effect on the self-reaction of *t*-butyl radicals has been reported.[12] In *n*-pentane, $\eta = 0.35$ cP, $2k_t = 5.4 \times 10^9\ M^{-1}\ sec^{-1}$; in cyclohexane, $\eta = 0.92$, $2k_t = 2.2 \times 10^9$; in *n*-tridecane, $\eta = 1.3$, $2k_t = 1.8 \times 10^9$. That is, $2k_t$ is approximately proportional to $\eta^{-0.8}$ in reasonable agreement with the simple theory of diffusion-controlled bimolecular reactions. Similarly, $2k_t$ has been found to be inversely proportional to viscosity in the free radical polymerization of methyl methacrylate.[90] For radicals generated in pairs in solvent cages, the efficiency of escape from the cage has been related both to[91] $\eta^{-1.0}$ and to $\eta^{-0.5}$.[92]
90. C. H. Bamford and S. Brumby, *Chem. Ind.* (*London*), 1020 (1969).
91. A. P. Stefani, G. F. Thrower, and C. F. Jordon, *J. Phys. Chem.*, **73**, 1257 (1969); F. E. Herkes, J. Friedman, and P. D. Bartlett, *Int. J. Chem. Kinet.*, **1**, 193 (1969); K. Chakrovorty, J. M. Pearson, and M. Szwarc, *ibid.*, **1**, 357 (1969).
92. T. Koenig, *J. Am. Chem. Soc.*, **91**, 2558 (1969); W. A. Pryor and K. Smith, *ibid.*, **92**, 5403 (1970).
93. M. C. Sauer, Jr., and I. Mani, *J. Phys. Chem.*, **72**, 3856 (1968).
94. J. A. Howard and K. U. Ingold, *Can. J. Chem.*, **45**, 785 (1967).
95. D. Behar, G. Czapski, J. Rabani, L. M. Dorfman, and H. A. Schwarz, *J. Phys. Chem.*, **74**, 3209 (1970).
96. R. L. McCarthy and A. MacLachlan, *Trans. Faraday Soc.*, **57**, 1107 (1961).
97. A. Beckett and G. Porter, *Trans. Faraday Soc.*, **59**, 2038 (1963).
97a. See also S. A. Weiner, *J. Am. Chem. Soc.*, **96**, 6978 (1971); P. B. Ayscough and M. C. Brice, *J. Chem. Soc.* (*B*), 491 (1971); P. B. Ayscough, R. C. Sealy, and D. E. Woods, *J. Phys. Chem.*, **75**, 3454 (1971); C. E. Burchill and P. W. Jones, *Can. J. Chem.*, **49**, 4005 (1971).
98. C. H. Bamford, W. G. Barb, A. D. Jenkins, and P. F. Onyon, *The Kinetics of Vinyl Polymerization by Radical Mechanisms*, Academic Press, New York, 1958.
99. A. F. Revzin, *Russ. Chem. Rev.*, **35**, 73 (1966).
100. J. Hughes and A. M. North, *Trans. Faraday Soc.*, **60**, 960 (1964).
101. S. W. Benson and A. M. North, *J. Am. Chem. Soc.*, **81**, 1339 (1959).
102. A. M. North and G. A. Reed, *Trans. Faraday Soc.*, **57**, 859 (1961).
103. Reference 7, Chap. 5.
104. R. L. Strong, *J. Am. Chem. Soc.*, **87**, 3563 (1965).
105. P. Pagsberg, H. Christensen, J. Rabani, G. Nilsson, J. Fenger, and S. O. Nielsen, *J. Phys. Chem.*, **73**, 1029 (1969).
106. A. Treinin and E. Hayon, *J. Chem. Phys.*, **50**, 538 (1969).
107. E. Hayon and M. Simic, *J. Am. Chem. Soc.*, **92**, 7486 (1970).
108. D. J. Carlsson and K. U. Ingold, *J. Am. Chem. Soc.*, **89**, 4885, 4891 (1967).
109. G. Dobson and L. I. Grossweiner, *Trans. Faraday Soc.*, **61**, 708 (1965).
110. S. A. Weiner and L. R. Mahoney, *J. Am. Chem. Soc.*, **94**, 585, 1412 (1972); and in press. However some unhindered phenoxy radicals have quite small $2k_t$ values (see text).
111. E. J. Land and G. Porter, *Trans. Faraday Soc.*, **59**, 2027 (1963).
112. J. R. Roberts and K. U. Ingold, *J. Am. Chem. Soc.*, **93**, 6686 (1971); **94**, 2159 (1972); and unpublished results.
113. P. T. Frangopol and K. U. Ingold, *J. Organomet. Chem.*, **25**, C9 (1970).

114. In the gas phase $2k_t = 2.5 \times 10^{10}$ M^{-1} sec^{-1} for the combination of $CH_3S\cdot$ radicals [D. M. Graham, R. L. Mieville, R. H. Pallen, and C. Sivertz, *Can. J. Chem.*, **42**, 2250 (1964)]. In benzene, $2k_t = 7 \times 10^8$ M^{-1} sec^{-1} for 2,4,6-tri-*t*-butyl-1-thiyl (S. A. Weiner, private communication). In water, $2k_t = 2.6 \times 10^9$ M^{-1} sec^{-1} for $HO_2CCH(NH_2)C(CH_3)_2S\cdot$ [J. W. Purdie, H. A. Gillis, and N. V. Klassen, *Chem. Commun.*, 1163 (1971)].
115. C. D. Cook and B. E. Norcross, *J. Am. Chem. Soc.*, **78**, 3797 (1956); **81**, 1176 (1959).
116. A. Hubele, M. Suhr, and U. Heilmann, *Chem. Ber.*, **95**, 639 (1962).
117. J. E. Bennett, *Nature*, **186**, 385 (1960).
118. A. I. Brodskii, V. D. Pokhodenko, and L. N. Ganyuk, *Rocz. Chem.*, **38**, 105 (1964); A. I. Brodskii, V. D. Pokhodenko, V. A. Khizhnyi, and N. N. Kalibabchuk, *Dokl. Akad. Nauk SSSR*, **169**, 339 (1966), and references cited therein.
119. R. Stebbins and F. Sicilio, *Tetrahedron*, **26**, 291 (1970).
120. K. U. Ingold, *Acc. Chem. Res.*, **2**, 1 (1969).
121. G. A. Russell, *J. Am. Chem. Soc.*, **79**, 3871 (1957).
122. Reference 78, Chap. 5.
123. H. Wieland and K. Roth, *Chem. Ber.*, **53**, 210 (1920).
124. A. R. Forrester and R. H. Thomson, *Nature*, **203**, 74 (1964).
125. K. Adamic, D. F. Bowman, T. Gillan, and K. U. Ingold, *J. Am. Chem. Soc.*, **93**, 902 (1971); D. F. Bowman, J. L. Brokenshire, T. Gillan, and K. U. Ingold, *ibid.*, **93**, 6551 (1971); D. F. Bowman, T. Gillan, and K. U. Ingold, *ibid.*, **93**, 6555 (1971).
126. J. L. Brokenshire, J. R. Roberts, and K. U. Ingold, *J. Am. Chem. Soc.*, in press.
127. M. Simic, P. Neta, and E. Hayon, *J. Phys. Chem.*, **73**, 4214 (1969).
128. T. L. Staples, J. Jagur-Grodzinski, and M. Szwarc, *J. Am. Chem. Soc.*, **91**, 3721 (1969).
129. L. S. Marcoux, R. N. Adams, and S. W. Feldberg, *J. Phys. Chem.*, **73**, 2611 (1969); R. F. Nelson and S. W. Feldberg, *ibid.*, **73**, 2623 (1969).
130. J. Spanswick and K. U. Ingold, *Can. J. Chem.*, **48**, 546, 554 (1970).
131. M. Szwarc, *Acc. Chem. Res.*, **2**, 87 (1969); M. Szwarc, *Prog. Phys. Org. Chem.*, **6**, 323 (1968); L. M. Dorfman, *Acc. Chem. Res.*, **3**, 224 (1970). See also S. Bank and B. Bockrath, *J. Am. Chem. Soc.*, **93**, 430 (1971); D. Carson and B. J. Tabner, *J. Chem. Soc.* (*B*), 1560, 1565 (1970).
132. P. D. Bartlett and T. Funahashi, *J. Am. Chem. Soc.*, **84**, 2596 (1962).
133. R. H. Schuler, *J. Phys. Chem.*, **68**, 3873 (1964).
134. A. G. Davies and B. P. Roberts, *J. Chem. Soc.* (*B*), 17 (1967).
135. P. G. Allies and P. B. Brindley, *Chem. Ind.*, 319 (1967).
136. M. S. Khloplyankina, A. L. Buchachenko, M. B. Neiman, and A. G. Vasil'eva, *Kinet. Katal.*, **6**, 394 (1965).
137. I. T. Brownlie and K. U. Ingold, *Can. J. Chem.*, **45**, 2427 (1967).
137a. R. L. Wilson, *Trans. Faraday Soc.*, **67**, 3008 (1971).
138. J. K. Kochi and R. V. Subramanium, *J. Am. Chem. Soc.*, **87**, 4855 (1965). (See also Chapter 11.)
139. D. L. Struble, A. L. J. Beckwith, and G. E. Gream, *Tetrahedron Letters*, 4795 (1970).
140. For the polystyryl radical $k \sim 2 \times 10^6$ M^{-1} sec^{-1}.[141] The relative reactivity of manganese(III), cerium(IV), and copper(II) in oxidizing secondary alkyl radicals is ~1:12:350.[142]
141. W. I. Bengough and T. O'Neill, *Trans. Faraday Soc.*, **64**, 1014 (1968).
142. E. I. Heiba and R. M. Dessau, *J. Am. Chem. Soc.*, **93**, 524 (1971).
143. See, e.g., J. F. Garst and F. E. Barton, *Tetrahedron Letters*, 587 (1969). (See also Chapter 9.

144. R. Blackburn and A. Charlesby, *Trans. Faraday Soc.*, **62**, 1159 (1966).
145. J. K. Thomas, *J. Phys. Chem.*, **71**, 1919 (1967).
146. K. U. Ingold and B. P. Roberts, *Free-Radical Substitution Reactions*, Wiley, New York, 1971.
147. G. H. Williams, *Homolytic Aromatic Substitution*, Pergamon Press, London, 1960.
148. S. W. Benson, *Thermochemical Kinetics*, Wiley, New York, 1968.
149. D. M. Golden and S. W. Benson, *Chem. Rev.*, **69**, 125 (1969).
150. M. G. Evans and M. Polanyi, *Trans. Faraday Soc.*, **34**, 11 (1938).
151. N. N. Semenov, *Some Problems in Chemical Kinetics and Reactivity*, Princeton University Press, Princeton, N.J., 1958.
152. H. S. Johnston, *Adv. Chem. Phys.*, **3**, 131 (1960).
153. T. Kagiya, Y. Sumida, and T. Inoue, *Bull. Chem. Soc. Japan*, **42**, 2422 (1969).
154. Johnston's[152] more complicated expression gets around this difficulty in spite of having only one adjustable parameter.
155. H. S. Johnston and C. Parr, *J. Am. Chem. Soc.*, **85**, 2544 (1963).
156. S. W. Benson and W. B. DeMore, *Ann. Rev. Phys. Chem.*, **16**, 397 (1965).
156a. A. A. Zavitsas, *J. Am. Chem. Soc.*, **94**, 2779 (1972).
157. It is worth noting that this is not possible for a highly reactive radical. The fast reaction of this radical with the substrate precludes a buildup in its concentration to the point where termination by a bimolecular self-reaction becomes important.
158. J. A. Howard and K. U. Ingold, *Can. J. Chem.*, **44**, 1119 (1966).
159. J. A. Howard and K. U. Ingold, *Can. J. Chem.*, **42**, 2324 (1964).
160. Chain transfer agents are not confined to hydrogen donors. Carbon tetrachloride, disulfides, and sulfur are also effective.
161. Reference 7, Chap. 4.
162. R. A. Gregg and F. R. Mayo, *Discuss. Faraday Soc.*, **2**, 328 (1947).
163. Solvent effects on radical reactivity can be very pronounced, see, e.g., the effect of solvents on the relative reactivities of hydrocarbons toward chlorine atoms.[164]
164. G. A. Russell, *J. Am. Chem. Soc.*, **80**, 4987, 4997, 5002 (1958).
165. A. S. Rodgers, D. M. Golden, and S. W. Benson, *J. Am. Chem. Soc.*, **89**, 4578 (1967).
165a. J. E. Packer, D. B. House, and E. J. Rasburn, *J. Chem. Soc.* (*B*), 1574 (1971).
166. R. F. Bridger and G. A. Russell, *J. Am. Chem. Soc.*, **85**, 3754 (1963).
167. J. A. Meyer, V. Stannett, and M. Szwarc, *J. Am. Chem. Soc.*, **83**, 25 (1961).
168. J. A. Kerr and A. F. Trotman-Dickenson, *Prog. React. Kinet.*, **1**, 105 (1961).
169. M. Cher, C. S. Hollingsworth, and F. Sicillio, *J. Phys. Chem.*, **70**, 877 (1966).
170. R. P. Buckley and M. Szwarc, *Proc. Roy. Soc.* (*London*), **A240**, 396 (1957).
171. R. P. Buckley, F. Leavitt, and M. Szwarc, *J. Am. Chem. Soc.*, **78**, 5557 (1956).
172. I. B. Burkley and R. E. Rebbert, *J. Phys. Chem.*, **67**, 168 (1963).
173. C. Walling and B. B. Jacknow, *J. Am. Chem. Soc.*, **82**, 6108, 6113 (1960).
174. C. Walling and W. Thaler, *J. Am. Chem. Soc.*, **83**, 3877 (1961).
175. K. U. Ingold, *Pure Appl. Chem.*, **15**, 49 (1967).
176. C. Walling and M. J. Gibian, *J. Am. Chem. Soc.*, **87**, 3361 (1965).
177. G. A. Russell, A. Ito, and D. G. Hendry, *J. Am. Chem. Soc.*, **85**, 2976 (1963).
178. D. D. Davis, W. Braun, and A. M. Bass, *Int. J. Chem. Kinet.*, **2**, 101 (1970).
179. H. O. Pritchard, J. B. Pyke, and A. F. Trotman-Dickenson, *J. Am. Chem. Soc.*, **77**, 2629 (1955).
180. G. A. Russell and C. DeBoer, *J. Am. Chem. Soc.*, **85**, 3136 (1963).
181. M. L. White and R. R. Kuntz, *Int. J. Chem. Kinet.*, **3**, 127 (1971).
181a. F. B. Wampler and R. R. Kuntz, *Int. J. Chem. Kinet.*, **3**, 283 (1971).

182. K. Schwetlick and S. Helm, *Tetrahedron*, **22**, 793 (1963).
183. J. L. Brokenshire, A. Nechvatal, and J. M. Tedder, *Trans. Faraday Soc.*, **66**, 2029 (1970).
184. J. A. Howard, K. U. Ingold, and M. Symonds, *Can. J. Chem.*, **46**, 1017 (1968).
185. D. L. Allara, T. Mill, F. R. Mayo, H. Richardson, and K. Irwin, *Div. Pet. Chem. Preprints*, **16** (1), B31 (1971).
186. D. D. Tanner, D. Darwish, M. W. Mosher, and N. J. Bunce, *J. Am. Chem. Soc.*, **91**, 7398 (1969).
187. K. D. King, D. M. Golden, and S. W. Benson, *Trans. Faraday Soc.*, **66**, 2794 (1970).
188. H. R. Anderson, H. A. Scheraga, and E. R. Van Artsdalen, *J. Chem. Phys.*, **21**, 1258 (1953).
189. S. W. Benson and J. H. Buss, *J. Chem. Phys.*, **28**, 301 (1958).
190. C. H. Bamford, A. D. Jenkins, and R. Johnston, *Trans. Faraday Soc.*, **55**, 418 (1959).
191. R. A. Jackson and D. W. O'Neill, *Chem. Commun.*, 1210 (1969).
192. Fluorine atoms[193,194] and at least some cyclopropyl radicals[195] are even less selective.
193. P. S. Fredericks and J. M. Tedder, *J. Chem. Soc.*, 144 (1960).
194. G. C. Fettis, J. H. Knox, and A. F. Trotman-Dickenson, *J. Chem. Soc.*, 1064 (1960).
195. H. M. Walborsky and J.-C. Chen, *J. Am. Chem. Soc.*, **92**, 7573 (1970).
196. The trifluoromethyl radical is more reactive than methyl both in abstractions and in additions.[197,198] Thus, for cyclohexane k/H for attack by $\cdot CF_3$ has been estimated to be $\sim 10^3$ M^{-1} sec^{-1} at 25°, which is about an order of magnitude greater than the value for $\cdot CH_3$ radical attack. This is believed to be due to a lower activation energy for abstraction by $\cdot CF_3$.
197. P. P. Infelta and R. H. Schuler, *J. Phys. Chem.*, **73**, 2083 (1969).
198. G. Bullock and R. Cooper, *Trans. Faraday Soc.*, **66**, 2055 (1970).
199. J. A. Howard and K. U. Ingold, *Can. J. Chem.*, **47**, 3809 (1969); **48**, 873 (1970).
200. J. A. Howard and S. Korcek, *Can. J. Chem.*, in press.
201. S. Korcek, J. H. B. Chenier, J. A. Howard, and K. U. Ingold, *Can. J Chem*, in press.
202. J. A. Howard and S. Korcek, *Can. J. Chem.*, **48**, 2165 (1970).
203. S. S. Friedrich, E. C. Friedrich, L. J. Andrews, and R. M. Keefer, *J. Org. Chem.*, **34**, 900 (1969).
204. G. A. Russell, private communication.
205. C. Walling, A. L. Rieger, and D. D. Tanner, *J. Am. Chem. Soc.*, **85**, 3129 (1963).
206. R. E. Pearson and J. C. Martin, *J. Am. Chem. Soc.*, **85**, 3142 (1963).
207. The most powerful electron-releasing and electron-withdrawing substituents in common use in reactions of this type are CH_3O and NO_2 for which the σ^+ constants are -0.78 and $+0.79$, respectively. Thus, for a ρ^+ value of -1.0 these substituents change the rate, relative to the base hydrocarbon, by 0.78 and 0.79 log units. For this ρ^+ value the *entire range* of substituent effects is given by $\log(k_{p\text{-}CH_3O}/k_{p\text{-}NO_2}) = 1.57$.
208. J. A. Howard, unpublished results.
209. See, e.g.: J. A. Howard and K. U. Ingold, *Can. J. Chem.*, **41**, 1744 (1963); G. A. Russell and R. C. Williamson, Jr., *J. Am. Chem. Soc.*, **86**, 2357 (1964); R. L. Huang and K. H. Lee, *J. Chem. Soc.* (*C*), 932, 935 (1966); B. R. Kennedy and K. U. Ingold, *Can. J. Chem.*, **45**, 2632 (1967); K. H. Lee, *Tetrahedron*, **25**, 4357, 4363 (1969); E. P. Change, R. L. Huang, and K. H. Lee, *J. Chem. Soc.* (*B*), 878 (1969); W. A. Pryor, U. Tonellato, D. L. Fuller, and S. Jumonville, *J. Org. Chem.*, **34**, 2018 (1969); W. D. Totherow and G. J. Gleicher, *J. Am. Chem. Soc.*, **91**, 7150 (1969);

Yu. L. Spirin, *Russ. Chem. Rev.*, **38**, 529 (1969); T. P. Low and K. H. Lee, *J. Chem. Soc.* (*B*), 535 (1970); H. Sakurai, A. Hosomi, and M. Kumada, *J. Org. Chem.*, **35**, 993 (1970); S. S. Friedrich, L. J. Andrews, and R. M. Keefer, *ibid.*, **35**, 944 (1970).
210. M. Sakurai and A. Hosomi, *J. Am. Chem. Soc.*, **89**, 458 (1967).
211. C. Walling and J. A. McGuinness, *J. Am. Chem. Soc.*, **91**, 2053 (1969).
212. C. K. Ingold, *Structure and Mechanism in Organic Chemistry*, 2nd ed., Cornell University Press, Ithaca, N.Y., 1969.
213. F. Minisci, G. P. Gardini, and F. Bertini, *Can. J. Chem.*, **48**, 544 (1970), and references cited therein.
214. The intramolecular reaction is the well-known Hoffman–Löffler reaction.
215. N. C. Deno, W. E. Billups, R. Fishbein, C. Pierson, R. Whalen, and J. C. Wyckoff, *J. Am. Chem. Soc.*, **93**, 438 (1971).
216. L. N. Shishkina and I. V. Berezin, *Zh. Fiz. Khim.*, **43**, 912 (1969).
217. I. V. Berezin, N. F. Kazanskaya, and N. N. Ugarova, *Zh. Fiz. Khim.*, **40**, 766 (1966).
218. M. Onyszchuk and C. Sivertz, *Can. J. Chem.*, **33**, 1034 (1955); C. Sivertz, *J. Phys. Chem.*, **63**, 34 (1959).
219. R. J. Cvetanovic and E. W. R. Steacie, *Can. J. Chem.*, **31**, 158 (1953).
220. G. C. Fettis and A. F. Trotman-Dickenson, *J. Chem. Soc.*, 3037 (1961).
221. J. C. Amphlett and E. Whittle, *Trans. Faraday Soc.*, **64**, 2130 (1968).
222. N. A. Gac, D. M. Golden, and S. W. Benson, *J. Am. Chem. Soc.*, **91**, 3091 (1969).
223. K. D. King, D. M. Golden, and S. W. Benson, *J. Am. Chem. Soc.*, **92**, 5541 (1970).
224. D. M. Golden, R. Walsh, and S. W. Benson, *J. Am. Chem. Soc.*, **87**, 4053 (1965).
225. K. U. Ingold, *Can. J. Chem.*, **41**, 2816 (1963).
226. J. H. T. Brook, *Trans. Faraday Soc.*, **53**, 327 (1957).
227. (a) P. J. Wagner, *J. Am. Chem. Soc.*, **89**, 2503 (1967); (b) D. R. G. Brimage, R. S. Davidson, and P. F. Lambeth, *J. Chem. Soc.* (*C*), 1241 (1971).
228. J. A. Howard and K. U. Ingold, *Can. J. Chem.*, **40**, 1851 (1962); **41**, 1744, 2800 (1963); **42**, 1044 (1964); **43**, 2724 (1965).
229. D. V. Gardner, J. A. Howard, and K. U. Ingold, *Can. J. Chem.*, **42**, 2847 (1964).
230. L. R. Mahoney and M. A. DaRooge, *J. Am. Chem. Soc.*, **92**, 4063 (1970).
231. L. R. Mahoney, *Angew. Chem., Int. Engl. Ed.*, **8**, 547 (1969).
232. I. T. Brownlie and K. U. Ingold, *Can. J. Chem.*, **44**, 861 (1966); **45**, 2419 (1967).
233. J. A. Howard, W. J. Schwalm, and K. U. Ingold, *Adv. Chem. Ser.*, **75**, 6 (1968).
234. M. P. Godsay, G. A. Harpell, and K. E. Russell, *J. Polym. Sci.*, **57**, 641 (1962).
235. R. A. Gregg and F. R. Mayo, *J. Am. Chem. Soc.*, **75**, 3530 (1953).
236. C. Walling, *J. Am. Chem. Soc.*, **70**, 2561 (1948).
237. G. Henrici-Olivé and S. Olivé, *Fortschr. Hochpolym. Forsch.*, **2**, 496 (1960–1961).
238. M. Anbar and P. Neta, *Int. J. Appl. Radiat. Isot.*, **18**, 493 (1967).
239. P. Neta, G. R. Holdren, and R. H. Schuler, *J. Phys. Chem.*, **75**, 449 (1971).
240. R. Wander, P. Neta, and L. M. Dorfman, *J. Phys. Chem.*, **72**, 2946 (1968).
241. P. Neta and L. M. Dorfman, *J. Phys. Chem.*, **73**, 413 (1969).
241a. P. Neta, R. W. Fessenden, and R. H. Schuler, *J. Phys. Chem.*, **75**, 1654 (1971).
242. R. O. C. Norman and R. J. Pritchett, *J. Chem. Soc.* (*B*), 926 (1967).
243. C. R. E. Jefcoate and R. O. C. Norman, *J. Chem. Soc.* (*B*), 48 (1968).
244. B. Cercek, *J. Phys. Chem.*, **72**, 3832 (1968).
245. G. E. Adams, B. D. Michael, and E. J. Land, *Nature*, **211**, 293 (1966).
246. G. E. Adams and B. D. Michael, *Trans. Faraday Soc.*, **63**, 1171 (1967).
247. E. J. Land and M. Ebert, *Trans. Faraday Soc.*, **63**, 1181 (1967).
248. A. Wigger, W. Grünbein, A. Henglein, and E. J. Land, *Z. Naturforsch.*, **24b**, 1262 (1969).
249. W. A. Pryor, J. P. Stanley, and M. G. Griffith, *Science*, **168**, 181 (1970); W. A.

Pryor and J. P. Stanley, *Intra-Sci. Chem. Rep.*, **4**, 99 (1970); W. A. Pryor and R. W. Henderson, *J. Am. Chem. Soc.*, **92**, 7234 (1970); W. A. Pryor and M. G. Griffith, *ibid.*, **93**, 1408 (1971).
250. W. A. Pryor and J. P. Stanley, *J. Am. Chem. Soc.*, **93**, 1412 (1971).
251. It is interesting to note that the relative reactivities of these three alcohols toward the $CH_3\cdot$ radical in water at 25° are 1:2.7:16 (with $k_{CH_3OH} = 0.22\ M^{-1}\ sec^{-1}$).[145] Extrapolation of higher-temperature (130–250°) gas-phase measurements with the $CD_3\cdot$ radical to 25° gives relative reactivities of 1:3.2:25 (with $k_{CH_3OH} = 10\ M^{-1}\ sec^{-1}$).[252]
252. A. A. Herrod, *Chem. Commun.*, 891 (1968).
253. C. Rüchardt, *Angew. Chem., Int. Ed.*, **9**, 830 (1970). An unusual reactivity-selectivity relationship for halogenation of 1- and 2-adamantyl radicals has been reported by I. Tabushi, Y. Aoyama, S. Kojo, J. Hamuro, and Z. Yoshida, *J. Am. Chem. Soc.*, **94**, 1177 (1972).
254. F. G. Edwards and F. R. Mayo, *J. Am. Chem. Soc.*, **72**, 1265 (1950).
255. F. W. Evans, R. J. Fox, and M. Szwarc, *J. Am. Chem. Soc.*, **82**, 6414 (1960).
256. J. F. Bunnett and C. C. Wamser, *J. Am. Chem. Soc.*, **88**, 5534 (1966).
257. D. L. Brydon and J. I. G. Cadogan, *J. Chem. Soc.* (*C*), 819 (1968).
258. W. C. Danen and D. G. Saunders, *J. Am. Chem. Soc.*, **91**, 5924 (1969).
259. W. C. Danen and R. L. Winter, *J. Am. Chem. Soc.*, **93**, 716 (1971); W. C. Danen, T. J. Tipton, and D. G. Saunders, *ibid.*, **93**, 5186 (1971).
260. H. R. Ward, R. G. Lawler, and R. A. Cooper, *Tetrahedron Letters*, 527 (1969); *J. Am. Chem. Soc.*, **91**, 746 (1969).
261. A. R. Lepley and R. L. Landau, *J. Am. Chem. Soc.*, **91**, 748, 749 (1969).
262. R. G. Lawler, H. R. Ward, R. B. Allen, and P. E. Ellenbogen, *J. Am. Chem. Soc.*, **93**, 789 (1971).
263. B. Cercek and M. Kongshaug, *J. Phys. Chem.*, **74**, 4319 (1970).
264. Chlorine abstraction from aromatic acid chlorides by the tri-*n*-butyltin radical also exhibits a positive ρ value [H. G. Kuivila and E. J. Walsh, Jr., *J. Am. Chem. Soc.*, **88**, 571 (1966)].
265. F. D. Greene and N. N. Lowry, *J. Org. Chem.*, **32**, 875 (1967).
266. P. D. Bartlett, R. E. Pincock, J. H. Rolston, W. G. Schindel, and L. A. Singer, *J. Am. Chem. Soc.*, **87**, 2590 (1965).
267. F. D. Greene and N. M. Lowry, *J. Org. Chem.*, **32**, 882 (1967).
267a. See also L. J. Altman and R. C. Baldwin, *Tetrahedron Letters*, 981 (1972).
268. A. A. Zavitsas, *J. Org. Chem.*, **29**, 3086 (1964).
269. C. Walling and V. P. Kurkov, *J. Am. Chem. Soc.*, **89**, 4895 (1967).
270. C. Walling and A. Padwa, *J. Org. Chem.*, **27**, 2976 (1962).
271. A. A. Zavitsas and S. Ehrenson, *J. Am. Chem. Soc.*, **87**, 2841 (1965).
272. H. G. Kuivila, *Acc. Chem. Res.*, **1**, 299 (1968), and references cited therein.
273. P. B. Chock and J. Halpern, *J. Am. Chem. Soc.*, **91**, 582 (1969).
274. J. K. Kochi and J. W. Powers, *J. Am. Chem. Soc.*, **92**, 137 (1970).
275. Y. Nagai, I. Shiojima, N. Kasaburo, and H. Matsumoto, *Yuki Gosei Kagaku Kyokai Shi*, **26**, 999, 1004 (1968).
276. A. G. Davies, K. U. Ingold, B. P. Roberts, and R. Tudor, *J. Chem. Soc.* (*B*), 698 (1971).
277. S. Korcek, G. B. Watts, and K. U. Ingold, *J. Chem. Soc., Perkin II*, 242 (1972).
278. A. G. Davies, D. Griller, B. P. Roberts, and R. Tudor, *Chem. Commun.*, 640 (1970).
279. J. L. Brokenshire and K. U. Ingold, *Int. J. Chem. Kinet*, **3**, 343 (1971).
280. J. Spanswick and K. U. Ingold, *Int. J. Chem. Kinet.*, **2**, 157 (1970).

281. J. F. Harris, Jr., *J. Am. Chem. Soc.*, **84**, 3148 (1962); *J. Org. Chem.*, **31**, 931 (1966).
282. M. Szwarc and J. H. Binks, "Theoretical Organic Chemistry," Int. Union of Pure and Applied Chem., paper presented to Kekulé Symposium, Chem. Soc. (London), 1958, Butterworths, 1959, p. 262, and references cited therein.
283. A. Streitwieser, *Molecular Orbital Theory for Organic Chemists*, Wiley, New York, 1960.
284. C. Walling, in *Molecular Rearrangements*, de Mayo, Ed., Wiley, New York, 1963.
285. R. Kh. Freidlina, *Adv. Free Radical Chem.*, **1**, 211 (1965).
285a. See also J. W. Wilt, Chapter 8.
286. F. D. Greene, M. L. Savitz, H. H. Lau, F. D. Osterholtz, and W. N. Smith, *J. Am. Chem. Soc.*, **83**, 2196 (1961).
287. F. D. Greene, M. L. Savitz, F. D. Osterholtz, H. H. Lau, W. N. Smith, and P. M. Zanet, *J. Org. Chem.*, **28**, 55 (1963).
288. C. Walling and A. Padwa, *J. Am. Chem. Soc.*, **83**, 2207 (1961); **85**, 1597 (1963).
289. See, however, (a) K. W. Watkins and L. A. Ostreko, *J. Phys. Chem.*, **73**, 2080 (1969); (b) L. Endrenyi and D. J. LeRoy, *ibid.*, **70**, 4081 (1966); K. W. Watkins, *J. Am. Chem. Soc.*, **93**, 6355 (1971).
290. F. F. Rust, *J. Am. Chem. Soc.*, **79**, 4000 (1957).
291. T. Mill and G. Montorsi, paper presented to the Division of Petroleum Chemistry, ACS Meeting, Los Angeles, March 28–April 2, 1971, *Preprints*, **16** (1), B23.
292. For isopropyl ether $k_{inter} \approx 0.1\ M^{-1}\ sec^{-1}$, and for benzyl ether $k_{inter} \approx 30\ M^{-1}\ sec^{-1}$.[199]
293. L. Kaplan, *J. Am. Chem. Soc.*, **88**, 4531 (1966).
294. L. Kaplan and D. Ehrenkranz, private communication.
295. Rough estimates of rate constants can be made in a few cases where the yields of rearranged and unrearranged products were measured as a function of the concentration of some H donor, for example, an added thiol or, in the case of radicals formed by acyl radical decarbonylation, the aldehyde concentration itself.

$$R\dot{C}O \longrightarrow R\cdot + CO \xrightarrow{RCHO(R''SH)} RH + R\dot{C}O(R''S\cdot)$$

$$\hookrightarrow R'\cdot \xrightarrow{RCHO(R''SH)} R'H + R\dot{C}O(R''S\cdot)$$

296. H. Wieland, *Ber.*, **44**, 2553 (1911); M. S. Kharasch, A. Fono, and W. Nudenberg, *J. Org. Chem.*, **15**, 763 (1950); M. S. Kharasch, A. C. Poshkus, A. Fono, and W. Nudenberg, *ibid.*, **16**, 1458 (1951); P. D. Bartlett and J. Cotman, Jr., *J. Am. Chem. Soc.*, **72**, 3095 (1950).
297. J. A. Howard and K. U. Ingold, *Can. J. Chem.*, **47**, 3797 (1969).
298. L. K. Montgomery and J. W. Mott, *J. Am. Chem. Soc.*, **89**, 3050 (1967).
299. C. R. Warner, R. J. Strunk, and H. G. Kuivila, *J. Org. Chem.*, **31**, 3381 (1966).
300. T. A. Halgren, M. E. H. Howden, M. E. Medof, and J. D. Roberts, *J. Am. Chem. Soc.*, **89**, 3051 (1967).
301. S. J. Cristol and R. V. Barbour, *J. Am. Chem. Soc.*, **90**, 2832 (1968).
302. P. J. Krusic, J. P. Jesson, and J. K. Kochi, *J. Am. Chem. Soc.*, **91**, 4566 (1969).
303. V. M. A. Chambers, W. R. Jackson, and G. W. Young, *Chem. Commun.*, 1275 (1970).
304. L. K. Montgomery, J. W. Matt, and J. R. Webster, *J. Am. Chem. Soc.*, **89**, 923 (1967); L. K. Montgomery and J. W. Matt, *ibid.*, **89**, 934 (1967).
305. See, e.g., M. Julia and M. Maumy, *Bull. Soc. Chim. France*, 434 (1966), 1603 (1968), 2415, 2427 (1969); R. C. Lamb, P. W. Ayers, and M. K. Toney, *J. Am. Chem. Soc.*, **85**, 3483 (1963); C. Walling and M. S. Pearson, *ibid.*, **86**, 2262 (1964);

J. F. Garst, P. W. Ayers, and R. C. Lamb, *ibid.*, **88**, 4260 (1966); R. C. Lamb, P. W. Ayers, M. K. Toney, and J. F. Garst, *ibid.*, **88**, 4261 (1966); C. Walling and A. Cioffari, *ibid.*, **92**, 6609 (1970), and references cited therein.

306. C. Walling, J. H. Cooley, A. Ponaras, and E. J. Racah, *J. Am. Chem. Soc.*, **88**, 5361 (1966).

307. The cyclization rate constant of 10^5 sec^{-1} has been utilized to estimate a rate constant of 4×10^7 M^{-1} sec^{-1} for the reduction of the 5-hexenyl radical by an ethylenediaminechromium(II) complex in aqueous dimethylformamide.[274]

308. D. L. Struble, A. L. J. Beckwith, and G. E. Gream, *Tetrahedron Letters*, 3701 (1968).

309. This rate constant has been used to estimate a rate constant of 5×10^7 M^{-1} sec^{-1} for the oxidation of the primary alkyl radical with copper(II).[139]

310. J. W. Wilt, S. N. Massie, and R. B. Dabek, *J. Org. Chem.*, **35**, 2803 (1970).

311. See, e.g., J. C. Chottard, M. Julia, and J. M. Salard, *Tetrahedron*, **25**, 4967 (1969).

312. The rate constant for the addition of ethyl radicals to octene-1 in the gas phase at this temperature is similar [D. G. L. James and T. Ogawa, *Can. J. Chem.*, **43**, 640 (1965)].

313. J. K. Crandall and D. J. Keyton, *Tetrahedron Letters*, 1653 (1969).

314. R. D. Rieke and N. A. Moore, *Tetrahedron Letters*, 2035 (1969).

315. W. A. Thaler, A. A. Oswald, and B. E. Hudson, Jr., *J. Am. Chem. Soc.*, **87**, 311 (1965).

316. D. B. Denney, R. M. Hoyte, and P. T. MacGregor, *Chem. Commun.*, 1241 (1967).

317. D. M. Golden, *Int. J. Chem. Kinet.*, **1**, 127 (1969).

318. It is interesting to note that in the reaction of *t*-butoxy radicals with tri-*n*-butylphosphine, product studies by Buckler[319] have shown that the phosphoranyl radical decomposes by β-scission:

$$(CH_3)_3CO\dot{P}(n\text{-Bu})_3 \longrightarrow (CH_3)_3C\cdot + OP(n\text{-Bu})_3$$

318a. However, relative and absolute rate measurements have now been made on the α- and β-scission of a variety of other phosphoranyl radicals. See, for example, A. G. Davies, D. Griller, and B. P. Roberts, *Angew. Chem., Int. Ed. Engl.* **10**, 738 (1971); idem, *J. Chem. Soc., Perkin II*, in press; G. B. Watts, D. Griller, and K. U. Ingold, *J. Am. Chem. Soc.*, in press; W. G. Bentrude and T. B. Min, *ibid.*, **94**, 1025 (1972); W. G. Bentrude, E. R. Hansen, W. A. Khan, and P. E. Rogers, *ibid.*, **94**, 2867 (1972); A. G. Davies, R. W. Dennis, D. Griller, and B. P. Roberts, *J. Organometal. Chem.*, in press.

319. S. A. Buckler, *J. Am. Chem. Soc.*, **84**, 3093 (1962).

320. C. Walling and A. Padwa, *J. Am. Chem. Soc.*, **85**, 1593 (1963).

321. Although Walling and Padwa reported values of A_β/A_α and of E_β/E_α these data have not been included because the ratios of the preexponential factors show an impossibly wide variation.

322. F. D. Greene, *J. Am. Chem. Soc.*, **81**, 2688 (1959).

323. J. K. Kochi, *J. Am. Chem. Soc.*, **84**, 1193 (1962).

324. P. Kabasakalian and E. R. Townley, *J. Am. Chem. Soc.*, **84**, 2723 (1962).

325. W. Braun, L. Rajbenbach, and F. R. Eirich, *J. Phys. Chem.*, **66**, 1591 (1962).

325a. D. Griller and B. P. Roberts, *Chem. Comm.*, 1035 (1971).

326. J. A. Kerr and A. C. Lloyd, *Q. Rev. (London)*, **22**, 549 (1968).

CHAPTER 3

The Decompositions of Peroxides and Azoalkanes

THOMAS KOENIG
Department of Chemistry, University of Oregon, Eugene, Oregon

I. INTRODUCTION

The presence of the peroxy or azo function in a molecule invariably leads to the suspicion that the compound should be a free radical initiator. While there is certainly a basis for this suspicion, the chemistry of peroxides goes

$$\text{R—O—O—R} \longrightarrow 2\text{R—O}\cdot \quad (1)$$
$$\text{R—N=N—R} \longrightarrow 2\text{R}\cdot + \text{N}_2 \quad (2)$$

considerably beyond those processes which begin with the simple homolyses indicated (equation 1 or 2).[1] A notable example of a nonradical peroxide reaction is the Criegee[2] rearrangement which is now known to be the result of a highly concerted heterolysis of the O—O bond with simultaneous carbon-to-oxygen migration of a substituent group (R′) to give the corresponding oxonium–acylate ion pair (3).

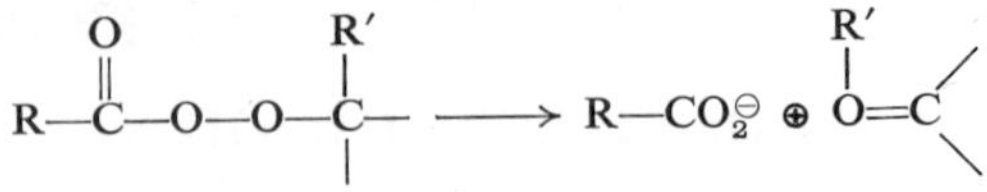

(3)

This mode of perester decomposition is favored by polar solvents[3] and electron-withdrawing substituents in the anion fragment.[4] Thus, *t*-butyl pertosylate has a half-life of only about one hour in methanol at 4.5°.[5] The nature of the migrating group (R′) also has a powerful effect on the stability of *p*-nitroperbenzoates.[4] Cases where the Criegee and normal homolysis reactions have fortuitously similar rates have also been reported.[6–11] The ionic reaction can usually be caused to be the favored one by increasing the solvent polarity.

The formally identical process in the diacyl peroxide series is the Leffler carboxy inversion.[12,13] This reaction was originally discovered for unsymmetrically substituted benzoyl peroxides but is now known to contribute to

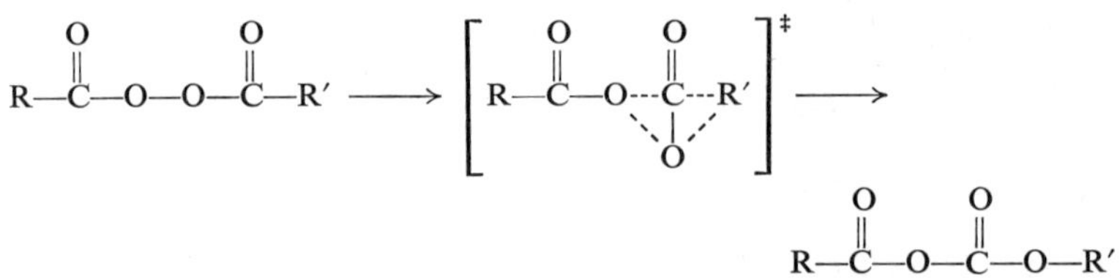

(4)

the overall reaction of simpler diacyl peroxides where the R′ group is secondary alkyl[14–16] or better cation. Again this reaction is favored by polar solvents and electron-withdrawing substituents in the potential anion fragment.[14]

The rearrangements above (equations 3 and 4) are unimolecular although they probably do depend on solvation energy to make them competitive with the homolytic reactions. (Tests for these reactions in the gas phase do not appear to have been made as yet.) Almost any O—O bond can be caused to undergo a heterolysis under the influence of acid catalyses.[17,18] The sulfuric acid-catalyzed decomposition of cumene hydroperoxide[19] is the basis of a commercial source of phenol. The key step here is the carbon-to-oxygen migration of the aryl group, which is analogous to the Criegee process. The uncatalyzed thermal decomposition[20] of cumene hydroperoxide takes an entirely different course. The sulfuric acid-catalyzed decomposition of *t*-butyl hydroperoxide can also give rise to such rearrangement products.[21] Many diacyl peroxides will undergo carboxy inversion in the presence of Lewis acid catalysts like boron trifluoride[22] or antimony pentachloride.[23] Rüchardt and Schwarzer[6,24] have provided some interesting examples of acid-catalyzed perester fragmentations when a very stable carbonium ion is formed through decarboxylation (5).

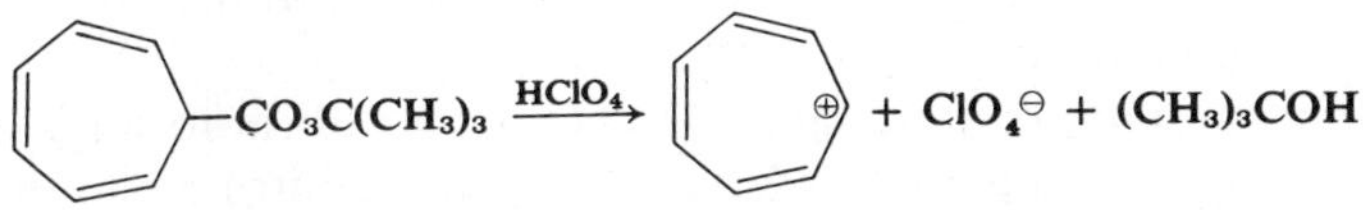

(5)

Base-catalyzed decompositions of peroxides are also known. *t*-Butyl performate[25] undergoes such a nonradical fragmentation which increases dramatically in rate as the solvent is frozen. This is the perester example of the general[26] process (6) of fragmentation by abstraction of a proton α to the O—O bond with formation of a carbonyl compound (carbon dioxide) and the alkoxide ion. *t*-Butyl peroxide has been reported[27] to be inert to base,

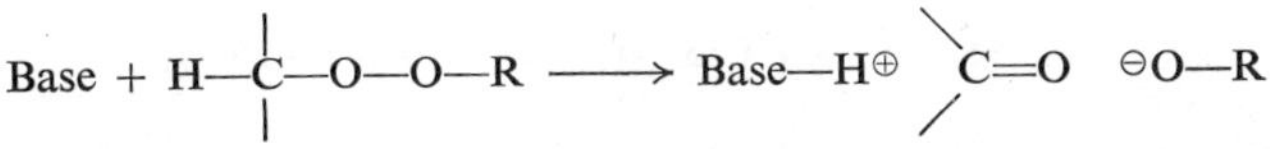

(6)

suggesting the necessity for the presence of an α-proton. However, acetyl peroxide undergoes an acetate ion-catalyzed isomerization to acetoxy acetic acid. It is supposed[28] that the acetate ion acts as a strong base in the solvents used.

Another large class of nonradical bimolecular reactions of peroxides comprises those involving some type of nucleophile. This general type of reaction (7) occurs with substances such as organometallics,[29] amines,[30] phenols,[31] sulfides,[32] olefins,[33] aromatics,[34] and trivalent phosphorus compounds.[35] Radicals are also formed to various extents but the products are

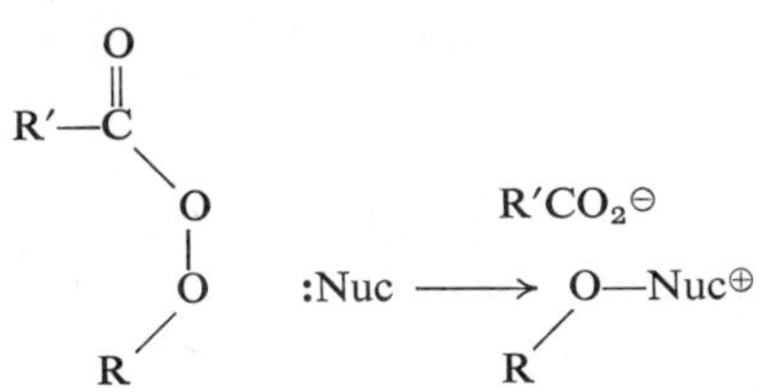

(7)

generally those which are expected from the ion-pair intermediate shown.

Relatively less work has been expended toward investigating nonradical reactions of simple azo compounds, although these certainly exist. The isomerization of azoalkanes to the corresponding hydrazones[36] is one example. This reaction can also be acid catalyzed.[36a] Dibenzoyldiimide is

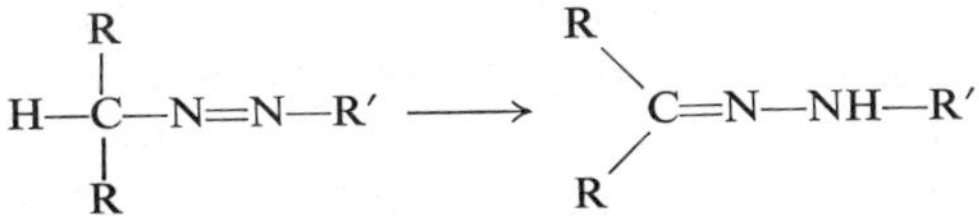

(8)

known to undergo radical decomposition in competition with a nonradical cyclization.[37] Benzoylphenyldiimide is a source of diazenyl anions in the presence of strong base,[37a] Diels-Alder reactions of azodicarboxylate with dienes are well known.

All of these results are important in studies of the radical-forming reactions of azo compounds and peroxides because they may contribute to the overall disappearance of the initiator to an extent which varies with temperature. This can cause serious errors in the apparent activation parameters for the overall decomposition. The simple demonstration of *some* free radical formation is not sufficient to ensure that all of the products are formed from such species. Additional criteria are required which allow one to estimate what overall fraction of the reaction is occurring through radicals.

II. INDUCED DECOMPOSITION

A second general problem, which arises after contributions from nonradical paths are ruled out, is the extent to which a potential initiator is destroyed by *chain* reactions. The decomposition of hydroperoxides is almost always by a chain process involving abstraction of the hydroperoxy hydrogen atom.[38] Direct evidence for the existence of the tetraoxy and trioxy[39] combination

$$\text{R—O—O—H} \longrightarrow \text{R—O}\cdot + \cdot\text{OH} \tag{9}$$

$$\text{R—O—O—H} + \text{RO}\cdot \longrightarrow \text{R—O—O}\cdot + \text{R—OH} \tag{10}$$

$$2\text{R—O—O}\cdot \rightleftharpoons \text{R—O—O—O—O—R} \longrightarrow \text{O}_2 + 2\text{RO}\cdot \tag{11}$$

$$\text{R—O}\cdot + \text{R—O—O}\cdot \rightleftharpoons \text{R—O—O—O—R} \tag{12}$$

products shown has recently been found. The cleavages of these O—O bonds are extremely rapid by comparison with other peroxides but consistent with the low O—H bond dissociation energy of the hydroperoxide.[40]

Peroxides and azoalkanes, having hydrogen atoms α to oxygen or nitrogen, also have an obvious reactive site for attack by a radical generated from a previous decomposition. Such chain processes have been implicated in the overall destruction of ethyl and isopropyl peroxide in both solution[41] and

$$\text{H—}\overset{|}{\underset{|}{\text{C}}}\text{—O—O—R} + \cdot\text{R}' \longrightarrow \,\rangle\!\cdot\text{C—O—O—R} \longrightarrow \,\rangle\text{C=O} + \cdot\text{OR} \tag{13}$$

gas phase.[42] In the gas phase the evidence for this step comes from studies of the products of reaction of the peroxide with radicals generated from an independent source. Temperature extrapolation allows the estimation of the degree to which these steps contribute to the overall destruction. In solution, this step is inferred from the chain transfer constant for the peroxide.

The rates of decomposition of azomethane are reduced by added nitric oxide[43] in the gas phase. This is an indication of the existence of a contribution from a short chain process. Abstraction of the α-hydrogen atom has again been suggested.[44] The same possibility has been discussed for azoethane

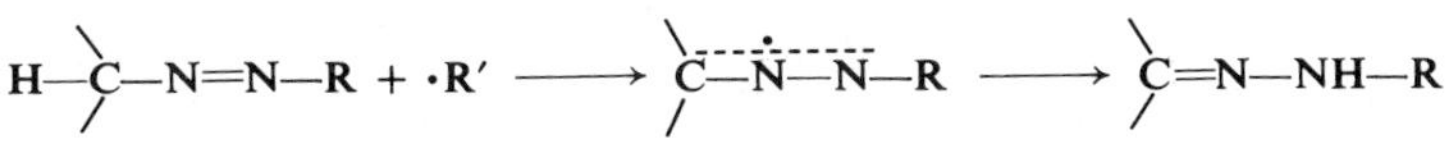 (14)

on the basis of product studies.[45] However, the ability of the azyl radical, formed from 2,2′-azopropane, to propagate chains is open to question.[46] The hydrazone has been detected[46a] in the thermolysis products but its origin has not been carefully studied.

Early studies[47] of the rates of decomposition of benzoyl peroxide showed the existence of a considerable solvent dependence. The rates were found to be reduced when scavengers were added and the rate law was greater than first order in some cases. These observations are consistent with the chain sequence shown. The order of the induced term in the overall rate law depends

Per. $\longrightarrow$ 2RO·			(15)
RO· + S—H $\longrightarrow$ R—OH + S·			(16)
S· + Per. $\longrightarrow$ R—O—S + RO·			(17)
S· + S· $\longrightarrow$ S—S	$\frac{3}{2}$ Order		(18)
S· + RO· $\longrightarrow$ R—O—S	1 Order		(19)
RO· + RO· $\longrightarrow$ Per.	$\frac{1}{2}$ Order		(20)

on the nature of the propagation and termination steps. (The three shown are not a complete list of the possibilities.)

The propagation step can involve a number of mechanistically different reactions. In diethyl ether, the rates are fast and first order and the induced decomposition product (1-ethoxyethyl benzoate) is formed in 95% yield.[48] This product, isolated from reaction with oxygen-18 labeled peroxide, shows very little randomization of the oxygen atoms.[49] Thus, one of the mechanisms for the propagation step can be depicted as a direct S_H2 displacement on the O—O bond. Similar results have been obtained with cyclohexenyl and trityl

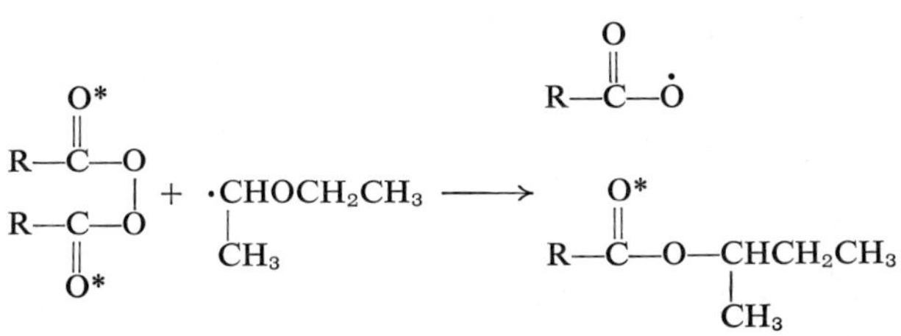 (21)

radicals.[50] The reaction constant for the latter reactions with *meta*- and parasubstituted benzoyl peroxides is rather large (+1.45) and the rates are fairly solvent dependent[50a] so that charge separation is important in the transition state.

The decomposition of benzoyl peroxide in carbon tetrachloride solution gives a significant (18%) yield of *p*-trichloromethylbenzoic acid. This has been suggested[51] to be the result of a *concerted* attack of the trichloromethyl radical on the aryl nucleus with simultaneous intramolecular displacement by the α-carbon of the ring on the O—O bond (22). The resulting α-lactone would certainly be expected to rearrange quickly to the observed acid. The corresponding acids have also been reported as products in the decomposition of

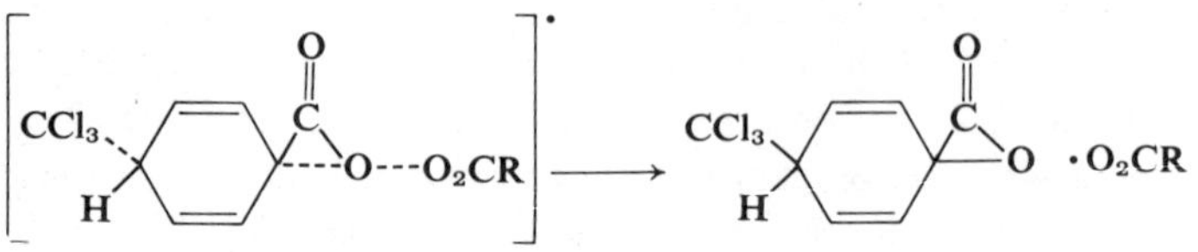 (22)

naphthoyl peroxide.[52] Cyclohexyl-substituted benzoic acids are formed in 75% yield in the decomposition of acetyl–benzoyl peroxide in cyclohexane.[53] High yields of carbon dioxide and methane are also observed. These results suggest the formation of the acetoxy radical and identify the benzoylperoxy oxygen atom as the site of induced attack. These results argue against the stepwise alternatives which have been proposed[54] for formation of these products. The latter involve the intermediacy of the adduct cyclohexadienyl radical which could transfer a hydrogen atom. The absence of any measurable deuterium isotope effect rules against hydrogen transfer as the rate-determining step.

There are several examples of intramolecular S_H2 displacements on peroxy functions. The decomposition of *t*-butyl peroxide in the gas phase[55] gives 3% isobutylene oxide. In neat liquid the yield is 35%.[56] The chain-propagating steps are β-hydrogen atom abstraction and displacement (23). A formally identical process has been proposed for the decomposition of neat *t*-butylperoxy isobutyrate.[57] Here, the α-lactone is not isolated but its polymer is

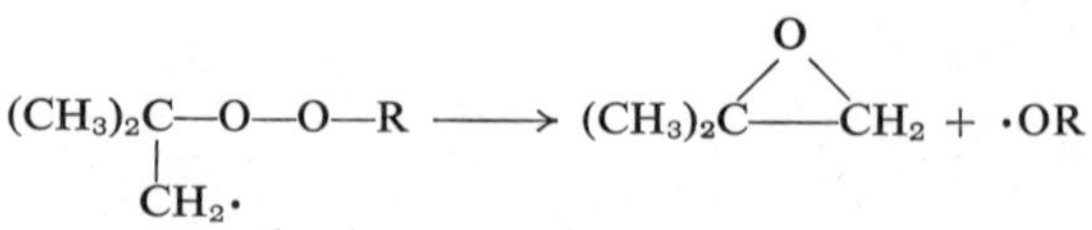 (23)

found in 76% yield. Malonate diperesters[57,58] give similar products, which provides good evidence for the displacement step. The self-polymerization of an α-lactone has also been verified[59] recently. Lactone products, which are found[60] in peroxides containing olefinic side chains, also suggest intramolecular S_H2 displacement after solvent radical addition to the double bond [instead of abstraction (24)].

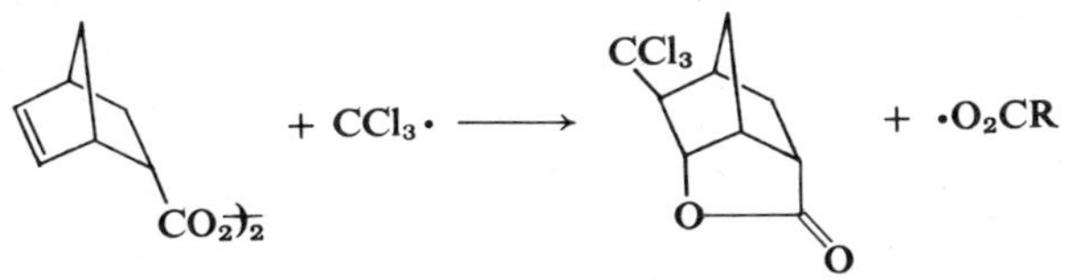

(24)

In view of these results it is somewhat surprising that radical bromination of *p*-tolyl peroxides and peresters gives the corresponding *p*-bromomethyl compounds without destruction of the O—O bond.[61] Rüchardt and Schwarzer[130] have provided evidence that the formation of the α-lactone which would be observed from the cycloheptatrienyl perester (5) is rather slow. The rate constant for the S_H2 step must be reduced for highly delocalized α radicals. Also, peresters of acrylic acids can be polymerized[62] so that the displacement step is not faster than polymerization.

The decomposition of peresters with an acetylenic side chain[63] showed the existence of another type of induced reaction. This process is the cleavage of the perester function when a radical center is β to the carbonyl carbon atom. This type of reactivity was indicated for cinamoyl peresters also. Previous

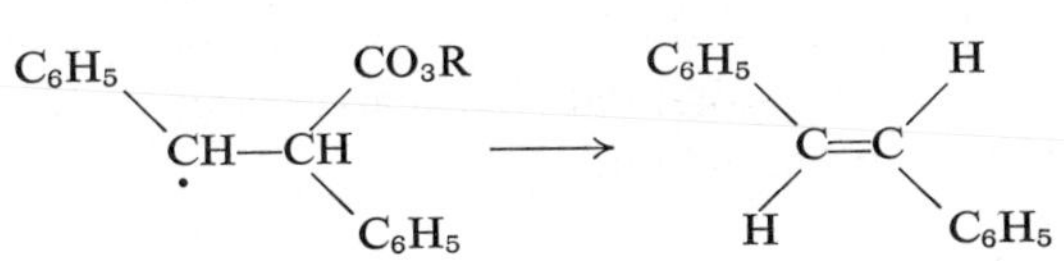

(25)

studies of the di-*t*-butylperoxy 1,2-diphenylsuccinates[64] have provided evidence for the rapid fragmentation step (25).

Huyser and his colleagues[65] have shown that direct hydrogen atom transfer can be a contributor to the destruction of a peroxide molecule (26).

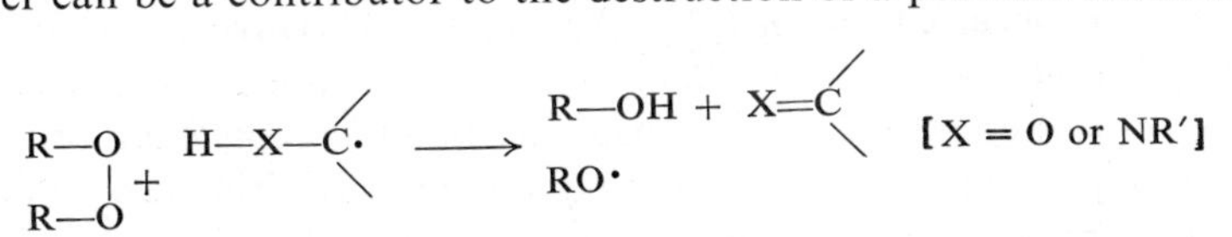

(26)

The rates of decomposition of *t*-butyl peroxide are fast in solvents made up of primary and secondary alcohols and amines but not in ethers or tertiary amines. A similar process has been invoked to explain the high yields of enol ethers, carbon dioxide, and *t*-butanol from perester decompositions in diisopropyl ether solvent.[66,67] Induced decomposition of *t*-butyl peracetate also appears to contain a contribution from this pathway.[68,69]

Close analogies to such S_H2 displacement reactions in azo compound decomposition do not exist. Products, such as tetramethyl hydrazine[70] from photolysis of azomethane, indicate that these initiators are destroyed, in part, by radical addition to the N═N bond to give the corresponding hydrazyl radical. Such processes have been demonstrated for solution also.[70a]

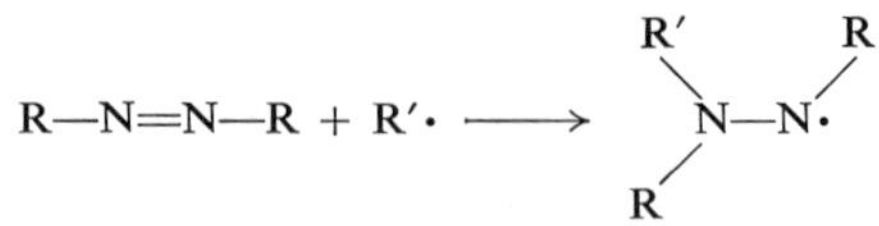
(27)

As with nonradical side reactions, small but variable contributions from one or more of these radical-induced reactions can seriously affect the magnitude of the activation parameters. Therefore some care should be taken in minimizing the likelihood of such contributions. The best chances for observing the simple unimolecular cleavage in solution are in pure nonpolar solvents which contain some scavenger and in as high dilution as feasible. Induced decomposition may be discernible from reductions in rate with dilution or with addition of a scavenger. Lamb et al.[71] have developed a method which allows the determination of first-order rates of disappearance of initiators in very dilute solution by following the consumption of a colored scavenger, such as galvinoxyl, which is added in slight excess. This method works well in some cases but is limited to fairly low temperatures by the instability of scavengers. Styrene[117,123] is frequently used as an inhibitor at higher temperatures. Even when the precautions are taken, it is difficult to be certain that none of these potential side reactions contributes to the overall disappearance of the potential initiator.

III. UNIMOLECULAR INITIATOR DECOMPOSITION

A considerable body of knowledge now exists concerning the nature of initiation reactions. Much of it deals with the identification of those reactions which are nonconcerted and involve the stepwise cleavage of an O—O or C—N bond to give a short-lived intermediate. These are distinguished from concerted processes wherein more than one bond is *completely* broken or made in a single unit process. The same products may ultimately be derived from either type of process though this is not necessarily the case. Concerted breaking of an old bond along with making of a new one can lead to a significant change in products. The latter type of concerted reaction is also referred to as anchimeric assistance. Figure 1 shows a schematic energy reaction coordinate profile which demonstrates the conceptual definition of the concerted process used here; the *complete* making or breaking of *more than one* bond by a process in which the starting material and product are separated by a single maximum in free energy. This is distinct from other definitions of concertedness which use *perturbation* of more than one bond or force constant as a basis. All reactions of molecules larger than diatomic involve

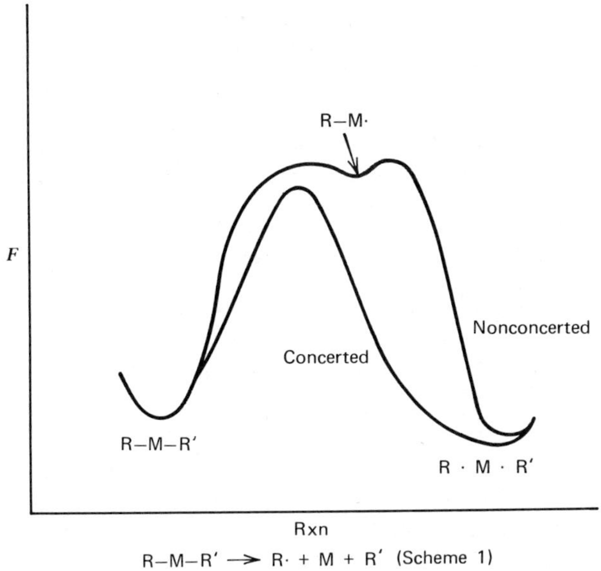

Fig. 1. Energy-reaction coordinate.

transition states in which more than the single reacting bond are perturbed to some extent. The ultimate distinction between stepwise and concerted processes always reduces to the existence of possible alternatives to reaction in the forward direction when the stepwise situation pertains. Thus, in the strict sense, the only time the stepwise process can be ruled out is when an infinitely effective scavenger for the intermediate fails to divert any of the reaction away from its normal course.

A. Nonconcerted Decompositions

A generalized outline of the kinetically important steps in a nonconcerted initiator decomposition is shown as Scheme 1. The definitive characteristic

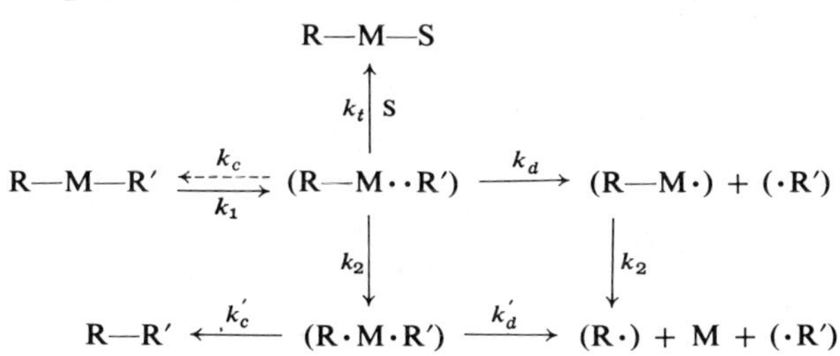

M = N_2: azoalkane

M = CO_2 or carbonyl compound: peroxide

S = scavenger

Scheme 1

of this mechanism is the presence of the general intermediate (R—M·) which may fragment further (k_2) losing a small molecule (M). The scheme can be applied equally well to azoalkanes and peroxides. It can also be applied to both gas-phase and solution reactions with the distinction that the solution reactions have the possibility of recombination (k_c and k_c') since separation (k_d and k_d') is not infinitely rapid. Trapping of the intermediate (k_t) is a formal possibility for either medium.

1. Gas-Phase Decomposition of Azoalkanes and Peroxides

The classic example of the simple nonconcerted homolytic cleavage mechanism is that of *t*-butyl peroxide.[72] Induced reactions are minimal in the gas phase. The *t*-butoxy radical is easy to trap since the β-scission step in this case requires 11–16 kcal/mole.[72a,99] Thus, almost any hydrogen donor will divert this intermediate away from acetone to *t*-butanol. Such results comprise a standard example of scavenging of the intermediate, which is the best method for demonstrating stepwise reactions. Benson and O'Neal[73] have advocated thermochemical additivity schemes in dealing with gas-phase unimolecular reactions. The hypothesis is that reactions involving the same type of bond should have the same activation energies except when there is electron delocalization or relief of strain in the intermediate formed. The activation energies for simple alkyl peroxides should thus be the same since only the O—O bond is being broken in the rate-determining transition state. Table 1 contains a selected list of activation parameters for these reactions.

TABLE 1

Activation Parameters—Peroxides R—O—O—R

R	k_{rel}[a]	E_a	Log A	E_a Solution	Ref.
CH_3	1	36.9	15.6	—	76
CH_3CH_2	1.8	37.3	16.1	35.2[b]	77
$CH_3(CH_2)_2$	1.1	36.5	15.4	—	78
$(CH_3)_2CH$	0.4	37.1	15.4	36.5[b]	77
$(CH_3)_3C$	0.3	37.4	15.6	35,[b] 37.5[c]	55, 56
CH_3CO	1.6×10^3	29.5	14.25	32.0[d]	79
CH_3CH_2CO	1.2×10^3	30.0	14.44	31.1[d]	80
$CH_3(CH_2)_2CO$	1.5×10^3	29.6	14.28	—	80

[a] Calculated at 80° using activation parameters listed.
[b] 50% Styrene–benzene. Reference 81.
[c] Cumene. Reference 71.
[d] Toluene. Reference 175.

The low values for the diacylperoxides can be rationalized by a small resonance stabilization (28) of the carboxy radical.[74] The constancy of the other values supports the hypothesis.[75]

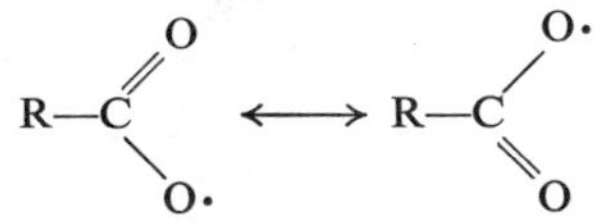

(28)

Similar lines of reasoning have been advanced in discussing the gas-phase decompositions of azoalkanes. Activation parameters are listed in Table 2. The steady reduction in E_a down the series parallels the corresponding C—H bond dissociation energy. An Evans–Polanyi[81] plot is shown as Figure 2. The isopropyl compound gives some hydrazone[46a] and its activation parameters are thus unreliable. The near unit slope of the line is as expected for essentially complete breaking of one of the C—N bonds in the transition state and zero activation energy for recombination. The loss of nitrogen from an alkyl diazenyl radical is exothermic[82] and no examples of trapping of such a species are known as yet. Some authors[83] choose to represent these same reactions as concerted processes. The observed β-secondary deuterium kinetic isotope effects do not allow a choice in the one gas-phase reaction in which they have been measured (3,3′-azopropene[88]). Recent studies[88a] of

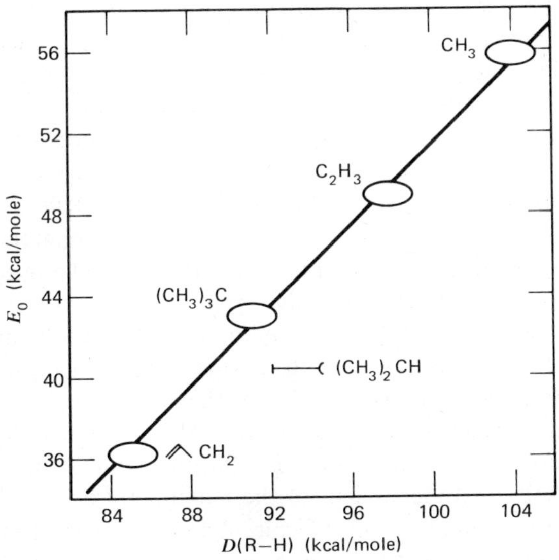

Fig. 2. Azoalkane activation energy-dissociation correlation. From reference 88.

TABLE 2

Gas-Phase Activation Parameters for Azoalkanes, R—N = N—R

R	k_{rel}[a]	Log A	E_a	DR—H[b]	Ref.
CF_3	1.0	16.2	55.2	106	85
CH_3	15.5	17.3	55.4	104	43
CH_3CH_2	42.7	16.0	51.2	98	84
$(CH_3)_2CH$	1.4×10^4	13.8	40.9	94.5	86
$(CH_3)_3C$	4.0×10^4	17.2	43.0[c,d]	91	87
$CH_2{=}CH{-}CH_2$	1.2×10^8	15.5	36.1	85	88
$C_6H_5CH_2$	1.5×10^7	14.1	35.0	85	89

[a] Calculated at 200° using activation parameters listed.
[b] In kcal/mole. Reference 92.
[c] E_a = 43.9 kcal/mole in diphenyl ether–isoquinoline (90:10) solution. Reference 91.
[d] E_a = 43.9 kcal/mole, cetane solution, 170–200°. Reference 90.

unsymmetrical alkyl allyl-azo compounds have strengthened the case for the stepwise process.

2. *Solution Phase*

The very close agreement in the rates and even activation parameters for the solution and gas-phase decompositions of 2,2′-azoisobutane (Table 2), *t*-butyl peroxide, and acetyl peroxide (Table 1) has been taken as justification for ignoring solvent effects for radical-forming reactions. This generalization has been shown to be incorrect, in a pure sense,[93,94] for *t*-butyl peroxide in solvents where induced decompositions are not important. The activation enthalpy varies from a low of 31 kcal/mole in acetonitrile to a high of 40.8 kcal/mole in cyclohexane. There is not very much difference in the rates because of compensation in the apparent activation entropies. Such correlations between $\Delta H^\ddagger$ and $\Delta S^\ddagger$ have also been observed with azobisisobutyronitrile[95,96] as well as phenylazotriphenylmethane.[97] A problem, which always exists with the interpretation of such isokinetic relationships, is the inherent correlation between the errors[94,96] in these parameters. However, the fragmentation-abstraction for *t*-butoxy radical has been shown to be subject to definite[99] solvent effects. Heats of solution of nitroxide radicals[100] also indicate that solvent (independent of the recombination problem) must play some role in the formation of radicals. Admitting this, it is still useful to consider the role of the solvent as constant and use reasoning similar to that for the gas phase. (See also Chapter 20 for further discussion.)

a. Dialkyl Peroxides and Arylazotriphenylmethanes. The decomposition of *t*-butyl peroxide in dilute solution in most solvents is relatively uncomplicated by induced or ionic reactions. The fragmentation-abstraction competition furnishes good evidence for the presence of the *t*-butoxy radical and the stepwise nonconcerted mechanism. The products include acetone and *t*-butanol in ratios which vary with the hydrogen-donating ability of the solvent.[71] The rates are essentially constant in contrast to the product variation. The activation energies for initiation of styrene polymerization for the rest of the alkyl peroxides of Table 1 seem consistent with the same mechanism although product studies are not available.

One of the features of the nonconcerted mechanism which is peculiar to the liquid phase is the possibility of reversible decomposition of an initiator as a result of the "cage effect" (see Chapter 4). If solvation is truly constant, then this reversibility should show up as a reduced rate of disappearance of the initiator with an increased solvent viscosity. The isolation of *t*-butyl peroxide as a product in the decomposition of the corresponding hyponitrite makes it certain that such a combination can occur, although it is not very effective at normal fluidities[101] (10% in isooctane at 45°). Walling and Waits[102] have shown that the rate of disappearance of *t*-butyl peroxide is reduced in Nujol compared with *t*-butylbenzene ($k_{rel} \sim 2.5$–4). The magnitude of the reduction indicates a minimum of 60% recombination in Nujol ($\eta = 2.6$ cP at 125°).

Pryor and Smith[103] have examined the possibility of using such rate reductions as a criterion for mechanisms of initiator cleavage. The relationship used is

$$\frac{1}{k_0} = \frac{1}{k_1} + \text{const}\sqrt{\frac{\eta}{A\eta}} \qquad \text{(c.-1)}$$

The reciprocal rate constants for decomposition of a number of initiators give approximately linear correlations with the square root of the viscosity. Such correlations allow extrapolation to zero viscosity and thus give an estimate of k_1 (Scheme 1), the unimolecular rate constant of initiator homolysis. Such results are shown as Figure 3 for *p*-nitrophenylazotriphenylmethane (NAT) and azocumene. The contrast is quite suggestive of nonconcerted cleavage for the former and concerted cleavage for the latter. Such a nonconcerted process would explain why so little cage product[104,105] (nitrobenzene indirectly) is formed in the NAT reaction. The phenyl diazenyl radical must lose nitrogen more slowly than it diffuses in normal solvents. Reversible decomposition of arylazothiolates[106] and sulfinates[107] has been proposed previously. The phenyl diazenyl radical has also been implicated in the reactions of phenyldiimide.[108] CIDNP polarization of the nitrogen nuclei of products of decomposition of *N*,*N*′-diphenyl triazene has also been

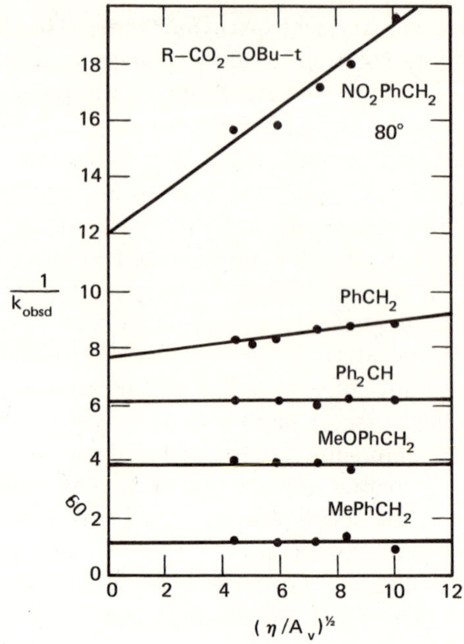

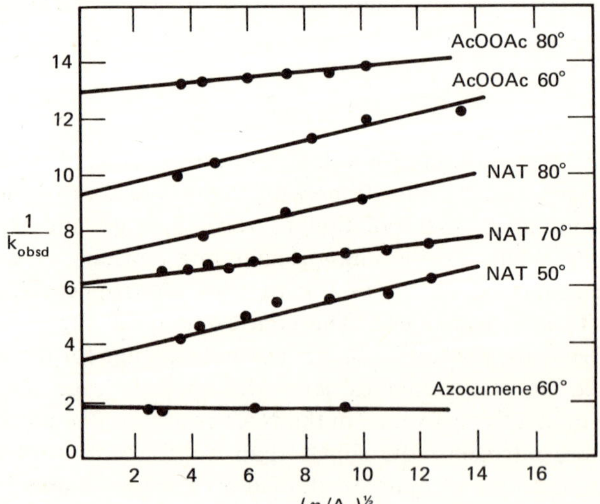

Fig. 3. Rate constant-viscosity correlations. From reference 103.

detected.[109] This latter method should prove especially valuable as a test for nonconcerted azo compounds which react rapidly enough to permit it.

Seltzer and co-workers[195,204] have measured α-secondary deuterium, carbon-13, and nitrogen-15 kinetic isotope effects in the decomposition of α-phenylethylazomethane. The results shown by the diagram (29) prove that the transition state is very unsymmetrical. They do not, however, answer the question of the existence of an alkyldiazenyl intermediate. Studies of the rate–viscosity dependence could provide some information on this question.

$k_H/k_D = 1.12$ $\quad$ $k_H/k_D = 0.99$

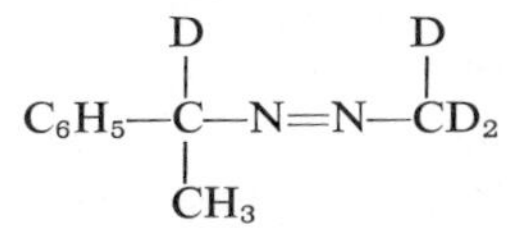

(29)

$k_{N^{14}}/k_{N^{15}} = 1.013$ $\quad$ $k_{C^{12}}/k_{C^{13}} = 1.0068$

b. Benzoyl Peroxide. The nonconcerted mechanism for the unimolecular (noninduced) decomposition of benzoyl peroxide has been demonstrated many times through the interception of the benzoyloxy radical by conventional scavengers including moist iodine,[110] galvinoxyl,[111] and vinyl monomers.[112] The latter studies indicate that the activation energy for decarboxylation of this intermediate must be more than 10 kcal/mole. The rate constant for this β-scission process is assigned a value of 10^4 sec^{-1} in DeTar's[113] computer analysis of the entire reaction scheme in benzene solution. Randomization of oxygen atoms has recently been demonstrated for benzoyl peroxide,[114] recovered after partial reaction. This result is consistent with the reversibility although the quantitative amount of recombination seems low.

The observation of CIDNP polarization in the *thermolysis*[115,115a] of this peroxide in cyclohexanone produces some problems with respect to interpretation. Emission is observed in the proton NMR of the benzene product. This is consistent with a singlet pair containing the phenyl radical. However, the slow rate of decarboxylation of the benzoyloxy radical implies that no phenyl radicals should be formed while in the vicinity of its geminate partner. Fischer argues that a small amount of the concerted reaction contributes to the decomposition.[115] He cites the isolation of small amounts of phenyl benzoate as support for his claim. (Esters are formed in aromatic hydrocarbon solvents by[116] benzoyloxylation of the solvent, which should not be confused with their formation in carbon tetrachloride[110] or cyclohexane.[116a]) There are a number of alternatives to the conclusion of concerted thermal reaction. These results are as indicative of the caution with which CIDNP effects must be viewed as they are of the mechanism of the homolysis reaction.

The radical decompositions of benzoyl peroxides[117] and *t*-butyl perbenzoates respond to polar substituents in the meta and para positions in a fairly strong manner. The relative rate constants can be correlated with

Hammett σ constants with ρ values of -0.4 and -0.7, respectively. These ρ values have been explained in terms of variations of ground-state stabilities which depend, in part, on the dipole–dipole repulsion between the two oxygen atoms of the peroxy linkage. The likelihood of some reversibility in the decompositions of these molecules makes the exact magnitudes of these reaction constants questionable. The rates of decomposition[117a] of para-substituted phenylazotriphenylmethanes are all very similar, although there is considerable variation in activation parameters. The reversibility may be more important here.

c. Acetyl Peroxide. The decomposition of acetyl peroxide furnishes an interesting demonstration of the challenge presented by attempting to distinguish concerted from nonconcerted mechanisms. The small reductions in both E_a and A factor for the gas-phase reaction (compared with dialkyl peroxides) are consistent with either possibility. The gas-phase products in the presence of toluene carrier are methane and carbon dioxide. In isooctane solution, the rates are not much different and the products are nearly the same except for small amounts of methyl acetate and ethane. The increasing yields[118] of these products with increasing viscosity make geminate combination a likely origin. Oxygen-18 randomization has been observed to increase in rate, with increasing solvent viscosity,[119a] while disappearance rate decreases. These results are again consistent with the stepwise scheme.

Koenig[119] has proposed a method of quantitative analysis of such results using the following relationships:

$$\frac{1}{y} - 1 = \frac{k_0}{k_s} = \frac{k_d}{k_c} + \frac{k_2}{k_c} = \sqrt{\frac{\text{const}}{\eta}} + \frac{k_2}{k_c} \qquad \text{(c.-2)}$$

Here k_0 is the observed rate constant for destruction of initiator, k_s is the observed rate constant for randomization, y is the yield of combination product, k_d is the *formal* rate constant for separation, k_c is the formal rate

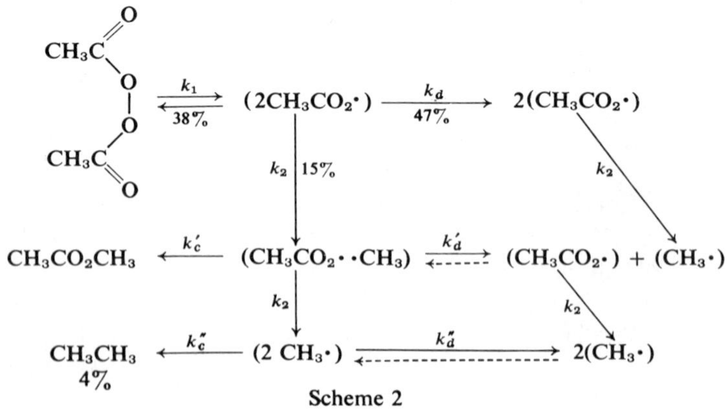

Scheme 2

constant for combination, and η is solvent viscosity. This analysis gives the results shown in Scheme 2. These results are identical to those of Pryor and Smith[103] but differ from Martin and Dombchick[118a] in that all of the geminate pairs are allowed the possibility of diffusive separation. The latter workers included diffusion of only the methyl radicals. This makes the viscosity effect on k_0 hard to accommodate. On the other hand, Scheme 2 must include the possibility of some combination of methyl–acetoxy pairs which are formed from separated acetoxy–acetoxy pairs (dotted arrows). Neither scheme is completely satisfactory.

Experiments with conventional trapping agents gave misleading results. The reduced yields of carbon dioxide in the presence of moist iodine[111] turned out to be due to hydrolysis to peracetic acid. Using DPPH[111] or carbon-14 acetate[28] ion led to other induced reactions. Reports of spin trapping[117b] of the acetoxy species also are likely to be results of such bimolecular reactions. Cyclohexene and other simple olefins give appreciable amounts of saturated esters and may be the result of the desired scavenging.[111,111a]

TABLE 3

Activation Parameters for Diacyl Peroxides, RCO_2—O_2CR

R	Solvent	$\Delta H^‡$	$\Delta S^‡$	k_{rel}[a]	Ref.
CH_3	Toluene	31.3	+11.6	(1.00)	175
CH_3CH_2	Toluene	30.4	+9.8	1.45	175
C_6H_5[b]	Benzene	33.3	+12.5	0.037	47
$C_6H_5(CH_2)_4$—	Carbon tetrachloride	30.4	+8.9	0.93	178
Ortho-C≡CC_6H_5 benzoyl	Chlorobenzene	24.0	−2.7	25.0	177
Ortho-CH=CHC_6H_5[c] benzoyl	Chlorobenzene	20.2	−8.7	275.0	173
$C_6H_5CH{=}CHC(CH_2)_2$—	Propylene carbonate	20.7	−11.3	34.0	178
	Carbon tetrachloride	26.2	−0.3	1.4	178
$(CH_3)_2CH$	Isooctane	26.6	+5.3	12.3	175
Cyclopropyl-	*p*-Cymene	29.1	+3.5	0.39	176
trans-4-*t*-Butylcyclohexyl	Carbon tetrachloride	19.5	−16.1	18.0	168

[a] Calculated at 80° using activation parameters listed.

[b] Activation parameters for the unimolecular reaction calculated from data of reference 47.

[c] Unsymmetrical benzoyl peroxide.

The constancy in activation parameters for acetyl, propionyl, and δ-phenylvaleryl peroxides (Table 3) compared with the increased yields of geminate combination products (ethane, 5% from acetyl vs. ethylene and butane, 35–40% from propionyl) provides further evidence for the *general* applicability of the stepwise scheme. The rate of decarboxylation of the propionyloxy radical is expected to increase with respect to diffusion giving more primary ethyl–ethyl pairs. δ-Phenylvaleryl peroxide gives 17.5% ester and 21.5% 1,8-diphenyloctane in carbon tetrachloride.[122]

The possibility of a mixed mechanism with a contribution from a concerted reaction, involving a transition state with significant methyl radical character, is ruled out by the small magnitude and viscosity sensitivity of the α-deuterium isotope effects ($k_H/k_D < 1.015$).[119a] These small values are compatible with the slight amount of decarboxylation (in isooctane, Scheme 2) which contributes to the reaction in the forward direction. The α-secondary deuterium kinetic isotope effect on the decarboxylation of the acetoxy radical was estimated as 1.03 per deuterium. This small value is consistent with this reaction being exothermic so that the transition state resembles the acetoxy radical more than it does methyl. This is in contrast to the isotope effect observed for β-scission of cumyloxy radicals ($k_H/k_D = 1.15$).[148] The carbon-13 kinetic isotope effects[120] can also be accommodated by such composite schemes.

d. Peresters. Table 4 summarizes the activation parameters reported for *t*-butyl peresters which would be expected to undergo nonconcerted decomposition. There are some rather serious discrepancies in the values for the same compound in different solvents or by different workers. However, there is again a remarkable consistency in most of the values listed. Confirmation of the nonconcerted nature of the reaction for the acetyl compound comes from decomposition rate reduction and oxygen-18 randomization rate increases with increasing viscosity, and from isolation of the perester from decomposition of the corresponding hyponitrite.[104] α-Secondary deuterium kinetic isotope effects[154] are again small ($k_H/k_D < 1.01$). Carbon-13 isotope effects have been interpreted in terms of the concerted reaction for this compound but this was before the viscosity results were available.[135]

The same criteria (except isotope effects) have been applied to the decomposition of the perbenzoate. The oxygen-18 randomization is much lower here than in the acetyl case.[119] CIDNP polarization has been reported for this compound[136] though the phenyl *t*-butyl ether ("cage" product) is not formed. The activation volume for the decomposition of this compound is large (+13 cc/mole) and has been interpreted in support of the reversible cleavage.[137]

TABLE 4

Activation Parameters for Nonconcerted Peresters—$RCO_3(CH_3)_3$

R	Solvent	$\Delta H^{\ddagger}$	$\Delta S^{\ddagger}$	k_{rel}[a]	Ref.
CH_3	Cumene	39.4	+24.6	(1.0)	124
	Chlorobenzene	38.0	+17.0	0.16	123
$CH_3(CH_2)_8$—	Nitrobenzene	35.3	+14.0	1.1	125
$CH_3(CH_2)_{12}$—	Nitrobenzene	35.0	+13.0	1.0	125
$C_6H_5(CH_2)_2$—	Chlorobenzene	35.0	+13.0	1.0	126
$C_6H_5C(CH_3)_2CH_2$—	Cumene	34.6	+12.0	1.0	127
(C_6H_5)—$(CH_2)_3$—	Chlorobenzene	36.0	+16.0	1.3	126
1-Norbornyl	Cumene	35.9	+14.8	0.8	128
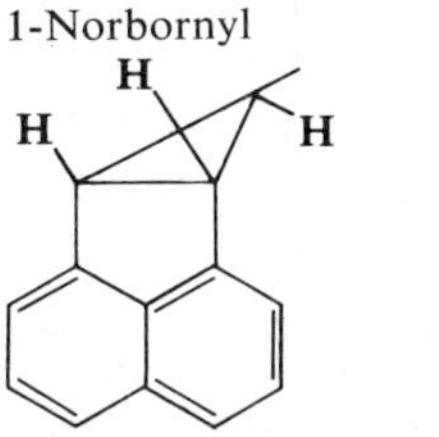	Cumene	37.1	+17.5	0.6	129
	Chlorobenzene	36.7	+15.0	0.3	130
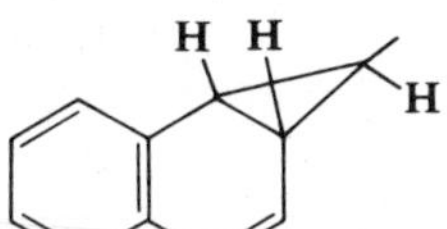	Chlorobenzene	35.3	+12.0	0.5	130
Cyclobutyl	Ethylbenzene	34.4	+13.3	2.8	131
$(CH_3)_3CO$	Chlorobenzene	35.4	+14.6	1.4	132
C_6H_5	Diphenyl ether	37.5	+16.1	0.2	133
	Chlorobenzene	33.5	+ 7.8	1.0	123

[a] Calculated at 100° using activation parameters listed.

B. Concerted Peroxide Fragmentations

1. Peresters

In 1958, Bartlett and Hiatt[123] published a paper which showed that radical formation from *t*-butyl peresters covered a rate range of over 10^5. These authors pointed out the direct relationship which exists between the stability of the decarboxylated radical and the rate of disappearance of the perester leading to it. In addition, a correlation between the decrease in the entropy of activation and the number of internal rotations which must be restricted, in a transition state which utilizes some of the stabilization of the developing carbon radical (Table 5), was noted. In spite of the numerous sources of possible error in the measurement of activation parameters for perester decomposition in solution, the implications of this entropy–enthalpy rela-

TABLE 5

Activation Parameters for Concerted Peresters—$RCO_3C(CH_3)_3$

R	Solvent	$\Delta H^‡$	$\Delta S^‡$	k_{rel}[a]	Ref.
p-$NO_2C_6H_4CH_2$—	Chlorobenzene	29.8	4.4	0.14	138
$C_6H_5CH_2$—	Chlorobenzene	27.9	2.0	(1.00)	138
p-$CH_3OC_6H_4CH_2$—	Chlorobenzene	25.7	0.0	10.0	138
$C_6H_5CH{=}CH{-}CH_2$—	Chlorobenzene	23.5	−6.0	13.0	123
$C_6H_5CH{-}CH{=}CH_2$	Chlorobenzene	23.0	−1.0	3.5×10^2	123
$CH_3CH{=}CHCH_2$	Cumene	25.8	−2.5	24.0	139
$CH_3{-}C{\equiv}C{-}CH_2$	p-Cymene	29.8	5.9	0.4	139
$C_6H_5{-}C{\equiv}C{-}CH_2$	p-Cymene	28.9	4.6	0.8	139
$C_6H_5{-}C{\equiv}CCH(CH_3)$	p-Cymene	26.7	3.2	11.0	139
$(C_6H_5)_2CH$	Cumene	25.0	0.5	37.0	57
	Chlorobenzene	24.3	−1.0	50.0	123
2.3-Diphenyl-cyclopropenyl	Ethylbenzene	23.6	2.0	7.6×10^2	140
$(C_6H_5)_3C$	Cumene	24.1	4.9	1.3×10^3	141
ROCO	Chlorobenzene	24–27.9	2.5–6.8		143–144
$(CH_3)_3CO_2CO$	Chlorobenzene	25.5	5.1	$1.7 \times 10^{+3}$	66
Cl	Cumene	29.1	10.5		161

[a] Calculated at 60° using activation parameters listed.

tionship have been rather well borne out by the results of subsequent studies. The cylindrical symmetry of the acetylenic group implies less rotational restriction than for an olefinic substituent, and indeed the observed activation entropies are less negative.[139] Rotation of the aryl groups in the trityl perester is already sterically restricted in the starting material and a more positive activation entropy is not surprising.[141]

The activation entropies of most tertiary alkyl peresters (Table 6) are lower than the 11 cal/mole-deg reported originally by Bartlett and Simons[142] for t-butyl perpivalate. Subsequent investigations of the kinetics of this compound in cumene gave similar rate constants but significantly altered activation parameters with $\Delta S^‡$ as low as −2.6 cal/mole-deg.[146,147] Only one internal rotation should be lost on going to the transition state for this compound (O—C—O—O). The low values and the variability of the activation parameters for the same compound, reported by different workers, detract from the quantitative use of the activation parameter criterion for the type of mechanism which pertains. However, the activation parameters for the 1-norbornyl case stand out and suggest that this compound reacts by the nonconcerted process. The transition state for the multibond reaction should

TABLE 6

Activation Parameters for Tertiary Peresters—$RCO_3C(CH_3)_3$

R	Solvent	$\Delta H^{\ddagger}$	$\Delta S^{\ddagger}$	K_{rel}[a]	Ref.
$(CH_3)_3C$	Chlorobenzene	30.0	+11.1	(1.00)	142
	Cumene	27.6	+3.6	0.9	146
	Cumene	25.5	−2.9	0.8	147
$C_6H_5C(CH_3)_2$	Chlorobenzene	26.1	+6.0	27.0	123
$(C_6H_5)_2CCH_3$	Chlorobenzene	24.7	+3.0	51.0	123
	Cumene	27.9	6.3	2.1	149
	Cumene	27.8	5.9	2.0	152
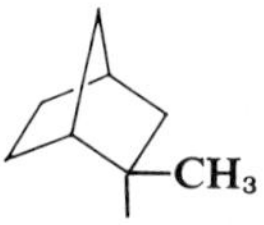	Cumene	30.3	9.3	0.2	149
	Cumene	29.8	7.8	0.1	152
trans-1,4-Dimethyl-cyclohexyl	Cumene	27.8	5.8	1.9	149
cis-1,4-Dimethyl-cyclohexyl	Cumene	27.5	5.1	2.1	149
cis-9-Decalyl	Cumene	26.6	4.7	6.8	151
trans-9-Decalyl	Cumene	26.4	4.2	7.2	151
1-Adamantyl	Cumene	27.9	4.9	1.1	146
	Cumene	27.6	3.7	0.9	147
	Cumene	29.8	11.1	0.7	128
1-Bicyclooctyl	Cumene	28.5	2.2	0.11	147
	Cumene	28.7	2.8	0.11	128
1-Norbornyl	Cumene	37.1	17.5	5.4×10^{-4}	128
	Cumene	35.9	14.8	8.7×10^{-4}	129
CH_3	Cumene	39.4	24.6	6.3×10^{-4}	124

[a] Calculated at 60° using activation parameters listed.

be relatively high in energy in this case because of the strain involved in planarizing the radical center. The other bridgehead compounds are classed as concerted by both the majority of the activation parameters and the fast relative rates. One might conclude from these results that the requirement for planarity around a radical center is less important for radicals than for cations. However, these relative rates give *direct* information about the transition states only and do not strictly imply anything about the corresponding decarboxylated carbon radicals.

Other criteria, applied to members of Tables 5 and 6, have led to conflicting conclusions. The rate–viscosity criterion[103] (Figure 3), applied to the phenylacetyl compounds, indicates that the *p*-nitro compound is undergoing nonconcerted decomposition since appreciable reduction in observed rate constants is noted as the solvent is changed from pentane to octadecane. Also, some oxygen-18 randomization was detected in samples of *t*-butyl perpivalate which were recovered after partial decomposition.[153] Either the oxygen-18 randomization occurs by a 1,3 sigmatropic shift or this perester must have a contribution from the stepwise process. On the other hand, appreciable α- or β-secondary deuterium isotope effects have been reported for both the phenylacetyl peresters[154] and *t*-butyl perpivalate,[154a] indicating significant perturbations in the C—H vibrations in the transition states. Such perturbations are expected of the concerted but not the nonconcerted mechanism. These conflicting conclusions point out the weaknesses involved in using a single criterion of mechanism and demonstrate the necessity of looking at more than one type of experiment.

The rate constants for decomposition of phenylacetyl peresters are fairly sensitive to the nature of the substituent in the meta or para position of the aryl nucleus.[138] These rate constants can be correlated with σ^+ values with a reaction constant of -1.09. Thus the transition state for the concerted reaction has been depicted as containing a significant degree of polar character. Here again a note of caution is needed since Pryor and Smith[103] have pointed out that the *curve* one obtains from these same data, plotted against σ values, can be interpreted in terms of concerted decomposition for the faster members but nonconcerted reaction for the *p*-nitro compound. Decomposition of monoperoxyoxalates[143,144] and α-heteroatom-containing peracetates[155] also responds to variation in substituent in the way expected of the charge-separated structure (Scheme 3).

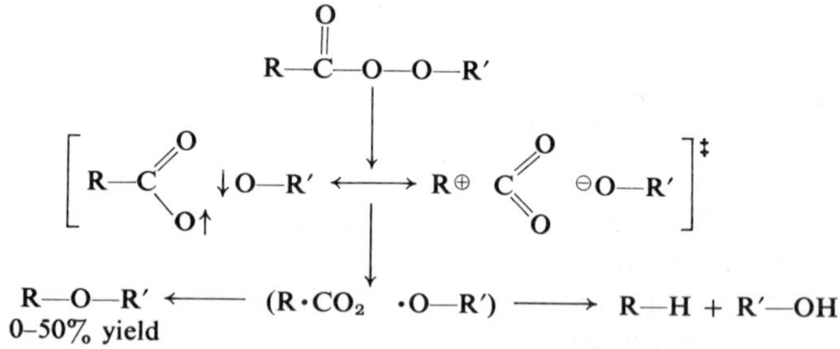

Scheme 3

Solvent effects in these reactions are unspectacular. The relative rates in isooctane compared with acetonitrile are all within a factor of 10. However, the activation volumes for the phenylacetyl compounds have been found to be extremely pressure sensitive, beginning low[156] and increasing with the pressure. Neuman and Behar[156] believe this dependence is the result of the solvent effect. At low pressures the increased volume of the transition state is offset by restriction of the solvent molecules. This latter effect becomes less effective as the pressure is increased. This rationalization is not necessarily in conflict with the small solvent effects at atmospheric pressure since these could be governed by an isokinetic relationship of the type discussed by Leffler[98] for azo compounds. Solvation effects could also be contributing to the low activation entropies of the tertiary radical peresters of Table 6.

The sensitivity shown by these compounds to polar substituents is again what one might expect from a transition state for the concerted process but not that for the stepwise process unless the intermediate acyloxy radical experiences very significant perturbations of its remaining bonds. Allowing such perturbations means that a stepwise transition state could hardly be distinguished from the concerted one. Trapping experiments are required to establish the intermediate in such cases.

Table 7 lists activation parameters for a series of "borderline" peresters.

TABLE 7

"Borderline" Peresters—$RCO_3C(CH_3)_3$

R	Solvent	$\Delta H^{\ddagger}$	$\Delta S^{\ddagger}$	k_{rel}[a]	Ref.
CH_3	Cumene	39.4	+24.6	(1.00)	124
	Chlorobenzene	38.0	+17.0	0.16	123
CH_3CH_2	Ethylbenzene	35.3	8.7	0.09	157
$(CH_3)_2CH$	Cumene	31.8	9.4	22.0	57
Exo-2-norbornyl	Cumene	30.9	6.1	16.0	158
	Chlorobenzene	30.9	6.7	22.0	158
Exo-2-norbornenyl	Cumene	33.0	11.0	10.0	159
Endo-2-norbornyl	Cumene	32.8	9.0	4.6	158
	Chlorobenzene[b]	36.8	20.0	4.0	158
Endo-2-norbornenyl	Cumene	32.6	9.0	1.6	159
Cyclohexyl	Chlorobenzene	31.3	8.6	32.0	158
$(CH_3)_3COO$	Chlorobenzene	31.8	7.1	6.3	160
$((CH_3)_3C)_2CH$	Cumene	29.9	7.9	150	148

[a] Calculated at 80° using activation parameters listed.
[b] Activation parameters were highly concentration dependent, suggesting induced reaction.

The chain-induced formation of the α-lactone has been suggested as an explanation for the low carbon dioxide yields and atypical activation parameters for the endo-2-norbornyl compound. In cumene, the products are more straightforward and carbon dioxide yields are high except for the endo-2-

norborn-6-enyl compound which gives some γ-lactone. The activation parameter criterion leads to the conclusion of the concerted mechanism. This does not explain the γ-lactone formation just mentioned nor the claim that the 2-norbornyl *t*-butyl ether products are not formed from "cage" reactions. Neuman and Behar[137] have isolated a 17% yield of cyclohexyl *t*-butyl ether for the cyclohexyl perester in cumene and have concluded that this and the cyclohexene are both products of the geminate pairs. The decomposition of the latter perester shows a low activation volume in cumene (+4 cc/mole) compared with *t*-butyl perbenzoate (13 cc/mole). This is proposed to be evidence that the decomposition is concerted in the former case. Each member of Table 7 should be examined using the rate–viscosity, oxygen-18, and activation volume measurements before definite conclusions are made concerning the exact nature of the decomposition mechanism.

2. *Diacyl Peroxides*

The diacyl peroxides corresponding to the peresters in Tables 5–7 give indirect evidence of the importance of the concerted mechanism since some free radicals and radical products can be detected in their relatively rapid reactions. The carboxy inversion product is also sometimes detected. In some cases, esters can be isolated without any evidence of either the carbonate anhydride or free radicals. Multiple mechanisms, with contributions from

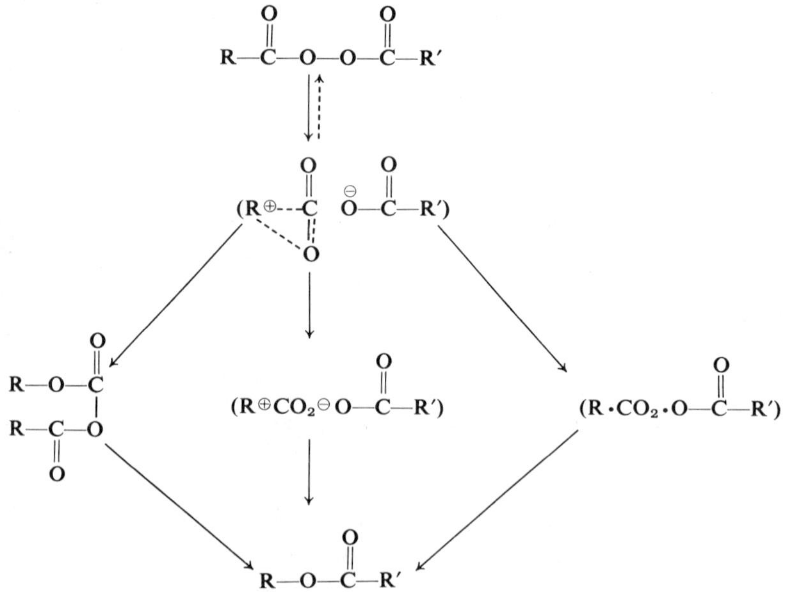

Scheme 4

more than one transition state, are one possible explanation. A more economical hypothesis[16] is that acyl or diacyl peroxides can give a tight ion pair (Scheme 4) which then is partitioned depending on the stabilities involved. While this hypothesis suffers from lack of precision in the definition of the ion pair, it does very well at rationalizing a very large body of otherwise contradictory data. Whatever the detailed explanation turns out to be, secondary and tertiary peroxides are prone toward carbonium ion-like reactions even when the cation center is at a bridgehead. Known or probable examples of such behavior are when R· of Scheme 4 is benzyl,[16,17,167] cyclopropylmethyl,[18] phenylethyl,[163] 2-propyl,[14–16,164] 2-butyl,[165] 1-phenyl-2-propyl,[166] epoxystyryl,[167] cyclohexyl,[168] 2-norbornyl,[60] *t*-butyl,[169] 1-apocamphyl,[170] and 1-triptocyl.[171]

C. Anchimerically Assisted Peroxide Cleavage

A number of examples of rapid decomposition of ortho-substituted *t*-butyl perbenzoates and benzoyl peroxides have been found. Table 8 summarizes

TABLE 8

Activation Parameters for Orthosubstituted Perbenzoates—

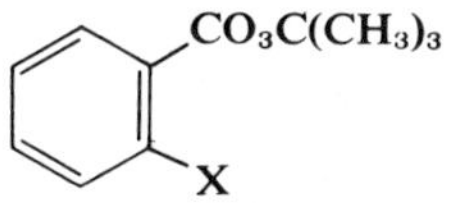

X	$\Delta H^{\ddagger}$	$\Delta S^{\ddagger}$	$k_{rel}^{60°}$	Ref.
—H	34.1	10.0	1.0	173
—I	28.0	−1.0	43.0	172
—CH=C(C_6H_5)$_2$	26.3	−5.0	67.0	173
—SCH_3	22.6	−5.0	140,000	172
—S—C_6H_5	23.0	−3.0	250,000	172
—C(CH_3)$_3$	34.2	+12.0	3.0	172
—SO_2CH_3	38.0	+19.0	0.4	134
—CH_2—S—C_6H_5	32.2	+ 7.2	5.0	174

some of the observed activation parameters. The products isolated from the decomposition of the ortho thiophenyl compound are those shown in Scheme 5.[172] The ortho vinyl compound gave dimers or telomers and the two lactones indicated.[173] The products and fast rates can be explained by transition states which involve formation of a partial bond between one of the peroxy oxygen atoms and the neighboring substituent as the O—O bond is broken. The extra bonding interaction should lower activation enthalpy as

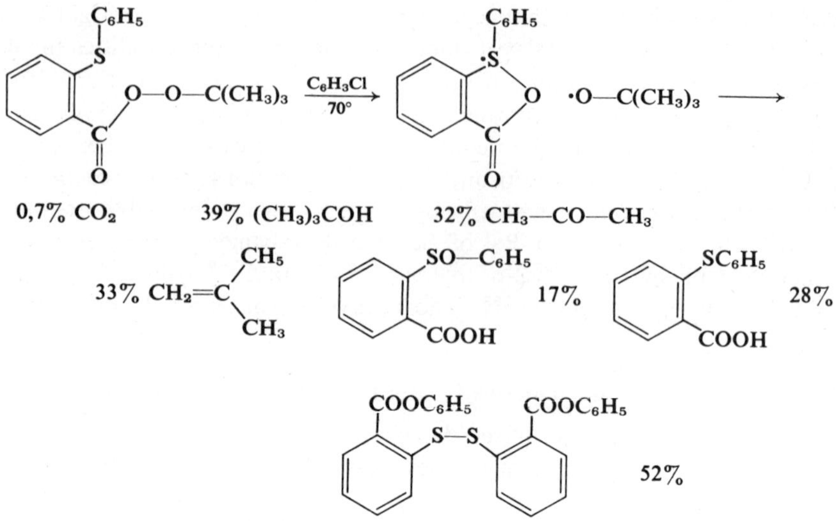

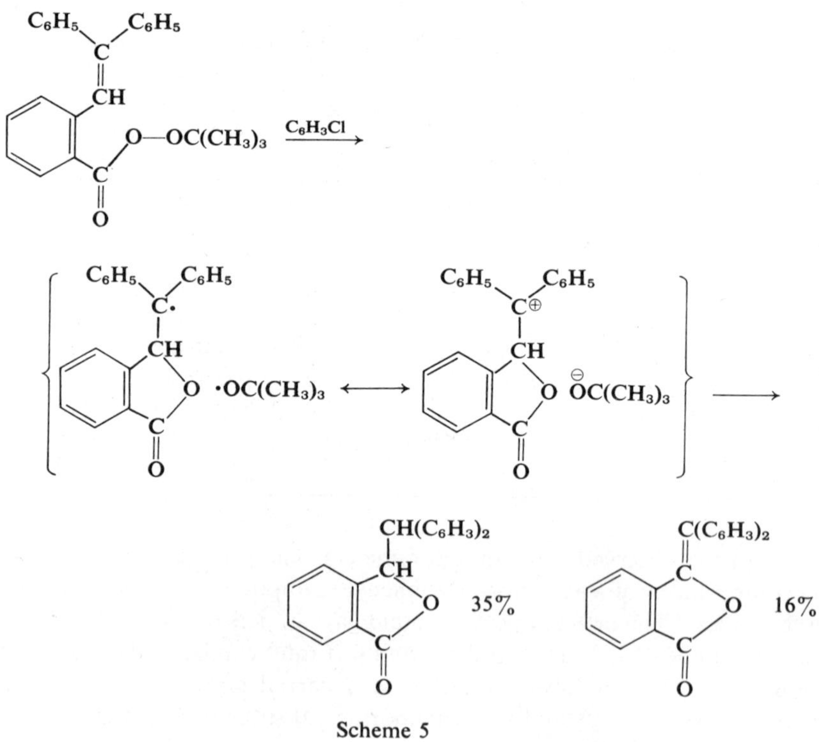

Scheme 5

observed. It would also require that the rotational restrictions reduce the entropy of activation, which also fits with the observations. Radical counting, by decolorization of galvinoxyl, and the isolation of products like *t*-butanol and acetone demonstrate that radicals are produced with reasonable efficiency. The failure of the dibenzothiophene and thioxanthone (Scheme 6) compounds to show the dramatic rate acceleration observed for the simple ortho sulfides indicates that the bond angles around the sulfur group are critical.[134] The transition states for the ortho thio compounds have thus been depicted as having a contribution from a valence shell-expanded sulfur atom with the S—R bond being directed out of the plane containing the benzoate group (Scheme 6).

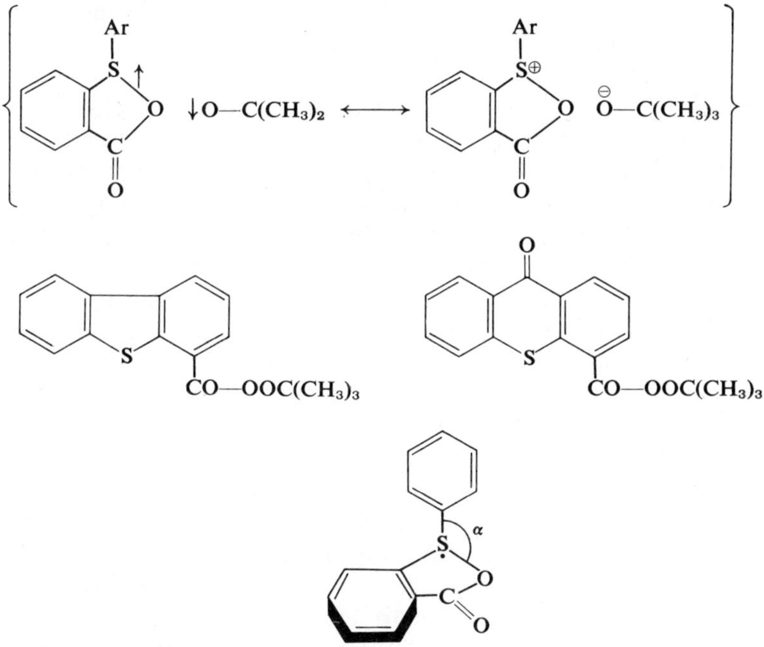

Scheme 6

Unusual rates and products have been found for some of the diacyl peroxide counterparts of the perbenzoates listed in Table 8. The decomposition of *O,O′*-diiodobenzoyl peroxide is rapid and gives some radicals capable of initiating polymerization.[179] The main products are the iodoso lactone and its polymer (60–75% yield). The decomposition of *o*-iodo-*p′*-nitrobenzoyl peroxide follows a similar course but with a slower rate constant. Bis-*β*-iodopropionyl peroxide also decomposes[180] rapidly but the products (iodine 32%, ethylene 29–42%) are not straightforwardly related to the aryl cases. Carboxy inversion also appears to play a role.

trans-γ-Benzylidenebutyryl peroxide[178] thermolyzes rather rapidly to give 45% radicals which can be scavenged by DPPH. The net rate acceleration, compared with δ-phenylvaleryl peroxide, is only 2–10, but this is due to the very large entropy decrease which is required for forming a cyclic transition state in the aliphatic compounds. Much larger accelerations were observed for ortho vinyl-substituted benzoyl peroxides.[173] At least one less free internal rotation is present in these starting materials. This unfavorable entropy factor is probably important in explaining the absence of any dramatic rate acceleration in peroxides containing simple olefinic linkages.[80,181]

Substituents in the para position of the ortho phenylthio group have a strong effect on the rate of decomposition of the corresponding perbenzoate. The reaction constant which correlates these data is -1.3,[134] indicating a significant degree of positive charge on the sulfur atom. The ortho styryl-substituted benzoyl[173] and γ-benzylidinebutyryl peroxides[182] also show this order of sensitivity to meta and para substitutions in the aryl group of the olefin. The transition states here have again included significant contributions from a charge-separated structure, as was the case for concerted fragmentation.

The solvent effects for the anchimerically assisted reactions are much larger than those noted for the fragmentation reaction. The ortho phenylthio compound showed a rate range of the order of 10^3 on going from cyclohexane to methanol. Salt effects were also demonstrated in this case. In contrast to the wide variation in rate, the efficiency of the thio perester in decolorizing galvinoxyl was relatively constant in all solvents. The rates for the ortho vinyl perbenzoates and benzoyl peroxides and γ-benzylidenebutyryl peroxide are also sensitive to solvent polarity but to a less dramatic extent.

These data are consistent with the generalized transition state with charge-separated contribution as indicated. They are also consistent with a process analogous to Scheme 4 in which the charge-separated structure is actually an intermediate which is partitioned between ionic and radical products in subsequent steps. Such a hypothesis more easily accommodates the low free radical efficiencies which approach zero as the stability of the leaving anionic group increases and is also able to unify the effects of neighboring olefinic and sulfide groups with their counterpart bimolecular reactions between olefins or sulfides and peroxides. The feasibility of such electron transfers has been demonstrated a number of times by the detection of free radicals from treatment of benzhydryl[183] or trityl halides[184] with hindered oxyanions. The CIDNP results (Chapter 6) on metallation reactions are also possibly related. The behavior of the recently studied[185] tetravalent sulfur compound with two oxy substituents could also be accommodated by this proposal.

A careful[52] study of the decomposition of naphthoyl peroxide produced the suggestion that this reaction, and perhaps that of benzoyl peroxide as well, may be subject to participation by the aromatic nucleus through the α-lactone structure (Scheme 7). The major products are α-chloronaphthalene

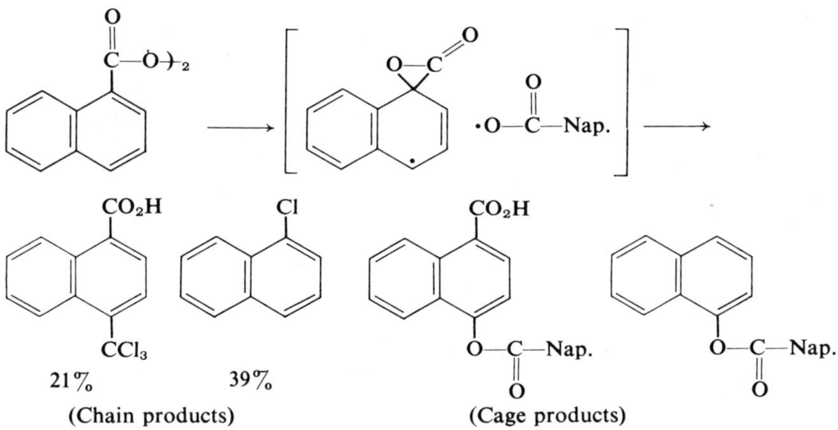

Scheme 7

(39%) and α-(4-trichloromethyl)naphthoic acid (see induced decomposition above for chain paths). The yields of the two naphthyl esters were not reduced by galvinoxyl or stilbene and thus were taken as "cage" products. The modest increase in rate for the naphthoyl compared with benzoyl peroxide and the formation of the 4-(α-naphthoyloxy)naphthoic acid can be rationalized by the α-lactone intermediate. It is possible that the reactions of benzoyl peroxides and peresters are themselves concerted in this sense. The low rate of decarboxylation of *p*-methoxybenzoate radical[112] could be a reflection of the increased importance of this type of structure. The formation of small amounts of phenyl benzoate and biphenyl from decomposition of benzoyl peroxide in cyclohexane[116a] could be explained in the same way. The problematic observation of CIDNP effects, mentioned earlier, also might be resolved with the proper coupling constants and *g* values for the cyclohexadienyl radical.

Activation parameters for several cinnamate and related peresters are listed in Table 9.[186–188] The activation entropies are suggestive of some rotational restrictions in the transition state. The products for the compounds[186,187] with phenyl groups cis to the peroxy functions give some coumarins but the similarity of the activation parameters for the trans compounds argues against neighboring aryl participation. An α-lactone-like structure can be envisaged here also. The similar values for the activation parameters of the α-phenylglycidates weaken this argument. However, the sum of the reported carbon dioxide and carboxylic acid yields for the latter was approximately 150%.

TABLE 9

Perester Decomposition—$RCO_3C(CH_3)_3$

R	$\Delta H^{\ddagger}$	$\Delta S^{\ddagger}$	Ref.
$C_6H_5CH{=}CCH_3$			
trans	30.0	1.0	187
cis	30.6	5.1	187
$C_6H_5C(CH_3){=}CCH_3$			
trans	29.3	2.9	186
cis	28.4	0.1	186
$C_6H_5CH{=}CC_6H_5$			
trans	32.0	8.0	187
cis	32.7	13.3	187
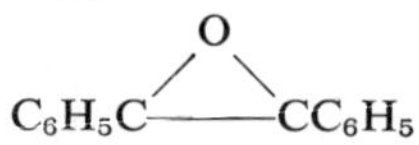			
trans	28.0	0.2	188
cis	26.8	0.4	188

Concerted fragmentation would have been the prediction for these compounds.

The decomposition[189] of β,β,β-triphenylpropionyl peroxide gives appreciable amounts of phenyl β-phenylcinnamate. Oxygen-18 labeling studies showed that 65% of the label is found in the alkyl oxygen atom of the product from carbonyl-labeled perester. Thus it has been suggested that this reaction

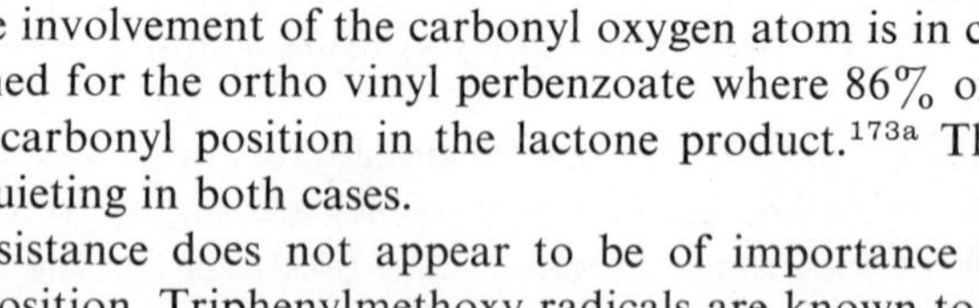

(30)

involves concerted bond formation between the carbonyl oxygen and the β-aryl group. The involvement of the carbonyl oxygen atom is in contrast to the results obtained for the ortho vinyl perbenzoate where 86% of the label remained in the carbonyl position in the lactone product.[173a] The lack of specificity is disquieting in both cases.

Anchimeric assistance does not appear to be of importance in dialkyl peroxide decomposition. Triphenylmethoxy radicals are known to rearrange by carbon-to-oxygen migration of an aryl group. However, peroxides which

$$Ar_3C{-}O{-}O{-}C(CH_3)_3 \longrightarrow Ar_3CO\cdot \longrightarrow Ar_2\dot{C}{-}OAr \qquad (31)$$

lead to such alkoxy radicals are unexceptional.[189a] Hydrogen donors can also trap these intermediates before rearrangement so that these are probably stepwise reactions. Very extensive investigation[7,127] of β-aryl perpropionate peresters, which would give the carbon radicals undergoing the analogous rearrangement, also showed no evidence of anchimeric acceleration.

D. Concerted Azoalkane Decomposition

Activation parameter comparisons provide good evidence for the perturbation of more than one C—N bond in the transition state for decomposition of a number of symmetrical azo compounds in solution. The unit slope of the line of Figure 2 means that a symmetrical and an unsymmetrical azo compound, having the alkyl group which is most stable as a radical in common, should both decompose with the same activation energy if the wholly nonconcerted mechanism applies. Table 10 contains data for a number

TABLE 10

Solution-Phase Activation Parameters for RN_2R and RN_2R^1

R (symmetrical)	R^1	k_{rel}[a]	Symmetrical E_a	Symmetrical $\Delta S^{\ddagger}$	Unsymmetrical E_a	Unsymmetrical $\Delta S^{\ddagger}$
$(CH_3)_2CCN$[96]	C_6H_5[190]	3.5×10^4	31.0	8.8	38.9	8.8
$(C_6H_5)_2CH$[191]	C_6H_5[192]	1.4×10^4	26.0	2.0	34.0	4.0
$(C_6H_5)_3C$[193]	C_6H_5[194]	—	Low[b]	—	27.0	5.0
$C_6H_5CHCH_3$[191]	CH_3[195]	9.4×10	32.6	7.0	38.6	14.0
$C_6H_5CHCH_3$[191]	$(CH_3)_2CH$[196]	1.2×10^2	32.6	7.0	36.5	9.0
$C_6H_5C(CH_3)_2$[197]	$(CH_3)_2CH$[196]	7.1×10^3	29.0	11.0	36.7	14.0

[a] Relative rate of symmetrical to unsymmetrical compound calculated at 100° using activation parameters listed and ignoring solvent effects.

[b] Too unstable to isolate. Reference 193.

of such pairs and in no case is this condition met. This demands that *at least* the symmetrical compounds must have transition states in which the stabilization (with respect to methyl) of both of the developing radicals and the nitrogen molecule is felt. These results leave open the question of mechanism for the unsymmetrical analogs. In view of the viscosity effects with NAT and phenylazotriphenylmethane,[103] it is attractive to propose that the arylazoalkanes are going by the nonconcerted route. Racemization of an optically active member of this series during thermolysis could provide an additional indication of reversibility as it has[201] for the photochemical decomposition.

The kinetic isotope effects for the α-phenylethylazoalkyl compounds provide a good confirmation of the conclusions reached from the activation parameters listed in Table 10. The α-secondary deuterium effects show that the transition state for the symmetrical compound is symmetrical[202] (k_H/k_D = 1.12). A small (k_H/k_D = 1.04) normal α effect for the isopropyl group and the large α effect for the phenylethyl group ($k_H/k_D \sim 1.12$) indicate that the transition state for the isopropyl compound is quite unsymmetrical, with the breaking of the phenylethyl C—N bond well in advance of the isopropyl group.[203] The methyl analog shows an α effect which is slightly inverse for the methyl hydrogen, indicating little or no methyl C—N bond[195] breaking in the transition state. β-Secondary deuterium,[202a] carbon-13,[204] and nitrogen-15[204] isotope effects confirm these findings in every case. The β-secondary deuterium isotope effect for the decomposition of ABN[205] also suggests a symmetrical transition state. Again, the demonstration of an unsymmetrical transition state does not directly answer the question of the possible intervention of an alkyldiazenyl radical intermediate. As discussed by Pryor and Smith,[103] racemization during thermolysis of the appropriate optically active azoalkane is a possible means of demonstrating the stepwise path. Controls ruling out other mechanisms for racemization (such as tautomerization) are necessary.

Table 11 lists some data which have recently become available for azo-

TABLE 11

Activation Parameters for 2,2¹-Disubstituted 2,2′-Azopropanes

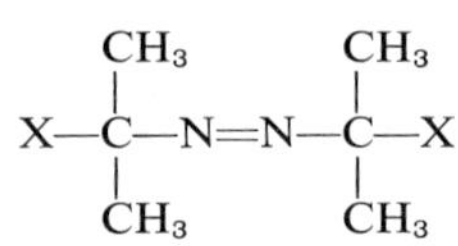

X	$\Delta H^{\ddagger}$	$\Delta S^{\ddagger}$	k_{rel}[a]	k_{rel}[b]	Ref.
CH_3	42.2	16.2	(1.00)	(1.00)	90, 91
$(CH_2)_2CH$—	37.8	12.4	75.0	—	91
N≡C	31.8	13.7	7.4×10^5	—	96
RO_2C	30.9	8.8	2.2×10^7	0.1[c]	198
C_6H_5	29.0	11.0	1.0×10^7	27.0[d]	197
C_6H_5—CH_2	36.4	4.1	8.6	—	199, 200
C_6H_5—O—	33.2	−2.5	29.0	1.5×10^4	199, 200
C_6H_5—S—	24.9	−12.6	2.5×10^4	—	199, 200

[a] Calculated at 80° using activation parameters listed and ignoring solvent effects.
[b] Relative rates for corresponding α-substituted *t*-butyl isobutyrates.
[c] Di-*t*-butylperoxy dimethyl malonate. Reference 58.
[d] Table 6.
[e] For the α-methoxy compound. Reference 155.

alkanes and peresters leading to α-substituted isopropyl radicals. The relative rates of the α-carboxy:α-methyl azo compounds are in the opposite order of the corresponding peresters. The ratios are of very different magnitudes for the α-phenyl and α-alkoxy groups. These comparisons emphasize the importance of the dipolar character of the transition state for the peresters. The activation parameters for the α-thiophenyl azo compound are suggestive of some highly ordered transition state. The products also do not correspond well to expectations for radicals. A concerted fragmentation to phenylisopropenyl sulfide, thiophenol, and 2-diazopropane could explain both.

The very small effects of *para* substituents on the decomposition of azocumenes (total range of rate ratio, 2.8)[206] also contrast with the polar character of the concerted perester reaction ($\rho = -1.09$). Rüchardt[207,208] has recently pointed out a major difference in the effect of varying ring size on the rates of decomposition of peresters compared with azobisnitriles (Figure 4). While it is possible to criticize the lack of recognition of possible changes in mechanism for the perester series, the absence of any significant rate maximum on going from the six- to the four-membered ring seems a firm conclusion. This ensures the departure from behavior expected for reactions involving significant planarization of the α-carbon atoms, behavior which is observed for the azo nitriles. The magnitudes of the α-secondary deuterium isotope effects for the phenylacetyl peresters[154] ($k_H/k_D = 1.05$) are smaller than those for the azoalkanes ($k_H/k_D = 1.12$).[202] This is also suggestive of a smaller degree of planarization for the perester transition state.

Another striking difference in the behavior of the two types of initiators is with respect to the ease of formation of bridgehead radicals. The relative rates (Table 6) for *t*-butyl, 1-adamantyl, 1-bicyclooctyl, and 1-norbornyl peresters parallel the results found earlier for the decarbonylation of the corresponding aldehydes.[209] The conclusion from both sets of data was that radicals could accommodate the degree of nonplanarity which is required of the adamantyl skeleton without any energetic expenditure. However, the decomposition of azoadamantane[90] was found to be much slower than 2,2′-azoisobutane. Rüchardt and colleagues[207] have found that all bridgehead azoalkanes, including unsymmetrical examples with one *t*-butyl group, decompose more slowly than the acyclic model. The concerted mechanism fits the observed rate ratios though activation parameters and product studies have not been reported as yet. There is a roughly linear correlation between the relative rates of the bridgehead perester decomposition (and azoalkanes) and relative rates of solvolysis of the corresponding arenesulfonates. The slope of the lines relating the radical and ionic reactions is only about 0.2, which suggests that the curvature of the potential surface for deviation from planarity for radicals is only 20% of that for cations. The fact that the 1-adamantyl perester is not slower than the pivaloyl compound and

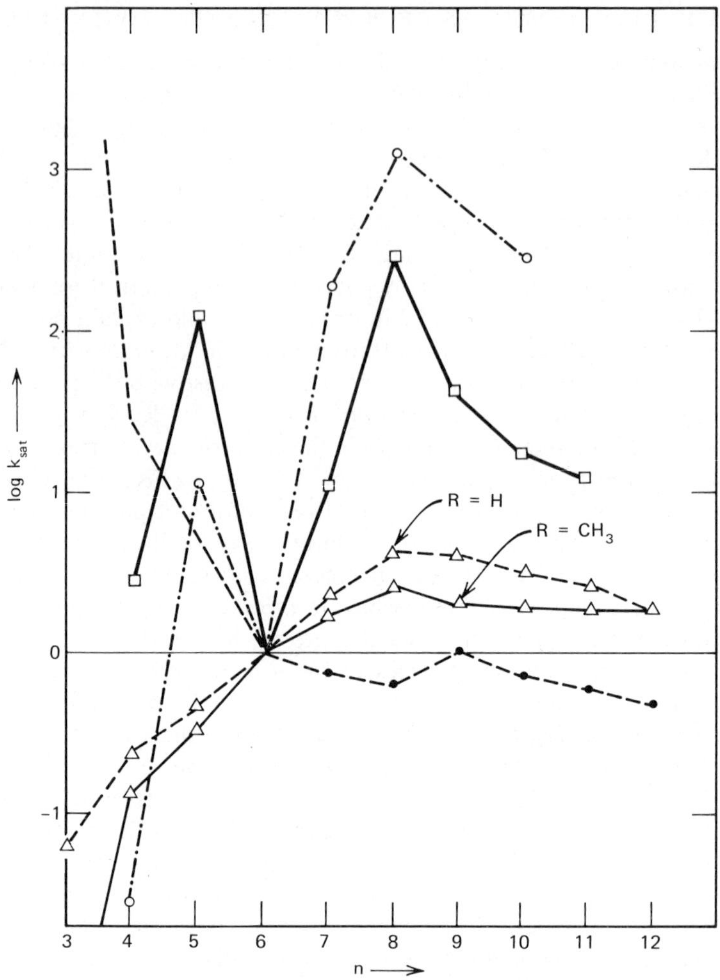

Fig. 4. Rate constant-ring size correlations. From reference 207.

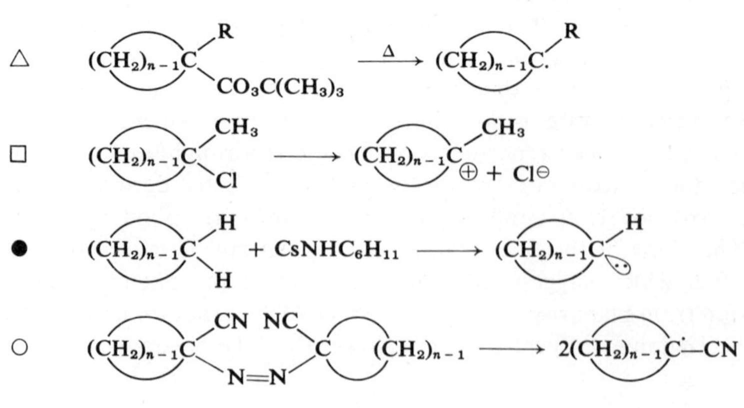

the possibility of a change in mechanism between the 1-bicyclooctyl and 1-adamantyl percarboxylates have not been adequately explained by this type of correlation. The conflicting conclusions which one would reach from comparisons of peroxides with azo compounds again point out that relative rates reflect differences in transition state stabilities which may respond to factors completely absent in the isolated radicals.

E. Anchimeric Acceleration and Steric Effects in Azoalkane Decomposition

Most tests for neighboring group effects in azo compound decomposition have been negative. Overberger and Ganier[210] have found that the rates of a number of β-phenyl-substituted azoethanes were unexceptional. Seltzer and Scheppele[211] have investigated the α-secondary deuterium kinetic isotope effects in the decomposition of *meso*- and *dl*-1,1,2,2-tetraphenylazoethane. The isotope effects were hardly reduced ($k_H/k_D = 1.10$) from those for the α,α'-diphenylazoethane ($k_H/k_D = 1.12$). The rates and activation parameters were also essentially unchanged. There was a definite difference in the rates for the meso and dl isomers of the tetraphenyl compound. This was taken to be a reflection of the steric effect on the relative stability of the starting materials. Larger effects attributable to steric strain in the ground states of azoalkanes[212] and azobisnitriles have been observed previously.[213]

The difference in the rates of solvolysis of exo- and endo-2-norbornyl arenesulfonates has been attributed to a type of anchimeric effect in which the $C_1—C_6$ bond assists the ionization of the exo compound. Berson and co-workers[214] investigated the possibility of a similar situation in the azoalkane decomposition. They found the products expected of the Wagner–Meerwein rearrangement for the 2-bornyl radical but showed that it was actually the result of a β-scission–readdition sequence. There was no evidence of any acceleration of the rate of this reaction. Rüchardt and Hinz[215] have reported a significant difference in the rates of exo,exo- and endo,endo-2,2-dimethyl-2,2′-azonorbornane ($k_{exo}/k_{endo} = 68$). The square root of this value is in qualitative agreement with the relative rates of the corresponding tertiary peresters (Table 6) and has been interpreted[207] as being a reflection of differences in the torsional strain of planarization of the radical centers. This interpretation is not quite consistent with the point of view that the transition state for concerted perester decomposition has relatively less planarization than the corresponding azo compound reaction (Figure 4). The square root of the rate ratio is correct if the concerted mechanism applies for the symmetrical azoalkanes. If the nonconcerted mechanism contributes at all, then the steric effect becomes larger. This points out the importance of careful attention to the effects of changing mechanism in radical reactions. The possibility of a small anchimeric effect for the 2-norbornyl radical still exists although it is certainly less important than in cation formation. Martin's group[91,216] has provided convincing evidence for some stabilization of transition states leading to the cyclopropylmethyl radical (Table 11) although it is certainly small compared with that for the corresponding cation.

The most dramatic demonstration of an acceleration in decomposition of azoalkanes is for the cis isomers. The cis structure has recently been demonstrated for perfluoroazomethane[217] which is apparently the slowest azoalkane (Table 2). *cis*-Azomethane[218] is observable by photoisomerization of the trans compound but rearranges back to the trans isomer without decomposition. However, the cis isomers of ABN, 2,2′-azoisobutane, and 2,2′-carboethoxy-2,2′-azopropane all decompose to radicals[219] in competition with isomerization to the trans material at room temperature. Porter and coworkers[201] have also reported radical formation from *cis*-phenylazo-(2-phenyl)butane at room temperature. The rate accelerations for the decompositions of the last-named four compounds, compared with their trans isomers, are extremely large. Further studies which allow a good understanding of the decompositions of the cis compounds and their relationship to the detailed mechanism of photodestruction of azoalkanes should prove very interesting.

REFERENCES

1. A. G. Davies, *Organic Peroxides*, Butterworths, London, 1961; E. G. E. Hawkins, *Organic Peroxides*, Van Nostrand, Princeton, N.J., 1961; A. V. Tobolsky and R. B. Mesrobian, *Organic Peroxides*, Interscience, New York, 1954; D. Swern, *Organic Peroxides*, Wiley, New York, 1970, 1971.
2. R. Criegee, *Chem. Ber.*, **77**, 722 (1944).
3. R. Criegee and R. Kaspar, *Ann.*, **560**, 127 (1948).
4. E. Hedaya and S. Winstein, *J. Am. Chem. Soc.*, **89**, 1661 (1967).
5. P. D. Bartlett and T. G. Traylor, *J. Am. Chem. Soc.*, **83**, 856 (1961).
6. C. Rüchardt, *Fortschr. Chem. Forsch.*, **6**, 251 (1966).
7. C. Rüchardt and H. Quadbeck-Seeger, *Chem. Ber.*, **102**, 3525 (1969).
8. T. Koenig and W. D. Brewer, *J. Am. Chem. Soc.*, **86**, 2728 (1965).
9. E. Hedaya, R. L. Hinman, L. Kibler, and S. Theodoropulos, *J. Am. Chem. Soc.*, 86, 2727 (1964).
10. E. Hedaya, R. L. Hinman, V. Schomaker, S. Theodoropulos, and L. M. Kyle, *J. Am. Chem. Soc.*, **89**, 4875 (1967).
11. R. Sheldon and J. Kochi, *J. Org. Chem.*, **35**, 1223 (1970).
12. J. E. Leffler, *J. Am. Chem. Soc.*, **72**, 67 (1950).
13. J. E. Leffler and C. D. Petropoulas, *J. Am. Chem. Soc.*, **79**, 3068 (1957).
14. F. D. Greene, H. P. Stein, C. C. Chu, and F. M. Vane, *J. Am. Chem. Soc.*, **86**, 2080 (1964).
15. R. C. Lamb and J. R. Sanderson, *J. Am. Chem. Soc.*, **91**, 5034 (1969).
16. C. Walling, H. P. Waits, J. Milovanovic, and C. Pappiannou, *J. Am. Chem. Soc.*, **92**, 4927 (1970).
17. P. D. Bartlett and J. E. Leffler, *J. Am. Chem. Soc.*, **72**, 3030 (1950).
18. H. Hart and R. Cipriani, *J. Am. Chem. Soc.*, **84**, 3697 (1962).
19. A. V. Tobolsky and L. R. Matlack, *J. Polym. Sci.*, **55**, 49 (1961).

20. M. S. Kharasch, A. Fono, and W. Nudenberg, *J. Org. Chem.*, **16**, 113 (1951).
21. J. O. Turner, *Tetrahedron Letters*, **No. 14**, 887 (1971); N. C. Deno, W. E. Billups, K. E. Kramer, and R. R. Lastomirsky, *J. Org. Chem.*, **35**, 3080 (1970).
22. R. Huisgen and W. Edl, *Angew. Chem.*, **74**, 588 (1962).
23. D. Z. Denney, T. M. Valega, and D. B. Denney, *J. Am. Chem. Soc.*, **86**, 46 (1964).
24. C. Rüchardt and H. Schwarzer, *Chem. Ber.*, **99**, 1871, 1879 (1966); *Angew. Chem.*, **74**, 251 (1962).
25. R. E. Pincock, *J. Am. Chem. Soc.*, **87**, 1274 (1965); R. E. Pincock and T. E. Kiovsky, *ibid.*, **87**, 4100 (1965).
26. N. Kornblum and H. E. DelaMare, *J. Am. Chem. Soc.*, **73**, 880 (1951); J. I. G. Cadogan, D. H. Hey, and W. A. Sanderson, *J. Chem. Soc.*, 4498 (1958). See also W. H. Richardson and R. S. Smith, *J. Am. Chem. Soc.*, **89**, 2230 (1967).
27. A. Tobolosky and R. Mesrobian, *Organic Peroxides*, Interscience, New York, 1954, p. 122.
28. T. Koenig and R. Wielesek, *J. Am. Chem. Soc.*, **91**, 2551 (1969).
29. S. O. Lawesson, S. Grönwall, and M. Anderson, *Ark. Kemi*, **17**, 457 (1961); S. O. Lawesson, G. Frisell, D. B. Denney, and D. Z. Denney, *Tetrahedron*, **19**, 1229 (1963); M. Okabo, K. Maruyama, and J. Osuki, *Bull. Chem. Soc. Japan*, **44**, 125 (1971).
30. C. Walling, *Free Radicals in Solution*, Wiley, New York, 1957; C. Walling and N. Indictor, *J. Am. Chem. Soc.*, **80**, 5815 (1958); J. C. Crano, *J. Org. Chem.*, **31**, 3615 (1966); M. A. Greenbaum, D. B. Denney, and A. K. Hoffman, *J. Am. Chem. Soc.*, **78**, 2563 (1956); S. Hasegawa, S. Kashino, and Y. Mukai, *Bull. Soc. Chem. Japan*, **44**, 69 (1971); H. Ohta and K. Tokumaru, *J. Chem. Soc.* (*D*), 1601 (1970).
31. D. B. Denney and D. Z. Denney, *J. Am. Chem. Soc.*, **82**, 1389 (1960); C. Walling and R. B. Hodgdon, *ibid.*, **80**, 228 (1958).
32. L. Horner and H. Brüggemann, *Ann.*, **635**, 22 (1960); L. Horner and E. Jürgens, *ibid.*, **602**, 135 (1957).
33. F. D. Greene, W. Adam, and J. E. Cantrill, *J. Am. Chem. Soc.*, **83**, 3461 (1961); C. Walling and L. Heaton, *ibid.*, **87**, 38 (1965); F. D. Greene, *ibid.*, **78**, 2250 (1956); J. R. Sheldon and C. Uzelmeir, *J. Org. Chem.*, **35**, 1576 (1970); M. Kobayashi, *Bull. Soc. Chem. Japan*, **43**, 1158 (1970).
34. E. M. Levi, P. Kovacic, and J. Gormish, *Tetrahedron*, **26**, 4537 (1970); R. Dannley and G. Corbett, *J. Org. Chem.*, **31**, 153 (1966); P. Kovacic and M. Kurz, *ibid.*, **31**, 2011 (1966); *J. Am. Chem. Soc.*, **87**, 4811 (1965).
35. G. Sosnovsky and D. J. Rawlinson, *J. Org. Chem.*, **34**, 3462 (1969); L. Horner and W. Jurgeleit, *Ann.*, **591**, 138 (1955); D. B. Denney, W. F. Goodyear, and B. Goldstein, *J. Am. Chem. Soc.*, **83**, 1727 (1961); W. Adam and A. Rios, *J. Org. Chem.*, **36**, 407 (1971); D. B. Denney and S. T. D. Gough, *J. Am. Chem. Soc.*, **87**, 138 (1965).
36. R. O'Conner, *J. Org. Chem.*, **26**, 4375 (1961); R. O'Conner and W. Rosenbrook, *ibid.*, **26**, 5708 (1961).
36a. R. G. Peterson and S. D. Ross, *Tetrahedron Letters*, **No. 23**, 18 (1960).
37. J. E. Leffler and W. Bond, *J. Am. Chem. Soc.*, **78**, 335 (1956); D. Mackay, U. Marx, and W. Waters, *J. Chem. Soc.*, 4793 (1964).
37a. R. W. Hoffman and G. Kuhn, *Chem. Ber.*, **100**, 1474 (1967); P. Huang and E. Kosower, *J. Am. Chem. Soc.*, **90**, 2354 (1968); G. Fraenkel and E. Pecchold, *Tetrahedron Letters*, **No. 54**, 4821 (1969); T. Tsuji and E. M. Kosower, *J. Am. Chem. Soc.*, **91**, 3395 (1970).
38. G. A. Russell, *J. Am. Chem. Soc.*, **79**, 3871 (1957); H. S. Blanchard, *ibid.*, **81**, 4548 (1959); P. D. Bartlett and T. G. Traylor, *ibid.*, **85**, 2407 (1963).
39. J. R. Thomas, *J. Am. Chem. Soc.*, **87**, 3935 (1965); P. D. Bartlett and P. Gunther, *ibid.*, **88**, 3288 (1966); P. D. Bartlett and R. Guaraldi, *ibid.*, **89**, 4799 (1967); P. G.

Thompson, *ibid.*, **89**, 4316 (1967); L. R. Anderson and W. B. Fox, *ibid.*, 4313 (1967).

40. S. W. Benson, *J. Am. Chem. Soc.*, **86**, 3922 (1964).
41. W. A. Pryor, D. M. Huston, T. R. Fiske, T. L. Pickering, and E. Ciuffarin, *J. Am. Chem. Soc.*, **86**, 4237 (1964); W. A. Pryor, *Tetrahedron Letters*, 1201 (1963).
42. P. L. Hanst and J. G. Calvert, *J. Phys. Chem.*, **63**, 104 (1969); M. J. Yee Quee and J. C. Thynne, *Trans. Faraday Soc.*, **64**, 1296 (1968).
43. W. Forst and O. K. Rice, *Can. J. Chem.*, **41**, 562 (1963).
44. W. Forst, *J. Chem. Phys.*, **44**, 2349 (1966).
45. H. S. Sandu, *J. Phys. Chem.*, **72**, 1857 (1968).
46. O. P. Straus, R. E. Berkeley, and H. E. Gunning, *Can. J. Chem.*, **47**, 3470 (1969); G. Geisler and J. Hoffman, *Z. Phys. Chem.* (*Frankfurt*), **57**, 318 (1968).

46a. S. G. Cohen and R. Zand, *J. Am. Chem. Soc.*, **84**, 586 (1962).

47. P. D. Bartlett and K. Nozaki, *J. Am. Chem. Soc.*, **68**, 1686 (1946); W. E. Cass, *ibid.*, **68**, 1976 (1946).
48. P. D. Bartlett and K. Nozaki, *J. Am. Chem. Soc.*, **69**, 2299 (1947); W. E. Cass, *ibid.*, **69**, 500 (1947).
49. D. B. Denney and G. Feig, *J. Am. Chem. Soc.*, **81**, 5322 (1959); E. H. Drew and J. C. Martin, *Chem. Ind.*, 925 (1959).
50. W. von E. Doering, K. Okamoto, and H. Krauch, *J. Am. Chem. Soc.*, **82**, 3579 (1960).

50a. H. Suehiro, A. Kanoya, H. Hara, T. Nakahama, M. Omori, and T. Komori, *Bull. Soc. Chem. Japan*, **40**, 678 (1967).

51. C. Walling and E. S. Savas, *J. Am. Chem. Soc.*, **82**, 1738 (1960).
52. J. E. Leffler and R. G. Zepp, *J. Am. Chem. Soc.*, **92**, 3713 (1970).
53. C. Walling and Z. Cekovic, *J. Am. Chem. Soc.*, **89**, 6681 (1967).
54. J. I. G. Cadogan, D. H. Hey, and P. G. Hibbert, *J. Chem. Soc.*, 3939 (1965); D. F. DeTar and C. Weis, *J. Am. Chem. Soc.*, **78**, 4296 (1956).
55. S. W. Benson and L. Batt, *J. Chem. Phys.*, **36**, 895 (1962).
56. E. R. Bell, F. F. Rust, and W. E. Vaughn, *J. Am. Chem. Soc.*, **72**, 337 (1950).
57. P. D. Bartlett and L. B. Gortler, *J. Am. Chem. Soc.*, **85**, 1864 (1963); N. A. Milas and A. Golubovic, *ibid.*, **80**, 5994 (1958).
58. L. B. Gortler and M. D. Saltzman, *J. Org. Chem.*, **31**, 3821 (1966).
59. P. D. Bartlett and R. Wheland, *J. Am. Chem. Soc.*, **92**, 6057 (1970).
60. H. Hart and F. Chloupek, *J. Am. Chem. Soc.*, **85**, 1155 (1963); R. C. Lamb, P. W. Ayers, and M. K. Toney, *ibid.*, **85**, 3483 (1963).
61. J. E. Leffler and M. M. Schwartz, *J. Am. Chem. Soc.*, **93**, 919 (1971).
62. V. Puchin, T. I. Yurzhenko, S. A. Vovonov, M. S. Britan, *Vysokomol. Soedin.*, **Ser. B**, 530 (1958); R. A. Bafford, E. R. Kamens, and O. L. Meglo, Ger. Offen. 1,909,662; *Chem. Abstr.*, **71**, 125, 246 (1969).
63. N. Maramoto, T. Ochaiai, D. Simamura, and M. Yoshida, *Chem. Commun.*, 717 (1968).
64. L. M. Bobroff, L. B. Gortler, D. J. Shan, and H. Wiland, *J. Org. Chem.*, **31**, 2678 (1966).
65, E. Huyser and C. J. Bredeweg, *J. Am. Chem. Soc.*, **86**, 2401 (1964); E. S. Huyser, C. J. Bredeweg, and R. M. Van Scoy, *ibid.*, **86**, 4148 (1964); R. L. Huang, T. Lee, and S. H. Ong, *J. Chem. Soc.* (*C*), 2522 (1969).
66. P. D. Bartlett, E. P. Benzing, and R. E. Pincock, *J. Am. Chem. Soc.*, **82**, 1762 (1960).

67. M. M. Martin, *J. Am. Chem. Soc.*, **83**, 2869 (1961).
68. E. Huyser, *Free Radical Chain Reactions*, Wiley, New York, 1970, p. 278.
69. C. Walling and J. C. Azar, *J. Org. Chem.*, **33**, 3888 (1968).
70. S. Cheng, J. Nimoy, and S. Tobey, *J. Phys. Chem.*, **71**, 3075 (1967); S. Tobey and J. Nimoy, *ibid.*, **70**, 867 (1966).
70a. J. K. S. Wan, L. D. Hess, and J. N. Pitts, *J. Am. Chem. Soc.*, **86**, 2069 (1964).
71. R. C. Lamb and J. G. Pacifici, *J. Am. Chem. Soc.*, **86**, 914 (1964); R. C. Lamb, J. G. Pacifici, and L. Spadafino, *J. Org. Chem.*, **30**, 3102 (1965); R. C. Lamb, W. E. McNew, Jr., J. R. Sanderson, and D. C. Lunney, *ibid.*, **36**, 174 (1971).
72. J. H. Raley, F. F. Rust, and W. E. Vaughn, *J. Am. Chem. Soc.*, **70**, 88, 1336, 2767 (1948); A. L. Williams, E. A. Oberright, and J. W. Brooks, *ibid.*, **78**, 1190 (1956).
72a. P. Gray and A. Williams, *Chem. Rev.*, **59**, 239 (1959); H. Hershenson and S. W. Benson, *J. Chem. Phys.*, **37**, 1889 (1962).
73. S. W. Benson and H. E. O'Neal, "Kinetic Data on the Gas Phase Unimolecular Reactions," National Bureau of Standards Data Series, NSRDS-NBS 21.
74. See, however, arguments by E. Hedaya, reference 10.
75. Reference 73 contains a wider list of the activation parameters showing the experiment at variations. See also P. Molyneaux, *Tetrahedron*, **22**, 2929 (1966).
76. Y. Takezaki and C. Takeuchi, *J. Chem. Phys.*, **22**, 1527 (1954).
77. C. Leggett and J. C. J. Thynne, *Trans. Faraday Soc.*, **63**, 2504 (1967).
78. E. J. Harris and A. C. G. Eggerton, *Proc. Roy. Soc.* (*A*), **168**, 1 (1938).
79. A. Rembaum and M. Szwarc, *J. Am. Chem. Soc.*, **76**, 5975 (1954).
80. A. Rembaum and M. Szwarc, *J. Chem. Phys.* **23**, 909 (1955).
81. M. G. Evans and M. Polanyi, *Trans. Faraday Soc.*, **34**, 11 (1938).
82. M. Page, H. O. Pritchard, and A. F. Trotman-Dickenson, *J. Chem. Soc.*, 3878 (1953).
83. B. G. Gowenlock, P. P. Jones, and D. R. Snelling, *Can. J. Chem.*, **41**, 1911 (1963).
84. H. C. Ramsperger, *J. Am. Chem. Soc.*, **49**, 912 (1927).
85. E. Leventhal, C. R. Simons, and C. Steel, *Can. J. Chem.*, **41**, 930 (1962); D. Clark and H. O. Pritchard, *J. Chem. Soc.*, 2136 (1956).
86. H. C. Ramsperger, *Proc. Natl. Acad. Sci.*, **13**, 849 (1927).
87. A. U. Blackham and N. L. Eatough, *J. Am. Chem. Soc.*, **84**, 2922 (1962).
88. B. H. Al-Sayer and R. J. Crawford, *Can. J. Chem.*, **48**, 2745 (1970).
88a. K. Takagi and R. J. Crawford, *J. Am. Chem. Soc.*, **93**, 5910 (1971).
89. C. Williams and A. S. C. Lawrence, *Proc. Roy. Soc.* (*L*), **A156**, 455 (1936).
90. M. Prochazka, O. Ryba, and D. Lim, *Collect. Czech. Chem. Soc.*, **33**, 3378 (1968).
91. J. C. Martin and J. W. Timberlake, *J. Am. Chem. Soc.*, **92**, 978 (1970).
92. J. A. Kerr, *Chem. Rev.*, **66**, 465 (1966).
93. E. Huyser and R. M. Van Scoy, *J. Org. Chem.*, **33**, 3524 (1968).
94. C. Walling and D. Bristol, *J. Org. Chem.*, **36**, 733 (1971).
95. F. M. Lewis and M. S. Matheson, *J. Am. Chem. Soc.*, **71**, 747 (1949); C. G. Overberger, M. T. O'Shaughnessy, and H. Shalit, *ibid.*, **71**, 2661 (1949); L. M. Arnett, *ibid.*, **74**, 2027 (1952).
96. R. C. Peterson, J. H. Markgraff, and S. D. Ross, *J. Am. Chem. Soc.*, **83**, 3819 (1961).
97. M. D. Cohen, J. E. Leffler, and L. M. Barbato, *J. Am. Chem. Soc.*, **76**, 4169 (1954); M. G. Alder and J. E. Leffler, *ibid.*, **76**, 1425 (1954).
98. J. E. Leffler and E. Grunwald, *Rates and Equilibria of Organic Reactions*, Wiley, New York, 1963, pp. 324–342.

99. C. Walling and P. J. Wagner, *J. Am. Chem. Soc.*, **86**, 3368 (1964).
100. W. G. Bentrude and A. MacKnight, *J. Am. Chem. Soc.*, **92**, 5259 (1970).
101. T. G. Traylor and H. Kiefer, *J. Am. Chem. Soc.*, **89**, 6667 (1967).
102. C. Walling and H. P. Waits, *J. Phys. Chem.*, **71**, 2361 (1967).
103. W. A. Pryor and K. Smith, *J. Am. Chem. Soc.*, **92**, 5403 (1970); **89**, 1741 (1967).
104. W. A. Pryor, J. T. Echols, and K. Smith, *J. Am. Chem. Soc.*, **88**, 1189 (1966).
105. W. G. Bentrude and J. L. Fu, *Tetrahedron Letters*, **No. 57**, 6033 (1968).
106. H. van Zwet and E. C. Kooyman, *Rec. Trav. Chim. Pays-Bas*, **86**, 1143 (1967).
107. C. G. Overberger and A. J. Rosenthal, *J. Am. Chem. Soc.*, **82**, 108 (1962).
108. H. C. Huang and E. M. Kosower, *J. Am. Chem. Soc.*, **90**, 2367 (1968).
109. E. Lippmaa, T. Pehk, and T. Saluveere, papers presented at the International Colloquium on CIDNP, General Assembly of the Belgian Chemical Society, Bruxelles, Belgium, March 1971.
110. G. S. Hammond and L. M. Soffer, *J. Am. Chem. Soc.*, **72**, 4711 (1951).
111. H. Shine, J. Waters, and D. Hoffman, *J. Am. Chem. Soc.*, **85**, 3613 (1963).
111a. J. C. Martin, J. W. Taylor, and E. H. Drew, *J. Am. Chem. Soc.*, **89**, 129 (1967).
112. G. Ayrey, *Chem. Rev.*, **63**, 645 (1963); J. C. Bevington, J. Toole, and L. Trossarelli, *Trans. Faraday Soc.*, **54**, 863 (1958); J. C. Bevington and T. D. Lewis, *ibid.*, **54**, 1340 (1958).
113. D. F. DeTar, *J. Am. Chem. Soc.*, **89**, 4058 (1967).
114. J. C. Martin and J. H. Hargis, *J. Am. Chem. Soc.*, **91**, 5399 (1969).
115. J. Bargon, H. Fischer, and U. Johnson, *Z. Naturforsch.*, **22a**, 1551 (1967); J. Bargon and H. Fischer, *ibid.*, **22a**, 1556 (1967); H. Fischer, papers reported at the 158th National Meeting of the American Chemical Society, Houston, Texas, Feb. 1970.
115a. E. Lippmaa, T. Pehk, A. Buchachenko, and S. Rybov, *Chem. Phys. Letters*, **5**, 521 (1970).
116. T. Nakata, K. Tokumaru, and O. Simamura, *Tetrahedron Letters*, **No. 34**, 3303 (1967); D. F. DeTar, R. Long, J. Rendleman, J. Bradley, and P. Duncan, *J. Am. Chem. Soc.*, **89**, 4051 (1967); J. Saltiel and H. C. Curtis, *ibid.*, **93**, 2056 (1971).
116a. C. G. Swain, L. Schaad, and A. J. Kresge, *J. Am. Chem. Soc.*, **80**, 5315 (1958).
117. C. G. Swain, W. H. Stockmayer, and J. T. Clarke, *J. Am. Chem. Soc.*, **72**, 5426 (1950).
117a. S. G. Cohen and C. H. Wing, *J. Am. Chem. Soc.*, **75**, 5504 (1954).
117b. E. G. Janzen and B. Blackburn, *J. Am. Chem. Soc.*, **91**, 4481 (1969).
118. W. Braun, L. Rajenbach, and F. R. Eirich, *J. Phys. Chem.*, **66**, 1591 (1962).
118a. J. C. Martin and S. A. Dombchick, "Oxidation of Organic Compounds," Vol. 1, *Advances in Chemistry Series, No. 75*, American Chemical Society, Washington, D.C., 1968, pp. 269–281.
119. T. Koenig, *J. Am. Chem. Soc.*, **91**, 2558 (1969); T. Koenig, J. A. Hoobler, and M. Deinzer, *ibid.*, **93**, 938 (1971); T. Koenig and M. Deinzer, *ibid.*, **90**, 7014 (1968); T. Koenig and M. Deinzer, *ibid.*, **88**, 4518 (1966).
119a. T. Koenig and R. Cruthoff, *J. Am. Chem. Soc.*, **91**, 2562 (1969).
120. M. J. Goldstein, *Tetrahedron Letters*, 1601 (1964).
121. J. Smid and M. Szwarc, *J. Am. Chem. Soc.*, **78**, 3315 (1956).
122. D. F. DeTar and C. Weiss, *J. Am. Chem. Soc.*, **78**, 4296 (1956).
123. P. D. Bartlett and R. R. Hiatt, *J. Am. Chem. Soc.*, **80**, 1398 (1958).
124. J. P. Lorand, Ph.D. thesis, Harvard University, Cambridge, Mass., 1964.
125. M. Trachtman and J. G. Miller, *J. Am. Chem. Soc.*, **84**, 4928 (1962).
126. M. Martin, *J. Am. Chem. Soc.*, **84**, 1986 (1962).
127. C. Rüchardt and R. Hecht, *Chem. Ber.*, **98**, 2460 (1965).

128. L. B. Humphrey, B. Hodgson, and R. E. Pincock, *Can. J. Chem.*, **46**, 3099 (1968).
129. T. G. Traylor, A. Sieber, H. Kiefer, and N. Clinton, *Intra-Sci. Chem. Rep.*, **3**, 289 (1969).
130. C. Rüchardt and H. Schwarzer, *Chem. Ber.*, **99**, 1861 (1966).
131. P. Lorenz, dissertation, University of Münster, private communication.
132. P. D. Bartlett and H. Sakurai, *J. Am. Chem. Soc.*, **84**, 3269 (1962).
133. A. T. Blomquist and I. Bernstein, *J. Am. Chem. Soc.*, **73**, 5546 (1951).
134. D. L. Tuleen, W. G. Bentrude, and J. C. Martin, *J. Am. Chem. Soc.*, **85**, 1938 (1963).
135. M. J. Goldstein and M. Yoshida, abstracts of papers presented at the 149th Meeting of the American Chemical Society, Detroit, Mich., April 1965, p. 54P.
136. S. V. Rykov and A. L. Buchachenko, *Dokl. Akad. Nauk SSSR*, **185**, 870 (1969).
137. R. C. Neuman and J. Behar, *J. Am. Chem. Soc.*, **91**, 6024 (1969).
138. P. D. Bartlett and C. Rüchardt, *J. Am. Chem. Soc.*, **82**, 1757 (1960).
139. M. Martin and E. B. Sanders, *J. Am. Chem. Soc.*, **89**, 3777 (1967).
140. C. Rüchardt and H. Schwarzer, *Chem. Ber.*, **99**, 1871 (1966).
141. P. D. Bartlett and J. P. Lorand, *J. Am. Chem. Soc.*, **88**, 3294 (1966).
142. P. D. Bartlett and D. Simons, *J. Am. Chem. Soc.*, **82**, 1753 (1960).
143. P. D. Bartlett, B. Gontarev, and H. Sakurai, *J. Am. Chem. Soc.*, **84**, 3102 (1962).
144. P. D. Bartlett, E. P. Benzing, and R. E. Pincock, *J. Am. Chem. Soc.*, **82**, 1762 (1960).
145. S. Seltzer and A. S. Zavitsas, *J. Am. Chem. Soc.*, **86**, 3836 (1964).
146. J. P. Lorand, S. D. Chodroff, and R. W. Wallace, *J. Am. Chem. Soc.*, **90**, 5266 (1968).
147. R. C. Fort and R. E. Franklin, *J. Am. Chem. Soc.*, **90**, 5267 (1968).
148. G. J. Abruscato and T. T. Tidwell, *J. Am. Chem. Soc.*, **94**, 672 (1972).
149. W. Schindel and R. E. Pincock, *J. Org. Chem.*, **35**, 1789 (1970).
150. F. Herkes, J. Friedman, and P. D. Bartlett, *Int. J. Chem. Kinet.*, 1, 193 (1969).
151. P. D. Bartlett, R. Pincock, J. H. Rolston, W. G. Schindel, and L. A. Singer, *J. Am. Chem. Soc.*, **87**, 2590 (1965).
152. P. D. Bartlett, G. N. Fickes, F. C. Haupt, and R. Helgeson, *Acc. Chem. Res.*, **3**, 177 (1970).
153. M. Goldstein and J. A. Hudson, *J. Am. Chem. Soc.*, **92**, 4120 (1970).
154. T. Koenig, J. Huntington, and R. Cruthoff, *J. Am. Chem. Soc.*, **92**, 5413 (1970).
154a. T. Koenig and R. Wolf, *J. Am. Chem. Soc.*, **91**, 2569 (1969); **89**, 2948 (1967).
155. C. Rüchardt and H. Boeck, *Chem. Ber.*, **104**, 577 (1971); C. Rüchardt and I. Mayer-Ruthardt, *ibid.*, 593 (1971).
156. R. C. Neuman and J. V. Behar, *J. Org. Chem.*, **36**, 654, 657 (1971).
157. W. A. Pryor, private communication.
158. J. M. McBride and P. D. Bartlett, *J. Am. Chem. Soc.*, **87**, 1727 (1965).
158a. P. D. Bartlett and R. E. Pincock, *J. Am. Chem. Soc.*, **84**, 2445 (1967).
159. M. Martin and D. C. DeJongh, *J. Am. Chem. Soc.*, **84**, 3526 (1962).
160. M. Martin, *J. Am. Chem. Soc.*, **83**, 2869 (1961).
161. P. D. Bartlett and H. Minato, *J. Am. Chem. Soc.*, **85**, 1858 (1960).
162. M. S. Kharasch and J. G. Bart, *J. Org. Chem.*, **16**, 150 (1951); T. Suehiro, S. Hibino, and T. Saito, *Bull. Soc. Chem. Japan*, **41**, 1707 (1968).
163. F. D. Greene, *J. Am. Chem. Soc.*, **77**, 4869 (1955).
164. D. B. Denney and N. Sherman, *J. Org. Chem.*, 3760 (1965); C. Walling, H. N. Moulden, J. H. Waters, and R. C. Neuman, *J. Am. Chem. Soc.*, **87**, 518 (1965).
165. M. S. Kharasch, J. Kuderna, and W. Nudenberg, *J. Org. Chem.*, **19**, 1283 (1955).
166. T. Kashiwagi, S. Kozuka, and S. Oae, *Tetrahedron*, **26**, 3617 (1970); D. F. DeTar and C. Weis, *J. Am. Chem. Soc.*, **79**, 3045 (1957).

167. J. A. Deyrup and U. Moyer, *J. Org. Chem.*, **34**, 1835 (1962).
168. H. Hart and H. Lau, *J. Am. Chem. Soc.*, **81**, 4897 (1959).
169. C. E. Boozer, C. Love, J. Motes, R. Turner, J. Toney, B. Moxey, W. Christian, and W. Stevens, U.S. Department of Commerce, Office of Tech. Serv. P. B. Rep., 154, 949 (1959); *Chem. Abstr.*, **59**, 2069*a* (1963).
170. M. S. Kharasch, F. Engelman, and W. H. Urry, *J. Am. Chem. Soc.*, **65**, 2428 (1963).
171. P. D. Bartlett and F. D. Greene, *J. Am. Chem. Soc.*, **76**, 1088 (1959).
172. J. C. Martin and W. G. Bentrude, *Chem. Ind.*, 192 (1959); W. G. Bentrude and J. C. Martin, *J. Am. Chem. Soc.*, **84**, 1561 (1960).
173. T. Koenig and J. C. Martin, *J. Org. Chem.*, **29**, 1520 (1964).
173a. T. Koenig and J. C. Martin, *J. Am. Chem. Soc.*, **86**, 1771 (1964).
174. T. H. Fisher and J. C. Martin, *J. Am. Chem. Soc.*, **88**, 3382 (1966).
175. J. Smid, A. Rembaum, and M. Szwarc, *J. Am. Chem. Soc.*, **78**, 3315 (1956).
176. G. Greig and J. C. Thynne, *Trans. Faraday Soc.*, **63**, 2196 (1967).
177. D. S. Huang and J. C. Martin, private communication.
178. R. C. Lamb, F. Rogers, G. Denn, Jr., and F. W. Voight, *J. Am. Chem. Soc.*, **84**, 2635 (1962).
179. W. Honsberg and J. E. Leffler, *J. Org. Chem.*, **26**, 733 (1962); J. E. Leffler, R. D. Falkner, and C. C. Petropoulos, *J. Am. Chem. Soc.*, **80**, 5435 (1958).
180. J. E. Leffler and J. S. West, *J. Org. Chem.*, **27**, 4191 (1962).
181. R. C. Lamb, W. E. McNew, J. R. Sanderson, and D. C. Lunney, *J. Org. Chem.*, **36**, 174 (1971); R. C. Lamb, P. W. Ayers, and M. K. Toney, *J. Am. Chem. Soc.*, **85**, 3483 (1963).
182. R. C. Lamb, L. P. Spadafino, R. G. Webb, E. B. Smith, W. E. McNew, and J. G. Pacifici, *J. Org. Chem.*, **31**, 147 (1966).
183. K. A. Bilevitch, N. Bubnov, and O. Okhlobystin, *Tetrahedron Letters*, **No. 31**, 3465 (1968).
184. K. Okamoto, Y. Matsui, and H. Shingu, *Bull. Soc. Chem. Japan*, **38**, 1844 (1965).
185. J. C. Martin and R. J. Arhart, *J. Am. Chem. Soc.*, **93**, 2339 (1971).
186. R. M. Fantazier and J. A. Kampmeier, *J. Am. Chem. Soc.*, **88**, 5219 (1966); **88**, 1959 (1966).
187. L. A. Singer and N. P. Kong, *J. Am. Chem. Soc.*, **88**, 5213 (1966); **89**, 5251 (1967).
188. A. Padwa and N. C. Das, *J. Org. Chem.*, **34**, 816 (1969).
189. D. B. Denney, R. Ellsworth, and D. Z. Denney, *J. Am. Chem. Soc.*, **86**, 1116 (1964).
189a. M. Kharasch, A. Poshkus, A. Fono, and W. Nudenberg, *J. Org. Chem.*, **16**, 1458 (1951).
190. B. W. Brooks, F. S. Dainton, and K. J. Ivin, *Trans. Faraday Soc.*, **61**, 1437 (1965).
191. S. G. Cohen and C. H. Wang, *J. Am. Chem. Soc.*, **77**, 2457 (1955).
192. S. G. Cohen and C. H. Wang, *J. Am. Chem. Soc.*, **77**, 3628 (1955).
193. H. Weiland, H. Hove, K. Borner, *Ann.*, **446**, 33 (1926).
194. S. G. Cohen and C. H. Wang, *J. Am. Chem. Soc.*, **75**, 5504 (1953).
195. S. Seltzer and F. T. Dunne, *J. Am. Chem. Soc.*, **87**, 2628 (1965).
196. C. G. Overberger and A. V. Digulio, *J. Am. Chem. Soc.*, **81**, 2154 (1959).
197. P. D. Bartlett and S. F. Nelson, *J. Am. Chem. Soc.*, **88**, 137 (1966).
198. K. Ziegler, W. Deparade, and W. Meye, *Ann.*, **567**, 141 (1950).
199. A. Ohno and Y. Ohnishi, *Tetrahedron Letters*, **No. 50**, 4405 (1969); A. Ohno, N. Kito, and Y. Ohnishi, *Bull. Chem. Soc. Japan*, **44**, 463, 467 (1971).
200. J. W. Timberlake and M. L. Hodges, *Tetrahedron Letters*, 4147 (1970).
201. N. Porter, M. Landis, and L. J. Marnett, *J. Am. Chem. Soc.*, **93**, 795 (1971).
202. S. Seltzer, *J. Am. Chem. Soc.*, **83**, 2625 (1961).

202a. S. Seltzer and E. Hamilton, Jr., *J. Am. Chem. Soc.*, **88**, 3775 (1966).
203. S. Seltzer, *J. Am. Chem. Soc.*, **85**, 14 (1963).
204. S. Seltzer and S. Mylonakis, *J. Am. Chem. Soc.*, **89**, 6584 (1967).
205. S. Rummel, H. Hubner, and P. Krumbieger, *Z. Chem.*, **7**, 351, 392 (1967).
206. P. Kovacic, R. R. Flynn, J. Gormish, A. H. Kappelman, and J. R. Sheldon, *J. Org. Chem.*, **34**, 3312 (1969); J. R. Sheldon, C. K. Liang, and P. Kovacic, *J. Am. Chem. Soc.*, **90**, 354 (1968).
207. C. Rüchardt, *Angew. Chem., Int. Ed.*, **9**, 830 (1970), and references cited therein.
208. C. Rüchardt, P. Lorenz, and E. Schacht, *Tetrahedron Letters*, **No. 32**, 2787 (1969).
209. D. E. Applequist and L. Kaplan, *J. Am. Chem. Soc.*, **87**, 2194 (1965).
210. C. G. Overberger and J. Ganier, *J. Am. Chem. Soc.*, **80**, 4561 (1958).
211. S. Seltzer and S. Scheppele, *J. Am. Chem. Soc.*, **90**, 358 (1968).
212. S. G. Cohen, S. J. Groszos, and D. B. Sparrow, *J. Am. Chem. Soc.*, **72**, 3947 (1950).
213. C. G. Overberger, W. F. Hale, M. B. Berenbaum, and A. B. Finestone, *J. Am. Chem. Soc.*, **76**, 6185 (1954).
214. J. A. Berson, C. J. Olsen, and J. S. Walia, *J. Am. Chem. Soc.*, **84**, 3337 (1962).
215. C. Rüchardt and J. Hinz, *Tetrahedron Letters*, 3095 (1970).
216. J. C. Martin, J. E. Schultz, and J. W. Timberlake, *Tetrahedron Letters*, 4629 (1967); C. G. Overberger and H. Biletch, *J. Am. Chem. Soc.*, **73**, 4880 (1951).
217. C. H. Chang, R. F. Porter, and S. H. Bauer, *J. Am. Chem. Soc.*, **92**, 5313 (1970).
218. R. F. Hutton and C. F. Steel, *J. Am. Chem. Soc.*, **86**, 745 (1964).
219. T. Mill and R. S. Stringham, *Tetrahedron Letters*, 1853 (1969).

CHAPTER 4

"Cage" Effects

THOMAS KOENIG
Department of Chemistry, University of Oregon, Eugene, Oregon

AND

HANSPETER FISCHER
Ciba-Geigy Chemical Company, Basel, Switzerland

I. INTRODUCTION

The existence of the phenomenon which has come to be broadly described as the "cage effect" was first recognized almost forty years ago in a paper by Frank and Rabinowitch.[1a] These authors pointed out that the quantum yields for photodissociation processes should be reduced in solution, compared with the gas phase, because of the constant state of collision between the virtually separating fragments and the surrounding solvent molecules. In a subsequent paper[1b] it was shown, for a mechanical model of a two-dimensional liquid, that collisions between two solute particles occur in sets which increase in size as the density of solvent particles is increased. Thus, at low density, the sets contained an average of about one collision between a pair of solute particles while at high solvent density they contained 10–100 such collisions. The reason for this increase in the size of the set is the reduction in the chance that one member of a pair, separating from a collision, will find a hole in the surroundings large enough to pass through without being reflected back into collision with its original partner, through collisions with the increasing density of solvent particles. Though relatively rare, density has been used specifically as a parameter in correlating ethane yields[2] from azomethane photolysis and in ^{18}F exchange with methyl fluoride.[3]

This mechanical picture of the solvent cage has been recognized to be of *possible* importance in thermal as well as photochemical processes many times since it was originally suggested. The term "cage effect" has been frequently mentioned with reference to the subsequent model which was put forth by Noyes.[4] This model recognizes three ways that a dimer (R—R′,

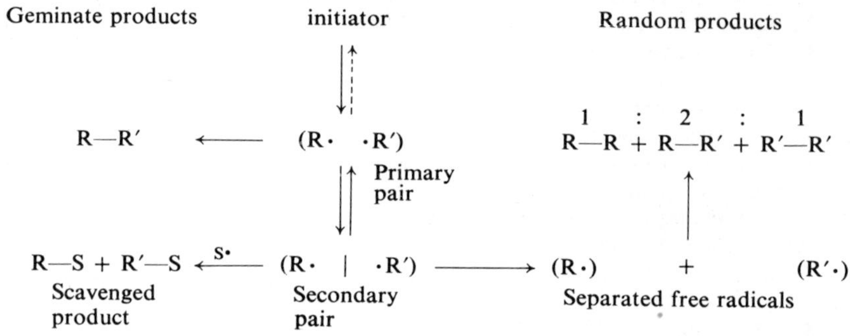

Scheme 1

Scheme 1) can form in solution. These processes are as follows: (1) from *primary pairs* (R′· ·R) which have never been separated (essentially equivalent to the original proposal of Frank and Rabinowitch); (2) from *secondary pairs* which have separated to some extent and returned; (3) from pairs formed from *free* radicals which have reached their statistical separations in the solution. The first two processes lead exclusively to the combination product formed from fragments originating within the same initiator molecule, whereas the third gives a random mixture of symmetrical and unsymmetrical dimers. Noyes therefore suggested that the *sum* of the first two processes be designated as *geminate* combinations to distinguish them from the random events which may lead to the same product. There are now many results which certainly reflect the existence of the phenomenon broadly designated as "cage effect," but the precise definition of the dimensions or other characteristics of this solvent property remains a problem today.

II. QUALITATIVE OPERATIONAL CRITERIA FOR GEMINATE EFFECTS

A. Quantum Yields

The original paper[1] by Frank and Rabinowitch dealt with differences in quantum yields for photodissociation in solution compared with the gas phase. Their arguments led to the prediction of a *reduction in the quantum yield in solution*, even when no predissociative state intervenes. Quantum yield variations are thus one possible operational criterion for demonstrating cage effects. The quantum yields for photodissociation of iodine or bromine molecules should be near unity[5] in the gas phase (at sufficiently high irradiation energy) while they are 0.14[6] and 0.22,[7] respectively, in carbon tetrachloride solution at 25° and 440 mμ irradiation wavelength. These reduced values for the liquids are certainly a demonstration of the cage effect concept. There are similar reductions in quantum yields for photodissociation of azo compounds in solution (0.25–0.0)[8–10] compared to the gas phase (1.0 at zero pressure).[11] However, at least one predissociative state has been implicated by the sensitivity of the quantum yields to the pressure of diluent gases.[10–12] Such observations render the reduced values in solution ambiguous as a criterion for geminate effects.

It is a very simple extension of the ideas of Frank and Rabinowitch to expect correlations of quantum yield with viscosity. Kodama[8] interpreted the variations in quantum yield for photodestruction of azomethane, which he observed with temperature (viscosity), as evidence for reversible fragmentation of only one of the C—N bonds. The quantum yields for photodestruction of azobiscyclohexylnitrile are independent of viscosity ($\Phi \sim 0.24$).[13] Isomerization (cis–trans) accompanies photodestruction of most azoalkanes[9–11,14,15] and the cis isomers are labile enough to give radicals at room temperature when the alkyl groups are tertiary.[14,15] Viscosity has a small effect on the cis–trans isomerizations of stilbenes[17] so that such effects are not unique for dissociations. The quantum yields for photodestruction of aryl disulfides have also been correlated with the viscosity of the medium.[17] Assuming solvation effects are constant, such observations are necessary but not sufficient to imply geminate recombination.

B. Racemization and Oxygen-18 Scrambling

A recent study[16] of the photolysis of optically active 2-phenylazo-(2-phenyl)butane (Scheme 2) provides a firmer basis for the existence of the cage effect in photolysis of azoalkanes. The most important observation was that the starting material, recovered after partial (40%) decomposition in hexadecane solution, was also partially racemized (26%). These authors also noted a linear dependence of the reciprocal of the quantum yields with the

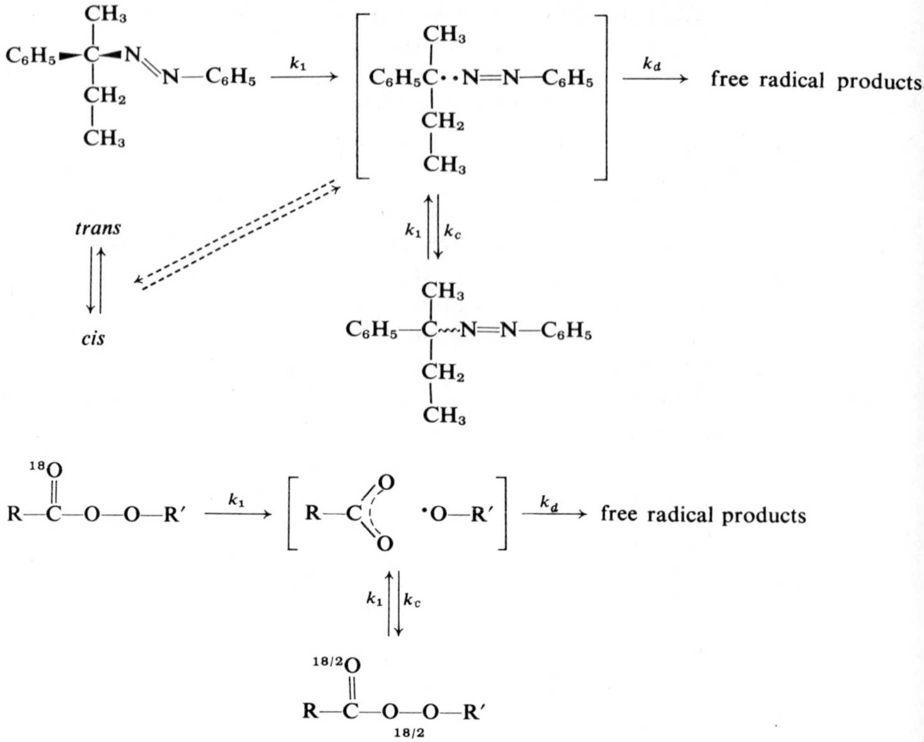

$$k_0 = k_1 \frac{k_d}{k_d + k_c} \qquad k_s = k_1 \frac{k_c}{k_d + k_c} \qquad k_1 = k_0 + k_s$$

Scheme 2

square root of the solvent viscosity. The linearity of this relationship allows an estimation of the quantum yield for photolysis at zero viscosity (0.062) where the fraction recombination should also be zero. This in turn allows an estimate of the fraction geminate recombination (52%) in hexadecane. The cis isomer was also formed in competition with photodestruction of the trans starting material. The former compound thermally isomerizes to the trans in competition with radical fragmentation. Examination of the optical activity of the cis isomer would be interesting since it might show the extent to which the cis–trans process is related to the formation of radicals.

Samples of carbonyl oxygen-18 labeled benzoyl peroxide, recovered after partial photolysis,[18] show some oxygen randomization. This isomerization could be taken to correspond exactly to the racemization process described above. This randomization also occurs during photolysis of *t*-butyl perbenzoate, *t*-butylperoxy cyclopropanecarboxylate but to a negligible extent with *t*-butyl peracetate.[19]

C. Rate–Viscosity Correlations

For thermal decompositions leading to radicals in solution, the original ideas of Frank and Rabinowitch translate into the expectation of reduced rate constants compared with the same process in the gas phase. Just such an order is found for acetyl peroxide[20,21] ($k_{gas}^{80°} = 1.5 \times 10^5\ sec^{-1}$; $k_{isooctane}^{80°} = 1.3 \times 10^{-5}\ sec^{-1}$). However *t*-butyl,[24c] propionyl,[20] and butyryl[20] peroxides all have gas-phase rate constants which are lower than their rate constants in solution at the same temperature. Solvation effects must therefore be important enough to override the small rate reductions expected of condensed phases. Pryor and Smith[21] have proposed that reductions in observed rate constants for destruction of initiators with increased solvent viscosity may still imply that recombination occurs (Chapter 3). Their method can accommodate solvation if it is a constant factor in every solvent. For peresters[19] and diacyl peroxides,[18] rate constants for oxygen-18 randomization can also be measured (Scheme 2). The observed values increased as the solvent viscosity increases, as expected for geminate recombination. The sum of the observed rate constant for destruction of a peroxide (k_0) and that for randomization of oxygen-18 label (k_s) should be constant and equal to the rate constant for unimolecular homolysis without recombination which the Pryor-Smith correlation gives. These conditions are well fulfilled for the case of acetyl peroxide,[18a] approximately so for *t*-butyl peracetate,[19b] and very poorly so for *t*-butyl perbenzoate.[19a] The deviations are always in the direction in which the reductions in k_0 predict more recombination than that counted by oxygen-18 scrambling. This could be true if combination occurs without randomization to some extent. On the other hand, Goldstein and Hudson[22] have provided evidence that oxygen-18 randomization occurs, in part, by a 3,3 sigmatropic shift at lower temperature. This implies that the rate of scrambling should over-estimate return. They also observed[22a] some oxygen-18 randomization in samples of *t*-butyl perpivalate, recovered after partial decomposition. Solvation effects and alternative mechanisms have thus clouded the interpretations placed on the rate-viscosity correlations.

D. Stable Product Yields

Studies of rates of the fragments produced in photolysis processes can give information about the separation–combination competition which is independent of complications surrounding the primary photochemical events. The photolysis of azomethane is a standard source of methyl *free* radicals. The chemical yield of ethane from this reaction is 3% at 1 mm and increases continuously as the pressure of added propane is increased.[2] In solution, the extrapolated yield at infinite viscosity is 100%.[59] This type of chemical yield behavior is the complement to that predicted for quantum yields when equivalence between density and viscosity is made. Thus, the observation of increasing yields of combination product, with increasing density or viscosity, is another possible operational criterion for demonstrating geminate combination. Yields of *t*-butyl peroxide,[24] *t*-butyl perbenzoate,[19a] and *t*-butyl peracetate,[19c] isolated from thermolysis of the corresponding hyponitrites (Scheme 3), have been found to respond to viscosity changes in a manner

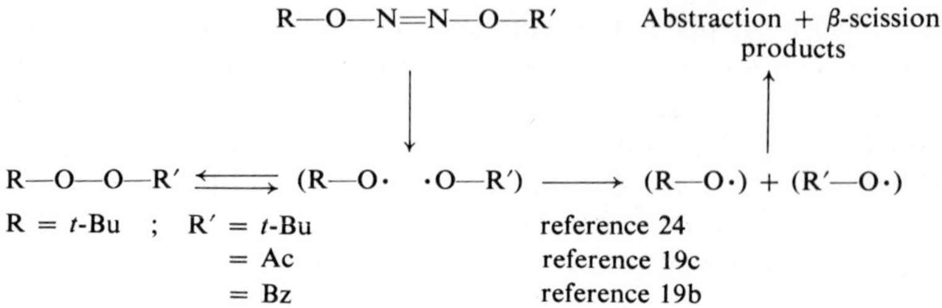

Scheme 3

consistent with this idea. Isolated yields are, of course, subject to the fewest complications with respect to interpretation.

E. Crossover Experiments

The case for geminate effects in azomethane photolysis is made considerably stronger by the results of crossover experiments. When a mixture of perdeuterio and protio azomethane is photolyzed in the gas phase, the isolated ethane is an essentially statistical mixture of the three possible dimers (1:2:1; C_2H_6:$C_2H_3D_3$:C_2D_6).[25] In isooctane solution, the 75% yield of ethane contains only the two symmetrical products. The very striking increase in yield of only the symmetrical dimers strongly supports the idea of geminate combination as the mechanism for the formation of this product. Similar results have been reported for the ethane and methyl acetate isolated from thermolysis of a mixture of protio and perdeuterio acetyl peroxide.[18c,19d] The photolysis of mixtures of different symmetrical diacyl peroxides in pentane does give some unsymmetrical dimer.[26] The yields of these random combination products are greatly reduced in decalin solution. When relatively unreactive radicals such as α-phenylethyl are involved, most (70%) of the dimer isolated from thermal decomposition of a mixture of protio and deuterio α,α′-phenylazoethane results from random combination, that is, large amounts of the unsymmetrically deuterated 2,3-diphenylbutanes are found.[27]

F. Scavenging Methods

As shown by Scheme 1 (and the α-phenylethyl radical results[27] above) free radicals, which are inert to the solvent molecules, may still end up as combination product. In these cases, adding reactive compounds to the solution should divert the *free* radicals away from dimer formation. Such compounds are called scavengers and can be divided into three groups. The first group is comprised of stable radicals such as 2,2-diphenylpicrylhydrazyl (DPPH),[28] nitroxides,[29] galvinoxyl,[30] Koelsch radical,[31] α,γ-bis(diphenylene)-β-phenylallyl,[32] or oxygen[33]. These molecules trap radicals by combination processes which are nearly as fast as combinations between reactive radicals. The second group reacts by fast hydrogen or halogen transfer and includes thiols,[34] cyclohexadiene,[36] dihydroanthracene,[37] iodine,[38] and bromine.[39] The third group adds radicals to form a more stable radical; quinones,[40] styrene,[41] and other vinyl monomers as well as higher aromatic hydrocarbons[42] are traditional examples. Recently, nitroso compounds[43] and nitrones[44] have been found to be excellent free radical trapping agents.

The yields of tetramethylsuccinonitrile from the thermolysis of azobisisobutyronitrile (ABN) falls from above 80% when no scavenger is present, to a constant value of 19% above 0.6 *M* *n*-butanethiol in carbon tetrachloride.[35] Similar decreases are observed in the 2,3-diphenylbutane yields from α,α'-azophenylethane when 2-methyl-2-nitrosopropane is added.[43] Thus nonzero limiting values for the yield of a dimerization product, as scavenger concentration is increased, is another method which has been used to infer geminate combination. The same inference can be drawn from the complement to combination product yield, free radical efficiency (1-yield geminate is free radical). The fraction of ABN molecules leading to initiation of styrene polymerization increases as the concentration of the monomer increases, reaching a limiting value of 70% above 2 *M* styrene.[41] The remaining 30% can be attributed to the reactions of the geminate pairs. Decolorization of galvinoxyl or other colored radical is another method of quickly estimating the efficiency of *free* radical formation. This may be done by colorimetrically observing the moles of scavenger decolorized by a deficient quantity of initiator[45] or by observing the zero-order disappearance of the scavenger in a large excess of initiator. For the latter method, the first-order rate constant for destruction of initiator and its concentration are needed to calculate the efficiency factor. Deviations from 2 moles of scavenger consumed per mole of initiator present have been attributed to geminate combination in many cases. The viscosity effects on this type of measurement have been examined recently for *t*-butylperoxy α-phenylisobutyrate,[46] azocumenes,[29] and azocymenes.[47] Polymerization efficiency in azobiscyclohexylnitrile has also been measured as a function of viscosity.[13] In all cases, the estimated fractions of free radical production decreased as the solvent viscosity was increased.

G. CIDNP Effects

The recent discovery of nuclear polarization in diamagnetic products produced from paramagnetic intermediates (CIDNP) has been interpreted in terms of the combination–separation competition for radical pairs with the added stipulation that the combination process proceeds exclusively through the electronic singlets. This forms the basis of the Closs[48]–Kaptein–Oosterhoff[48a] model for the effect which has gained wide acceptance and which is discussed in another chapter. The observation of a CIDNP effect for a given product is certainly good evidence for the intervention of a paramagnetic precursor for *some* of that product. However, this method suffers from the same ambiguity that exists for all the conventional criteria for geminate effects. In every case, there is at least a formal possibility of multiple mechanism with ionic or concerted as well as pairwise branches contributing to the formation of a given product. With CIDNP as with all of the other types of observables, quantitative analysis is required before it can safely be stated that all of a given reaction product has arisen by radical pairs. The quantitative results, discussed in the subsequent sections, are much less settling than the qualitative ones mentioned above.

III. PRIMARY–SECONDARY COMBINATIONS

All of the qualitative criteria mentioned so far deal only with the geminate effect and make no distinction between primary (cage) and secondary combination. *Quantitative* analysis of the effects of scavengers does suggest the possibility of demonstrating this distinction. Figure 1 shows the predictions

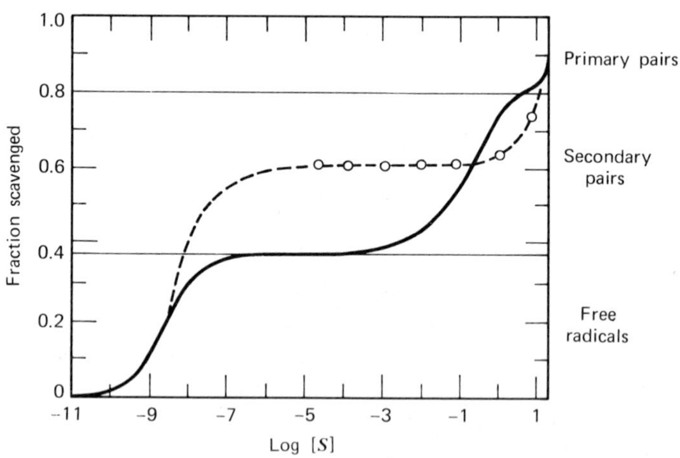

Fig. 1. Solid line: theoretical prediction (Noyes[4]). Dashed line: experimental R-R′, DPPH, Scheme 3 (Hammond et al.[48]).

made by Noyes[4] concerning this aspect of the problem. The three steps in the curve relating scavenger efficiency and scavenger concentration pertain to the three types of intermediates recognized in Scheme 1. The most serious attempt[49] at testing these predictions has been for the decomposition of the azobisnitriles (Scheme 4).

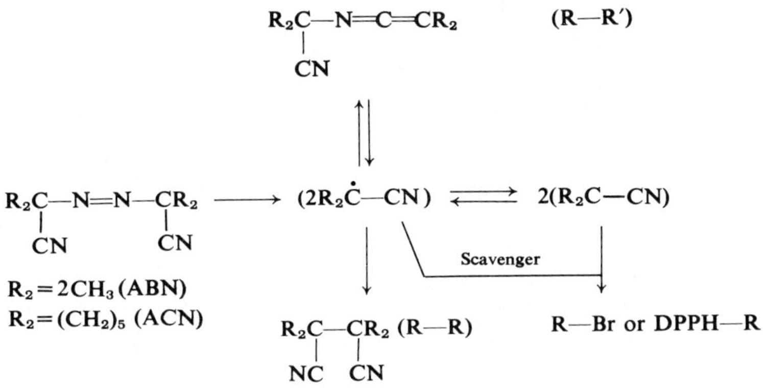

Scheme 4

The azo compounds decompose by the concerted breaking of both C—N bonds to give a pair of cyanoalkyl radicals separated by a nitrogen molecule. These radicals can dimerize to the stable succinonitrile (R—R) or the metastable ketenimine (R—R′), disproportionate, or diffuse. In the absence of scavengers, the pairs which reach statistical separation still react by the same processes (random combination) due to the absence of alternative routes for such kinetically stable free radicals. The eventual yields of R—R are thus high (~80%). The first step of Figure 1 is relatively easy to demonstrate and is actually the basis of the qualitative use of scavengers mentioned above. Addition of 0.6 *M* *n*-butanethiol reduces the yield of R—R from ABN to 19%. Very small concentrations of bromine or DPPH (in excess of ABN concentration) will accomplish this same reduction.

This high reactivity of the scavengers fulfills one of the requirements for demonstrating the distinction between primary and secondary recombinations. The effect of scavenger concentration on the yields of R—R from azobiscyclohexylnitrile (ACN) and the corresponding ketenimine (R—R′) have both been examined. The results for the cyclohexylketenimine using DPPH are shown by the dashed curve of Figure 1. The experimental values have been normalized for the 68%[49] yield of R—R which is obtained in the absence of scavenger. The remaining 32% of initiator results in disproportionation products. The experimental curve reflects the high reactivity of the stable radical scavenger (DPPH) with the cyanocyclohexyl radicals mentioned above. This is evident from the low concentration of scavenger at which the complete suppression of random recombination occurs. Secondly, the experimental curve rules out the possibility that any R—R product is formed from a competing concerted mechanism.[50] This conclusion is reached because of the clearly evident onset of the second step in the increasing efficiency of scavenging. It is important to recognize the fact that the majority of studies of the geminate effect have actually only inferred its existence by measuring the part of the reaction which is *not* occurring through the geminate processes. The above study stands virtually alone in providing direct indication of the existence of the geminate intermediate by a successful trapping experiment.

Figure 1 shows that the beginning of the interference with secondary combination processes is predicted[4] to occur at scavenger concentrations of around 10^{-2} *M*. The dependence on scavenger concentration is also predicted to be to the one-half power in this scavenger concentration range. In contrast, the second step for the experimental points does not begin until approximately 0.5 *M* scavenger. This high value and the absence of the predicted dependence on $(S)^{1/2}$ in the range 10^{-2}–10^{-1} *M* caused Hammond and his co-workers[49] to conclude that primary and secondary combination processes cannot be distinguished. They interpreted their results in terms of a model, which resembled the original proposal of Rabinowitch and Wood,[1] in which the second step of Figure 2 begins when there is an appreciable probability that a scavenger is one of the six molecules making up the solvent cage wall.

The formation of cyclohexyl acetate and smaller amounts of 1,2-diacetoxycyclohexane from the decomposition of acetyl peroxide in cyclohexene[18d] has been explained in just this way. The cyclohexene is proposed to act as a scavenger for the unstable acetoxy radical.[18] The evidence for supporting this conclusion is the constancy of the small amounts of methyl acetate and 1,2-diacetoxycylohexane detected with and without added scavenger (galvinoxyl). Both the absolute and relative rates of addition of acyloxy radicals to simple olefins must be considerably different from what one would expect based upon analogy with methyl affinities.[42a] Earlier workers had shown that the yield of ethane and methyl acetate in isooctane was not reduced by iodine.[42] However, the yield of C_{16} hydrocarbon dimer from decomposition of the diacyl peroxide of the C_9 carboxylic acid does show a clear reduction when 0.6 *M* iodine is present.[42b] The onset of the second step in the efficiency curve is again about 10–100 times higher scavenger concentration than predicted by the Noyes model.

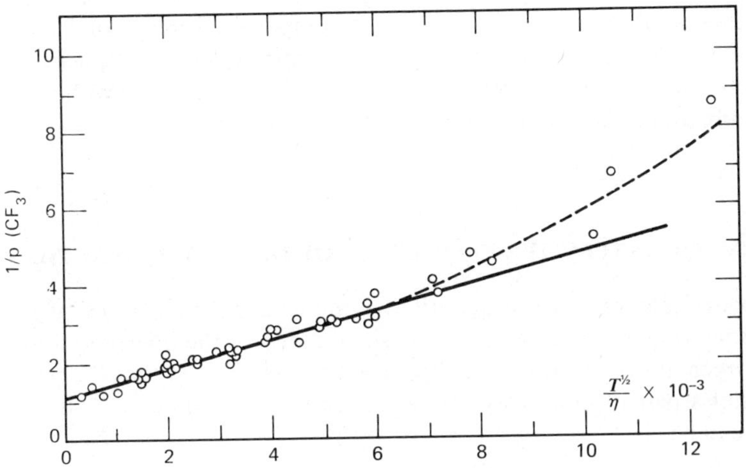

(a)

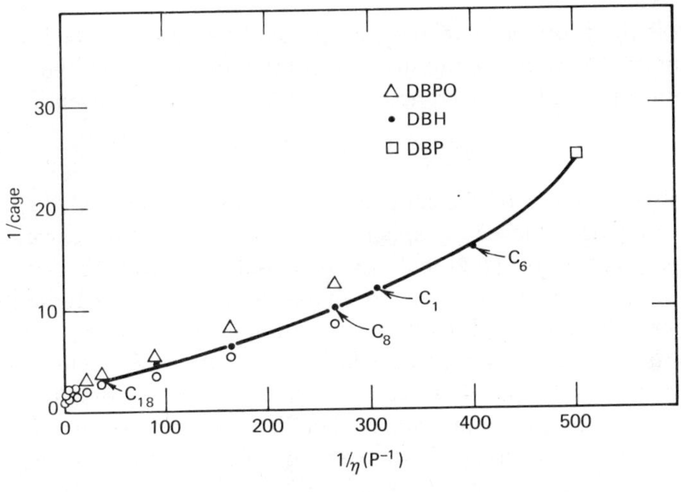

(b)

Fig. 2. (*a*) Reciprocal of perfluoroethane yield versus $T^{1/2}/$ (Szwarc[60]). (*b*) Reciprocal of *t*-butyl peroxide yields from DBH and DBPO (Kiefer and Traylor[24]).

Hamill and co-workers[52] have provided an alternative model for scavenger behavior which makes predictions similar to those of the Noyes version. These workers[53] have also provided some examples of γ-radiolysis reactions which appear to fit these predictions. However, the energetics of these reactions are rather different from those of the thermal decompositions of peroxides and azo compounds so that comparisons may not be justified. The scavenger effects on the decomposition of phenyl phenylsulfonyl diimide[54] have been interpreted in terms of the interference with secondary recombination. This system could bear reinvestigation.

IV. QUANTITATIVE EFFICIENCY OF GEMINATE REACTIONS

Noyes[55] has provided a quantitative model for the efficiency of geminate reactions which has served as the basis of many of the interpretations of the experimental investigations of this problem. He showed that the probability (β') that a pair of molecules separating from an encounter will subsequently react can be given as

$$\beta' = \frac{\alpha\beta}{1 - \beta + \alpha\beta} \tag{1}$$

where α is the probability of reaction per collision and β is the probability of reencounter. The total probability of recombination (β_0') may be taken as the product of probability for reencounter after the initiation event (β_0) times β'.

$$\beta_0' = \beta_0\beta' \tag{2}$$

The physical basis of the model rests simply on the final displacements which would be achieved by spherical particles which move with assumed velocities subject to the frictional drag of a viscous continuum. After reaching such points the further motions are taken as random. It is these initial displacements which are of the utmost importance according to this model. The separation r_0 is the distance achieved during the initiating event. This separation is calculated by assuming that the kinetic energy vectors are oriented by the nature of the event to give maximum displacement and may have magnitudes greater than average thermal energy. For example, in the photodissociation of an iodine molecule, r_0 was given as

$$r_0 = 2b + \frac{[m(h\nu - E)]^{1/2}}{3\pi\eta b} \tag{3}$$

where b is the diffusion radius, ν is the frequency of the absorbed photon, E is the dissociation energy of the process, and η is solvent viscosity. Thus β_0 can be written as

$$\beta_0 = \frac{2b}{r_0} = \frac{1}{1 + A_E/\eta} \qquad A_E = \frac{[m(h\nu - E)]^{1/2}}{6\pi b^2} \tag{4}$$

where the constants have been included in A_E.

The second displacement of this kind is achieved after subsequent encounters. Here the kinetic energy term is taken as $\frac{3}{2}kT$. This gives β as:

$$\beta = \frac{2b}{r_T} = \frac{1}{1 + A_T/\eta}, \qquad A_T = \frac{[(\frac{3}{2})mkT]^{1/2}}{6\pi b^2} \tag{5}$$

where the constants have been incorporated as A_T. The reciprocal of the total probability of geminate combination can now be written in terms of the product of the reciprocals of β_0 and β_0':

$$\frac{1}{\beta_0'} = 1 + \frac{(\alpha A_E + A_T)}{\alpha\eta} + \frac{A_T A_E}{\alpha\eta^2} \tag{6}$$

By the general line of reasoning one can see that an additional factor should be added, namely, the initial (R_0) separation which could be induced by the chemical structure as, for example, in the case of an azo compound where the departing nitrogen molecule might be expected to provide an initial displacement (R_0). This should be added as a simple multiplicative factor.

$$\frac{1}{\beta_0'} = \frac{R_0}{2b}\left[1 + \frac{(A_T + \alpha A_E)}{\alpha}\left(\frac{1}{\eta}\right) + \frac{A_T A_E}{\alpha}\left(\frac{1}{\eta}\right)^2\right] \tag{7}$$

A convenient form for analyzing the variable dependence of efficiency of geminate combination is through the reciprocal of the efficiency minus 1, which we define here as F.

$$F = (1/\text{cage efficiency}) - 1$$

For the Noyes model,

$$F = \left(\frac{R_0 - 2b}{2b}\right) + \left(\frac{R_0}{2b}\right)\left[\left(\frac{A_T + \alpha A_E}{\alpha}\right)\left(\frac{1}{\eta}\right) + \left(\frac{A_T \cdot A_E}{\alpha}\right)\left(\frac{1}{\eta}\right)^2\right] \tag{8}$$

This is a more general expression than Noyes wrote, but is used here since it shows the relationships between the reductions of it used by subsequent workers, which are all really derived from the ideas presented above. This equation shows the variables which are expected to pertain to the value of F. These variables are as follows:

a. The solvent fluidity ($1/\eta$).
b. The initial separation of the fragments due to intervening molecules (R_0).
c. The translational energy of the separating fragments (A_E).
d. The mass and radius of the fragments (A_E, A_T).
e. The probability per collision that the pair will react (α).

Other models have been proposed, including recent ones by Koenig[56] and

Stefani et al.[57] There is little basis for choosing between them at the present time and Noyes' is used here because it has been referred to more frequently.

A. Viscosity Dependence

Equation 8 implies that F should be linearly related to the fluidity $(1/\eta)$ except in the most fluid solvents, where the squared term becomes important. Relatively few tests of this dependence are available. The most careful was that by Booth and Noyes[58] in their studies of the photodissociation of I_2. The qualitative prediction of increasing quantum yield with increasing fluidity was borne out but the quantitative values of the quantum yields were too high at both the highest and lowest viscosities. The deviation at low viscosity was in spite of calculations which included the $(1/\eta)^2$ term; that is, the upward curvature in the F versus $(1/\eta)$ plot was greater than that *predicted* by equation 8.

Kodama[8,59] studied the effect of viscosity on the yields of ethane from azomethane photolysis. He found a fair region over which F was linear with $1/\eta$ but noted strong upward curvature in the high-fluidity region. Similar results have been obtained recently by Szwarc and co-workers[60] in the photolysis of perfluoroazomethane (Figure 2). Bartlett and co-workers[46] have followed Szwarc in their interpretation of the kinetically determined free radical efficiencies for *t*-butylperoxy α-phenylisobutyrate. The common observation in all of these studies is a linear dependence of F with fluidity out to 100 P^{-1} after which there is strong upward curvature. All of these workers came to the same essential conclusion concerning the greater than first-power dependence in highly fluid media. Namely, the highly fluid solvents acted to separate rather than to "cage" the pairs.

An additional observation of Noyes' study was the relatively small value for the quantum yield for iodine dissociation in carbon tetrachloride. This was interpreted as a point in favor of the original "cage" model of Rabinowitch since the latter is principally based upon momentum. The high molecular weight of the CCl_4 molecules should make them more effective in reflecting fragments back into collision with their geminate partners. This point has also been discussed with respect to a recent[63] computer simulation of a two-dimensional liquid which included intermolecular forces.

Figure 2 shows the overall viscosity behavior found for combination of *t*-butoxy radicals (Scheme 3 above).[24] This sigmoid curve seems to be typical for reactions which have been studied over wide enough viscosity ranges. The curve is roughly linear over the region between 10 and 0.5 cP. The curve also shows a strong downward turn at the very highest viscosities. Plots of F versus square root of fluidity are linear over the region 30–0.5 cP and the alternative models of Eirich et al.[64] and Koenig[56] have predicted square-root fluidity dependence. Szwarc[62] has argued that this type of observation is a reflection of the type of solvents used. The diffusion of small molecules in viscous solutions, made up of small amounts of high polymers dissolved in ordinary solvents, is not much different than that for the fluid solvent.[65] Long-chain hydrocarbons might, therefore, be expected to allow diffusive separation more easily than their Stokes-Einstein viscosities would predict. Thus the observation of good correlations of experimental data with one or another power of viscosity is not a very powerful argument for any particular model. Such models are helpful in providing a framework for comparison of the observations available. No adequate quantitative model is yet available for dealing with cage effects.

B. Intervening Molecule Effects

It is clear that the Noyes model places heavy emphasis on the initial distance of separation attained by the geminate pair before they begin their random walk. This implies that an initiator which would start the radical pair out with a positive separation, by splitting out a small molecule such as nitrogen or carbon dioxide, would produce a noticeable reduction in yield of combination product compared to an initiator which has the same geminate pair with no intervening species. For purposes of calibration, irradiation of I_2 with 4358 Å light in carbon tetrachloride ($\eta = 0.009$ P) at room temperature would lead to a pair of iodine atoms with an initial separation of approximately 1.2 Å, about the same separation expected by extruding the nitrogen molecule from N_2I_2. The theoretical (equation 8, $\alpha = 1$) quantum yield of such a pair is about 0.23 so F would be of the order of 0.3 If the iodine atoms could be formed with only $\frac{3}{2}kT$ of translational energy in the same solvent and at the same temperature, the initial separation would be about 0.2 Å. Taking this as a model for a decomposition leading to a pair with no intervening molecule, the quantum yield is calculated to be about 0.04 (essentially equal to F when it is so low). Thus the F value, which measures the diffusion–combination competition, is predicted to increase by a factor of 6 or 7 in the process which eliminates a nitrogen molecule.

There have been several tests for the effect of intervening small molecules. Hammond and co-workers[49] compared the decomposition of azobiscyclohexylnitrile with that of the corresponding ketenimine (Scheme 2 above). The possible geminate reactions for cyanocyclohexyl radicals include disproportionation and dimerization to either the stable succinonitrile (R—R) or the ketenimine (R—R′). The total efficiency of these three processes was 0.39 from the azo compound. The efficiency of reformation of the ketenimine cannot be measured directly. Assuming the ratio of the two dimerization processes for R—R′ is the same as that observed for the azo compound, the efficiency of formation of the stable dimer can be used to calculate reformation of R—R′. In this way, the total efficiency of the geminate processes in the ketenimine decomposition was estimated to be 0.73. The value of F for the azo compound (1.56) is thus 4.2 times that for the ketenimine (0.37). The effect of the intervening nitrogen molecule thus appears to be comparable in magnitude to the guess made above.

A much larger apparent difference was observed for the effect of the intervening nitrogen molecule in the decomposition of *O*-acetyl-*O'*-*t*-butyl hyponitrite compared with *t*-butyl peracetate (Scheme 3).[19b] The yield of the perester from the hyponitrite in a paraffin oil–carbon tetrachloride solution (viscosity ~17 cP at 32°) was 16%. At 0° the yield was estimated to be 20% at the same viscosity. Extrapolation, using these two numbers, predicts less than 5% geminate return at 130° and the same viscosity. The F value for the azo compound is about 20 times that estimated for the perester by oxygen-18 scrambling.

This factor seems too large compared with the model calculation above. The detailed mechanism of the hyponitrite decomposition is not known with certainty. It is possible that it is partly a concerted process which bypasses the acetoxy radical intermediate. On the other hand, it could be that the oxygen-18 scrambling of the perester, which was used in estimating its fraction return, gives high estimates due to contribution from 1,3 sigmatropic shift.[22] It is hard to see why this process would increase in rate with increased viscosity, but it is possible. More recent comparison of the benzoyl hyponitrite and perester gave results which suggested much less difference in the recombination efficiency for the nitrogen-separated and -unseparated pairs.[19a]

Kiefer and Traylor[24] found very little difference in the yields of *t*-butyl peroxide (DBP) from di-*t*-butyl hyponitrite (DBH) and di-*t*-butylperoxy oxalate (DBPO). Two carbon dioxide molecules intervene in the latter process while only one nitrogen intervenes in DBH. The relative rates of photochemical decomposition of DBP were measured in the same solvent series. The relative photochemical rates can be used to estimate fraction recombination if one such value is fixed, and these authors[24] did this by *assuming* that the geminate efficiency of DBP photolysis was the same as the measured yield of DBP from the other two sources in pentane. The recombination yields for the DBP photolysis in all the other solvents were then found to be not much different than those measured for DBH and DBPO. These workers concluded that there was no effect of an intervening nitrogen molecule. Their arguments would be strengthened considerably if the appropriate absolute quantum yields in the DBP photolysis were to be observed. The reductions in the rates of thermolysis of this peroxide at 125°, on going from *t*-butylbenzene to Nujol,[24b] imply more recombination than Traylor's prediction.

The separation due to an intervening molecule enters equation 8 explicitly as R_0. Values of R_0 greater than $2b$ should be reflected by positive intercepts in the F-fluidity correlations. Kodama[59] and Szwarc[40,60] found intercepts of the order of 0.04 to 0.10 in their correlations of ethane yield from azomethane and perfluoroazomethane photolysis. These were interpreted in terms of $(R_0/2b) - 1$. The effect of the intervening molecule would again appear to be very small. However, treating data on hyponitrite decompositions in the same way would give values which are too large to be reasonable.

There is yet another way in which to view the effect of the initial separation which has been utilized by Eirich[64] in his study of the yields of ethane and methyl acetate formed from the decomposition of acetyl peroxide. The first assumption necessary to this analysis is that the reaction is a totally stepwise process in which the peroxide first decomposes to a pair of acetoxy radicals. This intermediate can recombine or go on to react in one of the steps leading in the forward direction. One of these steps is diffusive displacement. There is no source of extra translational energy in the decomposition event so that the displacement distance which an "average" acetoxy radical achieves before decarboxylation is governed either by $(\frac{3}{2}kT)$ in Noyes' model or, as Eirich suggested, the mean-square displacement due to Brownian movement $[r_0 = (6Dt)^{1/2}]$. The latter quantity is calculated to be larger at viscosities greater than 1 cP for $t > 10^{-10}$ sec.

Using this initial separation distance in place of the Noyes frictional model changes the power in which $(1/\eta)$ enters but otherwise the ideas are the same. There would be one separation defined for the subsequently formed methyl–methyl pairs. More importantly, the magnitude of the sensitivity of the system to variation in viscosity is greatly changed because of the larger displacements. Indeed the yield of ethane from azomethane photolysis in 2-methyl-butane is 40% while for the *stepwise* acetyl peroxide decomposition it is 3% (in pentane). The ratio of the F factors is over 20. Extrapolations predict that both initiators would give quantitative yields of ethane (after normalization for methyl acetate) at the high viscosity limit. These ideas of Eirich's also explain the great increase in the geminate efficiency for the decomposition of propionyl compared with acetyl peroxide. An increase in the rate of decarboxylation of the propionyloxy radical is reasonable and would produce the combining fragments at shorter average separations. The possibility[66] that the results reflect size differences is decisively ruled out by the data of Section IV.D.

C. Effect of Reaction Energetics (A_E)

One of the predictions made by equation 8 (as well as by the original intuition of Rabinowitch) is that at constant viscosity, the value of F (and quantum yield) will rise as the wavelength of the absorbed light is reduced (A_E). The photodissociation of iodine is ideally suited for studies of this variable since all of the excess energy of the dissociation process must reside initially in some form of translation (due to the absence of internal modes). While the qualitative expectation of increased quantum yield with increasing A_E is borne out by the experimental results,[67] the quantitative agreement with the particular reduction of (8) used to interpret these results ($\alpha = 1$, $R_0 = 2b$) was disappointing. The dependence on the wavelength was much stronger than predicted by the model; that is, the theoretical value or power of A_E is too low. This gross feature could be related to the problem of the upward curvature of the F versus $1/\eta$ correlations in the low-viscosity media mentioned above.

In azomethane photolysis studies, Kodama[59] used a reduction of equation 8 which amounted to neglecting A_T ($A_T = 0$). He also used a modified form of A_E in which an effective energy (E') was deduced from the slope of the plot of F versus $1/\eta$. When molecules instead of atoms are involved in the fragmentation, it is not possible to know in advance how the excess energy will partition itself between internal and translational modes. Thus one loses the use of the exact $h\nu - E$ relationship above. However, the physical sense of Kodama's ideas is identical to that for iodine dissociation. The energy governing the initial separation is deduced to be 17 kcal/mole from the slope of the F–fluidity correlation. Such a factor (A_E) has also been discussed with respect to the reactions of ABN and the photolysis of dibenzyl ketone.[68] It has more often been neglected altogether. Szwarc[60] has interpreted the slopes of his F–fluidity correlations in terms of a reduction of equation 8 which neglects A_E altogether ($A_E = 0$, $A_T \neq 0$).

D. Mass (m) and Diffusion Radius (b) Effects

Equation 8 predicts that heavier radicals should give lower combination yields (higher quantum yields) as a result of the greater distances they would travel before the frictional drag of the solvent causes them to come to rest and they begin their random walk through the solution (either back together or further apart). On the other hand, an increase in radius of the diffusing particle (b) should lead to a decrease in quantum yield, so that the two effects tend to cancel one another. The slopes of the F versus fluidity ($1/\eta$) correlations should reflect the residual variations. The relative values of the quantum yields for photodissociation of I_2 and Br_2 in CCl_4 solution (0.14 and 0.22, respectively) are in the direction dictated by the order of diffusion radii[7] since the relative masses and A_E terms would reverse them.

The large increase in the efficiency of the geminate combination for radicals produced in propionyl compared to acetyl peroxide thermolysis has been attributed to the effect of size.[66] However, the very recent studies of Kochi and Sheldon[26] on peroxide photolysis provide a wider test for this basic idea. The quantum yields for carbon dioxide formation are near unity in every case so that all absorption leads effectively to a pair of alkyl radicals. Crossover experiments with mixtures of valeryl and butyryl peroxides gave only symmetrical dimers in decalin solution. The constancy of the geminate effects for the series ethyl through *n*-butyl (Table 1), measured under constant conditions, seems to rule out any significant effect of the size in question.

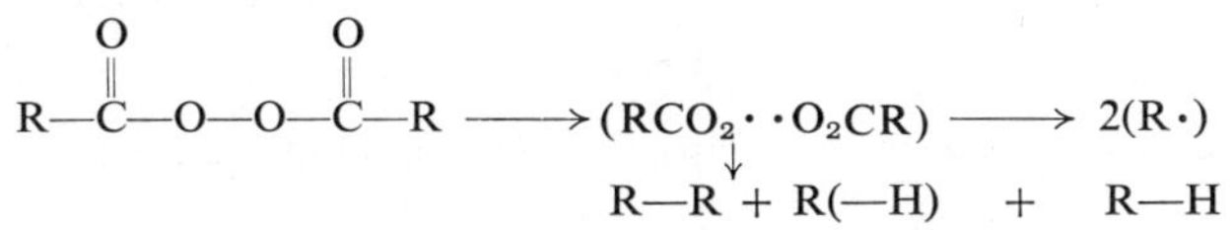

TABLE 1

Geminate Reactions of Alkyl Radical Pairs in Decalin at 30°[a]

Alkyl radical	Geminate %	F	ΦCO_2
$CH_3CH_2\cdot$	55	0.83	2.0
$CH_3CH_2CH_2\cdot$	57	0.75	2.1
$CH_3CH_2CH_2CH_2\cdot$	51	0.98	2.0
$CH_2{=}CH—CH_2—CH_2\cdot$	49	1.02	—
Cyclopropylmethyl·	50	1.00	2.0
$CH_3\dot{C}H—CH_3$	54	0.85	2.1
$CH_3—CH_2—\dot{C}H—CH_3$	57	0.73	—
Cyclobutyl·	61	0.64	—

[a] Data of Reference 26.

One comparison which is quite useful is that between the photolysis of azomethane and perfluoroazomethane. Here the mass terms change while the diffusion radius and R_0 could be expected to remain constant. If all other variables (including the fact[69] that the perfluoro compound is cis while the methyl compound is trans) are constant, then the slope of the F versus $1/\eta$ correlation is predicted $[(m/m')^{1/2}]$ to be two times larger for the heavier fluorinated radical pair. The slope for the fluorinated pair is indeed higher than that for the hydrocarbon but it is greater by more than a factor of 2.[60,61] This means that α or E' must also vary (depending on which reduction of equation 8 one chooses to adopt).

E. Variations in Radical Reactivity (α)

Szwarc's[40,60] reduction of equation 8 indicates that variations in the F–fluidity correlations not accounted for by size changes are related to α, the probability of reaction of the radical pair per collision. As just pointed out, the ratio of the slopes for the combination of methyl compared with trifluoromethyl pairs demands some such variation. The difference in entropy of activation is more than enough to explain the magnitude of this ratio when nonplanarity[70] of the trifluoromethyl radical is taken into account. The ratio of α's is predicted to be 4 if only the front side of the pyramidal trifluoromethyl radical is capable of reaction.

TABLE 2

Geminate Effects for Pairs Containing *t*-Butoxy Radicals

$R\cdot M\cdot OC(CH_3)_2$

R	M[a]	Solvent	$T°$	η[b]	Yield[c]	F[d]	Ref.
2-Norbornyl	CO_2	Benzene	80	—	0	∞	36
$(C_6H_5)_3C$	CO_2	Cumene	25	0.63	11.	9	44a
$(CH_3)_3CO$	N_2	Hexane	45	0.25	6	16	24
CH_3CO_2	N_2	Hexane–CCl_4	32	0.33	3	33	19c
$C_6H_5CO_2$	N_2	Hexane–CCl_4	32	0.33	9	10	19a
CH_3	N_2, CO_2	Hexane–CCl_4	32	0.33	9	10	19c
CH_3	CO_2	Pentane	30	0.21	27	2.7	71
CH_3	CO_2	Decalin	30	1.75	44	1.3	71
$(CH_2)_2CH—CH_2$	CO_2	Decalin	30	1.75	43	1.3	71
(Alkyl–alkyl)	CO_2	Decalin	30	1.75	60	0.6	26
$C_6H_5C(CH_3)_2$	CO_2	Cumene	60	0.49	38	1.6	46
$C_6H_5CH_2$	CO_2	Cumene	80	0.41	32	2.1	73c
C_6H_{11}	CO_2	Cumene	80	0.41	36	1.7	73b
C_2H_5OCO	CO_2	Cumene	25	0.63	49	1.0	72

[a] Intervening molecule(s).
[b] Viscosity in centipoise.
[c] Total percentage of geminate reaction.
[d] Equation 8.

Table 2 summarizes some of the observed yields of combination products for pairs containing one *t*-butoxy radical. There is again a relative constancy in the efficiency of geminate reactions of primary alkyl-*t*-butoxy for the pairs generated from perester photolysis in decalin at 30°. However, these values are certainly lower than those for the alkyl–alkyl pairs generated from diacyl peroxide cleavage (Table 1) and clearly much higher than combinations with other oxy radicals. Batt and Benson[24a] have deduced a rather low value for the rate constant for recombination of *t*-butoxy radicals in the gas phase. The low value is again due to an unfavorable entropy term. The unfavorable steric factor probably plays a role in the very low combination efficiency reported for *t*-butoxy–trityl pairs[44a] and the absence of any cage effect for 2-methyl-2-norbornyl-*t*-butoxy pairs.[35a]

Several groups[74–76] have recently measured termination constants for radicals of a variety of structures (Table 3). Carlsson and Ingold[76] found that the reciprocals of the observed termination constants for *t*-butyl radicals in pentane, cyclohexane, and tridecane correlate with $\eta^{0.8}$. For solvents thickened by small amounts of dissolved high polymer, no reduction in k_e with increased viscosity is observed. (North[65] has discussed this general result fully in his review on diffusion effects.) The sensitivity of termination constants of cyanocyclohexyl radicals to solvent viscosity has also been studied.[13] Comparisons of measured rate constants for cyanoisopropyl radicals in benzene and cyclohexane show changes which are entirely outside the range which could be expected from viscosity differences and imply some solvation effect. Recently measured heats of solution of a nitroxide radical also point to

TABLE 3

Solvent Effects on Termination Rate Constants (k_e) at 25°

Radical	Solvent	Viscosity (cP)	$k_e \times 10^{-9}$ M^{-1} sec^{-1}	Ref.
t-Butyl	Pentane	0.35	2.7	76
t-Butyl	Cyclohexane	0.92	1.1	76
t-Butyl	Cyclohexane	0.89	2.2	75
t-Butyl	*n*-Tridecane	1.30	0.9	76
t-Butyl	Benzene	0.61	0.73	75
n-Propyl	Cyclohexane	0.89	1.70	74b
n-Propyl	Benzene	0.61	0.21	74b
Cyanoisopropyl	Cyclohexane	0.89	4.0	75
Cyanoisopropyl	Benzene	0.61	0.94	75
Cyanoisopropyl	Hexafluorobenzene	—	0.48	75
Benzyl	Cyclohexane	0.89	2.0	74b
Benzyl	Benzene	0.61	1.8	74b

specific solvation by aromatics.[77] However, the free radical efficiencies for ABN photolysis and thermolysis were only slightly different in the two solvents (0.54 in benzene and 0.50 in cyclohexane for thermolysis). The approximately fivefold reduction in the termination rate constant for cyanoisopropyl radical in benzene compared with cyclohexane does not appear to have a counterpart in the cage efficiency. The rate of diffusion and combination for the "cage" pairs must be affected in similar ways by solvation.

Table 4 summarizes structural effects on measured termination rate constants (k_e). The most impressive feature of these data is their relative constancy. Comparison with the gas-phase values suggests that k_e is essentially

TABLE 4

Structure Effects on Termination of Radical Pairs

Radical	Solvent	Viscosity (cP)	$k_e \times 10^{-9}$ M^{-1} sec^{-1}	k_{gas}[a] $\times 10^{-9}$ M^{-1} sec^{-1}	Ref.
H	Propylene	—	(≥38)	20.0	78
I	Carbon tetrachloride	0.9	(8)	—	79
CH_3	Isooctane	—	(18.0)	20.0	2
CH_3	Cyclohexane	0.9	(4.5)	20.0	76
CCl_3	Cyclohexane	0.9	0.05	5.0	80, 81
CH_3CH_2	Isooctane	—	(4.7)	40.0	2
$CH_3CH_2CH_2$	Cyclohexane	0.9	1.7	—	74b
n-C_5H_{11}	Benzene	0.6	2.0	—	74
n-C_6H_{13}	Cyclohexane	0.9	1.1	—	76
Cyclohexyl	Cyclohexane	0.9	1.3	—	76
$(CH_3)_3C$	Cyclohexane	0.9	1.1	25.0	76
$C_6H_5CH_2$	Cyclohexane	0.9	2.0	—	74b
$C_6H_5CH_2$	Benzene	0.6	1.8	—	74b
$C_6H_5C(CH_3)_2$	Benzene	0.6	8.0	—	75
$(CH_3)_2CHC_6H_4C$-$(CH_3)_2$	Benzene	0.6	1.2	—	75
p-$BrC_6H_4C(CH_3)_2$	Benzene	0.6	0.23	—	75
$(C_6H_5)_3C$	Chloroform	—	3.6×10^{-6}	—	84
cy-$C_6H_{10}CN$	Benzene	0.6	2.0	—	75
$(CH_3)_2CCN$	Benzene	0.6	0.80	—	75
$(CH_3)_3CO$	—	—	1–0.1	0.2[b]	82, 83

[a] References compiled in 85.
[b] Estimated from thermodynamics of the reverse process. Reference 24a.
[c] Values in parentheses are calculated indirectly.

determined by diffusion. The constancy is thus surprising since one would expect some correlation with radius. Such a factor could be important for the series from H· to ethyl, but is no longer apparent for the larger radicals. Combination processes which have been studied in the gas phase have been found to occur without activation energy. Variations in rate constants for such processes are therefore thought to result from entropy effects. However, the effect of para substitution on the disappearance of cumyl radicals[75] is too large to be explained by entropy. Dimerizations of ketyl[75] radicals have also been found to be rather sensitive to polar substituent and solvent effects. It is not at all clear how these observations relate to geminate reactions and attempts at correlating the bimolecular termination rates with geminate efficiencies have not been very successful. Reversible complexation of the radicals by aromatic solvent molecules could explain both the constancy of the termination constants listed in Table 4 for alkyl radicals of varying size, and the variation in these values for the substituted benzylic radicals. Solvation effects are known to be important in the β-scission–abstraction competition for *t*-butoxy radicals.[77a] Also, reversible addition of benzoyloxy radicals to benzene seems clearly established.[87,88] It is possible that the process which competes with geminate combination in aromatic solvents is solvation of the nascent fragments rather than their separation.

V. SPIN MULTIPLICITY EFFECTS

The effects of spin multiplicity are not included explicitly in the Noyes or most other models for cage effects. Lyons' calculations[2] of rate constants for combination are the exception. A number of experimental investigations into this question have been carried out by comparing results of sensitized versus direct photolysis. The presumption underlying this type of experiment is that a known triplet sensitizer, which is effective in causing radical formation, gives triplet radical pairs. Direct photolysis is presumed to proceed by singlets. The triplet-sensitized decomposition of *N*-(1-cyanocyclohexyl)-pentamethylene–ketenimine [(R—R′) Scheme 4] gives only 8% succinonitrile in carbon tetrachloride; direct photolysis under identical conditions gave 24%. Thus the triplet pairs have only about one-third the probability of combining as do singlet pairs. A similar figure is derived from the yields of phenyl benzoate or *t*-butyl phenyl ether in the sensitized versus direct photolysis of benzoyl peroxide[87] or *t*-butyl perbenzoate (~3% vs. 8% phenyl benzoate in benzene at room temperature). Oxygen-18 randomization is also less important in the triplet-sensitized decompositions of these peroxides.[18]

Such relatively low efficiencies are not observed in sensitized decompositions of azo compounds.[86] The yields of combination products are almost identical with either direct or sensitized decomposition of azobiscyclohexylnitrile[86] or azocumene.[90] Engel and Bartlett[91] have recently shown that the azo compounds intercept singlets of the sensitizers and thus go through radical pairs of the same spin multiplicity as does the direct photolysis. There are other complications in the overall process but it is now clear that the earlier work on the photolysis of azo compounds provided no test of the spin multiplicity effect.

The reversal in the sign of the CIDNP polarization observed in the dimerization products of the direct photolysis of benzyl-9-azofluorene (singlet) compared with 9-diazofluorene (triplet) in toluene has been explained[92] in terms of the requirement of triplet–singlet crossing as a step prior to combination. Similar reversals in sign of the polarization have been reported recently in the direct versus sensitized photolysis of benzoyl peroxide.[93,94] The nearly statistical product distribution from the carbene insertion reaction also indicates that the combination of the triplet pairs, which are formed first, is relatively slow.

Two groups[95,68] have recently reported independent studies of the photofragmentation of dibenzyl ketone. The products are bibenzyls formed by random combination since crossover products are observed. Substituent effects indicate that the fragmentation involves a triplet state of the ketone. The quantum yields are not reduced by increasing the solvent viscosity. This is probably a result of the spin barrier rather than the translational excitation suggested. It would be interesting to see what type of CIDNP effect is observed in this case since the singlet–triplet crossing occurs before the radical pair is formed. Such results could serve as a valuable calibration point in the evaluation of the implications of CIDNP effects on reaction mechanisms.

VI. STEREOCHEMISTRY OF GEMINATE PROCESSES

A variety of spectroscopic results[96] all indicate that carbon radical (trivalent, neutral, carbon center) is either planar or slightly pyramidal with a low barrier to inversion. The question of the stereochemistry of a chemical reaction is only partly related to this conclusion since even for a strictly planar radical center, the racemization process requires relaxation with respect to the nearest neighbors. Free radicals have relatively long lifetimes and reactions involving these species generally lead to racemized products. On the other hand, geminate combinations of *radical pairs* are among the fastest of all possible chemical changes. The *possibility* of stereoselectivity in these reactions has been recognized for many years. The free radical chlorination 1-chloro-2-methylbutane gives 1,2-dichloro-2-methylbutane, which is completely racemic.[89] In contrast, the reaction of neutron-activated chlorine atoms with *meso*- or *d,l*-2,3-dichlorobutane gives radiochlorine incorporation with a significant amount of retention.[97] This has been interpreted in terms of recombination of alkyl–chlorine atom pairs in competition with rotational relaxation. Wiberg and co-workers[98] have suggested geminate combination with a large fraction of retention in the permanganate and chromic acid oxidations of optically active tertiary hydrocarbons.

Bartlett and McBride[99] showed that photolysis of *meso*- and *d,l*-azo-3-methyl-2-phenyl-2-butane at 77°K in a methylcyclohexane glass gave only one of the possible 3,4-diphenyl-2,3,4,5-tetramethylhexanes. These were assumed to be the meso and *d,l* compounds (respectively), which must be formed with high degrees of retention. In liquid media, around room temperature, none of these coupling products were formed when a scavenger was present. The steric bulk of the radicals involved here is apparently large enough to reduce the combination rate to a value too low to compete with the diffusion process.

Kopecky and Gillian[100] and Green[118] have independently shown that some retention of configuration does attend the geminate combination of α-phenylethyl radical with itself or with benzyl radicals. Thermolysis of the corresponding azo compounds in benzene was the source of the radical pair in both cases. The results of Greene's work are summarized in Scheme 5.

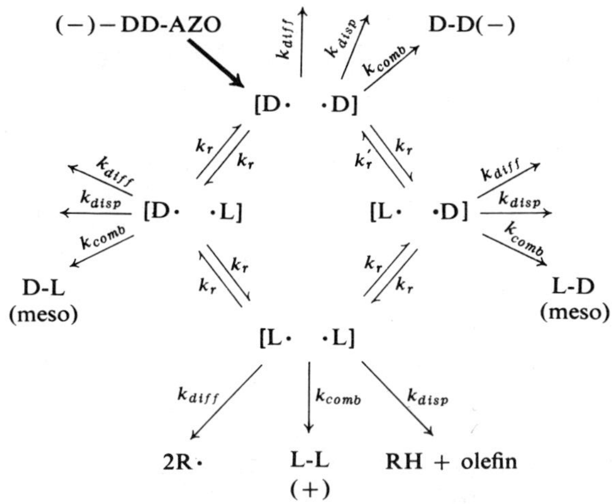

Scheme 5

Controls showed that there is no preference for meso over *d,l* product formation in random combinations. The total yields of the combination product drop from 88 to 28% when 2-nitroso-2-methylpropane is present to scavenge the free radicals. This figure is in good agreement with other estimates of geminate effect in this system. This 28% total yield was found to be composed of 31.3% S,S-(−) (double retention), 48.1% meso (one inversion), and 20.6% R,R(+) (double inversion) isomers. These results were interpreted in terms of the competition between combination, diffusion, and rotation of one of the optically active radicals 180° with respect to the other. This analysis gave $k_{\mathrm{rot}}/k_{\mathrm{comb}} = 15$ and $k_{\mathrm{diff}}/k_{\mathrm{comb}} = 2.5$. Kopecky and Gillian's[100] data give $k_{\mathrm{rot}}/k_{\mathrm{comb}} = 16$ at 110° in benzene for the α-phenylethyl–benzyl pair.

Greene[101] has investigated the decomposition of optically active peroxide of hydratropic acid. He found that a significant amount of ester was formed with retention of optical activity. The small amount of 2,3-diphenylbutane isolated from this reaction also had some net retention. However, the main decomposition path in this case is carboxy inversion so that quantitative interpretation is difficult. Similar results have recently been obtained by Oae[102] and co-workers for the thermal decomposition of the related bis-

(β-phenyl)isobutyryl peroxide. The latter workers also studied the photochemical destruction of both oxygen-18 labeled and optically active material. They obtained a significant amount of retention of both optical activity and oxygen atom identity in the ester product isolated. They interpreted their results in terms of several competing concerted processes. A more interesting possibility is that, like the rotational racemization, oxygen atom randomization is not instantaneously complete upon formation of a carboxy radical. The recognition of electron equivalence of the two oxygen atoms of such a species, when considered as an isolated system, is neither necessary nor sufficient to imply randomization in the combination product. Nuclear motions are the important motions for determining this result. The carboxy radical cannot be considered independent of its partner.

Walborsky and Chen[103] have found 37% retention in the disproportionation product from the decomposition of optically active 1-methyl-2,2-diphenylcyclopropanoyl peroxide. The radical center in such a system is strained so that the inversion (racemization) rate may be unusually slow.[104] Inversion has been reported in the reaction of a similar bromide with tin hydride.[105] The idea of a cage reaction is presented, in a complicated way, as the explanation for the latter results also.

If one assumes that the qualitative observation of polarized NMR transitions (CIDNP) in a product which *could* result from a geminate combination is sufficient to ensure that the entire reaction occurs by radical pairs, then there are some truly striking examples of retention for these reactions. The Meisenheimer[106] reaction of optically active *N*-(α-*d*-benzyl)-*N*-methylaniline-*N*-oxide gives the corresponding *O*-benzylhydroxylamine with 22–39% retention.[107] Conditions which allow the observation of CIDNP polarization have been found.[108] Oxygen scavenges 67% of the benzyl radicals so that the geminate product is formed with only 33% yield.[109] The cage process must therefore give the hydroxylamine with essentially complete retention. Similarly, some examples of the Stevens rearrangement,[110] which shows CIDNP effects,[111] also occur with greater than 78–80% retention of configuration.[112] Such high degrees of stereoselectivity leave little room for the existence of secondary combination processes.

It should be remembered that the implications of the qualitative observation of CIDNP may not be quite straightforward. For example, nuclear polarization is observed in the reaction of *n*-butyllithium with *sec*-butyl bromide[113] in the butene product. Metallation occurs with 20% retention.[114] However the substitution product is formed with over 90% inversion,[115] the same as is observed with the *sec*-butyl tosylate. This result requires a very sophisticated radical pair if indeed no contribution from the conventional backside nucleophilic displacement path actually pertains.

VII. ACTIVATION PARAMETERS

The reduction of equation 8 which is used by Szwarc[60] contains two types of temperature dependence. One is the $T^{1/2}/\eta$ factor which appears explicitly and which Szwarc considers a normalization factor. This allows plotting F values, obtained in different solvents (viscosities) and temperatures, on a single graph (such as Figure 2) using $T^{1/2}/\eta$ as the ordinate. Simple substitution of Arrhenius parameters for α and η (and setting A_E and $R_0 - 2b$ equal to zero) allows his reduction of equation 8 to be written as:

$$F = \frac{A_T}{A_\eta} \exp\left[-\frac{(E_\eta - E_\alpha)}{RT}\right] = \frac{k_d}{k_c} \tag{9}$$

Thus the slope of the plot of ln F versus $1/T$ should give the difference in the activation energy for viscous flow (E_η) and the combination process E_α. In phenomenological models, this slope is the difference of activation energy for the formal diffusion process and the combination process (i.e., E_η does not enter identically). DeTar and Weis,[117] Lamb and Pacifici,[32] and Bartlett[90] have used such slopes and E_η in discussing the combination process. More recent studies of Szwarc[60–62] have revealed rather striking constancy in these slopes for the ethane formation in azomethane photolysis in a wide variety of solvents. Free radical efficiencies in the decomposition of *t*-butylperoxy α-phenylisobutyrate[46] show constancy which is similar in magnitude. To be internally consistent with the Szwarc-Noyes model E_η enters explicitly and E_α, determined in this way, is proportional to E_η and becomes negative in fluid solvents. The interpretation for these observations has not yet been put forth.

Neuman and co-workers[73–73b] have initiated an extensive investigation of the effects of pressure on radical reactions. They have measured such effects for the combination–diffusion competition for *t*-butoxy radical pairs generated from di-*t*-butyl hyponitrite (DBH). The apparent activation volume difference ($\Delta V_d^\ddagger - \Delta V_c^\ddagger$) for this competition was highly pressure dependent. A fairly good correlation between the yields of combination product and the fluidity of the solvent (octane) at the various pressures was obtained. These authors suggested a value of -5 cc/mole for the activation volume for the combination process but the basis of this suggestion is weak. More recent work[73b] on the products of benzyl-*t*-butoxy pairs has led to a moderated point of view which recognizes the problem of the intervening nitrogen or carbon dioxide molecule.

The same group showed that the difference in activation volumes[73a] for the disproportionation–combination competition, which is possible for cyclohexyl-*t*-butoxy pairs, is essentially zero. This result was assumed earlier by Stefani[120] in his studies of the effects of temperature and solvent on this competition for ethyl radicals in liquids. There are small differences in the disproportionation–combination ratios which can be correlated with the square root of the internal pressure of the solvent. In isooctane the ratio is identical with that found for the gas phase. Sheldon and Kochi[26] have shown that there are no significant differences in these ratios in solution as compared with the gas phase for any of the extensive series of structurally different radical pairs which they produced from photolysis of peroxides and peresters. Table 5 contains a summary of such results.

The activation energy *difference* for this competition is certainly near zero. Several groups have proposed semiempirical equations for correlating these results on the basis of entropy.[122–124] Konar[121] has pointed out that these

TABLE 5

Disproportionation–Combination Ratios

$$R\cdot + R'\cdot \longrightarrow R(-H) + R'H$$

R	R′	Solvent	(k_d/k_c)	$(k_d/k_c)_{gas}$[h]
Ethyl	Ethyl	Glycol[a]	0.24	0.12
Ethyl	Ethyl	Pentane[b]	0.15	0.14
Ethyl	Ethyl	Decalin[b]	0.12	0.14
Ethyl	*t*-Butoxy	Decalin[c]	0.32	—
1-Propyl	1-Propyl	Decalin[b]	0.13	0.14
1-Butyl	1-Butyl	Decalin[b]	0.13	0.14
2-Propyl	2-Propyl	Decalin[b]	1.2	1.1
2-Propyl	*t*-Butoxy	Decalin[c]	2.4	—
2-Butyl	2-Butyl	Decalin[b]	1.1	1.5
Cyclobutyl	Cyclobutyl	Decalin[b]	0.5	—
t-Butyl	*t*-Butyl	Pentane[b]	7.2	4.5
t-Butyl	*t*-Butoxy	Decalin[c]	19.0	—
Cyanocyclohexyl	Cyanocyclohexyl	Chlorobenzene[d]	0.24	—
Cyclohexyl	*t*-Butoxy	Cumeme[e]	1.4	—
Cumyl	Cumyl	Cumene[f]	0.05	—
2-Carbomethoxy-2-propyl	2-Carbomethoxy-2-propyl	*o*-Dichlorobenzene[g]	0.6	—

[a] 65°, reference 120.
[b] 30°, reference 26.
[c] 30°, reference 71.
[d] 80°, reference 49.
[e] 80°, reference 73.
[f] 50°, reference 29.
[g] 100°, reference 126.
[h] Reference 127.

correlations are not very reliable and has suggested that there may be significant activation energies in both reactions. The competitions do not appear to be *exactly* temperature independent. The primary isotope effects on the disproportionation reaction are small for the reaction of ethyl–ethyl pairs[125] but this result is ambiguous since it is expected for a very exothermic reaction regardless of whether or not an activation energy is involved. Systematic investigation of isotope effects could provide some help in resolving the general problem of whether or not geminate reactions have associated activation energies. The preference for combination of cumyl radicals is significantly higher than the other tertiary radicals exhibit. This could be a result of a π complexation between the members of such pairs which directs the geminate process to this result. It would be interesting to see if unsymmetrically substituted pairs show evidence of such complexation.

REFERENCES

1. (a) J. Frank and E. Rabinowitch, *Trans. Faraday Soc.*, **30**, 120 (1934); (b) E. Rabinowitch and W. Wood, *ibid.*, **32**, 1381 (1936).
2. R. Lyons, *J. Am. Chem. Soc.*, **86**, 1907 (1964).
3. A. E. Richardson and R. Wolfgang, *J. Am. Chem. Soc.*, **92**, 3480 (1970).
4. R. M. Noyes, *J. Am. Chem. Soc.*, **77**, 2042 (1955); **78**, 5486 (1956).
5. J. Calvert and J. Pitts, *Photochemistry*, Wiley, New York, 1966, p. 226.
6. R. M. Noyes, *Z. Electrochem.*, **69**, 153 (1960).
7. R. L. Strong, *J. Am. Chem. Soc.*, **87**, 3563 (1965).
8. S. Kodama, *Bull. Soc. Chem. Japan*, **35**, 658 (1962); **39**, 1009 (1966).
9. R. F. Hutton and C. Steel, *J. Am. Chem. Soc.*, **86**, 745 (1969).
10. I. Abram, G. Milne, B. Solomon, and C. Steel, *J. Am. Chem. Soc.*, **91**, 1220 (1969).
11. W. C. Worsham and O. K. Rice, *J. Chem. Phys.*, **46**, 2021 (1967); E. C. Wu and O. K. Rice, *J. Phys. Chem.*, **72**, 542 (1968).
12. S. S. Collier, D. H. Slater, and J. G. Calvert, *Photochem. Photobiol.*, **7**, 733 (1968); D. H. Slater, S. S. Collier, and J. G. Calvert, *J. Am. Chem. Soc.*, **90**, 268 (1968).
13. J. P. Fisher, G. Mucke, and G. V. Schulz, *Ber. Bunsenges. Phys. Chem.*, **73**, 154 (1969).
14. T. Mill and R. S. Stringham, *Tetrahedron Letters*, **No. 23**, 1853 (1969).
15. N. Porter, M. Landis, and L. J. Marnett, *J. Am. Chem. Soc.*, **93**, 795 (1971).
16. D. Gegion, K. Muszkat, and E. Fischer, *J. Am. Chem. Soc.*, **90**, 12 (1968).
17. Y. Schaafsma, A. Bickel, and E. C. Kooyman, *Tetrahedron*, **10**, 76 (1960).
18. J. C. Martin and J. H. Hargis, papers presented at the 158th National Meeting of the American Chemical Society, Houston, Texas, Feb. 1970.

18a. J. C. Martin and S. A. Dombchick, "Oxidation of Organic Compounds," Vol. 1, *Advances in Chemistry*, Series No. 75, American Chemical Society, Washington, D.C., 1968, pp. 269–281.

18b. J. C. Martin and J. H. Hargis, *J. Am. Chem. Soc.*, **91**, 5399 (1969).

18c. J. C. Martin and J. W. Taylor, *J. Am. Chem. Soc.*, **89**, 6904 (1969).

18d. J. C. Martin, J. W. Taylor, and E. H. Drew, *J. Am. Chem. Soc.*, **89**, 129 (1967); H. J. Shine and J. R. Slayle, *ibid.*, **81**, 6309 (1959).

19. T. Koenig and J. A. Hoobler, unpublished work.
19a. T. Koenig, J. A. Hoobler, and M. Deinzer, *J. Am. Chem. Soc.*, **93**, 938 (1971).
19b. T. Koenig, J. Huntington, and R. Cruthoff, *J. Am. Chem. Soc.*, **92**, 5409 (1970).
19c. T. Koenig and M. Deinzer, *J. Am. Chem. Soc.*, **90**, 7014 (1968).
19d. T. Koenig and R. Cruthoff, *J. Am. Chem. Soc.*, **91**, 2562 (1969).
20. A. Rembaum and M. Szwarc, *J. Chem. Phys.*, **23**, 907 (1955).
21. W. A. Pryor and K. Smith, *J. Am. Chem. Soc.*, **92**, 5403 (1970), and references cited therein.
22. M. Goldstein and J. Hudson, *J. Am. Chem. Soc.*, **92**, 3119 (1970).
23. M. Goldstein and J. Hudson, *J. Am. Chem. Soc.*, **92**, 4120 (1970).
24. H. Kiefer and T. Traylor, *J. Am. Chem. Soc.*, **89**, 6667 (1967).
24a. L. Batt and S. Benson, *J. Chem. Phys.*, **36**, 895 (1962).
24b. C. Walling and H. P. Waits, *J. Phys. Chem.*, **71**, 2361 (1967).
24c. J. Raley, F. Rust, and W. Vaughn, *J. Am. Chem. Soc.*, **70**, 1336 (1948).
25. R. K. Lyon and D. H. Levy, *J. Am. Chem. Soc.*, **83**, 4290 (1961); R. E. Reberts and P. Ausloos, *J. Phys. Chem.*, **66**, 2253 (1962).
26. R. Sheldon and J. Kochi, *J. Am. Chem. Soc.*, **92**, 4395 (1970).
27. S. Seltzer and E. Hamilton, *J. Am. Chem. Soc.*, **88**, 3775 (1966).
28. C. H. Bawn and S. F. Mellish, *Trans. Faraday Soc.*, **47**, 1216 (1951).
29. P. D. Bartlett and S. F. Nelson, *J. Am. Chem. Soc.*, **88**, 139 (1966).
30. P. D. Bartlett and T. Funahashi, *J. Am. Chem. Soc.*, **84**, 2596 (1962).
31. C. F. Koelsch, *J. Am. Chem. Soc.*, **79**, 4439 (1957).
32. R. C. Lamb and J. G. Pacifici, *J. Am. Chem. Soc.*, **86**, 914 (1964).
33. G. S. Hammond, *J. Am. Chem. Soc.*, **72**, 3737 (1950); E. Janzen, F. T. Johnson, and C. C. Ayers, *ibid.*, **89**, 1176 (1967); P. D. Bartlett, R. Pincock, J. H. Rolston, W. G. Schindel, and L. A. Singer, *ibid.*, **87**, 2590 (1965).
34. G. S. Hammond, J. N. Sen, and C. E. Boozer, *J. Am. Chem. Soc.*, **77**, 3244 (1955).
35. P. D. Bartlett, G. N. Fickes, F. C. Haupt, and R. Hegelson, *Acc. Chem. Res.*, **3**, 177 (1970).
36. T. Koenig and J. C. Martin, *J. Am. Chem. Soc.*, **86**, 1771 (1964).
37. H. Shine, J. Waters, and D. Hoffman, *J. Am. Chem. Soc.*, **85**, 3613 (1963).
38. G. Hammond and L. M. Soffer, *J. Am. Chem. Soc.*, **72**, 4711 (1959); C. Walling and R. B. Hodgdon, *ibid.*, **80**, 228 (1958).
39. O. D. Trapp and G. S. Hammond, *J. Am. Chem. Soc.*, **81**, 4876 (1959).
40. M. Levy and M. Szwarc, *J. Am. Chem. Soc.*, **76**, 5982 (1954).
41. J. C. Bevington, *Trans. Faraday Soc.*, **51**, 1392 (1955); G. Ayrey, *Chem. Rev.*, **63**, 645 (1963).
42. M. Levy and M. Szwarc, *J. Am. Chem. Soc.*, **77**, 1949 (1955).
42a. J. Gresser, A. Rajenbach, and M. Szwarc, *J. Am. Chem. Soc.*, **83**, 3005 (1961).
42b. L. S. Silbert, D. Swern, and T. Asahara, *J. Org. Chem.*, **33**, 3670 (1968).
43. E. G. Janzen, *Accts. Chem. Res.*, **4**, 31 (1971).
44. E. Janzen and B. Blackburn, *J. Am. Chem. Soc.*, **91**, 4481 (1969); C. Lagercrantz and S. Forschult, *Acta Chem. Scand.*, **23**, 522, 708, 811 (1969); I. H. Leaver, G. C. Ramsay, and E. Suzuki, *Aust. J. Chem.*, **22**, 1891, 1899 (1969); S. Terabe and R. Konaka, *J. Am. Chem. Soc.*, **91**, 5655 (1969).
44a. P. D. Bartlett and J. Lorand, *J. Am. Chem. Soc.*, **88**, 3294 (1966).
45. R. C. Lamb, F. F. Rogers, G. C. Dean, and F. W. Voight, *J. Am. Chem. Soc.*, **84**, 2635 (1962).
46. F. E. Herkes, J. Friedman, and P. D. Bartlett, *Int. J. Chem. Kinet.*, **1**, 193 (1969).

47. W. F. Smith and K. L. Eddy, *Tetrahedron*, **26**, 1255 (1970).
48. G. Closs, *J. Am. Chem. Soc.*, **91**, 4552 (1969).
48a. R. Kaptein and R. Oosterhoff, *Chem. Phys. Letters*, **4**, 195, 214 (1969).
49. G. S. Hammond and H. P. Waits, *J. Am. Chem. Soc.*, **86**, 1911 (1964); C.-H. S. Wu, G. S. Hammond, and J. M. Wright, *ibid.*, **82**, 538 (1960).
49a. G. S. Hammond, C.-H. S. Wu, O. D. Trapp, J. Warkentin, and R. T. Keys, *J. Am. Chem. Soc.*, **82**, 5394 (1960).
50. J. C. Roy, J. R. Nash, R. R. Williams, and W. A. Hamill, *J. Am. Chem. Soc.*, **78**, 519 (1956); M. Talat-Erben and A. N. Isfendiyroĝlu, *Can. J. Chem.*, **37**, 1165 (1959).
51. A. Rembaum and M. Szwarc, *J. Am. Chem. Soc.*, **77**, 3486 (1955).
52. J. C. Roy, R. R. Williams, and W. H. Hamill, *J. Am. Chem. Soc.*, **76**, 3274 (1954).
53. H. A. Gillis, R. R. Williams, and W. H. Hamill, *J. Am. Chem. Soc.*, **83**, 17 (1961); J. R. Nash, R. R. Williams, and W. A. Hamill, *ibid.*, **82**, 5974 (1960).
54. C. G. Overberger and A. V. Rosenthal, *J. Am. Chem. Soc.*, **82**, 108 (1962).
55. R. M. Noyes, *Prog. React. Kinet.*, **1**, 129 (1961).
56. T. Koenig, *J. Am. Chem. Soc.*, **91**, 2558 (1969).
57. A. Stefani, G. Thrower, and C. F. Jordan, *J. Phys. Chem.*, **73**, 1257 (1969).
58. B. Booth and R. M. Noyes, *J. Am. Chem. Soc.*, **82**, 1868 (1960).
59. S. Kodama, *Bull. Soc. Chem. Japan*, **35**, 824, 827 (1962).
60. O. Dobis, J. M. Pearson, and M. Szwarc, *J. Am. Chem. Soc.*, **90**, 278 (1968).
61. K. Chakravorty, J. M. Pearson and M. Szwarc, *J. Am. Chem. Soc.*, **90**, 283 (1968); *Int. J. Chem. Kinet.*, **1**, 357 (1969).
62. M. Szwarc, papers presented at the 160th National Meeting of the American Chemical Society, Chicago, Illinois, Sept. 1970.
63. C. A. Emeis and P. L. Fehder, *J. Am. Chem. Soc.*, **92**, 2246 (1970).
64. W. Braun, L. Rajenbach, and F. Eirich, *J. Phys. Chem.*, **66**, 1591 (1961).
65. A. North, *Q. Rev.* (*London*), **20**, 421 (1966).
66. L. Herk, M. Feld, and M. Szwarc, *J. Am. Chem. Soc.*, **83**, 2998 (1961).
67. L. Meadows and R. M. Noyes, *J. Am. Chem. Soc.*, **82**, 1872 (1960).
68. W. K. Robins and R. H. Eastman, *J. Am. Chem. Soc.*, **92**, 6077 (1970).
69. C. H. Chang, R. C. Porter, and S. H. Bauer, *J. Am. Chem. Soc.*, **92**, 5313 (1970).
70. R. W. Fessenden and R. H. Schuler, *J. Chem. Phys.*, **43**, 2704 (1965).
71. R. Sheldon and J. Kochi, *J. Am. Chem. Soc.*, **92**, 5175 (1970).
72. B. A. Gontarev, H. Sakurai, and P. D. Bartlett, *J. Am. Chem. Soc.*, **82**, 3101 (1962).
73. R. C. Neuman and J. V. Behar, *J. Am. Chem. Soc.*, **91**, 6024 (1969); R. C. Neuman and E. S. Alhadeff, *J. Org. Chem.*, **35**, 3401 (1970).
73a. R. C. Neuman and R. G. Bussey, *J. Am. Chem. Soc.*, **92**, 2440, 7234 (1970).
73b. R. C. Neuman and J. V. Behar, *J. Org. Chem.*, **36**, 657 (1971).
73c. P. D. Bartlett and C. Rüchardt, *J. Am. Chem. Soc.*, **82**, 1756 (1960).
74. R. D. Burkhart, *J. Am. Chem. Soc.*, **90**, 273 (1968).
74a. R. D. Burkhart and J. C. Merrill, *J. Phys. Chem.*, **73**, 2699 (1969).
74b. R. D. Burkhart and J. C. Merrill, *J. Phys. Chem.*, **73**, 2703 (1969).
75. S. A. Weiner and G. S. Hammond, *J. Am. Chem. Soc.*, **91**, 987 (1969); T. Fujisuwa, B. Monroe, and G. Hammond, *ibid.*, **92**, 542 (1970); G. S. Hammond and S. A. Weiner, *Intra-Sci. Chem. Rep.*, **3**, 241 (1969).
76. D. J. Carlsson and K. U. Ingold, *J. Am. Chem. Soc.*, **90**, 7047 (1968).
77. W. G. Bentrude and A. K. MacKnight, *J. Am. Chem. Soc.*, **92**, 5259 (1970).
77a. C. Walling and P. J. Wagner, *J. Am. Chem. Soc.*, **86**, 3368 (1968).
78. M. Szwarc, *J. Phys. Chem.*, **68**, 389 (1962).
79. S. Levison and R. M. Noyes, *J. Am. Chem. Soc.*, **86**, 4525, 4529 (1964).

80. H. W. Melville, J. C. Robb, and R. C. Tutton, *Discuss. Faraday Soc.*, **14**, 150 (1954).
81. D. J. Carlsson, J. A. Howard, and K. U. Ingold, *J. Am. Chem. Soc.*, **88**, 4726 (1966).
82. S. A. Weiner and G. S. Hammond, *J. Am. Chem. Soc.*, **91**, 2182 (1969); M. Symons, *ibid.*, **91**, 5924 (1969).
83. D. J. Carlsson, J. A. Howard, and K. U. Ingold, *J. Am. Chem. Soc.*, **88**, 4725 (1966).
84. C. Walling, *Free Radicals in Solution*, Wiley, New York, 1957, pp. 531–533.
85. L. Bertrand, G. R. DeMare, G. Huybrechts, J. Olbregts, and M. Toth, *Chem. Phys. Letters*, **5**, 183 (1970).
86. J. R. Fox and G. S. Hammond, *J. Am. Chem. Soc.*, **86**, 4031 (1964).
87. T. Nakata, K. Tokumaru, and O. Simamaru, *Bull. Soc. Chem. Japan*, **43**, 3315 (1970); *Tetrahedron Letters*, 3303 (1967); C. Walling and M. J. Gibian, *J. Am. Chem. Soc.*, **87**, 3413 (1965).
88. J. Saltiel and H. C. Curtis, *J. Am. Chem. Soc.*, **93**, 2056 (1971).
89. H. C. Brown, M. S. Kharasch, and T. H. Chao, *J. Am. Chem. Soc.*, **62**, 3435 (1940).
90. S. F. Nelson and P. D. Bartlett, *J. Am. Chem. Soc.*, **88**, 143 (1966).
91. P. S. Engel and P. D. Bartlett, *J. Am. Chem. Soc.*, **92**, 5883 (1970).
92. G. L. Closs and A. D. Trifunac, *J. Am. Chem. Soc.*, **91**, 4554 (1969).
93. R. Kaptein, J. Hollander, D. Antheunis, and L. J. Oosterhoff, *J. Chem. Soc.* (*D*), 1687 (1970).
94. A. M. Trozzolo and S. Farenholz, *J. Am. Chem. Soc.*, **93**, 251 (1971).
95. P. S. Engel, *J. Am. Chem. Soc.*, **92**, 6074 (1970).
96. R. C. Fort and P. Schleyer, *Adv. Free Radical Chem.*, **1**, 284 (1966); M. Symons, *Nature*, **222**, 1123 (1969).
97. C. Wai, C. T. Ting, and F. S. Rowland, *J. Am. Chem. Soc.*, **86**, 2525 (1964); C. Wai and F. S. Rowland, *J. Phys. Chem.*, **74**, 434 (1970).
98. K. Wiberg and G. Foster, *J. Am. Chem. Soc.*, **83**, 423 (1961); K. Wiberg and A. Fox, *ibid.*, **85**, 3487 (1963).
99. P. D. Bartlett and J. M. McBride, *Pure Appl. Chem.*, **15**, 89 (1967).
100. K. R. Kopecky and T. Gillian, *Can. J. Chem.*, **47**, 2371 (1969).
101. F. D. Greene, *J. Am. Chem. Soc.*, **77**, 4689 (1955).
102. T. Kashiwagi, K. Fujmori, and S. Oae, *Tetrahedron*, **26**, 3639 (1970).
103. H. M. Walborsky and J. Chen, *J. Am. Chem. Soc.*, **93**, 671 (1971).
104. M. J. S. Dewar and J. M. Harris, *J. Am. Chem. Soc.*, **91**, 3652 (1969); J. Jacobus and P. Pensak, *Chem. Commun.*, 400 (1969); L. A. Singer and J. Chen, *Tetrahedron Letters*, **No. 14**, 939 (1971); T. Shono, M. Akashi, and R. Oda, *ibid.*, 1507 (1968).
105. L. J. Altman and B. W. Nelson, *J. Am. Chem. Soc.*, **91**, 5163 (1969); *Tetrahedron Letters*, 4891 (1970).
106. J. Meisenheimer, *Chem. Ber.*, **52**, 1668 (1919).
107. U. Schöllkopf, M. Putsch, and H. Schäfer, *Tetrahedron Letters*, 2515 (1964); U. Schöllkopf and H. Schäfer, *Ann.*, **683**, 42 (1965).
108. A. R. Lepley, P. M. Cook, and G. F. Willard, *J. Am. Chem. Soc.*, **92**, 1101 (1970).
109. John P. Lorand, R. Grant, P. Samuel, E. O'Connell, and J. Zaro, *Tetrahedron Letters*, **No. 46**, 4087 (1969).
110. T. S. Stevens, E. M. Creighton, A. B. Gordon, and M. MacNicol, *J. Chem. Soc.*, 3192 (1928).
111. A. R. Lepley, *J. Am. Chem. Soc.*, **91**, 1237 (1969).
112. E. F. Jenney and J. Druey, *Angew. Chem., Int. Ed.*, **1**, 155 (1962).

113. H. Ward, R. Lawler, and R. A. Cooper, *J. Am. Chem. Soc.*, **91**, 747 (1969).
114. R. L. Letsinger, *J. Am. Chem. Soc.*, **72**, 4842 (1950).
115. W. D. Korte, L. Kinner, and W. C. Kaska, *Tetrahedron Letters*, 603 (1970).
116. C. Walling and A. Lepley, *J. Am. Chem. Soc.*, **93**, 546 (1971).
117. D. F. DeTar and C. Weis, *J. Am. Chem. Soc.*, **79**, 3041 (1957).
118. F. D. Greene, M. A. Berwick, and T. C. Stowell, *J. Am. Chem. Soc.*, **92**, 867 (1970).
119. P. S. Dixon, A. P. Stefani, and M. Szwarc, *J. Am. Chem. Soc.*, **85**, 2551, 3344 (1963).
120. A. P. Stefani, *J. Am. Chem. Soc.*, **90**, 1694 (1968).
121. R. S. Konar, *Int. J. Chem. Kinet.*, **2**, 419 (1970).
122. J. N. Bradley, *J. Chem. Phys.*, **35**, 748 (1961).
123. A. F. Trotman-Dickenson, *Proc. Chem. Soc.* (*London*), 249 (1964).
124. P. Gray, R. Shaw, and J. C. J. Thynne, *Prog. React. Kinet.*, **4**, 90 (1967).
125. K. Watkins and H. Moser, *J. Phys. Chem.*, **70**, 1137 (1966); J. H. Current and B. S. Rabinowitch, *J. Chem. Phys.*, **38**, 783 (1963).
126. D. J. Trecker and R. S. Foote, *J. Org. Chem.*, **33**, 3527 (1968).
127. J. A. Kerr and A. F. Trotman-Dickenson, *Prog. React. Kinet.*, **1**, 107 (1961).

CHAPTER 5

Diradicals: A Case Study of Trimethylene

ROBERT G. BERGMAN

Division of Chemistry and Chemical Engineering, California Institute of Technology, Pasadena, California

I. INTRODUCTION

A large part of the goal of modern mechanistic chemistry has been the identification of generalizable and transferable aspects of the bond cleavage and re-formation process. It is now known that many very complex overall transformations begin with a dissociation reaction, and it has been found that dissociations in general can be classified into two broad areas: heterolytic (equation 1), in which the paired electrons of the chemical bond eventually reside on one atom, and homolytic (equation 2), in which each of the bound atoms escapes with one of the bonding electrons. The former process generates intermediates which are involved in reactions generally termed "ionic," and the latter is responsible for much of the free radical chemistry known today.

However, although much is known about the structure and reactions of complex ions and free radicals, chemical species which lie in between the completely bound and completely dissociated states are at present much less well understood. Because the lifetimes of such species are normally extremely short, one of the most profitable ways to examine them is to somehow slow the rates at which they move into the bound or dissociated manifold. This has typically been accomplished in two ways. The first is to carry out the dissociation in solvent, where there appears to be some delay in escape of partially dissociated fragments from the "solvent cage." Using this technique it has been possible to identify and study species now referred to as "ion pairs"[1] and "caged radical pairs"[2] (Chapter 4).

$$\mathrm{A{-}B} \rightleftharpoons \underset{\text{Ion pair}}{\mathrm{A^+B^-}} \longrightarrow \mathrm{A^+ + B^-} \tag{1}$$

$$\mathrm{A{-}B} \rightleftharpoons \underset{\text{Radical pair}}{\overline{\mathrm{A\cdot B\cdot}}} \longrightarrow \mathrm{A\cdot + B\cdot} \tag{2}$$

The second technique used to restrict such reactive species is to connect the dissociating atoms with a relatively inert molecular chain. This normally prevents complete escape of the dissociating atoms, and allows one to study reactions which are competitive with re-formation of the chemical bond. It is these intermediates which are most frequently referred to as "diradicals," and it is their structure and chemical behavior with which this chapter is concerned.

$$\overset{\frown}{\mathrm{A{-}B}} \rightleftharpoons \underset{\text{Diradical}}{\overset{\frown}{\mathrm{A\cdot\ B\cdot}}} \longrightarrow \text{products} \tag{3}$$

II. THEORETICAL DESCRIPTION OF DIRADICALS

One of the most useful parameters which characterizes a diradical species is the magnitude of the interaction between the "radical" electrons. From an experimental point of view, molecules with widely varying degrees of interaction have been recognized. At one extreme, we have species such as bisgalvinoxyl (**1**), in which the energy of interaction of the two odd electrons is

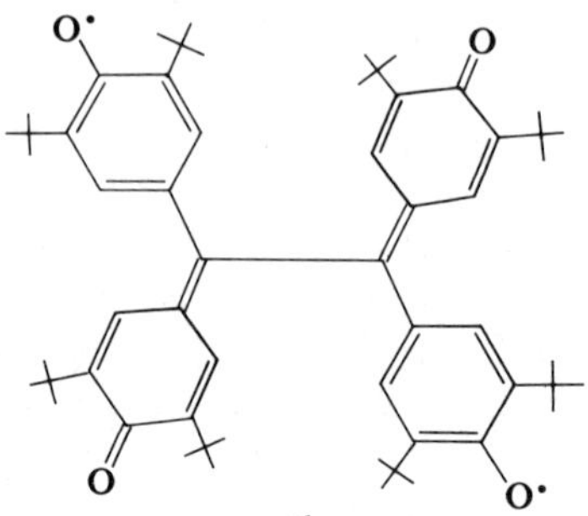

1

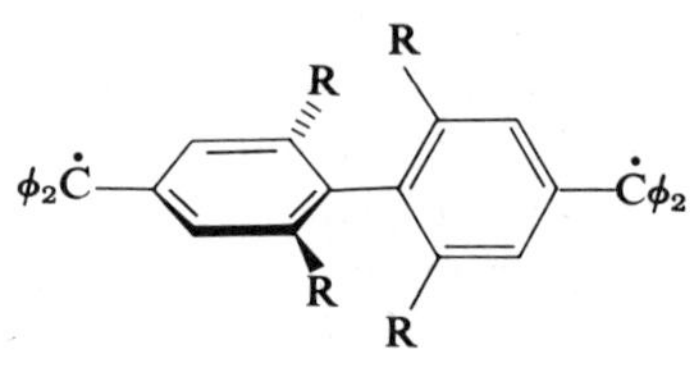

2

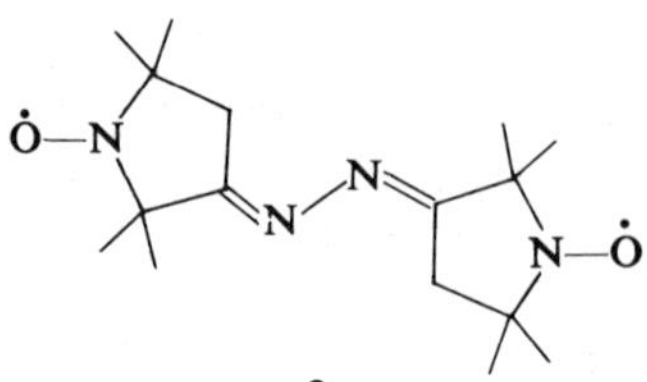

3

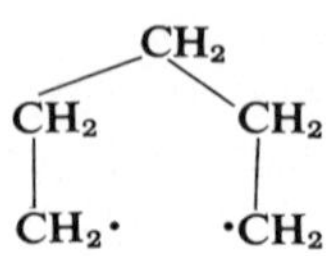

4

smaller than the hyperfine interaction between the odd electrons and nuclear spins. Such a molecule gives an electron spin resonance (ESR) spectrum characteristic of two doublet states.[3,7] A second case, in which electron–electron interaction is greater than hyperfine coupling, is represented by **3**, which shows only a well-defined triplet spectrum in the ESR.[4] Chichibabin's hydrocarbon (**2**) represents an intermediate case in which electron–electron interaction is greater than it is in **1**, but both the singlet and triplet are spectroscopically detectable.[5] Flexible systems such as **4**, formed by cleavage of a hydrocarbon ring, can in principle exist in conformations having either strong or weak electron–electron interaction.

An examination of the wave-mechanical picture of the interaction of two unpaired electrons is helpful in understanding the nature of available electronic states and also presents a useful picture of charge distribution in diradicals. Wilhite and Whitten[6] and Salem and Rowland[7] have presented molecular orbital descriptions of this two-electron interaction. Consider each of the electrons to be situated in similar chemical environments (Salem's "homosymmetric" case[7]), described initially by wave functions n_1 and n_2. It is possible to construct two molecular orbitals[6,7] using linear combinations of n_1 and n_2, one of which is symmetric (ψ_+) and the other asymmetric (ψ_-):

$$\psi_+ = \mathrm{N}(n_1 + n_2) \tag{4}$$

$$\psi_- = \mathrm{N}'(n_1 - n_2) \tag{5}$$

There are then three electronic configurations which must be considered. Double occupancy of ψ_+ represents the ground singlet configuration (S_0) when the two electrons are strongly bound. Double occupancy of ψ_- represents a doubly excited singlet (S_2) in the strong-binding situation. Placing one electron in ψ_+ and the other in ψ_- produces both an additional singlet (S_1 and a triplet (T) configuration.

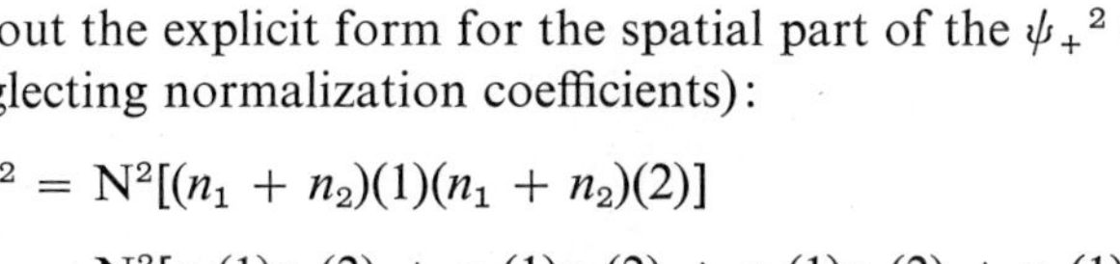

$\psi_+{}^2(S_0)$ $\psi_-{}^2(S_2)$ $^1(\psi_+\psi_-)(S_1)$ $^3(\psi_+\psi_-)(T)$

Writing out the explicit form for the spatial part of the $\psi_+{}^2$ configuration, we get (neglecting normalization coefficients):

$$\begin{aligned}\psi_+{}^2 &= \mathrm{N}^2[(n_1 + n_2)(1)(n_1 + n_2)(2)] \\ &= \mathrm{N}^2[n_1(1)n_2(2) + n_2(1)n_1(2) + n_1(1)n_1(2) + n_2(1)n_2(2)] \\ &= \psi_{\text{covalent}} + \psi_{\text{ionic}}\end{aligned} \tag{6}$$

It is evident that the first two terms of this wave function are covalent contributions to the two-electron interaction, and the last two terms are ionic contributions. This leads to one of the basic difficulties of molecular orbital theory: The ionic contributions generated for a two-electron bond are too high.[8] Furthermore, the situation becomes even worse as the AB bond begins to break and the overlap integral (S_{AB}) decreases in magnitude. Because of the ionic terms in equation 6, the energy of configuration $\psi_+{}^2$ rises much faster than does the true energy of the ground state as dissociation proceeds, leading to an erroneous description of the dissociation process.[7]

There are two ways to avoid this inaccurate picture. The first is to apply a valence bond (VB) (rather than molecular orbital) theory to the problem. In the valence bond approach, ionic terms can be left out of the ground singlet configuration, and there is no specific requirement for double occupancy of orbitals. It is therefore possible to construct one orbital for each electron (instead of one for each pair of electrons), and a state made up of such one-electron orbitals dissociates properly.[9] The other technique, which improves the molecular orbital picture of dissociation, mixes the $\psi_+{}^2$ configuration with the second configuration of the same symmetry, $\psi_-{}^2$, in such a manner that the ionic terms subtract. This is done by taking the negative linear combination

$$\lambda\psi_+{}^2 - (1 - \lambda^2)^{1/2}\psi_-{}^2$$

This is referred to as the configuration interaction (CI) approach. In the limit where the overlap integral goes to zero and the two configurations have equal energy ($\lambda = 1/\sqrt{2}$), we recover the purely covalent singlet:

$$\frac{1}{\sqrt{2}}(\psi_+{}^2 - \psi_-{}^2) = \frac{1}{\sqrt{2}}[n_1(1)n_2(2) + n_1(1)n_2(2)]$$

Similarly, $\psi_-{}^2$ is not an appropriate function for the doubly excited state at small S_{AB}, and is also represented by a linear combination of itself and the ground singlet $\psi_+{}^2$:

$$\psi_+{}^2(1 - \lambda^2)^{1/2} + \lambda\psi_-{}^2$$

Writing out the explicit form of S_1 (required to be symmetric with respect to exchange of electrons) shows that it is a completely ionic state:

$$^1(\psi_+\psi_-) = N[n_1(1)n_1(2) + n_2(1) + n_2(2)]$$

whereas the triplet is written explicitly as

$$^3(\psi_+\psi_-) = N[n_1(1)n_2(2) - n_2(1)n_1(2)]$$

and inspection of the terms in the wave function shows that this state is essentially completely covalent. The same spatial wave function appears in the three possible triplet configurations ($\alpha\alpha$, $\beta\beta$, and $\alpha\beta + \beta\alpha$).

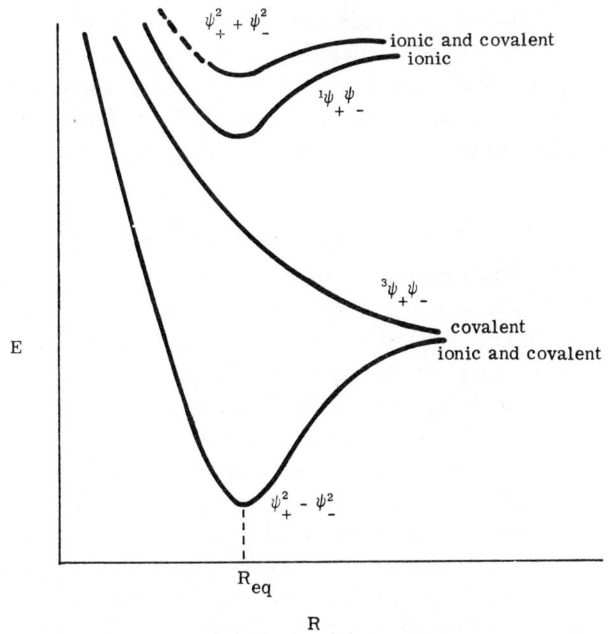

Fig. 1. Illustration of the dependence of potential energy on interorbital distance for the four state configurations generated in a two-electron system.

Illustrated in Figure 1 are the forms of typical potential functions for the states to be expected for the two-electron bonding problem; for example, the hydrogen molecule or an isolated covalent bond. The two lowest-lying states are the covalent triplet and mixed singlet; as Salem points out, which of these two is the ground state for the diradical will depend principally on the local energies of the two electrons, relative magnitudes of the overlap and exchange integrals between them, and the magnitude of the interorbital repulsion integrals.[7] One of the important lessons of the theoretical analysis, however, is that certain states are constructed from combinations of functions containing *both* covalent and ionic character. Thus it may often be meaningless to ask whether a given intermediate is a "diradical" or "zwitterion"; even from a molecular orbital viewpoint it may have the characteristics of both.

III. CASE HISTORY OF A DIRADICAL: TRIMETHYLENE

There now exists a very large number of reactions thought to proceed through diradical intermediates, and an exhaustive survey of these reactions is beyond the scope of this chapter. The following section, therefore, concentrates in some depth on the structure and chemistry of only one diradical—

trimethylene—which has been subjected to intensive scrutiny during recent years. Following that, a few other diradical intermediates are briefly discussed. It is hoped that the concepts and problems mentioned with regard to trimethylene will be of sufficient generality that the interested reader can obtain an understanding of phenomena extending beyond a single chemical system.

A. Early Experimental Work

The early 1930s was a period of great interest in fundamental free radical chemistry. Paneth had shown in 1929 that alkyl free radicals could be generated from metal alkyls, and that these intermediates had short but finite lifetimes.[10] In the years immediately following, Rice and his co-workers discovered that thermal decomposition of many organic molecules produces free radicals,[11] and in 1935 Rice and Herzfeld showed how free radical chain mechanisms could lead to the simple overall kinetic rate laws observed in the thermal decompositions of many organic compounds.[12] It was during this period that Chambers and Kistiakowsky carried out[13] the first extensive study of the kinetics of decomposition of cyclopropane (**5**). They found that the reaction was unimolecular and that propylene (**7**) was its sole product. They also suggested that the trimethylene diradical (**6**) might be an intermediate.[13] The conceptual history of trimethylene, therefore, is nearly as long as that of simple alkyl monoradicals.

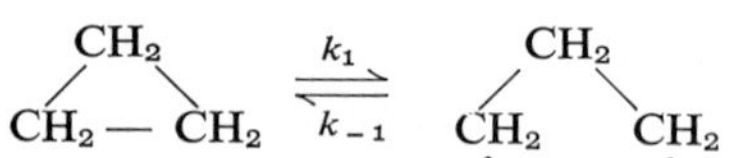

5 **6** **7**

Scheme 1

Since this initial study, the cyclopropane pyrolysis has assumed a central role in the field of reaction kinetics, due principally to its frequent use as a model for the testing of various theories of unimolecular reactions.[14] Little progress was made in elucidating the intimate details of the reaction mechanism, however, until Rabinovitch et al.[15] examined the thermal isomerization of trans-cyclopropane-d_2 (**8**). This molecule was found to undergo conversion to the cis isomer **9** somewhat more rapidly than it formed propylene-d_2, demonstrating that a pathway existed which was capable of producing rotation about at least one of the cyclopropane C—C bonds. This was exactly the

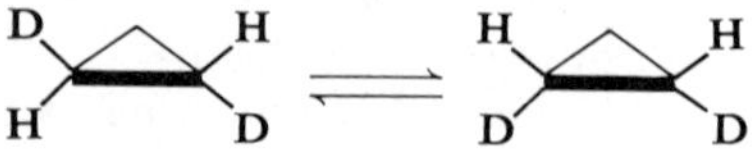

8 **9**

sort of behavior expected of trimethylene-d_2, and this experiment therefore constituted the first clear-cut evidence that a diradical intermediate was involved in the cyclopropane pyrolysis. Stereomutation has since been found to occur in other appropriately substituted cyclopropanes.[16]

B. Thermochemistry

An accurate evaluation of the energy required to interconvert cyclopropane and propylene with the proposed trimethylene intermediate is crucial to the understanding of the cyclopropane rearrangement mechanism. Most directly measurable are the overall activation energies (E_a) and frequency factors (A) for geometric (e.g., **8** → **9**) and structural (cyclopropane → propylene) isomerizations. A number of groups have obtained the parameters for the latter isomerization[17]; within experimental error the agreement is quite good and the high-pressure rate constant is given by:

$$k_s = A \exp(E_a/RT) = 10^{15.45} \exp(-65{,}600/RT)$$

Geometric isomerization occurs more rapidly; the rate expression is

$$k_g = 10^{16.0} \exp(-64{,}200/RT)$$

It is clear from these experiments that the energy required to "break" a cyclopropane C—C bond is considerably less than that required to break a corresponding bond in a normal alkane. This is attributed to destabilization of the cyclopropane ground state due to ring strain. However, it is of some interest that the difference between the activation energy for cyclopropane cleavage and that required to break, for example, the secondary–secondary bond in n-butane (18 kcal/mole[18]) is not as large as the total "strain energy" of cyclopropane (28 kcal/mole[18]). This raises the possibility that the transition state for three-membered ring opening is not completely strain free. The normal assumption that recombination of alkyl radicals takes place with essentially zero activation energy[19] therefore might not hold for trimethylene; that is, ring closure to cyclopropane (though formally a radical recombination process) could require some thermal activation.

In order to investigate this question, Benson and co-workers applied techniques developed to predict free radical and hydrocarbon thermochemistry to the trimethylene problem.[20a] The method and a number of examples are covered in Chapter 17 of Volume II, and so only its application to trimethylene is discussed here.

A possible schematic reaction coordinate is pictured in Figure 2; it refers to the simple mechanism outlined in Scheme 1. The activation enthalpy ΔH_1 is assumed to be approximated by the activation enthalpy for geometrical isomerization (64 kcal/mole[15]). In order to estimate the ring-closure barrier

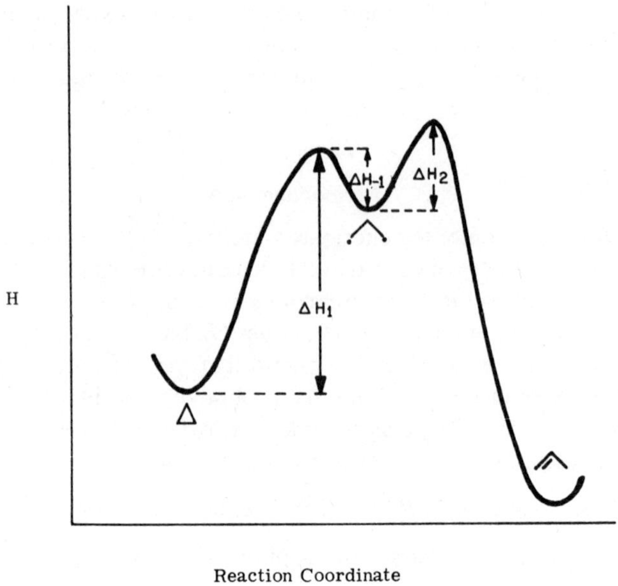

Fig. 2. Possible reaction coordinate for the isomerization of cyclopropane.

(ΔH_{-1}), therefore, one needs to know the heat of formation of trimethylene. This number is estimated in the following way: One can set up the simple thermodynamic cycle illustrated in Scheme 2. The ring-opening energy, ΔH°_{ro}, must be equal to the energy required to first hydrogenate cyclopropane to propane (ΔH_{hyd}) and then remove two hydrogen atoms from the primary

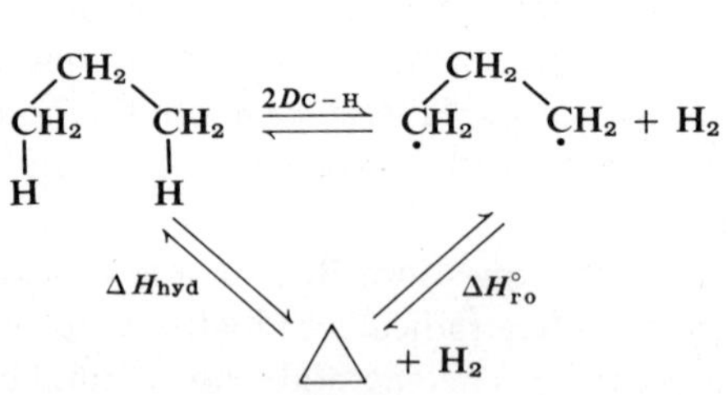

Scheme 2

carbons of propane; ΔH_{hyd} can either be measured directly or calculated using the difference in heats of formation of cyclopropane and propane. Assuming (vide infra) that the methyl C—H bond strengths in propane and propyl are identical and equal to 98 kcal/mole,[18] we can calculate ΔH°_{ro} as follows:

$$\begin{aligned}\Delta H^\circ_{ro} &= \Delta H_{hyd} + 2D_{C-H} - D(H_2)\\ &= \Delta H^\circ_{propane} - \Delta H^\circ_{cyclopropane} + 2D_{C-H} - D(H_2)\\ &= -25.1 - 12.7 + 196 - 104\\ &= 54 \text{ kcal/mole}\end{aligned}$$

(This is equivalent to calculating a heat of formation of +67 kcal/mole for trimethylene.[20a]) The value of ΔH°_{ro} calculated in this way is substantially (10 kcal/mole) lower than the 64 kcal/mole energy for ring opening measured kinetically.[16] This discrepancy is assumed to correspond to the activation energy for the reverse reaction.[20a] The estimate therefore predicts a substantial barrier to ring closure for trimethylene.

TABLE 1

Thermochemical–Kinetic Calculations of Ring-Closure Barriers in Some Possible Diradical Intermediates[20b]

Diradical	ΔH_{-1} (kcal/mole)
$\cdot CH_2CH_2CH_2\cdot$	10
$\cdot CH_2CH_2CH_2CH_2\cdot$	6.5
$CH_2{=}CH{-}\dot{C}H{-}CH_2\dot{C}H_2$	8.5
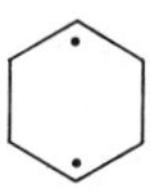	8.2
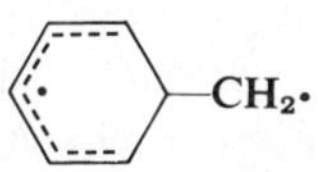	24
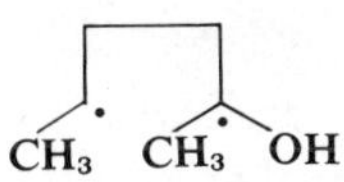	8
	8

Similar calculations can be carried out for other diradical species; the only information required is the activation energy for ring opening, the heats of formation of the appropriate cyclic hydrocarbon and the corresponding straight-chain analog, and the relevant C—H bond energies. Table 1 lists a number of barriers to ring closure calculated in this way.[20b] These and other calculated diradical ring-closure energies are remarkably similar—nearly all the values are between 6 and 10 kcal/mole, despite relatively significant differences in ΔH°_{ro} and in the strain energies of the ring-closed species. The one significant deviation is the diradical formed by 1,7 bond cleavage in norbornadiene, which is calculated to have a surprisingly large 24 kcal/mole barrier preventing ring closure.[20b]

It must be remembered that the critical assumption in these calculations is that the terminal C—H bond energy in a monoradical of general structure **11**

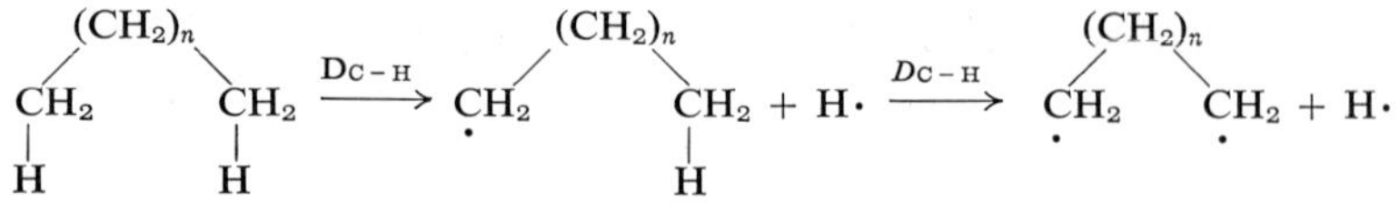

10 **11** **12**

is identical with the primary C—H bond energy in the hydrocarbon **10**.[20b] While one might perhaps make a case for trimethylene ($n = 1$) that the radical center at C-1 in **11** has a detectable effect on the C—H bond energy at C-3, it seems unlikely that such an effect would be appreciable in tetramethylene, where two methylene groups insulate the terminal centers from one another. In any case, it is amusing to extend this thermodynamic–kinetic estimation technique to one of its logical extremes, where $n = 0$. This corresponds to cis–trans isomerization in ethylene, and once again a thermodynamic cycle can be constructed[21] (Scheme 3). If we define the "diradical" as some form of ethylene with zero interaction between the two π electrons, a close approximation to its structure is probably the twisted singlet diradical **13**. Its heat of formation can be estimated[21] in a manner analogous to that used for other derivatives of **12**. This gives $\Delta H^\circ_\pi = +59$ kcal/mole. However,

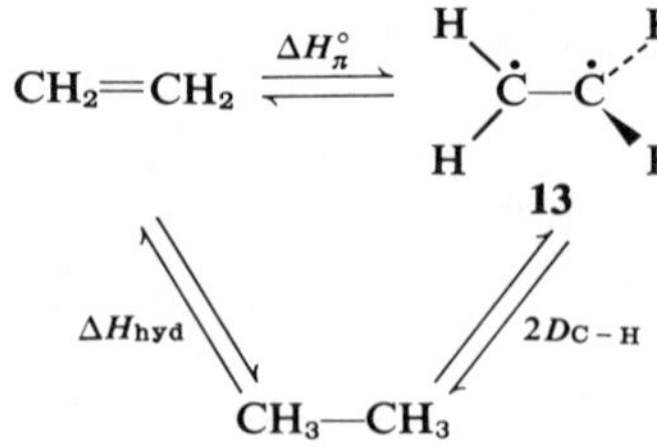

Scheme 3

the activation energy for cis–trans isomerization of 1,2-dideuterioethylene is 64 kcal/mole.[22] As with the hydrocarbon diradicals, then, this suggests that there is a substantial (5 kcal/mole) activation energy for the reverse reaction (**13** → ethylene)—which in this case is merely rotation of one of the methylene groups into the plane of the other.

It is difficult to believe that an appreciable activation energy is associated with such a simple, highly exothermic process. For this reason, Benson and O'Neal claim[21] that ΔH_π° is not the enthalpy associated with the ethylene → **13** process, but rather that associated only with the conversion of ethylene to the "flat" diradical **14**. In order to account for the discrepancy between the calculated ΔH_π° and the measured E_a for the ethylene isomerization, Benson and O'Neal postulate[21] that there is a 5 kcal/mole barrier to rotation in exactly the opposite direction: that is, for the conversion of **14** into the twisted conformation **13**. This barrier is suggested despite the reasonable

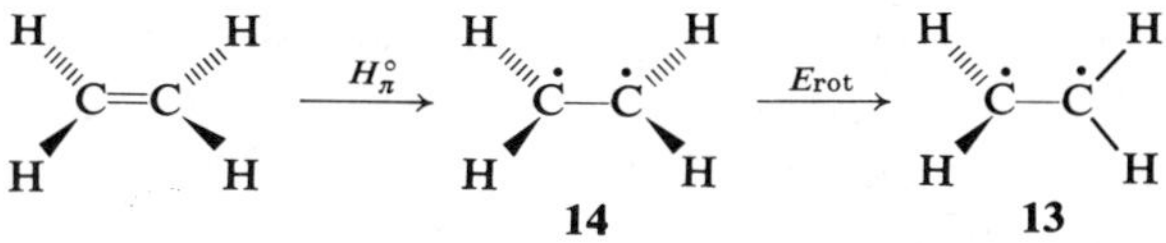

$$\Delta H^\circ_{\text{total}} = \Delta H_\pi^\circ + E_{\text{rot}}$$

presumption that **13**, in a staggered conformation, is sterically more favorable than the eclipsed **14**. The energy E_{rot} is accounted for[21] by reference to the fact that there is a 4 kcal/mole energy barrier to rotation about the "single" bond in butadiene, and the authors postulate that the same effect is present in planar diradical **14**, thus stabilizing it relative to **13**. We therefore have the interesting suggestion that **14**, which is *defined* as having no interaction between the two π electrons, has a residual resonance energy. This argument is not very much more satisfying than the presumption of an activation energy for the **13** → ethylene reaction. Only one other possible alternative is offered to explain the 5 kcal/mole discrepancy: "this may be . . . simply . . . resistance of the sp^2–sp^2 carbon to torsion." The physical source of this resistance is left unspecified.[21]

Theory provides support for the intuition that there should be no barrier for the reversion of **13** to ethylene. Rotation of the carbon atoms in ethylene has been extensively investigated by ab initio theoretical methods,[23] and the twisted diradical is predicted not to lie in an energy well. It is of some interest that ab initio theoretical techniques have also recently been applied to trimethylene. Nearly all of them predict little or no barrier to the lowest energy path for ring closure (see Section III.C). There seems to be a definite conflict, therefore, in the predictions of thermodynamics and theory concerning barriers to ring closure (or π bond formation) for both hydrocarbon and olefin diradicals, and the discrepancy always seems to be of nearly the same magnitude: 5–10 kcal/mole. It is probably important to emphasize, however, that as a predictive tool or model, thermodynamic–kinetic calculations are undeniably useful. A large number of rates of unimolecular reactions have been successfully calculated using the method,[20b,21] and they have made a distinct contribution to the diradical field by emphasizing the role that entropic effects play in restricting the rates of rotation about carbon–carbon bonds in substituted diradicals.[20a,36] A final decision on the existence of a barrier to ring closure in simple diradicals must therefore await the test of experiment.

C. Theoretical Studies

Little was known until recently about the calculated potential energy surface for ring opening of cyclopropane. In the last three years, however, stimulated perhaps by the success of orbital symmetry principles, an early and provocative paper by Hoffmann,[24] and the continuing improvement in the application of computational quantum chemical techniques to complex systems, a significant amount of theoretical work has appeared on this problem. Groups which have devoted considerable effort to the cyclopropane ring-opening question include those of Hoffmann, Salem, Hayes, Goddard, and Pople. An attempt will be made here to give a brief summary of some of the computational results.

Before 1968, there was almost no discussion in print concerning the various possible conformational alternatives for the ground state of trimethylene. Normally the intermediate was implicitly treated as though the terminal methylene groups were capable of relatively rapid rotation, and Benson's early attempts to estimate enthalpic and entropic barriers to the rotation contained no commitment as to the relative energies of various specific conformations of trimethylene.[20c] An interesting example of an attempt to begin discussion of this problem is the note by Smith suggesting that **15a** might be an intermediate in the cyclopropane → propylene reaction.[25] If Smith's **15a** corresponds to the basis orbital representation **15b**, this might be considered as an early suggestion of what has since come to be known as the "0,90"

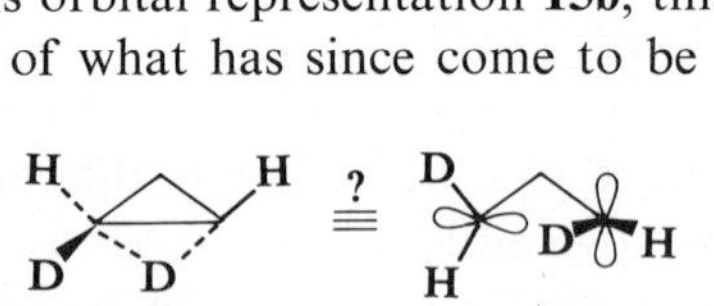

15a **15b**

conformation of trimethylene.

The problem was defined more explicitly by Hoffmann in 1968 in a paper reporting[24] some extended Hückel (EHT) calculations on trimethylene. Hoffmann calculated the total energy of trimethylene with respect to variation of the three most important degrees of freedom of the molecule: the central C—C—C angle and the rotations of the terminal methylene groups. He provided the terminology "0,0" for the conformation with both methylene groups in the plane of the carbons, "90,90" for the species with both H—C—H planes perpendicular to the C—C—C plane, and "0,90" for the conformation having one methylene in and one out of the molecular plane. Hoffmann's study led to the interesting prediction that there should be a potential energy

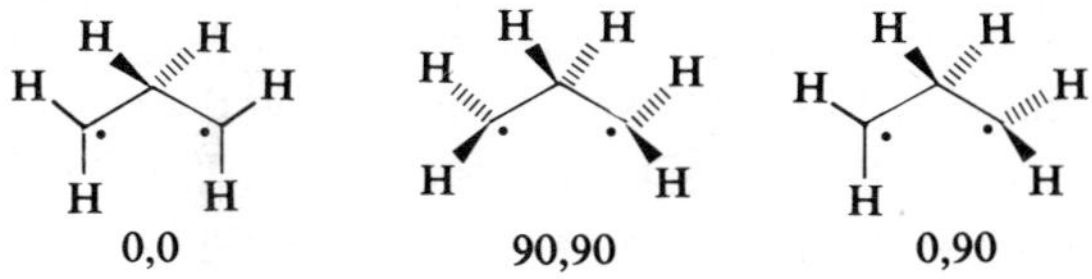

minimum for the 0,0 species at an energy of about 45 kcal/mole above that of ring-closed cyclopropane, and that this intermediate should have an energy some 4 kcal/mole lower than that of the 90,90 intermediate having the same central bond angle. 90,90 Conformations with smaller C—C—C angles were predicted to have lower energies than that of the 0,0 minimum, but did not constitute "resting places" on the potential energy curve. Most important was the prediction that the lowest-energy pathway between the 0,0 intermediate and the ground-state cyclopropane involved conrotatory motion of the terminal methylene groups, and this motion was expected to take place over a barrier height of *at most* 1 kcal/mole.

It is important to realize that the Hoffmann prediction of conrotatory closure for trimethylene is closely associated with the fact that configuration interaction is neglected in the EHT method. To a first approximation, the nonbonding molecular orbitals in the 0,0 species are, as pointed out in Section II of this chapter, linear combinations of the atomic p orbitals n_1 and n_2. However, in this somewhat more complicated system, the symmetric

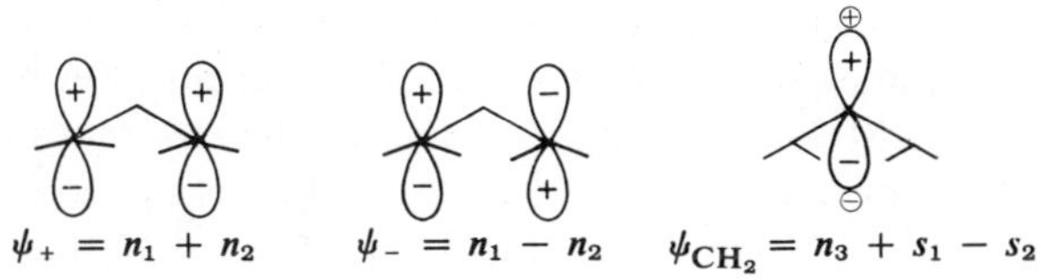

combination ψ_+ interacts strongly with a linear "pseudo-p" combination of carbon $2p$ and hydrogen $1s$ orbitals at the central methylene group,[24] which is also symmetric (ψ_{CH_2}). This raises the energy of the ψ_+ orbital, leaving the antisymmetric ψ_- orbital at lower energy. The double-occupancy requirement of molecular orbital theory then defines the ground singlet configuration as the one with ψ_- filled with two electrons and ψ_+ empty, and conrotatory ring closure is predicted. As indicated in Section II, however, configuration interaction strongly mixes the ground singlet ψ_-^2 with the doubly excited configuration ψ_+^2, especially if they are nearly coenergetic. To the extent that this mixing occurs, the stereochemical restrictions arising from the sole occupancy of ψ_-^2 will be relaxed.

This effect of configuration interaction was pointed out by Salem[26] in 1970, and in 1971 Hayes and Siu reported nonempirical SCF calculations including configuration interaction for 0,0 trimethylene and a number of its analogs.[27] These were aimed at quantitatively evaluating the relative contributions of ψ_+^2 and ψ_-^2 to the CI wave function, on the assumption that this would provide information about the propensity of the intermediate for stereospecific ring closure. Because configuration interaction is designed to decrease the contribution of ionic terms in the ground-state wave function (cf. Section II), a relatively equal contribution of ψ_+^2 and ψ_-^2 to the ground singlet will decrease both the stereospecificity of ring closure *and* the amount of ionic character in the intermediate. Thus intermediates with more "diradical" character by the quantum mechanical definition (i.e., low contribution of ionic terms to the ground singlet wave function) will behave nonstereospecifically, and intermediates with more "ionic" character will tend to behave stereospecifically. Hayes' calculations correspond to degrees of ionicity that one would predict intuitively: Trimethylene is predicted to have a large degree of diradical character and react relatively nonspecifically; the ring-opened intermediates from aziridine, ethylene oxide, and oxazirane are expected to be relatively ionic, and hence to react more stereospecifically.[27]

Whether these predictions are borne out even theoretically, however, can only be examined by calculating points along possible reaction coordinates on the multidimensional potential energy surfaces. Calculation of even a partially complete potential surface for a system with as many degrees of freedom as trimethylene is essentially out of the question, and so most workers have adopted the Hoffmann approach[24] of examining the behavior of the system as only certain intuitively important degrees of freedom are varied. One of the first studies was carried out by Buenker and Peyerimhoff, who used ab initio SCF–MO calculations to examine the energetics of stretching one of the bonds in cyclopropane[28] (this corresponds to Hoffmann's 90,90 reaction coordinate). These authors found that a two-determinant CI study gave a significant energy lowering compared with that obtained without CI, but including additional configurations produced less important further changes. Their results predict that the energy increase in going from 60 to 120° should be 47.45 kcal/mole, which is somewhat less than the activation energy measured for ring opening.

Siu et al. have reported similar ab initio SCF–MO calculations on trimethylene.[29a] They employed a less extensive CI technique, but examined the energies associated with rotation of the 90,90 geometry into the 0,90 form, as well as those associated with ring opening. Once again configuration interaction produced a significant effect on the latter energy; the difference between cyclopropane and the lowest-energy form of trimethylene is calculated to be 122 kcal/mole using single-determinant Hartree-Fock energies; CI lowers the energy of the opened form such that this energy drops to 32 kcal/mole.[29b] Their calculations also predict a 2.5 kcal/mole barrier for conversion of the 0,0 form into the 0,90 form in the singlet state and less than 1 kcal/mole for the triplet. For the 90,90 → 0,90 reaction the barrier is 1.5 kcal/mole for the singlet and less than 1 kcal/mole for the triplet. The authors suggest that because these numbers are relatively small, the barriers to concerted rotation (either conrotatory or disrotatory) of both methylene groups are probably even smaller and are beyond the accuracy of the calculation method.[29a]

A much more extensive investigation of the potential energy surface leading to and surrounding trimethylene is being carried out by Salem's group at Orsay. His preliminary results point out the pitfalls involved in making intuitive decisions about which sorts of structural deformations may be conveniently ignored in such calculations in order to reduce the complexity of the problem. Using an ab initio SCF–MO program written by Stevens, Salem first found that configuration **16**, which is a distorted form of the 0,90 intermediate having a partially inverted, nonplanar carbon at C-1, was significantly stabilized relative to its planar analog.[30] Shortly afterwards Jean and Salem reported that certain other conformations could also be stabilized by similar distortions.[31] Most striking was that the 90,90 species was calculated to be stabilized by no less than 6 kcal/mole by distortion of both methylene

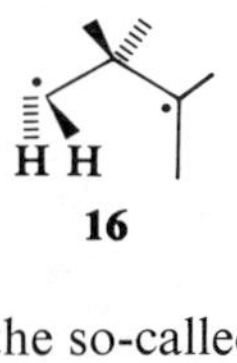

16

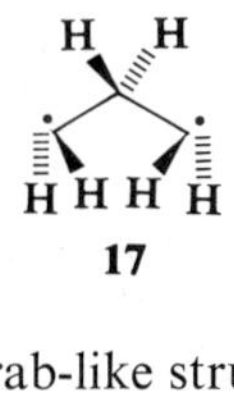

17

groups inwards, forming the so-called crab-like structure **17**. Jean and Salem attribute the stabilization to an increase in zwitterionic character in the diradical.[31] They also report[29c] that preliminary results predict no barrier to reclosure from **17**.

Finally, Hay et al. have applied a generalized valence bond (GVB) technique to the trimethylene problem.[32] This approach removes the inability of the Hartree-Fock method to handle dissociation reactions by "splitting" the doubly occupied orbital pairs $n_i(1)n_i(2)$ in the MO many-electron wave function into singlet-coupled pairs of the form $n_{ia}(1)n_{ib}(2) + n_{ib}(1)n_{ia}(2)$. These authors have carried out minimum basis set ab initio calculations for the GVB orbitals of various conformations of trimethylene allowing one

C—C pair to be split, but solving the full variational equations allowing each orbital to use all functions in the basis.

Most unique about these calculations is that they apparently provide an accurate value for the ring-opening energy—the minimum energy barrier is calculated to be 61 kcal/mole, in good agreement with the measured E_a of 64 kcal/mole. With planar end groups, the 0,0 intermediate is predicted to be more stable than the 0,90 or 90,90 structures. Hay and his co-workers have examined the barrier to concerted rotation of the methylene groups in the 0,0 species, and they find a weak preference for conrotation. Disrotatory conversion of the 0,0 into 90,90 conformation requires, at a fixed angle of 120°, 2 kcal/mole, whereas no barrier is required for conrotation. In striking agreement with the results of Jean and Salem, the GVB method also finds the distorted 90,90 ("crab") configuration **17** to be greatly stabilized (5 kcal/mole) relative to any of the planar-methylene structures. In this case, however, the stabilization is attributed to increased bonding between the two radical centers. This is caused by a change in shape of the radical-electron orbitals resulting from their attempt to remain as nearly orthogonal as possible to the other orbitals in the system. Hay et al. also have carefully examined the energetics involved in smoothly converting **17** back to cyclopropane. No barrier is predicted for this motion.[32]

In summary, although different calculational methods predict different absolute values for the ring-opening energy of cyclopropane, there is striking agreement in the general trends predicted for the behavior of the 0,0 and 90,90 intermediates. The 90,90 form nearly always rises monotonically in energy as the ring opens, and reaches no intermediate energy well. At some point near 110°, the 0,0 curve crosses the 90,90 and becomes the lower-energy form. Finally, the disrotatory and conrotatory modes are predicted to be nearly equienergetic for conversion of the 0,0 species back to cyclopropane, although the latter may be weakly favored. No significant barrier is predicted to exist for this motion (in contrast to the prediction made by the thermochemical–kinetic technique outlined in the previous section), even for the stable distorted "crab" configuration **17** examined by Salem's and Goddard's groups.

D. Chemical Behavior of Trimethylene

Although the energetics of the isomerization of cyclopropane to propylene were measured some time ago, systematic study of substituted cyclopropanes has begun only recently. As mentioned in Section III.A, the classic study of Rabinovitch, Schlag, and Wiberg demonstrated that cis–trans or "geometric" isomerization occurred in 1,2-dideuteriocyclopropane at a rate somewhat faster than "structural" isomerization to propylene.[15] Shortly afterward other studies demonstrated that both types of isomerization occurred in substituted cyclopropanes.[16] A large number of systems have since been examined, with the general result that alkyl substitution appears to lower the energy for ring opening by about 2 kcal/mole per alkyl group. Vinyl and phenyl substitution provide a conjugative-type stabilization of the intermediate diradical which is considerably greater; the former also provides an alternative mode of ring closure, and this process (**18** → **19**) is referred to as the vinylcyclopropane rearrangement. The evidence is quite good that nearly

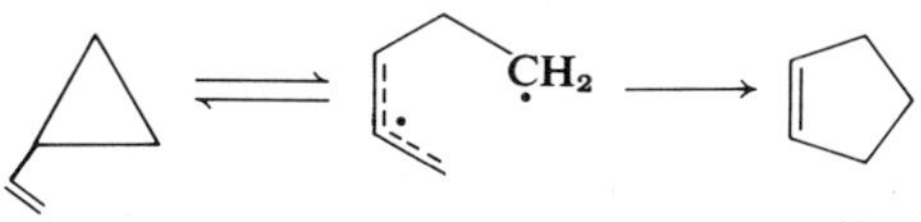

18 **19**

all of these reactions proceed through trimethylene diradicals; they have been reviewed[16] and are not treated in detail here.

The stereochemistry of ring closure of trimethylene, however, has not been reviewed in detail elsewhere. Because stereochemistry provides the most sensitive probe currently available into the symmetry and structure of diradicals, and because it relates rather closely to a number of questions posed by theoretical calculations on trimethylene, the stereochemical problem will be the primary focus of this section.

Information concerning the stereochemistry of ring closure of the trimethylene intermediate formed on cyclopropane pyrolysis is difficult to obtain. This is because the intermediate has an extremely short lifetime and therefore a low steady-state concentration, and because it is reversibly formed continuously during the pyrolysis. However, some information can be obtained by examining the overall stereochemical changes which take place at *both* carbons of a disubstituted cyclopropane. There are three overall processes possible (Scheme 4): inversion at one center and retention at the other (path *A*), inversion at both centers (path *B*), and stereochemical randomization at both centers (path *C*). The three modes of reaction can be associated with certain reasonable assumptions about the behavior of the intermediates which control them. Process *A* suggests that a 0,90 intermediate (**21**) is involved in the reaction. Process *B* would be predicted to occur for a diradical having an orbital symmetry bias (either conrotatory or disrotatory), because the principle of microscopic reversibility requires that the stereochemical pathway favored for destruction of the 0,0 intermediate **22** should also be favored in its formation. Path *C* would be produced by "freely rotating" diradical **23**; that is, one whose rate of rotation about C—C bonds is faster than that of ring closure.

The fact that geometrical isomerization occurs in substituted cyclopropanes[15,16] suggests that an intermediate such as **21** is at least accessible on

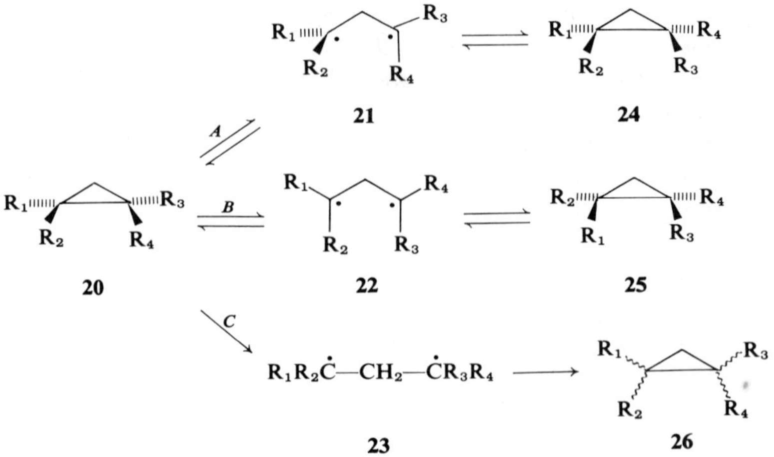

Scheme 4

the reaction energy surface. However, geometrical isomerization studies leave process *B* undetected, because this mode produces no cis–trans isomerization. Compounds **20** and **25** are, however, mirror images, and path *B* therefore would produce racemization ("optical" isomerization) of an optically active cyclopropane such as **20**. Occurrence of racemization at a rate faster than that of cis–trans isomerization would constitute evidence that path *B* (via intermediate **22**) was the lowest-energy route for ring opening.

This problem was independently investigated in substituted systems **27**, **28**, and **29** by our own group[33,35] and Berson's.[34] In both studies it was found that isomerization and racemization proceeded at competitive rates, indicating that there was no significantly large favorability associated with process *B*. The data can be fit to a partially randomizing mechanism (see below), or to a more complicated mixture of paths having different stereoselectivities. The results show, however, that neither a completely randomized intermediate (**23**) nor an in-place rotation of one group (path *A*) can be the sole process occurring.[34] Further, they permit no more than a fraction of the ring cleavages to produce *any* planar intermediate or transition state in these systems. As a working hypothesis, therefore, both groups fit their kinetic data to a simple mechanism in which each stereoisomeric cyclopropane undergoes ring cleavage between the most highly substituted carbons to give some nonplanar intermediate (suggested tentatively as a 90,90 structure by Carter and Bergman[33]) which can interconvert with stereoisomeric diradicals by rotation (Scheme 5). Application of the steady-state hypothesis to this mechanism allows calculation of the relative rate constants for rotation and cyclization of the postulated diradical intermediates. Berson and Balquist's data[34] give $k_{cyc}/k_{rot} \simeq 11$. This means that the experiment measures the contribution of

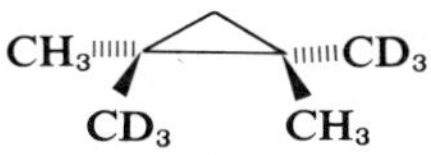

27

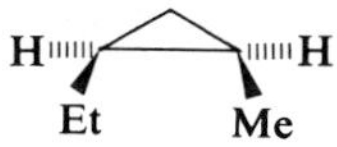

28

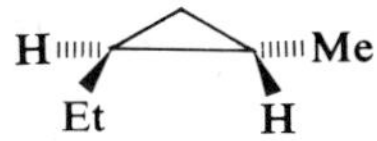

29

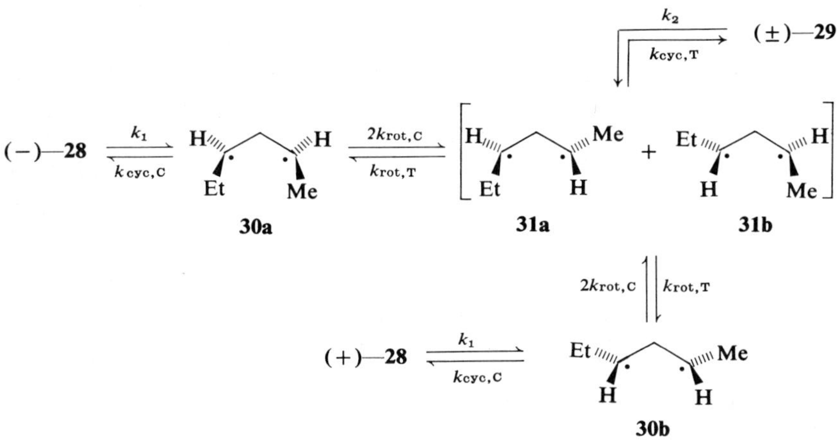

Scheme 5

the ineffective "no reaction" reaction in which 11 out of every 12 times compound **27** undergoes ring opening to the diradical and simply recloses without rotation to give the same optical isomer from which it arose.[34] Somewhat less precise results were obtained[33,35] from a kinetic analysis of the pyrolyses of optically active **28** and **29**, which was more complicated due to the lower symmetry of this system. The formation of essentially racemic **29** from **28** at short reaction times indicated that rotation about both C—C bonds in **30a** (and **30b**) occurred at nearly the same rate. This assumption allowed calculation of k_{cyc}/k_{rot} for both diradicals **30** and **31**. Within experimental error, rotation occurred about five times faster than cyclization in these intermediates. The difference in behavior of the intermediates formed from **27** and **28** is reasonable in view of the fact that both entropic and enthalpic hindrance to rotation should be considerably greater in the tetrasubstituted diradicals.[36,20a]

One of the principal difficulties associated with the study of trimethylene formed in cyclopropane pyrolyses is the fact that both ring opening and ring closure are going on simultaneously, and this seriously complicates the determination of the simple trimethylene reaction selectivity. For this reason, a number of workers have approached the problem from an alternative direction: generation of the presumed intermediate from a noncyclopropane precursor. The stereochemical results (which contrast markedly with data obtained in cyclopropane pyrolysis) fall into two groups: those occurring with predominant single inversion of stereochemistry, and those occurring with predominant double inversion.

A series of comprehensive investigations on cyclic azo compounds has been carried out by the research groups of Crawford and McGreer.[37] The experiments most relevant to the trimethylene stereochemistry problem[37a] are those involving the thermal decomposition of *cis*- and *trans*-3,5-dimethylpyrazolines (**32** and **33**). The major products of these decompositions are the stereoisomeric dimethylcyclopropanes, and the major pathway is apparent single inversion of stereochemistry in each case (cf. Scheme 6). Crawford and Mishra rationalize these observations by assuming that the pyrazolines decompose in the "envelope" conformation leading directly to 0,0 intermediates.[37a] Predominant conrotatory closure[24] then leads to overall single

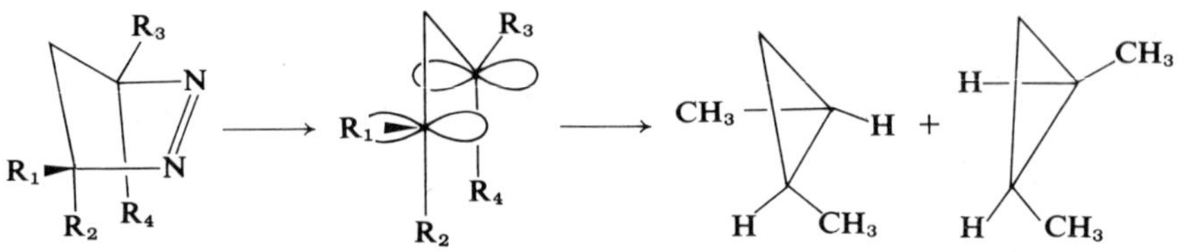

32, $R_1 = R_3 = CH_3$; $R_2 = R_4 = H$ ⟶ 66.1% 33.2%

33, $R_1 = R_4 = CH_3$; $R_2 = R_3 = H$ ⟶ 25.4% 72.6%

Scheme 6

inversion of stereochemistry. Other mechanisms are of course possible; one alternative is that the conversion of pyrazoline to cyclopropane occurs in only one step, as a reverse 2 + 2 reaction. Orbital symmetry considerations suggest[38] that this reaction might take place through a "twisted" transition

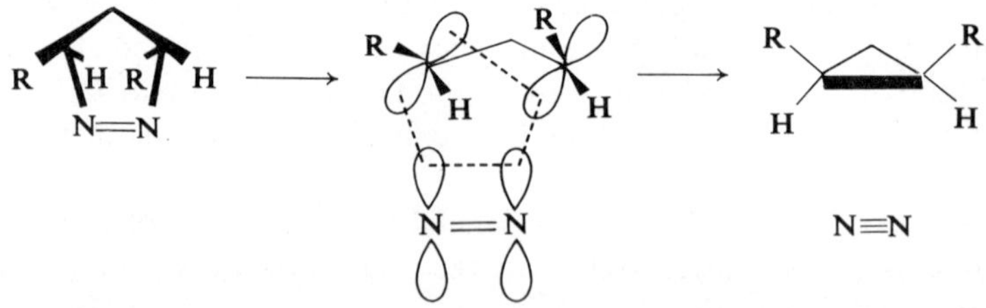

state, having a "mobius" arrangement of basis set orbitals,[39] so that the overall stereochemistry can be ${}_{\sigma}2_a + {}_{\sigma}2_s$, which is symmetry-allowed.[38] This predicts inversion of stereochemistry at one center, as is observed.

One of the most difficult mechanisms to rule out rigorously involves the possibility that only one C—N bond breaks initially, leading (in the case of *trans*-pyrazoline **33**) to diradical **34**. If the radical center at C-2 is now required to carry out a backside displacement of N_2 at C-4, a product of the correct stereochemistry is produced.[40] However, Crawford and his co-workers have carried out a number of elegant studies that provide support

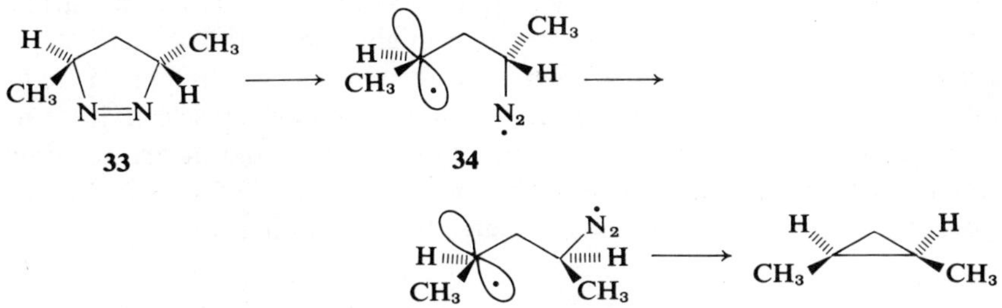

33 **34**

for a mechanism which involves simultaneous cleavage of both C—N bonds.[37] Although the cumulative decrease in rate[37j] in going from **35** to **35**-d_2 to **35**-d_4 is consistent with either simultaneous cleavage or a statistical mixture of single C—N cleavage at each of the two possible sites in the

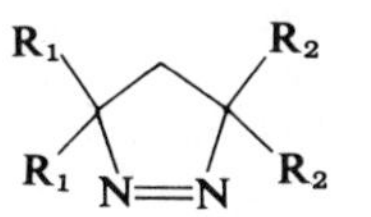

35, $R_1 = R_2 = H$
35-d₂, $R_1 = D, R_2 = H$
35-d₄, $R_1 = D, R_2 = D$

molecule, the magnitude of the effect per deuterium calculated assuming simultaneous cleavage is close to that observed by Seltzer[41a–d] in acyclic phenyl-substituted systems. For some time the simultaneous-cleavage ("three-fragment") mechanism appeared to be well established for acyclic systems due to this work. However, some recent observations of Takagi and Crawford on the thermal decomposition of a number of mono- and diolefinic azo compounds[41e] have caused Crawford to state that he is "convinced that two fragments [that is, the sequential cleavage process] are produced in the rate determining step" of these reactions.[41f] Further experiments are now under way in Crawford's laboratories to determine conclusively whether or not this is also the mechanism which operates in cyclic azo compound decompositions.[41f]

Crawford has reported labeling studies[37c] which demonstrated that the product-forming intermediate from 4-deuterio-3-methyl-1-pyrazoline has a plane of symmetry, although thermal decomposition of optically active **33** produces *trans*-1,2-dimethylcyclopropane with about 26% retention of optical purity.[37e] The latter result, however, only *requires* that 6% of the total product mixture in the dimethyl system is formed from chiral intermediates.

Some indication of the structural distortion which the 0,0 intermediate must be able to survive in order to account for pyrazoline decomposition stereochemistry is available in the decomposition[42] of bicyclic azo compounds **36a** and **36b**. The 0,0 intermediate generated here must have two of its carbons connected by a three-carbon bridge; one might expect this structural perturbation to destabilize the 0,0 pathway relative to the "retention" component of the mechanism. However, such a change in stereoselectivity is not observed—compounds **36a** and **36b** decompose with equal or greater single inversion selectivity than their monocyclic analogs **32** and **33**. This does not rigorously rule out the 0,0 intermediate, but it requires that it be rather

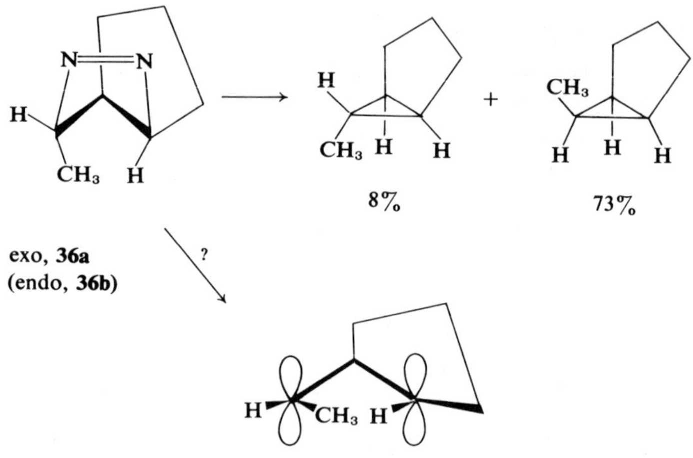

unusually insensitive to distortion effects, and this perhaps strengthens the case for an alternative mechanism.

Two cases have recently appeared in which other extrusion reactions have been used to generate possible trimethylene intermediates. Thietanonium salt **37a** and its cis isomer **37b** have been shown to react with butyllithium to give dimethylcyclopropanes, presumably via neutral tetravalent sulfur intermediates formed by attack of alkyllithium on the heteroatom, followed by extrusion of dialkyl sulfide.[43] It is striking that predominant single inversion is observed in the product formation, in qualitative agreement with the

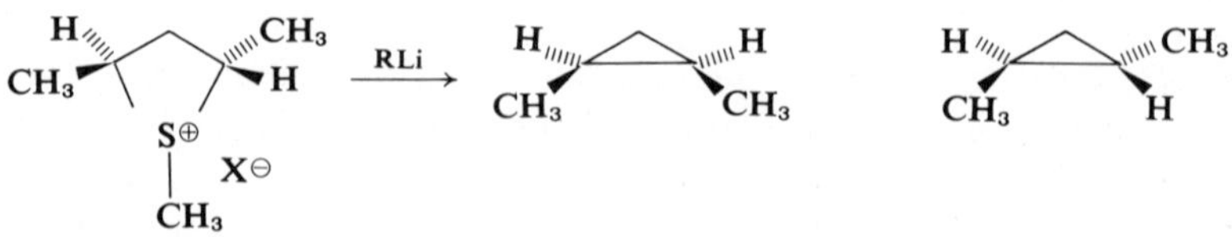

trans, **37a**
(*cis*, **37b**)

stereochemistry of decomposition of pyrazolines **32** and **33**. In an effort to determine the stereochemistry of a two-step mechanism involving displacement (i.e., **38** → **39** → **40**), Trost and co-workers[43] treated *meso*- and *dl*-2,4-dibromopentane with *n*-butyllithium, and found that this reaction gave predominant double *retention* of configuration. While this interesting result is not

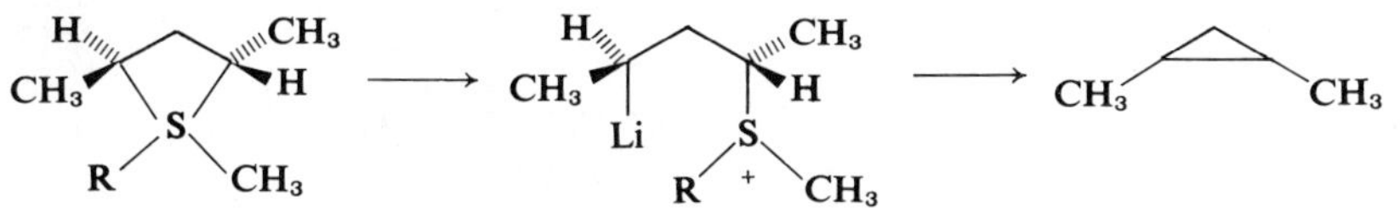

38 **39** **40**

extensively commented upon, the authors suggest that the similarity of the presumed intermediate to **39** indicates that the thietanonium salts must be reacting by a more concerted route, perhaps involving conrotatory ring closure of 0,0 dimethyltrimethylene diradicals.[43]

Similar preliminary results have been obtained[44] by Freeman and co-workers in the decomposition of *N*-nitrosoazetidines **41** and its cis isomer **42**. Reaction with thiosulfate presumably produces the short-lived diazene **43**, which again decomposes predominantly to *cis*-1,2-dimethylcyclopropane.

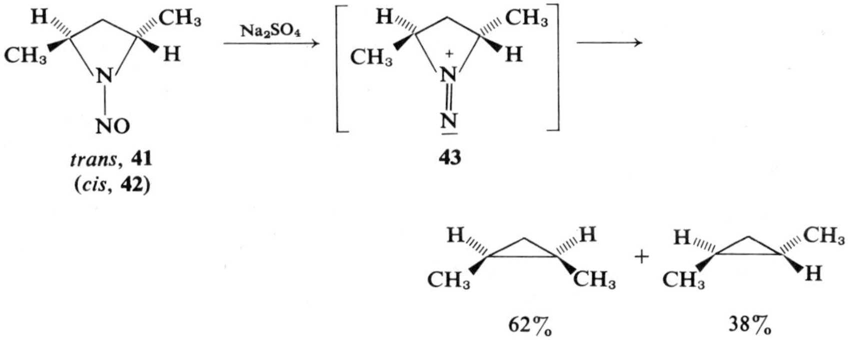

trans, **41**
(*cis*, **42**)

43

All of these monocyclic extrusion systems give somewhat different proportions of *cis*- and *trans*-1,2-dimethylcyclopropanes, but apparent single inversion of stereochemistry is observed predominantly in each case. The different proportions of the two cyclopropanes may be due to medium and temperature effects, or may reflect some essential difference in mechanism. However, it seems unlikely that leaving groups as different as N_2 and dialkyl sulfide would behave so similarly in displacement reactions such as that shown for diradical **34**, and this perhaps strengthens the case that 0,0 dimethyltrimethylene diradicals may be involved in these reactions.

However, sequential C—N cleavage is one of the mechanisms which has also been suggested to explain the stereochemistry of those azo compounds that undergo decomposition with *double* inversion.[40b,45] This mechanism is apparently observed only in the 2,3-diazabicyclo[2.2.1]hept-2-ene series (e.g., **44**); these molecules lose nitrogen and lead predominantly to bicyclo-[2.1.0]pentanes with a "bridge flip" (e.g., **47**). One way to account for this is to postulate[40b] that the first intermediate is nitrogen-containing diradical **45**;

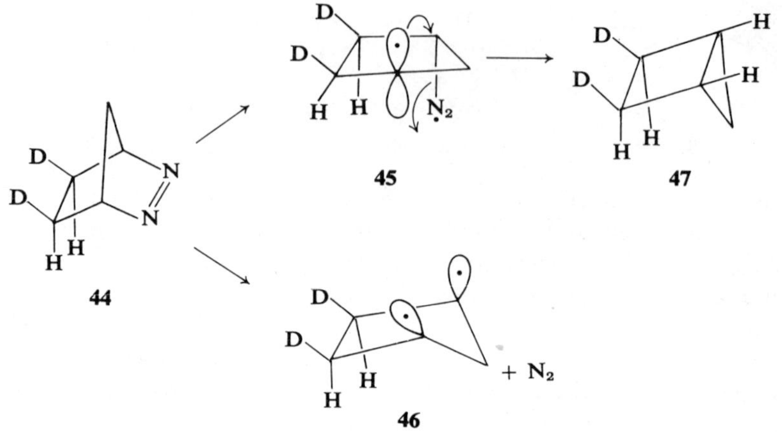

44 **45** **46** **47**

Scheme 7

structural factors now allow backside attack of the unpaired electron on the C—N bond only as pictured in Scheme 7, leading to predominant formation of **47**.

Allred and Smith, however, have suggested[45] an interesting alternative explanation for the double-inversion reaction, which they have observed in the 5-methoxy compound **48**. It is suggested that the leaving nitrogen molecule exerts a "recoil" force on carbons 1 and 4, generating **49** (or **46**) as the first-formed diradical intermediate. This species undergoes predominant ring closure, but can also equilibrate with stereoisomeric diradical **50**.

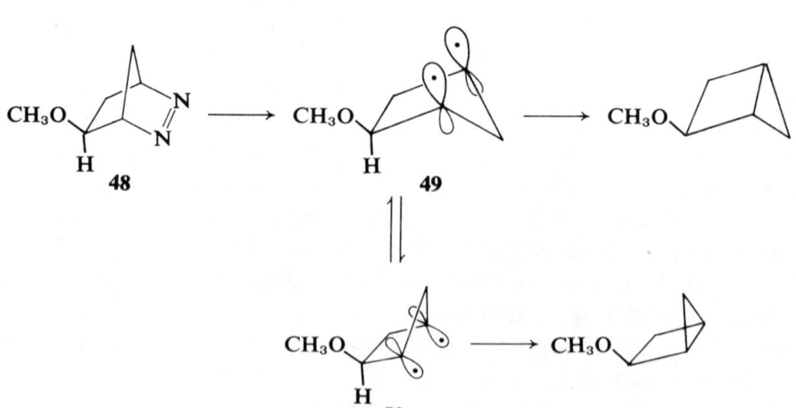

48 **49** **50**

The structural features which should cause the recoil effect to appear are not very clear. For example, one might imagine that the energy which causes recoil is essentially excess vibrational energy present in the diradical intermediate. However, the **44** → **46** reaction can be estimated to be nearly thermoneutral[18]; it is only after ring closure that exothermicity is released. Another interesting question concerns whether the **44** → **47** reaction is subject to the principle of microscopic reversibility. If so, one must presumably expect that the hypothetical reaction of N_2 with **46** (or **47**) to regenerate

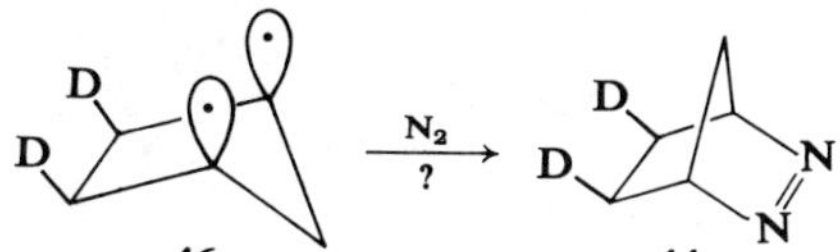

46 **44**

44 must also occur with double inversion. This would presumably require that nitrogen attack **46** from the sterically hindered underside of the ring envelope. It is difficult to understand why this should happen, and so the implication is that if the recoil effect is the cause of double inversion in the **44** → **47** reaction, the reverse of this process should perhaps *not* take place with double inversion.

It is of course impossible to observe the reaction of nitrogen with bicyclo-[2.1.0]pentane due to its endothermicity. However, approximations to this reaction have been observed; reactive unsaturated molecules such as 1,2-dicyanoethylene, diethyl azodicarboxylate, and benzyne have been found to add to strained bi- and tricyclic molecules; the stereochemistry of these reactions has been investigated.[46] Most informative perhaps is the reaction of bicyclo[2.1.0]pentane with reactive olefins. Compound **47** adds maleic an-

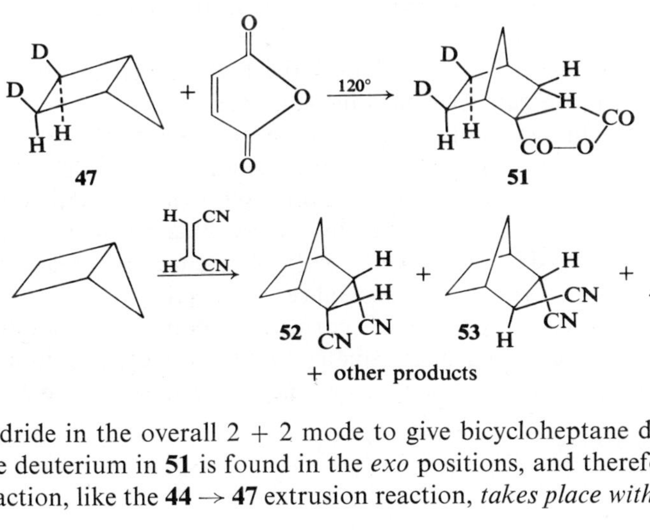

47 **51** **52** **53** **54**

hydride in the overall 2 + 2 mode to give bicycloheptane derivative **51**. All the deuterium in **51** is found in the *exo* positions, and therefore this addition reaction, like the **44** → **47** extrusion reaction, *takes place with double inversion* *of configuration*. Bicyclo[2.1.0]pentane also reacts with *cis*- and *trans*-dicyanoethylene. The stereochemistry about the reacting double bond, however, is scrambled in the products. These two pieces of information are nicely rationalized by the diradical mechanism[46] outlined in Scheme 8. Initial attack on the bicyclic system is assumed to occur with backside attack at C-1,

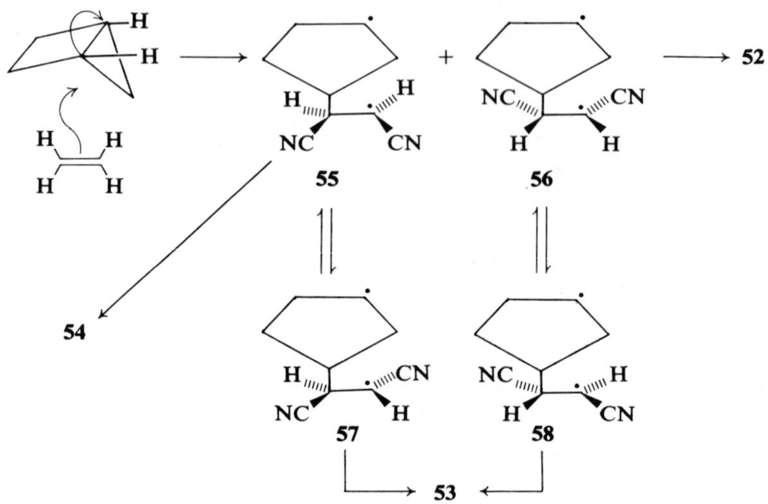

Scheme 8

generating intermediates **55** and **56**. These diradicals now can undergo rotation about the former C—C double bond, equilibrating with **57** and **58**, and are responsible for the bicyclic products observed. It is amusing that this mechanistic sequence is an approximate microscopic reverse of the **44** → **45** → **47** mechanism. It is tempting to associate the double inversion observed in both reactions with the same chemical phenomenon.

It is clear from the preceding discussion that the mechanism of decomposition of both monocyclic and bicyclic pyrazolines is incompletely understood. The sequential C—N cleavage mechanism probably provides rational interpretation of more diverse data than does any of the other mechanisms discussed; however, some data exist which mitigate against the sequential mechanism and in any case broad applicability of a particular mechanistic hypothesis does not necessarily mean it is correct. In addition, the causes of the differing behavior of similar intermediates formed on decomposition of cyclopropanes, diazenes, thietanonium salts, and pyrazolines remain to be elucidated.

E. Spin Correlation Effects

Up to this point we have limited the discussion of diradical chemistry to systems in which the intermediates are generated thermally. In nearly all such cases the assumption is made, either implicitly or explicitly, that the intermediates are formed and react in their singlet states. The evidence for such an assumption rests mainly on two grounds. First, because the ground states of essentially all precursors capable of generating diradicals in thermal reactions are singlets, the principle of spin conservation requires that the diradicals be generated, at least initially, also in singlet states.[47] The most likely way of reaching the triplet, then, is by intersystem crossing in the diradical (providing it has a triplet ground state or thermally accessible triplet excited state) and it seems reasonable, at least, that such intersystem crossing might be slower than the very fast rates which characterize other reactive modes available to the diradical. The second line of evidence comes from experiments intended to deliberately generate the intermediates as triplets. In nearly all cases, the presumed triplet diradicals behave quite differently from their thermally generated relatives.

The most common technique used for the deliberate generation of triplet diradicals is the photosensitized decomposition of diradical precursors. This is somewhat difficult to apply to cyclic hydrocarbons—although it appears to have been successful in at least one case[48]—and has therefore more often been applied to heterosubstituted molecules such as azo compounds and ketones. As an example, consider the photochemical decomposition[49] of 1,2,3-triazolines **59** and **60**. Direct irradiation leads to different mixtures of aziridines **63** and **64**, presumably via nitrogen-substituted diradicals **61** and **62**. Some, but not complete, stereochemical scrambling is observed in the formation of products, and this can be attributed to competitive ring closure and

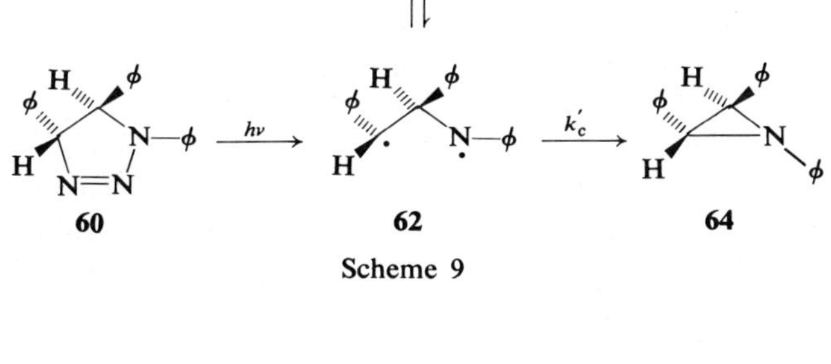

Scheme 9

rotation about the former ring C—C bonds in **61** and **62**. A kinetic analysis[35] of the mechanism outlined in Scheme 9, employing the product distribution from each isomer, allows calculation of the relative bond rotation–cyclization rate ratios for diradicals **61** and **62**. These numbers turn out to be $k_r/k_c = 0.43$ and $k_r'/k_c' = 0.56$. Both diradicals undergo ring closure faster than rotation about their C—C bonds.[35]

The use of benzophenone triplets to photosensitize the decomposition of **59** and **60** causes a dramatic change in product distribution; considerably more stereochemical scrambling is observed.[49] In this case, the triplet states of **61** and **62** are presumably the intermediates, and these must undergo spin interconversion to the singlet states either before or during the ring-closure process. The implication, therefore, is that the apparent rotation–cyclization rate ratios measured using the product distribution and kinetic analysis of Scheme 9 will be much larger in the photosensitized case. This turns out to be true; k_r/k_c now increases[35] to 6.7 for triplet **61** and $k_r'/k_c' = 10.0$ for triplet **62**.

Azo compounds apparently show similar effects. Bartlett and Porter have examined the thermal, direct photochemical, and sensitized photochemical decomposition of cyclic azo compounds **65** and **66**, which should decompose to give substituted tetramethylene (or 1,4) diradicals.[50] In this case, both

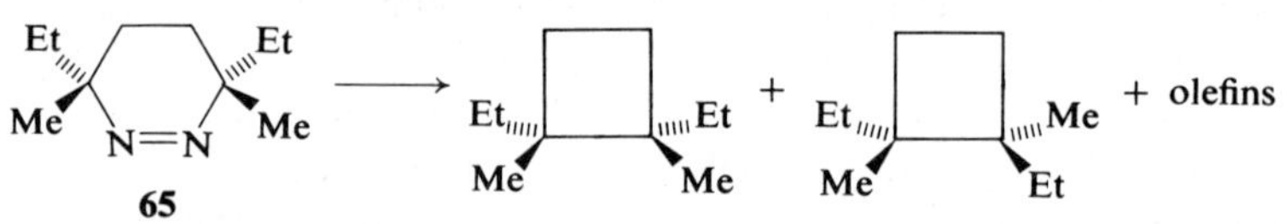

65
(*trans*, **66**)

direct irradiation and thermolysis give predominant retention of stereochemistry; the stereochemistry (as expected) is again scrambled to a higher degree in the photosensitized decomposition.[50] The photochemistry of pyrazolines **32–33**[51] and **36a–36b**[42] is complicated by a lack of knowledge as to how much of the single-inversion mechanism is present in the direct irradiation, but here, too, there appears to be an increase in stereochemical scrambling in the photosensitized decomposition.

These data appear to be well correlated by the assumption that triplet diradicals are the principal product-determining intermediates in the photosensitized decomposition of azo compounds.[50] However, this assumption requires that the rate for spin inversion in diradicals such as the one derived from **65** be somewhat slower than the rate of rotation about C—C bonds in these diradicals.[52] This is perhaps too slow a rate for systems in which the radical centers are weakly coupled and the singlet and triplet states are nearly coenergetic. Another explanation has therefore been suggested as a way of accounting for such "spin correlation effects." This interpretation is outlined in the next section,[52] following a short discussion of excess energy in diradicals.

F. Trimethylene with Excess Energy and an Alternative Explanation of Spin Correlation Effects

The difficulty of generating from different precursors "trimethylenes" which behave identically was discussed in some detail in Section III.D. While structural, medium, and spin effects will often account for the discrepancies in product distribution observed, there is an additional interesting phenomenon which should also be considered perturbing the reactivity of chemical species. This is the possibility that diradicals generated in different ways may have similar structures but different energy contents.[44,52]

In order to decide just what sort of chemical effects differing energy content should cause, we will first examine one or two situations in which it is clear that trimethylenes with different energies have been generated. The technique of "chemical activation" is the most useful way of producing and observing the effects of excess energy in relatively complex molecules, and Rabinovitch and others have reported a number of pioneering studies which bear on the problem of excess energy in cyclopropane thermal reactions.[53] In concept, the technique is a simple one: One attempts to produce a species in an exothermic reaction, and allow this molecule—which must initially contain the exothermicity of the reaction—a long enough lifetime to undergo further reaction before it gives up this excess energy and deactivates to its ground vibrational state. An example of such a reaction is the bimolecular association of methylene (CH_2) and ethylene.[54] The energetic situation is best

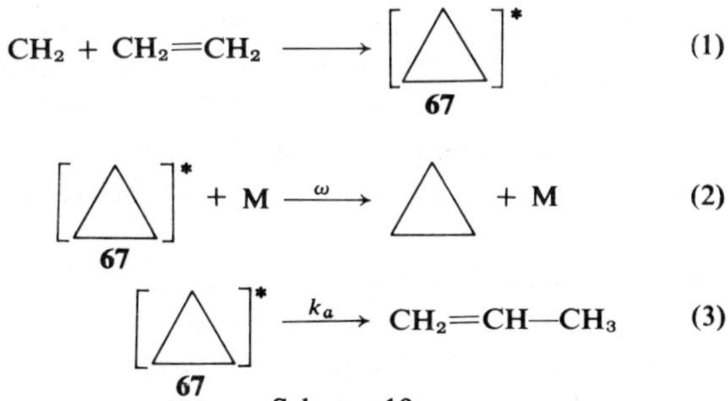

Scheme 10

illustrated with reference to Figure 3 and Scheme 10. The total energy of the vibrationally excited cyclopropane (**67**) which must be produced initially will be the sum of the heat of reaction 1 in Scheme 10 and any excess energy which the reactants contain. Because both the heat of formation of methylene and its excess energy (left over from photochemical generation from diazomethane

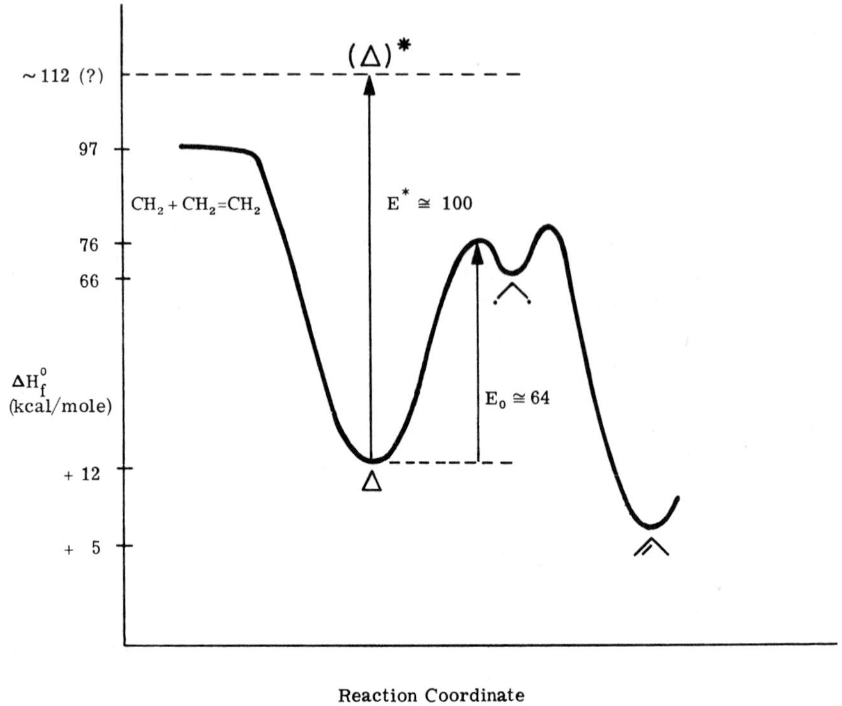

Fig. 3. Reaction coordinate diagram for the addition of methylene to ethylene and isomerization of the chemically activated cyclopropane produced.

or ketene) are very uncertain quantities, the total energy of the chemically activated cyclopropane produced in this reaction is not well known. However, Rabinovitch and Setser have estimated that **67** is formed initially about 100 kcal/mole above its ground vibrational state.[58] This is more than enough energy to allow the excited molecule to pass over the 64 kcal/mole barrier to propylene.

Deactivation of the excited molecule can occur on collision with "cooler" bath molecules in the surrounding gas (reaction 2, Scheme 10). Although it was once assumed that such deactivation, involving relatively complex molecules, took place in one collision, it seems now that the best models for energy transfer indicate each collision removes 5–10 kcal/mole of vibrational energy, depending on the efficiency of the bath gas.[55] This still means that only a few collisions are required to deactivate an excited molecule, and in order to be detected, chemical reaction of **67** (reaction 3) must take place at a rate competitive with that of collision (ω). The qualitative prediction, then, is that more isomerization of the energized molecule should occur as the pressure is

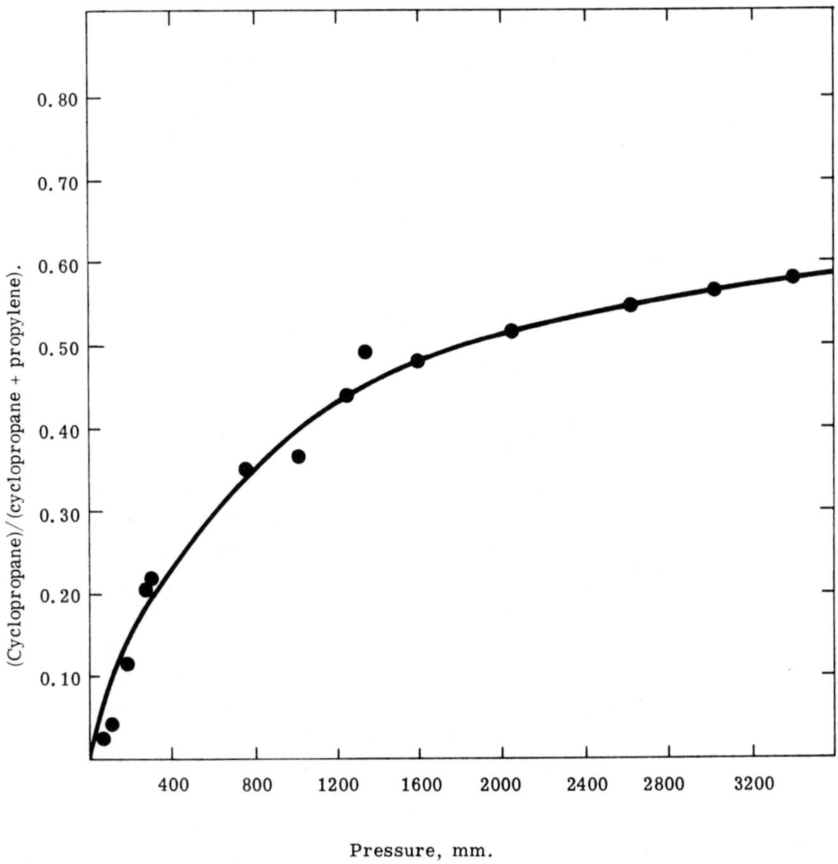

Fig. 4. Dependence on pressure of the ratio of cyclopropane to cyclopropane plus propylene in the addition of methylene to ethylene in the gas phase. Data taken from Reference 54b.

lowered. This is exactly what is observed; in the gas-phase addition of methylene to ethylene, the ratio of "stabilization product" cyclopropane (S) to "decomposition product" propylene (D) is highest at high pressures. As the pressure is lowered, the energized molecules survive longer before colliding with bath molecules that can remove their energy, and more of them isomerize to propylene.[54b] Thus the S/D ratio drops off as the pressure is lowered (Figure 4). From a kinetic analysis of the pressure dependence of S/D, it is possible to measure the rate constant k_a for isomerization of the excited species. It is interesting that comparison of these measurements with others which provide k_a for thermally activated cyclopropane indicates that the lifetime of the molecule produced in reaction 1 is about 6×10^{-3} that of the thermally activated species.[56] The large difference emphasizes the effect of the high internal energy of **67** on its rate of isomerization.

Theoretical models which allow calculation of these rate constants are reasonably well developed, and make use principally of quantum statistical techniques. An approach developed by Rice, Ramsberger, and Kassel (often referred to as the RRK model) led to an equation of the form

$$k_E = A\left[\frac{E^* - E_0}{E^*}\right]^{s-1}, \qquad E^* \geqslant E_0$$

where k_E is the rate constant for isomerization of an activated species with energy E, A is a constant which can be approximated by the Arrhenius frequency factor for thermally activated decomposition, E^* is the total energy of the activated molecule relative to its ground vibrational level, E_0 is the threshold energy for the subsequent reaction of the excited species (approximated by the Arrhenius activation energy E_a), and s is the number of so-called active degrees of freedom in the molecule.[57] Inspection of the equation shows that more complex molecules (large s) will undergo decomposition at slower rates, and large amounts of excess energy (large $E^* - E_0$) will produce faster decomposition. With no energy in excess of threshold ($E^* - E_0 = 0$), no decomposition will occur; as the energy E^* increases, the fraction in parentheses approaches 1.0, so that the limiting rate of decomposition at very high excess energies (or at very low threshold energies) is governed simply by the frequency factor; that is, it is the entropy-determined rate.

More recently, Rabinovitch has improved the calculational method by employing high-speed computers to do state counting for the energized molecule and its activated complex for decomposition.[58] This approach uses the Marcus–Rice (RRKM) extension of RRK theory; the equation for k_E here takes the form[59]

$$k_E = \frac{P_1{}^+ N_E{}^+}{h P_1^* N_E^*}$$

where k_E is once again the rate constant for decomposition of an activated molecule containing total energy E, $P_1{}^+$ is the product of partition functions for those degrees of freedom in the activated complex which are taken as adiabatic, P_1^* is the corresponding product of partition functions for the energized molecule, $N_E{}^+$ is the total number of internal vibrational and rotational energy levels for an activated complex of total internal energy E^+, and N_E^* is the total number of internal vibration and rotational energy levels for the activated molecule of total internal energy E. The N's are "energy densities," the counts of the number of energy states available to the activated complex and activated molecule with certain total internal energies. The observed rate constant k_a (Scheme 10) will be a weighted average of k_E values for the range of energies which characterize the excited molecule, which requires integrating over an energy distribution function for this species.

As indicated in the preceding discussion, generation of cyclopropane having a sufficiently high internal energy content (e.g., in an exothermic reaction such as the addition of methylene to ethylene[54]) will cause it to rearrange to propylene even in the absence of thermal activation. Somewhat more interesting from a strictly chemical point of view, however, are activated systems which have more than one channel available for further reaction. It is important to know just how the choice of various reaction modes in a given excited species depends on factors such as structure, amount of excess energy, number and type of vibrational degrees of freedom, and so forth. The chemistry of vibrationally excited, substituted cyclopropanes gives some qualitative information about this reaction selectivity.

A number of such molecules have been investigated (almost all of which have been produced in chemically activated form using methylene additions to appropriately substituted olefins), but only two will be discussed in detail here. The first is the chemistry of *cis*-1,2-dimethylcyclopropane,[55a,58,60] which is one of the products of the extrusion reactions discussed in the previous section. In Table 2 is listed the percentage of both cyclopropane and olefin products produced in the thermal decomposition of the disubstituted cyclopropane[61] and pyrazoline **32** (Section III.D) and in the addition of methylene to *cis*-2-butene at both high ($\geqslant$400 mm) and low ($\sim$0.5 mm) pressures (the values listed in Table 2 are those which arise only from singlet methylene). The large amounts of olefin formed in the high-pressure reaction result from direct insertion of methylene into the C—H bonds of the olefin; this is indicated by the very weak pressure dependence of the yields of these products.[60] Clearly, however, the vibrationally excited cyclopropane undergoes a great deal of stereochemical scrambling before it is deactivated at low pressures, and *trans*-2-pentene also arises entirely from the chemically activated molecule. These reactions disappear almost completely at pressures high enough to rapidly quench the excited cyclopropane.

Vinylcyclopropane, when activated thermally, rearranges over a 49.6 kcal/mole barrier, giving almost exclusively cyclopentene.[62] Three olefins are formed in less than 1% total yield, apparently only via cleavage of the substituted C—C bond followed by hydrogen shift (Table 3). As in the dimethylcyclopropane system, addition of methylene to butadiene at high pressure gives mainly vinylcyclopropane and C—H insertion products; at lower pressure, considerable isomerization of the excited initial adduct takes place, raising the yields of linear pentadienes.[63] It is interesting that at sufficiently low pressures, the cyclopentene formed from excited vinylcyclopropane is also chemically activated, and itself undergoes further reaction to cyclopentadiene and H_2. The percentage yield of cyclopentene therefore passes through a maximum as the pressure is lowered.

TABLE 2

Products Formed from Dimethyltrimethylene Diradicals Generated from Different Precursors

Reaction	Products (%)						Ref.
	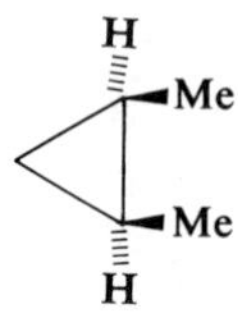	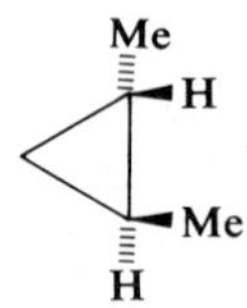		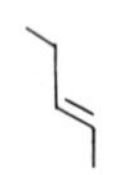	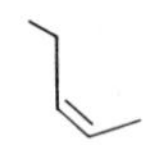	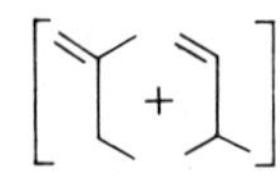	
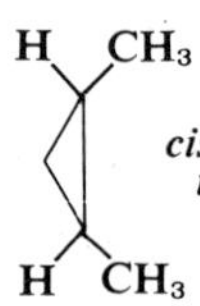 *cis* and *trans*, heat	60[b]		10	12	9	9[a]	61
32 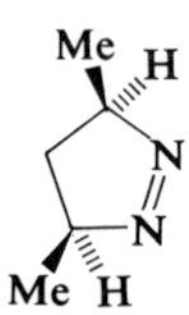, heat	33	66	0	0.68	0	0	37a
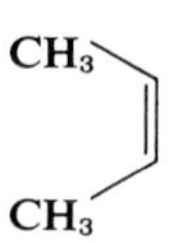 + CH_2, high pressure	45.0	0.6	16.5	0	37.5	0.4	60
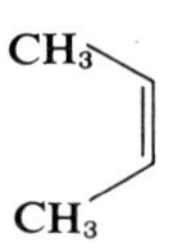 + CH_2, low pressure[c]	12	23	20	11	32	4	60

[a] Quoted as 2-methyl-1-butene.

[b] An equilibrium mixture of *cis*- and *trans*-cyclopropanes was pyrolyzed. Product distribution corresponds to 40% conversion of starting materials, but relative proportion of olefins is essentially invariant with time.

[c] About 1% of 1-pentene also formed.

"Hot" molecules are normally expected to react with less selectivity than "cooler" molecules. This expectation is borne out in the general results outlined in Table 3; for example, thermally activated vinylcyclopropane selects cyclopentene over olefin formation by a factor of about 50, and this selectivity is reduced to about 1 (correcting for insertion) when chemically activated vinylcyclopropane is generated from methylene and butadiene at low pressure. Because of complications from the triplet reaction and from inser-

TABLE 3

Products Formed from Vinyltrimethylene Diradical Generated from Different Precursors

	Products (%)						
Reaction	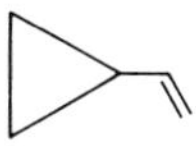		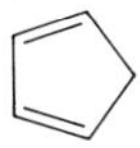		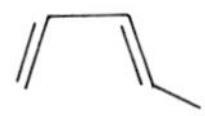		Ref.
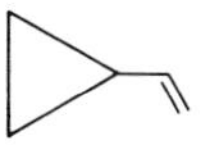, heat	—	96.02	0	1.16	1.18	1.64	62
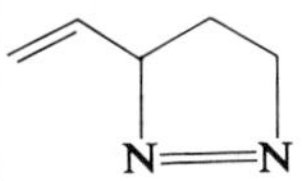, heat	98.91	0.91	0	0.04	0.07	0.07	64
 + CH_2, high pressure	77	4	0	4	7	4	63
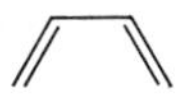 + CH_2, low pressure	26	12	9	13	18	18	63

tion, it is difficult to measure selectivities accurately enough in these systems to determine whether more energy *always* means less selectivity. However, theory is quite clear in its indication that this generality should be an oversimplification.[57,59] At high energies, rate constants should be entropy limited, and therefore high-energy chemical processes may not necessarily occur with complete randomness. In fact, it is possible that an increase in the energy of a chemically activated molecule will actually *increase* the selectivity of its subsequent reactions. Suppose, for example, a molecule B can produce C or D by rate constants k_C and k_D, and that when thermally activated ($M + B \rightleftharpoons M + B^* \rightarrow C + D$), $k_C/k_D \cong 1$—specifically because the path to C has both a high frequency factor and activation energy, and the path to D is characterized by lower values for these parameters. Thus we are assuming that $k_C/k_D = 1$ due to a compensation, in this case, in enthalpic and entropic effects. If B is now generated with a large amount of excess energy (e.g., in an exothermic reaction from *A*), the difference in rate constants caused by the difference between $\Delta H_C^\ddagger$ and $\Delta H_D^\ddagger$ will tend to damp out, leaving the difference in *A* factors as the determinant of selectivity. Thus the compensation is lost, and the pathway with the higher *A* factor will now be favored.

This point has recently been made in a detailed analysis of diradical reactions by Stephenson and Brauman.[52] These authors have attempted to account for the difference in selectivities of diradicals produced from different precursors, a problem outlined in some detail with regard to trimethylene in

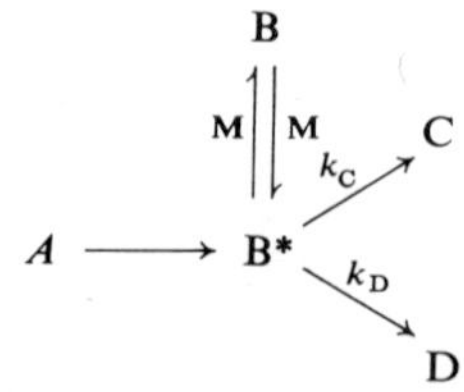

Section III.D of this chapter. Stephenson and Brauman addressed themselves principally to the behavior of substituted tetramethylene (or 1,4) diradicals, where similar problems exist. Specifically, there is some evidence that substituted singlet tetramethylenes generated by cyclobutane ring opening undergo ring closure and rotation about their carbon–carbon bonds at competitive rates, but when generated from azo compounds, the "same" diradicals appear to ring close much faster than they undergo carbon–carbon rotation. Furthermore, a number of triplet 1,4 diradicals (generated by photosensitized decomposition of six-ring azo compounds or in Norrish type II

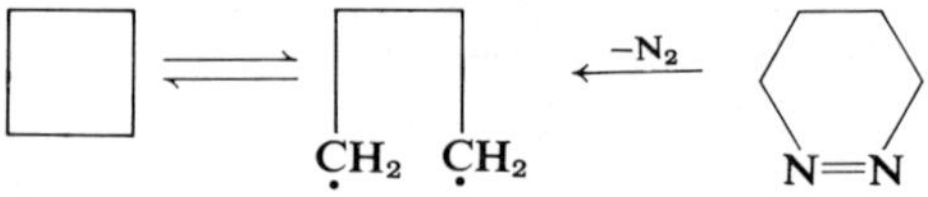

photocleavage of acyclic ketones[64]) undergo spin inversion back to the singlet (a step which presumably must precede or accompany ring closure) at rates which are *slower* than those associated with rotation about C—C bonds in the diradical. Since these rotation rates can be estimated, in substituted systems, as having values around 10^6–10^8 sec^{-1}, the rate associated with spin inversion and ring closure must be around 10^4–10^6 sec^{-1}. Although little experimental work is available upon which to base an expectation for this intersystem crossing rate, Stephenson and Brauman feel that 10^4–10^6 sec^{-1} is extremely slow for a system in which the two electrons are so weakly coupled and where the number of rotational and vibrational states is undoubtedly dense in both the singlet and triplet manifolds.[52]

As a result of this intuitive inference (which may, in fact, not be correct; vide infra), they argue that another interpretation of "spin correlation effects" is needed.[52] The explanation they suggest is perhaps best outlined with reference to Figure 5, which illustrates a possible reaction coordinate for the ring opening of a cyclobutane to a diradical and the cleavage of this diradical to two olefins. The authors point out that in the thermal decomposition of substituted cyclobutanes, the rates of ring closure, cleavage to olefins,

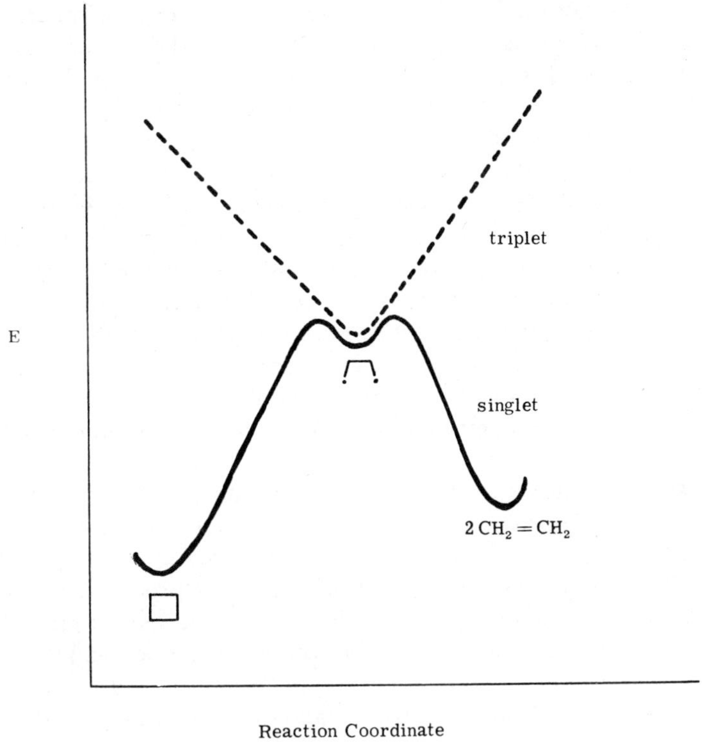

Fig. 5. Schematic cross sections through possible lowest singlet and triplet potential surfaces for cyclobutane decomposition.

and rotation about C—C bonds in the intermediate diradical all appear to be competitive.[65] Generation of similar diradicals from azo compounds such as **65** and **66** (Section III.E), however (by thermolysis or *direct* irradiation), gives a much larger degree of stereochemical retention, indicating that the rate of rotation in these intermediates is much slower than those for ring closure or cleavage. Photosensitized decomposition, as we have seen (Section III.E), normally proceeds with relatively greater amounts of stereochemical randomization, which is reminiscent of cyclobutane thermolysis behavior.

The authors suggest that these changes in rotation–cyclization ratios might be an excess energy or "hot molecule" effect.[52] Because the thermal ring opening of cyclobutane is very endothermic (Figure 5), the diradical intermediate is presumably generated without much excess energy. Thermochemical–kinetic calculations indicate that ring closure will be inhibited mainly by an activation energy barrier (however, see Section III.B) but that C—C rotation will be inhibited mainly by an entropic effect.[20b] The overall result is that the rates for ring closure and rotation will be comparable, but for different physical reasons. This, of course, is just the sort of "compensation effect" discussed in the paragraphs above, and one would expect that generation of the diradical in a vibrationally excited state should tend to favor the high-frequency-factor, high-barrier process—that is, ring closure. The authors argue[52] that azo compound decompositions (either photochemical or thermal) generate such vibrationally excited intermediates. In photochemical decomposition, the diradical carries a significant fraction of the energy of the irradiating photon, and even in the thermal decomposition (because the reaction leading from azo compound to diradical is nearly thermoneutral[18]) the diradical must be born with an energy above its ground vibrational state equal to the activation energy for decomposition (30–40 kcal/mole[37]). In each of these cases, the effect of the potential energy barrier to ring closure will be damped out, making this process correspondingly more favorable.

The decrease in stereospecificity associated with photosensitized decomposition is also explained with reference to Figure 5. In the regions above the deep minima in Figure 5, the triplet potential surface (dotted line) is strongly repulsive. The vibrationally excited triplet diradical produced in a photosensitized decomposition is presumed, therefore, to deactivate preferentially into the "central" or diradical region of the potential surface, and then undergo intersystem crossing to the singlet at this point. This accounts for the similarity in behavior between the triplet- and thermally generated diradicals.

Stephenson and Brauman explain the discrepancy in stereochemical behavior of trimethylenes generated from cyclopropanes and azo compounds by postulating that the 0,0 intermediate is located on a potential energy surface somewhat higher in energy than the stereochemically random diradical. The products are then accounted for by supposing that the excess energy generated in the pyrazoline decomposition allows the diradical to explore the higher-energy surface, which is not accessible to the "cooler" trimethylene generated from cyclopropane.[52]

Although the Stephenson-Brauman hypothesis is an interesting way of accounting for so-called spin correlation effects, there are a number of assumptions in the new explanation which remain to be tested, and some data now exist which indicate that the hypothesis may well be incorrect. One problem is that the stereochemistries of cyclobutane and six-ring azo compound decompositions have *not* been compared in identically substituted systems. It is possible that it is the bulky substituents present at the termini of the tetrasubstituted diradical formed from **65**, and not the presence of excess vibrational energy, which lowers k_{rot}/k_{cyc} in this system relative to that observed in thermal decomposition of 1,2-disubstituted cyclobutanes. This question can be solved by examining the stereochemistry of the decomposition of an appropriate disubstituted azo compound[66] such as **68**. A second problem with the hot molecule explanation is the requirement that there be two distinct trimethylene species, one with a propensity for conrotation and one without. As we have seen (Section II), sophisticated theoretical calculations tend to weaken the case for the existence of a trimethylene diradical which

TABLE 4

Products Formed in the Thermolysis of 2,3-Diaza-5,6-Dimethylnorbornenes and 2,3-Dimethylbicyclo[2.2.0]hexanes[69]

		Products (%)					
	Reactant	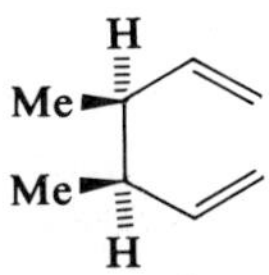	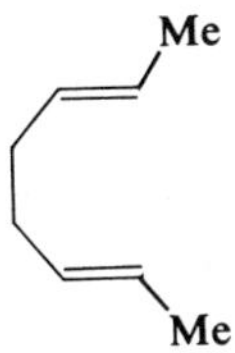	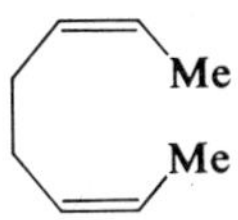	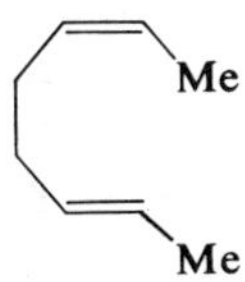	**70a**	**70b**
70a	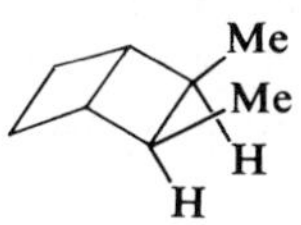	37.4	3.1	0	54.9	[a]	4.6
69a	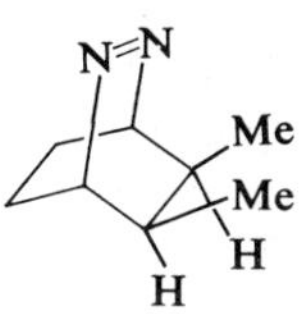	28.7	6.3	2.5	51.5	[b]	7.1
70b	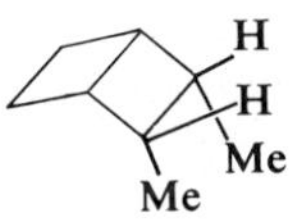	33.1	1.4	0	47.6	17.9	[a]
69b	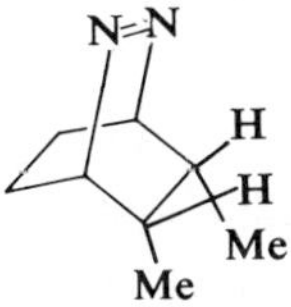	31.1	2.3	0	47.2	17.1	[b]

[a] These decompositions were carried to only 5% conversion; the percentages listed are the percentages of other compounds kinetically formed from **70a** and **70b**.

[b] The percentages listed have been calculated excluding the **70b** formed from **69b** (7.0%) and **70a** formed from **69a** (11.7%), in order to be able to compare precisely the azo compound and hydrocarbon thermolysis product distributions.

behaves stereospecifically. Thirdly, although there is evidence that the cyclopropanes formed on photolysis of pyrazolines contain excess energy,[67] collisional deactivation appears to be rapid relative to chemical reaction of the excited species even at pressures as low as 7 torr,[51] and *no* evidence for vibrationally excited cyclopropanes was found in the thermal decomposition of two typical pyrazolines even at pressures as low as 0.0014 torr.[68]

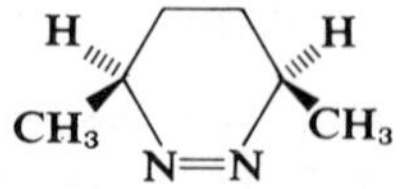

68

With regard to a comparison of product distributions from identically substituted six-ring azo compounds, some recent work carried out in Roth's group is perhaps informative.[69] The product distributions from azo compounds **69a** and **69b** have been compared with those from bicyclohexanes **70a** and **70b** (Table 4). The corrected (see Table 4) product percentages at low conversion formed from similarly substituted azo compound–hydrocarbon pairs are very similar. A larger percentage of ring closure relative to rotation (which in this case constitutes ring flip) in the intermediate 1,4-cyclohexanediyl diradicals generated from the azo compounds, which would be expected on the basis of the hot diradical hypothesis,[52] is *not* observed in this case.

In summary, if vibrationally excited intermediates are involved in the thermal decomposition of cyclic azo compounds, these intermediates lack certain of the characteristics which we might expect them to have based on observations made in more clear-cut cases of chemical activation, such as the addition of methylene to olefins. On the other hand, the need to postulate a very low rate of diradical intersystem crossing was the primary impetus for seeking a new explanation for the spin correlation effect; if this low rate is correct, there is nothing really unsatisfactory about the traditional interpretation. In any case, we may soon see an experimental resolution of this question.

IV. OTHER DIRADICALS

The primary focus of this chapter has been the trimethylene problem. In this context, however, we have discussed certain other diradicals, such as those formed from cyclobutane and vinylcyclopropane. Before concluding, brief mention should be made of a number of additional systems for readers interested in further study of the subject.

The Norrish type II reaction is the name given to the photoprocess which converts ketones into fragmentation products and cyclobutanols (e.g., **71** → **73**, **74**, **75**). This reaction has received very careful study in recent years, and has been shown to proceed via the intermediacy of substituted tetramethylene diradicals (e.g., **72**). A review of these studies has recently appeared.[70]

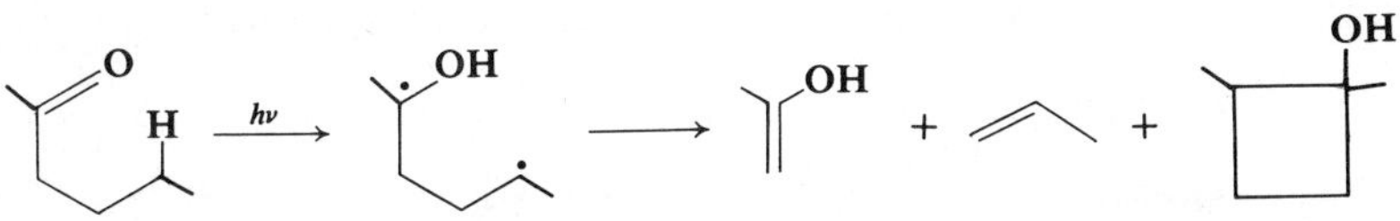

71 **72** **73** **74** **75**

Interesting studies of other tetramethylene systems are also in the literature. Bartlett's group has carried out extensive research on the behavior of substituted tetramethylenes formed during cycloaddition of halogenated olefins; this work has also been reviewed.[71] A number of groups have established that benzyne adds to both olefins and strained ring compounds via diradical intermediates.[72] In an elegant investigation paralleling their study of tetramethyltrimethylene stereochemistry, Berson's group has found[36] that optically active tetrasubstituted cyclobutane **76** undergoes thermolytic cleavage to olefin without detectable cis–trans isomerization or racemization, either because of the hindrance to rotation of the tetrasubstituted diradical centers in **77** or because **77** undergoes rapid rotation about the central C—C bond into a more stable conformation which can fragment but not cyclize.

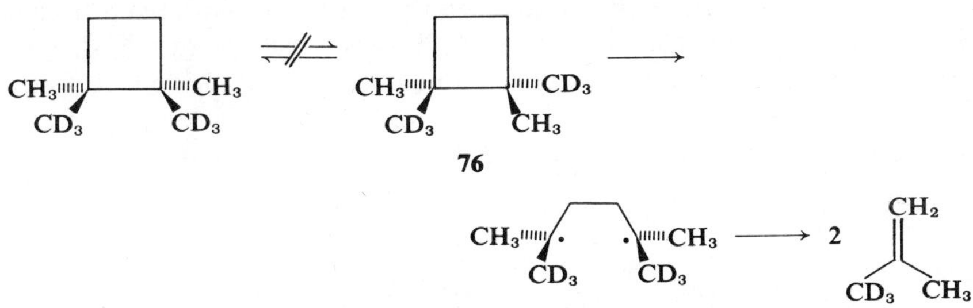

76

77

Some information is also available concerning diradicals generated from larger rings. Overberger's group has examined the decomposition of eight-membered cyclic azo compounds such as **78**; here "disproportionation" (to give **79**) as well as ring closure is observed.[73,74] The behavior of diradicals

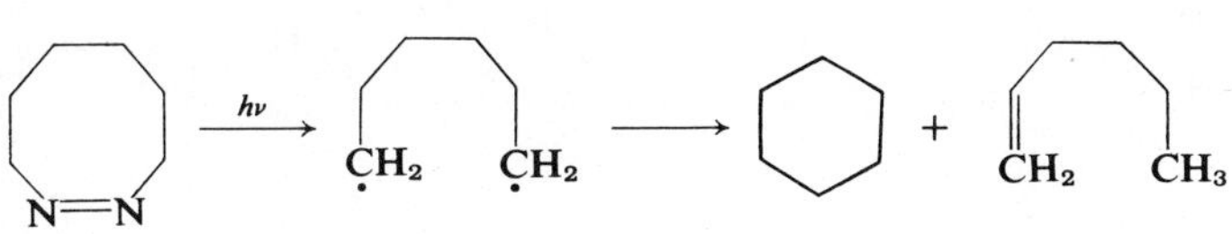

78 **79**

formed on cleavage of the peripheral bond in *p*-cyclophane (e.g., **80**) has been studied,[75] as has [76] 1,8-dehydronaphthalene (**81**). Both Berson's and Willcott's groups have reported painstaking investigations of diradicals formed during thermolysis of substituted bicyclo[2.2.1]heptyl systems.[77]

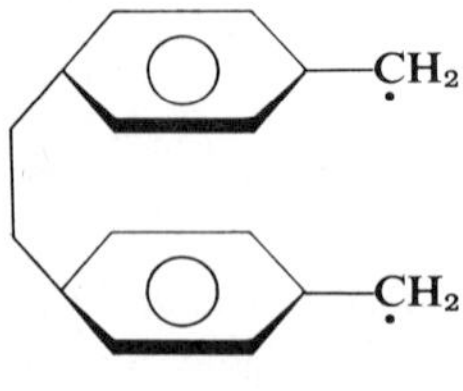

80

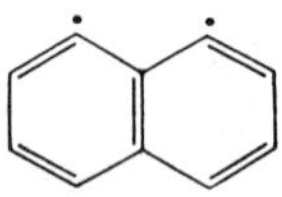

81

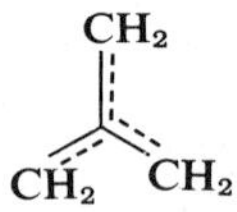

82

Unfortunately, a number of more complicated but interesting substituted trimethylene systems also had to be excluded from extensive discussion here due to space limitations. Among these were molecules which undergo complex secondary rearrangements, such as spiropentane ⇌ methylenecyclobutane.[78] Also neglected were diradicals having triplet ground states, but which may undergo thermal rearrangement through different intermediates or states than those detected by low-temperature electron spin resonance spectroscopy. An interesting example is trimethylenemethane **82**, which has been shown to exist in the (presumably planar) triplet state at low temperatures by ESR.[79] Stereochemical studies on the rearrangement of substituted, optically active methylenecyclopropanes, however, have shown[80] that rearrangement (e.g., **83** → **85**) proceeds with some retention of optical purity. At least some of the intermediates must therefore be chiral; "perpendicular" diradicals of type **84** have been suggested[80a] to explain the experimental data.

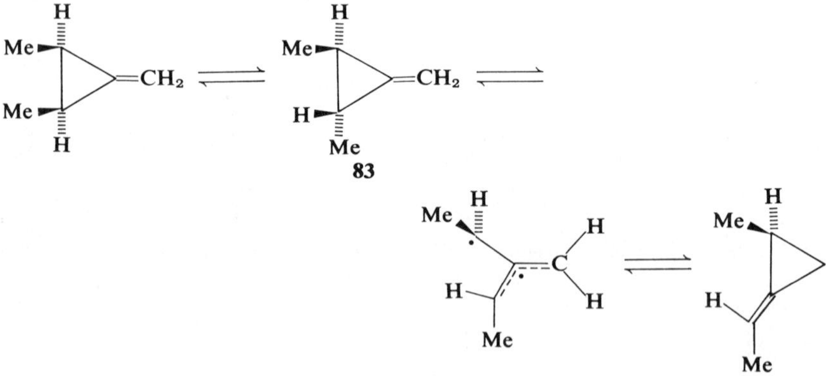

84 **85**

Similar observations have been made on the rearrangement of substituted 1,2-bismethylenecyclobutanes, which takes place at least partially via tetramethyleneethane diradicals.[81]

Although most of the trimethylene studies we have discussed have been limited to simply substituted species which can be generated in the gas phase, a number of investigations on more highly substituted systems have been carried out. The 1,2-diphenyltrimethylene diradical has been generated from diphenylpyrazolines[82] and from both racemic[83] and optically active[84] diphenylcyclopropanes. Several substituted pyrazoline decompositions have also been examined; the stereochemical results are strongly dependent on substitution, but single inversion is observed in certain instances.[37a–d] Finally, evidence has been obtained that cyclopropanes substituted with polar groups undergo ring opening in solution to intermediates with strong dipolar character; in certain instances the intermediates can be trapped by solvent.[85]

Finally, mention should be made of Huisgen's observations[86] on the isomerization of trisubstituted aziridines (e.g., **86**). Here, a very clear case can

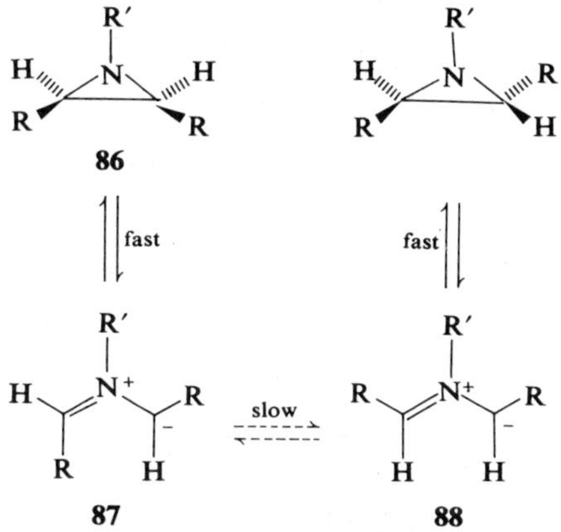

be made for orbital symmetry control of the reaction course, presumably because conjugation through the heteroatom ensures a true allyl-anion-like π system in the 0,0 intermediates **87** and **88**. Aziridine **86** opens in a stereospecific, conrotatory manner, giving **87**, which can be trapped with complete retention of stereochemistry (4 + 2 reaction) by dimethyl acetylenedicarboxylate. While **87** and **88** have not been generated by the azo compound route, similar intermediates containing sulfur as a heteroatom have; these also are most stable in the 0,0 form and undergo stereospecific conrotatory ring closure.[87]

ACKNOWLEDGMENTS

The author is grateful for stimulating discussions, during the course of the writing of this chapter, with Mr. Thomas Morton and Drs. William Goddard, Lionel Salem, John Brauman, L. M. Stephenson, David Golden, Gerhard Binsch, J. P. Freeman, Joseph J. Gajewski, Ronald Levin, and Jerome A. Berson. Thanks are also due to Professors Salem, Goddard, Stephenson, Binsch, Freeman, Gajewski, Wolfgang Roth, and F. H. Dorer for supplying data prior to publication, and to the Alfred P. Sloan and Camille and Henry Dreyfus Foundations, the National Science Foundation, and the Petroleum Research Fund of the American Chemical Society for financial support.

REFERENCES

1. See, e.g., D. Bethell and V. Gold, *Carbonium Ions*, Academic Press, London, 1967, pp. 152–158.
2. See, e.g., R. M. Noyes, *J. Chem. Phys.*, **18**, 999 (1950); **22**, 1349 (1954); *J. Phys. Chem.*, **65**, 763 (1961).
3. E. A. Chandross, *J. Am. Chem. Soc.*, **86**, 1263 (1964).
4. R. M. Dupeyre, H. Lemaire, and A. Rassat, *J. Am. Chem. Soc.*, **87**, 3771 (1965).
5. (a) A. E. Chichibabin, *Ber.*, **40**, 1810 (1907); (b) W. J. van der Hart and L. J. Oosterhoff, *Mol. Phys.*, **18**, 281 (1970).
6. D. L. Wilhite and J. L. Whitten, *J. Am. Chem. Soc.*, **93**, 2858 (1971).
7. L. Salem and C. Rowland, *Angew. Chem. Int. Ed.*, **11**, 92 (1972).
8. J. C. Slater, *Quantum Theory of Molecules and Solids*, Vol. 1, McGraw-Hill, New York, 1963, pp. 66–70.
9. J. C. Slater, *Quantum Theory of Molecules and Solids*, Vol. 1, McGraw-Hill, New York, 1963, p. 52.
10. F. Paneth and W. Hofeditz, *Ber.*, **62**, 1335 (1929).
11. (a) F. O. Rice, *J. Am. Chem. Soc.*, **53**, 1959 (1931); (b) F. O. Rice, W. R. Johnston, and B. L. Evering, *ibid.*, **54**, 3529 (1932); (c) F. O. Rice and A. L. Glasebrook, *ibid.*, **56**, 2381 (1934).
12. F. O. Rice and K. F. Herzfeld, *J. Am. Chem. Soc.*, **56**, 284 (1934).
13. T. S. Chambers and G. B. Kistiakoswky, *J. Am. Chem. Soc.*, **56**, 399 (1934).
14. (a) N. B. Slater, *Proc. Roy. Soc. (London)*, **A218**, 224 (1953). See also (b) E. W. Schlag and B. S. Rabinovitch, *J. Am. Chem. Soc.*, **82**, 5996 (1960); and (c) E. W. Schlag, Ph.D. dissertation, University of Washington, Seattle, Washington, 1958; (d) R. C. Golike and E. W. Schlag, *J. Chem. Phys.*, **38**, 1886 (1963); (e) K. J. Laidler, *Chemical Kinetics*, McGraw-Hill, New York, 1965, pp. 159–163.
15. B. S. Rabinovitch, E. W. Schlag, and K. B. Wiberg, *J. Chem. Phys.*, **28**, 504 (1958).
16. For a review, see H. M. Frey, *Ad. Phys. Org. Chem.*, **4**, 147 (1966).
17. (a) Reference 13; (b) H. O. Pritchard, R. G. Sowden, and A. F. Trotman-Dickenson, *Proc. Roy. Soc. (London)*, **A217**, 563 (1953); (c) E. S. Corner and R. N. Pease, *J. Am. Chem. Soc.*, **67**, 2067 (1945); (d) W. E. Falconer, T. F. Hunter, and A. F. Trotman-Dickenson, *J. Chem. Soc.*, 609 (1961).
18. S. W. Benson, *Thermochemical Kinetics*, Wiley, New York, 1968.
19. W. A. Pryor, *Free Radicals*, McGraw-Hill, New York, 1966, pp. 49–50.
20. (a) H. E. O'Neal and S. W. Benson, *J. Phys. Chem.*, **72**, 1866 (1968); (b) H. E. O'Neal and S. W. Benson, *Int. J. Chem. Kinet.*, **II**, 423 (1970); (c) S. W. Benson and P. S. Nangia, *J. Chem. Phys.*, **38**, 18 (1963); (d) S. W. Benson, *ibid.*, **34**, 521 (1961).

21. S. W. Benson and H. E. O'Neal, *NSRDS-NBS Monograph 21*, Feb. 1970, pp. 21–23, 349.
22. J. E. Douglas, B. S. Rabinovitch, and F. S. Looney, *J. Chem. Phys.*, **23**, 315 (1955).
23. (a) U. Kaldor and I. Shavitt, *J. Chem. Phys.*, **48**, 191 (1968); (b) W. A. Goddard and W. L. Hunt, unpublished results.
24. R. Hoffmann, *J. Am. Chem. Soc.*, **90**, 1475 (1968).
25. F. T. Smith, *J. Chem. Phys.*, **29**, 235 (1958).
26. L. Salem, *Chem. Commun.*, 981 (1970).
27. E. F. Hayes and A. K. Q. Siu, *J. Am. Chem. Soc.*, **93**, 2090 (1971).
28. R. J. Buenker and S. D. Peyerimhoff, *J. Phys. Chem.*, **73**, 1299 (1969).
29. (a) A. K. Q. Siu, W. M. St. John, and E. F. Hayes, *J. Am. Chem. Soc.*, **92**, 7249 (1970); (b) It may be unfair to quote a ring-opening energy for the CI case. As the authors point out, use of the molecular orbitals from the lowest triplet in the CI calculation gave a lower energy than use of those optimized in the singlet, and it is this lower energy which is taken as the CI energy of 120°, 0,0 trimethylene.[29a] The triplet MO's cannot be used for the 60°, 90,90 ground state. The only way to calculate a ring-opening energy, therefore, is to take the difference between the energy of trimethylene at 120° (calculated using triplet MO's in the CI approach) and that of ground-state cyclopropane (90,90 form, 60°, calculated using singlet MO's and no CI). The fact that the two energies are obtained using these different methods may account for the difference between the calculated and experimental values; (c) J. A. Horsley, Y. Jean, C. Moser, L. Salem, R. M. Stevens, and J. S. Wright, *J. Am. Chem. Soc.* **94**, 279 (1972).
30. L. Salem, *Bull. Soc. Chim. France*, 3161 (1970).
31. Y. Jean and L. Salem, *Chem. Commun.*, 382 (1971).
32. P. J. Hay, W. J. Hunt, and W. A. Goddard, *J. Am. Chem. Soc.*, **94**, 638 (1972).
33. W. L. Carter and R. G. Bergman, *J. Am. Chem. Soc.*, **90**, 7344 (1968).
34. J. A. Berson and J. M. Balquist, *J. Am. Chem. Soc.*, **90**, 7343 (1968).
35. R. G. Bergman and W. L. Carter, *J. Am. Chem. Soc.*, **91**, 7411 (1969).
36. J. A. Berson, D. A. Tompkins, and G. Jones, III, *J. Am. Chem. Soc.*, **92**, 5799 (1970)
37. (a) R. J. Crawford and A. Mishra, *J. Am. Chem. Soc.*, **88**, 3963 (1966); (b) R. J. Crawford and D. Cameron, *ibid.*, **88**, 2589 (1966); (c) R. J. Crawford and G. L. Erickson, *ibid.*, **89**, 3907 (1967); (d) R. J. Crawford and L. J. Ali, *ibid.*, **89**, 3908 (1967); (e) A. Mishra and R. J. Crawford, *Can. J. Chem.*, **47**, 1515 (1969); (f) D. E. McGreer, R. S. McDaniel, and M. G. Vinje, *ibid.*, **43**, 1389 (1965); (g) D. E. McGreer, N. W. K. Chiu, M. G. Vinje, and K. C. K. Wong, *ibid.*, **43**, 1407 (1965); (h) D. E. McGreer, N. W. K. Chiu, and M. G. Vinje, *ibid.*, **43**, 1398 (1965); (i) D. E. McGreer and W.-S. Wu, *ibid.*, **45**, 461 (1967); (j) B. H. Al-Sader and R. J. Crawford, *ibid.*, **46**, 3302 (1968).
38. R. B. Woodward and R. Hoffmann, *The Conservation of Orbital Symmetry*, Verlag Chemie, Academic Press, Weinheim, Germany, 1970.
39. H. E. Zimmerman, *J. Am. Chem. Soc.*, **88**, 1564, 1566 (1966).
40. (a) A similar mechanism has been suggested[40b] to explain the stereochemistry of decomposition of certain bicyclic azo compounds (vide infra); (b) W. R. Roth and M. Martin, *Ann.*, **702**, 1 (1967).
41. (a) S. Seltzer and S. G. Mylonakis, *J. Am. Chem. Soc.*, **89**, 6584 (1967); (b) S. Seltzer, *ibid.*, **83**, 2625 (1961); (c) S. Seltzer, *ibid.*, **85**, 14 (1963); (d) S. Seltzer and F. T. Dunne, *ibid.*, **87**, 2628 (1965); (e) K. Takagi and R. J. Crawford, *J. Am. Chem. Soc.*, **93**, 5910 (1971); (f) R. J. Crawford, private communication.
42. P. B. Condit and R. G. Bergman, *Chem. Commun.*, 4 (1971).

43. B. M. Trost, W. L. Schinski, and I. B. Mantz, *J. Am. Chem. Soc.*, **91**, 4320 (1969).
44. J. P. Freeman, D. G. Pucci, and G. Binsch, *J. Org. Chem.*, **37**, 1894 (1972).
45. A. L. Allred and R. L. Smith, *J. Am. Chem. Soc.*, **89**, 7133 (1967).
46. P. G. Gassman, *Acc. Chem. Res.*, **4**, 128 (1971).
47. N. J. Turro, *Molecular Photochemistry*, Benjamin, New York, 1965, p. 27.
48. J. A. Bell, *J. Am. Chem. Soc.*, **87**, 4966 (1965).
49. P. Scheiner, *J. Am. Chem. Soc.*, **88**, 4759 (1966).
50. P. D. Bartlett and N. A. Porter, *J. Am. Chem. Soc.*, **90**, 5318 (1968).
51. R. Moore, A. Mishra, and R. J. Crawford, *Can. J. Chem.*, **46**, 3306 (1968).
52. L. M. Stephenson and J. I. Brauman, *J. Am. Chem. Soc.*, **93**, 1988 (1971).
53. For a review, see B. S. Rabinovitch and M. C. Flowers, *Q. Rev.*, **18**, 122 (1964).
54. (a) G. B. Kistiakowsky and K. Sauer, *J. Am. Chem. Soc.*, **78**, 5699 (1956); (b) H. M. Frey, *ibid.*, **79**, 1259 (1957); (c) B. S. Rabinovitch, E. Tschiukow-Roux, and E. W. Schlag, *ibid.*, **81**, 1081 (1959).
55. (a) J. D. Rynbrandt and B. S. Rabinovitch, *J. Phys. Chem.*, **74**, 1679 (1970); (b) S. C. Chan, B. S. Rabinovitch, J. T. Bryant, L. D. Spicer, T. Fujimoto, Y. N. Lin, and S. P. Pavlou, *ibid.*, **74**, 3160 (1970); (c) J. H. Georgakakos, E. A. McAlduff, and B. S. Rabinovitch, *J. Chem. Phys.*, **52**, 2143 (1970).
56. J. A. Bell, *Prog. Phys. Org. Chem.*, **2**, 1 (1964).
57. S. W. Benson, *The Foundations of Chemical Kinetics*, McGraw-Hill, New York, 1960, pp. 211–266.
58. D. W. Setser and B. S. Rabinovitch, *Can. J. Chem.*, **40**, 1425 (1962).
59. R. A. Marcus and O. K. Rice, *J. Phys. Colloid Chem.*, **55**, 894 (1951).
60. J. W. Simons and B. S. Rabinovitch, *J. Phys. Chem.*, **68**, 1322 (1964).
61. M. C. Flowers and H. M. Frey, *Proc. Roy. Soc.* (*London*), **A260**, 424 (1961).
62. C. A. Wellington, *J. Phys. Chem.*, **66**, 1671 (1962).
63. H. M. Frey, *Trans. Faraday Soc.*, **58**, 516 (1962).
64. (a) R. G. W. Norrish, *Trans. Faraday Soc.*, **33**, 1521 (1937); (b) N. C. Yang and D.-H. Yang, *J. Am. Chem. Soc.*, **80**, 2913 (1958).
65. H. R. Gerberich and W. D. Walters, *J. Am. Chem. Soc.*, **83**, 4884 (1961).
66. Such a study is in progress (private communication from Professor L. M. Stephenson).
67. (a) F. T. Thomas, C. I. Sutin, and C. Steel, *J. Am. Chem. Soc.*, **89**, 5107 (1967); (b) F. H. Dorer, *J. Phys. Chem.*, **73**, 3109 (1959); (c) F. H. Dorer, *ibid.*, **74**, 1142 (1970).
68. F. H. Dorer, manuscript submitted for publication.
69. (a) W. R. Roth, private communication. The azo compound data have been updated from those reported earlier[69b]; (b) W. R. Roth and M. Martin, *Tetrahedron Letters*, 3865 (1967).
70. P. J. Wagner, *Acc. Chem. Res.*, **4**, 168 (1971).
71. (a) P. D. Bartlett, *Q. Rev.*, **24**, 473 (1970); (b) P. D. Bartlett, *Science*, **159**, 833 (1968).
72. I. Tabushi and R. Oda, *Tetrahedron Letters*, 3743 (1968); see also M. Jones, Jr. and R. H. Levin, *ibid.*, 5593 (1958); H. H. Wasserman, A. J. Solodar, and L. S. Keller, *ibid.*, 5597 (1968); L. Friedman, R. J. Osiewicz, and P. W. Rabideau, *ibid.*, 5735 (1968).
73. C. G. Overberger and J. W. Stoddard, *J. Am. Chem. Soc.*, **92**, 4922 (1970).
74. C. G. Overberger, J. W. Stoddard, C. Jaroslavsky, H. Katz, and J.-P. Anselme, *J. Am. Chem. Soc.*, **91**, 3226 (1969).
75. H. J. Reich and D. J. Cram, *J. Am. Chem. Soc.*, **91**, 3517 (1969).
76. C. W. Reese and R. C. Storr, *J. Chem. Soc.* (*C*), 760, 765 (1969).
77. (a) J. A. Berson and E. J. Walsh, *J. Am. Chem. Soc.*, **90**, 4732 (1968); (b) M. R. Willcott and C. J. Boriack, *ibid.*, **90**, 3287 (1968); **93**, 2354 (1971).

78. (a) J. J. Gajewski, *J. Am. Chem. Soc.*, **92**, 3688 (1970); (b) J. C. Gilbert, *Tetrahedron*, **25**, 1459 (1969); (c) W. R. Roth and K. Enderer, *Ann.*, **733**, 44 (1970).
79. (a) P. Dowd, A. Gold, and K. Sachdev, *J. Am. Chem. Soc.*, **90**, 2715 (1968); (b) J. A. Berson, R. J. Bushby, J. M. McBride, and M. Tremelling, *ibid.*, **93**, 1544 (1971).
80. (a) J. J. Gajewski, *J. Am. Chem. Soc.*, **90**, 7178 (1968); (b) W. von E. Doering and H. D. Roth, *Tetrahedron*, **26**, 2825 (1970).
81. J. J. Gajewski and C. N. Shih, *J. Am. Chem. Soc.*, **94**, 1675 (1972).
82. J. W. Timberlake and B. K. Bandlish, *Tetrahedron Letters*, 1393 (1971), and references cited therein.
83. (a) C. H. De Puy and L. B. Rodewald, *Tetrahedron Letters*, 2951 (1964); (b) E. W. Valyocsik and P. Sigal, *J. Org. Chem.*, **36**, 66 (1971).
84. R. J. Crawford and T. R. Lynch, *Can. J. Chem.*, **46**, 1457 (1968).
85. (a) E. W. Yankee and D. J. Cram, *J. Am. Chem. Soc.*, **92**, 6328, 6331, 6333 (1970); (b) D. J. Cram and A. Ratajczak, *ibid.*, **90**, 2198 (1968).
86. R. Huisgen and H. Mäder, *J. Am. Chem. Soc.*, **93**, 1779 (1971), and earlier papers.
87. R. M. Kellogg, S. Wassemaar, and J. Buter, *Tetrahedron Letters*, 4689 (1970).

CHAPTER 6

Chemically Induced Dynamic Nuclear Polarization

HAROLD R. WARD
Metcalf Research Laboratories, Brown University, Providence, Rhode Island

I. INTRODUCTION

Five years ago, a chemist who observed negative peaks in an NMR spectrum might well have attempted instrument adjustment or repair to return to a "normal" absorption spectrum. Today a substantial number of research groups around the world strive mightily to observe such "abnormal" spectra, to understand their bases, and to use the results in the clarification of reaction mechanisms. I will present here an outline of the rapid evolution during this interlude, will give a brief and qualitative summary of the currently accepted theory, and will discuss at rather greater length the mechanistic applications which have been made.

II. THE PHENOMENA AND THE NOMENCLATURE

The expectation that these unusual spectra might bestow some chemical benefit (beyond their values as spectral curiosities) is reasonable, since they are observed only from reaction products either during or shortly following their formation. Furthermore, in every case, the reaction in question can reasonably be believed, on independent evidence, to possess free radical character, so that the spectra would seem to have their origin in the process of a radical transmutation of reagent to product. In addition to emission lines, these spectra display, usually concurrently, absorption signals of unusually high intensities, reflections of the deflection of the populations of

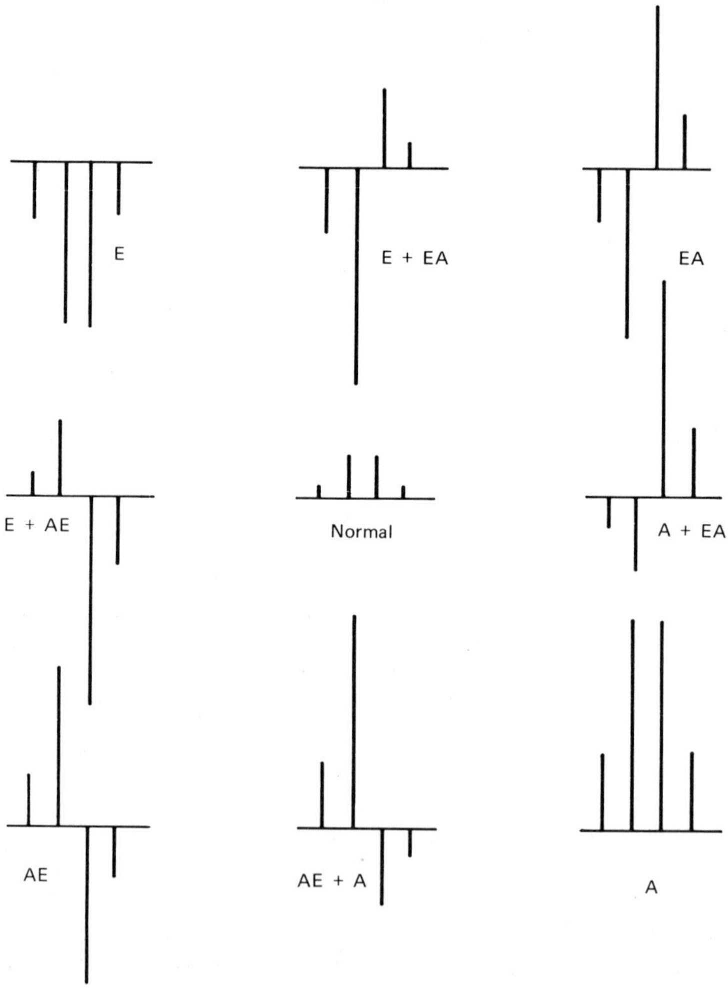

Fig. 1. Examples of multiplet effect (AE and EA) and net (A and E) polarization, appearing separately and in combination.

nuclear spin state levels from equilibrium in either direction, or the "polarization" of nuclei. Depending on the polarizing conditions, a multiplet of lines may show emission (E) or enhanced absorption (A) for each line (*net* polarization) or E for some lines and A for others. When both E and A appear in this way, the polarization is said to be of a *multiplet effect* type; the purest form of which, for a first-order spectrum, would have exactly as much E as A. The order with which each occurs with increasing magnetic field (during a spectral scan) will be a matter of particular interest; and this *phase* may be either EA or AE. Finally, the combination of net polarization and multiplet effects in a single spectrum is not uncommon. Examples of the extreme polarization types and their combinations are given in line spectra in Figure 1.

III. THEORY

A. Overhauser Model

Following the first observations[1,2] of nuclear polarization during chemical reaction, it was reasonable to search for whatever analogies chemical experience could provide, and the search revealed what appeared to be a closely related effect named dynamic nuclear polarization[3] (DNP). The similarities were convincing, since both E and A could be observed, and particularly because the polarizing mechanism depended on the interaction of nuclei with an unpaired electron (in a free radical or paramagnetic ion). DNP experiments produced polarization by pumping electron spin levels with the appropriate microwave frequencies and simultaneously observing the nuclear polarization which resulted from electron–nuclear cross-relaxation (Overhauser effect). It seemed a safe extension[1,4] to suggest that a chemical reaction might produce free radicals in which the electron spin levels were saturated, and that the Overhauser effect which results as the electron levels seek equilibrium population could explain the observed nuclear effects. Accordingly, the name of chemically induced dynamic nuclear polarization (CIDNP) was selected.[1] The details of this model have been presented in a review by Fischer and Bargon[5] and will not be reproduced here.

B. Radical Pair Model

Flaws in the Overhauser model became apparent shortly after its formulation, and no reasonable modifications were capable of bringing a respectable fit of theory with experiment. In particular, the magnitude of the observed polarizations was larger than allowed by the Overhauser model, and there seemed to be no way to account for the multiplet effect. Also, evidence accumulated that the polarization occurred during a product-forming step, and, in some cases, in reactions of radicals with near equilibrium electron spin populations.

A theory which explains the polarization by interactions of paired radicals was proposed by Kaptein and Oosterhoff[6] and independently by Closs[7] and has proved to be a far more satisfactory theory. The mathematical treatment of this theory is available in the original presentations,[6,7] in extensions,[8–12] and in review,[13] and it does not serve the purpose of this review to repeat it here. Instead, the classical model developed by Lawler will be presented, to give a qualitative idea of the spectral predictions, and to relate the various manifestations of polarization to the physical conception of radical reactions.

The effects giving rise to nuclear polarization are perhaps more easily visualized if one follows the peregrinations of a radical formed in solution as a partner in a pair. If the pair is formed by a thermal bond homolysis of a diamagnetic reagent, or by electron transfer between diamagnetic reagents, the pair should be in the singlet state, S, since no mechanism exists for changing the spin multiplicity in the 10^{-13} sec in which such acts occur. Conversely, the reaction of a triplet-state reagent will be expected to produce a triplet radical pair. The singlet and one of the triplet-state pairs (T_0) are shown in Figure 2, for the situation where the pairs are exposed to a strong external magnetic field (H_0) and the interaction within the pair (exchange

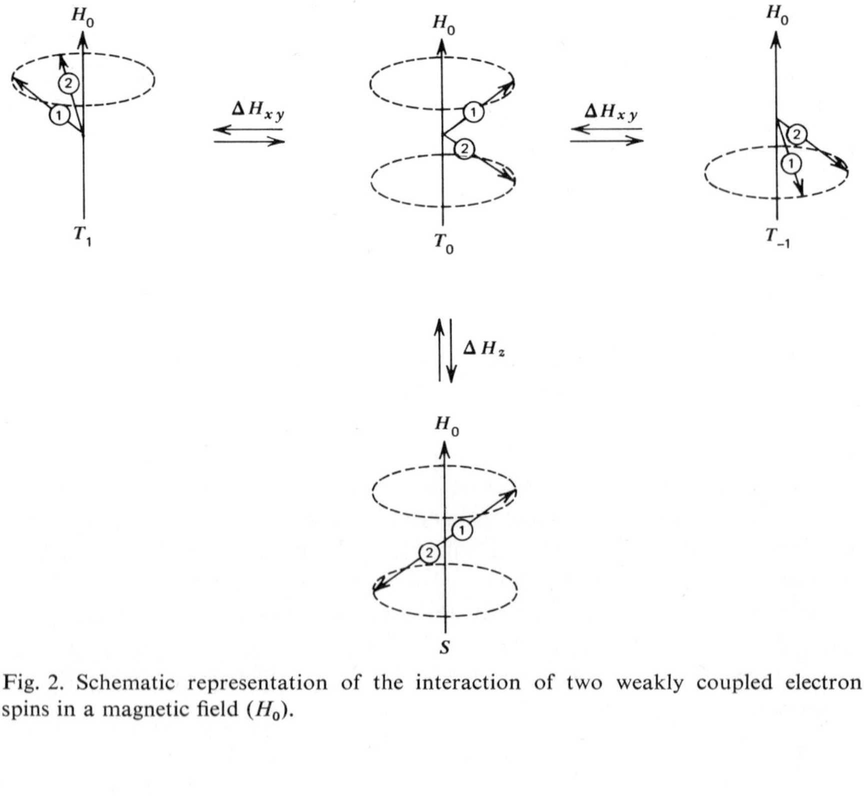

Fig. 2. Schematic representation of the interaction of two weakly coupled electron spins in a magnetic field (H_0).

interaction) is weak. The electrons will precess around the Z axis at a frequency which depends on the effective field which they experience, a combination of the external field and the hyperfine field from any nuclei with which they may be coupled, and on the g factor of the radical. It is apparent from Figure 2 that a difference in precessional frequency between radicals 1 and 2 will lead to a mixing between the S and T_0 states, and that, within a limited time period, the larger the frequency difference, the greater the mixing.

The paired radicals, confined as they are by surrounding solvent molecules, may well react, by combination or disproportionation, before diffusing apart (to "infinity"). The ratio of pairs reacting in this way to the total number of pairs formed is the cage effect. Following the guidance of Noyes,[14] we shall term such pairs as geminate and their progeny as geminate products. Since these products are unlikely to form in the triplet state, it is assumed that any geminate reaction which occurs will be from predominantly S pairs. In addition, S pairs can intersystem cross to T_0 or become separated into free radicals. On the other hand, T_0 pairs will either separate by diffusion or will intersystem cross to S. Since the inherent rate of diffusion is taken to be multiplicity independent, it follows that T_0 pairs will be longer lived than S pairs.

The relation between this basic idea of S–T_0 mixing and observable CIDNP spectra of the net E or A type (Figure 1) can be seen in an artificially simplified model system. Given two radicals of differing g factors formed in a singlet geminate pair, it is clear that, even in the absence of a hyperfine field, the pair will develop increased triplet character with the passage of time. If the radical of lower precessional frequency (lower g factor) is coupled to a proton with, say, a $+\frac{1}{2}$ spin (the lower nuclear level) by a negative hyperfine coupling constant ($a < 0$, as it would be for an α proton), the hyperfine field will oppose the applied field, and the precessional frequency will be decreased. The result is a greater frequency difference in the two radicals, a more rapid crossing to the triplet state, and a concomitant greater probability of finding radicals with $m = +\frac{1}{2}$ states free in solution. Had the radical with lower g factor instead been coupled to an $m = -\frac{1}{2}$ proton, again with a negative a, its precessional frequency would have been increased, the S–T_0 crossing of the pair decreased, and the probability of the formation of geminate product would remain relatively high. Emission would then be anticipated for geminate product and enhanced absorption for any free radicals which could be scavenged. A proper chemical example to which this reasoning can be directly applied is the thermal decomposition of acetyl trichloroacetyl peroxide in a carbon tetrachloride solution containing iodine.[15] The coupling product, 1,1,1-trichloroethane, shows a single emission line at $\delta 2.7$, and free methyl radicals are trapped by iodine to give methyl iodide, with enhanced absorption at $\delta 2.1$.

It is easy to extend these ideas to a similar pair with the proton this time implanted on the radical of higher g factor (once again with $a < 0$). Following through the same arguments will show that $m = -\frac{1}{2}$ will assist singlet–triplet mixing—giving E in the scavenged radicals, and $m = +\frac{1}{2}$ will remain in the geminate products, giving A. Of course, not *all* of $m = +\frac{1}{2}$ is found in geminate products, but even 1% spin selection is adequate to lead to spectral enhancements of 10^3.

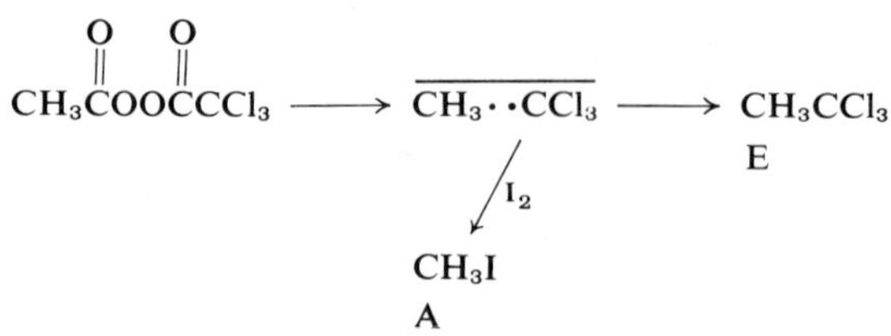

Another perturbation in the model system (proton on the low g-factor radical, $a < 0$) is to initiate the pair in the triplet state. Now a greater difference in precessional frequency leads to the reactive S state and the $-\frac{1}{2}$ state should predominate in free radicals, and the $+\frac{1}{2}$ state in geminate product, just the antithesis of the singlet pair result.

Finally, it is necessary to consider a pair formed by the random motion of two free radicals, a diffusive encounter pair. Statistically S and T_0 form equally, and a priori any S–T_0 mixing should be compensated exactly by T_0–S mixing, leading to no spin selection. However, recalling an earlier conclusion that T pairs are longer lived than S pairs, it can be seen that in the macrospin system T_0 pairs will dominate, and that spin selection in diffusive encounters should mimic T_0 geminate polarization.

Pure multiplet-effect polarization evolves from pairs of radicals of equal or near equal g factors (e.g., most unsubstituted alkyl radicals). Since $\Delta g = 0$, in the absence of other effects, only the hyperfine field is available to induce intersystem crossing. Obviously, if the hyperfine fields experienced by both radicals are equal, no mixing will occur. Equally obviously, in some pairs, the hyperfine fields will not be equal. A little thought reveals that it is those radicals with high (absolute) hyperfine fields that will, on the average, have a greater Δg compared to radicals with randomly chosen spin states. An ethyl radical paired with a (hypothetically) protonless radical of equal g factor provides a useful example. The negative a for the α and positive a for the β protons lead to a high value of $|am|$ for the inner nuclear levels, as shown in Figure 3. Given a positive nuclear spin–spin coupling constant (J), the overpopulation of these levels will give rise to an AE spectrum in scavenged free radicals. A proper example is seen in the decomposition of propionyl peroxide in solutions containing iodine,[16] with the AE clearly visible for the α-CH_2 quartet and β-CH_3 triplet of ethyl iodide (Figure 4).

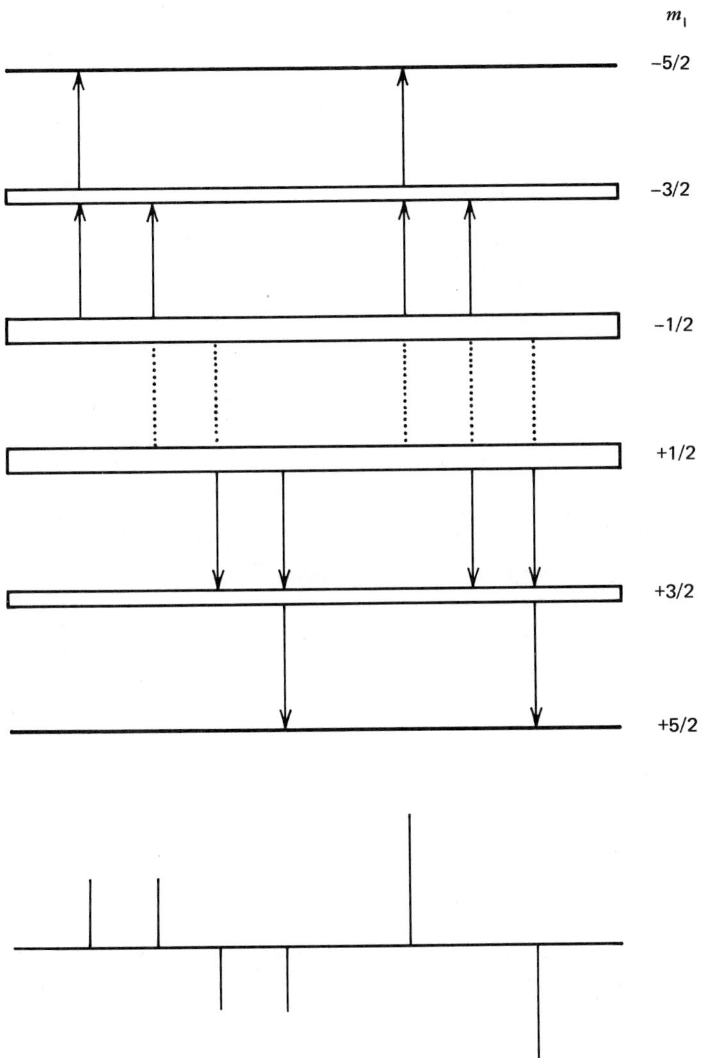

Fig. 3. Simplified nuclear spin energy level diagram for ethyl iodide formed from ethyl radicals which escaped a geminate pair reaction.

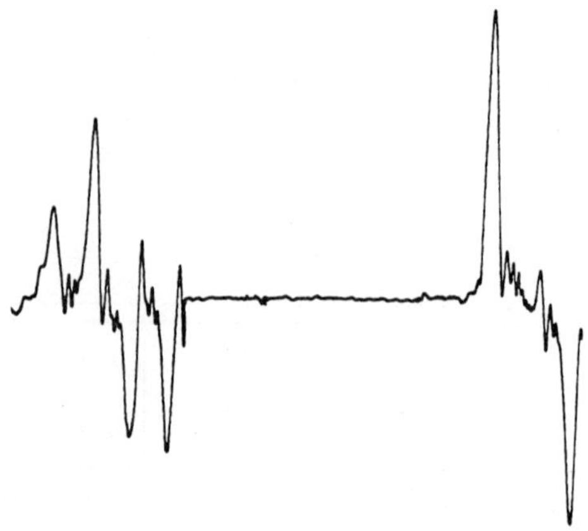

Fig. 4. CIDNP spectrum of ethyl iodide formed in the thermolysis of propionyl peroxide in *o*-dichlorobenzene solution containing iodine.

The complementary spin population must remain in geminate products, which, given a $J > 0$, will have an EA multiplet effect. The expected phase reversal occurs for pairs formed as triplets or by diffusive encounters, leading to EA for scavenged radicals.

These specific examples are generalized in Table 1 for geminate products. The tabulated predictions can be applied to scavenged radicals by simply

TABLE 1

Polarization Character of Geminate Products

	a_A	a_B	Birth	g	J_{AB}	H_A	H_B	
HA· + ·BH → HABH								
	−	−	S	A > B	+	A,EA	E,EA	(1)
	−	+	S	A > B	+	A,AE	A,AE	(2)
	−	−	T	A > B	+	E,AE	A,AE	(3)
	−	−	S	B > A	+	E,EA	A,EA	(4)
	−	−	S	A > B	−	A,AE	E,AE	(5)
R· + CH_BCH_A· → RCH_ACH_B								
	−	+	S	R > C	+	E,EA	A,EA	(6)
	+	−	S	R > C	+	A,EA	E,EA	(7)
	−	+	T	R > C	+	A,AE	E,AE	(8)
	−	+	S	C > R	+	A,EA	E,EA	(9)
	−	+	S	R > C	−	E,AE	A,AE	(10)

reversing the predicted phase (i.e., A,AE becomes E,EA). This qualitative treatment is no substitute for the quantitative calculations of spectral enhancements, and cannot predict the relative amount of net polarization and multiplet effect as a proper spectral simulation can. At the moment, absolute spectral enhancements contribute little to the understanding of reaction mechanisms. This will change as the quantitative aspects of the theory are refined, and a computer simulation of the polarized spectra will then almost certainly be *de rigueur* in CIDNP studies.

IV. MECHANISTIC APPLICATIONS

The mechanistic information that can be derived from CIDNP spectra falls into several classes:

a. Mere proof of radical ancestry for particular products.

b. The identity and characteristics of radical precursors.

c. The identity of pairs in which polarization occurs and the mode of pair formation.

d. Some information on radical pair separation and on radical lifetimes.

e. Potentially, the direct observation of unstable diamagnetic reaction intermediates.

Before a sensible discussion of CIDNP applications can be presented, it is essential to examine critically the creditability and applicability of each point. One attribute of CIDNP pervades all of the topics and must be kept constantly in mind as each topic is considered in turn: namely, that the spectral enhancements which can be observed are truly enormous. To reach the same level of enhancement by the use of signal averaging with a computer of average transients, for example, would require 10^4 to 10^6 spectral scans! This sensitivity offers attractive advantages, but also presents clear dangers of misinterpretation, as we shall see.

There is no reason to doubt that when sizable spectral enhancements are observed for a particular product, that product was formed, at least in part, from a radical intermediate in which the nuclei which show enhancements were coupled to the unpaired electron. (Intermolecular polarization, in which nuclei in one molecule are polarized by encounters with an unpaired electron in another molecule, is a known effect in dynamic nuclear polarization, and ought, in theory, to be observable in CIDNP systems, in the same sense that Overhauser effects ought to be possible. In fact, no intermolecular polarization has yet been reported, presumably because these cross-relaxation effects simply are much smaller than those of spin selection.) Unfortunately, even granting that the polarizations are intramolecular does not resolve all ambiguities. The formation of some portion of the product by a radical route does not exclude the concomitant formation by an ionic or a concerted path, and by far the most common criticism of the mechanistic utilization of CIDNP (especially in reactions not widely held to be radical) is that the polarization may arise from a chemically insignificant reaction, while the bulk of the product forms in a "traditional" manner. The point will be resolved when enhancement factors can be calculated and measured with reasonable confidence. Until then, the objection can be seen not to be generally important simply because it is unlikely that two parallel reaction paths, one of which is radical, will exist in the delicate energetic balance necessary to allow competition. Nonetheless, it is a precaution which must apply to every *individual* study.

The (inevitable) comparison of CIDNP with ESR on the points of sensitivity and latitude of information favors CIDNP on both accounts (for reactive radicals). The free radical concentrations in several systems which give good NMR intensities can be calculated at 10^{-6}–10^{-7} M, just at, or below, the detectability limit of ESR. In the only systems which have been investigated by both techniques,[17,18] either signal accumulation or drastic acceleration of reaction rates (by factors of 100–1000) was necessary for ESR detection. The inference of a radical precursor to some product of interest based solely on the ESR observation of the radical is viewed with suspicion because of the possibility that the radical may be quite independent of the path of product formation. In CIDNP at least this ambiguity is absent, since the effect arises from the product and not the intermediate. As an example, during the reaction of Grignard reagents with ketones, the ESR spectrum of the ketyl radical was observed,[19] and the addition product was supposed to follow by radical combination. Subsequent kinetic studies[20] showed the ketyl radical formation and decay rate to be unrelated to the rate of product formation, and established the independence of the two paths. The corresponding CIDNP studies would have inspected the NMR spectrum of the product for intensity irregularities. (Reactions of Grignards with a variety of ketones have shown no CIDNP spectra,[21] in situations which would in other respects seem suitable for their observation.) The identity of a radical intermediate obviously cannot be assigned as certainly by CIDNP as if it could be directly observed, but enough information is often available to allow selection among reasonable alternatives. Furthermore, for radicals which have not been (or cannot be) observed by ESR, g factors and hyperfine coupling constants can be determined by the values necessary for simulation of experimental CIDNP spectra.

The identities of the partners of a radical pair often are available from CIDNP studies, but can be obtained equally well from a careful examination of product structures. Chemical and structure studies usually do not reveal the spin multiplicity of the pair, however, and CIDNP is unique in its ability to assign predominant singlet or triplet character to a geminate pair or to distinguish between a singlet geminate encounter or a diffusive encounter. It is important to remember, however, that such assignments refer only to the dominant polarizing encounter, and the current, qualitative treatments have no way of apportioning the contributions of, say, a geminate and a diffusive encounter, when both occur. Such information may be available when quantitative CIDNP techniques are developed, but even then probably only for systems where the kinetics are well understood.

Most information on free radical lifetimes and on radical pair separations has been derived from the effect of scavenger concentrations and reactivities on product distributions. Since polarization intensities depend sensitively on pair lifetimes, the effect of added scavenger on spectral enhancements will provide an independent approach to the estimation of radical pair longevity.

Finally CIDNP should be particularly well suited to the detection of diamagnetic reaction intermediates, with lifetimes limited by a high probability for further reactions. If such an intermediate exists for about 0.5 sec or longer after its formation, the line widths (5 cycles or less) and enhancements should allow its direct observation. If subsequent reaction of the intermediate is rapid (less than $1/T_{1n}$), polarization will be detectable also in the reaction product, but at a lower intensity. This short listing of applications is almost certain to be incomplete, and should be expected to expand on the basis of firmer theory and continued experimentation.

The interpretation of the *total* absence of CIDNP as negative evidence for radical intermediates has been deliberately excluded since, at this stage of development, it is in high risk of error. Reaction rates, cage effects, nuclear spin relaxation times, and, in the case of diradicals, the exchange interactions must all be known to some precision in order to responsibly use the failure to observe enhancements as evidence for a nonradical reaction. However, the absence of a particular polarization, in the presence of CIDNP signals from other products of the same reactions, may in some cases be permissive evidence for the exclusion of solvent-separated radical pairs from the list of possible product progenitors.

V. EXPERIMENTAL TECHNIQUES

There is little doubt that, while the basic interest in CIDNP may arise from its novelty, the rather large number of research groups that have embraced the technique have done so because many CIDNP experiments are exceedingly easy to conduct (a characteristic which led to its serendipitous discovery). For many studies, a standard high-resolution NMR spectrometer (either 60 or 100 MHz) can be used with no modification, and, if a variable-temperature probe is available to allow control of reaction rates, a wide range of systems can be investigated. In the simplest approach, the conditions are chosen which give a reaction half-life of 30 sec to 30 min, and the spectrum of the reacting solution is scanned under standard slow-passage conditions. At the radical concentrations produced by rates in this range, the higher the reaction rate, the larger will be the observed enhancement (because of increasingly successful competition with T_{1n}). Of course, at the higher rates, less time is available for spectral scans, and only small regions of the spectrum may be examined during a single reaction. The resulting composite scans are adequate to observe the qualitative character of the polarization (e.g., E or A, EA or AE) but do not allow accurate quantitative comparisons between enhancements of different proton signals in the same reaction. In cases where the polarization is substantial and the kinetics are known, this difficulty is minimized, but more general solutions are desirable. These approaches can take two forms, either increasing the rate of scanning (Fourier transform spectroscopy) or by simulating steady-state conditions by the use of flow systems or by photolysis at relatively low light flux. Both the flow and photolytic systems require special apparatus or probe design which may be discouraging to the dilettante. When very rapid reactions are studied, the magnitude of the magnetic field in which reagents are mixed may determine the character of the polarization.[30,31] Reactions which lead to net enhanced absorption for particular protons when conducted in the spectrometer field may show net emission for the same protons if the reaction is completed at low fields and the products transferred to the spectrometer for observations. At this point in CIDNP development reactions at high field have greater mechanistic utility, since greater confidence can be placed in the theory developed for the high-field limit. Lower-field studies have potential, however, since it appears that they may reveal polarization which may not be observable at higher fields. Discussion in this chapter is confined entirely to reactions conducted at high fields.

The polarization of nuclei other than protons can be carried out on the same equipment that is used for normal absorption spectroscopy for these nuclei. For ^{13}C studies at the isotopic natural abundance, however, it appears to be necessary to noise decouple protons to increase the signal-to-noise ratio to a useful level.[24]

VI. CHEMICAL SYSTEMS

The majority of the CIDNP studies have been directed toward a better understanding of the phenomena and most chemical systems have been chosen to provide polarization examples rather than to test mechanistic ideas. This has been an entirely proper approach and has been moderately successful in the development of working theoretical models. It is certainly reasonable to expect greater mechanistic utilization in the future when more experiments designed for this purpose have been conducted, but even the early studies, sometimes fortuitously, have required new interpretations of reaction pathways, or have provided support for earlier mechanistic suggestions. These aspects will be emphasized in the following sections, in which summaries of CIDNP studies are presented, divided into related functional groupings.

A. Peroxide Decompositions

The thermal decompositions of diacyl peroxides have been selected as the first example of CIDNP spectra because they are chemically the least ambiguous. Kinetic and isotopic labeling studies[25] have shown that for *n*-alkyl diacyl peroxides (other than acetyl) the first step of oxygen–oxygen bond cleavage is followed rapidly by decarboxylation, thus generating two alkyl radicals in close proximity. The multiplicity of the pair so generated is certain to be largely singlet, and the cage effects (usually $\sim$0.5) are entirely suitable for spin selection. Further, the decompositions occur at a convenient rate at 100–200°, temperatures easily available in an NMR probe. Two *n*-alkyl radicals will react by combination and disproportionation, the latter giving olefins in which the vinyl protons were coupled to an unpaired electron in the radical, and by fortunate coincidence, with line positions in a spectral region which is clear of interference by solvents and reagents. Radicals which escape these pair reactions can be scavenged, and thus both portions of the nuclear spin selection can be observed directly. Very nice examples of such reactions were provided by Kaptein,[26] using hexachloroacetone as solvent (and scavenger), so that radicals escaping a geminate encounter were trapped as the alkyl chlorides. Our studies[16,27] have used alkyl iodides or iodine as scavenger agents (as adjuncts to alkyllithium reaction studies) and the polarization from one such system, the decomposition of lauroyl peroxide in solvents containing isopropyl iodide, is suitable for discussion and leads to the spectrum in Figure 5.

The polarization of the methylene protons in undecyl iodide (δ3.2) is a pure multiplet effect, with phase AE. The polarization of the vinyl protons of

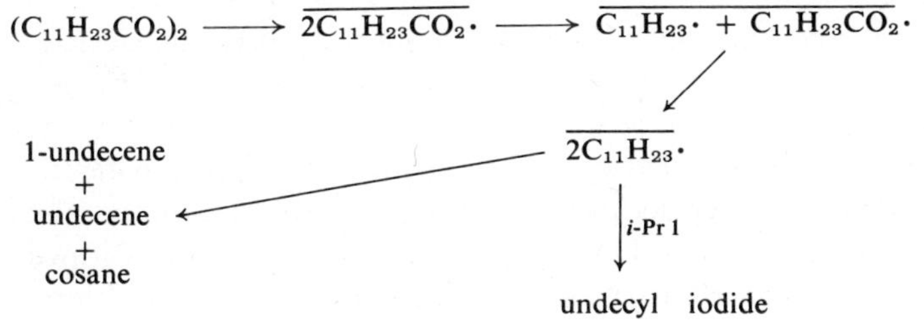

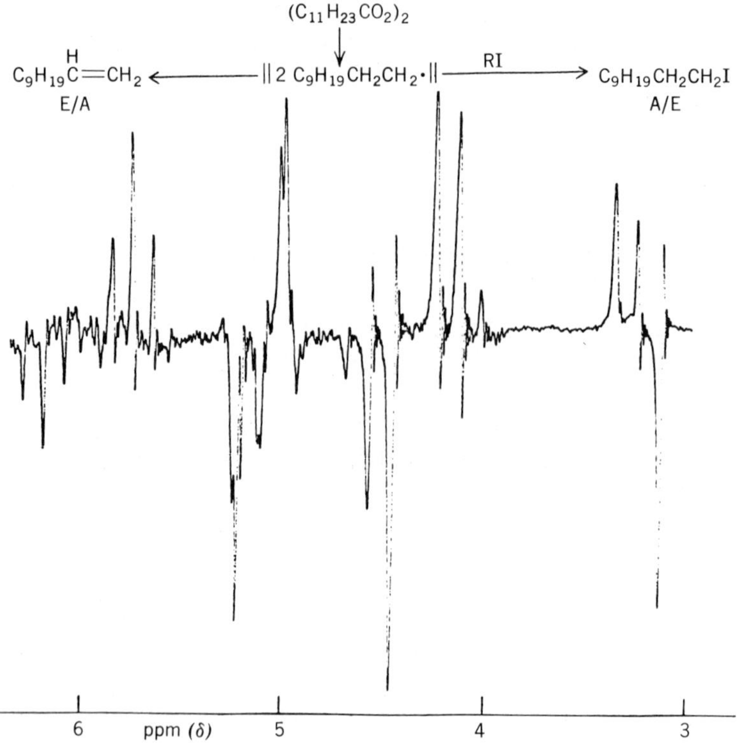

Fig. 5. Thermolysis (120°) of lauroyl peroxide in *o*-dichlorobenzene solution containing isopropyl iodide.

1-undecene (methylene δ5.0, methyne δ5.9), though more complex, is similarly of a multiplet type but, necessarily, of the opposite (EA) phase. Both are consistent with the predictions in Table 1, line 6, and with reasonable chemical intuition as to the mode of their formation. (The polarization of the methyne proton of isopropyl iodide at δ4.3 is discussed later.) The absence of any net polarization is probably the most mechanistically significant point in this study. The timing of the decarboxylation of the acyloxy radicals is uncertain, and only a lower limit can be set to its rate by the failure to trap these radicals by chemical scavenging. If, however, a pair of alkyl and acyloxy radicals (formed by a single decarboxylation) existed for 10^{-9}–10^{-10} sec, *even if there were no chemical reaction between them*, the Δg effect should be observed by net polarization in products of the succeeding alkyl radical pair. In fact, very minor amounts of ester do form in this reaction, but exhibit no polarization. Thus it is possible to conclude that the alkyl radical pair is formed in times less than approximately 10^{-10} sec, and that the ester forms within this time, or results from the independent path of carboxy inversion. (Notice that the criticisms of negative CIDNP evidence given earlier do not apply with the same force to the absence of polarization in one product when significant polarization can be observed in other products of the same reaction.)

The photolytic decomposition of the *n*-alkyl diacyl peroxides in scavenging solvents[28] gives results comparable to those from thermal decomposition when irradiation is direct, confirming the expectation that reaction occurs from an excited singlet state (or during its internal conversion). When the decomposition is photosensitized by aromatic ketones (known triplet sensitizers), the phase of the product polarization is inverted, good evidence for the formation of the geminate pair predominantly in the triplet state. Since the photolysis was carried out in a scavenging solvent, there is small probability that this polarization results from diffusive encounters.

Gratifyingly, in the exceptional case of acetyl peroxide thermolysis[26] (in hexachloroacetone) where there is independent evidence to support a longer-lived acetoxy radical, net polarization is observed for methyl acetate, methyl chloride, and ethane. Again the polarization of the geminate products (E) is opposite that of methyl radical (A) scavenged by chlorine donation. The net polarization of ethane, which cannot result (at high fields) from encounters between identical radicals, must have arisen from the acetoxy–methyl interaction, followed by decarboxylation and dimerization. The decomposition of acetyl peroxide (^{13}C enriched in the methyl positions) in hexachloroacetone[29] shows both multiplet and Δg effects in the ^{13}C spectrum for the ethane, methyl acetate, and methyl chloride which form, from which the g value of the acetoxy radical can be calculated at 2.0058 ± 0.0002. The short lifetime of the acetoxy radical has precluded direct ESR measurement of this value.

The mixed peroxides of trichloroacetyl with acetyl and propionyl[15] generate a geminate pair of a trichloromethyl ($g = 2.0091$) and an alkyl radical ($g = 2.0025$ for methyl) with a Δg of 0.007. As expected from such a large Δg, the polarization is large and can be clearly observed in slow reactions for periods of up to an hour. In the ethyl–trichloromethyl pair, the multiplet effect is totally obscured by the much larger net polarization from the high Δg.

The decomposition of a cyclic C_{12} diacyl peroxide (prepared in situ) provided the first example of polarization from a diradical.[29] Clearly the eight-carbon separation in the 1–10 diradical is adequate to allow nuclear spin selection. The theoretical prediction is that the spin selection will vanish at small radical separations (where the exchange interaction is large) but the exact point at which the effect will be observable is still undetermined. (However, see the section on azo compound decompositions.)

The mechanistic problems in the decomposition of branched-chain diacyl peroxides are more challenging. Products, in addition to those from normal radical reactions, are increased amounts of ester and substantial quantities of carbonate esters, formed by a rearrangement termed "carboxy inversion." Both rate and product distributions are sensitive to solvent polarity, with carboxy inversion products increasing with increasing polarity, while cage radical products and scavengable free radicals decrease. The simplest example, isobutyryl peroxide, has been examined by CIDNP in our laboratories[16] and by Walling et al.[30] with kinetic and product studies. Walling has proposed a general scheme (1) which encompasses this dual ionic and radical nature by the postulate of an intimate pair with both characteristics, and in which the radical description becomes better as the pair separates.

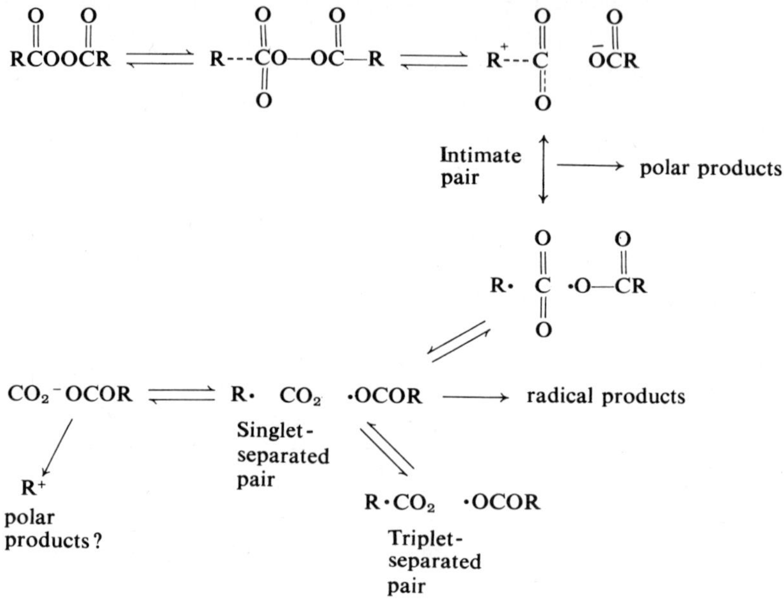

Scheme 1

CIDNP studies of isobutyryl peroxide decomposition in *o*-dichlorobenzene were necessarily carried out at higher temperature (110°) than were the kinetic studies (20–70°), which should slightly increase the proportion of "polar" products. The presence (and absence) of polarization in certain products allows several comments on the validity of Scheme 1.

$$(i\text{-PrCO}_2)_2 \longrightarrow \underset{\text{NP}}{i\text{-PrO}\overset{\text{O}}{\overset{\|}{\text{C}}}\text{O}\overset{\text{O}}{\overset{\|}{\text{C}}}i\text{-Pr}} + \underset{\text{NP}}{i\text{-PrO}\overset{\text{O}}{\overset{\|}{\text{C}}}i\text{-Pr}} + \underset{\text{E/A}}{\text{C}_3\text{H}_6} + \text{C}_3\text{H}_8 + \text{C}_6\text{H}_{14}$$

The strong E/A polarization of propene apparently arises from a geminate encounter of two isopropyl radicals, and is entirely analogous to the mode of primary peroxide decomposition (polarization is present in propane and 2,3-dimethylbutane, but is less easily observed because of spectral complexities and absorption overlap). It is highly unlikely that intersystem crossing occurs in pairs of isopropyl and isobutyryloxy radicals, since no net polarization occurs in any product. Further, it is unlikely that such pairs ever reach a separated state since, from such a state, net polarization from spin selection *should* occur. (When benzoyloxy–alkyl geminate pairs are formed, this net polarization is quite strong.) The *total* absence of polarization in the ester and carboxy inversion product is difficult to certify because of partial line position overlap with the strongly enhanced signals from propene. However, if polarization in these products is present, it must be very weak. This does not exclude some radical character in the "intimate pair" in Scheme 1, but it does require either that there is a very small hyperfine coupling between the isopropyl proton and an unpaired electron or that this pair exists for a very short time (or both). There is no evidence for polar product formed by recombination of a separated (radical) pair.

The decomposition of aroyl peroxides (specifically benzoyl) was the first system[1] in which CIDNP was observed and recognized. The reaction has been repeated both thermally[24,27,31,32] and photolytically[23,22] in several laboratories, and has certainly received the most careful study of any of the CIDNP systems. Early interpretations[1] of polarization observed in the benzene formed have been revised to consistency with the spin selection model,[34] and it is now most probable that the principal polarizing interaction is a geminate encounter between a benzoyloxy and a phenyl radical giving the

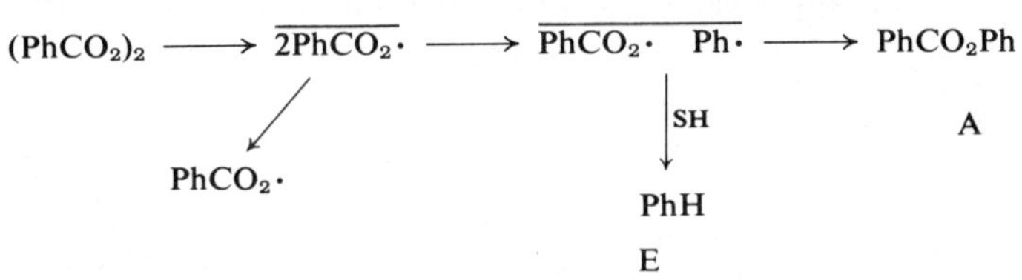

expected emission for the phenyl protons in benzene and enhanced absorption for the ring protons in the *O*-phenyl group. Such a geminate pair was postulated earlier[35] to explain the inability to prevent the formation of about 5% phenyl benzoate by addition of large amounts of scavenger. The decomposition by direct photolysis gives the same polarization, but triplet sensitizers lead to a triplet radical pair, and show the characteristic inverted polarization phase.[28,33] The detection of singlet energy transfer[28] from certain hydrocarbon sensitizers is a very nice demonstration of the ability of CIDNP to identify the spin multiplicity of radical pairs in photosystems.

Carbon-13 polarization has been observed[24] (at natural abundance) during benzoyl peroxide decomposition at 110°, and has been shown to provide some information beyond that available from the proton spectrum. Predictions of ^{13}C polarization are complicated by uncertainties in signs and magnitudes of the appropriate hyperfine coupling constants althought the spectrum is generally explicable in terms of a benzoyloxy–phenyl radical pair. The substituted ring carbons and the carbonyl carbon in phenyl benzoate show strong enhanced absorption, and benzene and carbon dioxide give complementary emission lines.

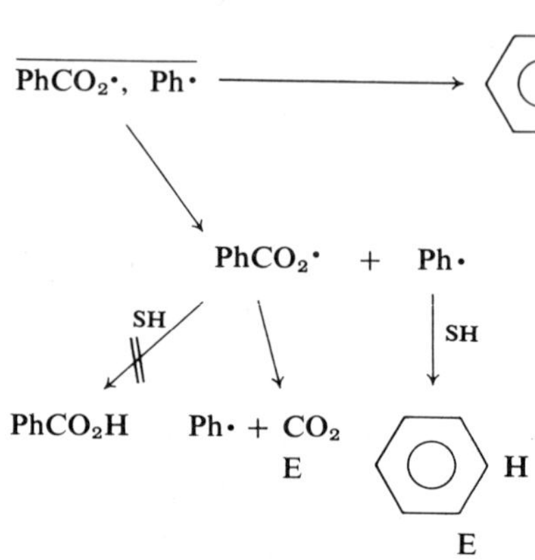

Since benzoic acid shows no polarization it is improbable that it forms from a hydrogen abstraction by benzoyloxy radical, and induced decomposition is a more likely source. Proton spectra could not have given this information because no appreciable proton hyperfine interaction exists in the benzoyloxy radicals. Even with the large polarization expected from the substantial ^{13}C hyperfine coupling and the longer ^{13}C T_1, the successful observation of polarized spectra in this case required proton decoupling and spectra accumulation.

Radicals which escape the geminate pair provide, via abstraction reactions, a convenient extra-cage radical preparation. If lauroyl peroxide is decomposed in a solution containing 2-iodopropane, for example, iodine abstraction forms unpaired isopropyl radicals (and undecyl iodide with polarization displayed in Figure 5).[16] There is no evidence that the act of abstraction produces any nuclear polarization[36] (for the sufficient reason that it is too rapid) and isopropyl radicals will form with the nuclear spin state population of the isopropyl iodide. Since the abstraction of iodine by the undecyl radical is rapid,[37] the isopropyl radical will be the dominant free radical and, since it is formed in concentrations of about 10^{-6} M, it will be consumed by diffusive encounters. Combination products show the appropriate multiplet effect and isopropyl radicals which do not react carry the complementary polarization. As these radicals abstract iodine from isopropyl iodide (an identity reaction)

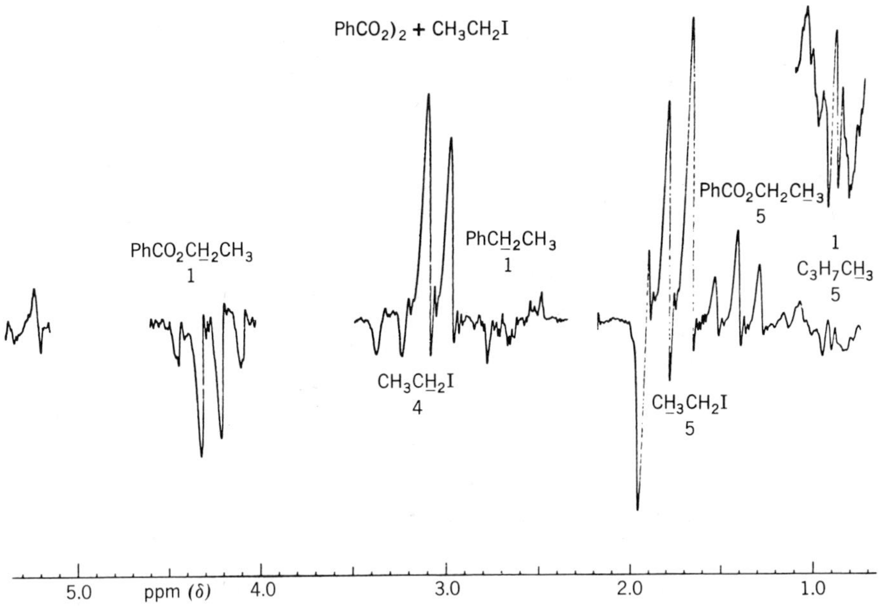

Fig. 6. Thermolysis (130°) of benzoyl peroxide in *o*-dichlorobenzene solution containing ethyl iodide.

they introduce an EA polarization into the reagent 2-iodopropane (methyne proton δ4.5, Figure 5).

Aroyl peroxide decomposition in mixed solvents has been used by Lehnig and Fischer[38] to generate mixtures of free radicals and to study polarization in unsymmetrical combination products. The method proves to be particularly useful in the formation of pairs with known Δg, so that spectra can be simulated for comparison with experimental observations. Alternatively, given confidence in the simulation, relative *g* factors can be calculated by spectrum fitting.

When the trapping agent is sufficiently reactive, as with iodoethane in a benzoyl peroxide decomposition, reasonable evidence of trapping one partner of a geminate pair can be obtained.[16] The polarization of cross-coupling products (in this example, ethyl benzoate; Figure 6) is indicative of a geminate pair spin selection and seems to result from a sort of pair substitution of ethyl

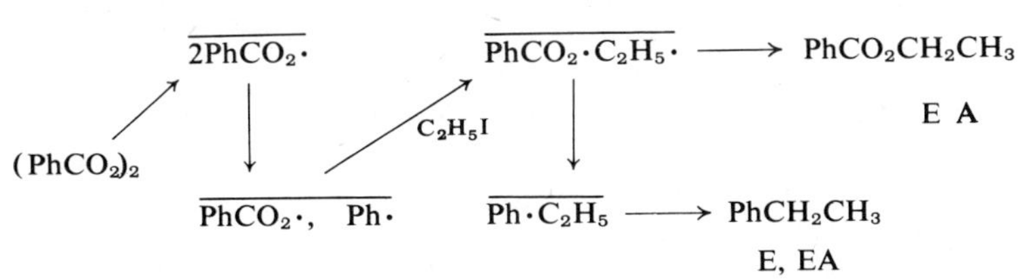

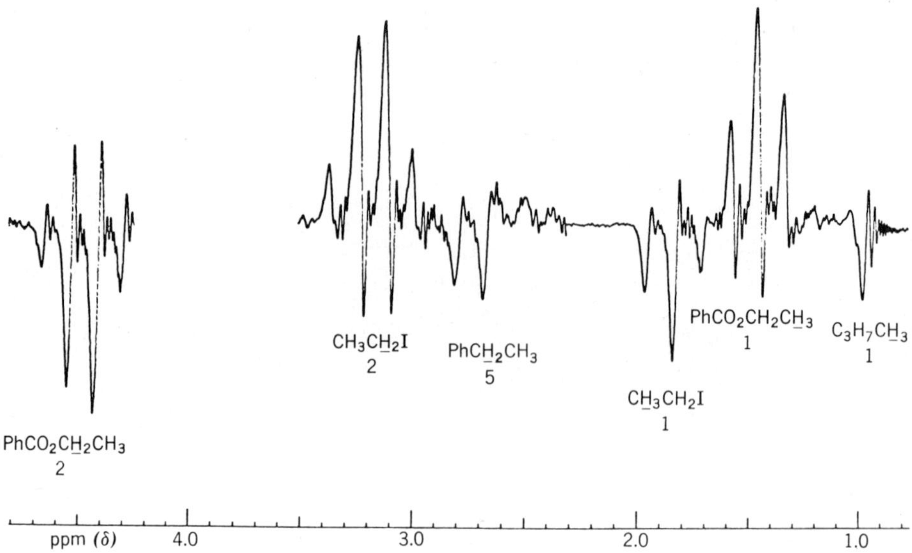

Fig. 7. Thermolysis of benzoyl propionyl peroxide in *o*-dichlorobenzene solution containing iodine.

for phenyl, while maintaining singlet character. The reduction of the iodobenzene polarization with increasing iodoethane concentration is nicely consistent with this model. The best evidence for this idea comes from the exact correspondence of the ethyl benzoate and ethylbenzene polarization from the trapping experiment and from a benzoyloxy–ethyl singlet pair generated directly from the mixed peroxide (Figure 7).

$$\mathrm{Ph\overset{O}{\overset{\|}{C}}O O\overset{O}{\overset{\|}{C}}CH_2CH_3} \longrightarrow \overline{\mathrm{PhCO_2\cdot,\ CH_3CH_2\cdot}}$$

In both reactions the ethylbenzene–methylene proton polarization (E + EA) suggests formation in a geminate combination, following benzoyloxy decarboxylation, rather than a diffusive encounter of phenyl and ethyl (AE) or a concerted bond cleavage to form singlet phenyl–ethyl pairs (EA). This evidence that benzoyloxy decarboxylates in a "cage" lifetime indicates that a concerted cleavage of benzoyl peroxide to form benzoyloxy–phenyl pairs (which form phenyl benzoate) is unnecessary. The enhanced absorption observed in the ^{13}C spectrum of biphenyl[24] in benzoyl peroxide thermolysis constitutes a similar proof. (The difference in the ethyl iodide polarization displayed in Figures 6 and 7 results from the possibility of a substantial number of diffusive encounters possible in the former case, but not in the latter.)

B. Organometallic–Alkyl Halide Reactions

The mechanistic application of CIDNP to peroxide systems is necessarily to the details of radical interactions, since there is no serious reason to doubt that radicals are the principal intermediates in peroxide decompositions. For alkyl halide–organometallic reactions, CIDNP can be used to satisfy the more basic question of whether radicals are produced at all. The suggestion of radical character in alkyllithium–alkyl halide reactions has been made periodically in the past,[39] but the dominance of ionic mechanology was so complete that products continued to be explained by SN_2 and E_2 processes. One of the most common of this type of reaction (because of its use in the Gilman titration) is that between benzyl chloride and an alkyllithium which, in the case of ethyllithium, gives an EA multiplet for all the alkyl protons

$$PhCH_2Cl + C_2H_5Li \longrightarrow \overline{PhCH_2\cdot,\quad C_2H_5\cdot} \longrightarrow PhCH_2CH_2CH_3$$

in *n*-propylbenzene (Figure 8).[40,41] Clearly, at least some portion of the reaction proceeds by electron transfer to form a benzyl–ethyl geminate pair.

The reaction of *p*-fluorobenzyl chloride with *n*-butyllithium leads to a fluorine multiplet effect of opposite phase for *n*-pentylfluorobenzene and *p*-*p*′-difluorobibenzyl.[42]

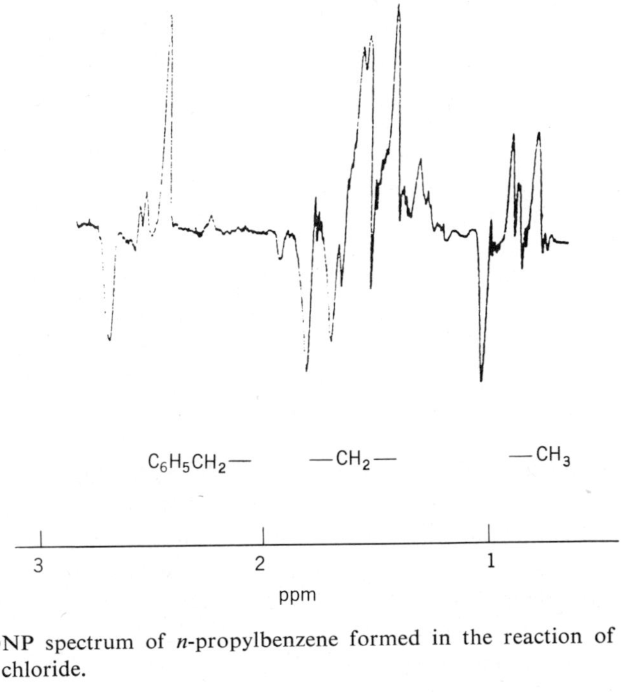

Fig. 8. CIDNP spectrum of *n*-propylbenzene formed in the reaction of ethyllithium with benzyl chloride.

The addition of an α-Cl, beyond a substantial rate acceleration, does little to change the character of the reaction, and the α-chloropropylbenzene from the benzal chloride–ethyllithium reaction also shows the EA multiplet, now

$$PhCHCl_2 + C_2H_5Li \longrightarrow \overline{Ph\dot{C}HCl, \quad \cdot C_2H_5} \longrightarrow PhCHClCH_2CH_3$$

combined with net polarization from the Δg introduced by spin orbit coupling with chlorine in the α-chlorobenzyl radical.[43] The closely related reaction of 2,2-dichloropropane with ethyllithium gives polarization which is generally explicable in a similar way.[43] The same reaction with *gem*-dichlorocyclopropanes,[43,44] however, has an added dimension, since ring opening to allenic products is possible. The polarization of the vinyl protons of the allene is

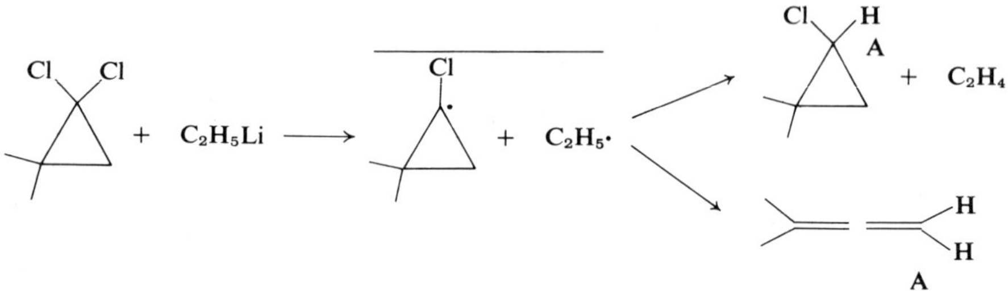

proper for formation from α-chlorocyclopropyl radicals which avoid geminate combination, but the details of the radical-to-allene conversion are unknown.

The reaction of primary and secondary alkyl chlorides (excepting allyl and benzyl chlorides) with alkyllithium reagents is too sluggish for CIDNP purposes but alkyl bromides react at a suitable rate[2] and the combination and disproportionation products exhibit a pure multiplet effect suitable for singlet geminate reactions. In these reactions, run in hydrocarbon or hydrocarbon–ether solvents, the alkyllithium reagents certainly are associated. Since these reactions have been shown kinetically[43–48] to occur directly between the associated form of the alkyllithium (hexamer or tetramer) and the halide, there is reason to question whether electron transfer would lead to a "free" or a complexed radical (from the alkyllithium). Among the variety of parameters that appear to influence this complexation, perhaps the most interesting is the degree of aggregation of the alkyllithium. This aggregation can be varied from hexameric to tetrameric by the addition of diethyl ether[49] to a hydrocarbon solution of *n*-butyllithium. When the ratio of ether to alkyllithium (R_m) reaches 0.5, the transition of the hexamer to a disolvated tetramer is complete. When the region of vinyl proton absorption is examined during reactions of *n*-butyllithium with *t*-butyl bromide as a function of R_m (Figure 9), a remarkable transformation is observed.[41] At low R_m, only

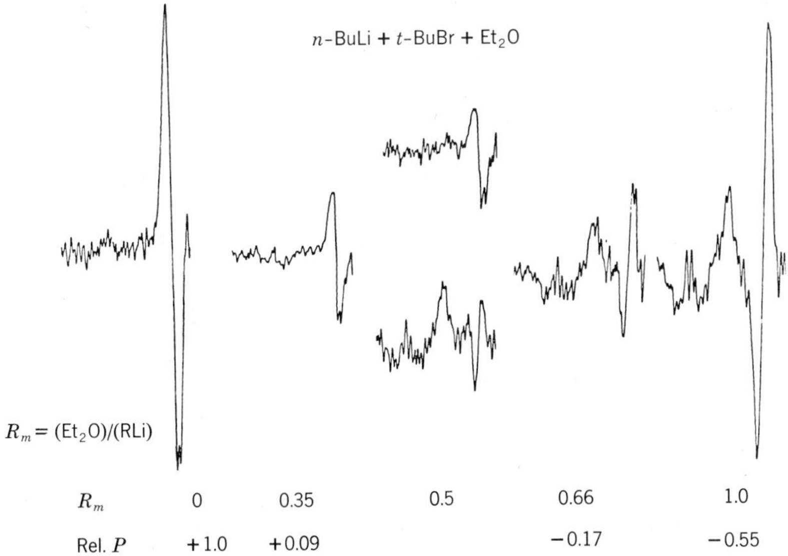

Fig. 9. Vinyl proton polarization variation, as a function of R_m, from reactions of *n*-butyllithium with *t*-butyl bromide.

diffusive polarization from isobutylene is observable. At $R_m = 0.5$, the isobutylene polarization switches phase (to that characteristic of geminate spin selection) and 1-butene polarization appears. At higher R_m both olefins show strong geminate polarization. Apparently, when the hexameric *n*-butyllithium transfers an electron to *t*-butyl bromide, the *n*-butyl radical remains associated, and the only spin selection occurs in diffusive encounters of *t*-butyl radicals. In the tetrameric form, perhaps because of increased steric crowding, both radical species are freed to lead to geminate polarization. When triethylamine serves as a radical trap in an *n*-butyllithium, 1-bromobutane reaction, AE multiplet polarization is observed in *N*,*N*-diethylvinylamine,[50] the product of disproportionation of the α-amino radical in a diffusive encounter.

Reactions of alkyl iodides with alkyllithiums[51–53] proceed at an appropriate rate without the inducement of ether addition, and show the expected polarization for geminate combination products. The facile iodine transfer between free radicals and alkyl iodides gives an extra dimension to the CIDNP spectra, however, since polarization is observed in the reagent iodide. These polarizations are of the multiplet type in every case, but, depending on the structure of the reagents, may be of either AE or EA phase. The radical derived from the reagent iodide will react in both geminate and diffusive encounters, as shown for ethyllithium and isopropyl iodide (Scheme 2,

Figure 10), and the dominant spin selection will depend on cage effects and the relative efficacies in the two types of encounters.

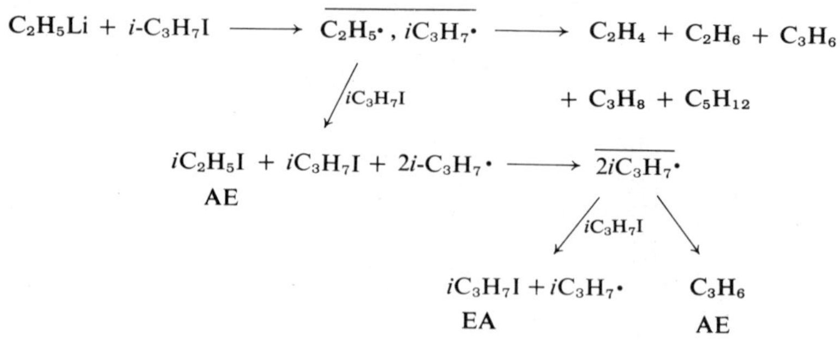

Scheme 2

In every case the iodide derived from the alkyllithium shows the AE phase on escape from a geminate encounter. Each pair which does not react produces two radicals from the reagent iodide (by iodine abstraction), and, given a cage effect of less than 0.5, these radicals will react more often in diffusive than in geminate encounters. Experimentally, when the alkyl moieties of the reagents are not the same, the reagent iodide develops an E/A phase during reaction, as though determined in a diffiusive encounter. This appears to demand that diffusive encounters are within a factor of 2 of the geminate encounter in spin selection efficiency.

Whether lithium atom transfer occurs in these systems is still uncertain, although CIDNP spectra give it scant support. In the majority of systems, no signal can be observed for a product alkyllithium, and in the few cases where signals are observed, they show no polarization. The absorption from the α-methylene protons (in ethyl- or butyllithium) can be seen to decrease smoothly during reaction with no intensity distortion. The report[53] that this absorption is no longer detectable, while the polarization of the alkyl iodide protons persists, is adequately explained by the difference in detection limits which results from the enhancement in only the iodide protons. When the alkyllithium–alkyl iodide reactions are run at low temperatures in ethereal solvents, conditions known to favor halogen–metal exchange over coupling and disproportionation, even at rates suitable for CIDNP, no polarization is observed.[41]

The alkyllithium–alkyl halide systems are one of the few classes of reactions for which both CIDNP and ESR results are available. The reactions cannot be run in precisely the same way for both types of investigations, since the inherently lower sensitivity of ESR requires a higher radical concentration. The addition of ether or TMEDA in molar equivalence to the lithium reagent

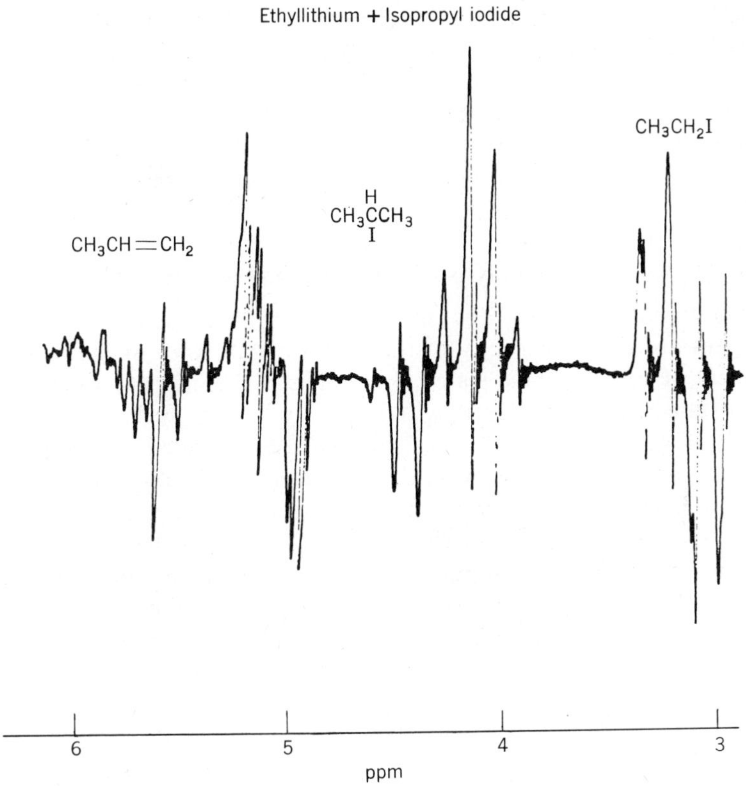

Fig. 10. Reaction of ethyllithium and isopropyl iodide.

and the use of a flow system meets this radical concentration requirement, but prevents an exact comparison with the CIDNP experiments which were run in hydrocarbon solution, with trace amounts of ether added to accelerate certain lethargic reactions. Unfortunately, the two available ESR studies[17,18] make no net contribution to the problem of the degree of radical character, since the conclusions are, in one study, that "only a small fraction of the R′I (or RLi) consumed is being converted to radicals observable by ESR spectroscopy"[18] and in the other study that "free radical formation reactions are the only initiating reactions of alkyllithium compounds with alkyl halides."[17]

There are brief reports of reactions of alkyl halides with two other organometallic systems. The reaction with Grignard reagents[54] shows multiplet polarization in disproportionation products. The polarization differs from that shown by the organolithium systems in suggesting, by its phase, that the spin selection occurs in diffusive encounters. Reactions of sodium naphthalene and alkyl sodium reagents with alkyl iodides lead to polarization when the reactions are conducted at low magnetic fields.[55] An explanation of this low-field polarization involves T_{-1}–S mixing in radical pairs, and suggests that reactions which do not show polarization effects at high field (perhaps because of a high exchange interaction) may usefully be examined at low field.

C. Photolysis of Aldehydes and Ketones

Unquestionably two of the more engaging uncertainties in carbonyl photolyses are the multiplicity of the reacting state and, in the case of involvement of triplet excited states, the timing of the intersystem crossing to ground-state products. CIDNP, with its ability to distinguish between singlet and triplet geminate pairs, appears to be ideally suited to answer these questions for the reactions where two radicals (i.e., not a biradical) are produced. Ironically, CIDNP may not be applicable to the most fashionable systems, the type II cleavages (which involve 1,4 biradicals), but has already been shown to give useful information on their intermolecular analog, carbonyl photoreduction. Closs[32,56,57] has reported a number of benzophenone and benzaldehyde photolyses in alkylbenzene solutions in which polarization is observed from the coupling product. It has been firmly

$$Ph_2CO + PhCH_3 \xrightarrow{h\nu} \overline{Ph_2\dot{C}OH,\ \ PhCH_2\cdot} \longrightarrow \underset{\displaystyle \mathbf{1}}{Ph_2\underset{\displaystyle OH}{\underset{|}{C}}CH_2Ph}$$

established, largely by the efficacy of triplet quenchers, that the triplet state of benzophenone precedes the radical pair. It was still possible to ask whether the pair forms with spin conservation, or, perhaps, with an intersystem crossing which converts electronic into vibrational excitation and may aid in the transfer of the hydrogen atom to form the ketyl radical. The polarization does not support a singlet pair. Instead, for the benzylic protons of **1**, enhanced absorption is observed, which, since the benzyl *g* factor is lower than that of the ketyl radical, is expected for a triplet geminate *or* a diffusive encounter product. Polarization phase alone cannot distinguish between the two. Rubin[58] reports that all benzophenone ketyl radicals may be trapped by hydrogen transfer to camphorquinone, which appears to favor reaction via diffusive encounters. It is not clear, however, that the minor amount of triplet geminate product which is expected ($\sim 1\%$) would have been detected in Rubin's study. In any case, it seems certain that intersystem crossing to form product occurred *after* the formation of the radical pair. Photo-CIDNP results from quinone photoreduction[59] can be explained in analogous ways.

Closs has used the variation in g factors[56c] which results from halogen substitution in benzaldehyde and benzophenone and in toluene to provide a known and controllable Δg in otherwise similar geminate pairs, and finds substantial agreement between simulated and experimental polarizations. Remarkably, the ketyl radicals produced by laser flash of benzophenone and benzaldehyde in hydrocarbon solvents have been shown by Atkins and McLauchlan to show full ESR emission.[60] There is little doubt that the nuclear and the electron polarizations in the nearly identical systems are related, but the details of the relation are not yet definitely established.[60b]

The photolysis of benzaldehyde[57,61] (especially in inert solvents) shows polarization for both aldehyde (A) and benzoin (E) protons. The triplet state was shown to precede reaction by piperylene quenching, and changes in solvent did not influence the polarization. The irreversible formation of a triplet geminate pair of benzoyl and hydroxybenzyl radical, as suggested by Closs and Paulson,[57] provides a suitable rationalization of the polarization.

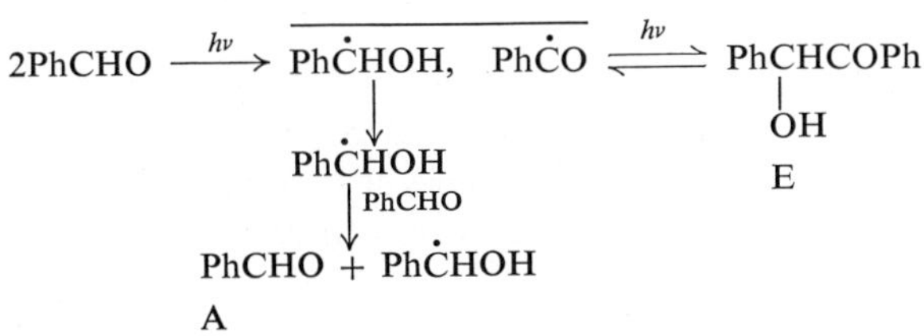

The benzoin formed by geminate combination should show E for the benzyl proton [g (benzoyl) < g (hydroxybenzyl)] and the hydroxybenzyl radicals escaping reaction may re-form benzaldehyde by donation of a hydrogen to another benzaldehyde in solution. Photolysis of benzoin gives the same polarization pattern, suggesting a reversion of the triplet geminate pair to benzoin. All these data, however, are equally consistent with product formation by diffusive encounter, and of themselves, do not demand reaction in the triplet geminate encounter, although such reaction seems probable.

D. Molecular Rearrangements

CIDNP studies have confirmed the intermediacy of radical pairs in a number of rearrangements where radical character was in question, most especially in the rearrangement of ammonium[63,66] and sulfonium[62,64,65,67–69] ylids (the Stevens rearrangement). Intramolecularity and retention of configuration in the rearrangement have, until recently, been interpreted as the result of either a concerted reaction or an intimate ion pair intermediate. However, the rearrangement of **2** generated in situ by Lepley[70] gave a clear

EA multiplet for the benzylic proton in **3**, apparently the result of spin selection in the singlet geminate pair. (The α-nitrogen has no apparent effect

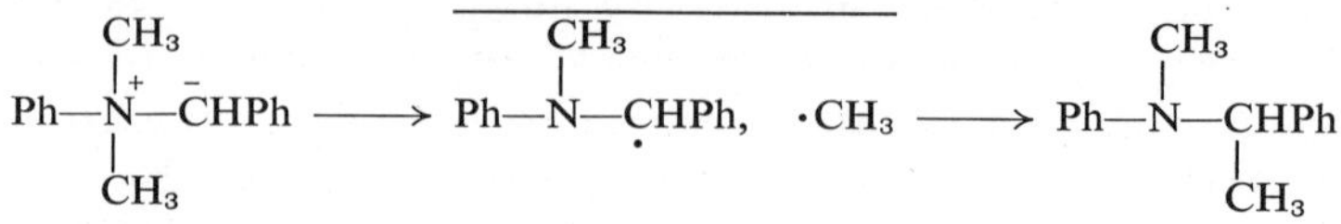

2 **3**

on the g value, and Δg in this pair is ~ 0.) The rearrangement of the phenacyl ammonium ylid **4**, reported by Schöllkopf et al.,[64] shows product polarization for the benzyl protons in **5** which also is consistent with spin selection in a

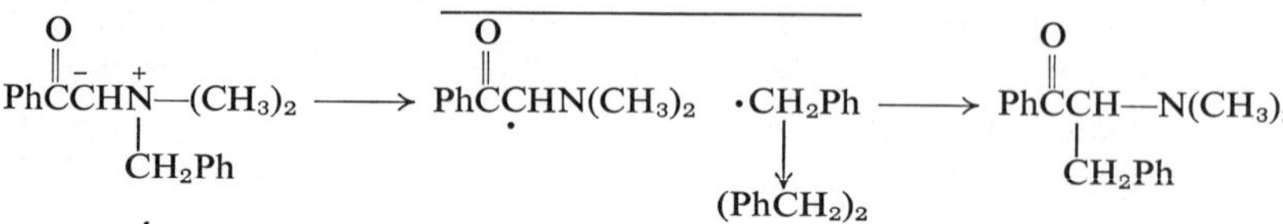

4 **5**

geminate singlet encounter. However, in the methyl analog of this ylid, where the stereochemistry of the rearrangement could be observed, product was formed with 95% retention of configuration, a result which may seem at first sight to be inconsistent with a radical pair mechanism as the only reaction path. A simultaneous but separate concerted reaction would provide a

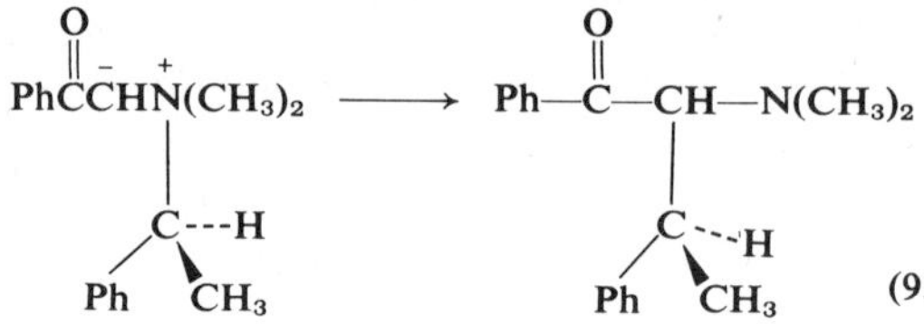

(95% retention)

device for the maintenance of stereochemical integrity but, through orbital symmetry insight, should give product of inverted configuration. An intimate ion pair, the other alternative for a competitive reaction, remains as a possibility, but may present no advantage over the pure radical route in explaining stereoselectivity. The singlet pairs formed by these rearrangements differ from the others discussed in this review in that no molecules or ions intervene between the partners of the pair. (In the reactions in the first two sections, pairs are separated by one or two molecules of carbon dioxide, or by lithium halide.) It is reasonable that these intimate pairs might, in large part, combine before racemization occurs. Very little spin selection should result from combination of such pairs since their existence must be ephemeral and the exchange interaction should remain high. The formation of 5% bibenzyl in the rearrangement of **4** requires that some pairs must dissociate and a corollary is that some pairs must be separated by a few solvent molecules and then recombine. It is probable that virtually all of the polarization is in the (racemized) products of these pairs. It is surprising that the retention is as high as 95%, but, since the stereochemistry and the polarization were not studied in the same system, and the enhancement factor was not measured, the radical rearrangement as a unique path remains a strong possibility.

Fortunately, in the sulfur analog of the Stevens rearrangement, Baldwin et al.[68] have investigated both stereochemistry and polarization in the same system. The 36% retention of configuration in the product is fully consistent

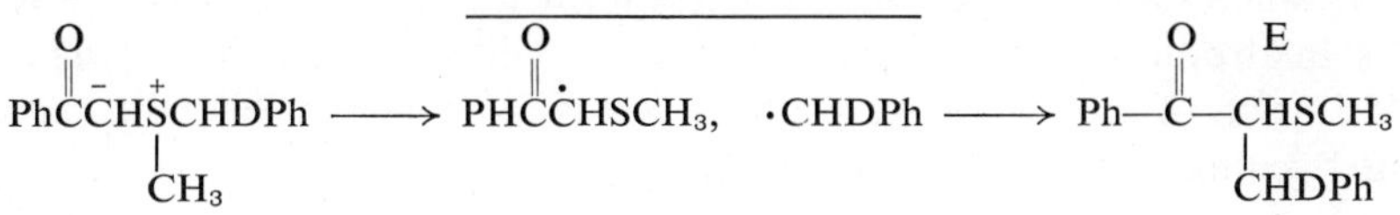

A

with the radical pair mechanism. Other reports of polarization in sulfur ylid rearrangements give, in each case, the type of spin selection expected from singlet geminate pairs.

The formally related Meisenheimer rearrangement of tertiary amine oxides[71–73] appears, from at least two examples, also to form singlet pairs and the *N*-methyl(s) and benzyl protons in **6** and **7** show strong emission during the rearrangement. (Emission from both partners in the radical pair results

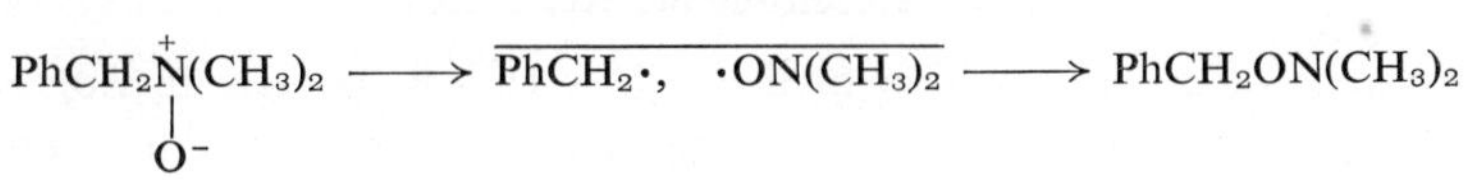

6

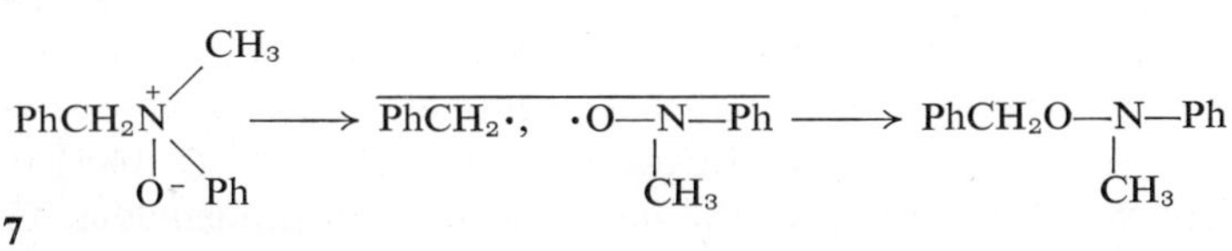

7

from a negative hyperfine coupling in the benzyl radical and a positive one in the nitroxyl radical.) Similar radical pairs and CIDNP spectra result from the Martynoff rearrangement of nitrones.[74]

The thermal rearrangement of sulphenates to sulfoxides bears strong resemblance to both the Stevens and Meisenheimer rearrangements, and there is no reason to believe that the emission signal from the benzylic protons in **8** does not derive from spin selection in the analogous radical pair.

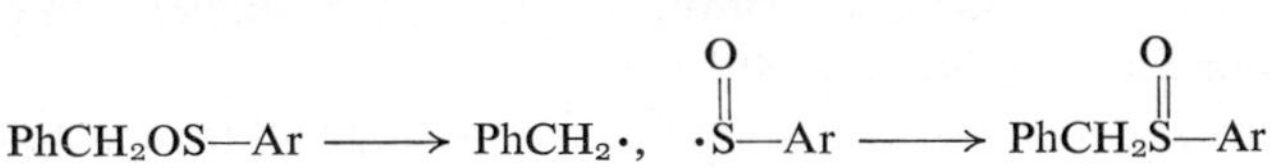

8

This polarization is interpreted rather differently by Jacobus,[75] who attributes the polarization to a thermal decomposition of the sulfoxide **8** to a radical pair which, on recombination, shows emission. He has carried out the proper experiment to test this hypothesis by examining the spectrum of **8** under rearrangement conditions and finds no polarization. This negative evidence is disregarded by Jacobus who makes the curious assertion that the smaller amount of **8** in the rearrangement system will give relatively more reaction than a more concentrated sample of **8** at the same temperature. The cancellation effect that he observes is, in fact, the normal, expected result of the buildup of product, and the spin selection as represented above.

The sole example of a nonylid rearrangement which gives CIDNP is the 1,3 sigmatropic rearrangement reported by Baldwin and Brown.[76] The EA polarization of the methylene protons in **9** is quite compatible with the suggested geminate pair intermediate.

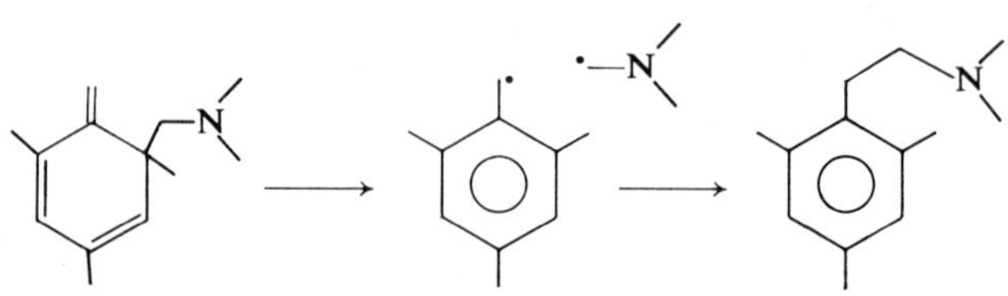

9

E. Miscellaneous Reactions

A number of systems which give rise to CIDNP spectra do not fall into any of the preceding categories and are included here for completeness. Of these, some are formally related by the common characteristic loss of nitrogen during radical generation. The thermal decomposition of azo compounds has been used by Closs[77] to generate singlet geminate pairs, for comparison with the same pairs formed by hydrogen abstraction by the (triplet) diphenylmethylene which results from the decomposition of diphenyldiazomethane. The opposite phase of the polarization of the coupling product in the two

$$PhCH_2{-}N{=}N{-}CHPh_2 \longrightarrow \overline{PhCH_2\cdot,\quad Ph_2CH\cdot}$$

$$Ph_2CN_2 \xrightarrow{h\nu} Ph_2C: \xrightarrow{PhCH_3} \overline{Ph_2CH\cdot\ PhCH_2\cdot}$$

cases was the first confirmation of the general idea of singlet–triplet mixing as the underlying source of CIDNP.

The thermal decomposition of the azo compound **10** is reported by Berson et al.[78] to lead to the triplet trimethylenemethane **11**. The emission signals observed from the dimerization products of **11** have been neatly explained by Closs[79] to be a result of T_{-1}–S mixing in the triplet biradicals formed during stepwise reaction. This model predicts that the only observable polarization from such biradical reactions will be emission and that singlet biradicals will lead to no polarization.

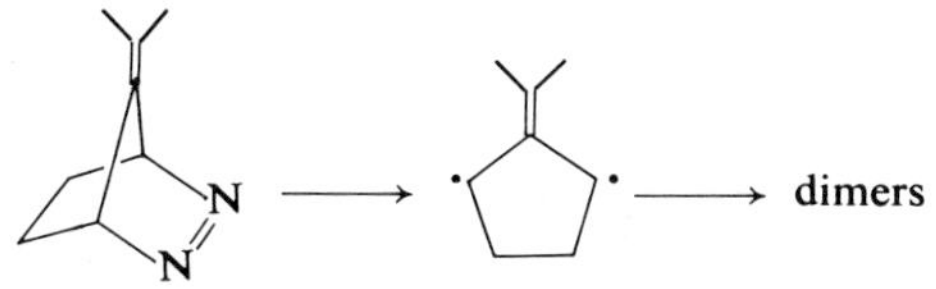

10 **11**

The photolysis of methyl diazoacetate in carbon tetrachloride[80] leads to the insertion product **12**, for which emission is observed in the methyne proton.

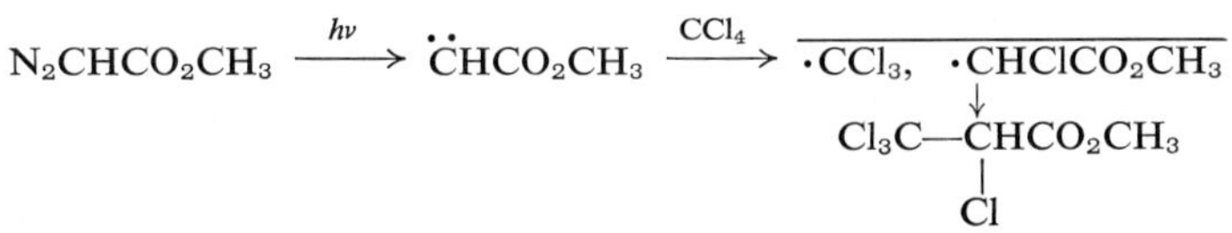

12

Since the trichloromethyl radical has a higher *g* factor than its partner radical, the pair would appear to have been formed by abstraction by a singlet carbene.

The loss of nitrogen from **13** in a carbon tetrachloride solution gives strong emission from chloroform, weaker emission from the methyl protons of propene, and enhanced absorption for the methyls of isopropyl chloride,[81] all of which is consistent with spin selection in a diffusive encounter between isopropyl and trichloromethyl radicals. The disproportionately strong emission from chloroform is a result of an unusually long T_1.

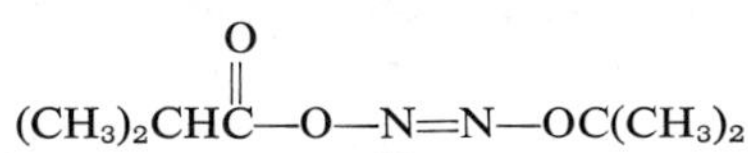

13

The reaction of amine oxides with acetic anhydride in benzene solution leads to radical pair generation and thus to CIDNP spectra.[82,83] In the case of 4-picoline-*N*-oxide[83] an intermediate *N*-acetate was proposed, and its rearrangement by bond homolysis was suggested. Iwamura[83] reports emission from the methylene of **14**, and E,EA for the methylene and A,EA for the methyl of 4-ethylpyridine. Curiously, the analogous reaction of 2-picoline-*N*-oxide[82] does not give polarization in the product corresponding to **14**, and shows polarization from ethane and methyl acetate which appear to result from a geminate acetoxy–methyl pair. The route to such a pair, and the difference in the 2- and the 4-picoline oxide reaction is not obvious.

The reduction of aryl diazonium salts gives emission from aromatic protons[84,85] but the identity of the radical pair leading to polarization has not been established.

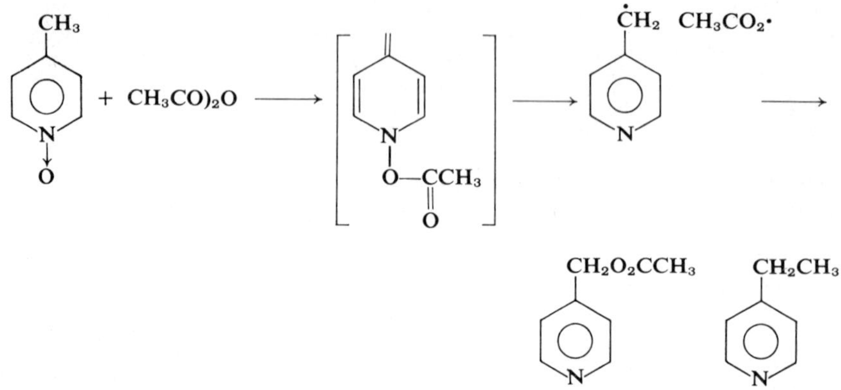

The stepwise addition of two 1-cyano-1-methylethyl radicals (from AIBN decomposition) to nitrosobenzene[86] or to nitrones[87] leads to emission signals from one set of methyl protons, the expected polarization from a diffusive encounter as the second addition occurs.

VII. CONCLUSIONS

Proper interpretation of CIDNP spectra, based on the theory at its present evolutionary stage, allows the detection of radical pair processes, the assignment of the multiplicity of these pairs, and the identity of the pair partners and may enable calculations of *g* factors and hyperfine fields of the individual radicals. It is not yet possible to predict absolute enhancements of spectral intensities, largely because the diffusion of radicals in solution and the variations of exchange interactions during the pair lifetime are both poorly understood. It follows that the degree of radical character in a given reaction cannot be determined directly from the CIDNP spectrum. Despite these uncertainties it is clear that a substantial body of information can be taken from the enhanced spectra, and that mechanistic applications have only just begun.[88]

REFERENCES

1. J. Bargon, H. Fischer, and U. Johnsen, *Z. Naturforsch.*, **22a**, 1552 (1967); J. Bargon and H. Fischer, *ibid.*, **22a**, 1556 (1967).
2. H. R. Ward, *J. Am. Chem. Soc.*, **89**, 5517 (1967); H. R. Ward and R. G. Lawler, *ibid.*, **89**, 5518 (1967).
3. K. H. Hausser and D. Stehlik, *Adv. Magnetic Resonance*, **3**, 79 (1967).
4. R. G. Lawler, *J. Am. Chem. Soc.*, **89**, 5519 (1967).
5. H. Fischer and J. Bargon, *Acc. Chem. Res.*, **2**, 110 (1969).
6. R. Kaptein and L. J. Oosterhoff, *Chem. Phys. Letters*, **4**, 214 (1969).
7. G. L. Closs, *J. Am. Chem. Soc.*, **91**, 4552 (1969).
8. H. Fischer, *Chem. Phys. Letters*, **4**, 611 (1970).
9. G. L. Closs and A. D. Trifunac, *J. Am. Chem. Soc.*, **92**, 2183 (1970).
10. F. J. Adrian, *J. Chem. Phys.*, **53**, 3374 (1970).
11. H. Fischer, *Z. Naturforsch.*, **25a**, 1957 (1970).
12. M. Tomkeiwicz and M. Cocivera, *Chem. Phys. Letters*, **8**, 595 (1971).
13. R. G. Lawler, *Acc. Chem. Res.*, in press.
14. R. M. Noyes, *J. Am. Chem. Soc.*, **77**, 2042 (1955).
15. P. Livant, unpublished observation.
16. R. A. Cooper, R. G. Lawler, and H. R. Ward, *J. Am. Chem. Soc.*, in press.
17. H. Fischer, *J. Phys. Chem.*, **73**, 3834 (1969).
18. G. A. Russell and D. W. Lanson, *J. Am. Chem. Soc.*, **91**, 3967 (1969).
19. C. Blomberg and H. S. Mosher, *J. Organomet. Chem.*, **13**, 519 (1968).
20. J. Billet and S. G. Smith, *J. Am. Chem. Soc.*, **90**, 4108 (1968).
21. T. A. Marzilli, unpublished results.
22. H. R. Ward, R. G. Lawler, H. Y. Loken, and R. A. Cooper, *J. Am. Chem. Soc.*, **91**, 4928 (1969).
23. M. Lehnig and H. Fischer, *Z. Naturforsch.*, **24a**, 1771 (1969).
24. E. Lippmaa, T. Pehk, A. L. Buchachenko, and S. V. Rykov, *Chem. Phys. Letters*, **5**, 521 (1970).
25. For a recent comprehensive review, see R. C. P. Cubbon, *Prog. React. Kinet.*, **5**, 29 (1970).
26. R. Kaptein, *Chem. Phys. Letters*, **2**, 261 (1968).
27. H. R. Ward, R. G. Lawler, and R. A. Cooper, *Tetrahedron Letters*, 527 (1969).
28. R. Kaptein, J. A. den Hollander, D. Antheunis, and L. J. Oosterhoff, *Chem. Commun.*, 1687 (1970).
29. R. Kaptein, Ph.D. thesis, University of Leiden, 1971.
30. C. Walling, H. P. Waits, J. Milovanovic, and C. G. Pappiaonnou, *J. Am. Chem. Soc.*, **92**, 4927 (1970).
31. S. V. Rykov and A. L. Buchachenko, *Dokl. Akad. Nauk SSSR*, **185** 870 (1969); S. V. Rykov, A. L. Buchachenko, and V. I. Baldin, *Zh. Strukt. Khimi*, **10**, 928 (1969); A. L. Buchachenko, S. V. Rykov, A. V. Kessenikh, and G. S. Bylina, *Dokl. Akad. Nauk SSSR*, **190**, 839 (1970).
32. G. L. Closs and A. D. Trifunac, *J. Am. Chem. Soc.*, **92**, 2186 (1970).
33. A. Trozzolo and S. R. Farenholtz, *J. Am. Chem. Soc.*, **93**, 251 (1971).
34. B. Blank and H. Fischer, *Helv. Chim. Acta*, in press.
35. D. F. DeTar, *J. Am. Chem. Soc.*, **89**, 4058 (1967).
36. F. Gerhart and G. Ostermann, *Tetrahedron Letters*, 4705 (1969); F. Gerhart, *ibid.*, 5061 (1969).
37. For an estimate of iodine abstraction rate see reference 77a.

38. M. Lehnig and H. Fischer, *Z. Naturforsch.*, **25a**, 1957 (1970).
39. D. Bryce Smith, *J. Chem. Soc.*, 1603 (1956).
40. J. S. Wishnok, unpublished results.
41. R. A. Cooper, Ph.D. thesis, Brown University, Providence, R.I., 1971.
42. J. W. Rakshys, Jr., *Chem. Commun.*, 578 (1970).
43. H. Y. Loken, Ph.D. thesis, Brown University, Providence, R.I., 1971.
44. H. R. Ward, R. G. Lawler, and H. Y. Loken, *J. Am. Chem. Soc.*, **90**, 7359 (1968).
45. A. I. Shatenstein, E. A. Kovrizhnykh, and V. M. Basmanoua, *Kinet. Katal.*, **7**, 953 (1966).
46. F. S. Dyachkovskii and A. E. Shilov, *Kinet. Katal.*, **4**, 919 (1963).
47. J. F. Eastham and G. W. Gibson, *J. Am. Chem. Soc.*, **85**, 2171 (1963).
48. W. Glaze and R. West, *J. Am. Chem. Soc.*, **82**, 4437 (1960).
49. H. L. Lewis and T. L. Brown, *J. Am. Chem. Soc.*, **92**, 4664 (1970).
50. A. R. Lepley, *J. Am. Chem. Soc.*, **90**, 2710 (1968).
51. H. R. Ward, R. G. Lawler, and R. A. Cooper, *J. Am. Chem. Soc.*, **91**, 746 (1969).
52. A. R. Lepley and R. L. Landau, *J. Am. Chem. Soc.*, **91**, 748 (1969); A. R. Lepley, *ibid.*, **91**, 749 (1969).
53. A. R. Lepley, *Chem. Commun.*, 69 (1969).
54. H. R. Ward, R. G. Lawler, and T. A. Marzilli, *Tetrahedron Letters*, 521 (1970).
55. J. G. Garst, R. H. Cox, J. T. Barbas, R. D. Roberts, J. I. Morris, and R. C. Morrison, *J. Am. Chem. Soc.*, **92**, 5761 (1970); J. F. Garst and R. H. Cox, *ibid.*, **92**, 6389 (1970).
56. (a) G. L. Closs and L. E. Closs, *J. Am. Chem. Soc.*, **91**, 4550 (1969); (b) G. L. Closs and A. D. Trifunac, *ibid.*, **91**, 4554 (1969); (c) G. L. Closs, C. E. Doubleday, and D. R. Paulson, *ibid.*, **92**, 2185 (1970).
57. G. L. Closs and D. R. Paulson, *J. Am. Chem. Soc.*, **92**, 7229 (1970).
58. M. Rubin, *Tetrahedron Letters*, 3931 (1969).
59. K. Marlyama, H. Shindo, and T. Mariyama, *Bull. Chem. Soc. Japan*, **44**, 585 (1971).
60. (a) P. W. Atkins, I. C. Buchanan, R. C. Gurd, K. A. McLauchlan, and A. F. Simpson, *Chem. Commun.*, 513 (1970); (b) P. W. Atkins, R. C. Gurd, K. A. McLauchlan, and A. F. Simpson, *Chem. Phys. Letters*, **8**, 55 (1971).
61. M. Cocivera and A. M. Trozzolo, *J. Am. Chem. Soc.*, **92**, 1772 (1970).
62. U. Schöllkopf, G. Ostermann, and J. Schossig, *Tetrahedron Letters*, 2619 (1969).
63. A. R. Lepley, *Chem. Commun.*, 1460 (1969).
64. U. Schöllkopf, U. Ludwig, G. Ostermann, and M. Patsch, *Tetrahedron Letters*, 3415 (1969).
65. R. W. Jemison and D. G. Morris, *Chem. Commun.*, 1226 (1969).
66. D. G. Morris, *Chem. Commun.*, 1345 (1969).
67. U. Schöllkopf, J. Schossig, and G. Ostermann, *Ann.*, **737**, 158 (1970).
68. J. E. Baldwin, W. F. Erickson, R. E. Hackler, and R. M. Scott, *Chem. Commun.*, 576 (1970).
69. H. Iwamura, M. Iwamura, T. Nishida, M. Yoshida, and J. Nakayama, *Tetrahedron Letters*, 63 (1971).
70. A. R. Lepley, *J. Am. Chem. Soc.*, **91**, 1237 (1969).
71. A. R. Lepley, P. M. Cook, and G. F. Willard, *J. Am. Chem. Soc.*, **92**, 1101 (1970).
72. H. Iwamura, M. Iwamura, T. Nishida, and I. Muira, *Bull. Chem. Soc. Japan*, **43**, 1914 (1970).
73. G. Ostermann and U. Schöllkopf, *Ann.*, **737**, 170 (1970).
74. D. G. Morris, *Chem. Commun.*, 221 (1971).
75. J. Jacobus, *Chem. Commun.*, 709 (1970).
76. J. F. Baldwin and J. E. Brown, *J. Am. Chem. Soc.*, **91**, 5103 (1969).

77. G. L. Closs and L. E. Closs, *J. Am. Chem. Soc.*, **91**, 4549 (1969); G. L. Closs and D. R. Paulson, *ibid.*, **92**, 7227 (1970).
78. J. A. Berson, R. J. Bushby, J. M. McBride, and M. Tramelling, *J. Am. Chem. Soc.*, **93**, 1544 (1971).
79. G. L. Closs, *J. Am. Chem. Soc.*, **93**, 1546 (1971).
80. M. Cocivera and H. D. Roth, *J. Am. Chem. Soc.*, **92**, 2573 (1970).
81. T. Koenig and W. R. Mabey, *J. Am. Chem. Soc.*, **92**, 3804 (1970).
82. H. Iwamura, M. Iwamura, T. Nishida, and I. Miura, *Tetrahedron Letters*, 3117 (1970).
83. H. Iwamura, M. Iwamura, T. Nishida, and S. Sato, *J. Am. Chem. Soc.*, **92**, 7474 (1970).
84. A. G. Lane, C. Rüchardt, and R. Werner, *Tetrahedron Letters*, 3213 (1969).
85. A. Rieker, P. Niederer, and D. Leibfritz, *Tetrahedron Letters*, 4287 (1969).
86. H. Iwamura, M. Iwamura, M. Tamura, and K. Shiomi, *Bull. Chem. Soc. Japan*, **43**, 3638 (1970).
87. H. Iwamura and M. Iwamura, *Tetrahedron Letters*, 3723 (1970).
88. For recent additional references see also H. R. Ward, *Acc. Chem. Res.*, **5**, 18 (1972); R. G. Lawler, *ibid.*, **5**, 25 (1972); "CIDNP," G. L. Closs and A. R. Lepley, Eds.; Interscience, New York, 1972.

CHAPTER 7

Reactivity, Selectivity, and Polar Effects in Hydrogen Atom Transfer Reactions

GLEN A. RUSSELL
Department of Chemistry, Iowa State University, Ames, Iowa

I. INTRODUCTION

The attack of an atom or radical upon a carbon–hydrogen bond is believed to proceed via attack on the hydrogen giving rise to a more or less linear transition state (Scheme 1)

$$R\text{—}H + X\cdot \longrightarrow [R\text{------}H\text{------}X]^{\ddagger} \longrightarrow R\cdot + HX$$

Scheme 1

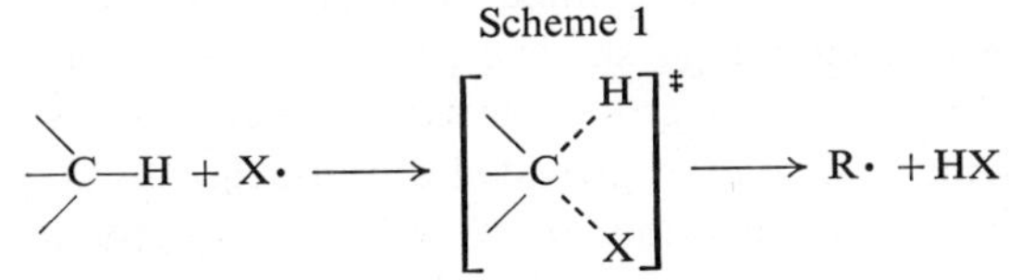

Scheme 2

Scheme 2, involving attack on carbon to yield the intermediate or transition state shown, is excluded by a number of experimental considerations. The primary deuterium or tritium isotope effect which varies greatly with the nature of the attacking radical or atom is not consistent with the addition–elimination Scheme 2.[1] Moreover, when X· is D· no hydrogen–deuterium exchange is observed (ESR spectroscopy) in attack on methylene groups (Scheme 3).[2]

In the following sections various factors affecting the stability of the transition state of Scheme 1 are analyzed as a function of the structure of R, the

$$D\cdot + CH_2(CO_2H)_2 \begin{cases} \longrightarrow \cdot CH(CO_2H) + HD \\ \not\longrightarrow \cdot CD(CO_2H) + H_2 \end{cases}$$

Scheme 3

structure of X, and the nature of the solvent. Evidence for strong polar effects is developed which requires that still a third mechanism for substitution be considered (Scheme 4).[3]

$$X\cdot + RH \longrightarrow RH\cdot^{+} + X^{-} \longrightarrow R\cdot + HX$$

Scheme 4

Scheme 4, involving an ionic or a dipolar intermediate, is discarded on the basis that it does not predict the variation in selectivity observed when various radicals or atoms are allowed to react with a series of carbon–hydrogen bonds in inter- or intramolecular processes.[4] Instead, polar contributions to the transition state for Scheme 1 are advocated as an explanation of the polar effects.

II. TECHNIQUES

The most general method for measuring reactivity and selectivity, and hence the polar nature and steric requirements, of free radicals or atoms is to measure the relative rate constants in hydrogen atom transfer reactions. In some cases absolute rate data are available for hydrogen abstraction steps. Absolute rate data are, of course, a true measure of reactivity. However, a more accurate picture of selectivity can often be obtained from more approximate competitive techniques. Competition of a radical or atom for two hydrocarbons each containing a single type of hydrogen atom, or competition for different hydrogen atoms within the same molecule for a given radical or atom, can provide precise data provided a well-defined free radical reaction is being studied. Halogenation reactions such as photochlorination,[1,5] photobromination,[6,7] chlorination with *t*-butyl hypochlorite,[8] or bromination with bromotrichloromethane,[6,9] have yielded competitive rate data toward the chlorine atom, bromine atom, trichloromethyl radical, or *t*-butoxy radical.

$$X\cdot + \begin{cases} R_1H \xrightarrow{k_1} R_1\cdot + HX & (1) \\ R_2H \xrightarrow{k_2} R_2\cdot + HX & (2) \end{cases}$$

$$R_1\cdot, R_2\cdot + X\text{—}Y \longrightarrow R_1Y, R_2Y + X\cdot \quad (3)$$

It is, of course, necessary to establish that the yields of R_1Y and R_2Y in these processes are true measures of k_1 and k_2. Serious complications arise if reactions 1–3 are reversible or if reaction 4 becomes important.

$$R_1\cdot + R_2H \rightleftharpoons R_2\cdot + R_1H \tag{4}$$

Such possibilities can be investigated experimentally, for example, by the use of a deuterated or tritiated reactant.[1]

$$X\cdot + R_1H \xrightarrow{k_H} R_1\cdot + HX \tag{1'}$$

$$X\cdot + R_2D \xrightarrow{k_D} R_2\cdot + DX \tag{2'}$$

If the ratios HX/DX and R_1Y/R_2Y are equal, then serious complications from reverse reactions, or hydrogen atom transfer involving secondary radicals, can be eliminated.

Competitive selectivity data can be used to establish the mechanism of other radical substitution processes. In this manner it was possible to establish that *N*-bromosuccinimide bromination proceeds via bromine atom attack[10–12] and that *N*-chloroamides usually react via a chlorine atom mechanism.[13]

Table 1 summarizes some data reported by Tanner, by Ingold, and by

TABLE 1

Products of Free Radical-Initiated Reactions

Chlorinating species	% Attack: Cl—CH_2 ↑	—CH_2 ↑	—CH_2 ↑	—CH_3 ↑	Attacking radicals	Ref.
Cl_2, 34°	7	24	51	18	Cl·	14
t-BuOCl, 40°	21	20	44	15	*t*-BuO·	14
$C_6H_5ICl_2$, 40°	5	21	61	12	Cl·	14
Cl_2O, 40°	22	20	51	8	ClO·	14
HOCl, CCl_4/-H_2O, 40°	7	23	44	26	Cl·	14
Cl_2, 80°	5	21	52	23	Cl·	14
SO_2Cl_2, 80°	6	21	43	30	Cl·, SO_2Cl·	14
$(CH_3)_2NCl$/-H_2SO_4, HOAc	4	10	77	8	$(CH_3)_2NH\cdot^+$	15, 16
Cl_2/H_2SO_4—HOAc	5	23	46	26	Cl·	15, 16
$(CH_3)_2NBr$/-H_2SO_4—HOAc[a]	5	10	79	7	$(CH_3)_2NH\cdot^+$	16
Br_2/H_2SO_4—HOAc[a]	24	26	50	0.5	Br·	16

[a] Bromination.

Minisci which suggest that the nature of the attacking radical is different in several substitution (chlorination) processes.

Once it has been established that a given mechanism predominates for one halogenation species it does not follow that for other substrates the same attacking radical predominates. Thus, the $SO_2Cl\cdot$ radical is involved in the sulfuryl chloride chlorination of alkanes. However, in chlorination in the presence of aromatic substrates or solvents, both molecular chlorine and sulfuryl chloride give selectivity patterns which are apparently the result of the chlorine atom complexed with the aromatic ring.[17]

$SO_2Cl\cdot + C_6H_6 \longrightarrow SO_2 +$ 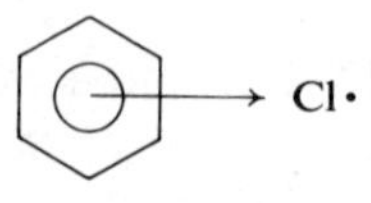 (5)

Another example of the effect of substrate are the 1°-:2°-:3°-hydrogen reactivities of Table 2.[18] Here it is obvious that at least in reaction with the more reactive 2°- and 3°-hydrogen atoms that $C_6H_6ICl_2$ propagates the chain via $C_6H_5ICl\cdot$ attack. Further examples of a specific solvent effect (complex formation) in reactions of chlorine atoms are given in Table 3 where data

TABLE 2

Relative Reactivities in Chlorination Reactions

	Relative reactivities		
Chlorinating species	1°-H	2°-H	3°-H
Cl_2, 40°	1	4	5
t-BuOCl, 40°	1	8	44
$C_6H_5ICl_2$, 40°	1	21	368

are expressed as per hydrogen reactivity. Table 3 also demonstrates the use of relative reactivity data to identify the abstracting radical species

For the hydrogen atom, methyl, trideuteriomethyl, trifluoromethyl, or phenyl radicals the substitution products resulting from attack on carbon–hydrogen bonds are complex and competitive techniques other than isolation of substitution products must be used. One of the first techniques developed was to allow the methyl radical (from acetyl peroxide thermolysis) to compete between an alkane and carbon tetrachloride.[23] Since the value of k_c is independent of the structure of RH, a series of k_{RH}/k_c can be converted into a table of relative k_{RH} values. More recently this technique has been applied to

$$CH_3\cdot + R_1H \xrightarrow{k_1} CH_4 + R_1\cdot \quad (6)$$

$$CH_3\cdot + CCl_4 \xrightarrow{k_c} CH_3Cl + CCl_3\cdot \quad (7)$$

$$CH_3\cdot + R_2H \xrightarrow{k_2} CH_4 + R_2\cdot \quad (8)$$

phenyl radicals,[24,25] and to substituted phenyl radicals,[26] obtained by the thermolysis of the arylazotriphenylmethanes.

The use of standard deuterium-containing substrates or of deuterated radical precursors has been a valuable technique. In this way an extensive

TABLE 3

Relative Reactivities of Hydrogen Atoms in *n*-Alkanes in Photochlorinations

Conditions	Attacking radical species	Relative reactivity per hydrogen: CH_3— ↑	—CH_2— ↑	—CH_2— ↑	—CH_2—R ↑	Ref.
Cl_2, 20°, neat	$Cl\cdot$	0.3	1.0	0.9	0.9 n-C_3H_7	19
Cl_2, 20°, C_6H_6	$Ar \rightarrow Cl\cdot$	0.1	1.0	1.2	1.1 n-C_3H_7	19
Cl_2, 20°, CS_2	$CS_2Cl\cdot$	0.04	1.0	1.3	1.1 n-C_3H_7	20, 21
Cl_2, 65°, neat	$Cl\cdot$	0.3	1.0	0.9	0.9 C_2H_5	22
105°, CCl_4	$CCl_3\cdot$	Low	1.0	0.6	0.6 C_2H_5	22
125°, CCl_4, CCl_2=CHCl, ROOR	CCl_3CH—$ClCCl_2\cdot$	<0.03	1.0	0.8	0.8 C_2H_5	22
125°, C_2Cl_4, ROOR[a]	$C_6H_{13}C(Cl)_2$—CCl_2	1.2	1.0	0.2	0.2 n-C_2H_5	22
Cl_2, 20° neat	$Cl\cdot$	0.3	1.0	0.9	0.9 n-C_4H_9	20, 21
Cl_2, 20°, CS_2	$CS_2Cl\cdot$	0.1	1.0	1.0	0.9 n-C_4H_9	20, 21
t-BuOCl, 20°	*t*-BuO·	0.1	1.0	0.8	0.7 n-C_4H_9	
CCl_4, Bz_2O_2, 98°	$CCl_3\cdot$	0.03	1.0	0.7	0.6 n-C_4H_9	20, 21
CCl_3SO_2Cl, Bz_2O_2, 98°	$CCl_3\cdot$	0.03	1.0	0.7	0.6 n-C_4H_9	20, 21
CCl_3SCl, 98°	$CCl_3S\cdot$	0.1	1.0	1.0	1.0 n-C_4H_9	20, 21
NCS, Bz_2O_2, 98°	$Cl\cdot$	0.3	1.0	0.9	0.8 n-C_4H_9	20, 21
PCl_5, Bz_2O_2, 98°	$PCl_4\cdot$	0.04	1.0	1.2	1.2 n-C_4H_9	20, 21

[a] Products are $C_6H_{13}CCl{=}CCl_2$ plus HCl·.

series of relative reactivities toward the trideuteriomethyl radical, prepared by the photolysis of perdeuterioacetone, has been obtained.[27,28]

$$CD_3COCD_3 \xrightarrow{h\nu} 2CD_3\cdot + CO \tag{9}$$

$$CD_3\cdot + CD_3COCD_3 \xrightarrow{k_D} CD_4 + CD_3COCD_2\cdot \tag{10}$$

$$CD_3\cdot + R_1H \xrightarrow{k_1} CD_3H + R_1\cdot \tag{11}$$

$$CD_3\cdot + R_2H \xrightarrow{k_2} CD_3H + R_2\cdot \tag{12}$$

Trifluoromethyl radicals (from the photolysis of hexafluoroacetone or hexafluoroazomethane) have usually been allowed to compete between the alkane and deuterium.[28,29]

$$CF_3\cdot + D_2 \xrightarrow{k_D} CF_3D + D\cdot \tag{13}$$

$$CF_3\cdot + R_1H \xrightarrow{k_1} CF_3H + R_1\cdot \tag{14}$$

$$CF_3\cdot + R_2H \xrightarrow{k_2} CF_3H + R_2\cdot \tag{15}$$

Two reliable competitive techniques have recently been developed for hydrogen atoms in solution. Photolysis of deuteriomercaptans in the presence of a series of alkanes allowed a table of k^D values to be established.[30]

$$RSD \xrightarrow{h\nu} RS\cdot + D\cdot \tag{16}$$

$$D\cdot + RSD \xrightarrow{k_S} D_2 + RS\cdot \tag{17}$$

$$D\cdot + R_1H \xrightarrow{k_1} HD + R_1\cdot \tag{18}$$

$$D\cdot + R_2H \xrightarrow{k_2} HD + R_2\cdot \tag{19}$$

Similarly, photolysis of *t*-butylperoxyformate in the presence of a series of hydrogen donors and one standard deuterium donor (e.g., thiophenol-d_1) yielded a series of k^H values.[31]

$$(CH_3)_3COOCHO \xrightarrow{h\nu} H\cdot + (CH_3)_3CO\cdot + CO_2 \tag{20}$$

$$H\cdot + R_1H \xrightarrow{k_R} HD + R\cdot \tag{21}$$

$$H\cdot + R_1H \xrightarrow{k_1} H_2 + R_1\cdot \tag{22}$$

$$H\cdot + R_2H \xrightarrow{k_2} H_2 + R_2\cdot \tag{23}$$

Another general technique utilized for methyl, trichloromethyl, or a variety of polymer radicals has been to allow the radical to compete between addition to a standard olefin and a series of hydrocarbons.

$$R\cdot + R'CH{=}CH_2 \xrightarrow{k_0} RCH_2\dot{C}HR' \tag{24}$$

$$R\cdot + R_1H \xrightarrow{k_1} RH + R_1\cdot \tag{25}$$

$$R\cdot + R_2H \xrightarrow{k_2} RH + R_2\cdot \tag{26}$$

In the case of methyl radicals[32] and trichloromethyl radicals[33] the analytical difficulties are serious and this technique does not appear to be fruitful. In polymerization the ratio of k_1/k_0 is the chain transfer constant of R_1H and it can be readily determined from the degree of polymerization in those polymerizations wherein the ratio $R_1H/R'CH{=}CH_2$ is reasonably constant and the rate of initiation can be held constant.[34] Styrene, vinyl acetate, and tetrafluoroethylene have been extensively studied as substrates for chain transfer during polymerization.

The fluorine atom has been difficult to work with experimentally, except for a few intramolecular competition reactions.[35] However, a recent study has compared hydrogen abstraction and addition to acetylene by near thermal fluorine atoms.[36]

A general technique in free radical chain processes is to measure under steady-state conditions the relative value of the propagation rate constant and the square root of the termination rate constant (for bimolecular termination). This technique has been applied extensively in studies of peroxy radical reactivities. Under normal circumstances the following mechanism applies in an autoxidation reaction.

$$\text{catalyst} \xrightarrow{k_i} R\cdot \tag{27}$$

$$R\cdot + O_2 \xrightarrow{\text{fast}} ROO\cdot \tag{28}$$

$$ROO\cdot + RH \xrightarrow{k_p} ROOH + R\cdot \tag{29}$$

$$2ROO\cdot \xrightarrow{k_t} \text{nonradical products} + O_2 \tag{30}$$

If it can be proved experimentally that the rate of reaction is independent of oxygen pressure, it follows that termination reactions involving R· radicals are not important. The values of k_p and k_t both change with the structure of R·. One approach to this problem has been to measure k_t, and hence k_p, by the rotating sector technique.[37] Another approach has been to perform competitive oxidations between pairs of hydrocarbons. If the yields of R_1OOH and R_2OOH, or the disappearances of R_1H and R_2H, can be measured, two experiments will yield the values of k_{11}/k_{12} and k_{22}/k_{21}, by application of the copolymerization equation.[38]

$$R_1OO\cdot + R_1H \xrightarrow{k_{11}} R_1OOH + R_1\cdot \qquad (31)$$

$$R_1OO\cdot + R_2H \xrightarrow{k_{12}} R_1OOH + R_2\cdot \qquad (32)$$

$$R_2OO\cdot + R_2H \xrightarrow{k_{22}} R_2OOH + R_2\cdot \qquad (33)$$

$$R_2OO\cdot + R_1H \xrightarrow{k_{21}} R_2OOH + R_1\cdot \qquad (34)$$

Still a third technique has been to perform autoxidations in the presence of cumene hydroperoxide (R_COOH). This allows the ratios of $k_{RH}{}^C/(k_t{}^C)^{1/2}$ to be measured from which the relative values of $k_{RH}{}^C$ are readily obtained.[39]

$$R_COO\cdot + RH \xrightarrow{k_{RH}{}^C} R_COOH + R\cdot \qquad (35)$$

$$R\cdot + O_2 \longrightarrow ROO\cdot \qquad (36)$$

$$ROO\cdot + R_COOH \longrightarrow ROOH + R_CCOO\cdot \qquad (37)$$

$$2R_COO\cdot \xrightarrow{k_t{}^C} \text{nonradical products} \qquad (38)$$

A fourth technique which has been productively applied has been to consider only those autoxidations when it can be deduced that the termination rate is constant.[40] This can often be achieved by studying the rates of oxidation of hydrocarbon mixtures. The relative reactivities obtained are for reaction 39 and can reflect changes in both the reactivity of R_XH and $R_XOO\cdot$ as R_X is varied.

$$R_XOO\cdot + R_XH \xrightarrow{k_X{}^X} R_XOOH + R_X \qquad (39)$$

Direct measurement of hydrogen abstraction products and bimolecular coupling or disproportionation products is a common technique in nonchain processes. It has been applied for methyl radicals (measurement of methane and ethane)[41] and methoxy and ethoxy radicals (measurement of the methanol or ethanol and formaldehyde or acetaldehyde).[42] In the case of *t*-butoxy radicals the hydrogen abstraction step and the unimolecular decomposition to acetone and methyl radical are often compared by measurement of yields of acetone and *t*-butoxy alcohol.[43] However, the ratio k_R/k_d appears to be

$$(CH_3)_3CO\cdot + RH \xrightarrow{k_R} (CH_3)_3COH + R\cdot \qquad (40)$$

$$(CH_3)_3CO\cdot \xrightarrow{k_d} CH_3COCH_3 + CH_3\cdot \qquad (41)$$

quite solvent sensitive[44] and the direct comparison of the loss of R_1H and R_2H in a competitive process seems to be preferred. It should be mentioned that the direct or indirect competitive technique for the *t*-butoxy radical generated from *t*-butyl hypochlorite may be invalidated by the occurrence of a chlorine atom chain.[45]

A number of semiquantitative studies have been performed wherein approximate relative reactivities have been calculated from isolated products of coupling (42–44). This technique has been applied to the products obtained

$$X\cdot + R_1H \xrightarrow{k_1} HX + R_1\cdot \tag{42}$$

$$X\cdot + R_2H \xrightarrow{k_2} HX + R_2\cdot \tag{43}$$

$$R_1\cdot + R_2\cdot \longrightarrow R_1{-}R_1,\ R_1{-}R_2,\ R_2{-}R_2 \tag{44}$$

from a single substrate with methyl and/or acetoxy radical attack[46–48] or for methoxy radical reaction.[49] Dimethylamino radicals (from tetramethyltetrazene) have been allowed to compete for two substrates.[50] Reaction 4 is a serious complicating factor (as has been shown for acetyl peroxide decompositions in mixed solvents)[51] and corrections should be made.

In a few cases electron spin resonance spectroscopy has been used to measure the point of attack of a radical on a hydrogen-donating substrate. This technique has been applied in a number of hydroxyl radical reactions (45–47).[52,53]

$$H_2O_2 + Ti^{III} \longrightarrow HO\cdot + Ti^{IV} + OH^- \tag{45}$$

$$HO\cdot + CH_3CH_2CO_2H \longrightarrow \cdot CH_2CH_2CO_2H \text{ (Major)} \tag{46}$$

$$HO\cdot + CH_3CH_2CO_2H \longrightarrow CH_3\dot{C}HCO_2H \text{ (Minor)} \tag{47}$$

III. RELATIVE REACTIVITIES OF 1°-, 2°-, AND 3°-HYDROGEN ATOMS IN ALKANES AND ARALKYL HYDROCARBONS

Some commonly used C—H bond dissociation energies (*D*) are listed in Table 4.[28,54] We note from Table 4 that a methyl or chlorine substituent imparts about equal resonance stabilization to an alkyl radical and that a phenyl substituent is much more effective in stabilizing the radical. We also note a leveling effect of substituents on radical stabilizing since a single methyl (or chlorine) substitution in methane decrease *D*(C—H) by 6–8 kcal whereas the third methyl (or chlorine) substitution reduces *D* by only 3–4 kcal/mole.

Evans and Polanyi were among the first to point out that factors which stabilize the radicals also stabilize the transition state in a hydrogen atom transfer reaction (Figure 1).[55] The attack of methyl radicals on a series of aliphatic C—H bonds has been shown to follow the relationship:[56]

$$E^{\ddagger} = 0.5D(\text{C—H}) - 38.5$$

TABLE 4

Bond Dissociation Energies (kcal/mole)

R—H	$D(\pm 1)$
CH_3—H	104
CH_3CH_2—H	98
$(CH_3)_2CH$—H	94
$(CH_3)_3C$—H	91
$C_6H_5CH_2$—H	85
$ClCH_2$—H	96
Cl_2CH—H	94
Cl_3C—H	90

As we will see, a direct relationship between $E^{\ddagger}$ and D(C—H) or ΔH is not universally observed. However, for the alkanes the relationship does appear to apply when steric effects are not important, and we will begin our discussion on this topic. Table 5 lists the 1°-:2°-:3°-hydrogen relative reactivities observed for alkanes toward a variety of radicals. Also listed are log A and $E^{\ddagger}$ for attack of the radicals on methane where known.

In Table 5 we note excellent qualitative agreement between selectivity

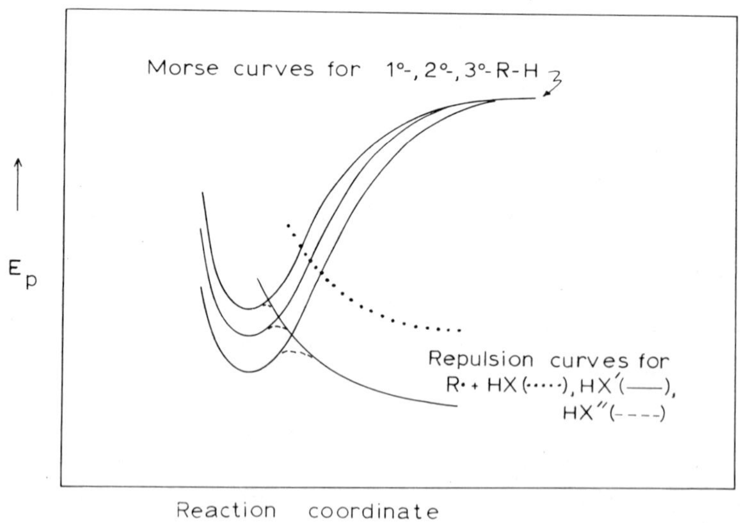

Fig. 1. Potential energy diagram in the manner of Evans and Polanyi for the attack of atoms X·, X·, and X′ on 1°-, 2°-, and 3°-carbon-hydrogen bonds.

TABLE 5

1°-, 2°-, and 3°-Hydrogen Atom Relative Reactivities in Alkanes, R = Simple Alkyl Group

Radical (X·)	Conditions	CH_3—H ↑	:RCH_3 ↑	:R_2CH_2 ↑	:R_3CH ↑	*D*(H—X)[a]	$E^‡(CH_4)$[a]	Log $A(CH_4)$[b]	Ref.
F·	25°, gas	0.5	1	1.2	1.4	135	1.2	14	57. 58
Cl·	25°, liquid	—	1	3.5	4.2	—	—	—	59
Cl·	100°, liquid	—	1	2.0	3.0	—	—	—	59
Cl·	100°, gas	0.004	1	4.3	7.0	103	3.8	13.4	60
"Cl·"	25°, 4 *M* C_6H_6	—	1	5	20	—	—	—	61
H·	25°, liquid	—	1	4.8	40	103	12	14	30
C_6H_5·	60°, liquid	—	1	9.0	44	104–112	—	—	24
p-$NO_2C_6H_4$·	60°, liquid	—	1	11	50	—	—	—	26
$(CH_3)_3CO$·	40°, liquid	—	1	12	44	102	11.5	12	44
HO·	25°, gas	0.04	1	7	45	118	3.7	12.5	62
CH_3·	164°, gas	0.09	1	10	80	104	17.3	13.1	28
CF_3·	164°, gas	0.06	1	8.3	90	105	11	12	28
"Cl·"	25°, 12 *M* CS_2	—	1	25	225	—	—	—	61
Br·	150°, gas	0.002	1	80	1700	87	18	14	63
I·	325°, gas	0.006	1	18	—	71	34	117	64
CN·	100°, gas	0.25	1	—	—	111	—	—	65
$SbCl_4$·	78°, CCl_4	—	1	45	—	—	—	—	66

[a] In gas phase, kcal/mole.
[b] Cm^3/(mole-sec).

(1°:2°:3°) and reactivity ($E^{\ddagger}$). There is some agreement between $E^{\ddagger}$ and ΔH but reactions with similar ΔH's (Cl·, H·, CF_3·) can give vastly different values of $E^{\ddagger}$ and of selectivity. Figure 1 illustrates this point by the dotted lines indicating a lowering in $E^{\ddagger}$ for X″· relative to X′· although ΔH_r remains constant. Such a lowering in $E^{\ddagger}$ can occur, for example, when different polar effects are involved for X· and X′·, for example, **1**.

$$\mathrm{R{:}H{\cdot}X} \longleftrightarrow \mathrm{R\overset{+}{\cdot}H{:}X^-}$$

1

Figure 1 also illustrates the case (X· and X′·) where ΔH_r has an effect on both $E^{\ddagger}$ and the 1°:2°:3° selectivity. Here other factors such as polar and steric differences between X· and X′· are constant or not important. Figure 1 demonstrates that for a family of symmetrical Morse curves for 1°-, 2°-, and 3°-carbon–hydrogen bonds, reactions with higher $E^{\ddagger}$ involve more bond breaking in the transition state. This concept is often referred to as the Hammond postulate,[67] and a more conventional potential energy diagram is given in Figure 2. Figure 1 emphasizes that the degree of bond breaking, and hence relative reactivities, cannot be predicted qualitatively from the ΔH_r. We will presently see that relative values of D(R—H) will not necessarily qualitatively predict relative reactivities toward a given radical, even though such a relationship holds in Figure 1 where a series of symmetrical Morse curves are involved.

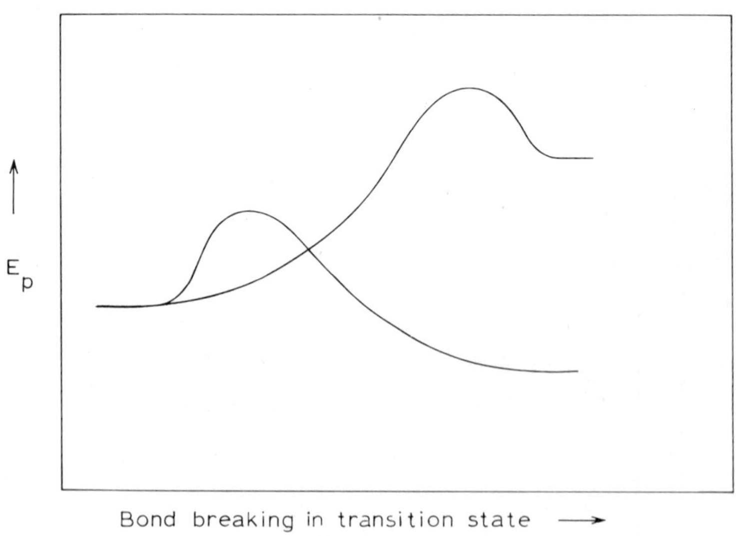

Fig. 2. Potential energy diagram indicating the interrelationship of bondbreaking, energy of activation, and exothermicity in the manner of Hammond.[67]

TABLE 6

Relative Reactivities of 1°-, 2°-, and 3°-Benzylic Hydrogen Atoms, Liquid Phase

Radical	Condition	$C_6H_5CH_3$ ↑	$C_6H_5CH_2CH_3$ ↑	$C_6H_5CH(CH_3)_2$ ↑	Log k, 40°[a]	$E^{\ddagger}$[a]	Ref.
Cl·	40°, infinite dilution	1	2.5	5.5	9–10.5	1	69
t-BuO·	40°	1	3.2	6.8	3.7	5.6	44
"Cl·"	40°, 4 *M* C_6H_6	1	3.6	8.1	—	—	69
C_6H_5·	60°	1	4.6	9.7	—	—	24
p-$NO_2C_6H_4$·	60°	1	4.7	28.5	—	—	26
CH_3·	65°	1	4.1	12.9	2.3	7.5	70
$(CH_3)_2N$·	151°	1	—	~12	—	—	51
Polyvinyl acetate	60°	1	3.9	13	—	—	71
2°-Alkyl-OO·	30°	1	9	14 }	−0.4	10	72
3°-Alkyl-OO·	30°	1	9	18 }			39
Polystyrenyl	60°	1	8	20	−3.7	~15	34
Polyacrylonitrile	60°	1	9.2	21	—	—	73
Br·	77°	1	25	57	5.5	7.2	7, 74
H·	40°	1	32	177	—	—	31
CCl_3·	40°	1	50	260	0.15	11.6	6

[a] For toluene, k in cm^3/mol-sec; $E^{\ddagger}$ in kcal/mol.

Using the Polanyi equation in the form, $\Delta E^{\ddagger} = -\alpha\Delta(\Delta H_r)$ for methane and ethane leads to the following values of α: $CH_3\cdot$, 0.44; $CF_3\cdot$, 0.40; H·, 0.45; HO·, 0.22; ROO·, 0.63; F·, 0; Cl·, 0.45; Br·, 0.66; I·, 0.89; and ·CN, 0.34.[68] Many radicals and atoms give values of α close to 0.4 although the highly reactive fluorine atom yields a transition state with little bond breaking ($\alpha = 0$) and iodine atoms yield a transition state with extensive bond rupture ($\alpha = 0.9$).

Table 6 extends our reactivities series to 1°-, 2°-, and 3°-benzylic hydrogen atoms. Although *D*(C—H) is not accurately known, it appears that the situation is qualitatively similar to the one depicted in Figure 1, that is, a series of symmetrical Morse curves.

In Table 6 there is some agreement between ΔH_r and $E^{\ddagger}$ for toluene. However, there is rather poor agreement between ΔH_r and $E^{\ddagger}$ where the nature of the attacking radical is varied. When the radical is held constant for a series of aralkanes it has been found that reactivity toward peroxy radicals,[75] or toward trichloromethyl radicals,[76] can be correlated by molecular orbital calculations of ΔE_{π} between the hydrocarbon and the benzylic radical.

Table 7 demonstrates that the relative reactivities of 1°, 2°, and 3° hydrogen atoms in alkanes parallel that of their benzylic analogs. Changes in reactivity by substitution of phenyl for alkyl are presented in a more graphical form for several radicals in Table 8.

In Table 8 the effect of alkyl substitution for hydrogen is seen in the horizontal rows, the effect of phenyl substitution for hydrogen in the vertical columns, and the effect of phenyl substitution for methyl in the vertical or diagonal directions. We note that there is a leveling effect in substitution of

TABLE 7

Relative Reactivities of Methyl, Methylene, and Methine Groups in Alkanes and Aralkanes, R = Methyl or 1°-Alkyl

	Relative reactivity per hydrogen			
Radical	$RCH_3/C_6H_5CH_3$	$R_2CH_2/C_6H_5CH_2R$	$R_3CH/C_6H_5CHR_2$	Ref.
Cl·, 40°	1.0:1.3	4.3:3.3	6.0:7.3	69
t-BuO·, 40°	1.0:10	12:32	44:69	44
$C_6H_5\cdot$, 60°	1.0:7.7	7.8:35	37:74	24
p-$NO_2C_6H_4\cdot$, 60°	1.0:6.0	11:28	50:170	26
H·, 40°	1.0:8.3	5:27	40:150	32
Br·, 80°	1.0:64,000	220:1,600,000	19,400:3,800,000	63

TABLE 8

Effect of Structure on Reactivity of Alkyl and Aralkyl Hydrogen Atoms[a]

	CH_3—H	CH_3CH_2—H	$(CH_3)_2CH$—H	$(CH_3)_3C$—H
Cl·, 40°	~0.004	1.00	4.3	6.0
C_6H_5·, 60°		1.00	7.8	37
Br·, 77°	~0.0007	1.00	220	19,400
	$C_6H_5CH_2$—H	$C_6H_5CH(CH_3)$—H	$C_6H_5C(CH_3)_2$—H	
Cl·, 40°	1.3	3.3	7.3	
C_6H_5·, 60°	7.7	35	74	
Br·, 77°	64,000	1,600,000	3,800,000	
	$(C_6H_5)_2CH$—H	$(C_6H_5)_2C(CH_3)$—H		
Cl·, 40°	2.6			
C_6H_5·, 40°	59	157		
Br·, 77°	1,100,000	4,200,000		
	$(C_6H_5)_3C$—H			
Cl·, 40°	9.5			
C_6H_5·, 40°	350			
Br·, 77°	6,400,000			

[a] References 26, 69, 6, and 74.

methyl or phenyl for hydrogen. The incremental rate increase falls off sharply as the second and third substitutions are made. It is also noted that activation by phenyl substitution relative to methyl substitution is much more effective for bromine atom attack than for chlorine atom attack. For chlorine atom attack where the transition state involves little bond breaking, alkyl and phenyl substituents activate to about the same extent. Here, inductive effects of the substituents are important and resonance effects are minimized since the incipient free radical is not well developed. Although phenyl can stabilize a radical better than a methyl substituent (by about 10 kcal/mole when substituted in methane) the inductive effect of a methyl stabilizes the electron-deficient transition state (structure **1**) whereas the phenyl would destabilize inductively by electron withdrawal. In the case of bromination, where the carbon–hydrogen bond is extensively broken in the transition state, the resonance-stabilizing effect of the phenyl substituent overshadows that of the methyl substituent.

Figure 3 illustrates how this situation might appear in a potential energy diagram where the dotted line indicates the stabilization of the transition

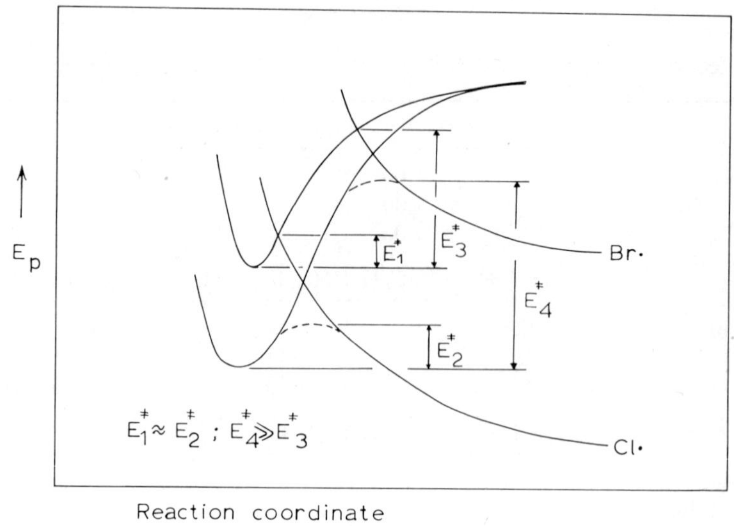

Fig. 3. Possible explanation of the reversal in the relative reactivities of toluene and cyclohexane toward chlorine and bromine atoms based on inductive polar effect consideration.

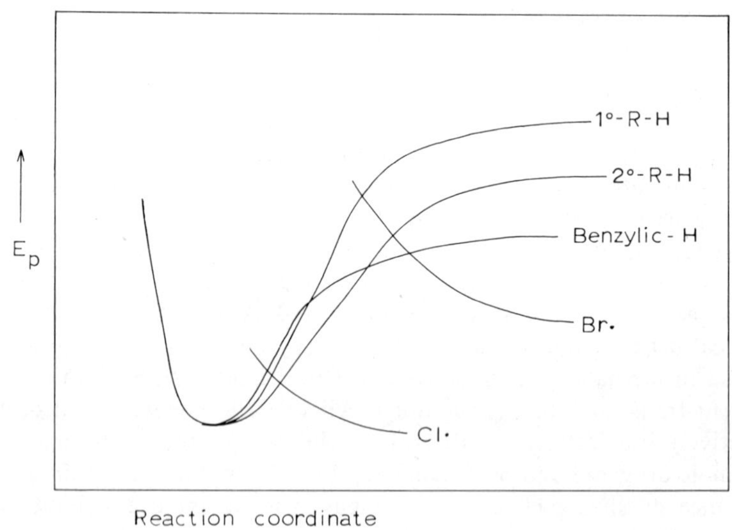

Fig. 4. Possible explanation of the reversal in the relative reactivities of toluene and cyclohexane toward chlorine and bromine atoms.

state with alkyl substituents by the inductive effect. An alternate explanation is that the benzylic resonance stabilization does not develop quantitatively in a similar fashion to alkyl hyperconjugative stabilization; that is, a series of symmetric Morse curves are not involved. If the benzylic resonance stabilization can only develop when the carbon–hydrogen bond is extensively broken, a situation similar to Figure 4 might result. As Figures 3 and 4 suggest, it is possible to select two substrates which will demonstrate a reversal in selectivity when the degree of bond breaking in the transition state is varied widely.

In Figure 4 the Morse curve has been drawn so that at very little bond breaking the $E^\ddagger$ of 1°-aliphatic and that of 1°-benzylic bonds are similar.[77] This would lead to a situation where $E^\ddagger$ (1°-aliphatic or 1°-benzylic) > $E^\ddagger$ (2°-aliphatic). With extensive bond breaking the reactivities would be controlled by the ΔH_r and $E^\ddagger$ (1°-benzylic) > $E^\ddagger$ (2°-aliphatic). In Table 9 are listed the relative reactivities of toluene and cyclohexane toward a series of radicals and atoms. The relative reactivities given by the radicals of low electron affinity, such as hydrogen, methyl, and particularly phenyl, suggest that the explanation pictured in Figure 4 is more nearly correct. On the other hand, the data for phenyl and *p*-nitrophenyl, or the complexed chlorine atom in benzene or carbon disulfide solution, suggest the polar inductive effect as outlined in Figure 3. Both effects seem to be important.

TABLE 9

Relative Reactivities of Cyclohexane to Toluene (= 1.00)

Attacking species	Relative reactivity of cyclohexane: per Molecule	per Hydrogen	Ref.
Bromine, 80°	~0.06	~0.015	5, 12
Peroxy, 65°	0.06	0.014	78
Polystyrenyl, 60°	0.20	0.05	34, 73
Methyl, 100°, gas	1.80	0.45	79
Phenyl, 60°	4.0	1.0	25
Methyl and/or acetoxy, 100°, liquid	6.4	1.6	24
t-Butoxy, 40°	6.0	1.5	80
p-NO_2-Phenyl, 60°	6.8	1.7	26
Hydrogen, 40°	2.4	0.6	31
Chlorine, 40°	7.6	1.9	5
Chlorine, 4 *M* C_6H_6, 40°	10.0	2.5	69
Chlorine, 12 *M* CS_2, 40°	7.2	1.8	69

TABLE 10

Relative Reactivities of Haloalkanes

		$E^‡$	Log *A*	Ref.
(a)	Br·, CH_4	17.3	13.2	81
	CH_3F	14.8	12.9	82
	CH_2F_2	15.3	12.5	82
	CHF_3	21.1	12.3	82
	CH_3Cl	14.5	13.6	81
	CH_3Br	16.0	13.7	81
	$CHCl_3$	9.3	12.3	81
	CHF_3	23.5	13.46	82
(b)	CH_3·, CH_4	14.2	11.8	83
	CH_3F	11.8	11.2	84
	CH_2F_2	10.4	11.0	84
	CHF_3	10.2	10.4	85
	CH_3Br	10.1	12.5	27
	CH_2Br_2	8.7	12.2	27
	CH_3Cl	9.4	11.8	27
	CH_2Cl_2	7.2	11.3	27
	$CHCl_3$	5.8	10.8	27
	$(CH_3)_4Si$	10.8	12.0	86
	$(CH_3)_3SiF$	11.5	12.0	86
	$(CH_3)_2SiF_2$	12.1	12.0	86
	CH_3SiF_3	12.6	12.1	86
(c)	Cl·, CH_4	3.9	10.4	87
	CH_3Cl	3.4	10.7	87
	CH_2Cl_2	5.5	11.6	87
	$CHCl_3$	6.5	11.6	87
(d)	CF_3·, CH_4	10.3	11.7	29
	CH_3Cl	10.6	12.1	88
	CH_3Br	9.9	11.6	88
	CH_3I	7.5	11.6	88
	$(CH_3)_4Si$	7.3	11.9	89
	$(CH_3)_3SiF$	9.5	12.4	89
	$(CH_3)_2SiF_2$	10.5	12.3	89
	CH_3SiF_3	11.7	12.0	89

With functional groups more pronounced polar effects in $E^{\ddagger}$ and in relative reactivity can be discerned. Table 10 lists some pertinent $E^{\ddagger}$ data for the halomethanes while Table 11 presents relative reactivity data toward some common radicals.

From Tables 10 and 11 it is seen that quite different reactivity series are observed for highly electron-deficient atoms or radicals ($Cl\cdot$, $CF_3\cdot$, $Br\cdot$) than for radicals with a low electron affinity ($CH_3\cdot$, $C_6H_5\cdot$). These results can be rationalized in terms of transition state **1** and potential energy diagrams similar to Figures 1 and 3. However, before making these correlations it would appear appropriate to (1) examine the polar effect in the absence of steric effects by use of the Hammett $\rho\sigma$ correlations, and (2) to establish a comprehensive scale of relative reactivities toward a radical where polar effects are not important.

TABLE 11

Relative Reactivity of Chloroalkanes toward Several Radicals

Conditions	$CH_2Cl—H$ ↑	$CHCl_2—H$ ↑	$CCl_3—H$ ↑
$Cl\cdot$, 125°, vapor[87]	1.00	1.15	0.19
Polytetrafluoroethyl·, 125°[90]	1.00	1.6	2.9
$CF_3\cdot$, 125°, vapor[91]	1.00	5.0	12.0
$C_6H_5\cdot$, 40°, liquid[24]	1.00	10.0	140.0

IV. POLAR EFFECTS IN SUBSTITUTED TOLUENES

The polar effect (i.e., structure **1**) has been invoked to explain numerous directive effects observed in aliphatic systems. Thus methyl radicals attack propionic acid preferentially at the *alpha* position, $k_\alpha/k_\beta = 7.8$ (per hydrogen)[46,47] whereas chlorine prefers to attack at the *beta* position, $k_\alpha/k_\beta = 0.03$ (per hydrogen).[92–95] In an investigation of *t*-butyl derivatives a semiquantitative relationship was observed between relative reactivity and the polar effect of the substituent as evidenced by the pK_a of the corresponding acid.[96]

A plot of the logarithm of relative rate (Table 12) versus ΔpK_a gave a satisfactory straight line indicating the electron-deficient character of the transition state for chlorine atom attack.

In the case of meta- and para-substituted toluenes it has been observed that very small directive effects exist for atoms or radicals not expected to have a pronounced polar effect. When treated by the Hammett $\rho\sigma$ equation it is observed that $\rho = -0.1$ for $H\cdot$,[97] $C_6H_5\cdot$,[24] *p*-$CH_3C_6H_4\cdot$,[26] and $CH_3\cdot$.[98]

TABLE 12

Chlorination of *t*-Butyl Derivatives [X-C(CH_3)$_3$]

Substituent (X)	Relative rate (80°)	$pK_a^{XCH_2CO_2H} - pK_a^{C_6H_5CH_2CO_2H}$
$(CH_3)_3C$—	1.2	0.7
C_6H_5—	1.0	0
p-$NO_2C_6H_4$—	0.5	−0.4
C_6H_5CO—	0.4	−0.8
ClCO—	0.19	−1.5
CN—	0.17	−1.7

Numerous radicals with an appreciable electron affinity show a pronounced polar effect in reaction with these toluenes. Figure 5 summarizes data for chlorine atom attack wherein a somewhat better agreement with σ^+ is achieved than with the original Hammett σ constants.[99] This is expected, at least when there is considerable bond breaking in the transition state because of the similarity of structures **2** and **3**.[100] The somewhat low value of ρ

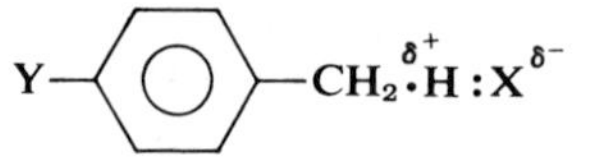

2

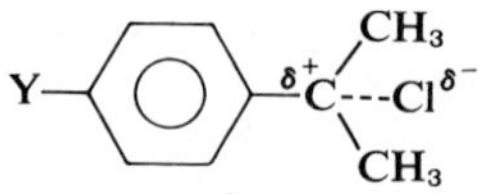

3

(−0.7) indicates the low degree of bond breaking in the transition state. For bromine atoms, which have an electron affinity similar to chlorine, ρ is −1.4,

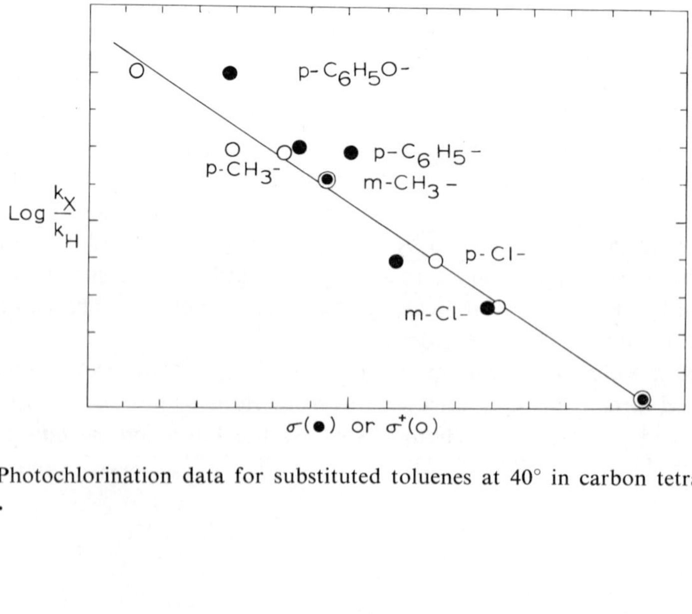

Fig. 5. Photochlorination data for substituted toluenes at 40° in carbon tetrachloride solution.

TABLE 13

Benzylic Attack of Radicals and Atoms

Abstracting species	ρ	σ or σ^+	Ref.
$(CH_3CH_2)_3Si\cdot$	+0.3	σ^*	101
$H\cdot$, 40°	−0.1	—	97
$C_6H_5\cdot$, 60°	−0.1	—	24
p-$CH_3C_6H_4\cdot$, 60°	−0.1	—	26
$CH_3\cdot$, 100°	−0.1	—	98
p-$BrC_6H_4\cdot$, 60°	−0.3	σ^+	26
$HO_2CCH_2\cdot$, 130°	−0.7	σ^+	102
$(CH_3)_3CO\cdot$, 40°	−0.4	σ^+	45, 103
$(C_6H_5)_3CO\cdot$, 22°	−1.1	σ^+	104
$(CH_3)_3COO\cdot$, 30°	−0.6	σ^+	39
p-$NO_2C_6H_4\cdot$, 60°	−0.6	σ^+	27
$Cl\cdot$, CCl_4, 40°	−0.7	—	99
$SO_2Cl\cdot$, CCl_4, 40°	−0.6	σ	106
$SO_2Cl\cdot$, CCl_4, 40°	−0.5[a]	σ	106
$SO_2Cl\cdot$, CCl_4, 80°	−0.5[b]	σ	107
$Cl_3CSO_2\cdot$, 80°	−0.5[b]	σ	108
p-$ClC_6H_4C(=O)OO\cdot$	−0.8[b]	—	109
$C_6H_5CO_2\cdot$, 80°	−0.6[c]	σ^+	110
$(CH_3)_2N\cdot$, 136°	−1.1	σ^+	111
$Br\cdot$, CCl_4, 80°	−1.4	σ^+	11, 12
$(CH_2)_5NH\cdot^+$, 30°	−1.4	σ^+	112
$CCl_3\cdot$, C_6H_5Cl, 50°	−1.5	σ^+	9
$CCl_3\cdot$	−0.5[d]	—	101
1-Cyano-1-cyclohexyl, 110°	−0.4[e]	—	113

[a] Substituted ethylbenzenes.
[b] Substituted benzaldehydes.
[c] Substituted dibenzyl ethers.
[d] Substituted aryldimethylsilanes.
[e] Substituted thiophenols.

indicative of extensive bond breaking. Table 13 lists a compilation of ρ values for various radicals.

In the original analysis of the polar effect in benzylic attack it was predicted that for a radical with a high degree of electron affinity, ρ should increase with bond breaking and the Hammett relationship should change from an σ constant to an σ^+ constant correlation.[100] Substrates with increased reactivity should possess transition states with lower degrees of bond breaking. This prediction appears to be at least partially supported by the data of Table 14.

A further example of the effect of bond breaking on the magnitude of the polar effect is the increase in ρ (σ^+ correlation) from -0.9 for attack on phenols by *t*-butoxy radicals (65°),[126] to -1.1 for polystyrenyl radicals (65°),[126] -1.6 for peroxy radicals (65°),[123] and -2.8 for diphenylpicrylhydrazyl (30°, benzene).[130]

Solvent effects on the value of ρ have been reported in chlorination and bromination of substituted toluenes. Thus ρ varies in chlorination from -0.7 (CCl_4, 40°) to -1 (C_6H_6, 60°), -0.8 (C_6H_5Cl, 80°), and -0.5 ($C_6H_5NO_2$, C_6H_5CN, 80°).[131] For bromination it is reported that ρ is -1.4 (CCl_4, 80°), -1.7 (C_6H_5Cl, 80°), -1.6 (C_6H_5Br, 80°), -1.1 ($C_6H_5CF_3$, 60°), -0.7 (C_6H_6, 60°), and -0.3 ($C_6H_5NO_2$, 80°).[131]

Reaction of a series of substituted *t*-butylbenzenes with sulfuryl chloride at 70° in chlorobenzene solution gave a rather poor $\rho\sigma$ correlation ($\rho = -0.44$).[132] Much better agreement was achieved by use of a Kirkwood-Westheimer charge–dipole interaction model,[133] that is, a field-effect model. A series of 1-substituted adamantanes gave relative reactivity data toward the trichloromethyl radical which could be correlated by the field-effect model by $\sigma^\ddagger$ constants in the Taft equation[134] ($\rho^* = -0.40$).[135] For bromination at 40° this system yields $\rho^* = -0.6$.[136] From the Kirkwood-Westheimer treatment the fraction of ionic character in the transition state, that is, structure **1**, can be estimated to be 0.17 for bromination of adamantane 0.12 for trichloromethyl radical attack on adamantanes, and 0.056 for chlorine atom complexed by chlorobenzene attacking the 1°-hydrogens of *t*-butylbenzene.[136]

Reaction of *t*-butoxy radicals with a series of substituted anisoles at 45° gave a better correlation with σ ($\rho = -0.4$) than σ^+.[105] Apparently the direct conjugation possible in structure **2** is not important in **4**.

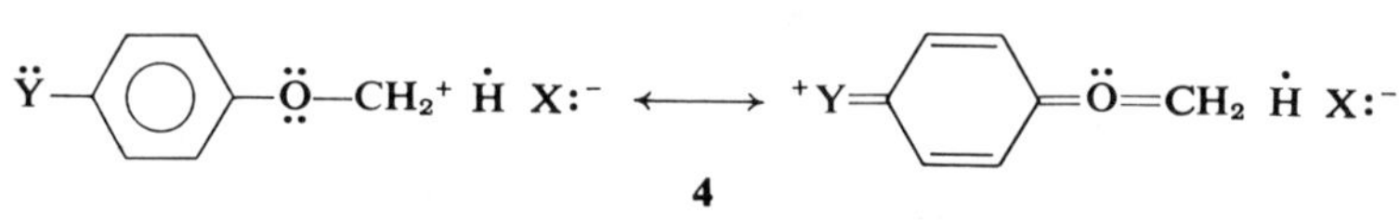

4

Polar effects of α substituents have been examined quantitatively in a few cases. In bromination at 77° a σ_p^+ correlation with $\rho = -2.5$ is observed for α-substituted toluenes.[137] The extreme reactivity of benzyl methyl ethers is thus connected with the stability of transition state structures such as **5**.

$$C_6H_5CH{=}\overset{+}{O}CH_3\ \dot{H}\ Br{:}^-$$

5

The small values of ρ for meta- and para-substituted ethers is rationalized by the fact that electron supply to the carbon free radical by the alpha oxygen

TABLE 14

ρ Values in Hammett $\rho\sigma$ Correlations

Substrates	ρ	Relative reactivity of unsubstituted substrate	Ref.
	Bromine atom		
Toluenes, 80°, C_6H_6	$-1.4\ (\sigma^+)$	1	11
Toluenes, 80°, CCl_4	$-1.3\ (\sigma^+)$	—	12
Diphenylmethanes, 77°	$-1.0\ (\sigma)$	18	114
Neopentylbenzenes, 70°	$-0.75\ (\sigma^+)$	—	115
Ethylbenzenes, 80°	$-0.7\ (\sigma^+)$	25	116
Allylbenzenes, 70°	$-0.8\ (\sigma^+)$	26	117
Deoxybenzoins, 80°	$-0.9\ (\sigma)$	—	118
Cumenes, 70°	$-0.4\ (\sigma)$	58	119
Benzyl methyl ethers, 80°	$-0.35\ (\sigma^+)$	159	120
Benzyl methyl ethers, 40°	$-0.4\ (\sigma^+)$	—	121
α,α-Dimethoxytoluenes, 80°	$-0.4\ (\sigma^+)$	—	120
Dibenzyl ethers, 80°	$-0.1\ (\sigma^+)$	—	122
Benzhydryl methyl ethers, 80°	0	High	116
	Peroxy radical		
Toluenes, 30°	$-0.6\ (\sigma^+)$	1	39
Cumenes, 60°	$-0.4\ (\sigma^+)$	18	99
Cumenes, 30°	$-0.3\ (\sigma^+)$	18	39
Benzyl phenyl ethers, 60°	0	11	99
Phenols, 65°	$-1.5\ (\sigma^+)$	—	123
2,6-Di-*t*-Bu-Phenols, 140°	$-1.0\ (\sigma^+)$	—	124
Phenylmethylanilines, 65°	$-1.6\ (\sigma^+)$	—	125
Diphenylamines, 65°	$-0.9\ (\sigma^+)$	—	125
Benzaldehydes, 30°	-0.8	—	109
	t-Butoxy radical		
Toluenes, 40°	$-0.4\ (\sigma^+)$	—	103
Phenols, 122°	$-1.2\ (\sigma^+)$	—	126
Benzyl methyl ethers, 40°	$-0.4\ (\sigma)$	—	127
Dibenzyl ethers, 110°	-0.5	—	110
	Trichloromethyl radical		
Toluenes, 55°	$-1.5\ (\sigma^+)$	—	9
Neopentylbenzenes, 70°	$-0.9\ (\sigma^+)$	—	115
Ethylbenzenes, 80°	$-0.5\ (\sigma^+)$	—	128
Cumenes, 70°	$-0.7\ (\sigma^+)$	—	119
Allylbenzenes, 70°	$-0.6\ (\sigma)$	—	117, 119
Benzyl methyl ethers, 80°	$-0.4\ (\sigma^+)$	—	120
α,α-Dimethoxytoluenes, 80°	$-0.2\ (\sigma^+)$	—	128
Benzhydryl methyl ethers, 80°	$-0.1\ (\sigma^+)$	—	128
Benzaldehydes, 80°	$-0.7\ (\sigma)$	—	129

atom is so sufficient that ring substituents can contribute little more stabilization. For α-substituted toluenes in reaction with peroxy radicals a $\sigma_p{}^+$ correlation with $\rho = -1.2$ is observed.[138] It is hard to imagine that in these cases only a polar effect is involved. The *alpha* substituent will certainly have an appreciable effect on the stability of the benzylic radical. Thus, toward the phenyl radical, a species with low electron affinity, the relative reactivities of Table 15 are observed.

Groups such as cyano or acyl activate alpha hydrogen atoms to attack by neutral radicals such as phenyl by stabilizing the incipient radical, for example:

$$R\dot{C}HC{\equiv}N \longleftrightarrow RCH{=}C{=}N\cdot$$

$$R\dot{C}HC({=}O)R' \longleftrightarrow RCH{=}C({-}O\cdot)R'$$

However, toward the bromine atom, where partial electron transference in the transition state has occurred, these substituents actually deactivate. It has been argued that for an attacking species such as bromine atom, complete electron transference could occur (48, 49)[3] but this seems unlikely in the gas phase or with nonpolar solvents.[4]

$$R{-}H + X\cdot \xrightarrow{\text{slow}} R\cdot H^+ + X^- \tag{48}$$

$$R\cdot H^+ + X^- \xrightarrow{\text{fast}} R\cdot + HX \tag{49}$$

TABLE 15

Relative Reactivities of Alpha-Substituted Toluenes

Substituent	$C_6H_5\cdot$, 60°[a]	$Br\cdot$, 77°[b]
None	1.0	1.0
CH_3—	4.5	25
C_6H_5—	7.7	18
CH_2=CH—	10	—
NO_2—	—	0.01
CN—	11	0.2
HO_2C—	3.0	0.27
Br—	2.6	0.2
Cl—	3.1	0.5
$C_6H_5CO_2$—	1.9	1.4
C_6H_5CO—	4.4	—
C_6H_5O—	5.7	46
HO—	6.7	—
CH_3S—	8.2	58

[a] Reference 24.
[b] Reference 114.

V. RELATIVE REACTIVITIES OF ORGANIC SUBSTRATES TOWARD A NONPOLAR RADICAL

It is of interest to examine the directive effects of some of the common functional groups in a homolytic substitution process that is free of the polar effect. The most extensive reactivity series presently known is for the phenyl radical, which displays medium selectivity and low polar requirements.[24,25] Tables 16–19 list some pertinent results.[139,140]

TABLE 16

Relative Reactivity toward Phenyl Radicals, 60°, per Hydrogen Atom

X=	$(CH_3)_3CX$ ↑	CH_3X ↑	CH_2X_2 ↑	CH_3CH_2X ↑	$(CH_3)_2CHX$ ↑	$C_6H_5CH_2X$ ↑
H—	1.0	—	—	1.0	9	9
CH_3—	1.2	1.0	9	9	47	41, 24[a]
C_6H_5—	1.2	9.0	69	41	87	69
α-Naph	—	6.0	—	23	—	—
β-Naph	—	2.5	—	27	—	—
CH_2=CH—	—	10–15	59	32	141	32
HO_2C—	0.6	2.9	—	—	—	27
CH_3O_2C—	—	2.9	9	—	—	—
$C_6H_5CO_2$—	0.2	0.3	—	2.0	3.0	17
C_6H_5CO—	0.5	1.1	7.5	31	24	40
RCO—[b]	—	2.8	15[c]	31	59	40, 46[d]
O_2N—	—	1.6	—	8.0	19	—
NC—	—	2.6	65	17	67	105
Cl—	1.1	2.3	23	—	—	28
Br—	1.0	—	68	8.6	28	23
HO—	0.7	4.3	—	—	—	60
C_6H_5O—	0.5	3.1	4.5	14	28	51
RO—[b]	—	4.7	6[c]	33	89	29
ROO—[b]	1.3	3.0	—	—	—	—
RSO—[b]	—	0.7	—	—	7	37
RSO_2—[b]	—	—	—	11	—	—
C_6H_5S—	2.3	13	81	26	88	225
RS—[b]	—	17	—	54	108	74
RSS—[b]	1.4	—	—	—	64	20
NCS—	—	2.2	—	10	—	—
R_2N—[b]	—	33	—	145	—	161

[a] X = $CH_2C_6H_5$.
[b] Symmetrical substrates.
[c] R = CH_3.
[d] X = $C_6H_5CH_2C(=O)$—.

TABLE 17

Relative Reactivity of Additional Functionally Substituted Methanes, CH_3X, toward the Phenyl Radical, 60°

X=	Reactivity per hydrogen atom
Alkyl	1.0
$(CH_3)_3Si$—	2.4
$C_6H_5Si(CH_3)_2$—	2.3
$(CH_3)_3SiSi(CH_3)_2$—	7.4
$(CH_3)_3SiOSi(CH_3)_2$—	1.9
$(CH_3)_3Sn$—	4.3
CH_3Hg—	5.1
$CH_3N{=}N$—	Very low
$CH_3OC({=}O)O$—	0.7
$(CH_3O)_2P({=}O)O$—	0.2
$(CH_3O)_2P({=}O)$—	1.2[a]
$(CH_3O)_3SiO$—	4.0
CH_3COCO—	1.2
o-$C_6H_4(CO)_2N$—	8.5
$C_6H_5CON(CH_3)$—	13
$C_6H_5CH{=}CH$—	14
$CH_3C{\equiv}C$—	15
$C_6(CH_3)_5$—	17
Phenyl	9 (69)[b]
2-Pyrrole	30
5-Methyl-2-pyrrole	24
3-Pyrrole	9
1,2,5-Trimethylpyrrole	156[a]
N-Pyrrole	10
N-Indole	3
2-Indole	35
N-Carbazole	17
N-Imidazole	<0.2
N-Pyrazole	4
3-Pyrazole	5
1-Benzotriazole	<0.2

(continued)

Table 16 displays many reasonable correlations. The relative abilities of phenyl and methyl substituents (X) to activate carbon–hydrogen bonds show a ninefold difference in CH_3X, 7.5-fold in CH_2X_2, 4.5-fold in CH_3CH_2X, 1.9-fold in $(CH_3)_2CHX$, and 1.7-fold in $C_6H_5CH_2X$. As the reactivity of the system with X = alkyl increases, the ability of the phenyl substituent to activate still further falls off. With many of the functional group substituents

TABLE 17 (cont.)

X=	Reactivity per hydrogen atom
2-Benzotriazole	<0.2
2-Benzimidazole	9
2-Benzoxazole	6
4-Thiazole	6
5-Thiazole	10
2-Benzothiazole	10 (97)[b]
1-Furan	2
2-Furan	2 (140)[b]
2-Benzofuran	19
3-Benzofuran	2
2-Thiophene	15 (210)[b]
3-Thiophene	8
2-Benzothiophene	14
3-Benzothiophene	14
Ferrocene	9 (125)[b]
2-Pyridine	6 (26)[b]
3-Pyridine	7
4-Pyridine	9
2,4,6-Trimethylpyridine	37[a]
5-Methyl-2-pyrazine	6
3-Pyridazine	33
4-Pyridazine	4
2-Pyrimidine	5
4-Pyrimidine	5
5-Pyrimidine	7
4-Methyl-1-tetrazine	0.3
3-Methyl-2-quinoxaline	6
7-Methyl-6-quinoxaline	19
4,6-Dimethyl-2-*sym*-triazine	16
N-Phenoxazine	65
N-Phenothiazine	45

[a] Total reactivity of molecule.
[b] R_2CH_2.

we see either activation or deactivation relative to X=alkyl. Thus, phenoxy substituent activates CH_3X (threefold), deactivates CH_2X_2 by 50%, activates CH_3CH_2X by 1.5-fold, deactivates $(CH_3)_2CHX$ by 60%, and activates $C_6H_5CH_2X$ by 25%. The cyano group is a puzzling substituent since its activation (relative to methyl) varies from 2.6-fold (CH_3X) to sevenfold (CH_2X_2), to twofold (CH_3CH_2X), to 1.5-fold [$(CH_3)_2CHX$], to 2.5-fold

TABLE 18

Relative Reactivity of Cyclic Compounds toward Phenyl Radicals at 60°

Compound	Relative reactivity per Molecule	Relative reactivity per α-Hydrogen
1°-Aliphatic	—	1.0
Ethylene oxide	Very low	—
Cyclopentane	104	10
Cyclopentene	370	~90
Cyclopentanone	138	25
1,1-Dichlorocyclopentane	98	—
Tetrohydrofuran	165	31
Tetrahydrothiophene	307	65
Tetramethylene sulfone	22	—
Butadiene sulfone	150	37
3-Pyrroline	950	240
Succinimide	34	—
Succinic anhydride	30	—
Ethylene carbonate	Very low	—
Indene	266	133
Indan	310	72
Fluorene	350	175
Phthalide	60	30
Oxindole	150	75
Indole	Very low	—
1,3-Dioxolane	180	—
2-Phenyl-1,3-dioxolane	320	—
Acenaphthalene	370	92
1,3-Indanedione	80	40
Cyclohexane	108	9
Cyclohexanone	128	18
Cyclohexene	440	~100
1,4-Cyclohexanedione	235	30
Tetrahydropyran	79	—
1,4-Dioxane	95	12
Pentamethylene sulfide	217	38
1,4-Thioxane	160	20
sym-Trioxane	21	3.5
Tetralin	480	110
9,10-Dihydroanthene	1400	350
Xanthene	1300	600
Thioxanthene	510	260
Anthrone	800	400

(continued)

TABLE 18 (cont.)

Compound	Relative reactivity per Molecule	per α-Hydrogen
Piperazene	325	40
N,*N*-Dimethylpiperazine	700	—
1,4-Diazabicyclo[2·2·2·]octane	200	17
N-Methylmorpholine	516	—
Cycloheptane	230	16
Cyclooctane	290	18

TABLE 19

Relative Reactivity toward the Phenyl Radical at 60° of Compounds Containing 3°-Hydrogen Atoms, Aldehydic Hydrogen Atoms, or Hydrogen Bonded to a Heteroatom

Compound	Relative reactivity, per Hydrogen
1°-Aliphatic	1.00
Chloroform	320
Triphenylmethane	350
Diphenylsilane	390
Triphenylsilane	480
Tetraphenyldisilane	1100
Diphenylamine	230
Indole	Very low
Diphenylphosphine	6000
Benzal chloride	117
1,1-Diphenylethane	160 (α)
Diphenylacetonitrile	440
Phenol	43
Benzaldehyde	194
Trimethylacetaldehyde	630
Methyl formate	11 [a]
N,*N*-Dimethylformamide	55 [a]

[a] Total reactivity of molecule.

($C_6H_5CH_2X$). There may be complications from groups which can exist in tautomeric structures (e.g., carbonyl, cyano, nitro). However, the reason for the discrepancies may be more fundamental. The ability of a substituent to influence the bond dissociation energy of an alpha carbon–hydrogen bond may well vary with the other substituents on the carbon atom. For example. Rüchardt has pointed out that the difference in bond dissociation energies (ΔD) between CH_3—X and $(CH_3)_3C$—X is a function of X.[141] Ground-state nonbonded interactions must be important as illustrated by the data of Table 20.

TABLE 20

Differences in Bond Dissociation Energies[142,143]

X	ΔD (kcal/mole)[a]
H	13
$C(CH_3)_3$	12
$CH(CH_3)_2$	11
C_6H_5	9
C_2H_5, CH_3	8
I, Br	7
Cl	5
F, OH, OCH_3, NH_2	1–2

[a] $D_{CH_3-X} - D_{(CH_3)_3C-X}$

Thus, as illustrated previously, even without contributions from polar factors, the relative reactivities of carbon–hydrogen bonds depend on other factors besides ΔH_r. Moreover, the effect of a given substituent on ΔH_r may depend on the system in which that substituent is introduced.

VI. OTHER EXAMPLES OF POLAR EFFECTS

In Table 21 have been gathered reactivity data for the phenyl radical, which has little propensity to enter into polar structures such as **1**,[24] as well as the bromine atom[114] and the *t*-butylperoxy radical,[138] both of which display pronounced polar effects. The reactivities observed toward these latter species show little resemblance to the "normal" reactivity series as measured by phenyl radicals, which are apparently controlled by bond dissociation energies. Most substituents such as alkoxide, halo, or ketonic groups activate slightly toward phenyl attack. Alkoxy groups are powerful activating groups toward bromine atoms and peroxy radicals, while halo substituents deactivate. All of these observations are readily understood in terms of resonance structure **1**.

TABLE 21

Relative Reactivity of Br·, C_6H_5·, and *t*-BuOO· toward α-Substituted Toluenes

	Relative reactivities, per hydrogen		
Substrate	C_6H_5·	Br·	*t*-BuOO·
$(C_6H_5)_3CH$	39	102	28
$(C_6H_5)_2CHCH_3$	18	66	7.3
$C_6H_5CH_2CN$	12	0.2	1.1
$C_6H_5CH(CH_3)_2$	9.5	58	18
$(C_6H_5)_2CH_2$	8	18	9.4
$C_6H_5CH_2OH$	7	—	6.7
$C_6H_5CH_2OC_6H_5$	4.5	46	11
$C_6H_5CH_2CH_3$	4.5	25	11
$C_6H_5CH_2COCH_2C_6H_5$	4.5	—	5
$C_6H_5CH_2OCH_2C_6H_5$	3	159	33
$C_6H_5CH_2Cl$	3	0.5	0.9
$C_6H_5CH_2Br$	2.5	0.2	0.7
$C_6H_5CH_2O_2CC_6H_5$	2	—	0.95
$C_6H_5CH_2O_2CCH_3$	2	1.4	0.75
$C_6H_5CH_3$	1.0	1.0	1.0
C_6H_5CHO	21	—	95

The ability of heteroatoms to stabilize an incipient radical when partially developed, as in phenyl radical attack, undoubtedly involves the delocalization depicted in structure **6**.

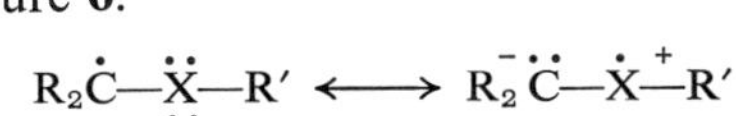

6

From Table 21 it is easy to gain the impression that toward heteroatom radicals, polar effects are more important than the bond dissociation energies of the carbon–hydrogen bonds.

Toward substrates without polar groups, that is, the alkanes, radicals with high and low electron affinities give similar reactivity scales. It does not seem that polar effects contribute significantly to the well-known 3° > 2° > 1° reactivity sequence. One quantitative correlation in this area links $E^{\ddagger}$ for hydroxyl radical attack with $E^{\ddagger}$ for methyl radical attack on a series of alkanes in the vapor phase by equation 50.[62]

$$E^{\ddagger}(\cdot OH) = 0.77E^{\ddagger}(CH_3\cdot) - 5.7 \text{ kcal/mole} \qquad (50)$$

In Table 22 are collected intramolecular reactivity data for a series of *n*-alkyl derivatives.

TABLE 22

Substitution in *n*-Propyl and *n*-Butyl Derivatives

Reagents	R—	% Attack CH_2 ↑	% Attack CH_2 ↑	% Attack CH_3 ↑	Attack on R	Ref.
Br_2, 160°, gas	CH_3—	49	49	1	1	144
Cl_2, 75°, gas	CH_3—	35	35	15	15	144
Cl_2, 68°, liquid	CH_3—	32	32	18	18	145
Cl_2, 68°, 13 *M* CS_2	CH_3—	42	42	8	8	145
Cl_2, 0°, 13 *M* CS_2	CH_3—	49	49	1	1	145
Cl_2, 75°, gas	$ClCH_2$—	26	46	19	10	144
Cl_2, 34°, liquid	$ClCH_2$—	24	51	18	7	144
SO_2Cl_2, 70°, liquid	$ClCH_2$—	22	47	24	7	146
Cl_2, 11 *M* CS_2, 0°	$ClCH_2$—	20	70	6	4	145
Br_2, 160°, gas	$ClCH_2$—	22	54	1	23	144
t-BuOCl, 40°	$ClCH_2$—	20	44	16	21	147
Cl_2O, 40°	$ClCH_2$—	19	51	9	22	14
$(CH_3)_2NCl$, 4 *M* H_2SO_4, 25°	$ClCH_2$—	10	78	7	6	15, 16
Cl_2, 60°, liquid	$BrCH_2$—	27	49	21	3	148
F_2, 20°, gas	FCH_2—	22	28	42	<8	35
Cl_2, 78°, gas	FCH_2—	22	48	19	12	35
Br_2, 146°, gas	FCH_2—	8	81	1	9	35
SO_2Cl_2, 70°, liquid	Cl_2CH—	13	48	37	2	146
Cl_2, 80°, liquid	Cl_2CH—	11	47	39	3	149
t-BuOCl, 40°	Cl_2CH—	8	39	23	30	147
SO_2Cl_2, 80°, liquid	Cl_3C—	8	42	50	—	146
Cl_2, 16°, liquid	F_3C—	0	45	55	—	150
SO_2Cl_2, 70°, liquid	Cl_3Si—	16	49	35	—	151
SO_2Cl_2, 70°, liquid	CH_3SiCl_2—	22	53	25	—	152
Cl_2, 20°, liquid	ClCO—	3	54	43	—	19
Cl_2, 20°, 8.5 *M* C_6H_6	ClCO—	2	70	29	—	19
Cl_2, CCl_4, liquid	HO_2C—	8	52	40	—	152
Cl_2, HF, liquid	HO_2C—	3	49	49	—	152
t-BuOCl, 30°	HO_2C—	21	46	33	—	153

(continued)

TABLE 22 (cont.)

Reagents	R—	% Attack $\underset{\uparrow}{CH_2}$—	$\underset{\uparrow}{CH_2}$—	$\underset{\uparrow}{CH_3}$	Attack on R	Ref
Cl_2, 25°, liquid	CH_3O_2C—	13	61	24	2	154
$(CH_3)_2NHCl^+$, Fe^{2+}, 25°	CH_3O_2C—	—	86	14	—	155, 156
t-BuOCl, 40°	CH_3CO_2—	30	43	25	—	147
Cl_2, 40°	CH_3CO_2—	25	20	49	—	93
Cl_2, 80°, liquid	Cl_3CO_2—	3	50	47	—	157
Cl_2, 25°	Cl—	12	48	30	—	154
Br_2, 25°	Cl—	40	60	Trace	—	154
$(Et)_2NHCl^+$, Cu^+, 25°	Cl—	17	51	32	—	154
t-BuOCl, 40°	Cl—	48	41	11	—	147
t-BuOCl, 40°	NC—	22	44	40	—	147
Cl_2, 40°	NC—	0	69	31	—	94
SO_2Cl_2, 70°	Cl—	0	60	40	—	146
t-BuOCl, 40°	O_2N—	13	33	54	—	147
Cl_2, 75°, gas	$ClCOCH_2$—	27	50	20	3	144
Cl_2, 20°, 8.4 *M* C_6H_6	$ClCOCH_2$—	14	72	14	0	19
Cl_2, 75°, gas	$FCOCH_2$—	22	57	20	1	144
Br_2, 160°, gas	$ClCOCH_2$—	23	55	1	21	144
Br_2, 160°, gas	$FCOCH_2$—	18	57	1	24	144
Cl_2, 100°, gas	$CH_3O_2CCH_2$—	27	51	19	23	158
Br_2, 150°, gas	$CH_3O_2CCH_2$—	23	60	1	16	158
Me_2NHCl^+, Fe^{2+}, 25°	$CH_3O_2CCH_2$—	13	80	7	0	15
Cl_2, 80°	$ClSO_2CH_2$—	4	35	61	0	149
Cl_2, 80°	$Cl_3CO_2CH_2$—	5	36	49	0	149
Br_2, 160°, gas	CF_3CH_2—	7	90	2	1	144
Cl_2, 75°, gas	CF_3CH_2—	17	61	21	1	144
Cl_2, 100°, gas	$CH_3CO_2CH_2$—	27	51	19	3	158
Cl_2, 106°, gas	$CF_3CO_2CH_2$—	20	55	23	3	158
Br_2, 153°, gas	$CF_3CO_2CH_2$—	9	86	2	3	158
SO_2Cl_2, 80°	Cl_3SiCH_2—	24	53	18	4	151
Cl_2, 40°, CCl_4	$C_6H_5CH_2$—	19	28	10	42	69
Cl_2, 40°, C_6H_6	$C_6H_5CH_2$—	22	29	3	45	69
Cl_2, 50°, gas	CH_3OCH_2—	7	43	15	35	69

The ability of a substituent to direct the attack of a radical with a high electron affinity to a remote part of the molecule increases the order, $—CH_2Cl$, $—CH_2Br$, $—CH_2F$, $CH_3SiCl_2—$ < $—CHCl_2$, $—SiCl_3$, $CH_3CO_2—$, Cl— < $—Cl_3C$ < —COCl, $—CO_2H$, $Cl_3CO_2—$, $—CF_3$. A positive-charged radical or substrate accentuates this effect; for example, compare Cl· and $(CH_3)_2NH^{\dot{+}}$. Protonation of butyric acid in hydrofluoric acid solution shows considerable deactivation at the α position in chlorine atom attack.[152] Protonation of the substrate and attack by a radical cation ($R_2N^{\dot{+}}$) show an extremely strong polar effect in that the site of attack is at the methylene group, the furthest removed from the site of protonation (Table 23).

TABLE 23

Chlorination of Hexanoic Acid Derivatives, % Attack at 25°

Reagents	R—	—CH_2— ↑	—CH_2— ↑	—CH_2— ↑	—CH_2— ↑	—CH_3 ↑	
$(CH_3)_2NHCl^+$, Fe^{2+}, H_2SO_4	MeO_2C—	—	0.7	6.3	87.3	5.7	155
$(CH_3)_2NHCl^+$, Fe^{2+}, $HOAc/H_2SO_4$	MeO_2C—	—	5	14	78	4	155
$(CH_3)_2NHCl^+$, Fe^{2+}	ClCO—	2.7	8.6	27.2	36.6	25	156
Cl_2, CCl_4	HO_2C—	4	13	28	37	18	159
$((CH_3)_2CH)_2NHCl^+$, $h\nu$, H_2SO_4	HO_2C—	0	0	0	93	1	160
Cl_2, 90% H_2SO_4	HO_2C—	4	7	40	7	42	159
Cl_2, neat	ClCO—	—	8	29	41	21	161
Cl_2, C_6H_6	ClCO—	—	3	12	42	5	161

Some very unusual product distributions have been reported in photochlorinations of carboxylic acids in 90% sulfuric acid solution which may reflect the intramolecular reactions of a species such as **7**.

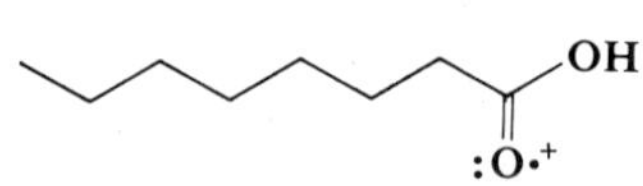

7

Octanoic acid derivatives (Table 24) illustrate these unusual directive effects. The attack at the γ-methylene position for **7** is not unexpected since a similar directive effect is noted in the Hofmann–Löffler–Freytag reaction (20°)[162] and in intramolecular reactions of alkoxy radicals (0°) **8a, b.**[163] The preferred cyclic transition state involves five atoms plus the hydrogen being attacked. Similar directive effects are noted in the Barton[164] and Yang[165] reactions.

TABLE 24

Chlorination of *n*-Octanoic Acid, % Attack at 25°[159,160]

	HO_2C—	CH_2—	CH_2—	CH_2—	CH_2—	CH_2—	CH_2—	CH_3
Cl_2, CCl_4, $h\nu$		1.5	9	15	15	19	24	17
Cl_2, 90% H_2SO_4, $h\nu$		5	8	31	10	2	3	41
$((CH_3)_2CH)_2NHCl^+$, $h\nu$, H_2SO_4		0	0	0	5	14	80	1

Quantitative relationships between reactivity and the σ^* polar substituent constants of Taft have been achieved in several instances. The Taft equation (51)[134] is given below:

$$\log \frac{k_{\text{substituted}}}{k_{\text{unsubstituted}}} = \rho^*\sigma^* + r^*\sigma^r + \delta E_s \tag{51}$$

where ρ^*, r^*, and δ are proportionality constants characterizing the polar (σ^*), resonance (σ^r), and steric (E_s) effects of the substituents. Liquid-phase chlorination of aliphatic esters, acids, and nitriles gives $\sigma^* = -0.2$.[166] Some other applications of the Taft equation are summarized in Table 25.

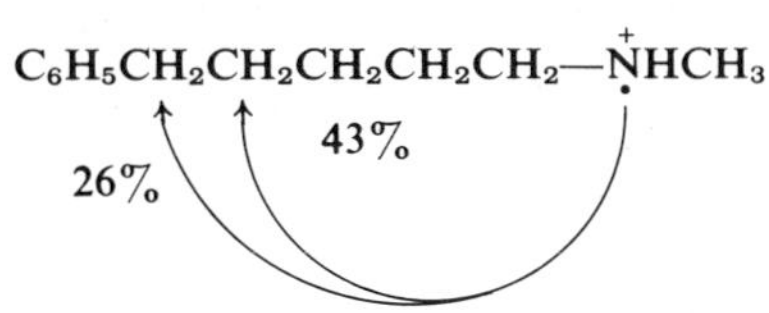

8a

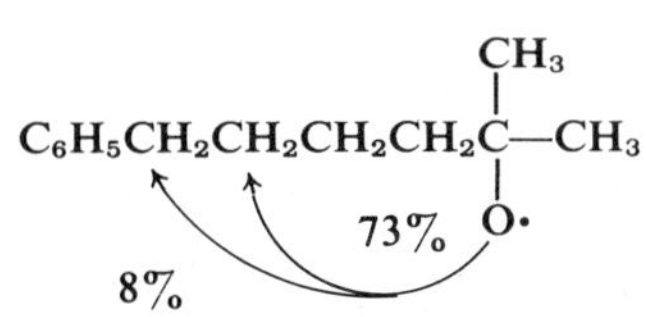

8b

Afanas'ev has applied the Taft equation with $\delta = 0$ by assuming r^* to be equal to $\rho^* = 2.33$ for a standard reaction, chain transfer in the telomerization of ethylene.[68] Using known values of σ^* he could calculate a series of σ^r values which he further related to spin densities in the fully developed free radicals. When applied to the available data for monosubstituted methanes, methylene chloride, and chloroform he obtained satisfactory correlations with methyl radicals ($\rho^* = r^* = 1.5$), polyvinyl acetate radicals ($\rho^* = 1.0$, $r^* = 1.2$), phenyl radicals ($\rho^* = 1.1$, $r^* = 0.9$), trifluoromethyl radicals

TABLE 25

Taft Equation Applied to Free Radical Reactions

Radical	Substrate	Conditions	ρ^*	Ref.
$CCl_3\cdot$	Adamantanes	40°	−0.4	135
Br·	Adamantanes	40°	−0.6	136
$R(CH_2CH_2)_3\cdot$	HCH_xCl_{3-x}	70–100°	+2.33	167
$(C_2H_5)_3Si\cdot$	$ClCCl_2X$	80°	+0.3	101
Cl·	HCH_2X	100°	−1.00	168
Cl·	HCH_2X	20°	−1.25	169
Cl·	HCH_2X	40°	−0.9	170
Cl·	HCH(Cl)X	20°	−0.8	169
Cl·	$HCCl_2X$	20°	−0.5	169
$(CH_3)_3CO\cdot$	HCH_2X	40°	−0.7	171
$(CH_3)_3CO\cdot$	Cyclohexanes	135°	−0.4	172

($\rho^* = 0.56$, $r^* = 1.55$), hydroxyl radicals ($\rho^* = 0.01$, $r^* = 1.6$), iodine atoms ($\rho^* = 1.7$, $r^* = 3.1$), and chlorine atoms ($\rho^* = -3.54$, $r^* = 1.3$).[68] However, the generality of this approach is limited since the activating (or deactivating) effect of a substituent is known to depend on the system in which it is substituted, for example, the effect of methyl substitution in the series methane, ethane, propane, isobutane.

Reversal in the preferred point of attack of an atom or radical is often a dramatic illustration of the polar effect. The preferred α attack by alkyl or phenyl radicals in propionic acids or nitriles is sharply contrasted by the preferred β attack of chlorine atoms (Table 26).

Four isolated but model examples of a reversal in directive effect are qualitatively summarized below:

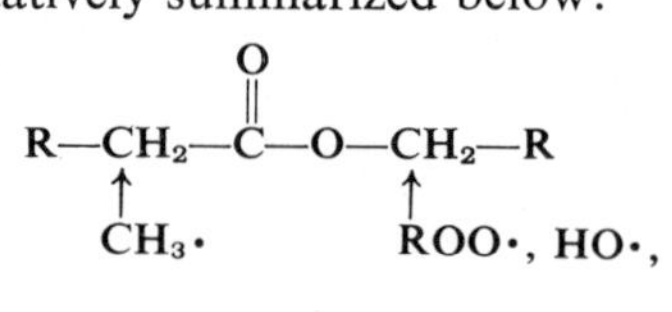

9 (ref. 46,53)

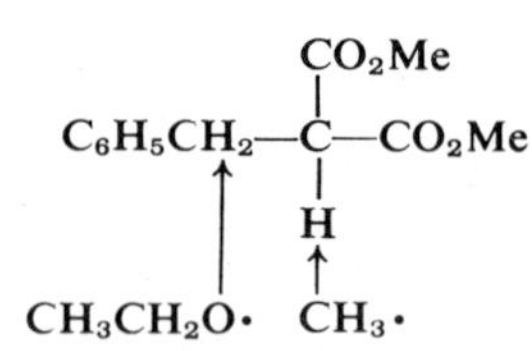

10 (ref. 176,177)

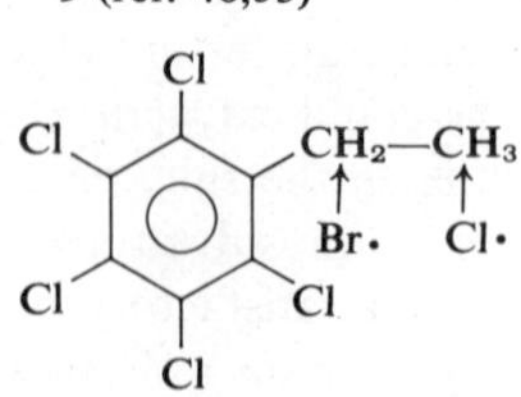

(ref. 178)

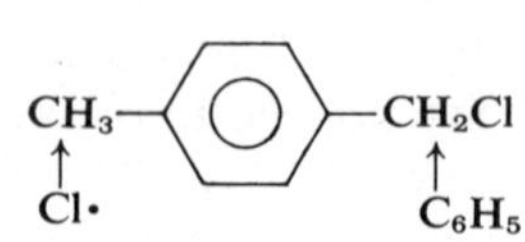

(ref. 24,179)

TABLE 26

Attack of Atoms or Radicals upon Propionic and Isobutyric Acids

Radical	Relative reactivity per hydrogen atom: X—	—CH_2— ↑	—CH_3 ↑	Ref.
$CH_3\cdot$, 100°	HO_2C	7.8	1.0	173
$Cl\cdot$, 80°	HO_2C	0.03	1.0	92
$HO\cdot$, 25°	HO_2C	~1	1.0	53
$C_6H_5\cdot$, 60°	NC	17	1.0	24
$Cl\cdot$, 60°	NC	0.5	1.0	174
$Cl\cdot$, 8 *M* C_6H_6, 60°	NC	0.9	1.0	174
$Cl\cdot$, 12 *M* CS_2, 40°	NC	1.06	1.0	174
	X—	—CH═ ↑	═$(CH_3)_2$ ↑	
$HO\cdot$, 25°	CH_3O_2C	~1	1.0	53
$CH_3\cdot$, 100°	ClCO	12.4	1.0	47
$Cl\cdot$, 80°	ClCO	1.0	1.0	175
$CH_3\cdot$, 100°	HO_2C	>25	<1	46
$C_6H_5\cdot$, 60°	NC	67	1.0	24
$Cl\cdot$, 60°	NC	2.7	1.0	174
$Cl\cdot$, 8 *M* C_6H_6, 60°	NC	6.6	1.0	174

VII. DEUTERIUM ISOTOPE EFFECT

Carbon–hydrogen and carbon–deuterium bonds have a zero-point energy difference of nearly 1.5 kcal/mole because of the effect of mass on stretching and bending frequencies.[180] A portion of this energy difference can be lost in the transition state in which these bonds are ruptured. The energy difference is associated with bending and stretching vibrational modes. If only a stretching vibration is lost in the transition state, a primary isotope effect of approximately 7 is expected. If the transition state is not symmetrical, for example, if it possesses little bond breaking or a great amount of bond making, the stretching vibration will be only partly lost and a greatly reduced isotope effect will result. Extremely reactive atoms or radicals, with low E_a, will involve a transition state with little bond breaking and with a low k^H/k^D ratio. As the reactivity of the radical or the substrate is changed, k^H/k^D will increase with E_a until the symmetrical transition state with approximately 50% bond breaking and making is reached. Table 27 demonstrates this point.

TABLE 27

Hydrogen–Deuterium Isotope Effects

Radical	Temperature (°C)	Substrate(s)	k^H/k^D	Ref.
Cl·	0	Methane	12.1	181
Cl·	127	Ethane	2.5	182
Cl·	−15	Isobutane (3°)	1.4	1
Cl·	77	Toluene	1.3	183
Cl·, 6 *M* toluene	77	Toluene	2.1	183
Br·	77	Toluene	4.9	184
Br·	77	Ethylbenzene (α)	2.7	184
Br·	77	Cumene (α)	1.8	184
Me_2N·	136	Toluene	4.0	111
C_6H_5·	60	Toluene	4.5	24
C_6H_5·	60	Acetone	4.2	24
CH_3·	164	Toluene	7.9	185
$(CH_3)_3CO$·	40	Toluene	5.5	45
Cumylperoxy	30	Cumene (α)	8.0	39
Cumylperoxy	60	Cumene (α)	5.5	186
t-Butylperoxy	30	Cumene (α)	20	39
CF_3·	164	$CHCl_3$	32	187
Cl·	127	$CHCl_3$	2	182

For the high values of k^H/k^D in Table 27, for example, 32 for CF_3·, the reaction mode must correspond to more than bond stretching, and one or more of the bending modes in the reactant must be lost in the transition state.

VIII. STERIC EFFECTS

Attack of free radicals on carbon–hydrogen bonds is not particularly sensitive to steric hindrance to approach of the attacking radical, or to relief of strain in the conversion of a tetrahedral carbon atom to a trigonal geometry in acyclic systems. Ring strain is important and a very low reactivity is observed for hydrogens at certain bridgehead positions and in the cyclopropyl ring (Table 28).

The chlorination of a series of branched-chain hydrocarbons revealed little tendency for steric acceleration in the reactivity of 3°- or 2°-hydrogens; see Chart 1. Presumably, bond breaking is not far enough advanced in the transition state for steric relief of strain to be important although there may well be a hint of acceleration for the 3° position in 2,3,4-trimethylpentane. The data of Chart 1 do indicate a steric hindrance to approach by the chlorine atoms to the 3°-hydrogen atoms of 2,4-dimethylpentane, 2,2,4-trimethylpentane, and the 2°-hydrogen atom of 2,3,4-trimethylpentane, **11**.[193] Some

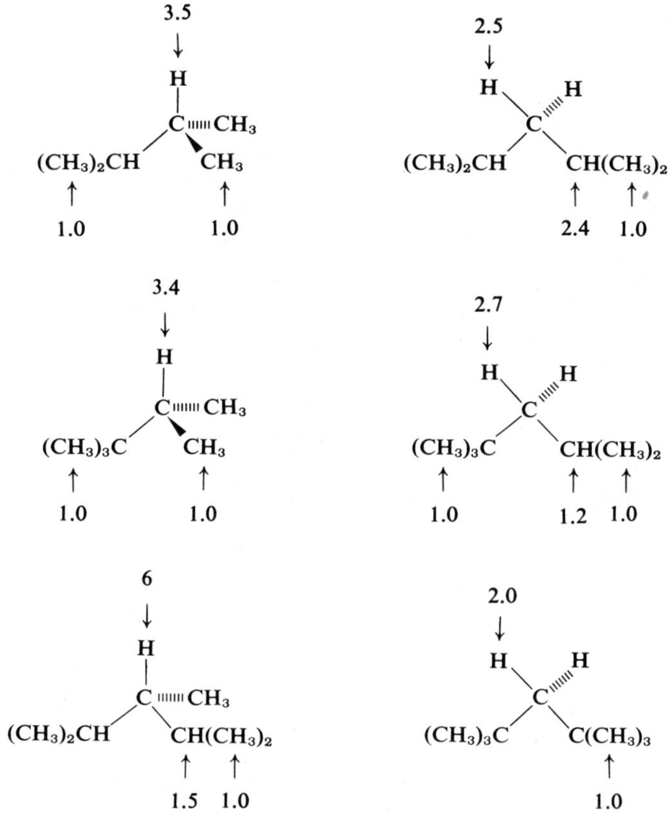

Chart 1. Relative reactivities of hydrogen atoms in a series of branched-chain hydrocarbons toward the chlorine atom at 40°.[59,193]

lowering of the reactivity of the 1°-hydrogen atoms in 2,2,4,4-tetramethylpentane has also been observed. A low reactivity has been observed for the 3° positions of isooctane in autoxidation, presumably because of the steric effect demonstrated in structure **11**.

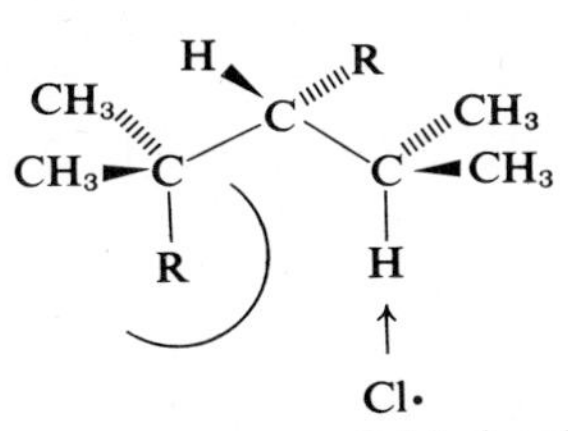

11

R = H or CH_3

TABLE 28

Relative Reactivities of Carbon–Hydrogen Bonds toward the Chlorine Atom, 0°

Substrate	Relative reactivity	Ref.
Cyclohexane	1.0	
Cyclopentane	0.95	188
Cyclobutane	0.84	188
Cyclopropane	0.048	188
Bicyclo[2,1,1]hexane		189
C-1	<0.2	
C-2	2.3	
C-5	<0.1	
Bicyclo[2,2,1]heptane		190
C-1	<0.3	
C-2	1.4	
C-7	0.1	
Bicyclo[2,2,2]octane		191
C-1	~3	
C-2	~0.5	
Adamantane		174, 192
C-1	1.4, 3.5,[a] 7.0[b]	
C-2	0.7, 1.0,[a] 1.4[b]	

[a] Cl·, 4 *M* benzene.
[b] Cl·, 12 *M* CS_2.

Chart 2 summarizes the measured reactivity (relative to carbon tetrachloride) and that calculated for phenyl radical attack using hydrogen atom relative reactivities of 0.0117 for *t*-butyl hydrogen atoms, 0.0098 for all other 1°-hydrogen, 0.091 for 2°-hydrogen, and 0.43 for 3°-hydrogens, values obtained in reactions of simpler hydrocarbons.[24] Again, low reactivities of the 3°-hydrogen atoms of 2,4-dimethylpentane and isooctane are indicated. Isooctane has a low reactivity in autoxidation, presumably because the 3°-hydrogen is shielded from peroxy radical attack.[194–196] Methyl radicals at 100° appear to attack preferentially the 1°-hydrogen atoms of isooctane.[197]

Steric factors have an important effect on benzylic radical stabilization, that is, steric inhibition of resonance. Tables 29 and 30 demonstrate this point. Apparently, even in ethylbenzene the developing radical does not achieve maximum stabilization because the preferred conformation does not place

	$(CH_3)_2CHCH(CH_3)_2$	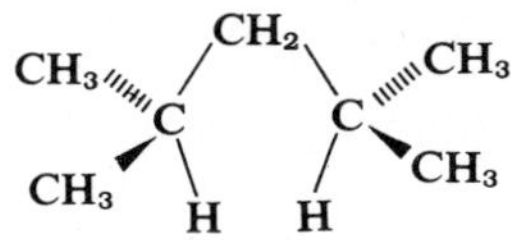	$(CH_3)_2CHCH_2CH_2CH(CH_3)_2$
Experimental	1.14	0.67	1.27
Calculated	0.98	1.16	1.34
	$(CH_3)_3CCH(CH_3)_2$	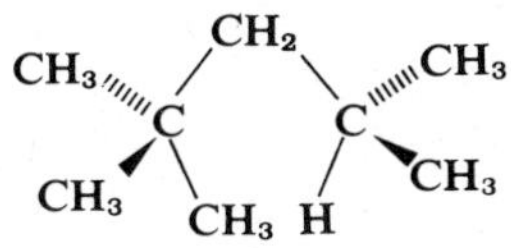	$(CH_3)_3CCH_2CH_2CH(CH_3)_2$
Experimental	0.71	0.34	0.81
Calculated	0.59	0.78	0.96

Chart 2. Relative reactivities (per molecule) of branched-chain hydrocarbons and carbon tetrachloride toward the phenyl radical at 60°.[24]

TABLE 29

Relative Reactivities (Total Benzylic) toward Radicals

	$C_6H_5\cdot$, 60°[24]	ROO·, 90°[40]	*t*-BuO·, 135°[43]	$CH_3\cdot$, 60°[70]	Br·, 40°[7]	Cl·, 40°[69]
$C_6H_5CH_3$	1	1	1	1	1	1
$C_6H_5CH_2CH_3$	3	5	2	3	14	2
Indan	11	49	—	11	49	6
Tetralin	16	89	10	15	47	5

the α-carbon–hydrogen bond in a coplanar arrangement with the π orbitals of the aromatic ring, **12**. In indan or for two of the α-hydrogen atoms of tetralin the stereoelectronic factors are better and an enhanced reactivity results.

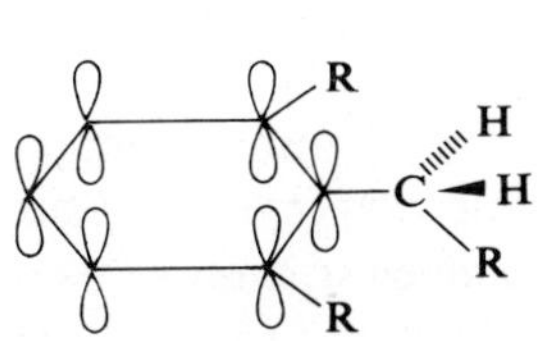

12

TABLE 30

Relative Reactivities of Aralkanes in Hydrogen Atom Transfer During the Telomerization of Carbon Tetrachloride and 1-Hexadecene[198]

Aralkane	Measure of transfer rate
Toluene	0.42
Ethylbenzene	1.30
Cumene	1.75
o-Cymene	0.4
m-Cymene	2.0
p-Cymene	2.5
p-Xylene	0.95
Isopropyl-*p*-xylene	0.78
2,5-Diisopropyl-*p*-xylene	0.74
1,2,4,5-Tetraisopropylbenzene	0.48
Hexaethylbenzene	0.68

The effect of preferred conformation becomes particularly important when attack on phenols, anilines, thiophenols, and so forth, is considered because of the possible delocalization of both the lone-pair electrons and the odd electron by the aromatic system, **13**.

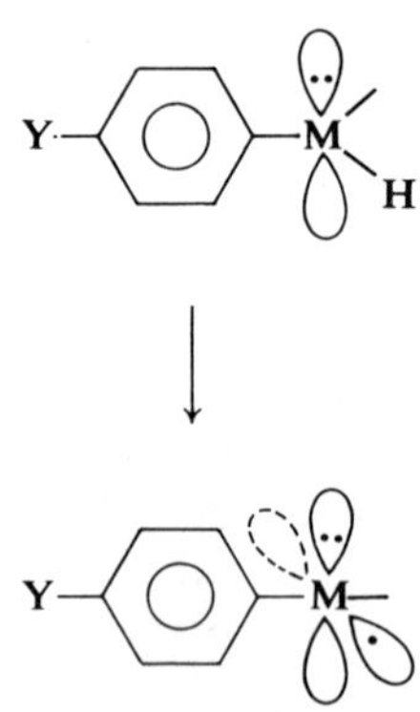

13

Geometric considerations are of utmost importance in determining the position of attack in intramolecular hydrogen atom abstraction reactions, some examples of which were given in a previous section. A further example is the reaction sequence of Scheme 4 in which the preferred point of attack ($y + 2$) varies from C-12 (49%) for $n + 2 = 14$, C-14 (66%) for $n + 2 = 16$,

C-11 to C-16 (17, 21, 18, 12, 5, 13%) for $n + 2 = 18$, and C-11 to C-16 (15, 20, 19, 19, 13, 8%) for $n + 2 = 20$.[199] Preferential attack at C-12 and C-14 in an α-cholestanol has also been achieved (Scheme 4).[200]

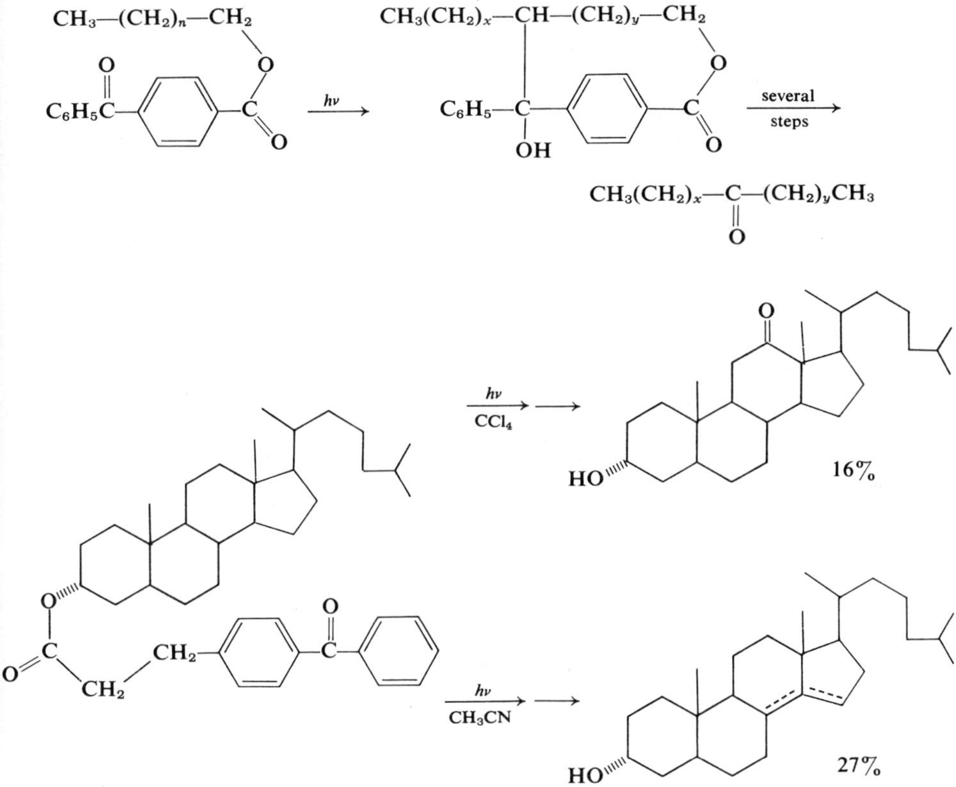

Scheme 4

IX. SOLVENT EFFECTS

As mentioned in previous sections, the chlorine atom is quite sensitive to certain solvent effects. With solvents such as SO_2, CS_2, C_6H_5I, or R_2S it can form new radical intermediates ($SO_2Cl\cdot$, $CS_2Cl\cdot$, $C_6H_5\dot{I}Cl$, $R_2\dot{S}Cl$) which are less reactive and more selective than the free chlorine atoms.[61] In the presence of aromatic solvents the chlorine atom becomes more selective in a fashion which parallels the basicity of the aromatic substance.[201] Apparently the chlorine atom is complexed by the aromatic ring and its reactivity and selectivity thus altered. As shown in Figure 6, such complexing is undoubtedly

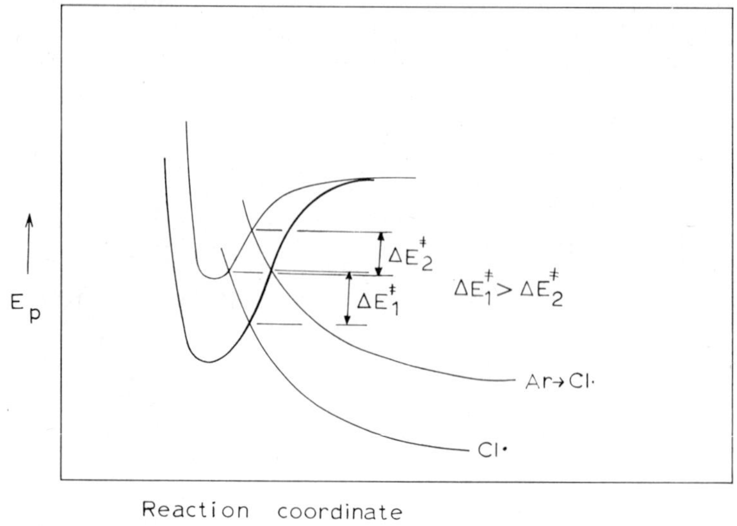

Fig. 6. Potential energy description of the aromatic solvent effect on the selectivity of chlorine atoms.

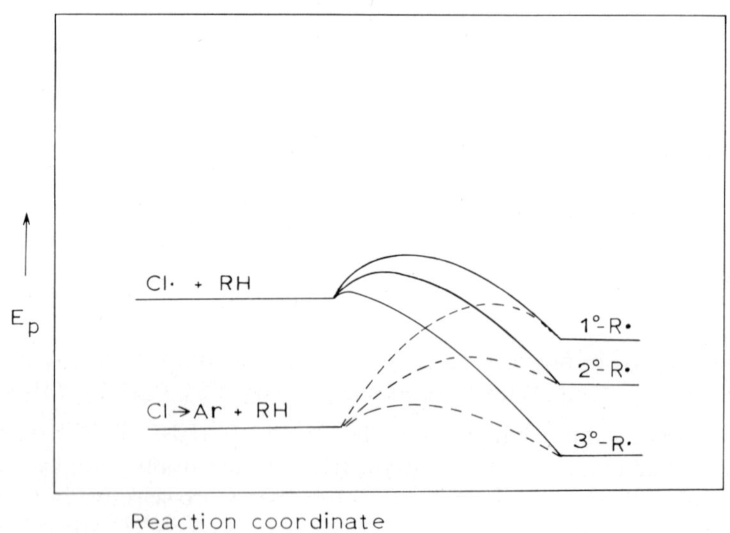

Fig. 7. Potential energy descriptions of the reactivity of free and complexed chlorine atoms.

more important for the free chlorine atom than for the transition state of the reaction. The point is further emphasized in Figure 7. Some typical aromatic solvent effects observed in the chlorine atom of 2,3-dimethylbutane are given in Table 31. For 2,5-dimethylhexane it was found that the increased selectivity in aromatic solvents of 3°- and 2°-hydrogen atoms over 1°-hydrogen atoms was the result of an increase in $E^{\ddagger}$ (Table 32).[174]

Aromatic solvents do not have much effect on the relative reactivities of hydrogen atoms which are controlled mainly by polar effects.[202] The more

TABLE 31

Relative Reactivities of 1°- and 3°-Hydrogen Atoms in 2,3-Dimethylbutane toward Chlorine Atoms at 55°[61]

Solvent, concentration	k_p/k_t
None	3.7
$C_6H_5NO_2$, 4.0 *M*	4.9
C_6H_5COCl, 4.0 *M*	6.4
$C_6H_5CF_3$, 4.0 *M*	6.9
$C_6H_5CO_2CH_3$, 4.0 *M*	10.2
C_6H_5Cl, 4.0 *M*	10.2
C_6H_5F, 4.0 *M*	10.3
Thiophene, 4.0 *M*	13.3
C_6H_6, 2 *M*	8.0
C_6H_6, 4 *M*	14.6
C_6H_6, 8 *M*	32
$C_6H_5CH_3$, 4 *M*	15.4
o-$C_6H_4(CH_3)_2$, 4 *M*	15
$C_6H_5CH_2CH_3$, 4 *M*	16.3
$C_6H_5Si(CH_3)_3$, 4 *M*	16.4
$C_6H_5OCH_3$, 4 *M*	18.4
p-$C_6H_4(CH_3)_2$, 4 *M*	18.6
$C_6H_5C(CH_3)_2H$, 4 *M*	20.3
m-$C_6H_4(CH_3)_2$, 4 *M*	22.4
$C_6H_5C(CH_3)_3$, 4 *M*	24
1,3,5-$C_6H_3(CH_3)_3$, 4 *M*	25
1-Chloronaphthalene, 4 *M*	33
C_6H_5I, 4 *M*	31
$(C_6H_5)_2S$, 2 *M*	24
$(C_6H_5)_2O$, 2 *M*	10
$C_6H_5COC_6H_5$, 2 *M*	8.8
$C_6H_5—C_6H_5$, 2 *M*	11.8

TABLE 32

Activation Parameters, Photochlorination of 2,5-Dimethylhexane, −35 to +60°

Solvent	$E_p^\ddagger - E_s^\ddagger$	$E_p^\ddagger - E_t^\ddagger$	Log (A_t/A_p)	Log (A_t/A_p)
Neat	0.01	0.32	+0.42	+0.32
4 *M* C_6H_6	0.80	1.7	+0.03	−0.22
6 *M* C_6H_5F	0.70	1.7	+0.20	−0.16
6 *M* C_6H_6	1.1	2.2	−0.06	−0.41
8 *M* C_6H_6	1.4	2.5	−0.12	−0.37
12 *M* CS_2	~2.0	—	~3.2	—

extensive bond breaking in the transition state for attack by complexed chlorine atoms would occur from attack on carbon–hydrogen bonds adjacent to many functional substituents (see Table 16). However, the more extensive bond breaking will also cause polar resonance structures for the incipient free radical to become more important and such structures are usually unfavorable when adjacent to an electron-withdrawing group. Table 33 illustrates this point.

There are definite trends wherein the complexing solvents favor attack on the bond of lowest dissociation energy, but these trends are small compared to the data of Tables 31 and 32. For molecules with polar substituents the effect of complexing solvents is often to reduce the difference in reactivities rather than to accentuate them. Thus, the relative reactivities of 1°-hydrogen atoms in alkanes and the hydrogen atoms in methylene chloride decrease

TABLE 33

Relative Reactivity of Various Carbon–Hydrogen Bonds toward Chlorine Atoms, 40° [202]

Substrate	Solvents		
	CCl_4	4 *M* $C_6H_5C(CH_3)_3$	12 *M* CS_2
2,3-Dimethylbutane, 1°	1	1	1
2,3-Dimethylbutane, 3°	3.9	30	200
Cyclohexane	2.7	6.9	20
Toluene	1.1	2.8	11
t-Butyl chloride	0.12	0.16	0.22
Methylene chloride	0.01	0.03	0.04
Chloroform	0.005	0.02	0.03

from 100/1 for the free chlorine atom to 33 for the chlorine atom in 4 *M t*-butylbenzene and 25-fold for the chlorine atom in 12 *M* carbon disulfide.[202] The combination of polar effects, temperature effects, and solvent effects can lead to many different results. For example, the complexing solvent actually changes the value of $E_\beta^\ddagger - E_\alpha^\ddagger$ in the chlorination of isobutyronitrile from −1.1 kcal/mole (neat) to +1.3 kcal/mole (8 *M* benzene).[174]

Small effects of solvents are observed in the rates of autoxidation reaction[203] and on the ratio of hydrogen atom abstraction and β-scission of *t*-butoxy radical (Table 34).[204] However, it is not obvious that the solvent effect occurs in the hydrogen abstraction step in either case.

TABLE 34

Solvent Effects in Chlorination of 2,3-Dimethylbutane, Autoxidation of Cyclohexene, and Reaction of *t*-Butoxy Radicals

Solvent	k_t/k_p, Chlorination at 55°[61]	Rate of oxidation of 2 *M* cyclohexene at 60°[203]	k_a/k_d for *t*-Butoxy radicals, cyclohexene, 70°[204]
CCl_4	3.5	0.036	12
CH_3CN	4.0	0.092	2.3
CH_3CO_2H	4.1	0.037	1.3
CH_3NO_2	3.3	0.088	—
$Cl_2C{=}CHCl$	3.6	—	4.9
Cyclohexene	3.6	0.030	—
C_6H_6	32(4 *M*)	0.034	7.6
C_6H_5Cl	10(4 *M*)	0.046	5.6
m-$C_6H_4Cl_2$	—	0.048	7.1

The exact nature of the small solvent effects reported in Table 34 is not understood. In the case of the oxidation reaction there is a rough correlation to the dielectric constant of the solvent.

X. OTHER EFFECTS

An often discussed effect on hydrogen atom reactivities is an activating effect by a neighboring group with nonbonding electron pairs which can interact with the developing radical center. In Table 16 there is no evidence that such an effect is important for groups such as Cl, Br, OR, or CO_2R.

There is some increased reactivity for phenyl *t*-butyl sulfide. Chlorination of bromocyclopentane does produce a large amount of *trans*-1-bromo-2-chlorocyclopentane, which has been interpreted in terms of the formation of a bridged radical, **14**.[205] Some activation at the 2 position is also noted for

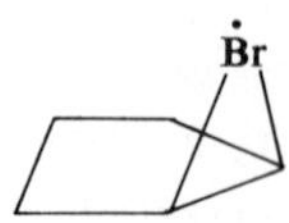

14

chlorocyclopentane and fluorocyclohexane but not for chlorocyclohexane, bromocyclohexane, or chlorocycloheptane (Table 35). Relative reactivities of the 2 and 3 positions, respectively, of the halocycloalkanes and 1-chloropentane at 78° in the gas phase, all relative to the 5 position in 1-chloropentane, are: 1-chloropentane, 2.1 and 3.7; chlorocyclopentane, 6.0 and 4.7; fluorocyclohexane, 4.3 and 3.0; chlorocyclohexane, 4.7 and 4.4; and chlorocycloheptane, 3.5 and 4.9. The experimental relative reactivities for the 2 position of the halocycloalkanes relative to a methylene group of the parent cycloalkane in the liquid phase for photochlorination at −30° are: bromocyclopentane, 0.5; chlorocyclopentane, 0.3; and chlorocyclohexane, 0.3.

It has been suggested that a bridged radical such as **14** is involved in the bromination of optically active isobutyl bromide to yield optically active (retention of configuration) 1,2-dibromo-2-methylbutane,[212] as well as optically active 1-chloro-2-bromo-2-methylbutane from the bromination of optically active isobutyl chloride (even though the chlorination of active isobutyl chloride yields inactive 1,2-dichloro-2-methylbutane).[213] It is also known that active α-deuterioethylbenzene on reaction with NBS yields only racemic α-phenethyl bromides.[214] However, it has been pointed out that bromination of 3-methylvaleronitrile yields an optically active 3-bromo-3-methylvaleronitrile although the cyano groups would not be expected to yield a bridged radical intermediate.[215] It has been suggested that the clean-cut bromination of *cis*-4-bromo-*t*-butylcyclohexane to yield *trans*-3-*cis*-4-dibromo-*t*-butylcyclohexane is also consistent with a bridged intermediate since the *trans*-4-bromo-*t*-butylcyclohexane yields a variety of products on photobromination.[216] The photobromination of erythro-α-bromo-β-deuteriobibenzyl to yield *meso*-α,β-dibromobibenzyl ($k_H/k_D = 0.5$) can also be explained by a bridged radical although other conformational explanations are preferred.[217] It was once thought that the *t*-butyl hypochlorite chlorination of *t*-butyl bromide to yield 1-bromo-2-chloro-2-methylpropane also proceeded by a bromine bridged intermediate.[218] However the observation that higher concentrations of *t*-butyl hypochlorite increasingly favor the unrearranged

TABLE 35

Products of Photochlorinations of Halocycloalkanes

Substrate	Substitution, %							Ref.
	C-1	C-2		C-3		C-4		
		Cis	Trans	Cis	Trans	Cis	Trans	
Bromocyclopentane, −30°, CCl_4	3.4	<1	43	23	30	—	—	205
Bromocyclopentane, −30°, CS_2	6	<1	45	21	29	—	—	205
Bromocyclopentane, −30°, CCl_4 [a]	9	<1	51	9	30	—	—	205
Chlorocyclopentane, 40°, CCl_4	5	<1	32	26	38	—	—	205
Chlorocyclopentane, 40°, CS_2	9	1	29	25	36	—	—	205
Chlorocyclopentane, 70°, gas	4	6	48	16	26	—	—	206
Methylcyclopentane, 104°, gas	20	15	20	37		—	—	206
Fluorocyclohexane, 78°, gas	9	1	45	15	17	8	14	207
Chlorocyclohexane, 40°, CCl_4	5	2	22	19	24	13	15	208
Chlorocyclohexane, 42°, CS_2	5	1	21	19	24	17	14	209
Chlorocyclohexane, 72°, gas	5	2	35	12	22	13	10	209
Bromocyclohexane, −30°, CCl_4	3	<1	15	26	20	13	24	208
Bromocyclohexane, −30°, CS_2	6	<1	17	21	21	17	19	208
Bromocyclohexane, −30°, CCl_4 [a]	16	<1	22	10	26	15	12	208
Chlorocycloheptane, 104°, gas	5	4	14	15	17	22	24	207
Methyl cyclohexanecarboxylate, 70°, CCl_4 [b]	5	7	18	14	31	13	13	210
Cyclohexanecarboxylic acid, 40°, C_6H_5Cl	4	9	9	24	24	18	12	211

[a] *t*-Butyl hypochlorite.
[b] Sulfuryl chloride chlorination.

1-chloro-2-bromo-2-methylpropane appears to demand Scheme 5, even at −80°.[219]

$$(CH_3)_2C(Br)\text{—}CH_2\cdot \longrightarrow (CH_3)_2C\text{=}CH_2 + Br\cdot$$

$$Br\cdot + CH_2\text{=}C(CH_3)_2 \longrightarrow BrCH_2\dot{C}(CH_3)_2$$

$$t\text{-BuOCl} + BrCH_2\dot{C}(CH_3)_2 \longrightarrow t\text{-BuO}\cdot + BrCH_2C(CH_3)_2Cl$$

Scheme 5

The high preference for *trans*-1,2-dihalides in all chlorinations of halocycloalkanes (but not for the 1,3- or 1,4-dihalides) has been attributed to a weak interaction between the halogen substituent and the free radical center which leads to a preferred conformation of the chlorocycloalkyl radical that is preferentially attacked by chlorine to yield the *trans*-1,2-dihalide.[206] Bromination of bromocyclopentane or bromocyclohexane at 60° in the liquid phase produces nearly exclusively the *trans*-1,2-dibromide, whereas bromination of chlorocyclohexane gives a spectrum of products with the 1-bromo-1-chlorocyclohexane predominant.[220] However, these results are claimed to be the consequence of reactions 52 and 53.[221,222] A further complication is the observation that reaction 54 occurs in the vapor phase[223] and 55 in the liquid phase.[224]

$$BrCH_2\dot{C}HCH_2CH_3 + HBr \xrightarrow{\text{slowly}} BrCH_2CH_2CH_2CH_3 + Br\cdot \qquad (52)$$

$$BrCH_2CH_2\dot{C}HCH_3 + HBr \xrightarrow{\text{rapidly}} BrCH_2CH_2CH_2CH_3 + Br\cdot \qquad (53)$$

$$RCH_2CH_2Br + HBr \longrightarrow RCH\text{=}CH_2 + 2HBr \qquad (54)$$

$$R_2C(Br)CH_3 + Br_2 \longrightarrow [R_2C\text{=}CH_2] \xrightarrow{-H^+} [R_2CH\text{=}CH_2] \xrightarrow{-Br^+} R_2C(Br)CH_2Br \qquad (55)$$

At low hydrogen bromide concentrations, for example, by use of NBS, the data of Table 36 are reported.[222]

In a similar fashion, the observed relative reactivities in photobromination of bromocyclohexane at 40° of $\alpha:\beta:\gamma:\delta = 0.3:3.3:0.5:1.0$ change to $\alpha:\beta:\gamma:\delta = 0.8:0.6:0.7:1.0$ when NBS is the brominating agent at 2% reaction.[222]

It appears that there is little anchimeric assistance by β substituents such as bromine* in hydrogen atom abstraction reactions but that substituents such as

* P. S. Skell and K. J. Shea have recently informed the author that the claims of references 221 and 222 can not be duplicated. They find that the ratio of 1,2-dibromide/1,3-dibromide in the liquid phase photobromination of 1-bromobutane is high under all conditions (as reported by Thaler[220]) and varies from ~4 at 60° to ~20 at −23°. Anchimeric assistance by a β-bromine atom is supported.[225]

TABLE 36

Liquid-Phase Halogenation of *n*-Butyl Halides, Relative Reactivities per Hydrogen Atom

Halogenating agent	X—	—CH_2— ↑	—CH_2— ↑	—CH_2— ↑	—CH_3 ↑	Ref.
Cl_2, 68°	Cl	0.2	0.5	0.5	1.0	145
Br_2, HBr, 60°	Cl	0.4	0.5	1.0	Low	220
Cl_2, 11 *M* CS_2, 68°	Cl	0.1	0.4	1.0	0.1	145
Cl_2, 60°	Br	0.1	0.4	1.0	0.5	220
Br_2, HBr, 60°	Br	0.1	6	1.0	Low	220
NBS, 40°	Br	0.3	0.3	1.0	Low	222
Br_2, 2% reaction, 40°	Br	0.5	0.5	1.0	Low	221
Br_2, 100°, vapor	Br	0.02	0.3	1.0	Low	223
Br_2, 100°, vapor, excess HBr	Br	Low	7	1.0	Low	223

2-bromo, 2-fluoro, 2-thio, and so on, can influence the conformation of a cycloalkyl free radical so as to produce a predominance of *trans*-1,2-dihalide. A bridged radical is one extreme picture of this interaction.[212,216,218] There is even some preference for methyl *trans*-2-chlorocyclohexanecarboxylate over its cis isomer in the chlorination of methyl cyclohexanecarboxylate but here the trans–cis ratio is higher for 3 attack than for 2 attack.[210,211] Polar effects between the dipole of the carboxyl substituent and the attacking chlorine molecule may be involved, for example, **15**. For cyclobutanecarboxylic acid the trans–cis ratio of monochlorination products vary from about 20 at the 2 position to about 3 at C-3.[226]

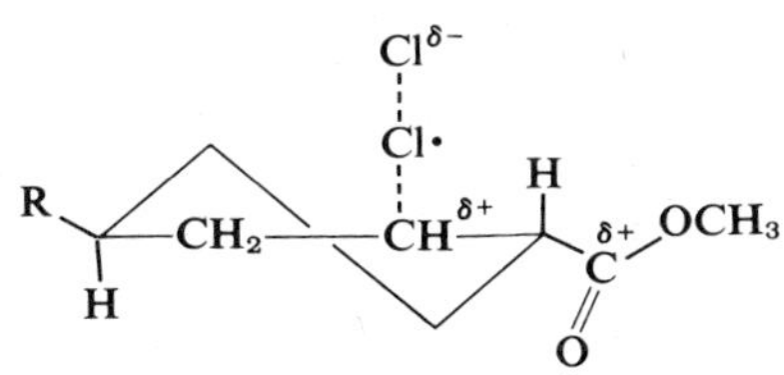

15

Conformational effects on relative reactivities have been mentioned previously under steric effects and in intramolecular processes such as the chlorination of long-chain carboxylic acids in sulfuric acid solution. A somewhat different effect is the observation that for long-chain alkanes, certain solvents apparently favor uncoiled conformations, which allows free radical attack at all methylene groups. Thus in chlorination of dodecane at 75°, 49% of the attack occurs at the C-4, C-5, and C-6 atoms and 10% at C-1.[227] In benzene the C-1 attack is drastically reduced, as expected (to 5%), but in addition the attack at the methylene positions now becomes nearly statistical (57% attack at C-4, C-5, C-6). It is argued that in benzene the conformations are uncoiled and that in a poor solvent, such as the fluoroalkanes, coiling of long chains should occur and attack at the center of the molecule would become quite difficult.[227] Another technique currently being investigated is the effect of adsorption of substrates. Table 37 summarizes some data for the photochlorination of *n*-octanoic acid at 25° where a significant increase in terminal attack is achieved by adsorption on alumina.[228] Chlorination of *n*-hexane in Linde 13X molecular sieves at 150° is reported to yield a high percentage of 1°-attack at low conversions ($k_{2^\circ-H} \approx k_{1^\circ-H}$).[229]

TABLE 37

Photochlorination of *n*-Octanoic Acid at 25°

	% Yield of monochlorooctanoic acids						
Conditions	2	3	4	5	6	7	8
Cl_2, CCl_4	2	9	15	15	19	24	17
Cl_2, C_6H_5Cl	—	3	14	24	30	26	3
t-BuOCl, CCl_4	6	7	14	17	22	28	6
Heterogeneous, 3% on alumina/CCl_4	1	3	5	8	20	30	33

ACKNOWLEDGMENTS

Some previously unpublished reactivity data were taken from the Ph.D. theses of Jerry Hunt, Joseph Schoeb, and Philip Haffley. This work was supported by a grant from the Petroleum Research Fund administered by the American Chemical Society.

REFERENCES

1. H. C. Brown and G. A. Russell, *J. Am. Chem. Soc.*, **74**, 3995 (1952).
2. T. Cole and H. Heller, *J. Chem. Phys.*, **42**, 1668 (1965).
3. J. Kenner, *Tetrahedron*, **3**, 78 (1958).
4. G. A. Russell, *Tetrahedron*, **5**, 101 (1959).
5. G. A. Russell and H. C. Brown, *J. Am. Chem. Soc.*, **77**, 4578 (1955).
6. G. A. Russell and C. DeBoer, *J. Am. Chem. Soc.*, **85**, 3136 (1963).
7. G. A. Russell and K. M. Desmond, *J. Am. Chem. Soc.*, **85**, 3139 (1963).
8. C. Walling and B. B. Jacknow, *J. Am. Chem. Soc.*, **82**, 6108 (1960).
9. E. S. Huyser, *J. Am. Chem. Soc.*, **82**, 391, 394 (1960).
10. G. A. Russell, C. DeBoer, and K. M. Desmond, *J. Am. Chem. Soc.*, **85**, 365 (1963).
11. R. E. Pearson and J. C. Martin, *J. Am. Chem. Soc.*, **85**, 354, 3142 (1963).
12. C. Walling, A. L. Reiger, and D. D. Tanner, *J. Am. Chem. Soc.*, **85**, 3129 (1963); C. Walling and A. L. Rieger, *ibid.*, **85**, 3135 (1963).
13. J. Adam, P. A. Gosselain, and P. Goldfinger, *Bull. Soc. Chim. Belges*, **65**, 523 (1956).
14. D. D. Tanner and N. Nychka, *J. Am. Chem. Soc.*, **89**, 121 (1967).

15. F. Minisci, G. P. Gardini, and F. Bertini, *Can. J. Chem.*, **50**, 1544 (1970).
16. J. Spanswick and K. U. Ingold, *Can. J. Chem.*, **50**, 546 (1970).
17. G. A. Russell, *J. Am. Chem. Soc.*, **80**, 5002 (1958).
18. D. D. Tanner and P. B. Van Bostelen, *J. Org. Chem.*, **32**, 1517 (1967).
19. P. Smit and H. J. den Hertog, *Rec. Trav. Chim. Pays-Bas*, **83**, 891 (1964).
20. B. Fell and L. H. Kung, *Chem. Ber.*, **98**, 2971 (1965).
21. F. Asinger and B. Fell, *Erdoel Kohle, Erdgas, Petrochem.*, **17**, 74 (1964).
22. F. F. Rust and C. S. Bell, *J. Am. Chem. Soc.*, **92**, 5330 (1970).
23. F. G. Edwards and F. R. Mayo, *J. Am. Chem. Soc.*, **72**, 1265 (1950).
24. R. F. Bridger and G. A. Russell, *J. Am. Chem. Soc.*, **85**, 3754 (1964).
25. W. A. Pryor, J. T. Echols, Jr., and K. Smith, *J. Am. Chem. Soc.*, **88**, 1189 (1966).
26. W. A. Pryor, K. Smith, J. T. Echols, Jr., and J. L. Fuller, *J. Org. Chem.*, **36**, 1753 (1972).
27. A. F. Trotman-Dickenson, *Q. Rev. (London)*, **7**, 198 (1953).
28. P. Gray, A. A. Herod, and A. Jones, *Chem. Rev.*, **71**, 247 (1971).
29. P. B. Ayscough, J. C. Polanyi, and E. W. R. Steacie, *Can. J. Chem.*, **33**, 743 (1955).
30. W. A. Pryor and J. P. Stanley, *J. Am. Chem. Soc.*, **93**, 1412 (1971).
31. R. W. Henderson and W. A. Pryor, *J. Am. Chem. Soc.*, **92**, 7234 (1970).
32. J. Gresser, J. R. Binks, and M. Szwarc, *J. Am. Chem. Soc.*, **81**, 5004 (1959).
33. E. C. Kooyman, R. van Helden, and A. F. Bickel, *K. Ned. Akad. Wet., Proc.*, **56B**, 75 (1953).
34. R. A. Gregg and F. R. Mayo, *Discuss Faraday Soc.*, **2**, 328 (1947).
35. P. S. Fredricks and J. M. Tedder, *Chem. Ind.*, 490 (1959).
36. R. L. Williams and F. S. Rowland, *J. Phys. Chem.*, **75**, 2709 (1971).
37. J. A. Howard, W. J. Schwalm, and K. U. Ingold, *Adv. Chem. Ser.*, **75**, 6 (1968),
38. G. A. Russell, *J. Am. Chem. Soc.*, **77**, 4583 (1955).
39. J. A. Howard, K. U. Ingold, and M. Symonds, *Can. J. Chem.*, **46**, 1017 (1968).
40. G. A. Russell, *Can. J. Chem.*, **78**, 1047 (1956).
41. W. A. Noyes, Jr., and L. M. Dorfman, *J. Chem. Phys.*, **16**, 788 (1948).
42. H. C. McBay and O. Tucker, *J. Org. Chem.*, **19**, 869 (1954).
43. A. L. Williams, E. A. Oberright, and J. W. Brooks, *J. Am. Chem. Soc.*, **78**, 1190 (1956).
44. C. Walling, *Pure Appl. Chem.*, **15**, 69 (1967).
45. C. Walling and J. A. McGuinness, *J. Am. Chem. Soc.*, **91**, 2053 (1969).
46. M. S. Kharasch and M. T. Gladstone, *J. Am. Chem. Soc.*, **65**, 15 (1943).
47. C. C. Price and H. Morita, *J. Am. Chem. Soc.*, **75**, 3686 (1953).
48. M. S. Kharasch, H. C. McBay, and W. H. Urry, *J. Org. Chem.*, **10**, 394 (1954).
49. H. C. McBay, O. Tucker, and A. Milligan, *J. Org. Chem.*, **19**, 1003 (1951).
50. K. M. Johnston, G. H. Williams, and H. J. Williams, *J. Chem. Soc. (B)*, 1114 (1966).
51. K. Schwetlick and S. Helm, *Tetrahedreon*, **22**, 793 (1966).
52. W. T. Dixon, R. O. C. Norman, and D. L. Buhey, *J. Chem. Soc.*, 3625 (1964).
53. P. Smith, J. T. Pearson, P. B. Wood, and T. C. Smith, *J. Chem. Phys.*, **43**, 1535 (1965).
54. T. L. Cottrell, *The Strengths of Chemical Bonds*, Academic Press, New York, 1954.
55. M. G. Evans and M. Polanyi, *Trans. Faraday Soc.*, **34**, 11 (1938).
56. A. F. Trotman-Dickenson, *Chem. Ind.*, 379 (1965).
57. G. C. Fettis, J. H. Knox, and A. F. Trotman-Dickenson, *J. Chem. Soc.*, 1064 (1964).
58. P. C. Anson, P. S. Fredricks, and J. M. Tedder, *J. Chem. Soc.*, 918 (1959).
59. G. A. Russell and H. C. Brown, *J. Am. Chem. Soc.*, **77**, 4031 (1955).

60. J. H. Knox and R. L. Nelsen, *Trans. Faraday Soc.*, **55**, 937 (1959).
61. G. A. Russell, *J. Am. Chem. Soc.*, **80**, 4987 (1958).
62. N. R. Greiner, *J. Chem. Phys.*, **53**, 1070, 1285 (1970).
63. J. M. Tedder, *Q. Rev.* (*London*), **14**, 336 (1960).
64. D. B. Hartley and S. W. Benson, *J. Chem. Phys.*, **39**, 132 (1963); P. S. Nangia and S. W. Benson, *J. Am. Chem. Soc.*, **86**, 2773 (1964); D. M. Golden, R. Walsh, and S. W. Benson, *ibid.*, **87**, 4053 (1965).
65. C. A. Goy, D. H. Shaw, and H. O. Prichard, *J. Phys. Chem.*, **69**, 1504 (1965).
66. P. Kovacic and J.-H. C. Chang, *J. Org. Chem.*, **36**, 3138 (1971).
67. G. S. Hammond, *J. Am. Chem. Soc.*, **77**, 334 (1955).
68. I. B. Afanas'ev, *Russ. Chem. Rev.*, **40**, 216 (1971).
69. G. A. Russell, A. Ito, and D. G. Hendry, *J. Am. Chem. Soc.*, **85**, 2976 (1963).
70. J. A. Meyer, V. Stannet, and M. Szwarc, *J. Am. Chem. Soc.*, **83**, 25 (1961).
71. S. Palit and S. K. Das, *Proc. Roy. Soc.* (*London*), **A226**, 82 (1954).
72. J. A. Howard and K. U. Ingold, *Can. J. Chem.*, **46**, 2661 (1968).
73. C. H. Bamford, A. D. Jenkins, and R. Johnston, *Trans. Faraday Soc.*, **55**, 418 (1959).
74. S. S. Friedrich, E. C. Friedrich, L. J. Andrews, and R. M. Keefer, *J. Org. Chem.*, **34**, 900 (1969).
75. D. G. Hendry and G. A. Russell, *J. Am. Chem. Soc.*, **86**, 2371 (1964).
76. J. D. Unruh and G. J. Gleicher, *J. Am. Chem. Soc.*, **91**, 6211 (1969).
77. M. Szwarc, "The Transition State," *Special Publ. 16*, The Chemical Society, London, 1962, p. 91.
78. K. U. Ingold, *Pure Appl. Chem.*, **15**, 49 (1967).
79. A. F. Trotman-Dickenson, *Discuss Faraday Soc.*, **14**, 230 (1953).
80. C. Walling and B. B. Jacknow, *J. Am. Chem. Soc.*, **82**, 6108, 6113 (1960).
81. G. C. Fettis, J. H. Knox, and A. F. Trotman-Dickenson, *J. Chem. Soc.*, 4177 (1960).
82. A. M. Tarr, J. W. Coomer, and E. Whittle, *Trans. Faraday Soc.*, **61**, 1182 (1965); P. Corbett, A. M. Tarr, and E. Whittle, *ibid.*, **59**, 1609 (1963).
83. G. A. Creak, F. S. Dainton, and K. J. Ivin, *Trans. Faraday Soc.*, **58**, 326 (1962).
84. G. O. Pritchard, J. T. Bryant, and R. L. Thommarson, *J. Phys. Chem.*, **69**, 664 (1965).
85. G. O. Pritchard and R. L. Thommarson, *J. Phys. Chem.*, **68**, 568 (1964).
86. T. N. Bell and A. E. Platt, *J. Phys. Chem.*, **75**, 603 (1971).
87. J. H. Knox, *Trans. Faraday Soc.*, **58**, 275 (1962); P. Goldfinger, *J. Chim. Phys.*, **55**, 234 (1958).
88. W. G. Alcock and E. Whittle, *Trans. Faraday Soc.*, **61**, 244 (1965).
89. T. N. Bell and U. F. Zucker, *J. Phys. Chem.*, **74**, 979 (1970).
90. L. O. Moore, *J. Phys. Chem.*, **75**, 2075 (1971).
91. W. G. Alcock and E. Whittle, *Trans. Faraday Soc.*, **62**, 664 (1966).
92. H. Magritte and A. Bruylants, *Ind. Chim. Belge*, **22**, 547 (1957).
93. A. Bruylants, M. Tits, and R. Danby, *Bull. Soc. Chim. Belges*, **58**, 210 (1949).
94. A. Bruylants, M. Tits, C. Dieu, and R. Gauthier, *Bull. Soc. Chim. Belges*, **61**, 266 (1952).
95. H. J. Den Hertog, B. de Vries, and J. van Brazt, *Rec. Trav. Chim. Pays-Bas*, **74**, 1561 (1955).
96. R. Van Helden and E. C. Kooyman, *Rec. Trav. Chim. Pays-Bas*, **73**, 269 (1954).
97. W. A. Pryor and R. W. Henderson, *J. Chem. Kinet.*, **4** (in press) (1972).
98. W. A. Pryor, U. Tonellato, D. L. Fuller, and S. Jumonville, *J. Org. Chem.*, **34**, 2018 (1969).

99. G. A. Russell and R. C. Williamson, Jr., *J. Am. Chem. Soc.*, **85**, 2357 (1964).
100. G. A. Russell, *J. Org. Chem.*, **23**, 1407 (1958).
101. Y. Nagai, K. Yamazaki, I. Shiojima, N. Kobori, and M. Hayashi, *J. Organomet. Chem.*, **9**, 21 (1967).
102. E. I. Heiba, R. M. Dessau, and W. J. Koehl, *J. Am. Chem. Soc.*, **91**, 138 (1969).
103. H. Sakurai and A. Hosomi, *J. Am. Chem. Soc.*, **89**, 458 (1967).
104. C. Walling and M. J. Gibian, *J. Am. Chem. Soc.*, **87**, 3361 (1965).
105. H. Sakurai, A. Hosomi, and M. Kumada, *J. Org. Chem.*, **35**, 993 (1970).
106. K. H. Lee, *Tetrahedron*, **25**, 4357 (1969).
107. K. H. Lee, *Tetrahedron*, **26**, 2041 (1968).
108. K. H. Lee, *Tetrahedron*, **26**, 1503 (1970).
109. C. Walling and E. A. McElhill, *J. Am. Chem. Soc.*, **73**, 2927 (1951).
110. R. L. Huang, K. H. Lee, and S. H. Ong, *J. Chem. Soc.*, 3336 (1962).
111. C. J. Michejda and W. P. Hoss, *J. Am. Chem. Soc.*, **92**, 6299 (1970).
112. R. S. Neale and E. Gross, *J. Am. Chem. Soc.*, **89**, 6579 (1967).
113. Y. Schaafsma, A. F. Bickel, and E. C. Kooyman, *Rec. Trav. Chim. Pays-Bas*, **76**, 180 (1967).
114. S. S. Friedrich, L. J. Andrews, and R. M. Keefer, *J. Org. Chem.*, **35**, 944 (1970).
115. W. D. Totherow and G. S. Gleicher, *J. Am. Chem. Soc.*, **91**, 7150 (1969).
116. R. L. Huang and K. H. Lee, *J. Chem. Soc.* (*C*), 935 (1966).
117. M. M. Martin and G. S. Gleicher, *J. Org. Chem.*, **28**, 3266 (1963).
118. T. P. Low and K. H. Lee, *J. Chem. Soc.* (*B*), 535 (1970).
119. G. J. Gleicher, *J. Org. Chem.*, **33**, 332 (1968).
120. R. L. Huang and K. H. Lee, *J. Chem. Soc.*, 5963 (1964).
121. K. H. Lee, *Tetrahedron*, **25**, 4357 (1969).
122. R. L. Huang, K. H. Lee, and M. S. Malhotra, *J. Chem. Soc.*, 5947 (1964).
123. J. A. Howard and K. U. Ingold, *Can. J. Chem.*, **41**, 1744 (1963).
124. K. U. Ingold, *J. Phys. Chem.*, **64**, 1636 (1960).
125. I. H. Brownlie and K. U. Ingold, *Can. J. Chem.*, **45**, 2419 (1967).
126. J. A. Howard and K. U. Ingold, *Can. J. Chem.*, **41**, 2800 (1963).
127. H. Sakurai, A. Hasomi, and K. Kumara, *J. Org. Chem.*, **35**, 993 (1970).
128. E. P. Chang, R. L. Huang, and K. H. Lee, *J. Chem. Soc.* (*B*), 878 (1969).
129. K. H. Lee, *Tetrahedron*, **24**, 4793 (1968).
130. J. S. Hogg, D. H. Lohmann, and K. E. Russell, *Can. J. Chem.*, **39**, 1588 (1961).
131. J. Hradil and V. Chvalousky, *Collect. Czech. Chem. Commun.*, **33**, 2029 (1968).
132. R. L. Harvey, G. J. Gleicher, and W. D. Totherow, *Tetrahedron*, **25**, 5019 (1969).
133. J. G. Kirkwood and F. H. Westheimer, *J. Chem. Phys.*, **6**, 506, 513 (1938).
134. R. W. Taft, Jr., in *Steric Effects in Organic Chemistry*, M. S. Newman, Ed., Wiley, New York, 1956, Chap. 7.
135. P. H. Owens, G. J. Gleicher, and L. M. Smith, Jr., *J. Am. Chem. Soc.*, **90**, 4122 (1968).
136. G. J. Gleicher, J. L. Jackson, P. H. Owens, and J. D. Unruh, *Tetrahedron Letters*, 833 (1969).
137. S. S. Friedrich, E. C. Friedrich, L. J. Andrews, and R. M. Keefer, *J. Org. Chem.*, **34**, 900 (1969).
138. J. A. Howard and S. Krocek, *Can. J. Chem.*, **48**, 2165 (1970).
139. Some data are from the unpublished theses of J. D. Hunt (1965) and J. H. Schoeb (1965), Iowa State University Library.
140. E. T. Strom, G. A. Russell, and J. H. Schoeb, *J. Am. Chem. Soc.*, **88**, 2004 (1966).
141. C. Rüchardt, *Angew. Chem., Int. Ed.*, **9**, 830 (1970).
142. S. W. Benson, *J. Chem. Educ.*, **42**, 502 (1965).

143. S. W. Benson, F. R. Cruickshank, D. M. Golden, G. R. Haugen, H. E. O'Neal, A. S. Rogers, R. Shaw, and R. Walsh, *Chem. Rev.*, **69**, 279 (1969).
144. H. Sing and J. M. Tedder, *J. Chem. Soc.*, 4737 (1964).
145. C. Walling and M. F. Mayahi, *J. Am. Chem. Soc.*, **81**, 1485 (1959).
146. H. C. Brown and A. B. Ash, *J. Am. Chem. Soc.*, **77**, 4014 (1955).
147. C. Walling and B. B. Jacknow, *J. Am. Chem. Soc.*, **82**, 6113 (1960).
148. W. Thaler, *J. Am. Chem. Soc.*, **85**, 2607 (1963).
149. L. Horner and L. Schläfer, *Justus Liebigs Ann. Chem.*, **635**, 31 (1960).
150. A. L. Henne and J. B. Hinkamp, *J. Am. Chem. Soc.*, **67**, 119 (1945).
151. Y. Nagai, N. Machida, H. Kono, and T. Migita, *J. Org. Chem.*, **32**, 1194 (1967).
152. J. Kollonitsch, G. A. Doldouras, and V. F. Verdi, *J. Chem. Soc. (B)*, 1093 (1967).
153. V. Shigemitsu, Y. Odaira, and S. Tsutsumi, *Bull. Chem. Soc. Japan*, **38**, 1450 (1965).
154. D. D. Tanner and M. W. Mosher, *Can. J. Chem.*, **47**, 715 (1969).
155. F. Minisci, R. Galli, A. Galli, and R. Bernardi, *Tetrahedron Letters*, 2207 (1967).
156. F. Minisci, R. Galli, and R. Bernardi, *Chem. Ind. (Milan)*, **49**, 594 (1967).
157. C. W. Gayler and H. M. Waddle, *J. Am. Chem. Soc.*, **63**, 3358 (1941).
158. H. Singh and J. M. Tedder, *J. Chem. Soc. (B)*, 608, 612 (1966).
159. N. C. Deno, R. Fishbein, J. C. Wyckoft, *J. Am. Chem. Soc.*, **92**, 5274 (1970).
160. N. C. Deno, K. Eisenhardt, R. Fishbein, C. Pierson, D. Pohl, H. Spinelli, R. C. White, and J. C. Wyckoff, XXIIIrd. International Congress of Pure and Applied Chemistry, Vol. 4, 1971, p. 155.
161. H. J. den Hertog and P. Smit, *Proc. Chem. Soc.*, 132 (1959).
162. R. S. Neale, M. R. Walsh, and N. L. Marcus, *J. Org. Chem.*, **30**, 3683 (1966).
163. C. Walling and A. Padwa, *J. Am. Chem. Soc.*, **85**, 1597 (1963).
164. D. H. R. Barton, J. M. Beaton, L. E. Geller, and M. M. Pechet, *J. Am. Chem. Soc.* **82**, 2640 (1960).
165. N. C. Yang and D. H. Yang, *J. Am. Chem. Soc.*, **80**, 2913 (1958).
166. J. Wantier and A. Bruylant, *Bull. Soc. Chim. Belges*, **72**, 222 (1963).
167. I. B. Afansas'ev and A. A. Beer, *Neftekhimiya*, **2**, 611 (1962).
168. W. Mack, *Tetrahedron Letters*, 4993 (1967).
169. T. Migita, M. Kosugi, and Y. Nagai, *Bull. Chem. Soc. Japan*, **40**, 920 (1967).
170. M. Kosugi, K. Takeuchi, and T. Migita, *Bull. Chem. Soc. Japan*, **42**, 3242, 3354 (1969).
171. M. Kosugi, K. Takeuchi, and T. Migita, *Bull. Chem. Soc. Japan*, **43**, 1535 (1970).
172. E. L. Patmore and R. J. Gritter, *J. Org. Chem.*, **27**, 4196 (1962).
173. F. C. Colton, Ph.D. thesis, University of Chicago Library, Chicago, Illinois, 1950.
174. P. Haffley, Ph.D. thesis, Iowa State University Library, 1967.
175. M. S. Kharasch and H. C. Brown, *J. Am. Chem. Soc.*, **61**, 2142 (1939).
176. H. C. McBay, O. Tucker, and A. Milligan, *J. Org. Chem.*, **19**, 1003 (1954).
177. M. S. Kharasch, H. C. McBay, and W. H. Urry, *J. Org. Chem.*, **10**, 394 (1954).
178. S. D. Ross, M. Markarian, and M. Nazzewski, *J. Am. Chem. Soc.*, **69**, 1914, 2468 (1947).
179. P. Beltrame, S. Carra, and S. Mori, *J. Phys. Chem.*, **70**, 1150 (1966).
180. K. B. Wiberg, *Chem. Rev.*, **55**, 713 (1958).
181. K. B. Wiberg and E. L. Motell, *Tetrahedron*, **19**, 2009 (1963).
182. G. Chiltz, R. Eckling, P. Goldfinger, G. Huybrechts, H. S. Johnson, L. Meyers, and G. Verbeke, *J. Chem. Phys.*, **38**, 1053 (1963).
183. G. A. Russell, *J. Am. Chem. Soc.*, **80**, 5002 (1958).
184. K. B. Wiberg and L. H. Slaugh, *J. Am. Chem. Soc.*, **80**, 3033 (1958).
185. M. Cher, C. S. Hollingsworth, and F. Sicilo, *J. Phys. Chem.*, **70**, 877 (1966).

186. G. A. Russell, *J. Am. Chem. Soc.*, **79**, 3871 (1957).
187. H. Carmichael and H. S. Johnson, *J. Chem. Phys.*, **41**, 1975 (1964).
188. C. Walling and P. S. Fredricks, *J. Am. Chem. Soc.*, **84**, 3326 (1962).
189. R. Srinivasan and F. I. Sonntag, *J. Am. Chem. Soc.*, **89**, 407 (1967).
190. E. C. Kooyman and G. C. Vegter, *Tetrahedron*, **4**, 382 (1958).
191. A. F. Bickel, J. Knotnerus, E. C. Kooyman, and G. C. Vegter, *Tetrahedron*, **9**, 230 (1960).
192. G. W. Smith and H. D. Williams, *J. Org. Chem.*, **26**, 2207 (1961).
193. G. A. Russell and P. G. Haffley, *J. Org. Chem.*, **31**, 1869 (1966).
194. J. P. Wibaut and A. Strang, *K. Ned. Akad. Wet., Proc.*, **B54**, 229 (1951).
195. C. E. Frank and A. V. Blackham, *Ind. Eng. Chem.*, **46**, 212 (1954).
196. F. F. Rust and D. O. Collamer, *J. Am. Chem. Soc.*, **76**, 1055 (1954).
197. A. L. J. Beckwith, *J. Chem. Soc.*, 2248 (1962).
198. E. C. Kooyman and A. Strang, *Rec. Trav. Chim. Pays-Bas*, **72**, 329 (1953).
199. R. Breslow and M. A. Winnik, *J. Am. Chem. Soc.*, **91**, 3083 (1969).
200. R. Breslow and S. W. Baldwin, *J. Am. Chem. Soc.*, **92**, 732 (1970).
201. G. A. Russell, *J. Am. Chem. Soc.*, **79**, 2977 (1957).
202. G. A. Russell, *J. Am. Chem. Soc.*, **80**, 4997 (1958).
203. D. G. Hendry and G. A. Russell, *J. Am. Chem. Soc.*, **86**, 2368 (1964).
204. C. Walling and P. J. Wagner, *J. Am. Chem. Soc.*, **86**, 3368 (1964).
205. G. A. Russell and A. Ito, *J. Am. Chem. Soc.*, **85**, 2983 (1963).
206. D. S. Ashton and J. M. Tedder, *J. Chem. Soc. (B)*, 1719 (1971).
207. D. S. Ashton and J. M. Tedder, *J. Chem. Soc. (B)*, 1724 (1971).
208. G. A. Russell, A. Ito, and R. Konaka, *J. Am. Chem. Soc.*, **85**, 2988 (1963).
209. D. S. Ashton and J. M. Tedder, *J. Chem. Soc. (B)*, 1031 (1970).
210. C. C. Price, C. D. Beard, and K. Akune, *J. Am. Chem. Soc.*, **92**, 5916 (1970).
211. J. C. Little, Y.-L. C. Tong, and J. P. Heeschen, *J. Am. Chem. Soc.*, **91**, 7090 (1969).
212. P. S. Skell, D. L. Tuleen, and P. D. Readio, *J. Am. Chem. Soc.*, **85**, 2850 (1963).
213. H. C. Brown, M. S. Kharasch, and T. H. Chao, *J. Am. Chem. Soc.*, **62**, 3435 (1940).
214. H. J. Dauben, Jr., and L. L. McCoy, *J. Am. Chem. Soc.*, **81**, 5404 (1959).
215. W. O. Haag and E. I. Heiba, *Tetrahedron Letters*, 3679 (1965).
216. P. S. Skell and P. D. Readio, *J. Am. Chem. Soc.*, **86**, 3334 (1964).
217. F. D. Greene, W. A. Remens, and J. W. Wilson, *J. Am. Chem. Soc.*, **79**, 1416 (1957).
218. P. S. Skell, R. G. Allen, and N. D. Gilmour, *J. Am. Chem. Soc.*, **83**, 504 (1961).
219. W. O. Haag and E. I. Heiba, *Tetrahedron Letters*, 3683 (1965).
220. W. Thaler, *J. Am. Chem. Soc.*, **85**, 2607 (1963).
221. D. D. Tanner, D. Darwish, M. W. Mosher, and N. J. Bunce, *J. Am. Chem. Soc.*, **91**, 7398 (1969).
222. D. D. Tanner, M. W. Mosher, N. C. Das, and E. V. Blackburn, *J. Am. Chem. Soc.*, **93**, 5846 (1971).
223. D. S. Ashton, J. M. Tedder, and J. C. Walton, *Chem. Commun.*, 1487 (1971).
224. G. A. Russell and H. C. Brown, *J. Am. Chem. Soc.*, **77**, 4025 (1955).
225. P. S. Skell, Special Publication No. 19, The Chemical Society, London, 1964, p. 131; P. S. Skell and K. J. Shea, *J. Am. Chem. Soc.*, **94**, 6550 (1972).
226. W. A. Nevill, D. S. Frank, and R. D. Trepka, *J. Org. Chem.*, **27**, 422 (1962).
227. D. J. Hurley, R. W. Rosenthal, and R. C. Williamson, *J. Org. Chem.*, **30**, 4314 (1965).
228. N. C. Deno, R. Fishbein, and C. Pierson, *J. Am. Chem. Soc.*, **92**, 1451 (1970).
229. E. Perry, *J. Org. Chem.*, **35**, 2053 (1970).

CHAPTER 8

Free Radical Rearrangements

JAMES W. WILT
Department of Chemistry, Loyola University, Chicago, Illinois

I. INTRODUCTION

The history of organic free radicals can be a long one if the early studies of supposed radicals like cyanogen and various alkyls are included.[1] But the monumental work of Moses Gomberg in 1900 on the triphenylmethyl radical is customarily considered the start of this story. It is therefore of interest to the present discussion that this classic work of Gomberg has recently been reinterpreted. As is discussed later herein (p. 446), the dimerization of triphenylmethyl does not afford hexaphenylethane, but rather a rearranged dimer. So rearrangement has been part of organic free radical chemistry from its birth, though this was unsuspected for the better part of this century.

In 1944 Urry and Kharasch discovered the neophyl radical rearrangement. This particular carbon radical rearrangement was the first to be demonstrated in solution. The neophyl system represents an active research area even today. From the mid-1940s to the present time, radical rearrangements appeared with increasing frequency. Now free radical rearrangements represent a major subdivision of the study of free radicals generally.

The subject of radical rearrangements has been reviewed in several texts and many review articles. Most importantly, the classic reviews of Walling[2] and Freidlina[3] should be cited as the wellspring of this present effort. Other texts and reviews will be given at the appropriate reference. It is not the purpose of this chapter to redo all these past reviews. Rather, the literature has been searched from 1962 to 1970 and beyond as far as is possible.[4] Studies earlier than 1962 are cited when pertinent, particularly if they were critically important to the presently held mechanism.

Molecular rearrangements, including those involving radicals, come in many guises. Some restriction must therefore be imposed upon the present discussion. Generally, the thermal rearrangement of organic monoradicals in solution forms the basic material of this chapter. Occasionally, some photochemically induced rearrangements are included also, as are a few "biradical" rearrangements.[5] The extent of coverage of these last two types of rearrangement is much less, however.

II. THE FREQUENCY AND NATURE OF RADICAL REARRANGEMENTS

It is commonplace that organic molecular rearrangements are most frequently cationic in nature, with fewer radical or anionic examples. At least one factor involved in this propensity for cationic rearrangement is the phenomenon of participation. Cationic rearrangements can be simultaneous (or to some degree concerted) with development of charge, and few or no reactions from unrearranged intermediates complicate matters. Moreover, as participation can arise from σ, n, or π electrons, the types of cationic rearrangement multiply fantastically.[6] On the other hand, radicals form without such participation in nearly every instance, although this statement is increasingly subject to scrutiny. Side reactions involving no rearrangement from the initial radical are therefore quite common, thus reducing both the number and types of radical rearrangements significantly. A major thrust to change this situation in radical rearrangement studies may be confidently predicted.

There remain the questions of why participation is notably absent in radical rearrangements, and why subsequent rearrangement of the initial radical is often so poorly competitive with nonrearrangement processes. Theoretical arguments have been employed to rationalize these observations, at least in cases involving the shift of some group to an adjacent radical site (1,2 shift). Particularly effective use of LCAO molecular orbital calculations some years ago[7] gave a rational basis for the more difficult nature of such radical rearrangements compared to their cationic analogs. Some of these results (simplified) are given in (1).

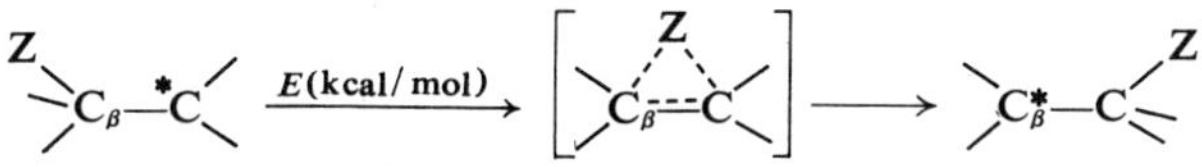

(1)

Z	Species	E/β[a]
Alkyl	Cation	-0.40
Alkyl	Radical	0.41
Alkyl	Anion	1.22
Aryl	Cation	-1.18
Aryl	Radical	-0.85
Aryl	Anion	-0.52

[a] β was not assigned a value, though dimensionally it is kcal/mole. Presumably, however, the more negative E/β, the more likely the rearrangement.

A more recent theoretical treatment of this topic utilized a more refined LCAO molecular orbital approach.[8] All of these calculations, though they differ in sophistication, are similar. They simply extend the naive Hückel molecular orbital treatment relating the stabilities of H_3^+, $H_3\cdot$ and H_3^- in triangular versus linear configurations.[9] This treatment is summarized in (2).

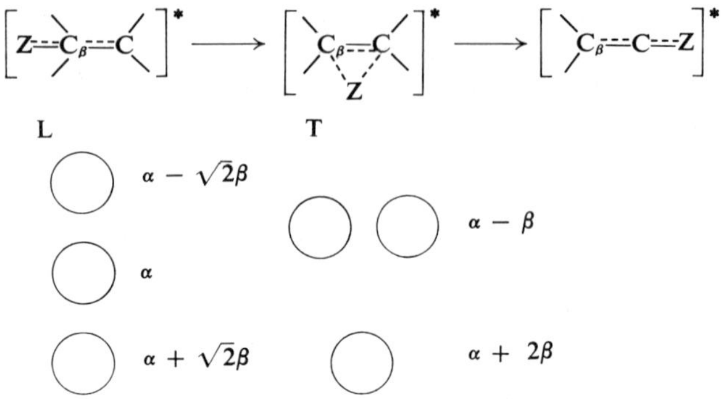

(2)

	Orbital energies, E		
Species	Linear (L)	Triangular (T)	$E_L - E_T$
Cation	$2\alpha + 2.83\beta$	$2\alpha + 4\beta$	-1.17β
Radical	$3\alpha + 2.83\beta$	$3\alpha + 3\beta$	-0.17β
Anion	$4\alpha + 2.83\beta$	$4\alpha + 2\beta$	0.83β

Three interacting atomic orbitals allow a triangular array (T) with one bonding molecular orbital and two less stable antibonding molecular orbitals. Cationic rearrangements bring the two electrons of the migrating group toward a "virtual" or empty orbital on the adjacent reactive center. Consequently, this rearrangement fills the bonding molecular orbital only and the transition state is totally bonding. Radical, and more so anionic, rearrangements of necessity have some antibonding character in the transition state because the orbital at the reactive center is not empty. As a consequence, the antibonding molecular orbitals must be populated to some extent. The comparisons in (2) between the linear (L), unrearranged state and the triangular one indicate roughly the degree of driving force for the three reactive species. The calculations indicate that the triangular state is more stable than the linear one for the cation, less stable for the anion, and about the same for the radical. Also, in those treatments that discussed it,[7,8] the relative ease of migration in the radical 1,2 shift was aryl > alkyl > hydrogen.

The theoretical rationales mentioned above have been mildly criticized[10] because they are based upon π approximations to (at times) totally σ electron systems. Particularly for a 1,2 hydrogen shift such an approximation may be less than apt.

Unfortunately, the powerful concept of orbital symmetry control[11] appears to be inapplicable to this topic. The 1,2 cationic shift is suprafacial and an allowed sigmatropic process.[12] Just how one should extend this treatment to radical (or anionic) analogs is unclear. A 1,2 suprafacial migration with retention of configuration in the migrating group as a radical across the ethylene π bond would be allowed (3); such a process for a cationic migrator across the ethylene π^* bond in a radical anion would not (3). In this latter case, antarafacial migration is obviously impossible, but suprafacial migration with inversion of configuration remains a possibility, for carbon at least.

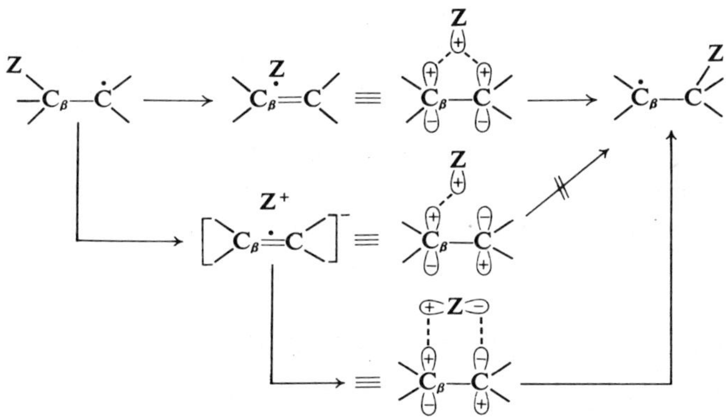

(3)

The theoretical rationale for 1,x shifts in radicals with $x > 1$ has not been studied very much. One interesting view[13] is that rearrangements of this type are most favorable when the σ bond of the migrating group lies colinearly with the radical center as in (4). Such a view is said to be supported by unreported approximate molecular orbital calculations. Obviously, this situation

(4)

is better met in larger-sized ring transition states; 1,2 shifts would be the worst possible in such a view. Another explanation has been advanced,[14] one based upon vibrational modes and frequencies in the radical. Basically, rearrangement will attend those cases where there is a high probability of an intramolecular vibration causing the appropriate geometry for migration. Heavier α substituents such as aryl or halogen, as well as an extension of the distance between the reaction site and the migrating group were considered to be favorable structural features for rearrangement.

Another factor affecting radical rearrangements should be mentioned, especially as it bears upon the second question above, namely, why subsequent rearrangement in radicals is often so poorly competitive with other processes. Cationic rearrangements are characterized by considerable driving forces at times. Such driving forces derive ultimately from the fact that cationic stabilities vary significantly, depending upon their structure and environment. Except for special cases that usually involve heteroatoms or extensive delocalization of the unpaired electron, the energetic difference between, for example, a primary and a tertiary pair of radicals is much less than that between the analogous cations. Indeed, the orthodox view that radical stabilities increase in the order primary, secondary, and tertiary has recently been questioned.[15] It would appear that this trend is correct only for endothermic transitions. Moreover, because cationic stabilities are subject to medium effects, their rearrangement can often be assisted by proper choice of reaction conditions. Radicals are less susceptible to such effects, however; hence the investigator has less maneuverability in his studies of potential radical rearrangements.

Chemistry remains an experimental science. However satisfying the above theoretical considerations may be, the search for radical rearrangements continues. The nature of science is such that it seems the search is all the more intense in those areas where theory predicts failure.

III. THE GENERAL TYPES OF RADICAL REARRANGEMENT

Any decision on the organization of radical rearrangements is of course a bit arbitrary. The organization used in this chapter, while close to that used by others,[16] is somewhat different. The present treatment seems to be adequate in that it groups all the rearrangements into six major classes and yet allows the development of a specific mechanism to cover nearly all cases in any given class.

a. *1,x Shifts of elements other than hydrogen*: This is an important radical rearrangement type. With the convention Z = migrating group, A = migration origin, and B = migration terminus,[17] the following general reactions (5) illustrate the types:

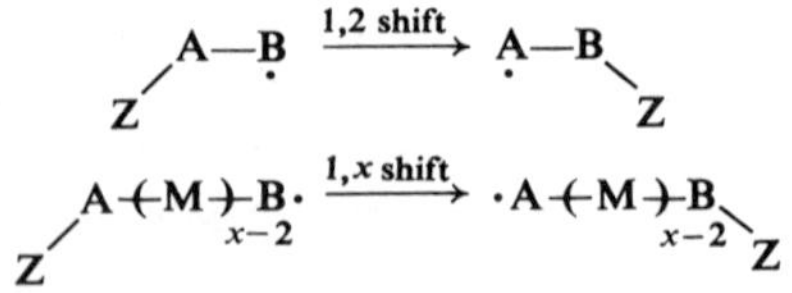

(5)

b. *Hydrogen abstractions*: Hydrogen abstractions, when intramolecular, are formally 1,*x* shifts. Because certain features of these shifts differ mechanistically from those in part a, the hydrogen shifts are grouped separately.

Their general type is given in (6).

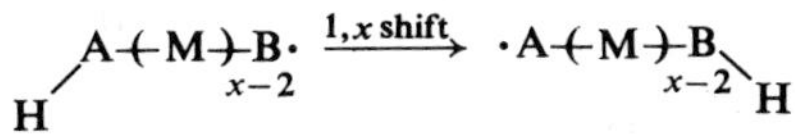

(6)

c. *Eliminations*[18]: In cyclic cases, ring opening of radicals leads to rearrangement. Some examples follow (7):

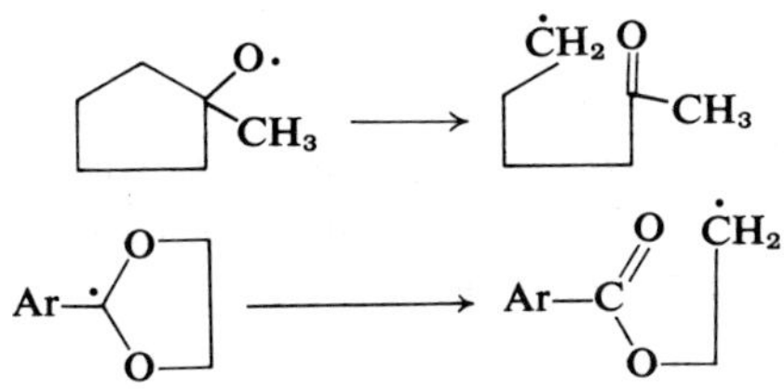

(7)

d. *Cyclizations*[18]: Essentially the reverse of the radical rearrangements in part c, these cyclizations are normally intramolecular radical additions to unsaturated functions, as shown in (8).

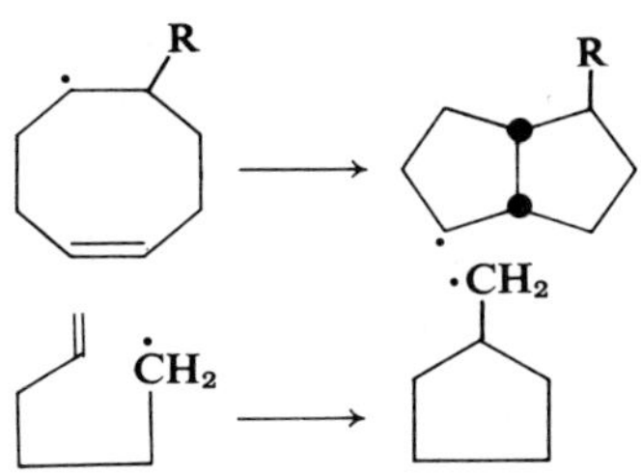

(8)

e. *Geminate and other recombinations*[19]: This type of radical rearrangement is gaining in interest. Rearrangements long thought to be electrophilic in nature now appear to proceed via radical–ion pairs. Again in terms of Z, A, and B, these may be formulated as in (9).

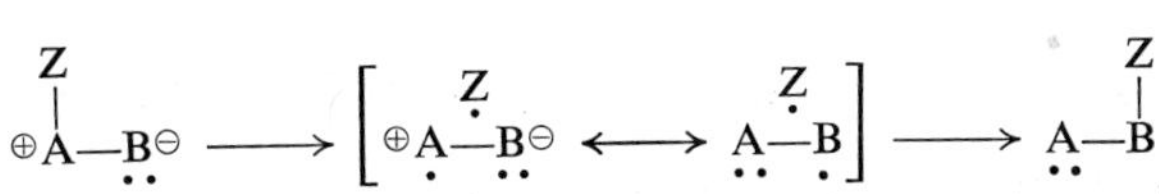

(9)

f. *Vinylic, allylic, and homoallylic rearrangements*: Because many studies have dwelled on these types of rearrangement, particularly the last, they will be discussed together under their own heading. Vinylic rearrangements (10) are commonly geometric isomerizations.

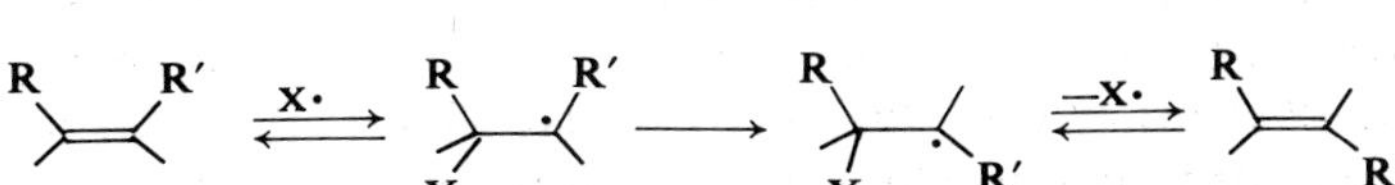

(10)

Allylic rearrangements are really still another type of 1,x shift, this time involving the double bond as the migrating group (11).

$$\begin{array}{ccc} [H_2C{=}CH{-}\dot{C}Cl_2 & \longleftrightarrow & H_2\dot{C}{-}CH{=}CCl_2] \\ \downarrow Y\cdot & & \downarrow Y\cdot \\ H_2C{=}\dot{C}H{-}CCl_2Y & & YH_2C{-}\dot{C}H{=}CCl_2 \end{array} \tag{11}$$

However, some allylic rearrangements actually involve an elimination–readdition sequence and are intermolecular, as, for example, the rearrangement shown in (12). Such rearrangements may arbitrarily be classed as

$$Ph{-}C(OR){=}CH_2 \xrightarrow[\Delta]{} Ph{-}C({=}O){-}CH_2R \tag{12}$$

"false" rearrangements because of their intermolecularity.[18] For completeness, however, such rearrangements are included in the present work. Homoallylic rearrangements are formally a type of elimination (13).

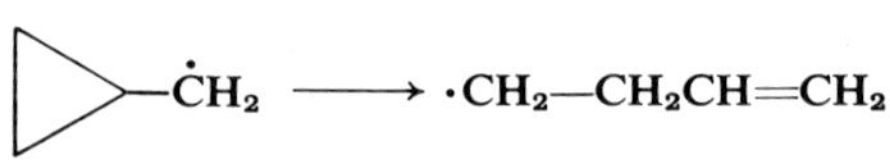

(13)

IV. DISCUSSION OF THE GENERAL TYPES OF RADICAL REARRANGEMENTS

A. 1,x Shifts

1. 1,2 Shifts (Vicinal Shifts) of Alkyl or Aralkyl groups

The chemical literature abounds in vain attempts to observe a vicinal shift of an alkyl group in a radical reaction in solution.[20] Even theory is against the process! The energy difference between the unrearranged radical and the half-migrated state (which approximates the transition state) has been calculated by LCAO–MO methods to be $+0.41\beta$ (unfavorable).[7] Still unobserved, such a rearrangement is probably the most sought after today. Good summaries of past efforts in the gas phase are also available.[21] But even here, the early views on alkyl (usually methyl) shifts in gas-phase radical reactions were usually amended later to alternative mechanisms. In solution work (with which this chapter is concerned), a very early observation in Kolbe electrolysis was interpreted as a methyl shift in the radical (14).[22]

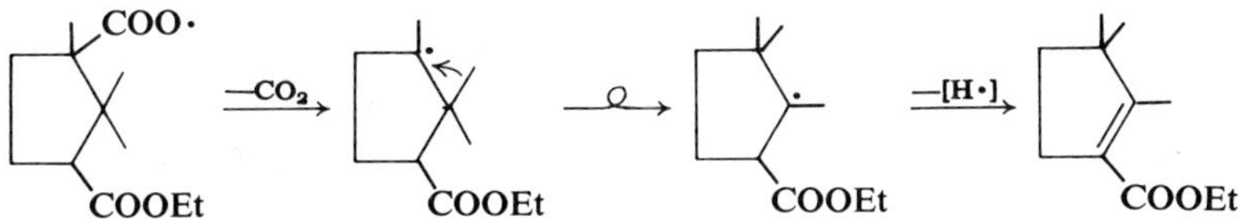

(14)

Far more likely, however, is a prior oxidation of the radical to a carbonium ion (the Hofer-Moest reaction), followed by the well-known neopentyl cation rearrangement.[23] An interesting attempt has been made recently to find such an alkyl radical shift caused by ring strain relief (15).[24]

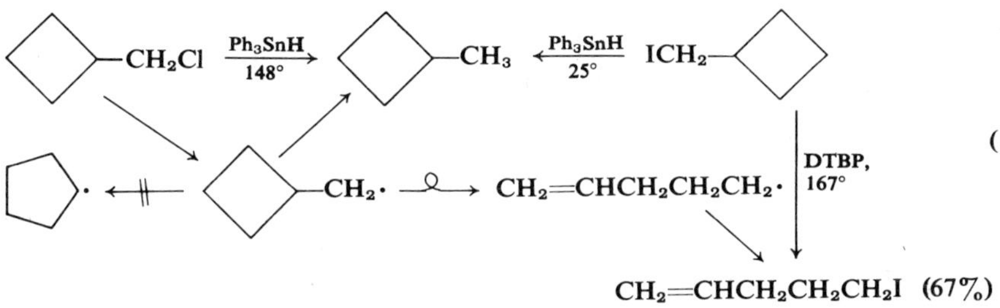

(15)

Although the cyclobutylcarbinyl radical did rearrange by elimination (see p. 406), it did not ring expand. Interestingly, the 1-phenyl-substituted radical did give phenylcyclopentane.[25] It is perhaps more likely, however, that this "rearrangement" was really elimination–readdition, as shown in (16).

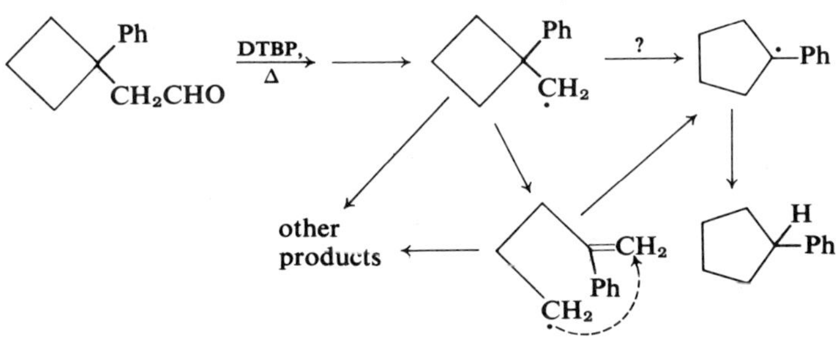

(16)

Products from all the above radicals were observed. But the significant percentage of olefinic products, the propensity for radical addition to form five-membered rings (see p. 420), and the benzylic nature of the adduct from such an addition all lend credibility to the pathway in (16) rather than to a direct ring expansion mechanism. Earlier, an analogous alkyl rearrangement in the

2-bornyl radical was demonstrated to be exactly such an elimination–readdition and not a direct alkyl shift (17).[26]

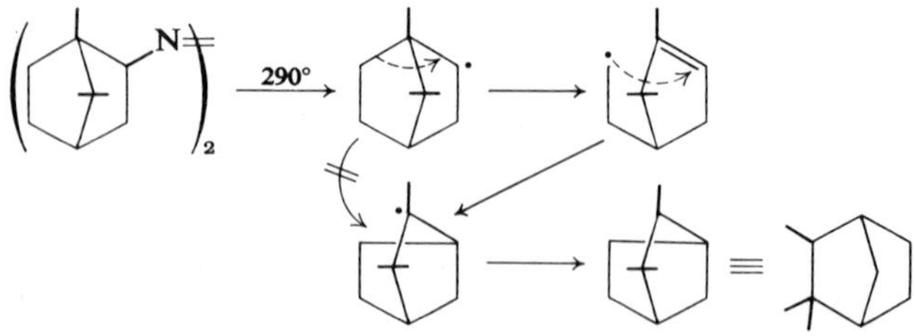

(17)

Notwithstanding all the above failures to observe vicinal alkyl shifts, there does remain such a possibility in a modification[27] of the Arndt-Eistert reaction (18).

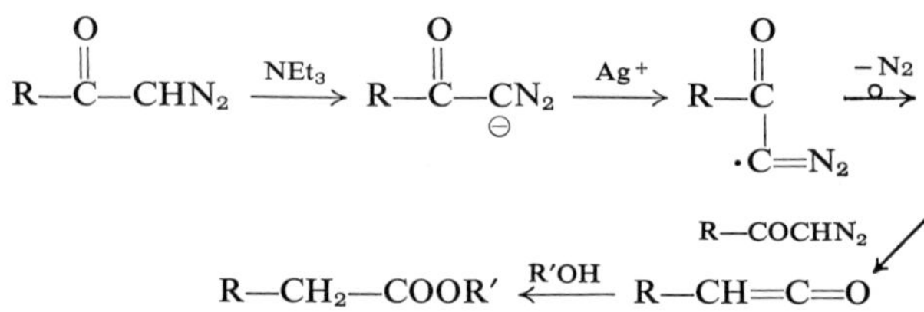

(18)

This interesting process, though used frequently, has apparently not been investigated since its discovery. Indeed, the rearrangement step of the Arndt-Eistert sequence, the venerable Wolff rearrangement, is still of uncertain mechanism![28] Then, too, an occasional process, usually photochemical, will apparently involve a vicinal alkyl shift. An example is the recent photochemical Beckmann rearrangement shown in (19).[29]

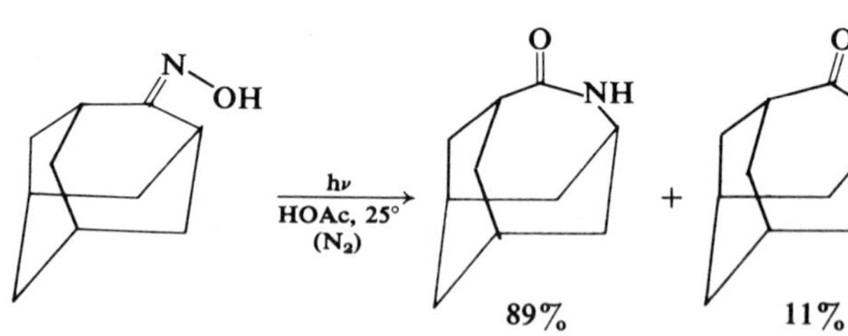

(19)

No fragmentation, a process typical for such oximes, was observed. The deleterious effect of oxygen, which drops the yield of lactam to 11%, may imply a radical mechanism. However, no definite mechanistic details are as yet available.

Recent work indicates that vicinal alkyl shifts in certain *diradicals* may occur and, indeed, may predominate over otherwise highly competitive processes. The cyclic perester shown in (20) decomposes thermally with methyl migration predominantly.[30]

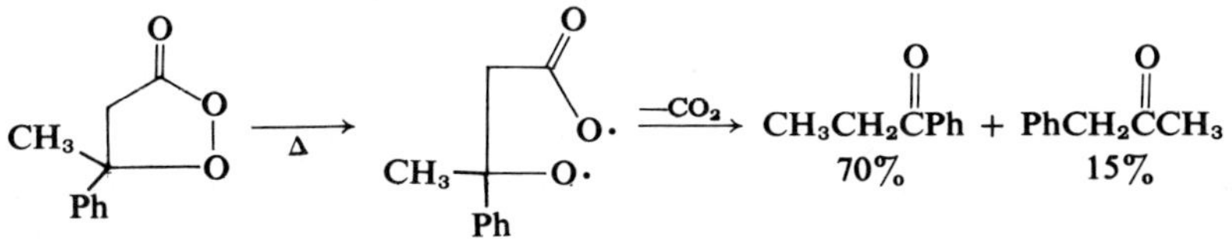

(20)

Other work[31] indicated a migration preference *i*-propyl > benzyl > ethyl > methyl > phenyl. This preference was manifested after the rate-determining step, however, as all the peresters decomposed at essentially the same rate. The above migration order is also the one associated with cleavage of groups from oxy radicals derived from hypochlorites (see p. 416). It would thus appear that bond breaking is advanced in this process beyond bond making. Therefore migratory ability is related to the stability of "Z" as a radical and not to the ability of "Z" to stabilize a bridged transition state by delocalization. Such lack of synchrony does not necessitate a prior dissociation step, however. Indeed, the smallish rate difference between *i*-propyl and phenyl (55:1) and the inversion observed[32] at the migration terminus (equation 21) indicate that rearrangement is a concerted push–pull process, with a pull from the departing carbon dioxide and a push from the developing carbonyl group.

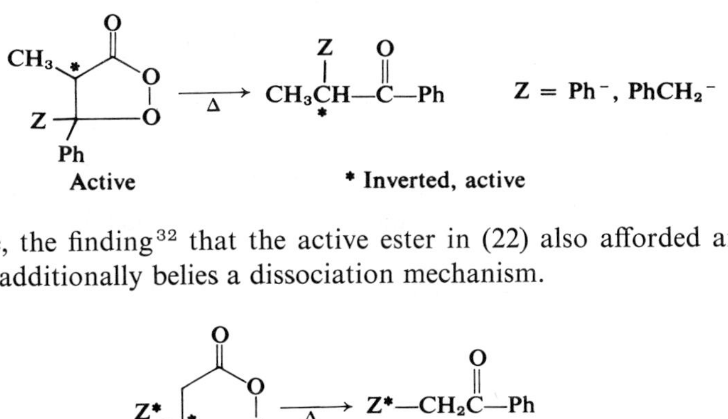

(21)

Likewise, the finding[32] that the active ester in (22) also afforded an active product additionally belies a dissociation mechanism.

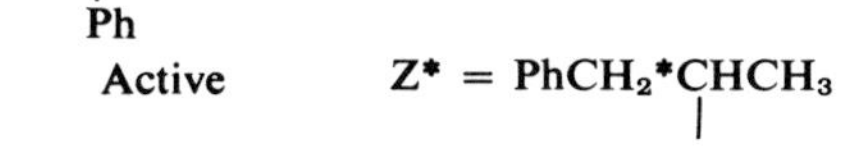

(22)

In another view, it has been suggested[33] that the transition state is quite polar (carbonium ion-like) and considerably assisted in its formation by the development of the carbonyl group (equation 23).

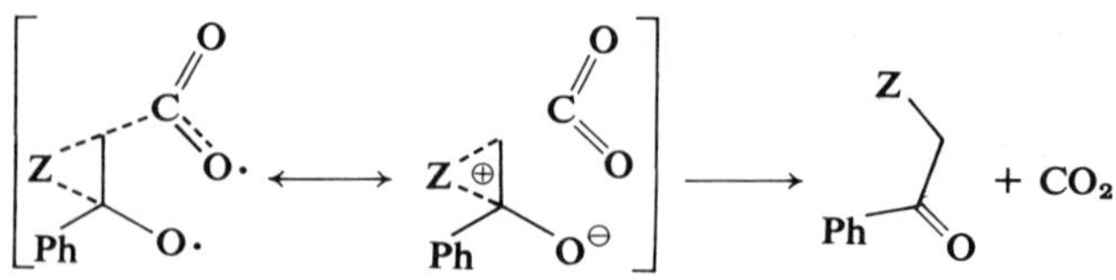

(23)

It is thus quite pertinent that photochemically produced diradicals also show this preference for alkyl versus phenyl migration (24), presumably via a triplet diradical.[34]

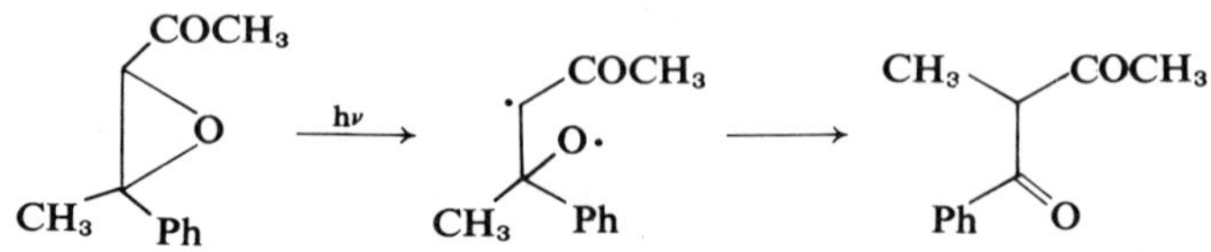

(24)

In this case, if a polar contributor to the transition state for rearrangement were invoked to explain the alkyl shift, intersystem crossing of the initial triplet to a singlet would be required. In any case, it remains unclear why an alkyl shift is preferred if the transition state has polar character. In bona fide polar rearrangements which pit alkyl against phenyl, for example, reaction 25, phenyl migration predominates.[35]

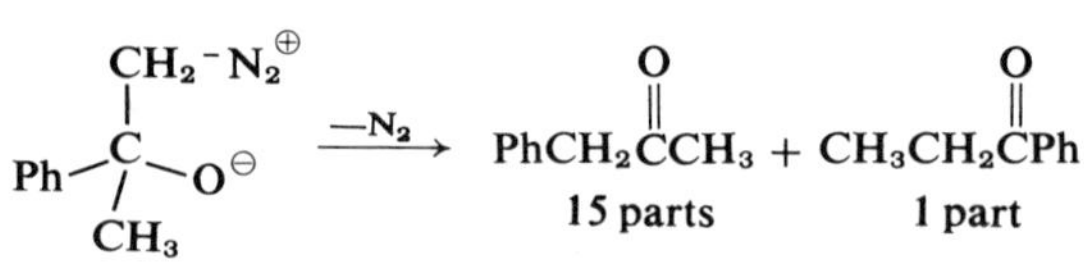

(25)

Migration of alkyl groups concertedly with elimination of carbon dioxide makes these rearrangements of peresters the first cases of such an assisted radical pathway. One may expect increased interest in such processes henceforth.

Moving to aralkyl group vicinal migration, a 1,2 shift of the benzyl group was sought in the reactions shown in (26).[36]

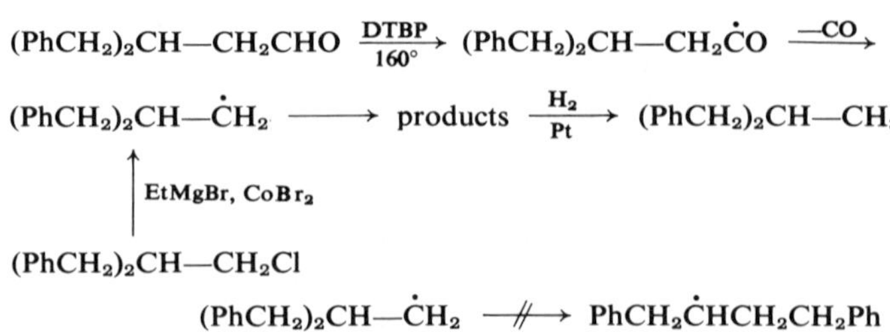

(26)

The work was based upon the reported 1,2 shift of benzyl in an analogous carbanionic rearrangement.[37] However, no such shift was observed in the radical processes. High yields of unrearranged hydrocarbon were obtained instead. As fate would have it, a later report[38] indicated that the basis of the study was itself inappropriate. In the carbanionic rearrangement (equation 27), radiochemical research established that the "1,2 shift" of benzyl was an intermolecular elimination–readdition process that bears no resemblance to the well-known intramolecular 1,2 phenyl shifts (see p. 346).

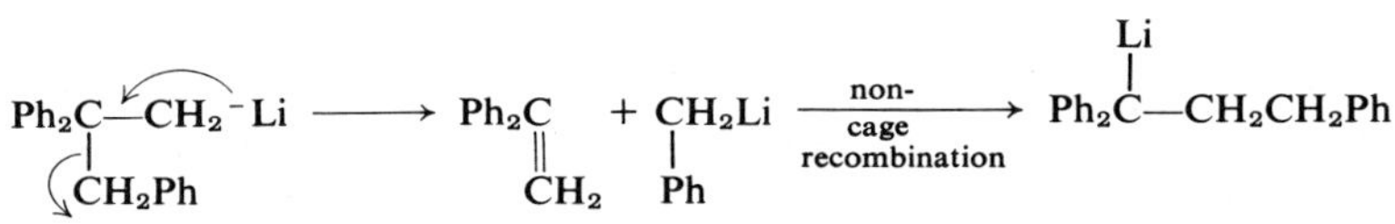

(27)

So, as with simple alkyl groups, no 1,2 benzyl shift is yet known.

2. 1,3 Shifts of Alkyl or Aralkyl Groups

An extremely interesting 1,3 shift (ring expansion) has been observed in unpublished work (28).[39a]

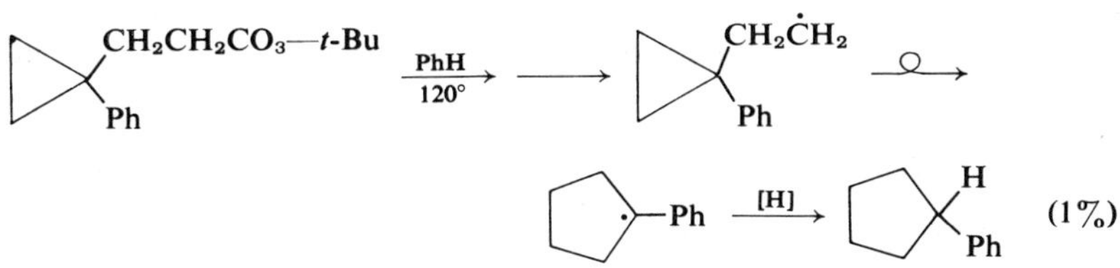

(28)

This unique result is difficult to rationalize in the light of no known 1,2 analog. One certainly waits for additional studies. That such a rearrangement is not beyond belief is indicated by the similar rearrangement shown in (29). The course of the rearrangement was said to involve intramolecular radical attack on the cyclopropyl ring.[39b]

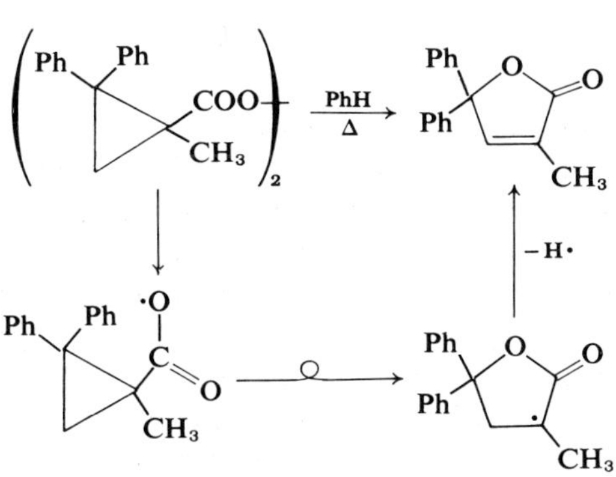

(26%) + other products (See p.399)

(29)

3. *1,2 Shifts of Aryl Groups*

The 1,2 shift of the phenyl group (equation 30) is the "neophyl" rearrangement, though the name "neophyl" perhaps more properly is applied to the β-phenylisobutyl group only.

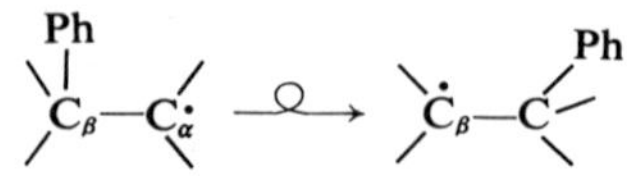

(30)

It ranks as perhaps the best known radical rearrangement and excellent coverage of the rearrangement is available.[40] The parent rearrangement was discovered in 1944 in a Kharasch-Grignard reaction.[41] It was the first demonstration of a carbon radical rearrangement in solution. The rearrangement in the parent or related systems has since been observed in a variety of other processes: peroxide-induced decarbonylation of aldehydes,[42–44] radical additions to olefins, [45] decomposition of azo compounds,[46] thermolysis of peroxy compounds,[47] Kolbe electrolyses,[48] the Hunsdiecker reaction,[49] peroxide attack on alkylbenzenes,[50] radical halogenations,[51] and formation of organometallic reagents.[52a,b]

The rearrangement is occasionally absent where it might have been expected (30a).[52c,d]

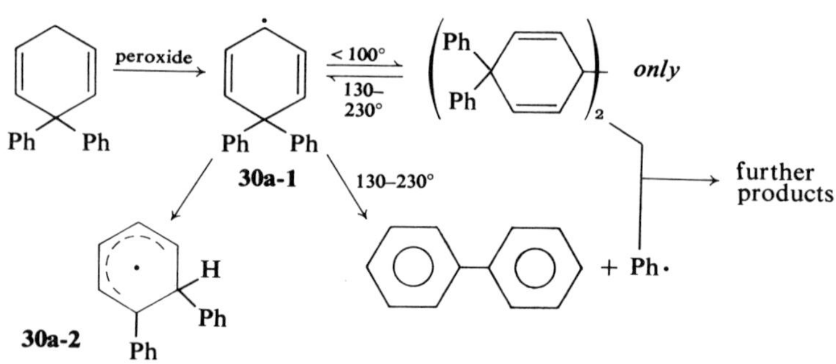

(30a)

Presumably, at the higher temperatures the drive to aromatize **30a-1** directly by phenyl radical loss is too competitive for a 1,2 phenyl shift to occur. In fact, an activation energy as high as 35 kcal/mole was suggested later[52d] for this potential rearrangement. At lower temperatures *either* process seems disadvantageous relative to simple dimerization, however, perhaps reflecting the allylic stability of **30a-1**.

Abundant evidence exists that the neophyl rearrangement involves discrete unrearranged and rearranged radicals, with bridged radicals being involved only as transition states or short-lived unstable intermediates.[53] Thus, the extent of rearrangement is inversely dependent upon substrate concentration,[54,55] and it is decreased if efficient chain transfer agents like mercaptans are present.[45,56] Recent supporting evidence for the absence of a bridged radical intermediate lies in the decarbonylation of the optically active aldehyde shown in (31) which gave racemic product.[57]

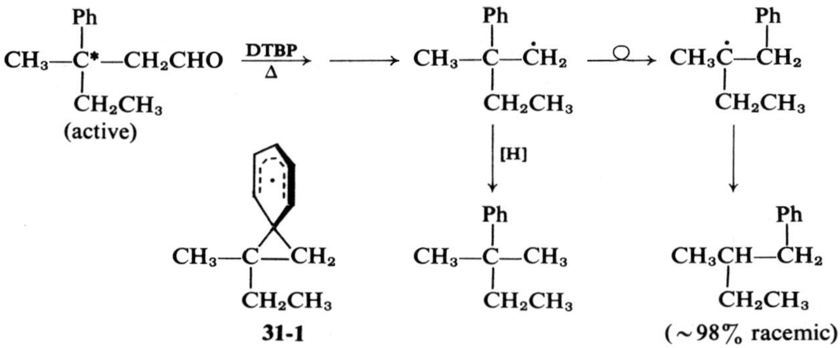
(31)

Had the bridged species **31-1** been a product-forming intermediate, an inverted product would have been expected. Whereas a substituent effect on the migrating ability of the aryl group has been found for decarbonylation (32), as well as for other reactions involving the neophyl radical,[58,47d] such an

$$(CH_3)_2\overset{\displaystyle Ar}{\overset{|}{C}}\text{—}CH_2CHO \xrightarrow[\Delta]{\text{DTBP}} (CH_3)_2CH\text{—}CH_2Ar \qquad (32)$$

$$\rho \cong +1 \qquad Ar = p\text{-}GC_6H_4\text{—}$$

electronic effect is not involved in the rate-determining step of the analogous perester decompositions, shown in (33).[47d]

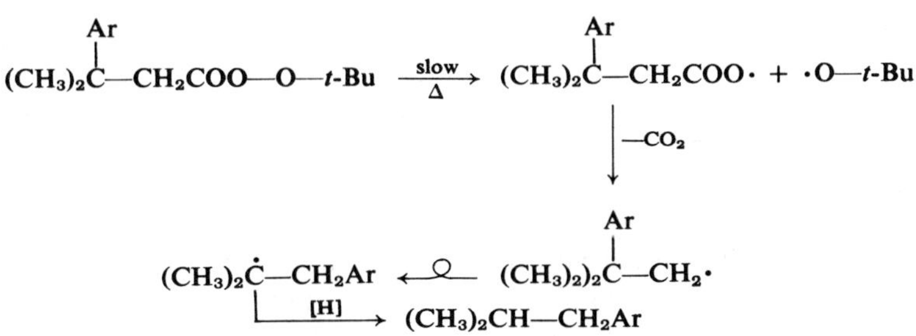
(33)

Indeed, these rates were hardly different from that for a decomposition of a nonneophyl substrate (cyclohexyl instead of aryl), as illustrated in Table 1.

Moreover, these reactions possessed the $\Delta H^{\ddagger}$ and $\Delta S^{\ddagger}$ parameters associated with nonconcerted decomposition of peresters.[59] The same conclusion

TABLE 1

Decomposition rates of *t*-butyl β-aryl perisovalerate esters

$$(CH_3)_2\overset{\displaystyle Ar}{\overset{|}{C}}—CH_2CO_3—t\text{-Bu} \xrightarrow[\Delta]{} (CH_3)_2CH—CH_2Ar$$

$G—C_6H_4(Ar)$	$10^4\,k$ (120°, Ethylbenzene or cumene)	% Rearrangement[a,b]
G = H	2.09	44
G = *p*-CH_3	2.43	35
G = *o*-OCH_3	2.45	73
G = *p*-OCH_3	2.72	19
G = *p*-F	1.92	25
G = *o*-Cl	1.92	71
G = *p*-Cl	2.03	47
G = *m*-Cl	1.98	78
G = *p*-Br	1.96	60
G = *p*-NO_2	2.26	96
Ar = Cyclohexyl	2.60	0

[a] In tetralin at 130°.
[b] Includes related olefin products.

was reached in earlier studies on the decomposition of azo compounds related to these peresters.[46a] So the consensus is that the neophyl rearrangement takes the following course (34):

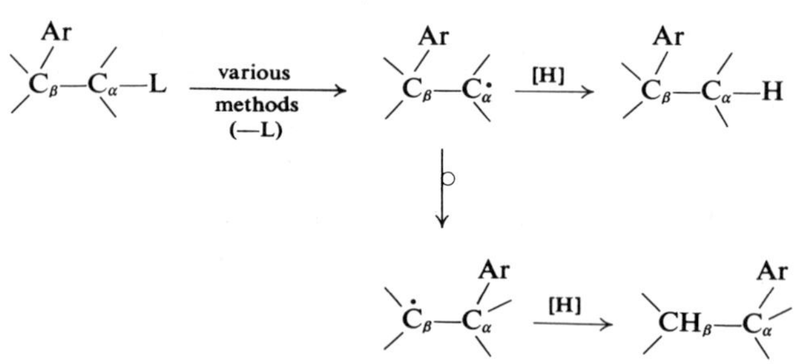

(34)

with the rearrangement step itself proceeding via a transition state (or short-lived intermediate) with some polar overtones, as in (35).

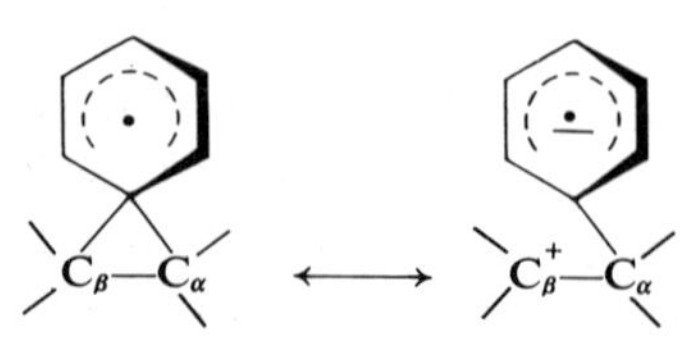

(35)

Generally, the rearrangement proceeds best when a tertiary or benzylic radical is formed and when steric compression at the β-carbon atom is relieved. For example, the energy difference between unrearranged and half-migrated radicals (the "bridging" energy) for $PhCH_2CH_2\cdot$, $Ph_2CHCH_2\cdot$, and $Ph_3CCH_2\cdot$ are -0.85β, -1.17β and -1.61β, respectively.[7b] Note the contrast with the 0.41β calculated for an alkyl shift. Because the β-phenylethyl radical itself rearranges,[60] however, such product radical stabilization and relief of steric strain are obviously not necessary for rearrangement. Nonetheless, these factors assist rearrangement, because β-phenylethyl itself rearranges only slightly (2–5% at 145–175°), whereas highly substituted analogs may rearrange essentially completely, for example, triphenylethyl.[61] A report[62] of 20% phenyl rearrangement from β-phenylethyl bromide in the Kharasch-Grignard reaction has been shown to be in error.[63] No rearrangement at this reaction temperature (~35°) in fact occurs. One further instance of rearrangement not attended by a gain in stability is given in (36).[64] Rearrangement amounted to 5% at 160° and 14% at 180°.

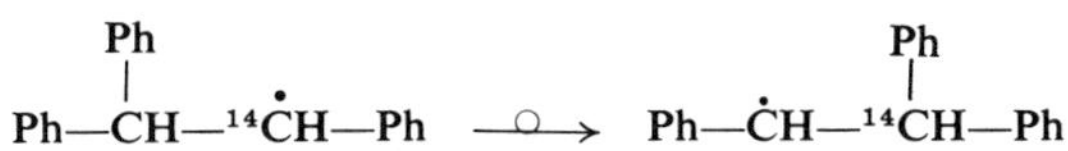

(36)

The ability of the aromatic migrator to achieve the proper orbital overlap for an Ar_1-3 transition state can be modified by ring size effects. As shown in Tables 2 and 3, this ability is lessened in smaller or in fused ring systems, leading to a decrease in rearrangement and/or an increase in competition from other processes.

LCAO–MO calculations readily rationalize the migration of the aromatic group in this rearrangement.[7] Although the estimates of its magnitude vary, some barrier to rearrangement is anticipated. In one view,[68] the aromatic migrator loses 2β kcal/mole stabilization in the Ar_1-3 transition state with only 1.46β kcal/mole of this loss recompensed by the pentadienyl radical so formed (37).

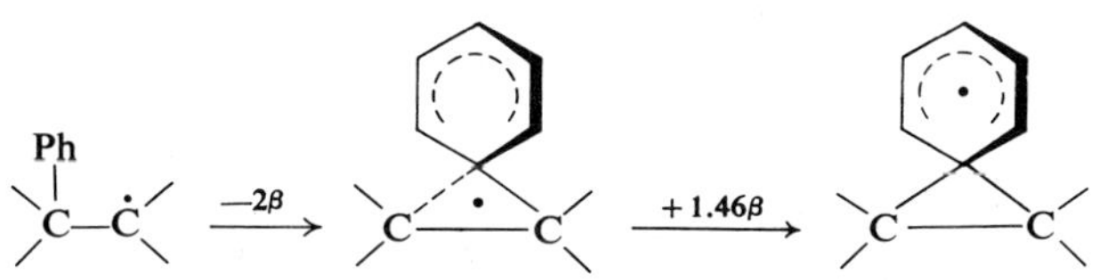

(37)

Attempts to detect this spiro radical intermediate via ESR spectroscopy at low temperature have been unsuccessful.[69] The energy barrier for the rearrangement of the neophyl radical itself was first estimated as 7–8 kcal/mole,[54a] but later work indicated that this estimate was too low.[54b] A calculation based on literature data[70] gives an Arrhenius activation energy for this

TABLE 2

Ring size effect: alicycles

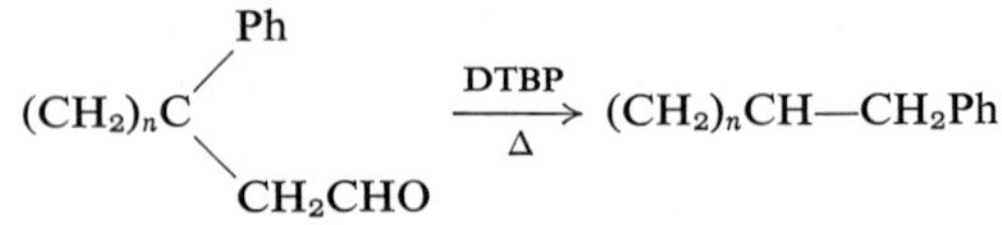

Ring size	n	% Rearrangement	Ref.
3	2	0[a]	25
4	3	4[a]	25
5	4	63	56a
6	5	89	56a
7	6	88	56b
"Neophyl"	—	64	54a

[a] Ring opening predominated.

TABLE 3

Ring size effect: benzoalicycles

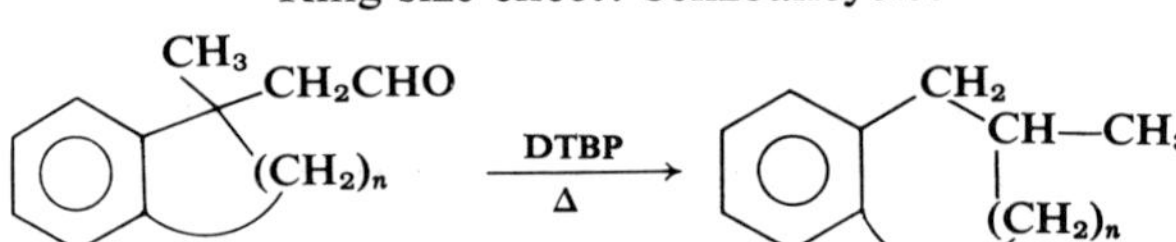

Ring size	n	% Rearrangement	Ref.
4	1	0[a]	65
5	2	40	54c
6	3	32	65
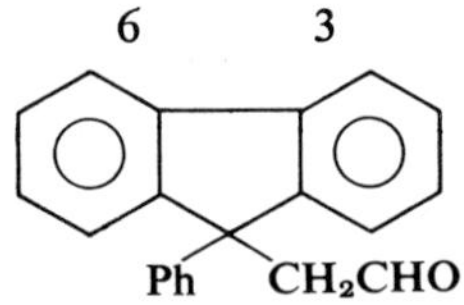		100[b]	66a
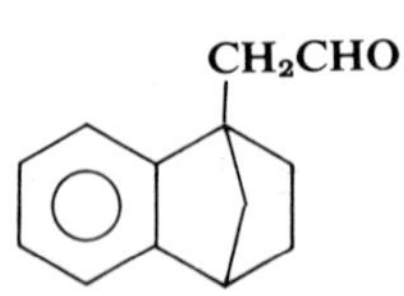		0[c]	67

[a] Decarbonylated with difficulty. The product was apparently a polymer.

[b] The product was (45% yield) 9-phenylphenanthrene (possibly via air oxidation of the 9,10-dihydro analog). Note that the rotationally restricted 9-phenyl group failed to migrate. Furthermore, parasubstituted 9-phenyl groups[66b] likewise did not migrate.

[c] The Ar_1-3 state was effectively precluded in this system.

case which is 2.9 kcal/mole higher, that is, 10–11 kcal/mole. A similar calculation recently reported for the rearrangement of the β-phenylethyl radical[71] gave a barrier of about 11 kcal/mole. This must be too low, however, because it was based upon the earlier (incorrect) value of approximately 7 kcal/mole for the barrier in the neophyl case. With the β,β,β-triphenylethyl radical, however, rearrangement is undoubtedly somewhat easier than in the neophyl radical itself. Early studies established this point,[61] and a recent restudy[72] of the reaction in (38) demonstrated that this rearrangement can even be competitive with a trapping reaction by triphenylmethyl radical.

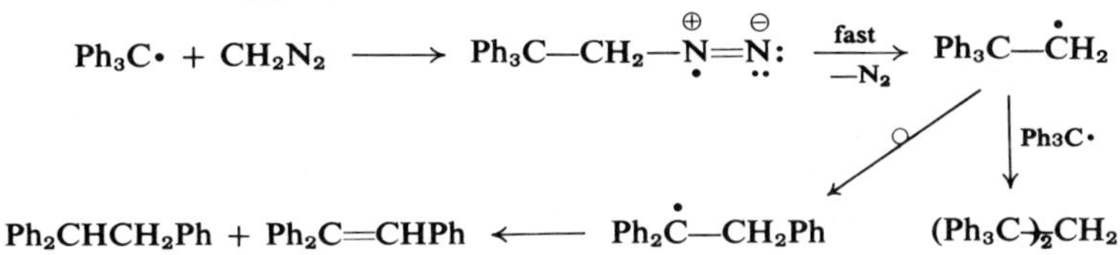

(38)

The rearrangement increased with dilution of the reactants, as would be expected by this scheme. The β,β,β-triphenylethyl radical can be trapped also by triphenyltin hydride (equation 39), a definitive study[73] that showed the presence of discrete unrearranged and rearranged radicals.[74]

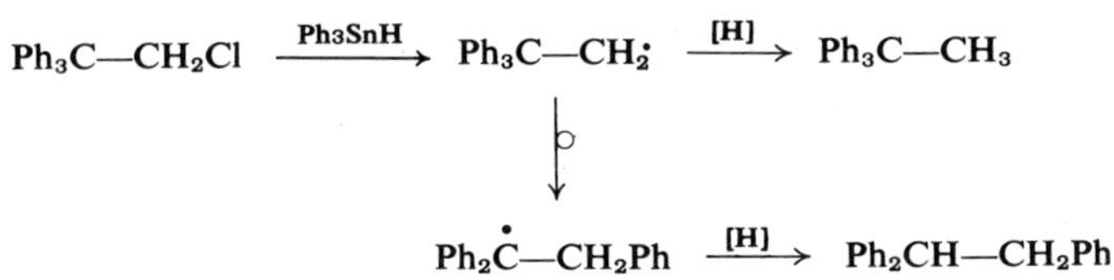

(39)

Rearrangement was again favored by lower tin hydride concentrations. Because reduction of halides with tin hydrides has rate data available,[75] a rate constant for the rearrangement in (39) can be extrapolated. The value obtained, $k = 5 \times 10^7\ \text{sec}^{-1}$ at 100°, indicates an energy of activation for the process of approximately 9 kcal/mole (using $PZ = 10^{13}\ \text{sec}^{-1}$).

An interesting 1,2 aromatic ring migration has been observed in the reduction of 9,10-dihydro-9,10-ethanoanthracenes with tri-*n*-butyltin hydride,[76] at

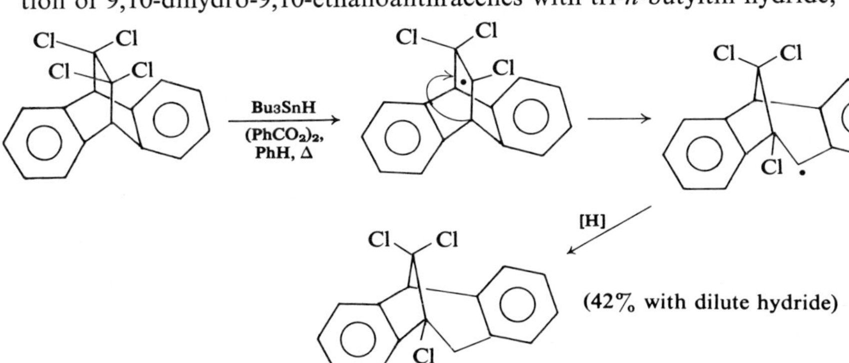

(40)

least under certain reaction conditions, as shown in (40). Similarly, addition of chlorine to benzonorbornadiene using nitryl or sulfuryl chloride gave some *exo,syn*-dichloride, and the rearrangement shown in (41) was suggested.[77]

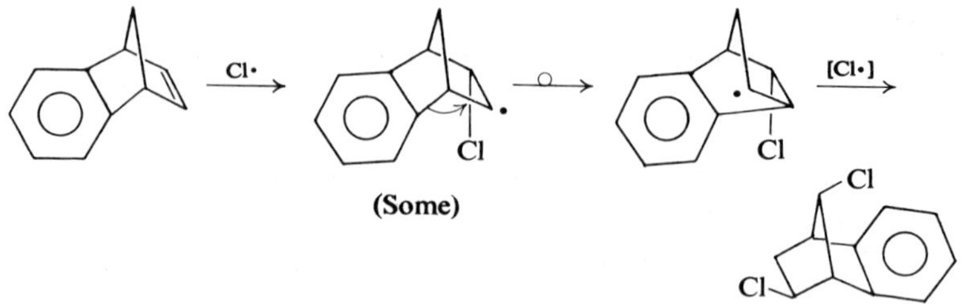

(41)

However, no rearrangement was detected in any of several radical additions involving other reagents to benzonorbornadiene.[78] In another recent study, no rearrangement was observed in the bromination shown in (42).[79]

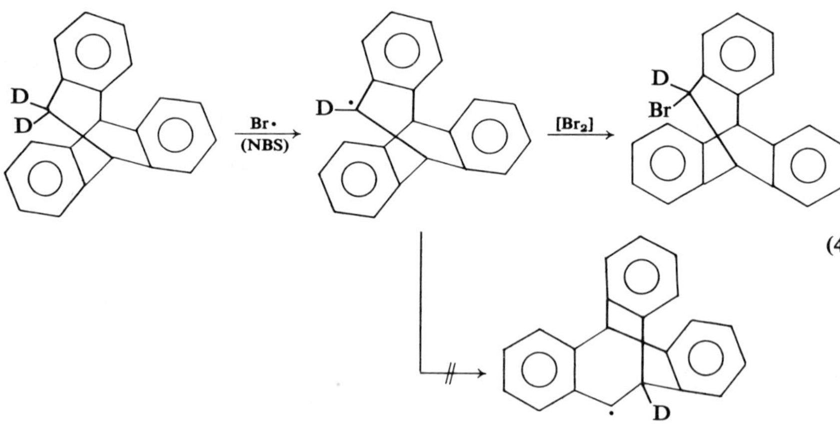

(42)

One might note that the cation related to the radical does rearrange. The unusual occurrence of rearrangement in (40) has been ascribed[76] to difficult

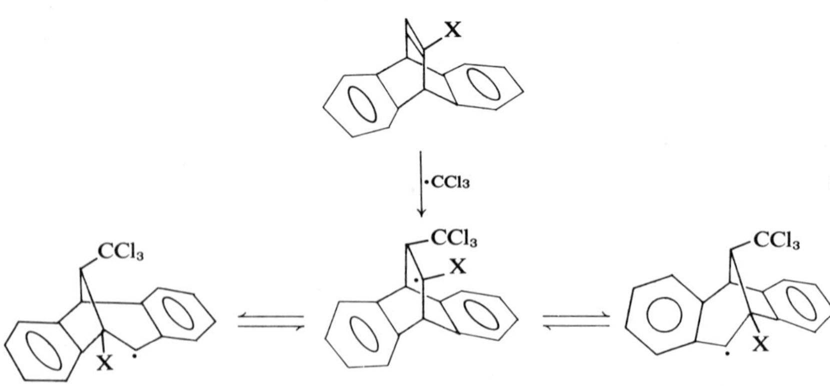

(43)

43-1 X = H, Cl, Br 43-2

chain transfer caused by the adjacent *gem*-dichloro function. In support of this view, no rearrangement is found in those cases where either no or only one β-chlorine is present.[80] In work pressing this point, the rearrangement in (43) was observed[81] to be complicated by *reversibility*, a unique result in the neophyl rearrangement area. Rearranged products increased with temperature and with lower concentrations of the chain transfer agent X–CCl_3, but product from **43-2** gained more than that from **43-1** under the latter condition, probably because exo chain transfer at the benzylic radical site of **43-2** is easier. It would be informative to approach this set of equilibria from the sides rather than the middle. In any event, the reconversion of *benzylic* radicals as shown is unusual.

Vicinal aryl shifts to or from sites other than carbon are known; the Wieland rearrangement[82] (reaction 44) is one such and actually represents one of the earliest free radical rearrangements in solution.

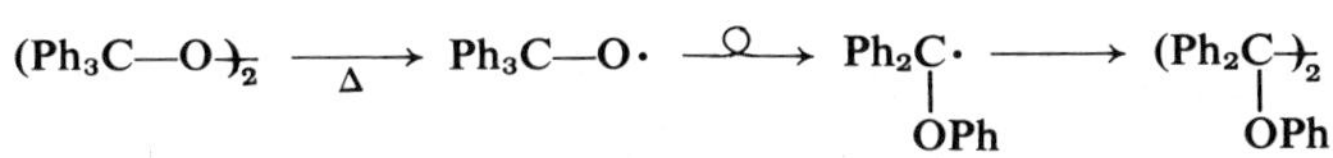

(44)

The early work has been much discussed,[83] but little recent work seems to have been done.

A somewhat related reaction is the Baeyer-Villiger oxidation. Though usually considered an ionic process,[84] reactions of the ketones shown in (45) with peracetic acid gave migratory aptitudes different from those obtained with trifluoroperacetic acid.[85]

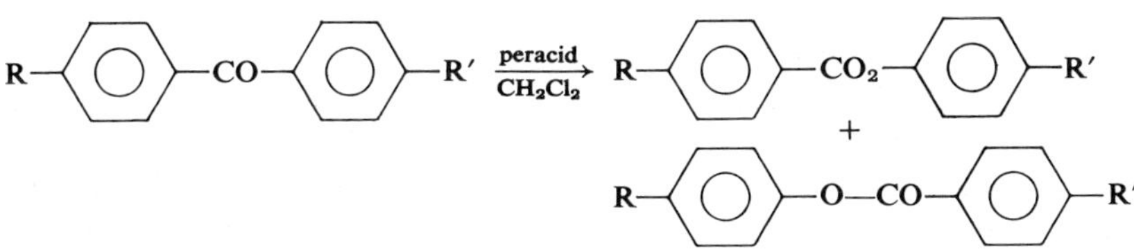

(45)

The migratory aptitudes in peracetic acid resemble radical arylation factors, especially in the first four entries. So a conceivable pathway is shown in (46). The acetoxy radical abstracts hydrogen as the aryl group migrates, presumably in a solvent cage because little carbon dioxide is obtained. Free acetoxy radicals readily decarboxylate.

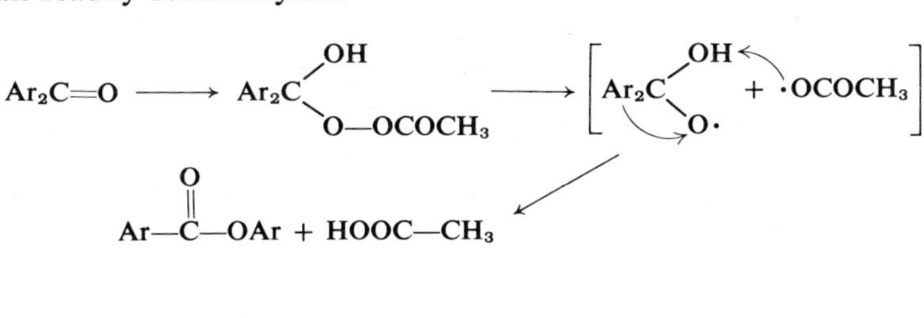

(46)

R	Migratory ability CF$_3$COOOH[a]	CH$_3$COOOH[b]
NO_2	1.00	3.20
t-Butyl	7.93	—
Cl	12.15	1.22
H	—	1.00
Br	13.94	2.14
CH_3	26.05	8.88
OCH_3	251	152.5

[a] $\rho = -0.77$ (using σ^+).
[b] Contained a catalytic amount of CF_3COOOH.

Because radical arylation data have been largely misinterpreted,[86] the fundamental basis of this study may be misplaced. It remains curious that little carbon dioxide was formed because methylene chloride is not an effective cage solvent.[87]

It is curious that vicinal migration of aryl groups from silicon to carbon is not known (equation 47).[88] The free-energy change associated with the trans-

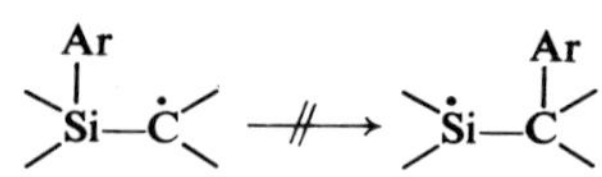

(47)

formation cannot be grossly unfavorable, however, because farther shifts of this kind (1,4 and 1,5) are known (p. 374).[89] Two features of α-silyl carbon radicals distinguish them from their γ- and δ-silyl analogs. First, there may be d_π–p_π delocalization of the unpaired electron into a silicon d_ε orbital as indicated ("backbonding") in (48). Such stabilization could reduce the driving

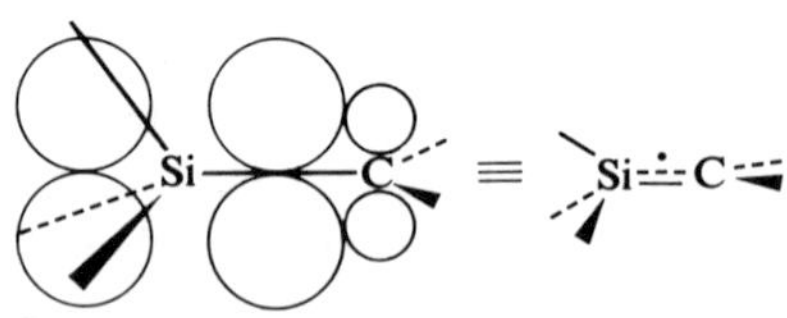

(48)

force for rearrangement. ESR data support such a view.[90] Secondly, the Ar$_1$-3 transition state would be severely strained (silacyclopropanes are unknown), and the strain might prohibit rearrangement via such a state. In the farther rearrangements known,[89] both of these antirearrangement factors are absent. Successful rearrangement in these instances adds strength, but not certitude, to these arguments for no rearrangement in the α-silyl radical.

The silicon analog of the Wieland rearrangement (p. 353) has been reported (49).[91] A recombination of radicals within a solvent cage seems favored. It is

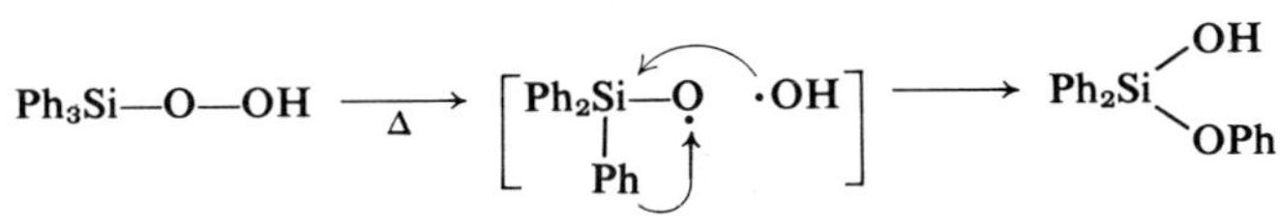

(49)

noteworthy that this α-silyl radical rearranges whereas the α-silyl carbon radical does not. It is conceivable that the *n* electrons on oxygen are delocalized into the d_ε orbitals of silicon and so prevent backbonding of the unpaired electron. This in turn allows rearrangement, although assistance by concomitant hydroxy radical attack on silicon undoubtedly plays a role as well.

A 1,2 shift of the phenyl group from oxygen to carbon (50) has been claimed,[92] although an earlier attempt to find this process failed.[93]

$$Ph—OCH_3 \xrightarrow[\Delta]{DTBP} Ph—O\dot{C}H_2 \longrightarrow \dot{O}CH_2—Ph \xrightarrow{[—H\cdot]} O{=}CH—Ph \qquad (50)$$

Other than for these reports, the rearrangement seems to have had little study. A related earlier claim for an *alkyl* shift of this type more probably is an elimination–readdition sequence (equation 51) because of the other products found.[94]

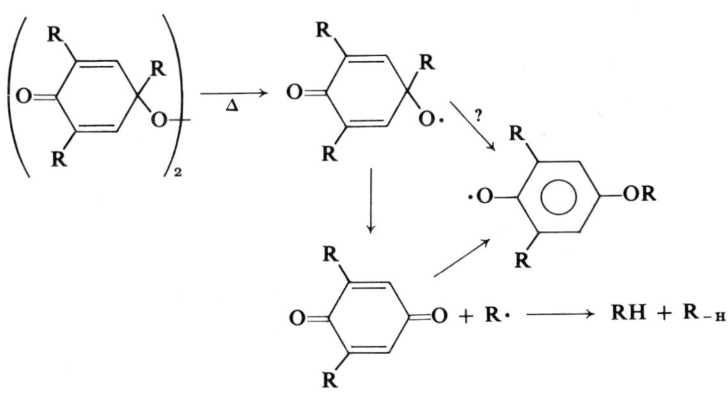

(51)

A recent attempt to find a neophyl-like rearrangement from nitrogen to carbon (52) has been described.[95] No migration was detected, although some *N*-methylcarbazole was observed. α-Amino radicals may have some stabilization as shown, thus reducing their tendency toward rearrangement.

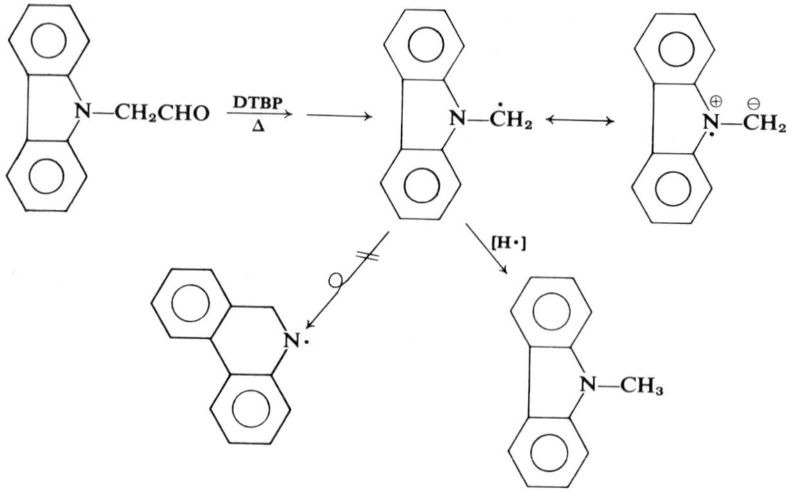

(52)

4. *1,2 Shifts of Vinyl Groups*

Vinyl groups seem able to rearrange in a vicinal manner. At elevated temperatures (~450°) iodination of alkenes led to rearranged products that were best explained via the vinyl shift shown in (53).[96]

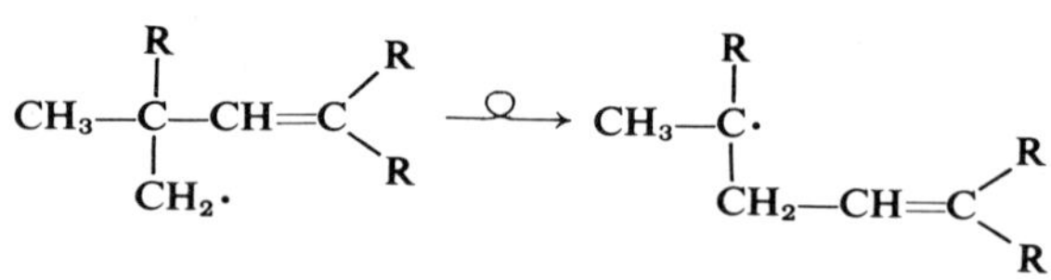

(53)

The rearrangement in (54) was likewise explained[97] by competitive shifts of a phenyl group in one case and a vinyl group in the other.

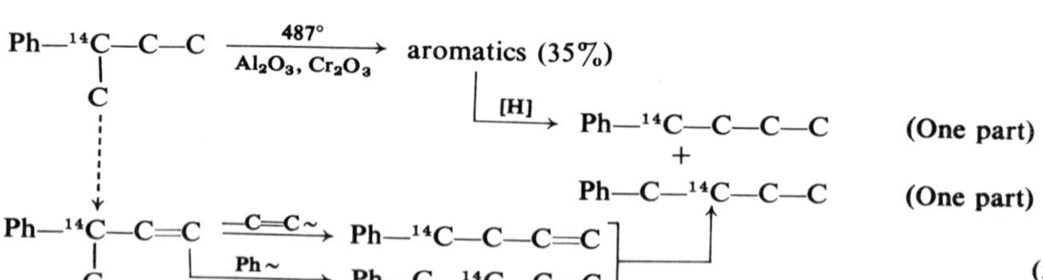

(54)

The catalyst produced no benzene and isobutylene from *t*-butylbenzene at 481°, so it was claimed to be nonacidic and one that did not induce cationic processes. Although intramolecular vinyl group migration does appear to be reasonable from such results (and others below), a cautionary note has been sounded.[98] In the absence of iodine and at sufficiently high temperatures, quasirearrangement of the vinyl group is in fact an elimination–readdition sequence, as shown in (55).

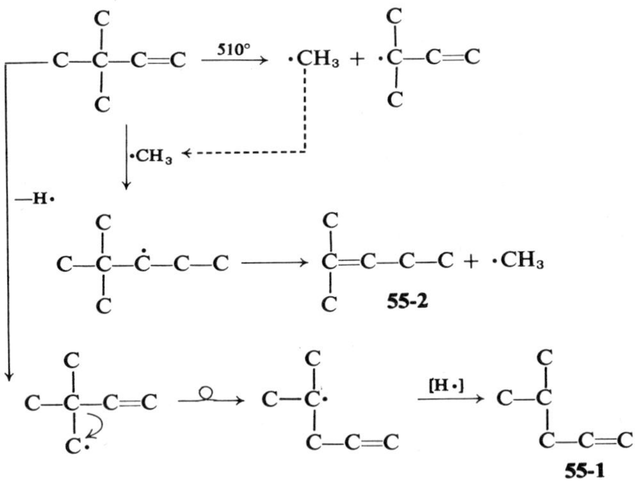

(55)

No product **55-1** was observed to accompany product **55-2** under reaction conditions where the conversion of the former to the latter was unlikely. The elimination–readdition sequence requires methyl radicals. Iodine traps these methyl radicals, so the bona fide rearrangement can be observed in the presence of such a trap. Lower temperatures also favor true rearrangement because methyl radical production is slower.

In solution, vicinal rearrangement of the vinyl group is well documented.[99] Although the secondary radical did not rearrange, the reverse rearrangement of the primary radical did occur (56).

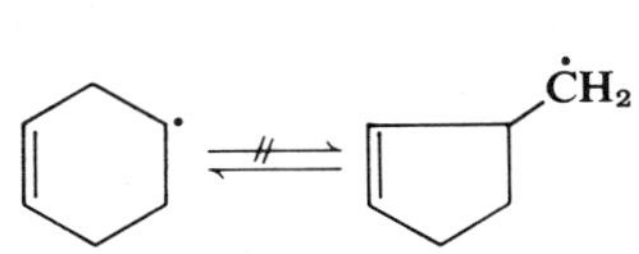

(56)

Possibly the strain energy differences between these species account for the one-sidedness of the process. Other examples where strain energy considerations precluded rearrangement are indeed known. The possibility of elimination–readdition instead of true rearrangement in (56)—always a bothersome possibility—was neatly disposed of by use of a chlorine atom donor (57).

Only 4-chlorocyclohexene was observed, so no elimination–readdition, or, for that matter no 3,4 σ electron pair migration, occurred.

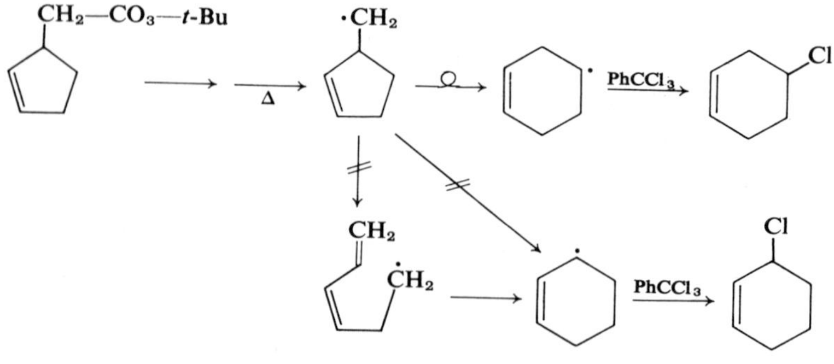

(57)

The question of vicinal vinyl group migration is intimately bound up with allylcarbinyl (homoallylic) radical rearrangements. Two distinctly different paths for rearrangement are possible (equation 58).

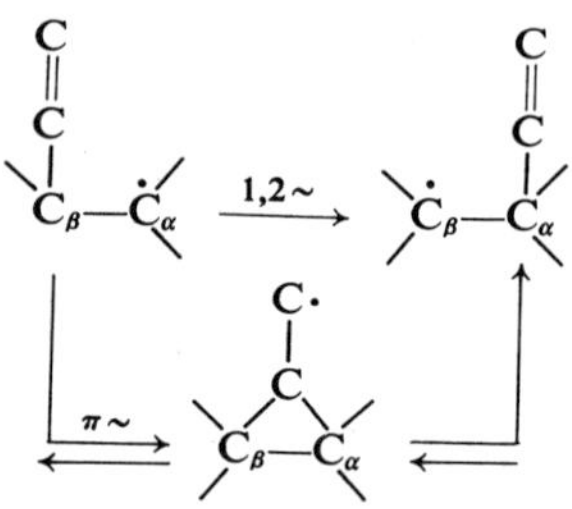

(58)

The first path involves a direct shift of a bond comprised of σ electrons in an sp^2–sp^3 carbon–carbon bond. The second path involves an interaction of the π electrons of the vinyl group and proceeds via a three-membered ring species somewhat analogous to that utilized in the neophyl rearrangement. A number of studies have implicated the second path preferentially. Among these have been studies of the decarbonylation of unsaturated aldehydes. Decarbonylation of the aldehyde shown in (59) gave rearranged product.[68]

$$\mathrm{C{=}C{-}\underset{\displaystyle |\atop\displaystyle C}{C}{-}C{-}CHO} \xrightarrow[\Delta \atop (-\mathrm{CO})]{\mathrm{DTBP}} \mathrm{C{-}C{-}C{-}C{=}C} + \mathrm{C{=}C{-}\underset{\displaystyle |\atop\displaystyle C}{C}{-}C} \qquad (59)$$

The ratio of the two products depended on the aldehyde concentration, so again discrete unrearranged and rearranged radicals were probably involved. The rearrangement itself could, a priori, take one or several paths (60).

The path requiring methyl migration (*a*) is unlikely and besides, no 2-pentene was observed. This olefin would be expected if the allylic radical **60-1** had intervened. Path *b* was supported by formation (less than 0.1%, however) of 1,2-dimethylcyclopropane. Such a product strongly implicates radical **60-2**.

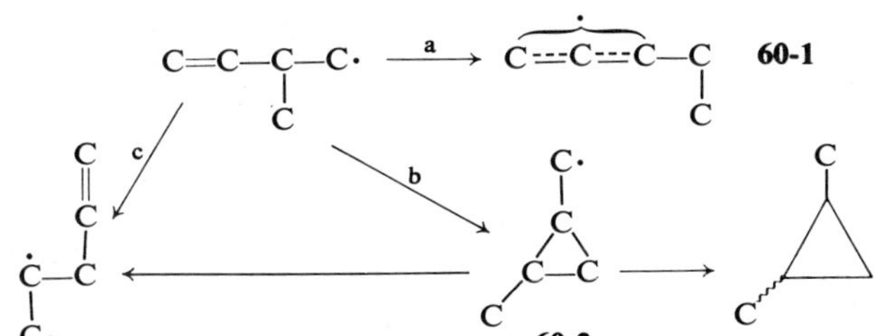

60-1

60-2

(60)

Indeed, a companion study (61)[100] showed that the four radicals depicted interconvert, most likely via radical **61-1**, which must possess a lifetime long

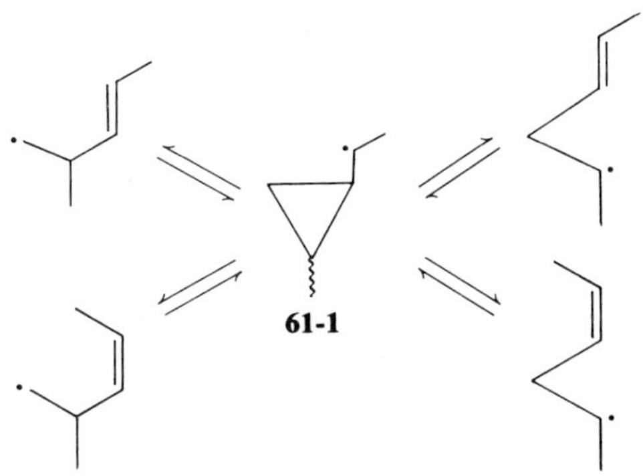

61-1

(61)

enough to allow rotations in bonds prior to collapse. This lifetime may not be long enough, however, to allow the formation of cyclic product in all cases. As is noted later (p. 403), ring opening of cyclopropylcarbinyl radicals is very fast ($k = \sim 10^8\ \text{sec}^{-1}$). Two studies of this point[101,102] using deuterated radicals further supported the intermediacy of the cyclic species. Such results clearly illustrate the similarity of these vinyl group rearrangements and the neophyl rearrangement. They argue against any direct 1,2 shift of a σ (sp^2–sp^3) bond. In a very complex process, the products resulting from the reaction in (62) have been rationalized via similar vinyl group rearrangements.[103]

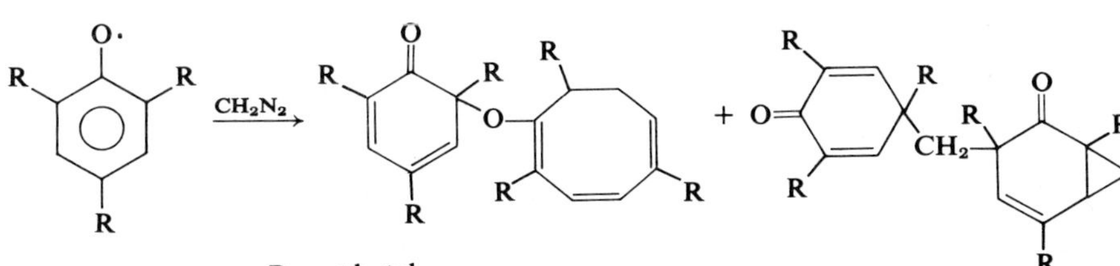

R = *t*-butyl

(62)

Likewise, the plethora of hydrocarbons formed by reduction of bullvalene bromides **62a-1** and **62a-2** have been similarly explained (62a).[103a]

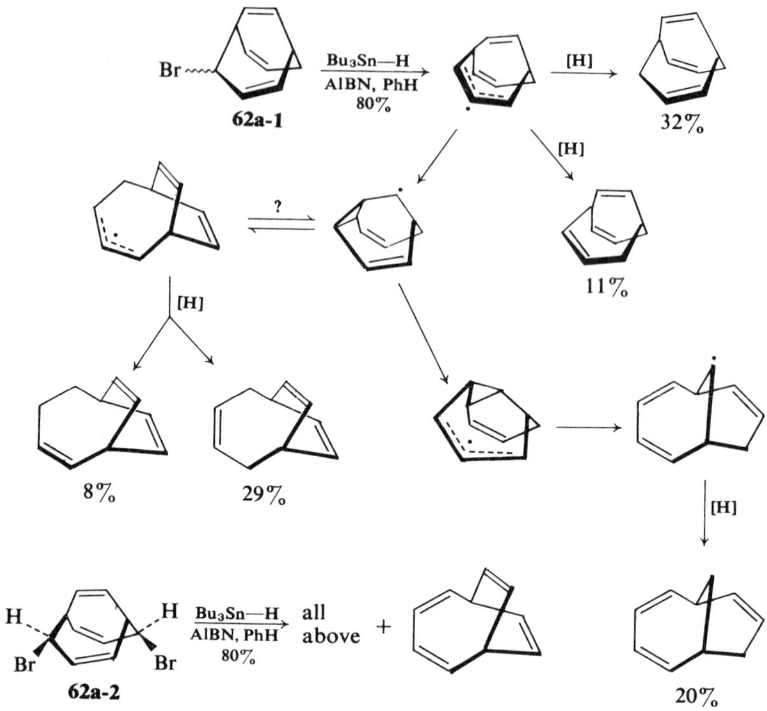

(62a)

5. *1,2 Shifts of Heteroatoms*

A recent study[104] reported the exceptionally clean vicinal migration of the acetoxy group (63).

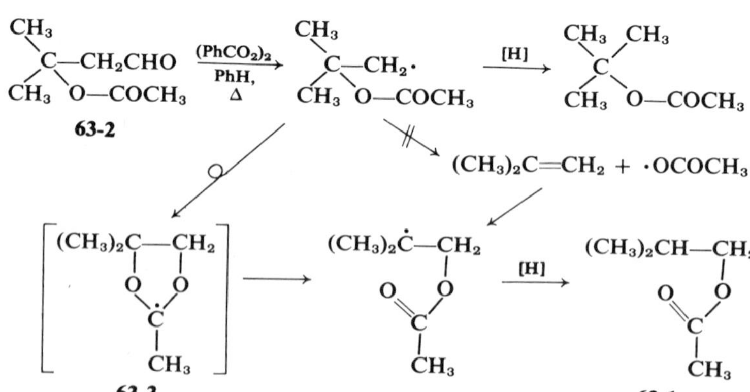

(63)

Added isobutylene was without effect in the process and the product composition showed less rearranged **63-1** (more unrearranged product) with more concentrated reactant **63-2**. The first point effectively precluded an elimination–readdition sequence as an alternative to the vicinal intramolecular shift via species **63-3**. The second point indicated that **63-3** itself is probably not an intermediate but rather a transition state. α-Ketal radicals are known species, however (see p. 412), and **63-3** may be a product-forming intermediate along with other, acyclic radicals in the sequence. Once more the conversion of a primary to a tertiary radical is apparently necessary. β-Acetoxy-*n*-butyraldehyde did not exhibit the rearrangement, which in its case involves a potential primary-to-secondary radical conversion. This vicinal migration of the acetoxy group occurs also in the addition reaction shown in (64), as well as in similar reactions.[105]

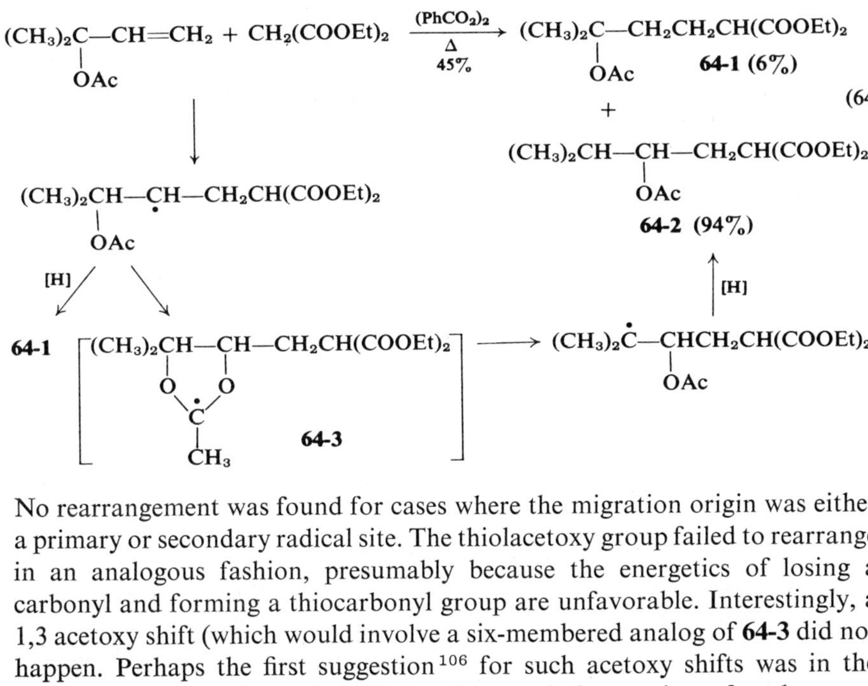

(64)

No rearrangement was found for cases where the migration origin was either a primary or secondary radical site. The thiolacetoxy group failed to rearrange in an analogous fashion, presumably because the energetics of losing a carbonyl and forming a thiocarbonyl group are unfavorable. Interestingly, a 1,3 acetoxy shift (which would involve a six-membered analog of **64-3** did not happen. Perhaps the first suggestion[106] for such acetoxy shifts was in the biradical pathway shown in (65) for the pyrolytic opening of cyclopropyl

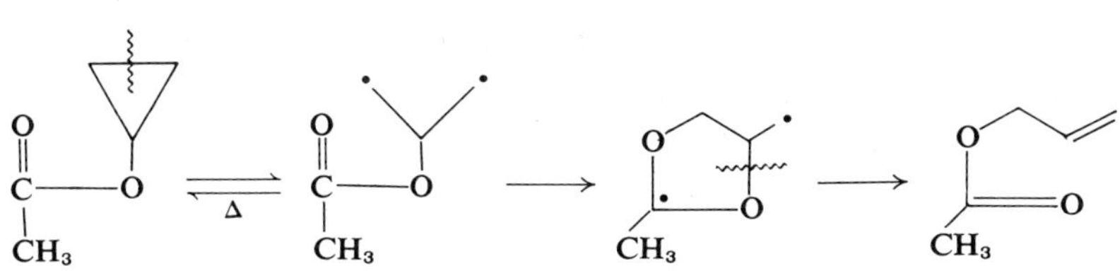

(65)

acetate. Such a pathway appeared better in accord with substituent effects than did a concerted quasi-Cope process.

An interesting vicinal migration of the thiophenyl group has been reported (equation 66). The interest lies in the better ability of this group (probably using the *n* electrons) to migrate than phenyl itself.[107]

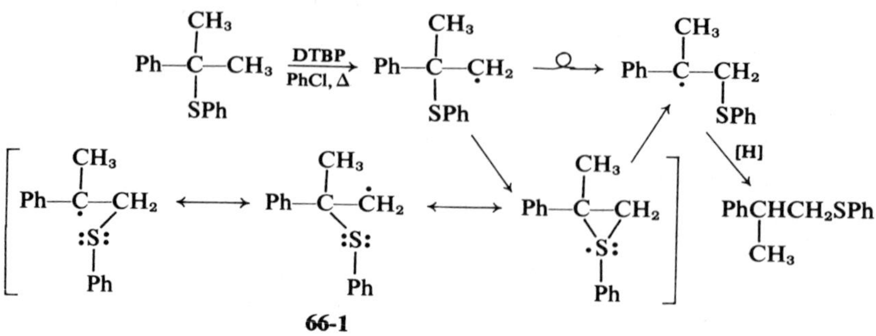

(66)

Possibly a bridged transition state **66-1**, available to sulfur because it can accommodate the expanded octet, is preferred over the Ar_1-3 transition state for phenyl migration. The latter does, after all, lose aromaticity.

The reported[108] migration of the trimethylsilyl group (equation 67) may occur similarly, although generation of the same radical in solution at about 130° led to no rearrangement.[109]

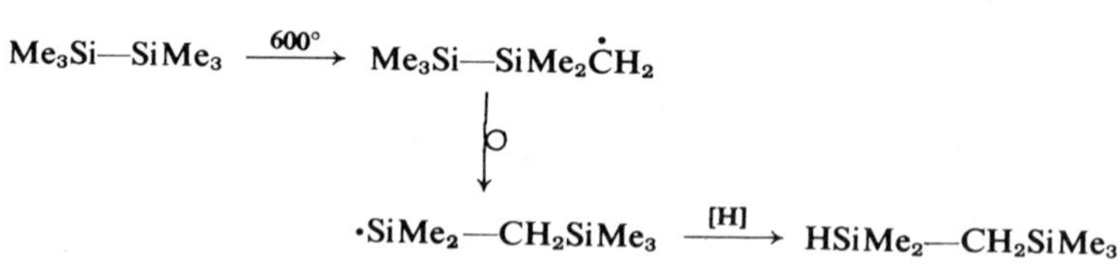

(67)

Far and away, the vicinal migration of chlorine (the Nesmeyanov rearrangement) is the most exemplified member of this class of rearrangements. The rearrangement has been thoroughly reviewed.[110] It also probably involves an expanded octet in its transition state, this time on chlorine rather than on sulfur as shown in (66). In 1951 two reports appeared on the 1,2 shift of chlorine in free radicals, one[111] correcting an earlier study[112] of an addition reaction (68), the other[113] reporting the amazing photochemical reaction of diazomethane with carbon tetrachloride and other trichloromethyl compounds (69), a reaction later shown to possess a fairly high quantum yield.[114]

$$Cl_3C{-}CH{=}CH_2 \xrightarrow[\text{peroxide or } h\nu]{HBr} Cl_2CH{-}CHCl{-}CH_2Br \qquad (68)$$

$$X{-}CCl_3 + CH_2N_2 \xrightarrow{h\nu} XCH_2{-}C(CH_2Cl)_3 \qquad (69)$$

$$X = {-}H, {-}Cl, {-}Br, {-}COOMe$$

From that time numerous examples of the rearrangement have been described. In general, the process involves the step shown in (70), although simpler cases (71) are also known (vide infra).

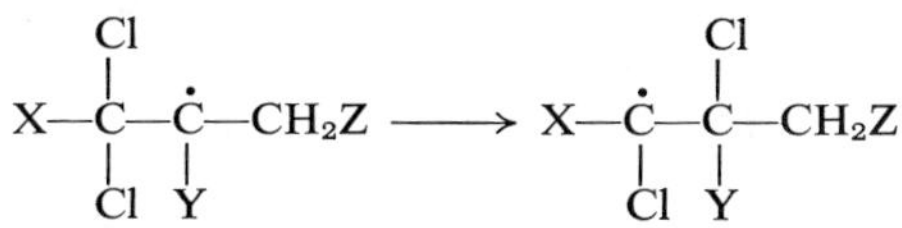

(70)

X = H, F, Cl, $—CH_3$; Y = H, Cl, Br, $—CH_3$; Z = H, Br, $—CCl_3$, —SR

$$Cl_2CH—\dot{C}H_2 \longrightarrow Cl\dot{C}H—CH_2Cl$$
$$Cl_3C—\dot{C}H_2 \longrightarrow Cl_2\dot{C}—CH_2Cl \qquad (71)$$

Some clarification of the much-discussed[115] mechanism of reaction 69 has appeared. The reactions shown in (72) have allowed the trapping of several intermediate radicals involved in the sequence.[116]

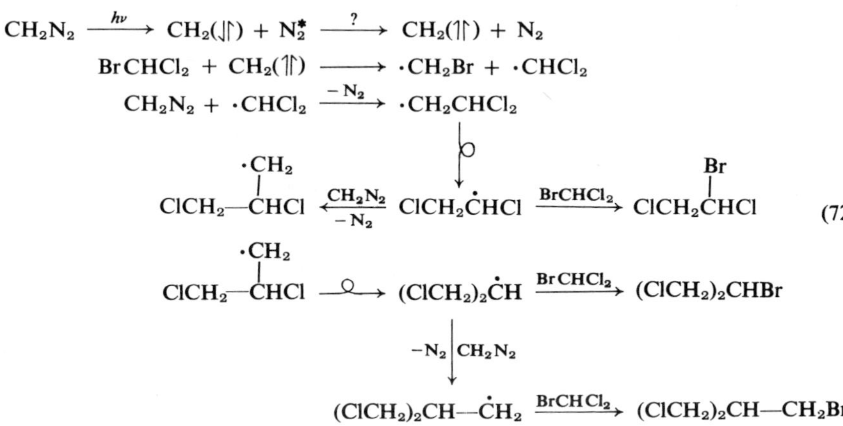

(72)

Other halides were similarly treated. The results are given in Table 4.

The spread of products indicates that the rates of vicinal chlorine migration and chain transfer with the reactant bromo compound are similar. Nonetheless, the rearrangement rate constant must be greater than that for chain transfer because the diazomethane concentration is considerably smaller than that of the bromo compound, which was the solvent.

Such trapping of individual radicals, together with the previously noted[114] high quantum yield in the reaction of diazomethane and carbon tetrachloride and the action of inhibitors,[113] strongly supports the chain sequence first proposed.[113] A recent restudy[117] of these β-chloro radicals produced in the alternative fashion shown in (73) confirmed their rearrangement.

Some C—Cl abstraction was apparent in this study also, because CH_3-$CCl_2—CCl_2CH_3$ and $CH_2{=}CCl_2$ accompanied the dimers. The latter product

TABLE 4

Reaction products from diazomethane and polyhalides

Halide	Products (% yields based upon diazomethane)		
$BrCHCl_2$	$BrCHClCH_2Cl$	$BrCH(CH_2Cl)_2$	$BrCH_2CH(CH_2Cl)_2$
	7	13	15
$BrCl_2CCH_3$	$ClCH_2CBrClCH_3$	$CH_3CBr(CH_2Cl)_2$	$CH_3C(CH_2Cl)_2CH_2Br$
	10	20	13
$BrCl_2CCH_2$-Cl	$BrClC(CH_2Cl)_2$	$BrC(CH_2Cl)_3$	$BrCH_2C(CH_2Cl)_3$
	7	13	18
$BrCCl_3$	$BrCl_2CCH_2Cl$	$BrClC(CH_2Cl)_2$	$BrC(CH_2Cl)_3$
5–15°	1	16	15
40–70°	13	26	18
	$BrCH_2C(CH_2Cl)_3$	—	—
5–15°	14[a]		
40–70°	0		

[a] Sole product with high concentration of diazomethane.

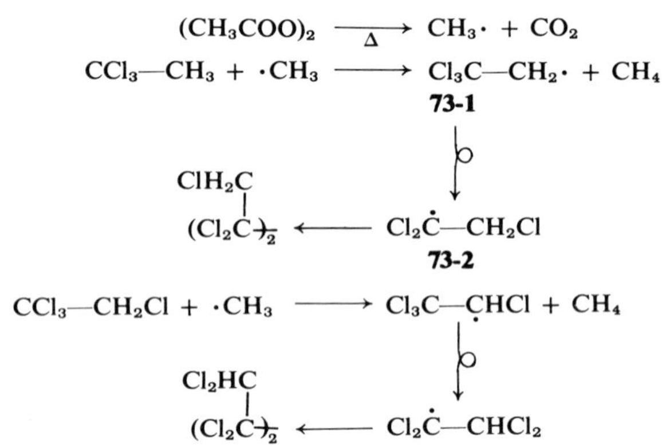

(73)

probably represents a competitive β elimination of a chlorine atom from **73-2** (or less probably, **73-1**). Such behavior is more common with β-bromo radicals (see p. 365), but it was suggested for β-chloro radicals as well, especially as the temperature increased.[116]

Similar vicinal chlorine migrations occur in the reaction of Group IV metalloid compounds with diazomethane (74).[118] The process seems to be

$$PhSnCl_3 \xrightarrow[(Cu)]{CH_2N_2} PhSn(CH_2Cl)_3 \quad (74)$$

nonradical in nature, however, and it may involve pentacoordinated metalloid species.

The reaction of diazo esters with polychloromethanes does not proceed analogously to that of diazomethane. Rather, a single vicinal shift of chlorine takes place (75).[119] The quantum yield (0.8–0.9) of the process, the lack

$$Cl_3C\text{—}Cl + N_2CH\text{—}COOR \xrightarrow{h\nu} Cl_2C\text{—}CHCl\text{—}COOR \qquad (75)$$
$$R = \text{—}CH_3, \text{—}CH_2CH_3$$

of inhibition, and the fact that dibromodichloromethane undergoes reaction 76 all indicate a distinctively different mechanism for this reaction vis-à-vis that using diazomethane.[120] A reaction of it analogous to (76) does not occur.

$$Br_2CCl_2 + N_2CH\text{—}COOMe \xrightarrow{h\nu} Br_2CClCHCl\text{—}COOMe \qquad (76)$$

A mild controversy centers on the reaction. Originally explained[119] as a radical chain process similar to (69), the low quantum yield and lack of inhibition of the reaction occasioned a change in viewpoint to a carbene insertion mechanism (77).[120] One recent study[121] concurred in the carbene

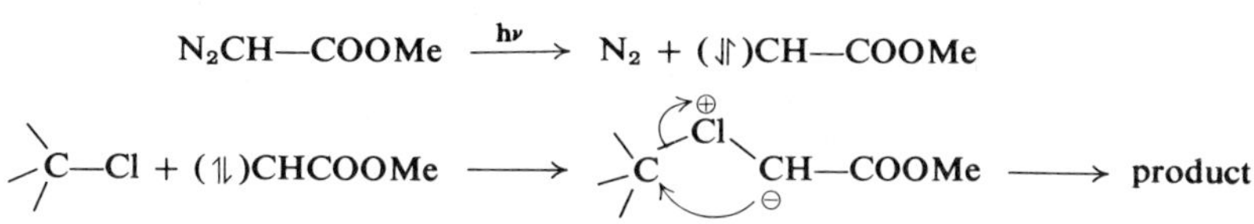
(77)

mechanism, largely because a bona fide radical process initiated by benzoyl peroxide gave oligomeric products from carbomethoxycarbene, some of which did not involve the polyhalomethane at all. Another recent study,[122] however, resurrected the original radical chain pathway. The reaction in (75) with the methyl ester exhibited chemically induced dynamic nuclear polarization (CIDNP) in the ester product. The slow decay of the NMR emission was taken as evidence for a chain process involving radicals. While the various studies need further development, it is probable that a radical chain process is present to some extent, but that the carbene pathway is concomitant and dominant in the reaction. Actually, this was claimed some time ago.[120]

The vicinal shift of bromine seems to be poorly competitive with its elimination. Thus, the β-bromo radical in sequence 78 loses a bromine atom rather than rearrange.[114]

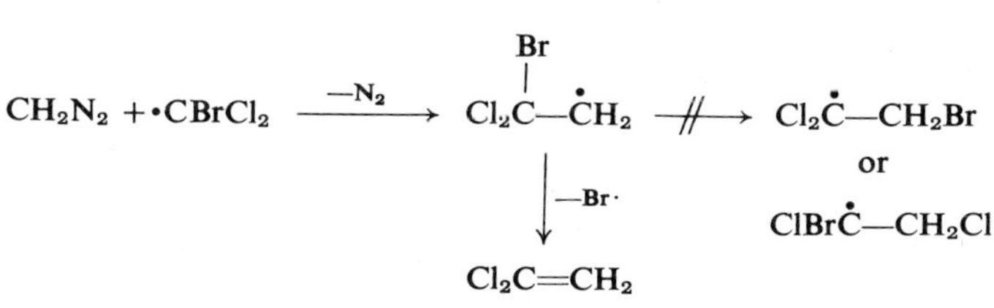
(78)

Analogous behavior occurs with iodoform. These reactions are more facile than that of diazomethane and carbon tetrachloride and, though unmeasured, their quantum yields are probably higher as well. Indeed, this preference for elimination characterizes the photochemical reactions of polybromo- and iodomethanes and diazomethane, as exemplified in (79), and distinguishes them from their counterparts using polychloromethanes.

$$CX_4 \xrightarrow[h\nu]{CH_2N_2} X_2C{=}CH_2 + CH_2X_2 \qquad (79)$$

$$X = Br, I$$

Parenthetically, the vicinal shift observed in reaction 76 renders unlikely any major pathway for that reaction that involves a β-bromo radical. The β elimination $Br{-}CH_2{-}CH_2\cdot \rightarrow Br\cdot + CH_2{=}CH_2$ has a $\Delta H = 5$ kcal/mole[123] and an activation barrier probably only slightly higher.[124] If 1,2 shifts of halogen have a comparable barrier to that in the neophyl rearrangement, then β elimination of bromine seems easier and its dominance over rearrangement is understandable.

Some studies did, nonetheless, claim vicinal bromine shifts, for example, reaction 80.[125]

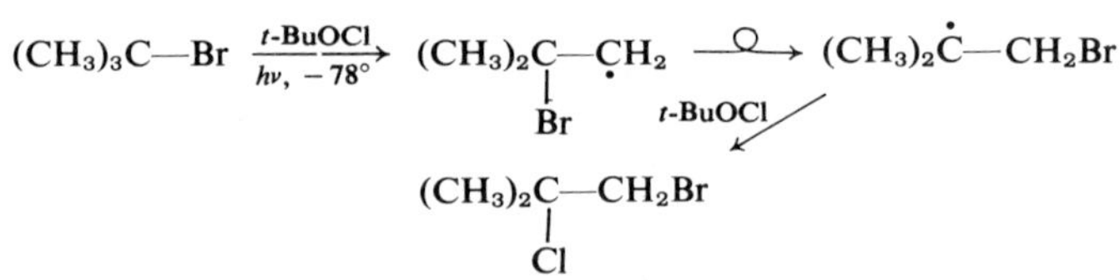

(80)

Other examples were also claimed.[126] Such reactions are, however, more likely elimination–readdition sequences. For example, bromine atoms are trappable in (80) using allene (81).[127,128]

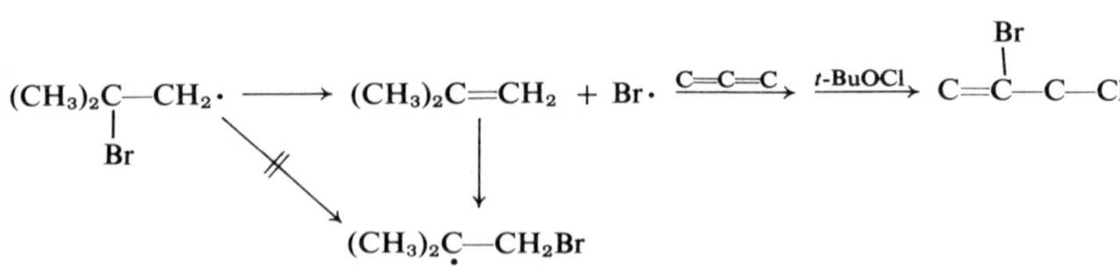

(81)

In this case β elimination was calculated to be six times faster than chain transfer. A more recent study[129] added further evidence for an elimination–readdition mechanism and against vicinal bromine shifts and the bridged bromo radicals[130] associated with such shifts. The product composition versus conversion in reaction 82 was in better keeping with elimination–readdition.

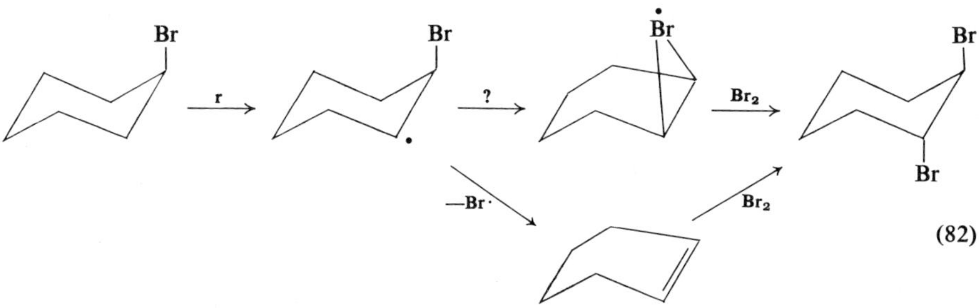

(82)

A radical rearrangement of silicon to sulfur is known (equation 83).[131] The rearrangement (and a similar one mentioned with it) derives from the strong Si—S bond formed at the expense of the weaker Si—Si bond.

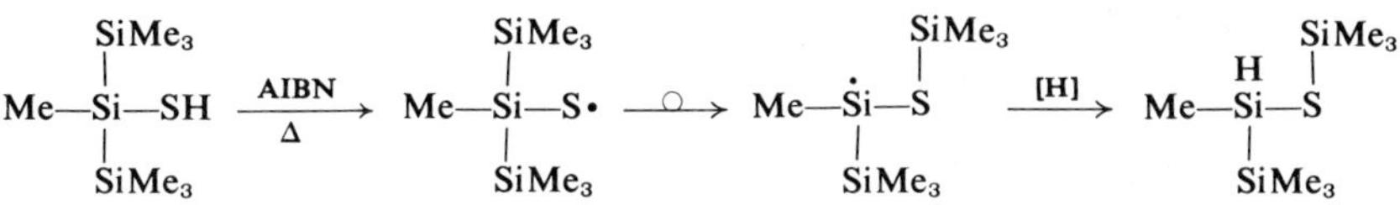

(83)

6. 1,3, 1,4, and 1,5 Shifts of Aryl Groups

No 1,3 phenyl shift was observed in the homolytic reactions noted in (84).[36]

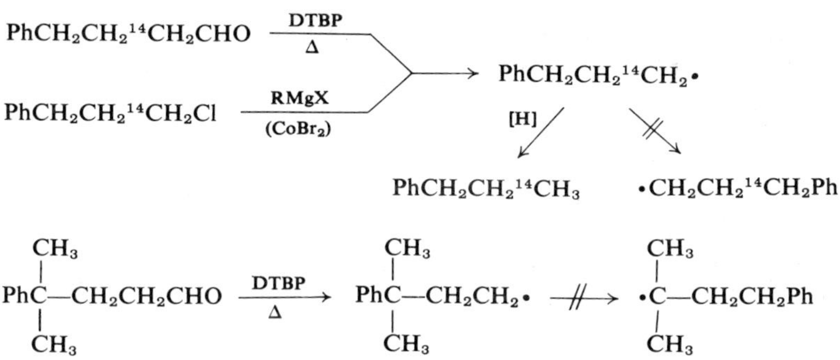

(84)

The limited rearrangement observed in the β-phenylethyl radical[60] (2–5%, see p. 349) via an Ar_1-3 transition state augurs ill for any rearrangement in the γ-phenylpropyl radical. Rearrangement in this instance would necessitate an Ar_1-4 transition state. Such a strained transition state is unfavorable, even in cationic rearrangements where the driving forces are greater.

1,4 Aryl shifts are well known. Both an Ar_1-5 and an Ar_2-6 mechanistic pathway have been established (the latter leads to cyclization) for the reaction shown in (85).[132]

The relatively strainless Ar_1-5 and Ar_2-6 transition states (or intermediates) make such rearrangements possible, when coupled of course with the

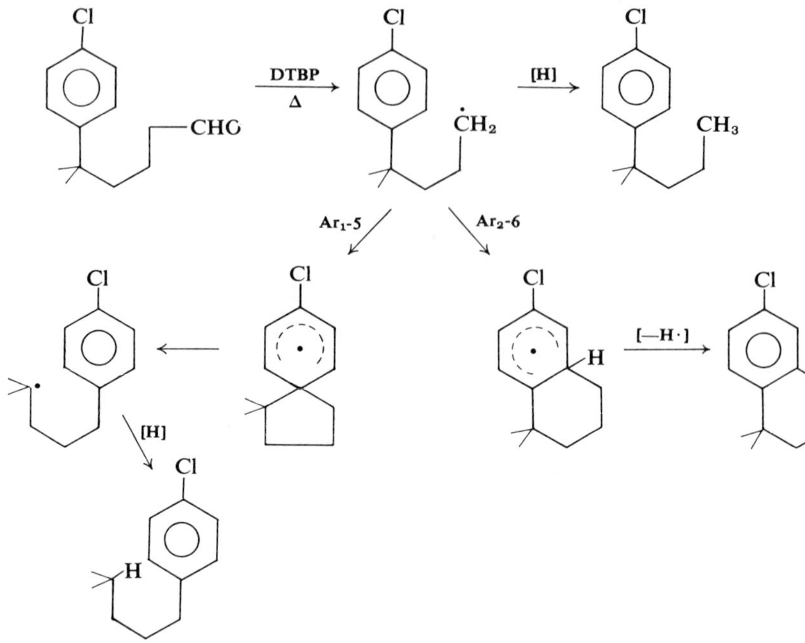

(85)

stability possible through electron delocalization into the aromatic ring. Such rearrangements occur readily to heteroatom termini as well (equations 86[133] and 87[134]).

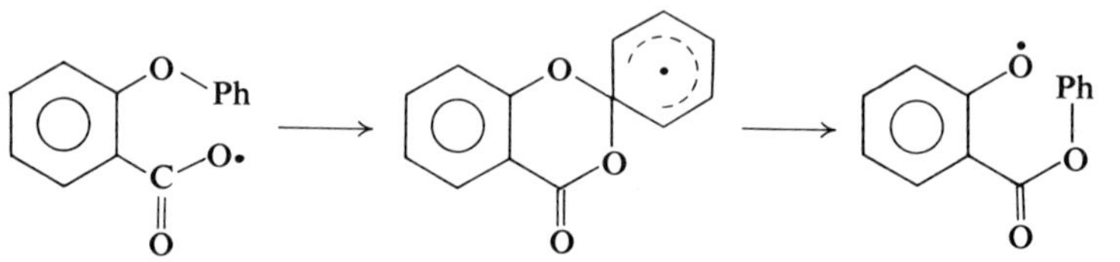

(86)

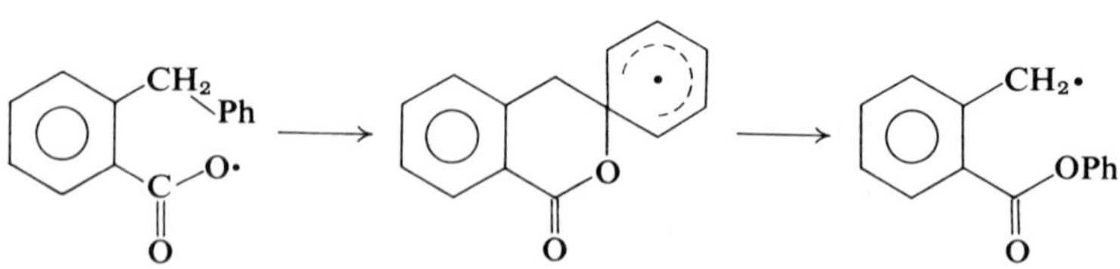

(87)

Formed at 100° from aldehydes and di-*t*-amyl peroxide, acyl radicals have been observed to rearrange in like fashion (equation 88).[135] This lower temperature (most decarbonylations are performed above 135° with DTBP)

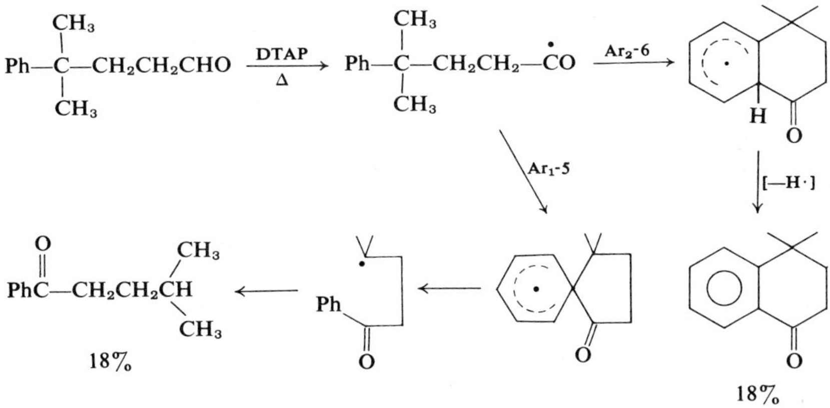

(88)

increases the lifetime of the acyl radicals and allows these processes. Actually, even Ar_2-5 cyclization is known (89).[61b,136]

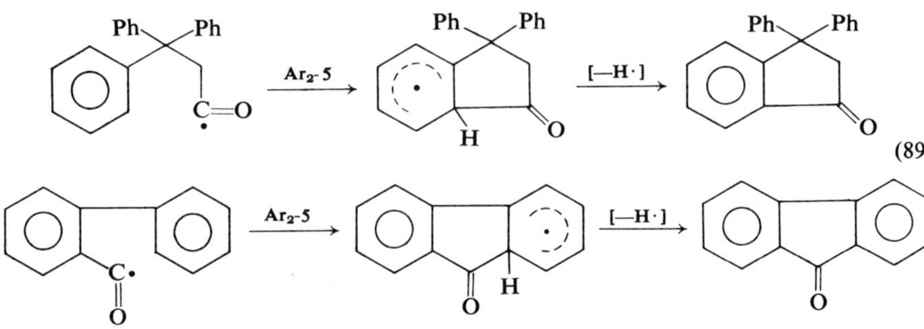

(89)

Such cyclizations to indanones and tetralones are not really germane to the 1,x shift category of rearrangements, but their formation under decarbonylation conditions along with 1,x-shifted products makes their brief mention excusable here. They are discussed in more detail later (p. 433 ff.).

As was noted in the neophyl rearrangement itself, a question could be asked if the 1,4 (or other 1,x) shift is simultaneous with radical formation to form a spiro radical precursor to all products, or do discrete unrearranged and rearranged species exist, connected via a spiro radical only as a transition state. Available evidence is meager in these cases, but copper(II) apparently oxidizes 4-phenyl-1-butyl and 5-phenyl-1-pentyl radicals over 10^3 liters/mole faster than these rearrange,[134] so the "discrete species" view seems preferable. Nevertheless, a longer lifetime for an Ar_1-5 spiro radical compared to an Ar_1-3 analog is expected and future work may implicate the former as an *intermediate* in some process. Indeed, an example may be known in an organosilyl case (see p. 375). That a spiro radical *can* serve as the precursor to both Ar_1-5 and Ar_2-6 products has, in fact, been recently reported (89a).[134a]

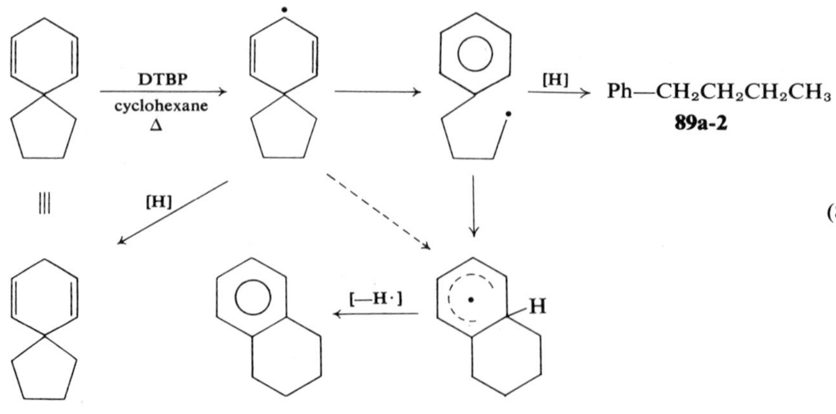

(89a)

Time (hr)	Temperature (°C)	% **89a-1**	% **89a-2**	% **89a-3**
2	200	10	35	26.5
60	140	9.5	22	19

These results unequivocally demonstrate that the spiro radical from hydrocarbon **89a-1** has a definite lifetime and need not be *just* a transition state in these 1,4 shifts. Whether the spiro radical is formed directly in reactions like (85) seems doubtful, however.

A type of 1,4 and 1,5 phenyl shift has been detected in substituted cyclohexyl and cyclohexylcarbinyl radicals, as shown in (90).[137]

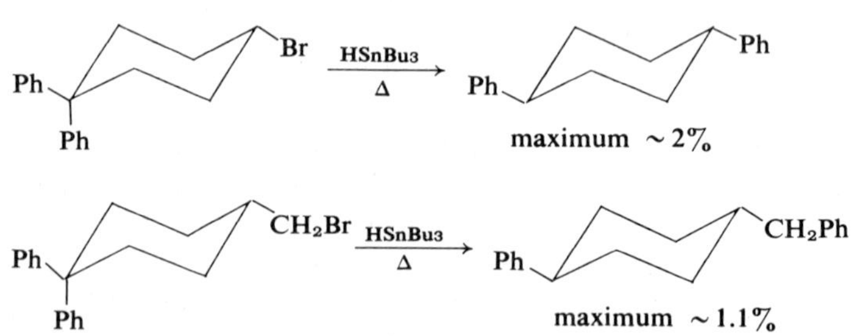

(90)

Analogous phenyl shifts were not found in three-, four-, or five-membered ring systems.[137,138] These "transannular neophyl" rearrangements are interesting because they are somewhat unexpected. The mild surprise is because analogous cationic examples are unknown in these systems. Normally, cationic rearrangements have considerably stronger driving forces and may show extensive rearrangement in cases where the corresponding radical process actually shows none (e.g., neopentyl). At first glance, the rearrangements in (90) occur in the face of some considerable strain (91).

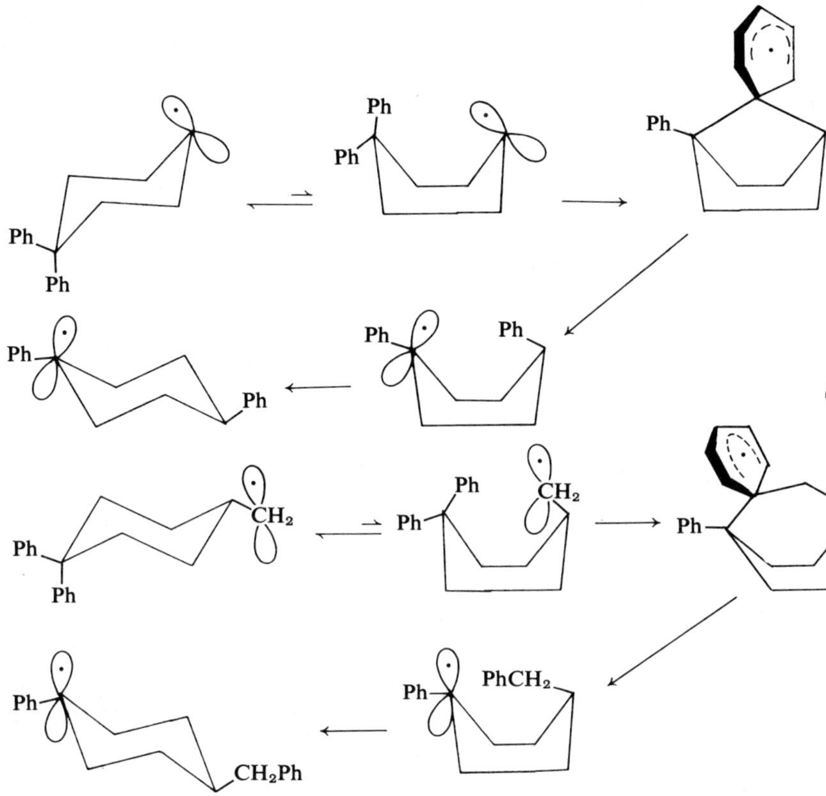

(91)

A strain energy difference of 11–12 kcal/mole has been estimated for the rearrangement. When compared with the neophyl rearrangement activation barrier of 10–11 kcal/mole, however, the difference is not too great and the rearrangements in (91) may be rationalized in this way. In the unsuccessful attempts at rearrangement in the smaller rings, the strain energy introduced with rearrangement is apparently simply too great, and other processes take precedence, namely, chain transfer with the tin hydride or ring opening (for the latter, see p. 398).

1,4 Aryl shifts are common in oxy radicals, such as those formed in Kolbe (92)[48] or Hunsdiecker (93)[139] reactions. In both these cases, the Ar_1-5 pathway for migration was established[48,140] by site retention in the product of an initial para substituent (an Ar_2-6 path effects a meta positioning of such a substituent). Likewise, decomposition of peroxides and peresters related to these systems has been studied and the 1,4 phenyl migration again detected.

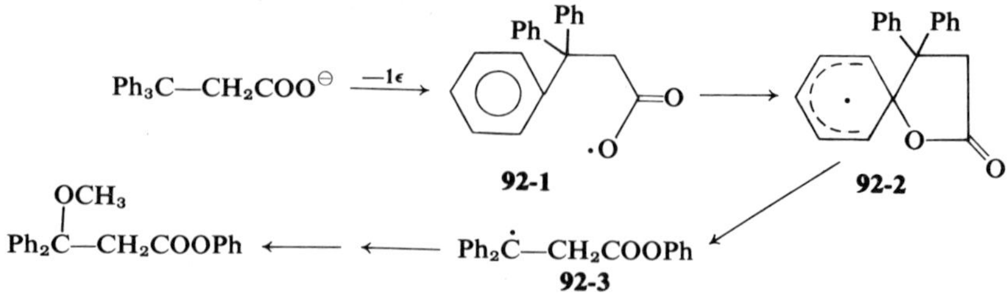

(92)

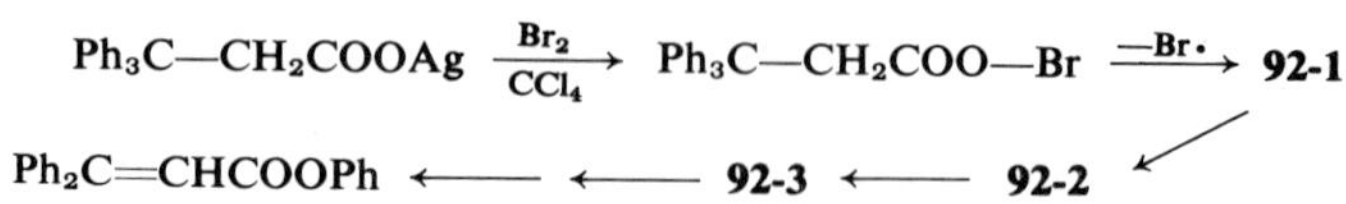

(93)

In the labeled peroxide shown in (94), no more than 60–65% of the rearranged ester could be derived from the free (carboxyl delocalized) radical.[47c]

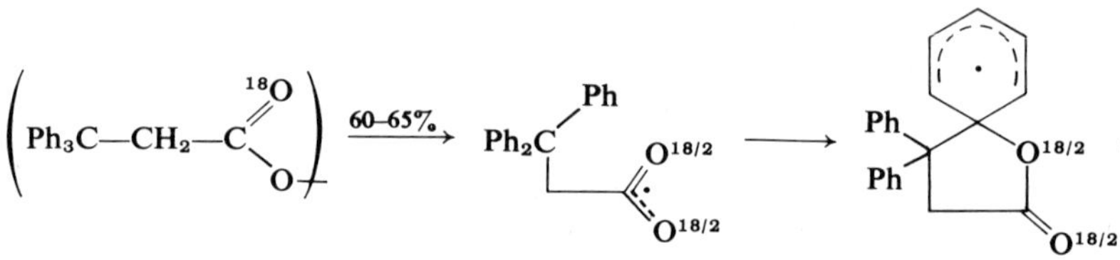

(94)

The remainder of the ester was formed by some path that involved phenyl–carbonyl oxygen interaction that kept the integrity of the labeled atom. Just what this interaction is remains unsolved, but it surely is associated with the close proximity of the aromatic ring to the carbonyl oxygen in the starting peroxide. The diphenyl peroxide analog, which has conformers without this proximity, failed to rearrange,[47a] although Kolbe electrolysis of the corresponding acid did afford 1,4 phenyl-shifted product.[141] The fact that peresters, as in (95), also rearrange[47b] is support for the view that the earlier-

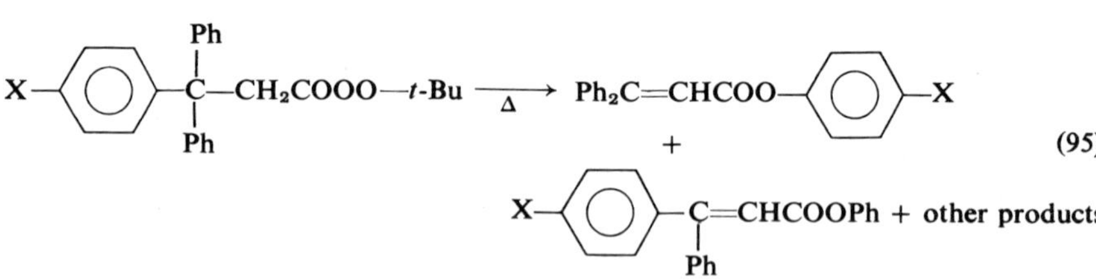

(95)

recorded Kolbe and Hunsdiecker examples are also bona fide radical rearrangements. Whereas perester decomposition is among the least suspect of radical processes, the mechanism of the Kolbe and Hunsdiecker reactions is far less certain, and alternative, nonradical pathways can be invented to explain the rearrangements observed in their case. The rates of decomposition of the peresters in (95) indicated simple O–O homolysis (no aryl participation). However, the data are qualitative and activation parameters are missing, so the possibility remains that such participation [see the ^{18}O study in (94)] might be present. A reaction constant ρ of -0.78 has been calculated using σ^+ values for the reaction in (95). The negative ρ is anticipated as acyloxy radicals are most certainly electron-seeking species. Site retention of X in (95) again supports an Ar_1-5 pathway.

The 1,4 aryl migration seen in these several studies is mimicked in others. A thorough study of the reaction in (96)[47e] indicated a nonradical pathway

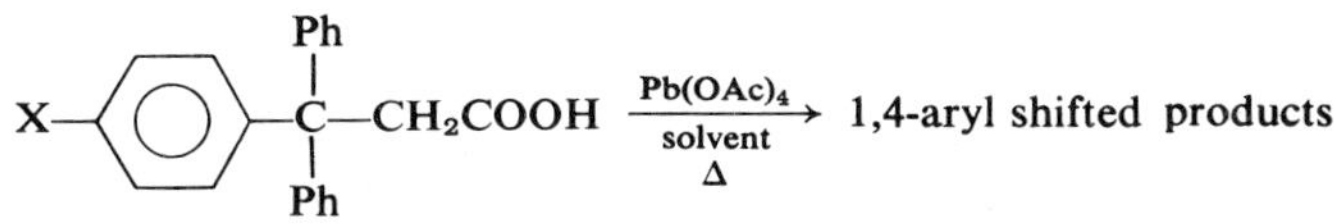

(96)

to the products. Though qualitatively similar to the perester decomposition in (95), the reaction in (96) gave a different distribution of carbon-to-oxygen rearrangement, decarboxylation, and cyclization; it had a different ρ value (-1.33 using σ^+) as well. Such results illustrate the need for careful, quantitative study of reactions that superficially resemble one another before mechanistic assessments are made.

The "anomalous" Hunsdiecker reaction shown in (93) was studied in substituted analogs and an Ar_1-5 pathway was thoroughly established.[140] The low yields of esters produced (less than 20%) and the complexity of the process precluded much mechanistic inference, but little difference was observed in migratory preferences in the aryl group. The values ranged from 10.3 for *o*-tolyl to 0.14 for *p*-fluorophenyl, relative to phenyl itself. Whereas no Ar_2-6 rearrangement pathway was detected in the β,β,β-triarylpropionic acid studies, the β,β-diphenyl analog gave a dihydrocoumarin product under Hunsdiecker reaction conditions,[142] clearly a result from an Ar_2-6 route, as shown in (97).[143]

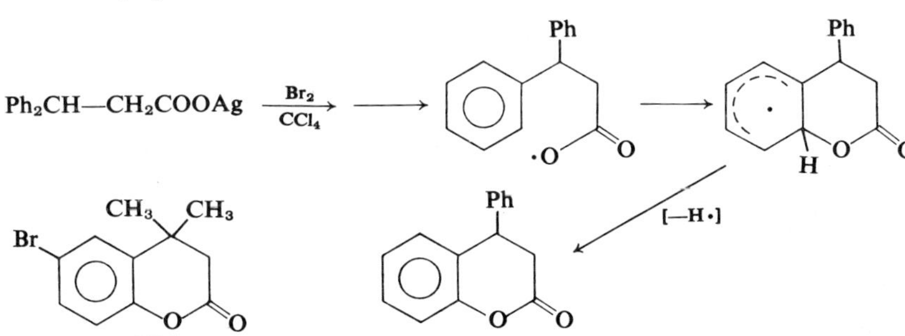

(97)

97-1

In fact, the Hunsdiecker reaction in the parent neophyl system also gave a dihydrocoumarin derivative, compound **97-1**.[49a] The difference in these last cases from the β,β,β-triphenyl case is probably the decreased interaction (of uncertain nature) between the aromatic ring and the carbonyl oxygen, as mentioned on p. 372. Such an interaction apparently favors the Ar_1-5 pathway and is best met in the triaryl compounds.

Although 1,5 phenyl shifts are recorded, as shown earlier in reactions 86 and 87,[144] an attempt to find such a path in the anomalous Hunsdiecker reaction (98) led to decarboxylative bromination (the "normal" Hunsdiecker reaction) instead.[145]

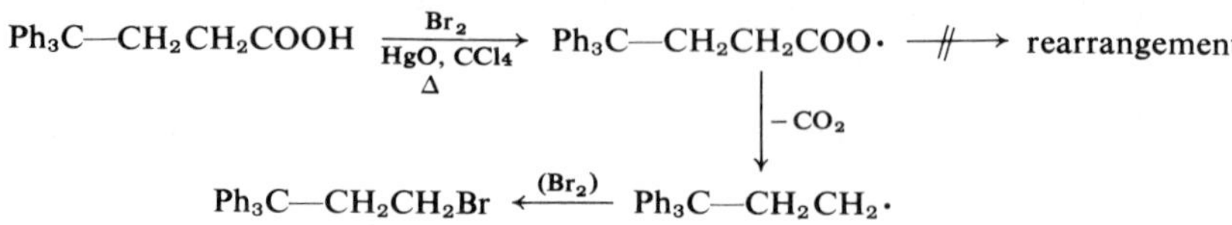

(98)

Because decarboxylation is some 16–18 kcal/mole more favorable for acyloxy rather than for aroyloxy radicals, the favorable instances of rearrangement in (86) and (87) possibly reflect the longer lifetime of the reactive species in those cases.

Both 1,4 and 1,5 phenyl shifts *from silicon to carbon* are known (99).[89]

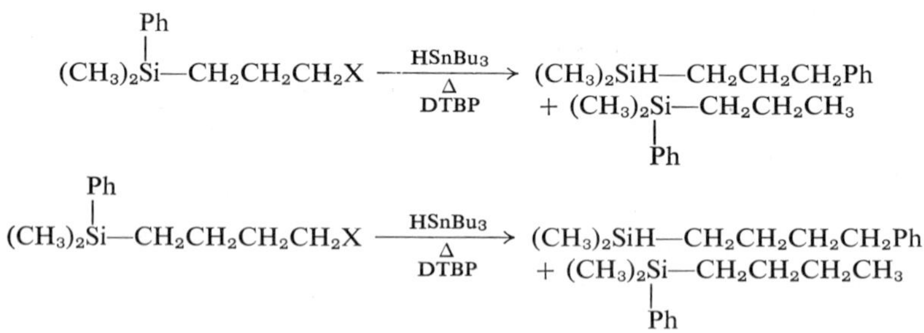

(99)

The extent of rearrangement depended on concentration, as shown in Table 5, and also on the halide.[146]

TABLE 5

Concentration and nature of halogen effects on Si → C phenyl rearrangement

	Percentage $(CH_3)_2SiH—(CH_2)_mPh$[a]			
$Ph(CH_3)_2Si—(CH_2)_mX$ *M*	1.0	0.1	0.01	0.001
$m = 3$, X = Cl	0	4.5	15	—
$m = 3$, X = Br	—	0	4	28
$m = 4$, X = Cl	0	2.5	21	—
$m = 4$, X = Br	—	0	1.5	16

[a] At 135° in benzene (no air) at molar ratios $RX:HSnBu_3:DTBP = 30:10:3$.

The rearrangement data fit the scheme in (100), illustrated for the Ar_1-5 pathway only.

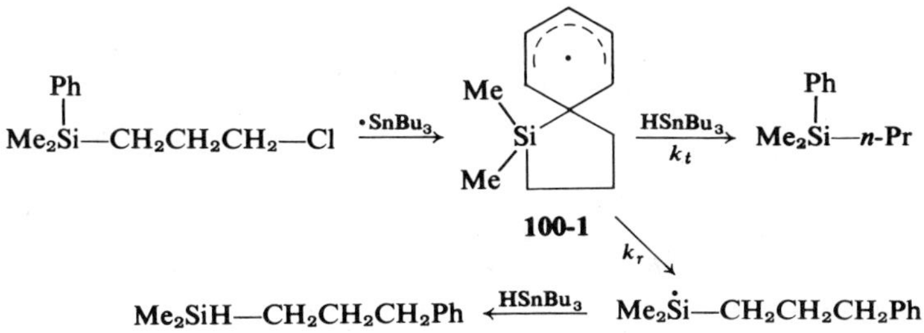

(100)

Such homolytic participation by phenyl to form **100-1** is unprecedented,[147] but it appears to be demanded by the fact that the corresponding bromide gives less rearrangement. In its case, the unrearranged radical is presumably formed, as shown in (101).

$$Me_2Si(Ph)—CH_2CH_2CH_2—Br \xrightarrow{\cdot SnBu_3} Me_2Si(Ph)—CH_2CH_2CH_2\cdot \qquad (101)$$

Because the two halides are postulated to form two different initial radicals, the rearrangement observed from each need not be the same. If radical **101-1** were formed from each halide, then the rearrangement percentages should have been the same because the fraction of rearrangement is k_r/k_t, which is proportional to $[HSn\text{–}Bu_3]^{-1}$, and radical **101-1** should have the same k_r and k_t regardless of its original source. As an additional point in favor of the participation in (100), the chloride was faster in the reduction than was $(CH_3)_3SiCH_2CH_2CH_2Cl$, whereas the bromide was not.[146,148] These rearrangements in γ- and δ-silyl carbon radicals are in obvious contrast to the lack of rearrangement in α-silyl carbon radicals (p. 354) and support the contention[88,90] that backbonding prevents the rearrangement in the latter.[149]

Radical migration of phenyl *from carbon to silicon* is also known, at least in certain cases. Though no vicinal phenyl shift of this type has been reported (equation 102),[150] recently the reverse of the rearrangements shown in (99) has been described (103).[151]

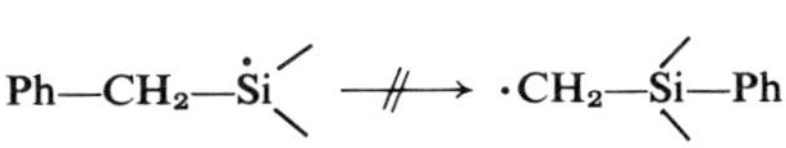

(102)

The retention of para X in product **103-1** is in agreement with this pathway. Also, as might be expected, the rearrangement to **103-1** is better when Y is CH_3 compared to H because the rearranged radical is secondary in the former. Cyclization to **103-2**, on the other hand, is favored by increased

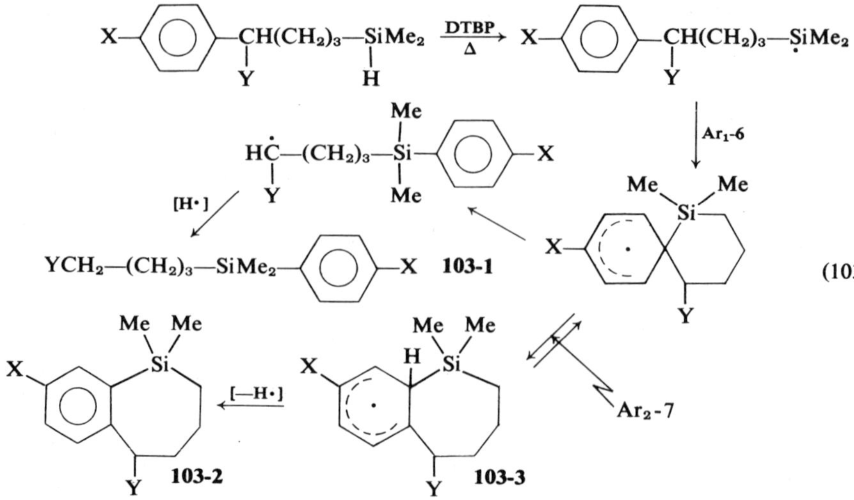

(103)

DTBP concentrations, presumably because the aromatization of the cyclized radical **103-3** is made easier via hydrogen transfer to DTBP-derived radicals. Cyclization, but no rearrangement, attends the lower homologs shown in (104).[151,152]

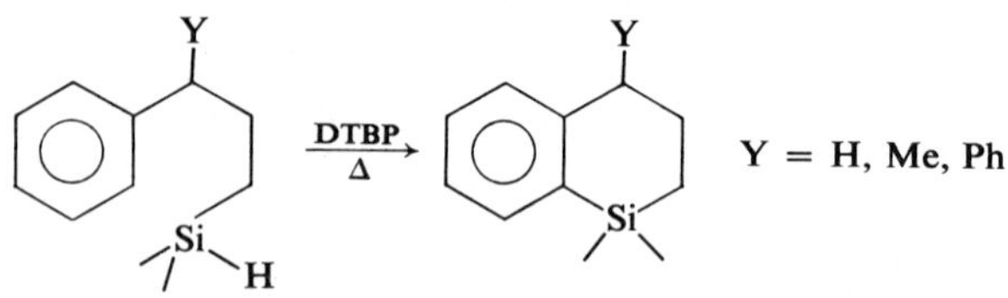

(104)

Homologs other than those of (103) and (104) gave neither cyclization nor rearrangement, so presumably only intermediates (or transition states) of the Ar_1-6 (for rearrangement) and Ar_2-6/7 (for cyclization) types are available for these reactions. A series of equilibria probably exists in these transformations, interconverting radicals as shown in (105).

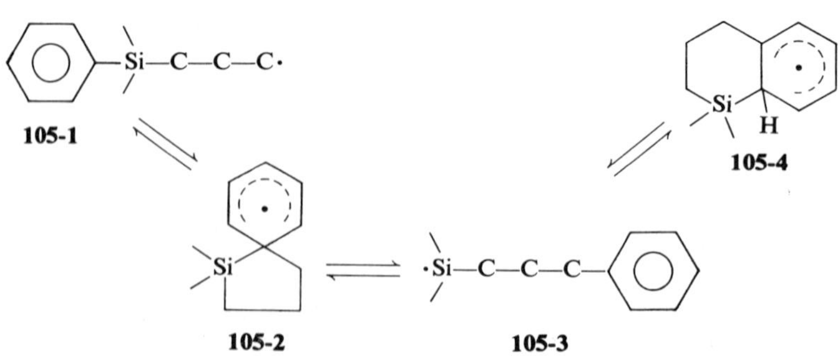

(105)

The conversion of **105-3** to **105-1** (undoubtedly via **105-2**) must be a bit unfavorable relative to closure to **105-4**, however, because no reversal of **105-3** to **105-1** has been reported under moderate reaction conditions. At 370°, however, the following uncatalyzed rearrangement occurred (106)[152]:

$$PhCH_2CH_2CH_2Si(CH_3)_2\text{—}H \xrightarrow[\Delta]{} CH_3CH_2CH_2Si(CH_3)_2\text{—}Ph \quad (106)$$

If this reaction involves radical **105-3**, then the change **105-3** to **105-1** may well be favored under such conditions, because **105-1** would be trapped out by the silane starting material whereas **105-4** would not. The fact that no cyclized product from **105-4** came about in the study using tri-*n*-butyltin hydride on the halides[89] similarly reflects the efficient trapping of both **105-1** and **105-3** by the hydride. The fact that no rearrangement occurred in (104) reflects the fast aromatization of *favored* **105-4** by DTBP-derived radicals, as mentioned.

7. *1,x Shifts of Heteroatoms*

A 1,3 shift of trimethylsilyl group has been claimed in the oxidation shown in (107).[153]

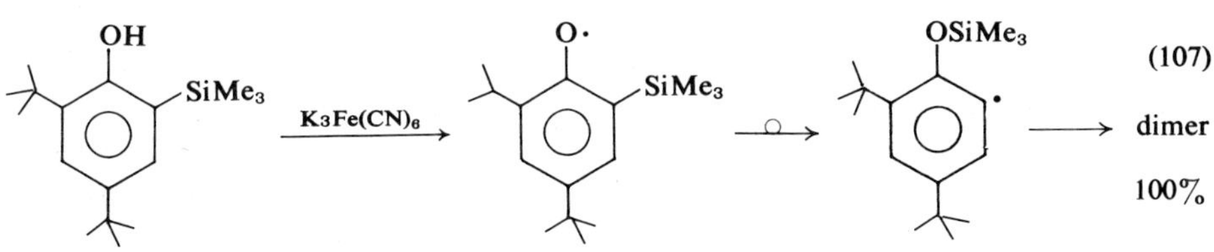

(107)

Similar rearrangements are known in thermal processes (250°). Although an intramolecular 1,3 shift seems strange a priori, the formation of the strong Si–O bond in the product radical (average bond energy 108 kcal/mole[154]) may militate for the process. It might be mentioned again at this juncture that a 1,3 shift of an acetoxy group did not materialize, though the vicinal shift is known (pp. 360–361). The attempt is given in (108).[105]

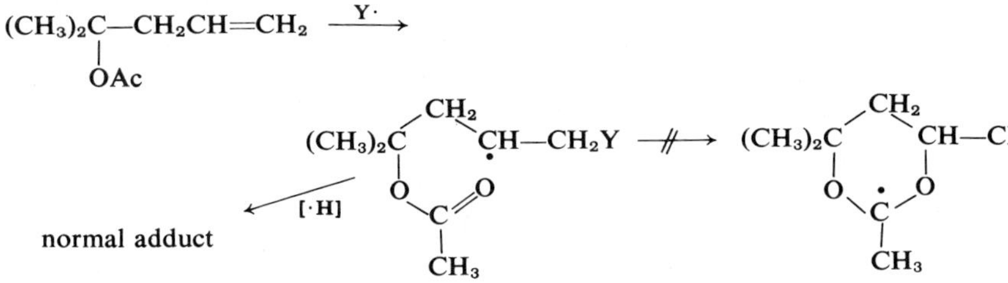

(108)

B. Hydrogen Abstractions

1. 1,2 Shifts

Early claims to the contrary, no bona fide intramolecular vicinal shift of hydrogen in a free radical has been substantiated, particularly in solution. But even in the gas phase, hydrogen atom transfers via transition states smaller than five-membered are not observed.[155] The status of the problem has been thoroughly documented.[156,157] Whether the failure to observe this rearrangement is explained by theory is also controversial. The triangular transition state for this vicinal migration would allow the least possible delocalization of the unpaired electron and therefore be the least favorable.[7,8] However, the π-electron model of Hückel used for such a view may be less wise for this hydrogen case where the electrons are σ.[10] If one adopts the viewpoint that intramolecular radical rearrangement is ideally radical displacement,[13] then the attacking center and the bond under attack should be colinear. *Any* 1,2 shift deviates strongly from this geometry. But with hydrogen especially, this colinearity factor seems important (vide infra). Lastly, the light mass of the system with hydrogen as the migrator lowers the probability of a properly oriented vibrational mode for rearrangement.[14]

2. 1,3 Shifts

Such hydrogen shifts have been claimed in solution, principally in radicals formed by thermal decomposition of peroxides.[158] Exactly analogous 1,3 hydride shifts have been invoked in cations.[159] In this case, the shift has been disputed[160] because isotopic labeling did not confirm the contention. So again the existence of such shifts in the radical case, while by no means disproved, remains a troubled matter. In recent studies, rearrangement was claimed in the reaction shown in 109.[161]

$$(CH_3CH_2CD_2{-}COO)_2 \xrightarrow[\Delta,\ -CO_2]{\text{solvent}} [CH_3CH_2\dot{C}D_2] \xrightarrow{\curvearrowright} \dot{C}H_2CH_2CHD_2 \qquad (109)$$

In carbon tetrachloride at 80° 1.3 ± 0.1% 1,3 hydrogen shift was apparently involved, as deduced from degradation of the *n*-propyl chloride formed. In the presence of the better chain transfer agent benzotrichloride at 150°, the 1,3 hydrogen shift was cut to 0.8 ± 0.1%. These results are in accord with expectations if the shift is considered to be a discrete process in competition with chain transfer. The 1,3 shift in (110),[162] which would be expected to be easier than that in (109), did not happen, however, and it is clear that 1,3

$$(CH_3)_2C(H){-}CH_2{-}\dot{C}H_2 \not\longrightarrow (CH_3)_2\dot{C}{-}CH_2{-}CH_2(H) \qquad (110)$$

hydrogen shifts have sensitive constraints with little definition of them available at present. A similar state of indecision exists for such shifts in the gas phase, though a definitive study has recently appeared.[155]

As part of an investigation of their chain inhibition of radical polymerization, aromatic nitro compounds were treated as shown in (111).[163]

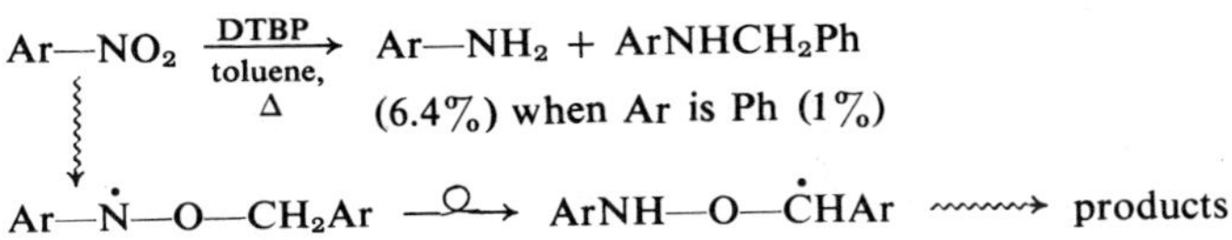

(111)

A 1,3 hydrogen shift was proposed in the reaction scheme. In view of the unsettled question of possible 1,3 shifts, and because no evidence was given for the proposed mechanism, this interpretation of the process is suspect.

An interesting 1,3 hydrogen shift has been advanced as part of the oxidation process undergone by the steroid in (112).[164]

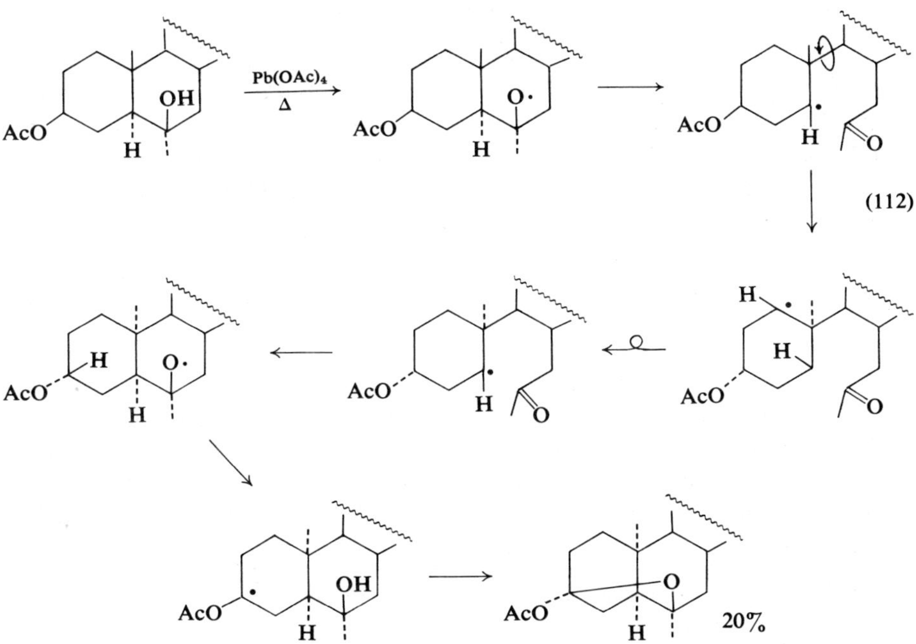

(112)

All sorts of radical processes are in this complex scheme and it might rank as one of the more complex on record! The shift of hydrogen was apparently substantiated in appropriately deuterated materials, although that work seems unpublished.

3. *1,4 Shifts*

An early indication that 1,4 hydrogen shifts were unfavorable arose in the cyclization shown in (113).[165] The fact that cyclization was preferred, even at

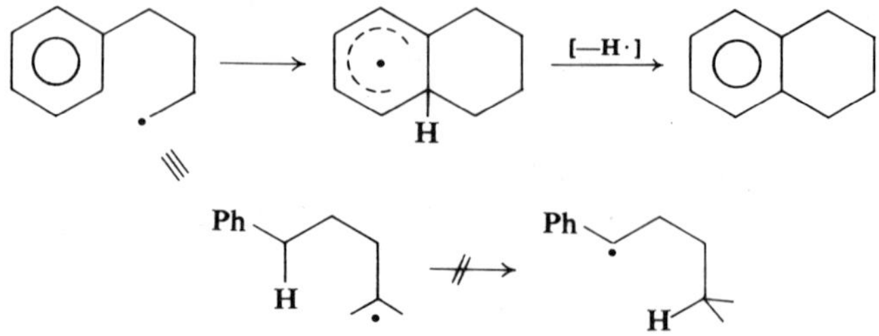

(113)

the expense of losing aromaticity in the transition state, over 1,4 hydrogen transfer to a benzylic radical, indicates that the barrier to the shift must be fairly high. One might expect the process to be better, therefore, at high temperatures in the gas phase. Indeed, under such conditions 1,4 hydrogen shifts apparently do occur.[155,166]

4. *1,x Shifts* ($x \geqslant 5$)

If indeed hydrogen transfer intramolecularly in a radical requires some approach to colinearity among the three atoms involved, then the occurrence of such transfer should be more frequent as one proceeds from 1,2 to 1,3 to 1,4, and so forth, shifts. With 1,5 hydrogen shifts (equation 114), a distorted six-membered ring transition state becomes possible which allows an approach to such colinearity.

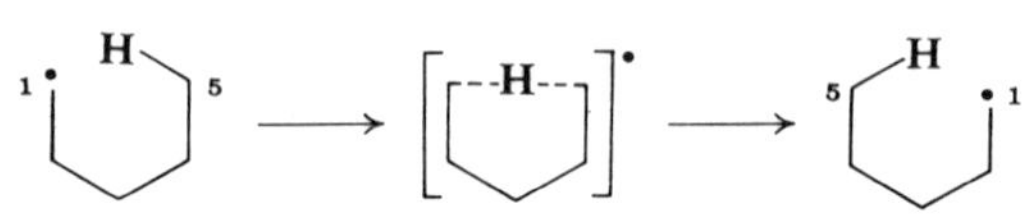

(114)

It is therefore somewhat expected that such shifts are very common. In fact, such 1,5 shifts dominate this area of organic radical rearrangements. Albeit that farther shifts should allow even more colinearity, such shifts are less common, although known. The probability that proper conformers would exist long enough to allow such farther shifts decreases with increasing chain length. Apparently this antirearrangement factor is crucial.

a. C to C 1,x Hydrogen Shifts. An early demonstration of this rearrangement is shown in (115).[167]

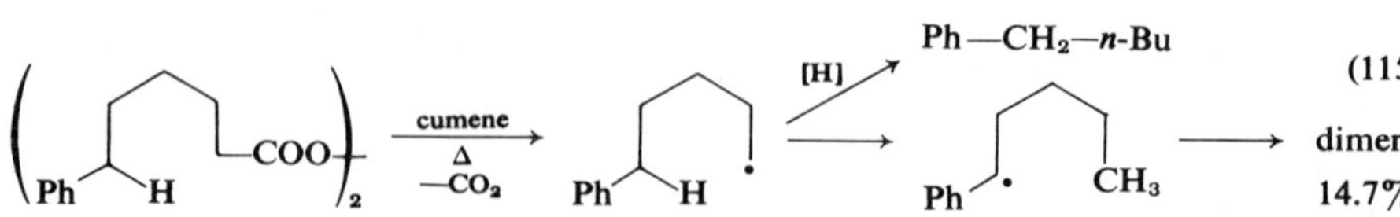

(115)

Neat reaction at 85° gave only 4.5% of the dimer. The increased yield of rearranged dimer in the presence of cumene indicated that the dimer did not arise from subsequent transformation of *n*-amylbenzene (116). Such reactions

$$Ph{-}CH_2{-}n\text{-}Bu \xrightarrow{r\cdot} Ph{-}\dot{C}H{-}n\text{-}Bu \longrightarrow \text{dimer} \qquad (116)$$

"outside the cage" should be minimized, not maximized, in the presence of the good hydrogen donor cumene. Perhaps the increase in rearrangement is a temperature effect (120° with cumene vs. 85° without). Other early examples of such shifts have been summarized.[168]

In telomerizations of ethylene, chain branching has been ascribed to 1,5 hydrogen shifts (117).[169]

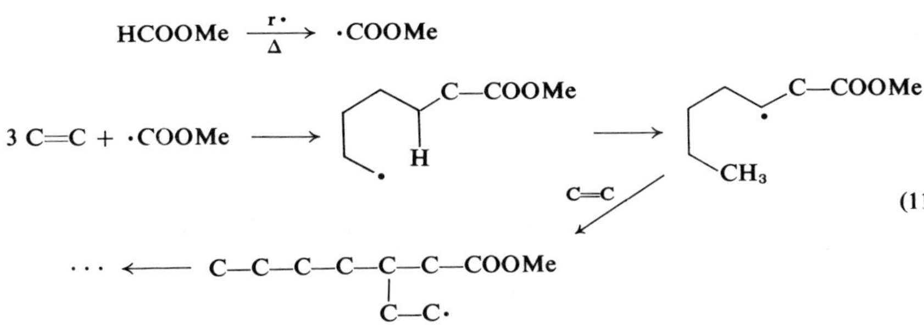

(117)

1,5 Hydrogen transfer accompanies certain radical additions to medium-sized cycloalkenes. For example, radical addition of carbon tetrachloride to cyclooctene proceeds as shown in (118).[170]

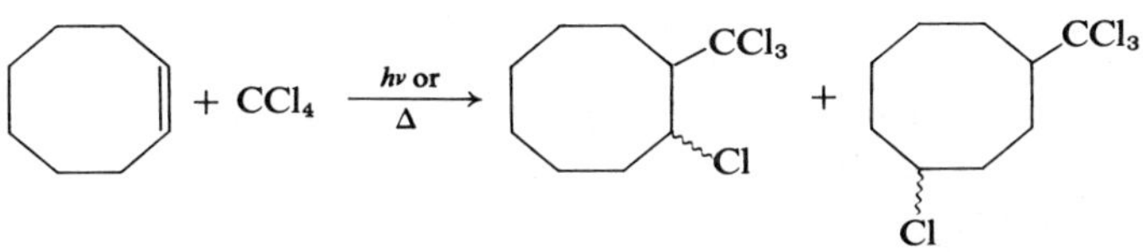

(118)

3.8–5% 58–68%
Cis–trans mixture

The better chain transfer agent bromotrichloromethane gave mostly vicinal addition in the photochemical reaction.[170,171] Likewise, trichloromethanesulfonyl chloride gave both vicinal (25%) and transannular (56%) products, in keeping with its intermediate chain transfer ability. The reaction pathway is summarized in (119).

The rate constants vary in the order $k_t'' > k_r > k_t' > k_t$. From assorted data, an energy banner of 18–19 kcal/mole was calculated for the transannu-

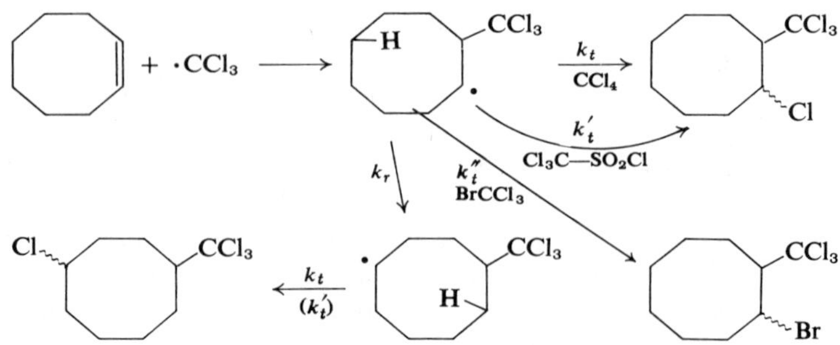

(119)

lar hydrogen shift. Other examples were described somewhat earlier (equation 120),[172] although the structure of the rearranged product was just surmised.

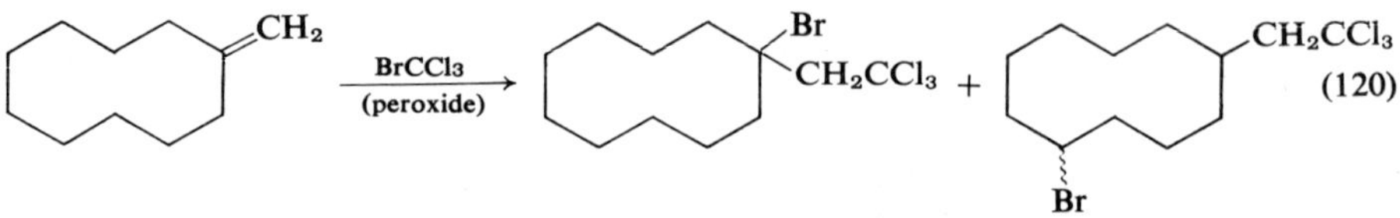

(120)

Finally, as part of a more complex scheme described later (p. 412), a rare intramolecular 1,4 hydrogen transfer (see p. 380) has been invoked for the radical addition of tetrahydrofuran to 1-octene (121).[173]

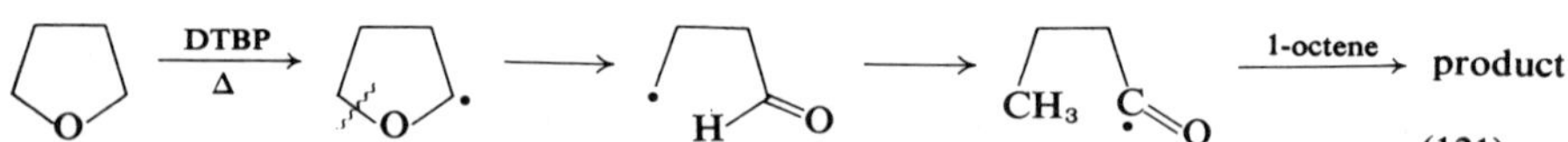

(121)

b. C to O 1,x Hydrogen Shifts. An extraordinary number of such hydrogen transfers are known. The area, particularly in steroids, has been reviewed.[174] Some gas-phase reactions exhibiting such shifts, as well as early solution studies involving this rearrangement, have also been surveyed.[175] The Barton reaction (equation 122)[176] is one of the well-known examples of this hydrogen transfer process.

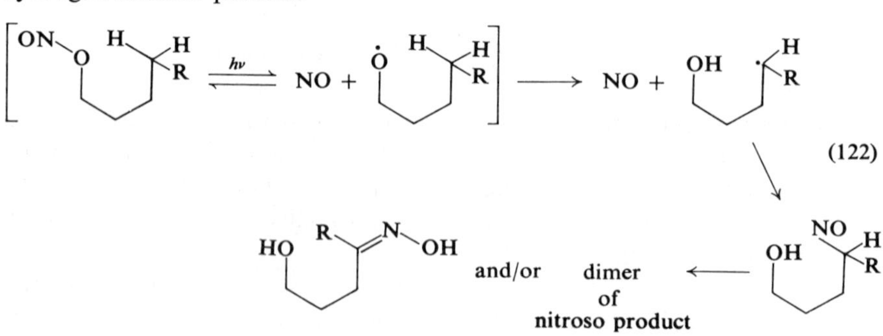

(122)

This photolytic cleavage–rearrangement process on nitrite esters has been used a great deal in steroid synthesis because it affords one of the few methods to effect substitution at angular methyl groups in such systems.[177] The primary homolysis is a reversible cage process,[178] but the products form from free ("out of cage") radicals.[179] In an extensive investigation of this process, several of its features became evident. Although the change shown in (123) is exothermic regardless of the value of n,[180] actually a sizable barrier

$$RCH_2(CH_2)_nCH_2O\cdot \longrightarrow R\dot{C}H(CH_2)_nCH_2OH \qquad (123)$$
$$\Delta H = -8.5 \text{ kcal/mole}$$

to transfer exists and the process quite rigidly demands a six-membered ring transition state, as shown in (124).[181]

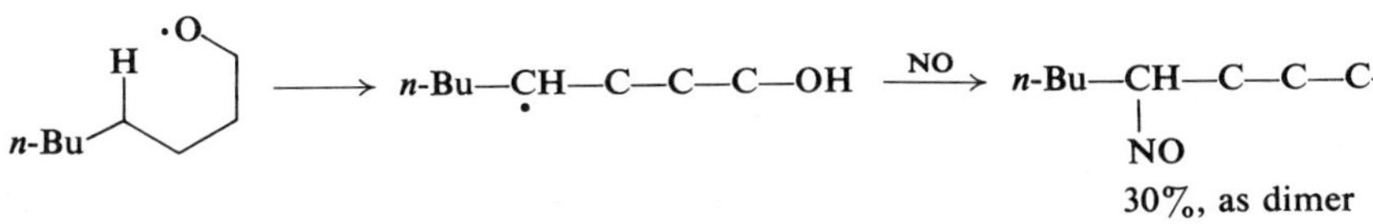

(124)

No such reaction occurs with smaller or larger possibilities. At times fragmentation occurs instead. Also, strongly bonded aromatic hydrogens resist transfer even though a six-membered ring transition state is available. These exceptions are shown in (125).

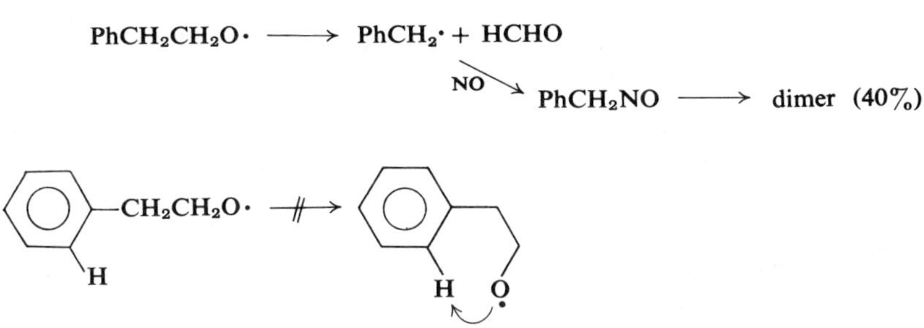

(125)

As might be anticipated, benzylic hydrogens transfer readily as do most secondary and tertiary hydrogens, provided that they are positioned appropriately for the six-membered ring transition state. If the hydrogen be primary, alternative reaction paths predominate.[182] Generally, nitroso dimers are the principal reaction products. Oxime formation seems, in fact, a special feature of the process in steroids.[178] A transannular example is also known (126).[183,184]

Another common class of 1,5 hydrogen transfer reactions is the decomposition of alkyl hypochlorites. A wide variety of products has been described since the early discovery of such decompositions,[185] but intramolecular

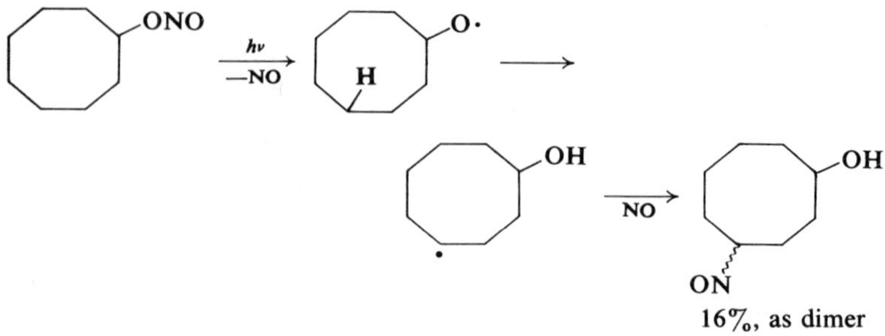

(126)

hydrogen transfer leading to chloro alcohols has been observed only recently.[186] These and other observations that crowded the scene rather quickly have been summarized.[187] A typical example of the process is given in (127).[186a,c]

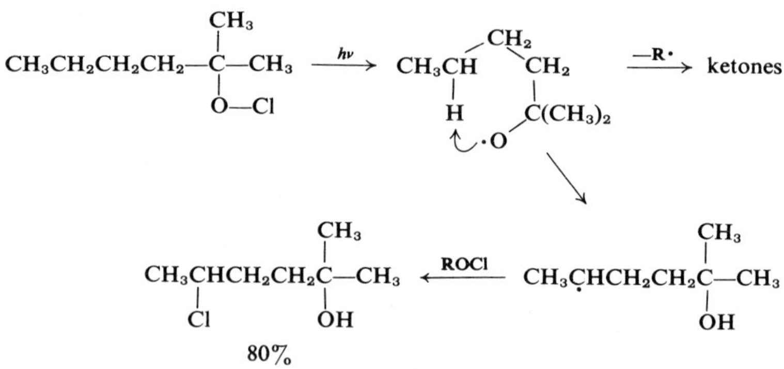

(127)

Because no significant change in chloro alcohol yield was observed with changes in alkyl hypochlorite concentration, the hydrogen transfer is intramolecular.[186b] Ease of 1,5 hydrogen transfer varies in the expected order: 3°-H > 2° − H > 1° − H, with relative values of 47:9:1 under certain conditions.[186c] Although most commonly found with tertiary alcohol hypochlorites, the process does occur in simpler cases (128).[188]

$$CH_3CH_2CH_2CH_2OCl \xrightarrow{h\nu} Cl—CH_2CH_2CH_2CH_2OH \quad (16\%) \qquad (128)$$

In most acyclic instances, 1,5 hydrogen transfer predominates. Only 1,6 hydrogen transfer seems at all competitive (129), and even when benzylic

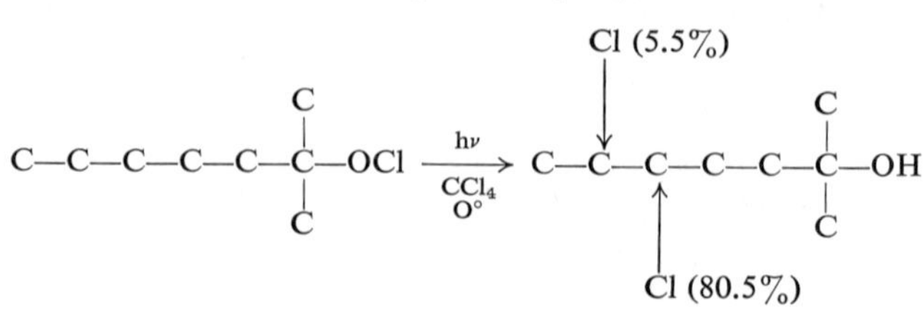

(129)

hydrogens are available for 1,6 transfer, 1,5 transfer is still preferred by 10:1.[186c]

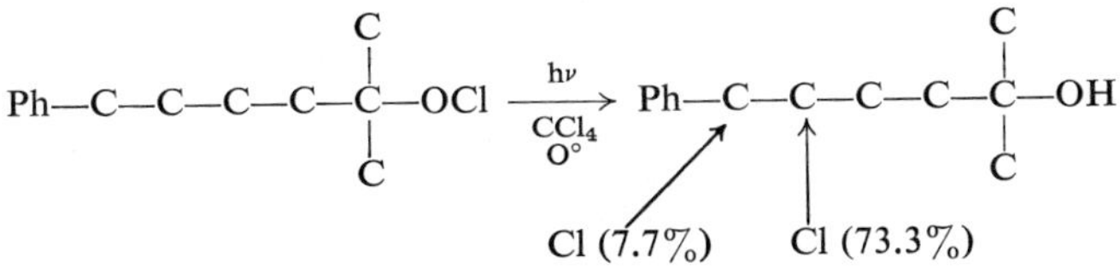

In a cyclic example, however, 1,6 hydrogen transfer was favored (equation 130).[189]

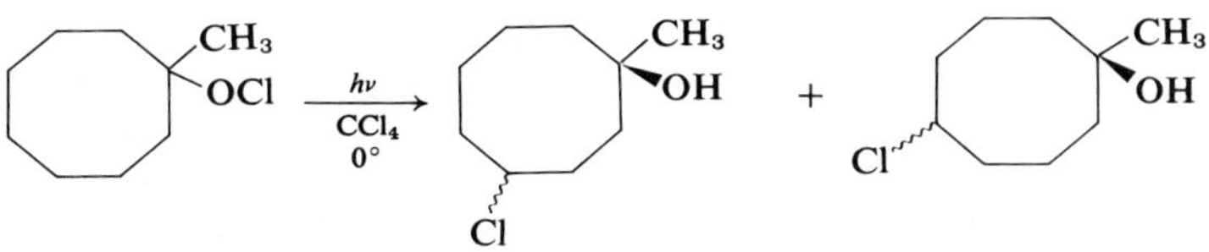

(130)

35% Trans/cis = 4/1

24% Trans/cis = 7/1

Because 1,5 hydrogen transfer from C-4 has a twofold statistical advantage, the hydrogens at C-5 actually have an increased reactivity of 1.4:1 (24:35/2). The stereochemistry of the chloro alcohol products may indicate configurational inversion upon chain transfer because of the bulk of the cyclic hypochlorite.[190]

A reaction of related interest is that of alkyl hypobromite decomposition. Treatment of alcohols with bromine and silver salts can give products analogous to those from hypochlorites. Though initially claimed to be an ionic process,[191] the examples shown in (131) and (132) indicated a radical stepwise mechanism.[192,193]

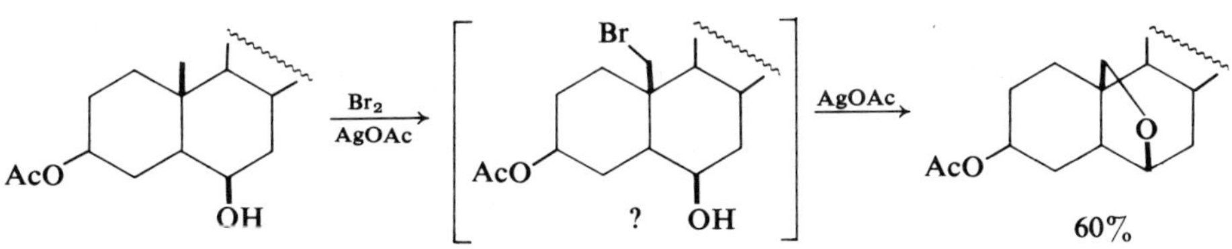

60%

(131)

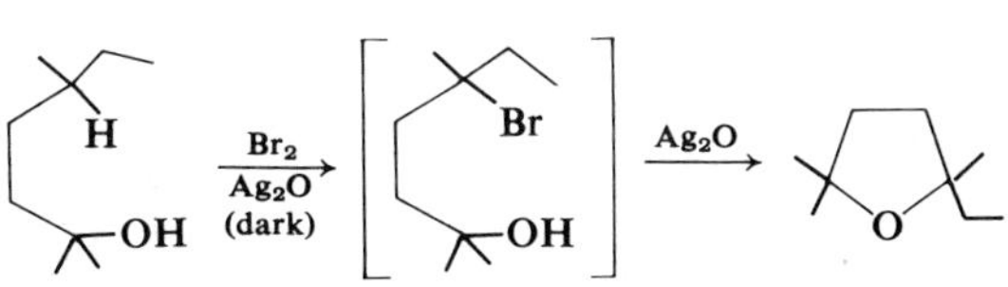

(132)

(+)S ? ~15% (±)

In (131), the hydrogen transfer from the angular methyl group mirrors known radical reactions (the Barton reaction). The production of racemic product in (132) is accommodated by an intervening radical intermediate, but not by a concerted (presumably ionic) process. There does remain, nonetheless, a possible path via a triplet $-O^+$ species.

Transannular or proximate hydrogen shifts also occur in oxidation of alcohols with lead tetraacetate (equation 133).

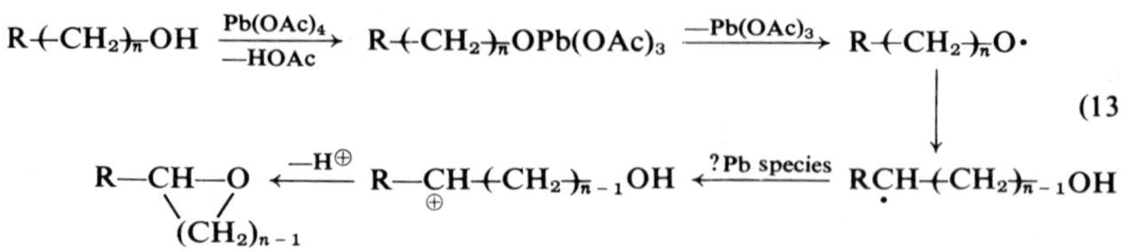

(133)

An analogous oxidation by silver ion may actually be involved in (131) and (132), bypassing the questioned bromo alcohol intermediates and leading directly to the cyclic ethers reported. A recent oxidation with lead tetraacetate indicated once again (see p. 385) that 1,6 hydrogen transfer can supersede a 1,5 shift (134).[194] As shown in (135), the proper geometry for 1,6 hydrogen transfer is missing in the epimeric alcohol, so 1,5 transfer again

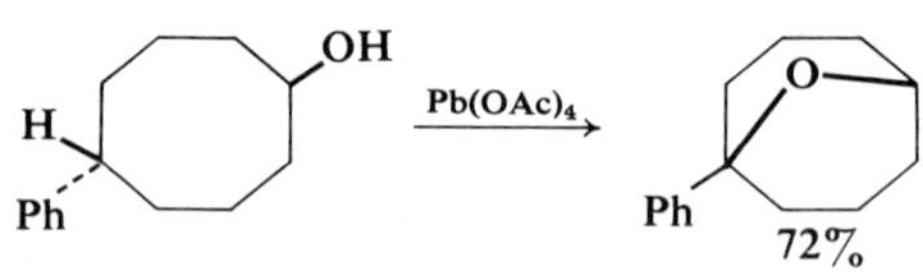

(134)

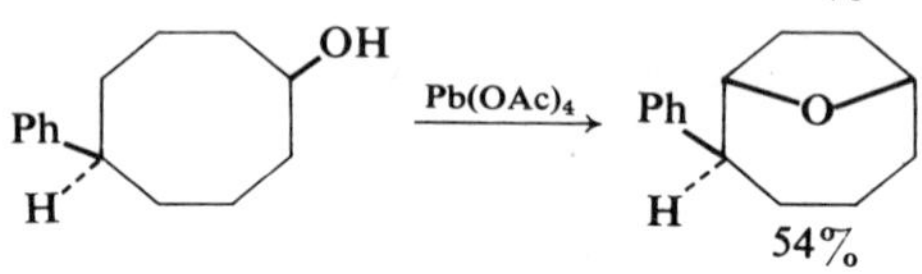

(135)

occurs. Another recent example shows what unbeatable geometry can do (136).[195]

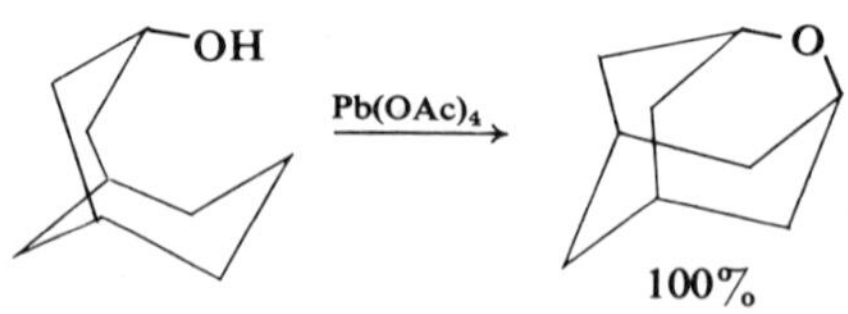

(136)

From the abundant data available, intramolecular hydrogen transfer appears easiest for oxy radicals in a chair conformation of a six-membered ring, as shown in (137).

If the distance between the oxygen radical site and the H—C— (the "O—C" distance) is 2.5–2.7 Å, intramolecular transfer of hydrogen predominates over

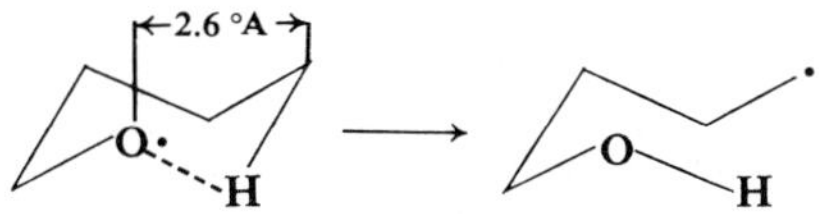

(137)

fragmentation or intermolecular reactions. As this distance approaches 2.8 Å the situation reverses.[174]

Intramolecular hydrogen transfer occurs upon irradiation of ketones possessing γ-hydrogen (equation 138).

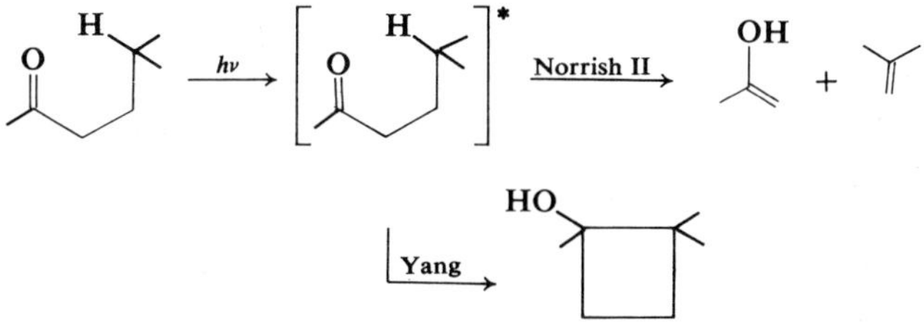

(138)

The usually predominant path (photoelimination or Norrish type II cleavage[196]) does not concern us here, but the other path does. It is the Yang rearrangement,[197] a reaction generally accepted as proceeding from the n,π^* excited state of the carbonyl compound. An early mechanism for the formation of cyclobutanols in this process was the concerted cycloaddition shown in (139).[198]

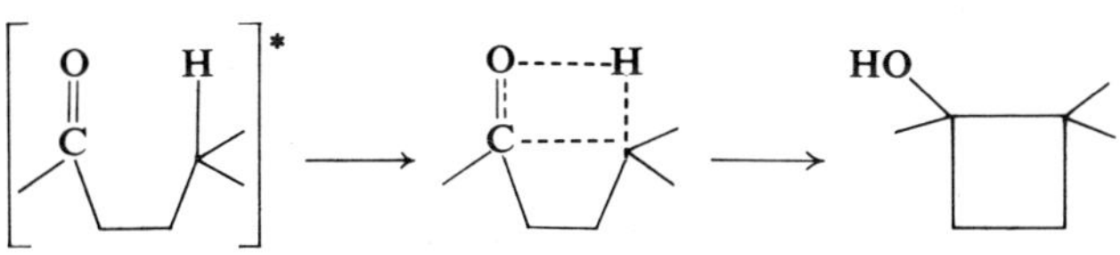

(139)

Such a concerted ${}_{\pi}2_s + {}_{\sigma}2_s$ process should be allowed photochemically.[11] However, a stepwise process was proposed to explain the reaction in (140).[199]

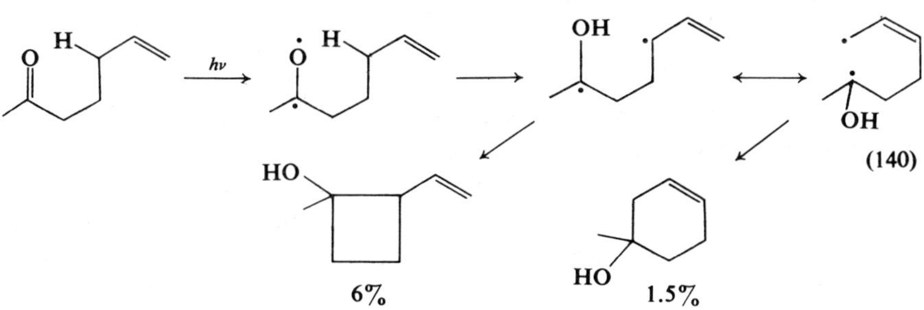

(140)

The similar conversion of α-diketones (141)[200a] is photosensitized by benzophenone and quenched by naphthalene. It thus should involve a triplet biradical intermediate—a point in favor of a stepwise pathway for the Yang rearrangement.

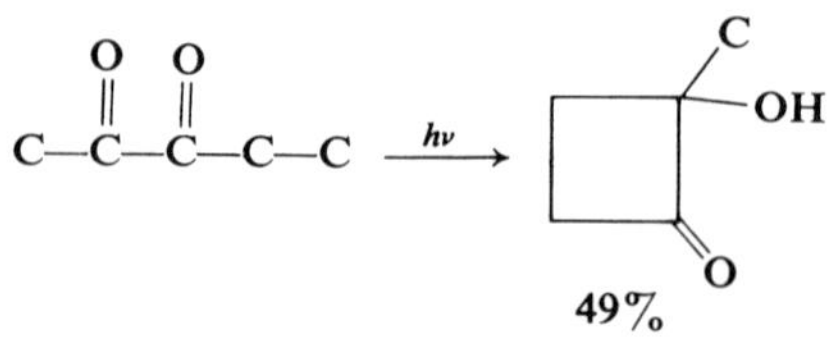

(141)

On the other hand, reaction of the optically active ketone shown in (142) gave a product with some activity retained.[201]

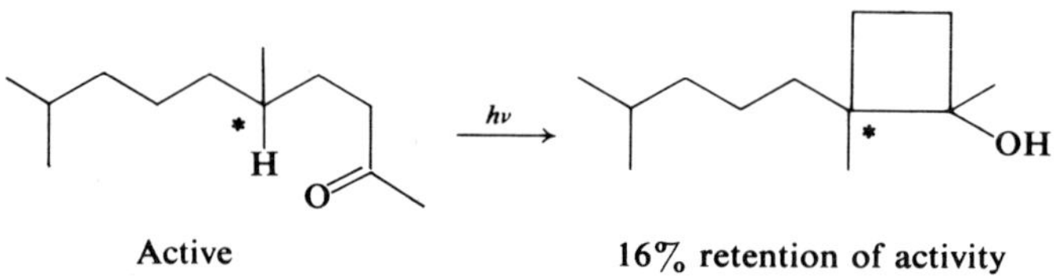

(142)

A concerted process should give 100% optically active product whereas a stepwise process might be expected to give racemic product. Perhaps both paths are followed. Nonetheless, a stepwise mechanism wherein ring closure of the biradical was comparable in rate to racemization at the chiral center would also allow the result in (142). Although it is still not conclusively established, recent investigators generally employ the stepwise mechanism in explaining their results, for example, reaction 143.[202]

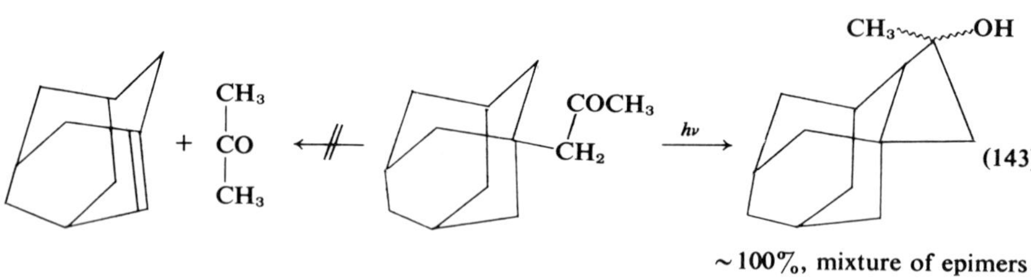

(143)

This and other cases of high yields of cyclobutanols[203] can be related to the intermediate radical's closing rather than cleaving to unstable (or highly energetic) olefins by photoelimination (reactions 144[204], and 145[200b])[205]. In cases where strain would be increased by cyclobutanol formation to an unlikely extent, an alternative rearrangement can occur, as shown for α,β-epoxy ketones (146).[206]

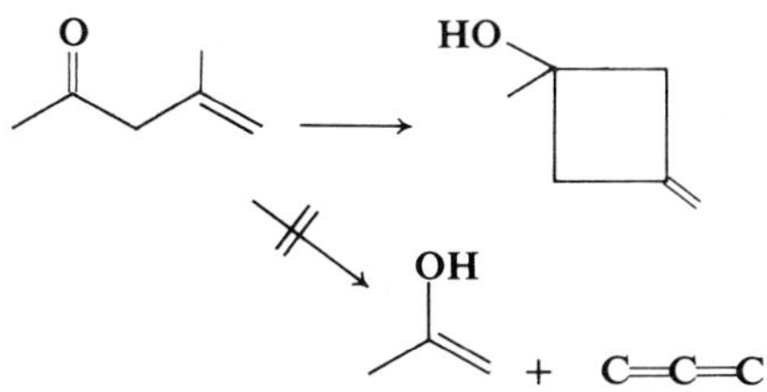

(144)

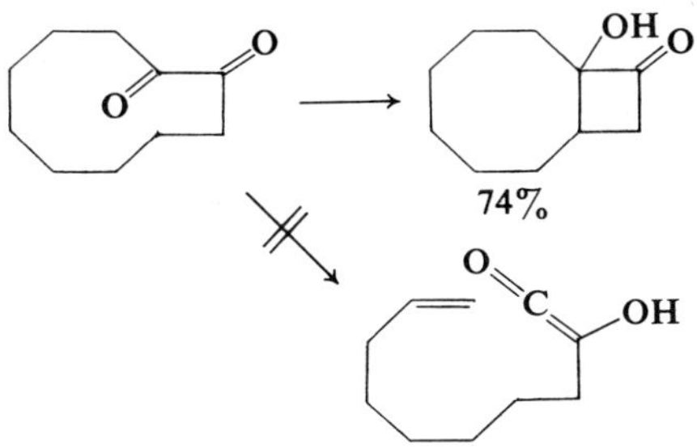

(145)

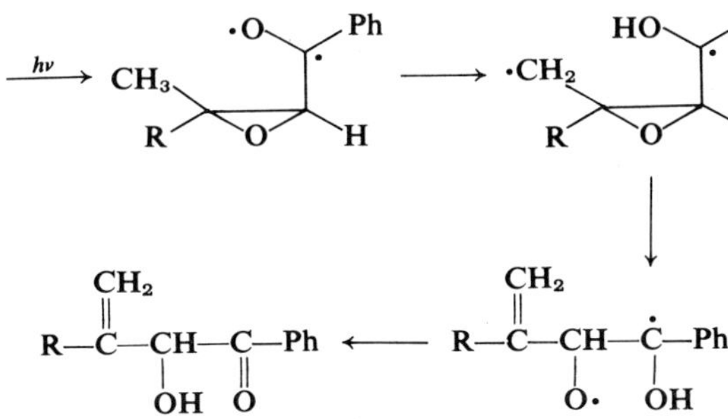

(146)

R = CH_3, 64%
R = *trans*-Ph, 25%

Proximate hydrogens other than γ can also be transferred (147).[207]

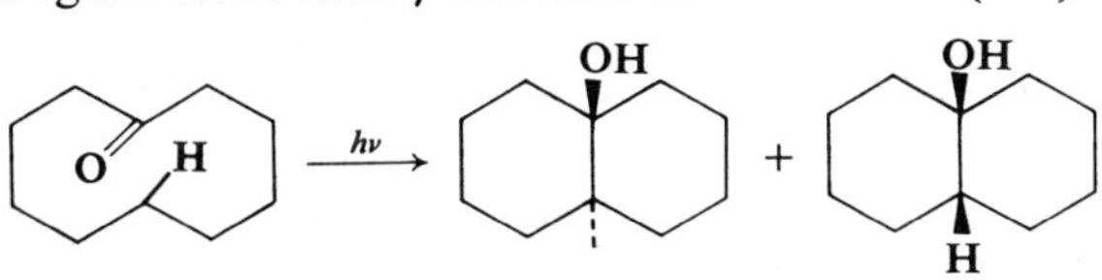

(147)

Notice that this decalol formation is quite different from the reaction observed with the α-diketone (145).

Recent interesting cases of preferred 1,6 hydrogen transfer are given in (148).[208]

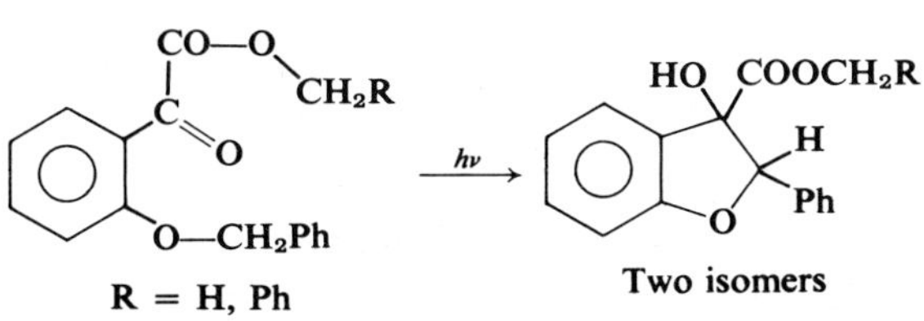

(148)

A spectacular 1,6 hydrogen "shift" has been recorded (149).[209]

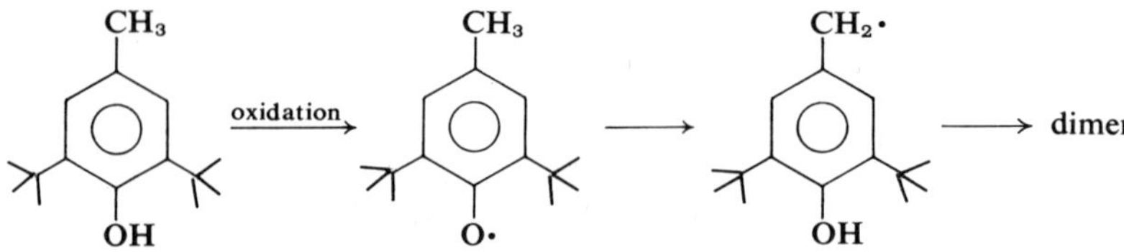

(149)

It seems clear that such a shift is really intermolecular benzylic hydrogen abstraction.

Lastly, many other examples of photochemical rearrangements of ketones are known, for example, reactions 150,[210] 151,[210] and 152.[211]

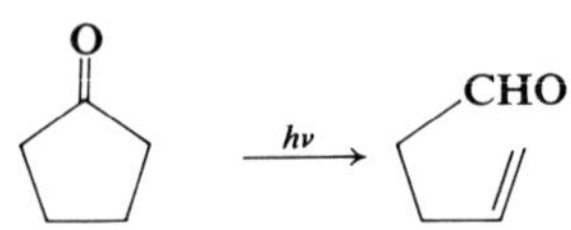

(150)

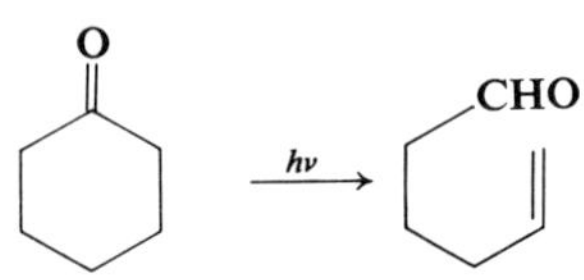

(151)

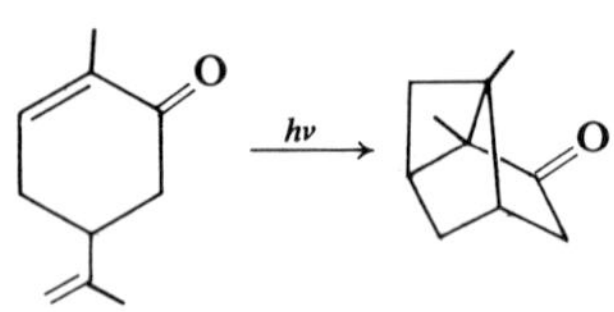

(152)

The interested reader is directed to more relevant reviews.[184]

c. C to N 1,x Hydrogen Shifts. In the elucidation of the structure of coniine, the rearrangement shown in (153) was discovered by Hofmann.[212]

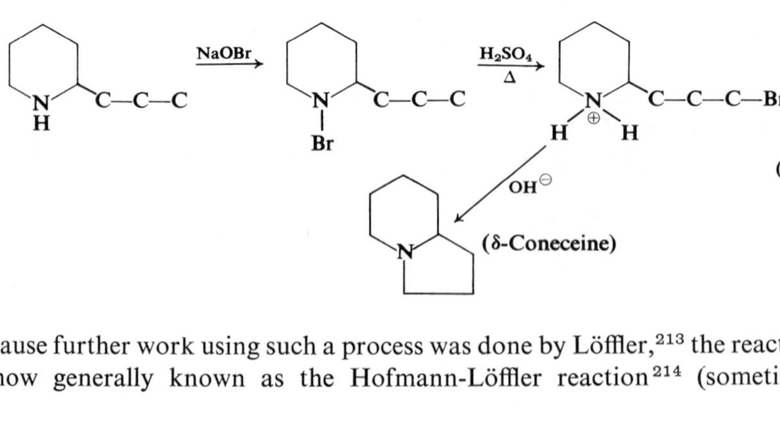

(153)

Because further work using such a process was done by Löffler,[213] the reaction is now generally known as the Hofmann-Löffler reaction[214] (sometimes

Freytag's name is added, too). The reaction is preparatively useful (154).[215]

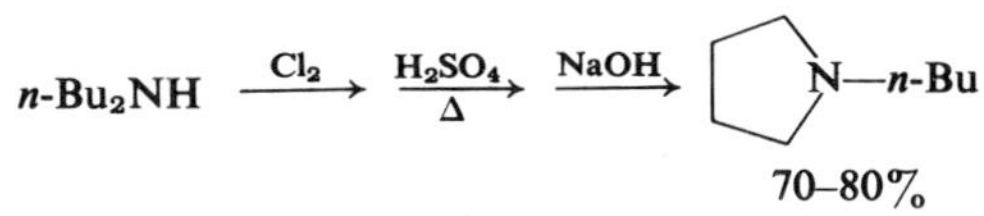

(154)

The generally accepted mechanism[216,217] can be illustrated on this amine (155) which is probably the most studied of them all.

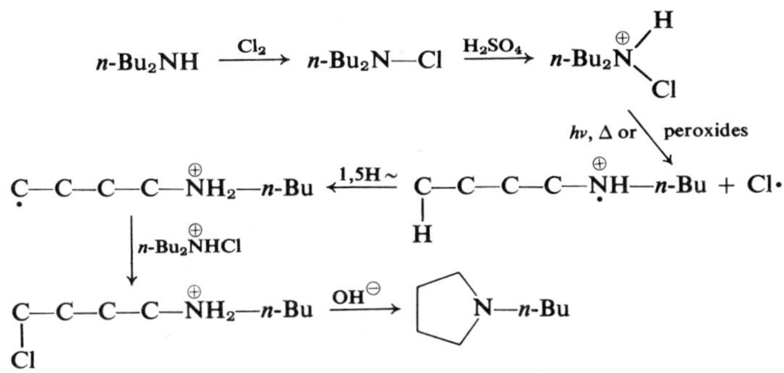

(155)

An early suggestion[217] that the highly acidic medium favors the reaction because intermolecular repulsion of the aminium radical cations should retard dimerization seems unnecessary. In fact, aminium radical cations do dimerize.[218] Recent work on the mechanism in (155)[219] indicated that increased acid strength is valuable because it reduces the concentration of *N*,*N*-dichloro-*n*-butylamine. It appears that this substance is the actual chain initiator (and not *n*-Bu_2N^+HCl as had been thought), but that too much of it retards reaction because it also gives rise to chain-terminating species. Neither acid nor light is sufficient for the process, though each is necessary (156).

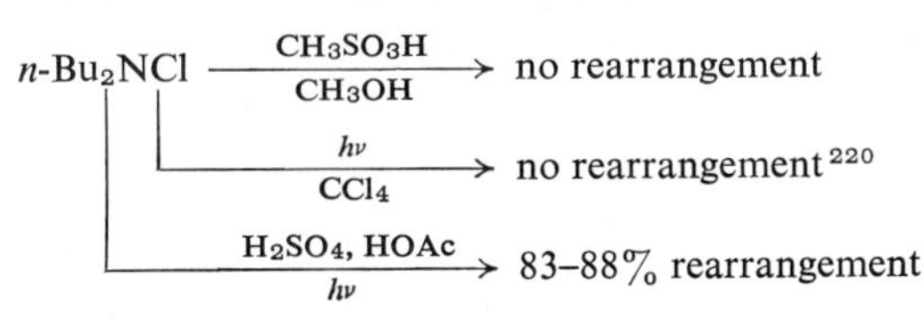

(156)

Use of trifluoroacetic acid and light is also an excellent method (157).

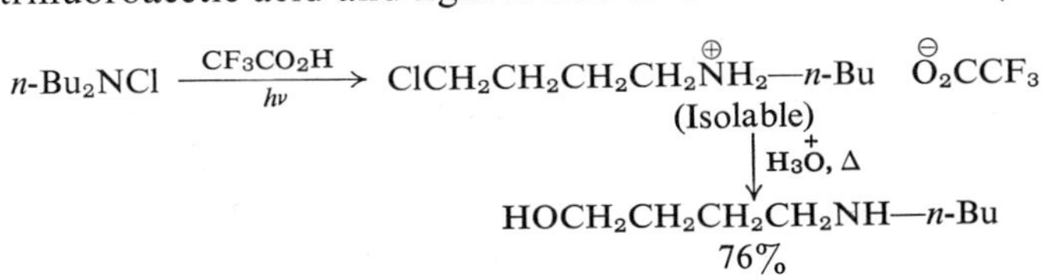

(157)

The data in (158) show the expected tendency toward 1,5 hydrogen transfer in the aminium radical cation.[221]

$$\mathrm{CH_3{}^5CH_2{}^4CH_2(CH_2)_3{-}\overset{\oplus}{N}HR(Cl)} \xrightarrow[h\nu,\ 25^\circ]{85\%\ \mathrm{H_2SO_4}} \underset{\textbf{158-1}}{\mathrm{CH_3CH_2{}^4CHCl(CH_2)_3\overset{\oplus}{N}H_2R}} + \underset{\textbf{158-2}}{\mathrm{CH_3{}^5CHClCH_2(CH_2)_3\overset{\oplus}{N}H_2R}} \quad (158)$$

R	% **158-1**	% **158-2**
CH_3	80	20
n-Hexyl	95.6	4.4

The presence of benzylic hydrogens can allow large amounts of 1,6 hydrogen transfer, however.[219b] One instance is given in (1959).

$$\mathrm{Ph{}^5CH_2{}^4CH_2(CH_2)_3{-}\overset{\oplus}{N}HCH_3(Cl)} \xrightarrow[\mathrm{HOAc}]{h\nu} \begin{array}{lll} 26\%\ \text{attack at C-5} & (38) \\ 43\%\ \text{attack at C-4} & (62) \\ \quad(\text{Yield}) & (\text{Ratio}) \end{array} \quad (159)$$

The 38% 1,6 hydrogen shift here is to be contrasted to the 7.7% such transfer in an analogous alkoxy radical (see reaction 129).

In the formation of bicyclic products, hydrogen transfer to the aminium radical site is allowed in both a seven- and six-membered ring transition state. In (160), both type products were observed at times.[222,223]

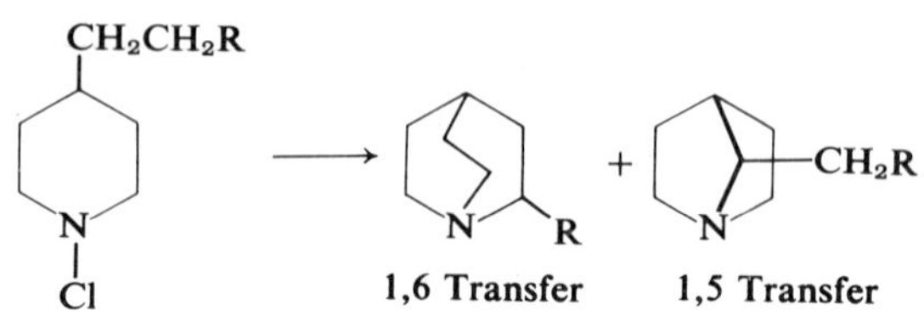

(160)

1,6 Transfer **160-1** 1,5 Transfer **160-2**

R	Yield (%)	**160-1** (%)	**160-2** (%)
H	21	50	50
CH_3	12	70	(Five other products were formed)

The simpler methyl derivatives were also studied (161). Clearly, however, the bicyclic ring strain engendered in 1,5 hydrogen transfer makes it less easy in these systems.

The increased tolerance of 1,6 hydrogen transfer in certain of the above reactions may be related to the linearity possible in a seven-membered ring transition state. The C—H—N angle can approach 180° in acyclic cases, compared to about 128° in a six-membered transition state. Even in cyclic ex-

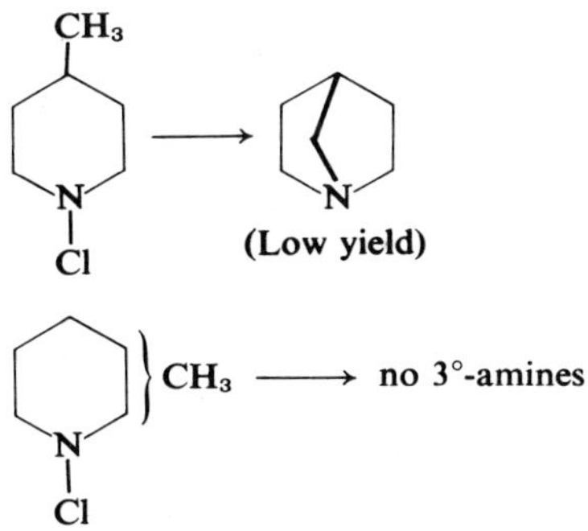

(161)

amples where colinearity may be impossible, an approach to linearity in the transfer step is again essential. In (162) 1,5 hydrogen transfer via a "strained" transition state still totally predominated over 1,4 transfer via a less strained one.[224,225]

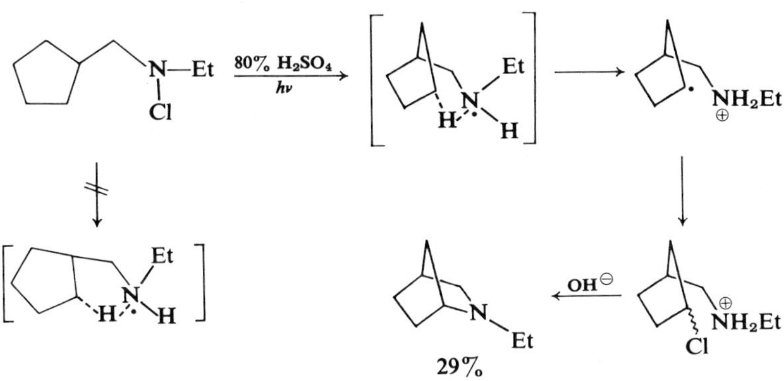

(162)

From inspection of stereomodels, the C—H—N angles were 145 versus 120° in the preferred and rejected transfers, respectively.

The Hofmann-Löffler reaction, contrary to common opinion, does proceed with primary amines.[226] Although *N*-chloro primary amines disproportionate readily in weak acid (reaction 163) to species then unable to achieve the Hofmann-Löffler reaction, this disproportionation is repressed in strong acid.

$$\text{R—NHCl} + \text{R}\overset{\oplus}{\text{N}}\text{H}_2\text{Cl} \rightleftharpoons \text{R—}\overset{\oplus}{\text{N}}\text{H}_3 + \text{R—NCl}_2 \qquad (163)$$

Initiation of the chain sequence may then be neatly accomplished with iron(II) salts (164).

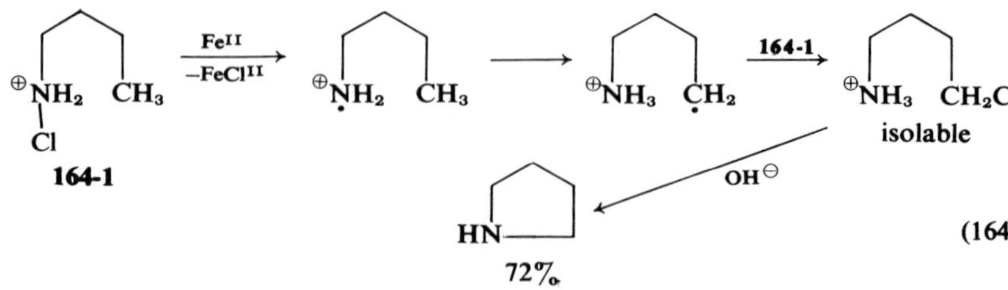

(164)

1,5 Hydrogen transfer is again preferred (165).

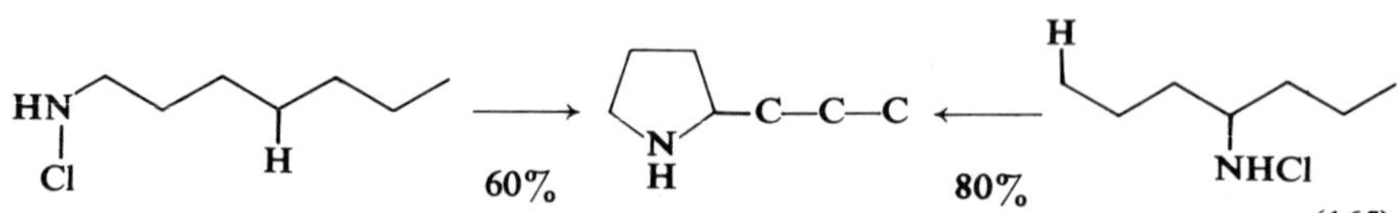

(165)

The *N,N*-dichloroamine is not an intermediate. Its use gave no pyrrolidine. Recall that the dichloro compound is an inhibitor when in too high a concentration (p. 391).

In recent years, a large number of 1,5 hydrogen transfers to nitrogen have been discovered in amine derivatives, for example, amides, imides, and sulfonamides. Photoinitiation is generally employed and the reactions have synthetic utility. One of the earliest such reactions involved *N*-chloroimides (166).[227]

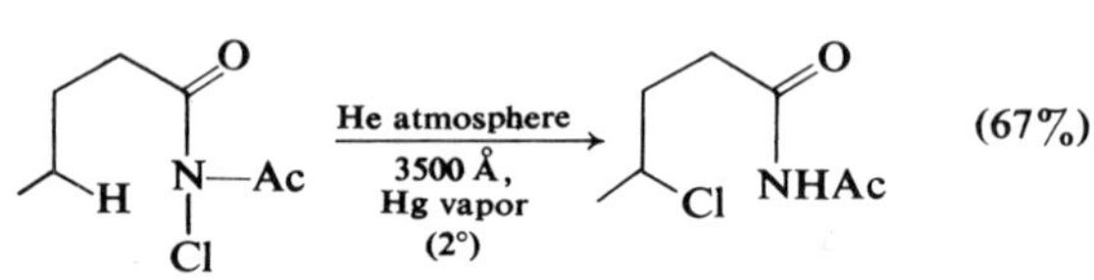

(166)

Use of either the long- or short-wavelength irradiation alone gave lower yields! Imides are not essential, however, as *N*-chloroamides behave similarly (167),[228] as do *N*-iodoamides.[229]

Accompanying the γ-lactone when possible is a δ-lactone, presumably formed by 1,6 hydrogen transfer. The presence of the carbonyl group may make 1,6 hydrogen transfer less difficult here. As previously mentioned, the C—H—N angle is always better for transfer[230] in a seven-membered ring transition state anyway. Air and cyclohexene had no effect on the *N*-chloroimide rearrangement[227] and it is thus probably not a chain reaction. A

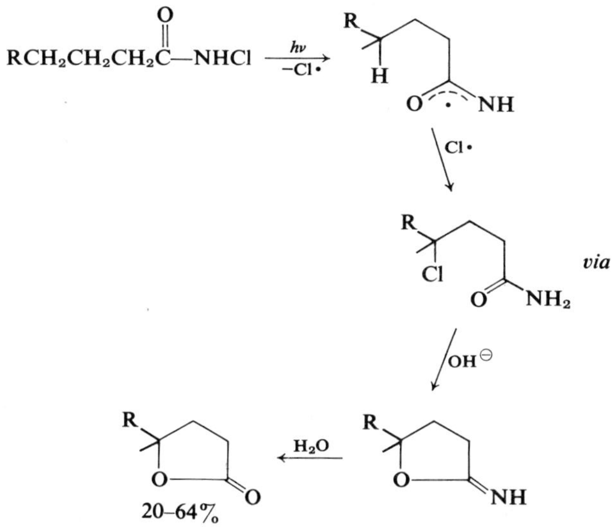

(167)

related rearrangement (168) of *N*-alkylamides is, contrariwise, claimed to be a chain sequence.[231]

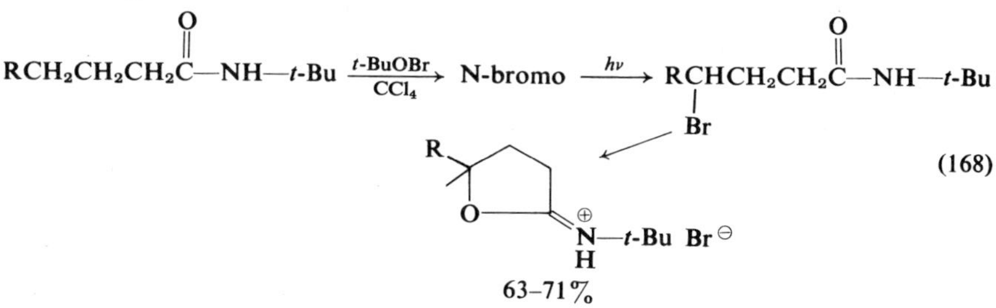

(168)

The reaction works best on *t*-butyl derivatives. For some reason, *N*-methyl, phenyl, or trityl derivatives are unusable and the parent amide is itself poor. A similar reaction of *N*-chloro analogs was reported along with the *N*-bromo ones. In the *N*-chloro cases, *N*-methyl *or N-t*-butyl derivatives may be used, but once more the parent amide failed. This is curious as such *N*-chloroamides do rearrange.[228]

Transannular hydrogen transfer in *N*-chloroamides in such reactions has been recently observed (168a).[231a]

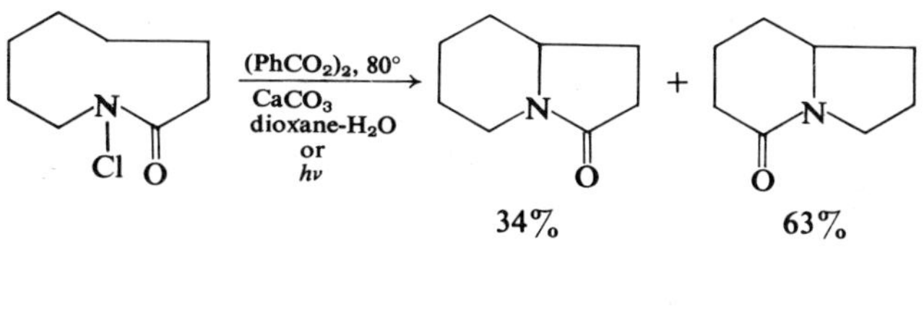

(168a)

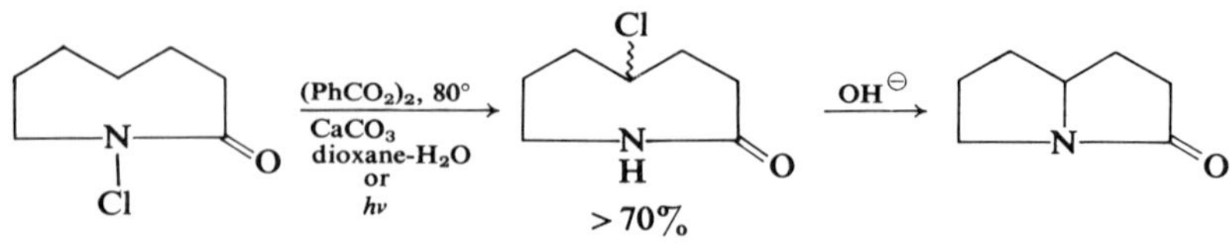

On occasion, however, the reaction took a different course (168b).

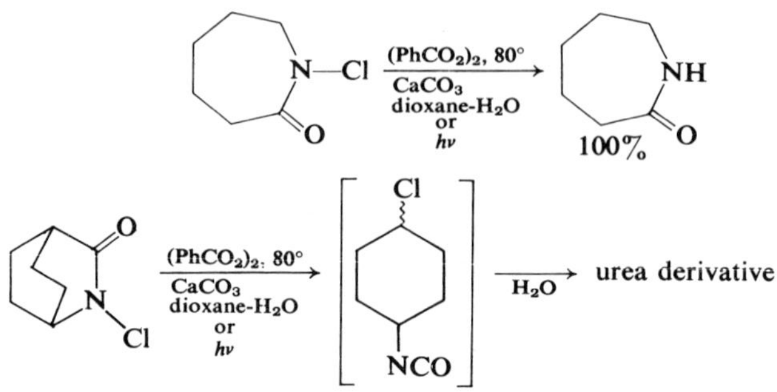

(168b)

Clearly the process is sensitive to reactant structure and any of the several paths in (168c) can be followed. Note again, however, the preference for 1,5 hydrogen transfer.

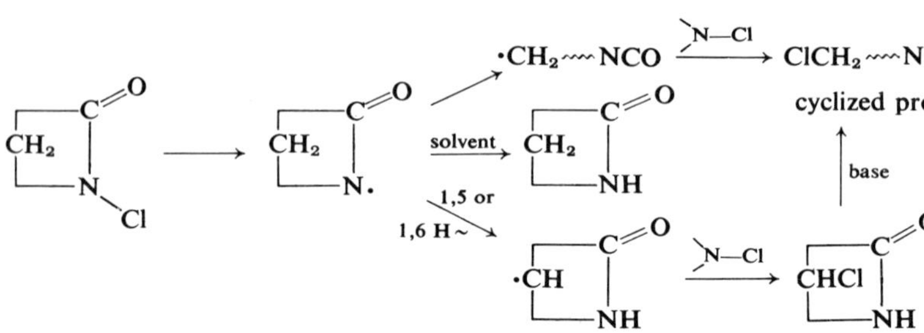

(168c)

N-Halosulfonamides join the list of rearrangement precursors. An early example is given in (169).[232]

$$\mathrm{RCH_2(CH_2)_3N(Cl){-}SO_2Ar \xrightarrow[\Delta]{85\%\ H_2SO_4} RCH(Cl)(CH_2)_3NHSO_2Ar} \qquad (169)$$

Alkanesulfonamides, especially *N*-*t*-butyl derivatives, behave analogously (170).[233–235] Both 1,5 and 1,6 hydrogen transfers can occur,[236]

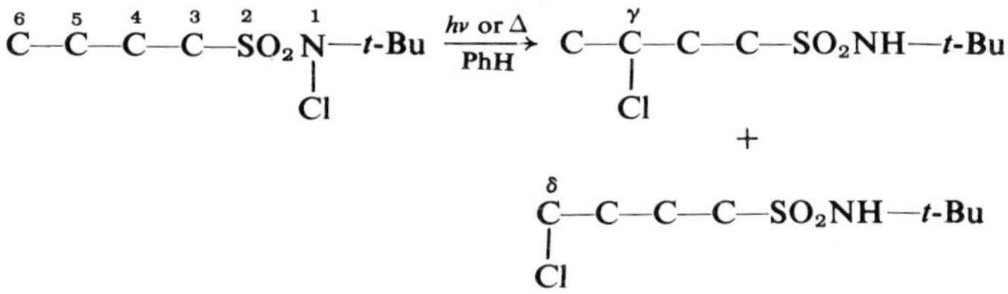

(170)

R—SO$_2$N(Cl)$^-$*t*-Bu	Conditions	γ-Chloro-sulfonamide (%)	δ-Chloro-sulfonamide (%)
R = *n*-Butyl	*hν*, PhH, 15°	50.8–67.1[a,b]	14.0–17.2
n-Amyl	*hν*, PhH, 15°	40.3–42.1	37.1–38.0
n-Amyl	*hν*, H$_2$O—HOAc—H$_2$SO$_4$	40.6–43.6	24.0–27.1

[a] The composition of product varied in all cases and was dependent on the *N*-chlorosulfonamide concentration.
[b] The remainder of product was unchanged reactant.

although the latter is somewhat decreased in acidic media.[237] The significance of this is unclear.

d. O to C 1,x Hydrogen Shifts. In a neat turnabout, the following 1,5 hydrogen transfer in a biradical has been advanced (reaction 171).[238]

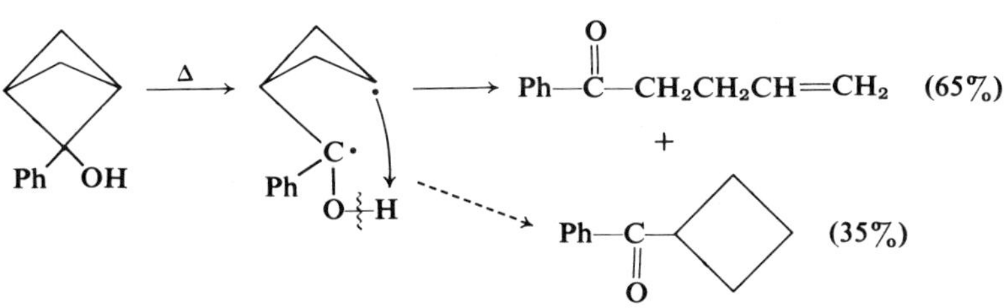

(171)

In reality, this is a "retro-Yang" rearrangement to the minor product. Actually, both 1,3[239] and 1,5 hydrogen shifts seem implicated in this process. Note the reactions of the deuterated alcohols in (172) and (173). First, in (172)

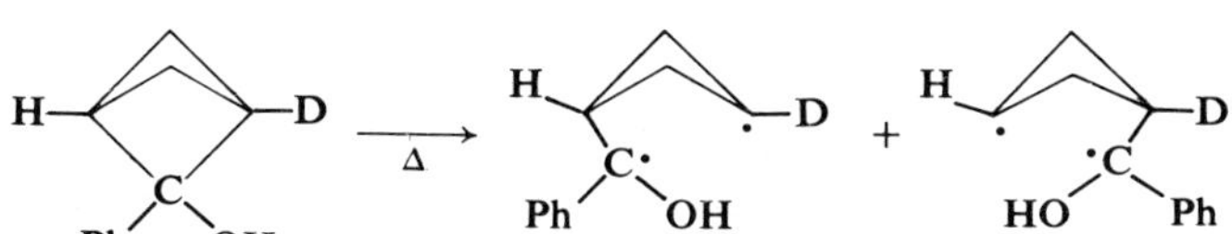

172-1 **172-2**

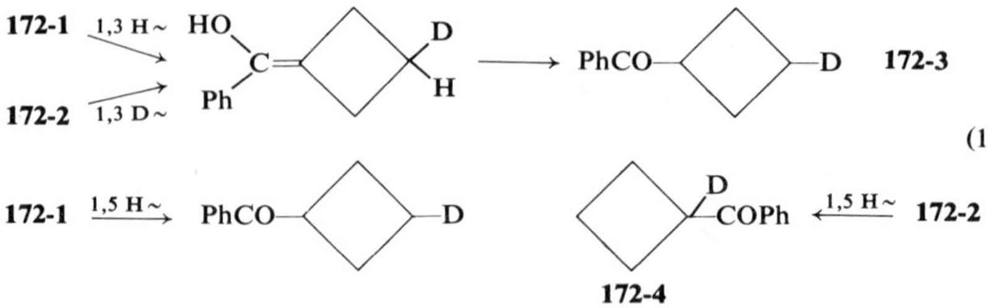

(172)

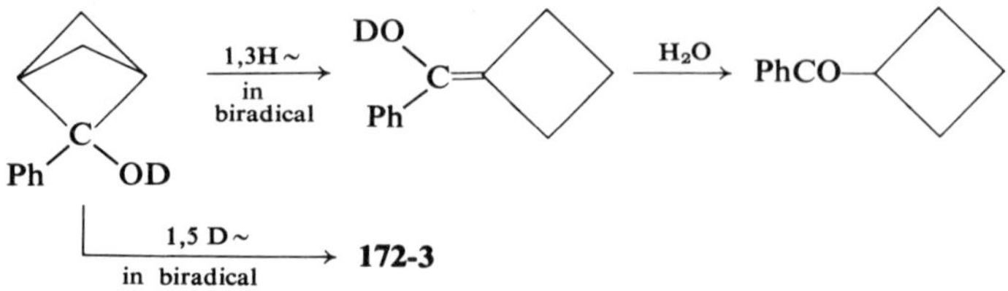

(173)

note that 1,3 hydrogen migration would produce all **172-3**, while 1,5 hydrogen migration would produce **172-3** and **172-4** in 1:1 ratio if no isotope effect were present. Furthermore, if all transfers were 1,5 then all the deuterium in (173) should be retained. The results in summary indicated that the rearrangement proceeded with some 60% 1,5 and 40% 1,3 hydrogen shifts.

C. Eliminations[240]

1. Without Formation of a Carbonyl or Related Group

a. Cyclopropyl Radicals. Cyclopropyl radicals resist ring opening even though relief of ring strain and an allyl radical should result. Some 30 kcal/mole of energy should be released by such an opening.[241] The mode of opening should probably be conrotatory, though no evidence is available. Indeed, it has been suggested that the mode may depend on "other factors of molecular structure."[242] In any case, stabilization of the transition state for cleavage appears to be poor and a number of reactions proceeding via simple cyclopropyl radicals do not give allylic products.[243] The activation energy at 186° and 50 mm for this ring opening has been calculated to be 19 kcal/mole.[244] Substituted cyclopropyl radicals behave similarly in several instances,[243] but two recent exceptions may be noted. In one,[137,138] reduction of the bromide in (174) produced olefin, presumably as shown.[245]

As indicated in Table 6, ring opening was favored by higher temperatures and more dilute solutions. If chain transfer with the tin hydride has a rate constant of approximately 10^6 M^{-1} sec^{-1},[75] then the barrier to the ring-opening reaction in the present case is approximately 11 kcal/mole. This lower barrier

TABLE 6

Ring opening of 2,2-diphenylcyclopropyl radical

Bromide (M)[a]	n-Bu_3SnH (M)	Temperature (°C)	% **174-1**	% **174-2**
0.30	0.33	78	88.5	11.5
0.10	0.11	78	69	31
0.30	0.33	150	27.5	72.5
0.10	0.11	150	13.5	86.5

[a] Reaction in benzene for 24 hr.

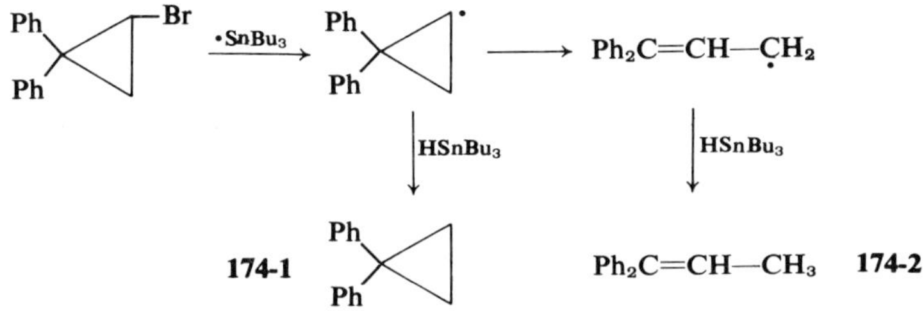

(174)

vis-à-vis the cyclopropyl one is most likely due to the stabilization afforded by the phenyl groups. In support of this last point is a similar cleavage reaction recently reported (equation 175).[39b,246] Here again it was stressed that the phenyl groups aided the ring cleavage.

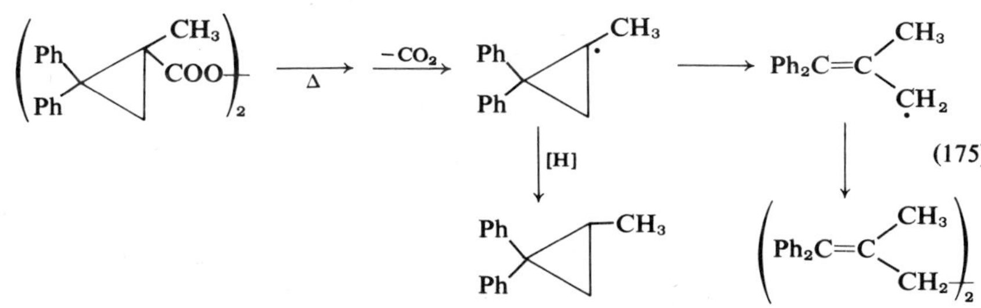

(175)

As an aside at this point, *cyclobutyl* radicals apparently fail to cleave,[247] and their rearrangement in any sense seems unknown. One mention exists of cyclobutyl radical formation from the cyclopropylcarbinyl radical,[248] but this, too, is otherwise unobserved.

b. Cyclopropylcarbinyl Radicals. The rearrangement of such radicals by ring opening leads to isomeric homoallylic radicals. These latter species themselves may rearrange via cyclopropylcarbinyl radical intermediates, leading to net vicinal migration of a vinyl group. So the present subject and the one dealing with 1,2 vinyl shifts (pp. 356–360) are intimately connected and can perhaps be considered two aspects of one more general situation. A well-established elimination process is known for cyclopropylcarbinyl radicals, as shown in (176)[135,249] and (177).[250]

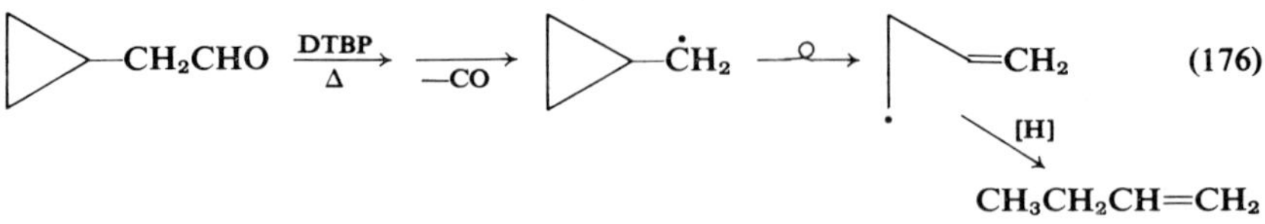

(176)

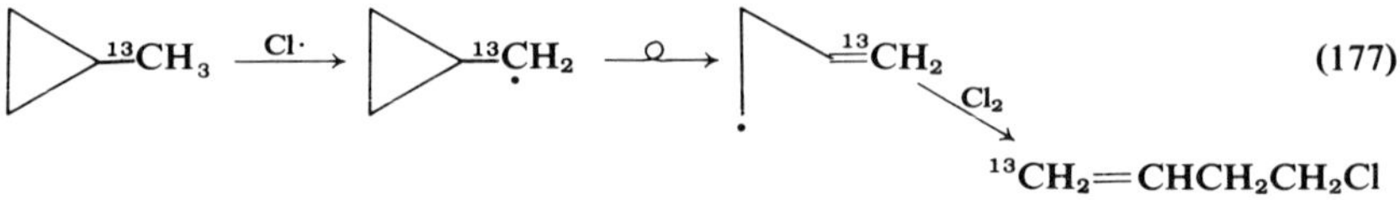

(177)

Such radicals can be produced in other ways as well, with rearrangement again commonly observed. In (178) rearrangement varied in an expected fashion with the nature of the chain transfer agent employed.[251]

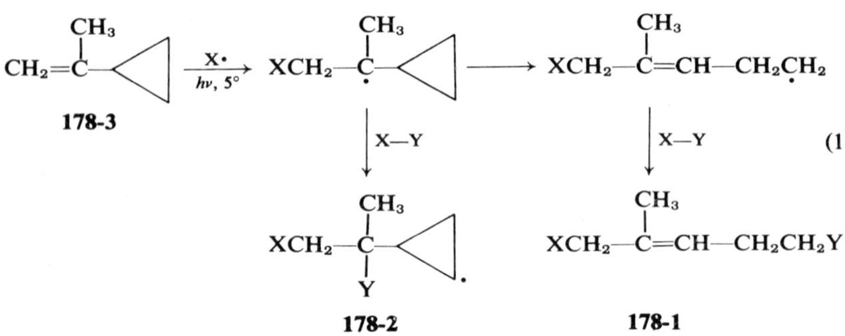

178-3

(178)

178-2

178-1

With X—Y = bromotrichloromethane, 68% of the ring-opened product **178-1** resulted. With X—Y = thiophenol, 57% of the vicinal adduct **178-2** was formed. With X—Y = methanethiol, the extent of rearrangement was variable, decreasing with increasing thiol concentration. Such variation has traditionally been viewed as evidence that two discrete radicals are the precursors to products, not one "nonclassical" species. Competition experiments between olefin **178-3** and the acyclic model 2,3-dimethyl-1-butene indicated the former to be 4.4–6.6-fold more reactive toward various radical addends. This (rather small) rate enhancement may mean some special stabilization exists for the cyclopropylcarbinyl radical intermediate in (178), as is the case for the cation. Other explanations are possible,[252] however, and the general absence of "specially stabilized" radicals augurs ill for their presence here.

A later study (179)[253] indicated that the elimination process produced trans olefinic products under conditions of kinetic control.

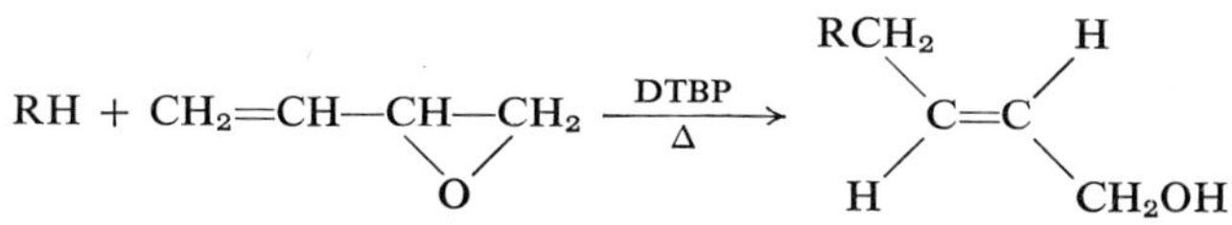

(179)

RH = cyclopentene, -hexene, and -octene.

By the Curtin-Hammett principle,[254] such specificity may indicate that cleavage from rotamer **180-1** in (180) has a lower energy barrier than cleavage from rotamer **180-2**.

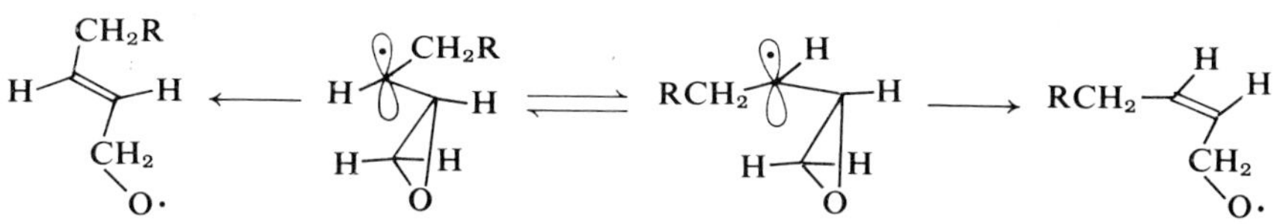

180-1 **180-2** (180)

It may also mean (but does not require) that cleavage from **180-1** be faster than rotation of **180-1** to **180-2**.[255] Presumably, cleavage of **180-2** to product is the more difficult because steric crowding becomes serious as the planar alkene product forms.

Similar additions under somewhat different conditions have also been reported (181).[256]

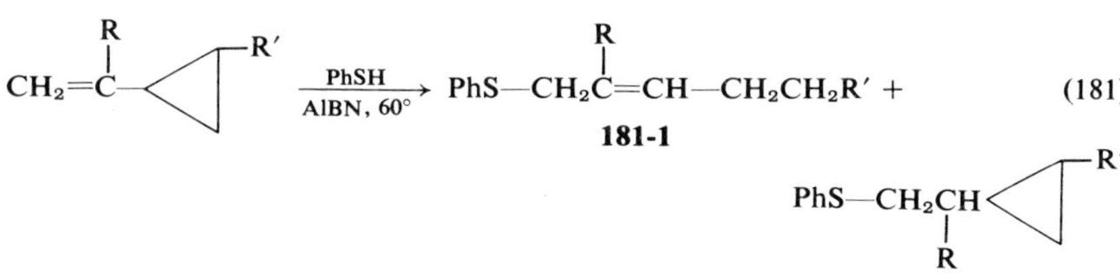

(181)

R	R′	% 181-1	% 181-2
H	H	40	60
CH_3	H	50	50
Ph	H	60	40
H	Ph	100	—

As the ring-opened radical intermediate was stabilized with R′ = phenyl or a variety of ester or amide functions, cleavage dominated the overall process. The elimination observed in these cases with thiophenol contrasts with the earlier study.[251] Possibly the higher temperature (60 vs. 5°) favored cleavage in the present case.

Radical polymerization of the alkene in (182) likewise involved ring opening.[257] Notice the contrast to ionic polymerization.

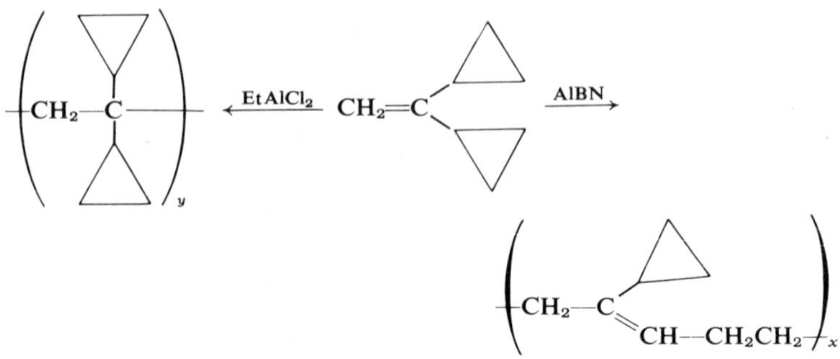

(182)

Electron spin resonance studies have been done on the cyclopropylcarbinyl radical.[258] The evidence supports two discrete radicals, **183-1** and **183-2**, with the former relatively stable to rearrangement below −140°. Above −120°, however, cleavage is very rapid and apparently irreversible (reaction 183).

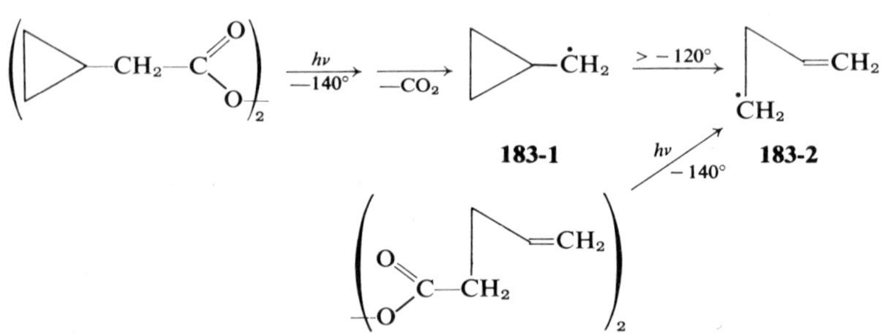

(183)

The radicals shown in (184) are also stable to cleavage under −140° but cleave rapidly (though observably) above −100°.[259]

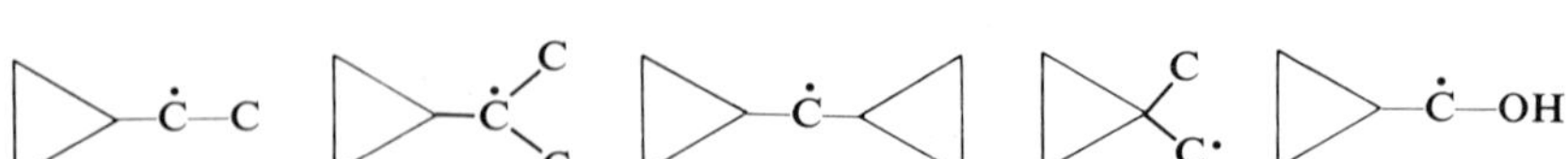

(184)

Curiously, the highly substituted example in (185) cleaved anomalously at $-126°$, albeit normally at a higher temperature.

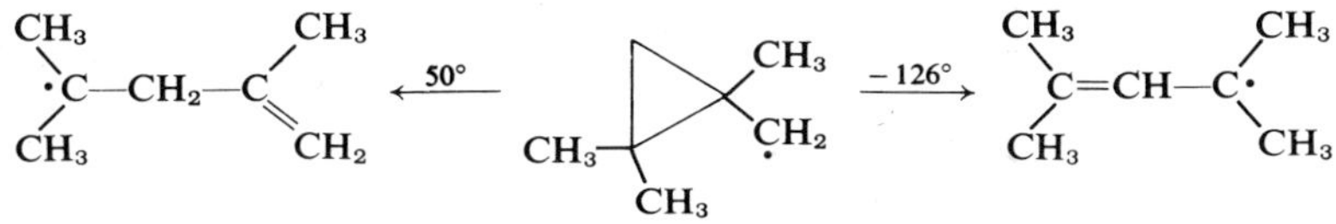

(185)

Although the rate constant for cleavage of such radicals at ordinary temperatures may be approximately 10^8 sec^{-1}, certain data indicate that cage recombination reactions can be several times faster.[260] For example, in (186) the yield of **186-1** and **186-2** ranged from 17 to 75%, depending on the viscosity of the solvent. However, the ratio of **186-1** to **186-2** was 5–10:1 and it did not depend on solvent viscosity. The relative constancy of this ratio and its independence of solvent viscosity imply that recombination of unrearranged radicals in the cage was severalfold (claimed to be sixfold) faster than cleavage.

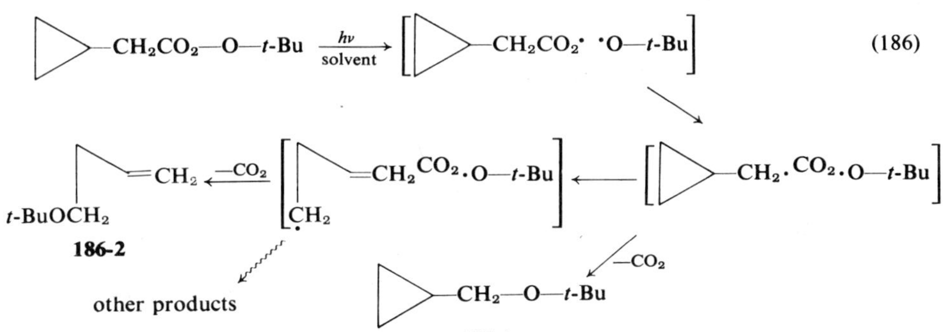

(186)

Other cyclopropylcarbinyl radicals may be trapped more or less effectively prior to cleavage (reactions 187, 188).[261]

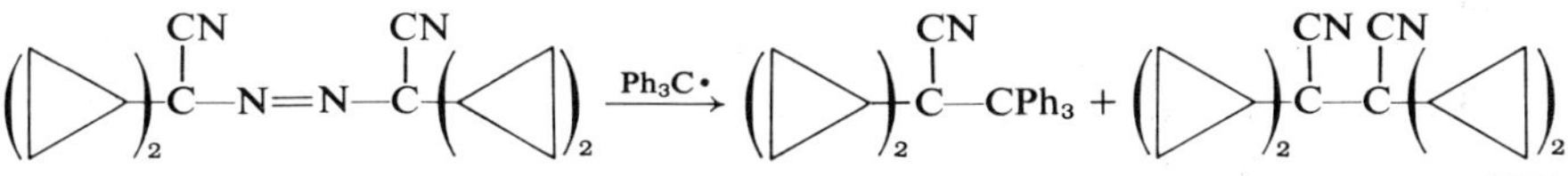

78% (187)

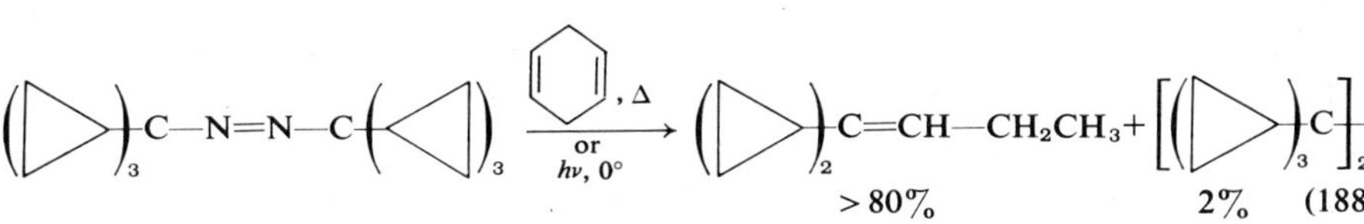

>80% 2% (188)

188-1

It is therefore seen that cleavage of cyclopropylcarbinyl radicals is subsequent to their formation. The faster rate of several reactions[262] that produce such radicals implies some special stabilization for them. Dimer **188-1** dissociates at 295° with a rate constant of 1.31×10^{-3} sec^{-1}, indicating a central carbon–carbon bond strength of about 45 kcal/mole. So cyclopropyl definitely stabilizes an adjacent radical center better than methyl does. 2,2,3,3-Tetramethylbutane possesses $D(C_2—C_3) = 67.5$ kcal/mole.[263]

(Phenyl)cyclopropylcarbinyl radical also cleaves under certain conditions (189).[264] Curiously, the opened radical apparently recloses under other conditions (190).[265]

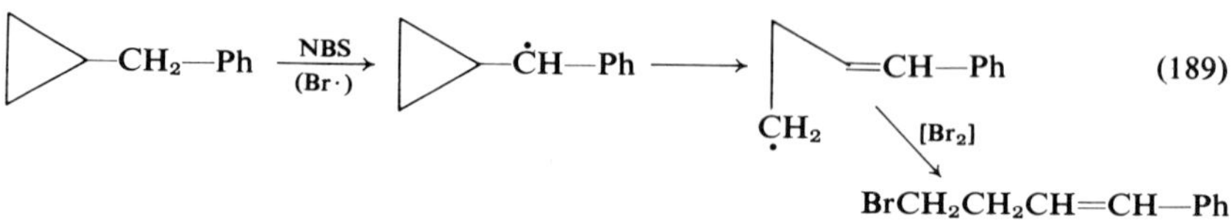

(189)

189-1

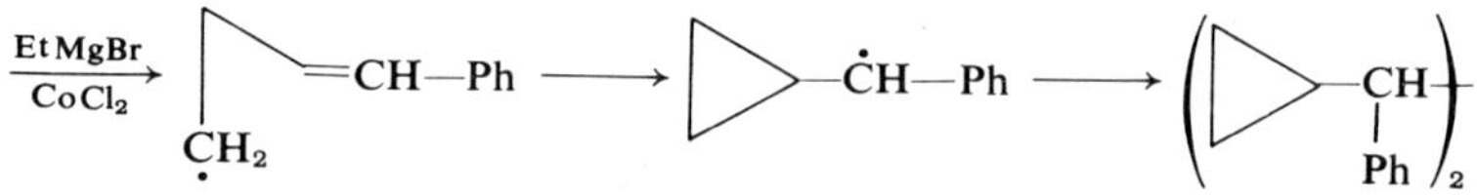

(190)

The Kharasch–Grignard process employed in (190) is, of course, the famous reaction that gave birth to radical rearrangements in solution (p. 346). There are many questions, however, about its detailed mechanism, particularly that involved in the formation of the dimers so frequently observed.[266] It is quite possible that the dimers form from organometallic intermediates distinctly different from those en route to the monomeric products. In any event, the opening in (189) and the closure in (190) perhaps indicate an equilibrium between the two radicals with chain transfer and dimerization rates determining the products formed.

Under still another set of conditions, the radical opened and reclosed to a 1,2-dialin.[267] The cyclization will be discussed later.

Another such example is found in the rearrangement shown in (190a).[267a] In this case it is curious that opening of radical **190a-1** proceeded preferentially to the primary allylic radical **190a-2** rather than to the secondary benzylic radical **190a-3**. A number of factors remain to be evaluated, however. The equilibria involved and the (perhaps) differing rates of chain transfer with the *N*-bromosuccinimide complicate any interpretation. So the curious opening remains such. An inductive destabilization, rather than a resonance stabilization, by the aromatic ring upon the transition state leading to

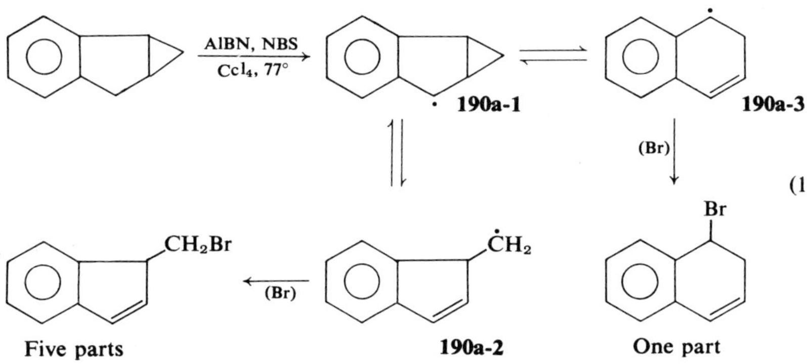

(190a)

190a-3 was suggested for the curious ring opening. However, it is worth noting that an *overlap control* mechanism for a somewhat similar ring opening (190b) has been advanced[267b] and it may apply here as well.

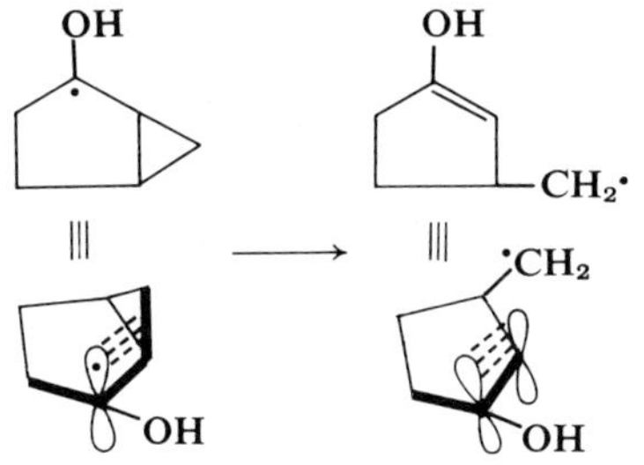

(190b)

(Diphenyl)cyclopropylcarbinyl radical, generated as in (191), rearranged to open-chain products, among them the C_{16} hydrocarbons shown.[268]

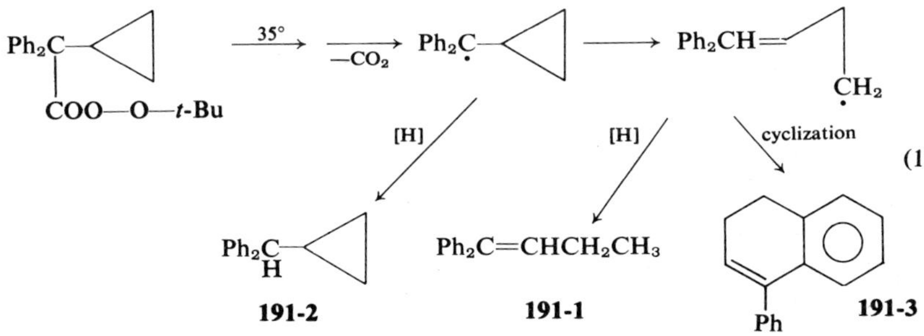

(191)

With the effective hydrogen donor 1,4-cyclohexadiene, the yields of products were: **191-1**, 45%; **191-2**, 14.6%; and **191-3**, 11%. In this system, rapid equilibration (reaction 192) between radicals has been demonstrated by deuterium labeling.

This stands in contrast to the parent radical case (p. 402). Understandably, the *gem*-diphenyl group definitely stabilizes both the opened and closed radi-

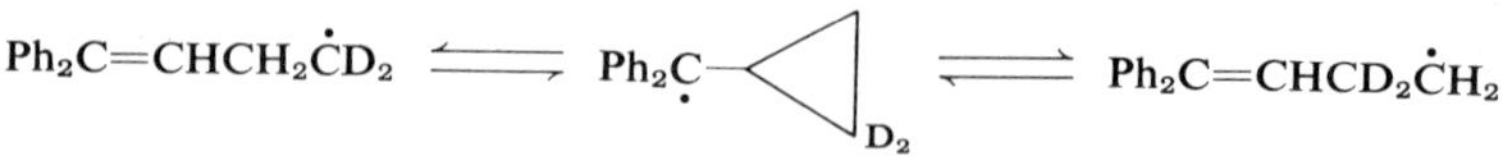

(192)

cals and presumably must lower the energy barrier between them. The reversibility of this system is shown also in (193). Here the open radical reclosed.

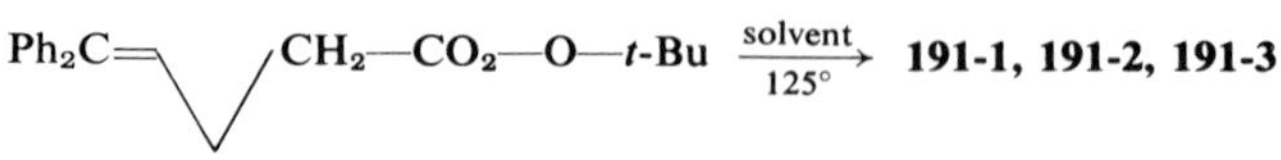

(193)

With hydrogen donors such as triethyltin hydride, 1,4-cyclohexadiene, and ether, the yields of cyclized **191-2** were 1.63, 0.3, and 0.6%, respectively. As might be expected, these studies indicate a greater stability for the open radical. The temperature difference must be involved in the different product spreads obtained in (191) and (193).

c. Cyclobutylcarbinyl Radical. As has been noted (p. 341), this radical did not ring expand. It did, however, rearrange by ring opening under certain conditions (15).[24] The barrier to ring expansion must indeed be formidable because the barrier to ring opening is itself calculated to be at least 15 kcal/mole.

d. Pinyl and Related Radicals. Addition of radicals to β-pinene **194-1** frequently leads to rearranged adducts (194).

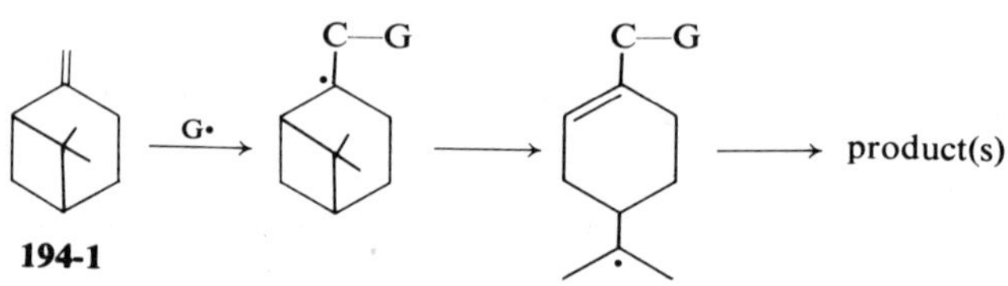

194-1

(194)

The process has been studied a great deal and summaries are available.[269] Recent examples of the reaction with β-pinene include the addition of cyanoacetic ester and related compounds (195),[270] long-chain aldehydes (196),[271] and tri-*n*-butyltin hydride (197).[272]

The aldehyde addition (196) deserves passing comment in that β-pinene apparently traps both the α-aldehyde and acyl radicals that form from aldehydes and di-*t*-butyl peroxide. No product resulting from addition of decarbonylated radicals $RCH_2\cdot$ to the pinene was reported. A detailed study of

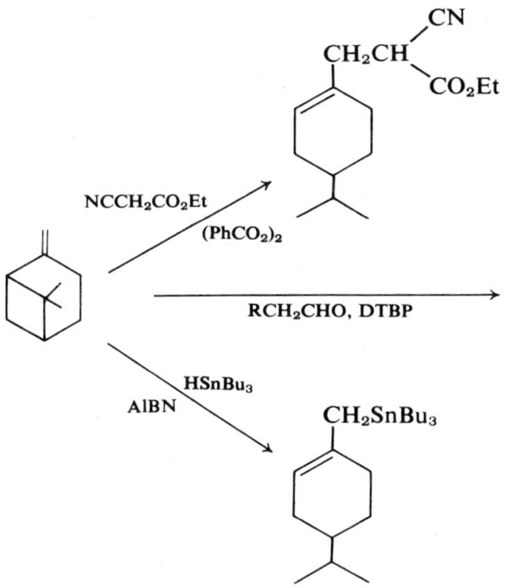

(195)

(196)

(197)

both α- and β-pinene, as well as sabinene, in such radical additions has recently been made.[273]

Several skeletal rearrangements are, in principle, possible in the radical addition reaction, as shown in (198).

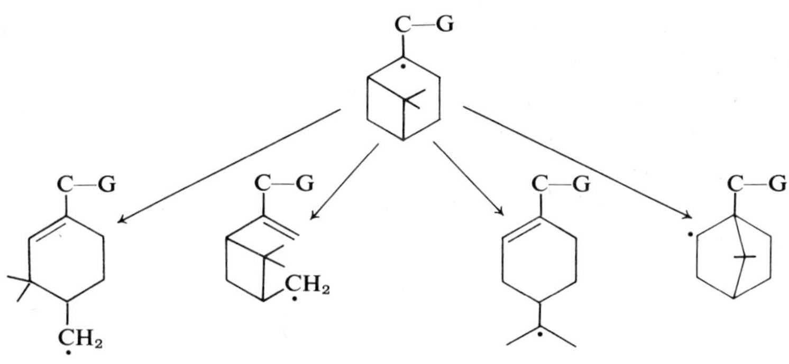

(198)

Why only one is observed becomes then an interesting question. The question is particularly interesting because *cationic* additions can (and do) follow more than one path. The usual answer to the question has been that only cleavage to **198-2** relieves ring strain *and* produces a tertiary radical as opposed to **198-4** which is a primary radical. Indeed, a release of perhaps 17–19 kcal/mole of strain energy could be expected in the cleavage to **198-2**,[274] and the greater stability of **198-2** compared to that of **198-4** should amount to another 7–8 kcal/mole (3° vs. 1° radicals).

Ring opening is not, however, always observed. With thiols (199), vicinal addition was claimed[275] and recently verified.[273]

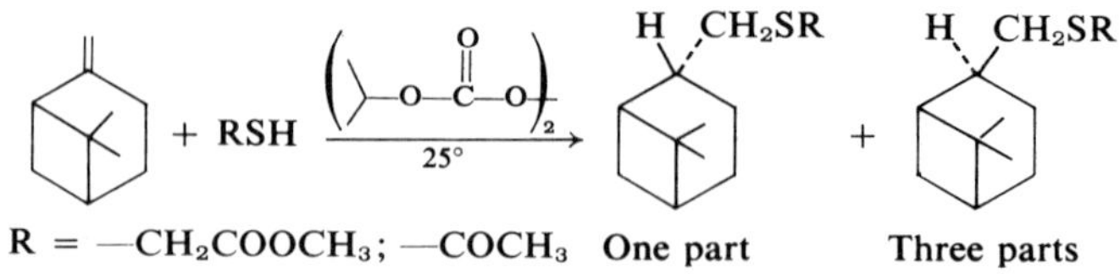

(199)

The major product was assumed to have the configuration shown because chain transfer should be sterically easier from the endo direction (200).

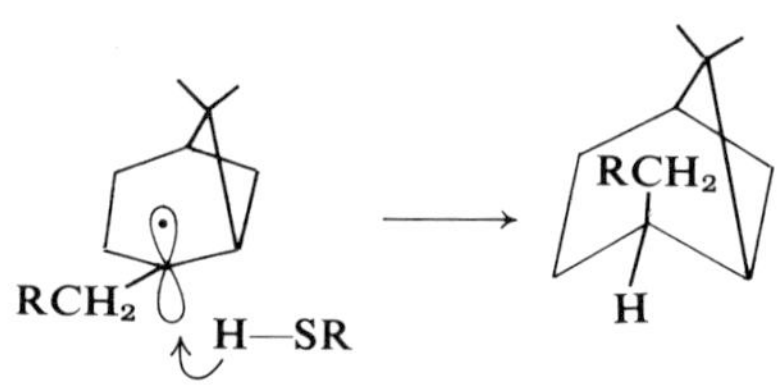

(200)

When di-*t*-butyl peroxide at a higher temperature (reflux) was used as the initiator instead of di-*i*-propyl peroxydicarbonate, ring opening again was found.[276] Clearly, the reaction products from such additions depend not only on release of ring strain but also on the temperature and the chain transfer ability of the hydrogen donor employed. That rearrangement to radical **198-1**, a Wagner-Meerwein rearrangement, does not occur is easily rationalized. Vicinal alkyl shifts in radicals are rare (if even known, see p. 340). Moreover, the delocalization possible in the cation 201, which does easily form in an analogous rearrangement, is unprecedented in radicals.

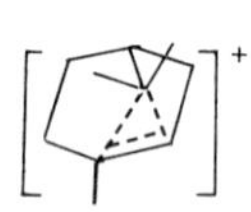

(201)

So while the cation may well be stabilized thereby in spite of the strain present, the radical is not. Rather, stabilization can result only through cleavage.

Fewer additions to α-pinene **202-1** have been reported. Early reports on the addition of trichlorosilane[277] and certain thiols[275a,278] did not give conclusive evidence for the structure(s) of the products. Recently, the reactions in (202) were reported in detail.[273]

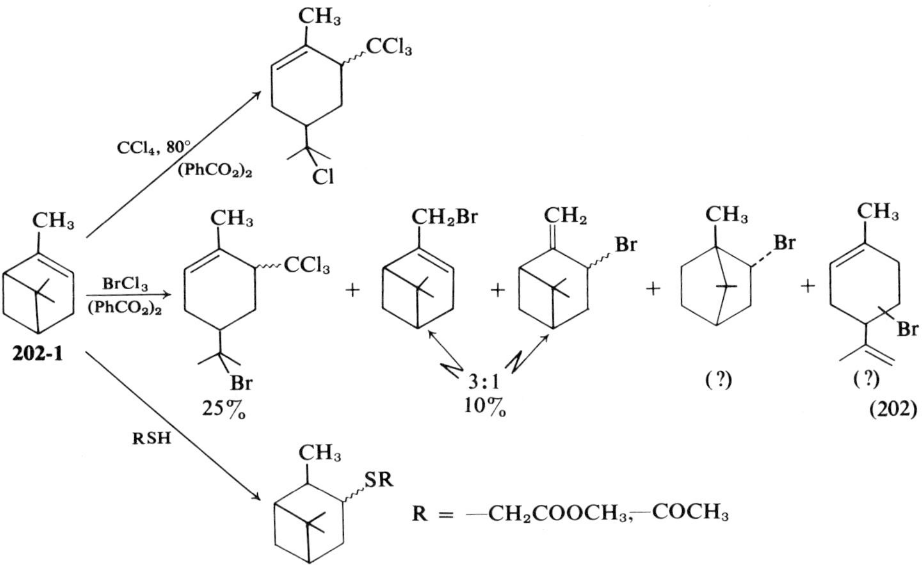

(202)

The carbon tetrachloride adduct was assigned a trans configuration (203) on the basis of its presumed mode of formation and spectral data.[279]

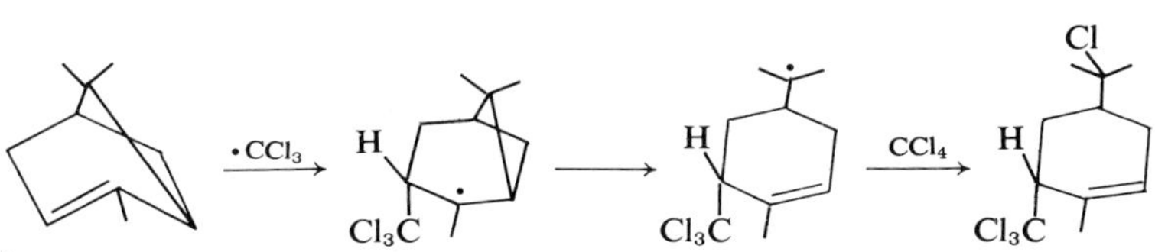

(203)

Probably an analogous structure would apply to the bromotrichloromethane adduct. The greater number of products in this latter reaction results from competitive processes: allylic bromination,[280] and possible *ionic* reactions caused by the hydrogen bromide produced by this allylic bromination reaction.[281] As with β-pinene, thiols gave vicinal addition.

Sabinene **204-1** afforded a cyclopropylcarbinyl-like radical upon such additions. Facile cleavage of such a radical is expected and, indeed, all addenda tried produced ring-opened products (204).[273]

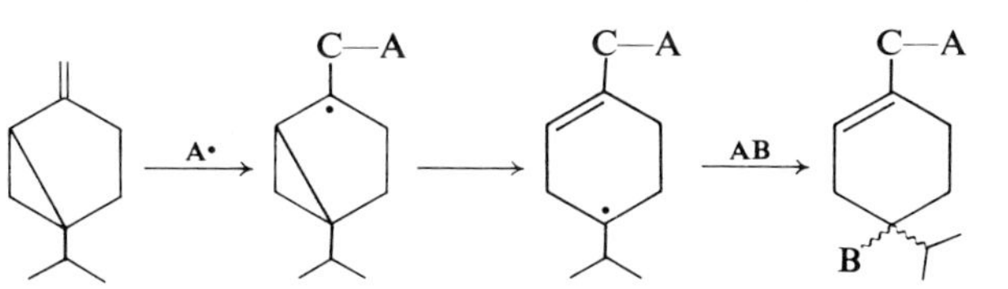

(204)

204-1 AB = CCl_3—Cl, CH_3OOCCH_2S—H, CH_3COS—H

Two adducts were formed in each case, presumably epimers, though this was not proved. Sabinene added methyl mercaptoacetate about one-hundredfold faster than did α-pinene and about sixteenfold faster than did β-pinene. Because earlier-cited studies on simpler cyclopropylcarbinyl radicals indicated that their cleavage was subsequent to formation, this increased reactivity of sabinene probably reflects its increased initial strain energy[283] relative to the pinenes rather than any simultaneity of addition and ring opening.

2. *With Formation of a Carbonyl or Related Group*

a. Oxiranyl Radicals. Perhaps the simplest example of this type of radical rearrangement is given in (205). One would expect a relatively facile cleavage

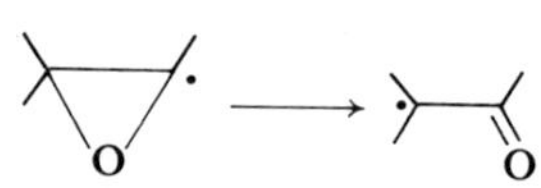

(205)

here because ring strain is relieved and a carbonyl group is formed. However as has been determined recently,[284] an adjacent carbonyl group lends no stabilization to radical sites, so some of the "facility" in the cleavage could be missing. Nevertheless, examples of this rearrangement are documented,[285] and a typical one is given in (206).[286]

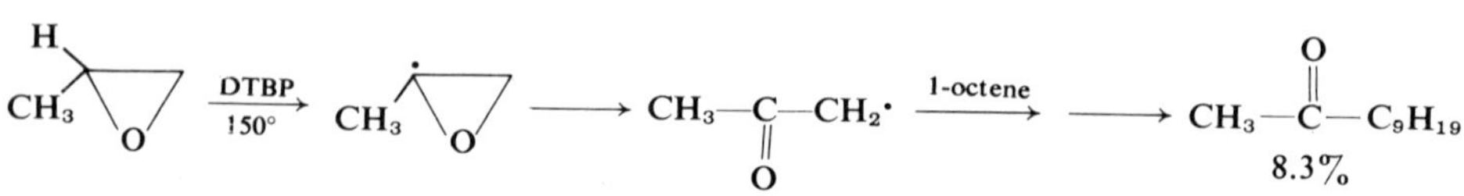

(206)

The opening encountered in this case may be due to the temperature. Oxiranyl radical itself does not open at low temperatures.[248] Another example of the rearrangement is given in (207).[287]

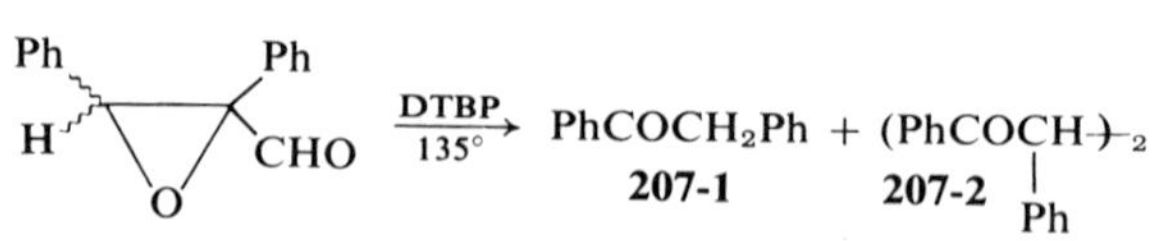

(207)

No other volatile products were observed. Use of the peresters 208 allowed the unrearranged radical to be trapped as *trans*-stilbene oxide (208-1).

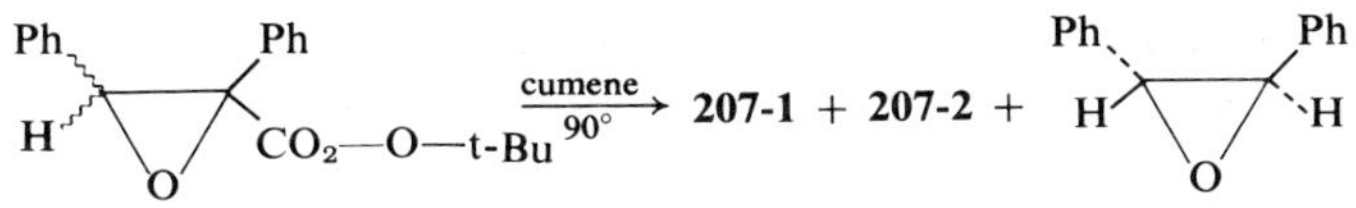

(208)

208-1

Product yields (%)

Ester	207-1	207-2	208-1
Cis	19	19	9
Trans	21	18	10

Obviously, rearrangement is not simultaneous with radical formation. At 90°, hydrogen abstraction by the unrearranged radical is competitive with the rearrangement, but not at 135°. In these oxiranyl radicals, cleavage is additionally favored by the formation of a benzylic (desyl) radical. The formation of only the trans oxide **208-1** may indicate a degree of tetrahedral character to the parent oxiranyl radical, positioning the large phenyl groups trans to each other during chain transfer. The low yield of oxide precludes extensive discussion, however.[288]

α,β-Epoxy ketones rearrange under photochemical conditions, apparently via a singlet biradical intermediate (equation 209).[289]

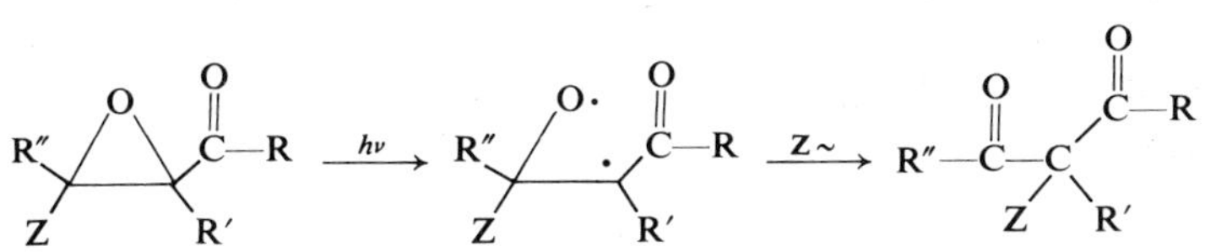

(209)

The migratory ability varies in the order Z = Ph_2CH— ~ $PhCH_2$— > H— > —CH_2— > CH_3— > Ph— and a *concerted* migration with some radical character has been suggested. The process actually resembles the biradical rearrangement involved in thermolysis of peroxylactones (p. 343). With radical

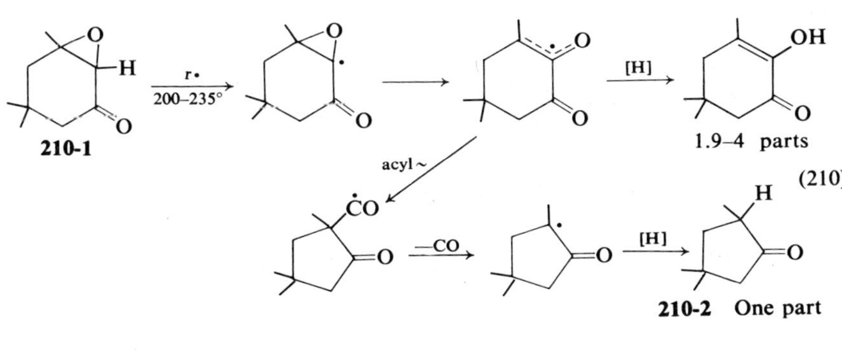

210-1

1.9–4 parts

(210)

210-2 One part

initiators, α,β-epoxy ketones bearing an oxiranyl hydrogen undergo yet another rearrangement, illustrated for isophorone oxide **210-1** in (210).[290]

The formation of **210-2** may be the result of an acyl migration via the path given in (211).[291]

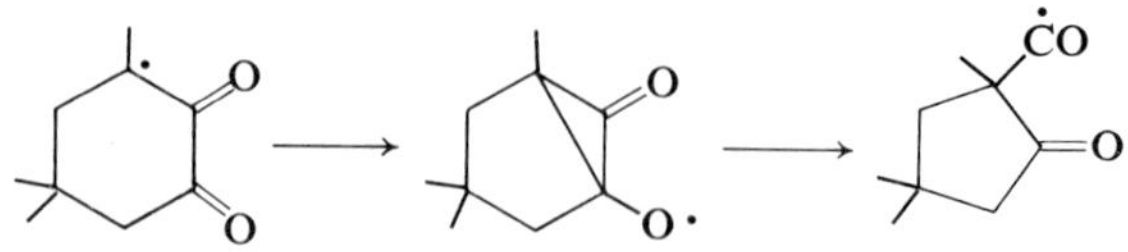

(211)

Such intramolecular additions of radical centers to carbonyl carbon do, in fact, seem established (see p. 479). Isophorone oxide yielded *only* **210-2** at 270° (dilute solution in decalin).[290,291] One is at somewhat of a loss to rationalize all these processes. This last thermal reaction is apparently a surface process. As such it could be difficult to clarify. Likewise, the reaction in (210) is not well behaved. Azo initiators fail to promote the transformation, for instance. It was suggested that radicals from azo compounds are nucleophilic radicals and therefore unsuccessful in this reaction, though it is not immediately obvious why this conclusion follows. Nonetheless, such is the grist for the mechanician's mill and one looks for more such challenging rearrangements.

b. Larger-Ring Ether Radicals. Cyclic ether radicals in general appear to cleave to ω-formyl radicals (equation 212). Most commonly, 1,x hydrogen transfer follows and the acyl radicals so formed are then trapped via addition to 1-alkenes.[292]

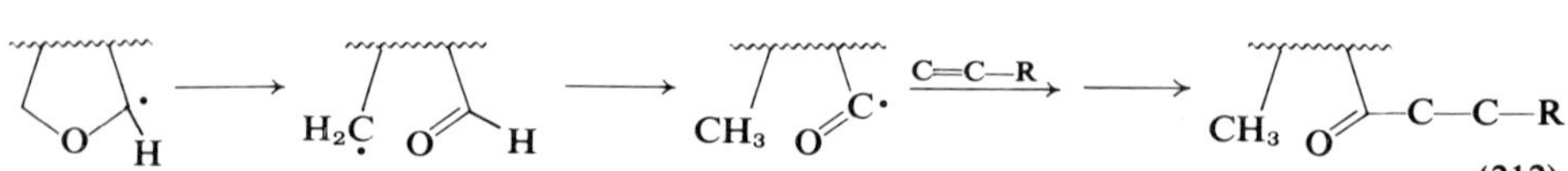

(212)

Such cleavage and/or 1,x hydrogen transfer does not always occur, however. The following phthalanyl radicals illustrate understandable exceptions (213).[293]

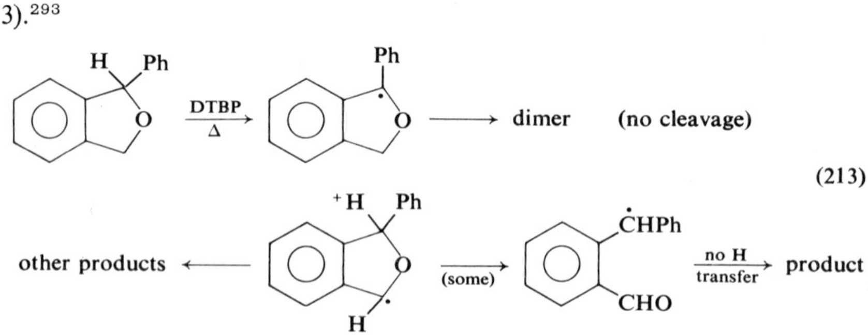

(213)

When such ring cleavage occurs in cyclic *acetals*, an interesting rearrangement results (214).[294]

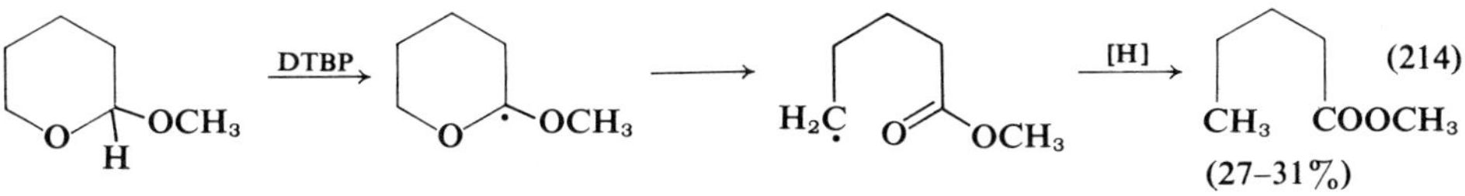

(214)

Other examples follow (215),[295] the last two showing a nice distinction between β cleavage at a secondary versus primary potential radical center.[296]

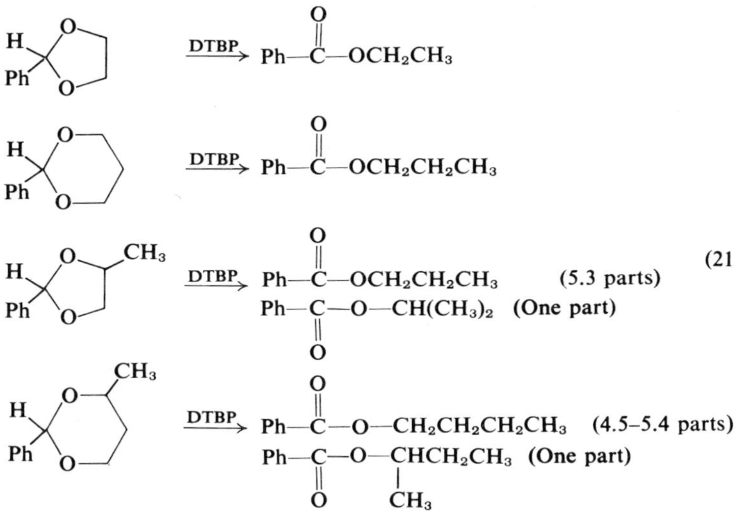

(215)

Each of the above cleavages involved conversion of a benzylic radical to (presumably) less stable primary or secondary radicals, an endothermic transition normally of some 15–20 kcal/mole. Ring opening is, however, entropically favored and a carbonyl group is formed as well. It is hence unlikely that these conversions have barriers this high. Indeed, the cleavages are chain reactions, albeit short ones (kinetic chain length of at most six).

A similar study was reported on the photochemical rearrangement of such acetals (216).[297]

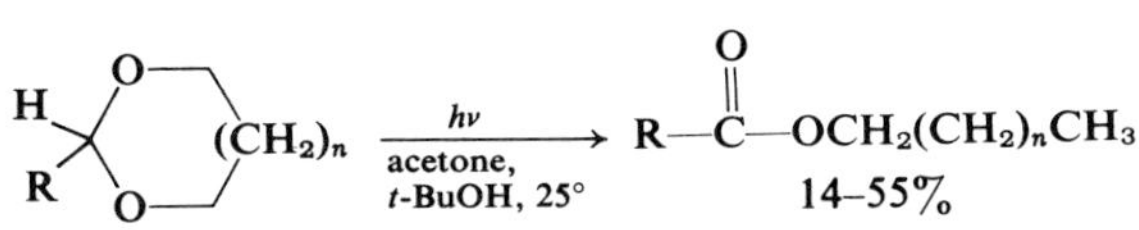

(216)

$n = 0$, R = n-C_5H_{11}, n-C_7H_{15}, n-C_9H_{19}, $PhCH_2$, $PhCH_2CH_2$
$n = 1$, R = n-C_7H_{15}, $PhCH_2CH_2$

These cleavages are much poorer in the absence of acetone, so probably the n,π^* triplet from acetone serves as the initial hydrogen atom abstractor. In accord with this view, oxygen can interfere with the reaction (217).[298]

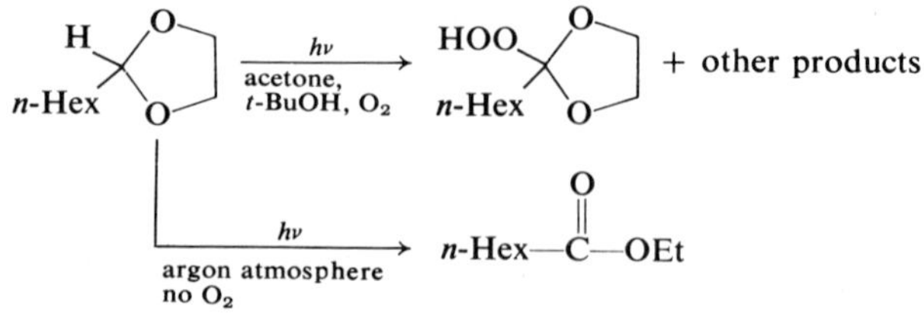

(217)

The parent compound, 1,3-dioxolane, has also been studied for rearrangement. Whereas its photochemical addition to 1-octene at 30° gave no rearrangement,[299] reaction with di-*t*-butyl peroxide at 160° led to a complex mixture of products, among them formate esters (218).[300]

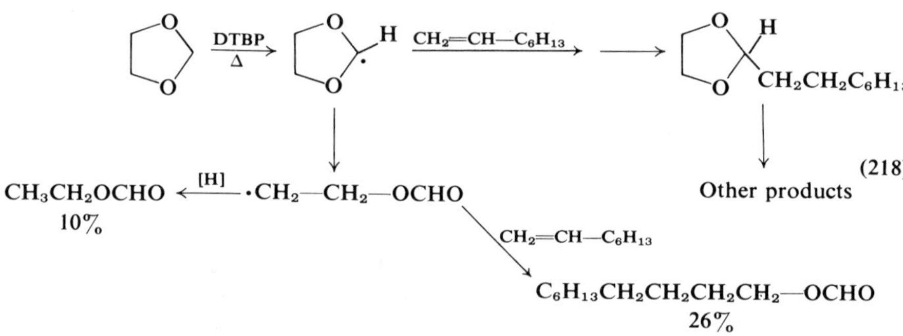

(218)

Other products (not shown) indicated several subsequent reactions as well. In addition, the other dioxolanyl radical (radical site at position 4) was also implicated.

c. Alcohol-Derived Carbon Radicals. Examples of this rearrangement are given in (219)[301] and (220).[302]

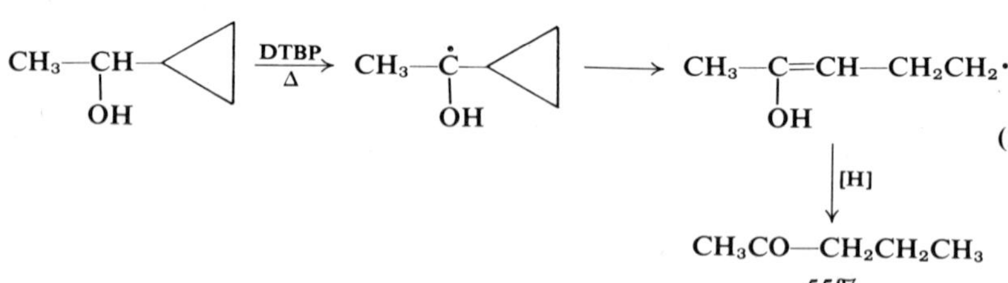

(219)

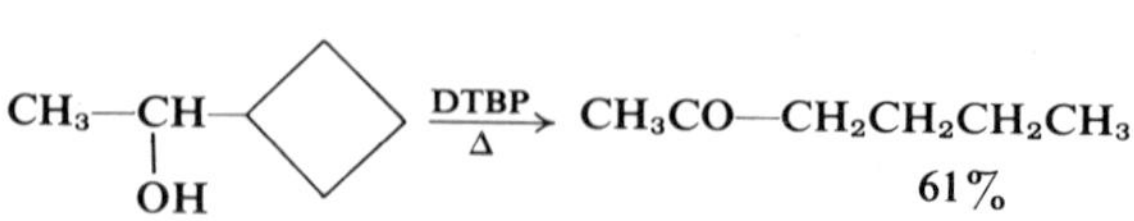

(220)

These rearrangements proceed via enols and involve no O—H cleavage. As evidence for this, the deuterated alcohol in (221) afforded only 3- and no 5-deuterio ketone.[301b]

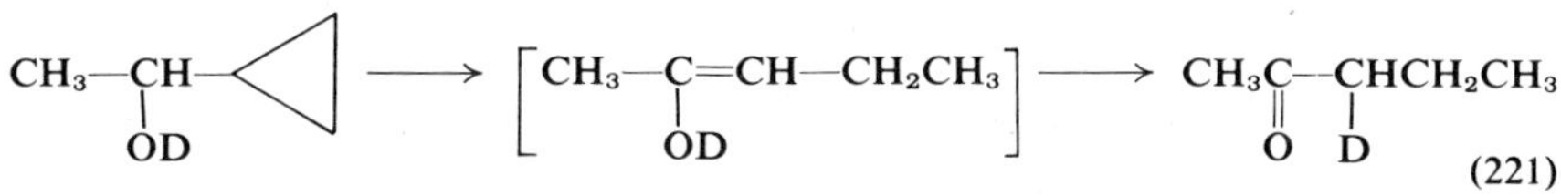

(221)

When the cleavage is attempted with aryl(cycloalkyl)carbinols, rearrangement is slower and a net oxidation process occurs instead (equation 222).

$$\text{Ar—CHOH—cyclo-R} \xrightarrow[\Delta]{\text{DTBP}} \text{Ar—}\overset{\text{O}}{\overset{\|}{\text{C}}}\text{—cyclo-R} \qquad (222)$$

Which reaction, rearrangement or oxidation, takes precedence depends on the stabilization of the α-hydroxy radical. In aryl cases, electron-supplying groups (+I and/or +R) tend to stabilize this species and to decrease the extent of rearrangement, as shown in Table 7. The effect, however, is a small one.[301,303]

TABLE 7

Ratios of Oxidation–Rearrangement Reactions in Aryl(cycloalkyl)carbinols

—Cyclopropylcarbinol	Oxidation/Rearrangement
Phenyl	2.34
p-Anisyl	3.76
p-Tolyl	2.84
p-Fluorophenyl	3.57
m-Trifluoromethylphenyl	1.86

Rearrangement of the alcohol in (223) produced the anticipated valerophenone preferentially.[301b]

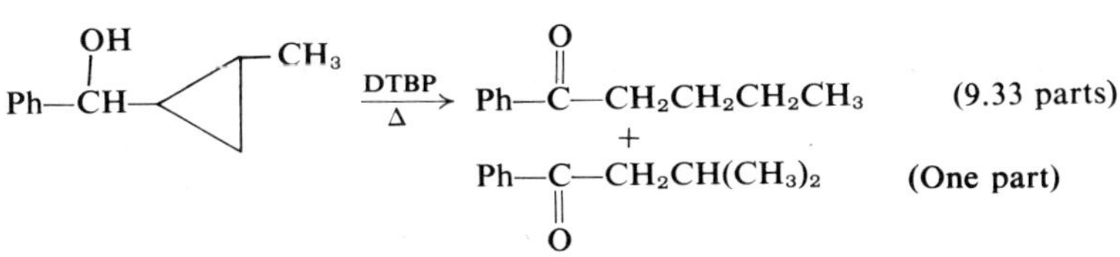

(223)

The result is the same as that observed in the cleavage of certain cyclic acetals (p. 413).

Alternatively, the rearrangements of these alcohols to ketones via enol radical intermediates can be mimicked by an entirely different process, as shown in (224).[304]

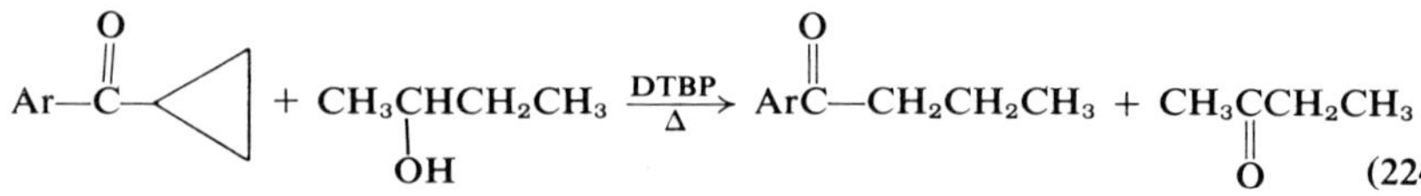

(224)

Although all steps are not definitely established,[305] the pathway of this process (reaction 225) resembles the analogous *photochemical* reaction of aryl ketones and alcohols. In fact, except for the chain initiation steps, the remainder of the chain sequence for each is probably the same, and, moreover, related to (223).

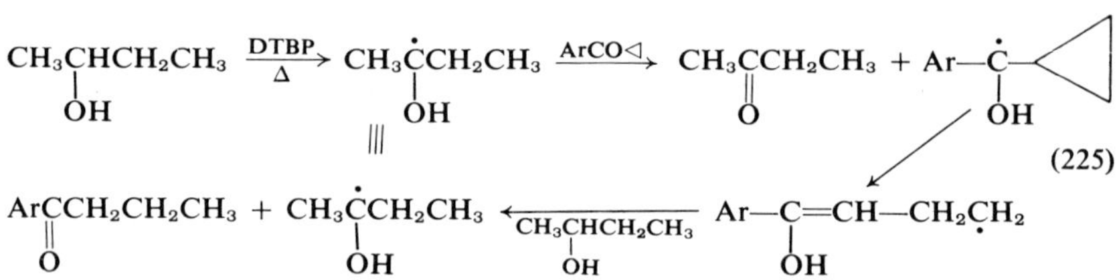

(225)

Thus, the rearrangement in (223) and the rearrangement of the corresponding ketone by the route in (224) gave the same ratio, essentially, of isomeric valerophenones. Analogous results have been obtained for cyclobutylcarbinols and cyclobutyl ketones.[302] Larger-ring homologs gave little cleavage, as might be anticipated because ring strain is less.

d. Alcohol-Derived Oxy Radicals. The rearrangement of oxy radicals by 1,*x* hydrogen transfer has been described already (pp. 382–390). In cyclic cases, an alternative rearrangement can occur as shown in (226).

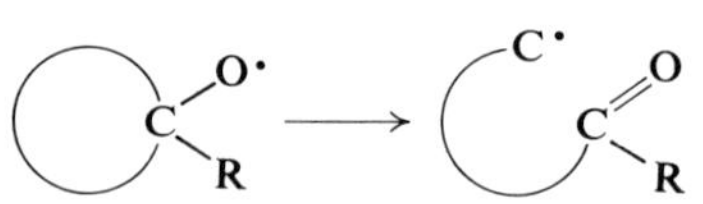

(226)

One of the more curious rearrangements of this type is that shown in (227).[306]

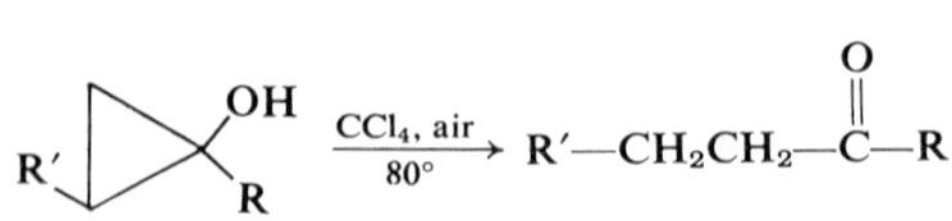

(227)

R = H, CH_3; R′ = H
R = H; R′ = *trans*-Ph

The cleavage failed to occur in cyclopropyl acetate esters or ketones. Reaction could be initiated by peroxides (but not by AIBN, see p. 412 for another such case) and it could be inhibited by antioxidants. Although the originally suggested mechanism was later recalled on the basis of substituent effects,[307] recently it has been advanced again,[308] largely on the basis of the rearrangements shown in (228).

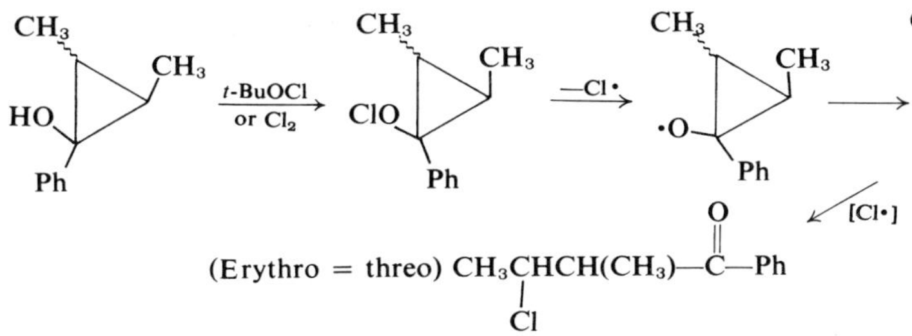

(228)

The lack of stereospecificity in the ring opening argues against electrophilic cleavage (which is specific). Because the intervention of an oxy radical formed via a hypochlorite ester is certainly possible here, it may therefore have been involved in the original cases (227) also.

The process in (226), aside from the above examples, is commonly performed in carbon tetrachloride under reflux or by weak ultraviolet irradiation at 0°, normally on hypochlorite esters formed in situ. The original example,[309] plus others in the patent literature,[310] involved *t*-cycloalkyl hypochlorites. Their rearrangement led to ω-chloro ketones, at times in good yield. Other examples followed with a rush,[186a,311–313] and some of these are collected in Table 8.

The poorer yields of chloro ketone in the larger ring sizes reflect the increased importance of intramolecular 1,*x* hydrogen transfer. This reaction transfers the radical site to a ring carbon and produces chloro alcohols instead. The rearrangement is sensitive to oxygen, in keeping with its possible chain length of more than 1000 (229).[186a,311]

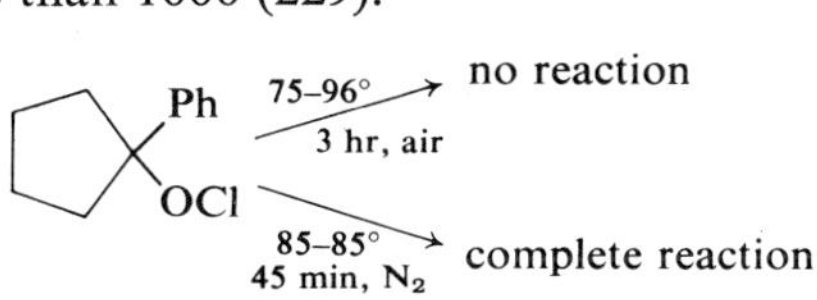

(229)

Dilute solutions also favor the process. At 0° in carbon tetrachloride, the yield of chloro ketone from 1-methylcyclohexyl hypochlorite rose from 74 to 97% as the hypochlorite concentration decreased from 1.9 to 0.001 *M*.[314] Olefins peculiarly suppress *intermolecular* hydrogen abstraction by oxy radicals and thereby allow the competitive fragmentation process a better chance.[314] This phenomenon has been carefully studied with acyclic hypo-

TABLE 8

Rearrangement of *t*-cycloalkyl hypochlorites

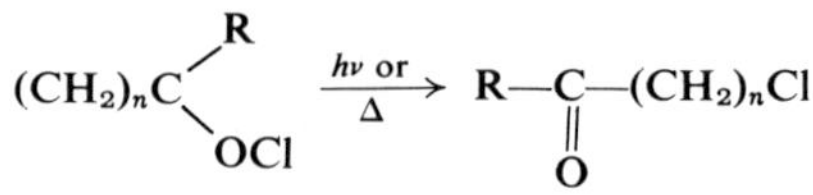

R	n	Yield of chloroketone (%)	Ref.
CH_3	4	90,52	309, 311
CH_3	5	59	311
CH_3	6	11	311
CH_3	7	1,4 to 8,15[a]	313
CH_3CH_2	5	68	186a
Ph	4	77	311
Ph	5	74	311
Ph	6	34	311
$(CH_3)_2CH$	4	94	186a
$(CH_3)_2CH$	5	13.5	186a
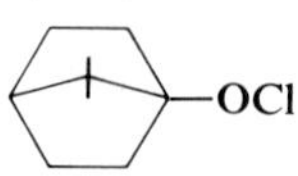		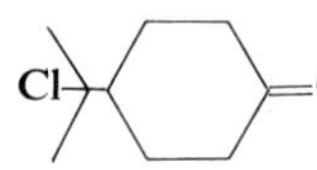, 95	186a

[a] The yield depended on conditions, being best upon thermolysis at 300°.

chlorite esters where such hydrogen abstraction can be serious. Although *intramolecular* hydrogen abstraction is the more serious side reaction in the present rearrangement, it would be nonetheless interesting to see what change, if any, an olefin would effect.

Lastly, generation of similar oxy radicals from other sources should also give rearrangement. A single example, reaction 230,[315] may perhaps suffice.

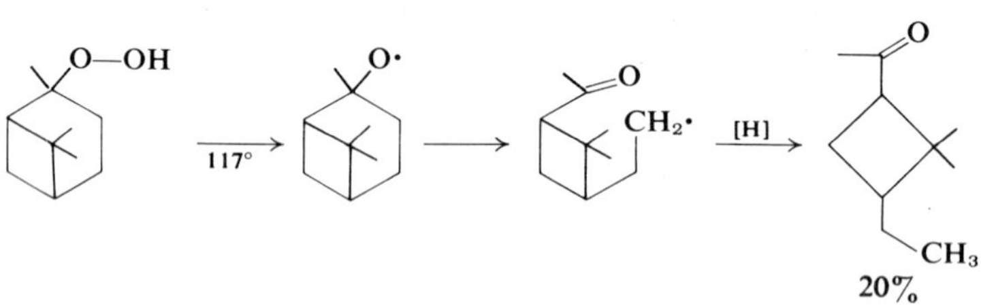

(230)

D. Cyclizations[316]

In recent years cyclization of radicals has become a major area of interest. Many examples are now known and a goodly number of them are synthetically useful reactions. Basically, the sequence shown in (231) is followed.

A few reviews are available.[317] They point out that cyclization of radicals was a conjectured phenomenon quite early in polymer chemistry, one

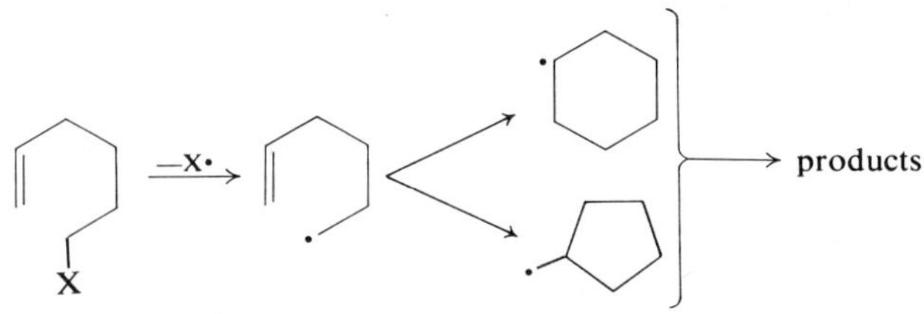

(231)

reference[318] being made shortly after the discovery of the prestigious neophyl rearrangement. Others detailed the radical cyclopolymerization of isolated dienes, as shown in (232).[319]

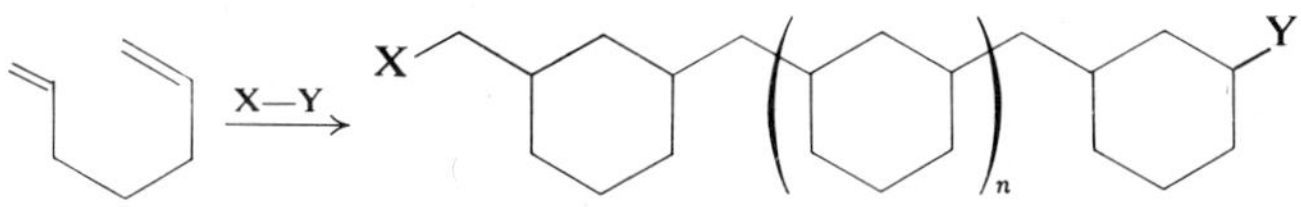

(232)

One of the problems in this area has been the identification of the cyclized product(s). In recent years it has become clear that the earlier reports of cyclohexyl products were erroneous. Cyclopentylcarbinyl products form instead or at least as serious competitors.

1. Cyclization to Small Rings

Claims have been made that γ elimination occurs in radical reactions to produce cyclopropanes (equations 233[320] and 234[321]).

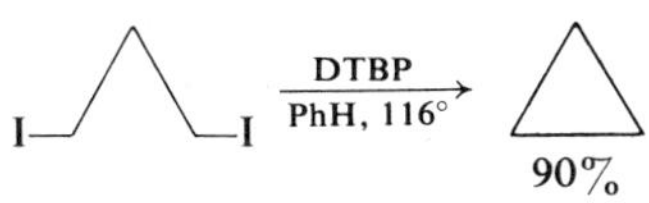

(233)

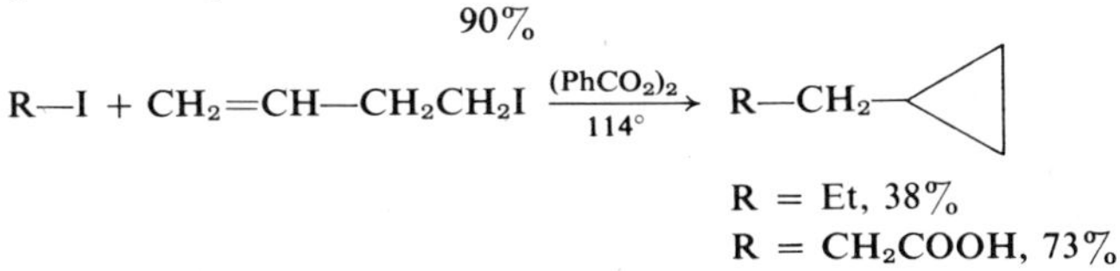

(234)

The first case is related in result (if not certainly in mechanism) to several early reports of cyclopropane formation from 1,3-dihalides.[322] Although no specific mechanism was suggested for the reaction in (234), the γ elimination in (233) has been estimated to be slightly exothermic.[323] This entices one to ascribe an S_H (homolytic displacement) mechanism to these closures. For the cases in (234) the scheme in (235) is possible, but not demanded.

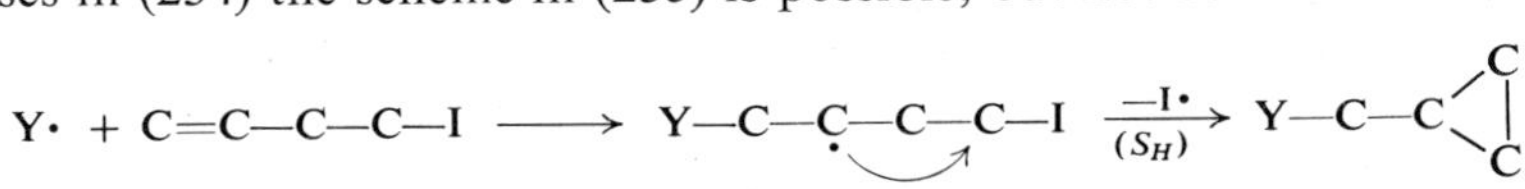

(235)

However, an obvious alternative pathway (236) seems absent because the reaction does not proceed if R = *t*-butyl or allyl. These should have been among the best cases if this mechanism held.

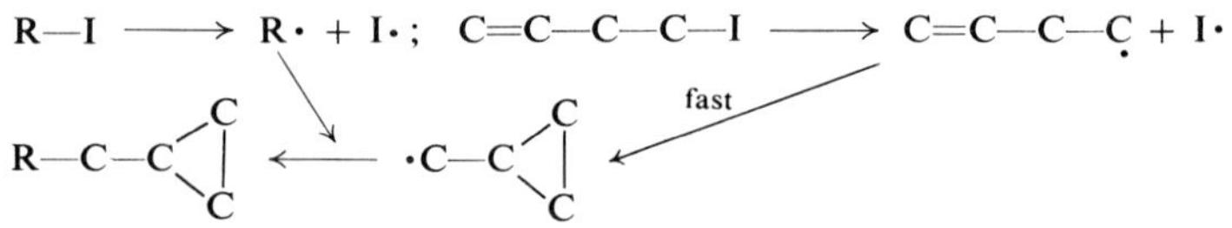

(236)

Although not specifically germane to the present section, further discussion of this possible S_H mechanism is perhaps warranted here. 1,5-Diiodopentane gave cyclopentane upon treatment with peroxides,[320a] leading to speculation that such radical displacements may be somewhat general. But this is not the case with the 4-iodobutyl radical. When generated from the diiodide with sodium naphthalene, this radical behaved as any "normal" primary alkyl radical. No evidence of an internal S_H process was obtained.[324] Similarly, the decomposition of the perester shown in (237) was no faster than the unbrominated analog.[325] No cyclopentane was found.

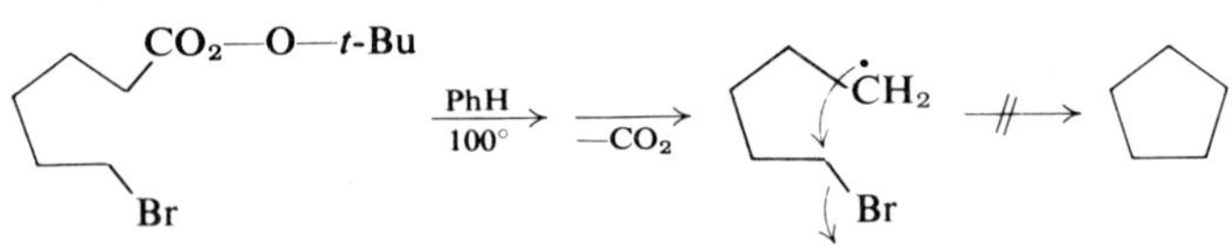

(237)

It would seem, therefore, that the S_H internal displacement was absent here as well. Obviously, much remains in question about the mechanism of the cyclizations in (233) and (234). But the conditions certainly indicate them to be bona fide radical rearrangements during their course.

2. Cyclization to Medium-Sized Rings.

Most of the reported cyclizations are in this category. Indeed, the question becomes one of selection because so many examples are known. The cyclization of the 5-hexen-1-yl radical **238-3** has been thoroughly investigated. Early studies in the gas phase[326] were followed by one on the decomposition in solution of the peroxide in (238).[327]

The rate of the decomposition was not enhanced by the presence of the double bond. The methylcyclopentane **238-1** predominated over cyclohexane **238-2** to the extent of 94 versus 6%, a point hard to understand at the time because the stabilities of the rings involved made the reverse the more likely. Because galvinoxyl effectively trapped the 5-hexen-l-yl radical **238-3**, ring closure of the latter had to be subsequently to its formation. This last makes the absence of any rate enhancement for the peroxide somewhat expected.

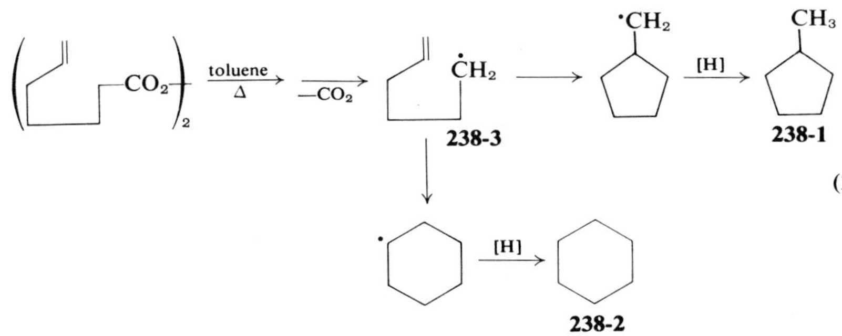

(238)

No evidence for reversibility in the ring closures was obtained. Radical **238-3** has been generated from the thiol in (239) using triethyl phosphite and an initiator.[328]

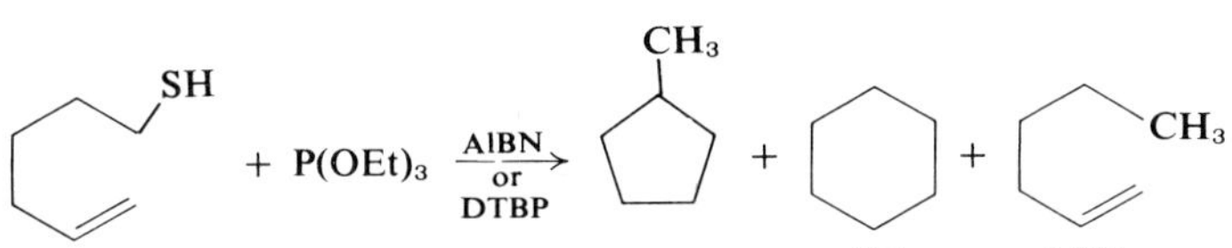

(239)

60°	43%	0%	15%
120°	50%	3%	13%

Again, cyclization to the five-membered ring product was favored, with the pathway in (240) as the probable route to radical formation.

$$\begin{aligned} R{-}SH + r\cdot &\longrightarrow R{-}S\cdot \\ R{-}S\cdot + (POEt)_3 &\longrightarrow R{-}S{-}\dot{P}(OEt)_3 \\ R{-}S{-}\dot{P}(OEt)_3 &\longrightarrow R\cdot + S{=}P(OEt)_3 \end{aligned} \qquad (240)$$

Interestingly, the thiol in (241) failed to cyclize at all.

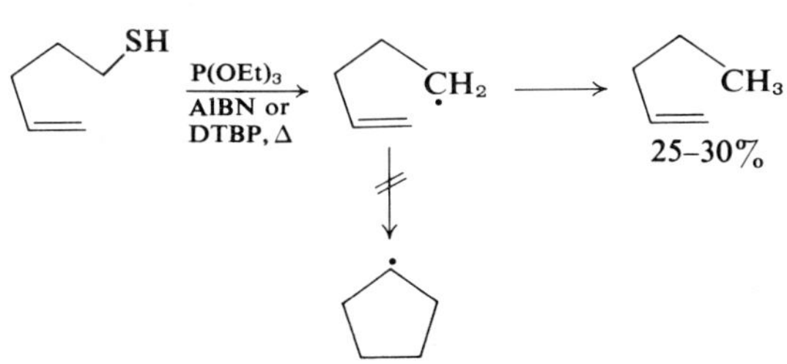

(241)

Ring closure somewhat analogous to that possible (but not found) in (241) can occur, however. As evidence for the homolytic nature of the reduction of halides by tin hydrides, the rearrangement in (242) was cited.[329]

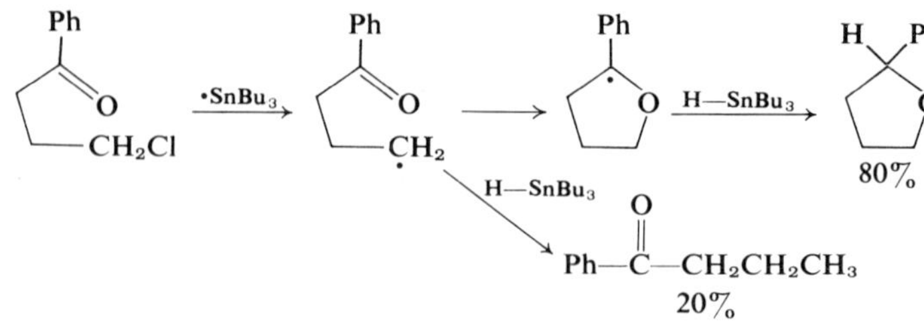

(242)

Here the ring closure is undoubtedly assisted by benzylic stabilization of the cyclized radical, a feature missing in (241).

In another instance, the thermal rearrangement of cyclopropyl acetate to allyl acetate was explained via such cyclization, this time in a diradical (243).[106]

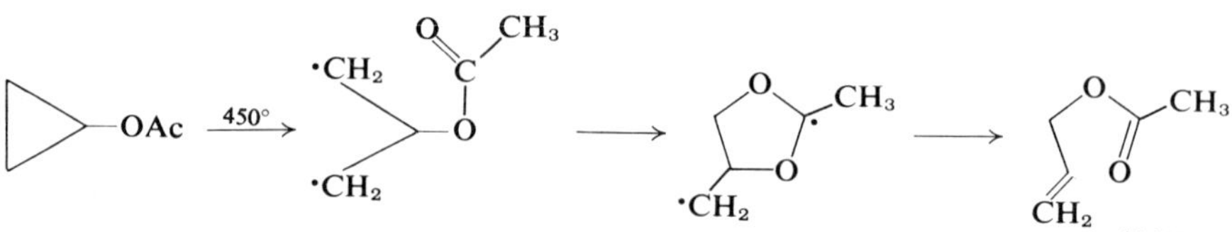

(243)

Although a concerted Cope-like pathway is also possible, the evidence based upon substituent effects on product composition makes the diradical pathway preferable.

Still other methods of generating the 5-hexen-1-yl radical **238-3** have been reported (reactions 244[330] and 245[331]).

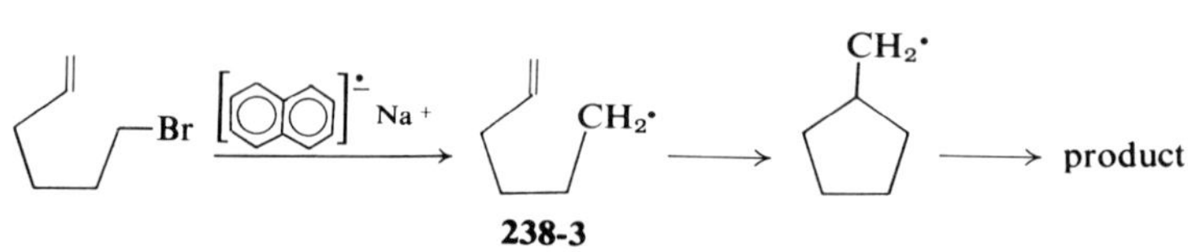

(244)

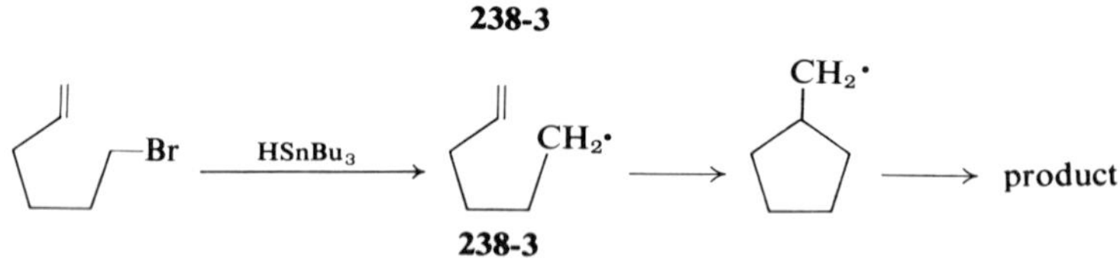

(245)

It is worth noting that the *anion* corresponding to **238-3** does not cyclize (for a similar situation, see the π route to norbornyl radical, p. 443). The detection of cyclopentyl products in the reactions shown in (246) therefore implicates radicals (and their attendant rearrangement) in the formation of the Grignard reagent. Such a situation exists in the neophyl system.[52a]

The oxygenation reaction presumably is *itself* a radical process, however, as is evidenced by the increased cyclization.[332] This interesting oxidative rearrangement has recently been studied in some detail.[333] No cyclization of the Grignard reagent in (246) occurred upon treatment with *t*-BuO—OMgCl (from *t*-BuMgCl + O_2). However, cyclization did occur upon oxygenation of the

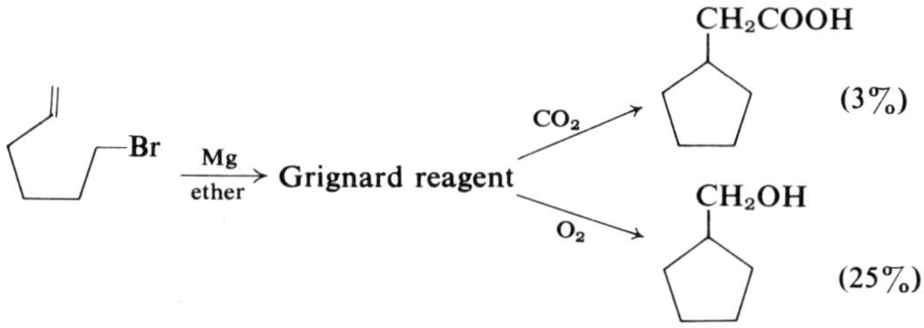

(246)

Grignard reagent and the extent of cyclization increased with decreased oxygen concentration in the solution. Although all steps are not totally clarified, the chain sequence in (247) is strongly supported by such facts.

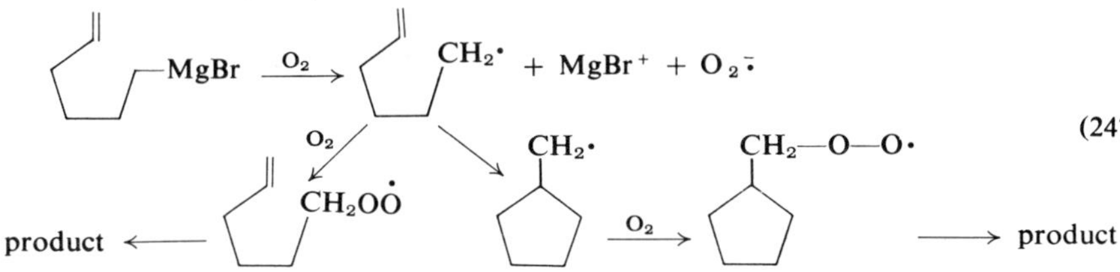

(247)

Quite analogously, the cyclization of radical **238-3** formed in (248) increased with decreased tin hydride concentration, as shown in the table.[331]

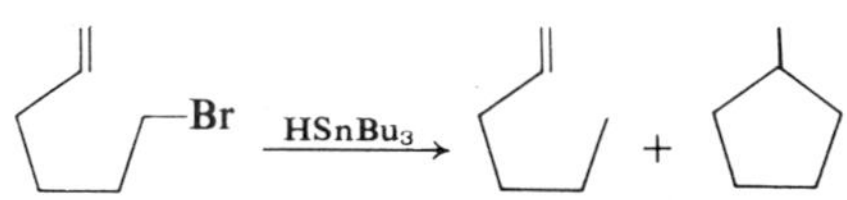

(248)

248-1 **248-2**

Bu_3Sn—H (*M*)	% Yield **248-1**	% Yield **248-2**
0.0160[a]	7	78
0.355[a]	49	42
0.0438[b]	38	50
0.526[b]	45.5	36.5

[a] At 40°.
[b] At 130°. The solvent was benzene and the bromide was in two- to tenfold excess. Initiation was by AIBN at 40°, DTBP at 130°.

Careful analysis of product composition versus tin hydride concentration led to calculations that chain transfer of **238-3** with the hydride was tenfold faster than cyclization. However, some "concerted cyclization" (249) also seemed to be implicated by this analysis.

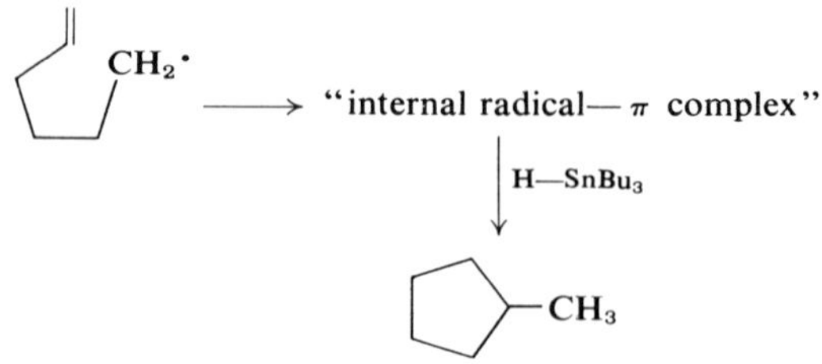

(249)

From data obtained by the rotating sector method, rate constants for reactions of halides with trialkyl- and triaryltin hydrides are known.[75] With the use of these rate constants, a rate constant for the cyclization of **238-3** (at 40°) equal to about 10^5 sec^{-1} can be calculated.[75] A recent study of the fluoride in (250) also allowed the calculation of this rate constant (upper limit of 7×10^5 sec^{-1}).[334]

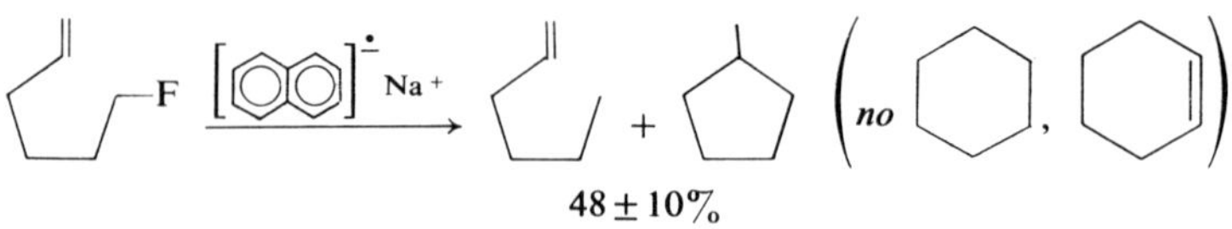

(250)

No "concerted cyclization" seemed present here because concentrated naphthalene sodium effectively quenched **238-3** without cyclization.[335]

Photochemical generation of radical **238-3** from the diacyl peroxide proceeds at −75° (equation 251).[69] At −35°, cyclization is very fast. Generation of radical **251-1** in a similar fashion led to no **238-3**, so the cyclization of the latter appears to be irreversible. All studies on the cyclization of this species support this contention.

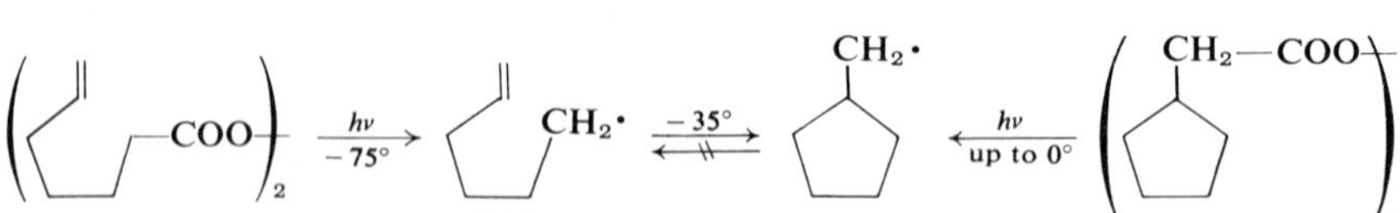

238-3 **251-1**

(251)

Another major type of radical cyclization is that of nonconjugated (isolated) dienes. Known earlier in polymerization, the initial demonstration elsewhere concerned the reactions in (252),[336] namely, radical cyclization to cyclohexyl derivatives.

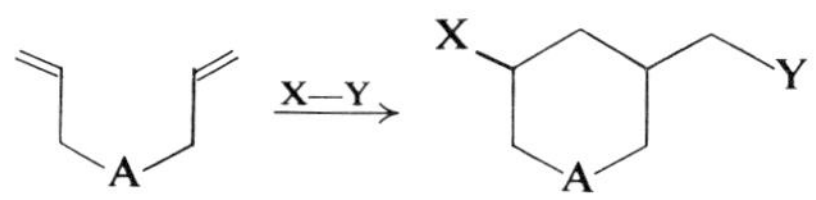

(252)

A = O, S
X—Y = Br—CCl_3, Cl—CCl_3, H—$SnBu_3$

The polymerizations described earlier[319] were, in effect, inhibited by use of these chain transfer agents, thereby allowing one-to-one addition. The initial assignment of ring size in these adducts was soon challenged, however. The cyclizations shown in (253)[337] and (254)[338] resulted either totally or at least partially in five-membered ring products. Clearly the situation was more complicated than had been originally thought.

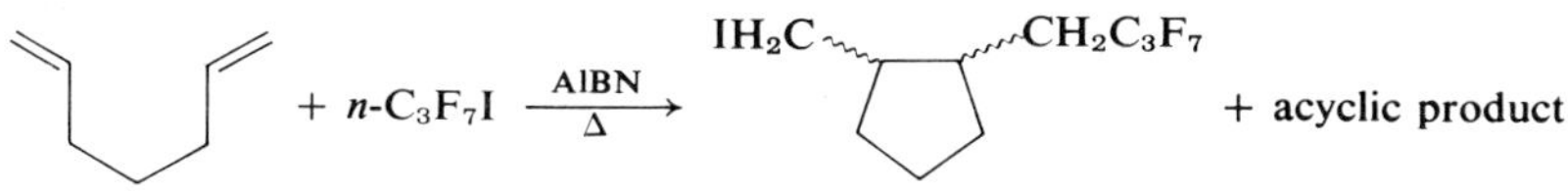

(253)

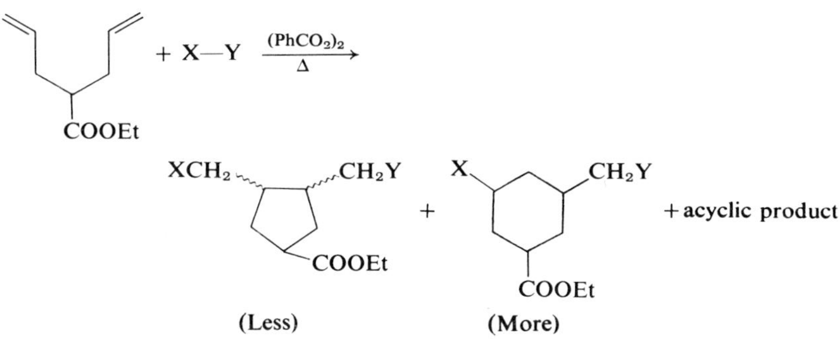

(254)

X—Y = H—$SnBu_3$, H—SAc, H—SCH_2COOEt, H—$P(O)(OEt)_2$

Both spectral and chemical evidence later indicated that the additions in (255) occurred as shown.[339] No cyclization to six-membered ring products was found.

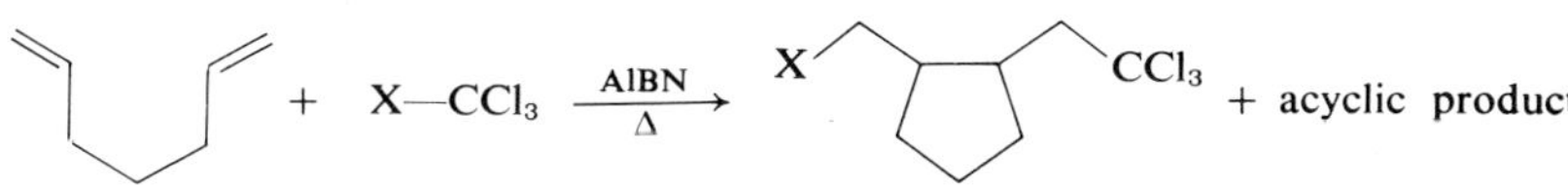

(255)

X = H, Cl, Br

Cis + trans

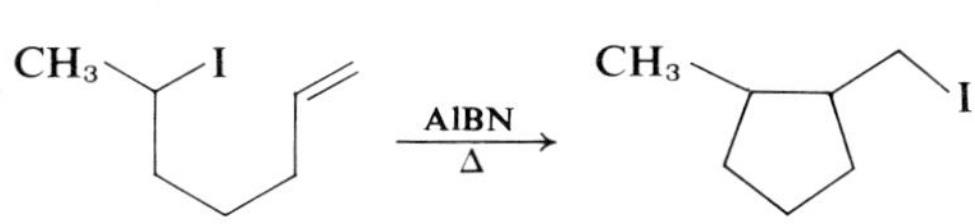

Trans/cis = 1/2.7

No cyclization occurred with 1,6-heptadiyne and 1-iodoperfluoropropane. Quite recently, the similar cyclization in (256) was reported.[340]

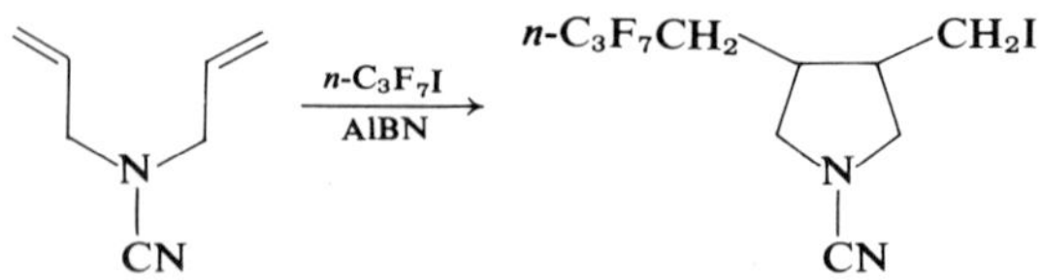

(256)

It may be seen that these reactions are general in 1,6 dienes and that they are synthetically useful. Their pathway may be summarized (257) as further examples of cyclopentylcarbinyl radical formation.

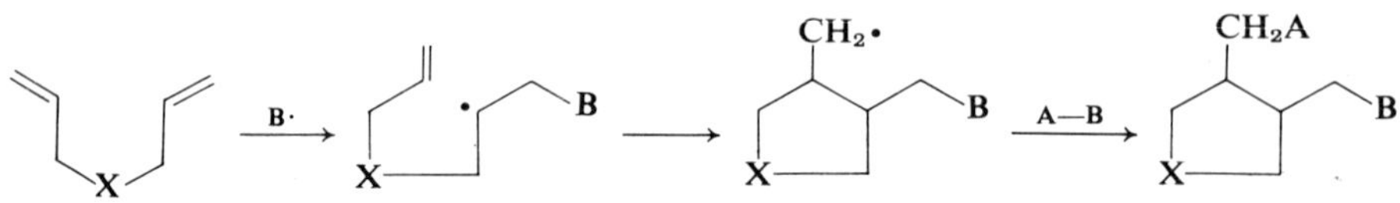

(257)

The formation of both cis and trans isomeric products in (255) may occur via the transition-state complexes shown in (258).

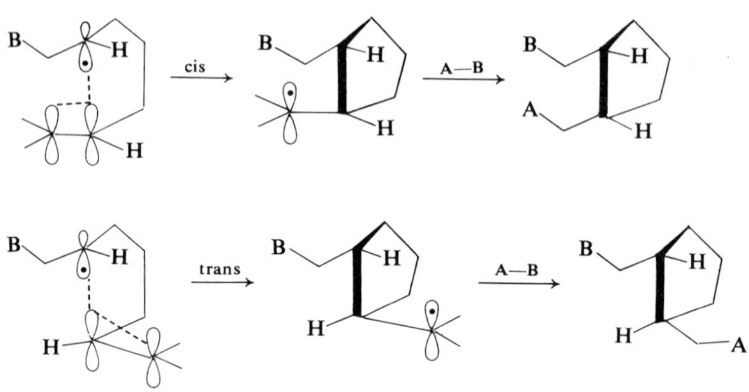

(258)

If so, such transition-state complexes must be reached early and bear little resemblance to products, because thermodynamic considerations favor the less crowded trans array. The *preference* for cis cyclization in (255) is perplexing.

Cyclizations of the present type are known also for heteroatom radicals such as thiyl and oxy radicals. The thiyl radical cyclizations are apparently reversible, a not unexpected feature of such radical additions. For example, a recent report gave the following data (258a)[340a]:

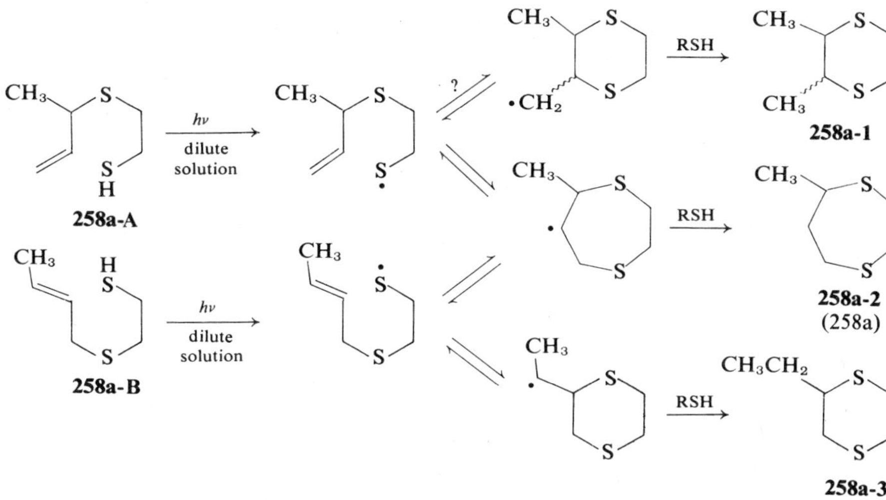

Reactant in cyclohexane	Temperature (°C)	Yield (%)	% **258a-2**	% **258a-3**
258a-A	80	47	3	95
258a-B	80	46	3	95
258a-A	−65	61	76	22
258a-B	−65	52	50	50

Obviously, at 80° thermodynamic control obtained and the two reactants gave identical product distribution, demonstrating the reversibility of the additions. Cyclization to the primary radical precursor to **258a-1** was never observed. Presumably the results at −65° represent kinetically achieved product distribution. The relative amounts of **258a-2** and **258a-3** formed under each type of control must be a function of conformational and electronic factors not precisely understood.

Oxy radicals likewise cyclize (reaction 258b).[340b]

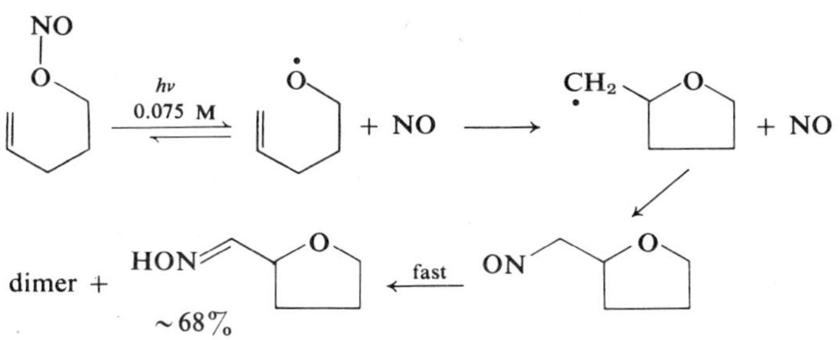

(258b)

Presumably the Barton reaction (p. 382) is precluded here by the absence of suitably positioned C—H functions.

Undoubtedly the most extensive series of studies on radical cyclizations deals with the general scheme given in (259).[341]

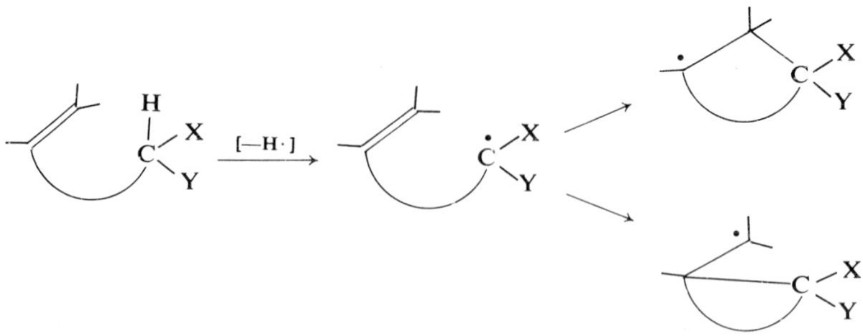

(259)

In these reactions a radical site is generated by hydrogen abstraction (as contrasted to addition or elimination reactions as in the cases above) and cyclization to a distant π center then follows. The usual reaction conditions employed a dilute solution of the substrate (to favor cyclization) in cyclohexane at 80° with benzoyl peroxide as the initiator. Chain lengths were not long and considerable peroxide was often required. Whereas the early results indicated cyclization to six-membered ring products,[342] more refined analytical procedures later established the formation of both five- and six-membered ring products (260).[343] Some results are given in Table 9.

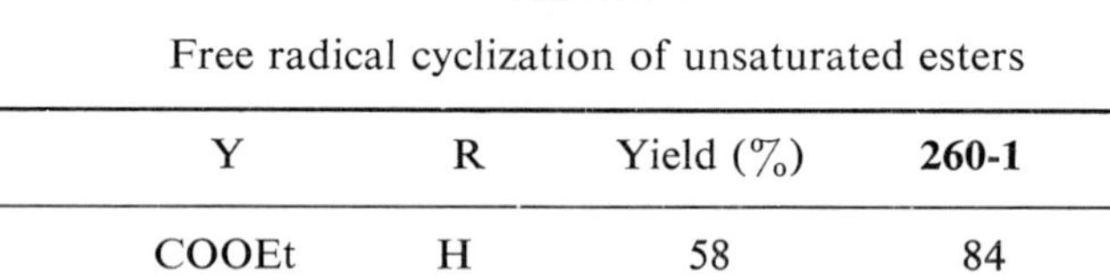

(260)

TABLE 9

Free radical cyclization of unsaturated esters

X	Y	R	Yield (%)	**260-1**	**260-2**
CN	COOEt	H	58	84	16
CN	COOEt	Me	90	100	0
COOEt	COOEt	H	55	40	60
COOEt	COOEt	Me	26	65	35
H	COOEt	H	30[a]	44	56[b]
H	COOEt	Me	25[a]	0	100[c]

[a] Di-*t*-butyl peroxide was the initiator.
[b] Trans/cis = 60/40.
[c] Trans/cis = 75/25.

Considerable variation on this theme can be achieved. A number of such varied cyclizations are given in the succeeding examples.

Example 1 (261).[344]

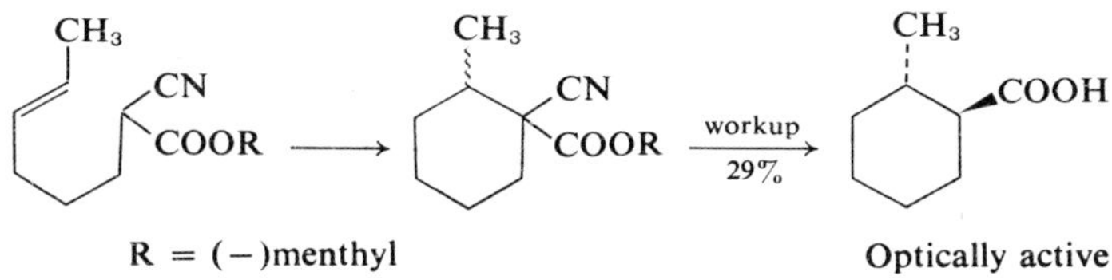

(261)

This is an example of asymmetric induction in such cyclizations. As such, it could have considerable potential value in synthesis.

Example 2 (262).[345]

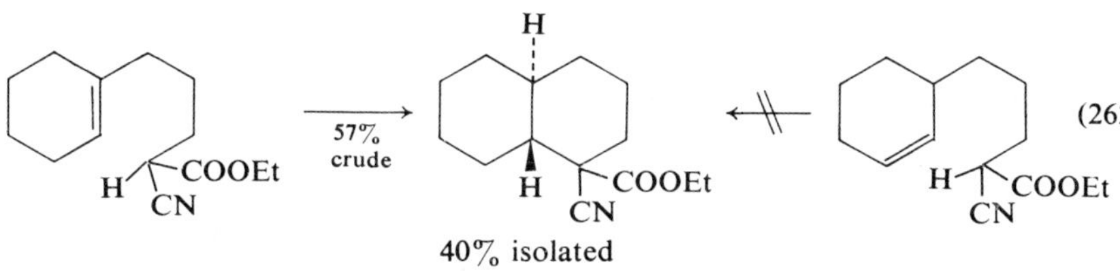

(262)

Interestingly, though the same product could form, cyclization of the isomeric cyano ester failed, mostly because the competitive hydrogen abstraction from the allylic tertiary position was too easy. Placement of a methyl group at the double bond also prevented cyclization (equation 263), most likely for steric reasons.

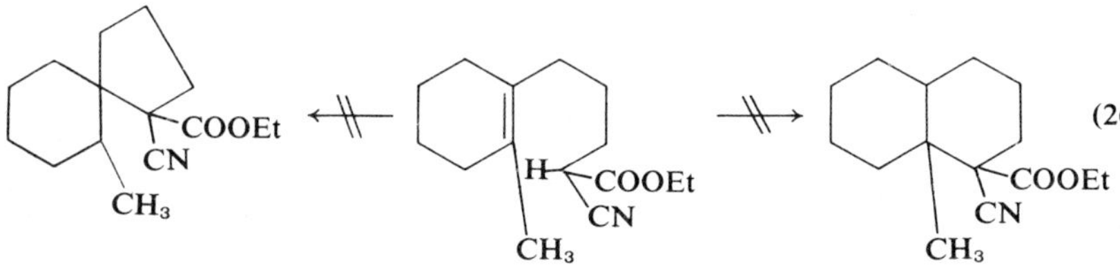

(263)

The contention that the strain involved in closure to a [5.4]spirane prevented cyclization in that direction is suspect. Such spiranes can be prepared by radical cyclization (see p. 440).

Example 3 (264).[346]

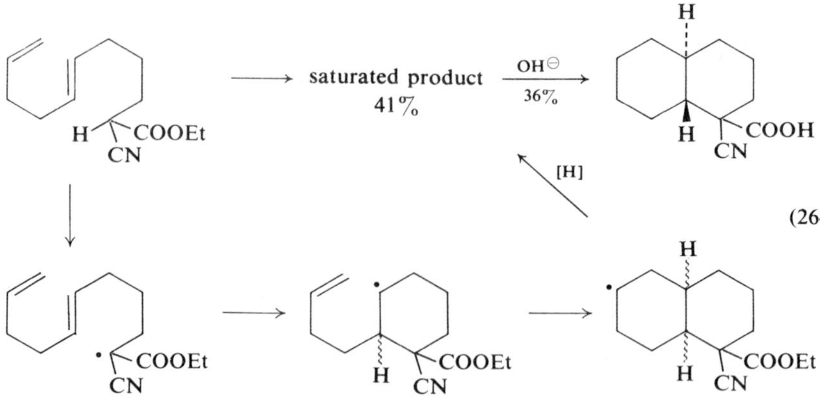

(264)

It is amazing that a process with such a labyrinthine pathway succeeds at all![347] The stereochemistry of ring closure in many of these cyclizations is unclear. Mixtures are undoubtedly produced and the isolated products (usually the higher-melting trans isomers) do not in themselves tell enough. The situation needs clarification.

Example 4 (265).[348]

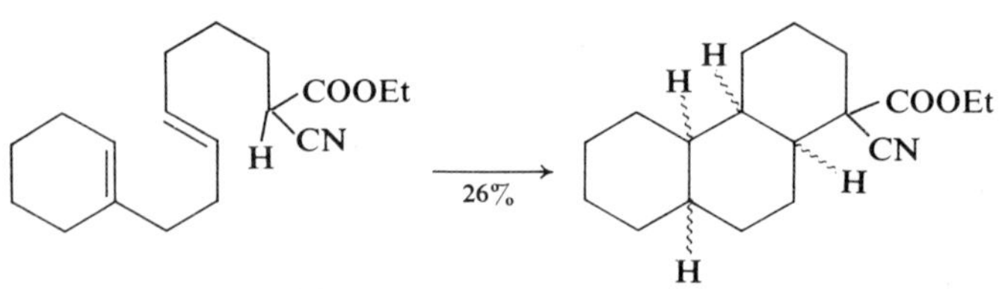

(265)

This one is equally amazing!

Example 5 (266).[349]

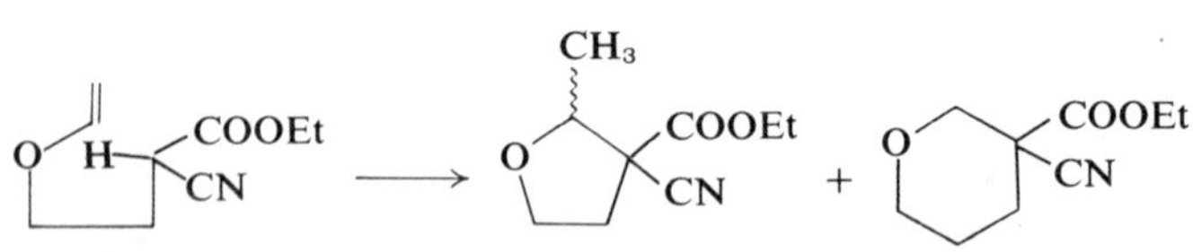

(266)

These heterocyclic systems can be prepared analogously to the carbocyclic cases. High dilution and higher temperatures are needed, however, because polymerization is a serious side reaction here. The yields of cyclized products varied in substituted cases from 40 to 60%.

Example 6 (267).[350]

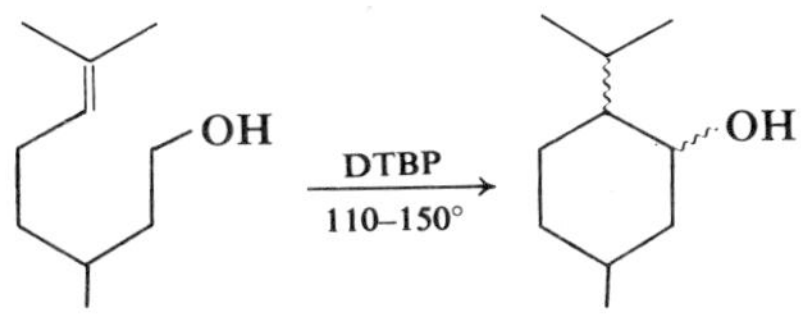

(267)

This cyclization of citronellol afforded a mixture of menthol (6.2%), neomenthol (4.8%), isomenthol (1.1%), and neoisomenthol (0.6%). The percentages suggest that the sterically less crowded products predominate. In fact, the composition is comparable to that obtained by the hydrogenation of thymol over Raney nickel at 200° and 20 atm.

Radical ring closure to five-membered ring products is usually, but not always, less competitive in these cyclizations. At times, such cyclizations fail (268).[351]

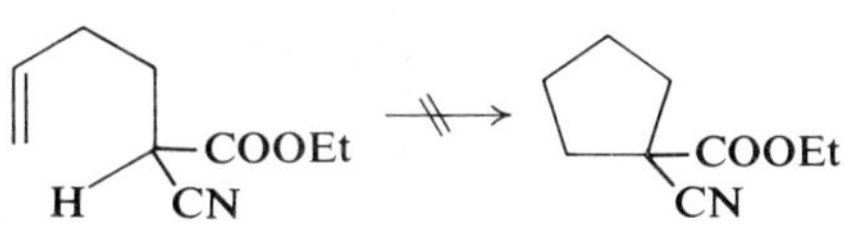

(268)

If the cyclization is assisted by the presence of an electron-enriched double bond, then it may occur (269).[351,352]

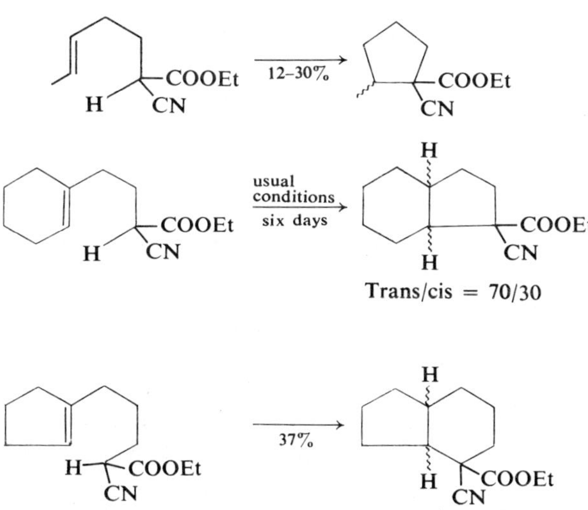

(269)

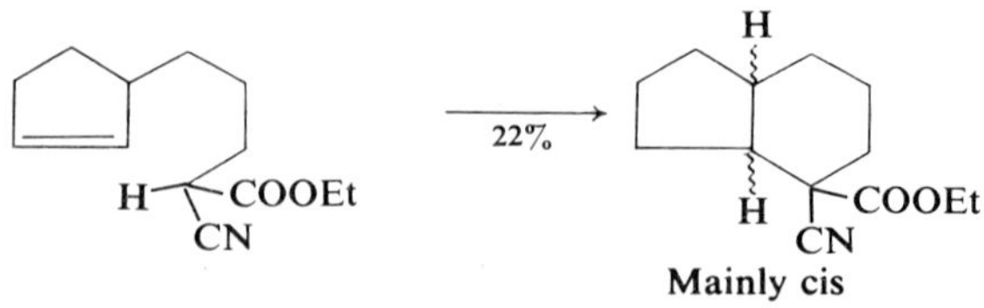

This facilitation by alkyl groups on the double bond may be polar in origin, stabilizing the intramolecular addition because the α-cyano ester radical is itself undoubtedly electron seeking. Again, the stereochemistry of these cyclizations has been deduced from solid derivatives isolated from rather complex reaction mixtures and via lengthy workup procedures. Some degree of uncertainty must be present.

The important observation has been made that these cyclizations are reversible *in certain cases*. Evidence exists that the simple case is irreversible (see p. 424). However, when the acyclic radical is stabilized by cyano and carboethoxy groups, reversibility has been demonstrated.[353] Under such conditions, product stability then becomes a controlling feature. Some indications of this fact are given in Table 10.

TABLE 10
Radical cyclization to 5- versus 6-membered ring products

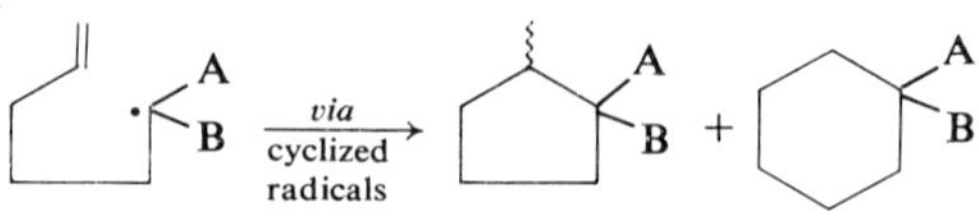

A	B	%	%	Ref.
H	H	100	0	334
Ph	H	92	8	354
COOEt	H	56	44	343
COOEt	CN	16	84	343

In cases where closure to both five- and six-membered ring radicals can occur, the kinetic preference is for the former, while the thermodynamic preference is for the latter. As the cyclizations change from irreversible (unstabilized acyclic radical) to reversible (stabilized acyclic radical) the control passes from kinetic to thermodynamic and the proportion of the stabler six-membered ring product increases. Other reaction features can also affect the product distribution, for example, temperature and solvent. Some of such data is collected in Table 11.[355]

TABLE 11
Effect of temperature and solvent on cyclization

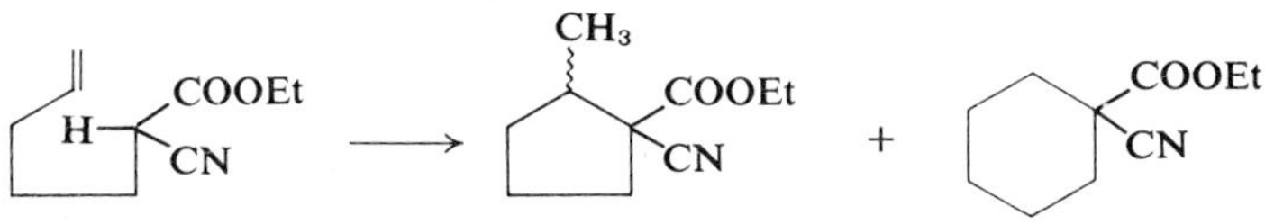

Photoinitiated in decalin		
°C	%	%
−70	50	50
25	35	65
105	29	71
Photoinitiated at 11°		
Solvent		
Alkanes	63–68	37–32
Cyclohexane + cumene (10%)	78	22
Toluene	90	10

Clearly cyclization to the stabler six-membered ring product is also favored by higher temperatures and poor hydrogen donor solvents. Good donor solvents trap out the kinetically favored five-membered ring product. Although the thermodynamic preference for the six-membered ring product reflects the well-known greater stability of this ring size vis-à-vis the five, the kinetic preference for the five-membered ring is less easily explained. One view of this preference is mentioned later (see p. 442).

3. *Cyclization to Hydroaromatics and Related Compounds*

Such cyclizations have been observed with a variety of radicals. Two early examples of acyl radical closure are illustrated in (270)[61b] and (271).[136]

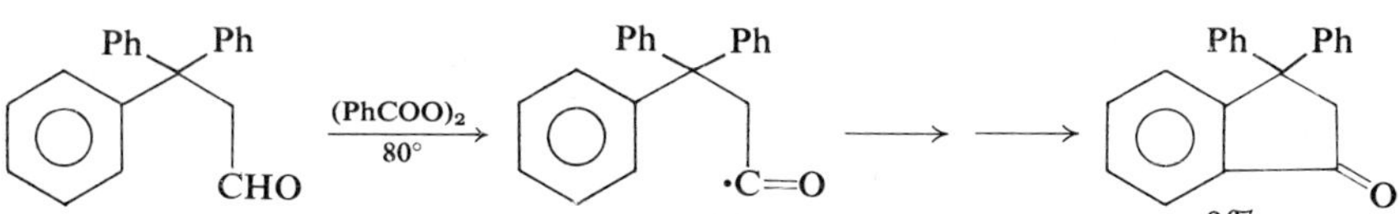

(270)

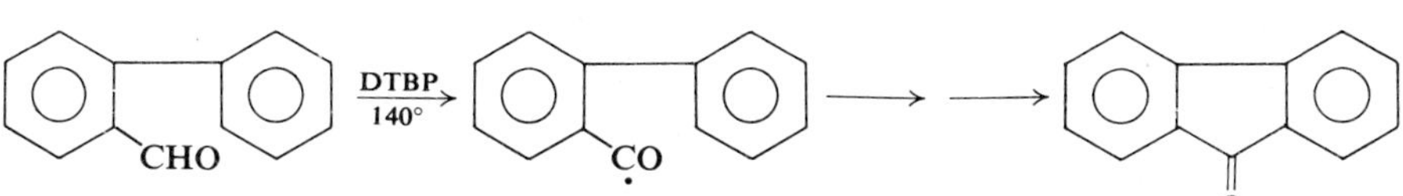

(271)

A more recent instance proceeded analogously (272).[135]

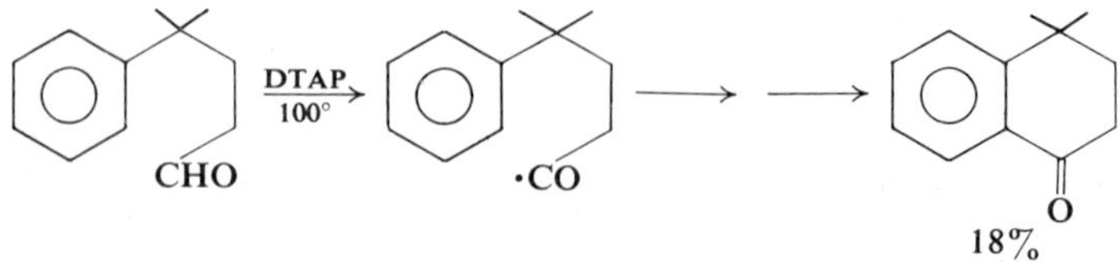

(272)

Other similar cyclizations were also described. These indanones and tetralones are formed in a chain-termination step (273) and their yield can be high (e.g., 90% on the basis of peroxide used).

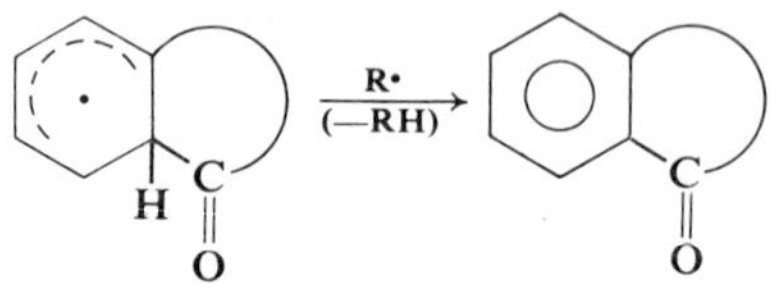

(273)

A comparison of the ease of closure shows that certain acyl radicals cyclize somewhat more readily than their alkyl counterparts (equation 274).

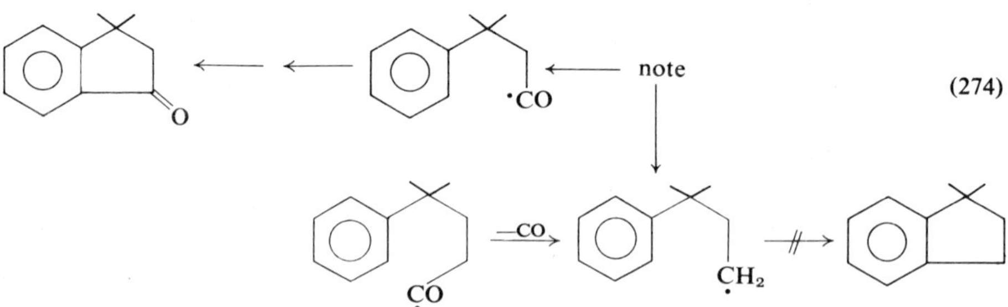

(274)

Perhaps an overlap of the π electrons of the carbonyl group in the acyl radical with the aromatic ring π system stabilizes the transition state for ring closure in this case. Other views are possible, however.[135]

Aralkyl radicals can, nonetheless, cyclize. Both indane and tetralin products result from such processes. Some simpler examples of the latter type of cyclization are given in (275),[165] (276),[356] and (277).[132]

These cyclizations are Ar_2-6 homolytic rearrangements. At times Ar_1-5 processes can compete (see p. 440). Although little has been reported on the aromatization step in the above cyclizations, it may occur via disproportionation, oxidation by other radicals present, or by oxidation during the processing of the reaction. Because Ar_2-5 cyclization involves more strain in its transition state, fewer examples of indane formation are available. Indanes do form, however, from alkylated benzenes at elevated temperatures (over

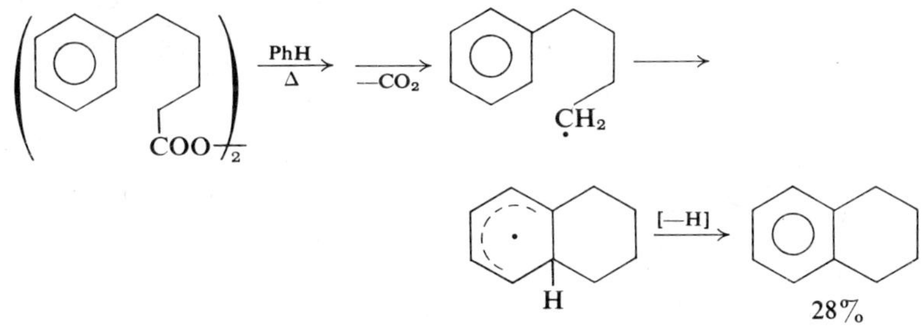

(275)

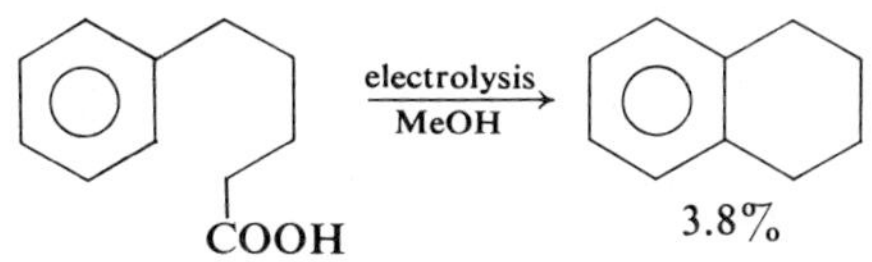

(276)

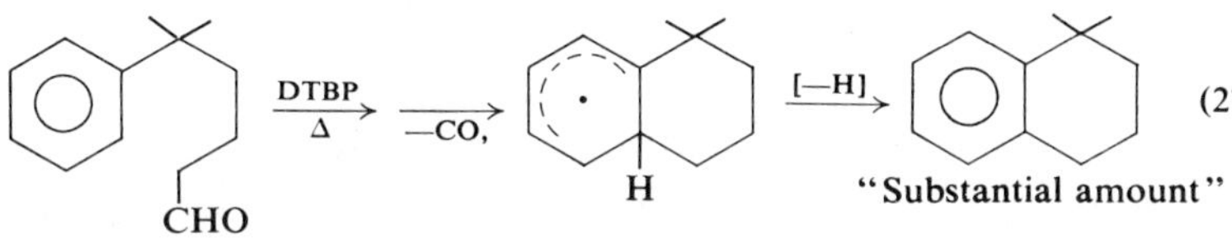

(277)

400°), presumably via radical cyclization.[357] Furthermore, lower-temperature cyclizations are known (278).[354]

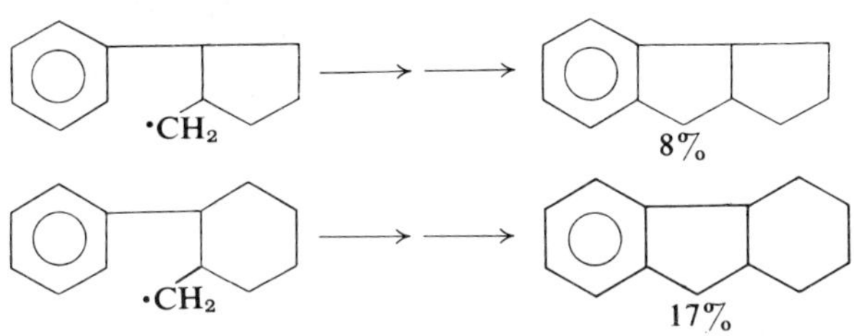

(278)

These products form with others in a complex reaction scheme from ω-phenyl-1-alkenes and di-*t*-butyl peroxide at 140–150°. The percentages represent composition, not yield. Cinnamyl and homocinnamyl radicals can cyclize under certain conditions (reactions 279[358] and 280[267]), although the cyclization of the former appears to be reversible, with the rate-determining step being the aromatization of the cyclized radical.

As an extension of the staggering number of cyclizations observed in acyclic and alicyclic cyano esters, such cyclizations can also occur via Ar_2-6 closure. Although the direct approach, as in (281), failed, an indirect one succeeded (reaction 282), at least when done under carefully controlled conditions.[359]

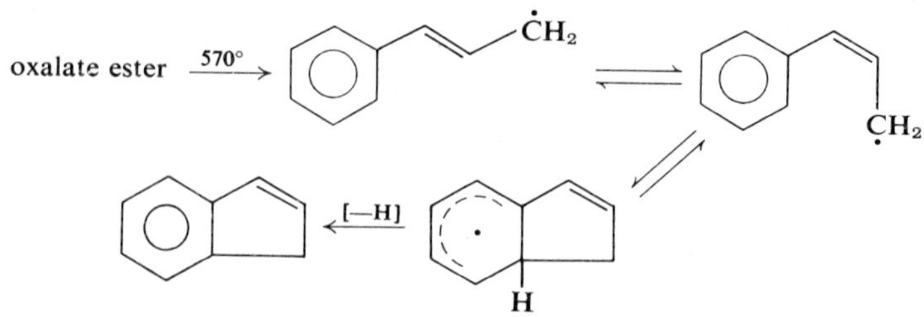

(279)

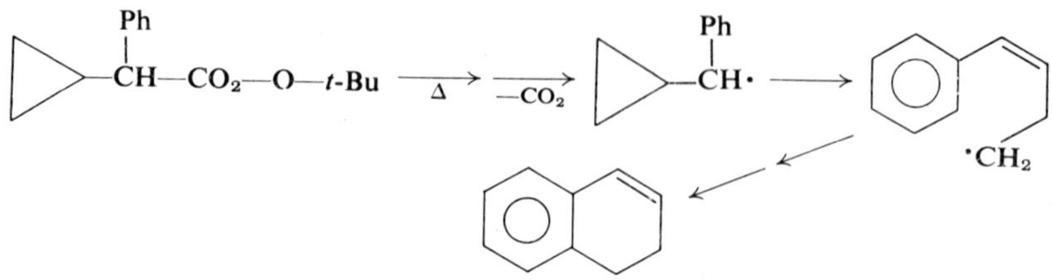

(280)

Cyclization in (282) failed when di-*t*-butyl peroxide was substituted for the benzoyl peroxide. Rather, simple vicinal addition of the cyano ester was

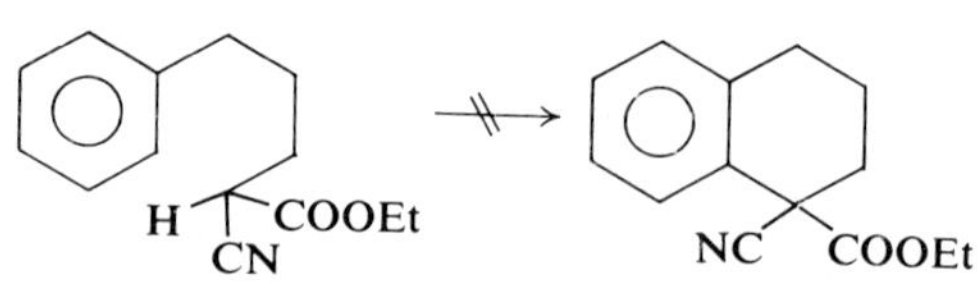

(281)

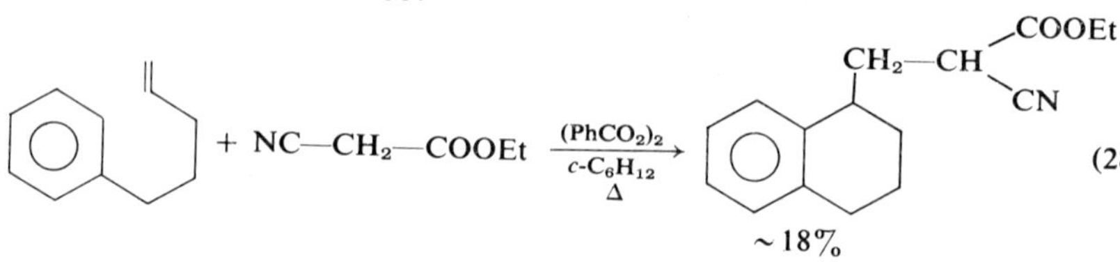

(282)

observed. Some of this occurred with benzoyl peroxide as well. The key step apparently is the aromatization of radical **283-2**, as shown in (283), before it can revert to the opened radical **283-1**, which affords vicinal adduct.

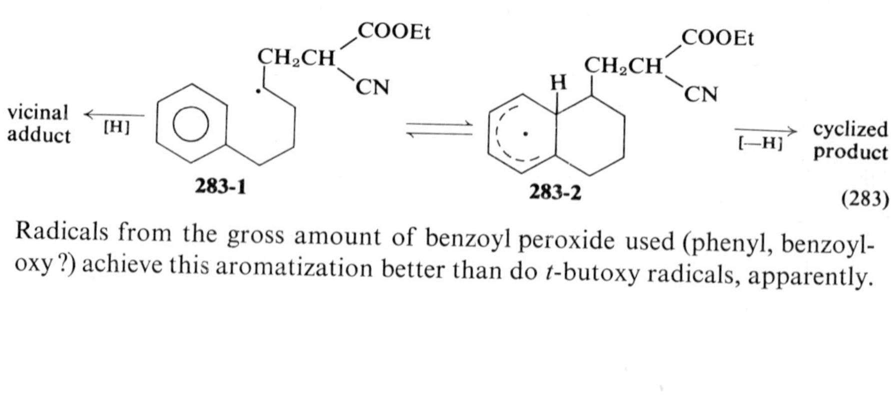

(283)

Radicals from the gross amount of benzoyl peroxide used (phenyl, benzoyloxy ?) achieve this aromatization better than do *t*-butoxy radicals, apparently.

The failure in (281) may, in fact, be due to a reduced aromatization rate (and not cyclization itself) because of steric hindrance near the reaction site. Indeed, the failure in (281) can be remedied under other conditions (284).[360]

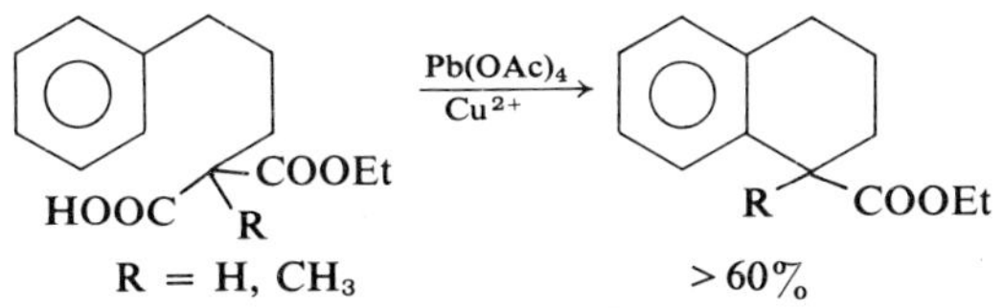

(284)

As in other cases of lead tetraacetate reactions (see pp. 373, 386), there is considerable doubt, however, that the radical character of (284) approaches that of these other cyclizations. When the initial radical is tertiary, cyclization also failed to occur (285), either for steric or electronic reasons.

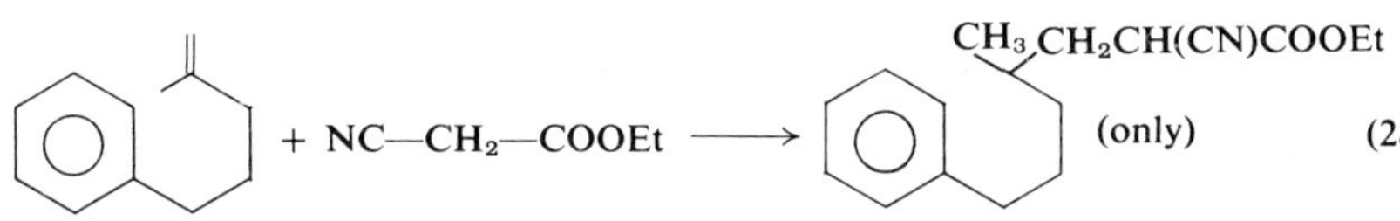

(285)

The Ar_2-6 closure can be embellished with a prior cyclization, as shown in (286).[361]

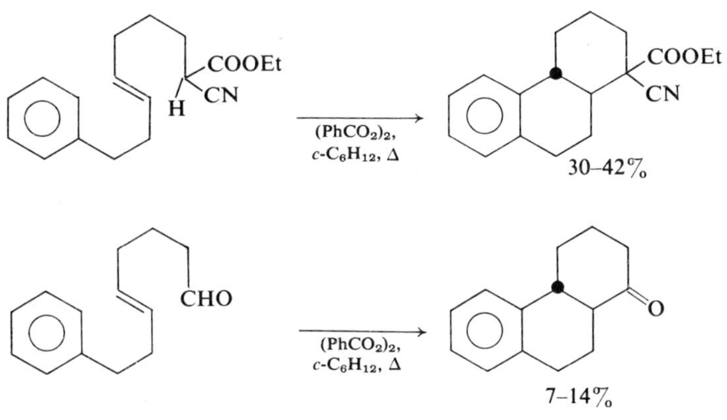

(286)

Spectral data supported the trans ring juncture in these products. However, as previously mentioned, isolation of such trans-joined products does not preclude other isomers as additional products.

4. *Bicyclizations*

Transannular radical addition to a double bond leads to a bicyclic product. Several studies of 1,5-cyclooctadiene in such a bicyclization have been reported. A brief note on its reaction with acetic anhydride in the presence of di-*t*-butyl peroxide indicated the course of the radical rearrangement as shown in (287).[362]

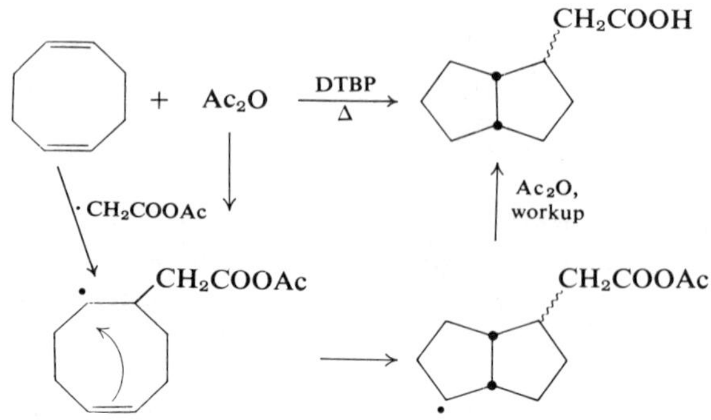

(287)

Two other, quite similar, studies involved other addenda (reactions 288[363] and 289[364]).

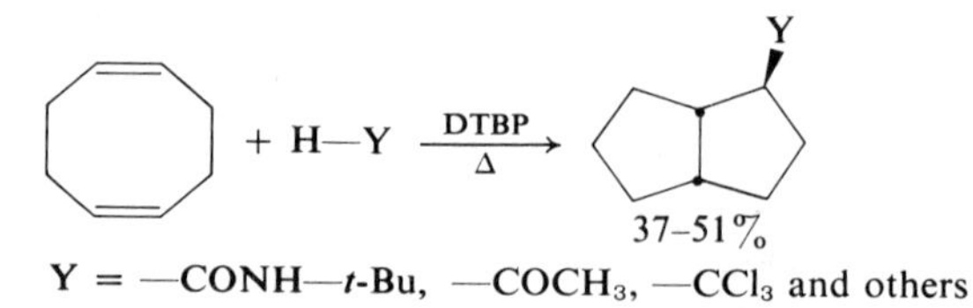

(288)

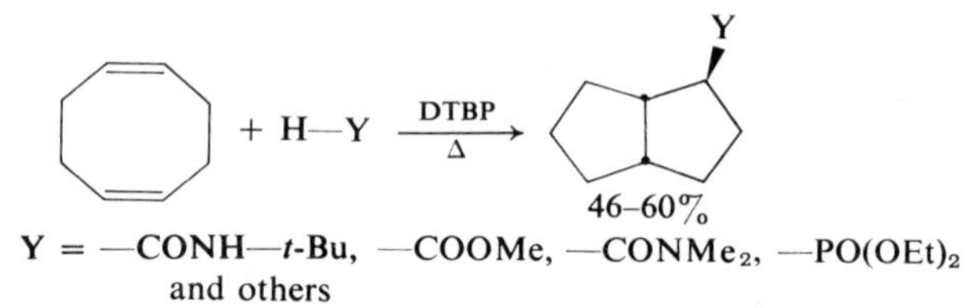

(289)

The cis ring juncture of the bicyclo[3.3.0]octane derivatives always found is essentially forced upon the reactant because a trans juncture would be considerably more strained.[365] The adduct with carbon tetrachloride (equation 290) has, moreover, both substituents apparently also stereo-electronically controlled in their orientation.[363]

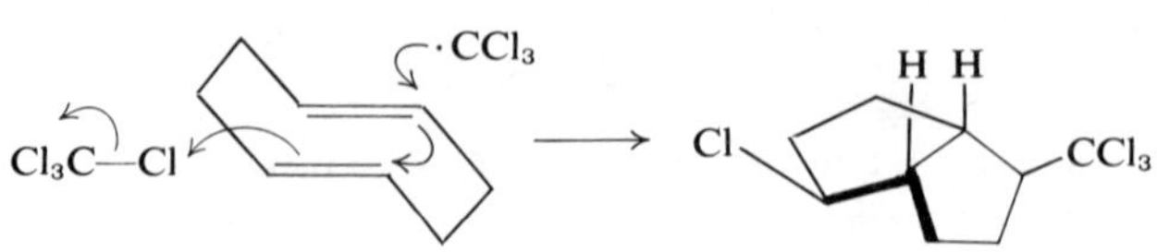

(290)

The trichloromethyl group appears from spectral data to be definitely exo. Because chain transfer should be easier from the less crowded *exo* side, the chloro function is probably also exo. It is of some interest in this regard that *ionic* bicyclization of this diene gave endo-2 derivatives (291)![366]

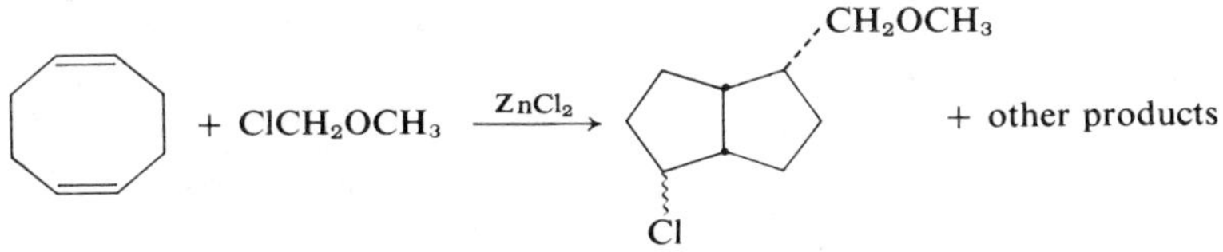

(291)

The case illustrated gave both 6-chloro epimers in comparable percentages. The difference between the radical and ionic bicyclizations has been explained as shown in (292).[366]

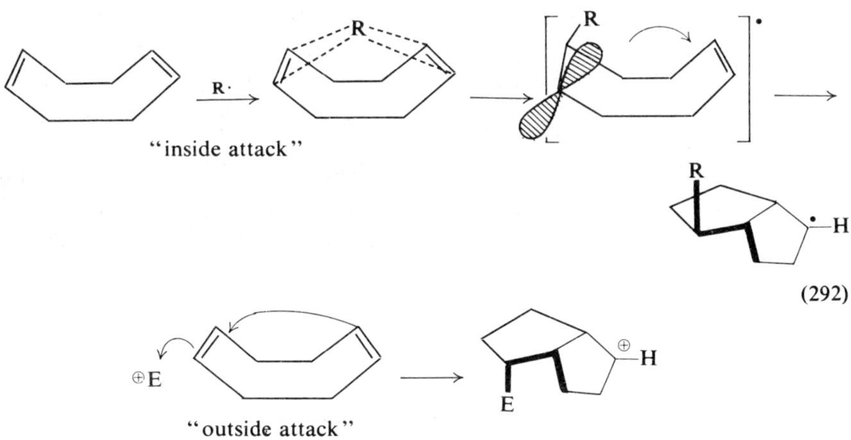

(292)

Radical attack is postulated to involve "inside attack" by the radical on the π system. Bicyclization of the intermediate radical then positions R exo. This is best shown through use of molecular models. On the contrary, electrophilic attack is postulated to occur on the "outside" of one double bond, and probably with participation by the other double bond, forcing E endo. Indeed, it is this participation that occasions the difference in the two additions. Because concerted addition of this type gives endo adducts, the radical additions are probably stepwise. A concerted addition mechanism [much as shown in (290)] had been speculated upon prior to the findings in the ionic cases.[363] Such a concerted pathway has also been advanced for the bicyclization of germacrene **292a-1** caused by the photochemically induced addition of several addenda (292a).[363a]

It is claimed that the specificity results from a concerned process utilizing a *trans*-decalin-type radical with synchronous C—X and C_5—C_{10} bond

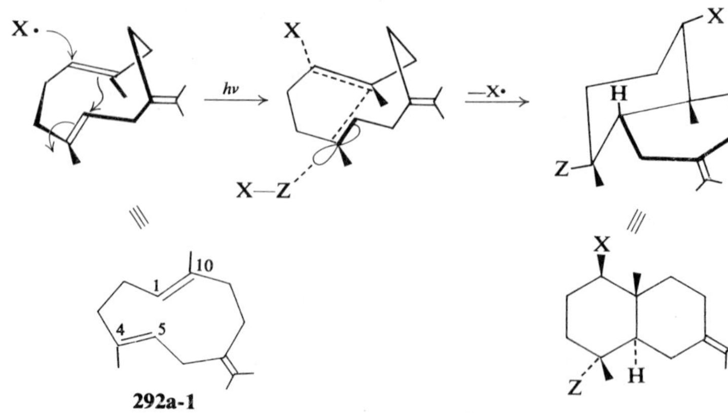

292a-1

(292a)

formation. Alternative attack by the radical center at C_{10} upon the transannular double bond at C_4 is presumably retarded by the greater strain present in the incipient bicyclo[6.2.0]decane system. The X—Z addenda used in (292a) were thiophenol, diphenyl disulfide, and carbon tetrachloride. Yields of adducts ranged from 32 to 51%.

Other suitable dienes should undergo bicyclization. Cases involving norbornadiene are numerous (see Section IV.F.3), but otherwise not too much seems to have been done. 4-Vinylcyclohexene underwent radical attack, but no bicyclization (293).[367]

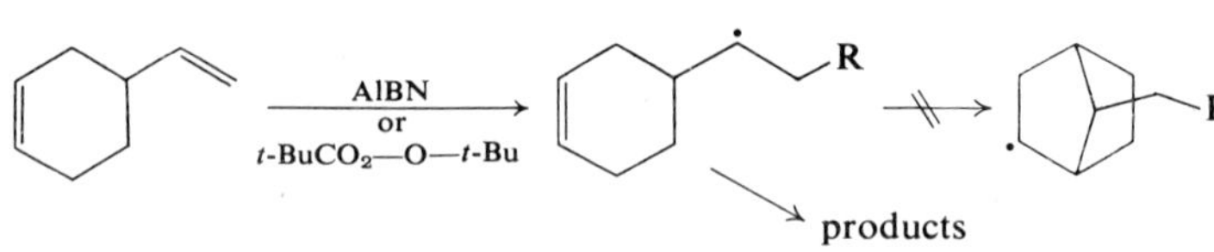

(293)

Two studies reported bicyclization of olefinic bromides upon reduction with tri-*n*-butyltin hydride. In the first, a spirane and both decalins resulted (294).[368]

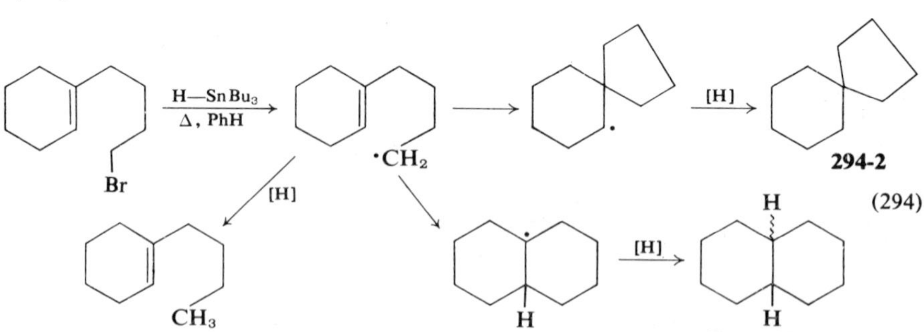

294-2

(294)

294-1

Trans **294-3**
Cis **294-4**

No evidence for radical disproportionation was found, as might be expected in the presence of the excellent chain transfer agent employed. The product composition varied with concentration, as shown in Table 12.

TABLE 12

Cyclization products from (4-bromo-1-butyl)cyclohexene

Product	% Composition[a]	
294-1	78[b]	43[c]
294-2	12	31
294-3	9	22
294-4	1.5	3.8

[a] Yields were quantitative.

[b] Values in this column were obtained when R—Br $M = 0.1$ and Bu_3SnH $M = 0.14$.

[c] Values in this column were obtained when R—Br $M = 0.02$ and Bu_3SnH $M = 0.028$.

The ratio of spirane to total cyclized product stayed constant (~0.85) over a considerable range of tin hydride concentrations. "Concerted" cyclization (see p. 424) thus appeared to be absent. Moreover, the ratio of cis and trans decalins also stayed constant, so the decalyl radical was presumably conformationally equilibrated under all the conditions employed. No conclusion was therefore possible on the stereochemistry of the actual cyclization step (see p. 430). This cyclization has recently been achieved as well from the peroxide shown in (295).[369]

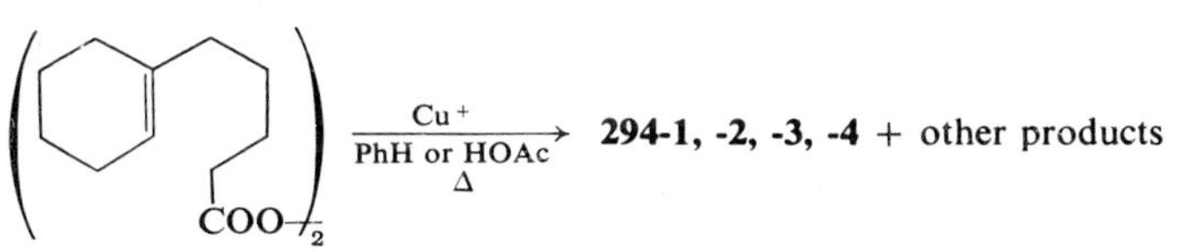

(295)

The similarity of product composition from this reaction *in benzene* to that in (294) led to the conclusion that this Cu^+-catalyzed process was also radical in nature. Use of acetic acid as solvent, however, changed the pathway to that of an ionic solvolysis. An intimate view of cyclization has been offered from these studies.[368] The initial stage was pictured as a vertical approach of the radical center within the plane of the π system to interact with the π^* olefinic orbital (296).

Such a viewpoint helps one to understand the preferential kinetic formation of cyclopentane instead of cyclohexane derivatives (the vertical orbital overlap is easier). From rate data available,[75] the rate constant for this cyclization

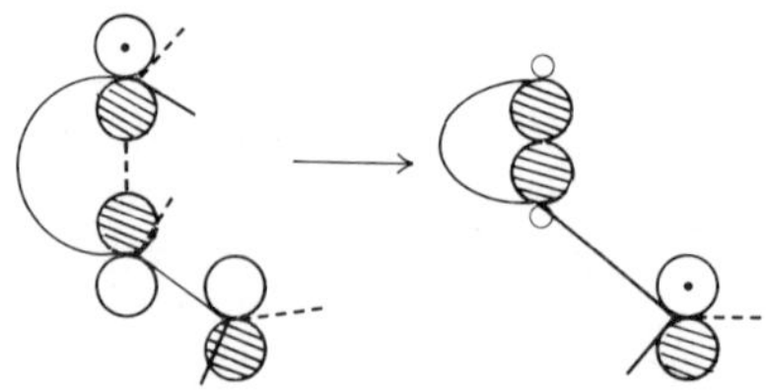

(296)

at 65° is approximately 4×10^4 sec^{-1}. Because intermolecular radical additions have $k = \sim 10^3\ M^{-1}$ sec^{-1}, this radical cyclization involves an "effective concentration" of double bonds of about 40 *M*!

The second study of reductive cyclization of bromides indicated that the "π route" to 2-norbornyl radical was open (297).[370]

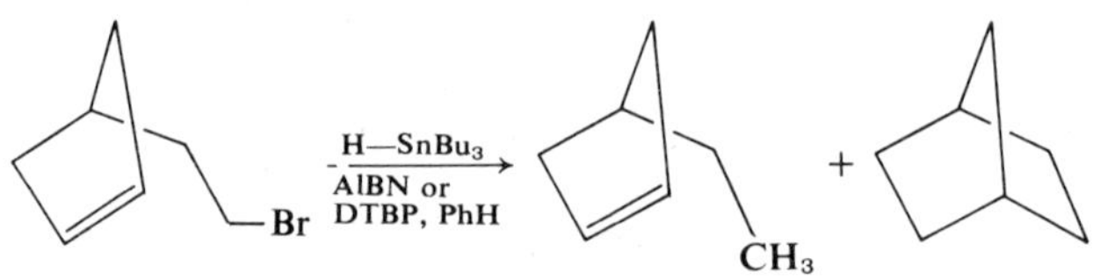

297-1 **297-2**

(297)

This cyclization occurred in spite of the considerable strain involved.[371] A tempting explanation for such an unexpected result centers on the earlier-mentioned mechanism for cyclization (296).[368] In the present instance this mechanism could appear as shown in (298).

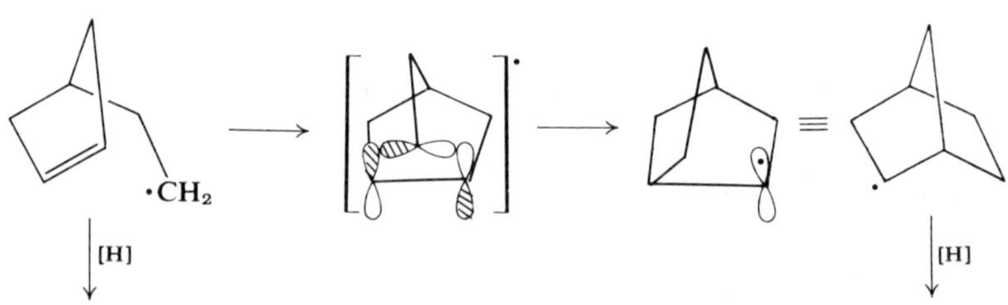

297-1 **297-2**

(298)

Here perhaps the interaction of the double bond and the radical center in the transition state could be symmetric with the π^* olefinic orbital, allowing considerable stabilization therefrom. In any case, cyclization is facile and increases with temperature and/or decreased tin hydride concentration, as shown in Table 13.

TABLE 13

Cyclization of 2-(Δ^3-cyclopentenyl)ethyl bromide

Bu_3SnH (M)[a]	% **297-1**	% **297-2**
0.100[b]	93	7
0.025[b]	80	20
0.100[c]	82	18
0.025[c]	59	41

[a] Alkenyl bromide in threefold excess.
[b] 93°, AIBN initiated.
[c] 130°, DTBP initiated.

The energy barrier to cyclization was calculated to be 14–16 kcal/mole, with a change in heat content (ΔH) of about -9 kcal/mole. 2-Norbornyl radical is a classical species devoid of extraordinary stabilization via 1,6 σ or other electronic delocalization.[372] That it forms so easily here indicates that the analogous ionic cyclization[373] *need not* involve nonclassicality in the 2-norbornyl cation either. In line with theoretical arguments (see Section II), the corresponding anion (at least as represented by Grignard and lithium organometallics) does not cyclize significantly (299).[374]

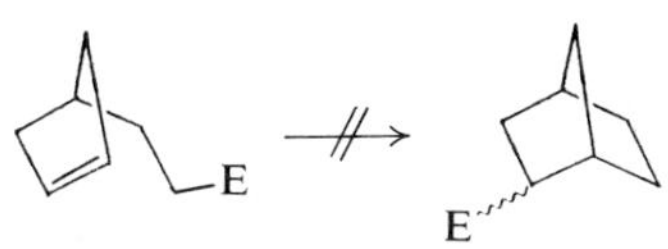

E = MgBr, Li (299)

5. *Cyclizations Involving Triple Bonds*

One of the stranger combinations of radical reactions was reported for the reaction of 1-heptyne and carbon tetrachloride (300).[375]

$$CH_3{+}CH_2{)_4}C{\equiv}CH + CCl_4 \xrightarrow[\Delta]{(PhCO_2)_2} CH_3{+}CH_2{)_4}CCl{=}CH{-}CCl_3 \quad 40\%$$

300-1

(300)

$+$ 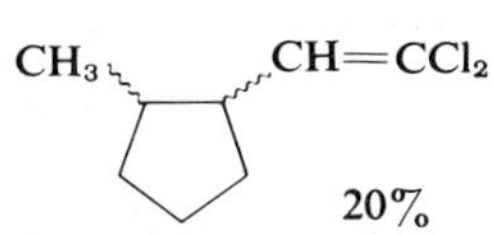

20%

300-2

$+ \; CH_3{+}CH_2{)_4}CCl{=}CHCl \quad 6\%$

300-3

An analogous reaction was observed with 6-methyl-1-heptyne. Use of bromotrichloromethane in place of carbon tetrachloride gave only the vicinal adduct in each case. The following sequence of reactions rationalizes the products formed (301).

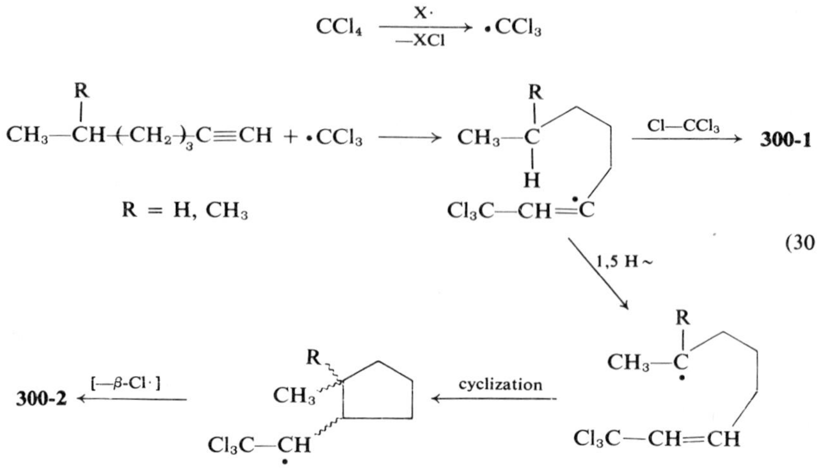

(301)

Product **300-3** apparently resulted from a slightly competitive chlorine atom chain. The better chain transfer agent, bromotrichloromethane, trapped out the initial radical prior to the hydrogen shift. The sequence in (301) combines two previously discussed radical rearrangements: the stereoelectronically favored 1,5 hydrogen shift (especially favored here because the vinyl radical is quite reactive and forms a strong bond between carbon and hydrogen), and a cyclization to a five-membered rather than a six-membered ring. The final β elimination of a chlorine atom, from which arises the chlorine atom chain that eventuates in **300-3**, is usually less favored than the vicinal shift (the Nesmeyanov rearrangement). However, β elimination of a chlorine atom is a known process, particularly observed when made irreversible by a chlorine atom trap, in this case the 1-alkyne.[376]

Similar closure resulted from propargylic esters (equation 302).[377]

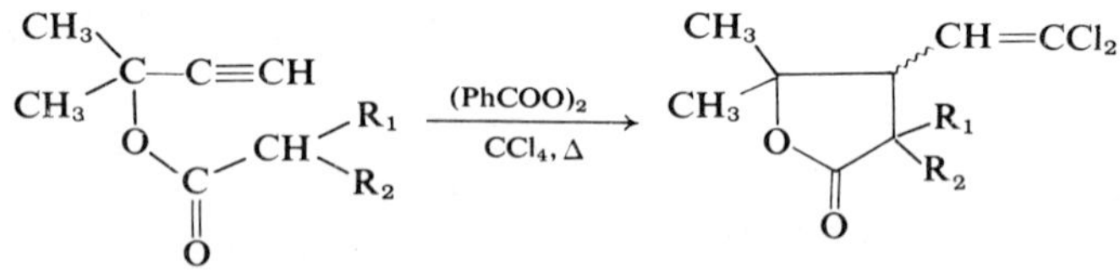

(302)

Here the hydrogen shift was easiest with benzylic or tertiary carbon–hydrogen bonds. Moreover, the subsequent cyclization must be more rapid than configurational racemization because optically active esters ($R_1 \neq R_2$) led to optically active products.

Apparent radical cyclization also occurred in the reaction shown in (303).[378]

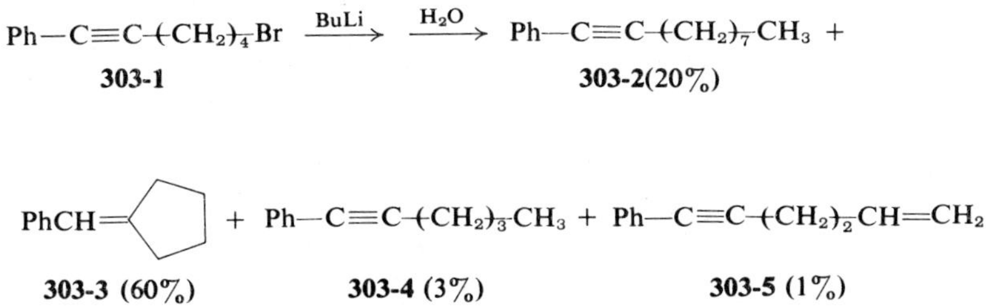

(303)

1-Phenylcyclohexene, if even formed, constituted less than 0.05% of the product. The conditions augur a carbanionic mechanism; indeed, somewhat similar carbanionic cyclizations were reported earlier.[379] However, use of deuterium oxide in processing afforded too little deuterated product for such a mechanism. Moreover, products **303-4** and **303-5** look suspiciously like radical disproportionation products. The presence of radicals (presumably as precursors to products, but this is less obvious now than then) was evidenced by chemically induced nuclear polarization (one of the first observations of this phenomenon). Therefore, the following scheme was suggested:

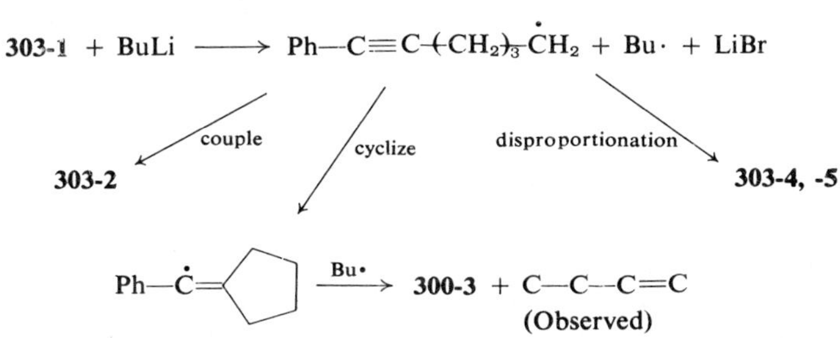

(304)

It is now well known that reactions of bromides and organolithiums involve radicals.[380]

Certain thiols have been reported to cyclize as well (reaction 305), although complete details are unavailable as yet.[381]

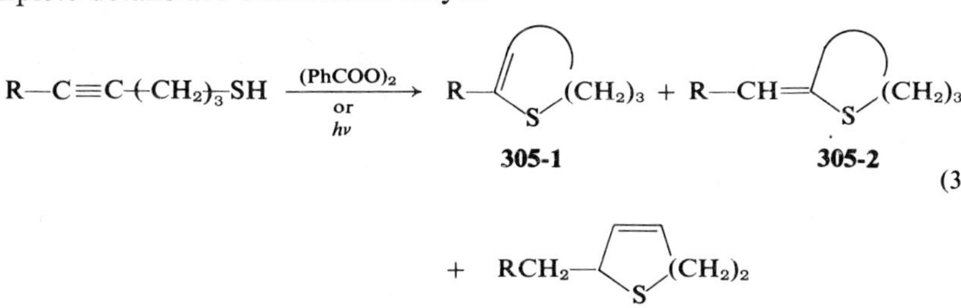

(305)

With R = H, **305-1** is the main product. When R = CH_3 or phenyl, product **305-2** predominates (75% with R = phenyl, light promoted), although all three products are present.

Although both thiyl and oxy radicals cyclize via addition to alkene linkages (p. 427), the aforementioned process (305) finds no analog with oxy radicals. The 4-pentynyloxy radical **305a-1** does not cyclize (305a).[381a]

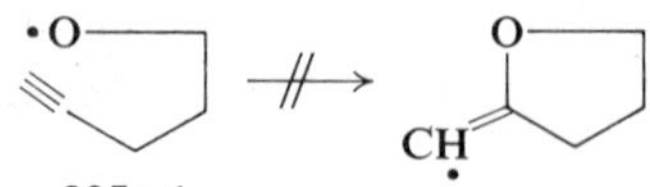

305a-1

(305a)

E. Geminate and Other Recombinations

1. "*Hexaphenylethane*" *and Related Compounds*

To the organic chemist interested in free radicals it conceivably could be less of a problem to believe the contention that silver chloride is a "green gas"[382] than to accept the fact that his septuagenarian science had doubtful parentage! Yet such is the case. Triphenylmethyl radical indeed dimerizes, not to hexaphenylethane,[383] but rather to the cyclohexadiene derivative shown in (306).[384]

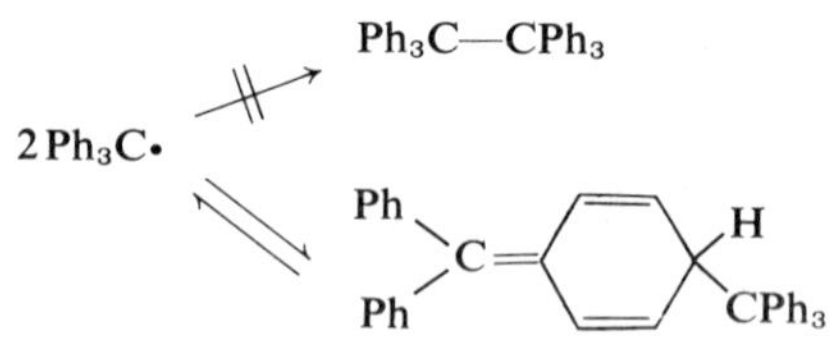

306-1

(306)

Structure **306-1** for the dimer was suggested long ago[385] but was not accepted. Now ultraviolet[384,386] and NMR *spectral* data, the latter both ^1H[384,386–388] and ^{13}C,[387] unequivocally support this early-assigned structure. Treatment of the radical–dimer system with antimony pentachloride led to *p*-trityl-triphenylcarbinol **307-1**, as illustrated (307).[388]

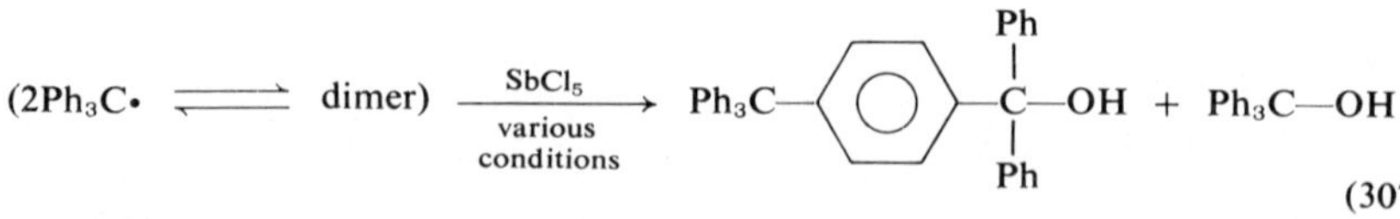

(307)

307-1 (19–44%) 43–72%

The possible formation of alcohol **307-1** by a Friedel-Crafts side reaction was eliminated and its formation from the dimer was established. This represents *chemical* evidence for the structure of the dimer. In (308) are given the Hückel molecular orbital π electron energies of the three most reasonable dimers of triphenylmethyl.[388]

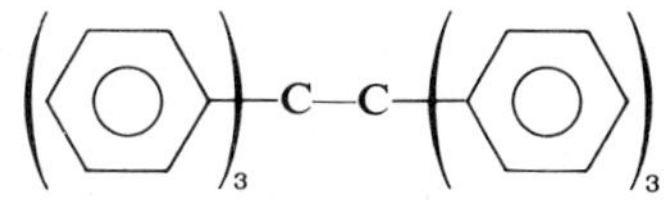

E_π 48.0000β

308-1

47.6137β

306-1

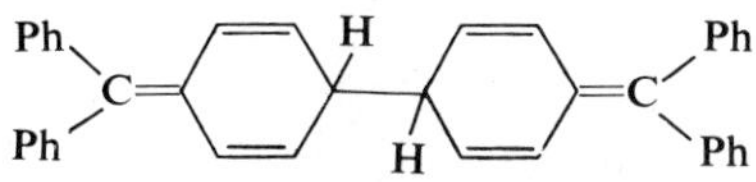

(308)

47.2274β

308-2

From the free valence density values for triphenylmethyl and the above E_π values, dimer **308-1** (hexaphenylethane) would seem the most probable. However, steric hindrance in **308-1** appears formidable and would cause a weakening in the central C—C σ bond. Contrariwise, dimer **306-1** would have a stronger σ bond with a slightly lower π electron delocalization stability ($\sim$7–8 kcal/mole). Apparently, in the balance between σ bond strength and E_π, dimer **306-1** is favored. So, after all these years, it seems hexaphenylethane is still unknown. Aside from a few bona fide examples,[389] all the detailed study of the "dissociation" of hexaphenylethane and its substituted analogs, all of the conclusions reached on the steric and resonance contributions to the dissociation constants, all the work of decades must be reinterpreted.[390] Quite so, a "landmark has fallen."[386]

True tetraphenylethanes are available by a new method (309).[384]

$$Ar_2CHOH \xrightarrow[\text{acetone, HCl, }25^\circ]{V^{2+}\text{ or }Cr^{2+}} \underset{10\text{–}20\%}{Ar_2CH\text{—}CHAr_2} \qquad (309)$$

Such compounds are difficult to dissociate thermally to radicals. The radicals so formed, however, then dimerize with rearrangement to produce the substances long thought to be the true tetraphenylethanes (310).

$$Ar_2CH\text{—}CHAr_2 \xrightarrow[\Delta]{} 2Ar_2CH\cdot \rightleftharpoons$$

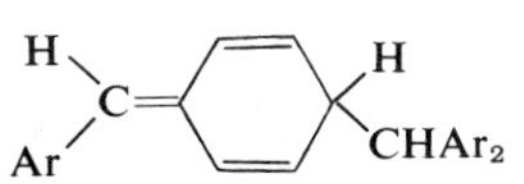

(310)

These last are easily dissociable, reforming the diarylmethyl radicals—but usually no tetraphenylethanes. The exceptions are some bulky diaryl radicals with *t*-butyl groups located at the meta or para ring positions.[390a] Although not yet proved, "true" hexaphenylethanes when available will undoubtedly undergo the same rearrangement. Much further work may be anticipated in this area.

2. *Ylid Rearrangements*

a. Stevens Rearrangement. This long-known rearrangement (311)[391] has been the subject of a recent burst of renewed interest.

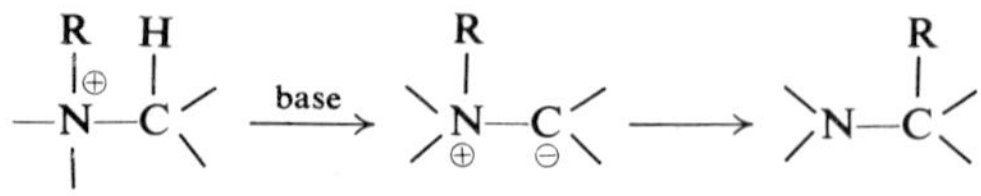

(311)

The rearrangement has been classified as an electrophilic 1,2 rearrangement,[17] although the several detailed mechanisms suggested over the years make it clear that the process is a complex one.[392]

Much of the recent work in this area has centered on the observance of chemically induced dynamic nuclear polarization for this process. Although the phenomenon of CIDNP is discussed extensively elsewhere in this volume (Chapter 6 in particular), a few words at this point are appropriate. CIDNP phenomena[393] are observed "for species which come from free-radical-generating reagents directly or by radical transfer."[394] Its occurrence in some process means that radicals are somehow involved in forming the products of the reaction. A caveat has been voiced, however, stressing that CIDNP may arise from paths unrelated to the main one.[395] At times NMR emission or enchanced absorption signals (the experimental indications of CIDNP) are strong and can be detected even though the radical pathway may be minor.

In the Stevens rearrangement, the reaction shown in (312) exhibited CIDNP (NMR emission) from the α-C—H in the amine product.[394]

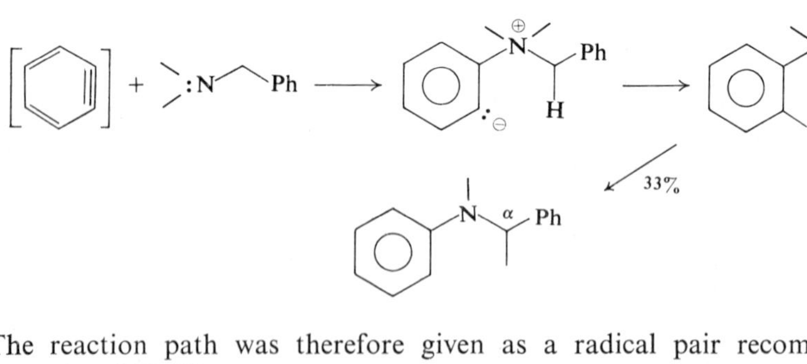

(312)

The reaction path was therefore given as a radical pair recombination (equation 313).[396]

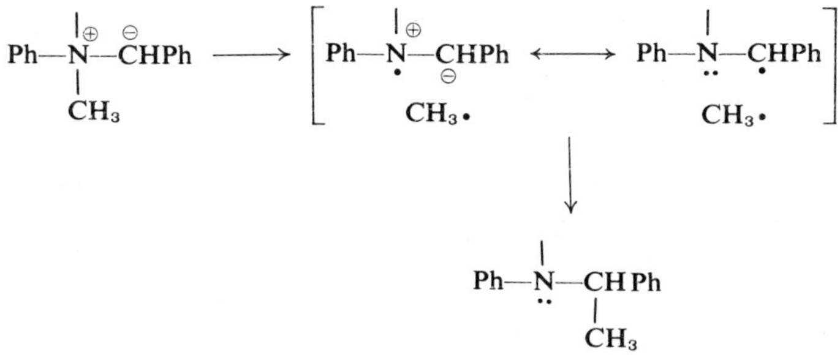

(313)

Use of another amine (314) led to a product showing NMR emission in *both* the α and β positions.[397]

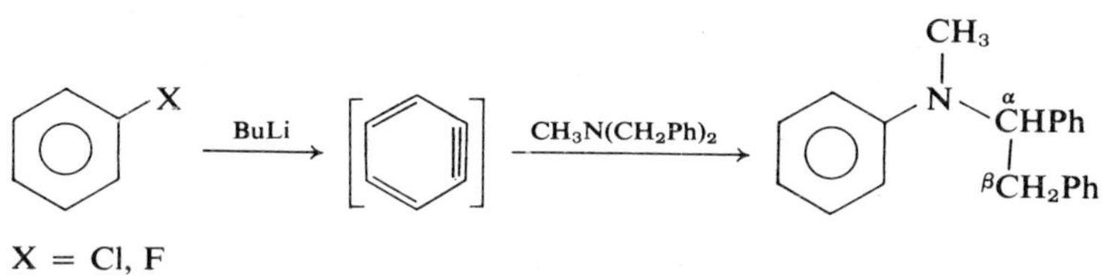

(314)

This result was interpreted as a *caged* radical pair recombination (geminate recombination). For the observed CIDNP, the combining radicals must have had the unpaired electron at that atom of each that eventually formed the α,β bond. However, the interpretation of just what CIDNP means is getting more complex (almost daily). Apparently *some* leakage from such a cage as above must occur for the phenomenon to be observed.[398] Also, just how "radical" the radicals need be is questionable (see p. 458).

In similar fashion, the reaction of the sulfonium ylid in (315) produced a variety of products, among them the Stevens-like rearranged product.[399]

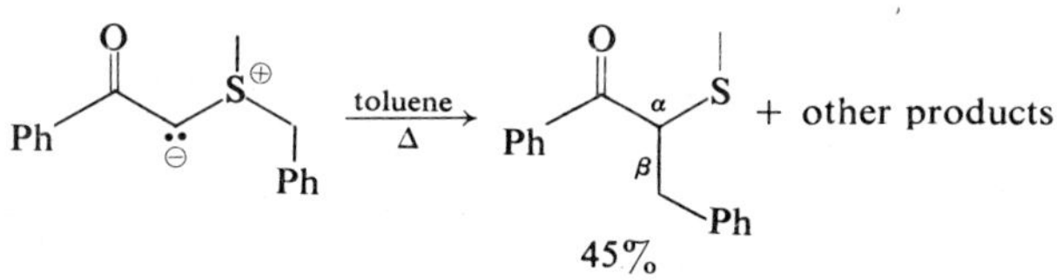

(315)

When done at 130° in diphenyl ether, CIDNP was observed in both the α and β protons. Crossover experiments indicated that the process was largely an intramolecular rearrangement (82 ± 6%). Stereochemical data from reaction of the optically active ylid shown in (316) indicated that the intramolecular process in this case proceeded with 44 ± 20% retention of configuration.

While not the only one possible, the most likely pathway again involved the

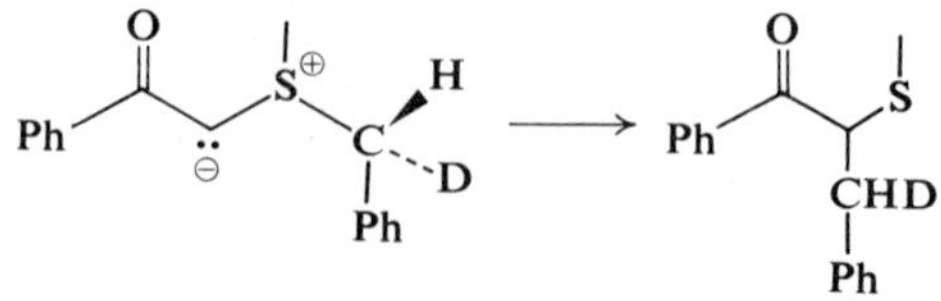

(316)

geminate recombination of a radical pair, as in (317). Such recombination would have to be faster than racemization of the chiral species involved.[400]

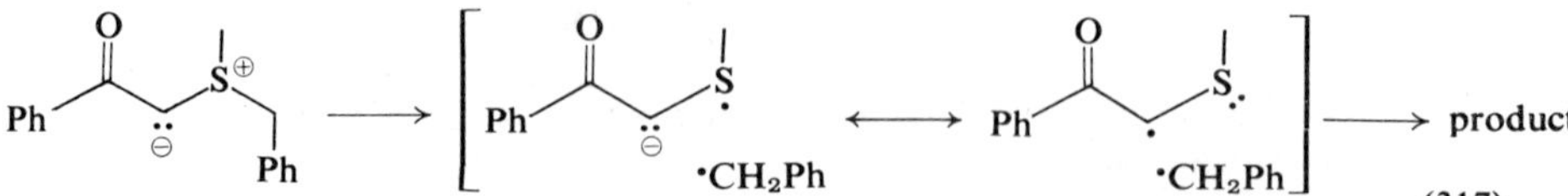

(317)

A number of other ylid rearrangements were suggested to proceed in analogous fashion.[401]

Another recent observation of CIDNP was in the process given in (318).[402]

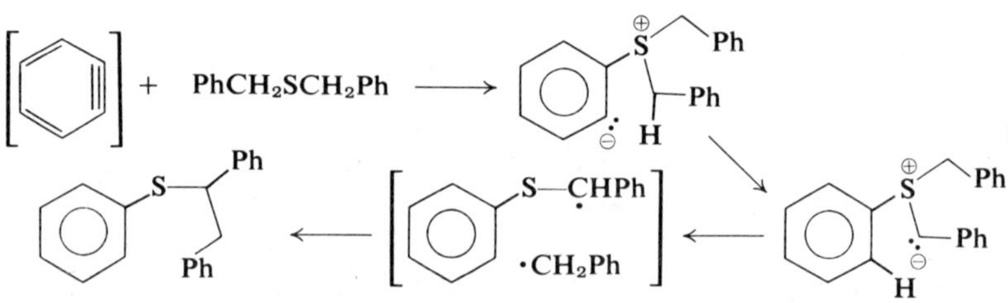

(318)

These authors noted that CIDNP has been found so far only for those ylid rearrangements for which orbitally symmetric pathways are forbidden. When so allowed (e.g., in allyl sulfonium ylids), these rearrangements are more facile and appear to proceed largely via a nonradical, concerted pathway.[402,403]

Although impressive in itself, the CIDNP phenomenon is not the only indication of radicals in the Stevens rearrangement. In the process given in (319), all three coupling products expected from radical intermediates were in fact observed.[404]

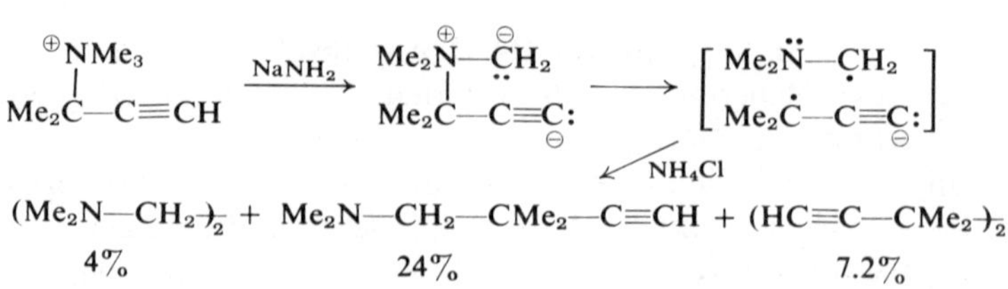

(319)

Such a finding complements an earlier[405] isolation of bibenzyl (5%) from a Stevens rearrangement of a benzylamine derivative. Likewise, the Stevens-like rearrangements of the aminimides in (319a) have been claimed to be *totally* radical in nature on the basis of radical-trapping experiments.[405a]

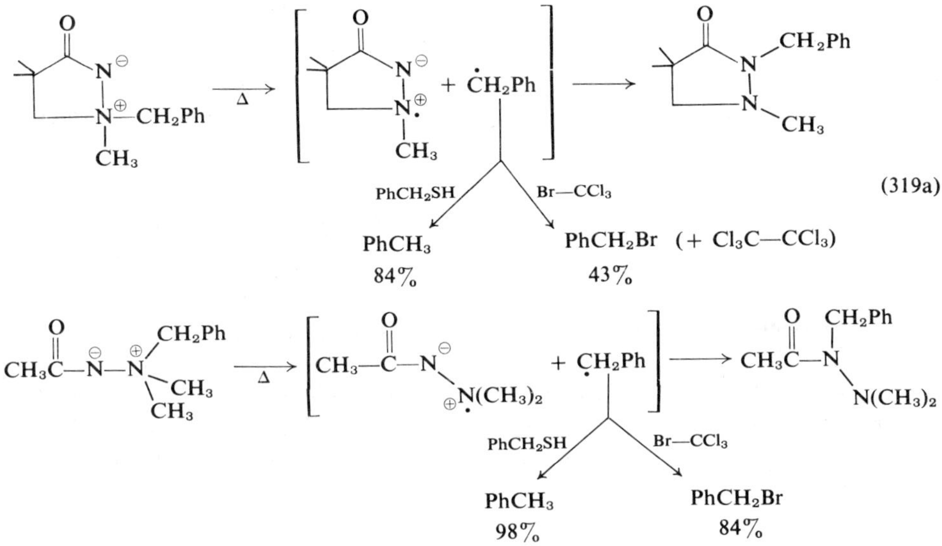

(319a)

Not all Stevens rearrangements show CIDNP. The reaction in (320) did not, for example,[406] and the products were more consistent with an ion pair mechanism suggested years ago (321).[407]

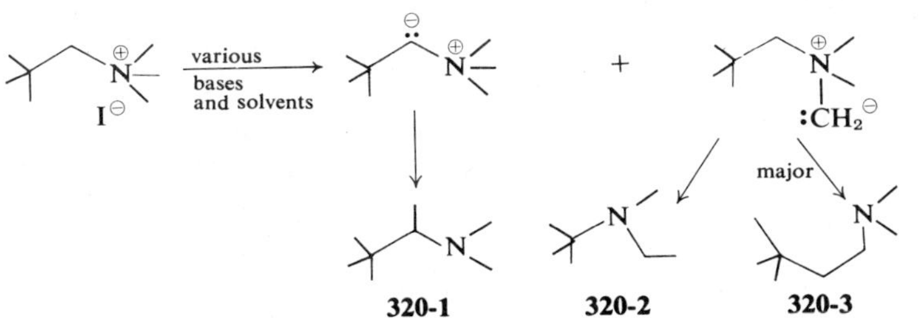

(320)

Neopentane was detected (possibly via leakage from the ion pair of the neopentyl anion followed by protonation by solvent). Also, carbanionic addition to the methylene immonium ion was detected.

In summary, radicals are clearly present at some stage in specific Stevens rearrangements. At present, what stage this is remains somewhat obscure, and it may even be subsequent to the actual rearrangement step. Contrariwise, the rearrangement may be one of a number of combined ion pair–radical pair processes. This has been recently suggested[408] and is discussed at greater length a bit later.

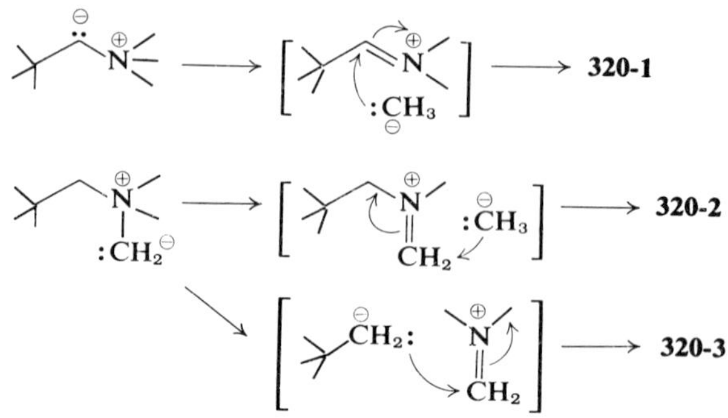

(321)

b. Meisenheimer Rearrangement. The thermal rearrangement of tertiary amine oxides to substituted hydroxylamines[409] is another example of a process formerly considered polar in mechanism and now considered radical. Below its rearrangement temperature, the amine oxide in (322) clearly dissociated via ESR measurements to the nitroxyl radical (the benzyl radical was not observed).[410]

$$\mathrm{Ph{-}N(CH_3)(\rightarrow O){-}CH_2Ph} \rightleftharpoons \mathrm{Ph{-}\dot{N}(CH_3)(\rightarrow O)} + \cdot\mathrm{CH_2Ph} \qquad (322)$$

The rearrangement in substituted cases (presumably by coupling of the two radicals produced as above) possessed a low ρ value ($\sim +0.9$)[410] and it was inhibited to a considerable (but not total) extent by oxygen.[411] Such indications of a radical process, possibly of the geminate cage recombination type, have been augmented by recent detection of CIDNP in Meisenheimer rearrangement products. The amine oxides in (323)[412] and (324)[413] exhibited CIDNP of their rearranged products.

$$\mathrm{PhCH_2{-}N(CH_3)_2 \rightarrow O} \xrightarrow{\Delta} \mathrm{(CH_3)_2N{-}OCH_2Ph} \qquad (323)$$

$$\mathrm{PhCH_2{-}N(CH_3)(Ph) \rightarrow O} \xrightarrow{\Delta} \mathrm{(CH_3)(Ph)N{-}OCH_2Ph} \qquad (324)$$

The suggested pathway involved geminate radical pair formation with recombination inside (20–40%) and outside (60–80%) the solvent cage (325).[412]

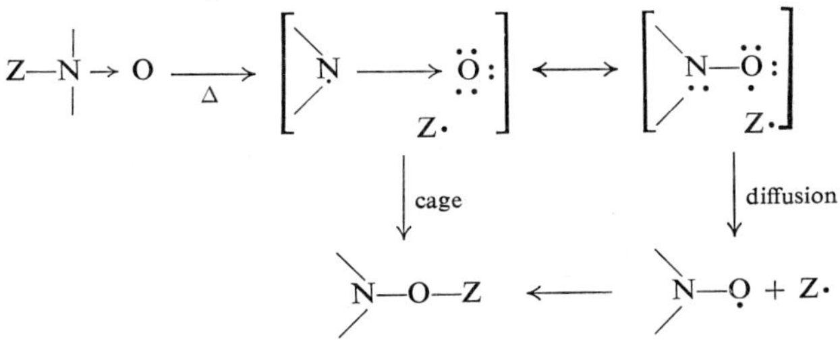

(325)

The reaction parameters ($\Delta S^* = +7.9$ eu, $\Delta H^* = 34.2$ kcal/mole)[414] seem in accord with this scheme, as does the partial but not total inhibition by oxygen[411] (a scavenger only for radicals *outside* the cage). CIDNP is exhibited by the benzylic protons of the product, as would be expected by this mechanism for at least the caged species.[415] The methyl protons also showed CIDNP, however, and these studies are the first to show such an effect for protons on groups *not* directly involved in the migration.

The Meisenheimer rearrangement is somewhat related to that of nitrones (326).[416]

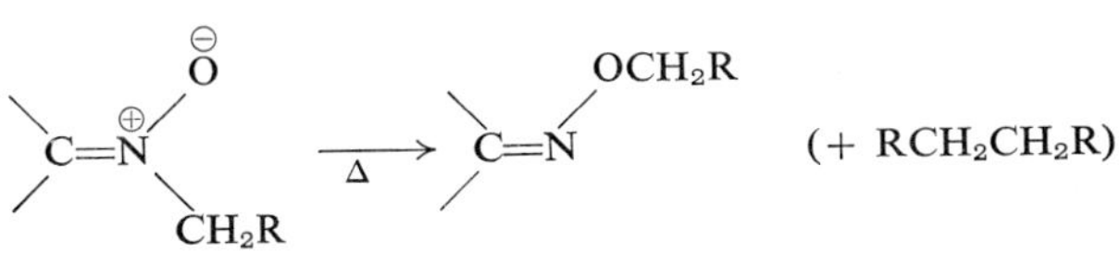

(326)

This latter process apparently proceeds by straightforward homolysis and recombination. The dimeric by-product indicates intermolecularity. In the cyclic amine oxides shown in (327), mixing experiments showed that the Meisenheimer rearrangement under certain conditions behaved quite comparably.[417]

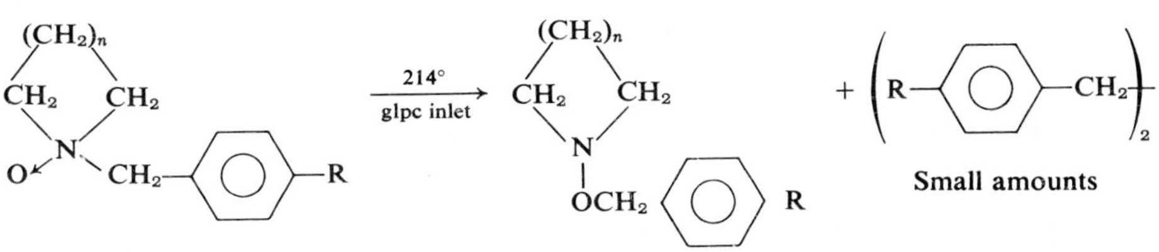

327-1, $n = 3$, R = H
327-2, $n = 2$, R = CH_3
(Mixed together)

327-3, $n = 3$, R = H
327-4, $n = 3$, R = CH_3
327-5, $n = 2$, R = CH_3
327-6, $n = 2$, R = H

(327)

A mixture of oxides **327-1** and **327-2** gave all possible combination products (**327-3** through **327-6**), as well as dimeric products. Of course, cage recombination seems somewhat remote a mechanistic possibility in a glpc inlet at 214°! So an intermolecular pathway is undoubtedly favored here. Again, the dimeric products attest to this route as well.

c. Wittig Rearrangement. This rearrangement of benzyl ethers with strong bases (328)[418] has joined the group of rearrangements that show CIDNP in the products.

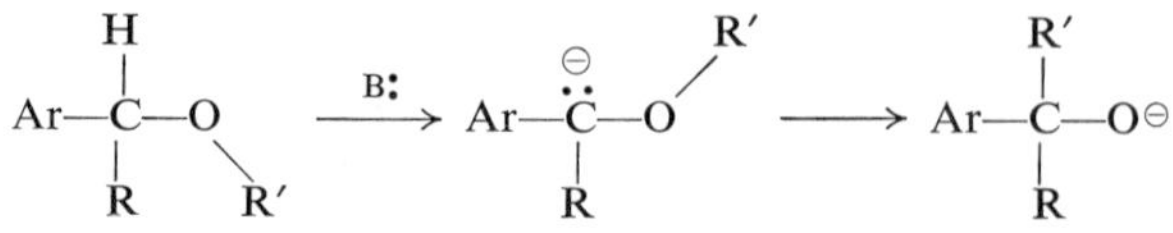

(328)

The reaction proceeds intramolecularly (>80%). When chiral migrators are involved, the product is largely racemized, although retention of configuration is somewhat favored.[419] The early suggested S_Ni mechanism is incorrect[420] and a dissociation–readdition mechanism has been generally accepted (equation 329).

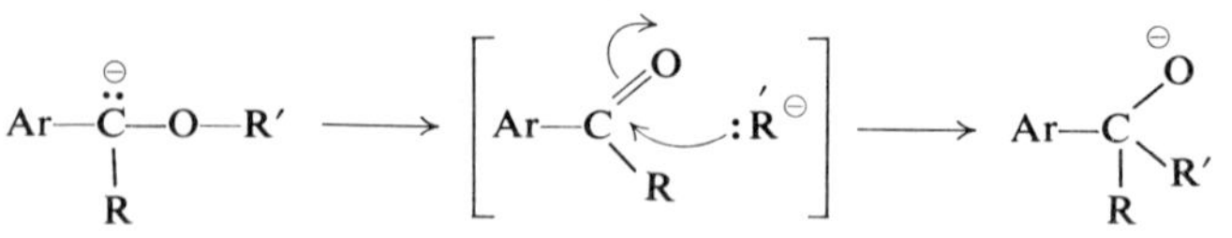

(329)

However, the Wittig rearrangement shown in (330) gave an alkoxide that exhibited CIDNP of the methinyl proton.[421]

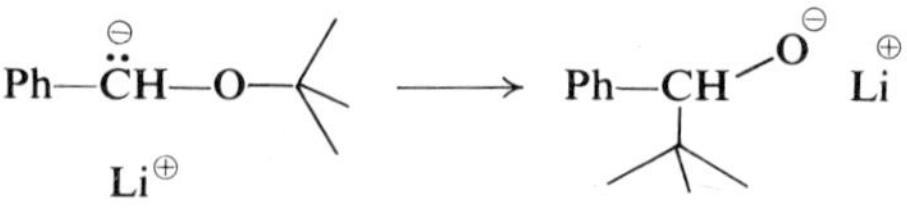

(330)

So, again, the presence of radicals is demanded and the actual mechanism could involve radical recombination. It is interesting to note that some net retention of configuration could be observed in such a process even though a chiral migrator itself became planar (331).

Although the migrator ($\cdot CR_1R_2R_3$) is planar, the radical pairs containing it are chiral. Whether total racemization or some net retention (or inversion) will occur depends on the relationship of k_1 and k_2. Apparently k_2 is slightly larger than k_1 in the Wittig rearrangement.

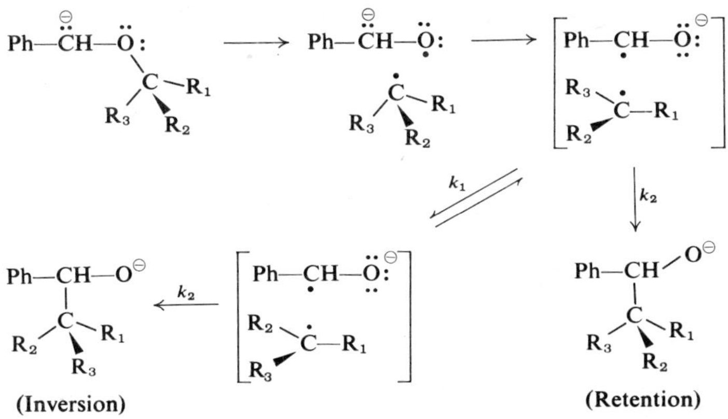

(331)

d. Miscellaneous Examples. The rearrangement of the sulfenic ester shown in (332) exhibited CIDNP of the benzylic methylene protons of the product.[422,395]

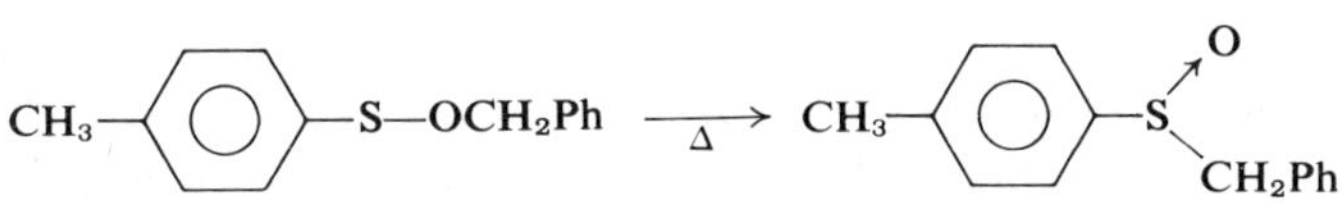

(332)

Although first rationalized as a radical process,[422] later work made an intramolecular, nonradical pathway more likely (though just what that pathway was was not stated).[423] Because the sulfoxide product is homolytically labile,[422,423] the CIDNP observed[395] was ascribed to such a homolysis *subsequent* to its rapid formation from the sulfenic ester—that is, an artifact! This led to the caveat previously mentioned (p. 448).[395]

Another somewhat related case of such a problem was the thermal rearrangement of 2-picoline oxide (333).

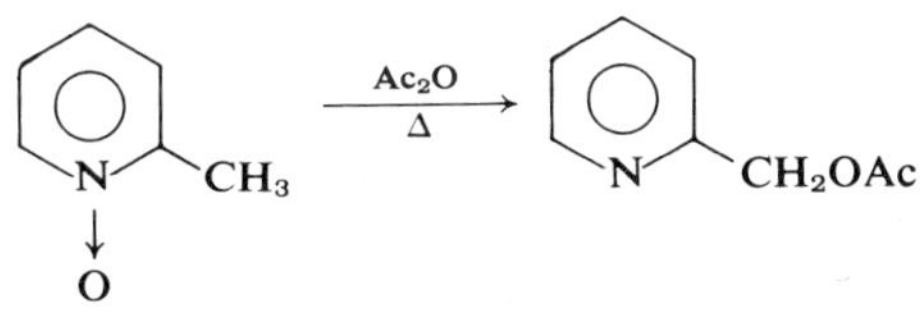

(333)

Radicals are definitely present during the process,[424] yet the reaction is insensitive to inhibitors.[425] So geminate radical recombination seemed a likely mechanistic pathway. Indeed, CIDNP was observed as the reaction proceeded.[426] But the NMR emission was due to by-products and *not* the pyridylcarbinyl acetate. A concerted pathway (nonradical) therefore has been advanced for this rearrangement.[426] A switch to 4-picoline oxide in this reaction (334) also switched the interpretation for it.[427]

Here CIDNP is again found, but this time it *is* due to the product **334-1**. So here a geminate radical recombination mechanism is better supported, although careful analysis of the NMR emission indicated that the radical path was accompanied by an ionic one. (Maybe there is but one path, one with

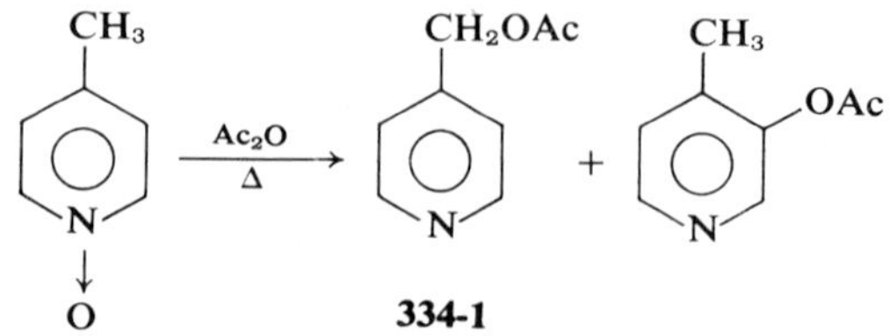

334-1

(334)

both radical and ion pair intervention; see below and p. 458.) It is educating to read the objections raised by a referee in a footnote to the report of this work. Their summary is that CIDNP *of itself* does not necessitate a radical mechanism for the reaction under study. Rather, a minor route, a subsequent homolysis, or impurities can cause the CIDNP. Although these artifacts are apparently not extant in (334), their possibility should make investigators wary in the future.

The acid-catalyzed rearrangement of nitroamines (335) apparently involves some geminate recombination radicals, at least in the formation of the nitroaniline products.[428]

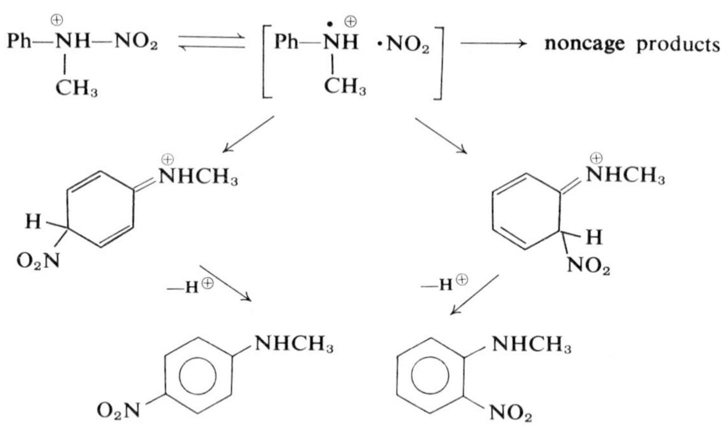

(335)

Such a conclusion was drawn from the fact that solvent viscosity determined both product yield and composition. Increased solvent viscosity impeded dissociation from the cage, so the yield of nitroanilines rose with such an increase. However, the migration of close partners in the cage from ortho to para positions was also impeded by such an increase. As a consequence, the ortho–para yield ratio also rose with viscosity.

A net 1,3 migration of arylazo groups has been reported, again most likely as the result of geminate radical recombination (336).[429]

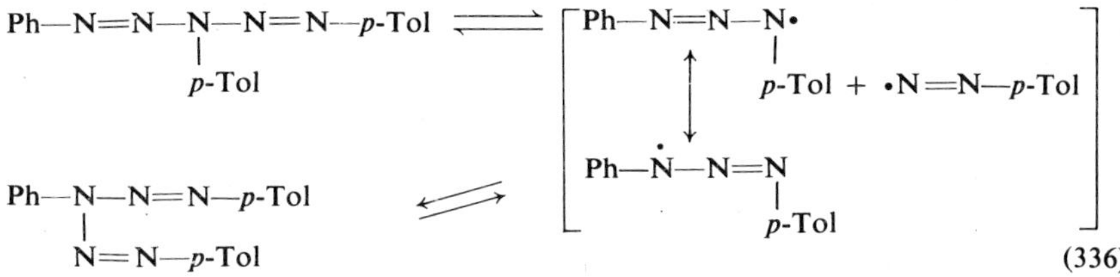

(336)

Other cases of this rearrangement (which is spontaneous upon production of the pentazenes) were also observed. The process is apparently intramolecular, from the results of mixing experiments. Recombination of radicals was implicated because CIDNP was observed in the *p*-tolyl methyl groups. Arylazo radicals had been suggested earlier[430] as reaction intermediates, and this rearrangement gives additional support to this view.

Finally, two examples of oxy radical involvement in such rearrangements may be mentioned. First, the photochemical rearrangement of aryl ethers (equation 337) is probably geminate radical recombination, at least in part.

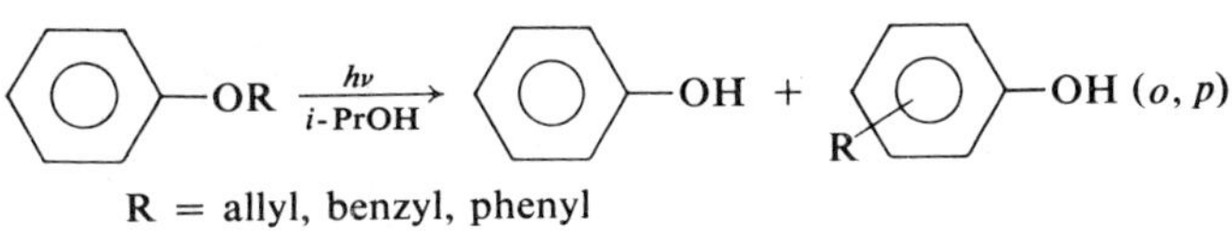

(337)

Although only para-substituted phenols were first reported,[431] ortho isomers are indeed formed as well.[432] The process is best viewed at present as a radical recombination, as shown in (338).[433]

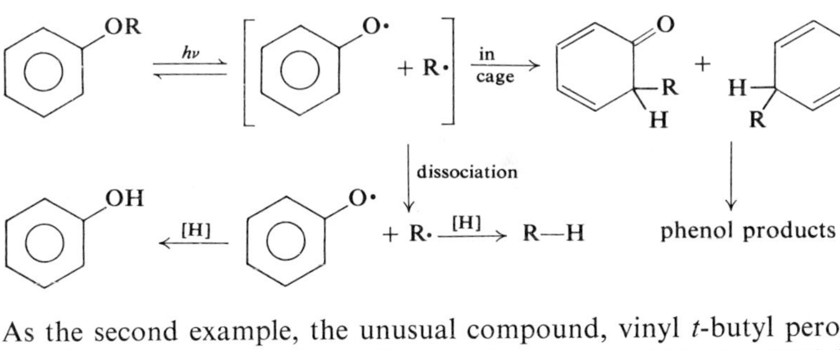

(338)

As the second example, the unusual compound, vinyl *t*-butyl peroxide, is reported to rearrange thermally to *t*-butoxyacetaldehyde (339).[434] No mechanism was claimed, but an obvious possibility is shown.

In closing this section, reference should be made to a novel mechanistic idea recently published.[408] In certain of the aforementioned processes (e.g., the Stevens and Wittig rearrangements), *both* ionic and radical pathways seem implicated. Ionic intermediates were indicated usually from substituent studies whereas radicals were invoked usually from CIDNP data. The recent mechanistic idea is that "no complete dichotomy can be demonstrated . . .

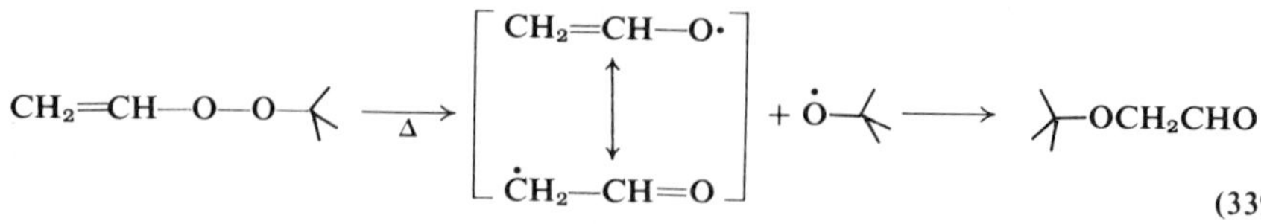

(339)

and, indeed, . . . they both probably proceed through the same rate-determining step. . . ." Fundamentally, the concept is that the division of the mechanism into ionic and radical paths occurs subsequently to passage over the highest energy barrier. As transition states have commonly been assigned ionic and radical contributors, so such contributors would now be assigned also to intermediates (e.g., caged pairs). In this view, the early stage of an ion–singlet radical pair in the solvent cage is modified quickly by solvent interaction or rotation to separated species, ions or radicals. The various "ionic" or "radical" products observed in such reactions then represent diversions of this initial caged pair at various stages of its metamorphosis.

F. Vinylic, Allylic, and Homoallylic Rearrangements

1. Vinylic Rearrangements

The term "vinylic rearrangement" can be a misnomer, but it blends well with the others of this section. Geometric isomerization by a radical pathway is the meaning usually ascribed to "vinylic rearrangement." Many examples of the process exist (340) and some of the earlier studies have been reviewed.[435]

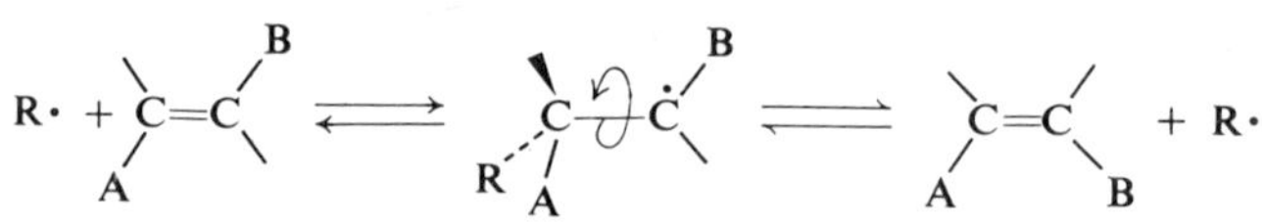

(340)

With respect to "true" vinylic rearrangements, it is of passing interest that vinyl radicals apparently can maintain some geometric identity. Photolysis of the isomeric iodides shown in (341), though not a rearrangement, is nevertheless interesting because the product distribution from each isomer was not the same.[436]

Though the isomerization of the excited vinyl radicals probably involves a barrier of less than 15 kcal/mole, this is sufficient to affect the product distribution. Not much more seems to have been reported in this area, at least on thermal processes. One senses that it is not in the mainstream of solution radical rearrangement research.

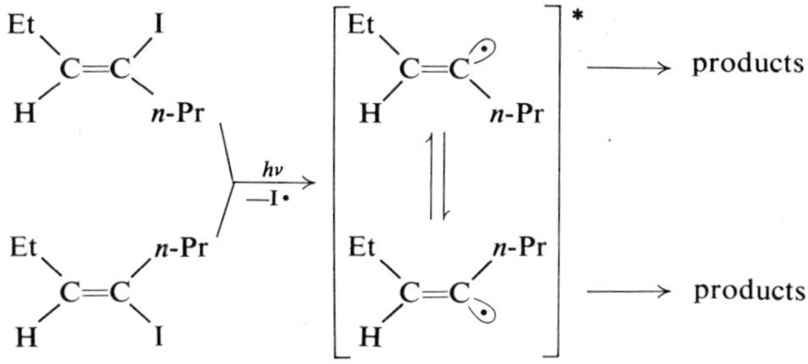

(341)

2. *Allylic Rearrangements*

Whenever an allylic radical is produced, rearrangement is possible. Many examples involve a reaction sequence as shown in (342) where hydrogen abstraction sets off the rearrangement. Such examples have been reviewed.[437]

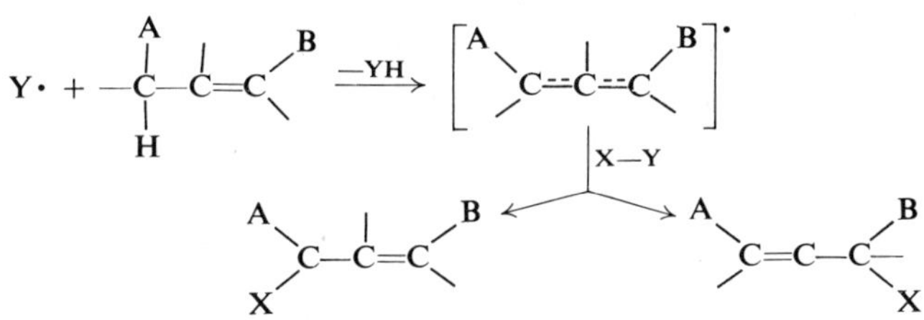

(342)

Less common are true allylic rearrangements (343)—true in the sense that actual isomerization occurs, generally via an intermolecular addition–elimination pathway.

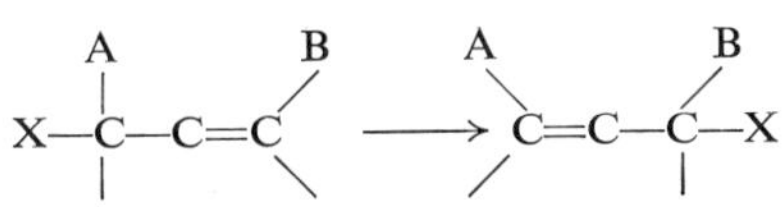

(343)

Some examples may be the bromine atom-catalyzed isomerization in (344)[438] and the thiyl radical-catalyzed isomerization in (345).[439]

$$Br\cdot + CH_2{=}CH{-}CH(Br)CH_3 \rightleftharpoons BrCH_2{-}\dot{C}H{-}CH(Br)CH_3 \rightleftharpoons BrCH_2CH{=}CHCH_3 + Br\cdot \quad (344)$$

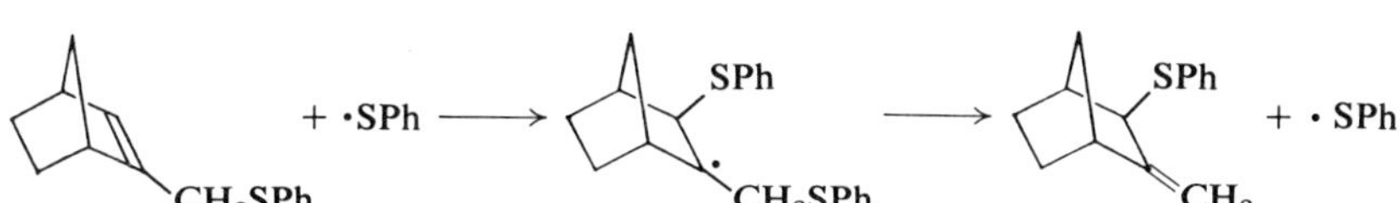

(345)

It is apparent that the addendum and leaving group must be identical and that it must be one of a special sort, namely, atoms or groups that both add and eliminate with facility, for example, Br–, I–, RS–. It is therefore of some interest that some recent work on photochemical and thermal rearrangements of allylic chlorides indicated a *chlorine* atom isomerization, for example, chloride **346-1** in (346).[80]

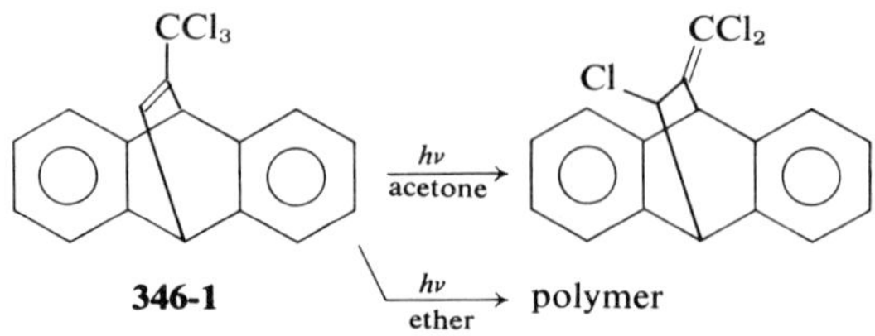

(346)

This would be an interesting, relatively uncommon, radical rearrangement, but some question exists about the nature of such photolyses, most especially when sensitized (e.g., by ketones). The photosensitized rearrangement in (347),[440] for instance, was rationalized as proceeding via vibrationally excited carbonium ion–halide ion pairs (348). A triplet biradical pathway (349) is also possible, however,[441] even though a peculiar 1,4 chlorine atom shift is required.

On the other hand, unsensitized photolysis of allylic chlorides is generally agreed to involve radicals,[442] but it seems that little synthetic value attaches

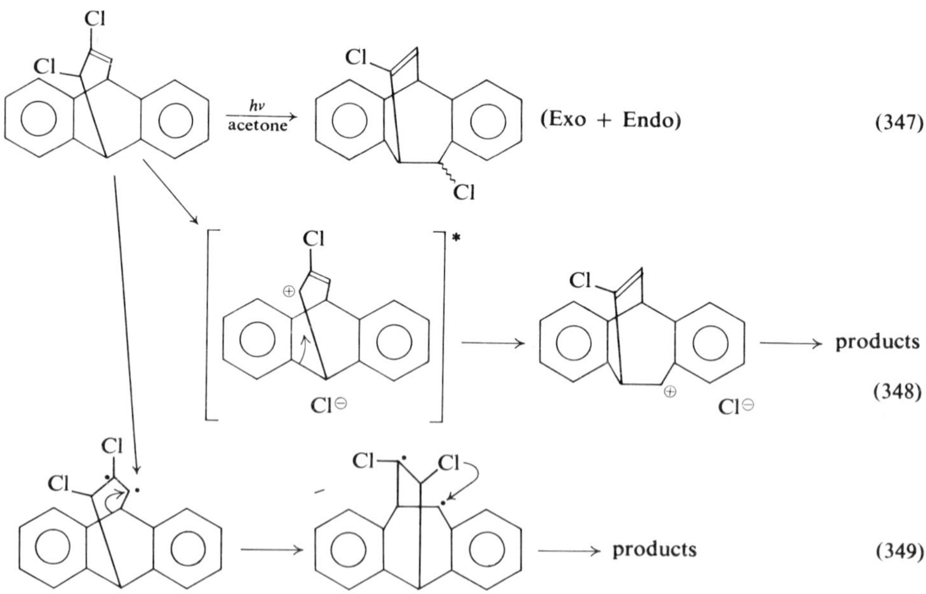

(347)

(348)

(349)

to the reaction. The situation has been confused a bit more in another study of chloride **346-1** (350).[443]

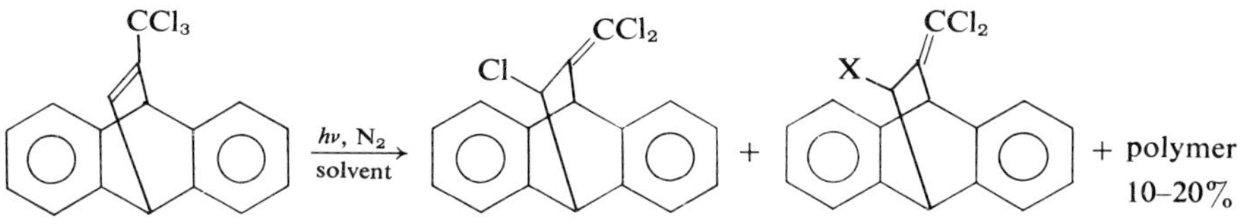

346-1 **350-1** **350-2**

(350)

The rate of disappearance of **346-1** is the same (by NMR analysis) in acetone, benzene, bromotrichloromethane, carbon tetrachloride, ether, or methanol–tetrahydrofuran. In bromotrichloromethane some **350-2** (X = Br) was formed, along with **350-1** (ratio **350-1**:**350-2** = 10:1). Also, in the methanol–tetrahydrofuran solvent, some **350-2** (X = OCH_3) was found. However, this apparently was a solvolysis product, presumably derived from **350-1**. Additional study showed that the bromide **350-2** did not form from the chloride **350-1** under reaction conditions. The solvent independence and the production of **350-2** (albeit in small amount) strongly suggest the radical pathway shown in (351), even in acetone. This last is in contrast to the ionic interpretation of (347).

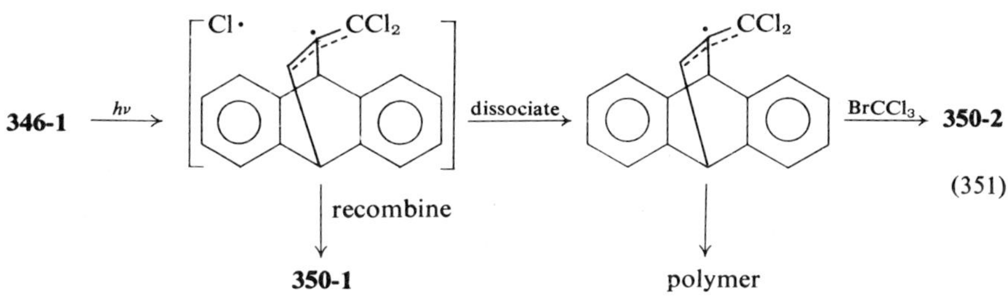

(351)

Thermal conversion of **346-1** to **350-1** was achieved also, with a rate definitely influenced by solvent (fastest in nitromethane, slowest in carbon tetrachloride). So here the ion pair mechanism was suggested for the process. Obviously, more work is needed to clarify the pathways involved in these fascinating rearrangements.

In a formal sense, the Claisen rearrangement[444] (the lesser known one) is also an allylic rearrangement (352).

$$\mathrm{Ph{-}C(OCH_3){=}CH_2 \xrightarrow[\Delta]{} Ph{-}C({=}O){-}CH_2CH_3} \qquad (352)$$

The following examples, shown in (353),[445] indicate that the rearrangement does not involve a configurationally retentive 1,3 shift (which would be disallowed thermally by orbital symmetry considerations, anyway) or a carbonium ion process.

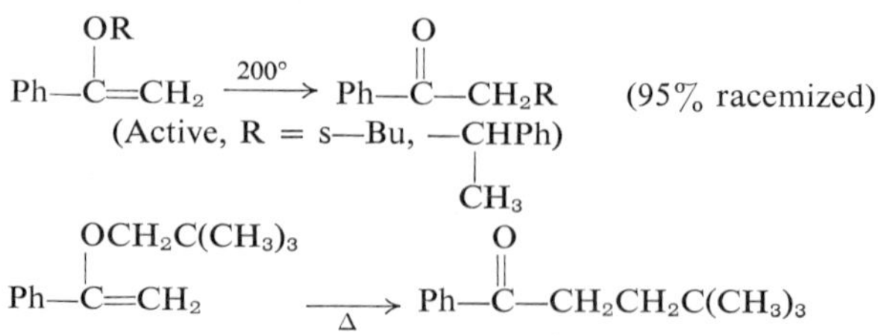

(353)

At 135°, di-*t*-butyl peroxide can initiate the reactions in (353), but no reaction occurred in its absence at this temperature. So a radical pathway (354) was suggested.[446]

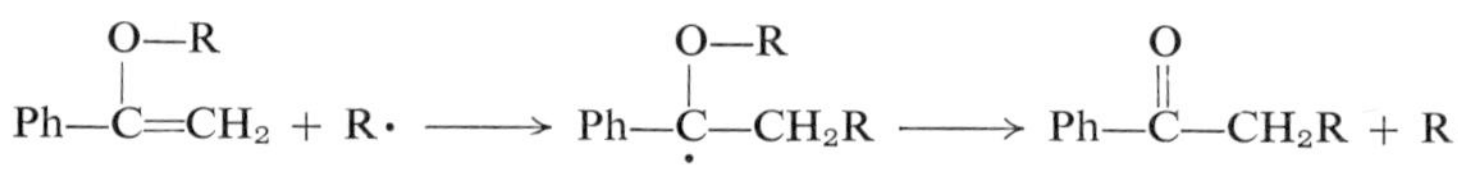

(354)

A *direct* transfer of R, as shown in (355), seems unlikely because chiral R groups are racemized in the product.

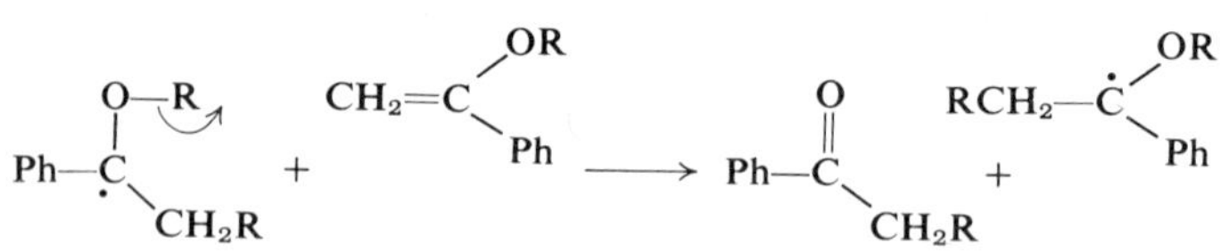

(355)

With the addition–elimination sequence thus more likely, it is curious that chain inhibitors like diphenylamine or hydroquinone did not interfere with the reaction.[445] An analogous pathway was made likely (by use of ^{13}C labeling) for the following rearrangements as well[447]:

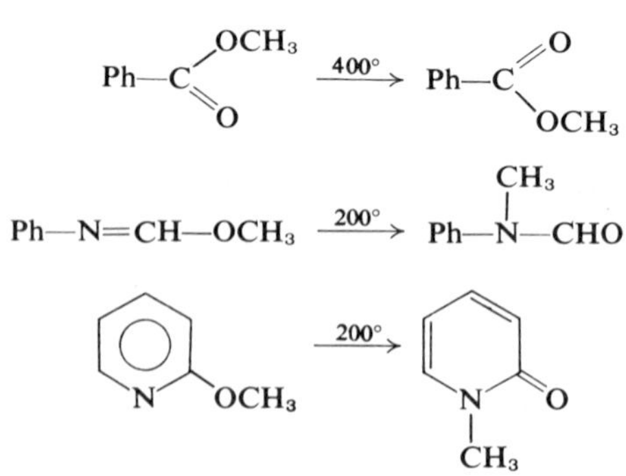

(356)

However, the Chapman rearrangement (equation 357), though superficially analogous to those in (356), differs from them in mechanism. The Chapman rearrangement is rather an intramolecular S_NAr process.[448]

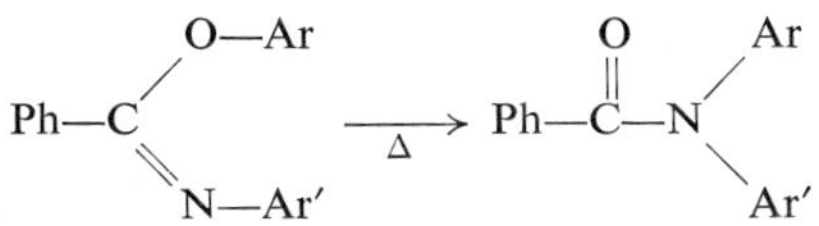

(357)

The object lesson here is that seemingly analogous processes may have quite different mechanisms and each must have its own pathway investigated. The Claisen rearrangement can be initiated by azobisisobutyronitrile, under which condition a chain reaction mechanism apparently again applies (358).[449]

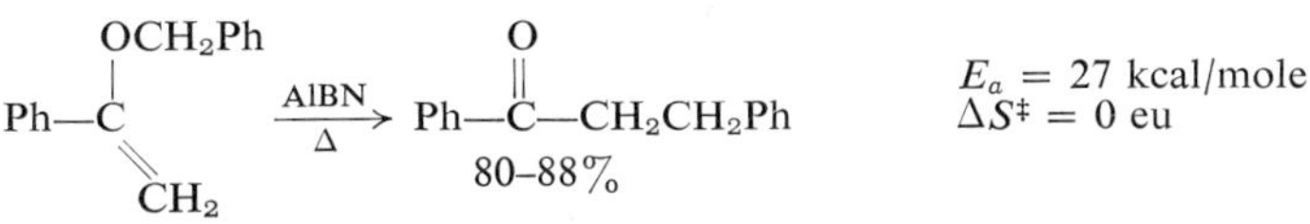

(358)

The chain length is only about 23, however, and this may explain the lack of inhibition of the reaction by chain inhibitors previously mentioned. The same process done thermally at higher temperatures in the absence of oxygen may involve this same chain path or a nonradical path using an eight-membered ring cyclic transition state.

A very interesting study of the photochemical Claisen rearrangement in (359) has been reported.[450]

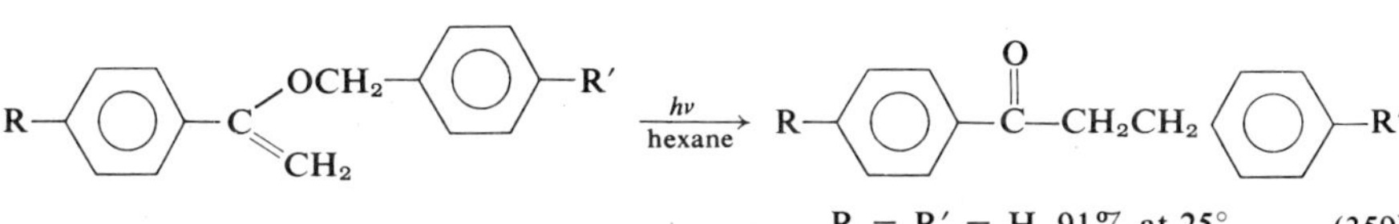

(359)

For a series of R groups, the quantum yields were 0.14–0.15 and very little mixing between different reactants during the process occurred. Under these *photochemical* conditions, the rearrangement was about 92% intramolecular and may proceed via the now allowed suprafacial 1,3 shift of the migrator.

Rearrangements related to the Claisen, (360)[451] and (361),[452] have also been shown to be radical chain mechanisms, though the chain lengths are short.

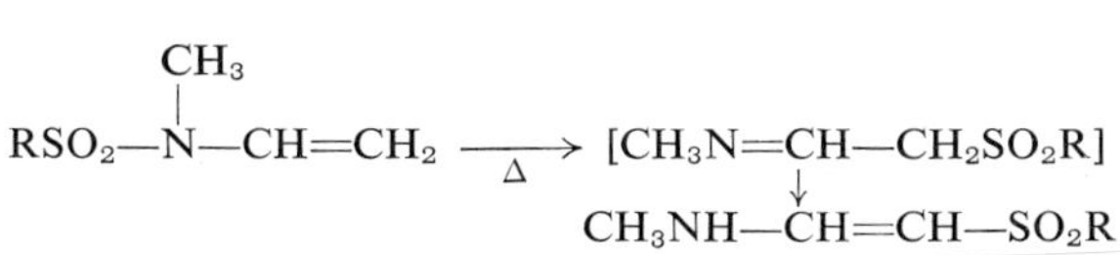

(360)

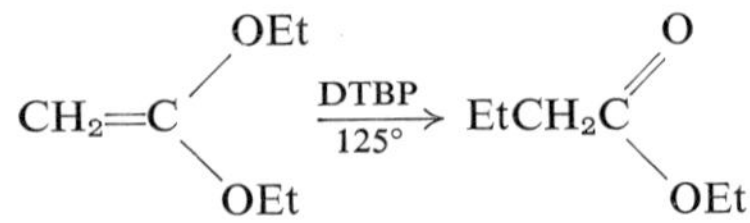

(361)

Another, very recent, example is given in (362).[453]

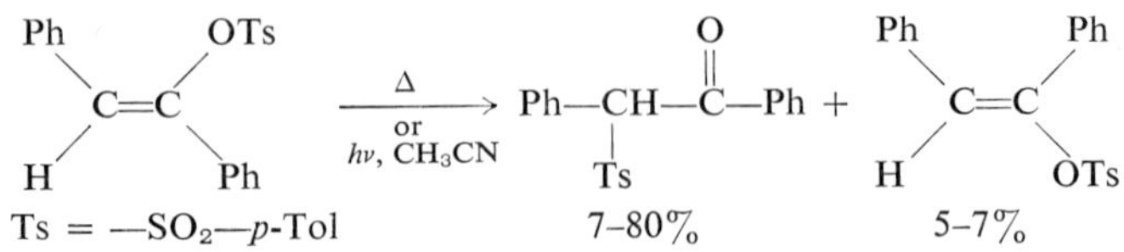

(362)

Although the yields are very subject to reaction conditions, the reaction pathway is clearly radical, being promoted by benzoyl peroxide and inhibited by hydroquinone. Both the geometric isomerization and the rearrangement were rationalized as shown in (363).

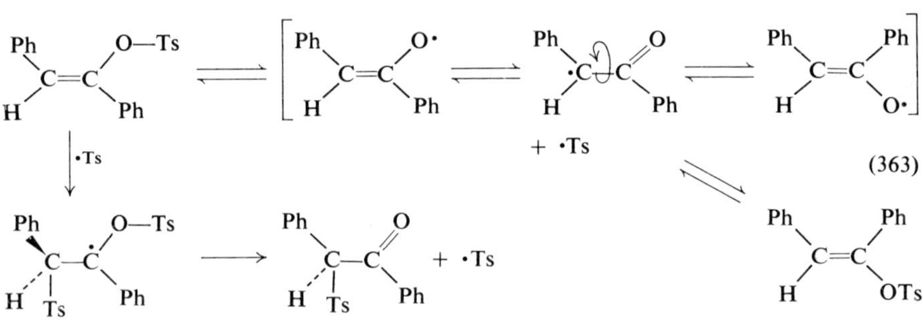

(363)

The cis enolic tosylate apparently did not allow radical addition; nor did $Ph_2C{=}CH{-}OTs$ or cyclo-$C_6H_{10}{=}CH{-}OTs$. Steric reasons were given, though the argument does not seem forceful. On the other hand, terminal alkene derivatives rearranged excellently (364).

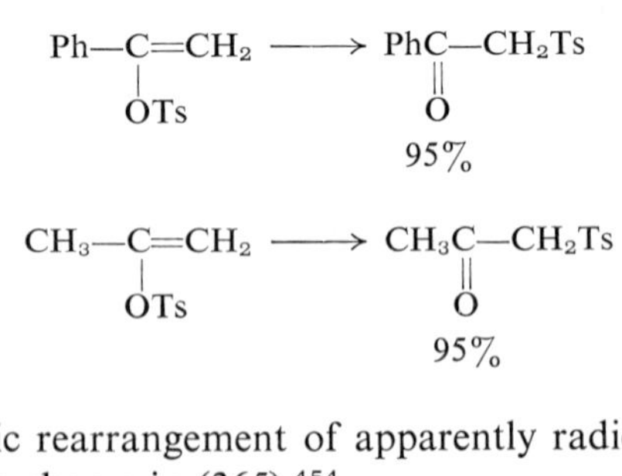

(364)

An extended allylic rearrangement of apparently radical nature occurred in the transformation shown in (365).[454]

An analogous rearrangement occurred with CH_3— in place of the —COOEt. Such distant migrations of trichloromethyl groups have been known for

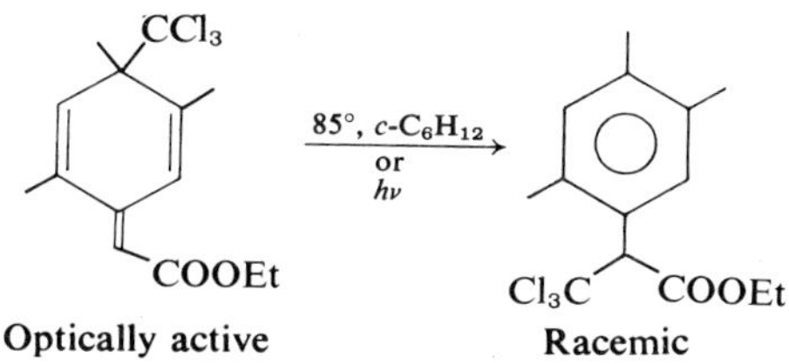

(365)

some time,[455] but the mechanism involved was unsettled. So facts such as: (1) the conversion showed an induction period; (2) the conversion could be inhibited by benzoquinone or thiophenol; and (3) the product was racemic, strongly implicate a radical pathway involving a chain sequence (366).[454]

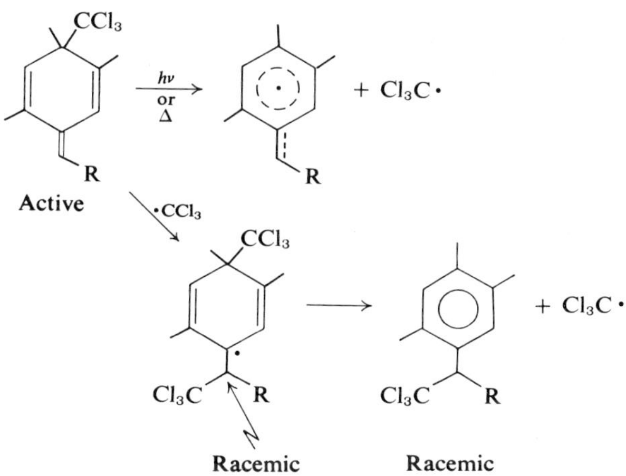

(366)

The possible recombination of radicals to form the product seems less likely because of the inhibition noted. Also, the racemized product could not arise solely because of subsequent enolization about an ester group as $R = CH_3$ allowed the reaction as well. Unfortunately, optically active ester product was unavailable, so its thermal stability to racemization could not be tested.

Finally, the Nesmeyanov rearrangement (p. 362 ff.) is often coupled with an allylic rearrangement. Some of these combined processes are shown in (367) and a summary is available.[456]

$$X{-}CCl_2{-}CBr{=}CH_2 \xrightarrow{h\nu} XClC{=}CCl{-}CH_2Br \qquad (367)$$

$$X = CH_3,\ Cl,\ F$$

The reader might enjoy the modest challenge of rationalizing these reactions in terms of a bromine atom chain carrier.

3. Homoallylic Rearrangements

Various kinds of rearrangements may be called homoallylic. Vicinal shifts of vinyl groups (p. 356) are one type, often proceeding via cyclopropylcarbinyl radical intermediates whose eliminative rearrangement (p. 400) completes the process.

But in this section, such rearrangements in bi- and polycyclic substrates will be emphasized. A large number of investigations in the norbornane and related ring systems involve rearrangement studies, most famously as cationic rearrangements. But a goodly number are radical processes—often tried to see if the cationic wizardry could be mimicked by the radicals. So the wealth of data alone merits a separate discussion of the homoallylic rearrangement in these cases.

Certain radical additions to bicyclic compounds lead to eliminative rearrangement (p. 406 ff.), but radical additions to norbornene and norbornene-related materials generally do not. Such additions are numerous and a good review of many of them through 1964 is available.[269a] The present account discusses newer work mainly. To set the stage, some of the earlier examples are given in the following equations (**368–371**).

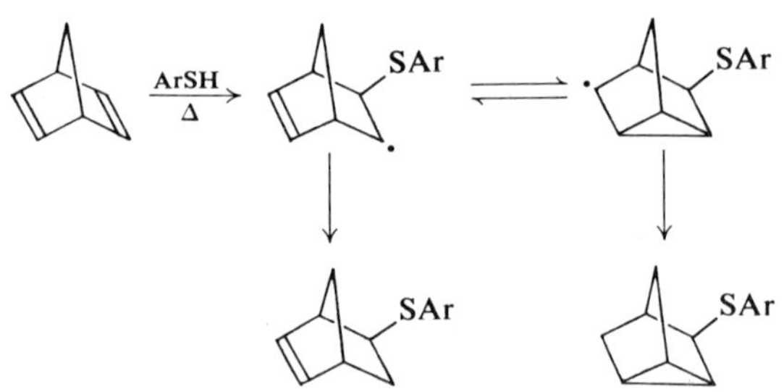

(368)[457]

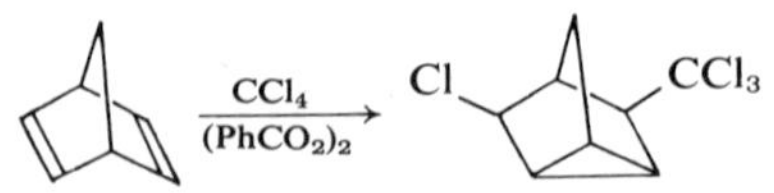

(369)[458]

In all these cases, as well as in others,[461] the bulk of the data supports the existence of two distinct radicals, **372-1** and **372-2** in (372), in equilibrium with one another. No good evidence for one, delocalized, "nonclassical" radical of this type exists.

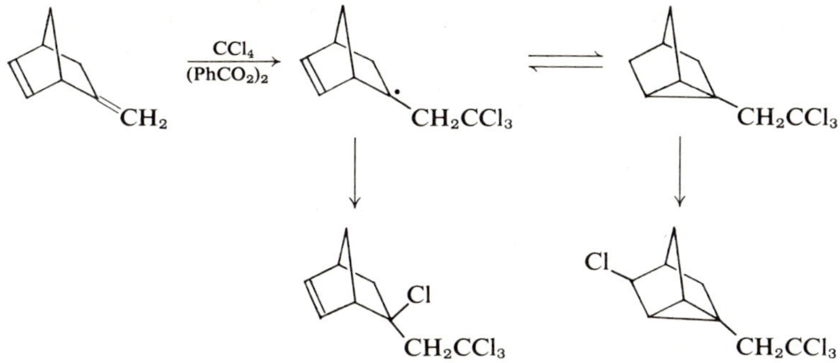

(370)[459]

(Stereochemistry unassigned)

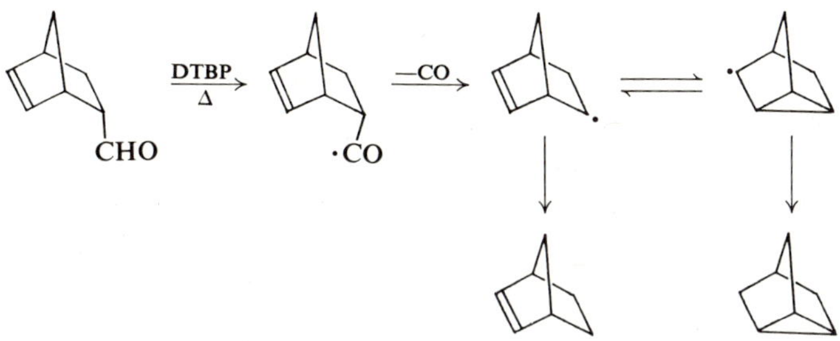

(371)[460]

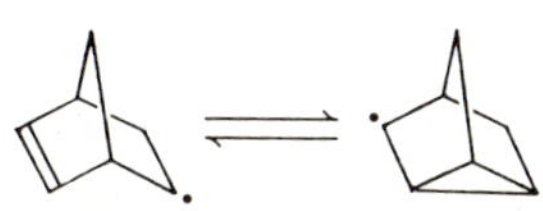

372-1 372-2

(372)

The composition of product that results from such radicals depends on whether or not the equilibrium in (372) has been established. From a norbornenyl reactant, good chain transfer agents or concentrated reactant will often favor more product from the parent radical, **372-1**. Normally, product from **372-2** is favored as equilibrium control is established. As a matter of fact, the equilibrium is rather quickly established, possibly in as little as 10^{-8} sec at 45°.[75] Two similar studies first demonstrated that radical **372-2**

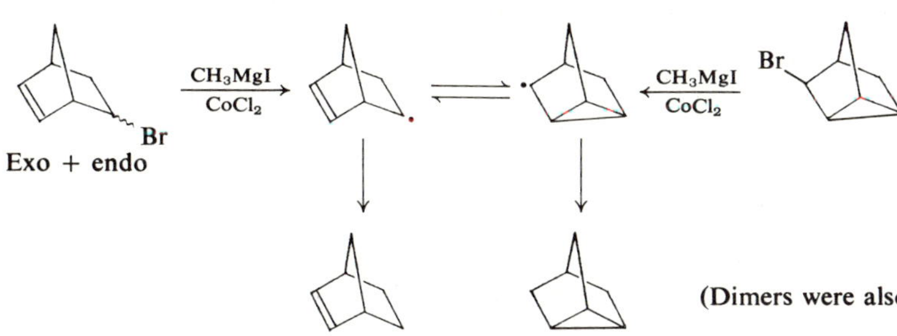

(373)

373-1 373-2

(Dimers were also formed)

could go to radical **372-1**. In the first (373),[462] the products from the bromides were formed comparably (**373-1**:**373-2** = 30:70), suggesting that the equilibrium in (372) had indeed been established and that **373-2** was favored. An earlier ratio of 45:55 in a decarbonylation study[460] probably reflects a different equilibrium composition at the higher temperature used there (135°) compared to that used here (35°). In the second (374),[463] both halides again gave both hydrocarbon products.

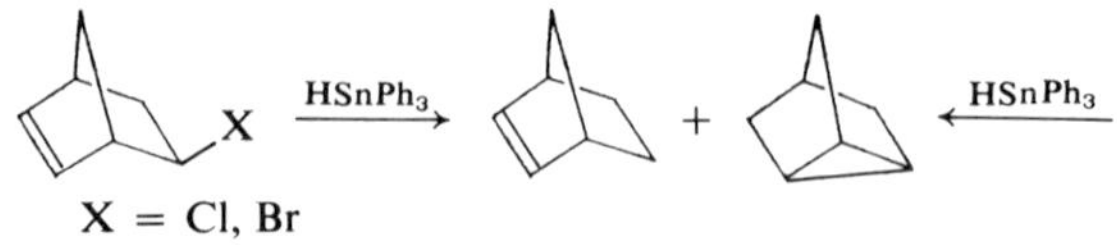

(374)

It is noteworthy that no special reactivity was shown by the norbornenyl halides, indicating that participation by the π electrons in the proximate double bond was not occurring. One might recall that such assistance is rare indeed in radical reactions, although a long-range case involving carbon may be known (p. 375).

Sodium biphenyl radical anion is a supremely effective reducing agent toward radicals. Using it, a recent study showed both the absence and presence (depending on conditions) of the equilibrium in (372). Some data from this study are collected in Table 14.[464]

At −58° the equilibrium in (372) was not established before the reactive reductant did its job. So the parent structure predominated in each case. By 25° even this reductant could not reduce the radicals prior to equilibration, thereby giving rise to similar product compositions in all cases. Note once again the favoritism toward nortricyclene.

Although not directly related to the above system, the reactions in (375) are of interest because again sodium biphenyl trapped an easily rearranged radical intermediate.[465]

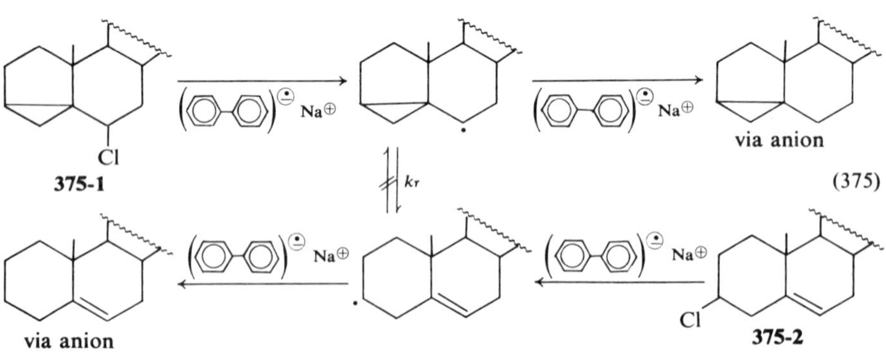

(375)

TABLE 14

Effect of conditions on the equilibrium **372-1** ⇌ **372-2**

Starting chloride (R–Cl)		Product composition (%)	
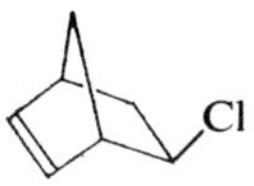	→	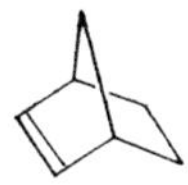 +	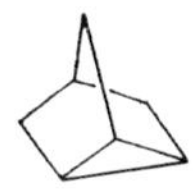
		64[a] (38)[b]	36 (62)
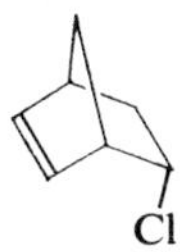	→	73 (35)	27 (65)
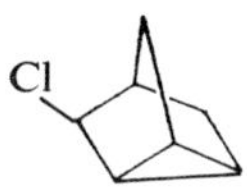	→	13 (35)	87 (65)

[a] Product compositions at −58°.
[b] At 25°.

The fact that chloride **375-2** was reduced without rearrangement suggests that under these conditions the cyclopropylcarbinyl–homoallylic rearrangement is not reversible. The reader may recall that the rearrangement is irreversible in simpler systems as well (see p. 402). From product composition data at various temperatures, the rearrangement rate constant k_r may be calculated to be 1.8×10^6 sec^{-1} at −20° and 1.25×10^8 sec^{-1} at 30°—fast indeed!

An elegant study recently interrelated yet another radical, **376-1**, with others in the norbornene–nortricyclene system (376).[466]

These same acetate esters were also produced from the chloromercuri acetate esters **376-3** and **376-4** upon treatment with sodium borohydride in tetrahydrofuran. Such a result may mean that a radical process is involved in this reaction also. Indeed, a nearly simultaneous and very similar report[74] strongly supports a radical pathway for the reduction of alkylmercuric halides with metal hydrides. The tripartite equilibrium in (376) is rapidly established, and the conversion of **376-1** to **376-2** has a rate constant at 20° of approximately 4×10^7 sec^{-1}.

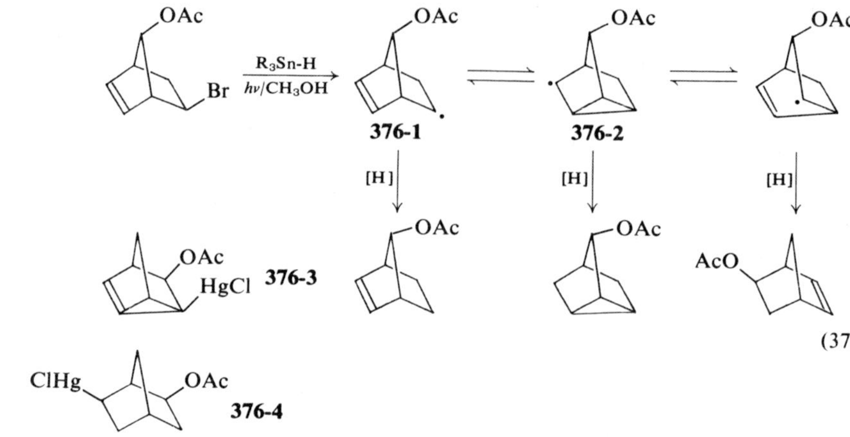

(376)

Rearrangements interconverting 5- and 7-substituted norbornenes (377) are rare in relatively unsubstituted cases, although (376) presents one instance.

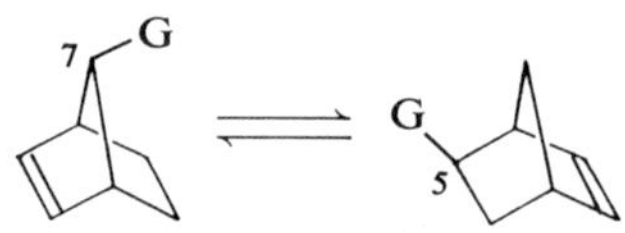

(377)

Two other recent examples are given in (378).[74]

Quite clearly, the transformation is in no way precluded. In the simple norbornenyl radical itself, the rearrangement would of course be undetectable without isotopic labeling (e.g., 377 with G = D).

Perhaps the earliest demonstration of the "5 ⇌ 7" rearrangement occurred in the exhaustive researches carried out on radical additions to hexachloronorbornadiene and related compounds. Conflicting structural assignments

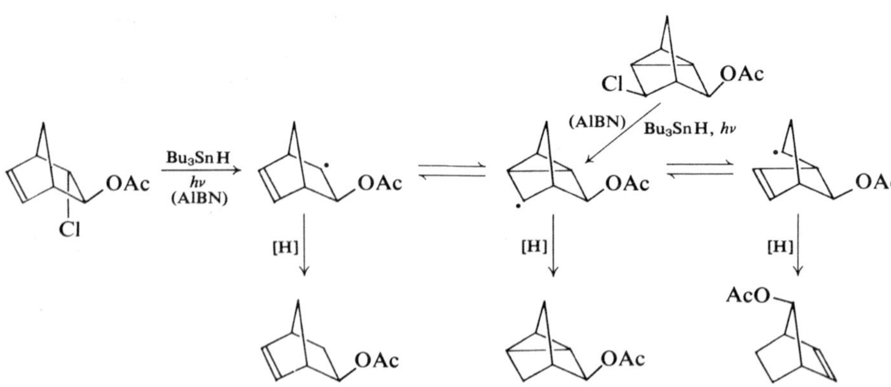

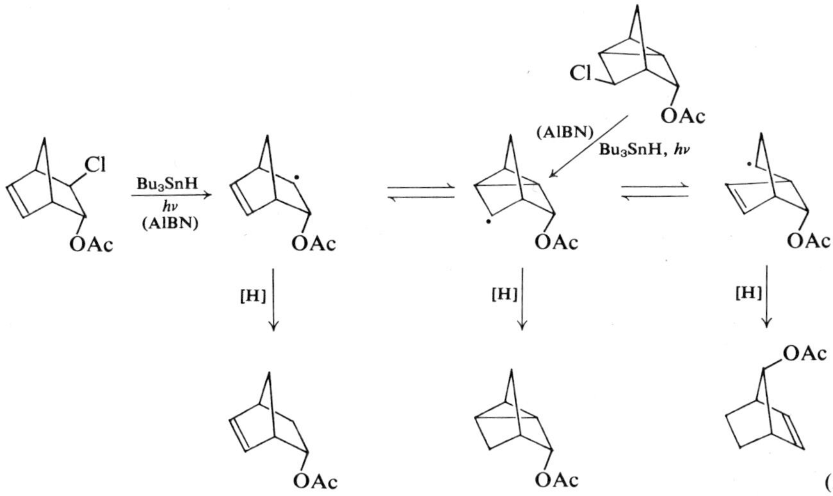

(378)

dominate this area. The derivatization procedures used sometimes turned out to be rearrangements themselves (!), so heavy reliance was placed on spectra and analogy. It reminds one in its complexity of the radical cyclization studies and the question of five- versus six-membered ring product formation (p. 420 ff.). The initial work centered on the process given in (379).[467]

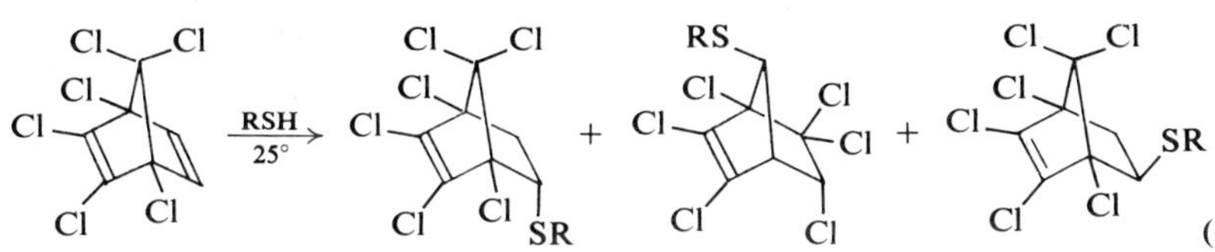

R = CH_3—, Et— **379-1** **379-2** **379-3** (379)

An earlier report claimed that **379-2** (R = Et) was the exo-1,2 adduct **379-3**.[468] But NMR spectral analysis supported **369-1** and **379-2** as the major products with a slight amount of **379-3**.[467] The insidious nature of things was illustrated by the oxidation of **379-1** and **379-2** to "their" sulfones, which were more easily handled derivatives (380).

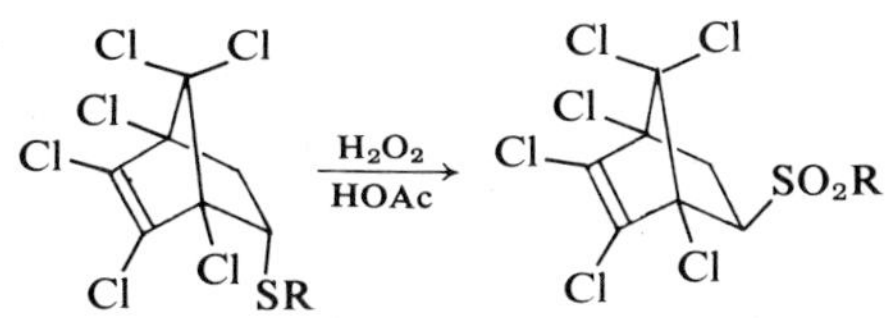

379-1

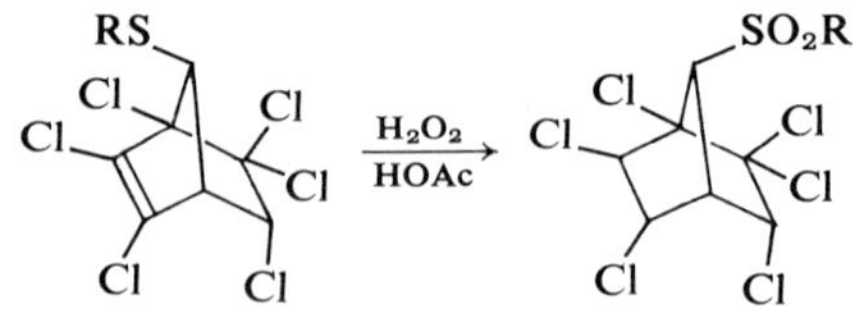

379-2

(380)

These epimerizations at C-5 and C-7 were at times avoidable when methanolic hydrogen peroxide was used. But little reliance was afterwards placed on such "proofs of structure." Whereas no mechanism was initially advanced for these additions, later work (381) brought one out.[469]

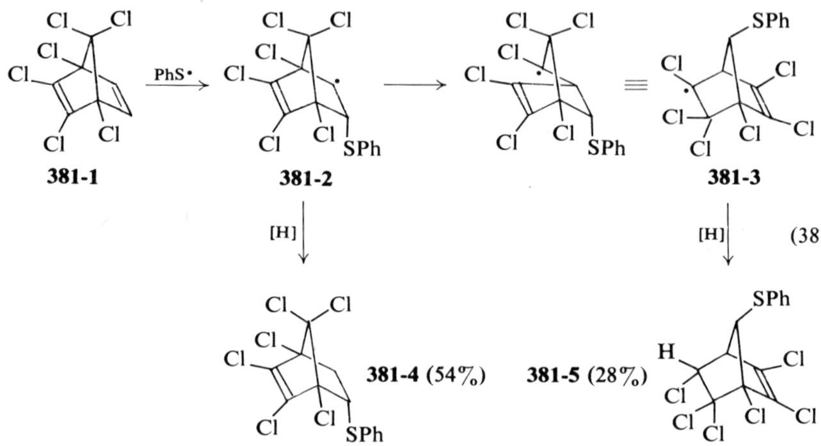

(381)

Although **381-4** was certified via independent, unequivocal synthesis, the structure of **381-5** was *based upon the mechanism*! Spectral data allowed an anti-7 epimeric structure for **381-5** as well. Indeed, the anti-7 epimer was claimed as the product.[470] The fundamental point affecting the configuration at C-7 is the initial direction of attack by the thiyl radical on **381-1**. If the attack be endo, a syn **381-5** will result; if exo, an anti epimer will form instead. Such attack is exo in norbornadiene itself,[457] but the 7-chloro function in **381-1** might favor endo attack in its case. Addition of bromotrichloromethane to **381-1** settled the question (382).[469]

(382)

Not only is the initial attack endo but the final chain transfer appears to be preferentially exo (as was the case in **381-5** as well). The syn nature of **381-5** was then later confirmed by the other contestants.[471]

An interesting aspect of the mechanism in (381) is the vicinal vinyl shift (**381-2** to **381-3**). Could the path really be via (383), as is now known in other cases?[74,466]

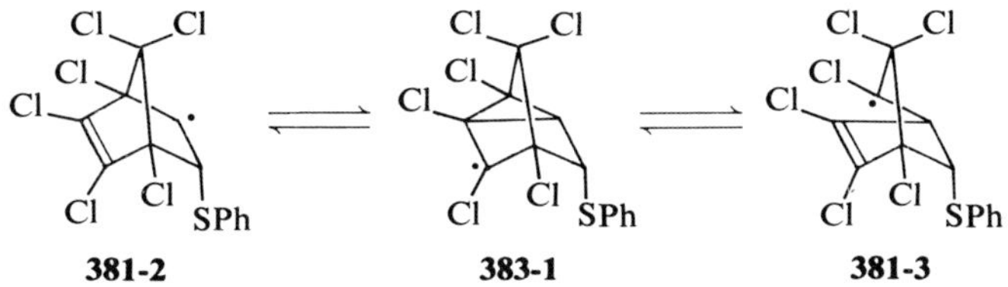

(383)

That such is probably the case was shown in other reactions (384).[472]

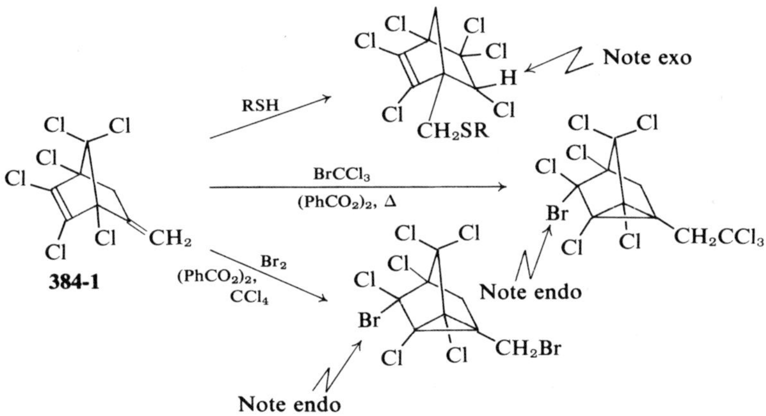

(384)

Undoubtedly, the three radicals shown (385) are available for product formation.

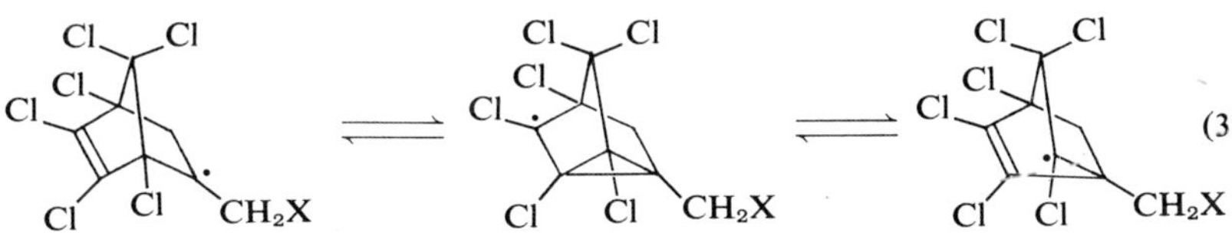

(385)

Just which products and in what proportions one gets depends on a number of factors, for example, the chain transfer ability of the addendum, the environment at the chain transfer site, whether or not the equilibria in (385) are established, and the influence of X on these equilibria. There seems to be no reason why such is not the case with the system in (383) as well.

Various other modified norbornadienes have been employed to develop these ideas further. In (386) the dichloro diene was used.[473]

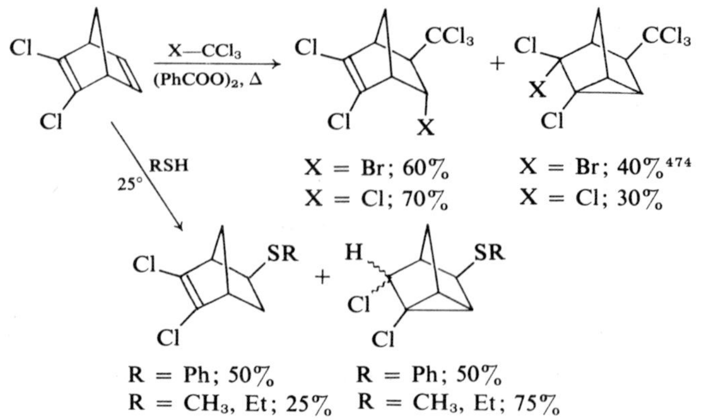

(386)

Clearly, in these cases, the equilibria in (387) would favor **387-1** at the expense of **387-2** because of the α-chloro function in the former (recall the stabilization of radical sites by α-chloro in the Nesmeyanov rearrangement, p. 362 ff.).

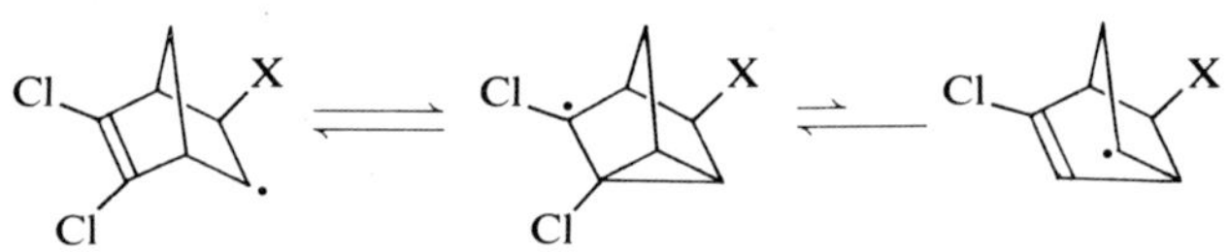

(387)

387-3 **387-1** **387-2**

As **387-2** can form apparently *only* via **387-1** (i.e., there is no direct vicinal vinyl shift converting **387-3** to **387-2**), only vicinal and nortricyclyl products—no C-7 substituted ones—are found here.

In the case of *hexa*chloronorbornadiene and thiols, the equilibria seemingly favor radicals **381-2** and **381-3** instead, because no nortricyclyl products resulted. A similar situation prevails with diene **384-1** in (384). In both of these instances, opening of nortricyclyl radical intermediates leads to α-chloro-stabilized radicals.

It is interesting to observe that bromotrichloromethane can trap out any one of the three radicals involved in all these studies. Thus, from hexachloronorbornadiene, trapping gave the vicinal adduct (equation 382). From diene **384-1**, trapping led to the nortricyclyl adduct (equation 384). Lastly, from diene **388-1**, the syn-7 adduct was obtained,[475] as shown in (388). Otherwise, results with **388-1** and other addenda indicated a gross similarity of this diene with hexachloronorbornadiene. Electronic effects (i.e., methoxy vs. chloro) was thus minor in these reactions, with steric factors clearly dominant in forcing the initial endo attack by radicals on those dienes with bulky C-7 substitutents.[476] Notwithstanding the importance of the C-7

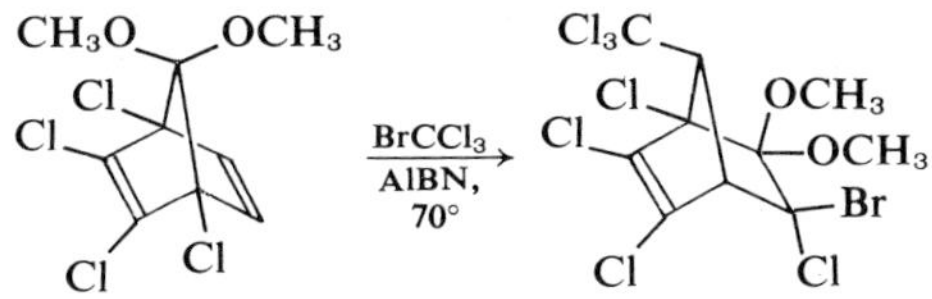

388-1 (388)

substituents, chlorines at C-1 and C-4 also play a role here. Whereas *di*chloronorbornadiene underwent initial exo attack as expected (386), *tetra*chloronorbornadiene **389-1** suffered some endo attack as well, even though no bulky C-7 substituent was present (389).[477]

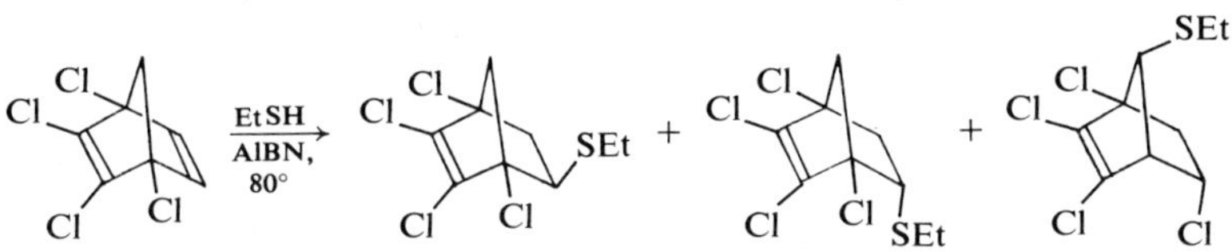

389-1 **389-2** (15%) **398-3** (21%) **389-4** (64%) (389)

The production of **389-4** necessitates the rearrangement in (390), as may be expected.

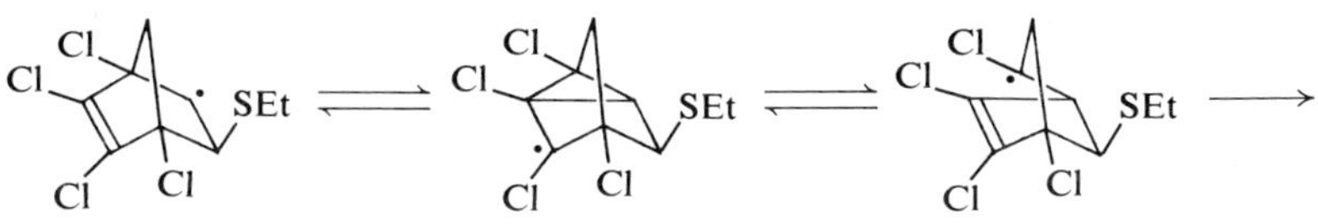

389-4 (390)

However, the fact that no syn-7 adduct resulted in the process means that the rearrangement in (391) did not occur, as may *not* have been expected.

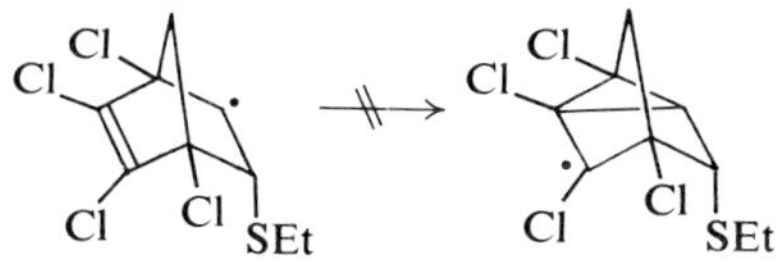

391-1 **391-2** (391)

Presumably, the *absence* of bulky C-7 substituents now does become important. Radical **391-1** must undergo chain transfer too rapidly with thiols to allow the rearrangement to **391-2**. In (390) the exo-2 thiyl group causes slower chain transfer and the rearrangement shown can occur. If so, then both

TABLE 15

Various Dienes vis-à-vis Exo–Endo Attack by Thiyl Radicals in Initial Addition

Diene	Preferred initial attack (%)
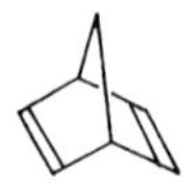	Exo (100)
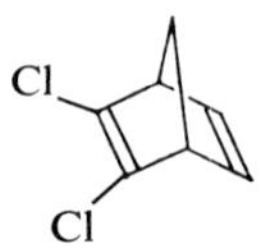	Exo (100)
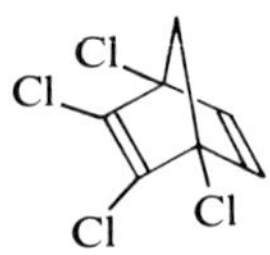	Exo (79)
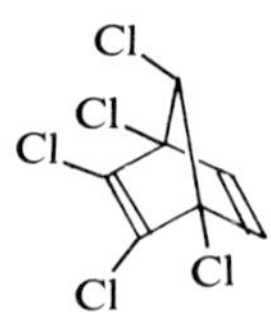	Exo (< 100)[477]
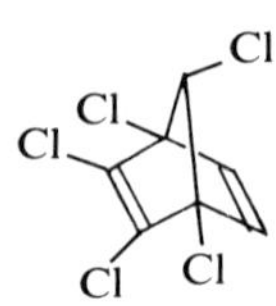	Endo (100)[477]
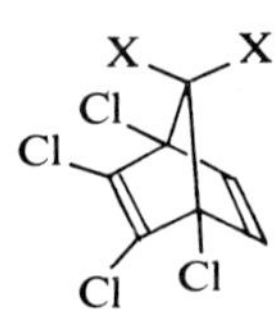 X = Cl, OCH_3	Endo (100)

389-2 and **389-4** come from the same process (exo attack) and amount to 79% of the product in (389). One may group the various dienes discussed vis-à-vis exo–endo radical attack in the initial addition by thiyl radicals. Such a grouping is given in Table 15.

The mind-boggling flippings uncovered in these additions make one suspicious of the implicit certitude in some earlier addition reactions, such as (392)[458] and (393).[478]

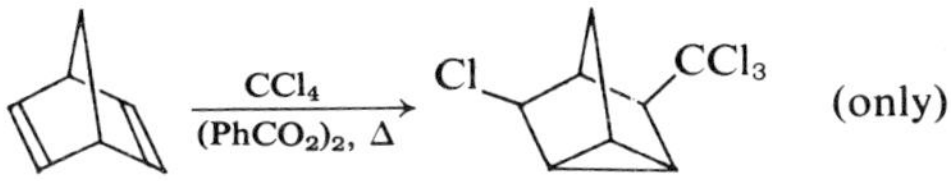

(392)

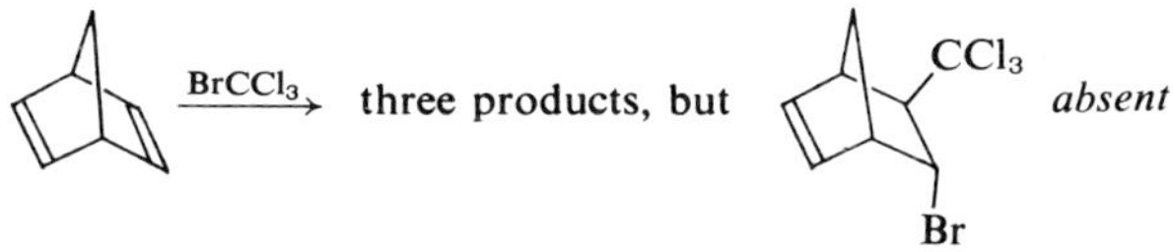

(393)

A reinvestigation[479] of both reactions indicated that the earlier conclusions were incorrect. In the first, chain transfer in fact gave considerable endo chloride (394).

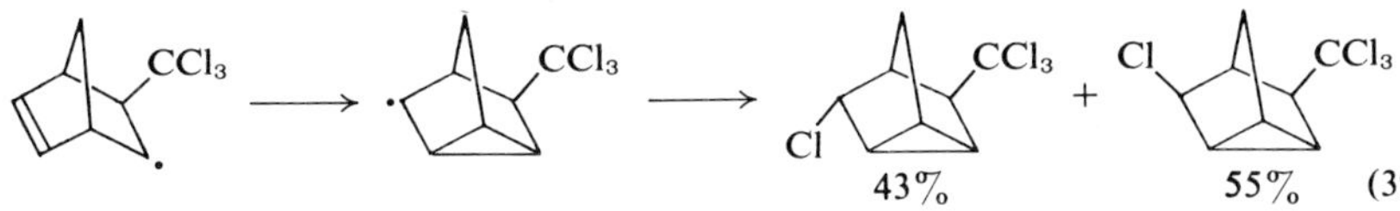

(394)

So nortricyclyl-type radicals show no particular exo preference in chain transfer, quite unlike norbornyl radicals. This result cleared up a minor mystery. Other addenda had given mixtures resulting from both exo and endo chain transfer with norbornadiene and the result[458] with carbon tetrachloride was anomalous.[480] Now it is not. What is now bothersome is that alkaline dehydrohalogenation of the adduct (mixture) gave only **395-1** (395)!

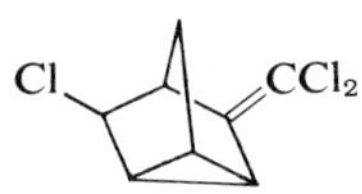

395-1

(395)

Somewhere in the elimination, epimerization of the endo chloride must occur.

In regard to (393), the culprit here was dissolved hydrogen bromide (often present in reactions with bromotrichloromethane). When this isomerization agent was removed from the product mixture, the vicinal adduct was detected (10%).[479]

In a somewhat different example, but related nonetheless, the following addition apparently involved a homoallylic rearrangement (396)[481]:

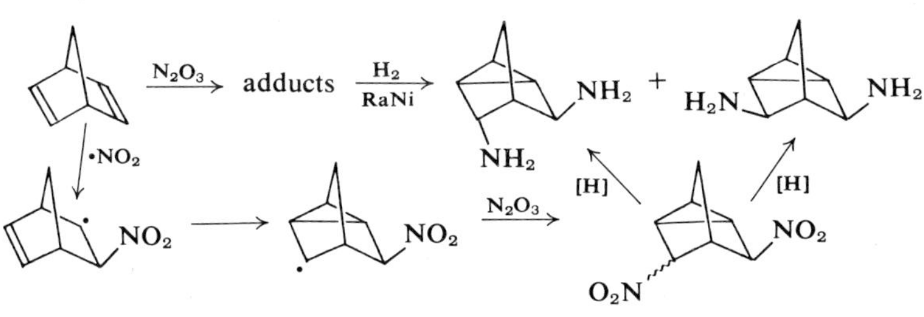

(396)

Radical additions to benzonorbornadiene can, on occasion, show such a homoallylic rearrangement. In this case the process involves a neophyl-like aromatic migration. Generally, as shown in (397), little rearrangement has been observed because active chain transfer agents have been employed.

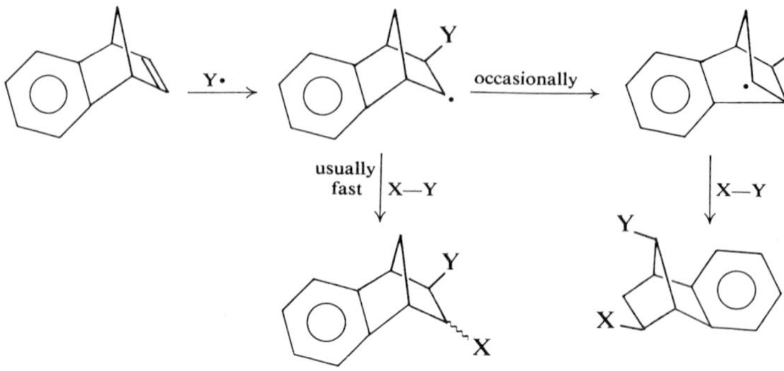

(397)

Incidentally, the final chain transfer act to give the vicinal adduct is mostly endo, and more so as Y is increased in size.[78,482] However, when Y is hydrogen, chain transfer is largely exo instead.[483] Obviously, steric factors are in control here.

Occasionally, rearrangement has been observed and an example has been given earlier (41). Another instance has very recently been reported (398).[484]

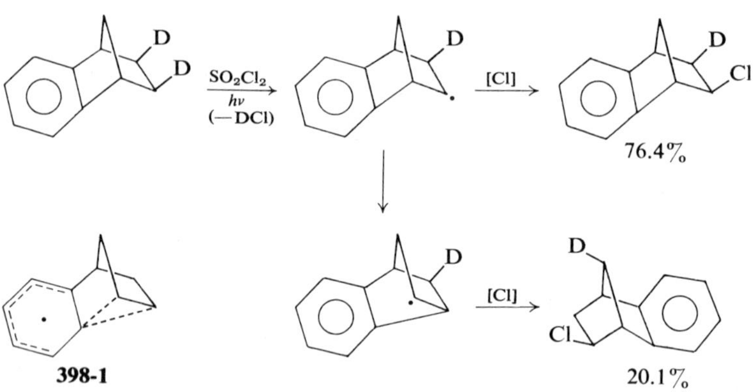

(398)

It might be noted, however, that in this case the process was initiated by abstraction, not addition. Radical rearrangement in the latter-type reactions involving benzonorbornadiene remain rare.[78] Perhaps in this system rearrangement is somewhat less easy because the transition involves a "phenonium"-type radical instead of a nortricyclyl radical as with norbornadienes. The latter clearly has some independent existence, whereas the former is postulated as a transition state only, as in **398-1**.

A homoallylic-type rearrangement of acyl groups completes the discussion in this area. For example, free radical acylations as in (399)[485] and (400)[486] definitely show that carbonyl *carbon* can be attacked by carbon radicals, an uncommon but known[487] reaction.

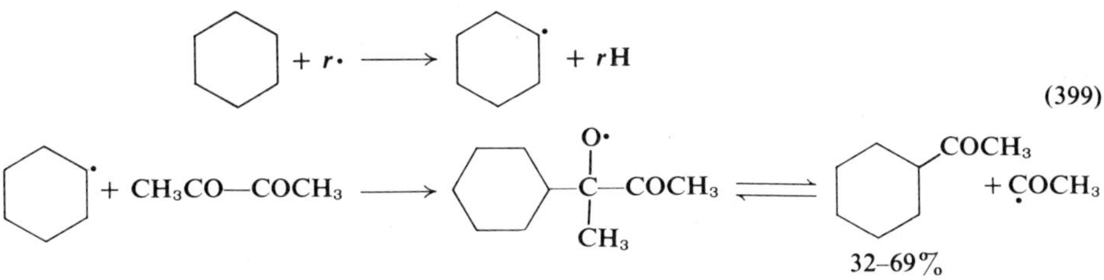

(399)

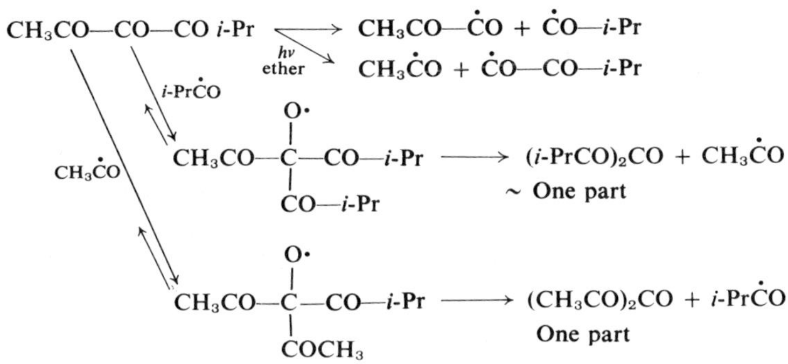

(400)

That this process can occur intramolecularly (and therefore be a rearrangement) is suggested by some additional studies, as shown in (401)[488] and (402).[489]

One might mention here another acyl migration, the isophorone oxide rearrangement (pp. 411, 412). Indeed, these migrations are 1,2 shifts, How-

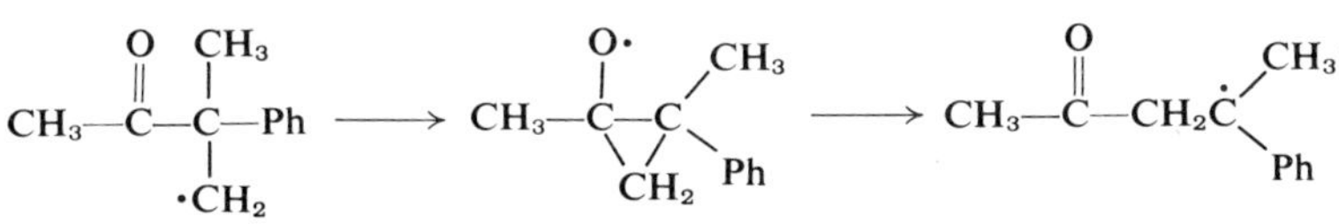

(401)

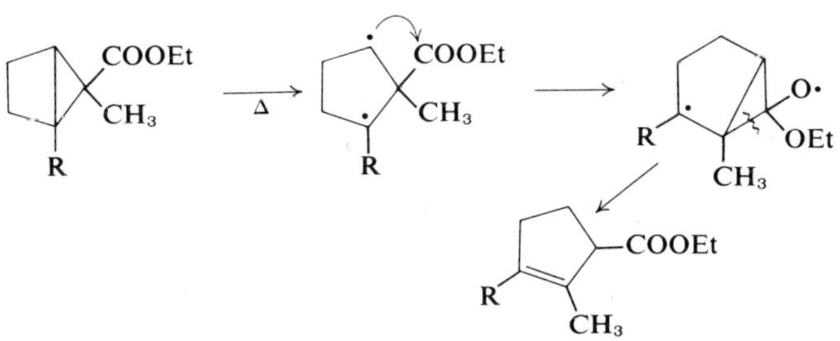

(402)

ever, as with the related 1,2 vinyl migrations, the shift actually involves a cyclized intermediate and they are perhaps better classified as homoallylic-type rearrangements akin to those in this section.

V. SUMMARY AND OUTLOOK FOR THE FUTURE

In the quarter-century since the Urry-Kharasch impetus, radical rearrangements in solution have multiplied in decent, but not embarrassing, numbers. By and large, the rearrangements discovered have mirrored cationic analogs. One can, in the opinion of the author, look for more of the same in the years ahead, particularly attempts to find the elusive 1,2 alkyl shift.

Whereas the first-generation radical rearrangements normally were self-evidently homolytic in mechanism, the new breed may not be. "Ionic" rearrangements such as the Stevens and the Wittig are now "something else," as CIDNP indicates. Future research will surely better answer what this "something else" is, as well as extend this knowledge to other rearrangements long held to be nonradical in mechanism.

Homolytic anchimeric participation by carbon must be waiting to be found. Because by definition such an effect involves rate data and because such data have been least available for radical rearrangement reactions, thus far the existence of this effect has eluded its hunters. Advances here will be made beyond doubt. Rate data are becoming available, as are better radical generating conditions (less "messy"). Together these aids will help to uncover the participation.

Nonclassical radicals, as ground-state species or as intermediates, are unknown (in rearrangements or in anything else). In every case, as discussed in several places in the foregoing, a presumed nonclassical radical intermediate was either dismissable by the data or at the least not demanded. There seems to be no reason why nonclassical radicals should be *so* forbidden. Admittedly, in theory there may be less *raison d'être* for them than for their cationic brethren, but some nonclassical radical awaits its discoverer somewhere—theory or no.

Free radical chemistry began in challenge to theory. It has matured within new theory. But the time has come to begin again, with some new Moses to lead the way.

REFERENCES AND NOTES

1. For texts on free radicals, see W. E. Bachmann, *Organic Chemistry*, 2nd ed., Vol. 1, H. Gilman, Ed., Wiley, New York, 1943, Chap. 6; W. A. Waters, *The Chemistry of Free Radicals*, 2nd ed., Oxford University Press, London, 1948; E. W. R. Steacie, *Atomic and Free Radical Reactions*, 2nd ed., Reinhold, New York, 1954; C. Walling, *Free Radicals in Solution*, Wiley, New York, 1957; W. A. Pryor, *Free Radicals*, McGraw-Hill, New York, 1966.

2. C. Walling, in *Molecular Rearrangements*, Vol. 1, P. deMayo, Ed., Interscience, New York, 1963, Chap. 7. Because citations to this reference are so frequent in the present chapter, they have been given as C. Walling, reference 2, with the appropriate pages listed.
3. R. Kh. Freidlina, *Advances in Free Radical Chemistry*, Vol. 1, G. H. Williams, Ed., Academic Press, New York, 1965, Chap. 6. This reference is again so often cited in this chapter that it has appeared as R. Kh. Freidlina, reference 3, with the appropriate pages listed.
4. Walling[2] covers through 1961 well, with a number of early 1962 references. Freidlina[3] is somewhat severe in what rearrangements should be covered, but she gives quite thorough coverage through 1964 in selected areas. Pryor[1] does not have much better coverage than Walling, even though his book is more recent. The radical rearrangement section is not as up to date as others.
5. There is no justification claimed for such extraneous inclusions other than personal interest. It should be stated clearly that in this chapter the blossoming field of photochemical (radical?) rearrangements is barely touched.
6. Compare P. D. Bartlett, *Nonclassical Ions*, Benjamin, New York, 1965.
7. (a) H. E. Zimmerman and A. Zweig, *J. Am. Chem. Soc.*, **83**, 1196 (1961); (b) H. E. Zimmerman, in *Molecular Rearrangements*, Vol. 1, P. de Mayo, Ed., Interscience, New York, 1963, pp. 394–399.
8. N. F. Phelan, H. H. Jaffe, and M. Orchin, *J. Chem. Educ.*, **44**, 626 (1967).
9. A. Streitwieser, Jr., *Molecular Orbital Theory for Organic Chemists*, Wiley, New York, 1961; J. D. Roberts, *Notes on Molecular Orbital Calculations*, Benjamin, New York, 1962; A. Liberles, *Introduction to Molecular Orbital Theory*, Holt, Rinehart and Winston, New York, 1966.
10. R. Kh. Freidlina, reference 3, p. 272.
11. R. B. Woodward and R. Hoffmann, *The Conservation of Orbital Symmetry*, Academic Press, New York, 1970.
12. R. B. Woodward and R. Hoffmann, reference 11, p. 118.
13. N. N. Semenov, *On Some Problems in Chemical Kinetics and Reactivity*, Vol. 1, translated by M. Boudart, Princeton University Press, Princeton, N.J., 1958, pp. 68–73.
14. L. S. Mayanz, *J. Phys. Chem. USSR*, **37**, 623 (1964).
15. C. Rüchardt, cited in footnote 37 of reference 419a.
16. The present organization is particularly close to those used by C. Walling[2] and R. Kh. Freidlina.[3]
17. The terminology of D. J. Cram, *Steric Effects in Organic Chemistry*, M. S. Newman, Ed., Wiley, New York, 1956, Chap. 5, p. 283.
18. Friedlina[3] chooses not to include such processes in detail. Her coverage emphasizes 1,x shifts and hydrogen abstractions.
19. The developments in this class are too recent for previous reviews. Specific areas in the class, for example, the ylid processes, have been reviewed, however. Citations will be made when appropriate.
20. For summaries of early work and references, compare C. Walling, reference 2, pp. 416–423; R. Kh. Freidlina, reference 3, pp. 263–267.
21. A. Fish, *Q. Rev. (London)*, **18**, 243 (1964); J. A. Kerr and A. C. Lloyd, *ibid.*, **22,** 549 (1968).
22. J. Walker, *J. Chem. Soc.*, **63**, 496 (1893); J. Walker and J. Henderson, *ibid.*, **67**, 337 (1895); J. Walker and W. Cormack, *ibid.*, **77**, 374 (1900).
23. For other such cases, see C. Walling, reference 2, pp. 420–421; J. A. Van Zorge, J. Strating, and H. Wynberg, *Rec. Trav. Chim.*, **89**, 781 (1970).

24. L. Kaplan, *J. Org. Chem.*, **33**, 2531 (1968).
25. J. W. Wilt, L. L. Maravetz, and J. F. Zawadzki, *J. Org. Chem.*, **31**, 3018 (1966).
26. J. A. Berson, C. J. Olsen, and J. S. Walia, *J. Am. Chem. Soc.*, **82**, 5000 (1960); **84**, 3337 (1962).
27. M. S. Newman and P. F. Beal, III, *J. Am. Chem. Soc.*, **72**, 5163 (1950).
28. G. Frater and O. P. Strausz, *J. Am. Chem. Soc.*, **92**, 6654 (1970), and references cited therein.
29. T. Sasaki, S. Eguchi, and T. Toru, *Chem. Commun.*, 1239 (1970).
30. F. D. Greene, W. Adam, and G. A. Knudsen, Jr., *J. Org. Chem.*, **31**, 2087 (1966). The preference for methyl migration varies with conditions, however. Photochemical decomposition of the perester in (20) gave $CH_3/Ph = 0.94$, while irradiation in the presence of the triplet quencher piperylene gave $CH_3/Ph = 0.45$; W. Adam and G. S. Aponte, *J. Am. Chem. Soc.*, **93**, 4300 (1971). These authors suggest that the first photochemical process represents an indiscriminate triplet process whereas the second represents a selective singlet reaction.
31. W. Adam and Y. M. Cheng, *J. Am. Chem. Soc.*, **91**, 2109 (1969).
32. W. Adam, Y. M. Cheng, C. Wilerson, and W. A. Zaidi, *J. Am. Chem. Soc.*, **91**, 2111 (1969).
33. C. Walling, H. P. Waits, J. Milovanovic, and C. G. Pappiaonnou, *J. Am. Chem. Soc.*, **92**, 4927 (1970).
34. H. E. Zimmerman, B. R. Cowley, C. Y. Tseng, and J. W. Wilson, *J. Am. Chem. Soc.*, **86**, 947 (1964).
35. H. O. House, E. J. Grubbs, and W. F. Gannon, *J. Am. Chem. Soc.*, **82**, 4099 (1960).
36. L. H. Slaugh, E. F. Magoon, and V. P. Guinn, *J. Org. Chem.*, **28**, 2643 (1963).
37. E. Grovenstein, Jr., and L. P. Williams, Jr., *J. Am. Chem. Soc.*, **83**, 2537 (1961).
38. E. Grovenstein, Jr., and G. Wentworth, *J. Am. Chem. Soc.*, **85**, 3303 (1963).
39. (a) M. P. Doyle, Department of Chemistry, Hope College, Holland, Michigan. Cited with permission. (b) H. M. Walborsky and J.-C. Chen, *J. Am. Chem. Soc.*, **92**, 7573 (1970).
40. C. Walling, reference 2, pp. 409–416; R. Kh. Freidlina, reference 3, pp. 249–260; W. A. Pryor, *Free Radicals*, McGraw-Hill, New York, 1966, pp. 266–273.
41. W. H. Urry and M. S. Kharasch, *J. Am. Chem. Soc.*, **66**, 1438 (1944). This study, done before the days of infrared, nuclear magnetic resonance, and gas–liquid chromatographic techniques, is a classic illustration of careful use of earlier analytical methods, e.g., indices of refraction. A similar study on the *p*-methylneophyl radical was done by W. H. Urry and N. Nicolaides, *ibid.*, **74**, 5163 (1952).
42. The initial study in solution was done by S. Winstein and F. H. Seubold, Jr., *J. Am. Chem. Soc.*, **69**, 2916 (1947). Since this first study a sizable number of others have been reported. Cf. R. Kh. Freidlina, reference 3, pp. 254–260 for tabular surveys.
43. The palladium-catalyzed decarbonylation of aldehydes proceeds differently. Far less rearrangement is observed in sensitive systems. Cf. J. W. Wilt and V. P. Abegg, *J. Org. Chem.*, **33**, 923 (1968).
44. Nonradical decarbonylation of aldehydes can be stereoselective and free of rearrangement. Cf. H. M. Walborsky and L. E. Allen, *Tetrahedron Letters*, 823 (1970), and references cited therein.
45. J. Weinstock and D. W. Lewis, *J. Am. Chem. Soc.*, **79**, 6243 (1957).
46. (a) C. G. Overberger and H. Gainer, *J. Am. Chem. Soc.*, **80**, 4561 (1958); (b) D. Y. Curtin and T. C. Miller, *J. Org. Chem.*, **25**, 885 (1960).
47. (a) W. Rickatson and T. S. Stevens, *J. Chem. Soc.*, 3960 (1963); (b) W. H. Starnes, Jr., *J. Am. Chem. Soc.*, **85**, 3708 (1963); (c) D. B. Denney, R. L. Ellsworth, and

D. Z. Denney, *ibid.*, **86**, 1116 (1964); (d) C. Rüchardt and R. Hecht, *Chem. Ber.*, **98**, 2460, 2471 (1965); (e) W. H. Starnes, Jr., *J. Am. Chem. Soc.*, **86**, 5603 (1964).

48. H. Breederveld and E. C. Kooyman, *Rec. Trav. Chim.*, **76**, 297 (1957).
49. C. E. Berr, Dissertation, University of California, Los Angeles, California, 1952, reported about 5% rearrangement in the reaction of silver β-phenylisovalerate with bromine. J. W. Wilt, *J. Am. Chem. Soc.*, **77**, 6397 (1955), reported no rearrangement in the same reaction. W. T. Smith, Jr., and J. T. Sellas, *Trans. Kentucky Acad. Sci.*, **16**, 72 (1955), claimed 52% rearrangement in the process. Clearly the study should be redone.
50. H. Pines and C. N. Pillai, *J. Am. Chem. Soc.*, **82**, 2921 (1960); M. Abramovici and H. Pines, *J. Org. Chem.*, **34**, 266 (1969).
51. In the gas phase by J. D. Backhurst, E. D. Hughes, and C. K. Ingold, *J. Chem. Soc.*, 2742 (1959); in solution by H. Meislich, J. Costanza, and J. Strelitz, *J. Org. Chem.*, **33**, 3221 (1968).
52. (a) C. Rüchardt and H. Trautwein, *Chem. Ber.*, **95**, 1197 (1962); (b) E. Grovenstein, Jr., and Y.-M. Cheng, *Chem. Commun.*, 101 (1970); (c) M. J. Perkins and P. Ward, *Tetrahedron Letters*, 2379 (1971); (d) D. J. Atkinson, M. J. Perkins, and P. Ward, *J. Chem. Soc. (C)*, 3240 (1971). It is worth noting that the rearrangement **30a-1** to **30a-2** was earlier suggested by D. M. Collington, D. H. Hey, and C. W. Rees, *ibid.*, 1026 (1968), but later retracted by D. H. Hey, G. H. Jones, and M. J. Perkins, *Chem. Commun.*, 1438 (1970).
53. The bridged "phenonium" ion involved in the analogous cationic rearrangement has been recently observed at −78° by NMR spectroscopy. Cf. G. A. Olah and R. D. Porter, *J. Am. Chem. Soc.*, **92**, 7627 (1970), and references cited therein. But no evidence for a comparably stable bridged radical exists.
54. (a) F. H. Seubold, Jr., *J. Am. Chem. Soc.*, **75**, 2532 (1953); (b) C. Rüchardt, *Chem. Ber.*, **94**, 2599 (1961); (c) J. W. Wilt and C. A. Schneider, *J. Org. Chem.*, **26**, 4196 (1961).
55. It is amusing that the differentiation between one nonclassical and two classical precursors to products has commonly been achieved by varying the concentrations of reactants. Because no nonclassical radical is known, it remains open to question how one would respond under these conditions.
56. (a) J. W. Wilt and Bro. H. Philip, *J. Org. Chem.*, **25**, 891 (1960); (b) J. W. Wilt, J. F. Zawadzki, and D. G. Schultenover, *ibid.*, **31**, 876 (1966).
57. C. Rüchardt and H. Trautwein, *Chem. Ber.*, **98**, 2478 (1965).
58. The decarbonylation reaction, C. Rüchardt, *Chem. Ber.*, **94**, 2609 (1961); the Kharasch-Grignard reaction, C. Rüchardt and H. Trautwein, *ibid.*, **96**, 160 (1963); the decomposition of *t*-butyl β-arylisopervalerate esters, C. Rüchardt and R. Hecht, reference 47d.
59. P. D. Bartlett and R. R. Hiatt, *J. Am. Chem. Soc.*, **80**, 1398 (1958).
60. L. H. Slaugh, *J. Am. Chem. Soc.*, **81**, 2262 (1959).
61. (a) D. Y. Curtin and M. J. Hurwitz, *J. Am. Chem. Soc.*, **74**, 5381 (1952); D. Y. Curtin and J. C. Kauer, *J. Org. Chem.*, **25**, 880 (1960).
62. W. B. Smith and J. D. Anderson, *J. Am. Chem. Soc.*, **82**, 656 (1960).
63. D. P. Thornhill and C. C. Lee, *Can. J. Chem.*, **41**, 2861 (1963).
64. W. A. Bonner and F. D. Mango, *J. Org. Chem.*, **29**, 29 (1964).
65. W. W. Pawlikowski, Jr., Dissertation, Loyola University, Chicago, Illinois, 1970.
66. (a) B. M. Vittimberga, *Tetrahedron Letters*, 2383 (1965); (b) P. N. Cote and B. M. Vittimberga, *J. Am. Chem. Soc.*, **93**, 276 (1971). In this report, decarbonylation of 9-(*p*-substituted-phenyl) fluorenylacetaldehydes possessed a reaction constant

$\rho = +0.426$, indicating that the polar contributor in (35) may be less important than previously thought. However, it remains to be shown that this system is related enough to the neophyl system to allow any quantitative comparisons between them.

67. J. W. Wilt, H. F. Dabek, Jr., J. P. Berliner, and C. A. Schneider, *J. Org. Chem.*, **35**, 2402 (1970).
68. L. K. Montgomery, J. W. Matt, and J. R. Webster, *J. Am. Chem. Soc.*, **89**, 923 (1967).
69. J. K. Kochi and P. J. Krusic, *J. Am. Chem. Soc.*, **91**, 3940 (1969).
70. Unpublished work by the author.
71. E. I. Snyder, *J. Am. Chem. Soc.*, **92**, 7529 (1970).
72. D. B. Denney and N. F. Newman, *J. Am. Chem. Soc.*, **89**, 4692 (1967). For earlier, similar studies, cf. W. Schlenk and C. Bornhardt, *Ann.*, **394**, 183 (1912); E. Müller, A. Moosmayer, and A. Rieker, *Z. Naturforsch.*, **186**, 982 (1963). For the somewhat similar reaction of the triphenylmethyl *cation* with diazomethane, see H. W. Whitlock, *J. Am. Chem. Soc.*, **84**, 2807 (1962).
73. L. Kaplan, *J. Am. Chem. Soc.*, **88**, 4531 (1966).
74. The parent neophyl radical itself, as well as other easily rearranged radicals, has been trapped in a similar reaction prior to rearrangement by G. M. Whitesides and J. SanFilippo, Jr., *J. Am. Chem. Soc.*, **92**, 6611 (1970).
75. D. J. Carlsson and K. U. Ingold, *J. Am. Chem. Soc.*, **90**, 7047 (1968).
76. B. B. Jarvis and J. B. Yount, III, *J. Org. Chem.*, **35**, 2088 (1970).
77. S. J. Cristol and G. W. Nachtigall, *J. Org. Chem.*, **32**, 3727 (1967).
78. M. M. Martin and R. A. Koster, *J. Org. Chem.*, **33**, 3428 (1968), and references cited therein.
79. S. J. Cristol and D. K. Pennelle, *J. Org. Chem.*, **35**, 2357 (1970).
80. B. B. Jarvis, *J. Org. Chem.*, **33**, 4075 (1968).
81. B. B. Jarvis, J. P. Govoni, and P. J. Zell, *J. Am. Chem. Soc.*, **93**, 913 (1971).
82. H. Wieland, *Ber.*, **44**, 2553 (1911). Because the structure of triphenylmethyl dimer has now been reassigned (p. 446), one is tempted to question the structures of *all* those compounds associated with early triphenylmethyl radical chemistry. Perhaps the peroxide in (44) should be reinvestigated also.
83. C. Walling, reference 1, p. 473; reference 2, p. 416; R. Kh. Freidlina, reference 3, p. 261.
84. P. A. S. Smith, in *Molecular Rearrangements*, Vol. 1, P. deMayo, Ed., Interscience, New York, 1963, pp. 577–589.
85. J. C. Robertson and A. Swelim, *Tetrahedron Letters*, 2871 (1967).
86. For a recent review, see D. H. Hey, in *Advances in Free Radical Chemistry*, Vol. 2, G. H. Williams, Ed., Academic Press, New York, 1965, pp. 47–86.
87. Methylene chloride is about one-half as viscous as carbon tetrachloride (4.4 vs. 9.5 mP at $\sim 21°$, *Handbook of Chemistry*, 9th ed., N. A. Lange, Ed., Handbook Publishers, Sandusky, Ohio, 1956, pp. 1658, 1660). Carbon tetrachloride is a fairly effective cage solvent, allowing only some 30% of the radicals from AIBN to escape its cage at 60° (C. Walling, reference 1, p. 79). Methylene chloride should be less effective, therefore, because less viscous solvents are usually poorer cage solvents.
88. J. W. Wilt and O. Kolewe, *J. Am. Chem. Soc.*, **87**, 2071 (1965); J. W. Wilt, O. Kolewe, and J. F. Kraemer, *ibid.*, **91**, 2624 (1969). For a recent review of organo-silicon radicals, see Y. Nagai, *Intra-Sci. Chem. Rep.*, **4**, 115 (1970).
89. J. W. Wilt and C. F. Dockus, *J. Am. Chem. Soc.*, **92**, 5813 (1970).

90. P. J. Krusic and J. K. Kochi, *J. Am. Chem. Soc.*, **91**, 6161 (1969).
91. R. L. Dannley and G. Jalics, *J. Org. Chem.*, **30**, 3848 (1965). See also R. L. Dannley and W. A. Ave, *ibid.*, **30**, 3845 (1965); R. L. Dannley and G. Farrant, *J. Am. Chem. Soc.*, **88**, 627 (1966); A. K. Shubber and R. L. Dannley, *J. Org. Chem.*, **36**, 3784 (1971).
92. M. F. R. Mulcahy, B. G. Tucker, D. J. Williams, and J. R. Wilmshurst, *Chem. Commun*, 609 (1965).
93. J. W. Wilt and M. P. Stumpf, *J. Org. Chem.*, **30**, 1256 (1965).
94. C. D. Cook, R. C. Woodworth, and P. Fianu, *J. Am. Chem. Soc.*, **78**, 4159 (1956).
95. B. M. Vittimberga and M. L. Herz, *J. Org. Chem.*, **35**, 3694 (1970).
96. L. H. Slaugh, R. D. Mullineaux, and J. H. Raley, *J. Am. Chem. Soc.*, **85**, 3180 (1963).
97. H. Pines and C. T. Goetschel, *J. Am. Chem. Soc.*, **87**, 4207 (1965).
98. H. Ladenheim and W. Bartok, *J. Am. Chem. Soc.*, **89**, 1786 (1967).
99. L. H. Slaugh, *J. Am. Chem. Soc.*, **87**, 1522 (1965). For a recent example, see E. C. Friedrich and R. L. Holmstead, *J. Org. Chem.*, **36**, 971 (1971).
100. L. K. Montgomery and J. W. Matt, *J. Am. Chem. Soc.*, **89**, 934 (1967).
101. L. K. Montgomery and J. W. Matt, *J. Am. Chem. Soc.*, **89**, 3050 (1967).
102. L. K. Montgomery and J. W. Matt, *J. Am. Chem. Soc.*, **89**, 6556 (1967).
103. E. Müller, R. Renner, and A. Rieker, *Tetrahedron Letters*, 891 (1968).
103a. H. P. Löffler, *Chem. Ber.*, **104**, 1981 (1971).
104. D. D. Tanner and F. C. P. Law, *J. Am. Chem. Soc.*, **91**, 7535 (1969).
105. J.-M. Surzur and P. Teissier, *Bull. Soc. Chim. France*, 3060 (1970).
106. C. H. DePuy, D. E. Zabel, and W. Wiedeman, *J. Org. Chem.*, **33**, 2198 (1968).
107. A. B. Terent'ev, *Izv. Akad. Nauk SSSR, Ser. Khim.*, 1258 (1965). For other, earlier work that established the generality of thioalkyl and thioaryl vicinal migration, see R. Kh. Freidlina, reference 3, pp. 269–271.
108. K. Shiina and M. Kumada, *J. Org. Chem.*, **23**, 139 (1958).
109. H. Sakurai, T. Kishida, A. Hosomi, and M. Kumada, *J. Organomet. Chem.*, **8**, 65 (1967).
110. R. Kh. Freidlina, reference 3, pp. 231–248. Much of the work has been done in Russia and this review is the most thorough coverage available.
111. A. N. Nesmeyanov, R. Kh. Friedlina, and V. I. Firstov, *Izv. Akad. Nauk SSSR, Otd. Khim. Nauk*, 505 (1951). It is worth mentioning that this α-chlorine shift does not occur from the —$SiCl_3$ group, C. A. Burkard and R. H. Krieble, *J. Am. Chem. Soc.*, **69**, 2687 (1947).
112. M. S. Kharasch, E. H. Rossin, and E. K. Fields, *J. Am. Chem. Soc.*, **63**, 2558 (1941).
113. W. H. Urry and J. R. Eiszner, *J. Am. Chem. Soc.*, **73**, 2977 (1951); **74**, 5822 (1952).
114. W. H. Urry, J. R. Eiszner, and J. W. Wilt, *J. Am. Chem. Soc.*, **79**, 918 (1957).
115. A recent study by H. D. Roth, *J. Am. Chem. Soc.*, **93**, 1527 (1971), indicates that *singlet* methylene is involved in the reaction of diazomethane and carbon tetrachloride, both in insertion and abstraction reactions. Via CIDNP studies, the products of the reaction were demonstrated to arise (in part) by radical recombination. Although not reported in the earlier work,[113,114] insertion products of one, two, and three methylene units into carbon tetrachloride are in fact formed—though the chain sequence originally proposed[113,114] still seems to be the major pathway. In fairness, the original report[113] did not preclude the presence of lower insertion products. Rather, it stated that, if formed, such products were present in quantities too small for isolation. Indeed, this seems so even in the recent study by Roth.

Subsequently, Roth [*J. Am. Chem. Soc.*, **93**, 4935 (1971)] has shown that *direct* photolysis of diazirine (a methylene source) produced *singlet* methylene whereas its *photosensitized* decomposition led to *triplet* methylene. The two species apparently differ in their atom abstraction selectivity toward deuteriochloroform. Triplet methylene abstracted deuterium while singlet methylene abstracted chlorine.

116. W. H. Urry and N. Bilow, *J. Am. Chem. Soc.*, **86**, 1815 (1964). The authors allow that some carbene insertion and radical initiation via $\cdot CH_2Br$ also are probable. Moreover, in light of Roth's work,[115] the spin nature of the methylene proposed in this work may be incorrect. It seems clear now that triplet methylene is not exclusively involved.
117. Y. A. Ol'dekop and R. V. Kaberdin, *Zh. Org. Khim.*, **6**, 1114 (1970).
118. K. Kramer and N. Wright, *Chem. Ber.*, **96**, 1877 (1963); D. Seyferth, *Chem. Rev.*, **55**, 1155 (1955).
119. W. H. Urry and J. W. Wilt, *J. Am. Chem. Soc.*, **76**, 2594 (1954).
120. J. W. Wilt, Dissertation, University of Chicago, Chicago, Illinois, 1954.
121. T. Migita, W. Ando, S. Kondo, H. Matsuyama, and M. Kosugi, *Nippon Kagaku Zasshi*, **91**, 374 (1970). The author thanks Mr. W. K. Chwang of Loyola University and Dr. Y. H. Liau, Department of Agricultural Chemistry, National Taiwan University, Taipei, Taiwan, for the translation of this article.
122. M. Cocivera and H. D. Roth, *J. Am. Chem. Soc.*, **92**, 2573 (1970).
123. C. Walling, reference 1, p. 241.
124. R. Barker and A. Maccoll, *J. Chem. Soc.*, 2839 (1963); P. I. Abell and R. S. Anderson, *Tetrahedron Letters*, 3727 (1964).
125. P. S. Skell, R. G. Allen, and N. D. Gilmour, *J. Am. Chem. Soc.*, **83**, 504 (1961).
126. Cf. J. G. Traynham and W. G. Hines, *J. Am. Chem. Soc.*, **90**, 5208 (1968).
127. W. O. Haag and E.-A. I. Heiba, *Tetrahedron Letters*, 3683 (1965).
128. For a claim of a bridged bromine radical in spite of all this, see P. S. Juneja and E. M. Hodnett, *J. Am. Chem. Soc.*, **89**, 5685 (1967).
129. D. D. Tanner, D. Darwish, M. W. Mosher, and N. J. Bunce, *J. Am. Chem. Soc.*, **91**, 7398 (1969). Even more recent studies lead to the same denial of bridged bromo radicals: D. D. Tanner, H. Yabuuchi, and E. V. Blackburn, *ibid.*, **93**, 4802 (1971); D. D. Tanner, M. W. Mosher, N. C. Das, and E. V. Blackburn, *ibid.*, **93**, 5846 (1971).
130. Such bridged bromine species as cations ("bromonium ions") are well known. See I. Roberts and G. E. Kimball, *J. Am. Chem. Soc.*, **59**, 947 (1937); G. A. Olah and J. M. Bollinger, *ibid.*, **89**, 4744 (1967); **90**, 947 (1968). One must wonder what the chances are that the *original* bromonium ion paper (actually a note) and the paper describing its observation via NMR spectroscopy some 31 years later would appear on the same page Another reason for α-bromo radical stability, one involving hyperconjugation, has been advanced by S. J. Cristol, M. A. Imhoff, and D. C. Lewis, *J. Org. Chem.*, **35**, 1722 (1970).
131. C. G. Pitt and M. S. Fowler, *J. Am. Chem. Soc.*, **90**, 1928 (1968).
132. S. Winstein, R. Heck, S. Lapporte, and R. Baird, *Experientia*, **12**, 138 (1956).
133. D. F. DeTar and A. Hlynsky, *J. Am. Chem. Soc.*, **77**, 4411 (1955).
134. J. K. Kochi and R. D. Gilliom, *J. Am. Chem. Soc.*, **86**, 5251 (1964).
134a. M. Julia and B. Mallassine, *Tetrahedron Letters*, 987 (1971).
135. W. H. Urry, D. J. Trecker, and H. D. Hartzler, *J. Org. Chem.*, **29**, 1663 (1964).
136. D. B. Denney and P. P. Klemchuk, *J. Am. Chem. Soc.*, **80**, 3289 (1958).
137. J. W. Wilt, R. B. Dabek, and K. Welzel, *Abstracts of Papers, Third Great Lakes Regional Meeting, American Chemical Society, Northern Illinois University, De Kalb, Illinois, June 1969*, p. 58; *J. Org. Chem.*, **37**, 425 (1972).

138. R. B. Dabek, Dissertation, Loyola University, Chicago, Illinois, 1970.
139. J. W. Wilt and D. D. Oathoudt, *J. Org. Chem.*, **21**, 1550 (1956); **23**, 218 (1958).
140. J. W. Wilt and J. L. Finnerty, *J. Org. Chem.*, **26**, 2173 (1961).
141. W. A. Bonner and F. D. Mango, *J. Org. Chem.*, **29**, 430 (1964).
142. U. K. Pandit and I. P. Dirk, *Tetrahedron Letters*, 891 (1963).
143. A troublesome feature of Kolbe, Hunsdiecker, and peroxide reactions—all seemingly involving acyloxy radicals—is that the three processes do not give the same products. A case in point is this β,β-diphenylpropionic acid. Its peroxide gave no rearrangement,[47a] the acid in Kolbe electrolysis produced Ar_1-5 rearranged product,[141] while the acid in the Hunsdiecker reaction afforded Ar_2-6 rearranged product.[142] Clearly the three reactions do not in fact utilize the same pathway. Intrusion of cationic processes (the Hofer-Moest in the Kolbe process and intramolecular displacement on an acyl hypobromite in the Hunsdiecker reaction) may distinguish these from the more accepted radical process from the peroxide.
144. Cf. also P. J. Bunyan and D. H. Hey, *J. Chem. Soc.*, **324**, 2771 (1962).
145. J. W. Wilt and J. A. Lunquist, *J. Org. Chem.*, **29**, 921 (1964).
146. C. F. Dockus, *M.S. thesis*, Loyola University, Chicago, Illinois, 1970; J. W. Wilt and C. F. Dockus, *Abstracts of Papers, 161st National Meeting of the American Chemical Society, Los Angeles, California, March 1971*, ORGN paper 163.

J. W. Wilt and C. F. Dockus, *J. Am. Chem. Soc.*, **92**, 5813 (1970), report the rearrangement data for the chlorides. Unfortunately these reported data have the rearrangement percentages for $m = 3$ and $m = 4$ reversed.
147. Such participation has been claimed for hetero and halogen functions, however.
148. Unpublished results of R. Niinemae and the author.
149. No certain data exist for the β-silyl radical case. However, treatment of β-(triphenylsilyl)ethyl bromide with organometallic reagents gave unrearranged products together with some Si—C cleavage products; A. J. Bourne, A. W. P. Jarvie, and A. Holt, *J. Chem. Soc. (C)*, 1740 (1970).
150. L. E. Nelson, N. G. Angelotti, and D. R. Weyenberg, *J. Am. Chem. Soc.*, **85**, 2662 (1963).
151. H. Sakurai and A. Hosomi, *J. Am. Chem. Soc.*, **92**, 7507 (1970). For a review of the reactions of silyl radicals in solution, cf. H. Sakurai, *Nippon Kagaku Zasshi*, **9**, 885 (1970).
152. H. Sakurai, A. Hosomi, and M. Kumada, *Tetrahedron Letters*, 1757 (1969).
153. G. A. Razuvaev, N. S. Vasileiskaja, and D. V. Muslin, *J. Organomet. Chem.*, **7**, 533 (1967).
154. J. D. Roberts and M. C. Caserio, *Basic Principles of Organic Chemistry*, Benjamin, New York, 1964, p. 1179.
155. E. A. Hardwidge, C. W. Larson, and B. S. Rabinovitch, *J. Am. Chem. Soc.*, **92**, 3278 (1970).
156. R. Kh. Freidlina, reference 3, pp. 215–219; C. Walling, reference 2, pp. 416–423; W. A. Pryor, *Free Radicals*, McGraw-Hill, New York, 1966, p. 273.
157. M. Lazar and O. Kysel, *Chem. Listy*, **60**, 192 (1966).
158. Such studies in solution have been done mainly by O. A. Reutov and co-workers. Cf. R. Kh. Freidlina, reference 3, pp. 219–223; C. Walling, reference 2, p. 423.
159. O. A. Reutov and T. N. Shatkina, *Tetrahedron*, **18**, 237 (1962).
160. C. C. Lee, J. E. Kruger, and E. W. C. Wong, *J. Am. Chem. Soc.*, **87**, 3985 (1965).
161. O. A. Reutov, G. M. Ostapchuk, K. Uteniyazov, and E. V. Binshtok, *Zh. Org. Khim.*, **4**, 1869 (1968).
162. Yu. G. Bundel, I. Yu. Levina, A. M. Krzhizhevskii, and O. A. Reutov, *Izv. Akad. Nauk SSSR, Ser. Khim.*, 2583 (1968).

163. R. A. Jackson and W. A. Waters, *J. Chem. Soc.*, 1653 (1960).
164. K. Heusler and J. Kalvoda, *Tetrahedron Letters*, 1001 (1963).
165. D. F. DeTar and C. Weiss, *J. Am. Chem. Soc.*, **78**, 4296 (1956).
166. W. A. Pryor, *Free Radicals*, McGraw-Hill, New York, 1966, pp. 280–282; C. Walling, reference 2, pp. 421–422; R. Kh. Freidlina, reference 3, pp. 221–223.
167. C. A. Grob and H. Kammüller, *Helv. Chim. Acta*, **40**, 2139 (1957).
168. R. Kh. Freidlina, reference 3, pp. 223–227.
169. R. Kh. Freidlina, reference 3, pp. 224–225. For recent work, cf. A. B. Terent'ev and R. Kh. Freidlina, *Izv. Akad. Nauk SSSR, Ser. Khim.*, 1854 (1966). For a review on such hydrogen shifts in telomerizations, see R. Kh. Freidlina, A. B. Terent'ev, M. Ya. Khorlina, and S. N. Aminov, *Zh. Vses. Khim. O-va. Im. D. I. Mendeleeva*, **11**, 211 (1966).
170. J. G. Traynham and T. M. Couvillon, *J. Am. Chem. Soc.*, **87**, 5807 (1965); **89**, 3205 (1967). This work corrects an earlier report of vicinal addition with these reactants; F. Reicheneder and H. Suter, Ger. Pat. 1,036,847 (Aug. 21, 1958); *Chem. Abstr.*, **54**, 22416*c* (1960).
171. J. G. Traynham, T. M. Couvillon, and N. S. Bhacca, *J. Org. Chem.*, **32**, 529 (1967), indicate that 3–10% transannular 1,5 hydrogen shift occurs in this reaction. The vicinal adduct is about one-to-one cis and trans.
172. M. Fisch and G. Ourisson, *Chem. Commun.*, 407 (1965). For a similar reaction between $BrCCl_3$ and longifolene, see G. Ourisson, *Proc. Chem. Soc.* (*London*), 281 (1964).
173. T. J. Wallace and R. J. Gritter, *J. Org. Chem.*, **26**, 5256 (1961).
174. K. Heusler and J. Kalvoda, *Angew. Chem.*, **76**, 518 (1964); *Angew. Chem., Int. Engl. Ed.*, **3**, 525 (1964). This excellent review is concerned principally with Barton reactions, hypoiodite decompositions, and alcohol oxidations with lead tetraacetate.
175. R. Kh. Freidlina, reference 3, pp. 227–230; C. Walling, reference 2, pp. 448–451; W. A. Pryor, *Free Radicals*, McGraw-Hill, New York, 1966, pp. 283–284.
176. D. H. R. Barton, J. M. Beaton, L. E. Geller, and M. M. Pechet, *J. Am. Chem. Soc.*, **82**, 2640 (1960); **83**, 4076 (1961). For a review of the Barton reaction, see R. H. Hesse, *Advances in Free Radical Chemistry*, Vol. 3, G. H. Williams, Ed., Academic Press, New York, 1969, pp. 83–137.
177. Cf. the synthesis of aldosterone acetate by D. H. R. Barton and J. M. Beaton, *J. Am. Chem. Soc.*, **82**, 4083 (1960).
178. P. Kabasakalian and E. T. Townley, *J. Am. Chem. Soc.*, **84**, 2711 (1962).
179. M. Akhtar and M. M. Pechet, *J. Am. Chem. Soc.*, **86**, 265 (1964).
180. P. Gray and A. Williams, *Chem. Rev.*, **59**, 239 (1959).
181. P. Kabasakalian, E. R. Townley, and M. D. Yudis, *J. Am. Chem. Soc.*, **84**, 2716 (1962).
182. P. Kabasakalian, E. R. Townley, and M. D. Yudis, *J. Am. Chem. Soc.*, **84**, 2718 (1962); P. Kabasakalian and E. R. Townley, *ibid.*, **84**, 2723, 2724 (1962).
183. P. Kabasakalian and E. R. Townley, *J. Org. Chem.*, **27**, 2918 (1962). The stereochemistry of the nitroso alcohol dimer was not determined.
184. Further examples and discussion may be found in R. O. Kan, *Organic Photochemistry*, McGraw-Hill, New York, 1966, p. 233 ff.
185. T. Sandmeyer, *Ber.*, **18**, 1767 (1885); **19**, 857 (1886); F. D. Chattaway and O. G. Backeberg, *J. Chem. Soc.*, **123**, 2999 (1923). These early studies involved thermal decompositions. Most recent studies use photoinduced decomposition.
186. (a) F. D. Greene, M. L. Savitz, H. H. Lau, F. D. Osterholtz, and W. N. Smith, *J. Am. Chem. Soc.*, **83**, 2196 (1961); F. D. Greene, M. L. Savitz, F. D. Osterholtz,

H. H. Lau, W. N. Smith, and P. M. Zanet, *J. Org. Chem.*, **28**, 55 (1963); (b) C. Walling and A. Padwa, *J. Am. Chem. Soc.*, **83**, 2207 (1961); (c) C. Walling and A. Padwa, *ibid.*, **85**, 1593 (1963); (d) M. Akhtar and D. H. R. Barton, *ibid.*, **83**, 2213 (1961); (e) J. S. Mills and V. Petrow, *Chem. Ind. (London)*, 946 (1961).

187. R. Kh. Freidlina, reference 3, pp. 228–229; C. Walling, reference 2, pp. 449–450; W. A. Pryor, *Free Radicals*, McGraw-Hill, New York, 1966, pp. 283–284.
188. E. L. Jenner, *J. Org. Chem.*, **27**, 1031 (1962). Cf. also reference 186b. The yields in such reactions can be raised to 60–90% if trichloroethylene is used as the solvent. See C. Walling and D. Bristol, *Abstracts of Papers, 161st National Meeting of the American Chemical Society, Los Angeles, California, March 1971*, ORGN paper 66.
189. A. C. Cope, R. S. Bly, M. M. Martin, and R. C. Petterson, *J. Am. Chem. Soc.*, **87**, 3111 (1965). Thermal decomposition gave similar results.
190. See, however, F. D. Greene, C. C. Chu, and J. Walia, *J. Org. Chem.*, **29**, 1285 (1964), for a discussion of product stereochemistry vis-à-vis a chain transfer mechanism.
191. R. A. Sneen and N. P. Matheny, *J. Am. Chem. Soc.*, **86**, 3905, 5503 (1964). The claim was based upon the failure to observe conversion of PhSH to PhSSPh during such a reaction. Such oxidative dimerization is not, however, an adequate test. Cf. C. Walling and A. Padwa, *ibid.*, **85**, 1597 (1963).
192. M. Akhtar, P. Hunt, and P. B. Dewhurst, *J. Am. Chem. Soc.*, **87**, 1807 (1965). This paper contains an excellent listing of earlier studies in hypohalite decompositions.
193. G. Smolinsky and B. I. Feuer, *J. Org. Chem.*, **30**, 3216 (1965). The enantiomeric alcohol is consistently misnamed in the paper.
194. A. C. Cope, M. A. McKervey, N. M. Weinshenker, and R. B. Kinnel, *J. Org. Chem.*, **35**, 2918 (1970).
195. M. Fisch, S. Smallcombe, J. C. Gramain, M. A. McKervey, and J. E. Anderson, *J. Org. Chem.*, **35**, 1886 (1970).
196. C. H. Banford and R. G. W. Norrish, *J. Chem. Soc.*, 1504 (1935). Cf. P. J. Wagner, *J. Am. Chem. Soc.*, **89**, 5898 (1967); *Tetrahedron Letters*, 1753 (1967).
197. N. C. Yang and D. H. Yang, *J. Am. Chem. Soc.*, **80**, 2913 (1958).
198. P. Buchschacher, M. Cereghetti, K. Schaffner, and O. Jeger, *Helv. Chim. Acta*, **43**, 367 (1960).
199. N. C. Yang, D. H. Yang, and A. Morduchowitz, *J. Am. Chem. Soc.*, **83**, 1017 (1963); P. Yates and A. G. Szabo, *Tetrahedron Letters*, 485 (1965). The photochemistry of alkanones with γ-hydrogen is complicated by the fact that *both* singlet and triplet n,π^* states are involved, and in different mechanisms. Aryl alkyl ketones, on the other hand, react via the triplet state only. The interested reader is referred to N. C. Yang and S. P. Elliott, *J. Am. Chem. Soc.*, **91**, 7550 (1969); N. C. Yang, S. P. Elliott, and B. Kim, *ibid.*, **91**, 7551 (1969).
200. (a) W. H. Urry and D. J. Trecker, *J. Am. Chem. Soc.*, **84**, 118 (1962); (b) W. H. Urry, D. J. Trecker, and D. A. Winey, *Tetrahedron Letters*, 609 (1962). Note that the hydrogen abstraction preferentially occurred from C-4, not C-8. This abstraction across the twinned carbonyl groups is curious.
201. I. Orban, K. Schaffner, and O. Jeger, *J. Am. Chem. Soc.*, **85**, 3033 (1963). Retention of activity has also been observed in the formation of cyclohexenols as in (140) by K. H. Schulte-Elte and G. Ohloff, *Tetrahedron Letters*, 1143 (1964). It is possible that stereospecificity results from a pathway using the singlet n,π^* state, whereas racemization results from the triplet n,π^* competitor; see reference 199.
202. R. B. Gagosian, J. C. Dalton, and N. J. Turro, *J. Am. Chem. Soc.*, **92**, 4752 (1970).
203. Such high yields are common in the Yang rearrangement of 20- or 11-keto steroids,

β,γ-unsaturated ketones, and α-diketones with γ-hydrogens. For a list of further references, see reference 202.
204. N. C. Yang and C. Thap, *Tetrahedron Letters*, 3671 (1966).
205. Note, however, that hydroxyketenes have been implicated in photoeliminations of alkyl phenylglyoxalates by E. S. Huyser and D. C. Neckers, *J. Org. Chem.*, **29**, 276 (1964).
206. H. E. Zimmerman, B. R. Cowley, C.-Y. Tseng, and J. W. Wilson, *J. Am. Chem. Soc.*, **86**, 947 (1964).
207. M. Banard and N. C. Yang, *Proc. Chem. Soc.* (*London*), 302 (1958). For abstraction of remote hydrogens in long-chain (C_{14}–C_{20}) compounds, see R. Breslow and M. A. Winnick, *J. Am. Chem. Soc.*, **91**, 3083 (1969); in steroids, R. Breslow and S. W. Baldwin, *ibid.*, **92**, 732 (1970); in an oriented substrate, R. Breslow and P. C. Scholl, *ibid.*, **93**, 2331 (1971).
208. S. P. Pappas, B. C. Pappas, and J. E. Blackwell, Jr., *J. Org. Chem.*, **32**, 3066 (1967); S. P. Pappas, J. E. Alexander, and R. D. Zehr, Jr., *J. Am. Chem. Soc.*, **92**, 6927 (1970).
209. R. Kh. Freidlina, reference 3, pp. 229–230.
210. R. Srinivasan, *J. Am. Chem. Soc.*, **81**, 1546, 2601, 2604 (1959).
211. G. Buchi and I. M. Goldman, *J. Am. Chem. Soc.*, **79**, 4741 (1957).
212. A. W. Hofmann, *Chem. Ber.*, **16**, 558, 583 (1883). The structure of δ-coneceine was determined by A. Lellmann, *Ann.*, **259**, 193 (1890).
213. K. Löffler and C. Freytag, *Chem. Ber.*, **42**, 3427 (1909); K. Löffler and S. Kober, *ibid.*, **42**, 3421 (1909).
214. M. E. Wolff, *Chem. Rev.*, **63**, 55 (1963). This is a general review of the reaction through June 1961.
215. G. H. Coleman, G. Nichols, and T. F. Martens, *Org. Syn.*, *Coll. Vol. III*, 159 (1955).
216. S. Wawzonek and P. J. Thelan, *J. Am. Chem. Soc.*, **72**, 2118 (1950); S. Wawzonek and T. P. Culbertson, *ibid.*, **81**, 3367 (1959).
217. E. J. Corey and W. R. Hertler, *J. Am. Chem. Soc.*, **82**, 1657 (1960).
218. G. S. Hammond and R. C. Neumann, Jr., *J. Am. Chem. Soc.*, **85**, 1501 (1963).
219. (a) R. S. Neale and M. R. Walsh, *J. Am. Chem. Soc.*, **87**, 1255 (1965); (b) R. S. Neale, M. R. Walsh, and N. L. Marens, *J. Org. Chem.*, **30**, 3683 (1965).
220. S. Wawzonek and J. D. Nordstrom, *J. Org. Chem.*, **27**, 3726 (1962).
221. S. Wawzonek and T. P. Culbertson, *J. Am. Chem. Soc.*, **82**, 441 (1960).
222. S. Wawzonek and T. C. Wilkinson, *J. Org. Chem.*, **31**, 1732 (1966).
223. The reaction with R = H was studied earlier by S. Wawzonek, M. F. Nelson, Jr., and P. J. Thelan, *J. Am. Chem. Soc.*, **73**, 2806 (1951); and R. Lukes and M. Ferles, *Collect. Czech. Chem. Commun.*, **20**, 1227 (1955). The latter claimed only 1,5 transfer product. Wawzonek explained[222] that the very volatile quinuclidine was lost in Lukes' study.
224. P. G. Gassman and D. C. Heckert, *Tetrahedron*, **21**, 2725 (1965). One wonders whether the transition state achieved *is* actually the more strained. Bicyclo[3.2.1]-octyl transition states seem, in fact, easily achieved. Cf. M. S. Newman, N. Gill, and B. Darré, *J. Org. Chem.*, **31**, 2713 (1966).
225. K. Löffler, *Chem. Ber.*, **43**, 2025 (1910) had earlier concluded that γ-hydrogens (1,4 transfer) were inactive toward abstraction in this reaction.
226. E. Schmitz and D. Murawski, *Chem. Ber.*, **99**, 1493 (1966).
227. R. C. Petterson and A. Wambsgans, *J. Am. Chem. Soc.*, **86**, 1648 (1964).
228. A. L. J. Beckwith and J. E. Goodrich, *Aust. J. Chem.*, **18**, 747 (1965).

229. D. H. R. Barton and A. L. J. Beckwith, *Proc. Chem. Soc.* (*London*), 335 (1963); D. H. R. Barton, A. L. J. Beckwith, and A. Goosen, *J. Chem. Soc.*, 181 (1965).
230. But which hereroatom actually abstracts the hydrogen is unknown.
231. R. S. Neale, N. L. Marcus, and R. G. Schepers, *J. Am. Chem. Soc.*, **88**, 3051 (1966). For a review of such reactions, cf. R. S. Neale, *Synthesis*, 1 (1971).
231a. O. E. Edwards, J. M. Paton, M. H. Benn, R. E. Mitchell, C. Watanatada, and K. N. Vohra, *Can. J. Chem.*, **49**, 1648 (1971). Earlier the process had been incorrectly rationalized via a nitrenium ion mechanism by O. E. Edwards, D. Vocelle, J. W. ApSimon, and F. Hague, *J. Am. Chem. Soc.*, **87**, 678 (1965).
232. G. H. Coleman, *Proc. Iowa Acad. Sci.*, **46**, 217 (1939). See also G. H. Coleman, C. C. Schulze, and H. A. Hoppens, *ibid.*, **47**, 264 (1940).
233. S. M. Verma and R. C. Srivastava, *Bull. Chem. Soc. Japan*, **40**, 2184 (1967).
234. M. Okahara, T. Ohashi, and S. Komori, *Tetrahedron Letters*, 1629 (1967); *J. Org. Chem.*, **33**, 3066 (1968).
235. R. S. Neale and N. L. Marcus, *J. Org. Chem.*, **34**, 1808 (1969).
236. Neale and Marcus[235] isolated only the γ-chlorosulfonamide from n-C_4H_9—SO_2—N(Cl)—t-C_4H_9 (75%), although spectral analysis indicated other t-butyl compound(s) were present in the reaction mixture. A similar reaction on the N-bromo analog apparently gave only γ-bromosulfonamide as the rearranged product.
237. The Japanese workers[234] state that "the formation of γ-chlorosulfonamide was greatly raised" in the acid medium. Actually, there was essentially no change. Rather, production of parent sulfonamide increased and δ-chlorosulfonanide formation decreased. They ascribe the decreased δ-chloro product to decreased intermolecular hydrogen abstraction in the acidic medium because of charge repulsion between the nitrogen radical cation intermediates. Aside from the known fact that such radical cations can dimerize,[218] photorearrangement of *neat* N-chlorosulfonamide in benzene, where intermolecular attack should be easier, actually gave less δ-chloro product than did most other concentrations studied.
238. A. Padwa and E. Alexander, *J. Am. Chem. Soc.*, **92**, 5674 (1970).
239. A 1,3 hydrogen shift in a biradical was also proposed by M. J. Jorgenson and T. J. Clark, *J. Am. Chem. Soc.*, **90**, 2188 (1968).
240. Such processes are perhaps not true rearrangements, as stated by R. Kh. Freidlina, reference 3, p. 213. They are discussed here, nonetheless, because such an opinion is not universal and because they are so interesting. For summaries of early studies, cf. C. Walling, reference 2, pp. 440–445; W. A. Pryor, *Free Radicals*, McGraw-Hill, New York, 1966, pp. 277–279.
241. J. D. Roberts and D. Schuster, *J. Org. Chem.*, **27**, 51 (1962).
242. E. M. Kosower, *An Introduction to Physical Organic Chemistry*, Wiley, New York, 1968, p. 233. Recently, M. J. S. Dewar and S. Kirschner, *J. Am. Chem. Soc.*, **93**, 4290 (1971), claim from MINDO/2 molecular orbital calculations that the radical would open in a disrotatory fashion. As these authors suggest, an experimental test between the conflicting predictions is obviously desirable.
243. C. Walling, reference 2, p. 442.
244. J. A. Kerr, A. Smith, and A. F. Trotman-Dickenson, *J. Chem. Soc.* (*A*), 1400 (1969).
245. Use of different aryl groups in diastereomeric bromides might allow a determination of the symmetry mode of cleavage because geometrically isomeric olefins could then form.
246. See also H. M. Walborsky, J.-C. Chen, and J. L. Webb, *Tetrahedron Letters*, 3551 (1964); J.-C. Chen, *ibid.*, 3669 (1971).

247. H. C. Brown and M. Borkowski, *J. Am. Chem. Soc.*, **74**, 1894 (1952).
248. C. Walling and P. S. Fredricks, *J. Am. Chem. Soc.*, **84**, 3326 (1962).
249. D. I. Schuster, Dissertation, California Institute of Technology, Pasadena, California, 1960.
250. E. Renk, P. R. Shafer, W. H. Graham, R. H. Mazur, and J. D. Roberts, *J. Am. Chem. Soc.*, **83**, 1987 (1961), and references cited therein.
251. (a) E. S. Huyser and J. D. Taliaferro, *J. Org. Chem.*, **28**, 3442 (1963); (b) for a recent, related study of the radical addition of methanethiol to bicyclo[3.1.0]-hexene-2, see P. K. Freeman, M. F. Grostic, and F. A. Raymond, *ibid.*, **36**, 905 (1971).
252. As the authors[251a] point out, the chain carriers in their study were all electron acceptors. The cyclopropylcarbinyl group in the transition state could, as a consequence, be somewhat electron deficient. Stabilization could then result from this cation-like character in the system. Unfortunately, a test of this notion using electron-donating acyl radical chain carriers was experimentally impracticable.
253. E. S. Huyser and L. R. Munson, *J. Org. Chem.*, **30**, 1436 (1965).
254. D. Y. Curtin, *Rec. Chem. Prog. Kresge-Hooker Sci. Libr.*, **15**, 111 (1954).
255. Cleavage of **180-1** may have a rate constant of about 10^8 sec^{-1} (the value obtained for another case, see p. 403). For such cleavage to be faster than rotation to **180-2**, the energy barrier to rotation would have to be over 8 kcal/mole.
256. I. S. Lishanskii, N. D. Vinogradova, A. M. Guloev, A. G. Zak, A. B. Zvyagina, O. S. Fomina, and A. S. Khachaturov, *Dokl. Akad. Nauk SSSR*, **179** (4), 882 (1968).
257. J. P. Kennedy, J. J. Elliott, and P. E. Butler, *J. Macromol. Sci., Chem.*, **2**, 1415 (1968).
258. J. K. Kochi, P. J. Krusic, and D. R. Eaton, *J. Am. Chem. Soc.*, **91**, 1877 (1969).
259. J. K. Kochi, P. J. Krusic, and D. R. Eaton, *J. Am. Chem. Soc.*, **91**, 1879 (1969).
260. R. A. Sheldon and J. K. Kochi, *J. Am. Chem. Soc.*, **92**, 4395, 5175 (1970).
261. J. C. Martin, J. E. Schultz, and J. W. Timberlake, *Tetrahedron Letters*, 4629 (1967).
262. C. G. Overberger and A. Lebovits, *J. Am. Chem. Soc.*, **76**, 2722 (1954); H. Hart and D. P. Wyman, *ibid.*, **81**, 4891 (1959). In the second reference there remains some question as to whether or not a radical mechanism is indeed involved.
263. The authors[261] place phenyl ahead of cyclopropyl in such stabilization because "hexaphenylethane" has $D(C—C) = 15$ kcal/mole. However, hexaphenylethane is actually unknown (see p. 446 ff.) so the use of this value is suspect.
264. (a) A. Maercker and J. D. Roberts, *J. Am. Chem. Soc.*, **88**, 1742 (1966); (b) the reaction was reported earlier[265] to give unrearranged product. Later the rearrangement was confirmed, J. G. Bennett, Jr., and S. C. Bunce, *J. Org. Chem.*, **35**, 4004 (1970).
265. J. G. Bennett, Jr., and S. C. Bunce, *J. Org. Chem.*, **25**, 73 (1960). The closure was reaffirmed at the same time that some of the original work was corrected.[264b] Reversible ring opening presumably can explain the curious result recently reported by E. C. Friedrich and R. L. Holmstead, *ibid.*, **36**, 971 (1971).
266. M. Tsutsui, *Ann. N.Y. Acad. Sci.*, **93**, 133 (1961); *Chem. Abstr.*, **57**, 16653 (1962).
267. D. Obitz, Thesis, University of Paris, Paris, France, 1967.
267a. E. C. Friedrich and R. L. Holmstead, *J. Org. Chem.*, **36**, 971 (1971).
267b. W. G. Dauben, L. Schutte, R. E. Wolf, and E. J. Deviny, *J. Org. Chem.*, **34**, 2512 (1969).
268. T. A. Halgren, M. E. H. Howden, M. E. Medof, and J. D. Roberts, *J. Am. Chem. Soc.*, **89**, 3051 (1967).
269. (a) D. I. Davies and S. J. Cristol, in *Advances in Free-Radical Chemistry*, Vol. 1,

G. H. Williams, Ed., Logos Press, London, 1965, pp. 156–157; (b) C. Walling, reference 2, pp. 440–441.

270. R. Lalande, M. Cazaux, and J. Moulines, *Ind. Chim. Belge*, 748 (1967).
271. A. Gaiffe and J. Castenet, *Compt. Rend. Acad. Sci., Ser. C*, **270**, 63 (1970).
272. J. Iyoda and I. Shiihara, *J. Org. Chem.*, **35**, 4267 (1970).
273. J. A. Claisse, D. I. Davies, and L. T. Parfitt, *J. Chem. Soc.* (*C*), 258 (1970). This article gives a particularly fine resumé of past work on radical additions to β-pinene.
274. The value 17–19 kcal/mole represents the difference in strain energies of cyclooctane (11.7 or 9.98 kcal/mole) and *cis*-bicyclo[4.2.0]octane (29.0 kcal/mole); P. v. R. Schleyer, J. E. Williams, and K. R. Blanchard, *J. Am. Chem. Soc.*, **92**, 2377 (1970).
275. (a) F. G. Bordwell and W. A. Hewett, *J. Am. Chem. Soc.*, **79**, 3493 (1957); (b) see also, J. R. Vincent and G. Etzel, U.S. Pat. 2,429,858 (1947); *Chem. Abstr.*, **42**, 1449 (1948); (c) C. M. Buess, C. N. Yiannios, and W. T. Fitzgerald, *J. Org. Chem.*, **22**, 197 (1957).
276. A. Gaiffe and J. Castenet, *Compt. Rend. Acad. Sci. Paris, Ser. C.*, **271**, 1012 (1970).
277. L. A. Goldblatt and D. M. Oldroyd, U.S. Pat. 2,533,240 (1950); *Chem. Abstr.*, **45**, 2262 (1951).
278. M. D. Sutherland and J. W. Wells, *J. Org. Chem.*, **21**, 1272 (1956); reference 275c; H. Behringer, *Ann.*, **564**, 219 (1949); reference 275a.
279. The conformational structure of Z given in reference 273 is difficult to interpret. It seems, nonetheless, that ring opening should place the trichloromethyl and chloro-*i*-propyl groups trans in the adduct.
280. Cf. E. S. Huyser, *J. Org. Chem.*, **26**, 3261 (1961).
281. *Ionic* addition of hydrogen bromide should afford bornyl bromide. Moreover, as certain ionic processes open α-pinene to limonene derivatives,[282] the latter are also possible. Nuclear magnetic resonance data implicate these among the products as well.
282. Cf. J. A. Berson, in *Molecular Rearrangements*, Vol. 1, P. deMayo, Ed., Wiley, New York, 1961, p. 185.
283. E.g., the strain energy of bicyclo[3.1.0]hexane is given as 33.91 kcal/mole. See the literature citation in reference 274.
284. K. D. King, D. M. Golden, and S. W. Benson, *J. Am. Chem. Soc.*, **92**, 5541 (1970). The C—H bond in acetone is as strong as in methane.
285. C. Walling, reference 2, pp. 444–445; E. C. Sabatino, *Diss. Abstr.*, **24**, 4408 (1964).
286. R. J. Gritter and T. J. Wallace, *J. Org. Chem.*, **26**, 282 (1961). Some 2% 5-hydroxy-2-hexanone was also observed. Indeed, at 125° it is the principal product. A mechanism involving radical displacement on carbon was used to explain its formation. While convenient and direct, this mechanism should be better established because such displacements are rare.
287. A. Padwa and N. C. Das, *J. Org. Chem.*, **34**, 816 (1969).
288. Nonetheless, more than 1% cis oxide was detectable, so some control was evident. But the implications of this control on the configuration of the oxiranyl radical are less clear.
289. C. S. Markos and W. Reusch, *J. Am. Chem. Soc.*, **89**, 3363 (1967).
290. W. Reusch and C. K. Johnson, *J. Am. Chem. Soc.*, **84**, 1759 (1962).
291. W. Reusch, C. K. Johnson, and J. A. Manner, *J. Am. Chem. Soc.*, **88**, 2803 (1966).
292. T. J. Wallace and R. J. Gritter, *J. Org. Chem.*, **26**, 5256 (1961).
293. R. L. Huang and H. H. Lee, *J. Chem. Soc.* (*C*), 929 (1966).
294. E. S. Huyser, *J. Org. Chem.*, **25**, 1860 (1960). L. P. Kuhn and C. Wellman, *ibid.*, **22**, 774 (1956), had shown earlier that acyclic acetals underwent radical cleavage to esters.

295. E. S. Huyser and Z. Garcia, *J. Org. Chem.*, **27**, 2716 (1962).
296. Though not germane to rearrangement studies, such cleavage in acetals of structure $CH_3CH(OR_1)(OR_2)$ has allowed a fascinating study of $R_1\cdot$ versus $R_2\cdot$ stability. Cf. E. S. Huyser and D. T. Wang, *J. Org. Chem.*, **29**, 2720 (1964).
297. D. Elad and R. D. Youssefyeh, *Tetrahedron Letters*, 2189 (1963).
298. H. E. Seyfarth, A. Hesse, and H. Pastohr, *Z. Chem.*, **9**, 150 (1969).
299. I. Rosenthal and D. Elad, *J. Org. Chem.*, **33**, 805 (1968).
300. R. Lalande, B. Maillard, and M. Cazaux, *Tetrahedron Letters*, 745 (1969).
301. (a) D. C. Neckers, *Tetrahedron Letters*, 1889 (1965); (b) D. C. Neckers, A. P. Schaap, and J. Hardy, *J. Am. Chem. Soc.*, **88**, 1265 (1966).
302. D. C. Neckers, J. Hardy, and A. P. Schaap, *J. Org. Chem.*, **31**, 622 (1966).
303. The α-hydroxy radicals perhaps induce the decomposition of DTBP to yield the ketone via

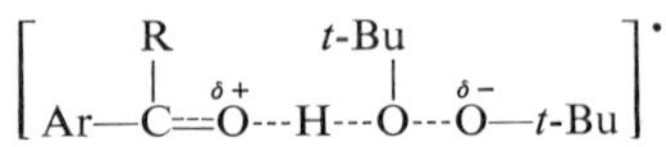

If so, then electron-rich Ar groups would facilitate the oxidation as well.[301]
304. E. S. Huyser and D. C. Neckers, *J. Am. Chem. Soc.*, **85**, 3641 (1963).
305. The nature of the hydrogen atom donation to the ketone from the α-hydroxy radical is particularly controversial. For discussion, see reference 301b.
306. C. H. DePuy, G. M. Dappen, and J. W. Hausser, *J. Am. Chem. Soc.*, **83**, 3156 (1961).
307. C. H. DePuy and F. W. Breitbeil, *J. Am. Chem. Soc.*, **85**, 2176 (1963).
308. C. H. DePuy, *Acc. Chem. Res.*, **1**, 33 (1968), specifically in footnote 42.
309. T. L. Cairns and B. E. Englund, *J. Org. Chem.*, **21**, 140 (1956).
310. B. E. Englund, U.S. Pat. 2,675,402 (1954); 2,691,682 (1954). Many additional examples with various R groups and ring sizes are given in these patents.
311. J. W. Wilt and J. W. Hill, *J. Org. Chem.*, **26**, 3523 (1961).
312. D. B. Denney and J. W. Hanifin, Jr., *J. Org. Chem.*, **29**, 732 (1964).
313. A. C. Cope, R. S. Bly, M. M. Martin, and R. C. Petterson, *J. Am. Chem. Soc.*, **87**, 3111 (1965). The low yield of chloro ketone here is caused by greatly increased hydrogen abstraction (transannular) in competition with cleavage.
314. C. Walling and A. Padwa, *J. Am. Chem. Soc.*, **84**, 2845 (1962).
315. G. S. Schmidt and G. S. Fisher, *J. Am. Chem. Soc.*, **76**, 5426 (1954).
316. Cyclization in the norbornenyl–nortricyclyl radical system, as well as in certain others, deserves a separate section (Section IV.F.3).
317. C. Walling, reference 2, pp. 445–446; M. Julia, *Rec. Chem. Prog., Kresge-Hooker Sci. Libr.*, **25**, 3 (1964); plenary lecture presented at the International Symposium on Free Radicals in Solution, University of Michigan, Ann Arbor, Michigan, Aug. 1966; *Pure Appl. Chem.*, **15**, 167 (1967); *Acc. Chem. Res.*, **4**, 386 (1971). These are good reviews of his work by one of the major contributors to this area.
318. C. Walling, *J. Am. Chem. Soc.*, **67**, 441 (1945).
319. For a review of this early work, cf. G. B. Butler, *J. Polym. Sci.*, **48**, 279 (1960).
320. (a) L. Kaplan, *J. Am. Chem. Soc.*, **89**, 1753 (1967); (b) L. Kaplan, *ibid.*, **89**, 4566 (1967).
321. L. Kaplan, *Chem. Commun.*, 754 (1968).
322. See M. S. Kharasch, M. Weiner, W. Nudenberg, A. Bhattacharya, T.-I. Wang, and N. C. Yang, *J. Am. Chem. Soc.*, **83**, 3232 (1961).
323. S. W. Benson, *J. Chem. Phys.*, **34**, 521 (1961).
324. J. F. Garst and J. T. Barbas, *J. Am. Chem. Soc.*, **91**, 3385 (1969). Cyclobutane *is*

formed, but not via an S_H process. See J. F. Garst, R. H. Cox, J. T. Barbas, R. D. Roberts, J. I. Morris, and R. C. Morrison, *ibid.*, **92**, 5761 (1970).
325. W. S. Trahanovsky and M. P. Doyle, *J. Org. Chem.*, **32**, 146 (1967). The rate similarity is of interest in the matter of 1,5 halogen-bridged radicals. Such bridged radicals form a portion of another chapter in this volume (L. Kaplan).
326. S. Arai, S. Sato, and S. Shida, *J. Chem. Phys.*, **33**, 1277 (1960); A. S. Gordon and S. R. Smith, *J. Phys. Chem.*, **66**, 521 (1962).
327. R. C. Lamb, P. W. Ayers, and M. K. Toney, *J. Am. Chem. Soc.*, **85**, 3483 (1963).
328. C. Walling and M. S. Pearson, *J. Am. Chem. Soc.*, **86**, 2262 (1964).
329. L. W. Menapace and H. G. Kuivila, *J. Am. Chem. Soc.*, **86**, 3047 (1964).
330. J. F. Garst, P. W. Ayers, and R. C. Lamb, *J. Am. Chem. Soc.*, **88**, 4260 (1966). For a review of this type of reduction, see J. F. Garst, *Acc. Chem. Res.*, **4**, 400 (1971).
331. C. Walling, J. H. Cooley, A. A. Ponaras, and E. J. Racah, *J. Am. Chem. Soc.*, **88**, 5361 (1966).
332. R. C. Lamb, P. W. Ayers, and M. K. Toney, *J. Am. Chem. Soc.*, **88**, 4261 (1966); cf. also M. E. H. Howden, J. Burdon, and J. D. Roberts, *ibid.*, **88**, 1732 (1960).
333. C. Walling and A. Cioffari, *J. Am. Chem. Soc.*, **92**, 6609 (1970).
334. J. F. Garst and F. E. Barton, II, *Tetrahedron Letters*, 587 (1969). The fluoride is slow enough to allow the kinetic study. Other halides are generally too fast. Again (see p. 420) no enhancement of rate due to π electron participation was observed.
335. Sodium naphthalenide is an efficient reducing agent. Its reaction rate with radicals is probably diffusion controlled. Thus, the ratio of its reaction rate constant with radical **238-3** as compared to that for cyclization of **238-3** is $1.6 \times 10^4\ M^{-1}$ [compare the value 10 M^{-1} with $(n\text{-}C_4H_9)_3SnH$].
336. W. S. Friedlander, *Abstracts of Papers, 133rd National Meeting of the American Chemical Society, San Francisco, California, April 1958*, paper 18N.
337. N. O. Brace, *J. Am. Chem. Soc.*, **86**, 523 (1964); *J. Org. Chem.*, **31**, 2879 (1966).
338. J. I. G. Cadogan, D. H. Hey, and A. O. S. Hock, *Chem. Ind.* (*London*), 753 (1964).
339. N. O. Brace, *J. Org. Chem.*, **32**, 2711 (1967).
340. N. O. Brace, *J. Polym. Sci.* (*A-1*), **8**, 2091 (1970); *J. Org. Chem.*, **36**, 3187 (1971).
340a. M.-P. Crozet, J.-M. Surzur, and C. Dupuy, *Tetrahedron Letters*, 2031 (1971).
340b. R. D. Rieke and N. A. Moore, *Tetrahedron Letters*, 2035 (1969).
341. The introductory paper discusses earlier cyclizations, e.g., the Pschorr reaction; M. Julia, J.-M. Surzur, L. Katz, F. LeGoffic, and C. James, *Bull. Soc. Chim. France*, 1106 (1964). See also reference 317.
342. M. Julia, J.-M. Surzur, and L. Katz, *Bull Soc. Chim. France*, 1109 (1964).
343. M. Julia and M. Maumy, *Bull. Soc. Chim. France*, 434 (1966).
344. M. Julia and M. Maumy, *Bull. Soc. Chim. France*, 2415 (1969).
345. M. Julia, J.-M. Surzur, L. Katz, and F. LeGoffic, *Bull. Soc. Chim. France*, 1116 (1964).
346. M. Julia, F. LeGoffic, and L. Katz, *Bull. Soc. Chim. France*, 1122 (1964).
347. As a process carried to an extreme, squalene could possibly yield a pentacyclic derivative in such a radical cyclization. It does not. Treatment with thiolacetic aid failed to cyclize the polyene; R. Breslow, E. Barrett, and E. Mohacsi, *Tetrahedron Letters*, 1207 (1962).
348. M. Julia and F. LeGoffic, *Bull. Soc. Chim. France*, 1129 (1964).
349. J.-M. Surzur and G. Torri, *Bull. Soc. Chim. France*, 3070 (1970).
350. E. Van Bruggen, *Rec. Trav. Chim. Pays-Bas*, **87**, 1134 (1968).
351. M. Julia and F. LeGoffic, *Bull Soc. Chim. France*, 1550 (1965).

352. M. Julia and F. LeGoffic, *Bull. Soc. Chim. France*, 1555 (1965).
353. M. Julia, M. Maumy, and L. Mion, *Bull. Soc. Chim. France*, 2641 (1967).
354. H. Pines, N. C. Sih, and D. B. Rosenfield, *J. Org. Chem.*, **31**, 2255 (1966).
355. M. Julia and M. Maumy, *Bull. Soc. Chim. France*, 2427 (1969).
356. P. J. Bunyan and D. H. Hey, *J. Chem. Soc.*, 1360 (1962). In hot methanol and pyridine, 4.5% of tetralin was obtained. No such product was observed by E. A. Evans and M. Whalley, *ibid.*, 3642 (1954), under these conditions.
357. H. Pines and J. T. Arrigo, *J. Am. Chem. Soc.*, **79**, 4958 (1957).
358. W. S. Trahanovsky and C. C. Ong, *J. Am. Chem. Soc.*, **92**, 7174 (1970).
359. M. Julia and J.-C. Chottard, *Bull. Soc. Chim. France*, 3691 (1968).
360. J.-C. Chottard, M. Julia, and J. M. Salard, *Tetrahedron*, **25**, 4967 (1969).
361. M. Julia, J.-C. Chottard, and J. J. Basselier, *Bull. Soc. Chim. France*, 3037 (1966); J.-C. Chottard and M. Julia, *ibid.*, 3700 (1968).
362. G. Pregaglia and G. Gregario, *Chim. Ind. (Milan)*, **45**, 1065 (1963).
363. R. Dowbenko, *J. Am. Chem. Soc.*, **86**, 946 (1964); *Tetrahedron*, **20**, 1843 (1964).
363a. T. W. Sam and J. K. Sutherland, *Chem. Commun.*, 970 (1971).
364. L. Friedman, *J. Am. Chem. Soc.*, **86**, 1885 (1964).
365. The strain energies of the *cis*- and *trans*-bicyclo[3.3.0]octanes are 12.8 and 18.8 kcal/mole, respectively. See the literature citation in reference 274.
366. I. Tabushi, K. Fujita, and R. Oda, *J. Org. Chem.*, **35**, 2376 (1970).
367. J. R. Shelton and E. E. Borchert, *Can. J. Chem.*, **46**, 3833 (1968).
368. D. L. Struble, A. L. J. Beckwith, and G. E. Gream, *Tetrahedron Letters*, 3701 (1968).
369. D. L. Struble, A. L. J. Beckwith, and G. E. Gream, *Tetrahedron Letters*, 4795 (1970).
370. J. W. Wilt, S. N. Massie, and R. B. Dabek, *J. Org. Chem.*, **35**, 2803 (1970).
371. One may say that the 2-norbornyl radical would possess a strain energy somewhere between that of norbornene (17.55 kcal/mole) and norbornane (27.2 kcal/mole).[274]
372. Reference 269a, Chap. 5.
373. R. G. Lawton, *J. Am. Chem. Soc.*, **83**, 2399 (1961); P. D. Bartlett and S. Bank, *ibid.*, **83**, 2591 (1961).
374. E. A. Hill, R. J. Thiessen, A. Doughty, and R. Miller, *J. Org. Chem.*, **34**, 3681 (1969). Intramolecular addition of certain organolithium reagents containing the *norbornene*-type double bond has, however, been reported by P. T. Lansbury and F. J. Caridi, *Chem. Commun.*, 714 (1970). At 100° in tetrahydrofuran a number of unsaturated Grignard reagents undergo cyclization as well. Cf. W. C. Kossa, Jr., T. C. Rees, and H. G. Richey, Jr., *Tetrahedron Letters*, 3455 (1971).
375. E. I. Heiba and R. M. Dessau, *J. Am. Chem. Soc.*, **88**, 1589 (1966); **89**, 3772 (1967).
376. R. Kh. Freidlina, reference 3, pp. 231–248.
377. E. I. Heiba and R. M. Dessau, *J. Am. Chem. Soc.*, **89**, 2238 (1967).
378. H. R. Ward, *J. Am. Chem. Soc.*, **89**, 5517 (1967).
379. S. A. Kandil and R. E. Dessy, *J. Org. Chem.*, **30**, 3857 (1965); *J. Am. Chem. Soc.*, **88**, 3027 (1966). These workers cyclized halides using *n*-butyllithium also. They suggested, interestingly, that similar cyclizations using organocopper reagents were probably radical in nature.
380. D. Bryce-Smith, *J. Chem. Soc.*, 1603 (1956), suggested this some time ago. The results of CIDNP studies confirmed it. Cf. H. R. Ward and R. G. Lawler, *J. Am. Chem. Soc.*, **89**, 5518 (1967); R. G. Lawler, *ibid.*, **89**, 5519 (1967).
381. J.-M. Surzur, C. Dupuy, M. P. Crozet, and N. Aimar, *Compt. Rend. Acad. Sci., Ser. C*, **269**, 849 (1969).
381a. R. D. Rieke and B. J. A. Cooke, *J. Org. Chem.*, **36**, 2674 (1971).
382. D. Davenport, *J. Chem. Educ.*, **47**, 271 (1970).

383. M. Gomberg, *Ber.*, **33**, 3150 (1900); *J. Am. Chem. Soc.*, **22**, 757 (1900).
384. H. Lankamp, W. T. Nauta, and C. MacLean, *Tetrahedron Letters*, 249 (1968). This muted article disguises the thunder within it very well, developing the past history of "hexaphenylethane" and the attendant doubts concerning its structure.
385. P. Jacobson, *Ber.*, **38**, 196 (1905).
386. W. B. Smith, *J. Chem. Educ.*, **47**, 535 (1970).
387. H. A. Staab, H. Brettschneider, and H. Brunner, *Chem. Ber.*, **103**, 1101 (1970).
388. H. Volz, W. Lotsch, and H.-W. Schnell, *Tetrahedron*, **26**, 5343 (1970).
389. Bis(9-phenylfluorenyl) is one true derivative of hexaphenylethane.[387] It gives no radicals by dissociation at 25°: W. Theilacker, H. Schultz, U. Baumgarte, H. G. Drössler, W. Rhode, F. Thater, and H. Uffmann, *Angew. Chem.*, **69**, 322 (1957). This dimer has had its structure confirmed by ^{13}C NMR spectroscopy; H. A. Staab, K. S. Rao, and H. Brunner, *Chem. Ber.*, **104**, 2634 (1971). The dimer of sesquixanthydryl radical,

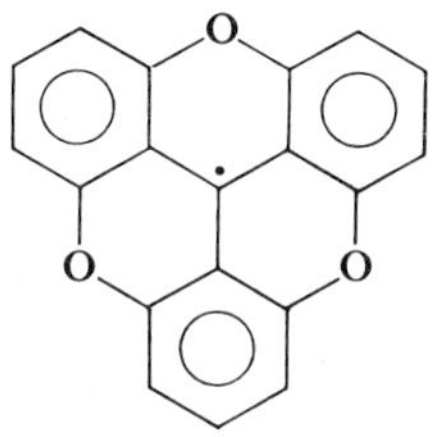

is another example. It, too, shows no tendency to dissociate; C. Rüchardt and I. Mayer-Ruthardt, *Chem. Unserer Zeit*, **3**, 41 (1969).
390. E.g., G. W. Wheland, *Advanced Organic Chemistry*, 3rd ed., Wiley, New York, 1960, pp. 763–791, gives an excellent summary of all this work. See also A. R. Forrester, J. M. Hay, and R. H. Thomson, *Organic Chemistry of Stable Free Radicals*, Academic Press, London, 1968, Chap. 2. This recent book came out just a bit too soon. Footnote 61 gives the new developments, however.
390a. F. Bölsing and K.-D. Korn, *Tetrahedron Letters*, 3865 (1971).
391. T. S. Stevens, E. M. Creighton, A. B. Gordon, and M. MacNicol, *J. Chem. Soc.*, 3193 (1928).
392. Recent reviews (but prior to CIDNP studies): D. J. Cram, *Fundamentals of Carbanion Chemistry*, Academic Press, New York, 1965, pp. 223–238; H. E. Zimmerman, *op. cit.* in reference 2, pp. 378–382.
393. H. Fischer and J. Bargon, *Acc. Chem. Res.*, **2**, 110 (1969), G. L. Closs, *J. Am. Chem. Soc.*, **91**, 4552 (1969); F. Gerhart, *Tetrahedron Letters*, 5061 (1969); R. Kaptein, *Chem. Commun.*, 732 (1971).
394. A. R. Lepley, *J. Am. Chem. Soc.*, **91**, 1237 (1969); A. R. Lepley, R. H. Becker, and A. G. Giumanini, *J. Org. Chem.*, **36**, 1222 (1971). For a related study, see A. R. Lepley and A. G. Giumanini, *ibid.*, **36**, 1217 (1971).
395. J. Jacobus, *Chem. Commun.*, 709 (1970).
396. Such a radical pair mechanism was first suggested by U. Schöllkopf and U. Ludwig, *Chem. Ber.*, **101**, 2224 (1968).
397. A. R. Lepley, *Chem. Commun.*, 1460 (1969). See also, U. Schöllkopf, U. Ludwig, G. Ostermann, and M. Patsch, *Tetrahedron Letters*, 3415 (1969).
398. Cf. R. Kaptein and L. J. Oosterhoff, *Chem. Phys. Letters*, **4**, 195, 214 (1969); G. L. Closs and A. D. Trifunac, *J. Am. Chem. Soc.*, **92**, 2183, 2186, 7227 (1970); G. L. Closs, C. E. Doubleday, and D. R. Paulson, *ibid.*, **92**, 2185 (1970).

399. J. E. Baldwin, W. F. Erickson, R. E. Hackler, and R. M. Scott, *Chem. Commun.*, 576 (1970). For a related study, see U. Schöllkopf, G. Ostermann, and J. Schosig, *Tetrahedron Letters*, 2619 (1969).
400. It would seem that this is precisely a study where the caveat[395] would apply. That is, the Stevens-like rearranged product might show CIDNP via a homolysis subsequent to its formation from the ylid.
401. An excellent list of references to recent mechanistic studies on ylid processes is given in reference 399.
402. H. Iwamura, M. Iwamura, T. Nishida, M. Yoshida, and J. Nakayama, *Tetrahedron Letters*, 63 (1971).
403. J. E. Baldwin and R. E. Hackler, *J. Am. Chem. Soc.*, **91**, 3646 (1969).
404. G. F. Hennion and M. J. Shoemaker, *J. Am. Chem. Soc.*, **92**, 1769 (1970). For another recent report of dimer formation in a Stevens rearrangement, see S. H. Pine, E. M. Munemo, T. R. Phillips, G. Bartolini, W. D. Cotton, and G. C. Andrews, *J. Org. Chem.*, **36**, 984 (1971).
405. U. Schöllkopf, U. Ludwig, G. Ostermann, and M. Patsch, *Tetrahedron Letters*, 3415 (1969).
405a. H. P. Benecke and J. H. Wikel, *Tetrahedron Letters*, 3479 (1971).
406. S. H. Pine, B. A. Catto, and F. G. Yamagishi, *J. Org. Chem.*, **35**, 3663 (1970).
407. T. S. Stevens, *J. Chem. Soc.*, 2107 (1930); T. S. Thomson and T. S. Stevens, *ibid.*, 55 (1932); E. F. Jenny and J. Druey, *Angew. Chem.*, *Int. Engl. Ed.*, **1**, 155 (1962).
408. C. Walling, H. P. Waits, J. Milovanovic, and C. G. Pappiaonnou, *J. Am. Chem. Soc.*, **92**, 4927 (1970).
409. J. Meisenheimer, *Ber.*, **52**, 1667 (1919); R. A. W. Johnstone, in *Mechanisms of Molecular Migrations*, Vol. 2, B. S. Thyagarajan, Ed., Interscience, New York, 1969, p. 249.
410. U. Schöllkopf, U. Ludwig, M. Patsch, and W. Franken, *Ann.*, **703**, 77 (1967).
411. J. P. Lorand, R. W. Grant, P. A. Samuel, E. O'Connel, and J. Zero, *Tetrahedron Letters*, 4087 (1969).
412. A. R. Lepley, P. M. Cook, and G. F. Willard, *J. Am. Chem. Soc.*, **92**, 1101 (1970).
413. G. Ostermann and U. Schöllkopf, *Ann.*, **737**, 170 (1970).
414. G. P. Schulman, P. Ellgen, and M. Conner, *Can. J. Chem.*, **43**, 3459 (1965).
415. G. L. Closs and L. E. Closs, *J. Am. Chem. Soc.*, **91**, 4550 (1969).
416. E. J. Grubbs, J. A. Villarreal, J. D. McCullogh, Sr., and J. S. Vincent, *J. Am. Chem. Soc.*, **89**, 2234 (1967); J. S. Vincent and E. J. Grubbs, *ibid.*, **91**, 2022 (1969). Very recently, CIDNP has been observed in a nitrone rearrangement by D. G. Morris, *Chem. Commun.*, 221 (1971).
417. N. Castagnoli, Jr., J. C. Craig, A. P. Melikian, and S. K. Roy, *Chem. Commun.*, 1327 (1970).
418. G. Wittig and L. Löhmann, *Ann.*, **550**, 260 (1942).
419. (a) For a recent discussion of this and other "carbanionic rearrangements", cf. U. Schöllkopf, *Angew. Chem.*, *Int. Engl. Ed.*, **9**, 763 (1970); (b) a trenchant review of the evolution of mechanistic thought in this area is given by S. H. Pine, *J. Chem. Educ.*, **48**, 99 (1971).
420. The suprafacial 1,2 sigmatropic shift with retention that would be necessary is disallowed by orbital symmetry considerations. The S_Ni' concerted pathway is allowed, however, and allyl migration in the Wittig rearrangement takes this path.[419]
421. P. T. Lansbury, private communication cited in footnote 39 in reference 419a.
422. E. G. Miller, D. R. Rayner, and K. Mislow, *J. Am. Chem. Soc.*, **88**, 3139 (1966).

423. E. G. Miller, D. R. Rayner, H. T. Thomas, and K. Mislow, *J. Am. Chem. Soc.*, **90**, 4861 (1968).
424. S. Oae, T. Kitao, and Y. Kitaoka, *J. Am. Chem. Soc.*, **84**, 3359 (1962).
425. V. Traynelis and R. F. Martello, *J. Am. Chem. Soc.*, **87**, 5710 (1965).
426. H. Iwamura, M. Iwamura, T. Nishida, and I. Mura, *Tetrahedron Letters*, 3117 (1970).
427. H. Iwamura, M. Iwamura, T. Nishida, and S. Sato, *J. Am. Chem. Soc.*, **92**, 7474 (1970).
428. W. N. White, H. S. White, and A. Fentiman, *J. Am. Chem. Soc.*, **92**, 4477 (1970), and earlier references cited therein.
429. J. Hollaender and W. P. Neumann, *Angew. Chem., Int. Engl. Ed.*, **9**, 804 (1970), For a related study on the thermal rearrangement of diazoaminobenzene, see L. F. Kasukhin, M. P. Ponomarchuk, and V. N. Kalinin, *Zh. Org. Khim.*, **6**, 2531 (1970).
430. J. F. Garst and G. S. Hammond, *J. Org. Chem.*, **23**, 98 (1958); W. A. Pryor and K. Smith, *J. Am. Chem. Soc.*, **89**, 1741 (1967).
431. M. S. Kharasch, G. Stampa, and W. Nudenberg, *Science*, **116**, 309 (1952).
432. D. P. Kelly, J. T. Pinhey, and R. D. G. Rigby, *Tetrahedron Letters*, 309 (1952).
433. H. J. Hageman, H. L. Louwerse, and W. J. Mijs, *Tetrahedron*, **26**, 2045 (1970).
434. E. Schmitz and O. Brede, *J. Prakt. Chem.*, **312**, 43 (1970).
435. C. Walling, reference 2, pp. 431–433.
436. R. C. Neumann, Jr., and G. D. Holmes, *J. Org. Chem.*, **33**, 4317 (1968), and references to earlier studies cited therein.
437. C. Walling, reference 2, pp. 433–438.
438. M. S. Kharasch, E. T. Margolis, and F. R. Mayo, *J. Org. Chem.*, **1**, 393 (1936). The ready isomerization of allylic halides, particularly bromides, as shown in (344) could make thermodynamic control predominant in radical bromination reactions involving allylic systems. Garnering reactivity data, and so on, from the product composition of such processes could therefore be unwise.
439. S. J. Cristol and R. Kellman, *J. Org. Chem.*, **36**, 1866 (1971).
440. S. J. Cristol, G. O. Mayo, and G. A. Lee, *J. Am. Chem. Soc.*, **91**, 214 (1969).
441. S. J. Cristol and G. A. Lee, *J. Am. Chem. Soc.*, **91**, 7554 (1969).
442. Cf. R. W. Phillips and D. H. Volman, *J. Am. Chem. Soc.*, **91**, 3418 (1969).
443. B. B. Jarvis and R. O. Fitch, *Chem. Commun.*, 408 (1970).
444. L. Claisen, *Ber.*, **29**, 2931 (1896).
445. K. B. Wiberg and B. I. Rowland, *J. Am. Chem. Soc.*, **77**, 1159 (1955).
446. C. Walling, reference 2, p. 448.
447. K. B. Wiberg, T. M. Shryne, and R. R. Kintner, *J. Am. Chem. Soc.*, **79**, 3160 (1957).
448. K. B. Wiberg and B. I. Rowland, *J. Am. Chem. Soc.*, **77**, 2205 (1955).
449. K. B. Wiberg, R. R. Kintner, and E. L. Motell, *J. Am. Chem. Soc.*, **85**, 450 (1963).
450. Y. Izawa and Y. Ogata, *J. Org. Chem.*, **35**, 3192 (1970).
451. F. W. Stacey, J. C. Sauer, and B. C. McKusick, *J. Am. Chem. Soc.*, **81**, 987 (1959).
452. E. S. Huyser, R. M. Kellogg, and D. T. Wang, *J. Org. Chem.*, **30**, 4377 (1965).
453. N. Frydman and Y. Mazur, *J. Am. Chem. Soc.*, **92**, 3203 (1970).
454. M. S. Newman and R. M. Layton, *J. Org. Chem.*, **33**, 2338 (1968).
455. K. von Auwers and W. Jülicher, *Chem. Ber.*, **55**, 2167 (1922); M. S. Newman and R. L. Tse, *J. Org. Chem.*, **21**, 638 (1956); M. S. Newman and J. A. Eberwein, *ibid.*, **29**, 2516 (1964).
456. R. Kh. Freidlina, reference 3, pp. 241–245.

457. S. J. Cristol, G. D. Brindell, and J. A. Reeder, *J. Am. Chem. Soc.*, **80**, 635 (1958).
458. D. J. Trecker and J. P. Henry, *J. Am. Chem. Soc.*, **85**, 3204 (1963). See the text (p. 477) for a later study where the real complexity of the process became apparent.
459. E. S. Huyser and G. Echegaray, *J. Org. Chem.*, **27**, 429 (1962).
460. J. W. Wilt and A. A. Levin, *J. Org. Chem.*, **27**, 2319 (1962).
461. S. J. Cristol and D. I. Davies, *J. Org. Chem.*, **29**, 1282 (1964); S. J. Cristol, T. W. Russell, and D. I. Davies, *ibid.*, **30**, 226 (1965).
462. D. I. Davies, J. N. Done, and D. H. Hey, *Chem. Commun.*, 725 (1966). The Kharasch-Grignard method used here is not among the less suspect of radical reactions. The dimers probably reflect a *nonfree* radical pathway (see p. 404). Their yield was small, however, 3.8–5.3%, so the bulk of the reaction is probably innocent enough.
463. C. R. Warner, R. J. Strunk, and H. G. Kuivila, *J. Org. Chem.*, **31**, 3381 (1966).
464. S. J. Cristol and R. W. Gleason, *J. Org. Chem.*, **34**, 1762 (1969).
465. S. J. Cristol and R. V. Barbour, *J. Am. Chem. Soc.*, **88**, 4262 (1966).
466. V. M. A. Chambers, W. R. Jackson, and G. W. Young, *Chem. Commun.*, 1275 (1970).
467. (a) J. A. Claisse and D. I. Davies, *Chem. Commun.*, 209 (1965); (b) C. K. Alden, J. A. Claisse, and D. I. Davies, *J. Chem. Soc. (C)*, 1540 (1966).
468. E. N. Prilezhaeva, V. A. Azovskaya, L. V. Tsymbal, E. N. Gur'yanova, G. Andrianova, and M. F. Shostakovskii, *Zh. Obsch. Khim.*, **35**, 39 (1965).
469. J. A. Claisse, D. I. Davies, and C. K. Allen, *J. Chem. Soc. (C)*, 1498 (1966). This is the second major paper in this series. Curiously, it appeared before the first one.[467b] Additional examples of reactions related to (381) may be found in D. I. Davies and P. Mason, *ibid.*, 288, 295 (1971), and R. Alexander, D. I. Davies, D. H. Hey, and J. N. Done, *ibid.*, 2367 (1971).
470. V. F. Bystrov, V. A. Azovskaya, N. P. Petukhova, A. U. Stepanyants, and E. N. Prilezhaeva, *Izv. Akad. Nauk SSSR, Ser. Khim.*, 318 (1966).
471. E. N. Prilezhaeva, V. A. Azovskaya, and N. P. Petukhova, *Zh. Org. Khim.*, **4**, 621 (1968).
472. C. K. Alden and D. I. Davies, *J. Chem. Soc. (C)*, 1017 (1967). Another example of interest is the recent report of R. Alexander and D. I. Davies, *ibid.*, 896 (1971), on the radical addition of hydrogen bromide to **384-1**.
473. C. K. Alden and D. I. Davies, *J. Chem. Soc. (C)*, 2007 (1967).
474. The unusual *smaller* amount of vicinal adduct with the normally more reactive chain transfer agent, bromotrichloromethane, may indicate a greater steric problem in chain transfer to it compared to carbon tetrachloride.
475. D. I. Davies and P. J. Rowley, *J. Chem. Soc. (C)*, 2245 (1967).
476. It is worth mentioning that C-7 substituents affect *cationic* additions to norbornenes as well; H. C. Brown and J. H. Kawakami, *J. Am. Chem. Soc.*, **92**, 201 (1970); H. C. Brown and K.-T. Liu, *ibid.*, **92**, 3502 (1970).
477. D. I. Davies and P. J. Rowley, *J. Chem. Soc. (C)*, 2249 (1967).
478. N. O. Brace, *J. Org. Chem.*, **27**, 3027 (1962).
479. D. I. Davies and L. T. Parfitt, *J. Chem. Soc. (C)*, 2691 (1967).
480. For a discussion, see reference 269a, pp. 170, 171.
481. M. L. Scheinbaum, *J. Org. Chem.*, **35**, 2785 (1970).
482. J. W. Wilt and P. J. Chenier, *J. Org. Chem.*, **35**, 1562 (1970), and references cited therein.
483. J. W. Wilt, G. Gutman, W. J. Ranus, Jr., and A. R. Zigman, *J. Org. Chem.*, **32**, 893 (1967).

484. S. J. Cristol and J. M. Sullivan, *J. Am. Chem. Soc.*, **93**, 1967 (1971).
485. W. G. Bentrude and K. R. Darnall, *J. Am. Chem. Soc.*, **90**, 3588 (1968).
486. W. H. Urry, M. H. Pai, and C. Y. Chen, *J. Am. Chem. Soc.*, **86**, 5342 (1964).
487. F. F. Rust, F. H. Seubold, and W. E. Vaughn, *J. Am. Chem. Soc.*, **70**, 4253 (1948); W. H. Urry, A. Nishihara, and J. H. Y. Liu, *J. Org. Chem.*, **32**, 347 (1967).
488. W. Reusch and C. L. Karl, *Abstracts of Papers, 155th National Meeting of the American Chemical Society, San Francisco, California, April 1968*, paper P79.
489. M. J. Jorgenson and T. J. Clark, *op. cit.* in reference 239.

CHAPTER 9

Electron Transfer Reactions of Organic Anions

By JOHN F. GARST
Department of Chemistry, The University of Georgia, Athens, Georgia

I. ELECTRON TRANSFER REACTIONS—DEFINITION AND SCOPE

Classifications of reactions, both elementary and complex, depend on somewhat arbitrarily chosen representations of reactants and products. Consider the postulated initial step of a reaction of an alkyllithium with an alkyl halide.

$$\text{R—Li} + \text{R—X} \longrightarrow [\text{R}\cdot\text{R}\cdot\text{Li}^+\text{Br}^-] \qquad (1)$$

$$\text{R:}^-, \text{Li}^+ + \text{R—X} \longrightarrow [\text{R}\cdot\text{R}\cdot\text{Li}^+\text{Br}^-] \qquad (2)$$

It is not clear that (1) represents an electron transfer (ET) reaction. A much better case can be made for (2): R:$^-$ loses one electron, R—X dissociates upon gaining one electron, and Li^+ is a spectator. Yet both represent the same reaction step, and justification for both points of view (and others) can be offered.

Instead of a definition of "electron transfer reaction," a classification of reaction representations is offered. These illustrate and define the kinds of reactions of concern here.

Electron attachment (EA):

$$e^- + A \longrightarrow A\cdot^- \qquad e^- + A\cdot \longrightarrow A{:}^-$$

Electron detachment (ED):

$$D\cdot^- \longrightarrow D + e^- \qquad D{:}^- \longrightarrow D\cdot + e^-$$

Dissociative electron attachment (DEA):

$$e^- + A\text{—}X \longrightarrow A\cdot + X{:}^- \qquad e^- + A\text{—}X\cdot^- \longrightarrow A{:}^- + X{:}^-$$

Associative electron detachment (AED):

$$D{:}^- + Y\cdot \longrightarrow D\text{—}Y + e^- \qquad D{:}^- + Y{:}^- \longrightarrow D\text{—}Y\cdot^- + e^-$$

The above can be regarded as components (half-reactions) of other ET processes, some examples of which follow:

ED–EA: $$D\cdot^- + A \longrightarrow D + A\cdot^-$$

ED–DEA: $$D\cdot^- + A\text{—}X \longrightarrow D + A\cdot + X^-$$

EA–AED: $$D{:}^- + Y\cdot + A \longrightarrow D\text{—}Y + A\cdot^-$$

This list is not complete, but extensions are obvious. For generality, the charges indicated should be regarded as relative rather than absolute. Other electronic multiplicities could be included, and there are further possible elementary classifications, such as dissociative electron detachment (DED) and associative electron attachment (AEA). In this framework, reaction 1 is DED–DEA; regarding Li^+ as a spectator, reaction 2 is ED–DEA.

But classification itself is unimportant. The important questions are as follows: Are reactions representable as ET really found as net reaction? as steps in reaction mechanisms? Are they significant in organic chemistry as a whole? What are the relevant structure–reactivity relationships? medium effects? other environmental effects? synthetic significance? and so on.

Electron transfer reactions are found, as both net reactions and steps in mechanisms, in thermal, photochemical, electrochemical, radiochemical, and biochemical systems. Structure–reactivity relationships, medium and other environmental effects, and synthetic applications of thermal, homogeneous ET reactions are still being actively explored, and generalities in these domains are not yet numerous nor firmly grounded.

In this chapter we consider selected aspects of thermal ET reactions of anions in systems which are initially homogeneous. Because they are frequently involved, we first review some of the chemistry of radical anions. Then we consider solvent and metal ion effects, which can be very large in some systems. Finally, we review some of the findings and interpretations for a number of reactions believed to proceed through electron transfer steps.

II. RADICAL ANIONS

A. Formation by Electron Attachment

Electron attachment to an unsaturated organic substrate sometimes gives a long-lived radical anion.

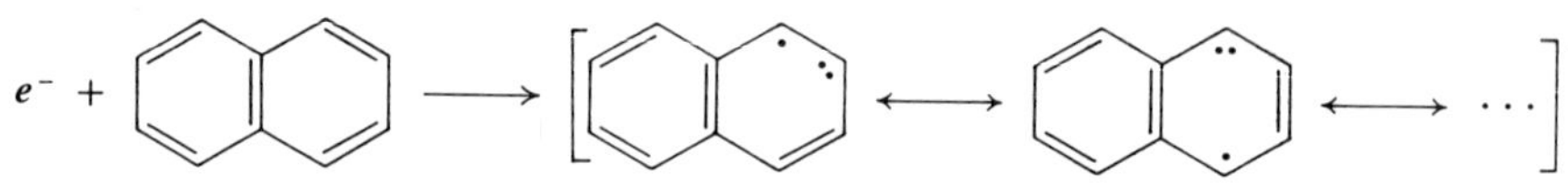

(3)

Naphthalene :Naph — Naphthalene radical anion :Naph·⁻

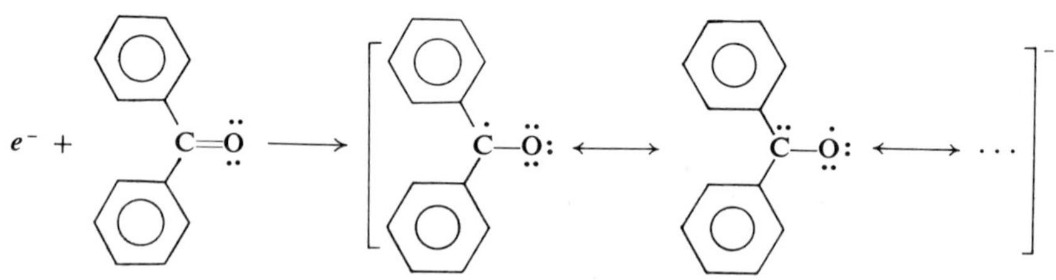

(4)

Benzophenone :Bzph — Benzophenone radical anion (a ketyl) :Bzph·⁻

Electron sources can be active metals, electrodes, populations of solvated electrons, radical anions, dianions, or other chemical species.

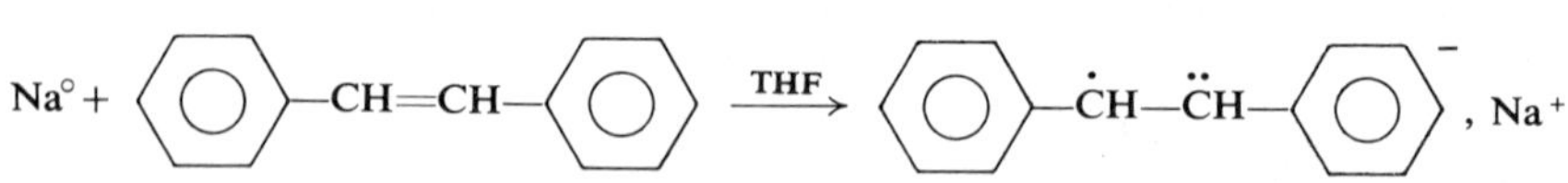

(5)

Stilbene — :Stil·⁻

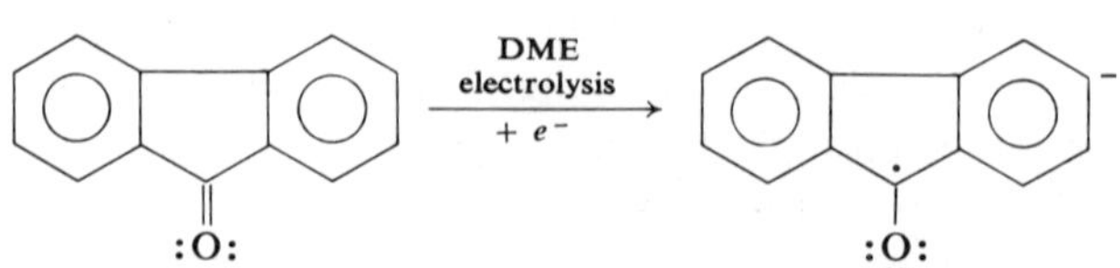

(6)

Fluorenone — :Fluo·⁻

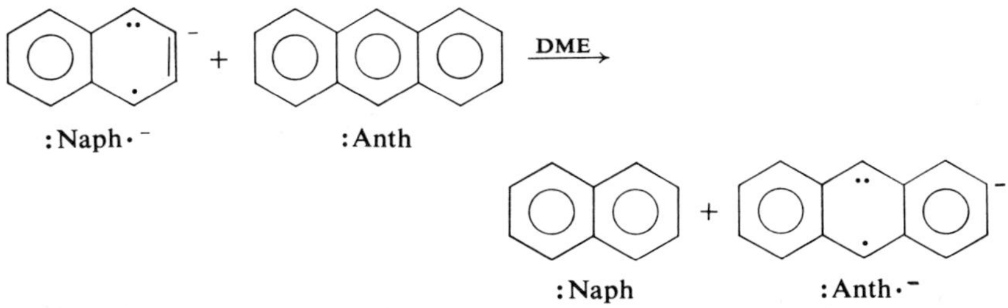

(7)

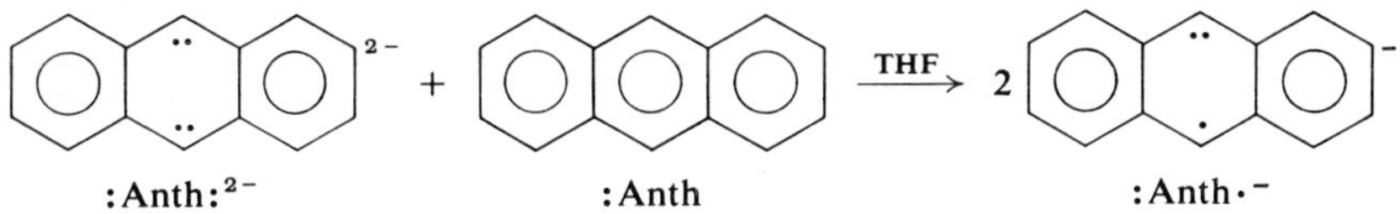

(8)

B. Solvents

Because many radical anions react with oxygen, water, and carbon dioxide, they are usually handled in rigorously dry and degassed solvents. Inert atmospheres may be used (nitrogen, argon), but evacuated vessels containing only solvent vapor are often preferable. The reactivities of alcohols, primary and secondary amines, ketones, esters, nitriles, and similar substances make them unsuitable solvents for the more reactive radical anions. Hydrocarbons may be sufficiently unreactive, but in them radical anion preparations are frequently unstable or insoluble. When placed in media of insufficient ion-solvating power, radical anion preparations decompose by ED–EA reactions with counterions.[6]

$$:\text{Naph} + \text{Na}\cdot \xrightarrow{\text{DME}} :\text{Naph}\cdot^{-}, \text{Na}^{+} \qquad \text{(Stable solution)} \tag{9}$$

$$:\text{Naph}\cdot^{-}, \text{Na}^{+} \text{ (solution in DME)} \xrightarrow{\text{DEE}} :\text{Naph} + \text{Na}\cdot \tag{10}$$

Diethyl ether (DEE) is clearly inferior to 1,2-dimethoxyethane (DME) in ion-solvating ability.

The more reactive radical anions are most often prepared in ethers. In addition, amines can be used as solvents, especially tertiary amines and pyridine. Even primary amines and alcohols will support very unreactive radical anions. Liquid ammonia has been used. *N,N*-Dimethylformamide (DMF) and dimethyl sulfoxide (DMSO) tend to be too reactive for many radical anions. Hexamethylphosphoramide (HMPA) has great ion-solvating power, a relatively high dielectric constant, and a low reactivity. Many radical anion preparations are stable in HMPA.

C. Types of Reactions

1. Typical Radical Reactions

Radical anions may undergo typical reactions of radicals. Sometimes they couple.

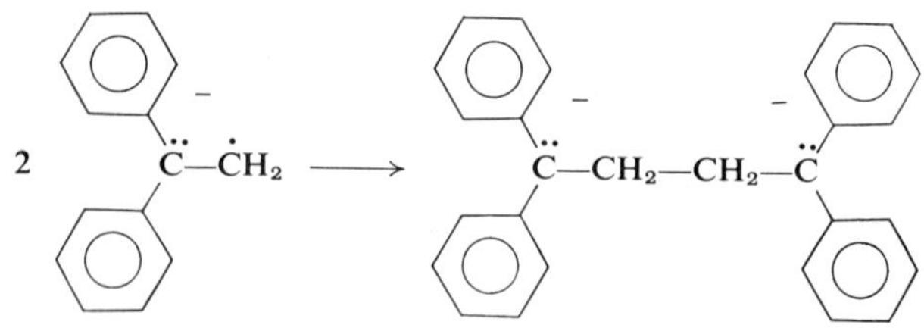

(11)[7]

But for many radical anions such reactions are unimportant, probably for thermodynamic reasons. The free radical properties of the stable radical anions can be summarized by saying that they are similar to those of stable (neutral) free radicals.

2. *Acid–Base Reactions*

Hydrocarbon radical anions, like :Naph·⁻, are believed to be substantially weaker bases than some carbanions which may appear to be closely related structurally. Consider the relative basicities of :Naph·⁻ and :NaphH⁻. These are measured by equilibrium constants K_b(:Naph·⁻) and K_b(:NaphH⁻), respectively, which are related by the following cycle to hydrogen atom dissociation constants K_H(HNaphH) and K_H(:NaphH⁻). In this cycle, all equilibrium constants refer to reactions as read from left to right or from top to bottom.

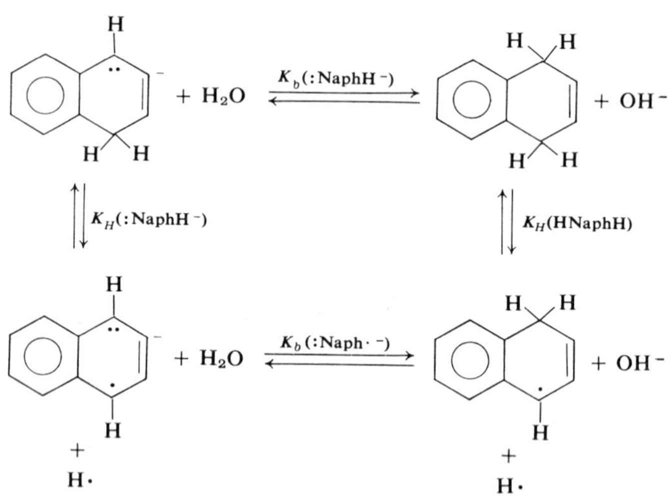

(12)

Clearly,

$$K_b(:\mathrm{Naph}\cdot^-) = K_b(:\mathrm{NaphH}^-)\left[\frac{K_H(\mathrm{HNaphH})}{K_H(:\mathrm{NaphH}^-)}\right] \qquad (13)$$

K_H(HNaphH) is expected to be smaller than K_H(:NaphH$^-$) for two reasons. The first is revealed by simple molecular orbital calculations[8]: The dissociation of H· from :NaphH$^-$ introduces additional cyclic delocalization (in :Naph·$^-$), whereas the dissociation of H· from HNaphH (giving ·NaphH) does not. The second reason to expect K_H(HNaphH) to be smaller than K_H(:NaphH$^-$) is not revealed by simple MO calculations, but it could be even more important than the first: Dissociation of H· from :NaphH$^-$ enlarges the skeleton over which the negative charge can be spread, whereas dissociation of H· from HNaphH accomplishes nothing similar. The consequent reduction in unbalanced electron–electron repulsion would increase K_H(:NaphH$^-$) relative to K_H(HNaphH). It is clearly predicted from these considerations, and equation 13, that K_b(:Naph·$^-$) < K_b(:NaphH$^-$), and this appears to be in accord with experiment.[8]

3. Electron Transfer Reactions

a. Simple Net ET Reactions with Iodine and Oxygen. Radical anions undergo various net ET reactions. For example, they can be titrated with iodine.[10]

$$2{:}\mathrm{Org}\cdot^- + I_2 \longrightarrow 2{:}\mathrm{Org} + 2I^- \qquad \text{(Net ED–DEA)} \qquad (14)$$

Radical anions often react with oxygen,[11–13] probably giving the superoxide directly, which may proceed to the peroxide.

$$\mathrm{Na}^+, {:}\mathrm{Bzph}\cdot^- + O_2 \xrightarrow{\text{THF}} {:}\mathrm{Bzph} + \mathrm{Na}^+, O_2\cdot^- \qquad \text{(ED–EA)} \qquad (15)^{11}$$

b. Competition among Nucleophilic, Typical Homolytic, and Electron Transfer Reaction Steps. The low reactivity of stable radical anions in typical radical reactions and their attenuated Lewis basicities support the idea that their ET reactions may be prevalent. One must view the three kinds of pathways as being competitive, so that low reactivity in two of these favors the third, provided that reactivity according to the third pathway is not similarly low.

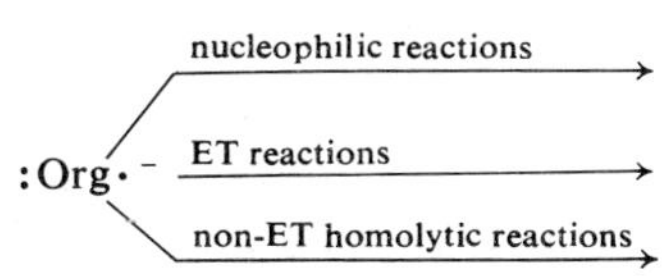

(16)

Questions surrounding this set of competitions pervade mechanistic investigations of reactions of radical anions. Similar competitions can be outlined for diamagnetic anions, and similar questions are raised.

c. Relationship of Stoichiometry to Mechanism. Because net, or overall, ET reactions do not necessarily have any ET steps in their mechanisms, there is the possibility of nomenclatural confusion. Consider the following mechanisms for reaction 14:

$$:\mathrm{Org}\cdot^- + \mathrm{I{-}I} \longrightarrow :\mathrm{Org} + \mathrm{I}\cdot + \mathrm{I}^- \quad \text{(ED–DEA)} \tag{17a}$$

$$:\mathrm{Org}\cdot^- + \mathrm{I}\cdot \longrightarrow :\mathrm{Org} + \mathrm{I}^- \quad \text{(ED–EA)} \tag{17b}$$

$$\cdot\mathrm{Org}:^- + \mathrm{I{-}I} \longrightarrow \cdot\mathrm{Org\text{-}I} + \mathrm{I}^- \quad (S_N2) \tag{18a}$$

$$:\mathrm{Org}\cdot^- + \cdot\mathrm{Org\text{-}I} \longrightarrow {}^-:\mathrm{Org\text{-}Org\text{-}I} \quad \text{(Homolytic association)} \tag{18b}$$

$${}^-:\mathrm{Org\text{-}Org\text{-}I} \longrightarrow :\mathrm{Org} + {}^-:\mathrm{Org\text{-}I} \quad \text{(Heterolytic dissociation)} \tag{18c}$$

$${}^-:\mathrm{Org\text{-}I} \longrightarrow :\mathrm{Org} + \mathrm{I}^- \quad \text{(Heterolytic dissociation)} \tag{18d}$$

Equations 18 represent a mechanism with no ET steps. If equations 18 were correct, would the reaction be an ET reaction? Of course it would be, overall, or net. But it would not have an ET mechanism. The term "electron transfer reaction" is often used in both senses. Context will distinguish between the possible meanings.

Only in relatively few cases have possible non-ET mechanisms been experimentally distinguished from possible ET mechanisms. With present experimental knowledge, Occam's razor finds more application in the field of electron transfer chemistry than it should.

III. SOLVENT AND METAL ION EFFECTS ON REACTIONS OF ANIONS

A. Introduction

Variations in solvent and metal ion can have dramatic effects on the chemistry of radical anions and carbanions. We noted earlier that sodium naphthalene will form in DME, but not in DEE, at room temperature. Shatenshtein and co-workers have investigated the equilibrium between metallic sodium and sodium biphenyl for a number of ether solvents,[14] and there are pronounced variations. Other examples are considered later.

$$\mathrm{Na}\cdot + :\mathrm{Biph} \rightleftharpoons \mathrm{Na}^+, :\mathrm{Biph}\cdot^- \tag{19}$$

B. Association and Solvation of Ions in Solvents of Low Dielectric Constant

The dielectric constants of ethers vary from about 2 (dioxane, DO) to about 7 (THF, DME, diglyme, and other "glymes").[15] In such media, association of ions to ion pairs and other aggregates is the rule.[1,16] For example, for sodium naphthalene in THF, K_d (25°) is about 10^{-7} M.[17]

$$\mathrm{Na}^+:\mathrm{Naph}\cdot^- \text{ (ion pair)} \underset{}{\overset{K_d}{\rightleftharpoons}} \mathrm{Na}^+ + :\mathrm{Naph}\cdot^- \tag{20}$$

For sodium anthracene in THF K_d (25°) is about $4 \times 10^{-6}\,M$, and for sodium biphenyl in DME it is about $5 \times 10^{-6}\,M$.[17] For oxyanions, and for related radical anions such as ketyls, K_d is probably much smaller yet in these solvents. These values are so small that the absence of a significant fraction of free ions in any solution of a practical concentration ($> 10^{-3}\,M$) is assured.

A brief consideration of Coulomb's law will illustrate how very large effects of solvent and metal ion variation could arise.

$$V = \frac{332 z_1 z_2}{Dr} \tag{21}$$

Here V is the potential energy, relative to separated charges, of a pair of charges z_1 and z_2 at a distance r from one another in a medium of dielectric constant D; V is in kilocalories per mole, r in angstroms, and z_1 and z_2 in units of the charge on a proton. Consider moving a pair of oppositely charged ions from a separation of 3 to 4 Å. In dioxane, this would require 14 kcal/mole of coulombic energy, but in DME it would take only 4 kcal/mole. If a process resulted in such a separation, going from reactants to products or from reactants to a transition state, the equilibrium or rate constant governing the process would be about 10^7 greater in DME than in dioxane, other factors being considered equal. This may be somewhat exaggerated because the estimate assumes the applicability of the full macroscopic dielectric constant difference at rather small ionic separations and because it assumes that Coulomb's law gives free energies. Nonetheless, the point is clear: One should expect the possibility of large effects of solvent variations, among ethers, on the chemical properties of organic anions, including radical anions. These effects extend, of course, beyond ethers to include other appropriate solvents.

In addition to generalized effects of solvent variation, there can be quite specific effects. Considerations of macroscopic dielectric constants do not fully represent solvent effects in such cases. In one analysis of specific ion solvation, the ions and solvent molecules are regarded as Lewis acid–base pairs. Alternatively, or additionally, ion–dipole interactions may be important in the associations illustrated in equations 22 and 23.

$$M^+ + :S \rightleftharpoons (M\text{—}S^+ \longleftrightarrow M^+:S) \tag{22}$$

$$\cdot Org:^- + S: \rightleftharpoons (\cdot Org\text{-}S:^- \longleftrightarrow \cdot Org:^-S:) \tag{23}$$

It can be considered that the interactions of the solvated ions with the remainder of the medium are given by classical equations involving gross solvent properties such as the dielectric constant.

There is no evidence that specific anion solvation (equation 23) is important for organic anions in ether solvents. Ethers are not Lewis acids of any appreciable strength and many of the organic anions of interest are large and delocalized besides, so that they are certainly not powerful Lewis bases. But alkali cations are small, and ethers are respectably strong Lewis bases. Therefore, it is entirely reasonable that specific cation solvation, but not specific anion solvation, should be important in the chemistry of organoalkali compounds in ethers.

C. Cation Solvation and the Chemistry of Anions

If we are interested in the chemistry of organic anions, why should we be concerned with specific solvation if it involves only the metal counterions? Because the systems often contain ion pairs or higher ionic aggregates. The free energy of anion–cation interaction can depend sensitively on the state of specific solvation of the cation. Since free energy differences govern chemical systems, cation solvation will affect the chemistry of the associated anions.

D. Physical and Chemical Properties of Anions—Solvent "Polarity"

Even the physical properties of anions are affected by solvation of associated cations. Garst and co-workers interpreted the shifts of the visible band of sodium benzophenone ketyl, as the solvent is varied, on this basis.[18] Hirota and Weissman suggested that these shifts merely reflect variations in the degree of aggregation (to ion pairs, ion quadrupoles, etc.) as the solvent is varied.[19] But this cannot be entirely correct for several reasons, one of which is that there is very little concentration dependence of the spectrum of sodium benzophenone in DME over a range through which both ion pairs and ion quadrupoles are believed to be present. This indicates that the ion pairs and ion quadrupoles have very similar visible spectra.

Another unlikely explanation for the spectral shifts with solvent is that there is a solvent-sensitive equilibrium between "solvent-separated" and "contact" ion pairs. It is unlikely that "solvent-separated" ion pairs are significant for sodium–oxyanion pairs in ethers. For the sterically hindered and highly delocalized galvinoxide ion, spectrophotometric studies suggest that no "solvent-separated" ion pairs are present, with any alkali counterion, in DME, THF, and so on.[20,20a]

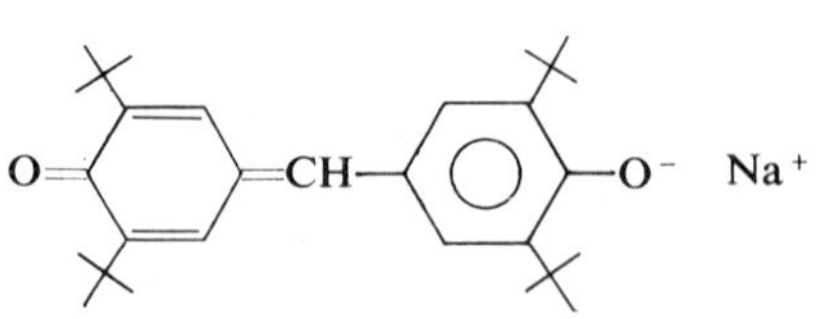

Sodium galvinoxide

Whatever the detailed mechanism through which the sodium benzophenone ketyl spectral shifts arise, they seem to reflect a solvent property we will simply term "polarity toward organosodium compounds." A number of chemical and physical properties follow the same solvent trend as the ketyl spectral shifts. According to this trend, the order of "polarity toward organosodium compounds" of the various ethers is

diglyme > DME > THF > MeTHF > dioxane > DEE

E. Specific Ion Pair Solvation

Hogen-Esch and Smid discovered evidence for different ion pair specific solvates ("solvent-separated" and "contact" ion pairs, or "loose" and "tight" ion pairs) in the electronic spectra of alkali fluorenides,[21] providing direct

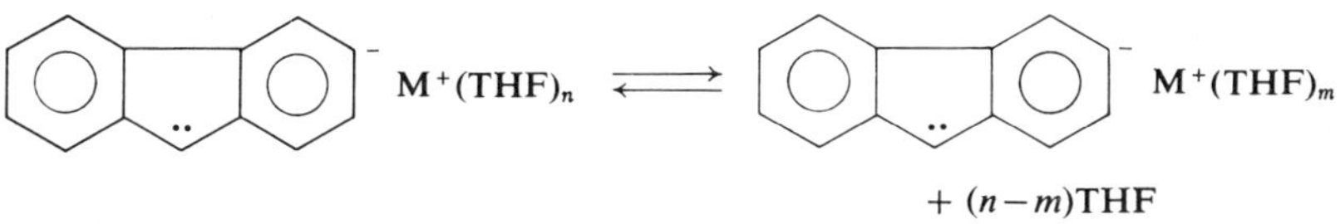

(24)

physical evidence for earlier speculations supported by indirect arguments.[22] Much progress has since been made in elaborating details of chemistry involving variations in specific cation solvation of organoalkali compounds in ethers.[1,22a,22b]

F. Disproportionation of Tetraphenylethylene Radical Anion, an Example of Large Solvent and Metal Ion Effects

An example of large solvent and metal ion effects on an anionic ET system is the disproportionation of tetraphenylethylene radical anion ($:TPE\cdot^{-}$).

$$2:TPE\cdot^{-}, M^{+} \rightleftharpoons :TPE:^{2-}, 2M^{+} + :TPE \qquad (25)$$

The symbol $:TPE\cdot^{-}, M^{+}$ does not represent an ion pair; $:TPE\cdot^{-}M^{+}$ would have that meaning here. Instead, where commas separate ions, the degree of aggregation is specifically unspecified. The apparent equilibrium constant for (25) is K_{app},

$$K_{app} = \frac{[:TPE:^{2-}, 2M^{+}][:TPE]}{[:TPE\cdot^{-}, M^{+}]^{2}} \qquad (26)$$

where $[:TPE\cdot^{-}, M^{+}]$ and $[:TPE:^{2-}, 2M^{+}]$ are the total concentrations of the ions $:TPE\cdot^{-}$ and $:TPE:^{2-}$ without regard to aggregation. Table 1 gives the variations of K_{app} with solvent and metal ion. It is seen that for sodium tetraphenylethylene, K_{app} spans a range of at least 10^{9}, going from HMPA to DEE.

TABLE 1

Solvent and Metal Ion Effect on the Disproportionation of Tetraphenylethylene Radical Anion[a]

M^+	HMPA[b]	Diglyme[c]	DME[c]	THF[c]	MeTHF[d]	DEE[e]
Li^+	—	—	0.8	1	—	—
Na^+	2×10^{-4}	—	0.2	20	>1000	>100,000
K^+	—	10	45	6000	—	—
Cs^+	—	300	2700	Large	—	—

[a] Values tabulated are K_{app} (equation 26). Concentrations are about $10^{-4}M$.
[b] From reference 24.
[c] From reference 25.
[d] From reference 26.
[e] From reference 26a.

G. Rationalization of Solvent and Metal Ion Effects

A qualitative, general view of solvent and metal ion effects is given here. Because highly specific ion–ion and ion–solvent interactions are treated as part of the general framework, and not taken specifically into account, there may be instances where this view fails. In spite of this, it is a useful approach. It is justified in detail elsewhere.[23]

Anions have differing cation affinities, differing abilities to attract cations. The cation affinity can be identified with the standard free energy change accompanying dissociation of a gas-phase ion pair. Anions with large, concentrated, sterically accessible charges have high cation affinities. Those with small, highly dispersed, or sterically inaccessible charges have low cation affinities.

Consider an equilibrium between anion A_1^- and anion A_2^-.

$$A_1^- + B \rightleftharpoons A_2^- + C \tag{27}$$

This could be a true chemical equilibrium, or it could be a virtual equilibrium between reactants and activated complexes. In the latter case, a rate constant rather than an equilibrium constant would be under discussion.

Let A_1^- have a larger cation affinity than A_2^-. Let K_{27}° be the equilibrium constant governing (27) for the unassociated ions in the gas phase. Ion association would shift the equilibrium toward A_1^-, the anion of larger cation affinity. Thus, if A_1^- and A_2^- were associated with counterions, K_{27} would be expected to be smaller than K_{27}°.

Now consider the aggregated system in solution. In ionic aggregates, the anions encumber cation solvation. Further, anions of higher cation affinity encumber the solvation of associated counterions more than do anions of lower cation affinity. Thus, more polar solvents will tend to shift equilibria toward anions of lower cation affinity, so that the metal counterions can be better solvated, lowering the free energy.

Metal ion effects can go in either direction. In solvents of low polarity, smaller cations are expected to shift equilibria toward anions of higher cation affinity, but in solvents of higher polarity, this effect can be reversed. For example, in MeTHF, Li^+ favors the disproportionation of stilbene radical anion, relative to Na^+, but in THF, Na^+ favors the disproportionation of :TPE$\cdot^-$, relative to Li^+.[25,27]

These principles apply successfully to the disproportionation of :TPE$\cdot^-$, where :TPE:$^{2-}$ is clearly the anion of higher cation affinity. More polar solvents shift the equilibrium to favor :TPE$\cdot^-$, the anion of lower cation affinity (Table 1).

Where the ions are free, not associated, and where there might be expected to be significant direct anion solvation, an appropriate modification of the approach outlined here would be required.

IV. SOME ELECTRON TRANSFER REACTIONS

A. ED–EA Reactions

1. Thermodynamics

Because either process (ED, EA) is the reverse of the other when the same species are involved, energy changes accompanying electron attachment and electron detachment are equal in magnitude but opposite in sign. Knowledge of the thermodynamics of EA for a list of compounds, using any constant electron source, is sufficient to establish ED–EA equilibrium constants.

$$\begin{array}{rcl@{\qquad}c}
:\mathrm{Org}_1 + [e^-] & \rightleftharpoons & :\mathrm{Org}_1\cdot^- & \Delta G_1^\circ \\
:\mathrm{Org}_2\cdot^- & \rightleftharpoons & :\mathrm{Org}_2 + [e^-] & -\Delta G_2^\circ \\
\hline
:\mathrm{Org}_1 + :\mathrm{Org}_2\cdot^- & \rightleftharpoons & :\mathrm{Org}_1\cdot^- + :\mathrm{Org}_2 & \Delta G_1^\circ - \Delta G_2^\circ
\end{array} \tag{28}$$

Similar relationships apply to other thermodynamic quantities. Table 2 gives some gas-phase electron affinities of some aromatic hydrocarbons and carbonyl compounds. In solution, effects of ion association and solvation must be taken into account.[1] These may, in some cases, invert the apparent electron affinities of a pair of compounds.

Electron affinities tend to increase with increasing extension of π networks. In addition, the presence of electronegative atoms and groups enhances electron affinities. The effects of carbonyl groups are illustrated above, and nitro groups have even more pronounced effects.

TABLE 2

Gas-Phase Electron Affinities[a]

Naphthalene	3.5 (kcal/mole)
Phenanthrene	7.1
Anthracene	12.7
Benz[*a*]anthracene	16.0
Benzaldehyde	9.7
Naphthaldehyde-1	17.2
Phenanthrene aldehyde-9	15.1

[a] Selected from references 28–30.

2. *Scope*

In a paper defining the scope of reactions of carbanions and nitranions with unsaturated organic compounds, Russell et al.[31] give many literature references to radical-producing net ET reactions. Many others are given by Bilevich and Okhlobystin.[5] The recognition of such processes can be traced to Schlenk, who reported that tritylsodium reacts with benzophenone to give trityl radical and benzophenone ketyl.[32]

$$Ph_3C:^-, Na^+ + Ph_2C{=}O \rightleftarrows Ph_3C\cdot + Ph_2\dot{C}{-}O^- \qquad (29)$$

In DMSO–*t*-BuOH mixtures containing *t*-BuOK and nitrobenzene, and in ethanol containing sodium ethoxide, Russell and co-workers detected radical anions from 39 substrates in the absence of oxygen.[31] These include relatively acidic hydrocarbons, nitriles, ketones, alcohols, phenols, thiophenols, mercaptans, nitroalkanes, and some miscellaneous compounds. Azobenzene, fluorenone, and other compounds (often aromatic) containing —N=N—, >C=C<, >C=N—, >C=O, and —N=O functionalities of various types were found to be suitable acceptors of electrons.

Alkyllithiums react with azobenzene, ketones, polynuclear aromatic hydrocarbons, diphenylacetylene, and so on, to yield radical anions corresponding to these acceptors.[31,33,34] Such reactions may be catalyzed by light.[35]

3. *Mechanisms*

It is probable that many base-catalyzed reactions of compounds such as those mentioned above involve initial ET. An example is the base-catalyzed condensation of nitrosobenzene and phenylhydroxylamine.[36] In ethanol containing sodium ethoxide, azoxybenzene is produced in up to 96% yields. Conventional heterolytic mechanisms can be written, but ESR measurements in a flow system reveal nitrosobenzene radical anion, which rapidly vanishes when the flow is stopped, following second-order kinetics. In DMSO–

t-BuOH, the nitrosobenzene radical anions do not rapidly decay to product. This is consistent with the following mechanism:

$$\longrightarrow 2\mathrm{PhNO}\cdot^- \rightleftarrows \underset{^-\mathrm{O}\quad \mathrm{O}^-}{\mathrm{PhN{-}NPh}} \xrightleftharpoons{\pm[\mathrm{H}^+]} \underset{\mathrm{HO}\quad \mathrm{O}^-}{\mathrm{PhN{-}NPh}} \longrightarrow \underset{\mathrm{O}^-}{\mathrm{PhN{=}\overset{+}{N}Ph}} + \mathrm{OH}^- \qquad (30)$$

In DMSO–*t*-BuOH the leftmost equilibrium remains stable, lying to the left. In ethanol, the remainder of reaction 30 occurs. Unfortunately, the results do not rule out the possibility that $PhN(OH)N(O^-)Ph$ is first formed through a conventional heterolytic condensation, followed by the rapid establishment of the equilibria of (30). This kind of question plagues investigations of mechanisms of ET reactions. Equation 31 represents an ED–EA rationale for the initial formation of $PhNO\cdot^-$.

$$\mathrm{PhNHOH} \xrightarrow{2\mathrm{OH}^-} \mathrm{PhNO}^{2-} \xrightarrow[\mathrm{ED{-}EA}]{\mathrm{PhNO}} 2\mathrm{PhNO}\cdot^- \qquad (31)$$

4. *Rates*

Thermoneutral and exothermic ED–EA reactions between radical anions and neutral species are usually very fast, most rate constants being in the range 10^7–10^{10} M^{-1} sec^{-1}.

$$:\mathrm{Bzph}\cdot^-\mathrm{Na}^+ + :\mathrm{Bzph} \xrightarrow[25^\circ]{\mathrm{DME}} :\mathrm{Bzph} + :\mathrm{Bzph}\cdot^-\mathrm{Na}^+ \qquad k = 10^8\ M^{-1}\ \mathrm{sec}^{-1} \qquad (32)^{37}$$

$$:\mathrm{Anth}\cdot^-\ (\text{free ion}) + :\mathrm{Anth} \xrightarrow{\mathrm{DMF}} :\mathrm{Anth} + :\mathrm{Anth}\cdot^- \qquad k = 5\times 10^8\ M^{-1}\ \mathrm{sec}^{-1} \qquad (33)^{38}$$

$$:\mathrm{Biph}\cdot^- + :\mathrm{Naph} \xrightarrow[25^\circ]{i\text{-}\mathrm{PrOH}} :\mathrm{Biph} + :\mathrm{Naph}\cdot^- \qquad k = 3\times 10^8\ M^{-1}\ \mathrm{sec}^{-1} \qquad (34)^{39,40}$$

$$:\mathrm{Biph}\cdot^- + :\mathrm{Anth} \xrightarrow[25^\circ]{i\text{-}\mathrm{PrOH}} :\mathrm{Biph} + :\mathrm{Anth}\cdot^- \qquad k = 6\times 10^9\ M^{-1}\ \mathrm{sec}^{-1} \qquad (35)^{39,40}$$

For various electron transfer reactions

$$:\mathrm{Arom}_1\cdot^- + :\mathrm{Arom}_2 \xrightarrow{i\text{-}\mathrm{PrOH}} :\mathrm{Arom}_1 + :\mathrm{Arom}_2\cdot^- \qquad (36)$$

the rate constants range from about 6×10^9 ($\Delta G^\circ = -15$ kcal/mole) to about 2×10^7 M^{-1} sec^{-1} ($\Delta G^\circ = +2.6$ kcal/mole).[40] The data offer limited support for Marcus' relationship among $\Delta G^\ddagger$, ΔG°, electrostatic work, and solvent parameters.[41]

Ion aggregation and accompanying solvent effects operate in ED–EA reactions.

$$(:\mathrm{Bzph}\cdot^-)_2(\mathrm{Na}^+)_2 + :\mathrm{Bzph} \xrightarrow[24^\circ]{\mathrm{MeTHF}} :\mathrm{Bzph} + (\mathrm{Bzph}\cdot^-)_2(\mathrm{Na}^+)_2 \qquad k < 10^6\ M^{-1}\ \mathrm{sec}^{-1} \qquad (37)^{42}$$

The difference in reactivities of free ions and ion pairs in such reactions is frequently distinct, but often not great.

It follows from these considerations that thermoneutral and exothermic ED–EA steps are probably not often rate determining for reactions whose overall rates are not very large. Ion aggregation, however, can affect this consideration.

5. Radical Anion Disproportionation

a. Gas Phase. In the gas phase for free anions, it can be expected that radical anion disproportionations (equation 38) will lie to the left at equilibrium.[43]

$$2{:}\mathrm{Org}\cdot^{-} \rightleftharpoons {:}\mathrm{Org}{:}^{2-} + {:}\mathrm{Org} \tag{38}$$

This is due to the uncompensated electron–electron repulsions in $:\mathrm{Org}:^{2-}$. Ion association can, however, provide appropriate compensation, so that the disproportionation of ion pairs to neutral triple ions is much less unfavored than the disproportionation of free ions, and it can even be favored, as in the cases of NaTPE[44] and sodium cyclooctatetraene.[44a–c] In both cases the free radical ions do not spontaneously disproportionate, but ion aggregates do.

$$2\mathrm{M}^{+}{:}\mathrm{Org}\cdot^{-} \rightleftharpoons \mathrm{M}^{+}{:}\mathrm{Org}^{2-}\mathrm{M}^{+} + {:}\mathrm{Org} \qquad K_{39} > K_{38} \tag{39}$$

b. Ion Association and the Disproportionation of Tetraphenylethylene Radical Anion. Consider the disproportionation of $:\mathrm{TPE}\cdot^{-}$ in some detail (equations 40).

$$\begin{array}{ccc}
2\mathrm{M}^{+} + 2{:}\mathrm{TPE}\cdot^{-} & \overset{K_{40a}}{\rightleftharpoons} & {:}\mathrm{TPE}{:}^{2-} + 2\mathrm{M}^{+} + {:}\mathrm{TPE} \quad (40a) \\
1/K_d({:}\mathrm{TPE}\cdot^{-}\mathrm{M}^{+}) \updownarrow & & \updownarrow 1/K_d({:}\mathrm{TPE}^{2-}\mathrm{M}^{+}) \\
\mathrm{M}^{+} + {:}\mathrm{TPE}\cdot^{-} + {:}\mathrm{TPE}\cdot^{-}\mathrm{M}^{+} & \overset{K_{40b}}{\rightleftharpoons} & {:}\mathrm{TPE}{:}^{2-}\mathrm{M}^{+} + \mathrm{M}^{+} + {:}\mathrm{TPE} \quad (40b) \\
1/K_d({:}\mathrm{TPE}\cdot^{-}\mathrm{M}^{+}) \updownarrow & & \updownarrow 1/K_d(\mathrm{M}^{+}{:}\mathrm{TPE}{:}^{2-}\mathrm{M}^{+}) \\
2{:}\mathrm{TPE}\cdot^{-}\mathrm{M}^{+} & \overset{K_{40c}}{\rightleftharpoons} & \mathrm{M}^{+}{:}\mathrm{TPE}{:}^{2-}\mathrm{M}^{+} + {:}\mathrm{TPE} \quad (40c)
\end{array}$$

Equations 41 follow from equations 40.

$$K_{40b} = K_{40a}\left[\frac{K_d({:}\mathrm{TPE}\cdot^{-}\mathrm{M}^{+})}{K_d({:}\mathrm{TPE}{:}^{2-}\mathrm{M}^{+})}\right] \tag{41a}$$

$$K_{40c} = K_{40b}\left[\frac{K_d({:}\mathrm{TPE}\cdot^{-}\mathrm{M}^{+})}{K_d(\mathrm{M}^{+}{:}\mathrm{TPE}{:}^{2-}\mathrm{M}^{+})}\right] \tag{41b}$$

Each of the ratios in brackets in equations 41 is expected to be large, the one in equation 41a being especially large. Thus, the equilibrium constants K_{40} should be progressively larger as one moves down equations 40, toward higher aggregation of ions. More polar solvents favor aggregation situations toward the top of equations 40, so that K_{app} would, in the limit, resemble K_{40a}. Less polar solvents favor situations toward the bottom of equations 40, so that K_{app} tends to resemble K_{40c}, which should be larger than K_{40a}. To a first approximation, one might ignore the solvent effect on K_{40a}; this is probably not well justifiable.[43] The data of Szwarc and co-workers then bear out the expectations:

In HMPA[24]: $K_{40a} = 2 \times 10^{-4}$ sodium counterion

In THF[45]: $K_{40b} = 3.3$ $K_{40c} = 400$ sodium counterion

A higher tendency toward disproportionation is calculated for :TPE$\cdot^-$ than for most other hydrocarbon radical anions, even if both :TPE$\cdot^-$ and :TPE:$^{2-}$ are treated as planar molecules.[43] It is an attractive hypothesis that :TPE:$^{2-}$ is not essentially planar, but that it consists of two linked benzhydryl anion units with their rough internal planes not coincident, but perhaps

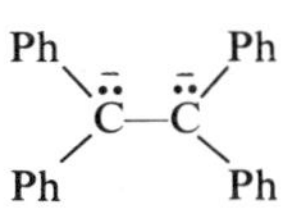
"Planar"

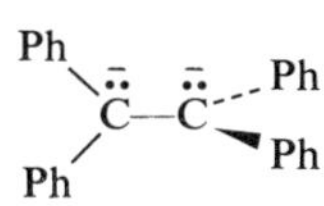
"Twisted"

(Probable "propellering" of phenyl groups through twists about the C—Ph bonds is understood.)

perpendicular.[1,26,42] The decrease in dianion free energy accompanying such a deformation from an essentially planar structure would be an additional factor favoring disproportionation. Such a deformation could explicitly account for the disproportionation enhancing effect of an α-methyl substituent on stilbene radical anion, and it finds at least rough analogy in similar deformations proposed for stilbene triplets.[26] However, there is no direct, commanding evidence relevant to the structure of :TPE:$^{2-}$.[46] The question is open.[46a]

c. Dianions as Intermediates in Radical Anion Reactions; Protonations of Radical Anions. Although :TPE$\cdot^-$ has a higher tendency to disproportionate than many radical anions, the same factors will tend to make disproportionation possible for all radical anions. For lithium naphthalene in dioxane, for example, :Naph:$^{2-}$ dominates even when :Naph is present.[47] Because of this, one cannot safely simply assume that a reaction of a radical anion with a reagent Z does not proceed through the related dianion (equation 42).

$$2\text{:Org}\cdot^-, \text{M}^+ \rightleftharpoons \text{:Org} + \text{:Org:}^{2-}, 2\text{M}^+ \xrightarrow{\text{Z}} \qquad (42)$$

An example is the reaction of perylene radical anions (:Pery$\cdot^-$) with *t*-BuOH in THF.[48] The reaction is second order in :Pery·⁻ and it is retarded by :Pery. The following scheme accounts for these facts:

$$2\text{:Pery}\cdot^{-}\text{Na}^{+} \rightleftarrows \text{Na}^{+}\text{:Pery:}^{2-}\text{Na}^{+} + \text{:Pery} \qquad (43a)$$

$$\text{Na}^{+}\text{:Pery:}^{2-}\text{Na}^{+} + t\text{-BuOH} \longrightarrow \text{protonation products} \qquad (43b)$$

In contrast to these results, the protonation of naphthalene radical anion by water in THF is first order in :Naph·⁻,[12] consistent with the following mechanism of protonation, first suggested by Paul et al.[49]:

$$\text{:Naph}\cdot^{-}\text{Na}^{+} + H_2O \longrightarrow \text{HNaph}\cdot + \text{NaOH} \qquad (44a)$$

$$\text{HNaph}\cdot + \text{:Naph}\cdot^{-}\text{Na}^{+} \longrightarrow \text{HNaph:}^{-}\text{Na}^{+} + \text{:Naph} \qquad (44b)$$

$$\text{HNaph:}^{-}\text{Na}^{+} + H_2O \longrightarrow \text{HNaphH} + \text{NaOH} \qquad (44c)$$

The differences between the two systems may be due to the different thermodynamics of disproportionation of the radical anions and the different acidities of the proton donors employed.

Szwarc has discovered several other systems in which a dianion, formed by radical anion disproportionation or a related process, is a key intermediate.[48]

B. Reactions of Naphthalene Radical Anion with Alkyl Halides[50]

1. Products and Stoichiometry

The overall reactions of alkali naphthalenes with alkyl halides can be regarded as sums of four subreactions: reduction, bimolecular reduction, monoalkylation, and dialkylation. In the following equations, counterions are omitted.

Reduction:

$$\text{R—X} + 2\text{:Naph}\cdot^{-} + [\text{H}^{+}] \longrightarrow \text{RH} + 2\text{:Naph} + \text{X}^{-} \qquad (45)$$

Bimolecular reduction:

$$2\text{R—X} + 2\text{:Naph}\cdot^{-} \longrightarrow \text{R—R} + 2\text{:Naph} + 2\text{X}^{-} \qquad (46)$$

(or RH + related olefin)

Monoalkylation:

$$\text{R—X} + 2\text{:Naph}\cdot^{-} + [\text{H}^{+}] \longrightarrow \text{RNaphH} + \text{:Naph} + \text{X}^{-} \qquad (47)$$

Dialkylation:

$$2\text{R—X} + 2\text{:Naph}\cdot^{-} \longrightarrow \text{RNaphR} + \text{:Naph} + 2\text{X}^{-} \qquad (48)$$

The alkylation products are mostly derivatives of 1,2- and 1,4-dihydronaphthalenes.[51,52]

The proportion of bimolecular reduction products is very sensitive to the halogen of the alkyl halide. Iodides give large yields (~70%), bromides give about 10%, chlorides about 1%, and fluorides 0%.

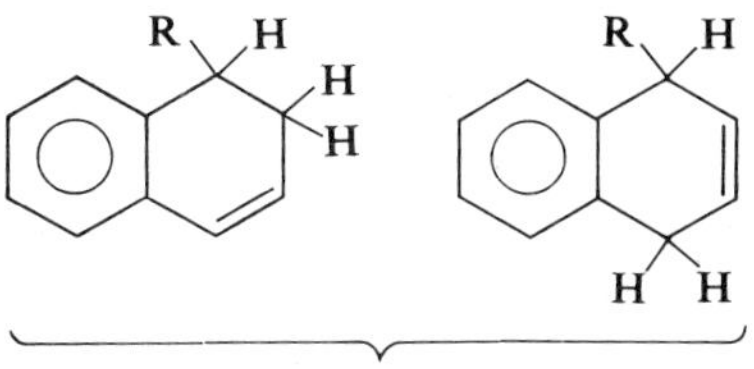

RNaphH

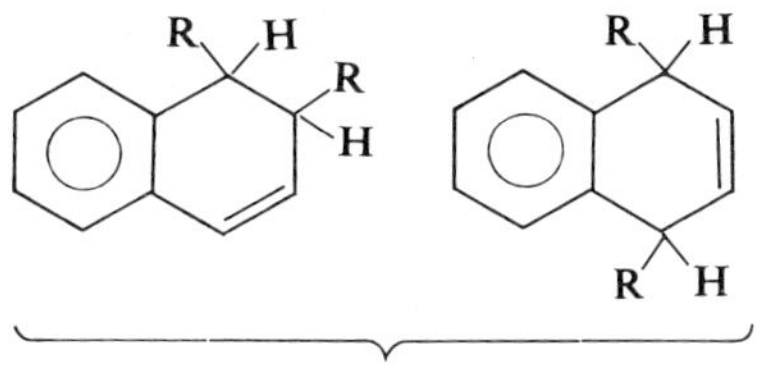

RNaphR

2. *Mechanisms of Formation of Reduction and Alkylation Products*

From 5-hexenyl halides, methylcyclopentane and 1-hexene are both formed, the latter predominating.[50,53,54] The cyclization suggests the intermediacy of 5-hexenyl radicals,[55] which cyclize to cyclopentylmethyl radicals with a rate constant of about 10^5 sec^{-1},[56] but the concomitant formation of larger quantities of 1-hexene demands that such intermediate radicals suffer some fate, leading eventually to 1-hexene, other than reaction with solvent, which is too slow compared to the cyclization to permit formation of 1-hexene by that route.[53,55]

When bimolecular reduction and dialkylation are eliminated or suppressed, the relative amounts of reduction and alkylation products formed are independent of the halogen through a series of primary alkyl halides.[50,57–59] This suggests that the halogens are not significantly bound into the transition states which determine the partitioning into these classes of products.

The evidence cited thus far, and other data, are accommodated by equations 49 (counterions omitted).

$$:\text{Naph}\cdot^- + \text{R—X} \longrightarrow :\text{Naph} + \text{X}^- + \text{R}\cdot \qquad \text{(ED–DEA)} \qquad (49a)$$

$$\text{R}\cdot \longrightarrow \text{R}'\cdot \qquad \text{(Radical cyclization or rearrangement, if applicable)} \qquad (49b)$$

$$\text{R}\cdot + :\text{Naph}\cdot^- \longrightarrow \text{R}:^- + :\text{Naph} \qquad \text{(ED–EA)} \qquad (49c)$$

$$\text{R}\cdot + :\text{Naph}\cdot^- \longrightarrow \text{RNaph}:^- \qquad (49d)$$

$$\text{R}:^- + CH_3OCH_2CH_2OCH_3 \longrightarrow \text{RH} + CH_3O^- + CH_2{=}CHOCH_3 \qquad (49e)$$

$$\text{RNaph}:^- + [\text{H}^+] \longrightarrow \text{RNaphH} \qquad (49f)$$

Where applicable, analogous equations apply for R′. This scheme accounts for the incomplete cyclization; reaction 49b competes with (49c) and (49d). The distribution of alkylation and reduction products is determined by the competition between (49c) and (49d). Halogens are not bound into reactants or products of either of these reactions, which is consistent with the observed lack of influence of the nature of the halogen on the reduction–alkylation product distribution.

The competition between (49b) and (49c, d) should be dependent on

[:Naph·⁻] since (49b) does not involve :Naph·⁻ whereas (49c) and (49d) do. This effect was documented for the reaction of 5-hexenyl fluoride.[54]

Sargent and Lux found that the amount of alkylation of sodium naphthalene by alkyl iodides *increases* in the order neopentyl < *n*-pentyl < *sec*-pentyl < *tert*-pentyl.[60] This would not be intelligible if the initial step of alkylation were an S_N2 reaction of :Naph·⁻ with RI, but it is easily understood in terms of the competition (49c) and (49d).

Bank and Bank interpreted the formation of Grignard reagents in reactions of sodium naphthalene with alkyl halides in THF containing $MgBr_2$ in terms of a modification of equations 49.[61]

$$\text{R—X} \xrightarrow{:\text{Naph}\cdot^-} \text{R}\cdot \xrightarrow{:\text{Naph}\cdot^-} \text{R}:^- \xrightarrow{\text{MgBr}_2} \text{RMgBr} + [\text{Br}^-] \qquad (50)$$

This amounts to evidence for the intermediacy of alkyl anions (alkylsodiums), since it is believed that $MgBr_2$ cannot compete with :Naph·⁻ for R·.[61] The method may be synthetically useful in the preparation of Grignard reagents which are otherwise difficult to prepare. It has been extended to the preparation of other organometallics.[62]

Tetrahydrofurfuryl halides act as their own carbanion traps. The opening

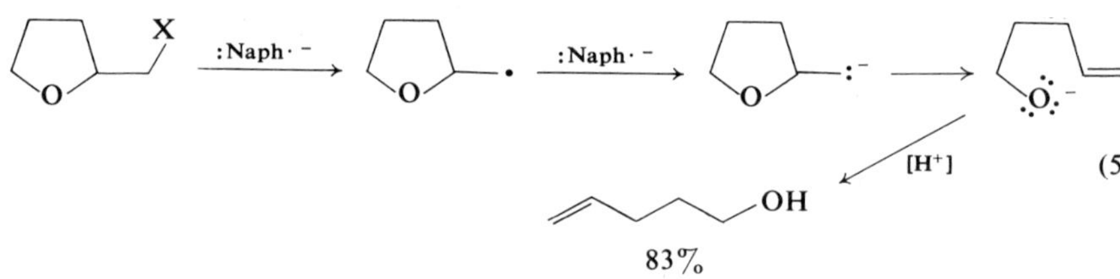

(51)

of tetrahydrofurfuryl radicals (equation 52) is thermodynamically unfavored, so this reaction cannot account for the observed ring opening.

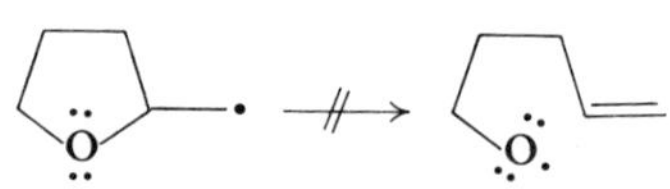

(52)

On the other hand, the facile fragmentation of carbanions $\text{RO}-\overset{|}{\underset{|}{\text{C}}}-\overset{|}{\underset{|}{\text{C}}}:^-$ is well known; this includes the opening of tetrahydrofurfuryl anions.[63] No methyltetrahydrofuran was detected in these reactions,[50,53] so tetrahydrofurfuryl radicals cannot react with solvent to a measurable extent. Instead, they must be scavenged by :Naph·⁻.

The second alkylation step is most likely a conventional S_N2 process,[50,60] though the evidence is scant.

$$\text{RNaph}:^- + \text{R—X} \xrightarrow[S_N2]{} \text{RNaphR} + \text{X}^- \qquad (53)$$

The mechanism of formation of alkyl dimers and disproportionation products (bimolecular reduction) is considered in a later section.

3. *Reactions of Naphthalene Radical Anion with Alkyl Radicals*

a. Rates. Garst and Barton determined the ratio $(k_{49c} + k_{49d})/k_{49b}$ to be about 2×10^4 M^{-1}.[54] Since $k_{49b} = \sim 10^5$ sec^{-1},[56] this leads to $(k_{49c} + k_{49d}) = \sim 2 \times 10^9$ M^{-1} sec^{-1}. Thus, reactions of alkyl radicals (at least primary ones) with sodium naphthalene in DME are nearly diffusion controlled. Vinyl[63a] and cyclopropyl[63b] radicals appear to react with sodium naphthalene at rates competitive with their rates of changes of configuration.

b. Competitive Pathways, ED–EA and Radical Combination. Such reactions take two competitive pathways, ED–EA (49c) and radical combination (49d). This situation is analogous to the conventional combination–disproportionation competition for reactions of alkyl radicals, the ED–EA reaction (which can be viewed as a sodium atom transfer) being likened to the disproportionation (hydrogen atom transfer). Equations 54 illustrate these comparisons.

(54a)

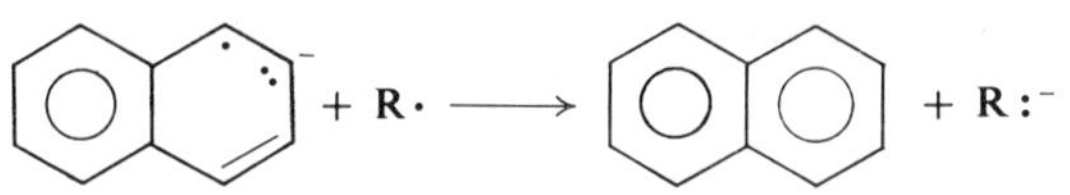

(54b)

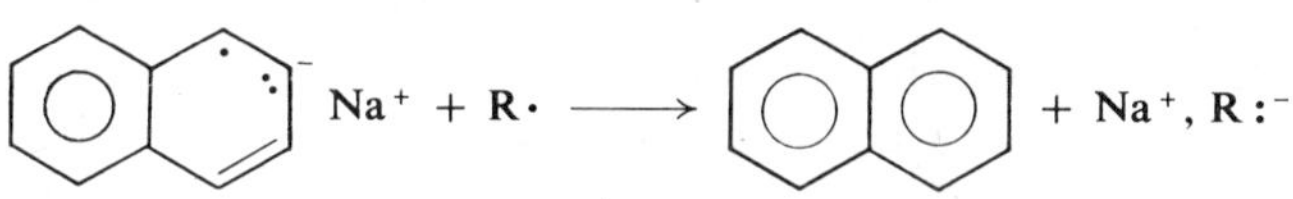

(54c)

The ratio k_{49c}/k_{49d} determines the partitioning of initially formed alkyl radicals into reduction and alkylation products. For primary alkyl radicals (5-hexenyl, pentyl, hexyl, 4-iodobutyl, 4-bromobutyl, 4-chlorobutyl, 5-iodopentyl, 5-bromopentyl, 5-chloropentyl) this ratio is well established as 1.4 in DME. For other radicals, k_{49c}/k_{49d} is less well established, but the data of Sargent and Lux[60] can be analyzed, with some assumptions, to give the following values:

	Neopentyl	*n*-Pentyl	*sec*-Pentyl	*tert*-Pentyl
k_{49c}/k_{49d}:	4.8	1.8	0.85	0.39

As an indication of the possible error in these values, 1.8 derived for neopentyl is to be compared with 1.4 established by more direct means for a variety of similar radicals. The trend is clear, and the values are probably reasonably correct.

Perhaps the similarities among the values of k_{49c}/k_{49d} are of more significance than the differences. The differences are small, consistent with very rapid competitive processes involving transition states which come early along the reaction coordinates. The trend could be a reflection of the thermodynamics of ED–EA in the kinetics of reaction 49c. Conventionally, tertiary carbanions are supposed to be less "stable" than primary ones, so that ED–EA (49c) might be inhibited for tertiary radicals. The larger amount of reduction for neopentyl suggests that a steric effect, inhibiting reduction less than combination, could be operative. However, the variations are so slight that they do not provide very firm grounds for such "explanations" as these.

Tetrahydrofurfuryl radicals apparently react with sodium naphthalene to give about 83% reduction.[50] This is slightly more than is obtained from the sterically similar cyclopentylmethyl radicals (~50–60%). This, again, could be a reflection of the thermodynamics of ED–EA; the alkoxy function in the tetrahydrofurfuryl radical would be expected to enhance the electron affinity of the radical, compared to cyclopentylmethyl. Again, the effect is too small to justify an authoritative claim that this expectation is borne out by experiment.

c. One-Step ED–EA versus Radical Combination Followed by Bond Heterolysis. While it is intuitively unattractive to imagine that net ED–EA reactions of alkyl radicals with :Naph·⁻ occur by a sequence involving radical combination and bond heterolysis, it is nonetheless satisfying that there is

$$\text{R}\cdot + :\text{Naph}\cdot^{-} \longrightarrow \text{R—Naph}:^{-} \longrightarrow \text{R}:^{-} + :\text{Naph} \qquad (55)$$

experimental evidence against this possibility. 1,4-Diiodo-, 1,4-dibromo-, and 1,4-dichlorobutanes react with sodium naphthalene to give the same relative amounts of reduction and alkylation products. If equation 55 applied, a halogen effect on the reduction–alkylation distribution would have been expected; it would have arisen through the competition of (56b) and (56c), where (56b), but not (56c), should be markedly affected by the nature of X.

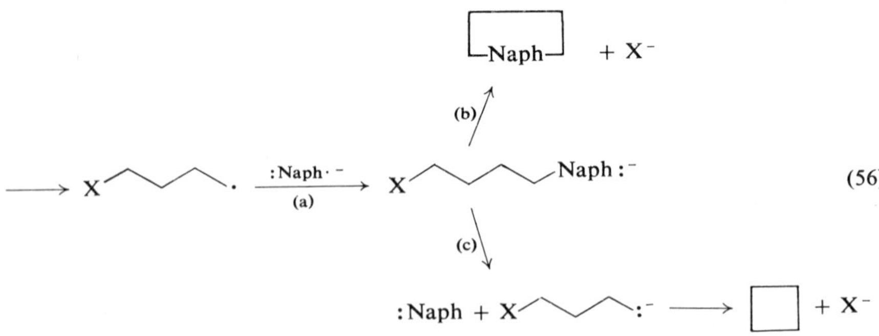

(56)

d. Solvent and Metal Ion Effects. There are small solvent and metal ion effects on the reduction–alkylation competition (Table 3). In DME and THF,

TABLE 3

Reduction–Alkylation Ratio for Primary R·[a]

M^+	DME	THF
Li^+	1.5[b]	1.2[b]
Na^+	1.4[b]	0.7[c]
Cs^+	0.7[c]	0.7[c]

[a] Based on reactions of alkali naphthalenes with 1,4-dihalobutanes and 1,5-dihalopentanes.[50]
[b] "Solvent-separated" ion pairs.
[c] "Contact" ion pairs.

the results form a dichotomy; where "solvent-separated" ion pairs predominate, k_{49c}/k_{49d} is about 1.4, but where "contact" ion pairs are believed to be the dominant species, the ratio is 0.7 (see Table 3). The small magnitudes of the effects again suggest that the transition states come very early along the reaction coordinates.

e. Other Radical Anions. Little definitive work has been done with radical anions other than :Naph·⁻; :Biph·⁻ seems to behave in an exactly parallel fashion[64]; :Anth·⁻ and :Bzph·⁻ give very little reduction product on reaction with alkyl halides.[65,66] Assuming that these reactions proceed according to a scheme such as (49), this suggests that these radical anions nearly always combine with alkyl radicals rather than undergo ED–EA reactions with them. It may be that the ED–EA reactions are thermodynamically unfavored in these cases; certainly they are less favored than for :Naph·⁻.

4. Mechanism of Formation of Bimolecular Reduction Products

Since neopentyl iodide reacts with sodium naphthalene to give bineopentyl in high yield (~70%), equations 57 seem inapplicable.[60] Neopentyl halides

$$\begin{array}{ccccccc} RX & \xrightarrow[(a)]{:Naph\cdot^-} & R\cdot & \xrightarrow[(b)]{:Naph\cdot^-} & R:^- & \xrightarrow[S_N2]{RX} & R{-}R \\ & & & & \Big\downarrow {\scriptstyle (d)\ DME} & {\scriptstyle (c)} & \\ & & & & RH & & \end{array} \quad (57)$$

are notoriously sluggish in conventional S_N2 reactions like (57c). On other, related grounds, Sargent and co-workers dismissed equations 57 and proposed equations 58.[67] This was before it was demonstrated that (58c) could not compete with (58d). Since k_{58d} is about $2 \times 10^9\ M^{-1}\ \text{sec}^{-1}$, since :Naph·⁻ is present in relatively high concentrations during most of the reaction, and since k_{58b} cannot be much greater than k_{58d}, scheme 58 loses a great deal of its attractiveness.

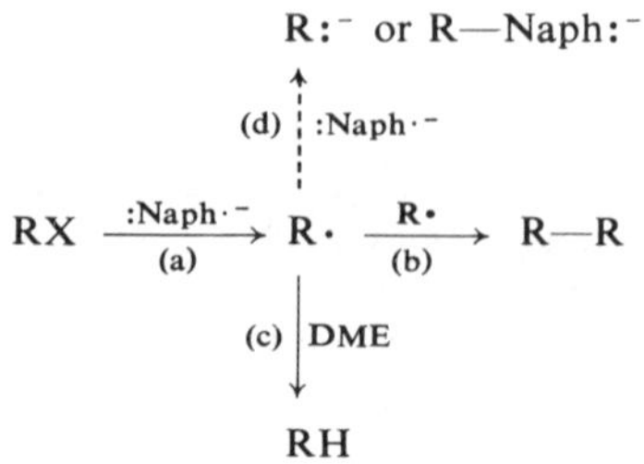

(58)

Scheme 58 has failed several specific tests. First, reactions of 1,4-diiodobutane and 1,5-diiodopentane give no products derived from dimerization or disproportionation of ω-iodoalkyl radicals. Instead, scheme 59 accounts for the products, where the reactions indicated with dashed arrows do not occur.

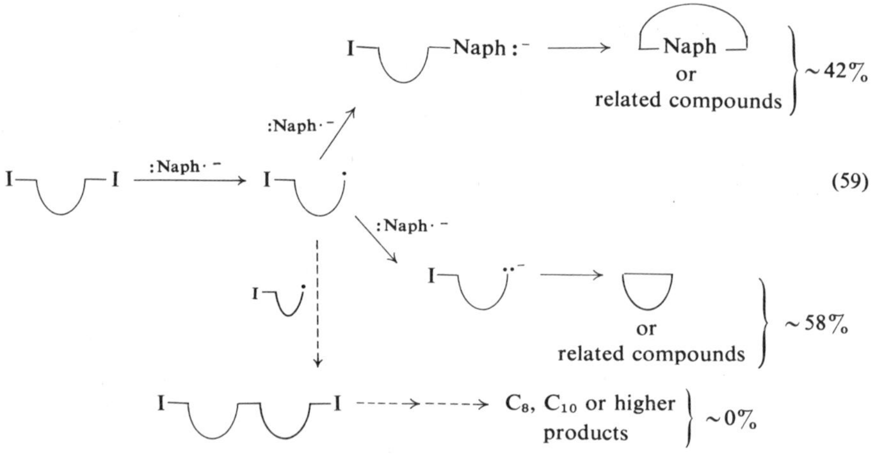

(59)

The reduction–alkylation ratio is very close to the 1.4 expected for a primary alkyl radical, indicating that scheme 59 is perfectly consistent with the data.[50,57–59] This consistency with the "normal" reduction–alkylation ratio would not have been expected if the ω-iodoalkyl radicals had bridged structures or if they underwent internal S_H reactions (equation 60).[58] If reaction 60 occurs, its rate constant is smaller than about 10^6 sec^{-1}.

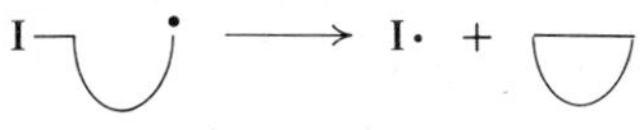
(60)

It is difficult to argue that 4-iodobutyl and 5-iodopentyl radicals are special in not coupling. A more satisfactory conclusion is that initially formed alkyl radicals do not couple in any reactions of sodium naphthalene with alkyl halides. Similar arguments apply to reactions of tetrahydrofurfuryl iodide with sodium naphthalene, from which no evidence of coupling of tetrahydrofurfuryl radicals was detected.[50]

The presence of $MgBr_2$ interferes with alkyl dimer formation when alkyl iodides react with sodium naphthalene in THF.[61] Interpreting the role of $MgBr_2$ as that of a carbanion trap, this interference is inconsistent with scheme 58.

Finally, reactions of alkyl iodides with sodium naphthalene give alkyl dimers which exhibit no CIDNP, provided the reactions are carried out in magnetic fields of thousands of gauss.[68] CIDNP would be expected if scheme 58 applied.

Thus, the evidence suggests that alkyl anions (alkylsodiums) are intermediates, but that products are not formed in S_N2 reactions of alkyl anions with alkyl halides. Rather, they seem most consistent with radical coupling–disproportionation, but the radicals involved are not the initially formed radicals. These requirements are met by scheme 49 with appended reaction 49g. This reaction has analogy in reactions of alkyllithiums with alkyl halides

$$R{:}^- + R\text{—}X \xrightarrow[\text{ED–DEA}]{} [R\cdot\ \cdot R] \longrightarrow R\text{—}R \text{ (or RH + olefin)} \qquad (49g)$$

(Sodium counterion omitted)

(see later). Although it is not clear precisely why (49g) would not give CIDNP, some of the related reactions, believed to proceed through a similar mechanism, also give no CIDNP. The CIDNP evidence weighs more heavily against scheme 58 than against scheme 49.

Scheme 49 [including (49g)] accounts for the halogen effects on the yields of alkyl dimers. Reaction 49g is slower, presumably, for halogens of lower atomic weight, but (49e), with no halogen bound to either reactant in any direct fashion, should be halogen independent, or nearly so.

5. *Kinetics of Radical Anion–Alkyl Halide Reactions*

Warhurst and co-workers studied the kinetics of reactions of some radical anions and dianions with alkyl and aryl halides.[71] Among radical anions derived from aromatic hydrocarbons, reasonable correlations of reaction rates with the thermodynamics of electron donation from the radical anions were found. In the few related reactions which have received detailed investigations (reactions of :Naph·⁻; to a lesser extent, reactions of :Bzph·⁻), the assumption of ED–DEA initial steps for these reactions has been supported. They are first order in each reactant, with pronounced solvent and metal ion effects on the rates. For reactions of benzophenone ketyl, more polar ethers give higher rates. This suggests that the ED–DEA transition states for these reactions have smaller cation affinities than :Bzph·⁻. In the ketyl, the negative charge may be rather highly localized on the electronegative oxygen atom. It may be more highly dispersed in the transition states, being shared between oxygen and halogen (and the rest of the system). Larger alkali cations lead to larger reaction rates, implying that the aggregation effect outweighs the solvation effect as the metal ion is varied.

Second-order rate constants for the reactions of alkali naphthalenes with 5-hexenyl fluoride in DME at 25° are given in Table 4.[72] The metal ion effect

TABLE 4

Second-Order Rate Constants for Reactions of Alkali Naphthalenes with 5-Hexenyl Fluoride in DME at 25°

Metal ion	k (M^{-1} sec^{-1})
Li^+	540×10^{-4}
Na^+	5.7×10^{-4}
K^+	0.05×10^{-4}
Cs^+	$<0.01 \times 10^{-4}$ [a]
Cs^+	$30{,}000 \times 10^{-4}$ [b]

[a] No reaction was detected up to 120 hr.

[b] Reactant was 5-hexenyl chloride.

is in the *opposite* direction from that for reactions of alkali benzophenone ketyls. One cannot deduce from this the relative cation affinities of reactant and transition-state anions, since the metal ion effect is the sum of aggregation and solvation effects which operate in opposite directions.

There is a possible practical application of the metal ion effect in this system. Sodium biphenyl is commonly used as a reagent to generate halide ions quantitatively from organic halogen compounds. The inorganic halides are then quantitatively determined.[73] For simple, "unactivated" alkyl fluorides, sodium biphenyl in DME reacts at about the same rate as sodium naphthalene; quantitative release of organically bound halogen would take many hours. But lithium biphenyl reacts about 100 times faster, so that it could prove to be the reagent of choice in such cases.

Another interesting consequence of the metal ion effect is the influence on multiple reduction. 1,6-Dichlorohexane is reduced by sodium naphthalene in DME to 1-hexene and cyclohexane (combined yield, 31%) and hexane (9%). The first two are products of two-electron reductions and the latter is the product of a four-electron reduction. Lithium naphthalene leads to a dominance of hexane (40%), with only 5% two-electron reduction products.[74] Scheme 61 provides a means of accounting for these results. It must be assumed that k_{61a} is less sensitive to metal ion variation than k_{61b}, or that they vary in opposite directions. This is entirely reasonable.

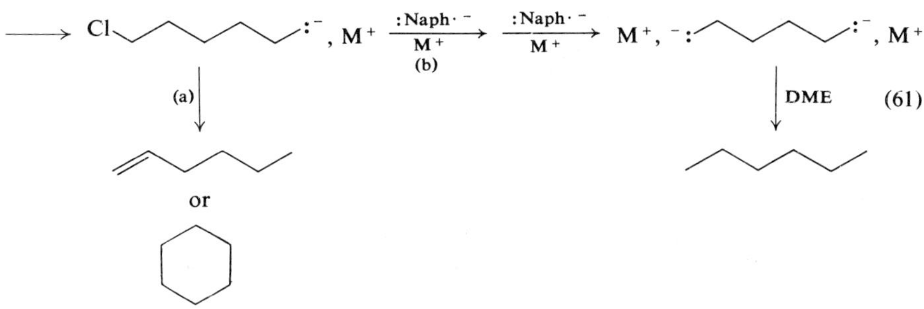

(61)

C. Anion Oxygenations

1. Oxidation of 2-Nitropropane

Many references to a voluminous literature of base-catalyzed oxidations are given in articles by G. A. Russell and co-workers.[75,76] In 1954 Russell noted that weakly basic solutions of 2-nitropropane are readily oxidized by molecular oxygen, but that strongly basic solutions are not. It appeared that this peculiarity must be connected with initiation of a chain process.[77,78] The correlation between ease of oxidation and spontaneous, base-catalyzed radical anion formation in the absence of oxygen[79] suggests that initiation involves ED–EA reactions of a conjugate base with the parent nitro compound. The following is a reasonable mechanism:

$$\mathrm{Me_2CHNO_2} \underset{\mathrm{H_2O}}{\overset{\mathrm{OH^-}}{\rightleftarrows}} \mathrm{Me_2C{=}NO_2^-} \qquad (62a)$$

$$\mathrm{Me_2C{=}NO_2^-} + \mathrm{Me_2CHNO_2} \longrightarrow \mathrm{Me_2CHNO_2\cdot^-} + \mathrm{Me_2\dot{C}NO_2} \qquad (62b)$$

$$\mathrm{Me_2\dot{C}NO_2} + \mathrm{O_2} \longrightarrow \mathrm{Me_2C(NO_2)OO\cdot} \qquad (62c)$$

$$\mathrm{Me_2CHNO_2\cdot^-} + \mathrm{O_2} \longrightarrow \mathrm{Me_2CHNO_2} + \mathrm{O_2\cdot^-} \qquad (62d)$$

As a higher concentration of base is employed, equilibrium 62a is shifted to the right, leaving less 2-nitropropane to participate in the initiation step 62b.

The conversion of $Me_2CH(NO_2)OO\cdot$ to products may follow equations 62e–62i.

$$Me_2C(NO_2)OO\cdot + Me_2C{=}NO_2^- \longrightarrow Me_2C(NO_2)OO^- + Me_2\dot{C}NO_2 \qquad (62e)$$

$$Me_2C(NO_2)OO^- + H_2O \rightleftharpoons Me_2C(NO_2)OOH + OH^- \qquad (62f)$$

$$Me_2C(NO_2)OOH + Me_2C{=}NO_2^- \longrightarrow Me_2C(NO_2)O^- + Me_2C(OH)NO_2 \qquad (62g)$$

$$Me_2C(NO_2)O^- \longrightarrow Me_2C{=}O + NO_2^- \qquad (62h)$$

$$Me_2C(OH)NO_2 + OH^- \longrightarrow Me_2C{=}O + NO_2^- + H_2O \qquad (62i)$$

Reactions 62c and 62e are propagation steps, and (62g–i) account for the ultimate products and stoichiometry. When the base concentration is very high, added nitrobenzene will initiate the reaction, playing the role ascribed to 2-nitropropane in reaction 62b.

2. Spin Conservation; Russell's General Chain Mechanism

Although oxygen does not react directly with the anion in the case of 2-nitropropane, it apparently does in some other anion oxidations. Citing a spin conservation rule, Russell has strongly criticized the view, recently prevalent, that anion oxidations involve one-step additions to oxygen.[75]

$$R{:}^- + O_2 \text{ (electronic triplet)} \longrightarrow ROO^- \text{ (singlet)} \qquad (63)$$

ED–EA is a much more attractive alternative. It has not been an easy matter, however, to establish (64), rather than (63), on experimental grounds. A

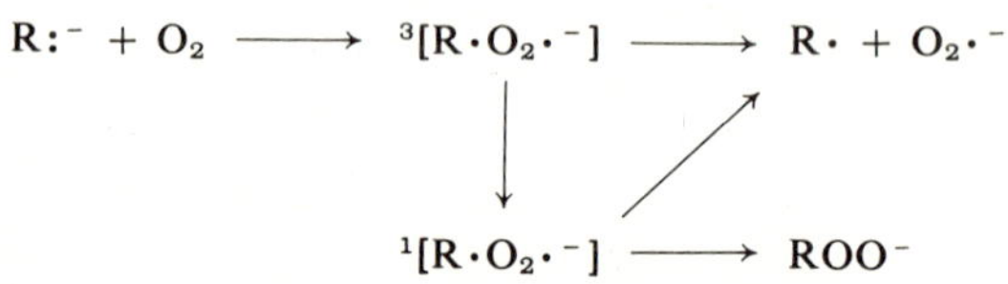

(64)

number of anion oxidations have been found to be catalyzed by nitroaromatics, suggesting chain mechanisms (equations 65). Strong similarities between the catalyzed and uncatalyzed reactions suggest that the uncatalyzed

$$R\cdot + O_2 \longrightarrow ROO\cdot \qquad (65a)$$

$$ROO\cdot + R{:}^- \longrightarrow ROO^- + R\cdot \qquad (65b)$$

processes are also chain reactions, probably initiated by reaction 64. Other processes, like the oxidation of triphenylmethane catalyzed by *t*-BuOK in HMPA or DMSO, show no evidence of being chain processes.[78]

3. *Hydroperoxide Salts*

Hydroperoxide anions (ROO^- or equivalent) are invariably invoked as intermediates in anion oxidations, but the final products are seldom hydroperoxides or their salts. An exception is the oxidation of triphenylmethane in HMPA.[78] In DMSO, however, the same oxidation gives triphenylmethoxide, presumably through the following reaction,

$$Ph_3COO^- + Me_2SO \longrightarrow Ph_3CO^- + Me_2SO_2 \qquad (66)$$

which resembles (62g) and a variety of other similar reactions.[80]

4. *Grignard Oxygenations*

The oxidation of Grignard reagents has commanded interest for many years Walling and Buckler confirmed the earlier suggestion of Porter and Steele (equations 67) with the finding that the hydroperoxide salts are the

$$RMgX + O_2 \longrightarrow ROOMgX \qquad (67a)$$

$$ROOMgX + RMgX \longrightarrow 2ROMgX \qquad (67b)$$

final products when Grignard reagents are slowly added to oxygen-saturated solvents at $-78°$.[81,82] Diphenylamine and butyraldehyde had no discernible effects on the reactions, suggesting that they do not occur through chain mechanisms like (65), but evidence for free radical intermediates in Grignard oxidations has since been presented.[83–85]

Lamb et al. found that 5-hexenylmagnesium bromide is oxidized to give about 25% cyclopentylcarbinol, but that hydrolyses and carbonations gave only about 5% cyclic products.[84] The latter were attributed to cyclization occurring during the formation of the Grignard reagent. The Russell kind of chain reaction (equations 65) would accommodate the data (equations 68).

$$R\cdot \longrightarrow R'\cdot \qquad (68a)$$

$$R\cdot + O_2 \longrightarrow ROO\cdot; \quad R'\cdot + O_2 \longrightarrow R'OO\cdot \qquad (68b)$$

$$\begin{aligned} ROO\cdot + RMgBr &\longrightarrow ROOMgBr + R\cdot; \\ R'OO\cdot + RMgBr &\longrightarrow R'OOMgBr + R\cdot \end{aligned} \qquad (68c)$$

Walling and Cioffari confirmed and extended these findings.[84a] They ruled out the possibility that cyclization was occurring during the second stage of the reaction (equation 67b) with the finding that 5-hexenylmagnesium bromide reacts with *t*-BuOOMgCl to give 5-hexen-1-ol but not cyclopentylcarbinol. Further, they found that slow oxidation gives somewhat more cyclic product than rapid oxidation, as predicted by the competitions of reactions 68. Unfortunately, there was also a temperature differential of 25° between the slow and rapid oxidations, so it is not clear that this is unambiguous support for reactions 68. Be that as it may, the evidence supports a radical chain mechanism. Although the data do not bear directly on the nature of initiation, they are consistent with reaction 69.

$$RMgBr + O_2 \longrightarrow R\cdot + BrMgO_2(Br^-Mg^+O_2\cdot^-) \qquad (69)$$

The presence of α-phenylethanol among the products of the oxidation of phenylmagnesium bromide in diethyl ether suggests radical attack on the solvent.[77] As a test for phenyl radical intermediates, the oxidation of phenylmagnesium bromide has been carried out in the presence of bromobenzene. Oxygen reacts relatively slowly with phenyl radicals,[85] and its presence *enhances* the yields of aromatic phenylation products.[86] Oxygen may serve to oxidize intermediate radicals as shown in (70). Therefore, a substituted benzene could act as a phenyl radical trap during an oxygenation of phenylmagnesium bromide.

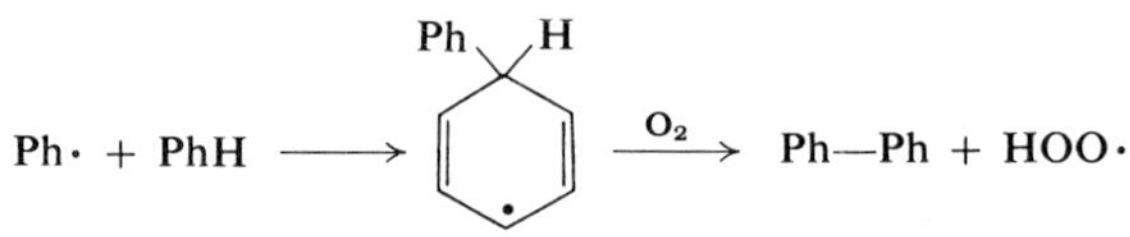

(70)

Biphenyl was formed during Grignard preparation, but it did not increase during oxygenation. Small quantities of bromobiphenyls were formed during preparation of phenylmagnesium bromide, but their concentration increased twentyfold or more during oxygenation. The ortho–meta–para isomer distribution was 58:29:13, comparing favorably with 55:30:15 found when bromobenzene reacts with phenylazotriphenylmethane at room temperature.[87] These results virtually require phenyl radical intermediates in the oxygenation.

5. Oxygenation of Koelsch's Anion; Solvent and Metal Ion Effects

There are solvent and metal ion effects in anion oxidations. A dramatic example is provided by the oxidation of the anion Koel:⁻, related to Koelsch's radical Koel·.[88]

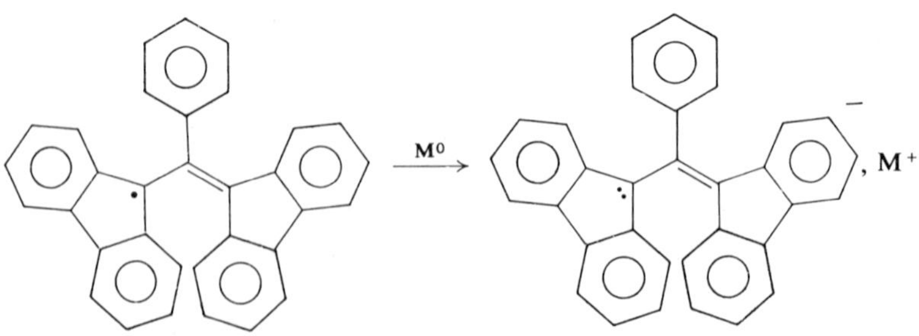

(71)

Koel· Koel:

The oxygenation of Koel:$^-$ in a variety of solvents yields Koel·, and the rate of disappearance of Koel:$^-$ is twice the rate of appearance of Koel·. While the stoichiometry is puzzling (perhaps there is an error, the most likely source

$$2\ \text{Koel}:^-,\ \text{M}^+ \xrightarrow{O_2} \text{Koel}\cdot + \text{other products} \tag{72}$$

for which would be in the extinction coefficients of Koel:$^-$ and Koel·), and the products are not entirely known, it is clear that Koel· does not enter into any chain process. The reactions are first order in Koel:$^-$ and in O_2 in the solvents (DME, THF) in which the kinetics could be conveniently followed. Thus, it is supposed that reaction 73 is kinetically isolated.

$$\text{Koel}:^-,\ \text{M}^+ + O_2 \longrightarrow \text{Koel}\cdot + \text{M}^+,\ O_2\cdot^- \tag{73}$$

Table 5 gives the pseudo-first-order rate constants (1 atm partial pressure of

TABLE 5

Rate of Koel:$^-$ Oxidation[a]

M^+	DMSO DMF	Pyridine	DME	THF	MeTHF	$(n\text{-Bu})_2O$, Et_2O Et_3N, benzene
Na^+	$<3 \times 10^{-4}$	—	0.34	0.65	>25	>25
K^+	$<3 \times 10^{-4}$	0.015	0.19	0.63	~ 15	>25
Cs^+	—	—	0.20	0.61	2.1	—

[a] Values tabulated are 10^3 times the pseudo-first-order rate constants (sec^{-1}) at 1 atm O_2. Initial concentrations of Koel:$^-$ were about 3×10^{-5} *M*.

oxygen) for several solvents and alkali counterions. Obviously less polar solvents enhance the rates of oxidation. Further, for NaKoel in DME, added sodium tetraphenylboride accelerates the reaction, indicating that ion pairs Na^+Koel:$^-$ are more reactive than free anions Koel:$^-$.[89] These observations are consistent with a transition state with a higher cation affinity than the reactant. Since in reaction 73 negative charge is being "poured from a large container (Koel:$^-$) into a small one ($O_2\cdot^-$)," this is reasonable if ED–EA has progressed to a considerable extent in the transition state. Although metal ions are not shown in (74), they must be incorporated (at least in the less polar solvents) as associated counterions or as bridges in the transition states.

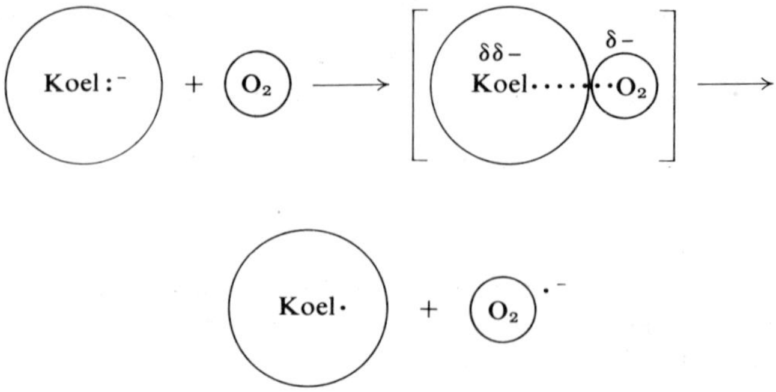

(74)

D. Associative Electron Detachment Reactions

1. One- and Two-Step Processes

Net AED reactions are most simply envisioned as one- or two-step processes.

$$\mathrm{R}\cdot + \mathrm{A}{:}^- \longrightarrow \mathrm{R{-}A} + e^- \tag{75}$$

$$\mathrm{R}\cdot + \mathrm{A}{:}^- \longrightarrow \mathrm{R{-}A}\cdot^- \tag{76a}$$

$$\mathrm{R{-}A}\cdot^- \longrightarrow \mathrm{R{-}A} + e^- \tag{76b}$$

Cases such as (76) may require a suitable electron acceptor to participate in (76b); the adduct radical anion may otherwise be stable. For many net AED reactions, the thermodynamics are favorable, even when there is no "receptacle" for the ejected electron.

2. Photolyses of Aryllithiums

One of the earliest recognized possible examples of AED occurs in the photolysis of aryllithiums in diethyl ether. Phenyllithium gives biphenyl; α-naphthyllithium gives exclusively 2,2′-binaphthyl.[90]

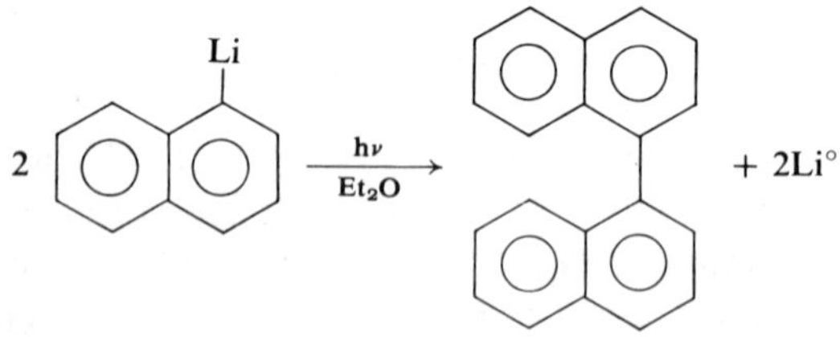

(77)

Mechanisms involving reactions of aryl radicals with solvent, followed by homolytic aromatic substitution on the resulting hydrocarbons, are ruled out by the specificity of the coupling and the low level of products derived from attack on the solvent. The most reasonable pathway invokes reactions of aryl radicals with aryllithiums.

$$Ar_2Li_2 \xrightarrow{h\nu} [Ar\cdot, Li\cdot, ArLi] \quad (78a)$$

$$[Ar\cdot, Li\cdot, ArLi] \longrightarrow Li\cdot + Li^+, Ar - Ar\cdot^- \quad (78b)$$

$$Li^+, Ar - Ar\cdot^- \longrightarrow Li\cdot + Ar - Ar \quad (78c)$$

Steps 78b and 78c are net AED, insofar as the reactant aryl radical and "anion" (or ArLi) are concerned. Although van Tamelen et al.[90] called upon the particular stability of biphenyl radical anion in accounting for their results, it should be noted that many AED reactions of simple species are highly exothermic, even when the radical anion which could be an intermediate is not stable with respect to ED to the medium.

3. *Gas-Phase Reactions*

Gas-phase AED reactions were first documented in the laboratory in 1966, though they had been postulated for many years.[91,92]

$$H\cdot + H{:}^- \longrightarrow H{-}H + e^- \qquad \Delta E = -88 \text{ kcal/mole} \quad (79)$$

$$H\cdot + {:}\ddot{\underset{..}{Cl}}{:}^- \longrightarrow H{-}\ddot{\underset{..}{Cl}}{:} + e^- \qquad \Delta E = -16 \text{ kcal/mole} \quad (80)$$

$${:}\ddot{\underset{..}{O}}\cdot^- + C_2H_4 \longrightarrow C_2H_4O + e^- \qquad \Delta E = -28 \text{ kcal/mole} \quad (81)$$

For each of these exothermic processes, the second-order rate constant is 10^{11}–10^{12} M^{-1} sec^{-1}. This seems typical for exothermic processes involving atomic reactants, but for more complex chemical systems, even exothermic processes are sometimes very slow. Slow exothermic reactions seem most likely when considerable atomic reorganization is required, for example, when more than one bond must be made or broken.

$${:}\ddot{\underset{..}{S}}\cdot^- + H_2 \longrightarrow H_2S + e^- \qquad \Delta E = -21 \text{ kcal/mole}, \quad k < 10^6\ M^{-1}\ \text{sec}^{-1} \quad (82)$$

4. *Alkylations of 2-Nitropropane with p-Nitrobenzyl Chloride and Related Reactions*

a. Carbon Alkylation through ET Mechanisms. Kornblum and co-workers were led to AED along a trail that began with observations of Haas and Bender[93] that only *p*-nitrobenzyl chloride, among nine para-substituted benzyl chlorides studied, gave any carbon alkylation of the sodium salt of 2-nitropropane. During studies of ambident anions, it was discovered that the fraction of carbon alkylation decreases substantially when better leaving groups than chloride are employed.[94]

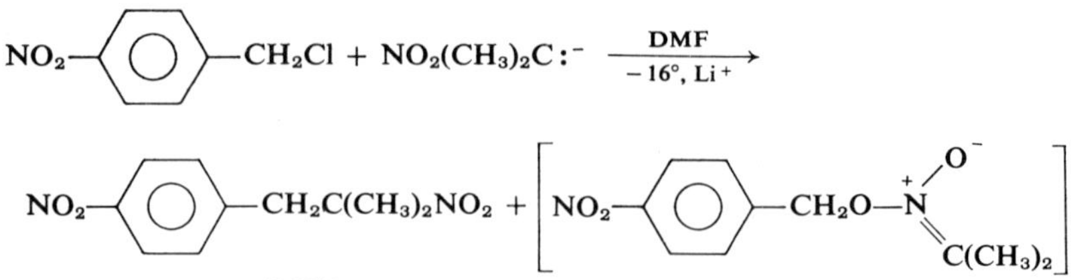

C (95%) O (1%)

(83)

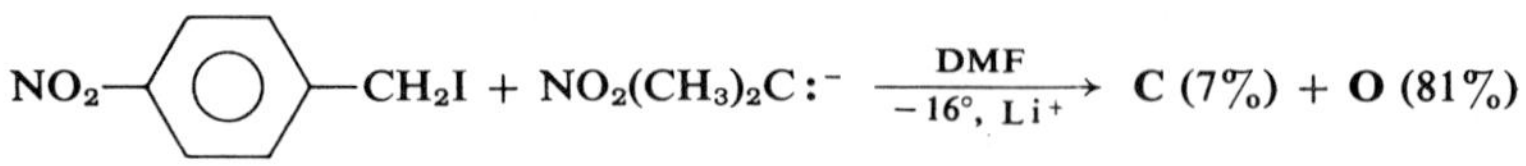

(84)

Kinetics suggests that carbon and oxygen alkylations are independent processes, oxygen alkylation representing the "normal" nucleophilic displacement, leaving carbon alkylation to arise in some other fashion. The rate constant for carbon alkylation with *p*-nitrobenzyl iodide is six times that for carbon alkylation with *p*-nitrobenzyl chloride, but the corresponding factor for oxygen alkylation is 900. The latter figure is reasonable for a nucleophilic displacement, but the former is unreasonably small for such a process.[95]

An initial net ED–DEA reaction, competing with nucleophilic displacement, is a reasonable hypothesis.[95] Consistent with this, the presence of

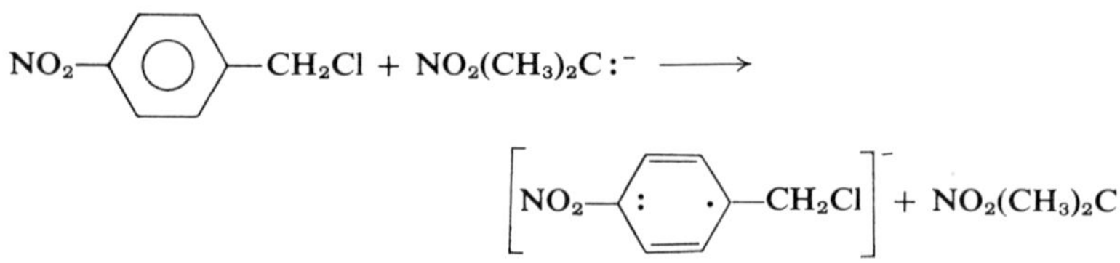

(85)

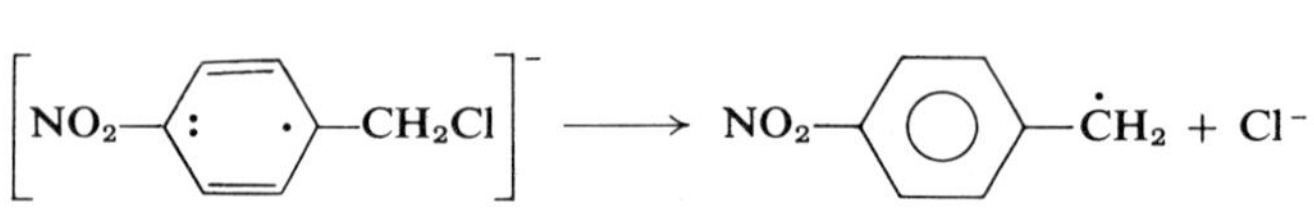

(86)

p-dinitrobenzene can almost completely suppress carbon alkylation. *p*-Dinitrobenzene should be a better electron sink than *p*-nitrobenzyl chloride, so that the former accrues electrons which would have gone, in its absence, to the latter, thus inhibiting reaction of *p*-nitrobenzyl chloride through the ET pathway.

It was proposed initially that carbon alkylation occurred through the coupling of *p*-nitrobenzyl and 2-nitro-2-propyl radicals, but this view proved

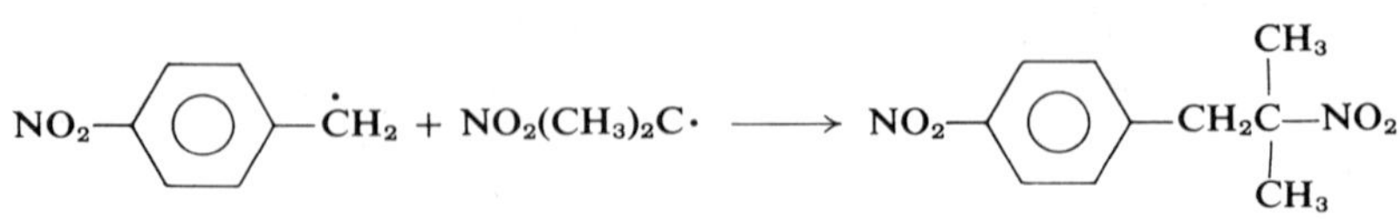

(87)

inadequate. Simultaneously, both Russell and Kornblum and their co-workers suggested that carbon alkylation arises in a chain reaction initiated by step 85. The propagating steps are (86), (88), and (89).[96,97]

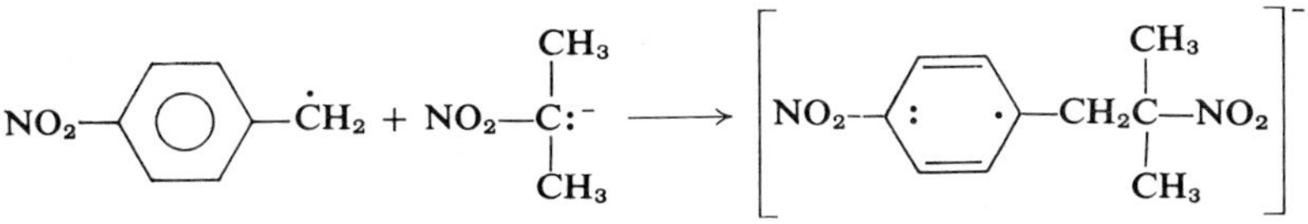

(88)

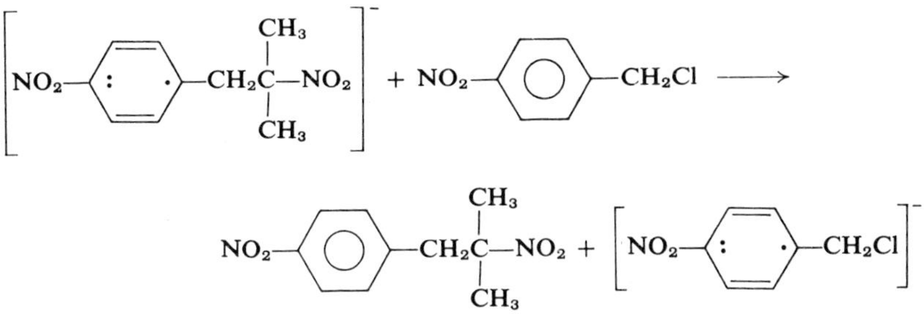

(89)

The most direct evidence for a chain reaction lies in the influence of scavengers. Oxygen suppresses the photochemically initiated reaction of the potassium salt of 2-nitropropane with *p*-nitrobenzyl chloride in ethanol.[97] No reaction occurs in this system in the dark, but the light-catalyzed reaction resembles closely the reaction in DMF, which occurs in the dark but which is accelerated by light. Kornblum and co-workers employed $CuCl_2$, in addition to nitroaromatics, as scavengers in studies of the reaction of *p*-nitrobenzyl chloride with the anion of a β-keto ester, 2-carbethoxycoumaran-3-one, a system which exhibits the same general features as that involving salts of 2-nitropropane.[98] $CuCl_2$ is an effective inhibitor, but $CuSO_4$ is not, supporting a ligand transfer mechanism for radical scavenging by $CuCl_2$.

It should be noted that reactions 88 and 89 are net AED for *p*-nitrobenzyl radical and the anion of 2-nitropropane.

b. Scope. Substitution through similar pathways extends to tertiary *p*-nitrobenzyl chlorides, for example, *p*-nitrocumyl chloride.[99] It also extends to leaving groups other than chloride and to "nucleophiles" other than 2-nitropropyl and β-keto ester anions. The α-nitro group of α,*p*-dinitrocumene is replaced in a reaction with thiophenoxide.[100] Neutral amines will serve as "nucleophiles."[101] Generally, these reactions are subject to catalysis by light and inhibition by oxygen and other radical scavengers. In addition, products specifically indicative of intermediate alkyl radicals (e.g., *p*-nitrobenzyl or derivatives) are sometimes found.[102]

c. Initiation and Propagation via Different Anions. In the chain mechanisms for these reactions, anions play two roles: (a) They react in initiation steps, and (b) they suffer AED reactions with intermediate radicals. In HMPA, 2-nitro-2-propyllithium is especially effective in the initiation; light is not required. The anions of various other added salts, not effective as initiators, can be more effective than 2-nitro-2-propyl anions in AED reactions with the intermediate radicals. Azide, benzene sulfinate, nitrite, and malonic ester anions are all effective agents in the AED steps, as is quinuclidine. The scope

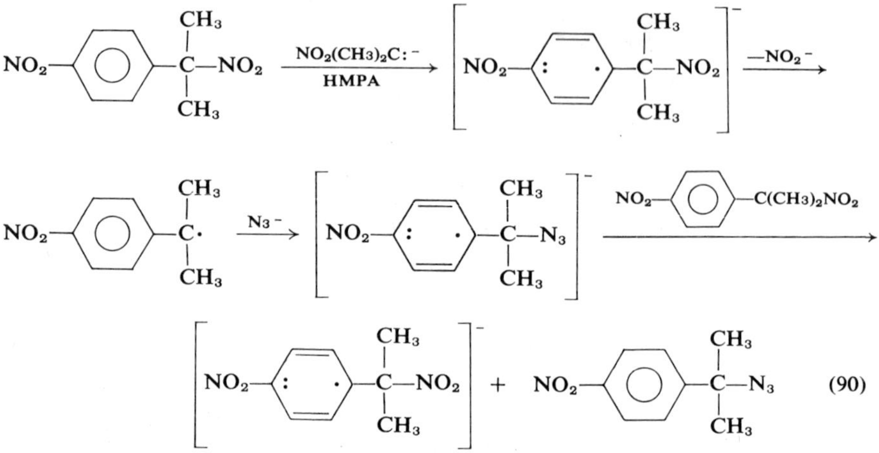

(90)

of such reactions has been widened to include α-nitro esters, ketones, and nitriles and α,α-dinitro alkanes as substrates reacting with salts of various nitroparaffins.[103–105]

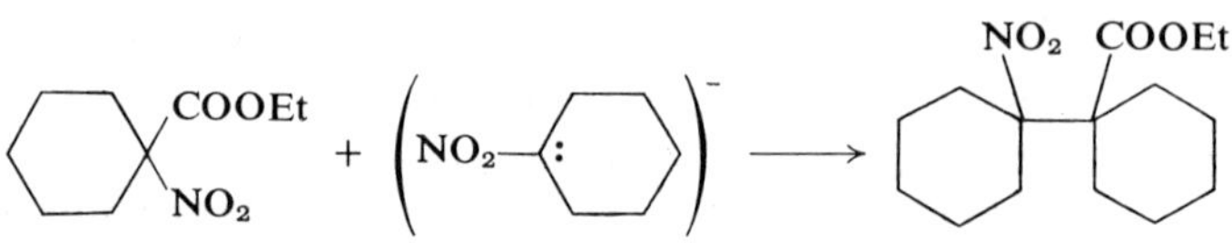

(91)

5. *Nucleophilic Aromatic Substitutions*

Kim and Bunnett have described another family of reactions, certain nucleophilic aromatic substitutions, which appear to proceed through the same kind of chain mechanism, involving AED.[106]

$$[e^-] + \mathrm{ArI} \longrightarrow [\mathrm{ArI}]\cdot^- \qquad \text{Initiation} \qquad (92a)$$

$$[\mathrm{ArI}]\cdot^- \longrightarrow \mathrm{Ar}\cdot + \mathrm{I}^- \qquad (92b)$$

$$\mathrm{Ar}\cdot + \mathrm{NH_2}^- \longrightarrow \mathrm{ArNH_2}\cdot^- \qquad \text{Propagation} \qquad (92c)$$

$$\mathrm{ArNH_2}\cdot^- + \mathrm{ArI} \longrightarrow \mathrm{ArNH_2} + [\mathrm{ArI}]\cdot^- \qquad (92d)$$

Such reactions can be promoted by solvated electrons in liquid ammonia. In addition to various halobenzenes, diphenyl ether is cleaved by a potassium–potassium amide solution, giving aniline (53%). Both potassium anilide and cyanomethylpotassium (from acetonitrile) also act as substituting agents, but in each case there is a variety of products.

6. *Alkyl Dimer Formation?*

It may be that AED reactions are more widespread in organic chemistry than has been realized. It is worth noting that for gas-phase free ions, reaction 93 is about 60 kcal/mole exothermic. Could some of the peculiarities sur-

$$CH_3\cdot + CH_3{:}^- \longrightarrow CH_3{-}CH_3 + e^- \qquad (93)$$

rounding alkyl dimer formation in reactions of alkyl iodides with sodium naphthalene or in reactions of alkyllithiums with alkyl halides be accounted for by invoking reaction 93 in a suitable mechanistic milieu?

E. Wurtz Reactions

1. *Mechanistic Uncertainty*

When high surface sodium is used in a reaction with an alkyl halide, organosodiums are generated at the expense of alkyl dimers, supporting the sequence shown in (94) for a Wurtz reaction.[107] There is evidence that at

$$R{-}X + 2M \longrightarrow RM + MX \qquad (M = \text{an alkali metal}) \qquad (94a)$$

$$RM + RX \longrightarrow R{-}R + MX \qquad (94b)$$

least some reactions 94b follow S_N2 pathways.[108,109] Others, however, clearly do not. The study of reactions 94b for alkylsodiums is very difficult if one desires homogeneous reaction conditions. Alkylsodiums tend to react with all solvents in which they might be soluble. The situation is even more acutely embarrassing for alkylpotassiums, alkylrubidiums, and alkylcesiums, but alkyllithiums present opportunities for controlled studies.

2. *Alkyllithium–Alkyl Halide Reactions*

In 1956 Bryce-Smith found that reactions of alkyllithiums with alkyl halides in the presence of cumene lead to substantial yields of bicumyl. This was interpreted in terms of radical intermediates.[110,111] Overwhelming additional evidence for reaction steps such as (95a) has since accumulated. Alkyl

$$RLi + RX \longrightarrow 2R\cdot + LiX \qquad (95a)$$

$$R\cdot + PhCHMe_2 \longrightarrow R\cdot + Ph\dot{C}Me_2 \qquad (95b)$$

$$2Ph\dot{C}Me_2 \longrightarrow PhCMe_2Me_2Ph \qquad (95c)$$

radical intermediates have been detected by ESR[112,113] and inferred from CIDNP,[114,115] stereochemical results,[109] unusual reactivity,[116] and the distribution of coupling products.[117]

Some insight into the reactions stems from CIDNP studies.[118] Although some ESR and CIDNP results suggest the possibility that halogen–metal interchange occurs through radical intermediates, perhaps reactions 96, no CIDNP has been detected in unreacted alkyllithiums during alkyllithium–

$$RLi + R'X \longrightarrow R\cdot + R'\cdot + LiX \tag{96a}$$

$$R\cdot + R'X \longrightarrow RX + R'\cdot \tag{96b}$$

$$R'\cdot + RLi \longrightarrow R'Li + R\cdot \tag{96c}$$

alkyl halide reactions, though it is detected in unreacted alkyl iodides. Further, under conditions in which metal–halogen interchange occurs, but not coupling, no CIDNP is observed at all. Only when coupling reactions accompany metal–halogen interchange is CIDNP observed. Ward and coworkers construe from this that metal–halogen interchange probably does not involve intermediate alkyl radicals.[118] However, negative CIDNP results cannot be regarded as being as convincing as positive ones at this time.

Assuming that alkyllithiums exist as hexamers or tetramers, the latter being favored by solvation in donor solvents (usually ethers or amines), Ward et al. ascribe CIDNP to reactions of aggregates with alkyl iodides. They

$$R_6Li_6 + R'I \longrightarrow [R_5Li_6IR\cdot R'\cdot]^s \tag{97}$$

$$R_4Li_4 + R'I \longrightarrow [R_3Li_4IR\cdot R'\cdot]^s \tag{98}$$

propose that the radicals of the electronic singlet geminate pairs formed in (97) and (98) are not associated with (complexed by) the residues of the alkyllithium clusters. Comparisons of the CIDNP from these reactions and from related peroxide decompositions support this view.

CIDNP is not observed for similar reactions of hexameric alkyllithiums with primary and secondary alkyl bromides, but it is found for corresponding reactions of tetrameric alkyllithiums. In reactions with allylic, benzylic, and tertiary bromides, it is found for both kinds of reactant aggregates.

Ward et al.[118] discuss these results in terms of "complexed" alkyl radicals. It is suggested that neither radical is released from the residue of the cluster when a hexameric alkyllithium reacts with a primary alkyl bromide. In other cases, like the reaction of *n*-butyllithium with *t*-butyl bromide in pentane, one

$$R_6Li_6 + R'Br \longrightarrow [R_6R'Li_6Br]^s \longrightarrow \underset{\text{(No CIDNP)}}{R{-}R' + R_5Li_6Br} \tag{99}$$

radical is "freed" and the remaining radical is "complexed." It must then be proposed that radical–radical reactions of complexed radicals do not give CIDNP.

$$(n\text{-Bu})_6Li_6 + t\text{-BuBr} \longrightarrow [(n\text{-Bu})_8Li_6Br\cdot t\text{-Bu}\cdot]^s \longrightarrow (n\text{-Bu})_6Li_6Br\cdot + t\text{-Bu}\cdot \quad (100a)$$

$$2t\text{-Bu}\cdot \rightleftharpoons [t\text{-Bu}\cdot t\text{-Bu}\cdot]^d \longrightarrow \text{products including } Me_2C{=}CH_2 \quad (\text{CIDNP}) \quad (100b)$$

$$(n\text{-Bu})_8Li_6Br\cdot + t\text{-Bu}\cdot \longrightarrow \text{products} \quad (\text{No CIDNP}) \quad (100c)$$

$$2(n\text{-Bu})_6Li_6Br\cdot \longrightarrow \text{products} \quad (\text{No CIDNP}) \quad (100d)$$

Here [ ]^d represents a nongeminate radical pair, formed from freely diffusing radicals.

It is difficult to assess such rationalizations. They seem to account for most of the data, with a few enigmas remaining or prompted by the nature of the rationalizations themselves. In any event, the finding of CIDNP characteristic of initially *singlet* associated radical pairs is direct support for reaction 95a, elaborated by consideration of aggregation as in equations 97–100.

3. *Reactions of Alkyl Iodides with Sodium Mirrors*

Some recent experiments with alkylsodiums add to the confusion. Reactions already mentioned are those of sodium naphthalene with alkyl iodides, for which the chemical evidence points to reaction 101 as the dimer-forming step, but for which no CIDNP has been detected which can be attributed to this

$$R:^-, Na^+ + RI \longrightarrow [R\cdot R\cdot NaI] \longrightarrow R{-}R + NaI \quad (101)$$

step. Perhaps it fails to generate CIDNP for reasons similar to those applicable to some of the alkyllithium–alkyl halide reactions. Garst and Cox have found CIDNP in the products of reactions of alkyl iodides with sodium mirrors, however.[119] They argue from the fact that methyl iodide gives CIDNP but 1,4-diiodobutane does not, that if the Morton sequence (equations 94) applies to both cases (intramolecularly to the 1,4-diiodobutane case), then reaction 94b must be responsible for the observed CIDNP, not reaction 94a. There is still confusion here, though. The phases of the CIDNP records, for both high- and low-field reactions, require polarization in the *collapse* of *triplet* or *nongeminate* radical pairs. This would be inconsistent with reaction 101 unless a large dominant polarization arose through escape of radicals from the geminate pairs and encounter of escaped radicals to form polarizing nongeminate pairs.

Wurtz reactions are still mechanistically unclear.

4. *ED–DEA versus S_N2 Steps*

Sauer and Braig proposed that organometallics containing simple alkyl groups react with alkyl halides through ET mechanisms whereas those containing allylic or benzylic groups proceed by S_N2 mechanisms.[109] It appears that this generality may hold for reactions of simple alkyl anions (alkylsodiums) and anions R–Naph:$^-$ as generated in reactions of sodium naphthalene with alkyl iodides (see earlier discussion). These two cases and the reaction of :Naph·$^-$ seem to correlate with the oxidation potentials of the anions involved. In the gas phase, the relevant oxidation potentials are :Naph·$^-$, 3.5 kcal/mole; CH_3:$^-$, about 23 kcal/mole; and R–Naph:$^-$, probably more than 23 kcal/mole.[120] The first two of these, with sodium counterions in DME, appear to react via ED–DEA and the last appears to undergo S_N2 reactions with alkyl iodides.

V. ACKNOWLEDGMENT

The preparation of this chapter was partially supported by a grant from the National Science Foundation.

REFERENCES

1. M. Szwarc, *Carbanions, Living Polymers, and Electron Transfer Processes*, Interscience, New York, 1968.
2. B. J. McClelland, *Chem. Rev.*, **64**, 301 (1964).
3. G. E. Coates, M. L. H. Green, and K. Wade, *Organometallic Compounds*, 3rd ed., Vol. I, Methuen, London, 1967, Chap. 1.
4. E. de Boer, *Adv. Organomet. Chem.*, **2**, 115 (1964).
5. K. A. Bilevich and O. Yu. Okhlobystin, *Russ. Chem. Rev.*, **37**, 954 (1968). This article, "Electron Transfer as an Elementary Act of Organic Reactions," gives 242 references largely in the area of electron transfer reactions and radical ions.
6. N. D. Scott, J. F. Walker, and V. L. Hansley, *J. Am. Chem. Soc.*, **58**, 2442 (1936).
7. T. L. Staples, J. Jagur-Grodzinski, and M. Szwarc, *J. Am. Chem. Soc.*, **91**, 3721 (1968), and earlier papers.
8. S. Bank and B. Bockrath, *J. Am. Chem. Soc.*, **93**, 430 (1971). These authors do not argue in terms of cycle 12, but the π energies[9] they give are applicable. One must note, however, that E_π(:NaphH$^-$) = E_π(·NaphH), E_π being the π energy, in simple MO calculations, when the coulomb integral α is taken as the zero of energy for each electron.
9. C. A. Coulson and A. Streitwieser, *Dictionary of π Electron Calculations*, Wiley, New York, 1961.
10. J. F. Garst, D. Walmsley, and W. R. Richards, *J. Org. Chem.*, **27**, 2924 (1962).
11. A. Le Berre and P. Goasguen, *Compt. Rend.*, **254**, 1306 (1962); I. I. Vol'nov, *Peroxides, Superoxides, and Ozonides of Alkali and Alkaline Earth Metals*, translated into English by J. Woroncow, A. W. Petrocelli, Ed., Plenum Press, New York, 1966, p. 99.
12. G. A. Russell and A. G. Bemis, *Inorg. Chem.*, **6**, 403 (1967).
13. V. L. Hansley, *Ind. Eng. Chem.*, **43**, 1759 (1951).
14. A. I. Shatenshtein, E. S. Petrov, and M. I. Belousova, *Org. React.*, **1**, 191 (1964); *J. Polym. Sci. C*, **16**, Part 3, 1729 (1967).
15. Reference 1, pp. 160–164. Properties of a number of ethers are given.
16. M. Szwarc, *Acc. Chem. Res.*, **2**, 87 (1969).
17. P. Chang, R. V. Slates, and M. Szwarc, *J. Phys. Chem.*, **70**, 3180 (1966).

18. J. F. Garst, C. Hewitt, D. Walmsley, and W. Richards, *J. Am. Chem. Soc.*, **83**, 5034 (1961); J. F. Garst, D. Walmsley, C. Hewitt, W. R. Richards, and E. R. Zabolotny, *ibid.*, **86**, 412 (1964).
19. N. Hirota and S. I. Weissman, *J. Am. Chem. Soc.*, **86**, 2538 (1964).
20. Unpublished spectrophotometric studies of J. F. Garst, J. G. Pacifici, and J. A. Pacifici, reported in part at the 152nd National Meeting of the American Chemical Society, New York, Sept. 1966, paper 149, Division of Organic Chemistry.
20a. Recent ESR studies from Hirota's group indicate that alkali ketyls derived from xanthone and fluorenone never exist as solvent-separated ion pairs over a wide range of temperatures in THF and DME. K. S. Chen, S. W. Mao, K. Nakamura, and N. Hirota, *J. Am. Chem. Soc.*, **93**, 6004 (1971).
21. T. E. Hogen-Esch and J. Smid, *J. Am. Chem. Soc.*, **87**, 669 (1965); **88**, 307 (1966), and later papers.
22. Some of the earlier arguments and speculations are given in reference 23, pp. 579 and 589.
22a. J. Smid, *Angew. Chem. Internat. Edit.*, **11**, 112 (1972).
22b. M. Szwarc, Ed., *Ions and Ion Pairs in Organic Reactions*, Wiley-Interscience, New York, 1972.
23. J. F. Garst, in *Solute–Solvent Interactions*, J. F. Coetzee and C. D. Ritchie, Eds., Dekker, New York, 1969, Chap. 8.
24. A. Cserhegyi, J. Jagur-Grodzinski, and M. Szwarc, *J. Am. Chem. Soc.*, **91**, 1892 (1968).
25. J. F. Garst and E. R. Zabolotny, *J. Am. Chem. Soc.*, **87**, 495 (1965).
26. J. F. Garst, J. G. Pacifici, and E. R. Zabolotny, *J. Am. Chem. Soc.*, **88**, 3872 (1966).
26a. J. F. Garst and R. S. Cole, *J. Am. Chem. Soc.*, **84**, 4352 (1962).
27. E. R. Zabolotny and J. F. Garst, *J. Am. Chem. Soc.*, **86**, 1645 (1964).
28. W. E. Wentworth, E. Chen, and J. E. Lovelock, *J. Phys. Chem.*, **70**, 445 (1966).
29. R. S. Becker and E. Chen, *J. Chem. Phys.*, **45**, 2403 (1966).
30. W. E. Wentworth and E. Chen, *J. Phys. Chem.*, **71**, 1929 (1967).
31. G. A. Russell, E. G. Janzen, and E. T. Strom, *J. Am. Chem. Soc.*, **86**, 1807 (1964).
32. W. Schlenk and R. Ochs, *Ber.*, **49**, 608 (1916); W. Schlenk and E. Bergmann, *Ann.*, **464**, 1 (1928).
33. H. W. Brown and R. C. Jones, *J. Chem. Phys.*, **36**, 2809 (1962).
34. J. E. Mulvaney, Z. G. Garlund, and S. L. Garlund, *J. Am. Chem. Soc.*, **85**, 3897 (1963).
35. H. J. S. Winkler, H. Winkler, and R. Bollinger, *Chem. Commun.*, 70 (1966).
36. G. A. Russell and E. J. Geels, *J. Am. Chem. Soc.*, **87**, 122 (1965); G. A. Russell, E. J. Geels, F. J. Smentowski, K.-Y. Chang, J. Reynolds, and G. Kaupp, *ibid.*, **89**, 3821 (1967).
37. F. C. Adam and S. I. Weissman, *J. Am. Chem. Soc.*, **80**, 1518 (1958).
38. P. A. Malachesky, T. A. Miller, T. Layloff, and R. N. Adams, *Proc. Symp. Exchange Reactions*, Upton, New York, 1965, p. 157.
39. S. Arai, D. A. Grev, and L. M. Dorfman, *J. Chem. Phys.*, **46**, 2572 (1967).
40. L. M. Dorfman, *Acc. Chem. Res.*, **3**, 224 (1970).
41. R. A. Marcus, *Ann. Rev. Phys. Chem.*, **15**, 155 (1964).
42. N. Hirota and S. I. Weissman, *J. Am. Chem. Soc.*, **86**, 2537, 2538 (1964).
43. N. S. Hush and J. Blackledge, *J. Chem. Phys.*, **23**, 514 (1955).
44. J. F. Garst, E. R. Zabolotny, and R. S. Cole, *J. Am. Chem. Soc.*, **86**, 2257 (1964).
44a. T. J. Katz, *J. Am. Chem. Soc.*, **82**, 3784 (1960); H. L. Strauss, T. J. Katz, and G. K. Fraenkel, *ibid.*, **85**, 2360 (1963).

44b. M. J. S. Dewar, A. Harget, and E. Haselbach, *J. Am. Chem. Soc.*, **91**, 7521 (1969).
44c. D. R. Thielen and L. B. Anderson, *J. Am. Chem. Soc.*, **94**, 2521 (1972).
45. R. C. Roberts and M. Szwarc, *J. Am. Chem. Soc.*, **87**, 5542 (1965).
46. A recent claim of evidence for essential planarity of :TPE:$^{2-}$ is based on spurious arguments and irrelevant data[46a]: D. H. Eargle, *J. Am. Chem. Soc.*, **93**, 3859 (1971).
46a. J. F. Garst, *J. Am. Chem. Soc.*, **93**, 6312 (1971).
47. A. I. Shatenshtein, E. S. Petrov, and M. I. Belouseva, *Org. React.* (*Tartu State University, Tartu, Estonian SSR*), **1**, 191 (1964).
48. M. Szwarc, *Acc. Chem. Res.*, **5**, 169 (1972), G. Levin, C. Sutphen, and M. Szwarc, *J. Am. Chem. Soc.*, **94**, 2652 (1972).
49. D. E. Paul, D. Lipkin, and S. I. Weissman, *J. Am. Chem. Soc.*, **78**, 116 (1956).
50. J. F. Garst, *Acc. Chem. Res.*, **4**, 400 (1971).
51. N. D. Scott and J. F. Walker, U.S. Pat. 2,150,039 (March 7, 1939); *Chem. Abstr.*, **33**, 4602[6] (1939).
52. D. Lipkin, G. J. Divis, and R. W. Jordan, *Preprints, American Chemical Society, Div. Petrol. Chem.*, **13**, D60 (1968).
53. J. F. Garst, P. W. Ayers, and R. C. Lamb, *J. Am. Chem. Soc.*, **88**, 4260 (1966); *Preprints, American Chemical Society, Div. Petrol. Chem.*, **13**, D65 (1968).
54. J. F. Garst and F. E. Barton, II, *Tetrahedron Letters*, 587 (1969).
55. R. C. Lamb, P. W. Ayers, and M. K. Toney, *J. Am. Chem. Soc.*, **85**, 3483 (1963).
56. D. J. Carlsson and K. U. Ingold, *J. Am. Chem. Soc.*, **90**, 7047 (1968).
57. J. F. Garst, J. T. Barbas, and F. E. Barton, II, *J. Am. Chem. Soc.*, **90**, 7159 (1968).
58. J. F. Garst and J. T. Barbas, *J. Am. Chem. Soc.*, **91**, 3385 (1969).
59. J. F. Garst and J. T. Barbas, *Tetrahedron Letters*, 3125 (1969).
60. G. D. Sargent and G. A. Lux, *J. Am. Chem. Soc.*, **90**, 7160 (1968).
61. S. Bank and J. F. Bank, *Tetrahedron Letters*, 4533 (1969).
62. S. Bank and J. F. Bank, *Abstracts of Papers, 162nd National Meeting of the American Chemical Society, Washington, D.C., 1971*, ORGN paper 5.
63. L. A. Brooks and H. R. Snyder, in *Organic Syntheses*, Collective Vol. III, E. C. Horning, Ed., Wiley, 1955, p. 698.
63a. G. D. Sargent and M. W. Browne, *J. Am. Chem. Soc.*, **89**, 2788 (1967).
63b. J. Jacobus and D. Pensak, *J. Chem. Soc.* (*D*), 400 (1969).
64. S. J. Cristol and R. V. Barbour, *J. Am. Chem. Soc.*, **88**, 4262 (1966); **90**, 2832 (1968).
65. J. T. Barbas, C. D. Smith, and J. F. Garst, unpublished observations.
66. J. F. Garst, *Polym. Preprints, American Chemical Society, Div. Polym. Chem.*, **11**, 8 (1970).
67. G. D. Sargent, J. N. Cron, and S. Bank, *J. Am. Chem. Soc.*, **88**, 5363 (1966).
68. Unpublished results from the author's laboratories. See references 69 and 70 for related work.
69. J. F. Garst, R. H. Cox, J. T. Barbas, R. D. Roberts, J. I. Morris, and R. C. Morrison, *J. Am. Chem. Soc.*, **92**, 5761 (1970).
70. J. F. Garst, F. E. Barton, II, and J. I. Morris, *J. Am. Chem. Soc.*, **93**, 4310 (1971).
71. D. J. Morantz and E. Warhurst, *Trans. Faraday Soc.*, **51**, 1375 (1955); E. Warhurst and R. Whittaker, *ibid.*, **62**, 707 (1966), and interim papers.
72. F. E. Barton, II, Dissertation, Univ. of Georgia, Athens, Georgia, 1969.
73. F. L. Benton and W. B. Hamill, *Ind. Eng. Chem., Anal. Ed.*, **14**, 449 (1942); L. M. Liggett, *Anal. Chem.*, **26**, 748 (1956).
74. J. T. Barbas, Dissertation, Univ. of Georgia, Athens, Georgia, 1971.
75. G. A. Russell, E. G. Janzen, A. G. Bemis, E. J. Geels, A. J. Moye, S. Mak, and E. T. Strom, "Oxidation of Hydrocarbons in Basic Solution," Chapter 10 of

Selective Oxidation Processes, Advances in Chemistry Series No. 51, American Chemical Society, Washington, D.C., 1965, pp. 112–171.

76. G. A. Russell, A. G. Bemis, E. J. Geels, E. G. Janzen, and A. J. Moye, "Oxidation of Carbanions," Chapter 14 of *Oxidation of Organic Compounds*, Vol. I, *Advances in Chemistry Series No. 75*, American Chemical Society, Washington, D.C., 1968, pp. 174–202.
77. G. A. Russell, *J. Am. Chem. Soc.*, **76**, 1595 (1954).
78. G. A. Russell, *Pure Appl. Chem.*, **15**, 185 (1967).
79. G. A. Russell and E. G. Janzen, *J. Am. Chem. Soc.*, **84**, 4153 (1962).
80. G. A. Russell and A. G. Bemis, *J. Am. Chem. Soc.*, **88**, 5491 (1966).
81. C. Walling and S. A. Buckler, *J. Am. Chem. Soc.*, **77**, 6032 (1955).
82. C. W. Porter and C. S. Steele, *J. Am. Chem. Soc.*, **42**, 2650 (1920).
83. M. E. Howden, J. Burdon, and J. D. Roberts, *J. Am. Chem. Soc.*, **88**, 1732 (1966).
84. R. C. Lamb, P. W. Ayers, M. K. Toney, and J. F. Garst, *J. Am. Chem. Soc.*, **88**, 4261 (1966).

84a. C. Walling and A. Cioffari, *J. Am. Chem. Soc.*, **92**, 6609 (1970).

85. G. A. Russell and R. F. Bridger, *J. Am. Chem. Soc.*, **85**, 3765 (1963).
86. R. T. Morrison, J. Cazes, N. Samkoff, and C. A. Howe, *J. Am. Chem. Soc.*, **84**, 4152 (1962).
87. A. C. Farrar, C. D. Smith, and J. F. Garst, unpublished results; A. C. Farrar, M.S. thesis, Univ. of Georgia, Athens, Georgia, 1972.
88. J. G. Pacifici, J. F. Garst, and E. G. Janzen, *J. Am. Chem. Soc.*, **87**, 3014 (1965). Footnote *a* of the table in this paper contains an error: The values tabulated therein are not the pseudo-first-order rate constants, but 10^3 times these.
89. J. G. Pacifici and J. F. Garst, unpublished observations.
90. E. E. van Tamelen, J. I. Brauman, and L. E. Ellis, *J. Am. Chem. Soc.*, **87**, 4964 (1965); **93**, 6141 (1971).
91. F. C. Fehsenfeld, E. E. Ferguson, and A. L. Schmeltekopf, *J. Chem. Phys.*, **45**, 1844 (1966).
92. E. E. Ferguson, *Acc. Chem. Res.*, **3**, 402 (1970).
93. H. B. Haas and M. L. Bender, *J. Am. Chem. Soc.*, **71**, 1767, 3482 (1949).
94. N. Kornblum, P. Pink, and K. V. Yorka, *J. Am. Chem. Soc.*, **83**, 2779 (1961); N. Kornblum and P. Pink, *Tetrahedron*, **19**, Suppl. 1, 17 (1963).
95. R. C. Kerber, G. W. Urry, and N. Kornblum, *J. Am. Chem. Soc.*, **86**, 3904 (1964); **87**, 4520 (1965).
96. N. Kornblum, R. E. Michel, and R. C. Kerber, *J. Am. Chem. Soc.*, **88**, 5662 (1966).
97. G. A. Russell and W. C. Danen, *J. Am. Chem. Soc.*, **88**, 5663 (1966).
98. N. Kornblum, R. E. Michel, and R. C. Kerber, *J. Am. Chem. Soc.*, **88**, 5660 (1966).
99. N. Kornblum, T. M. Davies, G. W. Earl, N. L. Holy, R. C. Kerber, M. T. Musser, and D. H. Snow, *J. Am. Chem. Soc.*, **89**, 725 (1967).
100. N. Kornblum, T. M. Davies, G. W. Earl, G. S. Greene, N. L. Holy, R. C. Kerber, J. W. Manthey, M. T. Musser, and D. H. Snow, *J. Am. Chem. Soc.*, **89**, 5714 (1967).
101. N. Kornblum and F. W. Stuchal, *J. Am. Chem. Soc.*, **92**, 1804 (1970).
102. N. Kornblum, G. W. Earl, N. L. Holy, J. W. Manthey, M. T. Musser, D. H. Snow, and R. T. Swiger, *J. Am. Chem. Soc.*, **90**, 6221 (1968).
103. N. Kornblum, R. T. Swiger, G. W. Earl, H. W. Pinnick, and F. W. Stuchal, *J. Am. Chem. Soc.*, **92**, 5513 (1970).
104. N. Kornblum, S. D. Boyd, and F. W. Stuchal, *J. Am. Chem. Soc.*, **92**, 5783 (1970).
105. N. Kornblum and S. D. Boyd, *J. Am. Chem. Soc.*, **92**, 5784 (1970).
106. J. K. Kim and J. F. Bunnett, *J. Am. Chem. Soc.*, **92**, 7463, 7464 (1970).

107. A. A. Morton and G. M. Richardson, *J. Am. Chem. Soc.*, **62**, 123 (1940); A. A. Morton, J. B. Davidson, and H. A. Newey, *ibid.*, **64**, 2240 (1942); A. A. Morton, J. B. Davidson, and B. L. Hakan, *ibid.*, **64**, 2242 (1942).
108. R. L. Letsinger, *J. Am. Chem. Soc.*, **70**, 406 (1948).
109. J. Sauer and W. Braig, *Tetrahedron Letters*, 4275 (1969).
110. D. Bryce-Smith, *J. Chem. Soc.*, 1603 (1956); *Bull. Soc. Chim. France*, 1418 (1963).
111. F. S. D'yachkovskii and A. E. Shilov, *Russ. Chem. Rev.*, **35**, 300 (1966), and works cited therein.
112. G. A. Russell and D. W. Lamson, *J. Am. Chem. Soc.*, **91**, 3967 (1969).
113. H. Fischer, *J. Phys. Chem.*, **73**, 3834 (1969).
114. H. R. Ward, R. G. Lawler, and R. A. Cooper, *J. Am. Chem. Soc.*, **91**, 746 (1969), and other works.
115. A. R. Lepley and R. L. Landau, *J. Am. Chem. Soc.*, **91**, 748 (1969), and other works.
116. The coupling of adamantyllithium with adamantyl iodide could not occur through a conventional S_N2 process. P. T. Lansbury, V. A. Pattison, J. D. Sidler, and J. Bieber, *J. Am. Chem. Soc.*, **88**, 78 (1966).
117. D. E. Applequist and D. F. O'Brien, *J. Am. Chem. Soc.*, **85**, 743 (1963), footnote 26.
118. H. R. Ward, R. G. Lawler, and R. A. Cooper, "The Role of CIDNP in a Mechanistic Investigation of Alkyllithium–Alkyl Halide Reactions," to be published in *Chemically Induced Magnetic Polarization*, G. L. Closs and A. R. Lepley, Eds., Wiley-Interscience, New York. Personal communication from H. R. Ward, 1971.
119. J. F. Garst and R. H. Cox, *J. Am. Chem. Soc.*, **92**, 6389 (1970).
120. The oxidation potential of R–Naph:$^-$ is judged to be considerably greater than that of CH_3:$^-$ on the basis of a cycle in which the difference in resonance energies of R–Naph· and R–Naph:$^-$ is the key consideration. The latter is more resonance stabilized than the former because in the latter π delocalization is accompanied by charge dispersion, whereas in the former it is not.

CHAPTER 10

Bimolecular Homolytic Substitution at Metal Centers

Alwyn G. Davies and Brian P. Roberts

Department of Chemistry, University College London, London, England

I. INTRODUCTION

Substitution reactions, generalized by equation 1, can conveniently be classified into three types on the basis of the nature of the electronic processes involved.

$$X + Y\text{—}Z \longrightarrow X\text{—}Y + Z \qquad (1)$$

In a *nucleophilic substitution*, the reagent X provides the electron pair for forming the new X—Y bond (equation 2).

$$\ddot{X} \curvearrowright Y\text{—}Z \longrightarrow X\text{—}Y + \ddot{Z} \qquad (2)$$

Reactions of this type, occurring at a saturated carbon center, have been studied extensively. Two mechanisms are distinguished which are independent of charge type (e.g., equations 3 and 4), but dependent on the number of molecules undergoing covalency change in the rate-determining stage.

$$S_N2 \qquad \ddot{X} \quad R{-}Z \longrightarrow \overset{\delta+}{X}{-}{-}{-}R{-}{-}{-}\overset{\delta-}{Z} \longrightarrow \overset{+}{X}{-}R + Y^{-} \qquad (3)$$

$$S_N1 \qquad R{-}Z \xrightleftharpoons{\text{slow}} Z^- + R^+ \xrightarrow{X} R{-}\overset{+}{X} \qquad (4)$$

In *electrophilic substitutions*, the electrons for forming the X—Y bond are derived from the substrate Y—Z.

$$\overset{+}{X} \quad Y{-}Z \longrightarrow X{-}Y + \overset{+}{Z} \qquad (5)$$

Studies dealing mainly with organometallic compounds have again shown the existence of bimolecular and unimolecular mechanisms.

$$S_E2 \qquad X \quad R{-}Z \longrightarrow \overset{\delta-}{X}{-}{-}R{-}{-}\overset{\delta+}{Z} \longrightarrow \overset{-}{X}{-}R + \overset{+}{Z} \qquad (6)$$

$$S_E1 \qquad R{-}Z \rightleftharpoons Z^+ + R^- \xrightarrow{X} RX^- \qquad (7)$$

In *homolytic substitution*, one electron of the X—Y bond is derived from the (radical) reagent, and one from the substrate molecule.

$$X\cdot \quad Y{-}Z \longrightarrow X{-}Y + Z\cdot \qquad (8)$$

Again, bimolecular and unimolecular mechanisms for this process can be envisaged.

$$S_H2 \qquad X\cdot \quad Y{-}Z \longrightarrow \overset{\delta\bullet}{X}{-}{-}{-}Y{-}{-}{-}\overset{\delta\bullet}{Z} \longrightarrow X{-}Y + Z\cdot \qquad (9)$$

$$S_H1 \qquad Y{-}Z \rightleftharpoons Z\cdot + Y\cdot \xrightarrow{X\cdot} XY \qquad (10)$$

Many examples can be quoted of substitution at a saturated carbon atom being achieved by an S_H1 process.* The activation energy for such reactions will normally be close to the bond dissociation energy.

S_H2 processes are known to occur at coordinatively unsaturated (sp^2 hybridized) carbon, when they usually involve an addition–elimination sequence [referred to below as S_H2 (stepwise)]. No examples, however, are known of S_H2 reactions occurring at an unstrained, four-coordinated (sp^3 hybridized) carbon center, although they can occur if the tetrahedral geometry at carbon is distorted, to change the hybridization and introduce "unsaturated" character.

The activation energy for the S_H2 reaction of deuterium atoms with methane

* An alternate symbolism, S_R2 and S_R1, is less commonly used.

at the carbon center has been calculated to be 37–40 kcal/mole,[1] and at the hydrogen center it is found to be 8–10 kcal/mole,[2] and hence under normal conditions the latter reaction is observed exclusively. This pattern is typical for other homolytic reagents: they react preferentially at peripheral monovalent hydrogen or halogen, and occasionally at bivalent oxygen or sulfur, but not at saturated carbon.

A rapid S_H2 reaction can, nevertheless, occur at a multivalent metallic center if it has accessible a low-lying unfilled orbital.

$$X\cdot + MR_n \longrightarrow XMR_{n-1} + R\cdot \qquad (11)$$

Thus, for example, a variety of radicals will bring about a rapid S_H2 reaction at the boron center in trialkyl boranes, where the $2p_z$ orbital is vacant, but not in the pyridine complexes of the boranes where this orbital is filled. Again, trialkylphosphines, which have vacant accessible $3d$ orbitals, react readily with similar radical reagents, whereas trialkylamines, which have no equivalent vacant orbitals, do not.

The basic reason for this high reactivity of organometallic compounds by an S_H2 mechanism is that most common radical reagents (e.g., $RO\cdot$, $ROO\cdot$, $RS\cdot$, $R_2N\cdot$, halogen atoms, or ketone triplets) involve an electronegative atom which will form a stronger bond to a metal than carbon does. When the reagent $X\cdot$ and the organometallic compound can combine to form the appropriate S_H2 transition state (or intermediate), reaction 11 is exothermic, the activation energy is low, and the reactions are fast. For example, if one accepts the approximate bond energy terms to be $\bar{D}$(B—C) = 83, $\bar{D}$(B—O) = 125, $\bar{D}$(H—C) = 99, and $\bar{D}$(H—O) = 110 kcal/mole, alkoxydealkylation at boron will be exothermic by 42 kcal/mole, but at hydrogen it will be exothermic by only 11 kcal/mole.

These homolytic substitution reactions which occur at a metallic center are discussed in this chapter.* Although they have been established only within the past few years, most of these reactions have been studied mechanistically, and examples are recognized in which the species $X\dot{M}R_n$ represents a transition state (at an energy maximum) or a metastable intermediate (at an energy minimum). The case for describing the second of these two processes as an S_H2 reaction is arguable,[4,5] but, for convenience, both will be referred to as S_H2 reactions, and, where necessary, the two will be differentiated by the terms S_H2 (synchronous) (equation 12) and S_H2 (stepwise) (equation 13).

$$X\cdot + MR_n \longrightarrow [X\dot{M}R_n]^{\ddagger} \longrightarrow XMR_{n-1} + R\cdot \qquad S_H2 \text{ (synchronous)} \qquad (12)$$

$$X\cdot + MR_n \longrightarrow X\dot{M}R_n \longrightarrow XMR_{n-1} + R\cdot \qquad S_H2 \text{ (stepwise)} \qquad (13)$$

* These reactions are reviewed briefly in reference 3, and more extensively in reference 4.

II. ALKYLPEROXYL RADICALS

The first organometallic S_H2 reactions to be investigated in detail were autoxidations which involve alkylperoxyl radicals, ROO·. If dilute solutions of alkyl derivatives of lithium, magnesium, zinc, cadmium, boron, or aluminum are treated with a liberal supply of oxygen, the corresponding alkylperoxy-metallic compounds are formed, usually in high yield.[6,7]

$$RM + O_2 \longrightarrow ROOM \qquad (14)$$

The alkyl compounds of a number of other metals (e.g., mercury,[6,7] tin,[6,7] antimony,[8] and bismuth[8]) give products which show the presence of some peroxidic material.

Early attempts to inhibit these autoxidations, with reagents, which were known to inhibit hydrocarbon autoxidation, were unsuccessful, and a polar mechanism was proposed for the reaction[9–11]:

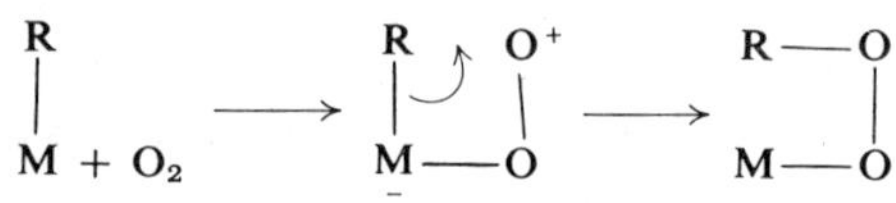

(15)

Subsequently it was shown that optically active 1-phenylethylboronic acid was autoxidized with at least 99% racemization (equation 16),[12] and that both the *exo*- and the *endo*-norborn-2-ylboron group reacted to give a mixture of 76% *exo*- and 24% *endo*-norbornylperoxyboron groups (equation 17).[13]

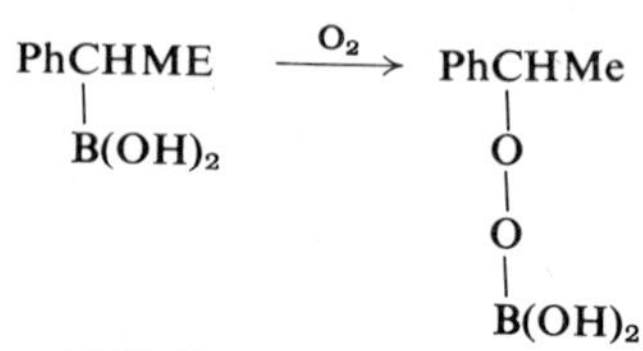

(16)

$[\alpha]_D$ −15.1° (C_6H_6)   0.0°

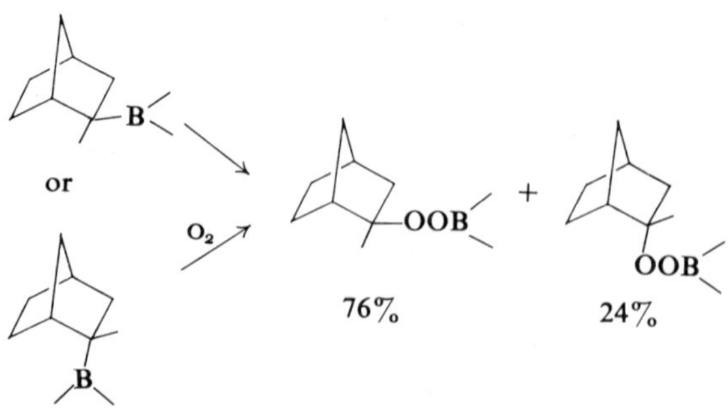

(17)

Norborn-2-ylmagnesium halides were autoxidized to give the same mixture of *exo* and *endo* isomers,[14] and diisopinocampheyl-*s*-butyl borane reacted with substantial loss of asymmetry in the *s*-butyl group.[15] The autoxidations of the organoboranes, and organic derivatives of zinc, cadmium, and aluminum, could be inhibited by powerful radical scavengers such as galvinoxyl, phenothiazine, or certain hindered phenols.[15,16]

Attempts to inhibit the autoxidation of Grignard reagents have not yet been successful, probably because the inhibitors which have been tried are rapidly removed by reaction with the Grignard reagents. Some further evidence that alkyl radicals are involved, however, is provided by the autoxidation of 5-hexenylmagnesium bromide.[17] With an excess of oxygen at low temperatures, 5-hexenylperoxymagnesium bromide is formed, but, if the supply of oxygen is restricted, or if the temperature is raised, a substantial amount of cyclopentylmethylperoxymagnesium bromide is formed by cyclization of the intermediate 5-hexenyl radical (equation 18).

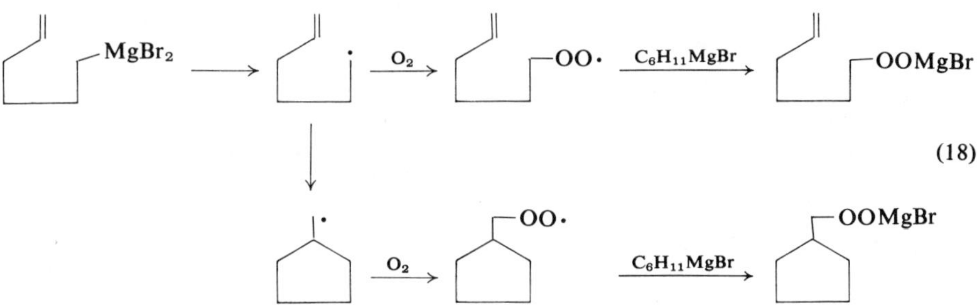

(18)

It was therefore proposed[12] that these reactions proceeded, not by a polar mechanism, but by a radical chain mechanism involving the S_H2 reaction of the alkylperoxyl radical at the metal center. This step must be much more rapid than the equivalent S_H2 reaction at the hydrogen centers in the alkyl groups or in the solvent; previous attempts at inhibiting the autoxidation had failed because the less efficient scavengers could not compete with this very fast reaction. The autoxidation of organometallic compounds would therefore be represented by a series of equations (19–22) similar to those which describe the autoxidation of hydrocarbons and should obey a similar rate law, that is, that of equation 23.

Initiation

$$\longrightarrow R\cdot \qquad (\text{Rate} = R_i) \tag{19}$$

Propagation

$$R\cdot + O_2 \longrightarrow ROO\cdot \tag{20}$$

$$ROO\cdot + MR \xrightarrow{k_p} ROOM + R\cdot \tag{21}$$

Termination

$$2ROO\cdot \xrightarrow{2k_t} \text{inactive products} \tag{22}$$

$$\frac{-d[O_2]}{dt} = k_p[ROO\cdot][MR] = \frac{k_p[MR]R_i^{1/2}}{(2k_t)^{1/2}} \tag{23}$$

This was confirmed (with oxygen overpressures in the range 150–760 mm) by a kinetic investigation of the autoxidation of a series of organoboron compounds in isooctane at 30°.[18–20] Rates of initiation (R_i) were determined by inhibiting the reactions with galvinoxyl or 2,6-di-*t*-butyl-4-methoxyphenol; values for the appropriate termination rate constants ($2k_t$) were taken from the literature. To reduce the rate of the uptake of oxygen by some trialkyl boranes to a measurable value, pyridine was added to convert them largely into the unreactive four-coordinate complexes. Rate constants for the S_H2 propagation step are given in Table 1.

TABLE 1

Rate Constants for Bimolecular Homolytic Alkylperoxydealkylation at Boron at 30°

	k_p (M^{-1} sec^{-1})				
Compound	R = *n*-Bu	R = *s*-Bu	R = *t*-Bu	R = $PhCH_2$	R = PhCHMe
R_3B	2×10^6	8×10^4	—	5×10^6	—
R_2BOBR_2	3×10^5	7×10^4	—	—	—
R_2BOOR	3×10^4	1×10^4	—	1×10^6	—
R_2BOR'	5×10^3	2×10^3	—	—	—
$(RBO)_3$	1×10^3	5×10^4	3×10^4	1×10^6	2×10^6
$RB(OR')_2$	Small	4	0.3	—	—

A similar mechanism was shown to apply to the autoxidation of triethyl borane in the gas phase.[21]

It will be seen from Table 1 that the values of k_p decrease as successive oxygen ligands are introduced at the boron center. This presumably results from the presence of p_π—p_π bonding from oxygen to boron, which must be broken to some degree when the alkylperoxyl radical attacks the boron to form the four-coordinate transition state or intermediate.

The alkylperoxydealkylation process is highly exothermic and the transition state should resemble the reactants rather than the products, with little bond fission and bond formation. In the trialkylboroxines, however, there is some indication of a parallel between the value of k_p and the stability of the displaced alkyl radical; the high value of k_p for tribenzylboroxine, in particular, may reflect the resonance stabilization of the benzyl radical. Again, the fact that diisobutyl-*t*-butyl borane reacts with oxygen to give mainly diisobutyl(*t*-butylperoxy)borane, may derive from the fact that the *t*-butyl radical is more stable than the isobutyl radical.[22]

Some steric effect also seems to be present. For example, the decrease in k_p on going from tri-*n*-butyl borane to tri-*s*-butyl borane, or from tri-*s*-butylboroxine to tri-*t*-butylboroxine, despite the usually accepted sequence of stabilities of the alkyl radicals ($R^p < R^s < R^t$), probably implies the incursion of steric strain in the four-coordinate transition state.

The autoxidation of trialkylphosphines (R_3P), which gives mainly the phosphine oxides (R_3PO) and phosphinates [$RP(O)(OR)_2$], is also a radical chain reaction, which can be inhibited by diphenylamine or quinol, but simple bimolecular homolytic alkylperoxydealkylation is probably not involved.[23] If the alkylperoxyl radical attacks at phosphorus to form a phosphoranyl radical, β-scission of the peroxide bond is easier than expulsion of an alkyl radical, and a phosphine oxide and alkoxyl radical are formed (equation 24).

$$ROO\cdot + PR_3 \longrightarrow ROO\dot{P}R_3 \longrightarrow RO\cdot + OPR_3 \qquad (24)$$

The alkoxyl radical will attack the phosphorus and the phosphoranyl radical which is formed may undergo α-scission resulting in an overall S_H2 (stepwise) alkoxydealkylation, or may undergo β-scission to give the phosphine oxide and alkyl radical.

$$RO\cdot + PR_3 \longrightarrow RO\dot{P}R_3 - \begin{cases} \xrightarrow{\alpha\text{-scission}} ROPR_2 + R\cdot \\ \xrightarrow{\beta\text{-scission}} R\cdot + OPR_3 \end{cases} \qquad (25)$$

Further reactions of these types lead to the formation of the products which are observed.

Lower down Group V, where the +v oxidation state becomes less stable, β-scission in the radical $ROO\dot{M}R_3$ is less likely, and the autoxidation of, say, trialkylbismuthines may proceed through homolytic alkylperoxydealkylation at the metal.[8]

Many compounds containing a metal-to-metal bond react with oxygen, and it is likely that these reactions proceed by chain mechanisms, where the chain-carrying species are metallyl and metalloperoxyl radicals. The evidence is strongest for the autoxidation of hexaethylditin. This is initiated at $-30°$ by ultraviolet light, and bis(triethyltin) peroxide can be identified as an unstable product.[24] The reaction is therefore probably propagated by the reactions:

$$Et_3Sn\cdot + O_2 \longrightarrow Et_3SnOO\cdot \qquad (26)$$

$$Et_3SnOO\cdot + Et_3SnSnEt_3 \longrightarrow Et_3SnOOSnEt_3 + Et_3Sn\cdot \qquad (27)$$

It is unfortunate that, whereas alkoxyl radicals can be generated in a number of different ways, sources of alkylperoxyl radicals are more limited, and reactions involving alkylperoxydealkylation have as yet been studied only in the context of the autoxidation reaction. A related reaction, however, is the direct reaction between oxygen and organoborane. This is the initiation process during the early stages of an autoxidation, and one reasonable mechanism by which it may occur is by an S_H2 reaction by triplet oxygen at the metallic center[19,25–27]:

$$O_2 + \;\rangle B{-}R \longrightarrow \dot{O}{-}OB\langle \; + R\cdot \qquad (28)$$

If iodine is present it prevents the radicals which are formed from initiating the chain reaction of equations 20 and 21, and the rate of disappearance of iodine, or of formation of alkyl iodide, may be interpreted as a measure of the rate of the initiation process.[26,27]

III. ALKOXYL RADICALS

A. Introduction

Alkoxyl and alkylperoxyl radicals show a similar type of S_H2 reactivity at hydrogen centers, and, as might be expected, they also show parallel behavior at metal centers. In the context of organometallic chemistry the behavior of alkylperoxyl radicals has been studied only as a component in the autoxidation process (see Section II), but alkoxyl radicals can be derived from a variety of sources. The principal processes which have been investigated are as follows:

1. The photolysis or thermolysis of peroxides or hyponitrites.
2. The homolysis of an alkyl hypochlorite.
3. The addition of a radical to an unsaturated carbonyl compound or epoxide.
4. The chemically induced decomposition of a peroxide.

The second and third of these reactions lead usually to a chain process, but the first reaction does not give rise to a chain, and under these conditions the concentrations of the alkyl radicals may be relatively high, allowing these species to be studied by the powerful technique of ESR.

The early suggestions of bimolecular homolytic alkoxydealkylation at metal centers arose from studies in the fields of organophosphorus and organoboron chemistry, and these remain the two metals which have received most attention. In 1962, Buckler reported that *t*-butyl di-*n*-butylphosphinite and tributylphosphine oxide were formed when di-*t*-butyl peroxide was decomposed in the presence of an excess of tri-*n*-butylphosphine at 130°.[28] Both products were suggested to come from the interaction of *t*-butoxyl radicals with the phosphorus center, involving, in the first case (equation 30) an S_H2 process.

$$t\text{-BuOO—}t\text{-Bu} \longrightarrow 2t\text{-BuO}\cdot \qquad (29)$$
$$t\text{-BuO}\cdot + \text{PBu}_3 \longrightarrow t\text{-BuOPBu}_2 + \text{Bu}\cdot \qquad (30)$$
$$t\text{-BuO}\cdot + \text{PBu}_3 \longrightarrow t\text{-Bu}\cdot + \text{OPBu}_3 \qquad (31)$$

In 1964, in an attempt to prepare chloromethylboron compounds, Matteson treated trimethylboroxine with *t*-butyl hypochlorite.[29] Under ultraviolet irradiation, a reaction occurred to give methyl chloride and an unstable oil which was apparently tri-*t*-butylboroxine, and he suggested that this product arose from a homolytic displacement at the boron center by the *t*-butoxyl radical (equation 32, R = Me).[29]

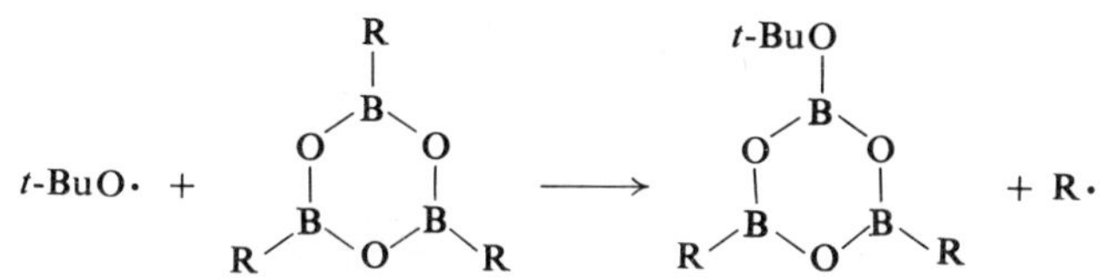
(32)

Similarly, Kruglova and Friedlina treated aryl- and aralkyl boroxines with di-*t*-butyl peroxide in boiling chlorobenzene, and obtained isobutene and boric oxide, which were thought to arise through thermal decomposition of *t*-BuOB groups formed again by reaction 32 (R = Ph, p-MeC_6H_4, or PhMeCH).[30]

B. Reactions Involving the Photolysis or Thermolysis of Peroxides

With the exception of some organic derivatives of the alkali metals, magnesium, and aluminum, most organometallic compounds derived from the main group metals do not react with di-*t*-butyl peroxide below its decomposition temperature. If the peroxide is photolyzed or thermolyzed, the alkoxyl radicals which are formed react at the metal center by an S_H2 process.

$$t\text{-BuOO}t\text{-Bu} \xrightarrow{h\nu} 2t\text{-BuO}\cdot \qquad (33)$$

$$t\text{-BuO}\cdot + \text{MR}_n \longrightarrow t\text{-BuOMR}_{n-1} + \text{R}\cdot \qquad (34)$$

This was established in 1969 for alkyl compounds of metals of Groups II, III, and V by Kochi and Krusic[31,32] and by Davies and Roberts,[33,34] who carried out the reactions in the cavity of an ESR spectrometer, and showed that the spectra of the displaced alkyl radicals could be observed. Thermolysis of di-*t*-butyl hyponitrite was used as an alternative source of *t*-butoxyl radicals for these organometallic compounds (e.g., Me_2Cd and Me_3Bi) which were photosensitive.[34]

The organometallic compounds which have been successfully studied by this technique are listed in Table 2. The extension of this technique to the determination of relative and absolute rate constants for the S_H2 processes is described in Section III.F.

Grotewold et al. studied the reaction between triethyl borane and di-*t*-butyl peroxide in the gas phase at 135°, and concluded that ethyl radicals were produced by the alkoxydealkylation reaction 34 ($MR_n = BEt_3$).[35]

Little work has been reported on the S_H2 reaction of alkoxyl radicals with organic derivatives of the Group IV metals, although the products from some reactions suggest that homolytic alkoxydealkylation may be occurring. For

TABLE 2

Alkylmetallic Compounds from Which the ESR Spectra of the Alkyl Radicals Have Been Observed, from the Reaction

$$t\text{-BuO}\cdot + MR_n \longrightarrow t\text{-BuOMR}_{n-1} + R\cdot$$

Group II	Et_2Zn
	Me_2Cd
Group III	Me_3B Et_3B Bu_3B i-Bu_3B s-Bu_3B neo-Pe_3B
	(n-$C_6H_{13})_3B$
	Bu_2BCl
	$(MeBO)_3$ $(BuBO)_3$ (i-$BuBO)_3$ (s-$BuBO)_3$ (t-$BuBO)_3$
	Me_3Al Et_3Al Bu_3Al i-Bu_2AlH Et_2AlOBu
	Et_3Ga
Group V	Me_3P Et_3P i-Pr_3P i-Bu_3P (cyclo-$C_6H_{11})_3P$ n-Bu_3P
	(n-$C_8H_{17})_3P$
	Et_3As Bu_3As
	Et_3Sb
	Et_3Bi

example, Vyazankin et al[36] studied the thermal decomposition of *t*-butyl peroxybenzoate in the presence of the compounds Et_4M (M = Si, Ge, and Sn). The silicon and germanium compounds gave the dehydrodimers $Et_3M(C_4H_8)MEt_3$, but tetraethyltin gave principally the compounds Et_3SnO*t*-Bu and $Et_3SnO\cdot CO\cdot Ph$. This reaction may involve an S_H2 reaction of the *t*-butoxyl and benzoyloxyl radicals at the tin center, but the possibility of the S_H2 attack of an $Et_3Sn\cdot$ radical at oxygen cannot be ruled out.

Neumann et al. showed that tributyl-*t*-butoxytin is formed when di-*t*-butyl peroxide is thermally decomposed in the presence of hexabutylditin,[37] and the S_H2 mechanism for the equivalent photolytic reaction of hexamethylditin has recently been confirmed by Watts and Ingold,[38] who observed the ESR spectrum of the displaced trimethylstannyl radical (equation 35, R = Me)

$$t\text{-BuO}\cdot + R_3SnSnR_3 \longrightarrow t\text{-BuOSnR}_3 + R_3Sn\cdot \tag{35}$$

Similarly, Torsell studied the thermal decomposition of di-*t*-butyl diperoxyoxalate in the presence of tetraethyllead and 2-methyl-2-nitrosopropane as a spin-trapping reagent.[39] *t*-Butyl(ethyl)nitroxide was detected, suggesting that *t*-butoxyl radicals had displaced ethyl radicals from lead.

C. Reactions Involving the Homolysis of Alkyl Hypochlorites

Walling and Kjellgren[40] studied the reaction of *t*-butyl hypochlorite with the phenyl derivatives of the metals Li, Mg, Zn, Cd, Hg, and Sn. In each case a major product was chlorobenzene, but the mechanisms of the reactions were not discussed.

The reactions between *t*-butyl hypochlorite and trialkyl boranes have been studied more thoroughly.[41–43] In the presence of an excess of borane, the yields of the alkyl chlorides according to equation 36 are usually good.

$$R_3B + t\text{-BuOCl} \longrightarrow R_2BOt\text{-Bu} + RCl \tag{36}$$

The reactions are self-initiated, and can be accelerated by di-*t*-butyl hyponitrite or ultraviolet light, and inhibited by galvinoxyl. When 1,1,2-trimethylpropyl hypochlorite is used, acetone and isopropyl chloride are also formed by the β-scission process:

$$i\text{-PrMe}_2CO\cdot \longrightarrow i\text{-Pr}\cdot + Me_2C{=}O \tag{37}$$

and if the reaction is carried out with *t*-butyl hypochlorite in a hydrocarbon solvent, *t*-butyl alcohol and the chloride derived from the hydrocarbon are formed by the competing process:

$$t\text{-BuO}\cdot + R'H \longrightarrow t\text{-BuOH} + R'\cdot \tag{38}$$

The reaction is therefore believed to follow a radical chain process:

$$t\text{-BuO}\cdot + BR_3 \longrightarrow t\text{-BuOBR}_2 + R\cdot \tag{39}$$

$$R\cdot + t\text{-BuOCl} \longrightarrow RCl + t\text{-BuO}\cdot \tag{40}$$

These reactions have been analyzed kinetically, and the results are presented in Section III.F.

D. Reactions Involving the Addition of Radicals to α,β-Unsaturated Carbonyl Compounds and Epoxides

Organometallic compounds may react with unsaturated carbonyl compounds by either 1,2 or 1,4 addition.[44,45]

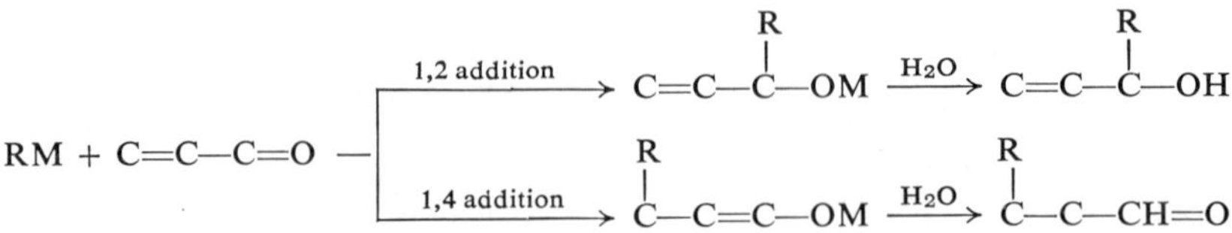

(41)

Polar organometallic compounds give largely 1,2 addition, probably by a heterolytic mechanism. If copper salts are present during the reaction of Grignard reagents, the proportion of 1,4 addition is increased, and radical anion mechanisms have been proposed.[46,47]

H. C. Brown and his co-workers have shown that the 1,4 addition of alkyl boranes to enones is a radical chain process.[48–50] Acrolein, methyl vinyl ketone, and their α-substituted derivatives of the structure $CH_2{=}CR'{\cdot}CO{\cdot}R$ react with trialkyl boranes at 25° by 1,4 addition, and the reactions can be inhibited by galvinoxyl. The corresponding β-substituted carbonyl compounds, $R'CH{=}CH{\cdot}COR$, do not react spontaneously, but their reaction can be initiated by benzoyl peroxide, acetyl peroxide, or oxygen. The following mechanism is therefore suggested:

$$R\cdot + CH_2{=}CHCH{=}O \longrightarrow RCH_2CH{=}CHO\cdot \tag{42}$$

$$RCH_2CH{=}CHO\cdot + BR_3 \longrightarrow RCH_2CH{=}CHOBR_2 + R\cdot \tag{43}$$

Acetylacetylene reacts in the same way in the presence of oxygen at 0°, azoisobutyronitrile at 40°, or under ultraviolet irradiation.[51]

$$R\cdot + CH{\equiv}CC(Me){=}O \longrightarrow RCH{=}C{=}C(Me)O\cdot \tag{44}$$

$$RCH{=}C{=}C(Me)O\cdot + BR_3 \longrightarrow RCH{=}C{=}C(Me)OBR_2 + R\cdot \tag{45}$$

The reaction between organoboranes and *p*-quinone to give the ring-alkylated quinol is also susceptible to inhibition by galvinoxyl,[52] and may involve a similar mechanism:

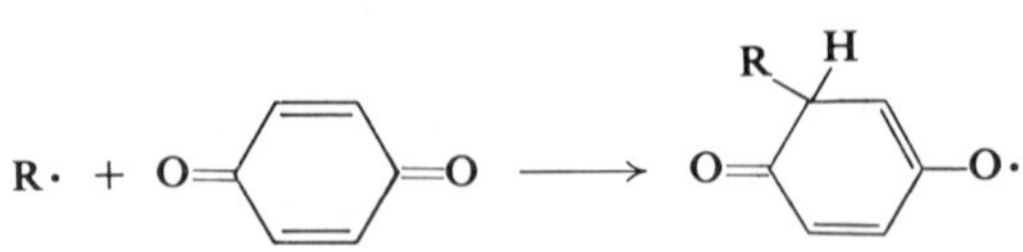

(46)

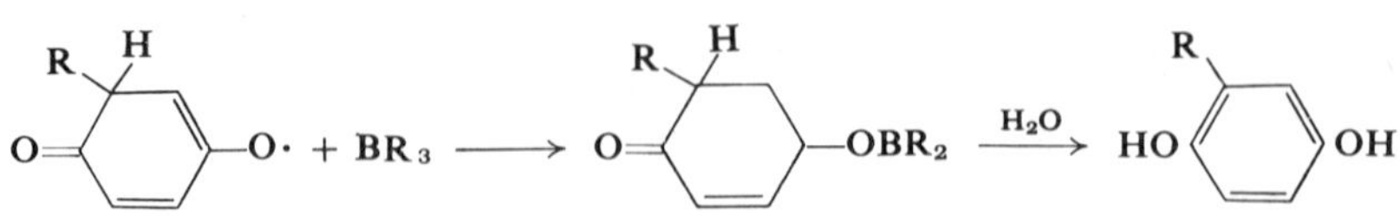

(47)

The addition of organoboranes to unsaturated carbonyl compounds results in a 3-carbon homologation. A similar reaction with unsaturated epoxides gives 4-carbon homologation.[53] Again, oxygen, or diacyl peroxides, or azoisobutyronitrile will initiate the reaction and galvinoxyl will inhibit it. Again, therefore, a radical chain mechanism is involved (equations 48–50).

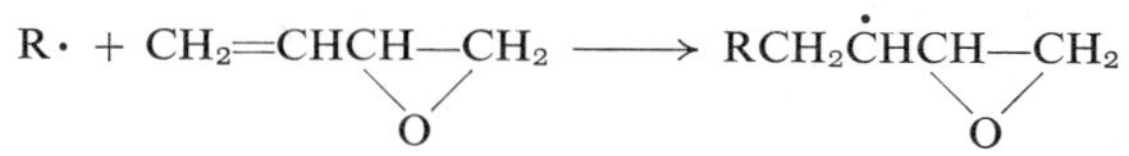

(48)

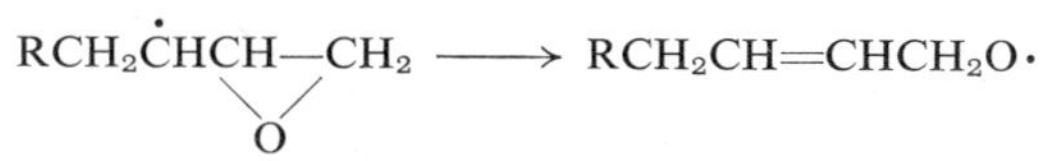

(49)

$$RCH_2CH{=}CHCH_2O\cdot + BR_3 \longrightarrow RCH_2CH{=}CHCH_2OBR_2 + R\cdot \quad (50)$$

E. Chemically Induced Homolysis of Peroxides

Many organometallic compounds are known to react with peroxides (particularly hydroperoxides) below the temperature at which unimolecular homolysis of the oxygen–oxygen bond occurs. There is strong evidence that some of these reactions involve hydroxyl, alkoxyl, or benzoyloxyl radicals, and it is a reasonable assumption, but as yet unproven, that these oxyl radicals react in turn with the organometallic compounds.

For example, trialkyl boranes (R_3B) react with neutral hydrogen peroxide to give boronic acids, $RB(OH)_2$, together with hydrocarbons and the compounds ROH, RCl (in carbon tetrachloride), and RI (in the presence of iodine). These derivatives are formed with some loss of stereospecificity in the alkyl group. One reasonable route for this cleavage involves the reactions 51 and 52.

$$R_3B + H_2O_2 \longrightarrow R_3\overset{-}{B}\underset{H}{\overset{+}{O}}OH \longrightarrow R\cdot + R_2BOH + \cdot OH \quad (51)$$

$$HO\cdot + R_3B \longrightarrow R_2BOH + R\cdot \quad (52)$$

A report that benzoyl peroxide reacts vigorously with trialkylboranes,[54] which might be interpreted in terms of a radical chain reaction, could not be confirmed.[55]

A related system is that in which α-phenylalkyl boranes react with hydrogen peroxide and ferrous salts.[56] The products of such reactions have been interpreted in terms of abstraction of hydrogen, by the $\cdot OH$ radical, from the α position of the benzylic group, but an S_H2 displacement at the boron center may also be involved.

F. Kinetics and Mechanism

Relative rate constants for homolytic alkoxydealkylation at metal centers have been determined by causing these reactions to compete with some other homolytic reaction, and measuring the relative concentrations of the respective intermediates (by ESR) or products (usually by glc). Absolute rate constants for the organometallic reactions can then be derived if those for the reference reactions are known.

The following four variants of the method have been used; as yet they have been applied only to boron compounds, but their potential extension to other metals is obvious.

1. By the ESR technique, *t*-butoxyl radicals are generated photolytically from di-*t*-butyl peroxide in the presence of, say, a trialkyl borane and cyclopentane in an ESR cavity (see Section III.B), and the relative concentrations of the radicals R· and C_5H_9· are determined from the superimposed spectra.[41,57]

$$t\text{-BuO}Ot\text{-Bu} \xrightarrow{h\nu} 2t\text{-BuO}\cdot \begin{cases} \xrightarrow[k_a]{BR_3} t\text{-BuOBR}_2 + R\cdot & (53a) \\ \xrightarrow[k_b]{C_5H_{10}} t\text{-BuOH} + C_5\dot{H}_9\cdot & (53b) \end{cases}$$

If it can be assumed that the rate constants describing the removal of these two radicals from the system are equal, it follows that:

$$\frac{k_a}{k_b} = \frac{[C_5H_{10}][R\cdot]}{[R_3B][C_5H_9\cdot]} \tag{54}$$

Values of k_b can be derived from information in the literature, hence absolute values of k_a can be calculated. This technique can readily be applied over a range of temperature, hence activation parameters can be derived.

2. This first method cannot be readily applied to, say, triphenyl borane, where no spectrum of the displaced radical has been detected, or to trineopentyl borane, where the spectrum is poor. In these cases, and more generally, the method of product analysis from a borane–hydrocarbon mixture is used.

A deficiency of *t*-butyl hypochlorite is caused to react with a mixture of, say, a trialkyl borane and cyclopentane, and the products (alkyl chloride) and cyclopentyl chloride) are analyzed by glc.[58] Trichloroethylene may be added to suppress any chlorine chain reaction which might occur.[59]

$$t\text{-BuOCl} \xrightarrow[C_5H_9\cdot]{R\cdot\text{or}} t\text{-BuO}\cdot - \begin{cases} \xrightarrow[k_a]{BR_3} t\text{-BuOBR}_2 + R\cdot \xrightarrow{t\text{-BuOCl}} RCl & (55a) \\ \xrightarrow[k_b]{C_5H_{10}} t\text{-BuOH} + C_5H_9\cdot \xrightarrow{t\text{-BuOCl}} C_5H_9Cl & (55b) \end{cases}$$

When the borane and cyclopentane are both present in large excess, the ratios of the two rate constants are given by:

$$\frac{k_a}{k_b} = \frac{[RCl][C_5H_{10}]}{[C_5H_9Cl][BR_3]} \tag{56}$$

3. This second method was unsuitable for triphenyl borane because of analytical difficulties. Triphenyl borane was therefore caused to compete with tributyl borane in benzene as solvent, and the relative rate constants were derived from the following expression[60]:

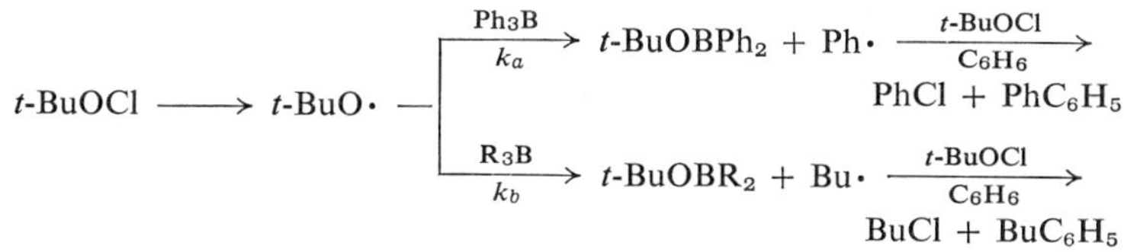

(57a)

(57b)

$$\frac{k_a}{k_b} = \frac{[\mathrm{Bu_3B}]([\mathrm{PhCl}] + [\mathrm{PhC_6H_5}])}{[\mathrm{Ph_3B}]([\mathrm{BuCl}] + [\mathrm{BuC_6H_5}])} \tag{58}$$

4. An alternative method was used to study the reaction of 1,1,2-trimethylpropyl hypochlorite;[41] here the *t*-alkoxyl radical undergoes unimolecular β-scission at a convenient and known rate, and this was used as the reference reaction.

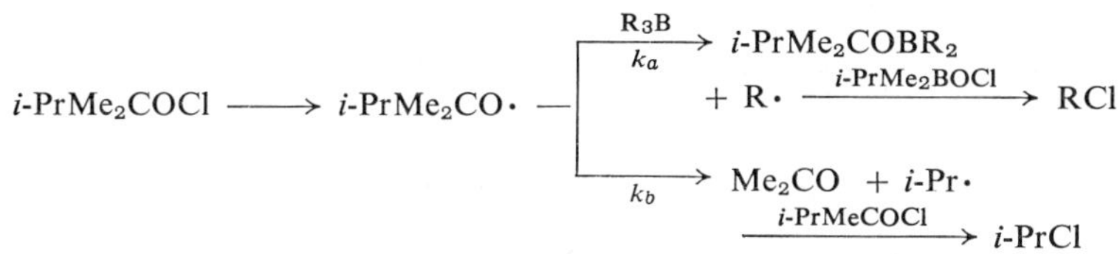

(59a)

(59b)

whence

$$\frac{k_a}{k_b} = \frac{[\mathrm{RCl}]}{[\mathrm{Me_2CO}][\mathrm{R_3B}]}$$

The reaction of *t*-butoxyl radicals with triethyl borane in the gas phase was studied by a similar method.[61]

These various methods give a consistent set of results which are listed in Table 3.

The rate constants for alkoxydealkylation, shown in Table 3, are from 3 to 500 times those for alkylperoxydealkylation, which are given in Table 1; as would be expected, the biggest differences involve the less reactive compounds, because the rates for the more rapid reactions are approaching the diffusion-controlled limit. In the series of boroxines, where steric hindrance to the approach of the *t*-alkoxyl radical would be expected to be small, there is some indication of reactivities following the order of stability of the displaced radicals (*n*-Bu < *s*-Bu < *t*-Bu). In the trialkyl boranes, where steric crowding would be larger than with the trialkylboroxines, reactivities (Et ~ *n*-Bu > *i*-Bu ~ *s*-Bu > neopentyl) are more in line with steric effects.

TABLE 3

Kinetic Parameters for Bimolecular Homolytic *t*-Alkoxydealkylation at Boron

R_3B	Method	Reference reactant	k ($T°$) (M^{-1} sec^{-1})	E_a (kcal/mole)	A (M^{-1} sec^{-1})	Ref.
Et_3B	4	—	1×10^7 (134.5)	—	—	61
n-Bu_3B	1	C_5H_{10}	3×10^7 (30)	0	3.5×10^7	57
n-Bu_3B	2	C_6H_{12}	1×10^7 (40)	0	6.1×10^6	58
n-Bu_3B	2	$Me_2CHCHMe_2$	5×10^6 (40)	—	—	58
n-Bu_3B	4	—	5×10^7 (40)	—	—	41
i-Bu_3B	1	C_5H_{10}	1×10^6 (30)	1.1	6.8×10^6	57
i-Bu_3B	2	C_6H_{12}	3×10^5 (40)	—	—	58
i-Bu_3B	2	C_5H_{10}	3×10^5 (40)	—	—	58
s-Bu_3B	1	C_5H_{10}	3×10^5 (30)	4.5	4.9×10^8	57
s-Bu_3B	2	C_6H_{12}	4×10^5 (40)	5.9	5.1×10^9	58
$(Me_3CCH_2)_3B$	2	C_6H_{12}	4×10^4 (40)	0.7	1.1×10^5	58
$(PhCH_2)_3B$	2	C_6H_{12}	2×10^7 (40)	—	—	58
Ph_3B	3	BBu_3	1×10^7 (25)	—	—	60
n-Bu_2BOnBu	4	—	6×10^5 (40)	—	—	60
$(MeBO)_3$	1	C_5H_{10}	2×10^6 (30)	2.8	2.0×10^8	57
(n-$BuBO)_3$	1	C_5H_{10}	6×10^5 (30)	2.2	2.4×10^7	57
(n-$BuBO)_3$	2	C_6H_{12}	3×10^5 (40)	—	—	58
(i-$BuBO)_3$	1	C_5H_{10}	1×10^6 (30)	2.3	5.8×10^7	57
(s-$BuBO)_3$	1	C_5H_{10}	6×10^5 (30)	—	—	57
(t-$BuBO)_3$	1	C_5H_{10}	3×10^6 (30)	2.6	2.3×10^8	57

Activation energies are always small, as would be expected for such exothermic reactions, and differences in rates are often due as much to variations in A as in E_a.

It is interesting that *t*-butoxyl radicals bring about the dephenylation of triphenyl borane at about the same rate as the dealkylation of tri-*n*-butyl borane[60]; this appears to be the first quantitative information on a radical chain reaction in which the phenyl radical is a chain carrier.

Attempts have been made by ESR spectroscopy to determine whether these homolytic alkoxydealkylation reactions proceed through an intermediate or a transition state, that is, whether the mechanism is S_H2 (synchronous) or S_H2 (stepwise) (see Section I).

$$RO\cdot + MR_n \longrightarrow [RO\dot{M}R_n] \longrightarrow ROMR_{n-1} + R\cdot \qquad (60)$$

All attempts to date to detect an intermediate where M is boron or aluminum have been unsuccessful.[31,61,62] Where M is phosphorus, however, an intermediate phosphoranyl radical, t-BuO$\dot{P}$R$_3$ ($R_3 = H_3$, MeH_2, Me_2H, and Me_3) can be observed, with a large phosphorus splitting.[31,62] Hyperfine splitting by the groups R_3 is consistent with a trigonal bipyramidal structure in which the singly occupied orbital is equatorially directed, and the t-BuO group is probably axial.

G. Related Oxygen Radicals

There are a number of reports of 2,4,6-trisubstituted phenoxyl radicals reacting with organometallic compounds (e.g., Na,[63] Li, Mg,[64] Al[64,65]), particularly Grignard reagents,[66] to give products derived from the corresponding organic radicals, for example,[67]

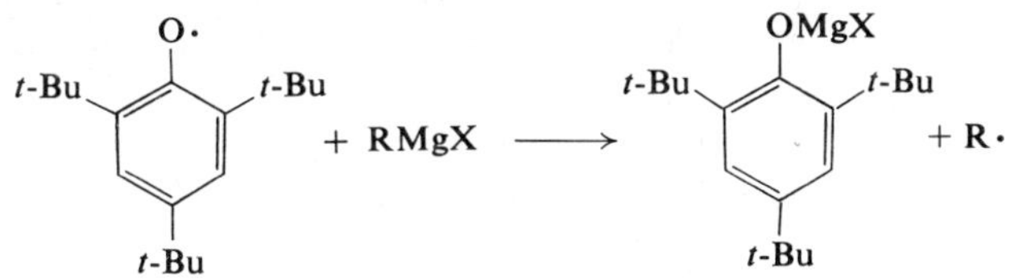

(61)

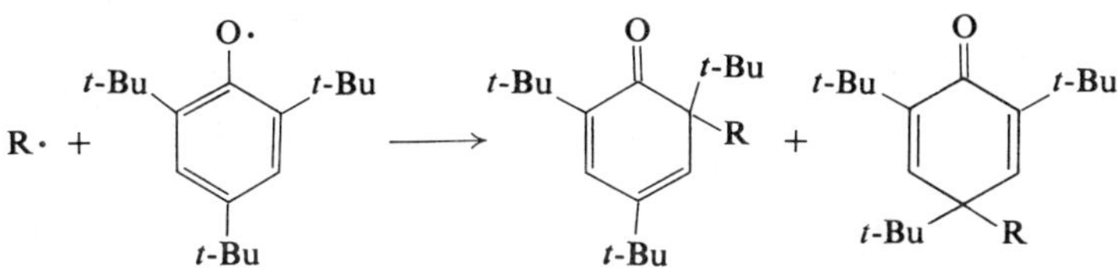

(62)

$$2R\cdot \longrightarrow R_2 \qquad (63)$$

An S_H2 process is a formal possibility for the initial step, but, in view of the steric hindrance which would be exerted by the ortho substituents, an electron transfer process is probably involved:

$$ArO\cdot + RMgX \longrightarrow ArO^- + \overset{+}{Mg}X + R\cdot \qquad (64)$$

Razuvaev et al. have described an interesting rearrangement of *ortho*-trimethylsilyl phenoxyl radicals, which can reasonably be represented as S_H2 aryloxydealkylation at the silicon center[68,69]; such a reaction would be facilitated by the high silicon–oxygen bond strength.

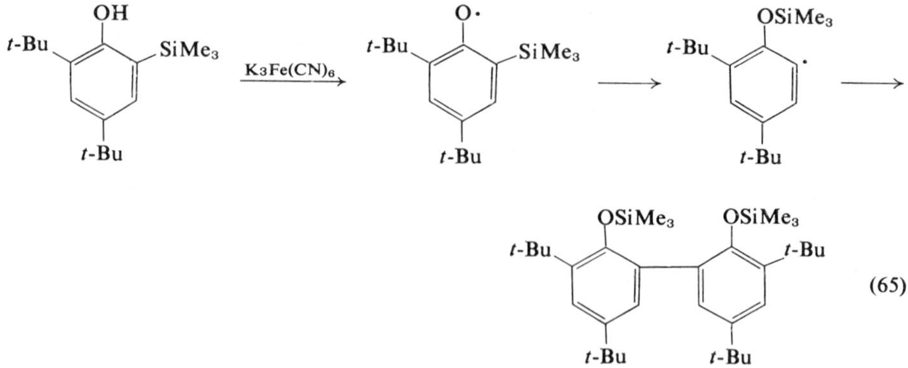

(65)

Nitroxyl radicals similarly react with Grignard reagents; for example, di-*t*-butylnitroxide and norborn-2-ylmagnesium bromide give *N*,*N*-di-*t*-butyl-*O*-norbornyl hydroxylamine in 40% yield.[70] Again, the initial step of the reaction could be represented either as an S_H2 displacement or an electron transfer process.

$$R_2NO\cdot + R'MgX \longrightarrow R_2NOMgX + R'\cdot \tag{66}$$

$$R'\cdot + R_2NO\cdot \longrightarrow R_2NOR' \tag{67}$$

The highly electronegative bis(trifluoromethyl)nitroxide behaves as a pseudohalide radical and displaces the halide from the compounds BBr_3, $AsBr_3$, AsI_3, $BiBr_3$, BiI_3, BiOI, and alkali metal halides.[71] These last reactions are clearly electron transfer processes, but others might well be represented as S_H2 reactions, for example,

$$(CF_3)_2NO\cdot + BBr_3 \longrightarrow (CF_3)_2NOBBr_2 + Br\cdot \tag{68}$$

A similar mechanism may also govern the reaction with tris(trifluoromethyl)arsine,[72]

$$(CF_3)_2NO\cdot + As(CF_3)_3 \longrightarrow (CF_3)_2NOAs(CF_3)_2 + \cdot CF_3 \tag{69}$$

$$(CF_3)_2NO\cdot + \cdot CF_3 \longrightarrow (CF_3)_2NOCF_3 \tag{70}$$

and may also be involved in the reaction between nitric oxide and organometallic compounds.[73–76]

H. Carbonyl Triplets

When a carbonyl compound is excited to its $n \rightarrow \pi^*$ triplet state, the non-bonding n electron which is promoted into the antibonding π^* orbital is concentrated largely on carbon, leaving the oxygen with a reactivity similar to that of an alkoxyl radical; for example, it will abstract hydrogen from a hydrocarbon.

$$R_2C{=}O \xrightarrow{h\nu} R_2\dot{C}{=}\ddot{O}\cdot \xrightarrow{R'H} R_2\dot{C}OH + R'\cdot \qquad (71)$$

Triplet carbonyl compounds would therefore be expected, like alkoxyl radicals, to react with organometallic compounds by an S_H2 process.

$$R_2C{=}O^* + MR'_n \longrightarrow R_2\dot{C}OMR'_{n-1} + R'\cdot \qquad (72)$$

Reactions of this type have been established for organoboranes both in the gas phase and in solution.[77–78]

The most direct evidence for this mechanism is that if a mixture of a ketone and an organoborane is irradiated in solution, the superimposed spectra of the two radicals which are formed (reaction 73a) can be observed.[77] The products which were identified result from the interaction of these radicals (reactions 73b, c, and d). With certain ketones, such as diisopropyl ketone, the Norrish I scission of the ketone adds the signals of the radical R· to the spectrum (reaction 73e). Results of experiments of this type are summarized in Table 4.

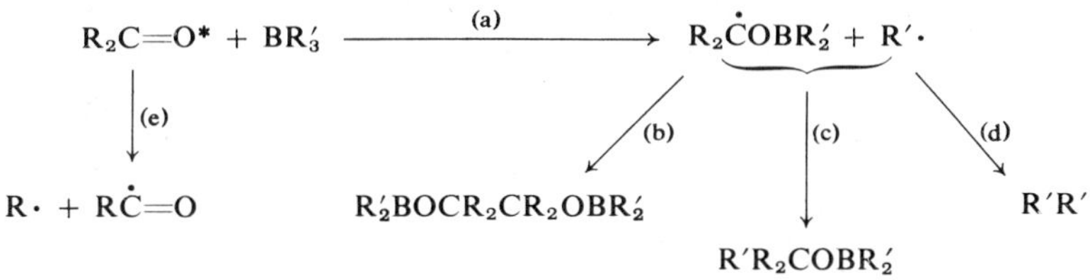

(73)

If a triplet quencher such as tetramethylethylene or piperylene is added to the mixture, the intensities of the spectra of the radicals (R·, R′·, and $R_2\dot{C}OBR'_2$) diminish, but the relative intensities stay constant.

Rate constants for the S_H2 reactions have been determined by measuring the concentrations of the appropriate radicals by ESR when the triplet is being removed from the system by (inter alia) the following competing processes:

1. The S_H2 process and the reaction with triplet quencher. The triplet is now being removed by the unimolecular reactions 74, and the bimolecular reactions 73a and 75.

$$R_2CO^* \longrightarrow \begin{cases} R_2CO \\ R_2CO + h\nu \\ R\cdot + R\dot{C}O \end{cases} \qquad (74)$$

$$R_2CO^* + Q \longrightarrow \begin{cases} R_2CO + Q \\ R_2CO + Q^* \\ \text{products} \end{cases} \qquad (75)$$

TABLE 4

Photolysis of Toluene Solutions of Mixtures of Aliphatic Ketones and Trialkyl Boranes

Trialkyl Borane	Ketone	Temperature (°C)	Radicals[a]
n-Bu_3B	Acetone or acetone-d_6	20 to −130	n-Bu· ~ X·
n-Pr_3B	Acetone or acetone-d_6	20 to −130	n-Pr· ~ X·
$(n\text{-}C_6H_{13})_3B$	Acetone or acetone-d_6	20 to −100	n-C_6H_{13}· ~ X·
i-Bu_3B	Acetone	0 to −110	i-Bu· ~ X·
s-Bu_3B	Acetone	−40 to −130	s-Bu· ~ X·
(neo-$C_5H_{11})_3B$	Acetone	20 to −70	None
n-Bu_3B	MeCOi-Pr	−30 to −110	X· ⩾ n-Bu·
n-Pr_3B	MeCOi-Pr	−30 to −110	X· ⩾ n-Pr·
i-Bu_3B	MeCOi-Pr	−90	X ~ i-Bu· ≫ i-Pr·
s-Bu_3B	MeCOi-Pr	−105	i-Pr· > s-Bu· ~ Y·
n-Pr_3B	i-Pr_2CO	−95	X· > n-Pr· ≫ i-Pr·
n-Bu_3B	i-Pr_2CO	−85	X· > n-Bu· > i-Pr·
i-Bu_3B	i-Pr_2CO	−95	i-Pr· ≫ i-Bu· ~ Y·
s-Bu_3B	i-Pr_2CO	−105	i-Pr·
n-Pr_3B	$(PhCH_2)_2CO$	+20 to −70	$PhCH_2$·
n-Bu_3B	$(PhCH_2)_2CO$	+20 to −70	$PhCH_2$·
n-Bu_3B	Et_2CO	−75	n-Bu· ~ Y·
n-Bu_3B	$(Cl_3C)_2CO$	−73	Cl_3C·
n-Pr_3B	Norcamphor	−20	n-Pr· ~ Y·
n-Bu_3B	Mesityl oxide	−70	n-Bu· ⩾ Y·
n-Bu_3B	3-Methyl-2-butene-2-one	−70	n-Bu· ⩾ Y·

[a] The radicals which were observed are listed in the order of decreasing concentration. The symbol X· represents an identified radical of the type $R_2\dot{C}OBR'_2$; Y· represents a similar boron-containing radical whose ESR spectrum could not be analyzed fully because of a poor signal-to-noise ratio.

Then

$$\frac{[R_2CO^*]_0}{[R_2CO^*]} = \frac{[R'\cdot]_0{}^2}{[R'\cdot]^2} = 1 + \frac{k_{75}[Q]}{k_{74} + k_{73a}[R'_3B]} \tag{76}$$

where the subscript 0 indicates the absence of quencher. If k_{75} is known, k_{73a} can therefore be determined.

2. The S_H2 process and the Norrish I reaction, whence

$$\frac{k_{73a}}{k_{73e}} = \frac{[R'\cdot]}{[R\cdot][R'_3B]} \tag{77}$$

3. The S_H2 process with two different boranes R'_3B and R''_3B, whence

$$\frac{k'_{73a}}{k''_{73a}} = \frac{[R'\cdot][R''_3B]}{[R''\cdot][R'_3B]} \tag{78}$$

Results are given in Table 5.

The photolysis of acetone in the presence of triethyl borane has also been studied in the gas phase. The borane reacts with the acetone triplets (reaction 73a) and thereby reduces the rate of their breakdown to give carbon monoxide via the acetyl radical (reaction 73e). From an analysis of the hydrocarbons and carbon monoxide produced, the value of k_{73a} at 20 and 126° was observed. The results are included in Table 5.

The rate constants for the S_H2 reactions of the acetone triplets (Table 4) are parallel with (and usually within a factor of 10 of) those for the *t*-butoxyl radical (Table 5): again, the reactivity appears to be determined more by steric than by electronic factors. The rates show little dependence on temperature, and activation energies are probably not larger than 2 kcal/mole.

TABLE 5

Rate Constants for the S_H2 Reaction of Acetone Triplets with Organoboranes

R_3B	k_{73a} (20°) (M^{-1} sec^{-1})
Et_3B	1×10^{8} [a]
n-Pr_3B	1.4×10^{7}
n-Bu_3B	7×10^{6}
n-Hex_3B	8×10^{6}
i-Bu_3B	8×10^{4}
s-Bu_3B	$< 10^{4}$

[a] 3.4×10^{8} M^{-1} sec^{-1} at 126°; reference 78.

IV. ALKYLTHIYL RADICALS

The reaction between thiols and organometallic compounds is a well-established route to alkylthiometallic compounds.[79,80]

$$RSH + MR'_n \longrightarrow RSMR'_{n-1} + R'H \tag{79}$$

For many metals these are doubtless polar processes, but in some cases there is a lack of correlation between the acidity of the thiol and its reactivity when compared with other Brønsted acids.[80]

The alkylthiodealkylation of organoboranes was assumed to be heterolytic,[81–84] but in 1964 Mikhailov and Bubnov found that trialkyl boranes which had been heated to 100° to free them from peroxides did not react with thiols below 150°.[85,86] The reactions could be initiated by hydrogen peroxide, or alkyl hydroperoxides, or air, or by broken porcelain which occluded air. Peroxides would also initiate the cyclization of β-mercaptoethoxyboranes (equation 80).

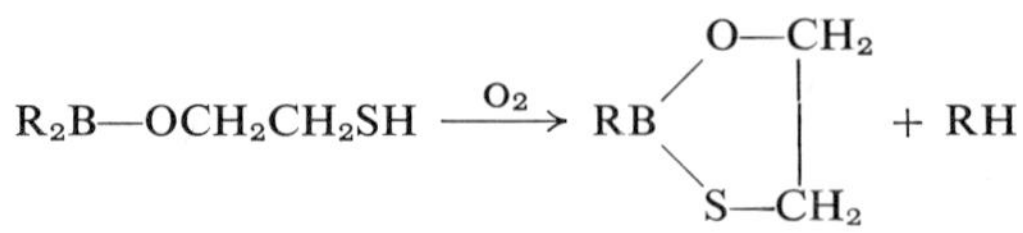
(80)

However, the obvious mechanistic deductions were not drawn from these observations.

The radical chain nature of the reactions was established in 1970 for organostibines and organobismuthines.[87] These react with benzenethiol to give the phenylthio compounds $R_nM(SPh)_{3-n}$, and the reactions can be accelerated by di-*t*-butyl hyponitrite, and sometimes inhibited by galvinoxyl, phenothiazine, or 2,2,6,6-tetramethylpiperid-4-one-1-oxyl.

The mechanism is therefore assumed to be a radical chain process, involving, as a product-forming step, the S_H2 reaction of the phenylthiyl radical at the metal center.*†

$$PhS\cdot + MR_3 \longrightarrow PhSMR_2 + R\cdot \quad (81)$$

$$R\cdot + PhSH \longrightarrow RH + PhS\cdot \quad (82)$$

$$RS\cdot + PR'_3 \longrightarrow [RS\dot{P}R'_3] \longrightarrow R\cdot + SPR'_3$$

A second, and a third, M—R bond can sometimes be cleaved by a similar mechanism.[87] An earlier report[88] that triphenylstibine and benzene-thiol did not react can probably be ascribed to the presence of some adventitious inhibitor.

The reaction of alkylthiyl radicals with trialkyl boranes has been studied more thoroughly.[92] If the disulfides RSSR (R = Ph, *t*-Bu, or *n*-Bu) are photolyzed in the presence of tri-*n*-butyl, triisobutyl, or tri-*s*-butyl borane (R'_3B), the ESR spectrum of the appropriate displaced butyl radical ($R'\cdot$) can

* When MR_3 = $SbPh_3$ or $BiPh_3$, these appear to be the first radical chain reactions to be identified to involve the $Ph\cdot$ radical as a chain-carrying species. The reaction of triphenyl borane with *t*-butyl hypochlorite (equation 57a) provides a further example.

† No equivalent alkylthiodealkylation is known for phosphines. Trialkyl- and triarylphosphines (and trialkyl phosphites), react with alkylthiyl radicals (from thiols or disulfides) to give the corresponding phosphorus(V) compounds.[89–91]

be observed. Quantum yields, however, are less than in the equivalent reactions of di-*t*-butyl peroxide.

$$\text{RSSR} \xrightarrow{h\nu} 2\text{RS}\cdot \tag{83}$$

$$\text{RS}\cdot + \text{BR}'_3 \longrightarrow \text{RSBR}'_2 + \text{R}'\cdot \tag{84}$$

Above about 35°, the displaced alkyl radical can react rapidly with diphenyl disulfide by an S_H2 process at sulfur, to regenerate the phenylthiyl radical, and a chain reaction can be induced. For example, with 2 mole % of di-*t*-butyl hyponitrite as initiator, diphenyl disulfide and tri-*n*-butyl borane react to give butyl phenyl sulfide in 76% yield after 115 min at 50°.

$$\text{PhS}\cdot + \text{BBu}_3 \longrightarrow \text{PhSBBu}_2 + \text{Bu}\cdot \tag{85}$$

$$\text{Bu}\cdot + \text{PhSSPh} \longrightarrow \text{BuSPh} + \text{PhS}\cdot \tag{86}$$

If tri-*n*-butyl borane and triisobutyl borane compete for the *n*-butylthiyl radical, the relative rate constants for the competing reactions are related to the intensities of the spectra of the two radicals by the expression 87.

$$\frac{k_n}{k_i} = \frac{[n\text{-Bu}\cdot]}{[i\text{-Bu}\cdot]} \times \frac{[i\text{-Bu}_3\text{B}]}{[n\text{-Bu}_3\text{B}]} \tag{87}$$

At 50°, k_n/k_i has the value of 16.9, and the activation energy for the S_H2 reaction of n-Bu_3B is 1.3 kcal/mole less than that for i-Bu_3B. Approximate absolute values for the rate constants were determined by causing the trialkyl borane and 1-octene to compete for reaction with the alkylthiyl radical derived from butanethiol, using di-*t*-butyl hyponitrite as an initiator.

$$\text{BuSH} \longrightarrow \text{BuS}\cdot \begin{cases} \xrightarrow{\text{BR}_3} \text{BuSBR}_2 + \text{R}\cdot \xrightarrow{\text{BuSH}} \text{RH} & \text{(88a)} \\ \xrightarrow{\text{C}_8\text{H}_{16}} \text{BuS}\dot{\text{C}}_8\text{H}_{16} \xrightarrow{\text{BuSH}} \text{BuSC}_8\text{H}_{17} & \text{(88b)} \end{cases}$$

If the thiol is present in deficiency, and the addition of the alkylthiyl radical to the olefin is essentially irreversible, it follows that

$$\frac{k_{88a}}{k_{88b}} = \frac{\log([\text{BR}_3]_0/[\text{BR}_3]_\infty)}{\log([\text{C}_8\text{H}_{16}]_0/[\text{C}_8\text{H}_{16}]_\infty)} \tag{89}$$

Values for k_{88a}/k_{88b} for four different organoboranes are given in Table 6, and if k_{88b} is taken to be approximately $4.3 \times 10^5\ M^{-1}\ \text{sec}^{-1}$ at 50°, the approximate absolute values of k_{88a}, which are listed, can be derived.

The butylthiyl radical is seen to have a reactivity toward organoboranes similar to that of the *t*-butoxyl radical, and again relative reactivities appear to be determined as much by steric as by electronic factors.

A number of other isolated observations have been recorded, which indicate that homolytic thiodealkylation is taking place.

TABLE 6

Relative and Absolute Rate Constants for Bimolecular Homolytic Butylthiodealkylation at Boron

Organoborane	k_{88a}/k_{88b} (50°)	k_{88a} (M^{-1} sec^{-1} at 50°)
n-Bu_3B	111	4.8×10^7
i-Bu_3B	10.4	4.5×10^6
s-Bu_3B	1.5	6.4×10^5
n-Bu_2BSn-Bu	0.8	3.4×10^5

Primary alkylmagnesium halides react thermally with di-*t*-butyl disulfide, but *t*-butylmagnesium chloride does not, and the fact that irradiation of their mixture with ultraviolet light gives the ESR spectrum of the *t*-butyl radical provides what is probably the most direct evidence of an S_H2 reaction at the magnesium atom in a Grignard reagent.[93]

$$t\text{-BuS}\cdot + t\text{-BuMgCl} \longrightarrow t\text{-BuSMgCl} + t\text{-Bu}\cdot \quad (90)$$

The rate of the reaction between benzenethiol and diphenylmercury in benzene at 35° is increased by di-*t*-butyl hyponitrite and reduced by phenothiazine, and must be, at least in part, a radical chain process.[94]

The reaction of elemental sulfur with organometallic compounds, under various conditions, may also contain a homolytic component. For example, tripropyl borane reacts with sulfur during 3.5 hr at 140–150° to give dipropyl(propylthiyl) borane in 38% yield,[95] and systems of this type are reported to be good initiators for the polymerization of chloroprene.[96]

V. DIALKYLAMINYL AND IMIDYL RADICALS

A. Dialkylaminyl Radicals

Little work has been reported on the reaction of dialkylaminyl radicals with organometallic compounds.

If dimethylaminyl radicals are generated, in the presence of tri-*n*-butyl borane by the photolysis of tetramethyltetrazene, the ESR spectrum of the butyl radical can be observed, and the principal products are dibutyl(dimethylamino) borane and octane, which are formed by reactions 92 and 93.[97]

$$Me_2NN{:}NNMe_2 \xrightarrow{h\nu} N_2 + 2Me_2N\cdot \quad (91)$$

$$Me_2N\cdot + BBu_3 \longrightarrow Me_2NBBu_2 + Bu\cdot \quad (92)$$

$$2Bu\cdot \longrightarrow \text{combination and disproportionation products} \quad (93)$$

The appropriate alkyl radical was also detected when the tetrazene was photolyzed in the presence of tri-*s*-butyl borane, oxybis(dibutyl borane), tri-*s*-butylboroxine, or triethylstibine.[97]

If the dimethylaminyl radical is obtained from *N*-chlorodimethylamine, the displaced alkyl radical abstracts chlorine to regenerate the dimethylaminyl radical, and a chain reaction is set up. The reaction with tri-*n*-butyl borane is self-initiating, and, in chlorobenzene, butyl chloride and di-*n*-butyl(dimethylamino) borane are formed in about 50% yield in a few minutes at room temperature.[98]

$$Me_2N\cdot + BBu_3 \longrightarrow Me_2NBBu_2 + Bu\cdot \tag{94}$$

$$Bu\cdot + Me_2NCl \longrightarrow BuCl + Me_2N\cdot \tag{95}$$

The remainder of the product consists of dibutyl chloroborane and butyldimethylamine, which are formed in a parallel polar reaction (equation 96), and, if galvinoxyl is added, the radical chain reaction is suppressed, and the chloroborane and *t*-amine are the only products.

$$Me_2NCl + Bu_3B \longrightarrow Me_2NBu + Bu_2BCl \tag{96}$$

N-Chlorodiethylamine and *N*-chloropiperidine are also reported to react with organoboranes to give the appropriate alkyl halide in 30–50% yield,[99] but chloramine itself, in aqueous alkaline tetrahydrofuran, reacts principally or completely by the polar route to give primary amine.[100]

In the presence of an excess of *N*-chlorodimethylamine, the second and third alkyl groups can be cleaved from the boron; this is now a purely homolytic reaction, with no polar component, and the ultimate products are butyl chloride and tris(dimethylamino) borane.[98]

No kinetic studies have yet been carried out on these reactions.

Alkylmagnesium halides also react readily with *N*-halogenoamines to give both alkyl halides and alkylamines.[101,102]

$$R_2NX + R'MgY \longrightarrow R'X \text{ and } R_2NR' \tag{97}$$

The epimeric norborn-2-ylmagnesium bromides react with *N*-chlorodimethylamine to give 2-chloronorbornane with substantial loss of stereospecificity[103]; again then, the reaction is, at least in part, a homolytic reaction involving the S_H2 process of equation 98.

$$Me_2N\cdot + RMgX \longrightarrow Me_2NMgX + R\cdot \tag{98}$$

$$R\cdot + Me_2NCl \longrightarrow RCl + Me_2N\cdot \tag{99}$$

The reaction between dimethylaminyl radicals (from tetramethyltetrazene) and methyl diphenylphosphinite has been studied, but an S_H2 reaction is not involved; the intermediate phosphoranyl radical undergoes β-scission, and the major products are *N,N*-dimethyldiphenylphosphinamide and methyldiphenylphosphine oxide.[104]

$$Me_2N\cdot + Ph_2POMe \longrightarrow Me_2N\dot{P}(Ph)_2OMe \longrightarrow Me_2NP(Ph)_2O + Me\cdot \quad (100)$$

$$Me\cdot + Ph_2POMe \longrightarrow Me\dot{P}(Ph)_2OMe \longrightarrow MePh_2PO + Me\cdot \quad (101)$$

B. The Succinimidyl Radical

N-Halogenoamides can react by both homolytic and heterolytic mechanisms, and it is often difficult to distinguish the two processes.

Allylic bromination is well established to be a free radical chain process.[105] Most of the evidence indicates that the bromine radical, and not the succinimidyl radical, abstracts hydrogen, the *N*-bromosuccinimide acting as a reservoir of bromine in low concentration (equations 102–104).

$$Br\cdot + HR \longrightarrow HBr + R\cdot \quad (102)$$

$$R\cdot + Br_2 \longrightarrow RBr + Br\cdot \quad (103)$$

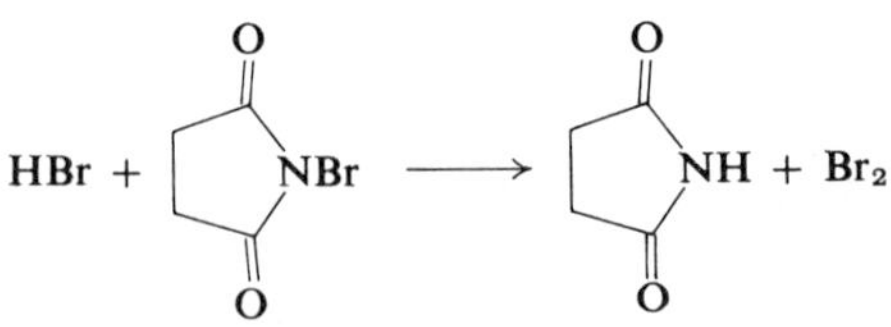

(104)

The fact that the succinimidyl radical did not act as a chain carrier was ascribed to its stabilization by resonance, but it now appears more likely that the unpaired electron is localized on the nitrogen in a largely sp^2 orbital[106]: *N,N*-bis(succinimidyl) is very stable to N-N scission, and the N—H bond strength in succinimide may be as high as 100 kcal/mole^{-1}. Abstraction of hydrogen in a weak CH bond should then be exothermic, but the only clear demonstration of the abstraction of hydrogen by the succinimidyl radical is in the oxidation of 1-phenylethanol to acetophenone by *N*-iodosuccinimide, which involves the step[107]:

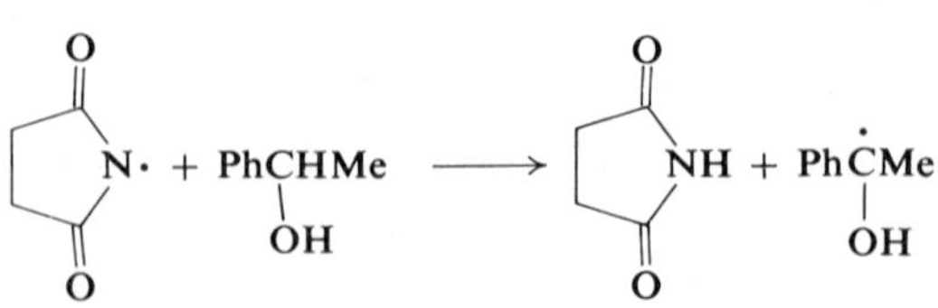

(105)

In the field of organometallic compounds, only the organotins have been studied in detail.[108] These react with *N*-halogenosuccinimides with the cleavage of one alkyl–tin bond according to equation 106.

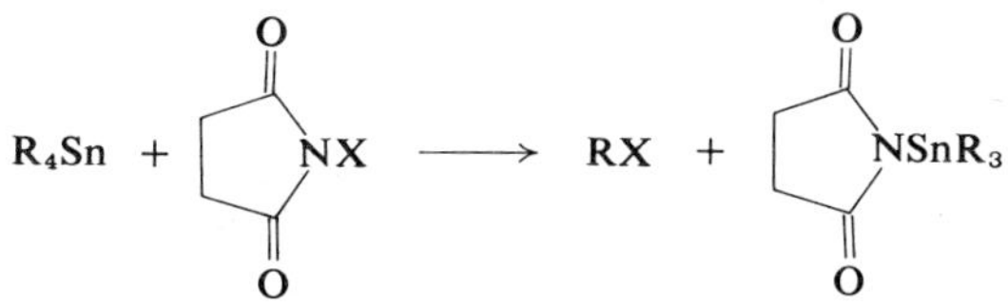

(106)

Benzyltributyltin reacts to give benzyl halides and *N*-tributylstannylsuccinimide exclusively,[108] and the cyclopropyl–tin bond in *cis*- or *trans*-trimethyl(2-phenylcyclopropyl)tin is cleaved with loss of stereospecificity.[109]

The reactions are accelerated by di-*t*-butyl hyponitrite and by phenylazotriphenylmethane and are inhibited by galvinoxyl or 2,6-di-*t*-butyl-4-methoxyphenol, and retarded by oxygen. The kinetics of the reactions of tetra-*n*-propyltin, tetra-*n*-butyltin, and benzyltri-*n*-butyltin in acetone were studied at 35°.[108] Termination was solely by the bimolecular interaction of succinimidyl radicals (reaction 109), except for the reaction of benzyltributyltin at low concentration of bromoimide, when the attack of benzyl radicals on *N*-bromosuccinimide became rate determining, and termination involved the interaction of two benzyl radicals (reaction 111).

The mechanism of the reaction is therefore as follows:

Propagation:

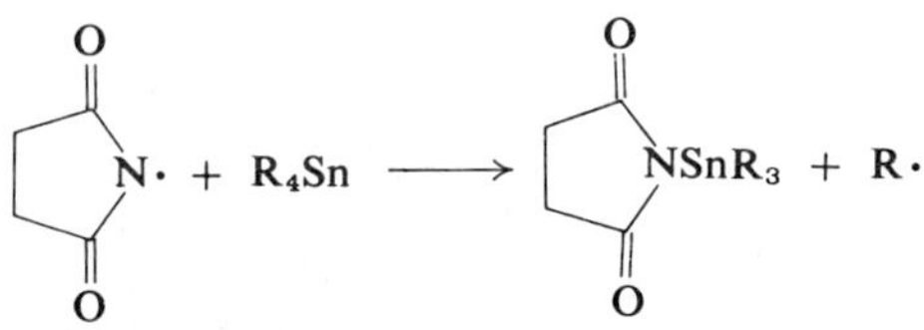

(107)

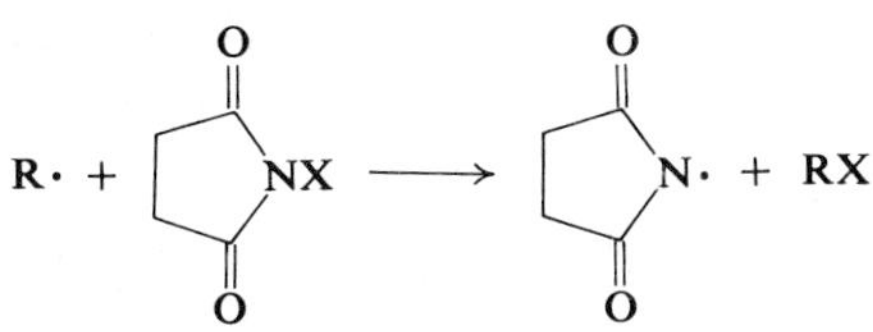

(108)

Termination:

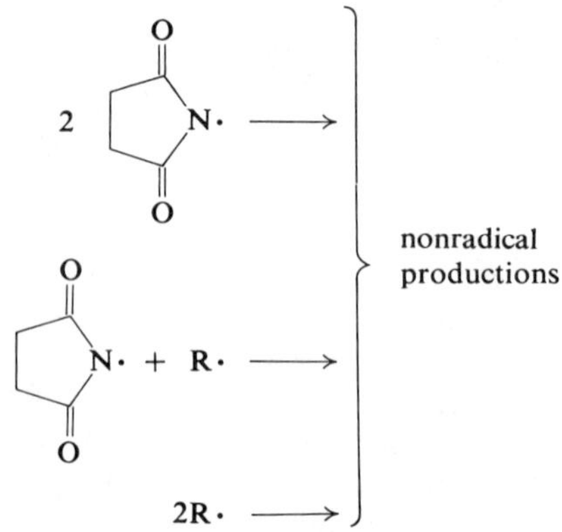

(109)

(110)

(111)

A reasonable value for the rate constant for the interaction of two succinimidyl radicals was assumed ($2 \times 10^9\ M^{-1}\ sec^{-1}$), and hence absolute rate constants for the S_H2 reaction of the succinimidyl radicals at the tin center (reaction 107) could be derived.* Values for further organotin compounds were then determined by the competition method. The results for reaction in acetone at 35° are given in Table 7.[108]

TABLE 7

Rate Constants for Bimolecular Homolytic Succinimidyl Dealkylation at Tin

R_4Sn	$k_{107}{}^a$ (35°) (M^{-1} sec^{-1})
Me_4Sn	$1.4 \times 10^{5\,I}$
Et_4Sn	$4 \times 10^{4\,I}$; $3.5 \times 10^{4\,B}$
Pr_4Sn	$4 \times 10^{3\,B}$
Bu_4Sn	$4 \times 10^{3\,B}$; $8 \times 10^{3\,B}$
s-Bu_4Sn	$8 \times 10^{2\,B}$
Bu_3SnCH_2Ph	$6 \times 10^{4\,B}$

[a] Determined in the reaction with (I) *N*-iodosuccinimide, (B) *N*-bromosuccinimide.

* The dimethylamino radical appears to be less reactive; no reaction could be induced between *N*-chlorodimethylamine and tetra-*n*-butyltin at 35°.[110]

The relative homolytic reactivity of the simple tetraalkyltin compounds, like that of the trialkyl boranes discussed in previous sections, appears to be determined principally by steric factors, but the high rate constant for the displacement of the benzyl group from tributylbenzyltin presumably reflects the resonance stabilization of the benzyl radical.

The aryl–tin bond in compounds such as tetraphenyltin,[111] dibenzyldiphenyltin,[112] or trimethyl-*p*-tolyltin[113] is cleaved rapidly and selectively by *N*-halogenosuccinimides to give the appropriate aryl halide, but the reactions do not appear to be subject to initiation or inhibition, and probably involve electrophilic attack by the halide on the aryl group.

Relatively little work has been reported on organic derivatives of other metals. Tetraalkyllead compounds react readily with *N*-bromosuccinimide according to equation 112.[114,115]

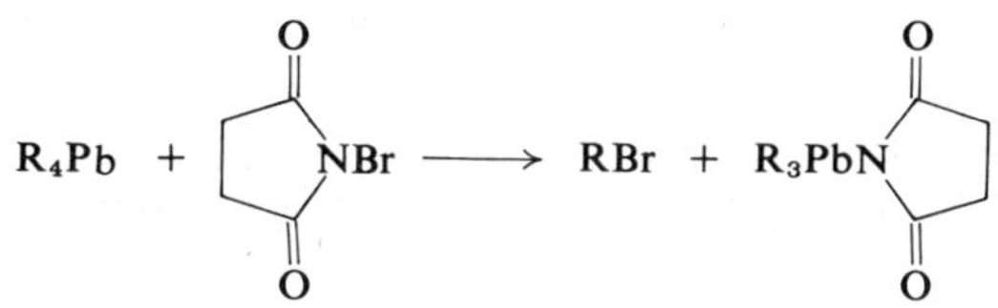

(112)

The mechanism of this reaction is not known, but the corresponding reaction of *N*-chlorosuccinimide occurs only in the presence of a source of free radicals, and is probably a radical chain process.[110] On the other hand, the dearylation of tetraaryl- or mixed alkylaryllead compounds[115,116] by *N*-bromosuccinimide probably involves electrophilic attack on the aromatic ring.

Dialkyl- and diarylmercury compounds similarly react with *N*-bromosuccinimide, and the relative reactivity of different aryl groups implies electrophilic attack.[117] *trans*-4-Methylcyclohexylmercuric chloride reacts with *N*-iodo- or *N*-bromosuccinimide to give only the *trans*-4-methylcyclohexyl iodide or bromide, but *N*-chlorosuccinimide reacts to give 93% of the *trans* and 7% of the *cis* chloride. This may be taken to imply the incursion of a homolytic component in the last reaction (equations 113 and 114).[118]

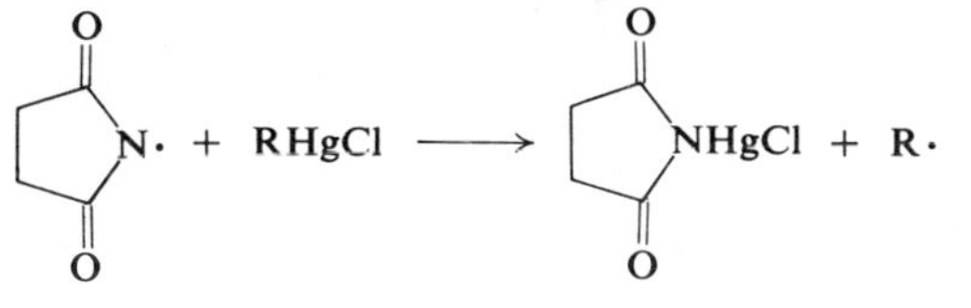

(113)

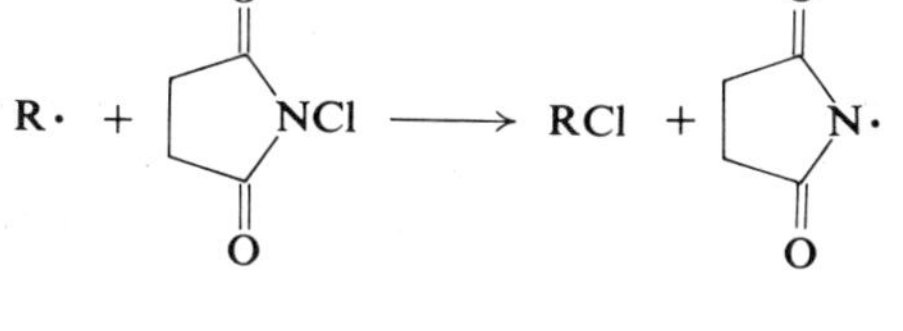

(114)

VI. ALKYL AND ARYL RADICALS

There are a number of isolated examples in the literature which indicate that alkyl radicals can bring about an S_H2 reaction at a metallic center, but little systematic work has been carried out in this field.

The early examples refer to organomercury compounds. If acetone is photolyzed in the gas phase in the presence of dimethylmercury, with either reagent labeled with CD_3 groups, products are obtained (CH_3D, CD_3H, CH_3CD_3) which are derived from both the radicals CH_3 and CD_3, and it was suggested that the methyl radical could react with dimethylmercury at both the hydrogen and the metal centers,[119,120] for example,

$$CH_3\cdot + CD_3HgCD_3 \begin{cases} \xrightarrow{\text{attack at Hg}} CH_3HgCD_3 + \cdot CD_3 \\ \xrightarrow{\text{attack at D}} CH_3D + \dot{C}D_2HgCD_3 \longrightarrow CD_2Hg + \cdot CD_3 \end{cases} \quad (115)$$

Similarly, the photolysis of azomethane in the presence of perdeuteriodimethylmercury in the liquid phase at 0° gave a methane fraction containing 12.5% of CD_3H, formed via the exchange of $\cdot CD_3$ and $\cdot CH_3$ radicals at mercury.[120]

Homolytic attack by alkyl radicals at mercury is also believed to occur in the decarboxylation of mercury(II) or mercury(I) carboxylates, which can be initiated by peroxides[121–126]

$$R\cdot + Hg(O\cdot CO\cdot R)_2 \longrightarrow RHg(O\cdot CO\cdot R) + RCO_2\cdot \longrightarrow R\cdot + CO_2 \quad (116)$$

and

$$R\cdot + (HgOCOR)_2 \longrightarrow RHgOCOR + RCO_2\dot{H}g \longrightarrow R\cdot + CO_2 + Hg \quad (117)$$

The S_H2 process is more firmly established for organoboron compounds.

Photolysis of azomethane or of acetone in the presence of triethyl borane in the gas phase between 5 and 90° gives products resulting from the attack of the methyl radical at the boron atom, and at the hydrogen of the α-methylene groups.[35,127,128]

$$CH_3\cdot + B(CH_2CH_3)_3 \begin{cases} \xrightarrow{\text{attack at B}} CH_3B(C_2H_5)_2 + C_2H_5\cdot & (118a) \\ \xrightarrow{\text{attack at H}} CH_4 + CH_3\dot{C}HB(C_2H_5)_2 & (118b) \end{cases}$$

The rate constants for these two reactions are described by the equations:

$$\log k_{118a} = 7.55 - \frac{5.20}{\theta} \qquad \text{whence} \quad k_{118a} = 5.9 \times 10^3\ M^{-1}\ \text{sec}^{-1} \text{ at } 27° \quad (119)$$

and

$$\log k_{118b} = 9.38 - \frac{7.78}{\theta} \qquad \text{whence} \quad k_{118b} = 5.3 \times 10^3\ M^{-1}\ \text{sec}^{-1} \text{ at } 27°* \quad (120)$$

* $\theta = 2.303RT$ in kilocalories per mole.

This value of k_{118b} for triethyl borane is to be compared with that for n-pentane (which has the same number of CH_2 groups) of 36 M^{-1} sec^{-1} at 27°.

The only other kinetic data which have been reported come from a study of the photolysis of hexafluoroacetone in the presence of tetramethyltin in the gas phase.[129,130] It was suggested that the trifluoromethyl radical displaced the methyl radical from the tin center (equation 122), and the rate constant for this process was found to be 1×10^4 M^{-1} sec^{-1} at 150°; the relative reactivity of other methylmetallic compounds followed the sequence $BMe_3 > SnMe_4 \gg GeMe_4 \sim SiMe_4$.

$$(CF_3)_2CO \xrightarrow{h\nu} CF_3\cdot \qquad (121)$$

$$CF_3\cdot + SnMe_4 \longrightarrow CF_3SnMe_3 + Me\cdot \qquad (122)$$

A number of other reactions have been reported, which have been, or may be, interpreted in terms of an S_H2 process by an alkyl radical at a metallic center. The equations for these reactions, and the appropriate references are noted below:

$$ArCH_3 + BBr_3 \xrightarrow{AIBN} ArCH_2BBr_2 + HBr \qquad (123)^{131,132}$$

$$CF_3I + Me_3SnSnMe_3 \xrightarrow{h\nu} Me_3SnCF_3 + Me_3SnI \qquad (124)^{133-135}$$

$$CF_2{=}CF_2 + Me_3SnSnMe_3 \xrightarrow{h\nu} Me_3SnCF_2CF_2SnMe_3 \qquad (125)^{136-138}$$

$$C_nF_{2n+1}I + R_4Pb \xrightarrow{h\nu \text{ or } \Delta} R_3PbC_nF_{2n+1} + RI \qquad (126)^{139}$$

$$CF_3I + P_4 \xrightarrow{220\text{—}220^\circ} (CF_3)_nPI_{(3-n)} \qquad (127)^{140}$$

$$CF_3I + (PPh)_4 \xrightarrow{h\nu \text{ or } \Delta} PhP(CF_3)_2 + PhP(CF_3)I \qquad (128)^{141}$$

$$CF_3I + Ph_2PPPh_2 \xrightarrow{h\nu \text{ or } \Delta} CF_3PPh_2 + Ph_2PI \qquad (129)^{142}$$

$$MeI + As(CF_3)_3 \xrightarrow{h\nu} MeAs(CF_3)_2 + CF_3I \qquad (130)^{143,144}$$

VII. HALOGEN RADICALS

Two factors tend to confuse the literature on the reactions of halogens with organometallic compounds. First, the halogen atoms are usually derived from halogen molecules, and these will often bring about a heterolytic cleavage of the carbon–metal bond to give the same products as from the homolytic reaction. Second, the rate of reaction at hydrogen in the alkyl groups may be as high as, or higher than, that at the metal center. The situation is rendered more difficult by the fact that no nonchain reactions are yet known that are amenable to study by ESR spectroscopy.

Most of the work which has been reported concerns alkyl derivatives of mercury, tin, and boron.

The reaction of a dialkylmercury or alkylmercury halide with a halogen[145,146] in a nonpolar solvent such as carbon tetrachloride or carbon disulfide can be accelerated by light or by the presence of peroxides, and strongly retarded by oxygen. This is consistent with a mechanism which involves in whole or in part a radical chain with the propagating steps shown in equations 131 and 132.[147–151]

$$X\cdot + HgRY \longrightarrow HgXY + R\cdot \qquad (131)$$

$$R\cdot + X_2 \longrightarrow RX + X\cdot \qquad (132)$$

This is confirmed by the reactions of alkylmercuric halides containing stereoisomeric groups R; *cis*- or *trans*-4-methylcyclohexylmercuric halide each reacts with iodine, bromine, or chlorine in carbon tetrachloride under nitrogen to give an approximately equimolar mixture of the *cis*- and *trans*-4-methylcyclohexyl halide, suggesting that the reactions involve a common 4-methylcyclohexyl radical.[152,153]

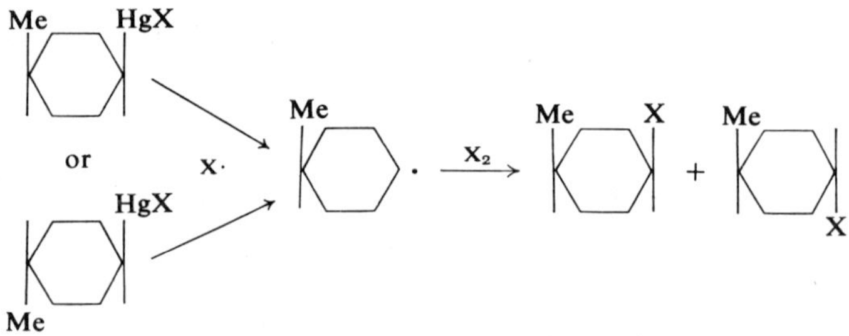

(133)

If more polar solvents (e.g., acetic acid, chloroform, or methanol) are used, bromination occurs with partial retention of configuration because of the incursion of a heterolytic component; under these conditions, the admission of air markedly increases the degree of retention because oxygen reacts with the alkyl radical and suppresses the homolytic reaction.

Similarly, optically active *s*-butylmercuric bromide reacts with bromine in carbon disulfide to give racemic *s*-butyl bromide, but if a small amount of alcohol is added to the solvent, partial retention of configuration occurs. At -40 to $-65°$, with pyridine as solvent, the reaction with bromine is wholly polar, and *s*-butyl bromide is obtained with complete retention of configuration.[154,155]

The behavior of organotin compounds is very similar. The initial proposal of a radical chain mechanism was made to account for the quantum yield greater than unity which was observed in the photolytic reaction between iodine and tetraethyltin in nonpolar solvents in the absence of oxygen.[156]

This was originally written as involving attack by the iodine atom at an α-carbon atom, but the alternative of S_H2 reaction at tin is much more likely:

$$I\cdot + SnEt_4 \longrightarrow ISnEt_3 + Et\cdot \quad (134)$$

$$Et\cdot + I_2 \longrightarrow EtI + I\cdot \quad (135)$$

The rate constant for the attack of the iodine atom was estimated to be approximately $10^5\ M^{-1}\ sec^{-1}$ at 25°.

Boué et al. have discussed the competition between the polar and the free radical reaction.[157] In the polar process, the relative reactivity of various tetraalkyltins, R_4Sn, follows a steric sequence in R (Me > Et > *n*-Pr > *i*-Pr), whereas homolytic bromodealkylation followed a sequence which was parallel to the strengths of the tin–alkyl bonds: Relative reactivities were *i*-Pr, 7.9; Et, 5.5; *n*-Bu, 4.3; *n*-Pr, 4.2; and Me, 1.0. This contrasts with the relative reactivities toward the succinimidyl radical, which are reported in Section V.B, namely, Me > Et > *n*-Pr, *n*-Bu > *s*-Bu.

The stereochemical studies which have been carried out all involve cyclopropyltin compounds.

It was found that *cis*- or *trans*-(2-methylcyclopropyl)trimethyltin reacted with iodine in methanol or acetic acid, or with bromine in acetic acid or chlorobenzene, to give the appropriate 2-methylcyclopropyl halide with retention of configuration.[158] Similarly, *cis*- or *trans*-(2-phenylcyclopropyl)-trimethyltin reacted with bromine in carbon tetrachloride to give 2-phenylcyclopropyl bromide with retention.[159] These secondary alkyl–tin bonds are therefore presumably cleaved by an S_E2 mechanism.

On the other hand, optically active (1-methyl-2,2-diphenylcyclopropyl)-trimethyltin reacted with bromine or iodine in carbon tetrachloride to give the corresponding cyclopropyl halide with racemization (equations 136 and 137), probably via the intermediate 1-methyl-2,2-diphenylcyclopropyl radical.[160,161] The 2-methyl-3,3-diphenylprop-1-ene which is formed in the latter reaction (equation 137) probably arises through β-scission of this intermediate radical.

The complication of competing reactions of halogen atoms at the metal center and at hydrogen is recognized most clearly in the reactions of organoboron compounds. The vacant 2*p* orbital on boron decreases the strength of

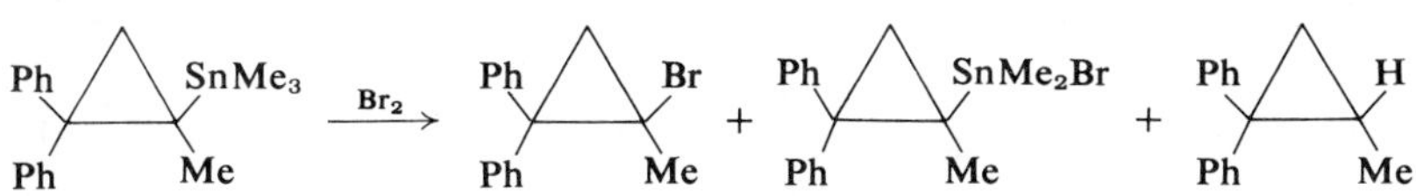

Yield		20%	30%	29%
Optical purity	87%	1.3%	87%	87%

(136)

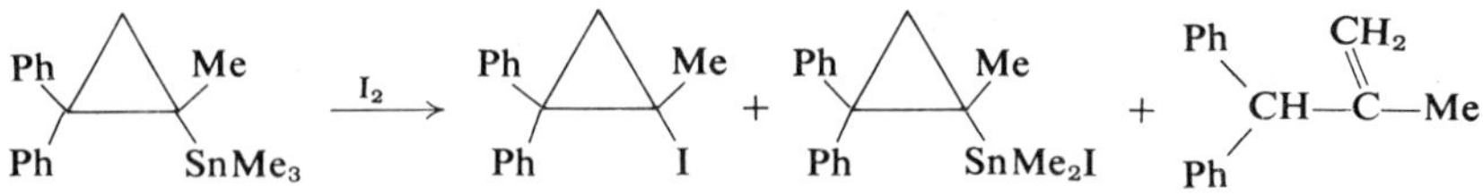

(137)

Yield	42%	41%
Optical purity 69%	2.3%	9%

the α-C—H bond by interacting with the singly occupied $2p$ orbital on carbon of the radical:

(138)

If one accepts Grotewold et al.'s value[162] of 80 kcal/mole for the strength of this α-C—H bond (discussed below), then the enthalpy changes in the reaction of the radical X at hydrogen (ΔH_H) and at boron (ΔH_B) in tributyl borane can be calculated from thermodynamic data given in reference 163; the results are given in Table 8.

TABLE 8

Enthalpy Changes for Homolytic Substitution at Hydrogen and at Boron

X	ΔH_H	ΔH_B	$\Delta H_B - \Delta H_H$
Cl	−23	−31	−8
Br	−7	−14	−7
I	+9	+6	−3
t-BuO	−26	−42	−16

These results are consistent with the observation that the attack of halogen atoms is less selective than that of alkoxyl radicals. The limited experimental information which is available suggests that chlorine and bromine atoms attack mainly the side chain, although there is also some evidence for the competing attack of bromine atoms at boron in trialkyl borane.[162]

At elevated temperatures, trialkyl boranes react with iodine to give the corresponding dialkyl iodoboranes in fair yield.

$$R_3B + I_2 \longrightarrow R_2BI + RI \qquad (139)$$

At 20°, the reaction can be initiated photolytically.[164]

There appears to be no evidence for attack at the alkyl group, but, if this were to contribute, it could lead to the same products by the sequence of reactions 140 and 141.

$$-CH_2-B\langle \;+\; I_2 \longrightarrow -CHI-B\langle \;+\; HI \tag{140}$$

$$-CHI-B\langle \;+\; HI \longrightarrow -CH_2I \;+\; IB\langle \tag{141}$$

Lissi and Sanhueza studied the reaction between triethyl borane and iodine in cyclohexane at 100–140°.[164,165] The reaction followed the rate law:

$$\frac{-d[I_2]}{dt} = k[Et_3B][I_2]^{1/2} \tag{142}$$

The following mechanism was proposed:

$$I_2 \underset{}{\overset{K}{\rightleftharpoons}} 2I\cdot \tag{143}$$

$$I\cdot + BEt_3 \xrightarrow{k_p} IBEt_2 + Et\cdot \tag{144}$$

$$Et\cdot + I_2 \longrightarrow EtI + I\cdot \tag{145}$$

This implies that $k = K^{1/2}k_p$, leading to approximate Arrhenius parameters for the S_H2 step of $E_a = 9.8$ kcal/mole, and of $\log A = 9.5$ (in $M^{-1}\,sec^{-1}$).

In the reaction of chlorine with organoboranes, only attack on the alkyl group, and no boron–carbon bond cleavage, appears to have been reported.[166–169] For example, when equimolar amounts of chlorine and trimethyl borane were allowed to react at $-95°$ for 1 hr, the material balance was as follows[168]:

$$Me_3B + Cl_2 \longrightarrow 0.31Me_2BCH_2Cl + 0.67HCl + 0.55Me_3B + \text{less volatile products} \tag{146}$$

The photochemical reaction between bromine and triethyl borane in the gas phase at 26° was studied by Grotewold et al.[162] The small amount of ethane and ethyl bromide which were formed were ascribed to the S_H2 reaction of the bromine atom at the boron center, but the major reaction was at the α-methylene group.

This was compared with the rate of the reaction of bromine atoms with cyclohexene and with toluene, and it was deduced that triethyl borane was at least five times as reactive as toluene at 26°, leading to an upper limit of E_a, for the above reaction, of 6.2 kcal/mole, and an estimate of $D(H{-}CHMeBEt_2)$ of 80 ± 3 kcal/mole. This is 14.5 kcal/mole less than $D(H{-}CHMe_2)$ in propane.

If the bromination is carried out in solution, and the hydrogen bromide is not removed, acidolysis of the α-bromoalkyl borane leads to the formation of the corresponding alkyl bromide.[170,171]

This activation of an α-C—H group apparently does not extend to the alkyldialkoxyboranes, perhaps because the 2*p* orbital on boron is already filled by overlap from the doubly occupied 2*p* orbitals on the oxygen atoms. Thus 2-phenylethyl(ethylenedioxy) borane and 1-phenylethyl(ethylenedioxy) borane react with bromine in carbon tetrachloride to give the products shown in equations 147 and 148, but the relative reactivity of these two boranes and of ethylbenzene is 0.30:0.63:1, respectively.[172]

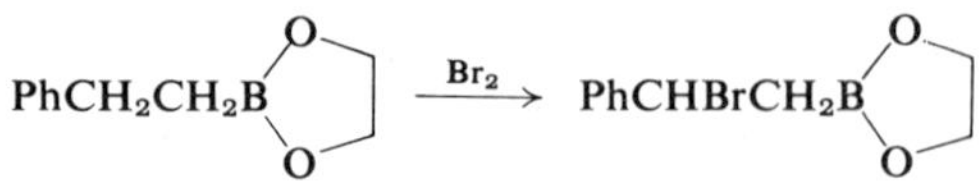

(147)

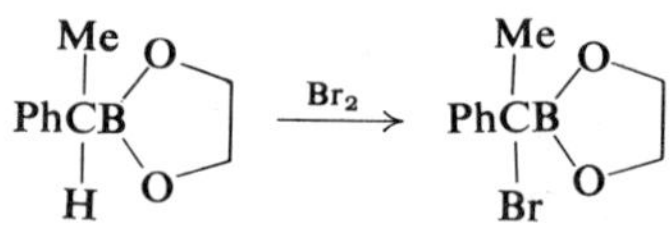

(148)

Similarly, 1-hexyl(dibutoxy) borane and 1-hexyl(ethylenedioxy) borane do not react with bromine in boiling carbon tetrachloride under irradiation with ultraviolet light, but 1-methylpentyl(ethylenedioxy) borane undergoes quantitative bromination in the α position.[173] Butoxy(1-hexyl)(methyl) borane consumes 2 molar equivalents of bromine to give a product which can be oxidized to 2-heptanone. A migration of the methyl group from boron to carbon is probably involved, as shown in equation 149.[173]

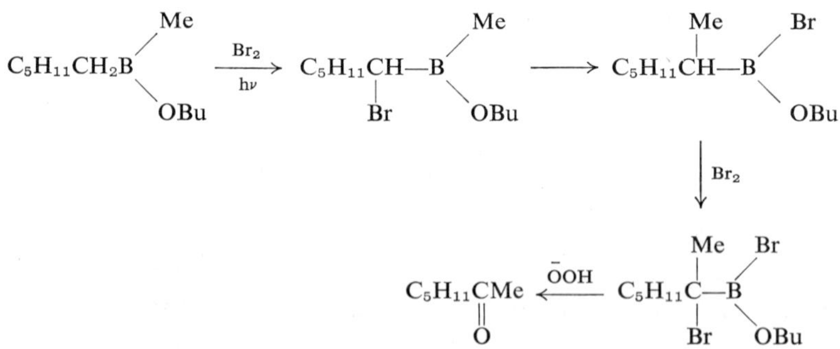

(149)

Rearrangements of this type are induced by nucleophilic reagents, and Brown and his co-workers have exploited this in organic synthesis.[174,175] For example, if triethyl borane is treated with bromine in the presence of water,

3-methyl-3-pentanol is obtained in 85% yield by the following sequence of reactions:

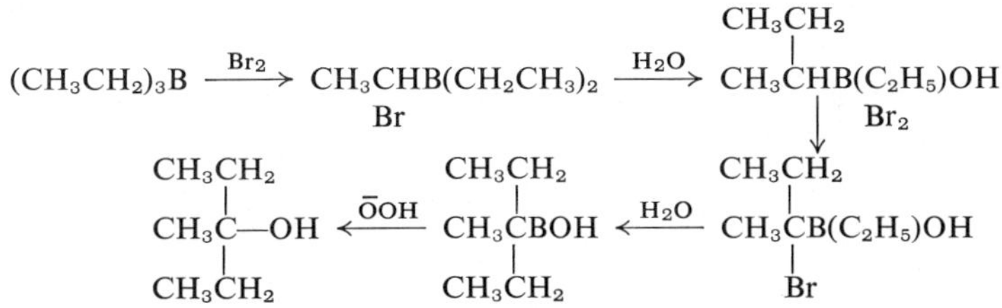

(150)

No clear example of homolytic displacement of an alkyl group from silicon appears to have been reported, but bimolecular homolytic iododesilylation at silicon appears to occur in the reaction between hexamethyldisilane and iodine in the gas phase at 175–250°.[176]

$$I_2 \rightleftharpoons 2I\cdot \tag{151}$$

$$I\cdot + Me_3SiSiMe_3 \longrightarrow ISiMe_3 + Me_3Si\cdot \tag{152}$$

$$Me_3Si\cdot + I_2 \longrightarrow Me_3SiI + I\cdot \tag{153}$$

The rate of the reaction is first order with respect to disilane, and half-order with respect to iodine. The S_H2 step (equation 152) is calculated to be exothermic by 2 kcal/mole, and

$$\log k_{152}\ M^{-1}\ \mathrm{sec}^{-1} = 8.23 \pm 0.50 - \frac{8.08 \pm 1.08}{\theta}$$

Some reactions which occur between pseudohalogens and organometallic compounds may involve S_H2 reactions at the metal center, but in no case is the evidence conclusive. For example, thiocyanogen reacts with the compounds R_nM (M = B, Cd, or Bi) by substitution to give $R_{n-1}MSCN$, but with R_nM (M = As or Sb) by addition to give $R_nM(NCS)_2$.[177] The former reactions may involve an S_H2 reaction by the NCS· radical, for example,

$$NCS\cdot + Ph_3B \longrightarrow Ph_2BSCN + Ph\cdot \tag{154}$$

$$Ph\cdot + (NCS)_2 \longrightarrow PhSCN + NCS\cdot \tag{155}$$

Similarly, chlorine azide reacts with the organometallic compounds R_nM (M = Hg,[178,179] Cd,[178] Al,[180] Ga,[180] or In[180]) to give the corresponding monoazides, $R_{n-1}MN_3$. Again a radical chain mechanism, now involving attack by the $N_3\cdot$ radical, is a possibility, for example,

$$N_3\cdot + Et_2Hg \longrightarrow EtHgN_3 + Et\cdot \tag{156}$$

$$Et\cdot + N_3Cl \longrightarrow EtCl + N_3\cdot \tag{157}$$

VIII. CONCLUSION

Bimolecular homolytic substitution reactions at a metal center were recognized only recently, but already (1971) they represent an important section of free radical chemistry. The reactions have been established to occur at some 15 metals (Li; Mg, Zn, Cd, Hg; B, Al, Ga; Si, Sn, Pb; P, As, Sb, Bi) and by eight classes of reagent [ROO·, RO·, RS·, R_2N·, $(CH_2CO)_2N$·, R·, halogen; $R_2C{=}O$*], and kinetic parameters are already available for the S_H2 process in many of these combinations. Some of the reactions (e.g., the addition of trialkyl boranes to α,β-unsaturated carbonyl compounds and expoxides) are beginning to find a place in routine organic synthesis.

It is tempting to try to predict the direction which work in this field will take in the next few years. As yet, almost no work has been carried out on derivatives of transition metals; this is a major field waiting to be developed. In conjunction with this, ESR spectroscopic studies will be extended to the identification of new intermediates, and of displaced organic radicals which extend beyond the simple alkyl radicals which have been observed up to now. The reaction of excited states with organometallic compounds is another large field of which only the fringe has been touched. Finally, the many new organometallic homolytic reactions which proceed very selectively and in high yield provide a valuable new context for studying the kinetic behavior of simple radicals quite apart from their S_H2 reactions which have been our interest here.

REFERENCES

1. E. Gorin, W. Kauzmann, J. Walter, and H. Eyring, *J. Chem. Phys.*, **7**, 633 (1939).
2. A. F. Trotman-Dickenson, *Adv. Free Radical Chem.*, **1**, 1 (1965).
3. A. G. Davies and B. P. Roberts, *Nat. Phys. Sci.*, **229**, 221 (1971).
4. K. U. Ingold and B. P. Roberts, *Free Radical Substitution Reactions*, Interscience, New York, 1971.
5. W. A. Pryor, *Nat. Phys. Sci.*, **231**, 181 (1971).
6. A. G. Davies, *Organic Peroxides*, Vol. II, D. Swern, Ed., Wiley, New York, 1971.
7. G. Sosnovsky and J. H. Brown, *Chem. Rev.*, **66**, 529 (1966).
8. A. G. Davies and S. C. W. Hook, *J. Chem. Soc.* (*C*), 1660 (1971).
9. C. Walling and S. A. Buckler, *J. Am. Chem. Soc.*, **77**, 6039 (1955).
10. T. D. Parsons, M. B. Silverman, and D. M. Ritter, *J. Am. Chem. Soc.*, **79**, 5091 (1957).
11. M. H. Abraham and A. G. Davies, *J. Chem. Soc.*, 429 (1959).
12. A. G. Davies and B. P. Roberts, *Chem. Commun.*, 298 (1966); *J. Chem. Soc.* (*B*), 17 (1967).
13. A. G. Davies and B. P. Roberts, *J. Chem. Soc.* (*B*), 311 (1969).
14. A. G. Davies and B. P. Roberts, *J. Chem. Soc.* (*B*), 317 (1969).
15. P. G. Allies and P. B. Brindley, *Chem. Ind.* (*London*), 319 (1967); 1439 (1968); *J. Chem. Soc.* (*B*), 1126 (1969).

16. A. G. Davies and B. P. Roberts, *J. Chem. Soc.* (*B*), 1074 (1968).
17. C. Walling and A. Cioffari, *J. Am. Chem. Soc.*, **92**, 6609 (1970); A. G. Davies and J. Eley, unpublished work.
18. K. U. Ingold, *Chem. Commun.*, 911 (1969).
19. A. G. Davies, K. U. Ingold, B. P. Roberts, and R. Tudor, *J. Chem. Soc.* (*B*), 698 (1971).
20. S. Korcek, G. B. Watts, and K. U. Ingold, *J. Chem. Soc.*, *Perkin II*, 242 (1972).
21. J. Grotewold, E. A. Lissi, and J. C. Scaiano, *J. Chem. Soc.* (*B*), 475 (1969).
22. A. G. Davies, D. G. Hare, and R. F. M. White, *J. Chem. Soc.*, 341 (1961).
23. M. B. Floyd and C. E. Boozer, *J. Am. Chem. Soc.*, **85**, 984 (1963).
24. Yu. A. Aleksandrov and B. A. Radbil', *Zh. Obschch. Khim.*, **36**, 543 (1966).
25. J. Grotewold, J. Hernandez, and E. A. Lissi, *J. Chem. Soc.* (*B*), 182 (1971).
26. M. M. Midland and H. C. Brown, *J, Am. Chem. Soc.*, **93**, 1506 (1971).
27. H. C. Brown and M. M. Midland, *Chem. Commun.*, 699 (1971).
28. S. A. Buckler, *J. Am. Chem. Soc.*, **84**, 3093 (1962).
29. D. S. Matteson, *J. Org. Chem.*, **29**, 3399 (1964).
30. N. V. Kruglova and R. M. Freidlina, *Izv. Akad. Nauk SSSR, Ser. Khim.*, 2044 (1965).
31. P. J. Krusic and J. K. Kochi, *J. Am. Chem. Soc.*, **91**, 3942 (1969).
32. J. K. Kochi and P. J. Krusic, *J. Am. Chem. Soc.*, **91**, 3944 (1969); **94**, 6033 (1972).
33. A. G. Davies and B. P. Roberts, *Chem. Commun.*, 699 (1969).
34. A. G. Davies and B. P. Roberts, *J. Organomet. Chem.*, **19**, P17 (1969).
35. J. Grotewold, E. A. Lissi, and J. C. Scaiano, *J. Chem. Soc.* (*B*), 1187 (1971).
36. N. S. Vyazankin, G. A. Razuvaev, and T. N. Brevnova, *Zh. Obshch. Khim.*, **35**, 2033 (1965).
37. W. P. Neumann, K. Rübsamen, and P. Sommer, *Chem. Ber.*, **100**, 1063 (1967).
38. G. B. Watts and K. U. Ingold, unpublished work.
39. K. Torssell, *Tetrahedron*, **26**, 2759 (1970).
40. C. Walling and J. Kjellgren, *J. Org. Chem.*, **34**, 1487 (1969).
41. A. G. Davies, D. Griller, B. P. Roberts, and R. Tudor, *Chem. Commun.*, 640 (1970).
42. A. G. Davies, B. P. Roberts, and R. Tudor, unpublished work.
43. T. Maki, unpublished work.
44. H. Gilman and R. H. Kirby, *J. Am. Chem. Soc.*, **63**, 2046 (1941).
45. J. J. Eisch, *The Chemistry of Organometallic Compounds*, Macmillan, New York, 1967.
46. T. Luong, T. Ngoe, and R. Henriette, *Compt. Rend. Acad. Sci., Ser. C*, **267**, 776 (1968).
47. H. O. House and W. F. Fischer, *J. Org. Chem.*, **33**, 949 (1968).
48. G. W. Kabalka, H. C. Brown, A. Suzuki, S. Honma, A. Arase, and M. Itoh, *J. Am. Chem. Soc.*, **92**, 710 (1970).
49. H. C. Brown and G. W. Kabalka, *J. Am. Chem. Soc.*, **92**, 712 (1970).
50. H. C. Brown and G. W. Kabalka, *J. Am. Chem. Soc.*, **92**, 714 (1970).
51. A. Suzuki, S. Nozawa, M. Itoh, H. C. Brown, G. W. Kabalka, and G. W. Holland, *J. Am. Chem. Soc.*, **92**, 3503 (1970).
52. M. F. Hawthorne and M. Reintjes, *J. Am. Chem. Soc.*, **86**, 951 (1964); **87**, 4585 (1965); D. J. Phillips, unpublished work.
53. A. Suzuki, N. Miyaura, M. Itoh, H. C. Brown, G. W. Holland, and E. Negishi, *J. Am. Chem. Soc.*, **93**, 2793 (1971).
54. J. R. Johnson and M. G. Van Campen, *J. Am. Chem. Soc.*, **60**, 121 (1938).
55. T. Maki, unpublished work.

56. D. J. Pasto, S. K. Arora, and J. Chow, *Tetrahedron*, **25**, 1571 (1969).
57. A. G. Davies, D. Griller, and B. P. Roberts, *J. Chem. Soc.* (*B*), 1823 (1971).
58. A. G. Davies, T. Maki, and B. P. Roberts, *J. Chem. Soc.*, *Perkin II*, 744 (1972).
59. C. Walling and J. A. McGuinness, *J. Am. Chem. Soc.*, **91**, 2053 (1969).
60. R. Tudor, unpublished work.
61. A. G. Davies and B. P. Roberts, unpublished work.
62. J. K. Kochi and P. J. Krusic, "Essays on Free-Radical Chemistry," *Chem. Soc. Spec. Publ. No. 24*, London, 1970, p. 147.
63. E. L. Altwicker, *Chem. Rev.*, **67**, 475 (1967).
64. A. Rieker, *Angew. Chem.*, *Int. Ed.*, **3**, 654 (1964).
65. E. Müller, P. Ziemek, and A. Rieker, *Tetrahedron Letters*, 207 (1964).
66. A. Rieker, E. Müller, and W. Beckert, *Z. Naturforsch*, **17b**, 718 (1962).
67. V. V. Ershov, A. A. Volod'kin, and M. V. Tarkhanova, *Izv. Akad. Nauk SSSR, Ser. Khim.*, 2470 (1967).
68. G. A. Razuvaev, N. S. Vasileiskaya, and D. V. Muslin, *J. Organomet. Chem.*, **7**, 531 (1967); *Dokl. Akad. Nauk SSSR*, **175**, 620 (1967).
69. G. A. Razuvaev, I. L. Khrazhanovskaya, N. S. Vasileiskaya, and D. V. Muslin, *Dokl. Akad. Nauk SSSR*, **177**, 600 (1967).
70. J. Gaasch, unpublished work, reported by G. M. Whitesides and J. San Filippo, *J. Am. Chem. Soc.*, **92**, 6611 (1970).
71. H. J. Emeléus, P. M. Spaziante, and S. M. Williamson, *Chem. Commun.*, 768 (1969).
72. H. Ang and K. F. Ho, *J. Organomet. Chem.*, **27**, 349 (1971).
73. M. H. Abraham, J. H. N. Garland, J. A. Hill, and L. F. Larkworthy, *Chem. Ind.* (*London*), 1615 (1962).
74. E. B. Baker and H. H. Sisler, *J. Am. Chem. Soc.*, **75**, 5193 (1953).
75. S. J. Brois, *Tetrahedron Letters*, 345, 805 (1964).
76. M. Inatome and L. P. Kuhn, *Adv. Chem. Ser.*, **42**, 183 (1964).
77. A. G. Davies, D. Griller, B. P. Roberts, and J. C. Scaiano, *Chem. Commun.*, 196 (1971); A. G. Davies, B. P. Roberts, and J. C. Scaiano, *J. Chem. Soc.* (*B*), in press.
78. M. V. Encina and E. A. Lissi, *J. Organomet. Chem.*, **29**, 21 (1971).
79. H. Gilman and J. P. Nelson, *J. Am. Chem. Soc.*, **59**, 935 (1937).
80. H. Gilman and H. L. Yale, *Chem. Rev.*, **30**, 281 (1942).
81. B. M. Mikhailov, V. A. Vaver, and Yu. N. Bubnov, *Dokl. Akad. Nauk SSSR*, **126**, 575 (1959).
82. B. M. Mikhailov and Yu. N. Bubnov, *Zh. Obshch. Khim.*, **31**, 160 (1961).
83. B. M. Mikhailov and F. B. Tutorskaya, *Zh. Obshch. Khim.*, **32**, 833 (1962).
84. L. H. Toporcer, R. E. Dessy, and S. I. E. Green, *J. Am. Chem. Soc.*, **87**, 1236 (1965).
85. B. M. Mikhailov and Yu. N. Bubnov, *Izv. Akad. Nauk SSSR, Ser. Khim.*, 2248 (1964).
86. B. M. Mikhailov, *Prog. Boron Chem.*, **3**, 313 (1970).
87. A. G. Davies and S. C. W. Hook, *J. Chem. Soc.* (*B*), 735 (1970).
88. H. Gilman and H. L. Yale, *J. Am. Chem. Soc.*, **73**, 2880 (1951).
89. C. Walling and R. Rabinowitch, *J. Am. Chem. Soc.*, **79**, 5326 (1957); **81**, 1243 (1959).
90. C. Walling and M. S. Pearson, *J. Am. Chem. Soc.*, **86**, 2262 (1964).
91. C. Walling, O. H. Besedow, and E. S. Savas, *J. Am. Chem. Soc.*, **82**, 2181 (1960).
92. A. G. Davies and B. P. Roberts, *J. Chem. Soc.* (*B*), in press.
93. B. P. Roberts, unpublished work.
94. J. M. Smith, unpublished work.
95. B. M. Mikhailov and Yu. N. Bubnov, *Izv. Akad. Nauk SSSR, Otd. Khim. Nauk*, 172 (1959); *Zh. Obshch. Khim.*, **29**, 1648 (1959).

96. Farbenfabriken Bayer Akv.-Ges., Br. Pat. 842,341 (1960); *Chem. Abstr.*, **55**, P9929*g* (1961).
97. A. G. Davies, S. C. W. Hook, and B. P. Roberts, *J. Organomet. Chem.*, **22**, C37 (1970).
98. A. G. Davies, S. C. W. Hook, and B. P. Roberts, *J. Organomet. Chem.*, **23**, C11 (1970).
99. J. G. Sharefkin and H. D. Banks, *J. Org. Chem.*, **30**, 4313 (1965).
100. H. C. Brown, W. R. Heydkamp, E. Bruer, and W. S. Murphy, *J. Am. Chem. Soc.*, **86**, 3565 (1964).
101. R. J. W. Le Fèvre, *J. Chem. Soc.*, 1745 (1932).
102. G. H. Coleman, *J. Am. Chem. Soc.*, **55**, 3001 (1933).
103. A. G. Davies, J. Eley, and B. P. Roberts, unpublished work.
104. R. S. Davidson, *Tetrahedron Letters*, 3029 (1968).
105. W. A. Thaler, *Methods in Free Radical Chem.*, **2**, 121 (1970).
106. E. Hedaya, R. L. Hinman, V. Schomaker, S. Theodoropulos, and L. M. Kyle, *J. Am. Chem. Soc.*, **89**, 4875 (1967).
107. T. R. Beeke and F. M. Howard, *J. Am. Chem. Soc.*, **91**, 3379 (1969).
108. A. G. Davies, B. P. Roberts, and J. M. Smith, *Chem. Commun.*, 557 (1970), and unpublished work.
109. K. Sisido, K. Ban, T. Isida, and S. Kozima, *J. Organomet. Chem.*, **29**, C7 (1971).
110. A. G. Davies, B. P. Roberts, and J. M. Smith, unpublished work.
111. E. J. Kupchik and T. Lanigan, *J. Org. Chem.*, **27**, 3661 (1962).
112. G. A. Razuvaev and U. Fetukova, *Zh. Obshch. Khim.*, **21**, 1010 (1951); *Chem. Abstr.*, **46**, 1479*f* (1952).
113. J. C. Maire, R. Prosperini, and J. van Rietschoten, *J. Organomet. Chem.*, **21**, P41 (1970).
114. B. C. Pant and W. E. Davidsohn, *J. Organomet. Chem.*, **19**, P3 (1969).
115. B. C. Pant, *J. Organomet. Chem.*, **24**, 697 (1970).
116. R. Kulmel and Z. Meibuer, *Z. Chem.*, **10**, 71 (1970).
117. G. A. Razuvaev and N. S. Vasileiskaya, *Dokl. Akad. Nauk SSSR*, **67**, 851 (1949); **74**, 279 (1950).
118. F. R. Jensen and J. Miller, unpublished results, reported by F. R. Jensen and B. Rickborn, in *Electrophilic Substitution of Organomercurials*, McGraw-Hill, New York, 1968, p. 89.
119. H. G. Oswin, R. Rebbert, and E. W. R. Staecie, *Can. J. Chem.*, **33**, 472 (1955).
120. R. E. Rebbert and P. Ausloos, *J. Am. Chem. Soc.*, **85**, 3086 (1963); **86**, 2068 (1964).
121. G. A. Razuvaev, Yu. A. Ol'dekop, and N. A. Maier, *Dokl. Akad. Nauk SSSR*, **98**, 613 (1954); *Zh. Obshch. Khim.*, **25**, 697 (1955).
122. Yu. A. Ol'dekop, N. A. Maier, V. N. Pshenichnyi, and Z. F. Izmailova, *Zh. Obshch. Khim.*, **40**, 308 (1970).
123. Yu. A. Ol'dekop, N. A. Maier, A. A. Érdman, and Yu. A. Dzhomidova, *Zh. Obshch. Khim.*, **40**, 637 (1970).
124. Yu. A. Ol'dekop, N. A. Maier, A. E. Érdman, and S. S. Stanovaya, *Zh. Obshch. Khim.*, **40**, 300, 305 (1970).
125. Yu. A. Ol'dekop, N. A. Maier, and Y. D. But'ko, *Zh. Obshch. Khim.*, **40**, 641 (1970).
126. Reviewed by L. G. Makarova and A. N. Nesmeyanov, *The Organic Compounds of Mercury*, North-Holland, Amsterdam, 1967, Chap. 10.
127. J. Grotewold and E. A. Lissi, *Chem. Commun.*, 21 (1965).
128. M. V. Encina and E. A. Lissi, *J. Organomet. Chem.*, **29**, 21 (1971).

129. T. N. Bell and A. E. Platt, *Int. J. Chem. Kinet.*, **2**, 299 (1970).
130. T. N. Bell and A. E. Platt, *Chem. Commun.*, 325 (1970).
131. A. K. Hoffmann and S. J. Groszos, U.S.Pat. 2,882,317 (1959); *Chem. Abstr.*, **53**, 16062*i* (1959).
132. Y. Ogata, Y. Izawa, H. Tomioka, and T. Ukigai, *Tetrahedron*, **25**, 1817 (1969).
133. H. C. Clark and C. J. Willis, *J. Am. Chem. Soc.*, **82**, 1888 (1960).
134. R. D. Chambers, H. C. Clark, and C. J. Willis, *Chem. Ind.* (*London*), 76 (1960).
135. R. A. Jackson, *Adv. Free Radical Chem.*, **3**, 231 (1969).
136. H. C. Clark and J. H. Tsai, *Chem. Commun.*, 111 (1965).
137. M. A. A. Beg and H. C. Clark, *Chem. Ind.* (*London*), 140 (1962).
138. H. C. Clark, J. D. Cotton, and J. H. Tsai, *Can. J. Chem.*, **44**, 903 (1966).
139. H. D. Kaesz, J. R. Phillips, and F. G. A. Stone, *Chem. Ind.* (*London*), 1409 (1959).
140. F. W. Bennett, H. J. Emeléus, and R. N. Haszeldine, *J. Chem. Soc.*, 1565 (1953).
141. M. A. A. Beg and H. C. Clark, *Can. J. Chem.*, **39**, 564 (1961).
142. M. A. A. Beg and H. C. Clark, *Can. J. Chem.*, **40**, 283 (1962).
143. H. J. Emeléus, R. N. Haszeldine, and E. G. Walaschewski, *J. Chem. Soc.*, 1552 (1953).
144. N. J. Friswell and B. G. Gowenlock, *Adv. Free Radical Chem.*, **2**, 1 (1967).
145. F. R. Jensen and B. Rickborn, *Electrophilic Substitution of Organomercurials*, McGraw-Hill, New York, 1968.
146. O. A. Reutov and I. P. Beletskaya, *Reaction Mechanisms of Organometallic Compounds*, North-Holland, Amsterdam, 1968.
147. J. Keller, Doctoral dissertation, University of California, Los Angeles, California, 1948, quoted by S. Winstein and T. G. Traylor, *J. Am. Chem. Soc.*, **78**, 2597 (1956).
148. G. A. Razuvaev and A. V. Savitskii, *Dokl. Akad. Nauk SSSR*, **85**, 575 (1952).
149. O. A. Reutov and I. P. Beletskaya, *Izv. Akad. Nauk SSSR, Otd. Khim. Nauk*, 1716 (1960).
150. I. P. Beletskaya, O. A. Reutov, and T. P. Guryanova, *Izv. Akad. Nauk SSSR, Otd. Khim. Nauk*, 2178 (1961).
151. I. P. Beletskaya, O. A. Reutov, and T. A. Azizyan, *Izv. Akad. Nauk SSSR, Otd. Khim. Nauk*, 223 (1962).
152. F. R. Jensen and L. H. Gale, *J. Am. Chem. Soc.*, **81**, 1261 (1959); **82**, 148 (1960).
153. F. R. Jensen and J. Miller, unpublished work, reported by F. R. Jensen and B. Rickborn, reference 145.
154. F. R. Jensen, L. D. Whipple, D. K. Wedegaertner, and J. A. Landgrebe, *J. Am. Chem. Soc.*, **81**, 1262 (1959); **82**, 2466 (1960).
155. E. V. Uglova, T. B. Svetlanova, I. P. Beletskaya, and O. A. Reutov, *Izv. Akad. Nauk SSSR*, 1151 (1968).
156. A. V. Savitskii and Y. K. Syrkin, *Dokl. Phys. Chem.*, **146**, 649 (1962).
157. S. Boué, M. Gielen, and J. Nasielski, *J. Organomet. Chem.*, **9**, 443, 461, 481 (1967).
158. P. Baekelmans, M. Gielen, and J. Nasielski, *Tetrahedron Letters*, 1149 (1967).
159. K. Sisido, K. Ban, T. Isida, and S. Kozima, *J. Organomet. Chem.*, **29**, C7 (1971).
160. K. Sisido, S. Kozima, and K. Takizawa, *Tetrahedron Letters*, 33 (1967).
161. K. Sisido, T. Miyanisi, T. Isida, and S. Kozima, *J. Organomet. Chem.*, **23**, 117 (1970).
162. J. Grotewold, E. A. Lissi, and J. C. Scaiano, *J. Organomet. Chem.*, **19**, 431 (1969).
163. A. Finch and P. J. Gardner, *Progress Boron Chem.*, **3**, 177 (1970).
164. E. A. Lissi and E. Sanhueza, *J. Organomet. Chem.*, **26**, C59 (1971).
165. E. A. Lissi and E. Sanhueza, *J. Organomet. Chem.* (1971), in press.
166. J. C. Perrine and R. N. Keller, *J. Am. Chem. Soc.*, **80**, 1823 (1958).

167. L. Zeldin and P. R. Giradot, *Abstracts, American Chemical Society Meetings*, **140,** 15-N (1961); U.S.Pat. 3,083,230 (1963); *Chem. Abstr.*, **59**, 10116*g* (1963).
168. R. Schaeffer and L. J. Todd, *J. Am. Chem. Soc.*, **87**, 488 (1965).
169. D. S. Matteson, *Prog. Boron Chem.*, **3**, 117 (1970).
170. C. F. Lane and H. C. Brown, *J. Am. Chem. Soc.*, **92**, 7212 (1970).
171. C. F. Lane and H. C. Brown, *J. Organomet. Chem.*, **26**, C51 (1971).
172. D. J. Pasto, J. Chow, and S. K. Arora, *Tetrahedron*, **25**, 1557 (1969).
173. D. J. Pasto and K. McReynolds, *Tetrahedron Letters*, 801 (1971).
174. C. F. Lane and H. C. Brown, *J. Am. Chem. Soc.*, **93**, 1025 (1971).
175. H. C. Brown and Y. Yamamoto, *J. Am. Chem. Soc.*, **93**, 2796 (1971).
176. S. J. Band and I. M. T. Davidson, *Trans. Faraday Soc.*, **66**, 406 (1970).
177. T. Wizemann, H. Müller, D. Seybold, and K. Dehnicke, *J. Organomet. Chem.*, **20**, 211 (1969).
178. K. Dehnicker, J. Strähle, D. Seybold, and J. Müller, *J. Organomet. Chem.*, **6**, 298 (1966).
179. A.-F. Shihada and K. Dehnicke, *J. Organomet. Chem.*, **26**, 157 (1971).
180. J. Müller and K. Dehnicke, *J. Organomet. Chem.*, **12**, 37 (1968).

CHAPTER 11

Oxidation—Reduction Reactions of Free Radicals and Metal Complexes

JAY K. KOCHI

Department of Chemistry, Indiana University, Bloomington, Indiana

I. INTRODUCTION

The oxidation and reduction of organic compounds can be effected by a variety of metal complexes.[1–3] A number of these reactions involve selective transformations of functional groups and therefore represent valuable tools for organic synthesis.[4–6] Basic to an understanding of these processes is the nature of the interaction between the organic substrate and the metal species.

Oxidations and reductions can proceed by essentially heterolytic or homolytic mechanisms, the differentiation between which is somewhat arbitrary since it is often difficult to assess the timing involved in making and breaking bonds. We make the distinction a purely operational one in which *discrete* one-electron changes are involved in a homolytic mechanism, and free radicals per se are intermediates. For example, the oxidation of ethylene to the glycol diacetate by cobalt(III) in acetic acid proceeds by several successive steps (equation 1); and the electron spin resonance (ESR)

$$\begin{aligned} Co^{III} + CH_2{=}CH_2 &\longrightarrow Co^{II} + CH_2\overset{+\cdot}{-}CH_2 \\ CH_2\overset{+\cdot}{-}CH_2 + HOAc &\longrightarrow \cdot CH_2{-}CH_2OAc + H^+ \\ \cdot CH_2CH_2OAc + Co^{III} + HOAc &\longrightarrow AcOCH_2CH_2OAc + Co^{II} + H^+ \end{aligned} \qquad (1)$$

spectrum of the cation-radical intermediate can be observed. In a number of other redox reactions, however, reactive intermediates are not readily detected, since stereospecific reactions can result within the coordination sphere and the changes in oxidation states of the metal nucleus cannot be followed.

In this chapter the mechanism involving the one-equivalent oxidation and reduction of free radicals by metal complexes is treated. Since electron transfer processes have been examined in inorganic systems for some time,[8] the mechanistic conclusions of those studies are used here as a basis for a discussion of the redox reactions of free radicals. The mechanism of representative oxidation and reduction reactions is then discussed with a framework of these elementary processes. No attempt is made to cover this subject in a comprehensive manner since a wide variety of examples are extant in the literature. Rather it is intended to treat redox reactions in as broad a scope as possible to allow electron transfer processes to be placed within the general context of reaction mechanisms.

II. OXIDATION–REDUCTION REACTIONS BETWEEN INORGANIC COMPLEXES

The well-known oxidation–reduction reaction in equation 2* represents

$$Cu^{1+} + Fe^{3+} \longrightarrow Cu^{2+} + Fe^{2+} \qquad (2)$$

one-equivalent changes since both half-reactions (Fe^{3+}–Fe^{2+} and Cu^{1+}–Cu^{2+}) involve a net transfer of one electron between the reduced and oxidized forms of the metal ions.

* In this and subsequent equations no attempt is made to include all the coordination (including solvent) around the metal or the charge on the ion unless pertinent to the discussion.

Second-order rate constants for electron transfer reactions encompass the inextensible range from very slow (months) to those that approach diffusion control ($> 10^8\ M^{-1}\ sec^{-1}$). Despite this broad spectrum in the rates, the majority of oxidation–reduction reactions involving one-equivalent changes occur by essentially one of two mechanisms, which are designated as *outer-sphere* and *inner-sphere* processes.[9] The lability of a metal complex (i.e., the readiness with which ligands in the first coordination sphere can be replaced) is important in redox processes since it can control how closely the metal nuclei approach each other and the ease with which electrons are transferred.

A. Outer-Sphere Electron Transfer

This mechanism takes place with the inner coordination spheres of both the oxidant as well as the reductant intact in the transition state. No ligand-to-metal bond is broken or made in this process as studies with substitution-inert complexes show. Thus, the reaction of hexacyanoferrate(II) with

$$Fe^{II}(CN)_6{}^{4-} + Ir^{IV}Cl_6{}^{2-} \longrightarrow Fe^{III}(CN)_6{}^{3-} + Ir^{III}Cl_6{}^{3-} \quad (3)$$

hexachloroiridate(IV) occurs with a second-order rate constant of $1.2 \times 10^5\ M^{-1}\ sec^{-1}$ at 25° despite the substitution inertness of both reactants.[10]

Outer-sphere electron transfer can be pictured as an electron "tunneling" through a potential barrier between regions of lowest energy. Such an electron transfer must not alter the total energy of the system, and an outer-sphere process is optimum when the oxidants and reductants are similar in bond energy and geometry. Since the total spin angular momentum must be constant during this exchange, it may impose a serious limitation in the rate, particularly for transition metals involving metastable high or low spin states. Finally the probability of electron transfer is unity if the reactants are sufficiently close and possess orbitals of suitable symmetry to allow for overlap (weakly adiabatic interactions). If the interaction is much less than 1 kcal/mole due to a lack of orbital overlap, the electron transfer is unlikely (i.e., a nonadiabatic reaction). The criterion of orbital overlap suggests that octahedral complexes with t_{2g} electrons (concentrated between ligands) may be more effective than those with e_g electrons (concentrated toward ligands). The addition of ligands with available orbitals should also facilitate electron transfer. A theory based on this model has been developed by Marcus,[11] who suggests the linear free energy relationship shown in equation 4. The theory also predicts that isotopic exchange k_{AB} between two complexes A

$$\Delta G^{\ddagger} = \text{constant} + \tfrac{1}{2}\Delta G_0 \quad (4)$$

and B is related to the equilibrium constant K_{AB} and the rates of self-exchanges k_{AA} and k_{BB} according to equation 5.

$$k_{AB} = (k_{AA}k_{BB}K_{AB}f_{AB})^{1/2} \quad (5)$$

B. Inner-Sphere Electron Transfer

This type of electron transfer proceeds via a bridged activated complex in which contact between the oxidant and reductant is maintained by a ligand which is bonded to both. For example, the oxidation of chromium(II) by $(NH_3)_5Co^{III}Cl$ has been shown to involve direct transfer of chlorine from the oxidant to the reductant.[12] This study was facilitated by the inertness of one

$$Cr^{II} + (NH_3)_5Co^{III}Cl \longrightarrow [(NH_3)_5Co\text{---}Cl\text{---}Cr]^{\ddagger} \longrightarrow (NH_3)_5Co^{II} + Cr^{III}Cl \quad (6)$$

$$(NH_3)_5Co^{II} \xrightarrow{\text{fast}} Co^{II} + 5NH_3$$

of the reactants $[(NH_3)_5Co^{III}Cl]$ and one of the products $[Cr^{III}Cl]$ to substitution, which allowed the use of radioactive chlorine as a tracer. The atom transfer shown in equation 6 is not always a necessary consequence of an inner-sphere process, but the bridging ligand is likely to be associated with the more substitution-stable product.

The extreme lability of chromium(II) to substitution (equation 7, $k > 10^8$ sec^{-1}) does not impose a serious kinetic restraint on the example of inner-

$$Cr^{II}(H_2O)_6 \rightleftharpoons Cr^{II}(H_2O)_5 + H_2O \quad (7)$$

sphere transfer given above. However, vanadium(II) is less labile and the rate of its oxidation can be limited by loss of ligand since in some cases it is the same as the rate of anation (equation 8).[13] The kinetics of electron transfer

$$V^{II}(H_2O)_6 + X^- \rightleftharpoons V^{II}(H_2O)_5X + H_2O \quad (8)$$

in these cases is independent of the concentration of the oxidant.

The bridging ligand has an important effect on the rates of these electron transfers, whereas the effect is minimal in outer-sphere processes. The relative reactivities of ligands to transfer, however, vary with the reductant. For example, the effectiveness of halides as bridging ions decreases in the order $X = I^- > Br > Cl^- > F$ in the reactions between chromium(II) and $Cr^{III}X$ or $(NH_3)_5Cr^{III}X$. However, the reverse trend in reactivity is observed in the Fe(II)—$(NH_3)_5Co^{III}X$, Eu^{II}—$(NH_3)_5Co^{III}X$, and Eu^{II}—$Cr^{III}X$ reactions.[14] The difference may be attributed to the relative stabilities of the transition states and their location along the reaction coordinate.[15] Various types of other ligands are also available for transfer including organic functional groups, particularly those with appropriate symmetry to provide overlap with the orbitals on the oxidant and reductant.

III. OXIDATION AND REDUCTION OF ORGANIC RADICALS

The one-equivalent oxidation and reduction of organic compounds can also be represented by appropriate half-reactions. For example, the oxidation of arenes affords cation radicals (equation 9) and reduction produces anion radicals (equation 10).

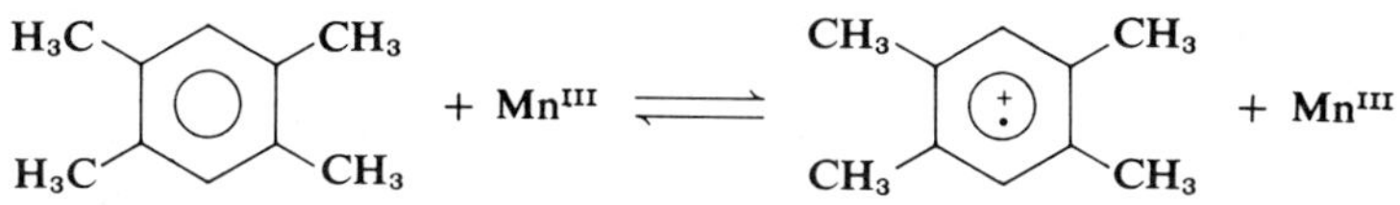
(9)

(10)

Since free radicals are required intermediates in such transformations, it is more convenient to initiate discussion using them as focal points. Thus, reactions 11 and 12 represent the standard reactions for formal oxidation and reduction, respectively, of an organic radical, $\dot{X}$. However, these reactions

$$\dot{X} \rightleftharpoons X^{+} + \varepsilon \tag{11}$$

$$\dot{X} + \epsilon \rightleftharpoons X\!:^{-} \tag{12}$$

are not necessarily reversible due to the metastability of carbonium ions and carbanions, especially if they are unencumbered with solvent.

The enhanced interest in the reactions of free radicals with metal complexes is due to their active role in various stoichiometric and chain processes proceeding via free radical intermediates such as oxidation, autoxidation, and polymerization.[16] The oxidation–reduction reactions between free radicals and metal complexes must be rapid in order to compete effectively with other facile homolytic processes. For example, the polyacrylamide radical[17] reacts with ferric and other salts with a second-order rate constant of about $10^3\ M^{-1}\ \text{sec}^{-1}$; and hydrogen atom, hydroxy, and hydroperoxy radicals undergo redox reactions with a number of metal complexes with rate constants in excess of $10^8\ M^{-1}\ \text{sec}^{-1}$.[18] The rates of these reactions have been studied by a variety of techniques developed to measure rate constants of rapid reactions. Physical measurements involving continuous and stop-flow methods, pulse radiolysis, and photolysis and time intermittency using a rotating sector have been employed, as well as chemical studies involving the inhibiting effect of an additive on a radical chain reaction and the method of competing reactions. The latter is particularly useful since relative rate constants can be obtained directly from the composition of the products.

A. Oxidation of Alkyl Radicals by Metal Complexes

The ionization potential of alkyl radicals measured in the gas phase[19] decreases in the order primary > secondary > tertiary. Estimates of the oxidation potentials of alkyl radicals in aqueous solution follow the same

$$R\cdot \longrightarrow R^+ + \varepsilon \qquad (13)$$

trend, but are obtained after a large correction is made for the heat of solution of the carbonium ion[20] (Table 1). Since the products of oxidation are usually

TABLE 1

Oxidation Potentials of Alkyl Radicals

		Ionization potential (eV)	
$R\cdot \rightarrow R^+$	$+\varepsilon$	H_2O	Gas
$CH_3\cdot \rightarrow CH_3^+$	$+\varepsilon$	2.9	9.98
$CH_3CH_2\cdot \rightarrow CH_3CH_2^+$	$+\varepsilon$	2.1	8.80
$CH_3CH_2CH_2\cdot \rightarrow CH_3CH_2CH_2^+$	$+\varepsilon$	2.1	8.69
$(CH_3)_2CH\cdot \rightarrow (CH_3)_2CH^+$	$+\varepsilon$	1.6	7.90
$(CH_3)_3C\cdot \rightarrow (CH_3)_3C^+$	$+\varepsilon$	1.3	7.43

alkenes and substitution products, the standard reactions for oxidation are more properly described as oxidative elimination (equation 14) and oxidative substitution (equation 15), respectively. The relative rates of oxidative

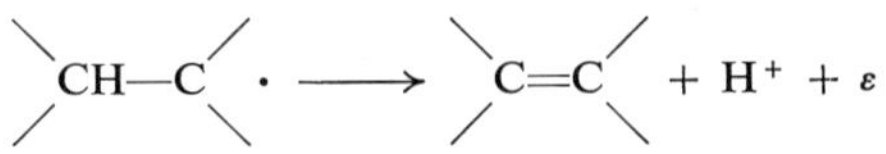
(14)

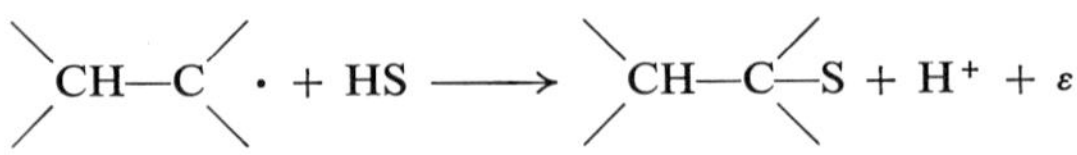
(15)

elimination k_e and oxidative substitution k_s of a series of simple alkyl radicals by various metal oxidants are compared in Table 2.

For a given alkyl radical these relative rates are surprisingly invariant with changes in the oxidant [copper(II) being the exception]. However, the insensitivity of the *product distribution* to the oxidant despite the large variations in the redox potentials [e.g., approximately 5 kcal/mole for copper(II) to 50 kcal/mole for cobalt(III) in water] suggests that a carbonium ion is formed as an intermediate. Furthermore, since there is no direct correlation between the oxidation potential of the metal and the *rate of oxidation* of isopropyl radical in isobutyric acid solutions (Table 3), an outer-sphere process is unlikely especially in view of the large differences in redox potentials. Since

TABLE 2

Relative Rates of Oxidative Elimination and Substitution of Alkyl Radicals by Metal Complexes

Oxidant M^n	Elimination/substitution (k_e/k_s)		
	$CH_3CH_2CH_2\cdot$	$(CH_3)_2CH\cdot$	$(CH_3)_3C\cdot$
Pb^{IV}	1	2	9
Co^{III}	1	2	5
Mn^{III}	1	1	3
Ce^{IV}	1	3	8
Ag^{II}	2	3	4
Cu^{II}	> 100	> 100	> 50

carbonium ions are apparently involved, these oxidations are described as *electron transfer processes* and the stoichiometry is given by equation 16.

$$R\cdot + M^n \longrightarrow [R(-H) + RS] + H^+ + M^{n-1} \quad (16)$$

The study of the oxidation of alkyl radicals by copper(II) complexes has provided considerable insight into the mechanism of these redox processes,[21] and are described in greater detail since the results are applicable to other

TABLE 3

Relative Rates of Oxidation of Isopropyl Radicals by Metal Complexes in Isobutyric Acid[a]

Oxidant M^n	E_0[b]	k_e/k_H	k_e[c]	k_e[CF_3CO_2H cat.][d]
Cu^{II}	0.15	9,400	500	900
Ce^{IV}	1.61	56	3	90
Mn^{III}	1.51	41	2	34
Co^{III}	1.82	20	1	17
Pb^{IV}	1.69[e]	48	2.5	—

[a] k_H is rate constant for hydrogen abstraction from isobutyric acid.
[b] Values in electron volts in water.
[c] Relative to cobalt(III) = 1.
[d] Catalysis by trifluoroacetic acid.
[e] Two-electron change.

systems. Electron transfer oxidation of alkyl radicals is generally effected by copper(II) acetate and other carboxylates.[22] For example,

$$PhCH_2CH_2\cdot + Cu^{II}(OAc)_2 \longrightarrow [PhCH{=}CH_2 + PhCH_2CH_2OAc] + H^+ + Cu^IOAc \quad (17)$$

Ligand transfer oxidation as described in equation 18 is usually carried out with X representing halide and pseudohalide complexes of copper(II).[23]

$$R\cdot + Cu^{II}X_2 \longrightarrow R{-}X + Cu^IX \quad (18)$$

The catalytic decomposition of peroxides affords a convenient method for the generation of free alkyl radicals and a study of their oxidation by copper(II) complexes.[24] In contrast to other homolytic decompositions of peroxides, the stoichiometries of the catalyzed reactions given by equations 19 and 20 are remarkably unequivocal:

$$(RCO_2)_2 \xrightarrow{Cu^I-Cu^{II}} CO_2 + RCO_2H + R_{ox} \quad (19)$$
$$R_{ox} = \text{alkene, alkyl ester}$$

or

$$(RCO_2)_2 + X^- \xrightarrow{Cu^I-Cu^{II}} CO_2 + RCO_2^- + R{-}X \quad (20)$$
$$X = \text{halide, pseudohalide}$$

These radical chain reactions are directly initiated by cuprous species and can be carried out at various temperatures. The mechanism of the catalytic reactions for the decomposition of diacyl peroxides is given by equations 21–25:

$$(RCO_2)_2 + Cu^I \longrightarrow RCO_2Cu^{II} + RCO_2\cdot \quad (21)$$

$$RCO_2\cdot \longrightarrow R\cdot + CO_2 \quad (22)$$

$$R\cdot + Cu^{II}O_2CR \longrightarrow R_{ox} + RCO_2H + Cu^I, \ldots \quad (23)$$

or

$$Cu^{II}O_2CR + X^- \rightleftharpoons Cu^{II}X + RCO_2^- \quad (24)$$

$$R\cdot + Cu^{II}X \longrightarrow R{-}X + Cu^I, \ldots \quad (25)$$

1. Electron Transfer Oxidation of Alkyl Radicals by Copper(II)

The oxidative elimination and substitution of alkyl radicals by copper(II) carboxylates often constitute parallel routes. Thus, benzylic, homoallylic, and β-arylethyl radicals generate mixtures of alkenes and alkyl acetates on oxidation by copper(II) complexes in acetic acid. On the other hand, simple alkyl radicals, such as *n*-butyl, *sec*-butyl, and *tert*-butyl radicals, with readily available β-hydrogens afford alkenes as principal (and often exclusive) products. At the other extreme, allylic radicals on oxidation yield only substitution products.[25]

Questions relevant to the electron transfer oxidation of alkyl radicals are: (a) the role of carbonium ions as intermediates, (b) the formation of alkenes by oxidative elimination, (c) the rates of oxidation of alkyl radicals by copper(II) acetate, (d) mechanisms for the oxidation of alkyl radicals by copper(II) acetate, (e) factors involved in oxidative elimination versus oxidative substitution, and (f) inner-sphere alkylcopper intermediates and solvolytic routes in electron transfer oxidation.

a. Carbonium Ions as Intermediates in Electron Transfer Oxidation. The *oxidation of neopentyl radicals* generated from *t*-butylacetyl peroxide by copper(II) acetate afforded only oxidation products which could be considered as arising from the *t*-amyl cation. Thus, 2-methylbutene-1 and -2 as well as *t*-amyl acetate were formed in acetic acid. The latter was replaced by *N*-*t*-amylacetamide when the oxidation was carried out in acetonitrile.[22]

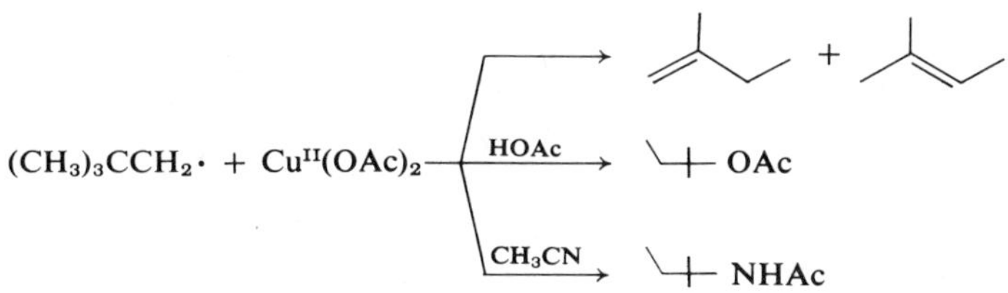

(26)

2-Methylbutenes and *t*-amyl acetate are common products of solvolysis or deamination of *t*-amyl and neopentyl systems in acetic acid.[26] Furthermore, the *N*-*t*-amylacetamide is reminiscent of *t*-amyl cations derived from the alcohol or alkene under much more acidic conditions required to generate carbonium ions in the Ritter reaction.[27] A similar behavior has been reported from the electrooxidation of *t*-amyl iodide in acetonitrile solutions.[28]

Cyclobutyl radicals generated from cyclobutanecarbonyl peroxide were oxidized by copper(II) acetate to cyclobutene and a mixture of cyclobutyl, allylcarbinyl, and cyclopropylmethyl acetates.[22] The composition of the mixture of homoallylic acetates was essentially independent of the oxidant and the solvent. *Allylcarbinyl* radicals under the same conditions afforded 1,3-butadiene and a mixture of homoallylic acetates. *Cyclopropylmethyl* radicals underwent a first-order rearrangement to allylcarbinyl radicals faster than they were oxidized by copper(II) acetate, and only those products observed in the oxidation of allylcarbinyl radicals were produced.

From the behavior shown by cyclobutyl and allylcarbinyl (cyclopropylmethyl) radicals toward copper(II) acetate, it was concluded that carbonium ions are important intermediates. One of the criteria applied, among others, is the invariance of the product composition with the nature of the oxidant, and the formation of the mixture of acetates from the oxidation of cyclobutyl radicals is attributed to the solvation of a C_4H_7 cation formed as an intermediate by electron transfer.

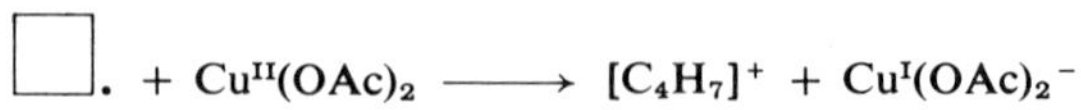
(27)

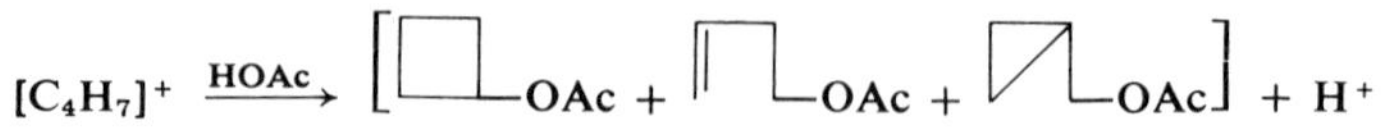
(28)

The distribution among cyclobutyl, allylcarbinyl, and cyclopropylmethyl acetates derived by solvolysis and deamination of various homoallylic derivatives is approximately the same as that observed in the oxidation of cyclobutyl radicals.[29] Furthermore, the acetolysis of bicyclobutane under conditions used in the catalytic decomposition of peroxide also gives the same mixture of isomeric C_4H_7 acetates.

The yields of products of *oxidative elimination*, such as cyclobutene and 1,3-butadiene from cyclobutyl and allylcarbinyl radicals, respectively, vary with the nature of the copper(II) oxidant. Ligands such as pyridine and α,α'-bipyridine facilitate oxidative elimination compared to oxidative substitution when such a competition exists. This dichotomy is also markedly influenced by solvent. An independent route for alkene formation that does not involve the same cationic intermediate presented for oxidative substitution is implied, and is discussed later.

The presence of cationic intermediates in the formation of substitution products can also be shown by deuterium labeling. Thus, *oxidation of β-anisylethyl radicals* by copper(II) acetate from the catalyzed decomposition of *p*-methoxyhydrocinnamoyl peroxide afforded β-anisylethyl acetate in excellent yields (> 90%), together with small amounts of *p*-methoxystyrene.[22] The α- and β-carbon atoms were scrambled during oxidative substitution, since β-anisylethyl-α,α-d_2 radicals yielded an equimolar mixture of β-anisylethyl-α,α-d_2 and β-anisylethyl-β,β-d_2 acetates. Even the unsubstituted *β-phenethyl radical* on oxidation gave β-phenethyl acetate in which the α- and β-carbon atoms were completely equilibrated, since the ester fraction isolated in 15% yield from the catalyzed decomposition of hydrocinnamoyl-β,β-d_2 peroxide consisted of both β-phenethyl-β,β-d_2 and α,α-d_2 acetates in equimolar amounts.

The loss of identity of the α- and β-carbon atoms during oxidative substitution of β-arylethyl radicals is attributed to the formation of bridged cationic intermediates since the corresponding β-arylethyl radicals do not rearrange under these conditions.

Complete equilibration of the α- and β-carbon atoms of the β-anisylethyl moiety during solvolysis of the tosylate derivative has been attributed to the intervention of the delocalized cation.[30] There is significantly less driving force

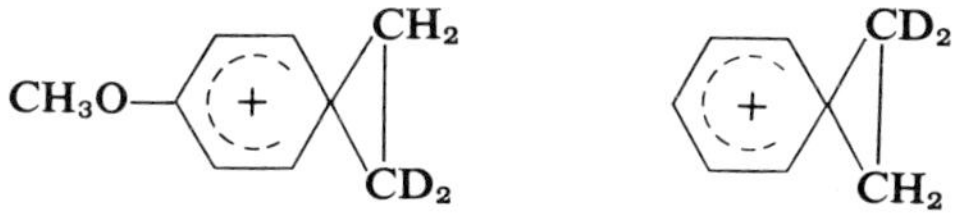

for the formation of the analogous ethylenephenonium ion in the solvolysis of phenethyl tosylate.[31] Convincing support for this ethylenephenonium ion has only recently been found through solvolytic studies in trifluoroacetic acid.[32]

b. Oxidative Elimination as a Synchronous Process. The more common route obtained in the oxidation of alkyl radicals by copper(II) acetate is elimination.[22] A discrete carbonium ion does not appear to be involved directly in this process, since oxidative elimination of alkyl radicals occurs with no rearrangement (of either the carbon skeleton or the hydrogens) diagnostic of other carbonium ion processes. For example, primary alkyl radicals afford only terminal alkenes, and isobutyl radicals generate only isobutylene (no butene-1 or -2). Furthermore, radicals such as cyclobutyl and 2-norbornyl afford significant amounts of cyclobutene and norbornene, respectively, whereas solvolysis or deamination of the corresponding tosylates and amines give little or none of these products.[33]

Neither β-anisylethyl-α,α-d_2 nor β-phenylethyl-β,β-d_2 radicals shows any evidence of rearrangement during oxidative elimination by copper(II) acetate to *p*-methoxystyrene or styrene, respectively. This is true in spite of

$$C_6H_5CD_2CH_2\cdot + Cu^{II}(OAc)_2 \longrightarrow C_6H_5CD{=}CH_2 + Cu^{I}OAc + DOAc \quad (29a)$$

$$p\text{-}CH_3OC_6H_4CH_2CD_2\cdot + Cu^{II}(OAc)_2 \longrightarrow p\text{-}CH_3OC_6H_4CH{=}CD_2 + Cu^{I}OAc + HOAc \quad (29b)$$

the extensive scrambling of the α- and β-carbon atoms observed in the products derived by oxidative substitution.

c. Rates of Oxidation of Alkyl Radicals by Copper(II) Acetate. Since alkyl radicals are the prime chain-carrying species in the copper-catalyzed decomposition of diacyl peroxides, in the presence of a hydrogen donor (HY) these radicals are reduced to alkane or suffer oxidation. The concentrations of copper(I) and copper(II) species remain invariant throughout the reaction.

$$R\cdot + HY \xrightarrow{k_H} RH + Y\cdot \quad (30)$$

$$R\cdot + Cu^{II} \xrightarrow{k_{ox}} R_{ox} + Cu^{I} + H^+ \quad (31)$$

The latter exists in acetonitrile–acetic acid solutions as a mixture of monomeric and dimeric copper(II) carboxylates, but it is only the monomeric form which oxidizes alkyl radicals.[34] Under these conditions, the dissociation constant K of the dimeric copper(II) species and the second-order rate

$$Cu_2^{II}(OAc)_4 \overset{K}{\rightleftharpoons} 2Cu^{II}(OAc)_2 \tag{32}$$

constants for oxidation k_{ox} and reduction k_H, are related to the measured excess of HY and the total copper(II) concentrations, Cu_t, according to equation 33.

$$\frac{(Cu_t)}{(SH)}\frac{RH}{R_{ox}} = 2K^{-1}\left(\frac{k_H}{k_{ox}}\right)^2 \frac{R_{ox}}{RH}(HY) + \frac{k_H}{k_{ox}} \tag{33}$$

Rates constants for oxidative elimination ($k_{ox} = k_e$) obtained in this manner are listed in Table 4.[34]

TABLE 4

Rate Constants for the Oxidation of Alkyl Radicals by Copper(II) at 57°

Alkyl radical	E (kcal/mole)	log A	log k_H (57°)	k_H/k_e	Log k_e
$CH_3CH_2\cdot$	5.9	8.1	4.2	—	—
$CH_3CH_2CH_2\cdot$	6.7	8.3	3.90	1.8×10^{-4}	7.64
$CH_3\dot{C}HCH_3$	6.3	8.3	4.15	2.8×10^{-4}	7.70
$CH_3CH_2CH_2CH_2\cdot$	5.4	7.9	4.35	2.0×10^{-4}	8.05
$(CH_3)_2CHCH_2\cdot$	6.5	8.7	4.42	5.3×10^{-4}	7.70
$CH_3CH_2\dot{C}HCH_3$	4.9	7.7	4.47	3.9×10^{-4}	7.88
$(CH_3)_3C\cdot$	4.3	7.5	4.70	9.2×10^{-5}	8.74

An alternative method for the determination of the rate of oxidation can be obtained from the irreversible cyclization of the 5-hexenyl radical, which proceeds with a first-order rate constant of 1×10^5 sec^{-1} at 25°C.[35] Cyclopentylmethyl and 5-hexenyl radicals were independently generated from

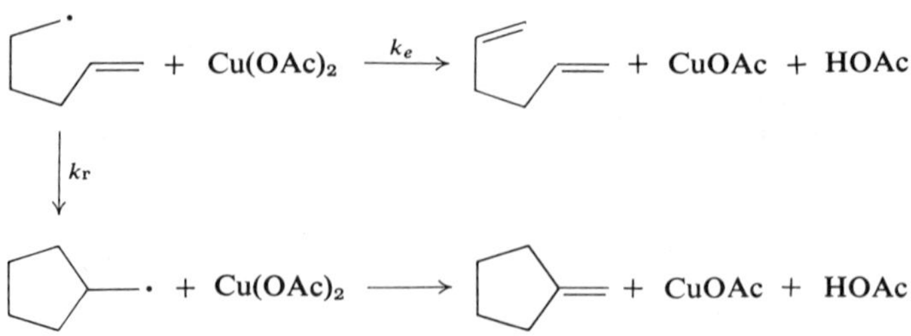

(34)

cyclopentylacetyl and 6-heptenoyl peroxides, respectively. The oxidation of 5-hexenyl radicals by copper(II) acetate under these conditions may be represented by the kinetic scheme shown in (34).[36] Since the isomerization is irreversible, the ratio of 1,5-hexadiene to methylenecyclopentane and methylcyclopentane may be expressed as

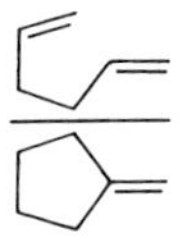

$$= \frac{k_e}{k_r}\,\mathrm{Cu^{II}(OAc)_2} \qquad (35)$$

The relative amounts of hexadiene and methylenecyclopentane varied linearly with the concentration of copper(II) acetate after correction for the monomer–dimer equilibrium. The value for k_e of $1.2 \times 10^6\ M^{-1}\,\mathrm{sec}^{-1}$ obtained in this manner compared reasonably well with the value of $k_e = 3.1 \times 10^6\ M^{-1}\,\mathrm{sec}^{-1}$ for the oxidation of *n*-butyl radicals under slightly different conditions using isobutyraldehyde as a hydrogen donor.

Oxidation of β-arylethyl radicals by copper(II) acetate afforded mixtures of β-arylethyl acetates and styrenes, the molar ratio of which represented the relative rates of oxidative substitution k_s and elimination k_e, respectively. These values of k_s/k_e varied with the ring substituent and the copper(II) oxidant. In particular, the amount of ester relative to styrene was optimum with β-anisylethyl radical. The ratio generally correlated with electron release at the para position of the aryl group. However, the latter conclusion

$$\mathrm{ArCH_2CH_2\cdot} + \mathrm{Cu^{II}(OAc)_2} \xrightarrow{k_s} \mathrm{ArCh_2CH_2OAc} + \mathrm{Cu^{I}OAc} \qquad (36)$$

$$\mathrm{ArCH_2CH_2\cdot} + \mathrm{Cu^{II}(OAc)_2} \xrightarrow{k_e} \mathrm{ArCH{=}CH_2} + \mathrm{HOAc} + \mathrm{Cu^{I}OAc} \qquad (37)$$

is not necessarily valid since the relative rates of oxidative substitution and elimination really only measure *differences* in polar effects of these two competing reactions.

In order to evaluate polar effects on oxidative substitution and elimination *separately*, the rate of oxidation of β-arylethyl radicals was measured relative to hydrogen transfer k_H to isobutyraldehyde. The yields of the substituted styrene as well as the β-arylethyl acetate were each determined relative to the

$$\mathrm{ArCH_2CH_2\cdot} + \mathrm{(CH_3)_2CHCHO} \xrightarrow{k_H} \mathrm{ArCH_2CH_3} + \mathrm{(CH_3)_2CH\dot{C}O} \qquad (38)$$

corresponding ethylbenzene at various concentrations of isobutyraldehyde and copper(II) applicable in equation 33. In this manner, it was possible to obtain values of k_e/k_H and k_s/k_H for each β-arylethyl radical.[22] Thus, the values of k_e and k_s listed in Table 5 could be evaluated separately from a common value of k_H.

TABLE 5

Polar Effects on the Rates of Oxidation of Substituted β-Phenethyl Radicals by Copper(II) Acetate

$Y\text{-}C_6H_4CH_2CH_2\cdot$	$k_e \times 10^{-6}$	$k_s \times 10^{-6}$	$(k_e + k_s) \times 10^{-6}$	k_e/k_s
m-Methoxy	—	—	—	50
H	1.6	0.050	1.7	33
p-Methyl	1.0	0.81	1.7	1.3
p-Methoxy	0.021	1.6	1.5	0.014

d. Mechanisms for the Oxidation of Alkyl Radicals by Copper(II) Acetate. Table 5 shows that the total rate of oxidation of β-arylethyl radicals given by $(k_e + k_s)$ *is remarkably constant for a series of substituents.* On the other hand, in varying the substituent from hydrogen to *p*-methoxy, the rate of oxidative elimination decreases by a factor of approximately 80, whereas the rate of oxidative substitution increases by about a factor of 30. At the same time the relative rates of elimination and substitution change by a factor of 2400. Therefore, *the rate-limiting step for oxidation precedes and is separate from the product-forming step.*

This conclusion militates against a mechanism in which elimination and substitution occur by two independent and parallel processes. It would be highly fortuitous for such concurrent reactions to proceed at widely varying rates, which compensate one another on ring substitution in such a fashion that the total rate remains invariant.

Two alternative mechanisms which are compatible with these kinetic constraints are a rate-limiting electron transfer (outer-sphere) process, or a rate-limiting reaction between the alkyl radical and copper(II) acetate. In both mechanisms, products are formed in a step subsequent to the formation of an intermediate generated in a rate-limiting step. Relative amounts of elimination and substitution products would then be determined by competing routes for decomposition of the intermediate. An alkyl cation could satisfy these requirements, since loss of a β proton and solvolysis are both well-established routes to alkenes and substitution products from carbonium ions generated as intermediates in a variety of reactions. However, if carbonium ions were direct precursors to both alkene and ester, the relative rates of oxidative elimination and substitution, k_e/k_s, would not be expected to vary with the nature of the copper(II) oxidant. However, in the majority of cases in which both oxidative elimination and substitution have been observed, the relative rates *are* dependent on the copper(II) species. Neopentyl radical is the outstanding exception. In this radical, however, rearrangement must precede the formation of oxidation products.

The influence of the copper(II) complex on the relative rates of oxidation elimination and substitution indicates that it must also be associated with the alkyl moiety in the product-determining step. The mechanism below, involving an alkylcopper intermediate, accommodates both of these important limitations.

$$R\cdot + Cu^{II}(OAc)_2 \rightleftharpoons RCu(OAc)_2 \quad (39)$$

$$RCu(OAc)_2 \xrightarrow{\text{elimination}} R(-H) + Cu^{I}OAc + HOAc \quad (40)$$

$$RCu(OAc)_2 \xrightarrow{\text{substitution}} [R^{+}Cu^{I}(OAc)_2] \xrightarrow{HS} RS + HOAc + Cu^{I}OAc \quad (41)$$

Substitution products are associated with heterolysis of the alkylcopper intermediate to copper(I) and a carbonium ion followed by solvation. The importance of this oxidative fragmentation is generally related to the cationic stability of the alkyl fragment and, thus, constitutes the more important process with cyclobutyl, *p*-methoxybenzyl, β-anisylethyl, and similar radicals. The properties of the resultant carbonium ion (rearrangement, addition, selectivity in solvation, elimination, etc.) would be essentially those observed in facile S_N1-$E1$ solvolyses of the corresponding alkyl tosylates and halides under comparable conditions.

The oxidation of β-arylethyl radicals illustrates the difference between elimination and substitution processes. Deuterium labeling has shown that the α- and β-carbon atoms are extensively scrambled in the β-arylethyl acetate, whereas no such scrambling occurs in the styrene. The importance of such a synchronous elimination of a β-hydrogen is also shown by the kinetic isotope effect in the oxidation of β-phenethyl-β,β-d_2 radical.[22]

$$C_6H_5CD_2CH_2\cdot + Cu^{II}(OAc)_2 \xrightarrow{k_s} [C_6H_5CD_2CH_2OAc + C_6H_5CH_2CD_2OAc] + Cu^{I}OAc$$

$$C_6H_5CD_2CH_2\cdot + Cu^{II}(OAc)_2 \xrightarrow{k_e} C_6H_5CD{=}CH_2 + Cu^{I}OAc + DOAc$$

The ratio of the second-order rate constants for oxidative elimination and substitution given by k_e/k_s was readily determined from the relative amounts of styrene and β-phenethyl acetate formed. This value was 33 for the oxidation of β-phenylethyl radical by copper(II) in 40 vol% acetonitrile–acetic acid, and 12 for β-phenylethyl-β,β-d_2 radical under equivalent conditions. The isotope effect in oxidative substitution is expected to be small and negligible, and the observed differences in rates are attributed solely to effects of deuterium substitution in oxidative elimination. On this basis, the kinetic isotope effect for the oxidative elimination of β-phenethyl radical is 2.8. The relatively small isotope effect observed is consistent with a high degree of electron transfer in the transition state for oxidative elimination. This conclusion is also in accord with the rather random loss of β-hydrogens during oxidation of *sec*-butyl radicals discussed previously.[22] Furthermore, the *magnitude* of the β-deuterium isotope effect for oxidative elimination is intermediate between values observed for the base-promoted elimination and for the unimolecular elimination of a variety of β-arylethyl derivatives.[37] *E*2 and *E*1 reactions commonly represent extremes of rupture of the β-hydrogen bond in the transition state for elimination.[38] A possible transition state for the loss of a β proton synchronously with electron transfer is represented schematically below:

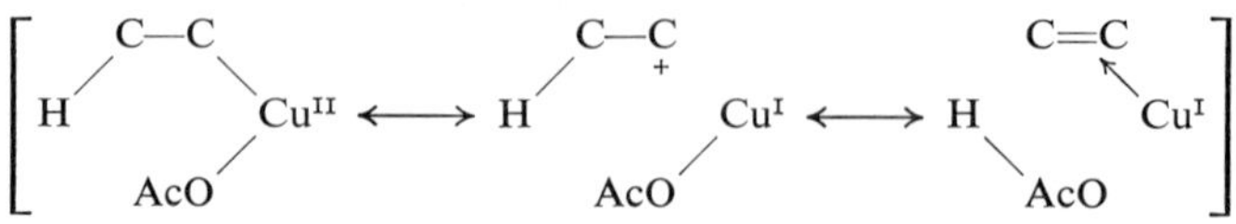

e. Oxidative Elimination versus Oxidative Substitution. The factors which influence the partitioning of the alkylcopper intermediate merit further discussion. There is no doubt that the stability of the carbonium ion formed by heterolysis influences the yield of substitution products, since excellent yields are obtained with those alkyl moieties which yield stable carbonium ions. However, the contrary is not valid, since primary and secondary alkyl radicals such as *n*-butyl and isopropyl, which do not produce stable cations, are oxidized as readily as (in fact, slightly faster than) the β-anisylethyl radical. Oxidative elimination is thus a kinetically (and energetically) equivalent route for decomposition of the alkylcopper intermediate.

Polar substituents exert diametrically opposed effects on oxidative substitution and elimination of β-arylethyl radicals. The rates shown in Table 5 do not strictly conform to a Hammett correlation, since oxidative elimination and substitution are mutually interdependent processes by a common alkylcopper intermediate. However, from the trend in rates, it is apparent that the substitution process is enhanced by electron-releasing substituents, and oxidative elimination is favored by electron-attracting substituents. Oxidative substitution is easily associated with a transition state in which the arylethyl moiety takes on cationic character. On the other hand, the charge requirements in oxidative elimination parallel those observed in proton transfers and base-promoted *E*2 reactions, previously noted in proton exchanges from substituted toluene and rates of elimination of β-arylethyl derivatives.[37,38]

The driving force for oxidative substitution can be enhanced in a remarkable way by simply replacing the acetate by a ligand less strongly coordinating, such as a perchlorate or a triflate.[36] Thus, the oxidation of simple alkyl radicals (e.g., *n*-butyl radicals) by copper(II) triflate proceeds with a high degree of carbonium ion character. The latter is indicated by the extensive rearrangement of the *n*-butyl to the *s*-butyl moiety, resulting in the formation of *cis*- and *trans*-butene-2 as well as *s*-butyl acetate and substantial amounts of *n*-butyl acetate. All four of these products were absent in the oxidation of

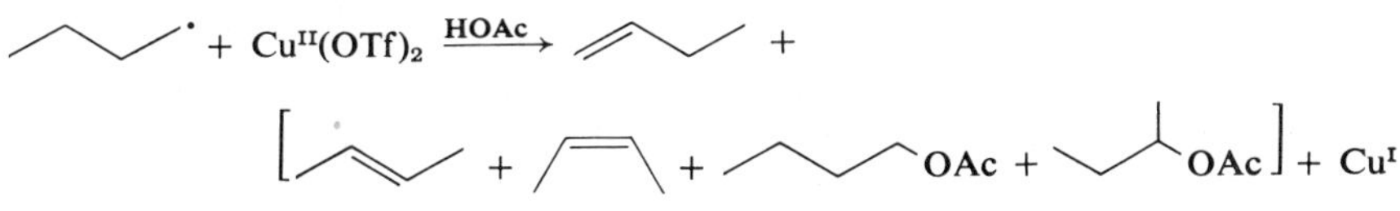

n-butyl radical by copper(II) acetate under the same conditions. The extent of the cationoid rearrangement of the n-butyl moiety has been cited by Jensen et al. to be a measure of the degree to which the carbonium ion is unencumbered in the solvolysis (see Table 6).[39]

TABLE 6

Rearrangement Accompanying Solvolysis of n-Butyl Derivatives

System	Reaction	Rearrangement (%)	Ref.
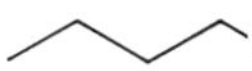 OTs	Solvolysis	0	40a
 N_2^+	Deamination	25	40b
 Hg^+	Demercuration	16	39
 $CuOTf^+$	Oxidation	35–50	36

The high degree of cationic driving force is also shown in the oxidation of isobutyl radicals by copper(II) triflate. The formation of butene-1, *cis*-butene-2, *trans*-butene-2, and *s*-butyl acetate as a result of methyl migration, and *t*-butyl acetate by hydrogen migration alludes to extensive cationoid rearrangements. Under the same conditions isobutyl radicals are oxidized by copper(II) acetate specifically to only isobutylene.

The solvolysis of primary alkyl tosylates has been investigated recently in trifluoroacetic, fluorosulfuric, and sulfuric acids.[41] The rather large α-d and β-d kinetic isotope effects observed in these media support the mechanism involving near-limiting solvolysis of isobutyl tosylate.[42] The internal competition between hydrogen and methyl migration in the anchimerically assisted solvolysis of isobutyl tosylate is also minimal in these highly ionizing and poorly solvating solvents. Extensive and competitive hydrogen and methyl migration in the deamination of isobutylamine[43] and in the oxidation of isobutyl radicals by copper(II) triflate indicate the involvement of the same high degree of cationoid transition states in these dissimilar processes.

The drastic changes induced in the course of oxidation of alkyl radicals by the replacement of an acetate by a triflate ligand on copper(II) are somewhat akin to large alterations in solvolysis effected by manipulating leaving groups and the solvent. In contrast, these large changes in the carbonium ion character of the oxidation process often occur in relatively poorly ionizing and highly nucleophilic solvents.

The comparison of the visible absorption spectra of copper(II) acetate and copper(II) triflate in acetic acid solutions shows clearly that the principal difference between these two oxidants is associated with the ionization of the ligand. Copper(II) acetate is present in acetic acid largely as an undissociated

$$Cu(OAc)_2 \rightleftharpoons CuOAc^+OAc^-$$

$$Cu(OTf)_2 \rightleftharpoons CuOTf^+OTf^- \rightleftharpoons Cu^{2+}2OTf^-$$

inner-sphere complex. On the other hand, copper(II) triflate like the perchlorate is highly dissociated and exists largely as an outer-sphere complex.[36]

In an analogous manner, the alkylcopper species formed as an intermediate in the oxidation of alkyl radicals by copper(II) acetate and triflate must also reflect these differences in the abilities of acetato and triflato to coordinate with the copper nucleus. Copper(II) complexes are substitution labile. For example, the rate of water exchange from the hexaaquocopper(II) ion is in excess of 10^9 sec^{-1}. The rates of anation of copper(II) complexes are also, no doubt, extremely fast and would not be expected to impose large kinetic limitations on the behavior of the metastable alkylcopper intermediates.

$$RCu(OTf)_2 \rightleftharpoons RCuOTf^+OTf^- \rightleftharpoons RCu^{2+}2OTf^-$$

$$RCu(OAc)_2 \rightleftharpoons RCuOAc^+OAc^-$$

We propose that a coordinately unsaturated species such as $RCuOTf^+$ and RCu^{2+} as well as $RCuOAc^+$ is directly responsible for the generation of carbonium ion intermediates, for example,

$$\begin{aligned} RCuOTf^+ &\longrightarrow R^+ + Cu^IOTf \\ RCuOAc^+ &\longrightarrow R^+ + Cu^IOAc \end{aligned} \tag{42}$$

The facility of the heterolyses represented in (42) is kinetically related to the dissociation constant of the alkylcopper intermediates, and is more important with triflato than acetato. Furthermore, other factors being equal, heterolysis is more favored by alkyl moieties most able to attain carbonium ion stability. For this reason, oxidations of alkyl radicals with copper(II) *acetate* usually circumvent the oxidative solvolysis route and proceed readily via oxidative elimination involving facile loss of β-hydrogen.

Oxidative substitution of primary alkyl radicals by copper(II) triflate does not proceed via an alkyl triflate intermediate, followed by subsequent solvolysis, for example,

$$R\cdot + Cu^{II}OTf^{+} \longrightarrow Cu^{I} + R\text{—}OTf \xrightarrow[HOAc]{} ROAc + HOTf$$

since *no rearrangement* was observed during acetolysis of *n*-propyl tiflate, whereas the oxidation of *n*-propyl radicals by copper(II) triflate in acetic acid affords substantial amounts of isopropyl acetate in addition to cyclopropane.[36] The latter is also produced during deamination of *n*-propylamine and is diagnostic of the formation of an *n*-propyl cationoid intermediate.[43] Finally, tetrafluoroborate derivatives of copper(II) cannot generate alkyl tetrafluoroborates as intermediates during oxidation of alkyl radicals. Nonetheless, oxidation of *n*-butyl radicals with copper(II) tetrafluoroborate affords essentially the *same mixture of rearranged butyl derivatives* as those derived from copper(II) triflate or perchlorate. An alkylcopper intermediate is the only species these oxidants have in common.

Various solvents and ligands can be employed to alter the relative amounts of oxidative elimination and substitution. Thus, acetonitrile and tetramethylene sulfone are solvents which are particularly good in enhancing the yields of oxidative substitution products. For example, the oxidation of β-anisylethyl radicals by copper(II) in acetonitrile–acetic acid solutions affords β-anisylethyl acetate in greater than 90% yield, whereas equivalent oxidations in benzene–acetic acid solutions generate *p*-methoxystyrene in excellent yields. A similar dichotomy has been noted with cyclobutyl radicals. Furthermore, the oxidation of neopentyl radicals by copper(II) in benzene is significantly slower than in acetic acid solutions.[22] These changes in rates of reaction may be partly attributed to a general solvent effect. Specific coordination to the copper species to generate a copper(I) complex which is a better leaving group in the heterolysis step may also be an important function of the solvent.

The ready oxidation of methyl radicals to methyl acetate merits separate attention since the methyl radical has a high ionization potential relative to other alkyl radicals. Furthermore, unlike other primary alkyl radicals, a pathway involving oxidative elimination is unavailable to methyl radical. Neopentyl radical, which is the exception, undergoes complete rearrangement via the carbonium ion to the *t*-amyl cation under these conditions.

The oxidation of methyl radical to methyl acetate by copper(II) acetate proceeds at approximately one-half the rate at which *n*-butyl radical is oxidized to butene-1.[36] The absence of *n*-butyl acetate among the products in the latter process indicates that its rate of formation must be at *least* ten times slower than the formation of methyl acetate. Mechanistically, such a conclusion would indicate that alkyl acetates are formed by a bimolecular displacement process. However, the oxidation of methyl radicals can be carried out under conditions in which the only available acetate is present as a complex with copper(II). The alkyl acetate, therefore, does not arise by a competing *second-order* process. However, the postulation of an alkylcopper intermediate which can be partitioned between alkene and alkyl acetate does accommodate these results. The methylcopper intermediate in contrast to other alkyl analogs has no access to β or even α elimination. The high yields of methyl acetate derived from the oxidation of methyl radicals by copper(II) acetate even in pure acetonitrile suggest that the acetate moiety is transferred internally, for example,

$$CH_3CuOAc^+ \longrightarrow [CH_3{}^+Cu^IOAc] \longrightarrow CH_3OAc + Cu^I$$

Such a pathway, which can be described as oxidative displacement, is unimportant with most alkyl radicals, since it can be readily circumvented by the competing oxidative elimination. The oxidation of methyl radicals by copper(II) triflate affords even higher yields of methyl acetate in acetic acid solution. Oxidation also proceeds rapidly in the absence of acetic acid, and solvation by acetonitrile and other nucleophiles is inferred.

f. Solvolytic Routes and the Mechanism of Electron Transfer Oxidation. The electron transfer oxidation of alkyl radicals by copper(II) complexes, thus proceeds via a metastable alkylcopper intermediate, which is partitioned between oxidative elimination and oxidative substitution. The anionic counterion exerts a dominant effect in the selection of pathways. Acetato ligand promotes oxidative elimination with most alkyl radicals. The importance of oxidative substitution increases as the alkyl moiety becomes capable of carbonium ion stabilization, and since it proceeds via a solvolysis of an alkylcopper intermediate the process is more properly described as oxidative *solvolysis*.

Oxidative solvolysis is also induced in cyclobutyl radicals by lithium perchlorate and an acetic acid–acetonitrile solvent. The effect induced by lithium perchlorate is analogous to that shown in the limiting solvolysis of a variety of alkyl derivatives, and has been described as the "special salt effect."[44] The analogy, however, is somewhat limited, since lithium perchlorate in solvolysis retards ion pair return and operates directly at the carbonium ion stage. Lithium perchlorate exerts a secondary effect in the example presented here. On the other hand, the oxidation of *n*-butyl radicals by copper(II) acetate is insensitive to the presence of lithium perchlorate. The difference between *n*-butyl and cyclobutyl radicals in this regard may be attributed to the relatively difficult ionization of the *n*-butylcopper intermediate to a *primary* butyl cation in contrast to the ready formation of a homoallylic cation. Under such circumstances the competition from the facile oxidative elimination to butene-1 far overshadows the cationic route. This behavior should be compared with that of copper(II) triflate itself which promotes oxidative solvolysis as the major route even with simple primary alkyl radicals. The following generalized scheme is proposed for electron transfer oxidation[36]:

Formation of an alkyl–copper complex:

$$R\cdot + CuX_2 \rightleftharpoons RCuX_2$$

Oxidative elimination:

$$RCuX_2 \longrightarrow R(\text{—}H) + HX + Cu^{I}X$$

Oxidative solvolysis:

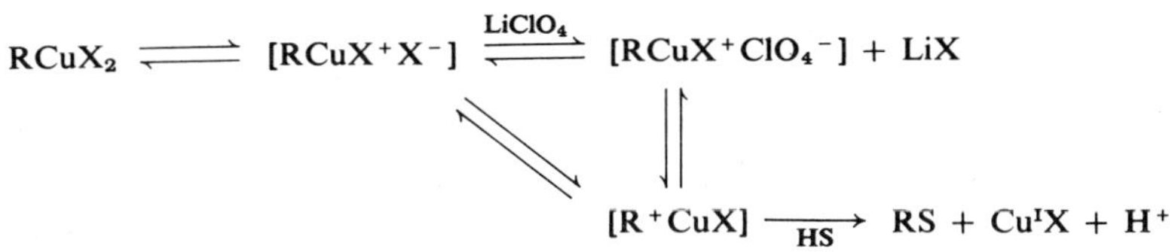

Oxidative displacement:

$$RCuX_2 \rightleftharpoons [R^+CuX_2^-] \longrightarrow R\text{—}X + Cu^{I}X$$

In this mechanistic scheme, the *dissociation of the ligand* from the alkyl-copper intermediate provides the driving force required for oxidative solvolysis to take place in competition with oxidative elimination during electron transfer oxidation of alkyl radicals by copper(II) complexes. Oxidative solvolysis induced by ionization has a formal analogy to the solvolytic demercuration of organomercurials reported by Jensen and Ouellette.[45] Thus, alkylmercury(II) acetates in acetic acid only undergo protonolysis. However, in

$$RHgOAc + HOAc \longrightarrow RH + Hg(OAc)_2$$

the presence of perchloric acid, facile demercuration occurs with the liberation of free mercury. In these cases, alkylmercurium species can be generated by ligand exchange. The S_N1 solvolysis shown in equation 43 exhibits great sensitivity to changes in structure of the alkyl moiety and to neighboring group participation. The latter are indicative of carbonium ions as intermediates or a high degree of carbonium ion character in the transition state. The driving force in these reactions is undoubtedly derived from the formation of Hg(0), and the following mechanism has been proposed:

$$RHgOAc + HClO_4 \longrightarrow RHg^+ClO_4^- + HOAc$$

$$RHg^+ \longrightarrow R^+ + Hg^0, \ldots \qquad (43)$$

Copper(I) plays a similar role as a leaving group in oxidation of alkyl radicals with copper(II). Interestingly, the acetolysis of the β-phenethylmercurinium ion leads to only 90% scrambling of the α and β positions due to an S_N2 component in the solvolysis.

2. *Ligand Transfer Oxidation of Alkyl Radicals by Copper(II)*

An inner-sphere process can also be formulated for the oxidation of alkyl radicals by metal complexes. Implicit in equation 44 is the direct

$$R\cdot + M^{n}X_{n} \longrightarrow R{-}X + M^{n-1}X_{n-1} \tag{44}$$

transfer of the ligand X from the metal reductant to the alkyl radical. However, an alternative formulation (equation 45) is also possible in which electron transfer (i.e., oxidation) precedes the transfer of the ligand. Organic systems

$$R\cdot + M^{n}X_{n} \longrightarrow [R^{+}M^{n-1}X_{n}] \longrightarrow R{-}X + M^{n-1}X_{n-1} \tag{45}$$

are particularly valuable in assessing this distinction, since carbonium ion intermediates show an unusual propensity for rearrangement, as studies of tight ion pairs in solvolysis of alkyl derivatives have shown.[46] Furthermore, since alkyl derivatives R—X are usually substitution stable, the transfer of ligand from the metal complex (whether by the direct or indirect mechanism) can be followed directly. These oxidations are described as *ligand transfer processes* and have also been studied by the catalytic decomposition of diacyl peroxides described in equations 21, 22, 24, and 25. The distinction between ligand transfer and electron transfer oxidation of alkyl radicals can be gained by considering: (a) rearrangements accompanying ligand transfer oxidation, (b) effect of solvent on the anation of copper(II) species, (c) carbonium ions as intermediates, and (d) dual mechanisms in ligand transfer processes. Finally, unifying principles in the mechanisms of these processes need consideration.

a. Rearrangement Accompanying Ligand Transfer Oxidations. The *oxidation of neopentyl radicals* by copper(II) chloride produced only neopentyl chloride. Control experiments showed that *t*-amyl chloride which was

$$(CH_3)_3CCH_2\cdot + Cu^{II}Cl_2 \longrightarrow (CH_3)_3CCH_2Cl + Cu^{I}Cl$$

deliberately added prior to the initiation of the decomposition of *t*-butylacetyl peroxide could be recovered quantitatively.[47]

The *oxidation of cyclobutyl and allylcarbinyl radicals* with copper(II) bromide occurred with essentially no rearrangement. The oxidation of cyclopropylmethyl radical by copper(II) bromide afforded predominantly cyclopropylmethyl bromide. The smaller amounts of allylcarbinyl bromide were attributable to the isomerization of the radical, the competition from which could be largely overwhelmed by increasing the concentration of copper(II) bromide.

The kinetic scheme presented above was based on a knowledge of this competing first-order rate constant for isomerization of 1×10^{8} sec^{-1} at

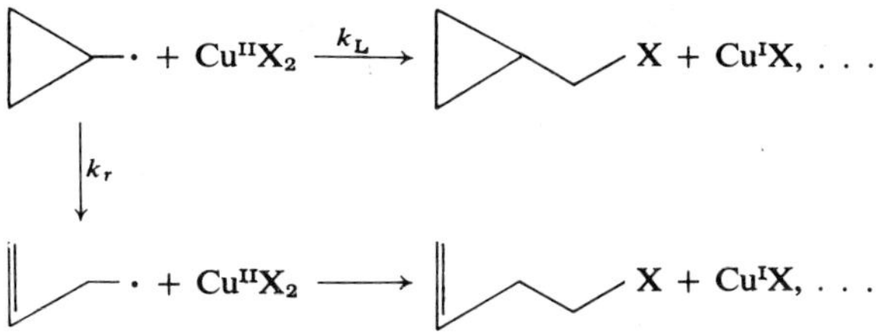

25°C. If it is assumed that the rate constants k_L for allylcarbinyl and cyclopropylmethyl radicals are approximately the same, the relative yields of products may be expressed as

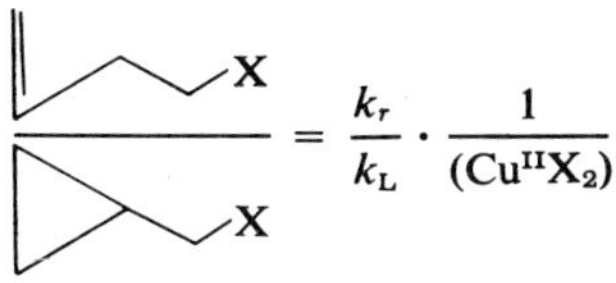

A value of $k_L(Br) = 4.3 \times 10^9\ M^{-1}\ sec^{-1}$ at 25°C was obtained. No isomerization of any of the alkyl bromides occurred under reaction conditions.

The oxidation of allylcarbinyl radicals by copper(II) *thiocyanate* proceeded without rearrangement. A similar treatment of cyclobutyl radicals also produced mainly cyclobutyl thiocyanate. The oxidation of cyclopropylmethyl by copper(II) thiocyanate afforded a mixture consisting largely of cyclopropylmethyl and allylcarbinyl thiocyanate, the relative amounts of which were directly determined by the concentration of copper(II) thiocyanate. A quantitative determination of the latter led to a value of the second-order rate constant k_L for thiocyanate transfer of $3.6 \times 10^8\ M^{-1}\ sec^{-1}$.[1] Small but discrete amounts of cyclobutyl thiocyanate were also formed. Scrutiny of the reaction showed, however, that no alkyl isothiocyanates were formed or isomerized under the conditions of the reaction.

The oxidation of each of these radicals by copper(II) *chloride*, in contrast to copper(II) bromide or thiocyanate, afforded a mixture of alkyl chlorides representing extensive rearrangement of the structure of the parent radical. Thus, the oxidation of cyclobutyl and allylcarbinyl radicals by copper(II) chloride in acetonitrile both produced unique mixtures of all three isomeric chlorides. *The composition of each mixture was independent of the concentration of the copper(II) chloride.*[47]

The composition of the products resulting from the oxidation of cyclopropylmethyl radical was highly dependent on the concentration of copper(II) chloride. At very low concentrations of copper(II) the isomeric mixture was the same as that obtained from the oxidation of allylcarbinyl radical. It could be shown from studies carried out at higher concentrations of copper(II) chloride that the oxidation of cyclopropylmethyl radicals also produced a unique mixture of isomeric chlorides. The composition of this mixture was also independent of the concentration of copper(II) chloride. Kinetic analysis of the type described for bromide and thiocyanate led to a second-order rate constant for chlorine transfer of k_L (Cl) $= 1.1 \times 10^9 \ M^{-1} \ \text{sec}^{-1}$.[23]

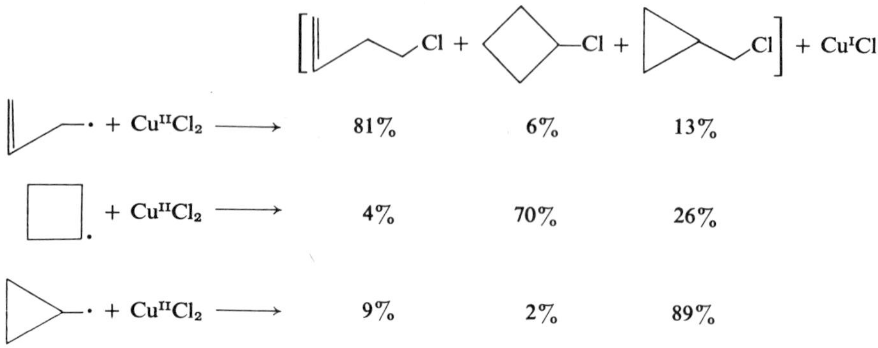

The oxidation of each C_4H_7 radical by copper(II) chloride in acetonitrile, thus produced a *unique* mixture of isomeric chlorides which was independent of the concentration of copper(II) chloride. Moreover, the predominant isomer in each mixture was that resulting from the intact structure of the precursor radical. Careful study proved that no rearrangement of the products occurred subsequent to their formation. The observed distribution of isomeric chlorides, thus resulted from a kinetically controlled oxidation process.

b. Effect of Solvent on the Oxidation of Cyclobutyl Chloride by Copper(II) Chloride. The foregoing studies showed that oxidation of cyclobutyl radicals in acetonitrile produced a mixture of C_4H_7 chlorides, the composition of which was independent of the concentration of copper(II) chloride. The medium was varied, therefore, in order to determine the effect of solvent on the distribution of isomeric chlorides and its relationship to the mechanism.

In fact, cyclobutyl chloride was the only product of oxidation of cyclobutyl radicals by copper(II) chloride when the reaction was carried out in pyridine. The reaction was slightly less selective in dimethyl sulfoxide (DMSO) and hexamethylphosphoric triamide (HMPA). Acetic acid, acetonitrile, and *N,N*-dimethylformamide, despite their different chemical and physical properties, induced approximately the same amount of rearrangement. The stability of cyclobutyl chloride to isomerization was established in each solvent under these conditions.

The specificity of the oxidation of cyclobutyl radicals by copper(II) chloride was enhanced by the addition of lithium chloride. Thus, the selectivity in the formation of cyclobutyl chloride was raised from 83% in the presence of 0.040 *M* copper(II) chloride to 95% on the addition of 0.106 *M* lithium chloride in DMSO solutions. The specificity could also be deliberately *lowered* by the use of 1,4-dioxane or 1,2-dimethoxyethane as the solvent.

c. Carbonium Ions as Intermediates in Ligand Transfer. The facile oxidation of alkyl radicals by various copper(II) halides and pseudohalides involves the transfer of a ligand from the copper(II) oxidant to the alkyl radical. Free carbonium ions as such are not intermediates since the transfer of chloride, bromide, iodide, thiocyanate, azide, and cyanide can be effected in protic media such as acetic acid without direct intervention of the external nucleophile.[23,47] However, cationoid intermediates are not necessarily precluded since the subtleties of reactions occurring from ion pairs can exclude direct participation of solvent, as studies of the solvolysis of alkyl derivatives have shown.[44,46]

A much more sensitive probe for the participation of carbonium ions along the reaction coordinate is provided by the study of alkyl moieties prone to cationic rearrangement. Three such systems have been employed in this study: neopentyl, homoallylic, and β-arylethyl radicals. A comparison of these alkyl radicals shows that the contribution from a cationic pathway varies with both the alkyl moiety as well as the copper(II) oxidant. Thus, ligand transfer oxidation of neopentyl radical occurs with *no evidence of rearrangement*, since *t*-amyl derivatives are absent.

On the other hand, cyclobutyl, allylcarbinyl, and cyclopropylmethyl cations are more susceptible to rearrangement. The predominant product obtained from the oxidation of each of these radicals by copper(II) chloride in acetonitrile was that isomeric chloride resulting from the radical with the structure intact. That is, cyclobutyl radical afforded principally cyclobutyl chloride, allylcarbinyl radical gave largely allylcarbinyl chloride. Even cyclopropylmethyl radical produced mainly cyclopropylmethyl chloride, provided the copper(II) chloride concentration was sufficiently high. In every case, the by-product from the oxidation of each isomeric radical was a mixture of the other two isomeric chlorides.

Oxidation of each radical by copper(II) bromide and by copper(II) thiocyanate gave similar results, although the amount of rearrangement varied appreciably with the copper(II) oxidant. The extent of rearrangement accompanying ligand transfer oxidation of each radical by various copper(II) complexes is summarized in Table 7.

Cyclobutyl cations are implicated as the precursors for the rearranged products, since the distributions of the isomeric by-products from the ligand transfer oxidations of all the radicals are characteristic of the results obtained in solvolytic studies[48] of these systems. Furthermore, any isomerization of the alkyl radical precursors has been taken into account in these results.

TABLE 7

Rearrangement Resulting from Ligand Transfer Oxidation of Homoallylic and Neopentyl Radicals by Copper(II) Complexes

		Homoallylic radical (%)			Neopentyl radical (%)
$Cu^{II}X_2$	Relative rate	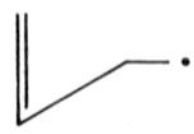		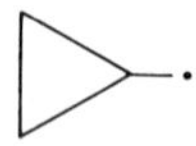	
$Cu^{II}Br_2$	12	0	0	< 1	—
$Cu^{II}Cl_2$	3	20	50	15	0
$Cu^{II}(NCS)_2$	1	0	5	3	—

The contribution of a polar substituent effect in promoting the cationic pathway is shown by comparing the *ligand transfer oxidation of β-phenethyl and β-anisylethyl radicals.* Isotopic labeling allows the examination of carbonium ion intermediates in this system, since complete equilibration of the α- and β-carbons invariably results from a phenonium ion intermediate. Solvolysis studies show that an even greater driving force for cation formation in the form of the bridged anisonium ion is provided by the presence of a

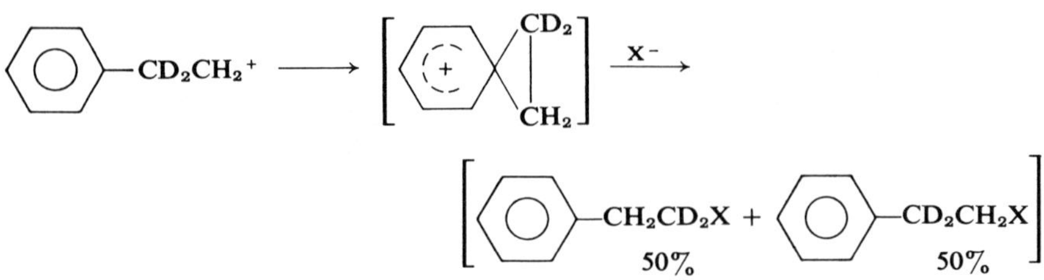

TABLE 8

Contribution from the Cationic Path in the Ligand Transfer Oxidation of β-Arylethyl Radicals in Acetonitrile

	Rearrangement (%)	
CuX2 oxidant	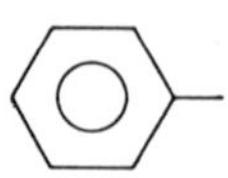 $-CD_2CH_2\cdot$	CH_3O-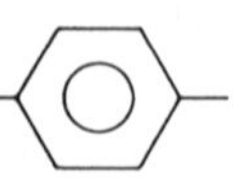$-CH_2CD_2\cdot$
$Cu^{II}Cl_2$	0	54
$Cu^{II}Br_2$	0	12

p-methoxy group. Rearrangement accompanying ligand transfer oxidation of β-phenethyl and β-anisylethyl radicals is summarized in Table 8. No spontaneous rearrangement of either β-phenethyl or β-anisylethyl radical occurs under these conditions.

d. Dual Mechanisms in Ligand Transfer Oxidation. These studies show that two modes of oxidation occur during ligand transfer oxidation of alkyl radicals by copper(II) halides and pseudohalides. One pathway involves *direct conversion* of the alkyl radical (R·) to the substitution product (RX). The other more *indirect route* generates a carbonium ion of sufficient integrity somewhere along the reaction coordinate to undergo complete equilibration of the α- and β-carbon atoms of the β-arylethyl moiety and extensive scrambling of the carbon atoms in the cyclobutyl precursors. Stabilization of the alkyl cation in these systems plays an important role in the indirect route, since the neopentyl and β-phenethyl analogs show no tendency to afford carbonium ion intermediates under the same conditions.

The results of *ligand transfer oxidation* of the series of neopentyl, cyclobutyl, and β-arylethyl radicals given here are in sharp contrast to the *electron transfer oxidation* of the same radicals previously examined under comparable conditions. In every example pertaining to the electron transfer oxidation of these alkyl radicals, *oxidative solvolysis invariably led to complete cationic rearrangement of the alkyl moiety.* Thus, electron transfer oxidation of neopentyl radicals afforded only products derived from the *t*-amyl cation. Similarly, oxidative substitution of the β-phenethyl radical by copper(II) acetate led to complete equilibration of the α- and β-carbon atoms. Finally the oxidation of cyclobutyl radicals under the same conditions produced the same extensively rearranged mixture of isomeric cyclobutyl acetates derived by other cationic processes.

The partial cationic rearrangement observed during the *ligand transfer oxidation* of certain alkyl radicals and the absence of rearrangement in others indicate that at least two processes are operative. The extent of rearrangement, furthermore, is invariant with the concentration of the copper(II) halide or pseudohalide, but highly dependent on the structure of, and the ligand associated with, the copper(II) nucleus. These results are inconsistent with competition from a kinetically first-order process. It was proposed[47] that the *direct* and *indirect* pathways in ligand transfer oxidation are represented by independent simultaneous second-order processes. The direct process is described as *atom transfer* and the indirect route is represented as *oxidative displacement*. These mechanisms will be elaborated further.

The Role of Oxidative Displacement in Ligand Transfer Processes. Ligand transfer oxidation proceeding via a pathway represented as *oxidative displacement* is given by equations 46 and 47, in which copper(II) chloride is used for illustrative purposes. The formation of a carbonium ion intermediate

$$R\cdot + Cu^{II}Cl_2 \rightleftharpoons RCuCl_2 \quad (46)$$

$$RCuCl_2 \rightleftharpoons [R^+Cu^{I}Cl_2] \longrightarrow R{-}Cl + Cu^{I}Cl \quad (47)$$

by this route is accommodated by a metastable alkylcopper species akin to that presented earlier in connection with electron transfer processes.

According to this formulation, the formation of carbonium ion intermediates leading to ligand transfer is promoted by ionization of the alkylcopper intermediate. Evidence for ionization of copper(II) complexes is obtained from an examination of the absorption spectra. Thus, the visible absorption spectrum of copper(II) chloride shows that it exists in dioxane as an outer-sphere complex.[47] In strong contrast, inner-sphere chlorocopper(II) species are prevalent in acetonitrile.[49]

$$CuCl_2 \xrightleftharpoons{\text{dioxane}} Cu^{2+}2Cl^-$$

$$CuCl_2 \xrightleftharpoons{CH_2CN} Cu^{2+}2Cl^-$$

The formation constants of chlorocopper(II) have not as yet been measured in dioxane. However, aqueous solutions form a reasonable approximation since the visible absorption spectrum of copper(II) chloride is virtually the same in dioxane as in water. The formation constants of inner-sphere chlorocopper(II) species also differ markedly in acetonitrile and in water as shown in Table 9.[50]

TABLE 9

Formation Constants of Inner-Sphere Chlorocopper(II) Species in Acetonitrile, Methanol, and Water

	Formation constant		
Equilibrium	CH_3CN	CH_3OH	H_2O
$Cu^{II} + Cl^- \rightleftharpoons Cu^{II}Cl^+$	$K_1 = 10^{9.7}$	$10^{4.2}$	1.0
$Cu^{II}Cl^+ + Cl^- \rightleftharpoons Cu^{II}Cl_2$	$K_2 = 10^{7.9}$	$10^{6.5}$	0.2
$Cu^{II}Cl_2 + Cl^- \rightleftharpoons Cu^{II}Cl_3^-$	$K_3 = 10^{7.1}$	—	0.4
$Cu^{II}Cl_3^- + Cl^- \rightleftharpoons Cu^{II}Cl_4^{2-}$	$K_4 = 10^{3.7}$	—	0.01

The ionization of the metastable chloroalkylcopper intermediate should be promoted by dioxane much like the chlorocopper(II) species themselves. Copper(II) complexes are, in general, highly substitution labile and anation should not place kinetic restraints on the reactivity of the alkylcopper species.

$$RCuCl_2 \xrightleftharpoons{\text{dioxane}} RCuCl^+Cl^-$$

The coordinatively unsaturated cationic species are subject to heterolysis of the alkyl–copper bond in a manner reminiscent of the previously described oxidative substitution in electron transfer oxidation and demercuration of

$$RCuCl^+ \longrightarrow R^+Cu^ICl, \ldots$$

alkylmercurinium cations.

Extensive cationic rearrangement of the cyclobutyl moiety during ligand transfer oxidation of cyclobutyl radicals by copper(II) chloride in dioxane is consistent with this formulation. A qualitative comparison of the importance of the cationic route in acetonitrile and dioxane is presented in Table 10. The

TABLE 10

Promotion of Oxidative Substitution by Solvent in Ligand Transfer Oxidations

Radical	$Cu^{II}X_2$	Solvent	Cationic rearrangement (%)
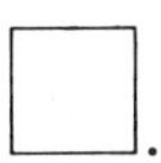	$Cu^{II}Cl_2$	CH_3CN	~ 50
	$Cu^{II}Br_2$	CH_2CN	0
	$Cu^{II}Cl_2$	Dioxane	~100
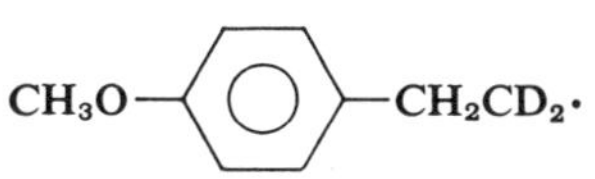	$Cu^{II}Cl_2$	CH_3CN	54
	$Cu^{II}Cl_2$	Dioxane	94
	$Cu^{II}Br_2$	CH_3CN	12
	$Cu^{II}Br_2$	Dioxane	18

same phenomenon can be evaluated quantitatively by examining the ligand transfer oxidation of isotropically labeled β-anisylethyl radical. The magnitude (both absolute and relative) of the solvent effect shown in Table 10 is more marked with copper(II) chloride than bromide.

The similarity between the oxidative displacement pathway in ligand transfer oxidation and oxidative solvolysis in electron transfer is shown by the oxidation of cyclobutyl radicals with copper(II) chloride in aqueous acetonitrile. The isomeric cyclobutyl alcohols formed under these conditions represent oxidative solvolysis. The composition of this mixture of isomeric alcohols is strikingly similar to the mixture of alkyl chlorides obtained from the ligand transfer oxidation of cyclobutyl radicals by copper(II) chloride in dioxane. The transition states for the reaction of the cyclobutyl moiety with water or chloride during these oxidative processes are no doubt related to similar alkylcopper species.

The Role of Atom Transfer in Ligand Transfer Processes. The alternate route available in ligand transfer oxidation is represented as *atom transfer* in equation 48. The transition state for atom transfer is free radical in nature,

$$R\cdot + Cu^{II}Cl_2 \longrightarrow [R\text{---}Cl\text{---}CuCl]^{\ddagger} \longrightarrow R{-}Cl + Cu^{I}Cl \qquad (48)$$

and polar effects are small. Thus, even alkyl radicals with electron-withdrawing α substituents such as cyano, carbonyl, and halogen are readily oxidized by copper(II) chloride.[51] The same radicals are, by and large, inert to electron transfer oxidants. Furthermore, the microscopic reverse process is represented by the direct transfer of a halogen atom from an alkyl halide to chromium(II), and a kinetic study of the reduction of substituted benzyl halides shows no polar effect (vide infra). The atom transfer mechanism is thus an example of a more general inner-sphere process involving a bridged activated complex described earlier.

e. Rates of Ligand Transfer Oxidation of Alkyl Radicals. The ligand transfer oxidation of alkyl radicals by copper(II) halides and pseudohalides is characterized by extremely high rates. Competitive kinetic studies based on the isomerization of 5-hexenyl and cyclopropylmethyl radicals provide results which are in remarkable agreement with one another, considering a variation in rates of a factor of 10^3. The second-order rate constants k_L for the ligand transfer oxidation of alkyl radicals by copper(II) thiocyanate, chloride, and bromide are summarized in Table 11.

TABLE 11

Rates of Ligand Transfer Oxidation by Copper(II) Halides and Pseudohalides

Radical	Copper(II) oxidant	k_r/k_L	k_L[a]	k_L (relative)
5-Hexenyl	$Cu^{II}(NCS)_2$	3.9×10^{-4}	2.6×10^{8}	1
Cyclopropylmethyl	$Cu^{II}(NCS)_2$	2.7×10^{-1}	3.6×10^{8}	
5-Hexenyl	$Cu^{II}Cl_2$	$<4 \times 10^{-4}$	$>2 \times 10^{8}$	3
Cyclopropylmethyl	$Cu^{II}Cl_2$	9.2×10^{-2}	1.1×10^{9}	
5-Hexenyl	$Cu^{II}Br_2$	$<4 \times 10^{-4}$	$>2 \times 10^{8}$	12
Cyclopropylmethyl	$Cu^{II}Br_2$	2.3×10^{-2}	4.3×10^{9}	

[a] At 25°C assuming $k_r = 1 \times 10^5$ sec^{-1} for 5-hexenyl radical and 1×10^8 sec^{-1} for cyclopropylmethyl radical, in units of M^{-1} sec^{-1}.

The values of k_L, however, represent a composite of a number of individual rate constants k_i for various copper(II) species. For example, the various

$$k_L(CuCl_2) = k_1(CuCl^+) + k_2(CuCl_2) + k_3(CuCl_3^-) + k_4(CuCl_4^{2-})$$

chlorocopper(II) species represented in Table 9 are present in appreciable amounts in acetonitrile solutions. At this juncture it is impossible to separate the specific rate constant k_i by which each of these chlorocopper(II) species oxidizes an alkyl radical. If the composite value of k_L listed in Table 11 includes all of these species, it is reasonable to expect that some values of k_i (particularly for $CuCl_4^{2-}$) may actually be at the diffusion-controlled limit since even the value of k_e for oxidative elimination by copper(II) acetate is in the range 10^6–10^7 M^{-1} sec^{-1}.

f. Atom Transfer versus Oxidative Displacement in Ligand Transfer Processes. In ligand transfer oxidations, the atom transfer mechanism (equation 48) is generally the energetically more favorable process and usually represents the major course of reaction. The alternative pathway involving oxidative displacement (equations 46 and 47), by virtue of the formation of carbonium ion intermediates, is more applicable to alkyl radicals capable of forming stabilized carbonium ions. Although solvents such as dioxane promote oxidative displacement, the effect is apparent in the ligand transfer oxidation of only radicals such as cyclobutyl and β-anisylethyl, in which the competition between atom transfer and oxidative substitution is already delicately balanced. Thus, oxidative displacement cannot be promoted in neopentyl and phenethyl radicals by solvent changes, and only atom transfer prevails.

For a given alkyl radical, the relative importance of atom transfer and oxidative displacement pathways also depends on the ligand involved. For example, the ligand transfer oxidation of cyclobutyl radicals by copper(II) bromide occurs only by atom transfer. On the other hand, as much as 50% of the oxidation of cyclobutyl radicals by copper(II) chloride takes place by oxidative displacement. Copper(II) thiocyanate occupies an intermediate position. The same general trend of $Cu^{II}Br_2 > Cu^{II}(NCS)_2 > Cu^{II}Cl_2$ in atom transfer processes is also established with cyclopropylmethyl and allylcarbinyl radicals.

The competition between atom transfer and oxidative displacement is summarized below:

Mechanisms of the Ligand Transfer Oxidation

Atom transfer (direct):

$$R\cdot + Cu^{II}X_2 \longrightarrow [R\text{---}X\text{---}CuX]^{\ddagger} \longrightarrow R\text{—}X + Cu^{I}X$$

Oxidative displacement (alkylcopper intermediate):

$$R\cdot + Cu^{II}X_2 \rightleftharpoons RCuX_2$$

$$RCuX_2 \rightleftharpoons [RCuX^{+}X^{-}] \rightleftharpoons [R^{+}Cu^{I}X_2^{-}]$$

$$[R^{+}Cu^{I}X_2^{-}] \longrightarrow R\text{—}X + Cu^{I}X$$

3. *A Unified View of Ligand Transfer and Electron Transfer Oxidations of Alkyl Radicals*

In a more general sense, the oxidation of alkyl radicals by copper(II) oxidants transcends the relatively arbitrary classification into electron transfer and ligand transfer categories, which are based largely on stoichiometry. For example, a distinction between the analogous metastable alkylcopper intermediates leading to (1) oxidative solvolysis in electron transfer and (2) oxidative displacement during ligand transfer cannot be clearly made. In each case, the ionization of the ligand plays an important role in the formation of carbonium ion intermediates.

A more unified and general classification of the mechanisms of oxidation of free radicals by metal complexes may be made by considering the position at which the metal complex is attacked by the free radical. For example, within the present context the reaction of alkyl radicals with copper(II) complexes at the ligand site leads to atom transfer, a process which is largely free radical or homolytic in nature. On the other hand, attachment of the alkyl radical to the copper(II) nucleus results in an alkylcopper intermediate which can be partitioned in a variety of ways including (1) oxidative elimination to alkenes, (2) oxidative displacement to alkyl derivatives from the ligand, and (3) oxidative solvolysis by participation of the solvent.

Central to this mechanistic classification is the ability of the free radical to discriminate among several sites on the metal complex. As a working hy-

TABLE 12

Hard and Soft Acid–Base Classification of the Oxidation of Alkyl Radicals by Cupric Complexes

$Cu^{II}(X_2)(L)_4$			Alkyl radical
	X^- (anion)	L (neutral)	
Perchlorato	ClO_4^-	Water	Methyl
Triflato	$CF_3SO_3^-$	Acetonitrile	Ethyl
Fluoroborato	BF_4^-	1,4-Dioxane	*n*-Butyl
Trifluoroacetato	$CF_3CO_2^-$	Pyridine	Isopropyl
Acetato	$CH_3CO_2^-$	2,2′-Bipyridine	*sec*-Butyl
Chloro	Cl^-	Phenanthroline	*t*-Butyl
Thiocyanato	SCN^-	Acetic acid	Benzyl
Bromo	Br^-	Dimethyl sulfoxide	Allyl
Iodo	I^-	Dimethylformamide	

pothesis for further study such a distinction may be based on the hard and soft acid–base classification proposed by Pearson and others.[52]

Ligands and alkyl radicals are listed in Table 12 qualitatively *in order of descending hardness*. The horizontal dashed line represents the border region above which alkyl radicals react with copper(II) complexes by attachment to the copper nucleus. Below the dashed line, reaction occurs primarily on the ligand and an atom transfer (inner-sphere) mechanism prevails. Attack on ligand and attachment to copper are competitive with the chloro ligand which lies in the borderline region.

Alkyl radicals can be similarly placed on a hardness scale, which increases from benzyl and allyl radicals to tertiary, secondary, and primary alkyl radicals and finally to methyl radical itself. The mode of reaction between a given alkyl radical and a copper(II) complex would then be determined by their relative positions on these scales. The extension of these concepts to the oxidation and reduction of other free radicals and metal systems would be desirable.

B. Reduction of Alkyl Radicals

The reduction of alkyl radicals generated from alkyl halides and reducing agents such as the aromatic anion radicals has been presented in Chapter 9, and both inner- and outer-sphere electron transfer processes can be described.

$$R\text{—}X \xrightarrow{\varepsilon} R\cdot + X^- \xrightarrow{\varepsilon} R:^- + X^- \qquad (49)$$

The reduction of alkyl radicals by transition metal ions such as chromium(II), titanium(III), and iron(II) can also be studied during reduction of diacyl peroxides and alkyl hydroperoxides.[53] Thus, the reduction of diacyl peroxide

$$(RCO_2)_2 + Cr^{II} \longrightarrow RCO_2Cr^{III} + RCO_2\cdot \qquad (50a)$$

$$RCO_2\cdot \longrightarrow R\cdot + CO_2 \qquad (50b)$$

$$R\cdot + Cr^{II} \xrightarrow[H_2O]{} R\text{—}H + Cr^{III} \qquad (51)$$

by chromium(II) proceeds according to equations 50 and 51. The facile reduction of *t*-alkyl hydroperoxides occurs by a similar series of one-equivalent reductions, for example,

$$\begin{aligned} R(CH_3)_2COOH + Cr^{II} &\longrightarrow R(CH_3)_2CO\cdot + Cr^{III}(OH) \\ R(CH_3)_2CO\cdot &\longrightarrow (CH_3)_2CO + R\cdot, \ldots \end{aligned} \qquad (52)$$

In both reactions the reduction to alkane depends on the efficiency of chromium(II) in the interception of alkyl radicals. The latter can be estimated from the competition with the irreversible cyclization of the ω-hexenyl radical.[54] If the reaction of ω-heptenoyl peroxide is carried out in the

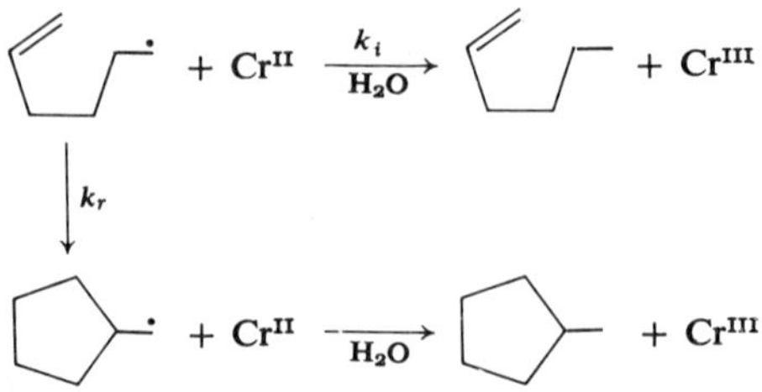

presence of a measured excess of chromium(II), the ratio of the rate constant for reduction k_i and for cyclization k_c is given by:

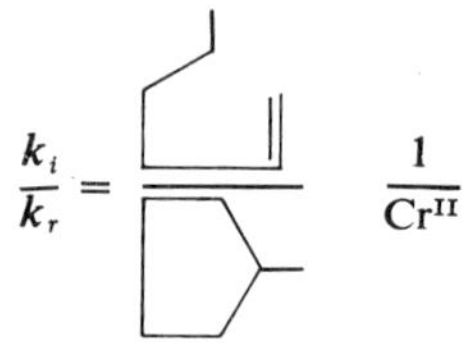

If a value of 1×10^5 sec^{-1} is used for k_r,[35] the rate constant k_i for reduction is 4×10^7 sec^{-1} at 25°C. The latter, however, is not sufficiently rapid to compete effectively with the rearrangement of the cyclopropylcarbinyl radical, and the reduction of a cyclopropylacetyl peroxide produces butene-1 principally.

The reduction of alkyl radicals by chromium(II) proceeds via an inner-sphere complex which is hydrolytically unstable. If the reactions are carried

$$R\cdot + Cr^{II} \xrightarrow{k_i} RCr^{2+} \qquad (51a)$$

$$RCr^{2+} + H_2O \longrightarrow R{-}H + Cr^{III}(OH) \qquad (51b)$$

out in aprotic media the alkylchromium species can be examined over longer periods and its salts have been isolated in crystalline form in some cases.[55] The rates of substitution of these organochromium species are more akin to those of other substitution-inert chromium(III) complexes and not to the labile chromium(II) complexes.[56]

The chromous reduction of alkyl radicals by an inner-sphere mechanism is consistent with the reduction of inorganic oxidants with this reagent discussed earlier. The reduction of peroxides and alkyl halides by chromium(II) also involve inner-sphere transfers (vide infra). Outer-sphere reductants are generally not sufficiently reactive to effect reduction of organic substrates, and conclusions have not yet been reached on the reduction of radicals by these reagents. Thus, it is not clear whether an inner-sphere process is necessary for the reduction of alkyl radicals by transition-metal complexes.

Reduction of radicals may occur by a third mechanism involving direct transfer of hydrogen from the reducing agent to the alkyl radical. For example, the reduction of alkyl halides by trialkyltin hydrides occurs by a radical chain process in which the propagation step involves such a transformation (see Chapter 25). A similar process appears to be involved in the

$$R\cdot + HSn^{IV}R_3 \longrightarrow R{-}H + \cdot Sn^{III}R_3, \ldots$$

decomposition of alkylmercuric hydrides formed as intermediates in the borohydride reduction of organomercurials, for example,[57]

$$R\cdot + RHg^{II}H \longrightarrow R{-}H + RHg^{I}$$

$$RHg^{I} \longrightarrow R\cdot + Hg, \ldots$$

These reactions represent an inner-sphere transfer of a hydrogen atom and are formally related to a large number of chain transfer reactions involving abstraction of hydrogen from a variety of donors (see Chapters 7 and 14). However, hydrogen transfer in the latter cases is not usually considered as an oxidation–reduction process despite the change in formal oxidation states involved, for example,

$$Cl\cdot + (CH_3)_3CH \longrightarrow HCl + (CH_3)_3C\cdot$$

$$(CH_3)_3CO\cdot + PhCH_3 \longrightarrow (CH_3)_3COH + PhCH_2\cdot$$

$$CH_3CH_2\cdot + CH_3SH \longrightarrow CH_3CH_3 + CH_3S\cdot$$

There are a variety of other metal hydrides known, and those associated with the transition metals are particularly noteworthy.[58] Hydrogen transfer to radicals has not been extensively examined in these systems, although participation of these processes in heterogeneous systems is well known in the dehydrogenation effected particularly by the noble-metal catalysts.[59]

Reduction of radicals by atom transfer is also possible. Thus, alkylperoxy and alkoxy radicals are reduced by oxygen atom transfer to trivalent phosphorus compounds (see Chapter 22). For example, the radical chain autoxidation of phosphines involve reactions such as

$$RO_2\cdot + PR_3 \longrightarrow RO\cdot + OPR_3$$

$$RO\cdot + PR_3 \longrightarrow R\cdot + OPR_3, \ldots$$

Thiyl radicals are similarly reduced by sulfur atom transfer.

C. Free Radical–Metal Complexes

In the foregoing discussion, the oxidation and reduction of free radicals by metal complexes represent examples in which the transfer of an electron from the reductant to the oxidant is more or less complete. However, in those cases in which redox reactions are not observed, a molecular or association complex may still be formed due to paramagnetic interactions, since many metal complexes possess unfilled or partially filled orbitals.[60] The potential importance of such complexes lies in the possibility of stabilizing free radicals and affecting their reactivity. Thus, the ability of transition-metal ions to catalyze the combination of methyl radicals has been attributed to the formation of a transient alkyl radical complex with the metal, which is stable against abstraction but long-lived enough to allow combination.[61] The reversible formation

$$CH_3\cdot + M^n \rightleftharpoons CH_3M^n$$

$$2CH_3M^n \longrightarrow CH_3CH_3 + 2M^n, \ldots$$

of a σ-bonded complex between benzyl radical and chromium(II) has been described.[62] Furthermore, cobalt complexes which are able to transfer a methyl ligand reversibly between cobalt(II) and cobalt(III) oxidation states

$$PhCH_2\cdot + Cr^{II} \rightleftharpoons PhCH_2Cr^{2+}$$

have been reported recently.[63]

$$Co^{II}(tfen) + CH_3Co^{III}(acen) \rightleftharpoons Co^{II}(acen) + CH_3Co^{III}(tfen)$$

The extent to which charge transfer is involved in a free radical complex with the metal nucleus complicates its definition, and it is often unclear whether such an entity should be alternatively described as a two-electron ligand with a metal in an unusual oxidation state, for example,

$$R\cdot M^n \qquad R^+M^{n-1} \qquad R:^-M^{n+1}$$

A series of related metal complexes in unusual oxidation states has been prepared by successive one-equivalent oxidations and reductions. These compounds are indeed usually characterized by having a ligand in which an electron can be extensively delocalized.[64]

Our interest here is directed mainly toward species in which the free radical retains much of its character. Recently, the ESR spectrum of a species derived from the oxidation of *sec*-butyllithium by copper(II) or iron(III) diisopropylsalicylate or acetylacetonate in solution has been detected.[65] The g factor (2.0834), absence of splitting by the copper nucleus, and the small values of the proton hyperfine splitting suggested to the authors the structure below,

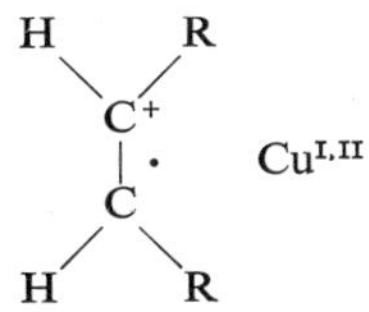

in which the unpaired electron is concentrated in a molecular orbital on carbon.

The hyperfine splitting by the silver nucleus has been observed in a variety of silver–alkenyl species ($g = 2.0017$) formed by γ-irradiation of polycrystalline silver–alkene complexes.[66] The same cyclohexadienylsilver species is obtained from the irradiation of either 1,3- or 1,4-cyclohexadienesilver perchlorate, and the proton hyperfine splitting constants are essentially the

same as those for the uncomplexed cyclohexadienyl radical. The magnitude of the silver hyperfine spitting indicates that there is only a small interaction of the hydrocarbon radical with the metal nucleus. Except for the electrostatics, the situation is somewhat akin to ion pairs formed between alkali metal ions and aromatic anion radicals.[67]

Complexes of stable free radicals with metals have also been described for semiquinones and nitroxides.[68] Recently the ESR spectrum of a complex between alkylperoxy radicals and cobalt(III) prepared from the reaction of *t*-butyl hydroperoxide and cobalt(II) acetylacetonate has been observed.[69] The similarity of the g factors for the complex with that of the uncomplexed alkylperoxy radical indicates minimal spin-orbit coupling associated with the metal nucleus. Analogous species appear to be formed from cobalt(II) acetylacetonate and *t*-butyl peroxyoxalate, and the g factor suggested an alkoxy radical complex. A phenoxy complex showing splitting by the cobalt nucleus has also been described. Such complexes no doubt play an important role in cobalt-catalyzed autoxidation of hydrocarbons, but their importance has not been fully assessed. The formation of hydroxy and hydroperoxy radical complexes with titanium(IV) has also been discussed.[70]

IV. CATALYTIC AND STOICHIOMETRIC PROCESS INVOLVING OXIDATION–REDUCTION REACTIONS

The variety of oxidation–reduction reactions of organic substrates induced by metal complexes either in a catalytic or stoichiometric manner makes a systematic discussion difficult. For this reason, we have arbitrarily chosen in the following section representative *classes* of reactions or reactants in order to illustrate some of the mechanistic principles described earlier.

A. Catalyzed Reactions of Peroxides

One-equivalent redox reactions of peroxides are represented by equations 53 and 54.

$$\text{R—O—O—R} + \varepsilon \longrightarrow \text{R—O}\cdot + \text{OR}^- \tag{53}$$

$$\text{R—O—O—H} \longrightarrow \text{R—O—O}\cdot + \text{H}^+ + \varepsilon \tag{54}$$

1. Hydrogen Peroxide

The mild oxidizing action of hydrogen peroxide is considerably enhanced in the presence of certain metal catalysts.[71] The most widely used catalyst is iron(II) and the combination is known as Fenton's reagent. It is now generally agreed[72] that the mechanism is given by equations 55–59, which is a slight modification of that originally proposed by Haber and Weiss.[73] Much evidence has been presented in support of a free radical chain process in-

$$\text{Fe}^{\text{II}} + \text{H}_2\text{O}_2 \longrightarrow \text{Fe}^{\text{III}} + \cdot\text{OH} + {}^-\text{OH} \tag{55}$$

$$\text{Fe}^{\text{III}} + \text{H}_2\text{O}_2 \longrightarrow \text{Fe}^{\text{II}} + \text{HO}_2\cdot + \text{H}^+ \tag{56}$$

$$\text{Fe}^{\text{II}} + \cdot\text{OH} \longrightarrow \text{Fe}^{\text{III}} + {}^-\text{OH} \tag{57}$$

$$\text{Fe}^{\text{III}} + \text{HO}_2\cdot \longrightarrow \text{Fe}^{\text{II}} + \text{O}_2 + \text{H}^+ \tag{58}$$

$$\text{HO}\cdot + \text{H}_2\text{O}_2 \longrightarrow \text{H}_2\text{O} + \text{HO}_2\cdot \tag{59}$$

volving hydroxyl radicals as transient intermediates. However, reaction 58 would be expected on energetic grounds to be much more facile than reaction 56. A catalytic cycle is possible via reactions 55, 58, and 59 without including reaction 56. In the reaction of metal ions with alkyl hydroperoxides, discussed later, the reaction analogous to that in equation 58 is slow, and alkyl hydroperoxides are generally insensitive to iron(III).[74]

The synthetic applications of Fenton's reagent to a wide variety of organic compounds have been reviewed recently.[75] Hydroxy radicals formed by reaction 55 are the important intermediates insofar as the organic substrates are concerned, although the extent to which they are associated with the iron species also present in solution is not clear. The mechanism of these oxidations falls into two categories: chain and nonchain reactions. A typical chain reaction is represented in the oxidation of alcohols by the following steps[76]:

$$\text{Fe}^{\text{II}} + \text{H}_2\text{O}_2 \longrightarrow \text{Fe}^{\text{III}}\text{OH} + \text{HO}\cdot \tag{55}$$

$$\text{R}_2\text{CHOH} + \text{HO}\cdot \longrightarrow \text{R}_2\dot{\text{C}}\text{OH} + \text{H}_2\text{O} \tag{60}$$

$$\text{Fe}^{\text{III}}(\text{OH}) + \text{R}_2\dot{\text{C}}\text{OH} \longrightarrow \text{R}_2\text{CO} + \text{Fe}^{\text{II}} + \text{H}_2\text{O}, \ldots \tag{61}$$

The last step (61) represents the rereduction of iron(III) and may be slow with certain alcohols. Termination may then occur via

$$R_2\dot{C}OH + HO\cdot \longrightarrow R_2CO + H_2O$$

$$2R_2\dot{C}OH \longrightarrow [R_2C(OH)\text{-}]_2 \quad \text{or} \quad R_2CHOH + R_2CO$$

or

$$Fe^{II} + HO\cdot \longrightarrow Fe^{III}(OH)$$

If oxidations occur by nonchain routes, at least 2 equivalents of iron(II) are required for each mole of substrate oxidized. Needless to say, only a catalytic amount of iron(II) is required if a chain process pertains. The addition of small amounts of copper(II) will increase the efficiency of a chain reaction[77] due to its enhanced reactivity toward carbon-centered radicals, for example,

$$R_2\dot{C}OH + Cu^{II} \longrightarrow R_2CO + Cu^{I} + H^+$$

$$Cu^{I} + Fe^{III} \longrightarrow Cu^{II} + Fe^{II}, \ldots$$

Fenton's reagent is used for the oxidation (hydroxylation) of aromatic hydrocarbons to the corresponding phenols,[72,78] Biphenyls formed via dimerization of the hydroxycyclohexadienyl radical intermediates are by-products in these reactions. Competition between these reactions is dependent on the iron(III) concentration since phenols are formed via oxidation of the intermediate by iron(III).

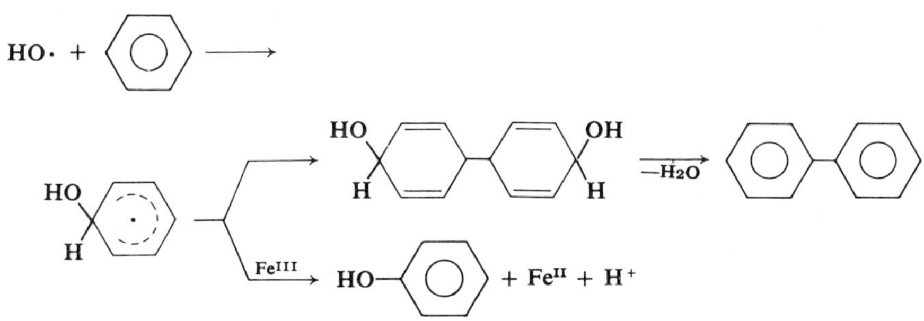

Another interesting use of Fenton's reagent is the conversion of hydrocarbons to carboxylic acids by reaction with hydrogen peroxide and carbon monoxide in the presence of iron(II).[79] This reaction has been used in the preparation of β-hydroxyisovaleric acid from *t*-butyl alcohol and carbon monoxide. The acyl radical intermediate is more prone to oxidation by

$$(CH_3)_3COH + CO + H_2O_2 \xrightarrow{Fe^{II}} (CH_3)_2C(OH)CH_2CO_2H$$

iron(III) than the alkyl radical precursor, and in the absence of carbon monoxide, dimers are formed. Analogous nonchain oxidative couplings have been

$$(CH_3)_3COH + H_2O_2 \xrightarrow{Fe^{II}} [(CH_3)_2C(OH)CH_2—]_2$$

applied to acids, nitriles, amides, amines, and ketones.[80] Additive dimerization results in the presence of butadiene and other vinyl monomers,[81] for example,

$HO\cdot +$ 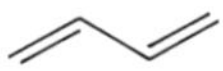 $\longrightarrow HO—C_4H_6\cdot \longrightarrow [HO—C_4H_6—]_2$

Hydroxyl radicals are also intermediates in the reaction of titanium(III) and H_2O_2.[82] This system is also capable of hydroxylation, but unlike Fenton's reagent only nonchain processes are possible since titanium(IV) is not usually an oxidant. A variety of interesting electron spin resonance (ESR) studies of radicals generated from this system has been reviewed recently.[83]

Ceric ion reacts with hydrogen peroxide to give the hydroperoxyl radical.[84] This system is not an effective reagent for hydroxylation since there appears to be no ready production of hydroxyl radicals by reaction with cerium(III).

$$Ce^{IV} + H_2O_2 \longrightarrow Ce^{III} + HO_2\cdot + H^+ \tag{62}$$

$$Ce^{III} + H_2O_2 \xrightarrow{slow} Ce^{IV}(OH) + HO\cdot \tag{63}$$

Other metal ions decompose hydrogen peroxide via reactions analogous to (55) and/or (56). For reactions analogous to (56) to be effective, the oxidant generally should be potent enough to oxidize water. Other metal systems studied include Cu^{I}/Cu^{II},[85] Mn^{II}/Mn^{III},[86] Co^{II}/Co^{III},[87] and Ag^{II}.[88] Many of these electron transfer reactions are thought to proceed via inner-sphere complexes of the metal and hydrogen peroxide.[86,89] Indeed, stable complexes are known for some of these metals. Thus, the stable cobalt(III)–hydroperoxide complex has been isolated[90] as the $Co(en)_3{}^{3+}$ salt. The analogous rhodium complex, as well as a ferric–hydroperoxide complex, has also been described.[91] It is significant that all known stable hydroperoxide complexes

$$\left.\begin{array}{r}[(CN)_5CoOOCo(CN)_5]^{6-} + H^+ \\ [Co(CN)_5OH]^{3-} + O_2\end{array}\right] \longrightarrow [Co(CN)_5OOH]^{3-}$$

$$[RhH(CN)_4H_2O]^{2-} + O_2 \longrightarrow [Rh(O_2H)(CN)_4H_2O]^{2-}$$

contain the metal ion in a higher oxidation state, probably because hydrogen peroxide is a strong oxidizing agent but only a mild reducing agent. Except with strong oxidizing agents, reactions represented in equation 55 are more facile than those typified by equation 56.

2. *Alkyl Hydroperoxides*

The metal-catalyzed decomposition of alkyl hydroperoxides is undoubtedly the most common pathway for catalysis of liquid-phase autoxidations by transition-metal complexes (see Chapter 12, "Autoxidation"). Reactions 64 and 65 are analogous to reactions 55 and 56, respectively, for hydrogen peroxide.

$$RO_2H + M^{n-1} \longrightarrow RO\cdot + M^n + {}^-OH \quad (64)$$

$$RO_2H + M^n \longrightarrow RO_2\cdot + M^{n-1} + H^+ \quad (65)$$

If a particular metal ion is only capable of effecting one of these reactions, a stoichiometric but not a catalytic decomposition of alkyl hydroperoxides would be expected. However, a catalytic reaction is feasible if there is an alternative pathway for regenerating the particular metal complex in an oxidation state which readily reacts with the alkyl hydroperoxide. In general, alkyl hydroperoxides like hydrogen peroxide are strong oxidizing agents but only mild reducing agents, and reaction 64 is faster than 65. Hence, there may be more favorable pathways available for regenerating M^{n-1} from M^n than reduction by the alkyl hydroperoxide, and several possibilities are given later.

When the metal complex is a strong oxidizing agent reaction 65 predominates. The best example of this type of oxidant is lead(IV). The reaction between lead tetraacetate and alkyl hydroperoxides has been studied by several groups.[92–95] The stoichiometric requirements are one $Pb(OAc)_4$ for every 2 moles of alkyl hydroperoxide, and the reaction is kinetically first order in $Pb(OAc)_4$ and second order in alkyl hydroperoxide.[93,95] High concentrations of alkylperoxy radicals are produced[94,95] which combine to form oxygen and

$$Pb^{IV}(OAc)_4 + 2RO_2H \longrightarrow Pb^{II}(OAc)_2 + 2RO_2\cdot + 2HOAc \quad (66)$$

alkoxy radicals. The mechanism, however, is more complicated than many authors have appreciated. First, the oxidation of the alkyl hydroperoxide by $Pb(OAc)_4$ must be distinguished from the radical-induced decomposition. To eliminate the latter, the hydroperoxide solution should be added slowly to the $Pb(OAc)_4$ solution since they react on mixing. Second, complications may arise from reactions between $Pb(OAc)_4$ and radicals derived from the solvent by hydrogen transfer with alkoxy radicals. Third, the fate of the intermediate lead(III) species must be considered, particularly with regard to decarboxylation of acetate. The kinetics have been reconciled with a mechanism involving the initial formation of a one-to-one complex of lead(IV) with alkyl hydroperoxide followed by a rate-determining reaction of the complex with a second molecule of the hydroperoxide.[95]

$$Pb^{IV}(OAc)_4 + ROOH \rightleftharpoons Pb^{IV}(OAc)_3(O_2R) + HOAc \quad (67)$$

$$Pb^{IV}(OAc)_3(O_2R) + ROOH \longrightarrow RO_2\cdot + Pb^{III}(OAc)_3, \ldots \quad (68)$$

Cerium(IV) resembles lead(IV) in its reactions with alkyl hydroperoxides, and these reactions have been used for producing high concentrations of alkylperoxy radicals for ESR studies.[96] The stoichiometry of the reaction has been reported to require one cerium(IV) for each alkyl hydroperoxide according to equation 69. However, it has been shown[97] that cerium(III) can

$$RO_2H + Ce^{IV} \longrightarrow RO_2\cdot + Ce^{III} + H^+ \tag{69}$$

be reoxidized to cerium(IV) in the presence of alkylperoxy radicals, although probably at a much slower rate. Other strongly oxidizing metals which might be expected to behave in a similar manner are silver(II) and thallium(III).

When the metal ion is a strong reducing agent, reaction 64 predominates. Chromous ion, chromium(II), reduces alkyl hydroperoxides to the corresponding alcohols.[53] The stoichiometric reaction requires two chromium(II) for each mole of the alkyl hydroperoxide and proceeds according to equations 70 and 71. The reaction cannot be catalytic since there is no route available

$$RO_2H + Cr^{II} \longrightarrow RO\cdot + Cr^{III}(OH) \tag{70}$$

$$RO\cdot + Cr^{II} \longrightarrow RO^- + Cr^{III} \tag{71}$$

for the regeneration of chromium(II). The reaction has been utilized for the preparation of benzylchromium ion in high yield as described earlier.

Copper(I) also reduces alkyl hydroperoxides to the corresponding alcohols[24] in a manner analogous to chromium(II). Indeed, the reaction of the Cu(I)/Cu(II) couple with peroxides has been studied extensively.[21–25] In contrast to the reaction with chromium(II), alkyl hydroperoxides react with copper(I) by a catalytic process since there are several routes for regenerating copper(I). Thus, mixtures of copper(II) and copper(I) catalytically reduce *t*-amyl hydroperoxide[24,98] in acetonitrile–acetic acid at 0°. Copper(I) is regenerated via electron transfer oxidation of the ethyl radicals formed by the fragmentation of *t*-amyloxy radicals. Copper(I) can also be regenerated by

$$\begin{aligned} CH_3CH_2C(CH_3)_2O_2H + Cu^I &\longrightarrow Cu^{II}OH + CH_3CH_2C(CH_3)_2O\cdot \longrightarrow \\ &\qquad (CH_3)_2CO + C_2H_5\cdot \\ C_2H_5\cdot + Cu^{II} &\longrightarrow C_2H_4 + Cu^I + H^+, \ldots \end{aligned} \tag{72}$$

electron transfer oxidation of radicals produced by hydrogen transfer from solvent by alkoxy radicals.[99] Thus, reactions carried out in hydrocarbon solvents will produce alkyl radicals which are oxidized by copper(II) at rates approaching diffusion control. Several authors[100] have suggested that copper(II) is reduced to copper(I) by alkyl hydroperoxides (73). Such a

$$ROOH + Cu^{II} \longrightarrow RO_2\cdot + Cu^I + H^+ \tag{73}$$

reaction seems unlikely, however, in view of the redox potential of the Cu^{II}/Cu^I couple. There are more favorable pathways for the regeneration of copper(I), especially in hydrocarbon solvents, which generally have not been considered as alternatives. Recently, Hiatt et al.[74] have reported that alkyl hydroperoxides were inert to cupric octoate alone at room temperature. Experimentally, the proof for the existence of reaction 73 is more difficult than for reaction 72, since the formation of alkylperoxy radicals alone does not eliminate the possibility of a radical-induced decomposition.

Ferrous complexes also reduce alkyl hydroperoxides (cf. Fenton's reagent). Direct evidence for the formation of alkoxy radicals in the aqueous phase is obtained by the reaction carried out in the presence of 1,3-butadiene. The products identified could best be explained by the following sequence of reactions[81,101]:

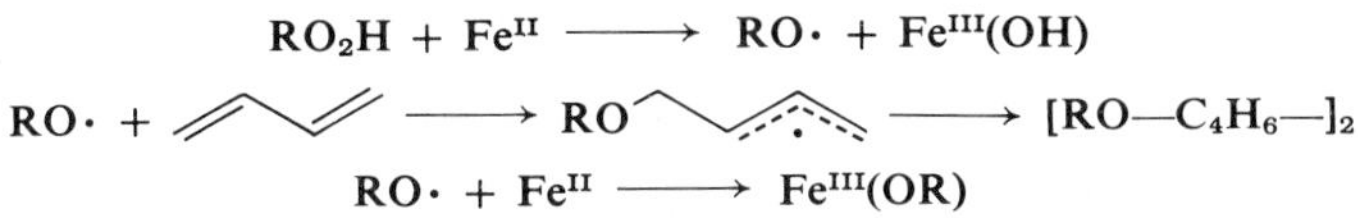

The reaction between ferrous salts and cumene hydroperoxide in aqueous solution was postulated on the basis of the kinetic parameters to proceed by dissociation of an unstable complex of the type (ROOH)Fe(II).[102] Some authors have assumed that iron(III) is reduced by alkyl hydroperoxides. However, there is no good evidence for this reaction as in the case of copper-

$$Fe^{III} + RO_2H \longrightarrow Fe^{II} + RO_2\cdot + H^+$$

(II), and Hiatt[74] states that alkyl hydroperoxides are inert to iron(III).

An interesting series of reactions has been developed based on the reaction between iron(II) and adducts of hydrogen peroxide and ketones. Cyclohexanone and hydrogen peroxide form a complex mixture of adducts,[103] which reacts with ferrous sulfate in methanol to produce the 5-(methoxycarbonyl)pentyl radical.[104] In the absence of reactive substrates, dimethyl dodecanedioate is formed by dimerization of these radicals.)

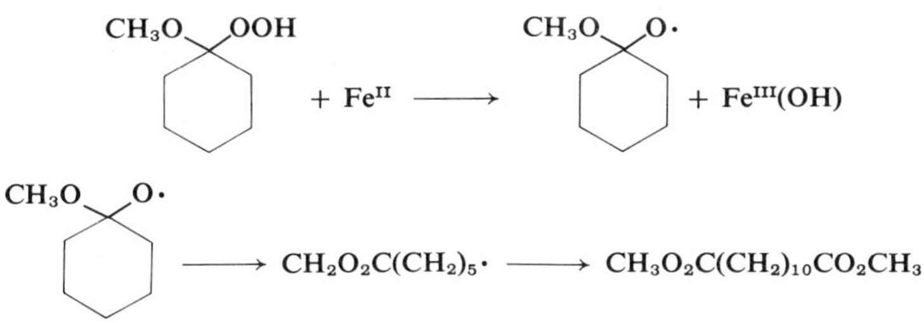

(74)

In the presence of butadiene, the 5-(methoxycarbonyl)pentyl radical generates an allylic adduct which dimerizes (equation 75) to a mixture of C_{20}

dicarboxylic acid esters in 65–75% yield.[104] A further modification[105] is accomplished by intercepting the allylic radicals with copper(II) as shown in equation 76, or the 5-(methoxycarbonyl)pentyl radical may also be intercepted by copper(II) as in equation 77. By the use of suitable olefins these

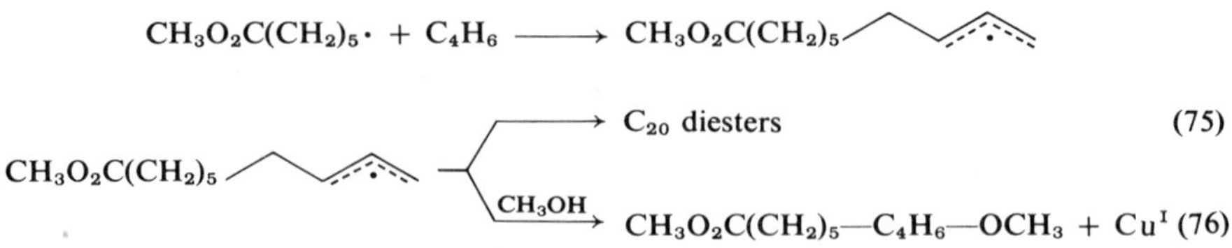

$$CH_3O_2C(CH_2)_5\cdot + Cu^{II} \longrightarrow CH_3O_2C(CH_2)_3CH{=}CH_2 + Cu^{I} \qquad (77)$$

reactions can be used for the preparation of a wide variety of polyfunctional long-chain molecules.[102] Copper complexes may be used for these reactions in the absence of Fe(II) since Cu(I) will also effect reaction 74. A further variation employs the ligand transfer oxidation of alkyl radicals by cupric halides. Thus, the reaction can be carried out in the presence of ferrous and cupric chlorides, and the 5-(methoxycarbonyl)pentyl radical formed as an intermediate can be efficiently intercepted via ligand transfer oxidation to give ω-chlorohexanoic acid methyl ester. These reactions have potential applica-

$$CH_3O_2C(CH_2)_5\cdot + Cu^{II}Cl_2 \longrightarrow CH_3O_2C(CH_2)_5Cl + Cu^{I}Cl$$

tion in the synthesis of many types of compounds.

The metal ion-induced decomposition of alkyl hydroperoxides can be employed as a method for introducing the alkylperoxy group into a variety of molecules.[75,101,106] Kharasch and co-workers discovered that in the presence of trace amounts of copper, cobaltous, or manganous salts, alkyl and aralkyl hydroperoxides react with substrates by replacing an activated hydrogen with an alkyperoxy group. Copper salts are usually the superior reagents for this reaction, which is analogous to the peroxy-ester reaction (vide infra).

$$2ROOH + R'H \xrightarrow{Cu^{II}\text{–}Cu^{I}} ROOR' + H_2O + ROH$$

Titanium(III), vanadium(II), and vanadium(III) are also known to reduce alkyl hydroperoxides in a manner analogous to chromium(II), iron(II), and copper(I) although these metals have not been studied as extensively.[21]

When the metal has two valence states of comparable stability, reactions 64 and 65 occur concurrently. Cobalt and manganese are the most effective autoxidation catalysts since they are able to induce the efficient catalytic (as opposed to stoichiometric) decomposition of alkyl hydroperoxides to radicals (see Chapter 12, "Autoxidation"). The contribution to the catalytic sequence of reactions is obtained by the addition of equation 78 to 79 to give the overall reaction 82. Equations 78 and 79 represent the reactions expected to prevail in aqueous solution. Outer-sphere electron transfer between metal ions and alkyl hydroperoxides without complex formation is expected to be favorable. Equations 80 and 81 represent the corresponding reactions in nonpolar solvents in which electron transfer probably proceeds via the formation of inner-sphere or covalent-bonded complexes. Aggregates of metal complexes are a further complication, particularly in hydrocarbon solvents.

Aqueous solution:

$$RO_2H + Co^{II} \longrightarrow RO\cdot + Co^{III}(OH) \qquad (78)$$

$$RO_2H + Co^{III} \longrightarrow RO_2\cdot + Co^{II} + H^+ \qquad (79)$$

Nonpolar solvents:

$$RO_2H + Co^{II} \longrightarrow RO\cdot + Co^{III}(OH)$$

or

$$RO_2Co^{II} \longrightarrow RO\cdot + Co^{II}O \qquad (80)$$

$$RO_2Co^{III} \rightleftharpoons RO_2\cdot + Co^{II} \qquad (81)$$

$$2RO_2H \longrightarrow RO_2\cdot + RO\cdot + H_2O \qquad (82)$$

The Co^{II}/Co^{III} and Mn^{II}/Mn^{III} redox systems are rather similar in their ability to decompose alkyl hydroperoxides. Cobalt compounds are generally the more effective, and consequently, have received more attention. The existence of alkylperoxy radicals can be shown by trapping with 1,3-butadiene.[101] Other evidence for the formation of alkylperoxy radicals is obtained from the formation of 4-peroxycyclohexadienones in the presence of phenolic inhibitors and by ESR studies.[107,108,110]

The investigation of these reactions has revealed two important conclusions which are relevant to all reactions between alkyl hydroperoxides and metal complexes. The first is the competition between metal ion–hydroperoxide reaction and the radical-induced decomposition of hydroperoxides. The second point concerns the relative rates of reactions 78 and 79 or 80 and 81, that is, which is the slower, rate-determining step.

The competition between metal ion-induced and radical-induced hydroperoxide decomposition was studied by Hiatt and co-workers[74] in the catalytic decomposition of *t*-butyl hydroperoxide by a variety of cobalt, iron, vanadium, cerium, and lead complexes. The time required for the complete decomposition of 0.1–0.2 *M* solutions of the hydroperoxide in chlorobenzene, by the addition of millimolar amounts of catalysts at 25°C, varied from a few minutes for cobalt carboxylates to several days for lead naphthenate. The products of these reactions consisted of approximately 86% *t*-butyl alcohol, 12% di-*t*-butyl peroxide, 0.5% acetone, and 93% oxygen (based on $\frac{1}{2}O_2$ from each alkyl hydroperoxide) and were *independent* of the catalyst or the rate of decomposition. Decompositions were strongly retarded by the addition of small amounts of materials which strongly complex the metal ions. The uniformity of products and the bimolecular rate expressions suggested a radical-induced decomposition of the usual type which was initiated by the interaction between the alkyl hydroperoxide and the metal complex. Results of decompositions carried out in hydrocarbon solvents (SH) agreed with the scheme proposed for the reaction in chlorobenzene, modified slightly by some radical attack on the more reactive alkanes. For *t*-

$$2RO_2\cdot \longrightarrow 2RO\cdot + O_2$$

$$RO\cdot + RO_2H \longrightarrow ROH + RO_2\cdot$$

$$RO\cdot + SH \longrightarrow ROH + S\cdot$$

butyl hydroperoxide such attack results in shorter kinetic chain lengths for the induced reaction, and lower yields of oxygen and di-*t*-butyl peroxide.

Competition between metal ion-induced and radical-induced decomposition of alkyl hydroperoxides is influenced by the relative concentrations of the metal complex and the alkyl hydroperoxide. At high concentrations of the hydroperoxide relative to the metal complex, alkoxy radicals will compete effectively with the metal complex for the alkyl hydroperoxide. Competition will also be affected by the relative amounts of the hydroperoxide and the solvent, particularly if the latter is readily attacked by alkoxy radicals. Hence, the contribution from the metal-catalyzed reaction is expected to predominate at low concentrations of the hydroperoxide and in reactive solvents. Generally, the contribution from the metal-induced decomposition to the overall reaction may be readily determined by carrying out the reaction in the presence of free radical inhibitors to minimize competition from the radical-induced decomposition.

Chuev et al.[108] observed that at high concentrations of alkyl hydroperoxide, the radical-induced decomposition competed with the cobalt acetate-catalyzed decomposition of *t*-butyl hydroperoxide in acetic acid or mixtures of benzene and acetic acid. Addition of 2,6-di-*t*-butyl-*p*-cresol retarded the rate of decomposition of the hydroperoxide but had no effect on the rate of reduction of Co(III). Kamiya et al.[111] observed that the initial rate of the cobalt acetate-catalyzed decomposition of tetralin hydroperoxide in mixtures of benzene and acetic acid decreased with decreasing initial concentrations of the hydroperoxide. The initial rate of the cobalt-catalyzed decompositions, when corrected for the contribution from the radical-induced decomposition by the addition of an inhibitor, was within experimental error equal to the limiting rate of the cobalt-catalyzed autoxidation of tetralin. Under these conditions (i.e., where the initial hydroperoxide concentrations are low) the cobalt-catalyzed decomposition of hydroperoxide represents the chain-initiating mechanism in the autoxidation reaction.

The relative concentration of M^n and M^{n-1} under steady-state conditions is related in part to the relative rates of reactions 78 and 79. On first inspecting the published data there appears to be two opposing schools of thought on which reaction, (78) or (79), is the slower rate-determining process. Closer inspection, however, reveals that authors favoring reaction 78 as the slow step have studied cobalt-catalyzed decompositions of alkyl hydroperoxides in polar solvents (or mixtures of nonpolar and polar solvents) under non-autoxidizing conditions. On the other hand, those favoring reaction 79 as the slow step have generally based their studies on cobalt-catalyzed autoxidations in hydrocarbon media. The differences in the results and consequently the conclusions may be explained by these significant differences in experimental conditions and may be attributed to a number of factors: (1) If the reaction between metal complex and alkyl hydroperoxide involves prior complex formation, it is expected to be much slower in polar coordinating solvents than in hydrocarbon solvents; (2) cobalt(III) compounds are known to exist as dimers or higher aggregates in solution. If only the monomeric species is the active catalyst the reaction with alkyl hydroperoxide would be faster in polar ionizing solvents; and (3) reaction 81 is a reversible reaction whereas reaction 80 is irreversible.

Dean and Skirrow[109] studied the cobaltous and cobaltic acetate-catalyzed decomposition of *t*-butyl hydroperoxide in acetic acid. They concluded from their results that reaction 79 was 250–300 times faster than reaction 78. Since cerium(IV), which has a lower redox potential than cobalt(III), is readily reduced by alkyl hydroperoxides[96] as given in equation 69, reaction 79 is also expected to be very rapid. The overall kinetic order of the reaction was greater than two, and a reversible formation of a complex between cobalt(II) and alkyl hydroperoxide was suggested to explain the kinetics. Complex formation is followed by reaction with another molecule of cobaltous acetate to produce radicals.

$$\mathrm{Co^{II} + RO_2H \longrightarrow [Co^{II}\cdot RO_2H]}$$

$$\mathrm{[Co^{II}\cdot RO_2H] + Co^{II} \longrightarrow Co^{III}(OH) + RO\cdot + Co^{II}}$$

Similarly, Richardson[107] showed that the reaction of cobaltous acetate with *t*-butyl hydroperoxide was first order in the hydroperoxide but of higher order (between one and two) in cobalt(II). He was able to study reaction 78 independently by carrying out the reaction in the presence of ethylenediaminetetraacetic acid which complexes strongly with cobalt(III). Under these conditions cobalt(II) was irreversibly oxidized to cobalt(III) via reaction 78. The rate of disappearance of the hydroperoxide was equal to the rate of cobalt(III) appearance, which followed the rate law:

$$\frac{d[\mathrm{Co^{III}}]}{dt} = k_2[\mathrm{Co^{II}}][t\text{-BuOOH}] + k_3[\mathrm{Co^{II}}]^2[t\text{-BuOOH}]$$

In another study[112] it was shown that the rate of decomposition of tetralin hydroperoxide catalyzed by cobaltous acetate in mixtures of acetic acid and benzene at 50° was proportional to the square of the cobalt concentration and first order in alkyl hydroperoxide. Kamiya et al.[111] also studied the decomposition of tetralin hydroperoxide catalyzed by cobaltous acetate in mixtures of benzene and acetic acid and concluded that reaction 78 was rate determining. However, the rate decreased as the molar ratio of acetic acid to benzene was increased, and at a given initial concentration of cobalt(II) and alkyl hydroperoxide, the rate was inversely proportional to the square of the acetic acid concentration.

Scott[113] investigated the reaction of cobalt(II) and cobalt(III) acetates with benzyl hydroperoxide and the following mechanism was suggested for the steady-state condition:

$$\mathrm{Co_2(OAc)_4(HO)_2} \rightleftharpoons \mathrm{2Co(OAc)_2(HO)}$$

$$\mathrm{Co(OAc)_2(HO) + PhCH_2O_2H} \rightleftharpoons \mathrm{2Co(OAc)_2(HO)(PhCH_2O_2H)}$$

$$\mathrm{Co(OAc)_2(HO)(PhCH_2O_2H) + Co(OAc)_2} \longrightarrow \mathrm{Co(OAc)_2(HO) + Co(OAc)_2 + PhCHO + H_2O}$$

Benzaldehyde, thus, results essentially by dehydration of benzyl hydroperoxide and free radicals as such are not important species. He suggests that the cobalt(III) is generated from cobalt(II) by reaction with alkylperoxy radicals and the reaction of benzyl hydroperoxide with the cobalt(III) dimer proceeds via a reaction of a binuclear $\mathrm{Co^{III}}$—$\mathrm{Co^{II}}$ complex with benzyloxy radicals.

In the catalyzed decomposition of alkyl hydroperoxides in chlorobenzene by cobalt abietate (the abietate salt was used because of its high solubility in hydrocarbons), divalent cobalt was oxidized instantaneously to trivalent cobalt on mixing the catalyst and the hydroperoxide.[114] The change in the cobalt species could be followed visually or spectrophotometrically, and was also observed by other workers.[115] The rate constants and the products of decomposition were independent of the oxidation state of the initial cobalt salt used.[116] These workers also studied the influence of the solvent on the valence changes of cobalt (acetate, abietate) during the catalyzed decomposition of a number of alkyl hydroperoxides, hydrogen peroxide, and peracids.[108,114] In solvents consisting of an organic acid (acetic, propionic, or butyric), mixtures of acids and chlorobenzene, or water, the catalyst is present almost completely in the divalent state during reaction. On the other hand, in chlorobenzene or benzene, the catalyst is present mainly in the trivalent state. The insensitivity of the rate of decomposition of the alkyl hydroperoxide to the oxidation state of the cobalt complex initially taken and the dramatic effect of solvents indicate large variations in the rate constants of reactions 78 and 79 [or (80) and (81)] in polar or nonpolar solvents. These results explain the apparent discrepancies obtained in other studies, but the reasons for the dramatic difference in the *relative* rates of these reactions have not been discussed at any length.

Reaction 79 is very rapid in aqueous solution[117] and it has been suggested that it takes place by an outer-sphere transfer between the ionic species (Co^{3+} and RO_2^-).[118] In a hydrocarbon medium in which the alkyl hydroperoxide does not dissociate appreciably, this reaction is expected to take place slowly, if at all. However, reaction 78 may proceed by an inner-sphere or covalently bonded complex, and hence should be facilitated by nonpolar and poorly coordinating solvents in which complex formation is facilitated. Exchange of ligands in cobalt(II) complexes, which are substitution labile, is rapid compared to cobalt(III) complexes, which are substitution inert.

The Mn^{II}/Mn^{III} system has not been studied in as much detail as the cobalt catalyst, but it appears to have many similarities. The same principles discussed above may apply to the manganese system, although reaction 83 appears to be much faster than (84).[119]

$$Mn^{II} + RO_2H \longrightarrow Mn^{III}(OH) + RO\cdot \qquad (83)$$

$$Mn^{III} + RO_2H \longrightarrow Mn^{II} + RO_2\cdot + H^+ \qquad (84)$$

Reaction 84 may be less favorable than the corresponding reaction with cobalt(III) due to the lower reduction potential for manganese(III). The kinetics of the reaction between manganous stearate and *n*-decyl hydroperoxide have been studied,[120] in which both the reaction between manganese-(II) and the hydroperoxide as well as the radical-induced decomposition of the hydroperoxide were observed.

The catalyzed decomposition of alkyl hydroperoxides by cobalt or manganese complexes discussed in the foregoing section is mostly relevant to autoxidation of organic substrates in which alkoxy radicals play an important role (see Chapter 12). In the absence of oxygen, moreover, the redox-catalyzed decomposition of hydrogen peroxide and alkyl hydroperoxides can be employed in a number of synthetic reactions.[121]

Rhodium and *iridium*, which are in the same group as cobalt in the periodic table, may also be expected to effect reactions, analogous to equations 78 and 79 and are supported by recent studies of autoxidations catalyzed by rhodium and iridium complexes.[122] Complexes of these metals rapidly decompose hydroperoxides in a catalytic reaction.[123]

Hydrogen peroxide and alkyl hydroperoxides are also capable of epoxidizing olefins in the presence of metal catalyst particularly those comprised of Groups IV, V, and VI. Oxygen transfer in these systems is effected hetero-

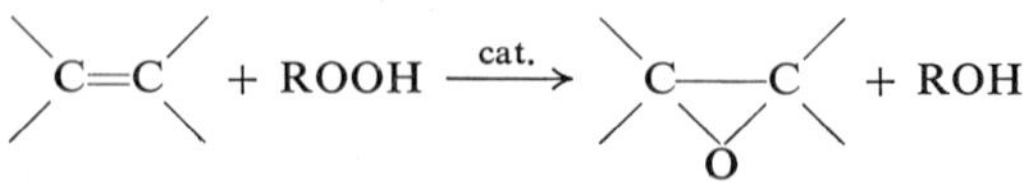

lytically without radical intermediates and the subject has been reviewed recently.[124]

3. *Peracids, Peresters, and Diacyl Peroxides*

Metal ions of variable valence catalyze the decomposition of peracids via redox reactions analogous to those with alkyl hydroperoxides.[114] There is, however, some ambiguity as to whether the one-equivalent reduction takes place by equation 85a or 85b. Thus, the formation of approximately 85%

$$RCO_2OH + M^{n-1} \longrightarrow RCO_2M^n + HO\cdot \qquad (85a)$$

$$RCO_2OH + M^{n-1} \longrightarrow RCO_2\cdot + M^n(OH) \qquad (85b)$$

carbon dioxide during the iron and cobalt-catalyzed decomposition of peracetic acid is consistent with the occurrence of reaction 85b.[125] However, reactions carried out in the presence of an acetate buffer labeled with ^{14}C in the carbonyl position showed that up to 30% of the carbon dioxide was derived from acetate.[126a] Since the exchange between peracetic acid and acetate ion is slow, the carbon dioxide must have arisen from the acetate ion itself. One possibility is as follows[126b]:

$$HO\cdot + CH_3CO_2^- \longrightarrow HO^- + CH_3CO_2\cdot, \ldots$$

The lability of the peroxidic proton is a source of this complication. For example, the catalytic decomposition of peracetic acid by manganese produces 90% carbon dioxide in acetic acid but only 24% in water. Large amounts of oxygen (60%) are produced in the aqueous medium. Carbon monoxide, which is also a product of these reactions, may arise by disproportionation of the acylperoxy radicals formed by oxidation of the peracid. In contrast, the iron-catalyzed decomposition of aroyl peracids in the presence

$$RCO_2OH + M^n \longrightarrow RCO_3\cdot + M^{n-1} + H^+ \qquad (86)$$

$$2RCO_3\cdot \longrightarrow 2RCO\cdot + O_2, \ldots$$

of styrene afforded polymers in which hydroxyl radicals were postulated to be responsible for initiation.[126c]

It has been shown that reaction 85 is a facile reaction with many metal complexes such as those containing cobalt(II) and manganese(II).[118,119] reaction 86 is usually not rapid and proceeds much more slowly than (85), if at all. Thus, in contrast to the reaction with alkyl hydroperoxides (see preceding section), the cobalt catalyst in the decomposition of peracetic acid[114] is present as cobaltic complexes in both chlorobenzene and acetic acid solvents and demonstrates that reaction 85 is faster than (86) in both solvents.

Tertiary alkyl peresters such as *t*-butyl peracetate like *t*-alkyl hydroperoxides and peracids are reduced by stoichiometric amounts of Cu(I). A catalytic reaction is possible when the alkoxy radical undergoes fragmentation

$$R'\text{—}CO_3C(CH_3)_2R + Cu^I \longrightarrow R'CO_2Cu^{II} + \cdot OC(CH_3)_2R \qquad (87a)$$

$$R(CH_3)_2CO\cdot + Cu^I \longrightarrow R(CH_3)_2COCu^{II} \qquad (87b)$$

(equation 88)[127] faster than reduction (equation 87b).

$$R(CH_3)_2CO\cdot \longrightarrow (CH_3)_2CO + R\cdot \qquad (88)$$

$$R\cdot + Cu^{II} \longrightarrow R(\text{—}H) + Cu^I + H^+, \ldots$$

Various substrates (RH) are selectively oxidized by *t*-butyl peresters, however, by catalytic amounts of transition-metal salts, particularly copper complexes.[128,129] It has been shown[24,25,98] that the relevant oxidation steps

$$RH + t\text{-Bu—}O_2Ac \xrightarrow{CuX_n} ROAc + t\text{-Bu—OH}$$

in these chain reactions are:

$$t\text{-Bu—}O_2Ac + Cu^I \longrightarrow Cu^{II}OAc + \cdot O\text{—}t\text{-Bu} \qquad (87a)$$

$$RH + \cdot O\text{—}t\text{-Bu} \longrightarrow R\cdot + HO\text{—}t\text{-Bu} \qquad (89)$$

$$R\cdot + Cu^{II}OAc \longrightarrow R\text{—}OAc + Cu^I, \ldots$$

The reaction is initiated by copper(I), although copper(II) may be employed directly if the reaction is carried out under conditions where the perester undergoes homolysis partially. The kinetic chain length of the reaction depends on the efficiency of the transfer step (89) since it must compete with the oxidation of copper(I) given in equation 87b. Alkenes, aldehydes, and ethers react readily with *t*-butoxy radicals (equation 89), and produce radicals (see Chapter 23) which undergo electron transfer oxidation with copper(II) sufficiently rapidly to promote an efficient catalytic process.

Unlike peracids there is no ambiguity as to the mode of reductive cleavage of peresters shown in equation 87a. The absence of carbon dioxide and the scavenging of alkoxy radicals[24,51b,98,127b] in the stoichiometric and catalytic reactions of peresters preclude the alternative fragmentation:

$$RCO_2OR' + Cu^I \longrightarrow RCO_2\cdot + Cu^{II}(OR')$$

Similarly *t*-alkyl hydroperoxides appear to undergo preferential cleavage of the alkoxy radical.[98,105,121]

$$ROOH + Cu^{I} \longrightarrow RO\cdot + Cu^{II}(OH)$$

The catalytic decomposition of diacyl peroxides was discussed earlier, and the mechanism is essentially the same as that presented for peresters.[130] Acycloxy radicals unlike most alkoxy radicals, however, are extremely unstable and readily liberate alkyl radicals on decarboxylation. Therefore, long kinetic chains usually result from the ready oxidation of alkyl radicals provided copper(II) is present in sufficiently high concentrations.

Aroyl peroxides are also catalytically decomposed by copper salts. Aroyloxy radicals decarboxylate more slowly than their alkyl counterparts and are capable of adding to arenes.[131] Oxidation of the cyclohexadienyl

$$(ArCO_2)_2 + Cu^{I} \longrightarrow ArCO_2\cdot + Cu^{II}O_2CAr$$

$$ArCO_2\cdot \longrightarrow Ar\cdot + CO_2$$

$$ArCO_2\cdot + C_6H_6 \longrightarrow ArCO_2{-}C_6H_6\cdot$$

adducts by copper(II) affords aryl benzoates in addition to arylbenzenes.

$$ArCO_2{-}C_6H_6\cdot + Cu^{II} \longrightarrow ArCO_2{-}C_6H_5 + H^+ + Cu^{I}, \ldots$$

$$Ar{-}C_6H_6\cdot + Cu^{II} \longrightarrow Ar{-}C_6H_5 + H^+ + Cu^{I}, \ldots$$

Aromatic oxygenation is also possible with alkoxycarboxy radicals generated in the copper-catalyzed decomposition of dialkyl peroxydicarbonates.[131]

$$ROCO_2{-}O_2COR + Cu^{I} \longrightarrow ROCO_2\cdot + Cu^{II}O_2COR, \ldots$$

B. Oxidation of Hydrocarbons with Metal Complexes

The oxidation of a variety of organic functional groups can be carried out selectively with inorganic reagents, and the mechanisms of these reactions have been reviewed at length.[1-3,132] In this section we are primarily interested in discussing oxidations involving one-equivalent changes in the breaking of a carbon–hydrogen bond.

1. Alkenes

Bawn and Sharp[133] studied the reaction between various olefins and cobalt(III) salts in polar media. The presence of oxygen or cobalt(II) had little effect on the reaction and the following rate expression was obtained from the data in aqueous acid

$$\frac{-d[Co^{III}]}{dt} = k_{obs}[Co^{III}]$$

where $k_{obs} = k[RH] + k'[H_2O]$. The term containing water was included since a plot of k_{obs} against olefin concentration gave a linear relation whose intercept corresponded to the value of the rate constant for the reduction of cobalt(III) by water. Reaction 90 was considered to be the rate-determining step followed by the further reaction of the radical cation with water and

$$RCH{=}CH_2 + Co^{III} \longrightarrow R\overset{+}{C}H{-}\dot{C}H_2 + Co^{II} \qquad (90)$$

cobalt(III) to form various products, such as aldehydes, ketones, and dienes. The reaction in nonaqueous solvents such as acetic acid is much slower, due partly to the dimeric nature of cobaltic acetate.[134] In trifluoroacetic acid, however, ethylene is rapidly and quantitatively oxidized by cobalt(III) to ethylene glycol ditrifluoroacetate.[135]

$$CH_2{=}CH_2 + Co^{III} \longrightarrow \overset{+\,\cdot}{CH_2{-}CH_2} + Co^{II}$$

$$\overset{+\,\cdot}{CH_2{-}CH_2} + CF_3CO_2H \longrightarrow CF_3CO_2CH_2CH_2\cdot + H^+$$

$$CF_3CO_2CH_2CH_2\cdot + Co^{III} \longrightarrow CF_3CO_2CH_2CH_2^+ + Co^{II}$$

$$CF_3CO_2CH_2CH_2^+ + CF_3CO_2H \longrightarrow CF_3CO_2CH_2CH_2O_2CCF_3 + H^+$$

Electron transfer from the alkene to cobalt(III) is the primary step, and the ESR spectra of radical cations have been observed in a number of cases.[136] Electron transfer has also been postulated during reaction of the outer-sphere reagent iron(III)(dipy)$_3$ with aminoalkenes such as vinylcarbazoles which form relatively stable radical cations.[137] Lead tetraacetate also reacts readily with a number of alkenes, but the reaction usually proceeds by electrophilic addition of a lead(IV) species followed by heterolysis of lead(II).[138] There is

$$\gt C{=}C\lt + Pb^{IV}(OAc)_3^+ \longrightarrow (AcO)_3Pb\overset{|}{\underset{|}{C}}{-}C^+\lt, \ldots \qquad (91)$$

ESR evidence, however, for an electron transfer process occurring between lead(IV) and particularly activated aromatic compounds in highly acidic media.[139] It is conceivable that in some cases equation 91 proceeds by two successive one-electron transfers,[140] although electrophilic addition by metal cations such as mercury(II) and thallium(III) is well established.

2. *Alkyl Aromatic Hydrocarbons*

The direct oxidation of alkylated aromatic hydrocarbons by metal salts has been extensively studied and is implicated in many important processes such as the metal-catalyzed autoxidation of *p*-xylene to terephthalic acid for which cobalt complexes are the preferred catalysts (see Chapter 12, "Autoxidation").

Dewar and co-workers[141] studied the manganic acetate oxidation of a number of aromatic ethers and amines in acetic acid. The inverse dependence of the rate on manganese(II) concentration was interpreted as a reversible electron transfer oxidation by manganese(III) to form the aromatic radical cation shown in equation 92a, followed by the rate-determining loss of a

$$\begin{aligned} ArCH_3 + Mn^{III} &\rightleftharpoons [ArCH_3]^{+} + Mn^{II} \\ [ArCH_3]\cdot^{+} &\xrightarrow{\text{slow}} ArCH_2\cdot + H^{+} \\ ArCH_2\cdot + Mn^{III} &\xrightarrow{\text{fast}} ArCH_2^{+} + Mn^{II} \\ ArCH_2^{+} + HOAc &\longrightarrow ArCH_2OAc + H^{+} \end{aligned} \tag{92a}$$

proton. The resultant benzylic radical is oxidized further by manganese(III) and the overall stoichiometry is given in equation 92b for *p*-methoxytoluene.

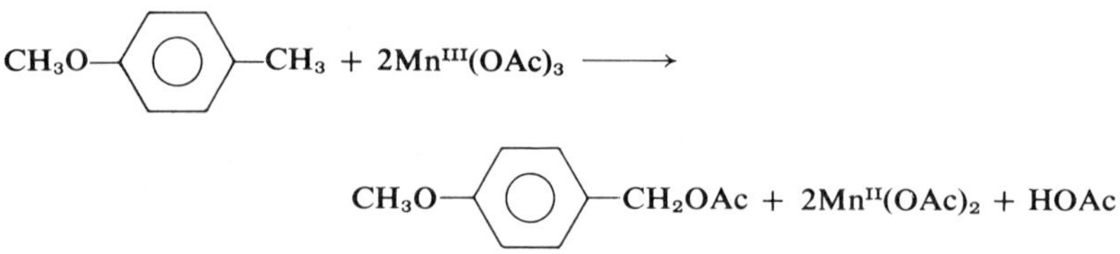

(92b)

The rate of oxidation is dependent on the ionization potential of the hydrocarbon[141] in agreement with the electron transfer mechanism.

Kooyman and co-workers[142] showed that no direct reaction between the arene and manganese(III) occurred in the manganic acetate oxidation of toluene. Products were satisfactorily explained by the participation of the carboxymethyl intermediate[143] analogous to those formed in the manganic acetate oxidation of olefins (vide infra).[144] A mixture of methyl benzyl acetates is produced from toluene as shown below:

$$\begin{aligned} Mn^{III}(OAc)_3 &\xrightarrow{HOAc} Mn^{II}(OAc)_2 + \cdot CH_2CO_2H \\ ArH + \cdot CH_2CO_2H &\xrightarrow{Mn^{III}} ArCH_2CO_2H \\ ArCH_2CO_2H + Mn^{III} &\longrightarrow ArCH_2\cdot + Mn^{II} + CO_2 + H^{+} \\ ArCH_2\cdot + Mn^{III} &\longrightarrow ArCH_2^{+} + Mn^{II} \\ ArCH_2^{+} + HOAc &\longrightarrow ArCH_2OAc + H^{+} \end{aligned}$$

The apparent duality of mechanism has been resolved by more recent studies,[145,146] in which the oxidation of aromatic hydrocarbons by manganic acetate has been shown to proceed by two competing paths: (a) a free radical mechanism resulting from interaction of the arene with $\cdot CH_2CO_2H$ radicals generated directly by thermolysis of manganic acetate, and (b) an electron transfer mechanism which is of importance in the oxidation of aromatic hydrocarbons having ionization potentials less than 8 eV.[146] Products of the reaction were consistent with the following schemes.[146]

1. Hydrocarbons such as toluene which have ionization potentials greater than 8 eV.[171b]

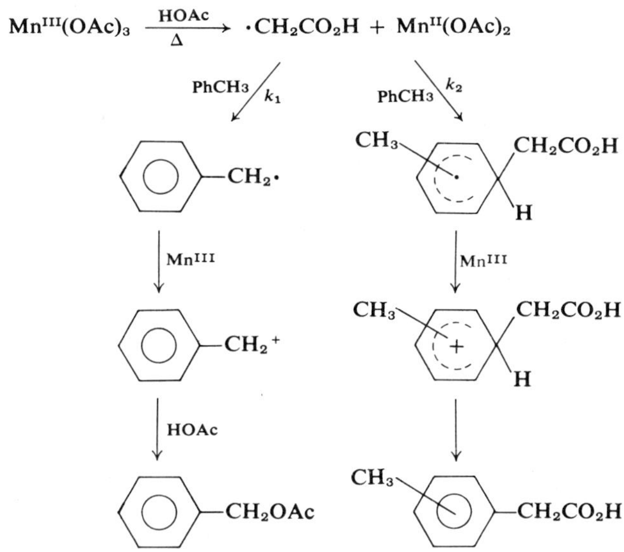

The mixture of arylacetic acids is subsequently decarboxylated (vide infra) by $Mn(OAc)_3$[147] to a mixture of xylyl acetates.

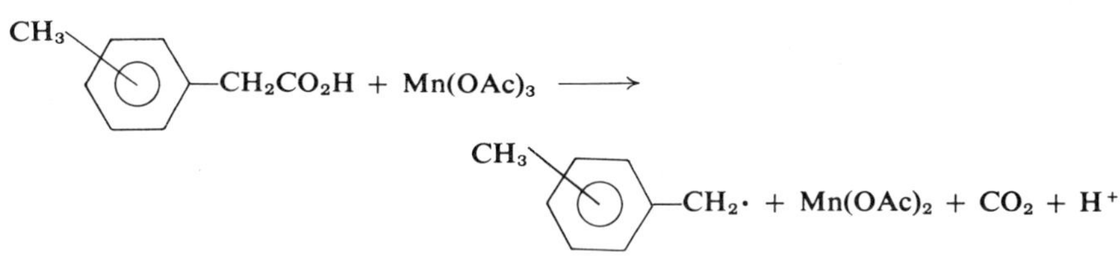

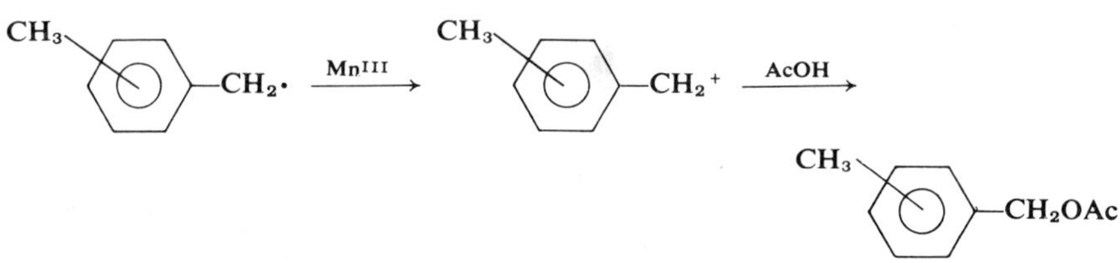

2. Hydrocarbons such as 2-methylnaphthalene which have ionization potentials less than 8 eV.[171b]

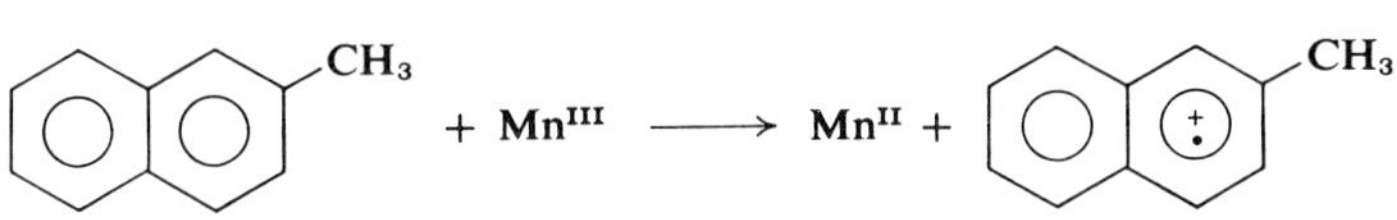

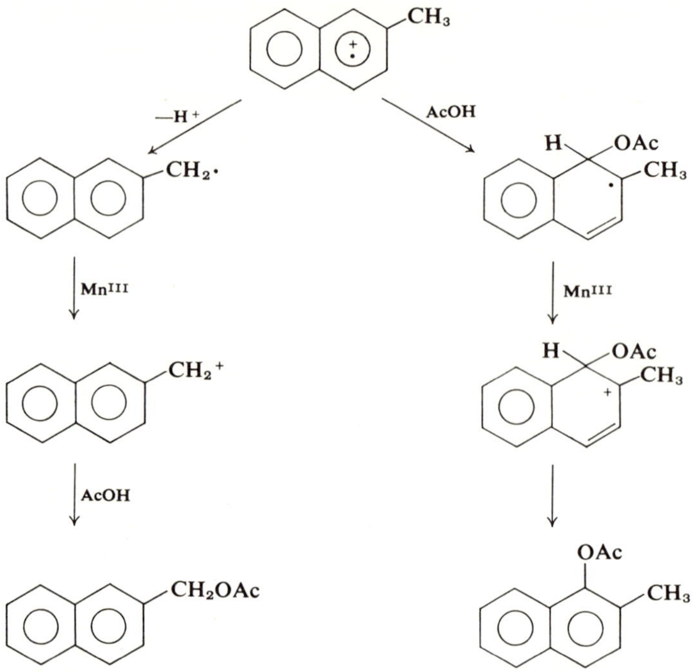

The route involving electron transfer can be suppressed by potassium acetate or manganous acetate or by carrying out the reaction under anhydrous conditions. Similar processes are also involved in oxidations with cerium(IV).[148]

Catalysis of these reactions by potassium bromide is observed,[143,145] Aralkyl bromides are formed as intermediates which undergo acetolysis to the corresponding acetates under the reaction conditions. The following mechanism was proposed, which involves ligand transfer oxidation by $Mn^{III}Br$.[145]

$$X{-}C_6H_4{-}CH_3 \xrightarrow[\text{HOAc}]{\text{Mn(OAc)}_3,\ \text{KBr}} X{-}C_6H_4{-}CH_2\cdot$$

$$X{-}C_6H_4{-}CH_2\cdot + Mn^{III}Br \longrightarrow X{-}C_6H_4{-}CH_2Br + Mn^{II}$$

$$X{-}C_6H_4{-}CH_2Br + HOAc \longrightarrow X{-}C_6H_4{-}CH_2OAc + HBr$$

Waters[149] and co-workers studied the oxidation of a series of alkylated aromatic hydrocarbons by cobaltic perchlorate in aqueous acetonitrile. The correlation between the reactivity of the arene and the ionization potential of the hydrocarbon is compatible with the formation of radical cations in an electron transfer process. Morimoto and Ogata[150] studied the autoxidation of toluene in acetic acid catalyzed by cobaltic acetate and concluded that reaction 93 was an important pathway for initiation. However, another kinetic study[151] showed that the rate of oxidation of substituted toluenes by

$$PhCH_3 + Co^{III} \longrightarrow PhCH_2\cdot + Co^{II} + H^+ \quad (93)$$

cobaltic acetate in acetic acid was rather insensitive to the C—H bond energy. The reaction was first order in toluene, second order in cobalt(III), and inverse first order in cobalt(II) [cf. the manganese(III) system above]. The apparent activation energy for the oxidation of toluene was 25 kcal/mole and the stoichiometry given in equation 94 was the same as that obtained for the oxidation of *p*-methoxytoluene by manganic acetate. A mechanism involving

$$PhCH_3 + 2Co(OAc)_3 \longrightarrow PhCH_2OAc + 2Co(OAc)_2 + HOAc \quad (94)$$

the formation of a radical cation via electron transfer oxidation of the aromatic ring is compatible with the evidence. In contrast to oxidations carried out with manganic acetate, the oxidation of alkylbenzenes by cobaltic acetate involves only electron transfer and no competition from other pathways involving alkyl radicals is apparent.[152]

The electron transfer mechanism is supported by Heiba and co-workers[153] who also observed the ESR spectra of radical cations when arenes were treated with cobaltic acetate in trifluoroacetic acid.[154] Cobaltic acetate is generally a better oxidant in trifluoroacetic acid than acetic acid[155] and benzene can be oxidized to phenyl trifluoroacetate in high yields.[152] In some cases the radical-cation intermediates can be observed for reactions carried out in acetic acid.[153] Further evidence for the radical-cation mechanism was obtained in the oxidation of *p*-methoxybenzyl phenyl sulfide.[153] The products were consistent with the following mechanism and could not be explained by a mechanism involving alkyl radicals.

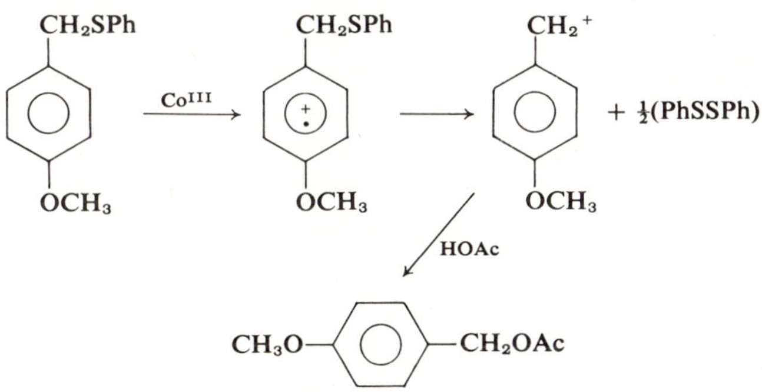

The addition of lithium chloride also leads to an enhancement of the reactivity of the cobaltic acetate,[153] and the acceleration is attributed to the formation of a cobaltic complex of relatively high oxidation potential and lower selectivity.

The direct oxidation of substrate by cobalt(III) probably plays an important role in the cobalt-catalyzed autoxidation of substituted toluenes which proceeds only when cobalt(III) is present.[150,156] No oxidation takes place if the cobalt is present only as cobalt(II). Oxidation via an electron transfer mechanism would explain the high selectivity shown by substituted toluenes in the cobalt-catalyzed autoxidations.[150] The large polar effect is incompatible with a rate-determining step involving hydrogen abstraction by alkylperoxy radicals. The requirement for high concentrations of cobalt catalyst is also unusual and suggests that the usual free radical chain mechanism is not the major reaction. A reaction involving electron transfer from the substituted toluene to the cobalt(III) must be operating in both the presence and absence of oxygen. For such a process to participate in autoxidation the subsequent steps require the reaction of the benzylic radical with oxygen to compete with further oxidation by cobalt(III).

$$\mathrm{PhCH_3} \xrightarrow{\mathrm{Co^{III}}} [\mathrm{PhCH_3}\cdot]^+ \longrightarrow \mathrm{PhCH_2}\cdot + \mathrm{H^+}$$

$$\mathrm{PhCH_2}\cdot \begin{cases} \xrightarrow{\mathrm{Co^{III}}} \mathrm{PhCH_2}^+ \\ \xrightarrow[\mathrm{O_2}]{} \mathrm{PhCH_2O_2}\cdot \end{cases}$$

Bromide ions have a pronounced synergistic effect on cobalt-catalyzed autoxidations[157,159–161] and the addition of an equimolar amount of hydrogen bromide to cobalt(II) acetate in acetic acid produces a catalyst which is quite active. Thus, *p*-xylene can be readily oxidized to terephthalic acid at atmospheric pressures and at temperatures as low as 60–100° under these conditions. Tetralin is oxidized rapidly at room temperature to α-tetralone. High concentrations of catalyst (approximately 0.1 *M*) are required for optimum rates of reaction. Other metal ions such as cerium and manganese showed the same effect, but to a more limited extent. Bromide is unique in its effect since none of the other halogens approach it in activity.

Hydrogen bromide enhances the rate of autoxidation of cumene,[157] which is best explained by the following reactions suggested by Rust and Vaughan.[158] Such a catalysis is insufficient to explain the properties of bromide since the

$$\mathrm{RO_2}\cdot + \mathrm{HBr} \longrightarrow \mathrm{RO_2H} + \mathrm{Br}\cdot$$

$$\mathrm{Br}\cdot + \mathrm{RH} \longrightarrow \mathrm{R}\cdot + \mathrm{HBr}$$

$$\mathrm{R}\cdot + \mathrm{O_2} \longrightarrow \mathrm{RO_2}\cdot, \ldots$$

autoxidation of alkylbenzenes proceeds at optimum rates only in the presence of both hydrogen bromide and cobalt. One function of the cobalt(II) is to maintain the concentration of hydrogen bromide at a sufficiently low level to prevent the acid-catalyzed rearrangement of hydroperoxides to phenols, which are inhibitors. A second function of the cobalt is to reconvert benzylic

$$Co(OAc)_2 + HBr \rightleftharpoons Co(OAc)Br + HOAc$$

bromides to bromine in order to keep the catalyst active. Thus, addition of cobalt(II) acetate to a warm solution of benzyl bromide in acetic acid immediately generates the intense blue color of cobalt acetate bromide.[157]

$$PhCH_2\cdot + Br_2 \longrightarrow PhCH_2Br + Br\cdot$$

$$Co^{III}Br + PhCH_2\cdot \longrightarrow Co^{II} + PhCH_2Br$$

$$Co(OAc)_2 + PhCH_2Br \longrightarrow Co(OAc)Br + PhCH_2OAc$$

In the absence of added bromide, the autoxidation of *p*-xylene catalyzed by cobalt or manganese salts proceeds with comparative ease to *p*-toluic acid. The efficient oxidation of *p*-xylene to terephthalic acid may only be achieved by using the cobalt or manganese catalyst in combination with an inorganic bromide. Bawn and Wright[160] studied the further autoxidation of *p*-toluic acid catalyzed by cobalt bromide and showed that reaction 95 occurred rapidly on mixing cobaltic acetate with lithium bromide in acetic acid. A trace of *t*-butyl hydroperoxide is necessary to initiate the $CoBr_2$-catalyzed

$$Co^{III} + Br^- \longrightarrow Co^{II} + Br\cdot \tag{95}$$

reaction at 90–100° and the oxidation of cobalt(II) by alkyl hydroperoxide provides the cobalt(III) for reaction 95.

The question arises[162] as to whether inner-sphere complexes of the aromatic hydrocarbon with cobalt(III) are involved in the electron transfer from the aromatic hydrocarbon. An investigation[163] was recently carried out on the oxidation of alkyl aromatic hydrocarbons by the heteropoly compound $K_5[Co^{III}O_4W_{12}O_{36}]\cdot H_2O$. Electron exchange between the cobalt(III) complex, which contains tetrahedral cobalt(III), and the corresponding cobalt(II) form proceeds via an outer-sphere mechanism. The oxidation products derived from xylenes and this oxidant in heterogeneous (hydrocarbon) and homogeneous (acetic acid) systems were also consistent with an aromatic radical cation as an intermediate formed by outer-sphere electron transfer. Inner-sphere electron transfer is eliminated by the stability and the inherent resistance to destruction of the tungstate framework which totally screens the cobalt(III) from direct interaction with the aromatic π system. An inner-sphere mechanism for electron transfer in the oxidation of alkyl aromatics is, thus, unnecessary since outer-sphere electron transfer is possible.

Although an electron transfer process for oxidations at benzylic positions is an attractive mechanism, it is not clear that hydrogen atom abstraction by the metal oxidant does not occur. Thus, it has been suggested that some oxidants, particularly those with oxygen ligands, such as chromate(VI) and permanganate(VII) are involved in hydrogen transfer processes, for example,[164,165]

$$R_3C\text{—}H + Mn^{VII}O_4^- \longrightarrow [R_3C\cdot Mn^{VI}O_4H] \longrightarrow R_3C\text{—}OMn^{V}O_3H^-, \ldots \quad (96)$$

Since a number of these conclusions are based on benzylic oxidations and involve outer-sphere reagents, an electron transfer process is possible. However, the latter is incompatible with the observation of a very large kinetic isotope (deuterium) effect and a low sensitivity to polar nuclear substituents in the MnO_4^- oxidation of γ-arylvaleric acids.[166a] The partial retention of configuration in the oxidation of the 4-methyl analog was accounted for by a cage recombination following hydrogen atom transfer (cf. equation 96).

The oxidation of toluene by ceric ammonium nitrate occurs preferentially at the side chain.[166b] Oxidation by hydrogen abstraction appears to be un-

$$PhCH_3 + Ce^{IV} \longrightarrow PhCH_2\cdot + Ce^{III} + H^+$$

likely for this reagent, in view of the absence of a kinetic isotope effect in the oxidation of cycloheptatriene-d_8, in which resonance stabilization of the tropenyl radical should optimize such a process (as well as a two-electron oxidation by removal of hydride).[166c] It has been proposed that oxidation of cycloheptatriene proceeds by addition of $NO_3\cdot$ followed by subsequent oxidation of the dienyl radical, although a case against electron transfer in the initial step is not convincing.[166b]

Hydrogen atom transfer has been proposed for reactions between inorganic ions in water, especially involving the Fe^{II}–Fe^{III} exchange. The situation, however, is far from clear due to the lability of the ligands,[168] as is the

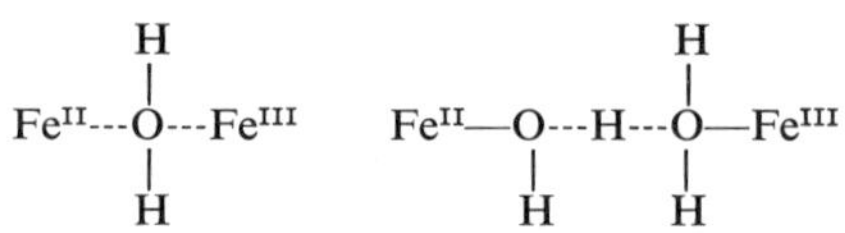

role of coordinated water as a bridging ligand in oxidation–reduction reactions.[9b]

3. *Saturated Hydrocarbons*

A number of inorganic oxidants such as chromic acid and permanganate are known to oxidize saturated hydrocarbons readily, and hydrogen atom transfer probably pertains.[164]

The stoichiometric oxidation of cyclohexane to a mixture of cyclohexanol, cyclohexanone, and adipic acid by cobaltic perchlorate in aqueous acetonitrile has been studied.[149] The overall rate of oxidation of cyclohexane is roughly equal to that of toluene under the same conditions. The oxidation of cyclohexane to cyclohexyl acetate by manganic acetate in acetic acid has been reported.[168] The direct oxidation of cyclohexane, which possesses no functional group able to complex a cobaltic ion, suggests that hydrogen transfer to the ligand is also possible with this oxidant, since an outer-sphere mechanism is prohibitively unfavorable energetically due to the high ionization potential of alkanes.

The mechanism of the oxidation of alkanes with chromic acid in aqueous acetic acid has been studied extensively, and hydrogen atom transfer has been proposed as the most reasonable first step[164]:

$$R_3C{-}H + H_2Cr^{VI}O_4 \longrightarrow [R_3C\cdot H_3Cr^{V}O_4] \longrightarrow R_3C{-}O{-}Cr^{IV}O_3H_3$$

$$R_3C{-}O{-}CrO_3H_3 + H_2O \longrightarrow R_3COH + H_4Cr^{IV}O_4$$

The chromium(IV) species, however, is a metastable intermediate and is involved further in one-equivalent redox reactions.[169] The hydrogen transfer shows carbonium ion character, and Rocek[165] has suggested that the intermediate (or transition state) should be considered in the same light as the "polar effect" in other hydrogen abstractions by free radicals (see Chapter 7).

$$[R_3C{-}HOCr^{VI}O_3H_2] \longleftrightarrow [R_3C\cdot H{-}OCr^{V}O_3H_2] \longleftrightarrow [R_3C^{+}H{-}OCr^{IV}O_3H_2]$$

C. Decarboxylation of Acids and Oxidation of Carbonyl Compounds

The oxidation of carboxylic acids represented by equation 97 is a useful source of alkyl radicals since the carboxyl radical is unstable.

$$RCO_2H \longrightarrow RCO_2\cdot + H^+ + \varepsilon \qquad (97)$$

$$RCO_2\cdot \longrightarrow R\cdot + CO_2 \qquad (98)$$

Decarboxylation (equation 98) is exothermic by about 15 kcal/mole, although it becomes less with increasing *s* character in the alkyl–carbonyl bond.[170] The decarboxylation carried out electrochemically is known as the Kolbe oxidation when alkyl dimers are formed, and as the Hofer-Moest reaction when oxidative substitution results.[171]

A number of metal oxidants such as lead(IV), silver(II), manganese(III), cobalt(III), cobalt(IV), and thallium(III) acetates are also capable of effecting decarboxylation. Lead tetraacetate has been widely used to effect oxidative substitution, (equation 99), oxidative elimination (equation 100), and halodecarboxylation (equation 101, X = Cl, Br).[172]

$$RCO_2H + Pb(OAc)_4 \longrightarrow ROAc + CO_2 + Pb(OAc)_2 + HOAc \qquad (99)$$

$$RCO_2H + Pb(OAc)_4 \xrightarrow{Cu^{II}} R({-}H) + CO_2 + Pb(OAc)_2 + 2HOAc \qquad (100)$$

$$RCO_2H + Pb(OAc)_4 + MX \longrightarrow R{-}X + CO_2 + Pb(OAc)_2 + MOAc + HOAc \qquad (101)$$

All of these reactions proceed via a radical chain process and are subject to inhibition by oxygen as well as initiation by peroxides or ultraviolet irradiation. A radical chain mechanism has been proposed[173]:

Initiation:

$$RCO_2H + Pb^{IV}(OAc)_4 \rightleftharpoons RCO_2Pb^{IV}(OAc)_3 + HOAc$$

$$RCO_2Pb(OAc)_3 \longrightarrow R\cdot + CO_2 + Pb^{III}(OAc)_3$$

Propagation:

$$R\cdot + RCO_2Pb^{IV}(OAc)_3 \longrightarrow ROAc + RCO_2Pb^{III}(OAc)_2 \quad (102)$$

$$RCO_2Pb^{III}(OAc)_2 \longrightarrow R\cdot + CO_2 + Pb^{II}(OAc)_2 \quad (103)$$

Oxidative substitution results from the oxidation of alkyl radicals by lead(IV). The oxidation equation 102 probably consists of two steps and involves a carbonium ion intermediate. Oxidation of alkyl radicals by lead(IV) is usually not as fast[174] as oxidation by copper(II), and a rapid catalytic reaction results from the following sequence of reactions when small amounts of copper(II) are added:

$$R\cdot + Cu^{II} \xrightarrow{\text{fast}} R(\text{—H}) + Cu^{I} + H^+$$

$$Cu^{I} + RCO_2Pb^{IV} \xrightarrow{\text{fast}} Cu^{II} + RCO_2Pb^{III}, \ldots$$

Moreover, the oxidation of alkyl radicals by Cu(II) under these conditions usually proceeds by oxidative elimination and alkenes are the primary products.[174]

Ligand transfer is also possible from halo lead(IV) species formed on metathesis and is a key step in a useful method for the preparation of alkyl halides.[175]

$$X^- + RCO_2Pb^{IV} \rightleftharpoons XPb^{IV}O_2CR$$

$$R\cdot + XPb^{IV}O_2CR \longrightarrow R\text{—}X + RCO_2Pb^{III}, \ldots$$

Lead(IV) carboxylates are rather unique in their ability to undergo radical chain decarboxylation, since lead(III) species are metastable (equation 103) and Pb(IV) is capable of oxidizing alkyl radicals (equation 102). The efficiency of the latter decreases from allylic, tertiary, secondary to primary alkyl radicals, and primary carboxylic acids generally undergo oxidative decarboxylation slowly in the absence of copper(II).[176]

Thallium(III) is not effective in the oxidation of alkyl radicals and thallium(III) carboxylates undergo only a nonchain decarboxylation when induced photochemically.[177] The fate of the alkyl radicals depends on the wavelength of the radiation.

$$(RCO_2)_3Tl^{III} \xrightarrow{h\nu} (RCO_2)_2Tl^{II} + R\cdot + CO_2$$

$$(RCO_2)_2Tl^{II} \xrightarrow{\text{fast}} RCO_2Tl^{I} + R\cdot + CO_2$$

Silver(II) is particularly effective in the decarboxylation of acids,[178] but in the absence of ligands such as pyridine it is metastable and disproportionates. However, silver(II) can be utilized in a catalytic manner by oxidation of silver(I) with peroxydisulfate.[178] Due to the low steady-state concentration of

$$\begin{aligned} 2Ag^{I} + S_2O_8^{2-} &\longrightarrow 2Ag^{II} + 2SO_4^{2-} \\ Ag^{II} + RCO_2H &\longrightarrow Ag^{I} + H^+ + RCO_2\cdot \longrightarrow R\cdot + CO_2 \\ R\cdot + Ag^{II} &\longrightarrow [R^+]_{ox} + Ag^{I}, \ldots \end{aligned} \tag{104}$$

silver(II), alkyl radicals are not efficiently oxidized (equation 104) by this method, except in the presence of copper(II). The efficient generation of alkyl radicals has been utilized in the homolytic alkylation of a variety of heteroaromatic bases in acidic media.[179]

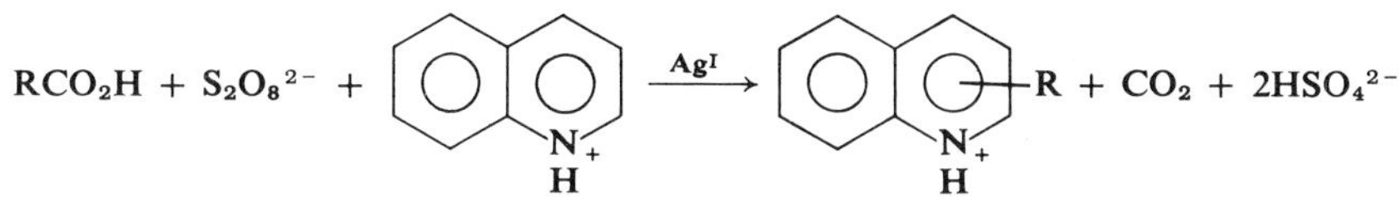

A chain mechanism is also not possible if the metal oxidant is constrained to a one-equivalent change. Thus, manganese(III), cobalt(III), cerium(IV), and silver (II) complexes decarboxylate acids by nonchain processes. The oxidative decarboxylation of propionic acid in aqueous perchloric acid obeys Michaelis-Menten kinetics, which are consistent with the following mechanism[180]:

$$\begin{aligned} RCO_2H + Co^{III}(H_2O)_6 &\rightleftharpoons RCO_2Co^{III}(H_2O)_5 + H_3O^+ \\ RCO_2Co^{III}(H_2O)_5 &\xrightarrow{\text{slow}} R\cdot + CO_2 + Co^{II}(H_2O)_5 \\ R\cdot + Co^{III}(H_2O)_6 &\xrightarrow{\text{fast}} ROH + Co^{II}(H_2O)_6 \end{aligned}$$

However, it has been suggested that the decarboxylation of acids by manganese(III) in the same medium proceeds by an outer-sphere reaction between hexaaquo manganese(III) and the carboxylate ion, and a similar mechanism has been proposed for the oxidation of alcohols.[181]

Decarboxylations have also been studied in carboxylic acids as solvents, and reactions in these media are best interpreted as inner-sphere processes involving homolysis of the metal–carboxylate bond.[176]

If pivalic, isobutyric, and *n*-butyric acids are taken as representative of tertiary, secondary, and primary acids, respectively, the rates of decarboxylation by lead(IV) vary significantly with the structure of the alkyl moiety in the manner: tertiary > secondary > primary. The trend, in more or less the same degree, has also been observed in the thermal decarboxylation of these acids under similar conditions by cobalt(III) and manganese(III). These oxidants are distinguished from lead(IV) in that a radical chain process is not involved in the decarboxylation. The comparison among these reagents is given in Table 13.

Homolysis of the carboxylatometal species is a common step in the mechanism of decarboxylation of acids by these metal oxidants. A multibond cleavage directly to carbon dioxide and alkyl radicals which occurs coincidentally with reduction has been postulated. Part of the driving force for such a simultaneous process is derived from the formation of carbon dioxide,

$$RCO_2M^{III} \longrightarrow R\cdot + CO_2 + M^{II} \quad \text{where} \quad M^{III} = Pb^{III}, Co^{III}, Mn^{III}$$

as it is in the thermolysis of some peresters.[182] As discussed by Bartlett and Hiatt,[182] the relative rates of homolysis would be determined by the stabilities of the various alkyl radicals. The increasing spread in selectivity from lead(III), cobalt(III), manganese(III) can be related to the increasing activation energy for the homolysis, and resultant increase in the extent to which the bonds are broken in the transition state.

Due to the chain character of the lead(IV) decarboxylation, the activation energy for homolysis is difficult to assess. No doubt, it is significantly less than for cobalt(III) carboxylate. The apparent activation energy for decarboxylation of cobalt(III) pivalate is 32 kcal/mole[183] and for manganese(III) pivalate it is 36 kcal/mole.[147] Qualitatively acids can be decarboxylated by lead(IV) at significantly lower temperatures than those necessary for decarboxylation by cobalt(III). Decarboxylation by manganese(III) is carried out at the highest temperatures and also shows the greatest selectivity.

TABLE 13

Relative Rates of Decarboxylation of Acids by Metal Oxidants

Oxidant $M(O_2CR)_n$	Temperature (°C)	Method[a]	Relative rate[b] Pivalic / *i*-butyric	Pivalic / *n*-butyric	Decarboxylation mechanism
Ag^{II}	60	Δ	1.7	2.0	Acyloxy
Pb^{IV}	30	$h\nu$	5	90	Multibond
Pb^{IV}	80	Δ	3	20[c]	Chain
Co^{III}	80	Δ	10	100[c]	Multibond
Mn^{III}	105	Δ	25	500[d]	Multibond
Ce^{IV}	30	$h\nu$	1.2	1.1	Acyloxy

[a] $h\nu$ = 3500 Å, Δ = thermal.
[b] By competition studies.
[c] At 100°.
[d] At 120°.

The absence of significant differences in rates of decarboxylation of these acids by silver(II) militates against the simultaneous fission of the alkyl–carbonyl bond of the type described above.[178] Decarboxylation by cerium(IV) is induced photochemically and also shows no selectivity.[184] In the latter case, sufficient energy is provided by radiation at 3500 Å to effect simple homolysis to acyloxy radicals, and driving force obtainable by multibond cleavage is not

$$RCO_2Ce^{IV} \xrightarrow{h\nu} RCO_2\cdot + Ce^{III}, \ldots$$

required. It has been concluded on this basis that the decarboxylation of acids by silver(II) is exothermic, as also indicated by the facility with which it takes

$$RCO_2H + Ag^{II} \longrightarrow RCO_2\cdot + H^+ + Ag^{I}, \ldots$$

place. An outer-sphere process may be involved.

The thermal oxidative decarboxylation of phenylacetic acid by ceric ammonium nitrate at 90°C affords a mixture of benzyl alcohol, benzaldehyde, and benzyl nitrate in aqueous acetonitrile solutions.[185] Ring-substituted derivatives follow a Hammett correlation with $\rho = -3.01 \pm 0.31$ using σ^+ substituent constants. The decarboxylation has been postulated to proceed via homolysis of an inner-sphere complex:

$$ArCH_2CO_2H + Ce^{IV} \underset{}{\overset{-H^+}{\rightleftharpoons}} ArCH_2CO_2Ce^{IV} \longrightarrow ArCH_2\cdot + CO_2 + Ce^{III}$$

The magnitude of ρ is significantly larger than even that for the thermolysis[186] of the *t*-butyl perester ($\rho = -1.20$) and may include a contribution from the preequilibrium step.

The oxidation of carboxylic acids can also proceed by loss of an α-hydrogen. Primary acids are particularly prone to this type of oxidation when treated with manganese(III) or cerium(IV).[143,144,148] Thus, carboxymethyl radicals are generated from acetic acid and add readily to alkenes to produce γ-lactones by the following mechanism[187]:

$$CH_3CO_2H + Mn^{III} \longrightarrow \cdot CH_2CO_2H + Mn^{II} + H^+$$

$$RCH{=}CH_2 + \cdot CH_2CO_2H \longrightarrow R\dot{C}H{-}CH_2CH_2CO_2H$$

$$R\dot{C}HCH_2CH_2CO_2H + Mn^{III} \longrightarrow$$ 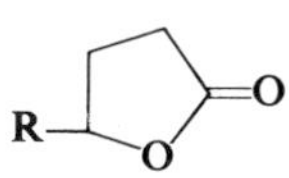 $+ Mn^{II} + H^+$ (105)

The formation of lactone in optimum amounts depends on the selective oxidation of the adduct radical (equation 105) in preference to the carboxymethyl radical which is deactivated by the electron-withdrawing carbonyl group.[51,105,188] Carbon-centered radicals are also produced from the oxidation of ketones, esters, and their derivatives.[189] It is possible that the oxidation proceeds via enolization,[142] for example,

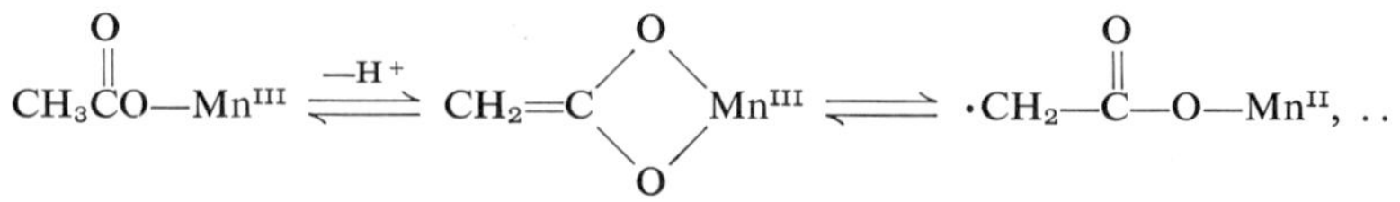

Carbonyl compounds are also oxidized in the α position by copper(II), iron(III), and manganese(III) complexes.[129,142,190] The oxidation of cyclohexanone by hexachloroiridate(IV) is particularly illuminating.[191] Kinetic data support an outer-sphere transfer from the enol to Cl_6Ir^{IV} since it proceeds with a rate constant of about $2 \times 10^3\ M^{-1}\ sec^{-1}$ at 25°, in contrast to the much slower aquation (approximately $10^{-6}\ sec^{-1}$ at 50°C). The ketocyclohexyl radical subsequently reduces a further molecule of Cl_6Ir^{IV} by a

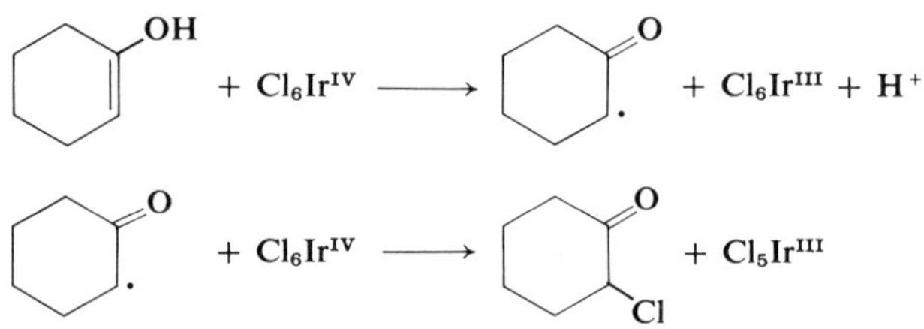

rapid inner-sphere path. Phenols are also oxidized by this reagent to quinones and diphenoquinones.[192] From the product of reduction, Cl_5Ir^{III}, it is deduced that the process occurs by two successive outer-sphere transfers.

Copper(II) chloride and bromide effect α-halogenation,[129] and autoxidations by manganese(III) have been suggested to proceed via α-ketoalkyl radicals.[142] Alternatively, phenols represent, of course, extreme forms of enols and an electron transfer process with outer-sphere reagents such as $Fe^{III}(CN)_6^{3-}$ is known,[193] as it is with thiols[194] and amines.[195]

The carbon–hydrogen bond in the aldehydic group is one of the weakest, and a number of transition-metal oxidants have been used to effect oxida-

$$RCHO \longrightarrow R\dot{C}O + H^+ + \varepsilon$$

tion.[196] However, aldehydes are prone to hydration as well as enolization and mechanistic studies using permanganate and chromate have been complicated due to competition from two-equivalent processes.[197]

The oxidation of aliphatic aldehydes by manganese(III) pyrophosphate is catalyzed by acid, but the rate is zero order in manganese(III). A rate-limiting formation of the enol is suggested,[198] and it would be similar to that involved in the autoxidation of ketones catalyzed by manganese salts.[142] The rapid oxidation of the enol is probably similar to the outer-sphere mechanism proposed for the oxidation of alcohols.[181]

A detailed kinetic study of the oxidation of benzaldehyde by cerium(IV) in aqueous acetic acid indicates that both 1:1 and 2:1 inner-sphere complexes are involved.[199] The large kinetic isotope effect ($k_H/k_D = 16$) is consistent

$$Ce^{IV} + RCHO \rightleftharpoons Ce^{IV}(RCHO)$$

$$Ce^{IV}(RCHO) + H^+ \longrightarrow Ce^{III} + R\dot{C}O \quad (106)$$

$$R\dot{C}O + Ce^{IV} \longrightarrow RCO_2H + Ce^{III} \quad (107)$$

with a rate-limiting homolysis (equation 106) followed by further one-equivalent oxidation of the acyl radical (equation 107). The formation of the latter is supported by the observation of triphenylcarbinol and carbon monoxide during oxidation of triphenylacetaldehyde.[200]

The rates of oxidation of aromatic aldehydes by cobalt(III) studied in aqueous acetonitrile solutions are first order in both reactants but show an inverse acid dependence.[201] The kinetic isotope effect also supports a direct transfer of hydrogen to the cobalt(III) but no decision could be reached as to whether an inner- or outer-sphere process is involved.

Traces of iron, copper, cobalt, and manganese salts catalyze the aerial oxidation of aldehydes. Initiation by direct interaction between the metal catalyst and the aldehyde was proposed as long ago as 1931 by Haber and Willstater.[202]

$$RCHO + Fe^{III} \longrightarrow R\dot{C}O + Fe^{II} + H^+$$

Bawn and co-workers have carried out detailed investigations of metal-catalyzed autoxidations of acetaldehyde[203] and benzaldehyde.[204] The mechanism of cobalt-catalyzed oxidation of benzaldehyde may be described by the following scheme:

$$R\dot{C}O + O_2 \longrightarrow RCO_3\cdot$$

$$RCO_3\cdot + RCHO \longrightarrow RCO_3H + R\dot{C}O$$

$$Co^{II} + RCO_3H \longrightarrow Co^{III} + RCO_2^- + \cdot OH \quad (108)$$

$$Co^{III} + RCHO \longrightarrow Co^{II} + R\dot{C}O + H^+ \quad (109)$$

$$Co^{III} + RCO_3H \longrightarrow Co^{II} + RCO_3\cdot + H^+ \quad (110)$$

The rate of chain initiation in the autoxidation of benzaldehyde catalyzed by cobalt acetate in acetic acid was equal to the rate of reaction of cobalt(III) with benzaldehyde in acetic acid, measured independently in the absence of oxygen. Moreover, the onset of absorption of oxygen coincided with the conversion of the cobaltous to the cobaltic oxidation state, and the catalyst was maintained largely in this state throughout the autoxidation. In separate experiments it was shown that the reaction of cobaltous species with perbenzoic acid (equation 108) is very rapid. Hence, the rate-determining step in the autoxidation is attributed to the reaction of benzaldehyde with cobaltic species to form the chain-initiating benzoyl radicals (equation 109). The reaction of cobalt(III) with perbenzoic acid (equation 110) plays no fundamental role in the autoxidation since it takes place at a much slower rate than reaction 109. The energy of activation for reaction 109 is 14.8 kcal/mole. In the cobalt- and manganese-catalyzed autoxidation of hydrocarbons,[118,119] the occurrence of a maximum in the concentration of cobalt(III) or manganese(III) coincides with the appearance of aldehydes among the products.

The rate of the cobalt-catalyzed autoxidation of benzaldehyde follows equation 111 up to catalyst concentrations of 2×10^{-4} *M*.[204,205] At con-

$$\frac{-d[O_2]}{dt} = k[\mathrm{PhCHO}]^{3/2}[\mathrm{Co(II)}]^{1/2} \qquad (111)$$

centrations of catalyst above 1.2×10^{-3} *M*, the rate is independent of catalyst concentration and longer induction periods are observed.[205] The scavenging of chain-initiating radicals by cobalt(II) is implied.

In the oxidation of benzaldehyde catalyzed by nickel(II) acetate the rate constants for reactions 112a and 112b were 4×10^{-3} and 1.8×10^{-2} liters/(mole-sec), respectively.[205] Hence, reaction 112a is the rate-determining

$$\mathrm{Ni^{II} + PhCO_3H \longrightarrow Ni^{III} + PhCO_2^- + \cdot OH} \qquad (112a)$$

$$\mathrm{Ni^{III} + PhCHO \longrightarrow Ni^{II} + Ph\dot{C}O + H^+} \qquad (112b)$$

step in the nickel-catalyzed autoxidation of benzaldehyde, in contrast to the cobalt- and manganese-catalyzed reactions. Consequently, the nickel remains mainly in the lower oxidation state throughout the reaction.

The oxidation of acetaldehyde[203,206] differs from the oxidation of benzaldehyde in that the acetaldehyde reacts with the peracetic acid to form acetaldehyde monoperacetate. In the presence of cobalt, manganese, or copper catalysts the acetaldehyde monoperacetate decomposes rapidly. Copper is less

$\mathrm{CH_3CHO + CH_3CO_2H} \rightleftharpoons$ 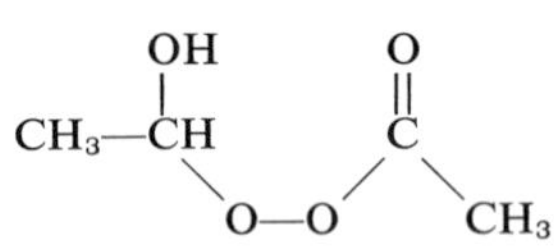

effective than manganese or cobalt as a catalyst for acetaldehyde autoxidation but all three catalysts follow the same general mechanism.[203] Direct reaction of acetaldehyde with the catalyst in its higher oxidation state constitutes the

$$\mathrm{Co^{III} + CH_3CHO \longrightarrow Co^{II} + CH_3\dot{C}O + H^+}$$

$$\mathrm{Co^{II} + CH_3CO_3H \longrightarrow Co^{III} + CH_3CO_2\cdot + OH^-} \qquad (113)$$

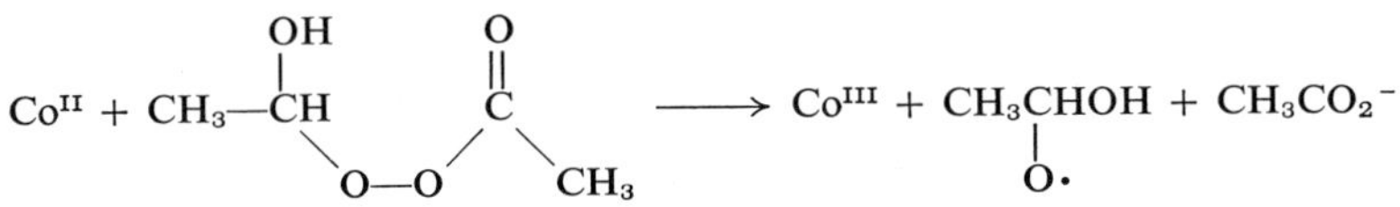
(114)

rate-determining step. Regeneration of cobalt(III) occurs by reaction 113 and/or 114. The oxidation of cobaltous acetate by peracetic acid given in equation 113 is very rapid and is the basis of a method for the preparation of cobaltic acetate.[207] Ozonation of cobaltous acetate, however, is the method of choice.[134]

Nikishin and co-workers[208] showed that the products of the stoichiometric oxidation of acetaldehyde are diacetyl, acetic acid, and manganous acetate.

$$CH_3CHO + Mn^{III}(OAc)_3 \longrightarrow CH_3\dot{C}O + Mn^{II}(OAc)_2 + HOAc$$
$$2CH_3\dot{C}O \longrightarrow CH_3COCOCH_3$$
$$CH_3\dot{C}O + Mn^{III} \longrightarrow CH_3\overset{+}{C}O + Mn^{II}, \ldots \qquad (115)$$

Electron transfer oxidation of acyl radicals represented in equation 115 competes with addition of oxygen in the metal-catalyzed autoxidations of aldehydes. The electron transfer step leads to the formation of acid anhydride by reaction with the carboxylic acid in the absence of water. Copper(II) appears to be one of the most effective oxidants for reaction 115 and a mixture of copper and cobalt acetates is effective in the conversion of acetaldehyde to acetic anhydride.[209]

$$Co^{III}(OAc)_3 + CH_3CHO \longrightarrow Co^{II}(OAc)_2 + CH_3\dot{C}O + HOAc$$
$$Co^{II}(OAc)_2 + CH_3CO_3H \longrightarrow Co^{III}(OAc)_3 + HO\cdot$$
$$CH_3\dot{C}O + Cu^{II}(OAc)_2 \longrightarrow CH_3\overset{+}{C}O + Cu^{I}OAc + AcO^-$$
$$Cu^{I}OAc + CH_3CO_3H \longrightarrow Cu^{II}(OAc)_2 + HO\cdot$$
$$CH_3\overset{+}{C}O + CH_3CO_2H \longrightarrow (CH_3CO_2)_2O + H^+$$

Oxidation of pivalaldehyde[208] by manganic acetate affords carbon monoxide, isobutane, and isobutylene, presumably by the following steps:

$$(CH_3)_3CCHO + Mn^{III}(OAc)_3 \longrightarrow (CH_3)_3C\dot{C}O + Mn^{II}(OAc)_2 + HOAc$$
$$(CH_3)_3C\dot{C}O \longrightarrow (CH_3)_3C\cdot + CO$$
$$(CH_3)_3C\cdot + (CH_3)_3CCHO \longrightarrow (CH_3)_3CH + (CH_3)_3C\dot{C}O$$
$$(CH_3)_3C\cdot + Mn^{III} \longrightarrow Mn^{II} + (CH_3)_3C^+ \longrightarrow (CH_3)_2C{=}CH_2 + H^+$$

The oxidation of aldehydes by metal complexes is a useful method for the generation of acyl radicals, and has been used for the synthesis of ketones by the radical chain addition of aldehydes to 1-alkenes with initiation by cobalt

$$RCHO + CH_2{=}CHR' \xrightarrow[O_2]{M^n} RCOCH_2CH_2R'$$

acetate in the presence of air.[210]

D. Reduction of Alkyl Halides

The reduction of alkyl halides by electron transfer from aromatic radical anions has been described in Chapter 9. An alternative process involving the inner-sphere transfer of halogen occurs mainly when the reducing agent is a metal or metal complex and derives part of its driving force from the strength of a metal–halogen bond. Thus, the reduction of benzyl chloride by chromium(II) to toluene has been shown to proceed by the following mechanism[62]:

$$PhCH_2Cl + Cr^{II} \xrightarrow{slow} PhCH_2\cdot + Cr^{III}Cl \qquad (116)$$

$$PhCH_2\cdot + Cr^{II} \xrightarrow{fast} PhCH_2Cr^{2+} \qquad (117)$$

$$PhCH_2Cr^{2+} + H_2O \longrightarrow PhCH_3 + Cr^{III} \qquad (118)$$

The rate-determining reduction step (equation 116) slows down with the trend I > Br > Cl of 555:124:1, and follows the order of bond-dissociation energies of the benzyl derivatives: 39, 51, and 60 kcal/mole, respectively. This pattern of reactivity and the activation parameters are similar to those observed for the inner-sphere transfer of halogen from a variety of substitution-inert cobalt(III) and chromium(III) halides.[9] Interestingly, polar substituents in the ring have little effect on the rate of reduction of benzyl halides by chromium(II). The absence of a significant polar effect in the ligand transfer of halogen to chromium(II) in the reduction of benzyl halides is in marked contrast to hydrogen abstractions from toluene by radicals such as *t*-butoxy, chlorine, bromine, and so on (see Chapter 7). The transition state for halogen transfer is more akin to that resulting from the thermally induced homolysis of the benzyl–halogen bond.[211]

Alkyl halides are generally inert to chromium(II), but its reactivity can be enhanced materially by the formation of $Cr^{II}(en)_2$ simply by adding ethylenediamine (en).[212] Under these conditions even primary alkyl chlorides are readily reduced to alkanes at room temperature. The reactivity of alkyl halides follows the order: tertiary > secondary > primary in a ratio of 100:10:1. Alkyl iodides are reduced 3×10^3 faster than the chlorides, and bromides have intermediate (4×10^2) reactivity. These trends together with the kinetics suggest that reduction of alkyl halides by $Cr^{II}(en)_2$ occurs by the same mechanism as that given in equations 116–118.

Alkylchromium species are intermediates in these reductions and form alkanes by subsequent hydrolysis in the aqueous medium. The rather large solvent isotope effect ($k_H/k_D = 4$) is attributed to the protonolysis step (118),[213] and is consistent with the inertness of the alkylchromium species to substitution.[55] Alkylchromium ions are stable for prolonged periods in the absence of protic sources.

Alkyl radicals formed as intermediates in the reaction between alkyl halides and chromium(II) complexes can be trapped by reactive hydrogen donors such as thiols.[214] The latter can be used to effect stereochemical changes in the product since the transition states for hydrogen abstraction and reaction with chromium(II) (followed by hydrolysis) are vastly different. For example, pure *cis*- or *trans*-4-*t*-butyl-1-methylcyclohexyl chloride affords the same mixture of 4-*t*-butyl-1-methylcyclohexane (*cis–trans* ~ 1.4), whereas the same reactions in the presence of butanethiol produced the *trans* isomer.[215]

A variety of β- and γ-substituted alkyl halides are reduced by chromium(II) complexes to alkenes and cyclopropanes, respectively. For example,[216]

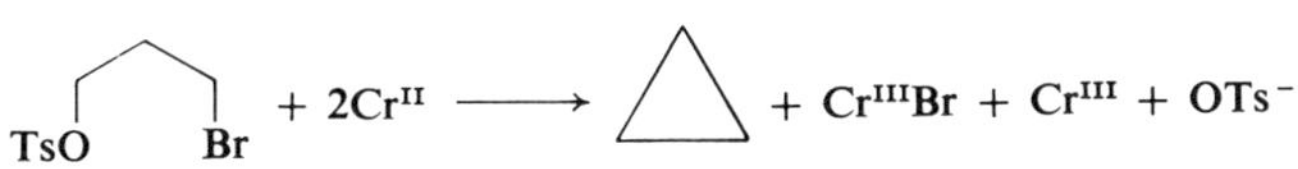

These reductive eliminations proceed by a mechanism similar to that for reductive protonations described above, except the presence of a β or γ substituent allows the intermediate to undergo syn or anti elimination as shown below[217]:

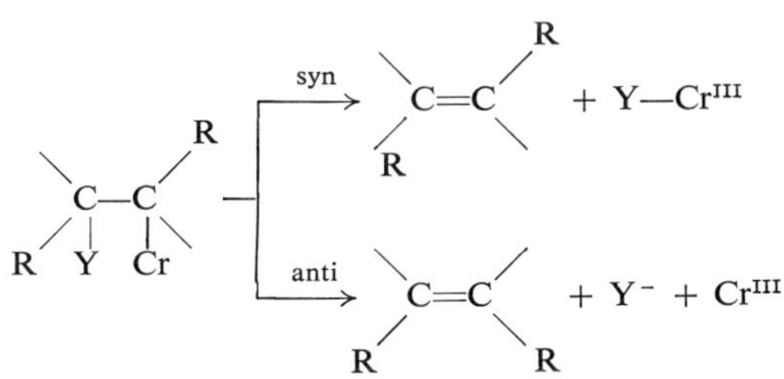

(119)

(120)

The relative rates of (119) and (120) depend on the substituent Y, and the bearing on the stereochemistry of reductive elimination has been discussed when Y = Br.

Geminal dihalides and polyhalomethanes react with chromium(II) by several rapid ligand transfer steps.[218] The α-haloalkyl radical formed in the first step is subject to α elimination and the resultant carbenoid species is reduced in a subsequent step, for example,

$$R_2CX_2 + Cr^{II} \longrightarrow R_2\dot{C}X + Cr^{III}X$$
$$R_2\dot{C}X + Cr^{II} \longrightarrow R_2C\colon + Cr^{III}X$$
$$R_2C\colon + Cr^{II} \xrightarrow{H^+} R_2\dot{C}H \xrightarrow{Cr^{II}} R_2CHCr^{+2}, \ldots$$

Polyhalomethanes are not reduced, thus, by a series of intermediates involving successive reductive protonations. The carbenoid intermediate [probably a complex with chromium(II)] has been trapped with unsaturated compounds.

The readily prepared *gem*-dibromocyclopropanes react under similar conditions to afford allenes in addition to reduced cyclopropanes. Rearrangement of cyclopropylidenes has been suggested.[219]

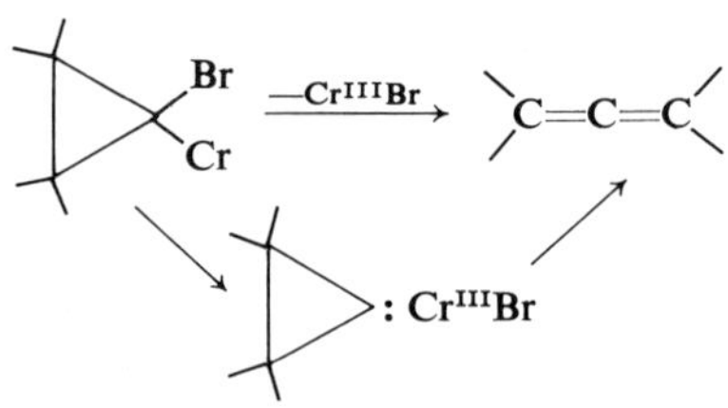

A number of low-spin cobalt(II) complexes also react with alkyl halides by an inner-sphere mechanism. Thus, the five-coordinate $Co^{II}(CN)_5^{3-}$ and Co^{II}(DMG)Py produce alkyl derivatives according to equations 121 and 122,[220,221] and the alkyl radical intermediate will initiate polymerization.[222]

$$L_5Co^{II} + R{-}X \xrightarrow{\text{slow}} L_5Co^{III}X + R\cdot \tag{121}$$
$$R\cdot + L_5Co^{II} \xrightarrow{\text{fast}} RCo^{III}L_5 \tag{122}$$

Outer-sphere processes have also been proposed on the basis of kinetic studies when *p*-nitrobenzyl halides are reduced by Co^{II}(salen)(M) in the presence of excess 1-methylimidazole (M).[223]

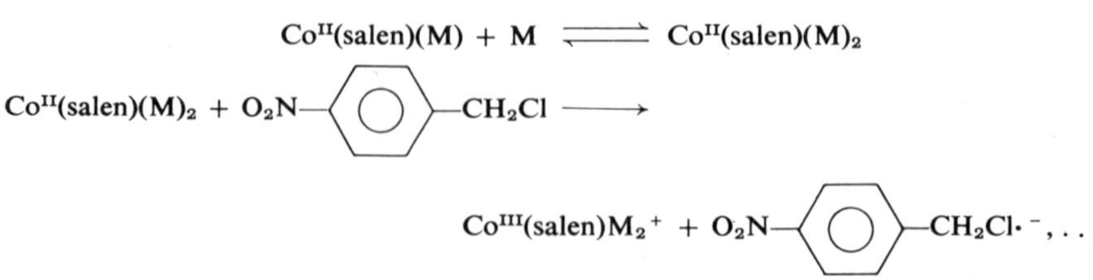

Outer-sphere transfer is promoted no doubt by the *p*-nitro group (which stabilizes the transient radical anion),[224] since other derivatives proceed via the inner-sphere mechanism. Finally, six-coordinate cobalt(II) complexes reduce alkyl halides only by an outer-sphere mechanism, and the stoichiometry is modified (to produce mainly R_2), since radical capture by the cobalt(II) species is precluded (although attack on the ligand is a possibility).

There are other routes by which organic halides can react with transition-metal complexes. Oxidative addition of alkyl halides particularly to d^8 metal complexes such as square planar iridium(I) is thought to entail S_N2 attack and free radicals as such are not intermediates.[225] Other coordinatively

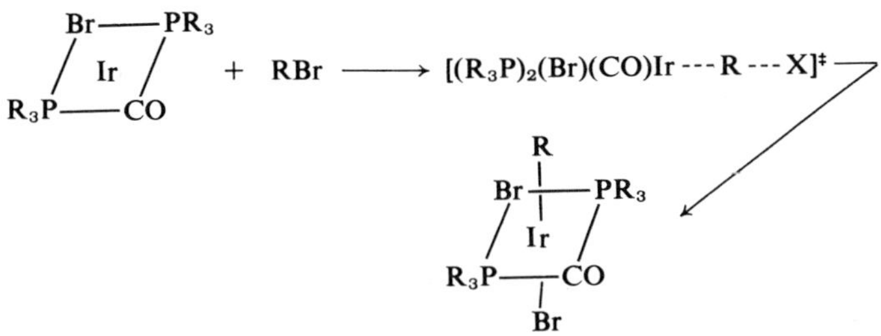

unsaturated complexes of neutral or anionic d^8 and d^{10} metal complexes show high nucleophilic reactivity toward carbon centers and a number of examples of S_N2 reactions with alkyl halides are known.[226] Thus, the reaction of the cobaloximine $Py(Co^I)_{Me}$ with alkyl halides proceeds by inversion of configuration.[227]

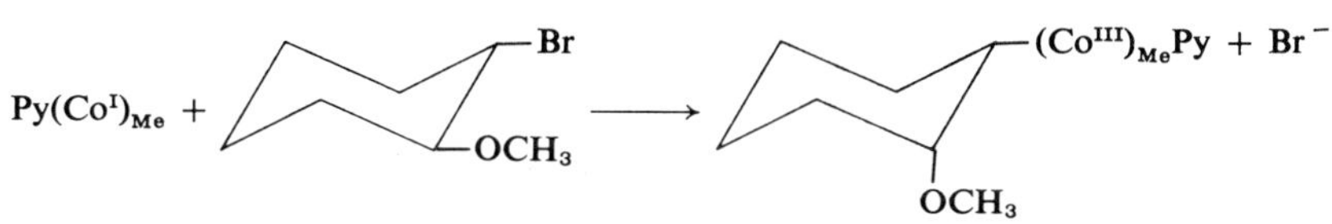

The formations of Grignard reagents and organolithium compounds are examples of the reduction of alkyl halides directly by metals. Partial loss of asymmetry during conversion of cyclopropyl halides has been attributed to the formation of alkyl radical intermediates,[228] since the cyclopropyl carbanion does not rearrange under these conditions, for example,

$$R\text{—}X \xrightarrow{M} [R\cdot + X:] \xrightarrow{M} RM + X:$$

It is not clear whether inner- or outer-sphere processes are involved since the reactions occurring on a metal surface offer only a limited number of probes. Recently the kinetics of the reaction of ethyl halides with atomically clean titanium surfaces have been examined,[229] but no conclusion was drawn about the initial transfer of halogen. Metal atoms generated in the gas phase by a method developed by Skell and co-workers should allow for more definitive studies.[230]

E. Metal-Catalyzed Addition Reactions

Redox reactions have been used effectively to generate free radicals for addition to unsaturated compounds. If the adduct radicals are capable of further redox reactions, a catalytic sequence is established. For example, the addition of chloroform to alkenes is catalyzed by copper and iron complexes and 1,1,3-trichloroalkanes are formed by the following catalytic reaction[231,232]:

$$Fe^{II} + HCCl_3 \longrightarrow Fe^{III}Cl + H\dot{C}Cl_2$$

$$H\dot{C}Cl_2 + CH_2{=}CHR \longrightarrow Cl_2CH{-}CH_2\dot{C}HR$$

$$Cl_2CH{-}CH_2\dot{C}HR + Fe^{III}Cl \longrightarrow Cl_2CH{-}CH_2CH(Cl){-}R + Fe^{II}, \ldots$$

The iron catalyst is continually recycled between iron(II) and iron(III) species as a result of ligand transfer of halogen. A copper–ethanolamine complex promotes a similar reaction between polyhaloalkanes and olefins.[233] The dichloromethyl radical is the chain-carrying species in this chain process,[231] in contrast to the trichloromethyl radical involved in the formation of the isomeric 1,1,1-trichloroalkane from the chain reaction initiated photochemically or with peroxides (see Chapter 13).

Copper chloride also catalyzes the addition of alkanesulfonyl chlorides to alkenes and vinyl monomers.[234] Addition to acetylenes[235] preferentially occurs by trans stereochemistry and constitutes a useful method for the preparation of β-chlorovinyl sulfones. In the absence of unsaturated compounds alkanesulfonyl chlorides are catalytically decomposed.[234]

$$RSO_2Cl + Cu^{I} \longrightarrow RSO_2\cdot + Cu^{II}Cl$$

$$RSO_2\cdot \rightleftharpoons R\cdot + SO_2$$

$$R\cdot + Cu^{II}Cl \longrightarrow RCl + Cu^{I}, \ldots$$

The latter proceeds by a mechanism which is similar to that of the copper-catalyzed chlorosulfonation of arenediazonium salts (vide infra).[236]

$$ArN_2^{+} + Cl^{-} + SO_2 \xrightarrow{Cu^{I}-Cu^{II}} ArSO_2Cl + N_2$$

The utilization of redox reactions between hydrogen peroxide or alkyl hydroperoxides and iron(II) in addition reactions has been discussed earlier. *N*-Chloramines and hydroxylamines are also reduced by iron(II) and titanium(III), and adducts of the amino radicals to alkenes can be observed

$$H_2NOH + Ti^{III} \longrightarrow H_2N\cdot + Ti^{IV}(OH)$$

directly by ESR.[237] Dialkylamino radicals produced by these redox systems form adducts according to the following steps[238]:

$$R_2NCl + Fe^{II} \longrightarrow R_2N\cdot + Fe^{III}Cl$$

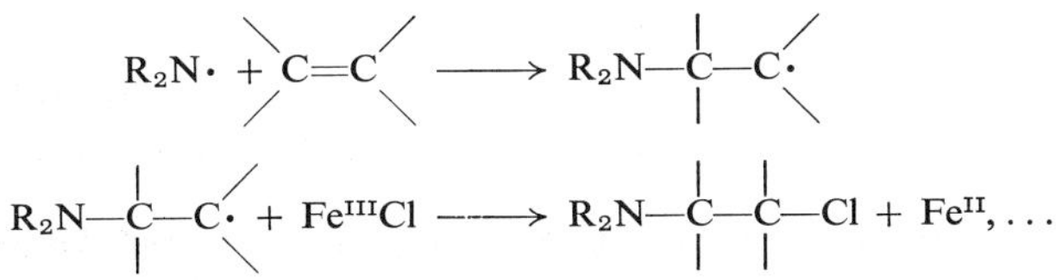

The reaction is reminiscent of the Hoffman–Leffler–Freytag reaction of *N*-chlorammonium ions.[239]

Homolytic aromatic aminations can be effected with a variety of dialkyl-*N*-chloramines and iron(II) and titanium(IV).[240] An interesting intramolecular amination has been effected with the chloramine below[241]:

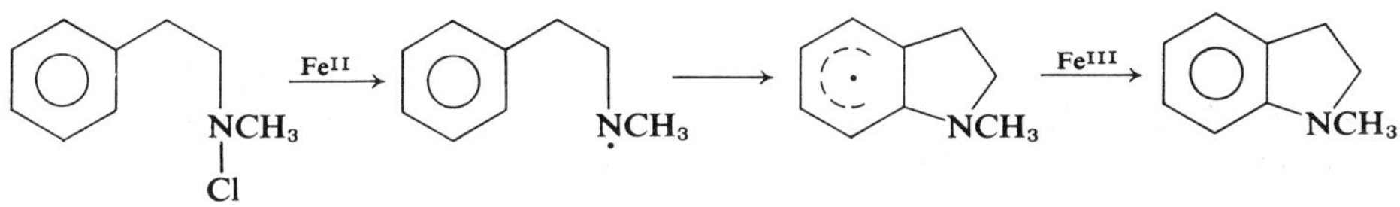

Oxaziridines also undergo similar reduction with metal salts to form amide-substituted alkyl radicals, for example,[242]

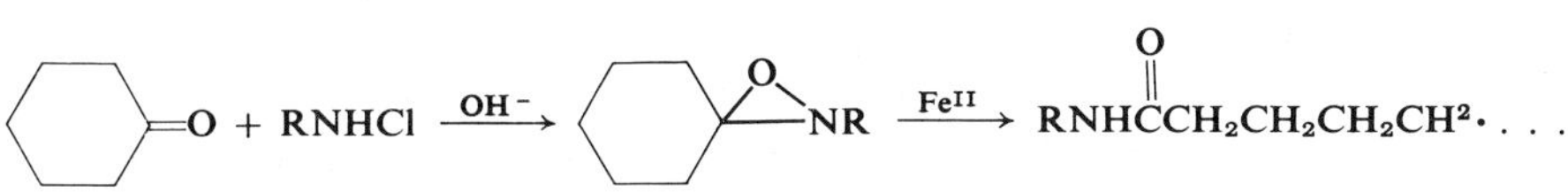

Azido radicals from the decomposition of hydrogen peroxide by iron(II) in the presence of azide can also be added to vinyl monomers.[243]

Aromatic hydrocarbons, particularly polycycles, are chlorinated by copper(II) chloride.[129] Halogen atom transfer has been suggested:

$$ArH + Cu^{II}Cl_2 \longrightarrow ArHCl\cdot + Cu^{I}Cl$$
$$ArHCl\cdot + Cu^{II}Cl_2 \longrightarrow ArCl + Cu^{I}Cl + HCl$$

Aryl iodides are formed in the presence of iodine or iodide salts.[244] This combination can be used to prepare α-chloro-β-iodoalkanes.[245]

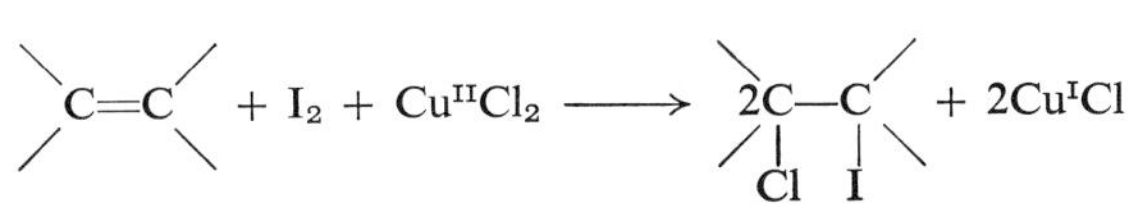

Chlorination of aliphatic hydrocarbons can also be effected by cupric chloride at high temperatures or photolytically.[246] Chlorine atoms derived from the homolysis of cupric chloride are probably responsible for attack on the organic substrate. The ambiguity in distinguishing between homolytic and

$$\mathrm{Cl\cdot + RH \longrightarrow R\cdot + HCl}$$
$$\mathrm{R\cdot + Cu^{II}Cl_2 \longrightarrow R{-}Cl + Cu^{I}Cl}$$

heterolytic mechanisms in halogenations carried out with metal halides such as $FeCl_3$ and $SbCl_5$ has been pointed out[247] and clarification is desirable.

F. Catalyzed Reactions of Arenediazonium Salts and Aromatic Compounds

Arenediazonium salts undergo a variety of radical chain reactions.[248] Various copper compounds have been used to promote the replacement of the diazo group. Classically, the Sandmeyer reaction was carried out in aqueous solution with stoichiometric amounts of cuprous halide necessitated by the precipitation of a diazo–copper complex.[249] When the reaction is carried out homogeneously in aqueous acetone solution, however, only catalytic amounts of copper salts are required.[250] Although the reaction is catalyzed by copper(I) species, copper(II) salts can be used since they are reduced by acetone in the presence of halide.[251]

$$\mathrm{2Cu^{II}Cl_2 + (CH_3)_2CO \longrightarrow 2Cu^{I}Cl + CH_3COCH_2Cl + HCl}$$

The one-equivalent reduction of the arenediazonium halides occurs readily as polarographic studies have shown[252] and can be induced by a mild reducing agent such as copper(I) or copper metal. The aryl radical resulting from the spontaneous fragmentation of the unstable arenediazo radical[253] subsequently undergoes ligand transfer oxidation by copper(II) halide to afford the Sandmeyer product and regenerate copper(I).[254]

$$\mathrm{ArN_2^+ + Cu^{I}Cl \longrightarrow ArN_2\cdot + Cu^{II}Cl}$$
$$\mathrm{ArN_2\cdot \longrightarrow Ar\cdot + N_2}$$
$$\mathrm{Ar\cdot + Cu^{II}Cl \longrightarrow ArCl + Cu^{I}, \ldots}$$

The rapid anation of copper(I) and copper(II), particularly in nonaqueous media, allows an adequate supply of halocopper(II) species to be maintained.

The Meerwein reaction results when the catalytic decomposition of the arenediazonium ion by copper complexes is carried out in the presence of unsaturated compounds. The aryl radicals are trapped as adducts which subsequently undergo ligand transfer oxidation.

$$\mathrm{Ar\cdot + {>}C{=}C{<} \longrightarrow Ar{-}\overset{|}{\underset{|}{C}}{-}\dot{C}{<}}$$
$$\mathrm{Ar{-}\overset{|}{\underset{|}{C}}{-}\dot{C}{<} + Cu^{II}Cl \longrightarrow Ar{-}\overset{|}{\underset{|}{C}}{-}\overset{|}{\underset{|}{C}}{-}Cl + Cu^{I}, \ldots}$$

The Meerwein reaction, thus, represents a homologation of the Sandmeyer reaction. Usually both types of products are formed, the relative amounts of which depend on the ease of addition of the aryl radical to the alkene and only to a limited extent on the selective oxidation of alkyl (adduct) radicals compared to aryl radicals.

Azobenzenes are formed by the decomposition of arenediazonium fluoroborates by copper powder or copper(I) in the absence of halides. A catalytic sequence involving aryl radicals in displacements with a diazocopper complex

$$\text{Ar}\cdot + \text{ArN}_2\text{Cu} \longrightarrow \text{ArN}_2\text{Ar} + \text{Cu}^{\text{I}}$$

has been proposed in nonaqueous media.[255] Substitution of arenes by aryl radicals is a more common route, and the Pschorr reaction is a useful intramolecular example.[256] Reduction is a common side reaction during the

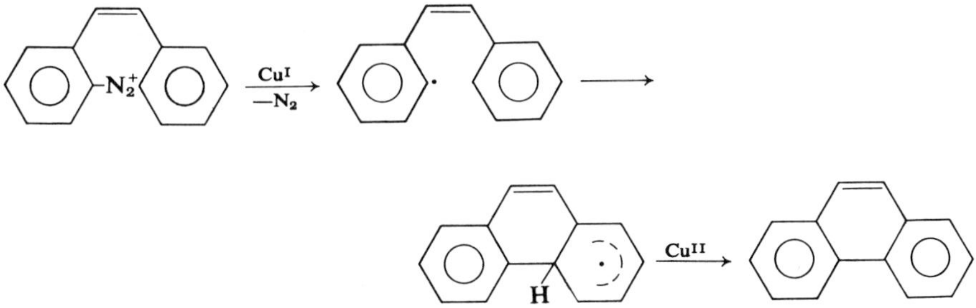

copper-catalyzed decomposition of arenediazonium salts, and results from hydrogen abstraction by aryl radical, particularly when an organic solvent or other hydrogen donors are employed.[248] The factors involved in the competition between arylation, cyclization, and reduction have been discussed.[257]

There are a number of other redox reactions of aromatic compounds which are promoted by metal complexes, particularly copper,[258] and they include the Ullman method of biaryl synthesis,[259] the oxidative decarboxylation of benzoic acids to phenols[260] and reductive decarboxylation to arenes,[261] the oxidative nitrosation of phenols,[262] the oxidative coupling of acetylenes,[263] and the autoxidation,[264] dimerization,[265] and polymerization[266] of phenols. The intermediacy of σ-bonded organocopper intermediate has been shown in some cases.[267] Evidence for or against the involvement of free radicals in these reactions is not compelling.[268] It is complicated by the possible formation of radical–metal complexes and may be obscured if attack (ligand transfer) of the radical on the organocopper intermediate is very rapid, as has been shown in a number of cases.

G. Redox Reactions of Grignard and Organometallic Reagents

Redox reactions of inorganic anions with metal complexes can proceed via a number of mechanisms involving electron transfer associated with substitution, dissociation, and/or displacement of ligands on the metal nucleus.[269]

Carbanions are generally unstable in protic media and under conditions in which they can be prepared they are highly associated with the metal counter-ion. Ion association and ion pairing must, therefore, be a part of any mechanism, and are implicit in the following discussion.

Outer-sphere electron transfers from carbanions are rare, although the oxidation of nitroalkane anions or enolate ions by copper(II) or iron(III)

$$(CH_3)_2CNO_2^- + Fe^{III} \longrightarrow (CH_3)_2CNO_2\cdot + Fe^{II}$$

may be examples,[270] since they are analogous to the autoxidation and oxidation of carbanions studied in depth by Russell and co-workers.[271] The dimeric product of the oxidation of nitroalkane anions does not arise via direct coupling of the radicals but by addition followed by oxidation:

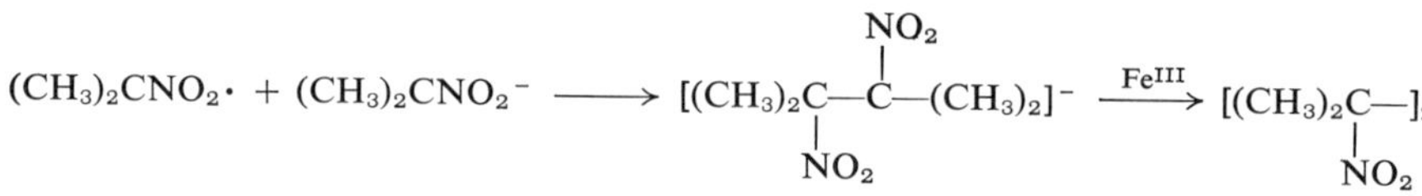

A similar step is a part of the mechanism proposed for the silver-induced nitration of nitroalkanes[272]:

$$(CH_3)_2CNO_2^- + Ag^I \longrightarrow (CH_3)_2CNO_2\cdot + Ag$$

$$(CH_3)_2CNO_2\cdot + NO_2^- \longrightarrow (CH_3)_2C(NO_2)_2\cdot^-$$

$$(CH_3)_2C(NO_2)_2\cdot^- + Ag^I \longrightarrow (CH_3)_2C(NO_2)_2 + Ag$$

An outer-sphere transfer may also be involved in the conjugate addition of dialkylcuprates to α,β-unsaturated carbonyl compounds, for example,[273]

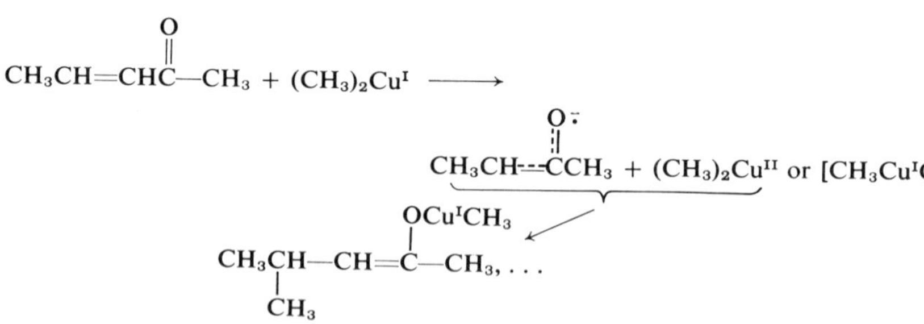

(123)

Alkylation at a carbon center by a homolytic process in this example is similar to the radical chain addition of alkyl boranes to the same substrate.[274] Furthermore, the subsequent reactions of the radical with the metal centers

(B in equation 124 and Cu in equation 123) are forms of ligand transfer processes. Bimolecular homolytic substitution reactions at metal centers are discussed more fully in Chapter 10.

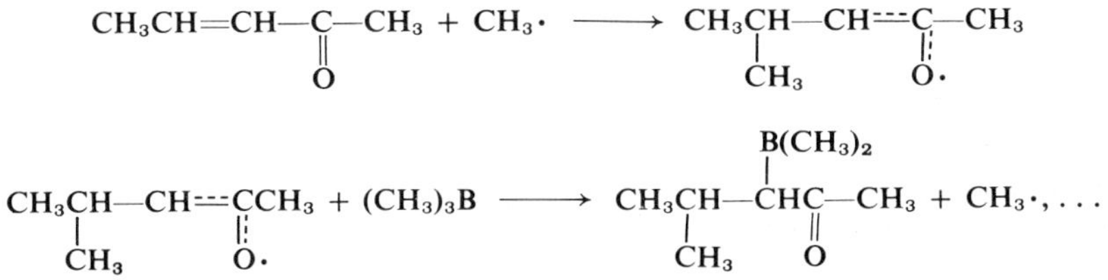

(124)

Association of the carbanion with the inner sphere of the metal nucleus should constitute a viable path for redox reactions.

$$R:^- + M^{n+} \longrightarrow R{-}M \longrightarrow R\cdot + M^{(n-1)+}$$

The homolysis of organometallic compounds, particularly those belonging to the main-group elements, has been reviewed.[275] Bond-dissociation energies for organometals of the transition series are generally undetermined. However, judging from the trend in the values known for homolysis of various alkyl metals of elements as far as the sixth period (Hg, Tl, Pb), it can be concluded that homolysis is not a readily accessible path for reactions carried out under relatively mild conditions, since it usually involves energies in excess of 30 kcal/mole. In addition, the tendency for organometals with methyl and alkyl derivatives containing no available β-hydrogen (including benzyl as in tetrabenzyltitanium) to be more stable than other analogous alkyls is not consistent with facile homolysis.[276] Thus, the simple trimethylsilylmethyl derivatives of iron (III and IV) and chromium (III and IV) as well as those of the related neopentyl metals have been prepared,[277] despite the fact that the trimethylsilylmethyl radical is appreciably stable and readily formed.[278] The stereospecific coupling of alkenyl groups during decomposition of organosilver and -copper compounds[279] is also indicative of the

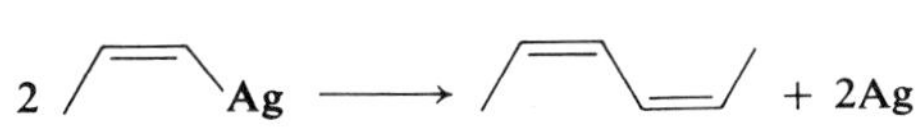

(125)

absence of free radicals as discrete intermediates, since vinylic radicals invert at rates greater than 10^{10} sec^{-1}. A variety of nonradical routes are available for the facile decomposition of organometals, including elimination, oxidative addition, and polynuclear processes. The stabilizing influence exerted by π bonding or polydentate ligands often observed in organometallic complexes is due less to the strengthening of the metal–alkyl bond than to occupation of the coordination sites around the metal necessary for elimination, since a variety of alkyl derivatives of substitution-inert complexes such as those of chromium(III), cobalt(III), and rhodium(III) are known.[280]

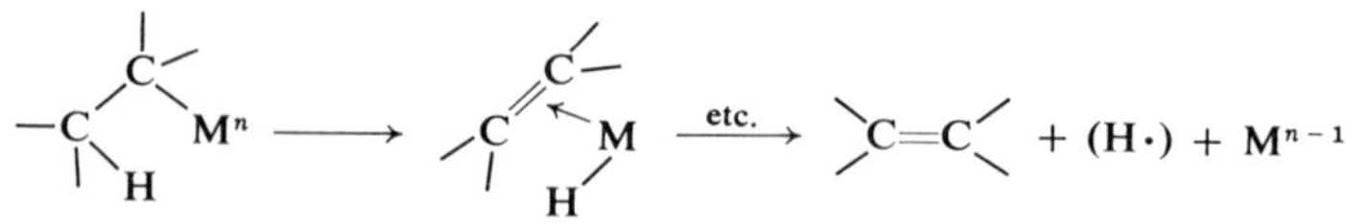

The situation described above for many organometallic complexes, however, is not unambiguous, since homolytic substitutions can occur by an addition–elimination mechanism involving changes in the inner sphere of the metal. For example, electron spin resonance studies[281] show that trialkylphosphines and phosphites form metastable phosphoranyl adducts with alkoxy and thiyl radicals, which undergo α and β homolysis as shown in equations 126 and 127. Homolytic substitution of organoboranes and other

$$R'O\cdot + PR_3 \longrightarrow [R'O\dot{P}R_3] \begin{cases} \xrightarrow{\alpha} R'OPR_2 + R\cdot & (126) \\ \xrightarrow{\beta} R'\cdot + OPR_3 & (127) \end{cases}$$

alkyl metals is also rapid, and the absence of adducts detectable by ESR[281] indicates that their lifetimes are exceedingly short or that the reaction is proceeding by direct homolytic displacement. In either case, the rise in the oxidation level of the metal (in the adduct or in the transition state) is relieved by homolysis of a ligand. A similar description applies to the reduction of alkylcopper intermediates in the oxidation of alkyl radicals by copper(II), the fragmentation of lead(III) species in oxidative decarboxylation by lead tetraacetate, and the metastability of thallium(II) carboxylates in the photolysis of thallium(III) derivatives discussed elsewhere in this chapter. Clearly, the metal, its oxidation state, and coordination are important factors to be considered in the homolysis of organometallic intermediates.[282] In this connection the applicability of the "rare gas rule" for the stability of organometallic intermediates is also noteworthy from the qualitative point of view.[283]

Grignard reagents, organolithium, and -aluminum compounds represent the readily available "carbanionoid" reagents, and a vast array of studies have been carried out with metal complexes in stoichiometric and catalytic reactions. The presence of free radical intermediates has often been inferred from the formation of dimeric products (in addition to alkanes and alkenes) and from hydrogen abstraction from the solvent.[284]

The reaction between Grignard reagents and alkyl halides in the presence of catalytic amounts of transition-metal complexes is of particular interest since there is extensive and varied evidence for the presence of alkyl radicals.[285] The coupling reaction 128 homogeneously catalyzed by silver salts has been

$$RMgX + R{-}X \xrightarrow{AgX} R_2 + MgX_2 \qquad (128)$$

reexamined in tetrahydrofuran, which has the advantage over the more commonly used diethyl ether as solvent since organometals are more stable in THF.[286] The rate of the coupling reaction is approximately first order in alkyl halide and the silver catalyst but is zero order in the Grignard reagent. The product-forming step involving decomposition of alkylsilver (equation 129)[286] is similar to that described in equation 125 for the more stable

$$\mathrm{RMgX + AgX \longrightarrow RAg + MgX_2}$$

$$\mathrm{2R{-}Ag \longrightarrow R_2 + 2Ag} \qquad (129)$$

$$\mathrm{Ag + R{-}X \longrightarrow R\cdot + AgX} \qquad (130)$$

$$\mathrm{R\cdot + Ag \longrightarrow R{-}Ag, \ldots} \qquad (131)$$

alkenylsilver[279] and does not involve free radical intermediates. Thus, when dissimilar alkyl groups are used (in the Grignard reagent and the alkyl halide), all three possible coupled products are formed, but only that alkyl group derived from the alkyl halide is selectively scavenged as an alkyl radical. Similarly, the catalytic reaction of *cis*-propenylmagnesium bromide and methyl bromide yields *cis*-butene-2, whereas the converse reaction between methylmagnesium bromide and *cis*-propenyl bromide leads to a mixture of *cis*- and *trans*-butene-2. According to the mechanism given in equations 129–

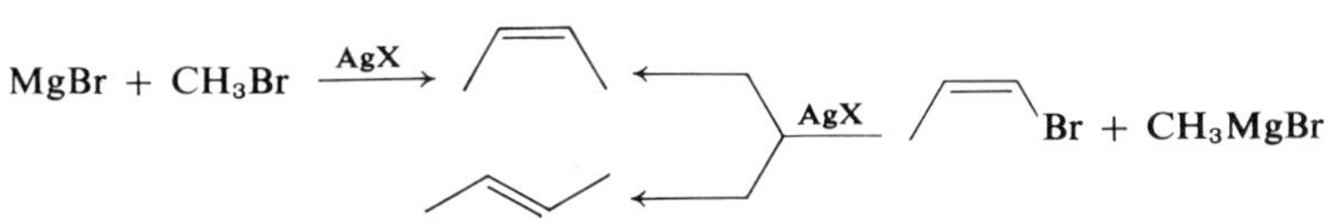

131, alkyl radicals are generated by the ligand transfer of halogen by reduced silver (equation 130). Thus, the relative reactivity of alkyl halides in the order: *t*-BuBr > *i*-PrBr > *n*-PrBr in a ratio of 20:3:1, is also similar to that in the analogous reduction by chromium(II) presented earlier.

The coupling reaction may proceed by mechanisms other than the redox process presented above. For example, the coupling reaction catalyzed by copper(I) proceeds by an entirely different mechanism.[287] Thus, the kinetics of coupling are first order each in copper(I), Grignard reagent, and alkyl halide. The reverse trend in the reactivity of various alkyl halides (*n*-PrBr > *i*-PrBr ≫ *t*-BuBr) suggests that a displacement process is involved, and there is no evidence for important radical intermediates. Catalysis by copper(I)

$$\mathrm{RMgX + Cu^{I}Br \longrightarrow RCu + MgXBr}$$

$$\mathrm{RCu + R'Br \longrightarrow R{-}R' + Cu^{I}Br, \ldots}$$

thus allows for specific cross-coupling to occur without mixing of the alkyl groups in the Grignard reagent and the alkyl halide.[288]

Other transition metal complexes are involved in a similar catalytic capacity to that described for silver and copper. It is, in most cases, not necessary to invoke homolysis of an organometallic intermediate to account for free radical intermediates except when the reaction is carried out under extreme conditions.[290] In those cases in which evidence for free radical intermediates from the Grignard component has been proposed, it is possible that they arise from side reactions involving the catalytic exchange of reactants.[276] Reduced forms of transition metals are also capable of dehydrogenating

$$\text{RMgX} + \text{R'X} \xrightleftharpoons{\text{MX}_n} \text{R—X} + \text{R'MgX}$$

ethers as well as alkanes, and attack on solvent per se is not necessarily indicative of free radical attack.[276,289]

REFERENCES

1. R. Stewart, *Oxidation Mechanisms*, Benjamin, New York, 1964.
2. K. B. Wiberg, Ed., *Oxidation in Organic Chemistry*, Academic Press, New York, 1965.
3. F. R. Mayo, Ed., *Oxidation of Organic Compounds*, *Advances in Chemistry Series 75–77*, American Chemical Society, Washington, D.C., 1968.
4. R. L. Augustine, Ed., *Reduction*, Dekker, New York, 1968.
5. R. L. Augustine, Ed., *Oxidation*, Dekker, Vol. I., New York, 1969; Vol. II., 1971.
6. L. J. Chinn, *Selection of Oxidants in Organic Synthesis*, Dekker, New York, 1971.
7. E.g., J. P. Collman, *Acc. Chem. Res.*, **1**, 137 (1968); J. Halpern, in *Homogeneous Catalysis*, *Advances in Chemistry Series 70*, American Chemical Society, Washington, D.C., 1968.
8. W. L. Reynolds and R. W. Lumry, *Mechanisms of Electron Transfer*, Ronald Press, New York, 1966; F. Basolo and R. G. Pearson, *Mechanisms of Inorganic Reactions*, Wiley, 1967, p. 454 ff; T. A. Turney, *Oxidation Mechanisms*, Butterworths, London, 1965.
9. H. Taube, *Adv. Inorg. Radiochem.*, **1**, 1 (1959); A. G. Sykes, *ibid.*, **10**, 153 (1967); J. Halpern, *Qt. Rev.*, **15**, 207 (1961).
10. B. M. Gordon, L. L. Williams, and N. Sutin, *J. Am. Chem. Soc.*, **83**, 2061 (1961).
11. R. A. Marcus, *J. Chem. Phys.*, **26**, 867, 872 (1957); **43**, 679 (1965); *J. Phys. Chem.*, **67**, 853 (1963); **72**, 891, 4249 (1968).
12. H. Taube and H. Myers, *J. Am. Chem. Soc.*, **76**, 2103 (1954).
13. P. R. Guenther and R. G. Linck, *J. Am. Chem. Soc.*, **91**, 3769 (1969); F. B. Baker, M. Orhanovic, and N. Sutin, *ibid.*, **89**, 722 (1967).
14. D. L. Ball and E. L. King, *J. Am. Chem. Soc.*, **80**, 1091 (1958); H. Diebler and H. Taube, *Inorg. Chem.*, **4**, 1029 (1965); J. H. Espenson, *ibid.*, **4**, 121 (1965); J. P. Candlin, J. Halpern, and D. L. Trimm, *J. Am. Chem. Soc.*, **86**, 1019 (1964); A. Adin and A. G. Sykes, *J. Chem. Soc.* (*A*), 354 (1968).
15. A. Haim, *Inorg. Chem.*, **7**, 1475 (1968).
16. W. A. Waters, *Chem. Soc.* (*London*) *Spec. Publ.*, **19**, 71 (1965); *Prog. Org. Chem.*, **5**, 12 (1961); *Mechanisms of Oxidation of Organic Compounds*, Methuen, London, 1964; *Pure Appl. Chem.*, **4**, 307 (1971); C. H. Bamford, A. D. Jenkins, and R.

Johnston, *Proc. Roy. Soc.*, **A239**, 214 (1957); J. S. Littler, *Chem. Soc.* (*London*) *Spec. Publ.*, **24**, 383 (1970).
17. E. Collinson, F. S. Dainton, B. Mile, S. Tazuke, and D. R. Smith, *Nature*, **198**, 26 (1963); *Discuss. Faraday Soc.*, **29**, 188 (1960); cf. also G. E. Adams and R. L. Wilson, *Trans. Faraday Soc.*, **65**, 2981 (1969).
18. E. T. Denisov, *Russ. Chem. Rev.*, **40**, 24 (1971).
19. F. P. Lossing and G. P. Semeluk, *Can. J. Chem.*, **48**, 955 (1970).
20. L. Eberson, *Acta Chem. Scand.*, **17**, 2004 (1963); K. Torssell, *Ark. Kemi*, **31**, 401 (1969).
21. J. K. Kochi, *Science*, **155**, 415 (1967); *Rec. Chem. Prog.*, **27**, 207 (1966).
22. J. K. Kochi, *J. Am. Chem. Soc.*, **85**, 1958 (1963); J. K. Kochi, A. Bemis, and C. L. Jenkins, *ibid.*, **90**, 4038, 4616 (1968).
23. J. K. Kochi and R. V. Subramanian, *J. Am. Chem. Soc.*, **87**, 1508 (1965); C. L. Jenkins and J. K. Kochi, *J. Org. Chem.*, **36**, 3095, 3103 (1971).
24. J. K. Kochi and H. E. Mains, *J. Org. Chem.*, **30**, 1862 (1965); J. K. Kochi and A. Bemis, *Tetrahedron*, **24**, 5099 (1968).
25. J. K. Kochi, *J. Am. Chem. Soc.*, **84**, 774, 3271 (1962).
26. H. C. Brown and M. Nakagawa, *J. Am. Chem. Soc.*, **77**, 3610, 3614 (1955); M. Silver, *J. Org. Chem.*, **28**, 1686 (1963); *J. Am. Chem. Soc.*, **83**, 3482 (1962); G. Fraser and H. Hoffman, *Chem. Commun.*, 561 (1967).
27. J. Ritter and P. Minieri, *J. Am. Chem. Soc.*, **70**, 4045, 4048 (1948).
28. L. Miller and A. K. Hoffman, *J. Am. Chem. Soc.*, **89**, 593 (1967).
29. R. Breslow, *Molecular Rearrangements*, P. deMayo, Ed., Interscience, New York, 1963, p. 254 ff.
30. S. Winstein, C. Lindegren, H. Marshall, and L. Ingraham, *J. Am. Chem. Soc.*, **75**, 147 (1953); H. C. Brown, R. Bernheimer, C. Kim, and S. Schepple, *ibid.*, **89**, 370 (1967); **87**, 2137 (1965); G. A. Olah, M. Comisarow, E. Namenworth, and B. Ramsey, *ibid.*, **89**, 711 (1967).
31. W. H. Saunders, S. Asperger, and D. Edison, *J. Am. Chem. Soc.*, **80**, 2423 (1958); C. C. Lee, G. Slater, and J. Spinks, *Can. J. Chem.*, **35**, 1417 (1957); **39**, 1510 (1961); *Tetrahedron*, **7**, 206 (1959); L. Eberson, J. Petrovich, R. Baird, D. Dyckes, and S. Winstein, *J. Am. Chem. Soc.*, **87**, 3504 (1965); M. Bentley and M. Dewar, *ibid.*, **90**, 1075 (1968).
32. J. Nordlander and W. Deadman, *J. Am. Chem. Soc.*, **90**, 1590 (1968); R. Jablonski and E. Snyder, *Tetrahedron Letters*, 1103 (1968); see also G. A. Olah and R. D. Porter, *J. Am. Chem. Soc.*, **93**, 6877 (1971).
33. S. Winstein and D. Trifan, *J. Am. Chem. Soc.*, **74**, 1147, 1154 (1952); E. J. Corey, J. Casanova, P. Vatakencherry, and R. Winter, *ibid.*, **85**, 169 (1963); R. Roberts, W. Bennett, and R. Armstrong, *ibid.*, **72**, 3121, 3329 (1950).
34. J. K. Kochi and R. V. Subramanian, *Inorg. Chem.*, **4**, 1527 (1965); *J. Am. Chem. Soc.*, **87**, 4855 (1965).
35. D. J. Carlsson and K. U. Ingold, *J. Am. Chem. Soc.*, **90**, 7047 (1968); R. C. Lamb, P. W. Ayers, and M. K. Toney, *ibid.*, **85**, 3483 (1963); C. Walling, J. H. Cooley, A. A. Ponaras, and E. J. Racah, *ibid.*, **88**, 5363 (1966); G. F. Garst and F. E. Barton, *Tetrahedron Letters*, 587 (1968); J. Eastham, *Chem. Commun.*, 139 (1969); J. K. Kochi and J. W. Powers, *J. Am. Chem. Soc.*, **92**, 137 (1970).
36. C. L. Jenkins and J. K. Kochi, *J. Am. Chem. Soc.*, **94**, 843 (1972).
37. A. Cockerill, S. Rottschaeffer, and W. H. Saunders, *J. Am. Chem. Soc.*, **89**, 901, 4985 (1967); **90**, 1775 (1965); L. Steffa and E. Thornton, *ibid.*, **89**, 6149 (1967); A. Streitwieser and J. Hammons, *Prog. Phys. Org. Chem.*, **5**, 60 (1965); D. Kevill

and J. Dorsey, *Chem. Ind.* (*London*), 2174 (1967); V. J. Shiner, Jr., W. Buddenbaum, B. Murr, and G. Lamaty, *J. Am. Chem. Soc.*, **90**, 418 (1968); M. G. Jones and J. L. Coke, *ibid.*, **91**, 4284 (1969); C. J. Kim and H. C. Brown, *ibid.*, **91**, 4289 (1969); P. v. R. Schleyer and C. L. Lancelot, *ibid.*, **91**, 4297 (1969).

38. W. H. Saunders, *Chemistry of Alkenes*, S. Patai, Ed., Interscience, New York, 1964, p. 149 ff; E. Thornton, *Solvolysis Mechanisms*, Ronald Press, New York, 1964, p. 207 ff.
39. F. R. Jensen, R. Ouellette, G. K. Knutson, D. Babbe, and R. Hartgerink, *Trans. N.Y. Acad. Sci., Ser.* (*II*), **30**, 751 (1968).
40. A. Streitwieser and W. D. Schaeffer, *J. Am. Chem. Soc.*, **79**, 6233 (1957); J. H. Bayless, F. D. Mendicino, and L. F. Friedman, *ibid.*, **87**, 5790 (1965); A. Streitwieser and W. D. Schaeffer, *ibid.*, **79**, 2888 (1957).
41. A. Diaz, I. L. Reich, and S. Winstein, *J. Am. Chem. Soc.*, **91**, 5635, 5637 (1969); P. C. Myhre and K. S. Brown, *ibid.*, **91**, 5639, 5641 (1969).
42. G. A. Dafforn and A. Streitwieser, *Tetrahedron Letters*, 3159 (1970).
43. G. J. Karabatsos, N. Hsi, and S. Meyerson, *J. Am. Chem. Soc.*, **92**, 621 (1970); G. J. Karabatsos, C. E. Orzech, J. L. Fry, and S. Meyerson, *ibid.*, **92**, 606 (1970); C. C. Lee and J. E. Kruger, *Tetrahedron*, **23**, 2539 (1967).
44. S. Winstein, B. Appel, R. Baker, and A. Diaz, *Chem. Soc. Spec. Publ.*, **19**, 109 (1965); see also C. A. Bunton, T. W. Del Pesco, A. M. Dunlop, and K.-U. Yang, *J. Org. Chem.*, **36**, 887 (1971).
45. F. R. Jensen and R. J. Ouellette, *J. Am. Chem. Soc.*, **83**, 4477, 4478 (1961); **85**, 363 (1963); see also G. A. Olah and P. R. Clifford, *ibid.*, **93**, 1261 (1971).
46. S. G. Smith and J. P. Petrovich, *J. Org. Chem.*, **30**, 2882 (1965); H. L. Goering et al., *J. Am. Chem. Soc.*, **92**, 7401 (1970).
47. C. L. Jenkins and J. K. Kochi, *J. Am. Chem. Soc.*, **94**, 856 (1972).
48. Cf. M. Hanack and A. J. Schneider, *Angew. Chem., Int. Ed.*, **6**, 666 (1967); R. Moss and F. Shulman, *Tetrahedron*, **24**, 2881 (1968); *J. Am. Chem. Soc.*, **90**, 2731 (1968); Z. Majerski, S. Borcic, and D. E. Sunko, *Tetrahedron*, **25**, 301 (1969); K. L. Servis and J. D. Roberts, *J. Am. Chem. Soc.*, **87**, 1331 (1965); P. v. R. Schleyer and G. W. Van Dine, *ibid.*, **88**, 2321 (1966); K. L. Servis and J. D. Roberts, *Tetrahedron Letters*, 1369 (1967); H. G. Richey and J. M. Richey, *J. Am. Chem. Soc.*, **88**, 4971 (1966). A common intermediate is not necessarily implied. See I. Lillien and L. Handloser, *ibid.*, **93**, 1682 (1971); *J. Org. Chem.*, **33**, 3841 (1968); G. A. Olah and C. H. Lin, *J. Am. Chem. Soc.*, **90**, 6468 (1968).
49. R. P. Eswein, E. S. Howald, R. A. Howald, and D. P. Keeton, *J. Inorg. Nucl. Chem.*, **29**, 437 (1967); see also R. D. Willet et al., *Inorg. Chem.*, **6**, 1666, 1885 (1967); C. Furlani and G. Morpurgo, *Theor. Chim. Acta*, **1**, 102 (1963); W. E. Hatfield and R. Whyman, *Transition Metal Chem.*, **5**, 47 (1969); F. Gutmann, *Coord. Chem. Rev.*, **2**, 239 (1967); D. W. Meek, W. E. Hatfield, R. S. Drago, and T. S. Piper, *Inorg. Chem.*, **3**, 841, 1637 (1964); D. Culpin, P. Day, R. Edwards, and R. J. P. Williams, *Chem. Commun.*, 450 (1965).
50. J. Bjerrum, *Maanedsbad*, **26**, 24 (1945); S. E. Manahan and R. T. Iwamoto, *Inorg. Chem.*, **4**, 1409 (1965); see also C. C. Hinckley, *ibid.*, **7**, 396 (1968); R. Meyers and R. D. Willet, *J. Inorg. Nucl. Chem.*, **29**, 1546 (1967).
51. J. K. Kochi and D. M. Mog, *J. Am. Chem. Soc.*, **87**, 522 (1965); J. K. Kochi and F. F. Rust, *ibid.*, **84**, 2785 (1962).
52. R. G. Pearson, *Science*, **151**, 172 (1966); *J. Am. Chem. Soc.*, **85**, 3533 (1963); S. Arhland, J. Chatt, and N. R. Davies, *Qt. Rev.* (*London*), **12**, 265 (1958); G. Klopman, *J. Am. Chem. Soc.*, **90**, 223 (1968).
53. J. K. Kochi and P. M. Mocadlo, *J. Org. Chem.*, **30**, 1134 (1965).

54. J. K. Kochi and J. W. Powers, *J. Am. Chem. Soc.*, **92**, 137 (1970).
55. R. G. Coombs, M. D. Johnson, and N. Winterton, *J. Chem. Soc.*, 7029 (1965); 177 (1966); *Chem. Commun.*, 251 (1965); R. P. A. Sneeden and H. P. Throndsen, *ibid.*, 509 (1965); H. H. Zeiss and R. P. A. Sneeden, *Angew. Chem., Int. Ed.*, **6**, 435 (1967); see also W. Schmidt, J. H. Swinehart, and H. Taube, *J. Am. Chem. Soc.*, **93**, 1117 (1971); M. Ardon, K. Woolmington, and A. Pernick, *Inorg. Chem.*, **10**, 2812 (1971).
56. M. Eigen and R. G. Wilkins, in *Mechanisms of Inorganic Reactions, Advances in Chemistry Series 49*, American Chemical Society, Washington, D.C., 1965, p. 55 ff.
57. G. M. Whitesides and J. San Fillippo, Jr., *J. Am. Chem. Soc.*, **92**, 6611 (1970); V. M. A. Chambers, W. R. Jackson, and G. W. Young, *Chem. Commun.*, 1275 (1970).
58. E. Muetterties, Ed., *Transition Metal Hydrides*, Dekker, New York, 1971; J. E. Bercaw, R. H. Marvich, and H. H. Brintzinger, *J. Am. Chem. Soc.*, **93**, 2045, 2046 (1971): **94**, 1219 (1972); J. Halpern, *Discuss. Faraday Soc.*, **46**, 9 (1968).
59. G. C. Bond, *Catalysis by Metals*, Academic Press, New York, 1962; J. J. Rooney, *Chem. Britain*, 242 (1966).
60. E.g., J. E. French and H. Taube, *J. Am. Chem. Soc.*, **91**, 6951 (1969); M. Z. Hoffman and M. Simic, *ibid.*, **92**, 5533 (1970).
61. J. K. Kochi and F. F. Rust, *J. Am. Chem. Soc.*, **83**, 2017 (1961).
62. J. K. Kochi and D. D. Davis, *J. Am. Chem. Soc.*, **86**, 5264 (1964).
63. A. Van den Bergen and B. O. West, *Chem. Commun.*, 52 (1971); XIV Int. Conf. Coord. Chem., Toronto, June 22, 1972, p. 196 absts.; cf. G. Anes et al., *ibid.*, 851 (1971).
64. G. N. Schrauzer, *Acc. Chem. Res.*, **2**, 72 (1969); N. A. Tokel, V. Katovic, K. Farmery, C. B. Anderson, and D. H. Busch, *J. Am. Chem. Soc.*, **92**, 400 (1970); E. I. Stiefel, L. E. Bennett, Z. Dori, T. H. Crawford, C. Simo, and H. B. Gray, *Inorg. Chem.*, **9**, 281 (1970); F. V. Lovecchio and D. H. Busch, 163rd ACS Meeting Boston, Mass., April 1972, p. INORG-147 absts.
65. H. J. M. Bartelink, H. K. Ostendorf, B. C. Roest, and H. A. J. Schepers, *Chem. Commun.*, 879 (1971); cf. G. Henrici-Olive and S. Olive, *ibid.*, 1482 (1969).
66. D. R. Gee, K. E. Russell, and J. K. S. Wan, *Can. J. Chem.*, **49**, 160 (1971); **48**, 2740 (1970); *J. Chem. Phys.*, **53**, 847 (1970).
67. N. Hirota, in *Radical-Ions*, E. T. Kaiser and L. Kevan, Eds., Interscience, New York, 1968, p. 35 ff.; K. S. Chen, S. W. Mao, K. Nakamura, and N. Hirota, *J. Am. Chem. Soc.*, **93**, 6004 (1971); P. Clopath and A. V. Zelewsky, *Chem. Commun.*, 47 (1971); N. M. Atherton, *Pure Appl. Chem.*, **4**, 469 (1971); H. Strehlow, *ibid.*, **4**, 525 (1971).
68. P. R. Eaton, *Inorg. Chem.*, **3**, 1268 (1964); B. M. Hoffman and T. B. Eames, *J. Am. Chem. Soc.*, **91**, 5168 (1969); K. M. Kuwata and Y. Shimizu, *Bull. Chem. Soc. Japan*, **42**, 864 (1969); D. G. Brown, T. Maier, and R. S. Drago, *Inorg. Chem.*, **10**, 2804 (1971); F. Leh and J. K. S. Wan, *Bull. Chem. Soc. Japan*, **45**, 308 (1972); *Can. J. Chem.*, **50**, 999 (1972); E. G. Janzen, *Anal. Chem.*, **44**, 113R (1972).
69. A. Tkac, K. Vesely, and L. Omelka, *J. Phys. Chem.*, **75**, 2575, 3580 (1971); see, however, K. U. Ingold, *J. Phys. Chem.*, **76**, 1385 (1972).
70. Y. Shimizu, T. Shiga, and K. Kuwata, *J. Phys. Chem.*, **74**, 2929 (1970).
71. J. H. Baxendale, *Adv. Catal.*, **4**, 31 (1952). For a review of free radical and electrophilic hydroxylation, see D. F. Sangster, in *The Chemistry of the Hydroxyl Group*, Part I, S. Patai, Ed., Wiley, London, 1971, p. 51; N. Uri, *Chem. Rev.*, **50**, 375 (1952); J. Weiss, *Adv. Catal.*, **4**, 343 (1952).
72. R. O. C. Norman and J. R. Lindsay-Smith, in *Oxidases and Related Redox Systems*,

Vol. 1, T. E. King, H. S. Mason, and M. Morrison, Eds., Wiley, New York, 1965, p. 131 ff.

73. F. Haber and J. Weiss, *Proc. Roy. Soc.*, **A147**, 332 (1934).
74. R. Hiatt, T. Mill, and F. R. Mayo, *J. Org. Chem.*, **33**, 1416 (1968); R. Hiatt, T. Mill, K. C. Irvin, and J. K. Castleman, *ibid.*, **33**, 1421 (1968); **33**, 1428 (1968); R. Hiatt and K. C. Irvin, *ibid.*, **33**, 1436 (1968); R. Hiatt, K. C. Irvin, and C. W. Gould, *ibid.*, **33**, 1430 (1968).
75. For a review see G. Sosnovsky and D. J. Rawlinson, *Organic Peroxides*, Vol. II, D. Swern, Ed., Interscience, New York, 1971, p. 269 ff.; Vol. III, 1972, p. 141 ff.
76. J. H. Merz and W. A. Waters, *J. Chem. Soc.*, 2427, S15 (1949); *Discuss. Faraday Soc.*, **2**, 179 (1947); I. M. Kolthoff and A. I. Medalia, *J. Am. Chem. Soc.*, **71**, 3777 (1949); D. L. Ingles, *Austral. J. Chem.*, **25**, 87, 97, 105 (1972).
77. C. Walling and S. Kato, *J. Am. Chem. Soc.*, **93**, 4275 (1971).
78. N. A. Milas, *J. Am. Chem. Soc.*, **59**, 543, 2342, 2345 (1937); H. Loebl, G. Stein, and J. Weiss, *J. Chem. Soc.*, 2074 (1949); *Nature*, **161**, 650 (1948); **166**, 1104 (1950).
79. D. D. Coffman, R. Cramer, and W. Mochel, *J. Am. Chem. Soc.*, **80**, 2882 (1958).
80. D. D. Coffman, E. L. Jenner, and R. D. Lipscomb, *J. Am. Chem. Soc.*, **80**, 2864 (1958).
81. M. S. Kharasch, F. S. Arimoto, and W. Nudenburg, *J. Org. Chem.*, **16**, 1556 (1951); **19**, 1977 (1954); D. D. Coffman and E. L. Jenner, *J. Am. Chem. Soc.*, **80**, 2872 (1958).
82. J. H. Baxendale, M. G. Evans, and G. S. Park, *Trans. Faraday Soc.*, **42**, 155 (1946); W. T. Dixon and R. O. C. Norman, *J. Chem. Soc.*, 3119 (1963); K. Gunther, W. G. Filby, and K. Eiben, *Tetrahedron Letters*, 251 (1971).
83. R. O. C. Norman, *Chem. Soc. (London) Spec. Publ.*, **24**, 117 (1970).
84. S. Baer and G. Stein, *J. Chem. Soc.*, 3176 (1953); E. Saito and B. H. J. Bielski, *J. Am. Chem. Soc.*, **83**, 4467 (1961).
85. H. Sigel, C. Flierl, and R. Gnesser, *J. Am. Chem. Soc.*, **91**, 1061 (1969), and following paper; see also A. Ya Syeher and Bui Ngok Tkho, *Russ. J. Phys. Chem.*, **44**, 1373 (1970).
86. C. F. Wells and D. Mays, *J. Chem. Soc. (A)*, 665 (1968).
87. J. G. Baxendale and C. F. Wells, *Trans. Faraday Soc.*, **53**, 80 (1957); C. F. Wells and M. Husain, *ibid.*, **67**, 760 (1971).
88. C. F. Wells and D. Mays, *Inorg. Nucl. Chem. Letters*, **4**, 43 (1968); **5**, 9 (1969).
89. W. A. Waters, *Discuss. Faraday Soc.*, **46**, 137, 158 (1968).
90. H. L. Roberts and W. R. Symes, *J. Chem. Soc. (A)*, 1450 (1968).
91. J. J. Lewis, D. H. Richardson, and D. A. Salter, *J. Chem. Soc.*, 2434 (1963).
92. For a general review see R. Criegee, in *Oxidations in Organic Chemistry*, K. Wiberg, Ed., Academic Press, New York, 1965, p. 302; R. Criegee, H. Pilz, and H. Flygare, *Ber.*, **72**, 1799 (1939).
93. D. Benson and L. H. Sutcliffe, *Trans. Faraday Soc.*, **55**, 2107 (1959).
94. P. D. Bartlett and P. Gunther, *J. Am. Chem. Soc.*, **88**, 3288 (1966).
95. P. D. Bartlett and G. Guaraldi, *J. Am. Chem. Soc.*, **89**, 4799 (1967); R. Sheldon, unpublished results.
96. W. T. Dixon and R. O. C. Norman, *Nature*, **196**, 891 (1962).
97. R. A. Sheldon and J. K. Kochi, *J. Am. Chem. Soc.*, **90**, 6688 (1968).
98. J. K. Kochi, *Tetrahedron*, **18**, 483 (1962).
99. C. Walling and A. A. Zavitsas, *J. Am. Chem. Soc.*, **85**, 2084 (1963).
100. B. G. Balkov, L. P. Skibida, and Z. K. Maizus, *Izv. Akad. Nauk SSSR*, **7**, 83, 1470, 1475 (1969); H. Berger and A. F. Bickel, *Trans. Faraday Soc.*, **57**, 1325 (1961);

H. Kropf, *Ann.*, **631**, 111 (1960); W. H. Richardson, *J. Am. Chem. Soc.*, **88**, 975 (1966).

101. M. S. Kharasch and A. Fono, *J. Org. Chem.*, **24**, 72 (1959); D. D. Coffman and H. N. Cripps, *J. Am. Chem. Soc.*, **80**, 2877, 2880 (1950).
102. J. W. L. Fordham and H. L. Williams, *J. Am. Chem. Soc.*, **72**, 4465 (1950); **73**, 1634 (1951); R. J. Orr and H. L. Williams, *Can. J. Chem.*, **30**, 985 (1952); F. Minisci, M. Cecere, R. Galli, and A. Selva, *Org. Prep. Proc.*, **1**, 11 (1969).
103. M. S. Kharasch and G. Sosnovsky, *J. Org. Chem.*, **23**, 1322 (1958).
104. M. S. Kharasch and W. Nudenberg, *J. Org. Chem.*, **19**, 1921 (1954).
105. J. K. Kochi and F. F. Rust, *J. Am. Chem. Soc.*, **84**, 3946 (1962).
106. M. S. Kharasch and A. Fono, *J. Org. Chem.*, **23**, 325 (1958).
107. W. H. Richardson, *J. Am. Chem. Soc.*, **87**, 247 (1965); *J. Org. Chem.*, **30**, 2804 (1965).
108. I. I. Chuev, V. A. Shushanov, M. H. Shchennikova, and G. A. Abakumov, *Kinet. Katal.*, **10**, 75 (1969); **11**, 426 (1970); M. K. Shchennikova, E. A. Kuz'mina, V. A. Shushanov, and G. A. Abakunov, *Dokl. Akad. Nauk SSSR*, **164**, 868 (1965).
109. M. H. Dean and G. Skirrow, *Trans. Faraday Soc.*, **54**, 849 (1958).
110. A. F. Bickel and E. C. Kooyman, *J. Chem. Soc.*, 3211 (1953).
111. Y. Kamiya, S. Beaton, A. Lafortune, and K. U. Ingold, *Can. J. Chem.*, **41**, 2020 (1963).
112. A. E. Woodward and R. B. Mesrobian, *J. Am. Chem. Soc.*, **75**, 6189 (1953); A. V. Tobolsky, D. J. Metz, and R. B. Mesrobian, *ibid.*, **72**, 1942 (1950).
113. E. J. Y. Scott, *J. Phys. Chem.*, **74**, 1174 (1970).
114. M. H. Shchennikova, E. A. Kuz′mina, O. D. Gernet, and L. P. Savinova, *Kinet. Katal.*, 467 (1968); M. H. Shchennikova, I. I. Chuev, and A. K. Meteles, *ibid.*, 847 (1968); A. K. Metelev, V. A. Shushanov, M. K. Shchennikova, and V. A. Guryanova, *ibid.*, **9**, 941 (1968).
115. E. Dyer, K. R. Carle, and D. E. Weiman, *J. Org. Chem.*, **23**, 1464 (1958).
116. M. H. Shchennikova, E. A. Kuz′mina, A. J. Kalinin, and A. N. Kornilina, *Kinet. Katal.*, 641 (1968).
117. J. A. Sharp, *J. Chem. Soc.*, 2026 (1957).
118. E. T. Denisov and N. M. Emanuel, *Russ. Chem. Rev.* (*Engl. Transl.*), 645 (1960).
119. N. M. Emanuel, E. T. Denisov, and Z. K. Maizus, *Liquid Phase Oxidation of Hydrocarbons*, Plenum Press, New York, 1967, Chap. VI.
120. M. S. Kharasch, W. Nudenberg, and G. J. Mantell, *J. Org. Chem.*, **16**, 524 (1951).
121. L. Boguslavskaya, *Russ. Chem. Rev.* (*Engl. Transl.*), **34**, 503 (1965); G. Sosnovsky and D. J. Rawlinson, reference 75, p. 153 ff.
122. V. P. Kurkov, J. Z. Pasky, and J. B. Lavigne, *J. Am. Chem. Soc.*, **90**, 4743 (1968). See also J. E. Lyons, *Chem. Commun.*, 562 (1971); L. W. Fine, M. Grayson, and V. H. Suggs, *J. Organomet. Chem.*, **8**, 219 (1970); E. W. Stern, *Chem. Commun.*, 736 (1970).
123. A. Fuji, R. Ugo, F. Fox, A. Pasini, and S. Cerini, *J. Organomet. Chem.*, **26**, 417 (1970).
124. R. A. Sheldon and J. K. Kochi, *Oxid. Combust. Rev.*, in press.
125. S. Havel, *Sb. Ved. Pr.*, *Vys. Sk.*, Pardubice **15(1)**, 95 (1967); *Chem. Abstr.*, **67**, 9336 (1967); C. E. H. Bawn and J. B. Williamson, *Trans. Faraday Soc.*, **47**, 735 (1951); M. J. Kagan and G. D. Lubarsky, *J. Phys. Chem.*, **39**, 837 (1935).
126. E. Koubek and J. O. Edwards, *J. Inorg. Nucl. Chem.*, **25**, 1401 (1963); cf. D. D. Tanner and S. A. Osman, *J. Am. Chem. Soc.*, **90**, 6572 (1968); L. Dogliotti and

E. Hayon, *J. Phys. Chem.*, **71**, 2511 (1971); W. Kern and R. Schultz, *Makromol. Chem.*, **13**, 210 (1954).

127. C. Walling, *Pure Appl. Chem.*, **15**, 69 (1967); C. Walling and P. J. Wagner, *J. Am. Chem. Soc.*, **86**, 3368 (1964); J. K. Kochi and P. J. Krusic, *Chem. Soc. (London) Spec. Publ.*, **24**, 152 (1970).
128. G. Sosnovsky and D. J. Rawlinson, *Organic Peroxides*, D. Swern, Ed., Wiley, New York, 1970, Chaps. X and XI; *Synthesis*, 1 (1972).
129. D. C. Nonhebel, *Chem. Soc. (London) Spec. Publ.*, **24**, 409 (1970).
130. R. Hiatt, reference 75, p. 799 ff.
131. M. E. Kurz and P. Kovacic, *J. Org. Chem.*, **33**, 1950 (1968); M. E. Kurz, P. Kovacic, A. K. Bose, and I. Kugajewsky, *J. Am. Chem. Soc.*, **90**, 1818 (1968); J. Saltiel and H. C. Curtis, *ibid.*, **93**, 2056 (1971).
132. R. G. R. Bacon, *Chem. Ind. (London)*, 19 (1962).
133. C. E. H. Bawn and J. A. Sharp, *J. Chem. Soc.*, 1854, 1866 (1957).
134. S. S. Lande, C. Falk, and J. K. Kochi, *J. Inorg. Nucl. Chem.*, **33**, 4101 (1971).
135. T. Bernath and J. K. Kochi, unpublished observations.
136. R. M. Dessau, *J. Am. Chem. Soc.*, **92**, 6356 (1970); A. Ledwith, *Acc. Chem. Res.*, **5**, 133 (1972).
137. F. A. Bell, R. A. Crellin, H. Fujii, and A. Ledwith, *Chem. Commun.*, 251 (1969).
138. R. M. Moriarty, in *Selective Organic Transformations*, B. S. Thyagarajan, Ed., Interscience, New York, 1972.
139. D. L. Allara, B. C. Gilbert, and R. O. C. Norman, *Chem. Commun.*, 319 (1965); D. H. Iles and A. Ledwith, *ibid.*, 498 (1968); cf. also C.-H. Wang, C. C. Sizman, and K. Stevenson, *J. Org. Chem.*, **35**, 2045 (1970).
140. R. E. Partch, *J. Am. Chem. Soc.*, **89**, 3662 (1967).
141. P. J. Andrulis, M. J. S. Dewar, R. Dietz, and R. L. Hunt, *J. Am. Chem. Soc.*, **88**, 5473 (1966); T. Aratani and M. J. S. Dewar, *ibid.*, **88**, 5479 (1966); P. J. Andrulis and M. J. S. Dewar, *ibid.*, **88**, 5483 (1966).
142. H. J. den Hertog and E. C. Kooyman, *J. Catal.*, **6**, 347, 357 (1966); R. van Helden and E. C. Kooyman, *Rec. Trav. Chim.*, **80**, 57 (1961).
143. R. E. van der Ploeg, R. W. de Korte, and E. C. Kooyman, *J. Catal.*, **10**, 52 (1968).
144. J. B. Bush and H. Finkbeiner, *J. Am. Chem. Soc.*, **90**, 5903 (1969); E. I. Heiba, R. M. Dessau, and W. J. Koehl, *ibid.*, **90**, 2706, 5905 (1969).
145. J. R. Gilmore and J. M. Mellor, *Chem. Commun.*, 507 (1970).
146. E. I. Heiba, R. M. Dessau, and W. J. Koehl, *J. Am. Chem. Soc.*, **91**, 138 (1969).
147. J. M. Anderson and J. K. Kochi, *J. Am. Chem. Soc.*, **92**, 2450 (1970).
148. E. I. Heiba and R. M. Dessau, *J. Am. Chem. Soc.*, **93**, 995 (1971); see also E. I. Heiba and R. M. Dessau, *ibid.*, **93**, 524 (1971).
149. T. A. Copper, A. A. Clifford, D. J. Mills, and W. A. Waters, *J. Chem. Soc. (B)*, 793 (1966); T. A. Copper and W. A. Waters, *ibid.*, 687 (1967).
150. T. Morimoto and Y. Ogata, *J. Chem. Soc. (B)*, 62 (1967); 1353 (1967).
151. K. Sakota, Y. Kamiya, and N. Ohta, *Can. J. Chem.*, **47**, 387 (1969).
152. R. Tang and J. K. Kochi, unpublished results.
153. E. I. Heiba, R. M. Dessau, and W. J. Koehl, *Preprints, Division of Petroleum Chemistry, American Chemical Society Convention, Minneapolis, April 1969*, p. A44; *J. Am. Chem. Soc.*, **91**, 6830 (1969).
154. R. M. Dessau, S. Shih, and E. I. Heiba, *J. Am. Chem. Soc.*, **92**, 412 (1970).
155. S. S. Lande and J. K. Kochi, *J. Am. Chem. Soc.*, **90**, 5196 (1968).
156. W. F. Brill, *Ind. Eng. Chem.*, **52**, 837 (1960); A. S. Hay, J. W. Eustance, and H. S. Blanchard, *J. Org. Chem.*, **25**, 616 (1960).

157. A. S. Hay and H. S. Blanchard, *Can. J. Chem.*, **43**, 1306 (1965).
158. F. F. Rust and W. E. Vaughan, *Ind. Eng. Chem.*, **41**, 2595 (1949).
159. D. A. S. Ravens, *Trans. Faraday Soc.*, **55**, 1768 (1959); Y. Kamiya, *Tetrahedron*, **22**, 2029 (1966); E. K. Fields and S. Meyerson, *Adv. Chem. Ser.*, **76**, p. 395 (ref. 10).
160. C. E. H. Bawn and T. K. Wright, *Discuss. Faraday Soc.*, **46**, 164 (1968); cf. also M. N. Malik, J. Hill, and A. McAuley, *J. Chem. Soc.* (*A*), 643 (1970).
161. Y. Kamiya, *Adv. Chem. Ser.*, **76**, p. 193 (ref. 10).
162. A. W. Chester, E. I. Heiba, R. M. Dessau, and W. J. Koehl, *Inorg. Nucl. Chem. Letters*, **5**, 277 (1969).
163. A. W. Chester, *J. Org. Chem.*, **35**, 1797 (1970).
164. R. Stewart, in reference 2, p. 36; K. B. Wiberg, reference 2, p. 119 ff.; K. B. Wiberg and A. S. Fox, *J. Am. Chem. Soc.*, **85**, 3487 (1963).
165. J. Rocek, *Tetrahedron Letters*, 135 (1962).
166. J. I. Braumann and A. J. Pandell, *J. Am. Chem. Soc.*, **92**, 329 (1970); W. S. Trahanovsky and L. B. Young, *J. Org. Chem.*, **31**, 2033 (1966); P. Müller, E. Katten, and J. Rocek, *J. Am. Chem. Soc.*, **93**, 7116 (1971).
167. D. L. Toppen and R. G. Linck, *Inorg. Chem.*, **10**, 2635 (1971).
168. H. L. Finkbeiner and J. B. Bush, U.S. Pat. 3,535,372 (1968). Cf. also *Discuss. Faraday Soc.*, **46**, 150 (1968); see also oxidation with Pb(IV), R. E. Partch, *J. Am. Chem. Soc.*, **89**, 3662 (1967); R. O. C. Norman et al., *J. Chem. Soc.* (*B*), 518 (1971).
169. J. Rocek and A. E. Radkowsky, *J. Am. Chem. Soc.*, **90**, 2986 (1968); P. M. Nave and W. S. Trahanovsky, *ibid.*, **92**, 1120 (1970); M. Rahman and J. Rocek, *ibid.*, **93**, 5455, 5462 (1971).
170. M. Szwarc, in *Peroxide Reaction Mechanisms*, J. O. Edwards, Ed., Interscience, New York, 1962, p. 153 ff.; T. G. Traylor, S. Sieber, H. Kiefer, and N. Clinton, *Intra-Sci. Chem. Rep.*, **3**, 289 (1969).
171. L. Eberson, in *The Chemistry of Carboxylic Acids and Esters*, S. Patai, Ed., Interscience, New York, 1969, Chap. 6; *Fort. Chem. Forsch.*, **21**, 1 (1971); C. K. Mann and K. K. Barnes, *Electrochemical Reactions in Nonaqueous Systems*, Dekker, New York, 1970; M. E. Peover, in *Electroanalytical Chemistry*, Vol. 2, A. J. Bard, Ed., Dekker, New York, 1967, p. 1 ff.
172. R. A. Sheldon and J. K. Kochi, *Org. React.*, **19**, 279 (1972).
173. J. K. Kochi, *J. Am. Chem. Soc.*, **87**, 3609 (1965); J. K. Kochi, J. D. Bacha, and T. W. Bethea, *ibid.*, **89**, 6538 (1967).
174. J. D. Bacha and J. K. Kochi, *Tetrahedron*, **24**, 2215 (1968); *J. Org. Chem.*, **33**, 83, 2746 (1968).
175. J. K. Kochi, *J. Org. Chem.*, **30**, 3265 (1965).
176. J. K. Kochi, R. A. Sheldon, and S. S. Lande, *Tetrahedron*, **25**, 1197 (1969).
177. J. Kochi and T. W. Bethea, *J. Org. Chem.*, **33**, 75 (1968).
178. J. M. Anderson and J. K. Kochi, *J. Org. Chem.*, **35**, 986 (1970); *J. Am. Chem. Soc.*, **92**, 1651 (1970).
179. F. Minisci, R. Bernardi, F. Bertini, R. Galli, and M. Perchinummo, *Tetrahedron*, **27**, 3575 (1971); F. Minisci, R. Mondelli, G. P. Gardini, and O. Porta, *ibid.*, **28**, 2403 (1972).
180. A. A. Clifford and W. A. Waters, *J. Chem. Soc.*, 2796 (1965); P. R. Sharan, P. Smith, and W. A. Waters, *J. Chem. Soc.* (*B*), 1322 (1968); 462 (1969); J. H. Hill and A. McAuley, *J. Chem. Soc.* (*A*), 1169 (1968).
181. C. F. Wells and C. Barnes, *Discuss. Faraday Soc.*, **46**, 193 (1968); *Trans. Faraday Soc.*, **64**, 3069 (1968); C. F. Wells and G. Davies, *ibid.*, **63**, 2737 (1967).
182. P. D. Bartlett and R. Hiatt, *J. Am. Chem. Soc.*, **80**, 1398 (1958), and related papers.

183. S. S. Lande and J. K. Kochi, *J. Am. Chem. Soc.*, **90**, 5196 (1968).
184. R. A. Sheldon and J. K. Kochi, *J. Am. Chem. Soc.*, **90**, 6688 (1968); cf. D. Greatorex and T. J. Kemp, *Trans. Faraday Soc.*, **67**, 1576 (1971).
185. W. S. Trahanovsky, J. Cramer, and D. W. Brixius, *Abstracts of Papers, 160th American Chemical Society Meeting, Chicago, Illinois, Sept. 1970*, ORGN paper 144.
186. P. D. Bartlett and C. Rüchardt, *J. Am. Chem. Soc.*, **82**, 1756 (1960).
187. E. I. Heiba and R. M. Dessau, *Discuss. Faraday Soc.*, **46**, 189 (1968).
188. H. C. Volger and W. Brackman, *Rec. Trav. Chim. Pays-Bas*, **85**, 1233 (1966).
189. E. I. Heiba and R. M. Dessau, *J. Am. Chem. Soc.*, **93**, 524 (1971).
190. E. T. Denisov, V. D. Komissarov, and D. I. Metelitza, *Discuss. Faraday Soc.*, **46**, 127 (1968); H. C. Brackman, *Rec. Trav. Chim. Pays-Bas*, **85**, 817 (1966).
191. R. Cecil, J. S. Littler, and G. Easton, *J. Chem. Soc.* (*B*), 626 (1970).
192. R. Cecil and J. S. Littler, *J. Chem. Soc.*, 1420 (1968).
193. L. Taimr and J. Pospisil, *Tetrahedron Letters*, 2807 (1971); A. I. Scott, *Q. Rev.*, **19**, 1 (1965).
194. K. B. Wiberg, H. Maltz, and M. Okano, *Inorg. Chem.*, **7**, 830 (1968).
195. C. A. Audeh and J. R. Lindsay-Smith, *J. Chem. Soc.* (*B*), 1281, 1741, 1745 (1971).
196. D. G. Lee, in reference 5, p. 81.
197. J. Rocek, in *The Chemistry of the Carbonyl Group*, S. Patai, Ed., Interscience, New York, 1965, p. 461 ff.
198. A. Y. Drummond and W. A. Waters, *J. Chem. Soc.*, 435 (1953); 497 (1955).
199. K. B. Wiberg and P. C. Ford, *J. Am. Chem. Soc.*, **91**, 125 (1969).
200. K. B. Wiberg and W. H. Richardson, *J. Am. Chem. Soc.*, **84**, 2800 (1962).
201. T. A. Cooper and W. A. Waters, *J. Chem. Soc.*, 1538 (1964).
202. F. Haber and K. Willstater, *Ber.*, **64**, 2844 (1931).
203. C. E. H. Bawn and J. B. Williamson, *Trans. Faraday Soc.*, **47**, 721, 735 (1951); C. E. H. Bawn, T. P. Hobin, and L. Raphael, *Proc. Roy. Soc.*, **A237**, 313 (1956).
204. C. E. H. Bawn and J. E. Jolley, *Proc. Roy. Soc.*, **A237**, 297 (1956); C. E. H. Bawn, *Discuss. Faraday Soc.*, **14**, 181 (1953).
205. F. Marta, E. Boga, and M. Matok, *Discuss. Faraday Soc.*, **46**, 173 (1968).
206. G. C. Allen and A. Aguilo, *Adv. Chem. Ser.*, **76**, p. 363 (ref. 10).
207. E. O. Koubek and J. O. Edwards, *J. Inorg. Nucl. Chem.*, **25**, 1401 (1963).
208. G. I. Nikishin, M. G. Vinogradov, and S. P. Verenchikov, *Bull. Acad. Sci. USSR*, 1698 (1969).
209. G. H. Twigg, *Chem. Ind.* (*London*), **476** (1966).
210. G. I. Nikishin, M. G. Vinogradov, and R. V. Kereselidge, *Bull. Acad. Sci. USSR*, 1570 (1967).
211. M. Szwarc, C. H. Leigh, and A. H. Sehon, *Proc. Roy. Soc.* (*London*), **A209**, 97 (1951); J. K. Kochi and D. D. Davis, *Nature*, **202**, 690 (1964).
212. J. K. Kochi and J. W. Powers, *J. Am. Chem. Soc.*, **92**, 137 (1970).
213. J. K. Kochi and D. B. Buchanan, *J. Am. Chem. Soc.*, **87**, 853 (1965).
214. D. H. R. Barton, N. K. Basu, R. H. Hesse, F. S. Morehouse, and M. M. Pechet, *J. Am. Chem. Soc.*, **88**, 3016 (1966).
215. R. E. Erickson and R. K. Holmquist, *Tetrahedron Letters*, 4209 (1969).
216. J. K. Kochi and D. M. Singleton, *J. Org. Chem.*, **33**, 1027 (1968).
217. D. M. Singleton and J. K. Kochi, *J. Am. Chem. Soc.*, **89**, 6547 (1967); **90**, 1582 (1968).
218. C. E. Castro and W. C. Kray, Jr., *J. Am. Chem. Soc.*, **88**, 4447 (1966).
219. H. Nozaki, T. Aratani, and R. Noyori, *Tetrahedron*, **23**, 3645 (1967).
220. J. Halpern and J. P. Maher, *J. Am. Chem. Soc.*, **87**, 5361 (1965); P. W. Schneider,

P. F. Phelan, and J. Halpern, *ibid.*, **91**, 77 (1969); P. B. Chock and J. Halpern, *ibid.*, **91**, 581 (1969).

221. G. E. Herberich and J. Schwarzer, *Angew. Chem.*, **82**, 883 (1970); *J. Organometal. Chem.*, **34**, C43 (1972).
222. S. Aoki, C. Shirafugi, and T. Otsu, *Makromol. Chem.*, **126**, 1 (1969).
223. L. G. Marzilli, P. A. Marzilli, and J. Halpern, *J. Am. Chem. Soc.*, **92**, 5752 (1970).
224. N. Kornblum, *Pure Appl. Chem.*, **4**, 81 (1971).
225. J. Halpern, *Acc. Chem. Res.*, **3**, 386 (1970); F. R. Jensen and B. Knickel, *J. Am. Chem. Soc.*, **93**, 6339 (1971); see, however, J. A. Osborn, *ibid.*, **94**, 4043 (1972).
226. R. B. King, *Acc. Chem. Res.*, **3**, 417 (1970); G. N. Schrauzer, *ibid.*, **1**, 97 (1968); E. Ochiai, K. M. Long, C. R. Sperati, and D. H. Busch, *J. Am. Chem. Soc.*, **91**, 3201 (1969); cf. also D. Seyferth et al., *Pure Appl. Chem.*, **6**, 135 (1971).
227. F. R. Jensen, V. Madan, and D. H. Buchanan, *J. Am. Chem. Soc.*, **93**, 5283 (1971); **92**, 1415 (1970).
228. H. M. Walborsky and M. S. Aronoff, *J. Organomet. Chem.*, **4**, 418 (1965); H. M. Walborsky, F. P. Johnson, and J. B. Pierce, *J. Am. Chem. Soc.*, **90**, 5222 (1968); J. L. Webb, C. K. Mann, and H. M. Walborsky, *ibid.*, **92**, 2042 (1970); H. W. Bodewitz, A. Blomberg, and F. Bickelhaupt, *Tetrahedron Letters*, 281 (1972).
229. J. F. Harrod and W. R. Summers, *J. Am. Chem. Soc.*, **93**, 5051 (1971).
230. P. S. Skell, *Pure Appl. Chem.*, **4**, 215 (1971); P. S. Skell and J. J. Havel, *J. Am. Chem. Soc.*, **93**, 6687 (1971).
231. M. Asscher and D. Vofsi, *J. Chem. Soc.*, 2261 (1961); 1887, 3921 (1963).
232. See also J. R. Lindsay-Smith, *J. Chem. Soc.* (*B*), 920 (1970); S. Murai, N. Sonoda, and S. Tsutsumi, *J. Org. Chem.*, **29**, 2104 (1964); Y. Mori and J. Tsuji, *Syn. Commun.*, **2**, 51 (1972).
233. D. J. Burton and C. J. Kehoe, *J. Org. Chem.*, **35**, 1339 (1970); **36**, 2596 (1971).
234. M. Asscher and D. Vofsi, *J. Chem. Soc.*, 4962 (1964).
235. Y. Amiel, *J. Org. Chem.*, **36**, 3691, 3697 (1971).
236. H. Meerwein, G. Dittmar, R. Gollner, K. Hafner, F. Mensch, and O. Steinfort, *Chem. Ber.*, **90**, 841 (1957).
237. W. E. Griffiths, G. F. Longster, J. Myatt, and P. F. Todd, *J. Chem. Soc.* (*B*), 530 (1967).
238. F. Minisci, M. Cecere, R. Galli, and R. Bernardi, *Chim. Ind.* (*Milan*), **49**, 946, 947 (1967); *J. Chem. Soc.* (*B*), 324 (1967).
239. F. Minisci, G. P. Gardini, and F. Bertini, *Can. J. Chem.*, **48**, 544 (1970); J. Spanswick and K. U. Ingold, *ibid.*, **48**, 547 (1970); R. S. Neale and N. L. Marcus, *J. Org. Chem.*, **33**, 3457 (1968), and earlier papers.
240. F. Minisci, R. Galli, and R. Bernardi, *Tetrahedron Letters*, 699 (1966), and earlier papers; *Synthesis*, in press (1972).
241. F. Minisci and R. Galli, *Tetrahedron Letters*, 2531 (1966).
242. F. Minisci, M. Cecere, and R. Galli, *Gazz. Chim. Ital.*, **98**, 358 (1968); see also G. Sosnovsky and D. J. Rawlinson, *Adv. Free Rad. Chem.*, **4**, 203 (1972).
243. F. Minisci and R. Galli, *Tetrahedron Letters*, 533 (1962).
244. W. C. Baird, Jr., and J. H. Surridge, *J. Org. Chem.*, **35**, 3436 (1970).
245. W. C. Baird, Jr., J. H. Surridge, and M. Boza, *J. Org. Chem.*, **36**, 2088 (1971).
246. L. G. Harrison and C. F. Ng, *Trans. Faraday Soc.*, **67**, 1801 (1971); J. K. Kochi, *J. Am. Chem. Soc.*, **84**, 2121 (1962).
247. P. Kovacic and J.-H. Chen Chang, *J. Org. Chem.*, **36**, 3138 (1971).
248. C. Rüchardt, E. Merz, B. Freudenberg, H.-J. Opgenorth, C. C. Tan, and R. Werner, *Chem. Soc.* (*London*) *Spec. Publ.*, **24**, 51 (1970).

249. C. Rømming and K. Waerstad, *Chem. Commun.*, 299 (1965).
250. J. K. Kochi, *J. Am. Chem. Soc.*, **79**, 2942 (1958).
251. J. Kochi, *J. Am. Chem. Soc.*, **77**, 5274 (1955).
252. R. M. Elofson and F. F. Gadallah, *J. Org. Chem.*, **34**, 854 (1969); J. K. Kochi, *J. Am. Chem. Soc.*, **77**, 3208 (1955).
253. E. M. Kosower, *Acc. Chem. Res.*, **4**, 193 (1971); J. I. G. Cadogan, *ibid.*, **4**, 186 (1971); R. M. Elofson and F. F. Gadallah, *J. Org. Chem.*, **34**, 3335 (1969); **36**, 1526 (1971).
254. S. C. Dickerman, K. Weiss, and A. K. Ingberman, *J. Org. Chem.*, **21**, 380 (1956).
255. J. I. G. Cadogan, in *Free Radicals in Solution*, Butterworths, London, 1967, p. 158 ff.
256. J. I. G. Cadogan, *Chem. Soc.* (*London*) *Spec. Publ.*, **24**, 71 (1970); R. M. Elofson and F. F. Gadallah, *J. Org. Chem.*, **36**, 1769 (1971).
257. A. H. Lewin and T. Cohen, *J. Org. Chem.*, **32**, 3844 (1967); A. H. Lewin, A. H. Dinwiddie, and T. Cohen, *Tetrahedron*, **22**, 1527 (1966).
258. R. G. R. Bacon and H. A. O. Hill, *Q. Rev.*, **19**, 95 (1965).
259. P. E. Fanta, *Chem. Rev.*, **64**, 613 (1964); M. Nilsson and O. Wennerstrom, *Tetrahedron Letters*, 3307 (1968).
260. W. W. Kaeding, H. O. Kerlinger, and G. R. Collins, *J. Org. Chem.*, **30**, 3754 (1965); **26**, 3144 (1961).
261. P. H. Leake, *Chem. Rev.*, **56**, 27 (1956).
262. K. Maruyama, I. Tanimoto, and R. Goto, *J. Org. Chem.*, **32**, 2516 (1967).
263. F. Sondheimer, Y. Amiel, and R. Wolovsky, *J. Am. Chem. Soc.*, **81**, 1771, 1772 (1959); A. A. Clifford and W. A. Waters, *J. Chem. Soc.*, 3056 (1963).
264. W. Brackman and E. Havinga, *Rec. Trav. Chim. Pays-Bas*, **74**, 939, 1021, 1070, 1100, 1107 (1955).
265. W. Kaeding, *J. Org. Chem.*, **28**, 1063 (1963).
266. H. Finkbeiner, G. F. Endres, A. S. Hay, and H. S. Blanchard, *J. Org. Chem.*, **31**, 549 (1966); G. D. Cooper and J. G. Bennett, Jr., *ibid.*, **37**, 441 (1972).
267. A. Cairncross, J. R. Roland, R. M. Henderson, and W. A. Sheppard, *J. Am. Chem. Soc.*, **92**, 3187 (1970); T. Cohen and R. A. Schambach, *ibid.*, **92**, 3189 (1970).
268. T. Cohen and A. H. Lewin, *J. Am. Chem. Soc.*, **88**, 4521 (1966).
269. Cf. M. Anbar, *Adv. Chem. Ser., Am. Chem. Soc.*, **49**, 126 (1965).
270. H. Volger and W. Brackman, *Rec. Trav. Chim. Pays-Bas*, **84**, 579, 1018, 1204, 1234 (1965); **85**, 817 (1966); G. A. Russell, *Chem. Soc.* (*London*) *Spec. Publ.*, **24**, 271 (1970); *Pure Appl. Chem. Suppl.*, **4**, 67 (1971).
271. G. A. Russell, E. G. Janzen, A. G. Bemis, E. J. Geels, A. J. Moye, S. Mak, and E. T. Strom, *Adv. Chem. Ser., Am. Chem. Soc.*, **51**, 112 (1965).
272. N. Kornblum, cited by H. Fever, *Tetrahedron Suppl.*, **1**, 107 (1964).
273. H. O. House, W. L. Respess, and G. M. Whitesides, *J. Org. Chem.*, **31**, 3128 (1966).
274. H. C. Brown and G. W. Kabalka, *J. Am. Chem. Soc.*, **92**, 713 (1970); **92**, 711 (1970).
275. N. J. Friswell and B. G. Gowenlock, *Adv. Free Radical Chem.*, **1**, 39 (1965); **2**, 415 (1966); J. D. Cox and G. Pilcher, *Thermochemistry of Organic and Organometallic Compounds*, Academic Press, New York, 1969; K. C. Bass, *Organomet. Chem. Rev.*, **1**, 391 (1966); A. G. Davies, *Angew. Chem., Int. Ed.*, **9**, 741 (1970); G. A. Razuvaev and V. N. Latyaeva, *Russ. Chem. Rev.*, **34**, 251 (1965).
276. M. Tamura and J. K. Kochi, *J. Am. Chem. Soc.*, **93**, 1483, 1485, 1487 (1971); **92**, 6656 (1970); *J. Organomet. Chem.*, **29**, 111 (1971); **31**, 289 (1971).
277. M. R. Collier, M. F. Lappert, and M. M. Trulock, *J. Organomet. Chem.*, **25**, C36 (1970); G. Yagupsky, W. Mowat, A. Shortland, and G. Wilkinson, *Chem. Commun.*, 1369 (1970); A. Jacot-Guillarmod, R. Tabucchi, and J. Porret, *Helv. Chim. Acta*, **53**, 1491 (1970); U. Giannini and U. Zucchini, *Chem. Commun.*, 940 (1968).

278. P. J. Krusic and J. K. Kochi, *J. Am. Chem. Soc.*, **91**, 6161 (1969).
279. G. M. Whitesides, C. P. Casey, and J. K. Krieger, *J. Am. Chem. Soc.*, **93**, 1379 (1971).
280. R. J. Cozens, K. S. Murray, and B. O. West, *Chem. Commun.*, 1262 (1970); D. A. White, *Organomet. Chem. Rev.*, **3**, 497 (1968).
281. J. K. Kochi and P. J. Krusic, *J. Am. Chem. Soc.*, **91**, 3944 (1969); **94**, 5641 (1972); T. Kawamura, unpublished results.
282. D. M. P. Mingos, *Chem. Commun.*, 165 (1972); R. S. Braterman and R. J. Cross, *J. Chem. Soc.*, Dalton, 657 (1972).
283. C. A. Tolman, *Chem. Soc. Revs.*, **1**, 222 (1972); R. Mason, *Pure Appl. Chem.*, **6**, 31 (1971).
284. M. S. Kharasch and O. Reinmuth, *Grignard Reactions of Nonmetallic Substances*, Prentice-Hall, Englewood Cliffs, N.J., 1954, p. 122 ff.
285. M. H. Abraham and M. J. Hogarth, *J. Organomet. Chem.*, **12**, 1, 497 (1968); D. I. Davies, J. N. Done, and D. H. Hey, *J. Chem. Soc.* (*C*), 1392, 2021, 2506 (1969); D. I. Davies, D. H. Hey, and M. Tiecco, *J. Chem. Soc.*, 7062 (1965); F. W. Frey, *J. Org. Chem.*, **26**, 5187 (1961); W. B. Smith, *ibid.*, **24**, 703 (1959); **26**, 4206 (1961); L. H. Slaugh, *J. Am. Chem. Soc.*, **83**, 2734 (1961); G. Costa, G. Mestroni, and G. Boscarato, *La Ricerca Sci.*, **7**, 315 (1964); R. O. C. Norman and W. A. Walters, *J. Chem. Soc.*, 950 (1957); C. C. Lee and D. P. Thornhill, *Can. J. Chem.*, **42**, 1502 (1964); Y. Obe and T. Matsuda, *J. Chem. Soc. Japan*, **89**, 298 (1968); V. D. Parker and C. R. Noller, *J. Am. Chem. Soc.*, **86**, 1112 (1964).
286. M. Tamura and J. K. Kochi, *J. Am. Chem. Soc.*, **93**, 1483 (1971).
287. M. Tamura and J. K. Kochi, *J. Am. Chem. Soc.*, **93**, 1485 (1971); **92**, 6656 (1970).
288. M. Tamura and J. K. Kochi, *Synthesis*, 303 (1971).
289. G. Kobrich and H. Buttner, *Tetrahedron*, **25**, 883, 2223 (1969); *J. Organomet. Chem.*, **18**, 117 (1969).
290. G. M. Whitesides, E. J. Panek, and E. R. Stedronsky, *J. Am. Chem. Soc.*, **94**, 232 (1972).

Subject Index

Immunopharmacology of Endotoxicosis

Immunopharmacology of Endotoxicosis

Proceedings of the
5th International Congress of Immunology
Satellite Workshop
Kyoto, Japan, August 27, 1983

Editors
M. K. Agarwal · M. Yoshida

W DE G

Walter de Gruyter · Berlin · New York 1984

Editors

M.K. Agarwal, M.Sc.; Ph.D., M.D.
Maître de Recherche au CNRS
Scientific Director: Laboratoire de Physik-Hormono-Réceptérologie
Faculté de Médecine Broussais Hôtel-Dieu
Université Pierre et Marie Curie
15, rue de l'Ecole de Médecine

F-75270 Paris Cédex 06
France

Masao Yoshida, M.D.
Professor
Department of Bacteriology
Iwate Medical University

19-1 Uchimaru, Iwate 020
Morioka
Japan

CIP-Kurztitelaufnahme der Deutschen Bibliothek

Immunopharmacology of endotoxicosis :
proceedings of the 5th Internat. Congress of Immunology
satellite workshop, Kyoto, Japan, August 27, 1983 / ed. M. K. Agarwal;
M. Yoshida. – Berlin; New York : de Gruyter, 1984

ISBN 3-11-009887-3

NE: Agarwal, Manjul K. [Hrsg.]; International Congress of Immunology
<05, 1983, Kyoto>

Library of Congress Cataloging in Publication Data

International Congress of Immunology Satellite Workshop
(5th : 1983 : Kyoto, Japan) Immunopharmacology of endotoxicosis.

Bibliography: p.
Includes indexes.
1. Endotoxins--Physiological effect--Congresses.
2. Endotoxins--Toxicology--Congresses.
3. Immunopharmacology--Congresses. I. Agarwal, M. K.
II. Yoshida, M. (Masao), 1925- . III. Title.
[DNLM: 1. Endotoxins--immunology--congresses.
2. Endotoxins--pharmacodynamics--congresses.
QW 630 I33 1983]
QP632.E4I54 1983 616.9'2 84-7650
ISBN 3-11-009887-3

 Printing: Gerike GmbH, Berlin. – Binding: Lüderitz & Bauer GmbH, Berlin. – Printed in Germany.

THIS BOOK IS DEDICATED

TO

PROFESSOR L. JOE BERRY

FOR HIS CONTRIBUTIONS IN THE FIELD OF ENDOTOXINS

SPANNING OVER HALF A CENTURY

PREFACE

Bacterial endotoxins have fascinated researchers for over a century. Their beneficial effects include nonspecific increase in resistance to various sorts of infections, induction of interferon, antitumour activity, adjuvanticity, immunogenicity and radioprotection. Their pyrogenic properties had been exploited for several centuries, but this use has since been abandoned. Equally impressive is their array of noxious properties, which include the Sanarelli-Shwartzman reaction, shock, microvascular coagulation, hemodynamic alterations and the depletion of carbohydrates, to mention only a few. No organ or cell type in the host is immune from the influence of endotoxins, but it is not clear whether these efforts are direct mediated and whether one site is affected or several sites simultaneously.

The purpose of this workshop was to bring together researchers from various disciplines working in the field of endotoxins. After surveying the immunopharmacological reactions evoked by the bacterial endotoxins, the influence of various pharmacological agents on endotoxin-mediated host reactions was discussed. Since the mechanism of action of these agents is sometimes quite well defined, it was hoped that insight could be gained into the manner of endotoxins reactivity in the host. This proved difficult, however, due to the diversity of experimental models. Finally, problem-oriented themes were chosen with the aim of arriving at a consensus as to the site and nature of endotoxin reaction.

It is hoped that we provided a forum for workers interested in a common problem to thrash out their differences in a congenial and relaxed atmosphere. If the workshop has helped redefine the problem in a clearer perspective, the goal of the organizers will have been accomplished.

The Editors

December 1983

ACKNOWLEDGEMENTS

Special thanks and appreciation are due to Professors Y. Yamamura and M. Hanaoka, President and General Secretary, respectively, of the 5th International Congress of Immunology, for their assistance regarding the opening and the publicity for this workshop.

The organizers and editors are especially indebted to Dr. M. Hirata for his painstaking and capable help in the organization and in the editorial work connected with this book.

Thanks are equally due to all the members of the Department of Bacteriology, Iwate Medical University, and to Ms. V. Braymer, Lecturer, Iwate Medical University, for their assistance during various phases.

For financial assistance we wish to thank the following:

Packard Instruments (France S.A.)

Falk Foundation e.V. (Dr. H. Falk)

Iwate Medical University (President Pr. K. Obara)

The Society for Promotion of Science of the Alumni Association of Iwate Medical University (Chairman Pr. M. Yoshida and Chief of the Academic Section Pr. S. Katsura).

Alumni Association of the Bacteriology Department, Iwate Medical University (President Pr. R. Kawana).

Members of Ichirokukai, Company of Odashima (President Mr. M. Odashima), and several other companies also contributed to the project.

Société Française d'Immunologie kindly provided some financial assistance to M.K.A. towards travel to Kyoto.

Finally, grateful appreciation is due to all those persons who contributed at different times in different ways although the space does not permit specific, individual mention.

CONTENTS

DISSOCIATION OF TISSUE LOCALIZATION OF ENDOTOXIN FROM ENDOTOXIN LETHALITY

George Lázár, Elizabeth Husztik[+], Susanna Ribárszki and Alexander Pintér
Institute of Pathophysiology, Institute of Medical Biology[+], University Medical School, Szeged, Hungary

Introduction

Correlations between reticuloendothelial activity and sensitivity to endotoxin are contradictory. The following substantiate the concept that reticuloendothelial system (RES) plays a pivotal role in resistance to endotoxin:
1) Soon after the injection of endotoxin there is deep depression in the reticuloendothelial activity, which persits until death. However, in survivors, RES function recovers and goes on to hyperfunctional state. Similar pattern of reticuloendothelial response was also observed in other types of experimental shock (1).
2) The blockade of the RES with inert, nonmetabolizable, colloidal materials sensitizes to endotoxin poisoning and abolishes the tolerance to the biological effects of endotoxin, including the lethal effect (2, 3).
3) Animals previously injected with one or more sublethal doses of endotoxin, exhibit hyperfunctional RES and increased tolerance to bacterial lipopolysaccharide (LPS) (4, 5).
Other studies, however, have indicate that the relationship between host RES activity and endotoxin sensitivity is more complex. It is true that endotoxin tolerant animals have hyperfunctional RES, but the classical RES stimulants such as BCG, zymosan, Corynebacterium parvum, particulate glucan

Immunopharmacology of Endotoxicosis

do not confer resistance of experimental animals to endotoxemia, but profoundly increase host susceptibility to endotoxin (6). Particulate glucan while increasing all parameters of RES renders the LPS nonresponder, C3H/HeJ, mice nearly as responsive as conventional mice (7). Substances not affecting RES activity such as Streptozotocin, potentiates the toxic effects of LPS in experimental animals (8); depression of RES activity by methyl palmitate, renders experimental animals highly resistant to lethal and supralethal doses of endotoxins (9).
In this study, to understand the correlations between reticuloendothelial activity and sensitivity to endotoxin, the vascular clearance and tissue distribution of ^{51}Cr-labelled endotoxin and endotoxin sensitivity have been studied in mice following treatment with gadolinium chloride (10, 11), sodium polyanethol sulphonate (12) and carrageenan (13). All of these substances, although having different physicochemical properties, significantly depress reticuloendothelial function.

Materials and methods

CFLP males weighing 30-35 g were maintained on a standard laboratory diet and tap water, ad libitum.
Gadolinium chloride (K. and K. Laboratories, Plainview, New York) was dissolved in 0.85% saline at a concentration of 2 mg/ml and was injected i.v. at a dose of 1 mg/100 g body weight 24 hours before testing; kappa, lambda, or iota carrageenan (Marine Colloids, Rochland) were dissolved in boiling 0.85% saline at a concentration of 10 mg/ml, and were injected i.p. at a dose of 5 mg/100 g body weight 24 hours before testing; sodium polyanethol sulphonate (Liquoid, Hoffman-La Roche, Basel) was dissolved in 0.85% saline at a concentration of 4 mg/ml, and was injected i.v.

at a doses of 3 or 2 mg/100 g body weight 1 hour before testing.
E. coli 026:B6 lipopolysaccharide B (Difco Lab., Detroit, lot 688839) was labelled with 50 µCi/mg ^{51}Cr-sodium chromate (Isotope Institute of Hungarian Academy of Sciences) by the method of Braude et al. (14). For organ-uptake studies, one hour after the i.v. injection of 250 µg ^{51}Cr-labelled endotoxin animals were killed and the radioactivities in the blood, liver, spleen, lung and bone marrow were determined in a well-type scintillation detector and the results expressed as a percentage of the injected dose.
For measurement of endotoxin sensitivity, mice were challenged i.p. with proportional doses of endotoxin (E. coli 026:B6 lipopolysaccharide B, Difco Lab., Detroit, lot 688839) and the number of survivors was recorded after 48 hours.
The significance of differences in the mean response between treatments was determined by the t test. The Chi squere test was used to determine the significance of differences between survivors after various treatments.

Results

Data in Table 1 show that animals were sensitized to LPS by pretreatment with either 2 mg (37.5% survival) or 3 mg (15% survival) polyanethol sulphonate per 100 g body weight compared to control (80% survival) treated only with LPS. Studies with ^{51}Cr-labelled LPS show that both doses of the polyanethol sulphonate caused the retention of radioactivity in the blood and reduced the hepatic uptake of the injected radioactivity. The uptake of ^{51}Cr-labelled LPS in the spleen and lung was significantly increased (Fig. 1).
The effect of carrageenan on endotoxin sensitivity and the distribution of ^{51}Cr-labelled endotoxin were very similar to those observed in mice pretreated with Liquoid. All forms

Table 1
Sensitization to LPS lethality by Liquoid

Treatment	Living/total (48 hr)	Survival (%)	Statistics
1. E	28/35	80	
2. L	19/20	95	2 vs 1 $p>0.05$
3. L + E	3/20	15	3 vs 1 $p<0.001$
4. L	20/20	100	4 vs 1 $p>0.05$
5. L + E	6/16	37.5	5 vs 1 $p<0.01$

E = endotoxin, 250 µg/10 g body weight i.p., one hr after the injection of Liquoid for lines 3 and 5; L = Liquoid, 3 mg/100 g body weight for lines 2 and 3 and 2 mg/100 g body weight for lines 4 and 5.

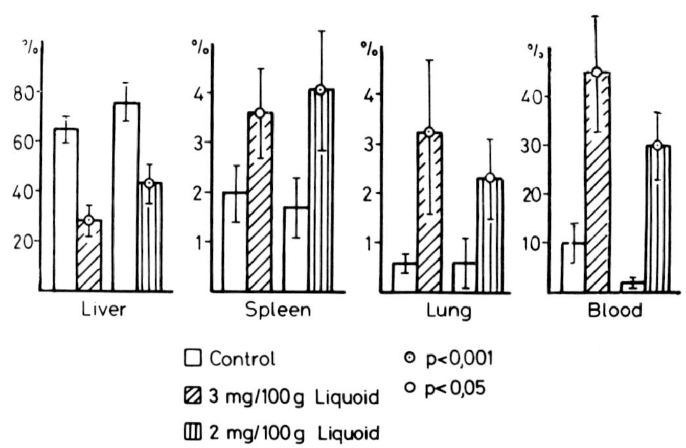

Fig. 1. Effect of Liquoid on the distribution of LPS. The radiocontents of the organs and blood were determined 1 hr after the i.v. injection of 250 µg LPS labelled with ^{51}Cr.

Table 2.
Sensitization to LPS lethality by carrageenan

Treatment	Living/total (48 hr)	Survival (%)	Statistics vs 1
1. E	30/50	60	
2. Kappa C	4/20	20	$p<0.01$
3. Lambda C	1/20	5	$p<0.001$
4. Iota C	3/20	15	$p<0.01$

E = endotoxin, 250 µg/10 g body weight i.p.; C = carrageenan, 5 mg/100 g body weight i.p. 24 hr before endotoxin administration.

of carrageenan aggravated endotoxin lethality. Only 20%, 5%, 15%, of the animals given kappa, lambda or iota carrageenans, respectively, survived endotoxin challenge compared to 60% survival in the control group given endotoxin

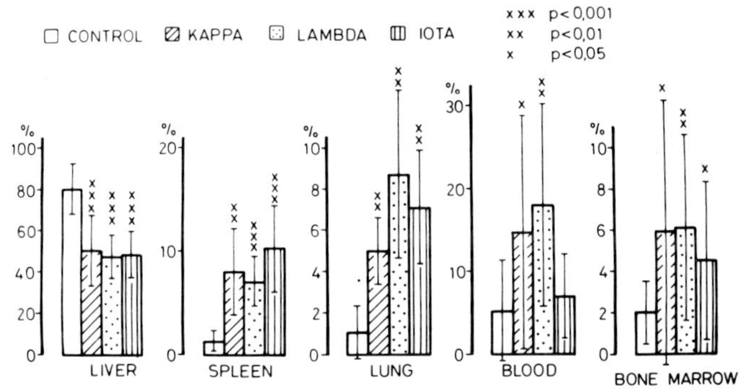

Fig. 2. Effect of carrageenan on the distribution of LPS. Carrageenan was injected i.p. at a dose of 5 mg/100 g body weight 24 hr before the i.v. injection of 250 µg LPS labelled with ^{51}Cr. The radiocontents of the organs and blood were determined 1 hr later.

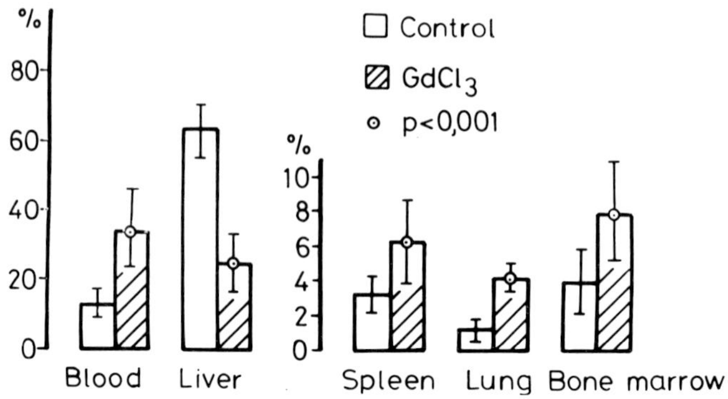

Fig. 3. Effect of gadolinium chloride on the distribution of LPS.

Gadolinium chloride was injected i.v. at a dose of 1 mg/ /100 g body weight 24 hr before the i.v. injection of 250 µg LPS labelled with ^{51}Cr. The radiocontents of the organs and the blood were determined 1 hr later.

alone. Carrageenans also caused the retention of ^{51}Cr-labelled LPS in the blood and at the same time reduced the hepatic and increased the extrahepatic uptake of endotoxin (Fig. 2).

Table 3.

Effect of gadolinium chloride on LPS sensitivity

Treatment	Living/total (48 hr)	Survival (%)	Statistics 2 vs 1
1. E	12/20	60	
2. Gd	11/20	55	$p>0.05$

E = endotoxin, 250 µg/10 g body weight i.p.; Gd = gadolinium chloride, 1 mg/100 g body weight i.v. 24 hr before endotoxin challenge.

In response to gadolinium chloride, the alterations in the reticuloendothelial function were very similar to those observed in mice pretreated with Liquoid or carrageenan (Fig. 3), but, contrary to the effect of these substances, gadolinium chloride did not alter endotoxin susceptibility of mice (Table 3).

Discussion

Present studies support earlier findings (15) that liver plays a major role in clearing endotoxin from the blood. This role of liver is even more pronounced during the state of endotoxin tolerance. Tolerant animals remove circulating endotoxin more rapidly than do normal animals, with a prompt and remarkable localization of LPS in the liver (15). However, endotoxin clearing capacity of the liver is not a crucial factor determining endotoxin sensitivity. Our studies with sodium polyanethol sulphonate, carrageenan, and gadolinium chloride indicate that whereas all of these reticuloendothelial depressants delayed vascular clearance of endotoxin and caused a great reduction in hepatic localization of endotoxin, only sodium polyanethol sulphonate and carrageenan enhanced endotoxin sensitivity - gadolinium chloride did not alter it. Other studies have indicate that the reticuloendothelial depressant trypan blue sensitizes animals to endotoxin lethality and methyl palmitate renders animals highly resistant to LPS, but none of them alter the hepatic uptake of endotoxin (16, 9). On the other hand, particulate glucan, which causes endotoxin hypersensitivity (6), greatly enhances the vascular clearance and the hepatic localization of endotoxin (9).

These studies dissociate vascular clearance and tissue localization of endotoxin from endotoxin sensitivity. RES active agents may have biological effects other that their influence on the RES granulopectic activity. Some of these biological effects such as metabolic disorders (17), blood coagulation (18, 19), labilization of lysosomal membrane (20), modulation of macrophage arachidonic acid metabolism (21), may unfavourably influence endotoxin shock and may represent critical lesions in endotoxicosis. On the other hand, some of these effects may favourably influence endotoxicosis. It is known that methyl palmitate reduces the extent of hepatic injury after endotoxin (22) or another toxicant (23), and the endotoxin-induced thromboxane production (21). Gadolinium chloride decreases the hyperfibrinogenaemic response to LPS (24) and may exert beneficial effects on other forms of shock such as the anaphylactic shock in mice (25).

Summary

The vascular clearance and tissue distribution of ^{51}Cr-labelled endotoxin and endotoxin sensitivity have been studied in mice following treatment with reticuloendothelial depressants, gadolinium chloride, sodium polyanethol sulphonate and carrageenan. It was found that all of these substances caused the retention of radioactivity in the blood and reduced the hepatic uptake of injected radioactivity with spillover into the extrahepatic organs. Our studies, however, indicate the dissociation of the tissue localization of endotoxin and endotoxin lethality; whereas sodium polyanethol sulphonate and carrageenan, aggravated, gadolinium chloride did not alter, endotoxin lethality.

References

1. Hershey, S:G.: in Kaley, G., Altura, B.M. (eds.) Micro-

circulation, Vol. 3., University Park Press, Baltimore, 1980.

2. Zweifach, B.W., Benacerraf, B., Thomas, L.: J. exp. Med. 106, 403-414 (1957).

3. Lázár, G., Sándor, Th., Karády, S., West, G.B.: Med. exp. 19, 287-292 (1969).

4. Zweifach, B.W., Thomas, L.: J. exp. Med. 106, 385-401 (1957).

5. Lázár, G., Sekiya, S., Agarwal, M.K.: J. Reticuloendothel. Soc. 22, 13-20 (1977).

6. Di Luzio, N.R., Al-Tuwaijri, A., Williams, D.L., Kitahama, A., Browder, W.: in Agarwal, M.K. (ed.) Bacterial Endotoxins and Host Response, Elsevier/North-Holland Biomedical Press, Amsterdam, New York, Oxford, 1980.

7. Lázár, G., Agarwal, M.K.: Biochem. Med. 28, 310-318 (1982).

8. Agarwal, M.K., Paillard, J.: in Agarwal, M.K. (ed.) Bacterial Endotoxins and Host Response, Elsevier/North--Holland Biomedical Press, Amsterdam, New York, Oxford, 1980.

9. Di Luzio, N.R., Crafton, C.G.: Proc. Exptl. Biol. Med. 132, 686-690 (1969).

10. Lázár, G.: J. Reticuloendothel. Soc. 13, 231-237 (1973).

11. Husztik, E., Lázár, G., Párducz, Á.: Br. J. exp. Path. 61, 264-230 (1980).

12. Lázár, G., Husztik, E., Somfay, A., Szabó, É.: Proc. of the Int. Congr. of Haematol. and Blood Transfusion, Budapest, 1982, p. 378.

13. Allison, A.C., Harrington, J.S., Birbeck, M.: J. exp. Med. 124, 141-153 (1966).

14. Braude, A,I., Carey, F.J.: J. clin. Invest. 34, 850-857 (1955).

15. Utili, R., Abernathy, Ch.O., Zimmerman, H.J.: Life Sci. 20, 553-568 (1977).

16. Noyes, H.E., McInturf, C.R., Blahuta, G.J.: Proc. Soc. Exptl. Biol. Med. 100, 65-68 (1959).

17. Agarwal, M.K., Hoffman, W.W., Rosen, F.: Biochim. Biophys. Acta 177, 250-258 (1969).

18. Van Aken, W.G., Vreeken, J.: Thrombos. Diathes. haemorrh. (Stuttg.) 22, 496-507 (1969).

19. Evensen, S.A., Jeremic, M.: Thrombos. Diathes. haemorrh. (Stuttg.) 19, 556-569 (1968).

20. Catanzaro, P.J., Schwartz, H.J., Graham, R.C.: Am. J.

Pathol. 64, 387-399 (1971).

21. Cook, J.A., Haluska, P.V., Wise, W.C.: Circulatory Shock 9, 605-617 (1982).

22. Di Luzio, N.R., Trejo, R.A., Crafton, C.G.: J. Reticuloendothel. Soc. 11, 637-653 (1972).

23. Lázár, G., Serra, D., Tuchweber, B.: Toxicol. appl. Pharmacol. 29, 367-376 (1974).

24. Lázár, G., Agarwal, M.K., Husztik, E.: Microbios Letters 21, 127-131 (1982).

25. Lázár, G., Husztik, E., Somfay, A.: Proc. of the 9th Int. RES Congr., Davos, 1982, p. 28.

SINUSOIDAL AND PARENCHYMAL CELLS AS TARGETS OF ENDOTOXIN EFFECTS ON THE LIVER

Riccardo Utili, Giovanni B. Gaeta, Augusto Andreana and Giuseppe Ruggiero.
Clinic of Infectious Diseases and Clinic of Tropical and Subtropical Diseases, University of Naples First Medical School, Naples, Italy.

Introduction

Among the critical events which follow endotoxin administration to experimental animals, alteration of liver functions represent an important and well documented effect (1). The liver plays a major role in clearing circulating endotoxin, and recent studies have confirmed that an injected dose of endotoxin can be almost completely recovered in pure Kupffer cell preparations (2). Moreover, since reticuloendothelial system (RES) blockade or stimulation may modify animal responsiveness to endotoxin, a mediation role of RES in endotoxicosis has been suggested (3).
However, evidence cumulates from our and others' groups that endotoxin may affect specific functions of liver parenchymal cells such as bile production (4, 5) and drug metabolizing activity (6). Moreover, pretreatment of rats with endotoxin induces refractoriness against the hepatotoxic effects of chlorpromazine (7) and erythromycin estolate (8), thus suggesting that endotoxin interacts with the hepatocellular targets of these agents and prevents the display of their toxicity. Evidence of localization of endotoxin in hepatocytes has also been provided (9) and a biliary route of excretion has been demonstrated (5).
Since endotoxin can interact with almost all mammalian cell membranes it was interesting to investigate whether endotoxin could affect independently

Immunopharmacology of Endotoxicosis

specific hepatocytic and sinusoidal cell functions and whether liver responsiveness to this agent could be modulated by an altered status of RES. By using in vitro models (isolated perfused rat liver; liver plasma membrane preparations) which allow to separate the role of humoral and cellular factors in the response to endotoxin stimulus, we studied 1) the effect of endotoxin administration on phagocytosis and bactericidal activity of hepatic RES versus circulating bacteria; 2) the interaction of endotoxin with liver plasma membranes; 3) the effect of an altered RES status on the endotoxin-induced cholestasis.

Methods and Materials

Endotoxin treatment: Male Sprague-Dawley rats (250 - 300 g) were injected intraperitoneally with increasing doses of E. coli endotoxin (055:B5, Difco), from 0.01 to 1 mg, over a 9-day period. This regimen induced endotoxin tolerance (10).

Liver perfusion studies: For bacterial phagocytosis studies, livers were isolated from controls and endotoxin-treated rats as described by Ruggiero et al. (11). The perfusing medium consisted of 5% serum from either control or endotoxin-treated animals in 100 ml Krebs-Ringer bicarbonate buffer (pH 7.4) supplemented with 200 mg glucose and 3 g bovine serum albumin (Sigma); in some experiments serum was omitted from the perfusate. Bacterial inoculum consisted of a nephelometric standardized suspension of E. coli (ATCC 25922, Difco) in phosphate buffered saline which provided an initial bacterial concentration in the perfusate of 2 x 10^7 cells/ml. The phagocytic function of RES was measured by following the bacterial clearance from perfusate at 10 min intervals for 30 min. At the end of the experiment recovery of viable bacteria was determined in the liver homogenate. In some experiments the liver was fixed by perfusion with 1.5% gluteraldehyde in cacodylate buffer (pH 7.4) and sections were prepared for electron microscopy observations. Phagocytosis was calculated by the percent of bacterial inoculum not recovered in the perfusate at the end of the experiment, after correction for the bactericidal activity of perfusate alone(11). The amount of phagocytized bacteria which were not recovered in the liver homogenate was considered to represent the amount of bacteria killed by the liver.

To study the effect of endotoxin on bile flow, livers were isolated from controls or from methyl palmitate (MP)-treated rats (200 mg/100 g b.w. suspended in 0.2% Tween 20 in 5% glucose, given i.v. 72 hrs before) and perfused as described by Utili *et al*. (4). Bile and perfusate flows were monitored throughout the 90 min experimental period.

Liver plasma membrane (LPM) preparation and enzyme assays: LPM were prepared from control and endotoxin-treated rats according to the procedure of Boyer and Reno (12), slightly modified by Utili *et al*. (13). Total ATPase was measured as described by Boyer and Reno (12). Na^+,K^+-ATPase was taken to be the difference between the total and Mg^{++}-ATPase. Adenylate cyclase (AC) was determined by measuring the conversion of ATP to cyclic AMP in the presence of an ATP-regenerating system (14). Addition of sodium fluoride (Fl) (10 mM) stimulated AC activity 6-7 fold. 5'-nucleotidase (5-NT) was measured as described by Song and Bodansky (15).

Results and Discussion

Effect of chronic endotoxin treatment on Kupffer cell function. The effect of chronic endotoxin administration *in vivo* on Kupffer cell function was studied by measuring the phagocytosis and bactericidal activity of perfused rat livers isolated from controls or endotoxin-treated animals, against circulating *E. coli* (Fig. 1). Endotoxin treatment enhanced both the clea-

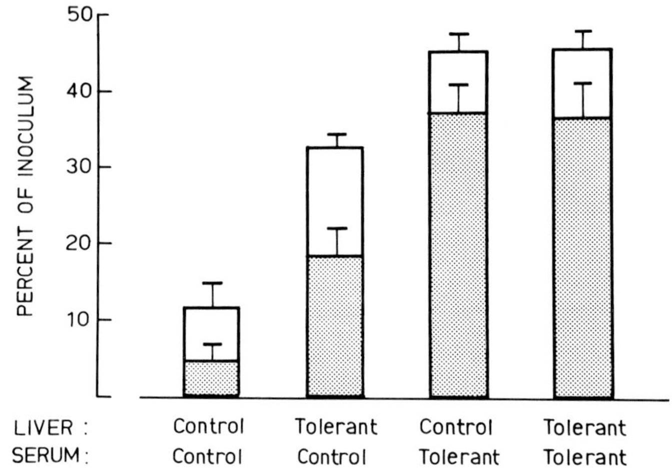

Fig. 1 - Phagocytosis and killing of viable E. coli by rat livers isolated from either controls or endotoxin-tolerant animals and perfused with a synthetic perfusate containing 5% serum from either controls or endotoxin-tolerant rats. The entire bar represents the amount of bacteria phagocytized; the shaded portion of each bar represents the intracellular killing. Values are means ± S.E. of four experiments for each group. Data are from ref. 11.

rance rate and the intracellular killing of bacteria. The intracellular localization of bacteria within Kupffer cells, as seen at electron microscopy, demonstrated that the bacterial clearance by the liver was the result of an active endocytic process (Fig. 2). The greatest activation of hepatic RES occurred when the serum from endotoxin-treated animals was present in the perfusate. Under these conditions, during the 30 min experimental time, 46% of the inoculum was phagocytized (12% in controls; $P<0.01$) and 38% was killed (5% in controls; $P<0.01$). This RES activation was also observed, albeit to a minor degree, when livers obtained from endotoxin-treated rats were perfused with serum from control animals. In these experiments phagocytosis was 33% and killing 19% ($P<0.05$ vs controls). Interestingly, in experiments where livers from endotoxin-treated animals were perfused with a serum-free perfusate, phagocytosis still occurred at

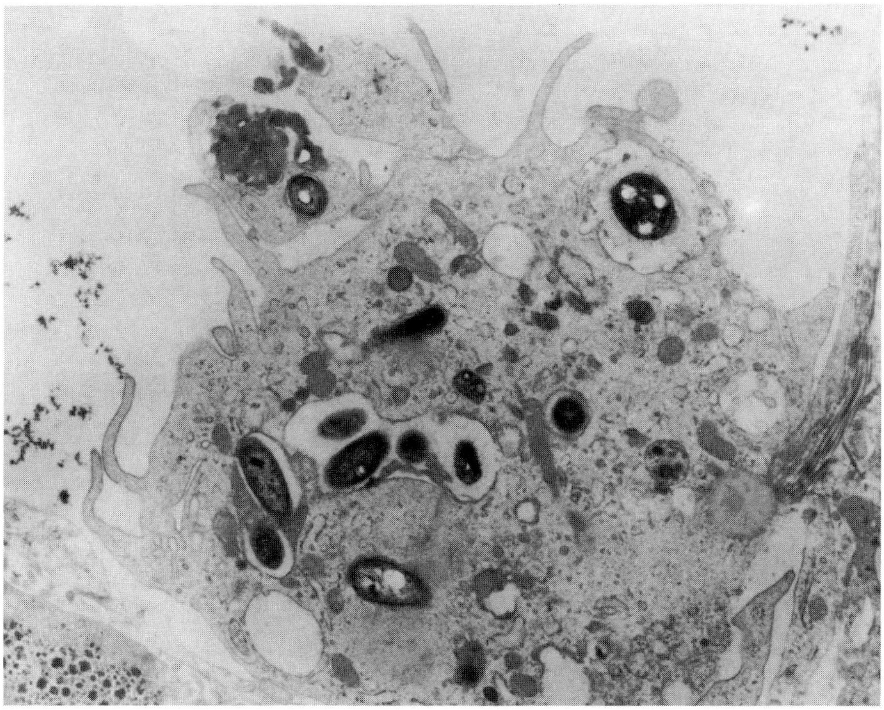

Fig. 2 - Intracellular localization of E. coli within endocytic vacuoles of an activated Kupffer cell after 30 min perfusion of isolated rat liver. Gaeta, G.B. and Wisse, E.; unpublished observations.

a significant greater extent than controls (38 ± 3% vs 15 ± 3%, respectively; P<0.05) and killing was 20% as opposed to no killing of controls. The data indicate that during endotoxin treatment humoral factors play an important role in the activation of hepatic RES. However a direct stimulation of Kupffer cells per se did occur independently of serum factors. It is suggested that a direct interaction of endotoxin with hepatic macrophages may enhance membrane movements, production of lysosomal enzymes and generation of superoxide, thus rendering these cells "activated" against bacteria.

Effect of endotoxin on hepatocytic membranes. The intracellular events which occur at the hepatocytic level after endotoxin administration involve alteration of bile production and organic anion excretion as well as changes in hepatic blood flow, energy production and carbohydrate and lipid metabolism (for review see 1). It is suggested that all these events could be triggered by a membrane perturbation which follows interaction with endotoxin. Due to its large amphiphatic structure, endotoxin may insert into the membrane bilayer; it is not clear however whether this occur at specific binding sites (receptors) which could recognize the lipid A moiety of endotoxin or through a non-specific affinity with the hydrophobic region of the membrane. To study the interaction of endotoxin with liver plasma membranes we evaluated the effect of endotoxin on LPM-bound enzymes with different location on the bilayer: Mg^{++}-ATPase and 5-NT on the external surface; Na^{+},K^{+}-ATPase a transmembrane enzyme, and basal and Fl-stimulated AC at the inner side. Membrane enzyme activities may vary with changes of their lipid environment and may serve as probes of endotoxin interaction.

TABLE

Effect of *E. coli* endotoxin on LPM-enzyme activities. Endotoxin was either added *in vitro* to LPM from untreated rats or given *in vivo* prior to LPM isolation (see Methods). Data are means ± S.D. of 3 to 5 experiments.

Endotoxin treatment		5'-NT	Mg^{++}-ATPase	Na^{+},K^{+}-ATPase	Basal AC	Fl-stim. AC
		(μmol Pi/mg prot/hr)			(pmol cAMP/mg prot/min)	
In vitro:	0	68 ± 2	72.7 ± 2.6	13.9 ± 1.5	66 ± 7	401 ± 19
(μg/ml)	20	67 ± 3	73.1 ± 1.0	6.8 ± 1.0*	63 ± 5	321 ± 20*
	40	69 ± 4	71.3 ± 2.0	6.3 ± 1.1*	64 ± 5	281 ± 15*
	80	67 ± 2	72.1 ± 1.8	4.3 ± 1.0*	65 ± 3	241 ± 22*
In vivo:		72 ± 3	76.3 ± 1.9	9.5 ± 1.8*	ND	ND

* $P < 0.05$

The table shows that endotoxin did not alter the activity of ectoenzymes (5'-NT and Mg^{++}-ATPase), nor the activity of the basal AC at the inner side, but significantly decreased both the Na^{+},K^{+}-ATPase and the Fl-stimulated AC activities (13, 14, 16). For both enzymes the inhibition was of a non-competitive type (13, 14). Since endotoxin is known to decrease membrane fluidity (17) the data are consistent with the view that interaction of endotoxin with the membrane occurs at the deep hydrophobic regions where it may restrict tha lateral diffusion of Na^{+},K^{+}-ATPase and the specific interaction of the Fl-stimulated regulatory unit with the catalytic AC component.

Effect of an altered hepatic RES activity in response to endotoxin. We verified whether the effect of endotoxin on bile secretion could by modified by altering the status of hepatic RES. Depression of RES activity was induced by methyl palmitate treatment. As shown in fig. 3,

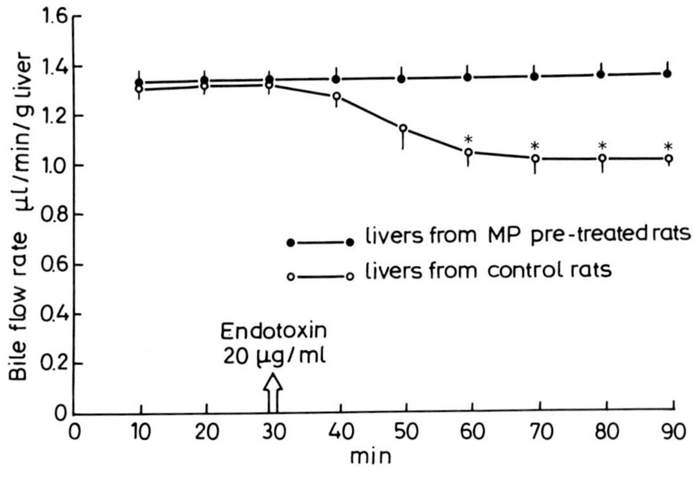

Fig. 3 - Effect of *E. coli* endotoxin on bile flow of livers from controls or methyl palmitate treated rats. Values are means ± S.E. of 4 experiments for each group. Data are from ref. 18.
* P<0.05

addition of endotoxin to the perfusate of control rat liver preparations caused a significant decrease of bile flow. In contrast, when the livers were isolated from MP-treated animals, endotoxin did not alter basal bile flow during all the experimental period. These results may suggest that hepatic RES mediates the display of biological effects of endotoxin on parenchymal cell functions. However, it cannot be excluded that MP, with its hydrophobic groups, may directly interfere with hepatocytic membrane, thus modifying the responsiveness to endotoxin (18).

The data presented provide strong evidence that, besides the activation of a number of humoral factors which modulate its effect (19), endotoxin is capable to interact directly with liver membranes and it can act independently on both parenchymal and non-parenchymal cells. However, communications between these two groups of cells can occur and may modulate the responsiveness of the liver to endotoxin. Binding of endotoxin to a cell triggers a number of intracellular events which may amplify or soften the signal. The knowledge of the nature of membrane alterations induced by endotoxin and of the intracellular events which follow, will help to clarify the multifaceted aspects of endotoxin effect on various biological targets.
(This work was supported in part by Consiglio Nazionale delle Ricerche)

References

1. Utili, R., Abernathy, C.O., Zimmerman, H.J.: Life Sciences 20, 553-568 (1977).
2. Praaning-Van Dalen, D.P., Brauwer, A., Knook, D.L.: Gastroenterology 81, 1036-1044 (1981).
3. Crafton, C.G., Di Luzio, N.R.: Am. J. Physiol. 217, 736-742 (1969).
4. Utili, R., Abernathy, C.O., Zimmerman, H.J.: Gastroenterology 70, 736-742 (1976).

5. Maitra, S.K., Rachmilewitz, D., Eberle, D., Kaplowitz, N.: Hepatology 1, 401-407 (1981).

6. Abernathy, C.O., Zimmerman, H.J., Utili, R: Res. Commun. Chem. Pathol. Pharmacol. 29, 193-196 (1980).

7. Utili, R., Abernathy, C.O., Zimmerman, H.J., Gaeta, G.B., Adinolfi, L., Lukacs, L.: Gastroenterology 80, 673-680 (1981).

8. Abernathy, C.O., Zimmerman, H.J., Lukacs, L., Utili, R., Gaeta, G.B.: IRCS Medical Sci. 8, 893-894 (1980).

9. Zlydaszyk, J.C., Moon, R.J.: Infect. Immun. 14, 100-105 (1976).

10. Nolan, J.P., Ali, M.V.: J. Med. 4, 28-38 (1973).

11. Ruggiero, G., Andreana, A., Utili, R., Galante, D.: Infect. Immun. 27, 798-803 (1980).

12. Boyer, J.L., Reno, D.: Biochim. Biophys. Acta 401, 59-72 (1976).

13. Utili, R., Abernathy, C.O., Zimmerman, H.J.: J. Infect. Dis. 136, 583-586 (1977).

14. Utili, R., Di Donato, A., Draetta, G., Paolisso, G., Abernathy, C.O., Illiano, G.: Experientia 38, 831-832 (1982).

15. Song, C.S., Bodansky, O.: J. Biol. Chem. 242, 694-699 (1967).

16. Utili, R., Abernathy, C.O., Zimmerman, H.J., Gaeta, G.B.: Proc. Soc. Exp. Biol. Med. 161, 554-558 (1979).

17. Rottem, S.: FEBS Letters 95, 121-124 (1978).

18. Gaeta, G.B., Utili, R., Adinolfi, L.E., Abernathy, C.O., Zimmerman, H. J.: In "Sinusoidal Liver Cells" D.L. Knook, E. Wisse eds. p. 471-472, Elsevier North Holland, Amsterdam, 1983.

19. Morrison, D.C., Ulevitch, R.J.: Am. J. Pathol. 93, 527-617 (1978).

LPS-INDUCED HYDROGEN PEROXIDE RELEASE FROM PERITONEAL MACROPAGES OF NORMAL AND IMMUNODEFECTIVE MICE

Hideyuki Kato,* Hideo Yaoita,* Tatsuo Saito-Taki, Masayasu Nakano
Department of Dermatology* and Department of Microbiology,
Jichi Medical School, Tochigi-ken 329-04, Japan

Introduction

It is well known that some of the anti-infectious, anti-parasitic and anti-tumoral properties of macrophages depend on their ability to release a large quantity of hydrogen peroxide after appropriate triggerings (1-8). Bacterial lipopolysaccharide (LPS), which is a constituent of the cell wall of gram-negative bacteria, is one of the stimulatory substances non-specifically enhancing the bactericidal or tumorcidal activity of macrophages (9-11). It has some ability to induce H_2O_2 production by macrophages, and the H_2O_2 released from the macrophages activated by LPS after the infection of bacteria may play an important role in bacterial killing.

On the other hand, there are some strains of mice which are immunologically defective and refractory to LPS. The best characterized strain of those possessing a defect in the responce to LPS is C3H/HeJ mice (12). The CBA/N strain is also shown to possess an X-linked defect resulting in the lack of a subset of mature B cells capable of responding to LPS (13). Both C3H/HeJ and CBA/N mice are very susceptible to Salmonella typhimurium (14, 15) and S. enteritidis (our unpublished observation), and their bactericidal function is relatively impaired. Beige mice with murine Chediak-Higashi syndrome are another immunodeficient model whose macrophages

Immunopharmacology of Endotoxicosis

exhibit delayed intracellular killing of bacteria (16). The susceptibility of these strains to infectious agents may be related to malfunction of macrophages.

In this paper, we deal with the H_2O_2 productivity of peritoneal macrophages obtained from normal mice and these immunologically defective strains of mice after the stimulation by endotoxins such as LPS, Boivin antigen, lipid A and lipid A-bovine serum albumin (BSA)-complex in vitro, and we demonstrate that macrophages of the defective strains are hampered in their extracellular production of H_2O_2.

Materials and methods

Mice

BALB/c, C3H/He, C3H/HeJ, C57BL/6, beige mutant of C57BL/6, (CBA/N female x BALB/c male)F1 and (beige female x C57BL/6 male)F1 mice, 8 to 12 weeks of age, were used. Parents of BALB/c, C3H/He and C57BL/6 mice were obtained from the Institute of Medical Science, University of Tokyo. LPS-nonresponder C3H/HeJ strain was obtained from the Kitasato University School of Medicine. CBA/N and beige mutant mice were gifts from the Division of Research Service, National Institute of Health, U.S.A. Since the CBA/N-defect is X-linked, CBA/N females mated to normal males (BALB/c in this series of experiments) produced F1 males which showed some defect in the function of B lymphosytes and female littermates which, although heterozygous for the defective X-chromosome, were phenotypically normal. The beige mutation in the homozygous condition (bg/bg) results in some functional defects in macrophages, but heterozygous F1 hybrid mice (bg/+) are phenotypically normal. These inbred mice and hybrid F1 mice were bred in our animal colony.

Macrophages

Mice were injected intraperitoneally with 2.5ml thioglycollate broth (Eiken Chemical Co. Ltd., Tokyo). The mice were killed 3 days after the injection, and then peritoneal exudate cells were harvested by rinsing the peritoneal cavities with Hanks' balanced salt solution (HBSS) (Nissui Seiyaku Co., Ltd., Tokyo) without anticoagulant. The cells obtained consisted of 70% macrophages and 30% other cells (majority of lymphocytes and few leucocytes). The cells were washed 3 times in HBSS and then suspended to a concentration of 1 x 10^6 cells/ml in HBSS containing 10% heat-inactivated (56°C for 30 min) fetal calf serum (FCS) (Grand Island Biological Co., U.S.A.). The cell suspension was poured into 16mm diameter plastic culture wells (1 ml/well) and incubated at 37°C for 3 hr in a 5% CO_2-95% air chamber in order to adhere the macrophages to the bottom surfaces of the wells. After the incubation, non-adherent cells were removed, and adherent cells were rinsed twice with warm (37°C) Earle's balanced salt solution (EBSS) without phenol red. The adherent cells were used as macrophages. The number of macrophages in a culture well was estimated by subtracting the number of non-adherent cells from the initial-number of cells. According to this estimation, there were 6.8-7.2 x 10^5 macrophages per culture well in our experiments, regardless of the strains used.

Culture

Macrophage monolayers in HBSS containing 10% FCS in the wells were cultured in the presence or absence of endotoxin at 37°C in a 5% CO_2-95% air chamber during a period of 1 to 24 hr.

Endotoxins

LPS was extracted from Salmonella typhimurium LT2 by the hotphenol-water procedure as described by Westphal (17) and then purified by repeated centrifugation at 100,000 x g for 60 min. Lipid A was prepared from the LPS by hydrolysis with 1%

acetic acid at 100°C for 4 hr. The lipid A was complexed with BSA by the method of Galanos et al. (18, 19). Boivin antigen was extracted from LT2 organisms by trichloroacetic acid (TCA) (20, 21). These endotoxins were dissolved in saline (mg/ml) and then diluted to adequate concentrations with HBSS. Since it was difficult to dissolve lipid A, the suspension was sonicated and the dispersion was added to the cultures.

Reagents

Phenol red (Sigma Chemical Co., St. Louis, Mo. U.S.A.) was dissolved in HBSS (0.1g/litter) and stocked in the cold. Before use, the phenol red solution (PRS) was kept at 37°C in a 5% CO_2-95% air chamber for 1 hr to allow it to equilibrate. Purified crystalline horseradish peroxidase (HRPO) (Sigma Chemical Co.) was dissolved in HBSS (165u/ml).

Measurement of H_2O_2 release from macrophages

Production of H_2O_2 by macrophages was quantified according to the method of Pick and Keisari (22). At the indicated time after the culture, macrophage monolayers were rinsed twice with warm EBSS, and then 1.55 ml PRS plus 0.05 ml HRPO solution was poured into each well where the macrophages remained. The culture plates were replaced in the 5% CO_2-95% air chamber. At 10, 20, 30 and 40 min after the incubation, 1 ml of supernatant from each culture was transferred to a 10 x 75 mm glass tube containing 10 μl of 1N NaOH. The absorbance of the samples were read at 610 nm. Standard curves were made using the same batch of PRS and H_2O_2 solutions, with final H_2O_2 concentrations of 10 to 1000 nmol. Amounts of H_2O_2 generated in 4 samples which had been obtained at 10, 20, 30 and 40 min, respectively, from the same group of cultures were summed up. The sum was taken as the total yield of H_2O_2/total period of incubation (100 min)/culture, and experimental values at the indicated time were calculated from the following formula;

$$\text{nmol. } H_2O_2\text{/25 min/culture} = \frac{\text{the sum}}{100} \times 25$$

The mean and its standard deviation were calculated from the values obtained from three independent experiments.

Results

Hydrogen peroxide production of BALB/c macrophages by LPS stimulation.

Prior to studying the hydrogen peroxide production of peritoneal macrophages obtained from genetically immuno-defective mice, we examined the ability of macrophages from normal mice. When macrophages of BALB/c mice, which were presumed to be immunologically normal, were incubated in the absence of LPS, the yield of H_2O_2 was quite low (Figure 1). However, in the presence of LPS (0.01-100μg/ml), the macrophages produced large amounts of H_2O_2. During 1 to 3 hr after the incubation with LPS, cultured macrophages continued to release a large quantity of H_2O_2, but at 6 hr, the production decreased. Ten μg/ml of LPS seemed to be optimal for inducing macrophage H_2O_2 production. One hundred μg/ml LPS was inferior to 10μg/ml LPS for the production, suggesting that an excess dose of LPS may have some toxic effect on the cultured macrophages. In the range from 0.01μg/ml to 10μg/ml, the yields seemed to be dose-dependent, since the more LPS was used, the greater production was observed.

Comparison of effects of endotoxins

Comparative studies on the effects of several kinds of endotoxins such as phenol-extracted LPS, TCA-extracted Boivin antigen, lipid A and artificial lipid A-BSA-complex were conducted by using BALB/c macrophage cultures. As shown in Figure 2, the cultures in the presence of LPS, Boivin antigen or lipid A-BSA-complex released large amounts of H_2O_2, while those in the presence of lipid A showed poor production.

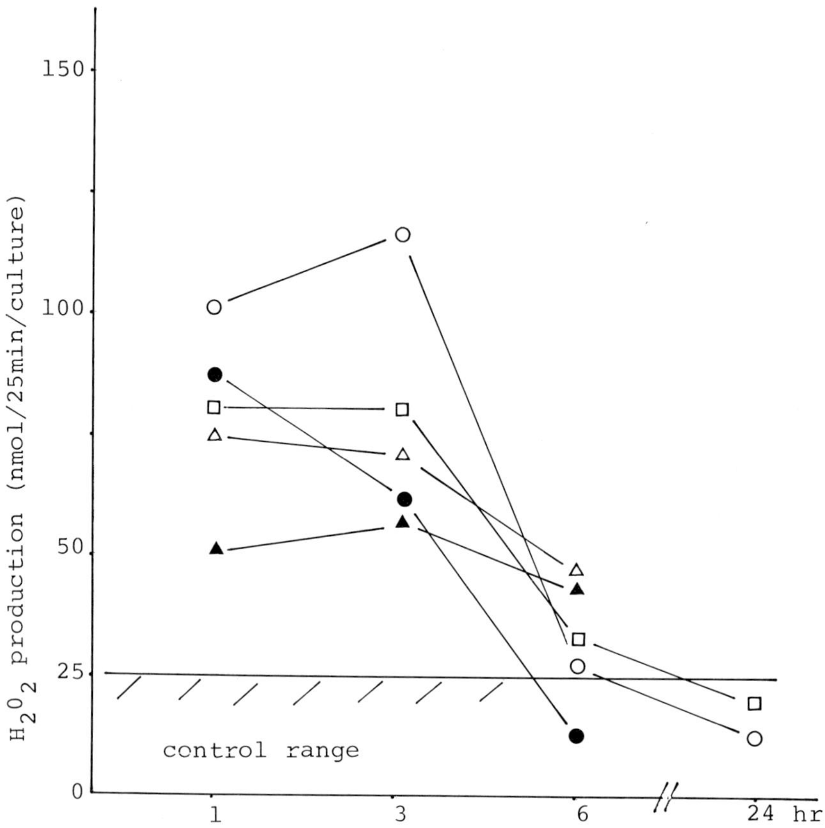

Figure 1. LPS-induced H_2O_2 production <u>in vitro</u> of peritoneal exudate macrophages obtained from BALB/c mice. Thioglycollate-induced macrophages were cultured in the presence or absence of LPS. At the times indicated on the abscissa, the amounts of H_2O_2 in the supernatants of cultures were determined. LPS added to cultures: ▲,0.01 μg/ml; △,0.1 μg/ml; □,1 μg/ml; ○,10 μg/ml; ●,100 μg/ml. ////: range of H_2O_2 yield in the supernatants of cultures in the absence of LPS.

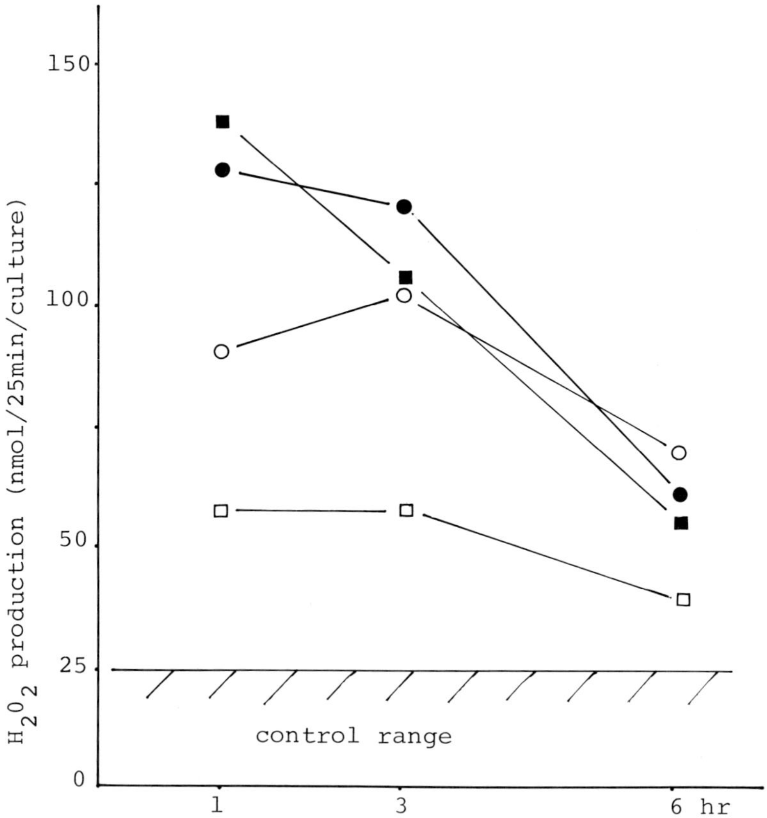

Figure 2. Endotoxin-induced H_2O_2 production *in vitro* of peritoneal exudate macrophages obtained from BALB/c mice. Thioglycollate-induced macrophages were cultured in the presence or absence of endotoxins. At the times indicated on the abscissa, the amounts of H_2O_2 in the supernatants were determind. Endotoxins (10 µg/ml) added to cultures: ○,LPS; ●,Boivin antigen; ■,lipid A-BSA-complex and □,lipid A. ///// : range of H_2O_2 yield in the supernatants of cultures in the absence of endotoxins.

These results indicate that Boivin antigen and the lipid A-BSA-complex as well as LPS have the ability to induce macrophages to produce an oxygen burst, and the lipid A part in the endotoxins is the active principle for the burst. The weak effect of lipid A itself in comparison with the lipid A-BSA-complex must be due to its insolubility.

Comparison of H_2O_2 production between C3H/He and C3H/HeJ macrophages.

The C3H/HeJ strain is a mutant from the C3H/He strain, and they possess the same genetic background except for responsiveness to LPS. It is known that C3H/HeJ mice are refractory to LPS, but C3H/He mice are responsive to LPS (12). However, the LPS-nonresponsive C3H/HeJ mice can respond to some kinds of endotoxins which have a protein moiety with lipid A (12). Therefore, it may be expected that HeJ macrophages can hardly produce H_2O_2 after the stimulation of LPS. Macrophages obtained from C3H/HeJ and C3H/He mice were cultured, and the H_2O_2 production was examined. As shown in Figure 3, He macrophages yielded large amounts of H_2O_2 after the stimulation of LPS, but HeJ macrophages did not. When the HeJ macrophages were cultured in the presence of Boivin antigen, the LPS-nonresponsive HeJ macrophages produced H_2O_2. These results indicate that in the production of H_2O_2, the C3H/HeJ macrophages are refractory to LPS, but Boivin antigen is effective on them.

Low H_2O_2 yields of CBA/N-defective macrophages

CBA/N-defective mice are low responders to LPS if their abilities are estimated by the proliferative response and polyclonal antibody production of B lymphocytes (23, 24). However, the effects of endotoxins on the CBA/N macrophages are quite obscure. We cultured the macrophages obtained from (CBA/N female x BALB/c male) F1 males or females, and endotoxin-induced H_2O_2 production of the cultures was examined.

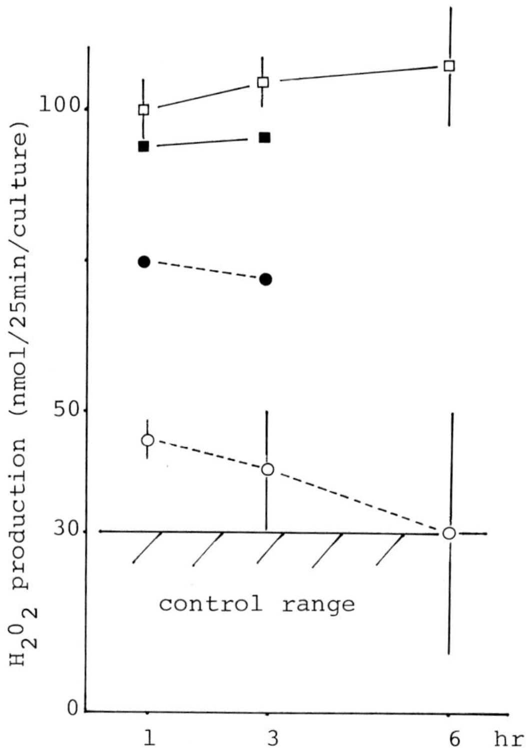

Figure 3. LPS- or Boivin antigen-induced H_2O_2 production _in vitro_ of peritoneal exudate macrophages obtained from C3H/HeJ or C3H/He mice. Thioglycollate-induced macrophages were cultured in the presence or absence of endotoxins. At the times indicated on the abscissa, the amounts of H_2O_2 in the supernatants were determined. Endotoxins (10 µg/ml) added to cultures: □—□,He cell cultures with LPS; ■----■,He cell cultures with Boivin antigen; ○—○,HeJ cell cultures with LPS and ●----●,HeJ cell cultures with Boivin antigen. ⫽⫽⫽ : range of H_2O_2 yield in the supernatants of cultures in the absence of endotoxins.

As shown in Figure 4, the H_2O_2 yields of the CBA/N-defective F1 male macrophages were obviously lower than those from phenotypically normal F1 female macrophages. Boivin antigen as well as LPS was unable to induce the oxygen burst normally in the CBA/N-defective macrophages. These findings suggest that this inability of the CBA/N-defective macrophages is not due to the "type" of endotoxins used for induction, but probably due to some functional deficiency in the macrophages themselves.

Defective H_2O_2 production of beige macrophages

Chediak-Higashi syndrome phagocytic cells are known to show decreased bactericidal capacity and decreased chemotaxis (16, 36). The basic defect is cellular, and it may cause abnormal mobility and/or fusion of intracellular organelles such as lysosomes. Beige mice are one of the available CHS animal models. Therefore, it seems worthwhile to examine whether or not the macrophages respond to endotoxins and release H_2O_2 into the culture medium. The macrophages of beige mice were cultured in the presence or absence of endotoxins, and the H_2O_2 production was assessed. As a control, the H_2O_2 yields of macrophages obtained from either F1 (bg/+) or C57BL/6 (+/+) were also determined. As shown in Figure 5, at 1 hr after the incubation, bg/bg macrophages could respond well to LPS, Boivin antigen and lipid A-BSA-complex and produced H_2O_2 to the similar extent as control macrophages. However, the H_2O_2 productivity of beige macrophages sharply dropped at 3 hr. Lipid A itself showed very low effects on either bg/bg macrophages or control ones. These results suggest that bg/bg macrophages can produce H_2O_2 in the early stage after endotoxin-stimulation, but the production does not continue as it does in normal macrophages.

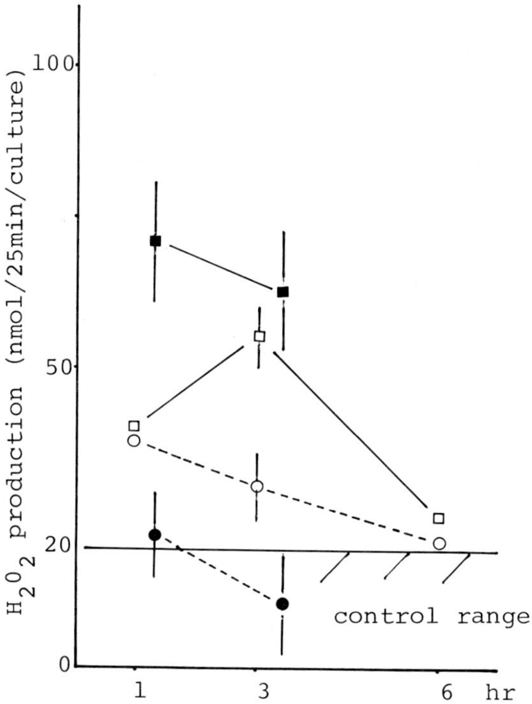

Figure 4. LPS- or Boivin antigen-induced H_2O_2 production *in vitro* of peritoneal exudate macrophages obtained from (CBA/N female x BALB/c male)F1 mice. Thioglycollate-induced macrophages of F1 male with CBA/N-defect (○and●) or phenotypically normal F1 female (□and■) were incubated in the presence or absence of LPS or Boivin antigen. At the times indicated on the abscissa, the amounts of H_2O_2 in the supernatants of cultures were determined. ○ and □ :LPS,10 μg/ml. ● and ■ :Boivin antigen, 10 μg/ml. ////: range of H_2O_2 yield in the supernatants of cultures in the absence of LPS or Boivin antigen.

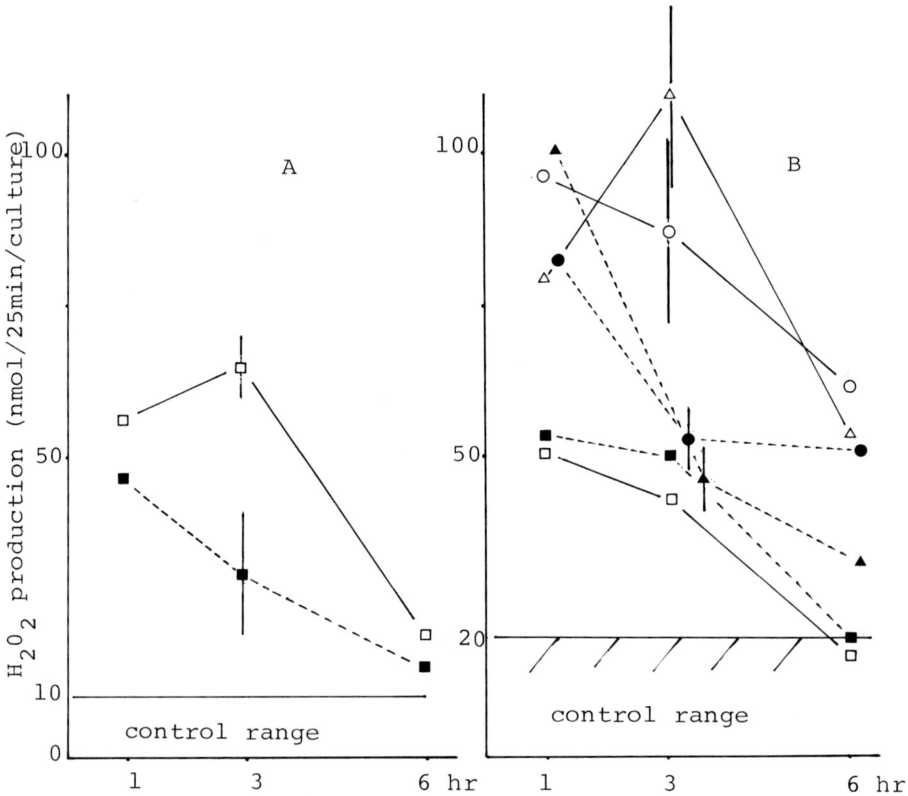

Figure 5. Endotoxin-induced H_2O_2 production *in vitro* of peritoneal exudate macrophages obtained from beige (*bg/bg*) or control (*bg/+* or *+/+*) mice. Thioglycollate-induced macrophages were incubated in the presence or absence of endotoxins (10 µg/ml). At the times indicated on the abscissa, the amounts of H_2O_2 in the supernatants of cultures were determined. In panel A, macrophages from *bg/bg* (■----■) or *bg/+* (□——□) were cultured with LPS. In panel B, macrophages from *bg/bg* (-----) or from *bg/+* (——) mice were cultured with Boivin antigen (○ or ●), lipid A-BSA-complex (△ or ▲) or lipid A (□ or ■). ///// : range of H_2O_2 yield in the supernatants of cultures in the absence of LPS or other endotoxins.

Discussion

Hydrogen peroxide is a key active oxygen product involved in host defense and inflammation. In vitro treatment of murine peritoneal macrophages with LPS resulted in the generation of hydrogen peroxide, unless the macrophages were derived from immunologically defective mice. Boivin antigen and lipid A-BSA-complex as well as LPS were also potent stimulators of the H_2O_2 production in cultured macrophages, while lipid A itself was not (Figures 1 and 2). These results suggest that the H_2O_2 production of macrophages is triggered by lipid A molecules in the endotoxins if it is given in a soluble form. It is believed that the lipid A is an active principle of endotoxins, though its endotoxic activity can hardly be shown due to its insolubility (18, 19). Our results are consistent with this concept.

There are great differences in the quantitative yield of H_2O_2 between the macrophages obtained from some kinds of immunologically defective mice and those from normal mice. We used three kinds of defective mice such as C3H/HeJ, CBA/N and beige strains. The cultured macrophages obtained from these mice showed quite poor production of H_2O_2 in response to LPS.

The C3H/HeJ strain has been reported to be unresponsive to LPS, but the mice can respond to endotoxins with a protein moiety (12). The HeJ macrophages are defective in their response to LPS with respect to macrophage-mediated tumor lysis, LPS-induced lethality, and production of prostaglandin and lymphocyte activating factor (25-31). Pabst and Johnston have already reported that levels of superoxide anion (O_2^-) released from LPS-primed HeJ macrophages by phorbol myristate acetate were low in comparison with those from normal C3Heb/FeJ macrophages (32). The low O_2^- formation in HeJ macrophages may participate in their decreased production of H_2O_2 in response to LPS, as these "cidal" reactive oxygen products are metabolically closely related to each other (33, 34). In

H_2O_2 generation, HeJ macrophages were unresponsive to LPS, but responsive to Boivin antigen (Fig. 3) which consists of endotoxic lipid, polysaccharide and protein (35). Therefore, HeJ macrophages seem to have receptors on their cell surfaces for endotoxins with a protein moiety.

The X-linked CBA/N gene, xid, seems to regulate negatively the endotoxin-induced H_2O_2 production since the production by macrophages from CBA/N-defective mice after endotoxin-stimulation was always poor (Fig. 4). Boivin antigen and lipid A-BSA-complex as well as LPS were also ineffective on the production of the defective macrophages. In this point, the defective H_2O_2 production in CBA/N macrophages is different from that in HeJ macrophages. Therefore, it may be thought that the inability of CBA/N macrophages to produce H_2O_2 is due to intracellular events in the cells, but not due to the lack of receptors on the cell surfaces for LPS.

The H_2O_2 production of macrophages obtained from beige mice with Chediak-Higashi syndrome were also affected to some extent by endotoxins (Fig. 5). Kinetics of the production in beige macrophages after endotoxin-stimulation was obviously different from those in HeJ and CBA/N macrophages. The beige macrophages showed a good yield of H_2O_2 in the early stage (at 1 hr after the culture) and the amounts were the same as that produced by normal macrophages. However, unlike that for normal macrophages, the production of beige macrophages could not continue till the later stages. Chediak-Higashi syndrome neutrophiles and monocytes possess decreased chemotactic responses and diminuated bactericidal activity, and the malfunctions of the Chediak-Higashi syndrome phagocytic cells are presumed to be caused by abnormal microtuble activity (36). Beige macrophages seem to have the ability to respond to endotoxins and generate H_2O_2 normally, but the continuation and/or release of H_2O_2 may be hampered by their abnormal microtuble activity.

In this paper, we clearly demonstrated that the HeJ, CBA/N and beige macrophages are defective in H_2O_2 productin after the stimulation of endotoxins. The susceptibility of these mice to some kinds of bacterial infections may be partially explained by the malproduction of H_2O_2 in their macrophages.

Summary

Hydrogen peroxide production of peritoneal exudate macrophages obtained from normal or immunologically defective mice were examined in vitro. Thioglycollate-induced peritoneal macrophages, which had been obtained from BALB/c, C3H/He, C57-BL/6, C3H/HeJ (LPS-nonresponder), (CBA/N female x BALB/c male) F1 male (CBA/N-defect) and F1 female (normal), beige (Chediak-Higashi syndrome) mice, were cultured for 1 to 24 hr in the presence or absence of endotoxins such as bacterial lipopolysaccharide (LPS), Boivin antigen, lipid A or lipid A-bovine serum albumin (BSA)-complex. Yields of H_2O_2 in the supernatants of the cell cultures were spectrophotometrically assessed by oxidation of phenol red in the presence of hydrogen peroxidase (Pick and Keisari's method).

Macrophages obtained from the mice with normal traits produced significant amounts of H_2O_2 after the stimulation by LPS, Boivin antigen or lipid A-BSA-complex. Lipid A itself was less effective than other endotoxins. The H_2O_2 production from C3H/HeJ macrophages by LPS was quite poor, while Boivin antigen could stimulate the macrophages to produce H_2O_2. Macrophages of CBA/N-defective mice could not produce sufficient amounts of H_2O_2 after the stimulations by these endotoxins. Beige macrophages were also defective in the ability of endotoxin-induced H_2O_2 production. The macrophages retained their ability to respond to endotoxins in the early stage (1 hr after the stimulation), but their H_2O_2 productivity was rapidly lost thereafter.

References

1. Nathan, C.F., Bruckner, L.H., Silverstein, S.C., Cohn, Z.A. : J. Exp. Med. 149, 84-99 (1979).
2. Nathan, C.F., Silverstein, S.C., Bruckner, L.H., Cohn, Z.A. : J.Exp. Med. 149, 100-113 (1979).
3. Nathan, C.F., Bruckner. L.H., Kaplan, G., Unkeless, J., Cohn, Z.A.: J. Exp. Med. 152, 183-197 (1980).
4. Murray, H.W., Junangbhanich, C.W., Nathan, C.F., Cohn, Z.A. : J. Exp. Med. 150, 950-964 (1979).
5. Murray, H.W., and Cohn, Z.A.: J. Exp. Med. 152, 1596-1609 (1980).
6. Murray, H.W., Nathan, C.F., Cohn, Z.A.: J. Exp. Med. 152, 1610-1624 (1980).
7. Dockrell, H.M., Playfair, J.H.L.: Infect. Immun. 39, 456-459 (1983).
8. Cohn, M.S., Taffet, S.M., Adams, D.O.: J. Immunol. 128, 1781-1785 (1982).
9. Peavy, D.L., Brandon, C.L.: Bacterial endotoxins and host response (Ed. by Agarwal, M.K.), 299-309, Elsevier/North-Holland Biomedical Press, Amsterdam (1980).
10. Hibbs, J.B.Jr.: Science 184, 468-471 (1974).
11. Currie, G.A., Basham, C.: J. Exp. Med. 142, 1600-1605 (1975).
12. Morrison, D.C., Ryan, J.L.: Adv. Immunol. 28, 293-450 (1979).
13. Scher, I.: Immunologic defects in laboratory animals 1 (Ed. by Gershwin, M.E., Merchant, B.), 163-190, Plenum Press, New York and London (1981).
14. O'Brien, A.D., Metcalf, E.S., Rosenstreich, D.L.: Cell. Immunol. 67, 325-333 (1982).
15. O'Brien, A.D., Scher, I., Campbell,G.H., MacDermott, R.P., Formal, S.B.: J. Immunol. 123, 720-724 (1979).
16. Brandt, E.J., Swank, R.T., Novak, E.K.: Immunologic defects in laboratory animals 1 (Ed. by Gershwin, M.E., Merchant, B.), 99-116, Plenum Press, New York and London (1981).
17. Westphal, O., Lüderitz, O.: Angew. Chem. 66, 407-417, (1954).
18. Galanos, C., Reitschel, E.T., Lüderitz, O., Westphal, O.: Eur. J. Biochem. 19, 143-152 (1971).
19. Galanos, C., Reitschel, E.T., Lüderitz, O., Westphal, O., Kim, Y.B., Watson, D.W.: Eur. J. Biochem. 31, 230-233 (1972).

20. Boivin, A., Mesrobeanu, J., Mesrobeanu, L.: Comt. Rend. Soc. Biol. 113, 490-492 (1933).

21. Kabat, E.A., Mayer, M.: Experimental immunochemistry, 2nd ed. Charles C. Thomas Publisher, Springfield Ill. (1961).

22. Pick, E., Keisari, Y.: J. Immunol. Methods 38, 161-170 (1980).

23. Scher, I., Ahmed, A., Strong, D.M., Steinberg, A.D., Paul, W.E.: J. Exp. Med. 141, 788-803 (1975).

24. Scher, I., Zaldivar, N.M., Moiser, D.E.: Microbiology-1977 (Ed. by Schlessinger, D.), 310-320, American Society for Microbiology, Washington, D.C. (1977).

25. Tagliabue, A., McCoy, J.L., Herberman, R.B.: J. Immunol. 121, 1223-1226 (1978).

26. Ruco, L.P., Meltzer, M.S.: J. Immunol. 121, 2035-2042 (1978).

27. Weinberg, J.B., Chapman, H.A.Jr., Hibbs, B.Jr.: J. Immunol. 121, 72-80 (1978).

28. Wahl, L.M., Rosenstreich, D.L., Glode, L.M., Sandberg, A. L., Mergenhagen, S.E.: Infect. Immun, 23, 8-13 (1979).

29. Ruco, L.P., Meltzer, M.S.: Cell. Immunol. 41, 35-51 (1978).

30. Rosenstreich, D.L., Vogel, S.N., Jacques, A.R., Wahl, L.M., Oppenheim, J.J.: J. Immunol. 121, 1664-1670 (1978).

31. Ryan, J.L., Glode, L.M., Rosenstreich, D.L.: J. Immunol. 122, 932-935 (1979).

32. Pabst, M.J., Johnston, R.B.Jr.: J. Exp. Med. 151, 101-114 (1980).

33. Badwey, J.A., Karnovsky, M.L.: Annu. Rev. Biochem. 49, 695-726 (1980).

34. Cohn, M.S., Ryan, J.L., Root, R.K.: J. Immunol. 127, 1007-1011 (1981).

35. Lüderitz, O., Staub, A.M., Westphal, O.: Bacteriol. Rev. 30, 192-255 (1966).

36. Katz, P., Roder, J.C., Zaytoun, A.M., Herberman, R.B., Fauci, A.S.: J. Immunol. 129, 297-302 (1982).

INDUCTION OF METALLOTHIONEIN IN MACROPHAGES: A MOLECULAR MECHANISM FOR PROTECTION AGAINST LPS-MEDIATED AUTOLYSIS.

STEVEN R. PATIERNO and DUANE L. PEAVY

Dept. of Pharmacology, University of Texas Medical School at Houston, Texas 77225; and the Dept. of Microbiology and Immunology, Baylor College of Medicine, Houston, Texas 77030

INTRODUCTION

Metallothioneins are a family of low-molecular weight, cysteine-rich, metal binding proteins whose synthesis is induced by certain heavy metals, corticosteroids and catecholamines. Originally thought to participate in detoxification of cadmium and mercury, metallothionein appears to have a much broader physiological role since its synthesis occurs in animals exposed to a variety of external stresses and to both chemical and bacterial toxins (For reviews see 1,2,3). In these latter circumstances, metallothionein may function in promoting the hepatic accummulation of zinc since this characteristic response to stress has recently been shown to occur subsequent to metallothionein induction[4,5]. The reason for hepatic accummulation of zinc during stress is not precisely understood; however, it is thought that the liver responds to stress by increasing its metabolic activity, particularly gluconeogenesis. Presumably, the increased metabolic rate results in an increased zinc requirement.

Metallothionein-dependent zinc accummulation occurs in the liver of animals following injection of bacterial lipopolysaccharide (LPS or endotoxin) and during gram negative infection. Interestingly, exogenous administration of zinc or corticosteroids induces hepatic metallothionein and protects against LPS lethality[6,7,8,9]. In the former case, metallothionein was synthesized by both parenchymal and non-parenchymal

Immunopharmacology of Endotoxicosis

cells[10]. Since many of the pathophysiological features of LPS toxicity result from direct interaction with cells of the mononuclear phagocyte system; stimulating release of pyrogens, prostaglandins, procoagulants, lysosomal, hydrolases and other factors possessing potent biological activity (For review see 11), we examined the potential role of metallothionein in protecting macrophages from the cytotoxic effects of LPS.

MATERIALS AND METHODS

Animals. Six to 8 week old C57BL/6 mice (Jackson Laboratories, Bar Harbor, ME) were housed in cages with barrier filters, fed standard laboratory chow and water and used within 6 weeks of arrival.

Endotoxin. LPS was extracted from formalin-fixed *Salmonella typhimurium* TV 119 using the hot phenol/water technique. Crude LPS was purified by differential centrifugation until spectrophotometrically free of RNA and protein. Purified LPS was lyophilized and stored at room temperature. Immediately before use, LPS was resuspended in sterile, distilled H_2O and was heated in a boiling water bath for 15 min. ^{3}H-LPS (sp. act. 8×10^5 cpm/μg) was a gift of Dr. Bobby J. Gormus, Delta Regional Primate Center, Covington, LA).

Steroids. Acetates of cortisone, corticosterone, desoxycorticosterone, estradiol, prednisolone and testosterone were obtained from Steraloids, Inc., Pawling, NY and Sigma Chemical Co., St. Louis, MO. Before use, each steroid was dissolved in sterile, distilled H_2O and sonicated for 10-15 min. with an ultrasonic cleaner (Mettler Electronics, Anaheim, CA).

Metals. Ultrapare, anhydrous chlorides of cadmium, calcium, lead (II), manganese, mercury (II), and zinc were purchased from Alfa Products,

Thiobol/Ventron Division, Danvers, MA. All metal salts were stored dessicated and were dissolved in sterile, distilled H_2O just before use.

Preparation of macrophage monolayers, treatment conditions and viability determination. A detailed description for collecting mouse peritoneal leukocytes and for preparing adherent cell monolayers has been reported previously[12]. Briefly, peritoneal leukocytes from normal mice were harvested by lavaging the peritoneum with 3 ml portions of complete tissue culture medium (TCM; Hanks' minimal essential medium supplemented with 5% aseptically collected, heat-inactivated, adult human serum; L-glutamine, gentamycin, sodium bicarbonate, and 1 u/ml heparin). Cells were washed twice, counted and resuspended in TCM without heparin @ 1.5×10^6 cells/ml. 0.3 ml of the leukocyte suspension was overlayed onto 22 x 22 mm, sterile glass coverslips and incubated @ $37^{\circ}C$ for 30 min. Coverslips were rinsed twice in TCM and were placed in 35 mm plastic culture dishes containing 1 ml TCM. Varying amounts of LPS, steroid hormones and metal salts were added in 0.1 ml for the indicated time periods and all cultures were incubated at $37^{\circ}C$ in a humidified culture chamber containing an atmosphere of 83% N_2, 10% CO_2 and 7% O_2. Pre-treatments with steroids or metals were performed as described above except that residual hormone and metal were removed by extensive washing with Hanks' balanced salt solution (HBSS, Microbiological Associates, Bethesda, MD) and coverslips were placed in new dishes containing 1 ml TCM and LPS. Metabolic inhibitors were included in designated cultures at final concentrations of 20 nM actinomycin C_1 (Boehringer/Mannheim, Mannheim, West Germany) or 1 μM cycloheximide (Boehringer/Mannheim). At termination, cell viability was determined by trypan blue exclusion.

Viable counts were based on observations of 200 mononuclear cells/coverslip. Results were expressed as the mean percent of viable cells obtained from three replicate cultures ± standard error of the mean (S.E.M.)

Measurement of metallothionein. Functional metallothionein was measured in macrophages in the following manner. For these experiments, Ham's F-12 medium without L-cysteine or $ZnSO_4$ (formulated by M.A. Bioproducts) was used instead of Hanks' MEM. Peritoneal leukocytes were harvested in the usual manner and monolayers prepared in 35 mm culture dishes. Cells were treated simultaneously with 10 μM $CdCl_2$, 10 μM $CdCl_2$ and 20 nM actinomycin C_1 or 0.1 ml HBSS and 10 μCi/ml (^{35}S)-cysteine (500 Ci/mmole; New England Nuclear, Boston, MA) and were incubated at $37^{o}C$ for varying time periods. At termination, all cultures were washed three times with 3 ml portions of TCM without serum. Macrophages were removed by gentle scraping into 2 ml TCM without serum with a rubber policeman, collected by centrifugation, resuspended in PBS lacking Ca^{++} and Mg^{++}, added to plastic vials, frozen and stored under liquid nitrogen.

Metallothionein was measured by a modified mercury binding assay which has been previously described in detail[13] and by the rate of incorporation of (^{35}S) cysteine into the metallothionein fraction. For the mercury binding assay, frozen cell samples were thawed and lysed by sonication. Mercury binding was determined by allowing 100 μl of each sample lysate to react with 50 μl of ^{203}Hg (1-2 nmoles of ^{203}Hg, 6 x 10^6 dpm) for 10 min. at room temperature. Non-metallothionein, protein-bound ^{203}Hg was precipitated by addition of cold trichloroacetic acid to a final concentration of 2.2%. Free ^{203}Hg was removed by centrifugation of the TCA supernatant through Sephadex G-10 minicolumns. The ^{203}Hg bound to metallothionein in the filtrates was counted directly in a Beckman Gamma 4000. The sensitivity and specificity of the ^{203}Hg-binding assay was verified using a homogeneous metallothionein standard from rat liver (cadmium/zinc metallothionein II) and other potential mercury-binding ligands such as glutathione. (^{35}S)-cysteine incorporation was measured by using the identical procedure except that ^{203}Hg addition was omitted. Filtrates were dissolved in 2 ml Beckman Ready-Solv HP and were counted in a Beckman LS 7500 liquid scintillation counter. Results were expressed as the ratio of control incorporation.

RESULTS

Administration of corticosteroids or stimulation of adrenal steroid synthesis by adrenocorticotropic hormone (ACTH) has been shown to have both prophylactic and therapeutic value in animals given lethal doses of LPS.[6,7] To determine if corticosteroids were exerting a direct effect on peritoneal adherent cells, a number of structurally-related, biologically-active steroids were added to normal macrophage monolayers that were subsequently incubated for 48 hrs in the presence or absence of LPS.

TABLE 1

STEROID PROTECTION AGAINST LPS TOXICITY

	CELL VIABILITY[a]					
	LPS[b]					Control
Steroid (μM)	0	0.1	1	10	100	(100 μM)
Cortisone	55	58	63	88	90	99
Corticosterone	38	65	85	90	92	92
Prednisolone	62	85	90	90	97	98
Desoxycortisone	38	40	47	62	80	98
Testosterone	55	53	60	54	60	96
Estradiol	55	60	52	61	57	97

[a]S.E.M. < 5 for all values
[b]50 μg/ml for 48 hrs.

0.1 μM cortisone acetate did not effect the viability of LPS-treated macrophages. However, 1 μM cortisone acetate produced a definite and statistically significant ($P < 0.025$) difference that increased in a concentration-dependent fashion. Acetates of corticosterone and prednisolone possessed activity similar to cortisone except that protection was observed over a wider range of concentrations. Desoxycorticosterone acetate was considerably less effective since significant protection only occurred using 100 μM. Estradiol acetate and testosterone acetate had no appreciable effect. Table 1 also illustrates that all steroids were non-toxic for normal macrophages since these substances did not affect cell viability even at the highest concentrations tested.

Intraperitoneal administration of zinc and manganese also protected mice from LPS toxicity;[8,9] therefore, these and other metal salts were examined for their ability to protect macrophages from LPS in vitro. LPS, 50 μg/ml, and varying amounts of divalent metal salts were added at culture initiation and cell viability was assessed after 48 hrs.

TABLE 2

METAL PROTECTION AGAINST LPS TOXICITY

Metal	CELL VIABILITY[a,b]				
(μM)	0	0.1	1	10	100
Zinc	99/50	n.d.[c]	99/65	98/85	60/55
Cadmium	98/35	97/43	96/67	92/38	n.d.
Mercury	97/40	95/39	93/71	30/21	n.d.
Manganese	98/53	99/50	97/78	96/85	80/80
Lead	99/45	95/49	90/42	81/42	n.d.
Calcium	98/49	97/45	91/46	83/36	75/29

[a]Metal/Metal + LPS
[b]S.E.M. < 6 for all cultures
[c]not done

As shown in Table 2, μM concentrations of zinc, cadmium, mercury and manganese protected macrophages from LPS toxicity in a concentration-dependent fashion. In all cases, protection reached an optimum after which the inherent toxicity of the metal presided. Lead and calcium had no protective effect at any concentration tested.

TABLE 3

TEMPORAL DEPENDENCE OF STEROID AND METAL PROTECTION

Treatment (10 μM)	CELL VIABILITY[a] Time of Addition (hrs)							
	-4	-2	-1	-0.5	0	+2	+4	+6
Cortisone	90	87	80	--	80	65	56	53
Zinc	--	95	85	79	83	72	52	49

[a]S.E.M. < 6 for all values
[b]Agents were added at the indicated hrs before or after 50 μg/ml LPS and remained in culture until termination.

The protective effects of steroids and heavy metals were examined further by altering the time of treatment in relation to LPS addition. Macrophages were markedly protected by 10 μM cortisone acetate when added during an interval of 1-4 hrs prior to LPS (Table 3). Cell viability also improved when both substances were added simultaneously; however, protection was considerably less impressive when cortisone acetate was added 2 hrs after LPS and no significant protection occurred after 4 and 6 hrs. Similar time-dependent effects were observed with zinc. Cortisone's protective effects were also examined by pre-treating macrophage monolayers for varying time periods. Coverslips were then removed, washed extensively in TCM and transferred to new culture dishes containing 1 ml TCM and 50 μg LPS. Pre-treatment of macrophages for 6 hrs with 10 μM cortisone acetate provided the same degree of protection that occurred when cortisone acetate was added simultaneously with LPS and remained in the medium for 24 hrs. Pre-treatment for 2 or 4 hrs provided similar results although protection under these circumstances was somewhat diminished (data not shown).

Since LPS associates with mammalian cell membranes and corticosteroids and zinc are known to stabilize biological membranes,[14] it was suspected that these agents may have altered LPS uptake and, therefore, its activity. However, using ^{3}H-LPS, no differences in uptake in normal, cortisone- or zinc-treated macrophages could be detected (data not shown). In addition, the morphological appearances of these cells were similar; again suggesting that LPS uptake was unaffected.

Since macrophages possess high affinity steroid receptors in their cytoplasm,[15,16] we became interested in their response to these agents with regards to RNA and protein synthesis. Brief exposure of macrophages to corticosteroids resulted in a rapid stimulation of RNA and protein synthesis. A 60 per cent increase in ^{3}H-uridine incorporation and a 40 per cent increase in ^{3}H-proline incorporation was observed 30 min. after addition of 10 μM corticosterone acetate. One μM prednisolone acetate and .01 μM dexamethasone acetate had similar activity (data not shown).

To determine if corticosteroids or metals protected macrophages through synthesis of one or more specific proteins, reversible inhibitors were added during the 2 hr pre-treatment period followed by removal by extensive washing of the monolayer. Inclusion of 20 nM actinomycin C_1, a concentration which inhibited ^{3}H-uridine incorporation in cortisone-treated macrophages by 78 per cent, or 1 μM cycloheximide, a concentration that inhibited ^{3}H-leucine incorporation by 89%, did not affect the viability of normal macrophages, but completely abolished the protective activity of cortison, zinc and cadmium (Table 4).

TABLE 4

DEPENDENCE OF STEROID AND METAL PROTECTION ON DE NOVO SYNTHESIS

Treatment[a]	CELL VIABILITY[b]	
	Control	LPS
None	99	58
Cortisone (10 μM)	98	89
Cortisone (10 μM) + actinomycin C_1	98	59
Cortisone (10 μM) + cycloheximide	97	50
Zinc (10 μM)	99	79
Zinc (10 μM) + actinomycin C_1	97	65
Cadmium (10 μM)	97	82
Cadmium (10 μM) + actinomycin C_1	96	59

[a]Monolayers were exposed to designated agents for 2 hrs @ 37°C, washed extensively and transferred to dishes containing 1 ml TCM and 50 μg LPS.

[b]S.E.M. < 4 for all values

The fact that protection against LPS toxicity by steroids and metals was dependent upon de novo RNA and protein synthesis, coupled with the knowledge that these agents can induce metallothionein, led us to investigate whether this protein was synthesized by macrophages. A standardized ^{203}Hg-binding assay, modified in this laboratory for use with small numbers of cultured cells, was employed. The incorporation of ^{35}S-cysteine into the metallothionein fraction was also measured simultaneously.

TABLE 5

SPECIFICITY OF METALLOTHIONEIN INDUCTION

Treatment	^{203}Hg binding (pmole/10^6 cells)[a]			
Time (hr)	2	4	8	12
None	0.08	0.10	0.07	0.08
Cadmium (10 μM)	0.16	0.12	0.52	0.40
Zinc (10 μM)	0.12	0.18	0.47	0.45
Calcium (10 μM)	0.14	0.22	0.12	0.14
Lead (10 μM)	0.09	0.11	0.14	0.12
Sodium (10 μM)	0.08	0.07	0.06	0.09

[a]Measured in TCA-soluble fitrates of Sephadex G-10 minicolumns.

As shown in Table 5, ^{203}Hg-binding activity, a measurement of functional metallothionein, doubled within 2 hrs and peaked after 8 hrs at a level 7.5 times the level of untreated controls. Zinc was equally effective as cadmium in inducing metallothionein; however, calcium, lead and sodium, cations that failed to protect against LPS in vitro, were ineffective inducers. In all cases, the degree of ^{35}S-cysteine incorporation into the metallothionein fraction, paralleled the increase in ^{203}Hg-binding activity, thus, verifying the protein as metallothionein.

We next attempted to determine whether protection was associated with metallothionein synthesis by measuring the level of this protein in macrophages under conditions allowing varying degrees of protection.

TABLE 6

CORRELATION BETWEEN METALLOTHIONEIN LEVELS AND PROTECTION

Treatment[a]	METALLOTHIONEIN[b] (pmoles/10^6 cells)	CELL VIABILITY
None	0.09	52
Cadmium (1 μM)	0.19	61
Cadmium (10 μM)	0.52	89
Cadmium (10 μM) + actinomycin C_1	0.20	58
Cortisone (1 μM)	0.05	49
Cortisone (3 μM)	0.05	52
Cortisone (10 μM)	0.21	62
Cortisone (33 μM)	0.34	65
Cortisone (100 μM)	0.44	73

[a]Monolayers were exposed for 2 hrs, washed and then incubated with LPS.

[b]Measured 8 hrs after treatment.

Table 6 demonstrates that 1 μM cadmium almost doubled the level of metallothionein and increased cell viability from 52 to 61 per cent. Ten μM cadmium raised the level of metallothionein to five times that of control and increased viability 37 per cent. Inclusion of actinomycin C_1 inhibited metallothionein synthesis and blocked induction of the protective effect. Addition of 1 or 3 μM cortisone acetate had no measurable effect on metallothionein levels and yielded no protection. However, 10, 33 and 100 μM cortisone acetate induced three-, four- and seven-fold increases in metallothionein while increasing the viability of macrophages only at these concentrations.

DISCUSSION

The results reported here indicate that agents that protected mice from LPS lethality such as cortisone, zinc and manganese also protected murine macrophages from LPS toxicity in vitro. Other anti-inflammatory steroids such as corticosterone, prednisolone and desoxycortisone were also active in this regard whereas testosterone and estradiol were inactive. Surprisingly, protection was also afforded by cadmium and mercury, two extremely toxic metals, within a narrow range of concentrations. Indeed, the concentrations of cadmium that gave protection have been shown by others to actually stimulate macrophages to increase their phagocytic activity[17] which theoretically should have potentiated LPS toxicity.[6] Macrophages were also protected by pretreatment with these agents followed by their removal by extensive washing indicating that a physical interaction with LPS was unlikely and that protection resulted from an intracellular response. Interestingly, delayed addition of these agents was only effective up to 4 hrs after LPS addition suggesting that there is a critical period after which LPS toxicity is irreversible. Protection in all cases was found to be dependent on de novo RNA and protein synthesis. The striking correlation between agents which protected macrophages from LPS toxicity and those agents which are known to induce metallothionein in vivo suggests that this protein may function intracellularly to inhibit LPS toxicity. This hypothesis is supported by the efficacy of pretreatment and the direct correlation between steroid concentration, the degree of protection and the level of metallothionein in macrophages.

Metallothioneins (MW 6000-10,000 daltons) are a family of metal binding proteins that occur in both parenchymal and nonparenchymal tissues of vertebrates, invertebrates and certain microorganisms. Of 61 amino acids, 20 are cysteine which participate in binding 7 g-atoms of metal per molecule of metallothionein. Originally thought to function primarily in chelation, transport and excretion of toxic metals, it is now clear that metallothionein has an important role in zinc and copper hemeostasis.[1,2,3,10,18]

The metallothionein content of several animal tissues, especially liver and kidney, can be increased by parenteral or dietary administration of certain metals such as Hg, Cd, Zn, Ag, Cu, Bi and Co and by parenteral administration of dexamethasone.[19] Metallothionein is also induced in animals by various stresses such as starvation, cold, strenuous exercise,[21] gram negative infection,[5] hepatectomy,[22] and by administration of LPS,[23] carrageenan[24] and alkylating agents such as carbon tetrachloride, iodoacetate, bromobenzene, diethylmaleate, and iodomethane.[25] Presumably, this response is the result of release of endogenous steroids or catecholamines from the adrenal glands. During endotoxemia, zinc accummulates in the liver bound to metallothionein which presumably was induced by endogenous steroids or catecholamines released from the adrenal glands in response to inflammatory stress.[5] Macrophages, however, also possess cytoplasmic steroid receptors making them additional targets for induction of metallothionein.[15] Thus, it appears that, at least for endotoxemia, the two cell types that are most likely to respond by increasing metabolic activity, i.e., hepatocytes and macrophages, are also capable of synthesizing metallothionein and do so under stressful conditions.

Considering that macrophages are one of the first cells to encounter gram negative bacteria or endotoxin, an interesting sequelae seems likely. Macrophages, upon interaction with bacteria or LPS, become activated and liberate leukocyte endogenous mediator by exocytosis or lysis. This substance, a known intermediate of inflammation, may act directly on hepatocytes and macrophages to stimulate metallothionein synthesis, and zinc accummulation, or more likely may stimulate the adrenals to release catecholamines and anti-inflammatory steroids. These then trigger metallothionein synthesis in both hepatocytes and macrophages.

At the molecular level, it is not yet clear what role metallothionein has; however, the involvement of zinc in various cellular processes may reflect its function. It is known that rapidly dividing tissue, i.e., tumor cells or regenerating liver, have a high zinc requirement.[26] Thus, it appears that, as in the case of hepatic zinc accummulation during stress, cells with relatively high metabolic activity require more zinc. It is thought that this is due to an increase in both amount and activity of replicative enzymes such as DNA polymerases and several metabolic enzymes such as aspartate transcarbamylase, leucine aminopeptidase, alcohol dehydrogenase, alkaline phosphatase and superoxide dismutase, all of which are zinc requiring apoenzymes. Induction of metallothionein may, therefore, be a mechanism for increasing the intracellular availability of zinc for enzyme activation. In addition, metallothionein probably directly donates zinc to the enzymes that require them.[27,28]

Metallothionein-mediated activation of superoxide dismutase supports the possibility that this protein may have a role in regulating the oxidative status of the cell. This may occur indirectly, via activation of superoxide dismutase which scavenges superoxide radicals; alternatively, this may occur by increasing the equilibrium concentration of zinc which is known to inhibit the autocatalytic process of lipid peroxidation. Additionally, induction of metallothionein may serve to increase the overall reducing capacity of the cell by providing an efficient sulfhydryl reservoir. While metallothionein does not seem to be a major detoxification mechanism for various alkylating agents,[29] the fact that metabolites of carbon tetrachloride are found associated with metallothionein indicates that this protein may interact with electrophilic reactive species.[30]

Finally, zinc is known to directly stabilize biological membranes[14] presumably by coordination of the polar phosphate head groups of membrane lipids. Thus, by increasing the intracellular concentration of zinc, metallothionein may stabilize both lysosomal and plasma membranes, thus preventing intracellular release of hydrolytic enzymes, inhibiting active exocytosis and preventing cell lysis and release of pharmacologically active substances.

Based on the probable role of metallothionein in zinc homeostasis, its involvement in metal redistribution during stress, its inducibility by steroids and catecholamines and its presence in macrophages and other cells capable of metabolic activation, we propose that metallothionein functions as an acute phase protein whose induction better enables cells to cope with metabolic stress.

REFERENCES

1. Brady, F.O.: TIBS. 7, 143-145 (1982.
2. Webb, M., and Cain, K.: Biochem. Pharmacol. 31, 137-142 (1982).
3. Cherian, M.G., and Goyer, R.A.: Life Sci. 23, 1-10 (1978).
4. Sobocinski, P.Z., Canterbury, W.J., and Mapes, C.A.: (Abstract) Fed. Proc. 36, 1100 (1977).
5. Sobocinski, P.Z., Canterbury, W.J., Jr., Mapes, C.A., and Dinterman, R.E.: Am. J. Physiol. 234, E399-E406 (1978).
6. Peavy, D.L., and Brandon, C.L.: Bacterial Endotoxins and Host Response, M.K. Agarwal, ed. Elsevier/North-Holland Biomedical Press. 299-309 (1980).
7. Spink, W.W.: Arch. Int. Med. 106, 433-442 (1960).
8. Snyder, S.L. and Walker, R.I.: Infect. Immun. 13, 998-1000 (1976).
9. Snyder, S.L., Walker, R.I., and Moniot, J.V.: Infect. Immun. 15, 337-343 (1977).
10. Sciortino, C.V., Failla, M.L., and Bullis, O.B.: Biochem. J. 204, 509-514 (1982).
11. Morrison, D.C., and Ulevitch, R.S.: Amer. J. Path. 93, 525-615 (1978).
12. Peavy, D.L., et al.: Infect. Immun. 21, 310-319 (1978).
13. Patierno, S.R., et al.: Life Sci. 32, 1629-1636 (1983).
14. Chvapil, M.: Life Sci. 13, 1041-1049 (1973).
15. Werb, Z., Foley, R., and Munck, A.: J. Exp. Med. 147, 1684-1694 (1978)
16. Werb, Z.: J. Exp. Med. 147, 1695-1712 (1978).

17. Koller, L.D., and Roan, J.G.: J. Reticuloendoth. Soc. 21, 7-13 (1977).

18. Menard, M.P., McCormick, C.C., Cousins, A.J.: J. Nutr. 111, 1353-1361 (1981).

19. Klaasson, C.D.: Toxicol. 20, 275-279 (1981).

20. Bremner, I., and Davies, N.T.: Biochem. J. 149, 733-738 (1975).

21. Oh, S.H., et al.: Am. J. Physiol. 234, E282-E285 (1978).

22. Ohtake, H., and Koga, M.: Biochem. J. 183, 683-690 (1979).

23. Suzuki, K.T., and Yamamwa, M.: Biochem. Pharmacol. 29, 2260 (1980).

24. Maitani, T., and Suzuki, K.T.: Biochem. Pharmacol. 30, 2353-2355 (1981).

25. Kotsonis, F.N. and Klaassen, C.D.: Tox. Appl. Pharm. 51, 19-27 (1979).

26. Kraker, A.J., and Petering, D.H.: Biol. Trace. Elem. Res. in press (1983).

27. Li, T., et al.: Proc. Nat. Acad. Sci. U.S.A. 77, 6334-6338 (1980).

28. Udom, A.O., and Brady, F.O.: Biochem. J. 187, 329-335 (1980).

29. Tobey, T., et al.: Canc. Res. 42, 2980-2984 (1982).

30. Cagen, S.Z., and Klaassen, C.D.: Tox. Appl. Pharm. 51, 107-116 (1979).

RES FUNCTION AND A ROLE OF PLASMA HDL IN ENDOTOXIN-POISONED MICE

Shuhei Sakaguchi, Hiroharu Abe, Osamu Sakaguchi

First Department of Hygienic Chemistry, Tohoku College of Pharmacy, Sendai, Miyagi 983, Japan

Introduction

It has generally been assumed that the resistance of animals to the harmful effects of endotoxins is depended on the detoxifying capacity of the reticuloendothelial system(RES), especially the Kupffer cells in the liver(1,2,3,4,5). Many investigators(2,4,6) have demonstrated that a ''blockade'' of the RES, such as saccharated iron oxide, throtrast, lead acetate, and trypan blue, increases the sensitivity of the animals to the lethal effect of endotoxin. On the other hand, RES stimulating agents, such as glucan, zymosan, and triolein, have been observed to enhance endotoxin senstivity(3,6,7). There are many paradoxical findings in numerous reports on the RES function following the administration of RES-stimulating or blocking agents. On the assuption that lead acetate sensitizes mice to endotoxin through the inactivation of sulfhydryl group , lead acetate may be considered to be a specific RES blockade dissimilar with the other agents(8).

Skarnes(9) demonstrated that endotoxin complexes with serum α_1-globulin having nonspecific esterase activity. Recently, Bøg-Hansen and his co-workers(10 ,11,12) surmised that HDL represents a transport protein for endotoxin from plasma to clearance organs or to other cellular targets, and that endotoxin interacts with plasma HDL apoprotein, forming a complex with a characteristic mobility.

Immunopharmacology of Endotoxicosis

The present work was carried out to observe the effect of endotoxin administration under RES depression by several agents and to determine the role of serum lipoprotein.

Materials and methods

Endotoxin. *Salmonella typhimurium* endotoxin(Westphal preparation obtained from Difco Detroit, MI, U.S.A) was used throughout this study.

Animals and treatment. Male ddy mice, 5 to 6 weeks old, weighing 22 to 25 g, were purchased from Shizuoka Experimental Animal Farme Co.(Japan) and maintained in our laboratory for one week before the experiments. Animals were injected intraperitoneally(i.p.) with 0.2 ml of saline solution containing an indicated dose of endotoxin. Control mice received i.p. 0.2 ml of saline only. Lead acetate(50 mg/kg) was injected intravenously(i.v.) into each mouse 10 min after endotoxin administration. An indicated dose(50 mg/kg, 100 mg/kg and 200 mg/kg) of 1.25% trypan blue(Wako Pure Chemicals Industries., Osaka) was injected i.v. into mice 24 hr before endotoxin administration. During the experimental period the mice were starved and had free access to water before being killed at desired intervals.

Lipid peroxide. Lipid peroxide in liver tissue homogenate (in 8.1% sodium dodecyl sulfate[SDS], containing 2% acetate buffer, pH 3.5) was estimated fluorometrically by thiobarbituric acid by a modification(13) of the method of yagi(14).

Glutathione(GSH)peroxidase[EC 1.11.1.19] and superoxide dismutase(SOD)[EC 1.15.1.1]. GSH peroxidase was assayed by the method of Stults(15). SOD was estimated by the method of Beauchamp and Fridovich(16). After centrifuged the SOD extracts obtained by the method of Sinet *et al*(17), SOD activity in the supernatant was assayed.

Superoxide anion. Superoxide anion $O_2^-\cdot$ was estimated by xanthin oxidase activity according to the method of Fried and Fried(18).

Isolation and electrophoresis of lipoprotein. Serum lipoprotein was isolated with the density gradient procedure by a modification of the method of Havel et al(19) using an RP 65 roter in a Hitachi 65 P-7 ultracentrifuge-separator at 15°. Electrophoresis of serum lipoprotein was performed by disk electrophoresis in 3.75% polyacrylamide gel which was done with a previously stained lipoprotein according to the method of Narayan(20).

HDL apoprotein. HDL isolated from serum according to the above mentioned method was subjected to electrophoresis on 7.5% polyacrylamide gel containing 8 M urea by the method on Kane (21).

Nonspecific heatstable arylesterase(NHAE). Serum from endotoxin-torelant mice 48 hr postintoxication was previously removed HDL by density gradient ultracentrifugation according to a modification of the method of Hatch and Lees(22). Then, we carried out partialy purification of NHAE by the following procedure.

The serum obtained from 40 mice was fractionated with ammonium sulfate to 33-67% saturation. The resulting precipitate was collected by centrifugation and dialyzed against water. The dialyzed solution was condensed, then chromatographed on Sephadex G-200 column(2.6 x 90 cm) equilibrated with 0.05 M phosphate buffer, pH 7.4. The column was eluted with the same buffer, and the protein fraction(Fraction 3) showing the esterase activity was collected, and condensed. The enzyme solution was placed onto a DEAE-sephadex A-50 column equilibrated with 0.0175 M phosphate buffer, pH 6.3, then the column was eluted with the phosphate buffer solutions(0.0175 M, pH 6.3-0.4 M, pH 5.2) by a stepwise procedure. The NHAE fraction (Fr.4) obtained was dialyzed aginst water, then dried in vacuo and used for the following experiments NHAE was assayed with β-naphtyl acetate reagent according to the method of Skarnes(2), and the esterase level was determined in the fractionated samples which had been heated at 60° for 25 min. The heated samples were followed by saturated, filtered solutin of β-naphthyl acetate in 0.1 M phosphate buffer(pH 7,4). Diisopropylphosphofluoridate(DFP) at a final concentration of 2 x 10^{-4} M was added in the substrate solution. The released β-naphthol was measured spectrophotometrically at 328 nm. Body temperature. The body temperature was measured using contact KN 90 thermometer(Natsume Seisakusho Co.Ltd.Tokyo) conected with continuous recorder.

Results

1) Effect of HDL and arylesterase(NHAE) administration.

HDL isolated from serum in preendotoxemic(normal) or 48 hr postendotoxemic(tolerant) animals was used for the experiment.

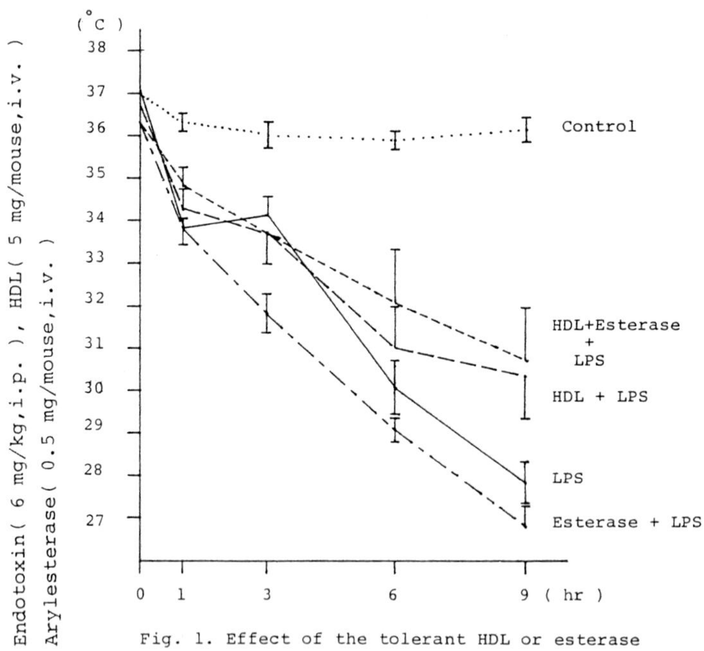

Fig. 1. Effect of the tolerant HDL or esterase (NHAE) administration on the body temperature of endotoxemic mice.

Effect of HDL and NHAE administration on body tempara-ture and blood glucose were observed in mice given endo-toxin(Fig. 1 and Fig. 2). As seen in figure, administ-ration of HDL from endotoxin-tolerant mice significantly increased the blood glucose level and body temperature of endotoxemic mice. There was no significant difference in the effect of injection of normal HDL on the blood glucose level of endotoxemic mice as compared with mice given endotoxin alone, but the tolerant HDL greatly elevated body temperature of the endotoxin-injected mice(Fig. 1). Thus, it was noted that HDL from tolerant mice exhibited a greater beneficial effect to protect mice against endo-toxicosis.

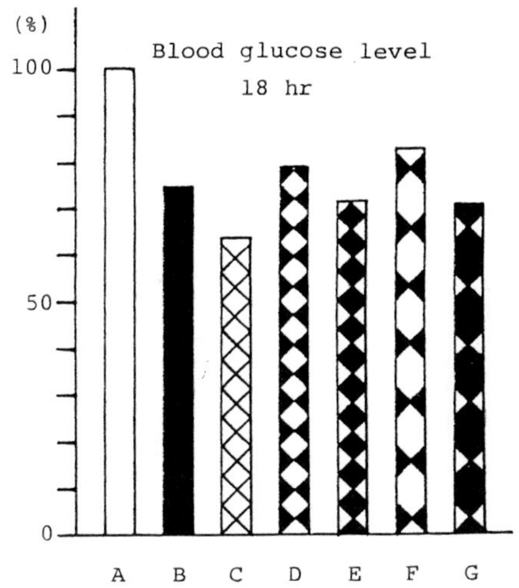

Fig.2. Effect of HDL and arylesterase(NHAE) administration on blood glucose level. A: Control mice. B: Endotoxin(6 mg/kg,i.p.) injected mice. C: NHAE(0.5 mg/mouse,i.v.) and endotoxin(6 mg/kg, i.p.) injected mice. D: HDL from normal mice(5 mg/mouse,i.v.) and endotoxin(6 mg/kg,i.p.) injected mice. E: HDL from normal mice(5 mg/mouse,i.v.), NHAE(0.5 mg/mouse), and endotoxin(6 mg/ kg,i.p.) injected mice. F: HDL from tolerant mice(5 mg/mouse, i.v.) and endotoxin(6 mg/kg,i.p.) injected mice. G: HDL from tolerant mice(5 mg/mouse,i.v.), NHAE(0.5 mg/mouse,i.v.), and endotoxin injected mice. Each bar represents the average of 6 mice, and express the percentage of the control(100 %).

The effect of administration of HDL and NHAE on the survival rate was observed in mice injected with endotoxin. Administration of NHAE alone was found to be harmful to the endotoxemic mice, but administration of NHAE only could not show toxic effect in the control mice. It is considered that by the administration of NHAE into mice given endotoxin, the esterase might act to partially hydrolyze LPS to form an active moiety.

The time course of change in survival rate was generally similar with the result on body temperature of mice(data are ommited). It seems ,therefore, most likely that the apoprotein of HDL from the endotoxin-tolerant mice is responsible for the in vivo detoxification of endotoxin.

2) Apoprotein A-1 and C-11 peptides in serum HDL.

Changes of apoprotein A-1 and C-11 peptides of HDL were analyzed in serum lipoprotein by the electrophoresis on 7.5% polyacrylamide gel disk containing 8 M urea 18 hr after endotoxin administration.

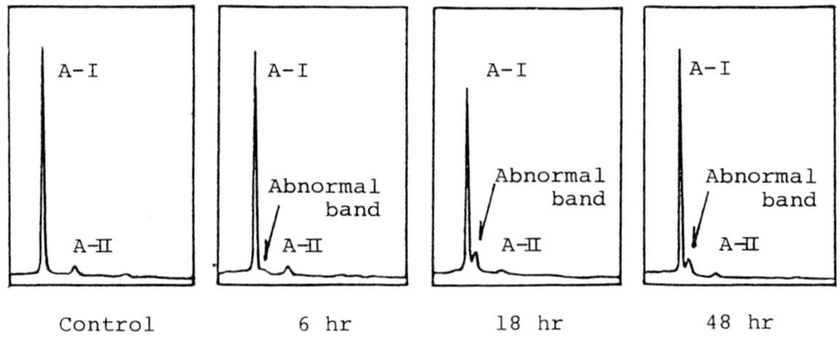

Fig.3. Time course of apoprotein A-I peptide in serum HDL lipoprotein in endotoxemic mice. HDl isolated from serum according to the above mentioned method was subjected to electrophoresis on 7.5 % polyacrylamide gel containing 8 M urea.

The electrophoretic pattern of apoprotein A-1 peptide in serum HDL lipoprotein in endotoxic mice is shown in Figure 3. As shown in figure, apoprotein A-1 peptide was decreased during endotoxicosis and divided into two peaks. The decline of apoprotein A-1 peptide level was maximum at 18 hr-24 hr postintoxication, thereafter the level showed a tendency of recorvery to normal range.

Analysis of apo C-11 peptide showed that the peptide level in endotoxin-poisoned mice was reduced 18 hr postintoxication as compared with that in the control.

3) Effect of lead acetate treatment.

Selye et al(8) found that sublethal intravenous administration of lead acetate increases the sensitivity of rats to endotoxins of different origins, about 100,000 times above normal, and that the highest mortality and most pronounced organ changes occured in the case of simultaneous injection of lead acetate and endotoxin. Therefore, we carried out the following experiments with simultaneous administration of lead acetate(50 mg/kg,i.v.) and endotoxin (6 mg/kg,i.p.) into mice. Because the mice were greatly sensitized to endotoxin by the lead treatment, all experiments were terminated by decapitation at 6 hr after endotoxin injection, and the liver or blood was used for analysis, unless otherwise noted.

[A]. Superoxide anion and lipid peroxide.

Superoxide anion $O_2^-\cdot$ was assayed by the derermination of xanthin oxidase activity. As shown in Figure 4, (a,b) the NBT reduction in liver of lead acetate plus endotoxin-treated mice was significantly greater than in the control. This result coincided with the findings(1) that the phagocytic activity of RES, as measured by the vascular clearence of colloidal carbon, shows a hypophagocytic state in lead acetate-endotoxin treated rats.

On the other hand, lipid peroxide levels 6 hr postintoxication were higher by about 1.5 fold in the livers in mice injected with endotoxin plus lead acetate than in those treated with endotoxin alone. This lipid peroxide formation could not be observed in C3H/HeJ mice(non responder strain) administered with endotoxin only, but these were a little formation in the mice injected with endotoxin plus lead acetate(data ommited).

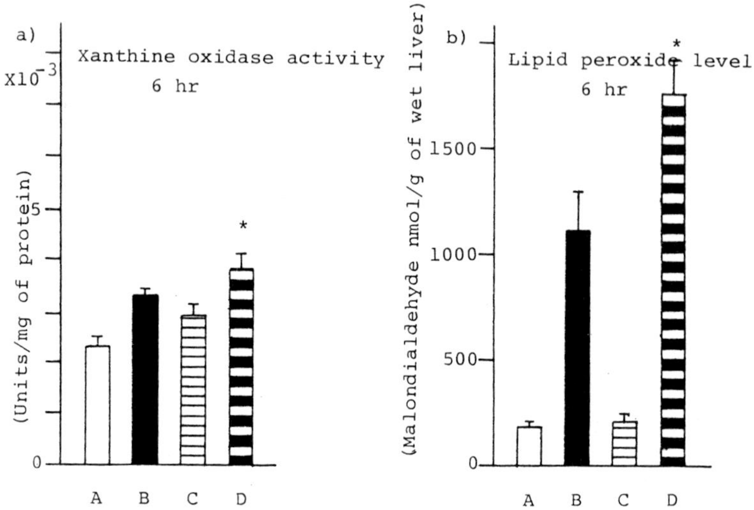

Fig.4. Effect of lead acetate on super oxide anion generation and lipid peroxide level in livers of endotoxin-poisoned mice. A: Control mice. B: Endotoxin(6 mg/kg,i.p.) injected mice. C: Lead acetate(50 mg/kg,i.v.) injected mice. D: Lead acetate(50 mg/kg,i.v.) and endotoxin(6 mg/kg,i.p.) injected mice. Mice in each group were starved for the indicated time before decapitation after injection of endotoxin. Each bar represents the mean ± S.E. of 6 mice. Asterisk(*) indicates significant difference from the value of endotoxin-poisoned mice(B) at $P<0.05$.

[B]. <u>Gltathione peroxide(GSH-Px) and superoxide dismutase(SOD) activities.</u>

GSH-Px is distributed for the destruction of hydrogen-peroxide(H_2O_2) and organic hydroperoxide compounds by which cellular membrane is greatly damaged. SOD plays an important role as a scanvenger of superoxide and other free radicals in the protection against cell damage.

As seen in Figure 5 (a,b), GSH-Px in the livers of endotoxin-lead treated mice exhibited about 30% less activity than that in the control 18 hr postintoxication , whereas the activity in endotoxin alone-poisoned mice significantly decreased than in mice treated with either lead or saline alone . The nonprotein SH level markedly decreased in the livers of lead plus endotoxin injected mice 6 hr postintoxication(data are omited). GSH plays an important role in tissues, especially in the oxidation-reduction system responsible for the inner tissue metabolism, and also protects SH enzyme and membran SH from free radicals.

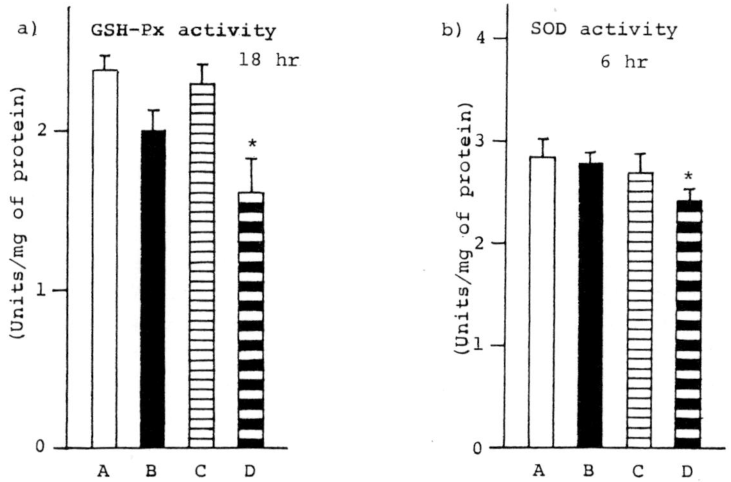

Fig.5. Effect of lead acetate on glutathioneperoxidase(GSH-Px) and superoxide dismutase(SOD) activities in livers of endotoxin-poisoned mice. A: Control mice. B: Endotoxin(6 mg/kg,i.p.)-injected mice. C: Lead acetate(50 mg/kg,i.v.)-injected mice. D: Lead acetate(50 mg/kg,i.v.) and endotoxin(6 mg/kg,i.p.)-injected mice. Mice in each group were starved for the indicated time before decapitation after injection of endotoxin.
Each bar for GSH-Px activity represents the mean $\pm$ S.E. of 12 mice, and SOD activity is the result from 6 mice. Asterisk(*) shows significant difference from the value of endotoixn-poisoned mice(B) at $P<0.05$.

Therefore, the decrease of GSH level resulted in a greater decrease in GSH-Px activity in the endotoxin-lead acetate treated mice. Also, SOD activity showed a significantly greater decrease in the livers of lead plus endotoxin treated mice than in mice injected with the lead or endotoxin alone. The GSH-Px or SOD inhibits the free radical formation against peroxidative attacks on lipids by preventing free radical formation.

4) Administration of lead acetate on serum lipoprotein.

Figure 6 shows serum lipoprotein disk gel electrophoretic patterns 18 hr postintoxication in endotoxin-poisoned mice injected(i.v.) together with lead acetate.

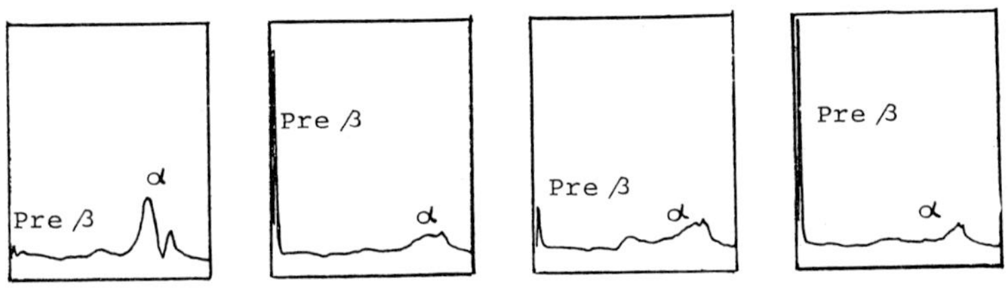

Control Endotoxin Lead acetate Lead acetate + Endotoxin

Fig.6. Analysis of polyacrylamide gel disk electrophoretic patterns of serum lipoproteins in lead acetate(50 mg/kg,i.v.) treated mice 18 hr after endotoxin administration.

As seen in Figure, there was an increase in the pre β lipoprotein in serum of lead plus endotoxin-treated mice. On the other hand, α_1-lipoprotein fraction was decreased markedly in the mice injected with lead and endotoxin than that in the control.

5) Effect of trypan blue administration in endotoxin-poisoned mice.

A) Trypan blue(TB)-pretreatment.

A 1.25% trypan blue(Wako Pure Chemicals Industries Ltd.,Osaka) was injected i.v. into mice 24 hr before endotoxin challenge at dose of 50 mg/kg, 100 mg/kg and 200 mg/kg.

Twenty four hours after TB treatment, endotoxin was injected i.p. into the mice. Control mice were injected i.p. with 0.2 ml of saline. The mice were fasted after endotoxin administration.

B) <u>Survival rate of TB pretreated mice.</u>

The percentage of survival after endotoxin administration in TB(200 mg/kg)-pretreated mice showed 95% mortality at 24 hr, and there was no survivor after 72 hr.
On the other hand, it is noticeable that TB(50 mg/kg)-endotoxin treated mice exhibited about 60% survival rate after 72 hr(Fig. 7)

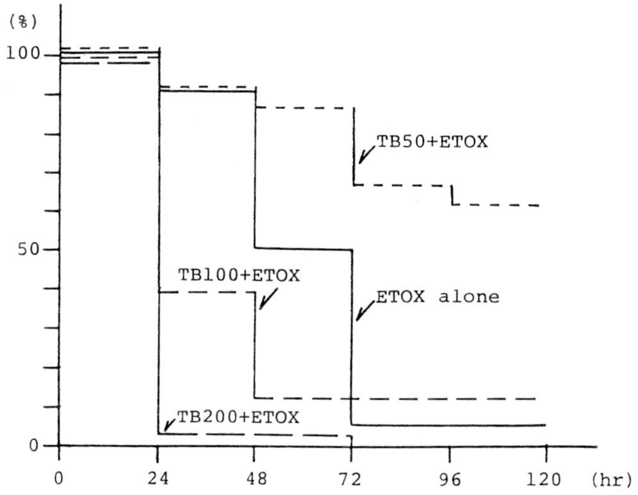

Fig.7. Effect of trypan blue pretreatment on endotoxin-induced lethality. Trypan blue(TB) was administered intravenously in a dosage of 50 mg/kg, 100 mg/kg, and 200 mg/kg into mice 24 hr prior to challennging with endotoxin(ETOX,18 mg/kg,i.p.).

C) Lipid peroxide formation in TB pretreated mice.

In previous report, we observed that endotoxin-poisoned mice exhibited a marked increase in lipid peroxide level in the liver compared with that in the control. The hepatic lipoperoxide level in mice livers injected with endotoxin 24 hr after pretreatment of 200 mg/kg of TB did not show a significant increase compared with that in the endotoxin alone-injected mice. However, it appears, generally, that TB(200 mg/kg) pretreated mice exhibited a tendency to increase in lipoperoxide level in the liver 6 hr after endotoxin administration rather than in the endtoxin alone. It is considered that the lipid peroxide level in endotoxemic mice may be decided by the mutual correlation between free radical generation and its scavengers in the experimental animals (Fig. 8).

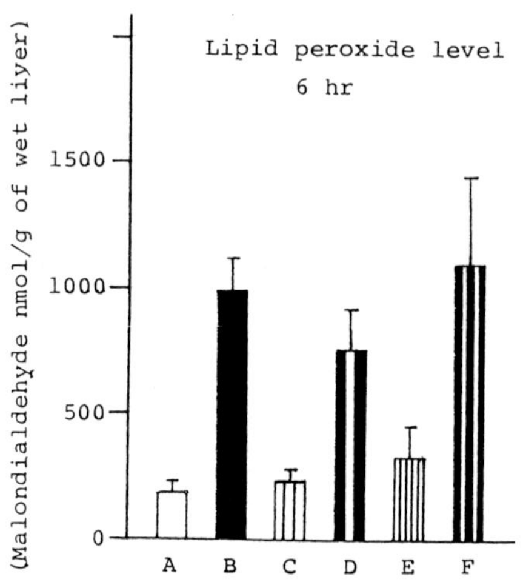

Fig.8. Lipid peroxide level in livers of trypan blue(TB)-pretreated mice after endotoxin administration. A: Control mice. B: Endotoxin(6 /mg/kg,i.p.) injected mice. C: TB(50 mg/kg,i.v.) pretreated mice. D: TB(50 mg/kg,i.v.) and endotoxin(6 mg/kg, i.p.) injected mice. E: TB(200 mg/kg,i.v.) pretreated mice. F: TB(200 mg/kg,i.v.) and endotoxin(6 mg/kg,i.p.) injected mice. Each bar represents the mean ± S.E. of 6 mice.

6) Polyacrylamide gel disk electrophoresis(PGDE) patterns of serum lipoprotein.

Skarnes(2) demonstrated that during detoxification in vitro endotoxin forms a complex with α_1-lipoprotein showing esterase activity. Bøg-Hansen et al(10.11.12) reported that intravenously injected endotoxin complexes in vitro with α_1-lipoprotein which supports the contention that endotoxin detoxification in the blood is important in vivo.

Ananlysis of PGDE patterns of serum lipoprotein was shown in mice 18 hr after endotoxin injection. Twenty four hours after the administration(i.v.) of TB(50 mg/kg and 100 mg/kg), the mice were injected(i.p.) with endotoxin, then decapitated at 18 hr postintoxication.

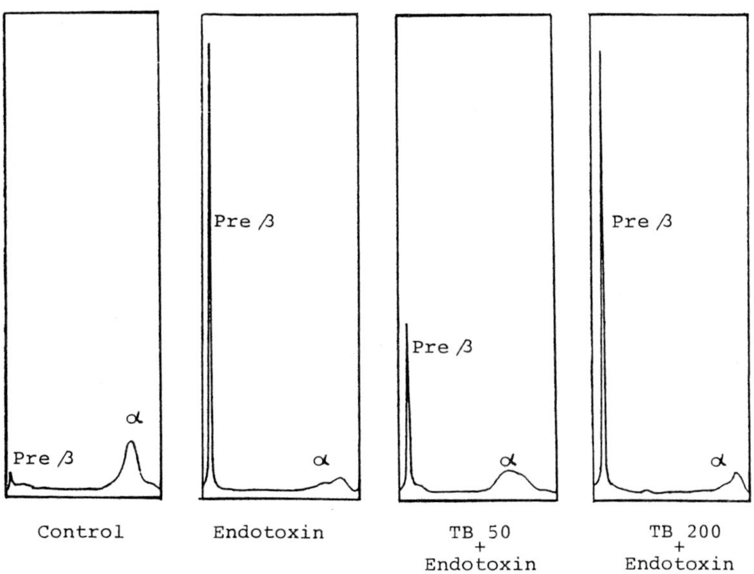

Fig.9. Analysis of polyacrylamide gel disk electrophoretic patterns of serum lipoproteins in trypan blue(TB)-pretreated mice 18 hr after endotoxin administration.

As shown in Figure 9, the mice treated with TB(200 mg/kg) plus endotoxin exhibited a greater peak of pre β-lipoprotein, while α_1-lipoprotein(HDL fraction) was reduced markedly in the mice. The mice treated with endotoxin and endotoxin plus TB(200 mg/kg) exhibited hyper-triglyceridemia 18 hr postintoxication. However, there was a decrease in pre β-lipoprotein level in the TB(50 mg/kg) pretreated mice but an increase in α_1-lipoprotein which shows a dissimilar mobility in the electrophoretic pattern. This finding showed the possibility that α_1-lipoprotein may interact with endotoxin under TB(50 mg/kg) pretreatment, and the detoxification may mainly accelerate after the uptake of endotoxin by the RES.

Discussion

Endotoxin administered in animals may impair the functional state of the RES either by direct damage to the RE cells or by indirect action on several endocrine, neural and vascular mechanisms. Although the gradual failure of the endotoxin-detoxifying mechanism of these cells during the progression of endotoxin shock may be related to a release of their lysosomal contents(5,23), the reason why activation or depression of RES increases susceptibility of animals to endotoxin remains to be uncertain.

RES stimulation by glucan preinjection in animals renders them more sensitive to endotoxin, but the treatment results in marked proliferation of the RE cells, and enhances the phagocytic capacity significantly(3,7). On the other hand, Selye <u>et al</u>(8) suggested that lead acetate may enhance endotoxin shock by blockade of RES. Trejo and DiLuzio(1) observed the functional defects in the RES induced by lead, and presented data suggesting an enhanced phagocytic state as evaluated by low carbon clearance and depressed capacity of the liver and spleen to detoxify endotoxin. Filkins and his co-workers, however, observed that no defects in either vascular clearance or hepatic detoxification of endotoxin were manifest after lead treatment(4,24).

Cook et al(25) described that pretreatment of rats with RES stimulant, glucan,markedly increases their sensitivity to endotoxin shock ,but this hyperphagocytic function, however, was abolished at 2 hr after endotoxin injection. They considered that altered glucoregulation appears to be a significant factor in the enhanced sensitivity of glucan-pretreated rats to endotoxin.

In this study the administration of endotoxin to mice rendered hypersensitive by lead acetate resulted in profound lipid peroxide formation in the liver 6 hr postintoxication, and GSH-Px and SOD activities was reduced in the mice liver, wherease superoxide anion generation was significantly increased in the livers of endotoxin plus lead acetate treated mice compared with that in mice given endotoxin alone.

The lethality of endotoxin was increased by preadministration with TB(200 mg/kg i.v.) in mice 24 hr before endotoxin, but pretreatment with TB(50 mg/kg i.v.) decreased the mortality of the poisoned mice to 40% at 72 hr postintoxication. In the mice liver treated with endotoxin plus TB(200 mg/kg), the lipoperoxide level was not significantly different from the level in endotoxin-poisoned mice. However, the treatment with TB(50 mg/kg) plus endotoxin decreased the lipoperoxide level in the liver as compared with 200 mg/kg of TB plus endotoxin.

The gradual failure of the endotoxin-detoxifying mechanisms of RE cells during the progression of endotoxin shock may be related to a release of their lysosomal contents, because the increase of lipoperoxide level can bring about membrane damage(26).

Skarnes(2) reported that intravenously injected endotoxin complexes in vivo with plasma α_1-lipoprotein(having arylesterase). Ulevith et al(27) reported evidence for an initial disaggregation by a lipoprotein followed by enzymatic detoxification. Freudenberg et al (28) described that endotoxin iteracts with plasma HDL forming a complex with characteristic electrophoretic mobility, and that some of HDL apoproteins may serve as coenzyme of LPS-inactivating enzyme different from arylesterase.

HDL fraction in mice injected with lead acetate plus endotoxin was markedly decreased as compared with that in the poisoned mice. Mice treated with TB(200 mg/kg) plus endotoxin exhibited a greater increase of pre β-lipoprotein, while α_1-lipoprotein was markedly decreased in the sera. However, there was a marked decrease in pre β-lipoprotein level in the TB (50 mg/kg)-endotoxin treated mice, while an increase in α_1-lipoprotein level was seen in the PGAE patterns in which HDL fraction showed a dissimilar mobility, suggesting the possibility that HDL may interact with LPS under TB(50 mg/kg) pretreatment and the detoxification may mainly accelerate after uptake of LPS by the RE cells. It was found that the injection of HDL from tolerant mice is more effectual to protect endotoxin lethality than from the normal mice. Tolerant HDL significantly increased the blood glucose level and survival rate of endotoxemic mice, however it was noticeable that administration of NHAE alone into mice was found to show harmful effect to the endotoxemic mice. Therefore it is considered that the NHAE alone-injection in endotoxemic mice might result in the formation of more active moiety from the endotoxin ,followed by binding with HDL subfraction for detoxification in Kupffer cells. Our finings may indicate the possibility that the subfraction in HDL represents a transport protein for LPS in plasma to Kupffer cells.

Recently Hoffman and Benditt(30) described that bacterial endotoxin is a potent inducer of the serum amyloid protein(apo-SAA) and that this apoprotein is distributed among high density HDL subfraction in serum 20 hr postintoxication; this response is regulated independently of other HDL apoproteins. The NHAE used in our study was partially purified in the serum from endotoxin-tolerant mice and free from HDL by density gradient ultracentrifugation according to the method of Hatch and Lee(22). We considered that HDL fraction in the serum from tolerant mouse must contain other subfractions dissimilar with those(apo A-1 and C-11 peptides) in normal HDL. The question of whether apo-SAA formed following endotoxin injection is identical with the subfraction of tolerant HDL must be futher investigated in detail .

Summary

The present study was carried out to observe the effect of endotoxin administration under RES depression by lead acetate or trypan blue(TB) and to determine the role of serum lipoprotein during endotoxemia.

The endotoxin challenge in mice rendered hypersensitive by lead acetate resulted in profound lipid peroxide formation in the liver 6 hr postintoxication. Endotoxin plus lead acetate-treated mice showed markedly depressed glutathione peroxidase and superoxide dismutase activities in these liver, whereas superoxide anion generation was significantly increased in the liver of endotoxin plus lead acetate-treated mice compared with that in endotoxin alone-treated mice. Serum HDL fraction in mice injected with lead acetate plus endotoxin was markedly decreased than in the group given endotoxin only.

The mice given endotoxin after treatment of TB(200 mg/kg) exhibited 95% mortality at 24 hr, while TB(50 mg/kg) plus endotoxin treated mice showed about 60% survival rate at 72 hr postintoxication. The mice treated with endotoxin plus TB (200 mg/kg) exhibited a tendency of increased lipoperoxide level in the liver 6 hr after endotoxin injection than in the endotoxin alone treated mice. However it was noted that the treatment with 50 mg/kg of TB plus endotoxin decreased less the lipoperoxide level as compared to that with 200 mg/kg TB plus endotoxin. The mice treated with TB(200 mg/kg) plus endotoxin indicated a greater peak of pre β-lipoprotein, while 0_1-lipoprotein(HDL) was reduced markedly in the mice sera. There was a decrease in pre β-lipoprotein level in the TB (50 mg/kg) pretreated mice but an increase in α_1-lipoprotein which shows a dissimilar mobility in the electrophoretic pattern.

In analysis of HDL apoprotein peptide in mice sera, apo C-ll level was reduced 18 hr postintoxication as compared to the control. The disk electrophoretic pattern of apoprotein A-1 peptide in serum HDL in poisoned mice was decreased during endotoxicosis and the peptide peak was divided into two peaks (main and abnormal bands).

Non specfic, heatlabile arylesterase(NHAE) from torelant mice sera was partially purified on DEAE-Sephadex A-50 column, and used for the experiment. The mouse injected with endotoxin plus NAAE showed a lower body temperature and greater mortality than the mouse injected with endotoxin alone. NHAE might partially hydrolyze LPS to form more active moiety in serum. However , the injection of endotoxin together with the HDL (from tolerant mice), resulted in less mortality than in the endotoxin alone-injected mice.

We surmise that apoprotein of HDL from tolerant mice may play an important role to transfer endotoxin into clearance organ.

References

1. Treijo,R.A, and DiLuzio : Proc.Exp.Biol.Med. 136, 889-893 (1971).
2. Skarnes,R.C. : J.Exp.Med. 132, 300-316(1970).
3. Cook,J.A.,Dougherty,W.J.,Holt,T.M. : Circ.Shock 7, 225-238 (1980)

4. Buchanan,B,J.,Filkins,J.P. : Am.J.physiol. 231,258-264 (1976).

5. Rangel,D.M.,Byfield,J.E.,Adominan,G.E.,Steven,G.H.,Fonkalsrud,E.W. : Surgery. 68, 503-511(1970).

6. Cooper,G.N.,Stuart,A.E. : Nature, 191, 294-295(1961).

7. Lemperile,G. : Proc.Soc.Exp.Biol.Med. 122, 1012-1015 (1966).

8. Selye,H.Tuchweber,B.,Bertók,L. : J.Bacteriol. 91, 884-896 (1966).

9. Skarnes,R.C. : J.Bacteriol. 95, 2031-2034(1968).

10. Bøg-Hansen,T.C.,Hertel•Wulff,B.,Back,U. : FEBS Letters, 104, 441-445(1979).

11. Bøg-Hansen,T.C.,Krog,H.H.,Back,U. : FEBS Letters, 93, 86-90(1978).

12. Bøg-Hansen,T.C.,Back,U. : Natural toxins, In Eaker,D.Wadstrom,T,(ed). Oxford,New York(1980).

13. Ohishi,N. : Saishin Igaku, 33, 660-663(1978). (in Japanese)

14. Yagi,K. : Biochem.Med. 15, 212-216(1976).

15. Stults,F.H.,Forstrom,J.W.,Chiu,D.T.Y.Tappel,A.L. : Arch. Biochem.Biophys. 183, 490-497(1977).

16. Beaucham,C.,Fridovich,I. : Anal.Biochem. 44, 276-287 (1971).

17. Sinet,P.M.,Lavelle,F.,Michelson,AM.,Jerome,H. : Biochim. Biophys.Res.Commun. 67, 904-909(1975).

18. Fried,R.,Fried,L.W. : Method of enzymatic analysis,vol.l. Verlag Chemic Weinheim.Academic Press Inc.,New York,London.

19. Havel,R.J.,Fielding,C.J.,Olivecrona,T.,Shore,V.G.,Fielding,P.E.,Egelrud,T. : Biochemistry 12, 1828-1833(1973).

20. Narayan,K.A.,Narayan,S.,Kummerow,F.A.: Nature 205, 246-248 (1965).

21. Kane,J.P. : Anal.Biochem. 53, 350-364(1973).

22. Hatch,F.T.,Lee,R.S. : Adv.Lipid.Res. 6, 1-68(1968).

23. Snyder,S.L.,Eklund,S.K.,Walker,R.I. : Can.J.Microbiol. 25, 1245-1259(1979).

24. Filkins,J.P. : J.Reticuloendothel.Soc. 22, 461-467(1977).

25. Cook,J.A.,Dougherty,W.J.,and Holt,T.M. : Circ.Shock 7, 225-238(1980).

26. Sakaguchi,S.,Kanda,N.,Hsu,C-C.,Sakaguchi,O. : Microbiol. Immunol. 25, 229-244(1981).

27. Uleuitch, R.J.,Johnston,A.R.,Weinstein,D.B. : J.Clin. Invest. 64. 1516-1524(1979).

28. Freudenberg,M.A.,Bøg-Hansen,T.C.,Back,U.,Galanos,C. : Infect.Immun. 28, 373-380(1980).

29. Hoffman,J.S.,Benditt,E.P. : J.Biol.Chem. 257, 10510-10517 (1982).

30. Hoffman,J.S., Benditt,E.P. : J.Biol.Chem. 257, 10518-10522 (1982).

NEUTRALIZING EFFECTS OF ANTI-SALMONELLA Re ANTIBODY ON ENDOTOXIC BONE MARROW REACTIONS AND INDUCTION OF PROCOAGULANT ACTIVITY

Masao Yoshida, Kazuaki Kudoh, Michimasa Hirata, Katsuya Inada and Masami Ogasawara

Department of Bacteriology, School of Medicine,
Iwate Medical University, Morioka, Iwate 020 Japan

Introduction

It has been known for a long time that antibody to endotoxin does not neutralize endotoxic activities. Since clarification of relationship between chemical structure of endotoxin and S-R mutation, the neutralizing effects of antibody to R-form on endotoxicity have been investigated (1-3). Especially, the protection of either DIC (disseminated intravascular coagulation) or lethality in endotoxemia is of clinical importance. We have studied bone marrow reactions due to endotoxin injection. The two reactions (4-6) were examined in the present study. (a) Leukocyte migration i.e. the decrease of nucleated cell numbers in bone marrow; this results from leukocytes migration into circulation. (b) Hemorrhage i.e. increase of red blood cell numbers in marrow; this results from hemorrhage due to injury of sinusoidal endothelial cells. In addition to the two reactions, (c) procoagulant activity (7-8) of marrow cells was also examined. We have proposed that this activity is one of the causes of DIC in endotoxicosis.
In the present paper, before endotoxin injection, anti-Re

Immunopharmacology of Endotoxicosis

antibody was pretreated and the three endotoxic reactions were examined. Preventing effects of the antibody on leukocyte migration and procoagulant induction due to endotoxicosis are evidenced. It was also investigated as to whether injurious (allergic) reaction was induced or not induced by antibody preinjection.

Materials and Methods

1. Preparation of antibody. a) Immunogen. Bacterial cells of *S. typhimurium* S(LT2), Rb (TV 161), *S. minnesota* Rb (R60), Rb (R345) and Re (R595) on nutrient agar plate were washed with saline, and heated at 100° C for 1 hr, followed by rewashing. Thereafter, 2 mg wet weight of bacterial cells were suspended in 1 ml saline. Two strains of *S. typhimurium* were supplied through the courtesy of Prof. J.Y. Homma (Institute of Medical Science, Tokyo). Three strains of *S. minnesota* were supplied through the courtesy of Prof. O. Lüderitz (Max-Planck Institut für Immunobiologie, Freiburg). b) Immunization. Immunization was carried out according to Johns (9); a male rabbit weighing 2-3 kg, was i.v. injected with 0.5, 1.0, 2.0 and 3.0 ml of the bacterial cells suspension at 4 day-intervals, a booster shot of 3.0 ml was then given, and 7 days after the final injection, sacrificed by blood-collection. The serum obtained was stored at -20° C until use.

2. Fractionation of immune-sera. Gel-filtration of immune-serum on Sephadex G-200 was carried out according to Braude (1). All procedures were carried out as aseptically as possible. The fractions of IgG, IgM and albumin were concentrated to the original volume and stored at -20° C and sterilized again with membrane filter before use.

3. Absorption of antibody. Lot 1 anti Re serum was absorbed with 10^9 of heat-killed cells of *S. minnesota* Re according to Young (10).

4. Passive hemagglutination. Antibody titers of the sera immunized were assayed by passive hemagglutination in which sheep red blood cells(SRBC) were sensitized with respective endotoxin-preparation. The method of Young (10) was used in the passive hemagglutination. a) Preparation of sera. The serum heated at 56° C for 30 min was absorbed with 1/2 volume of 50% SRBC at 37° C for 2 hr and kept at 4° C overnight to remove the heteroantibody against SRBC. b) Sensitization of SRBC with endotoxin. 500 μg/ml of endotoxin heated at 100° C for 1 hr was added to an equal volume of 10% SRBC, and incubated at 37° C for 2 hr, followed by washing with phosphate buffer saline (PBS). c) Procedure of hemagglutination. 0.2 ml of diluted serum was mixed with 0.2ml of 1% sensitized SRBC, and the reaction mixture was incubated at 37° C for 2 hr, followed by keeping overnight at 4° C. Thereafter, the titer was determined.

5. Preparations of endotoxins. S-LPS of *S. typhimurium* and Ra-LPS of *S. minnesota* were extracted by phenol water (11), and Rb-LPS of *S. typhimurium* was prepared by PCP method (12). S, Ra (R60), Rb (R345), Rc (R5), Rd1 (RT), Rd_2 (R4), Re(R 595) LPS and lipid A of *S. minnesota* were supplied through the courtesy of O. Lüderitz (Max-Planck Institut für Immunobiologie, Freiburg). *E. coli* O111:B4 LPS was purchased from Difco.

6. Preparation of bone marrow cells of mice. Male ddY mice (Shizuoka strain), at 10-11 weeks-old, weighing 30-35 g were used, and 5-6 mice were used for an experimental group. Two femora were taken from a mouse after being sacrificed for assays, and were cut into small pieces in citrate-saline (a mixture of 0.1M Na-citrate and 0.9% NaCl at 1:5 ratio, abbrev. C.S.), followed by preparation of cells suspension of bone marrow. The cell suspension was filtered with a piece of nylon-gauze, and washed by C.S. An aliquot of the suspension was used for the assay of two bone marrow reactions, while another part was used for the assay of procoagulant activity.

7. Bone marrow reactions. These were assayed using an electronic microcell-counter according to the method described previously (4-6).

8. Assay of procoagulant-induction of bone marrow cells. This assay was performed by the modified method (7) of Rickles (13). The cells suspension of bone marrow was adjusted to 4-8 x 10^7/ml, and subjected to sonication. A volume of 0.1 ml of mouse plasma used as the substrate was mixed with 0.1 ml of the cell-sonicate. The mixture was incubated at 37° C for 2 min, thereafter, 0.1 ml of 25 mM $CaCl_2$ which contained phospholipid(100μg/ml trostin, Chugai Pharmaceutical Co.), was added to the mixture. A clotting time was measured with a fibrometer. The mouse plasma was prepared by adding 1/10 volume of Na-citrate to one volume of mouse blood, and by centrifugation. The plasma was then stored at -20° C before use.

9. Neutralization in vivo by antiserum for endotoxic reactions. An aliquot of the antiserum or fraction of IgG or IgM was i.p. injected into a mouse; six hr later an endotoxin solution was i.v. injected, and 18 hr later the mouse was sacrificed for the assays.

10. Statistical analyses. Figures in Tables were represented with a mean ± standard error. The difference between two groups was checked by t-test.

Results

1) Hemagglutinations between antisera and various endotoxin-preparations. Anti-S and Rb sera of *S. typhimurium*, and anti-Ra, Rb and Re (lot 1 and 2) sera of *S. minnesota* were examined on passive hemagglutination with various endotoxins, as shown in Table 1. Antisera to *S. typhimurium* Rb, and *S. minnesota* Ra and Rb showed strong and broad

cross-reactivities for various endotoxins of both *S. typhimurium* and *S. minnesota*. Antiserum to *S. typhimurium* S-form showed also cross-reactivities for Rb of *S. typhimurium* and Rc of *S. minnesota*. The antiserum lot 1 to Re of *S. minnesota* showed weak cross-reactivities with *S. minnesota* S-form, Ra and Rb LPS, besides endotoxin of *E. coli* S-form, while the lot 2 showed no cross-reactivity with any endotoxin. No sera shows cross-reactivity with lipid A of *S. minnesota*.

Table 1 Passive hemagglutination

LPS \ Antisera		S. typhimurium		S. minnesota			
		S	Rb	Ra	Rb	Re 1	Re 2
S. typhimurium	S	1280	160	160	40	<20	<20
	Rb	640	640	640	640	<20	<20
S. minnesota	S	<20	160	160	320	40	<20
	Ra	<20	40	1280	160	20	<20
	Rb	<20	160	640	640	80	<20
	Rc	80	320	160	320	<20	<20
	Rd 1	<20	640	160	320	<20	<20
	Rd 2	<20	640	640	1280	<20	<20
	Re	<20	<20	<20	20	160	640
	lipid A	<20	<20	<20	<20	<20	
E. coli O111:B4		<20	<20	<20	20	40	<20

Figures represent reciprocals of final dilutions showing positive reactions.

2) Neutralizing effect of antiserum to *S. minnesota* Re on endotoxicities of the same endotoxin. Pretreatment of 0.8 ml of anti-Re serum lot 1 was given then 100 μg Re endotoxin was i.v. injected. The neutralizing effects of anti-Re serum on three reactions produced by Re endotoxin are shown in Table 2. On the 2nd line of Table 2, three positive endotoxic reactions were induced by injection of LPS only; decrease of nucleated cell number, increase of

RBC number and shortening of clotting time, compared to controls on the 1st line. As shown on the 4th line, however, both decrease of nucleated cell number and shortening of clotting time were inhibited by pretreatment of anti-Re serum lot 1. Normal serum did not inhibit any endotoxin reactions, as shown on the 3rd line. Since 2-ME treatment of this lot 1 serum did not decrease agglutinin-titer, it is likely that this serum contains mostly IgG antibody. Thus IgG antibody to Re could neutralize two endotoxic reactions. Increase of RBC number was not neutralized by pretreatment of anti-Re, but rather enhanced.

Table 2 Neutralizing effect of antiserum to S. minnesota on endotoxicities of Re endotoxin

Serum 0.8 ml i.p.	LPS 100 μg i.v.	Bone marrow reactions Nucleated cell number X 10^7/2 femora	RBC number	Procoagulant clotting time (sec)
-	-	2.85±0.06	0.80±0.05	181.2±4.4
-	+	1.89±0.19*	1.82±0.28*	130.2±7.6*
Normal serum	+	2.48±0.20	1.82±0.28	149.8±3.3
Antiserum lot 1	+	2.77±0.11**	2.26±0.23	169.5±5.5**

-: The same volume of saline was injected.
*: Difference from saline control was significant at $P<0.05$.
**: Difference from LPS control was significant at $P<0.05$.
i.p.: Intraperitoneal injection.
i.v.: Intravenous injection.

3) Neutralizing effect of anti-Re serum on endotoxicities of S-LPS of *S. minnesota*. Three reactions produced by 100 μg of S-LPS of *S. minnesota* were investigated as to neutralizing capacity of anti-serum lot 1, as shown in Table 3. In this case the leukocyte migration and the procoagulant induction of S-LPS were again inhibited by anti Re serum.

Normal serum also inhibited the leukocyte migration in this experiment, but such inhibition of normal serum was not observed in other same experiments.

Table 3 Neutralizing effect of anti-Re serum of endotoxicities of S-LPS of S. minnesota

Serum 0.8 ml i.p.	LPS 100 μg i.v.	Bone marrow reactions Nucleated cell number X 10^7/2 femora	RBC number	Procoagulant clotting time (sec)
–	–	4.05±0.20	0.70±0.11	198.3±5.4
-	+	2.15±0.36*	1.81±0.44	150.7±5.3*
Normal Serum	+	3.59±0.16**	2.06±0.14	167.8±6.4
Antiserum lot 1	+	4.00±0.23**	1.94±0.18	172.7±6.2**

Legends: See Table 2

4) Effect of antibody-absorption on its neutralizing capacity. To investigate as to whether the inhibiting effects of anti-serum mentioned above on two endotoxic reactions are dependent upon antibody in the serum or not, the similar experiment was performed with serum in which the majority of antibody was absorbed. The agglutinin-titer of the serum was decreased from 1:160 to 1:40 by absorption. In the experiment, inhibiting effect of the absorbed serum was not observed as to the three reactions examined. Thus it seems that the inhibiting effects of the anti-serum lot 1 are dependent upon the antibody in the serum and upon its neutralization to endotoxicities.

5) Neutralizing effect of anti-Re lot 2 serum on the activities of Re LPS. The antiserum lot 2 to Re of *S. minnesota* had an agglutinin-titer of 1:640, and this titer was decreased down to 1:80 by treatment of 2-ME. It is thus thought that the anti-Re lot 2 serum has mainly IgM antibody. Neutralizing capacity of this lot 2 serum was also

investigated on the endotoxicities of 100 μg Re-LPS, as shown in Table 4. The pretreatment of 0.4 ml of anti-Re lot 2 serum inhibited also the leukocytes migration and procoagulant-induction, even hemorrhage increase of RBC number produced by Re endotoxin. This result shows the anti-Re antibody which mainly consists of IgM also neutralized the activities of Re-LPS.

Table 4 Neutralizing effect of anti-Re lot 2 serum on the activities of Re LPS

Serum 0.4 ml i.p.	LPS 100 μg i.v.	Bone marrow reactions: Nucleated cell number $\times 10^7$/2 femora	Bone marrow reactions: RBC number	Procoagulant clotting time (sec)
-	-	2.73±0.13	0.06±0.02	247.8±9.9
-	+	1.41±0.17*	1.90±0.31*	149.4±3.4*
Normal serum	+	1.58±0.11	1.70±0.33	161.6±6.2
Antiserum lot 2	+	2.76±0.08**	0.79±0.14**	233.4±21.5**

Legend: See Table 2.

6) Neutralizing effects of antibody lot 2 to *S. minnesota* Re on activities of S-LPS of *S. typhimurium*. Anti-*S. minnesota* Re lot 2 serum was examined as to neutralizing the activities of S-LPS of *S. typhimurium*. Manifestation of leukocyte migration, hemorrhage and procoagulant were significantly inhibited by a pre-dose of 0.4 ml of anti-Re antibody (probably IgM). That is, 2.38±0.16 of nucleated number decreased by the endotoxin was restored to 3.19±0.12 by antibody; 4.20±0.38 of RBC number increased by the endotoxin was diminished down to 2.05±0.16 by antibody; 127.2 ±4.1 sec of clotting time shortened by endotoxin was restored to 164.8±3.5 by the antibody.

7) Neutralizing effects of antibody lot 2 *S.minnesota* Re on activities of S-LPS of *E. coli*. As shown in Table 5, anti-Re lot 2 serum neutralized even heterologous LPS, that of *E. coli* S-form. Hemorrhage reaction was however not inhibited by the antibody.

Table 5 Neutralizing effect of Re antibody lot 2 on activities of S-LPS of E. coli

Serum 0.4 ml i.p.	LPS 100 μg i.v.	Bone marrow reactions Nucleated cell number $X\ 10^7/$ 2 femora	RBC number	Procoagulant clotting time (sec)
-	-	3.18±0.16	0.59±0.12	190.6±5.1
-	+	1.85±0.05*	2.64±0.31*	103.4±4.9*
Normal serum	+	2.10±0.19	1.57±0.37	103.2±5.3
Antiserum lot 2	+	3.06±0.08**	1.64±0.37	161.0±7.7**

Legend: See Table 2.

8) Neutralization of IgM or IgG fraction of anti-Re serum. IgM and IgG fractions of anti-Re serum, and furthermore those of normal serum as control, were investigated on neutralizing capacity against activities of Re LPS. As shown in Table 6, 0.4 ml of IgM and IgG fractions of antiserum containing 1:320 and 1:640 antibody, were predosed to neutralize 100 μg of Re-LPS. Either IgM or IgG fraction of antiserum neutralized the manifestation of both leukocytes migration and procoagulant due to Re LPS, while neither IgM and IgG of normal serum neutralized these two reactions. IgM of antiserum tended to inhibit the increase of RBC number due to LPS. On the other hand, IgG of antiserum rather elevated the RBC number compared to the LPS-control (2nd line).

Table 6 Neutralization of IgM or IgG fraction of anti-Re serum

Fraction 0.4 ml i.p.		LPS 100 μg i.v.	Bone marrow reactions Nucleated cell number $X\ 10^7/2$ femora	RBC number	Procoagulant clotting time (sec)
-		-	4.28±0.15	0.91±0.14	191.9±5.1
-		+	1.50±0.28*	1.41±0.58	127.1±6.9*
Normal serum	IgM	+	1.57±0.19	0.92±0.26	131.0±10.3
	IgG	+	1.77±0.23	2.18±0.30	129.7±9.0
Antiserum	IgM	+	2.67±0.21**	0.88±0.18	163.6±8.8**
	IgG	+	2.49±0.09**	2.03±0.33	169.6±7.7**

Legend: See Table 2.

9) Neutralizing effect of IgM or IgG of anti-Re serum on activities of *E. coli*. 0.4 ml of the IgM or IgG fraction was pretreated and investigated as to neutralizing the activities of heterologous *E. coli* S-LPS. Either IgM or IgG of antiserum neutralized the activities of S-LPS of *E. coli* as to leukocytes migration and procoagulant induction although both IgM and IgG of normal serum did not give any modification for endotoxic reactions produced by *E. coli* S-LPS (data not shown).

10) Increase of RBC number in marrow due to injection of antigenic substances. As mentioned above, neutralization of anti-Re serum was indefinite only as to the increase of RBC number. Inhibition of the increase of RBC number was occasionally observed, especially when IgM antibody was preinjected, while even accelerations of increase of RBC number were often observed by predoses of other antisera. All the antisera used were prepared from rabbit. The

question has been thus raised as to whether a predose of rabbit serum alone can change the RBC number in the marrow. To solve the question, IgM and IgG of both normal rabbit serum and anti-Re serum were i.v. injected in the indicated volume and protein content, as shown in Exp. # 1 of Table 7. Increases of RBC number were observed in injections of both IgM and IgG of antiserum as compared to the control of Exp. # 1. Subsequently, the rabbit serum albumin (BSA) which did not have any activity of antibody was injected and investigated, since IgM and IgG had activities of antibody. Remarkable increase of RBC number was also obtained with injection of 30.4 mg/ 0.8 ml protein of BSA, compared to the control of Exp. # 2. As shown in Exp. # 3, 2.5% SRBC and 5 mg/ml of BSA increased also RBC numbers after a single injection. Sonicate of 2.5% SRBC increased also RBC number in mouse marrow as shown in Exp. # 4. During these experiments, it was assumed that these foreign proteins elevate RBC number in the marrow. Thus, sonicate of 2.5% RBC of C3H/ HeN mice was injected into C3H/HeN mice. However, no increase of RBC number was observed after injection of syngeneic RBC. From these results, it is likely that injection of allogeneic, and antigenic protein induces the increase of RBC number in the marrow.

Discussion

Strong and broad cross-reactivities of anti-R endotoxins presented in Table 1 show that lipopolysaccharide in the bacterial cell wall and LPS preparations can have some more deeper (S→R) and defective chemical structures in addition to the respective proper structure as termed. This has already been mentioned by McCabe (14) and Chedid (15).

Table 7 Increase of RBC number in marrow produced by single injection of various antigenic substances

Exp. #	Antigen (control)		Volume (ml)	Protein injected (mg)	Number of RBC
1.	Control		/	/	0.23±0.03
	Normal serum	IgM	0.4	1.6	0.18±0.02
		IgG	0.4	8.0	0.29±0.07
	Anti Re serum	IgM	0.4	3.2	0.42±0.14
		IgG	0.4	7.2	0.56±0.14
2.	Control		/	/	0.61±0.11
	Rabbit serum albumin		0.4	15.2	0.64±0.13
			0.8	30.4	1.29±0.18*
3.	Control		/	/	0.25±0.03
	2.5% SRBC		0.2	2.8	0.86±0.11*
	5 mg/ml BSA		0.2	1.0	0.79±0.11*
4.	Control		/	/	0.22±0.05
	Sonicate of 2.5% SRBC		0.2	2.8	1.11±0.09
5.	Control (C3H/HeN mice)		/	/	0.18±0.02
	Sonicate of 2.5% RBC of C3H/HeN mice		0.2	3.0	0.18±0.03

In Exp. #5, sonicate of RBC of C3H/HeN was i.v. injected into the syngeneic C3H/HeN mice.
*: The increase of RBC number was significant at<P 0.05 compared to respective control.

In the present study inhibition of endotoxicities by anti Re antibody has been intensively evidenced. These inhibitions of endotoxic reactions by antiserum was scarcely observed by the administration of normal serum. Absorption of antibody from anti-Re serum diminished the inhibition produced by antiserum. These results show that the inhibition results from neutralization by antibody. It has been thought by 2 ME-treatment that lot 1 anti Re serum has mainly IgG antibody and lot 2 antiserum has IgM in addition to a small quantity of IgG. Both lot 1 and 2 anti-serum had neutralizing capacity for endotoxic reactions. Both IgM and IgG fractions of antiserum had also the same capacity.

It is noteworthy that anti-Re antibody of *S. minnesota* can neutralize even the endotoxic reactions of *E. coli* S-form. These studies reported as to the significance of anti-R antibody have been noted for the past two decades. Tate (16) in 1966, and Chedid (15) in 1968 reported therapeutic significance of anti-R serum in animal experiments. McCabe (17) reported that patients with high-titer antibody to the rough form of gram-negative bacteria can be prevented from shock symptoms and death due to gram-negative injection. Braude (1)(3) reported that anti-J5 (Rc) of *E.coli* prepared from humans had a therapeutic effect on gram-negative septicemia. In the present paper these have been reconfirmed also by anti-Re antibody of *S. minnesota*.

The decrease of nucleated cells number in the marrow results from leukocytes migration into periperal circulation. Therefore this reflects essentially one of the protective host responses stimulated by endotoxin. However, these migrating leukocytes, especially granulocytes and macrophages, would carry procoagulant activity and cause DIC in periperal circulation, as proposed by Hirata in our laboratory. Therefore, the present study would give basic knowledge for the clinical control of DIC.

Since the active principle of the majority of endotoxic reactions is lipid A, anti-lipid A might be the best anti-serum to inhibit the reactions. However, anti-lipid A antibody has been shown to have no therapeutic effect on the endotoxin activities (9) (18).

This is not able to be explained at the present time. Anti-Ra and Rb sera of *S. minnesota* had stronger and broader cross-reactivities with various endotoxins than anti-Re serum, as shown in Table 1. Therefore, these antisera might have a comparable neutralizing activity with anti-Re serum. It should be stressed that the structure of Re-LPS is the most basic and common structure for many endotoxins of various bacterial genera.

Injected endotoxin would have been neutralized by the antibody before the expression of endotoxic reactions occur.

In the present study the antibody did not inhibit the hemorrhage reaction, but the hemorrhage producing activity of endotoxin must have been neutralized.

The present result also shows that a single injection of foreign proteins, such as rabbit antisera, can increase RBC number in the marrow. Thus, the neutralization of RBC reaction would become obscure.

We observed that the bone marrow reaction with hemorrhage was able to be induced also by allergic reaction of BSA (unpublished data). Therefore, the binding of the anti-Re antibody and endotoxin could cause the increase of RBC number in the marrow. The present study did not eliminate the possibility, but evidenced a more feasible possibility that a single injection of foreign and antigenic substances could cause the increase of RBC number. This seems to be a reason for obscure results as to neutralization of RBC reaction. Meanwhile, it is noteworthy that the injection of antigens increase the RBC number in the marrow, and it may be involved in immune response.

Acknowledgment

This work was supported by a grant-in-aid (00557109) for Fundamental Scientific Research from the Ministry of Education, Japan, 1980-1981. We appreciate the technical help of Miss N. Chiba.

References

1. Braude, A.I., Douglas, H. and Davis, C.E. J. Infect. Dis., 128, 157-164, 1973.

2. Ziegler, E.J., Douglas, H. and Braude, A.I. J. Clin. Invest., 52, 3236-3238, 1973.

3. Braude, A.I., Ziegler, E.J. and McCutchan, J.A. Schweiz. Med. Wschr., 108, 1827-1876, 1978.

4. Yoshida, M., Hirata, M. and Inada, K. Jap. J. Med. Sci. Biol., 26, 11-15, 1973.

5. Yoshida, M., Parant, F., Hirata, M. and Chedid, L. Jap. J. Med. Sci. Biol., 25, 143-147, 1972.

6. Yoshida, M., Hirata, M., Inada, K., Tsunoda, N., Mohri, H. and Matsui, S. In Bacterial Endotoxin and Host Response. ed by M.K. Agarwal, p 231-254, Elesevier/North Holland Biomedical press, 1980.

7. Hirata, M., Inada, K., Tsunoda, N. and Yoshida, M. In Bacterial Endotoxin and Host Response. ed. by M.K. Agarwal. p 255-272, Elesevier/North-Holland Biomedical Press, 1980.

8. Hirata, M., Tsunoda, N., Inada, K. and Yoshida, M. Jap. J. Med. Sci. Biol. 88, 30-31, 1980.

9. Johns, M.A., Bruins, S.C. and McCabe, W.R. Infect. Immun. 17, 9-15, 1977.

10. Young, L.S., Stevens, P. and Ingram, J. J. Clin. Invest., 56, 850-861, 1975.

11. Westphal, I., Luderitz, O. and Bister, F.Z. Naturforsch., 76, 148, 1952.

12. Galanos, C., Luderitz, O., Westphal, O. Eur. J. Biochem, 245-249, 1969.

13. Rickles, F.R., Levin, J., Hardin, J.A., Barr, C.F. and Conrad, M.E. J. Lab. Clin. Med., 89, 792-803, 1977.

14. McCabe, W.R. J. Immunol., 108, 601-610, 1972.

15. Chedid, L., Parent, M., Parant, F. and Bover, F. J. Immunol., 100, 292-301, 1968.

16. Tate, W.J., Douglas, H. and Braude, A.I. Ann. N.Y. Acad. Sci. 133, 746-762, 1966.

17. McCabe, W.R., Kreger, B.E. and Johns, M. N. Engl. J. Med. 287, 261-267, 1972.

18. Bruins, S.C., Stumacher, R., Johns, M.A. and McCabe, W.R. Infect. Immun., 17, 16-20, 1977.

MURINE IMMUNE RESPONSES TO THE SALMONELLA LIPOPOLYSACCHARIDE REGIONS

Emilio Jirillo*[+], Hiroshi Kiyono*, Suzanne M. Michalek*, Donato Fumarola**, Jerry R. McGhee*

*Departments of Microbiology and Pathology, Institute of Dental Research, Comprehensive Cancer Center, The University of Alabama in Birmingham, University Station, Birmingham, Alabama, USA

**Institute of Medical Microbiology, University of Bari, Medical School, Bari, Italy

[+]On sabbatical leave from the Institute of Medical Microbiology, University of Bari, Medical School, Bari, Italy

Introduction

Lipopolysaccharide (LPS), the endotoxic moiety of the outer membrane of gram-negative bacteria, exhibits many effects on the immune system of a susceptible host, including adjuvanticity, B cell mitogenicity, polyclonal response, and macrophage activation (1). It is also well known that LPS is immunogenic and acts as a T-independent antigen (2, 3). Several studies (4, 5, 6) have emphasized the importance of the O-polysaccharide chain (first signal), and the secondary role of lipid A (second signal) in the induction of anti-LPS responses. LPS sensitive C3H/HeN mice give good anti-LPS responses with specificity to the O-polysaccharide, while LPS insensitive C3H/HeJ mice do not produce significant anti-LPS responses. On the other hand, less is known regarding antibody responses to the R core and lipid A to injected LPS; however, it is clear that antibody responses to these regions can be induced by whole bacteria. In this regard, Galanos et al.(7) and Mattsby-Baltzer and Kaijser (8) induced anti-lipid A responses in rabbits by injection of acid

Immunopharmacology of Endotoxicosis

hydrolyzed Salmonella minnesota R595. These antibodies protected rabbits against toxic effects of lipid A (9, 10). Others have shown that rough bacteria can elicit antibody formation which reduces the severity of gram-negative infections (11, 12).

In the present study, evidence is presented that injection of C3H/HeN mice with purified LPS derived from either smooth or rough Salmonella, results in splenic plaque-forming cell (PFC) and serum antibody responses to all major LPS regions, while C3H/HeJ mice do not respond to these preparations. Since our past studies had shown that C3H/HeJ mice are high IgA responders to orally administered thymic-dependent antigens (13-16), it was also of interest to determine if this mouse strain, which is incapable of systemic responses to LPS, could form anti-LPS responses to Salmonella given by the gastrointestinal route. We present evidence that C3H/HeJ mice elicit PFC and antibody responses to the three LPS regions of the IgM, IgG and IgA isotypes. Furthermore, IgA anti-LPS antibody responses were favored and were considerably higher than those seen in identically treated C3H/HeN mice.

Materials and Methods

Mice. C3H/HeN (original breeders obtained from the National Institutes of Health, Bethesda, MD) and C3H/HeJ (original breeders obtained from Jackson laboratories, Bar Harbor, ME) mice were bred and maintained in the Core Facility for Immunocompromised Mice, the Comprehensive Cancer Center at the University of Alabama in Birmingham. All mice used in this study were 8 to 12 wk of age.

Bacterial Strains. Salmonella typhimurium LT-2 and Salmonella minnesota Ra(R60), Rb(R345) and Re(R595) were utilized in these studies.

LPS preparations. LPS was extracted from Salmonella typhimurium LT-2 (S-LPS) by a phenol water extraction method (17), while LPS from Salmonella minnesota R mutants, Ra, Rb, or Re (R-LPS) were prepared by the phenol-chloroform-petroleum ether (PCP) procedure of Galanos et al.(18). Free-lipid A was obtained by mild acid hydrolysis of Re-LPS (19) and alkaline lipid A by boiling free-lipid A in 0.25 N NaOH for 20 min (19).

Preparation of LPS-coated or TNP-coated sheep erythrocytes (SRBC). S-LPS (5 mg) was suspended in phosphate buffered saline (PBS) (5 ml) and boiled for 2 hr, while Ra-LPS, Rb-LPS and Re-LPS (3 mg) were suspended in distilled water (3 ml), boiled for 15 min, and, each preparation was added to 2% SRBC in PBS (45 ml) and incubated at 37°C for 30 min. Free-lipid A or alkaline-lipid A (800 µg) were added to a 2% suspension of SRBC in PBS and incubated for 30 min at 37°C (20). Coated erythrocytes were washed three times in PBS and used immediately or stored at 4°C. TNP-coupled SRBC (21) and horse erythrocytes (HRBC) were used for assessement of polyclonal B cell response.

In vivo immunization of mice with Salmonella LPS.Groups of mice were injected ip (0.1 ml) with 2.5 µg of S-LPS, Ra-LPS, Rb-LPS, or Re-LPS or 10 µg of alkaline lipid A or free-lipid A. On days 4, 7, 10, and 14 following immunization mice were bled and serum samples were collected for antibody assay. Animals were sacrificed and single spleen cell suspensions prepared for assessement of anti-LPS PFC responses (see below).

In vivo immunization of mice with killed Salmonella cells. Groups of C3H/HeJ and C3H/HeN mice were gastrically intubated (GI) with S.typhimurium LT-2 or S. minnesota Rb or Re (1×10^9 equivalent CFU in 0.25 ml of Hanks' balanced salt solution and 1.5% sodium bicarbonate). Mice were GI daily for three consecutive days/wk for 2 wk. One wk after the last GI, mice were boosted (iv) with either homologous bacterial cells (5 x 10^8 equivalent CFU/0.1 ml pyrogen-free saline) or with purified LPS (2.5 μg in 0.1 ml pyrogen-free saline). Four days following the final immunization, a portion of the animals were assessed for isotype and specificity of splenic anti-LPS PFC responses, using S-LPS, Rb-LPS and Re-LPS-coated SRBC. The remaining animals from each group were used for collection of serum and saliva 7 days after the final immunization for analysis of isotype and specificity of antibodies to LT-2, Rb or Re Salmonella by ELISA (see below). In another series , C3H/HeJ and C3H/HeN mice were immunized iv with heat-killed Salmonella LT-2, Rb, or Re cells (5 x 10^8 equivalent CFU/0.1 ml pyrogen-free saline) for 2 consecutive days/wk for 2 wk. Groups of mice were boosted iv 1 wk later with either homologous whole cells or purified LPS, as described above. The isotypes of splenic anti-LPS PFC responses were determined 4 days later.

Anti-LPS hemagglutination assay. Titers of anti-LPS antibodies in individual serum samples from mice immunized with LPS were determined by hemagglutination (HA) using SRBC coated with S-LPS, Ra-LPS, Rb-LPS, Re-LPS, free-lipid A or alkaline lipid A. The HA endpoints were recorded and then an optimal amount of goat anti-mouse IgM was added (22). Anti-LPS or anti-lipid A titers were expressed as the final dilution ($\log_2$) exhibiting agglutination after anti-SRBC HA antibody titers had been subtracted.

PFC assay. Splenic PFC were determined by the slide method of Cunningham and Szenberg (23). Appropriate numbers of spleen cells were mixed with guinea pig complement and SRBC coated with appropriate LPS preparations. IgG subclasses and IgA PFC were determined with optimal amounts of anti-mouse γ_1, γ_2, or α, as previously described (24).

Enzyme-linked immunosorbent assay (ELISA). The ELISA used in this study was a modification of a previously reported procedure (25, 26). Individual wells of EIA flat-bottom plates (Flow laboratories, Inc., McLean, VA) were coated with heat-killed Salmonella cells (200 µl; 5 x 10^8 CFU). Plates were incubated overnight at 37°C, washed extensively and 200 µl of 1% bovine serum albumin in 0.5% Tween-20-saline (Tween-saline) was added to block unreacted sites (90 min, 24°C). Plates were then washed five consecutive times with Tween-saline. Serial two-fold dilutions of serum and saliva were added to wells in quadruplicate (100 µl/well); one set of wells received Tween-saline only and served as control for comparison of absorbance readings with those containing saliva or serum samples from immune mice. Plates were incubated an additional 90 min (25°C) and washed four additional times with Tween-saline. Appropriate dilutions of alkaline phosphatase-labeled anti-mouse μ, γ, α chain sera were added to appropriate wells (100 µl/well). After overnight incubation at 4°C, plates were washed five additional times with Tween-saline, and substrate, freshly prepared by dissolving p-nitrophenylphosphate (1 mg/ml; Sigma 104 phosphatase substrate; Sigma Chemical Company, St. Louis, MO) in diethanolamine buffer was added to each well (100 µl/well). Plates were incubated at 25°C for 3 hr in the dark and the amount of color which developed was read at 405 nm with a Titertek Multiskan photometer (Flow).

Statistics. Results of splenic anti-LPS PFC responses (per 10^7 spleen cells) and HA titers ($\log_2$) were expressed as the mean ± SEM. ELISA values represent the mean ± SEM of quadruplicate determinations/sample. The significance of difference between means was determined by the Student's *t* test.

Results

Anti-LPS responses in mice systemically immunized with LPS. Splenic PFC responses to the three major regions of LPS in C3H/HeN mice given LPS (ip) is presented in Table 1. Maximum anti-O-polysaccharide responses, as determined with S-

LPS coated SRBC, were noted by day 4, while peak anti-R core and -lipid A PFC responses were observed at day 7 (Table 1). Serum HA antibody titers reflected this response pattern with peak anti-0-polysaccharide antibody responses at day 7 and anti-R core and -lipid A HA antibody responses by day 10 (Table 2). These results suggest that injection of S-LPS induces immune responses to all three major regions of the LPS molecule and initial responses are directed primarily to 0-polysaccharide and later responses to R core and lipid A.

We next determined the response pattern of C3H/HeN mice given R-LPS (Ra-, Rb- or Re-LPS). As expected, no splenic PFC responses to 0-polysaccharide were noted (Table 3) and maximum anti-R core and -lipid A PFC responses were seen on day 7. Highest PFC responses occurred with homologous antigen, e.g., mice immunized with Ra-LPS had higher anti-Ra-LPS than anti-Rb or Re-LPS responses. The finding that Re-LPS immunized mice had PFC responses detected by Ra- or Rb-LPS coated SRBC suggest that a significant anti-lipid A PFC responses are induced and can be detected with SRBC coated with these LPS preparations. Alternatively, Re-LPS may induce anti-keto-deoxyoctonate (KDO) responses, a moiety common to each of these R-LPS molecules. Maximum serum antibody responses to these R-LPS preparations were noted on day 10 (Table 4).

In a final series of experiments, C3H/HeN mice were injected with 10 µg of purified lipid A and splenic PFC responses assessed (Table 5). Maximum anti-lipid A PFC responses were noted on day 4. Anti-lipid A HA antibody titers in sera peaked on day 7 (Table 6). These results suggest that immune responses to purified lipid A occur earlier than are seen in mice given S- or R-LPS and may be related to greater accessibility of lipid A antigenic determinants. In data not shown, immunization of C3H/HeJ mice with each of these LPS preparations resulted in low or no splenic PFC or serum HA antibody responses.

Table 1

Kinetics of the splenic PFC responses to various <u>Salmonella</u> LPS preparations in C3H/HeN mice injected (ip) with 2.5 μg S-LPS[a]

Days after immunization	IgM anti-LPS PFC/10^7 spleen cells[b] (*Salmonella* LPS coated SRBC)					
	Smooth	Ra(60)	Rb(345)	Re(595)	alkaline lipid A	Free lipid A
Day 4	595±37	241±53	207±29	193±35	198±21	194±19
Day 7	315±22	503±87	613±61	580±23	587±25	511±15
Day 10	199±13	202±21	222±31	208±34	305±39	377±44
Day 14	83±11	89±14	53±7	57±5	59±11	63±8

[a]Groups of mice (24-36) were injected (ip) with 2.5 μg S-LPS and the level of splenic anti-LPS PFC responses were determined on the indicated day.

[b]Values are the mean ± SEM of 6-9 mice/time point and 2 determinations/sample.

Table 2

Kinetics of serum HA antibody responses to various *Salmonella* LPS preparations in C3H/HeN mice injected (ip) with 2.5 µg S-LPS[a]

Days after immunization	Serum anti-LPS HA titers ($\log_2$)[b] (*Salmonella* LPS coated SRBC)					
	Smooth	Ra(R60)	Rb(R345)	Re(R595)	Alkaline lipid A	Free lipid A
Day 4	6.5±1.2	3.7±0.9	6.7±1.3	2.2±0.8	2.7±0.9	2.3±0.6
Day 7	11.3±1.4	7.1±2.3	8.2±1.4	6.6±1.7	6.4±1.9	4.2±0.5
Day 10	6.3±1.3	10.2±2.1	13.1±0.5	9.8±1.3	10.1±0.9	8.1±1.1
Day 14	3.9±0.6	3.8±0.4	4.1±0.6	3.8±0.6	3.7±0.4	3.7±0.3

[a]Groups of mice (24-36) were injected (ip) with 2.5 µg S-LPS and serum anti-LPS HA titers were determined on the indicated day.

[b]Values are the mean ± SEM of 6-9 mice/time point and two determinations/sample.

Table 3

Kinetics of the splenic PFC responses to various Salmonella LPS preparations in C3H/HeN mice injected (ip) with 2.5 μg Ra-LPS, Rb-LPS, and Re-LPS[a].

Immunogen	Days after immunization	IgM anti-LPS PFC/10^7 spleen cells[b] (Salmonella LPS coated SRBC)					
		Smooth	Ra (R60)	Rb (R345)	Re (R595)	Alkal. lipid A	Free lipid A
Ra-LPS	Day 4	0.0	304±23	183±19	263±21	107±17	98±22
	Day 7	0.0	623±31	408±22	475±19	519±32	469±28
	Day 10	0.0	205±9	193±11	153±8	198±21	183±10
	Day 14	0.0	37±5	39±7	43±6	35±7	42±8
Rb-LPS	Day 4	0.0	102±26	204±11	67±13	183±10	194±11
	Day 7	0.0	357±24	606±11	369±28	393±9	387±8
	Day 10	0.0	102±12	198±13	97±10	186±9	169±10
	Day 14	0.0	33±5	0.0	0.0	0.0	0.0
Re-LPS	Day 4	0.0	201±29	173±21	204±34	223±13	225±16
	Day 7	0.0	410±11	403±9	781±17	802±25	793±33
	Day 10	0.0	153±12	171±8	183±13	367±14	339±12
	Day 14	0.0	38±5	43±6	65±11	87±11	101±8

[a]Groups of mice (24-36) were injected (ip) with 2.5 μg of the appropriate R-LPS.
[b]Values are the mean ± SEM of 6-9 mice/time point and two determinations/sample.

Table 4

Kinetics of the serum HA antibody responses to various LPS preparations in C3H/HeN mice injected with 2.5 μg Ra-LPS, Rb-LPS, and Re-LPS[a].

Immunogen	Days after immunization	Serum anti-LPS HA titers ($\log_2$) (Salmonella LPS coated SRBC)					
		Smooth	Ra (R60)	Rb (R345)	Re (R595)	Alkal.	Free lipid A
Ra-LPS	Day 4	0.0	6.3±0.8	5.6±0.5	2.2±0.4	3.7±0.5	2.3±0.3
	Day 7	0.0	8.3±0.9	6.2±0.4	3.8±0.5	7.1±0.3	6.7±0.4
	Day 10	0.0	11.1±1.1	8.3±0.8	7.6±0.5	9.2±1.2	8.2±0.8
	Day 14	0.0	7.2±0.3	5.8±0.4	6.7±0.3	6.5±0.3	6.4±0.7
Rb-LPS	Day 4	0.0	5.3±0.9	7.3±0.6	4.2±0.4	2.1±0.4	1.5±0.3
	Day 7	0.0	7.0±0.2	8.6±0.3	5.9±0.2	4.3±0.3	3.5±0.3
	Day 10	0.0	10.3±0.5	12.2±0.3	10.2±0.4	8.2±0.9	7.8±1
	Day 14	0.0	5.9±0.4	7.2±0.3	6.8±0.4	4.4±0.3	4.2±0.4
Re-LPS	Day 4	0.0	3.6±0.4	4.2±0.5	6.7±0.4	4.5±0.4	4.2±0.3
	Day 7	0.0	4.2±0.5	4.9±0.2	8.1±0.5	6.7±0.3	5.8±0.6
	Day 10	0.0	8.3±0.2	7.8±0.5	10.5±0.8	12.1±0.5	11.9±0.6
	Day 14	0.0	4.2±0.2	5.7±0.2	7.6±0.3	7.8±0.3	7.9±0.6

[a]Groups of mice (24-36) were injected (ip) with 2.5 μg of the appropriate R-LPS.
[b]Values are the mean ± SEM of 6-9 mice/time point and two determinations/sample.

Table 5

Kinetics of the splenic PFC responses to various *Salmonella* LPS preparations in C3/HeN mice injected (ip) with 10 μg alkaline lipid A or free lipid A[a].

Immunogen	Days after immunization	IgM anti-LPS PFC/10^7 spleen cells[b] (*Salmonella* LPS coated SRBC)					
		Smooth	Ra(R60)	Rb(R345)	Re(R595)	Alkaline lipid A	Free lipid A
Alkaline lipid A	Day 4	0.0	0.0	103±18	331±23	704±22	759±25
	Day 7	0.0	0.0	0.0	207±33	508±15	431±9
	Day 10	0.0	0.0	0.0	163±9	365±12	278±11
	Day 14	0.0	0.0	0.0	0.0	197±24	167±8
Free lipid A	Day 4	0.0	0.0	167±25	427±75	763±21	627±29
	Day 7	0.0	0.0	0.0	223±12	473±11	405±17
	Day 10	0.0	0.0	0.0	105±17	307±11	371±13
	Day 14	0.0	0.0	0.0	0.0	163±11	175±11

[a]Groups of mice (24-36) were injected (ip) with 10 μg of alkaline lipid A or free lipid A and the level of splenic anti-LPS PFC responses were determined on the indicated day.

[b]Values are the mean ± SEM ;f 6-9 mice/time point and 2 determinations/sample.

Table 6

Kinetics of serum HA antibody responses to various LPS preparations in C3H/HeN mice injected (ip) with 10 µg alkaline lipid A or free lipid A[a].

Immunogen	Days after immunization	Serum anti-LPS HA titers (log_2)[b] (*Salmonella* LPS coated SRBC)					
		Smooth	Ra(R60)	Rb(R345)	Re(R595)	Alkaline lipid A	Free lipid A
Alkaline lipid A	Day 4	0.0	0.0	0.0	0.0	2.1±0.6	1.5±0.3
	Day 7	0.0	0.0	0.0	3.7±0.5	12.3±0.3	11.9±0.5
	Day 10	0.0	0.0	0.0	0.0	6.2±0.3	3.9±0.2
	Day 14	0.0	0.0	0.0	0.0	4.2±0.3	3.9±0.2
Free lipid A	Day 4	0.0	0.0	0.0	0.0	3.4±0.5	2.2±0.2
	Day 7	0.0	0.0	0.0	4.1±0.6	11.7±0.2	11.5±0.5
	Day 10	0.0	0.0	0.0	0.0	5.4±0.5	5.2±0.4
	Day 14	0.0	0.0	0.0	0.0	3.5±0.4	3.4±0.3

[a]Groups of mice (24-36 mice) were injected (ip) with 10 µg of alkaline lipid A or free lipid A and the level of serum HA antibody responses were determined on the indicated day.
[b]Values are the mean ± SEM of 6-9 mice/time point and 2 determinations/sample.

It was also important to insure that responses in C3H/HeN mice were specific for LPS and were not due to polyclonal responses. Therefore, C3H/HeN mice immunized with appropriate doses of S- or R-LPS preparations or lipid A were assessed for TNP, SRBC and HRBC responses (data not shown). Although low anti-TNP and/or anti-SRBC PFC were noted, the response was clearly specific for LPS (data not shown).

Anti-LPS responses in C3H/HeN and C3H/HeJ mice orally immunized with S- or R-Salmonella.

Our past studies have clearly shown that C3H/HeJ mice are high IgA responders to orally administered thymic-dependent antigens (13-16). It was therefore of interest to determine if the C3H/HeJ mouse strain is capable of forming anti-LPS responses to orally administered gram-negative bacterial cells since this mouse is incapable of responding to systemically administered LPS. In preliminary studies, we compared immune responses to whole *Salmonella* cells given iv to C3H/HeN or C3H/HeJ mice. Repeated injections of bacteria resulted largely in IgM and IgG_1 and somewhat lower IgG_2 anti-LPS PFC responses in C3H/HeN mice. C3H/HeJ mice, on the other hand, exhibited low IgM and no IgA anti-LPS PFC responses (data not shown).

In the first experiment, groups of C3H/HeN and C3H/HeJ mice were given smooth *Salmonella* (LT-2) cells by GI and subsequently boosted (iv) with LT-2 cells or purified S-LPS. C3H/HeJ mice gave significantly higher PFC responses of the IgM and especially of the IgA isotype (Table 7). These PFC responses were directed to S-LPS (probably 0-polysaccharide chain) and to a lesser extent to Rb-LPS and Re-LPS. It should be noted that when mice were boosted systemically with S-LPS a dramatic increase in IgA anti-O-polysaccharide PFC responses was noted (Table 7).

Table 7

Splenic anti-*Salmonella* LPS PFC responses in C3H/HeN and C3H/HeJ mice orally administered with smooth *Salmonella* (LT-2)[a].

Mouse Strain	Antigen for boost	Anti-LPS PFC/10^7 spleen cells[b] (S-*Salmonella* LPS coated SRBC)			
		IgM	IgG_1	IgG_2	IgA
C3H/HeN	LT-2 cells	385±65	119±21	53±8	595±61
	S-LPS	475±67	163±17	92±11	797±54
C3H/HeJ	LT-2 cells	760±93	310±61	195±23	1192±32
	S-LPS	1310±72	1295±84	223±18	6671±23

[a]Groups of 8-10 mice orally administered as described in Materials and Methods.

[b]Values are the mean ± SEM of 2 determinations/sample.

Similar studies were carried out with Rb-*Salmonella* cells and groups of C3H/HeN and C3H/HeJ mice were GI with Rb cells and later boosted (iv) with Rb-*Salmonella* cells or purified Rb-LPS. C3H/HeJ mice again exhibited an elevated IgA splenic anti-LPS PFC response to Rb-LPS (Table 8). Interestingly, when mice were primed with Rb cells and boosted with Rb-LPS, the isotype enhancement was seven fold for IgM, five fold for IgG_1 and three fold for IgA. Nevertheless, IgA represented the highest isotype of anti-LPS PFC response seen.

Table 8

Splenic anti-Salmonella LPS PFC response in C3H/HeN and C3H/HeJ mice orally administered with rough Salmonella (Rb)[a].

Mouse Strain	Antigen for boost	Anti-LPS PFC/10^7 spleen cells[b] (Rb-Salmonella LPS coated SRBC)			
		IgM	IgG_1	IgG_2	IgA
C3H/HeN	Rb cells	104±15	102±8	0.0	573±21
	Rb-LPS	213±21	204±11	207±13	863±71
C3H/HeJ	Rb cells	636±73	1415±89	915±78	2421±63
	Rb-LPS	4298±193	6715±93	1315±71	8123±215

[a]Groups of 8-10 mice orally administered as described in Materials and Methods.

[b]Values are the mean ± SEM of 2 determinations/sample.

In a final series of experiments, groups of C3H/HeN and C3H/HeJ mice were orally immunized with Re-Salmonella cells and boosted (iv) with either Re cells or purified LPS. C3H/HeJ mice again exhibited elevated splenic anti-LPS responses and IgA was the major isotype seen (Table 9). In contrast, low IgA responses were seen in identically treated C3H/HeN mice in the absence of other isotype responses. Of interest was the observation that mice GI with Re-Salmonella and boosted with purified Re-LPS gave an anti-LPS PFC response which was similar in magnitude to that observed in C3H/HeJ mice injected with Re-bacteria (Table 9). These results suggest that purified Re-LPS and acid hydrolyzed Re-Salmonella exhibit similar potency for boosting PFC responses.

Table 9

Splenic anti-Salmonella LPS PFC responses in C3H/HeN and C3H/HeJ mice orally administered with rough Salmonella (Re)[a].

Mouse Strain	Antigen for boost	Anti-LPS PFC/10^7 spleen cells[b] (Re-Salmonella LPS coated SRBC)			
		IgM	IgG_1	IgG_2	IgA
C3H/HeN	Re cells	0.0	0.0	0.0	421±29
	Re-LPS	0.0	0.0	0.0	223±71
C3H/HeJ	Re cells	437±51	708±33	358±14	1129±34
	Re-LPS	587±81	757±73	405±61	1512±79

[a]Groups of 8-10 mice orally administered as described in Materials and Methods.

[b]Values are the mean ± SEM of 2 determinations/sample.

In data not shown, antibody levels in sera of C3H/HeJ mice orally immunized with S- or R-Salmonella were higher than seen in identically treated C3H/HeN mice. IgA represented the major isotype seen and levels were approximately three to four fold higher in C3H/HeJ than in C3H/HeN mice. IgA responses were also elevated in saliva of C3H/HeJ mice and four to ten fold higher responses were seen than in similarly treated C3H/HeN mice.

Discussion

Our results clearly suggest that systemic administration of S-LPS to mice leads to an immune response which is directed

to the three main regions of the LPS molecule. In this regard, we noted that mice injected with S-LPS formed splenic PFC and humoral antibody responses to O-polysaccharide and also to R core and lipid A. The anti-R core and -lipid A responses were specific since no anti-O-polysaccharide responses were observed in mice immunized with R-LPS. Moreover, the absence of significant polyclonal responses induced by the doses of LPS used confirm the specificity of these anti-LPS responses. The kinetics of anti-O-polysaccahride, -R core and -lipid A responses observed may result from several different mechanisms. In this regard, it is possible that, in the course of the anti-LPS response, the O-chain is removed from the intact LPS molecule with simultaneous liberation of R-LPS and induction of anti-R core and -lipid A responses. Alternatively, the anti-R core and -lipid A antibodies may be induced later in the immune response. We favor the former possibility since purified lipid A, free of polysaccharide, induced significant day 4 splenic PFC and day 7 serum antibody titers, a response pattern which was completely analogous to the anti-O polysaccharide response seen in S-LPS immunized C3H/HeN mice. Thus, we propose that lipid A remains "buried" in the molecule following antigen processing, and that significant digestion of the molecule may be required for exposure of lipid A antigenic sites.
Extensive studies by Galanos and his colleagues (7) and Mattsby-Baltzer and Kaijser (8) clearly indicate that the lipid A moiety bears antigenic sites; however, in their studies it was necessary to adsorb purified lipid A onto whole R595 bacteria in order to induce anti-lipid A responses in rabbits. On the other hand, Mita et al.(27) recently showed that anti-lipid A responses could be induced in mice given purified lipid A. Nevertheless, these latter workers reported very low anti-lipid A PFC responses with both low and high doses of antigen. Although the reason for this discrepancy is not known, we feel that the procedure used to coat

SRBC with lipid A is of importance. In our studies, we routinely employ both polyvalent (rabbit) anti-lipid A and monoclonal (mouse) anti-lipid A antibodies to test the degree of lipid A coating to SRBC. It should be pointed out that alkali treatment of free lipid A did not destroy the immunogenicity of this molecule. This finding confirms and extends the studies of Mita and coworkers (27), who also showed that both free lipid A and alkaline-lipid A induced anti-lipid A responses in mice. Finally, we suggest that lipid A may be more immunogenic when it is completely liberated from the remainder of the polysaccharide moiety. This supposition may provide an explanation for the presence of naturally occurring anti-lipid A antibodies in healthy subjects (28, 29) and for higher anti-lipid A titers in individuals with gram-negative infections (28, 29).

In studies with orally administered Salmonella, our data indicate that GI with either smooth S. typhimurium LT-2 or rough S. minnesota Rb or Re to C3H/HeJ mice, followed by iv boosting with homologous cells or purified LPS, results in significant humoral immune responses to the three major regions of the Salmonella LPS molecule. The IgA isotype was clearly favored, although good IgM, IgG_1, and less IgG_2 anti-LPS PFC responses were observed. Furthermore, since systemic immunization with purified S-LPS, Rb-LPS or Re-LPS induced good responses in C3H/HeN mice and negligible responses in the C3H/HeJ strain, we conclude that oral presentation of whole Salmonella cells clearly primes these mice for elevated anti-LPS responses and favors the IgA isotype.

The use of a panel of SRBC coated with S-LPS, Rb-LPS or Re-LPS allowed us to determine both the isotype and the major LPS regions to which the splenic anti-LPS response was directed. Mice GI with S-Salmonella and, subsequently boosted with the same antigen, elicited highest responses to S-LPS and lower responses to Rb-LPS and Re-LPS. Therefore, this response is largely directed to the O-polysaccharide. On the other hand, mice immunized with Rb-*Salmonella* and Re-*Salmonella*, failed to induce splenic immune responses to S-LPS coated SRBC. Additionally, Rb-immunized mice gave good responses to homologous LPS and lower responses to Re-LPS, while Re-immunized mice produced optimal splenic PFC responses to Re-LPS and poor responses to Rb-LPS. Therefore, we infer from these studies that GI of C3H/HeJ mice with S-*Salmonella* primes largely for anti-O-polysaccharide responses, while GI with Rb or Re-*Salmonella* primes the animals mainly for anti-R core or anti-lipid A PFC responses, respectively. The finding that purified S-LPS or Rb-LPS, when compared with whole *Salmonella* cells, induced higher splenic PFC responses in C3H/HeJ mice merits some discussion. In fact, according to Nixdorff and coworkers (30-32) the administration of LPS with hydrophobic molecules, e.g., phospholipids or lipoproteins favored the IgG_1 isotype response, while hydrophilic antigens, e.g., LPS and proteins, favored IgG_2 anti-LPS PFC responses (31, 32). In other words, both hydrophilic and hydrophobic derivatives act as adjuvants for enhancement of IgG subclass responses. Our results could also be ascribed to an adjuvant effect, since S-LPS, because of its long O-polysaccharide side chain would exhibit a hydrophilic character, providing a better immunogenic array for expansion of O-polysaccharide sensitized B cells. This could result in the expansion of IgA anti-LPS B cell clones in spleens of orally immunized C3H/HeJ mice. On the other hand, Rb-LPS, due to its hydrophobic nature may favor the enhancement of the IgG_1 response. However, we also suggest that purified Rb-LPS provides a better immunogenic array of R core determinants than occur on the *Salmonella* cell surface which generally facilitates IgA, IgG_1 and IgM anti-R core responses. A second possibility to explain this enhanced response may involve an adjuvant effect by lipid A. We do not favor this second possibility, since purified Re-LPS would induce higher splenic PFC responses if lipid A was inducing an adjuvant effect and this did not occur. We conclude that purified LPS provides a better immunogenic array for expansion of sensitized B cell clones in spleens of orally immunized C3H/HeJ mice.

Another interesting finding which emerged from these studies was that oral presentation of Salmonella to LPS responsive C3H/HeN mice largely results in an inability of these animals to form IgM and IgG subclass responses when systemically challenged. Generally, C3H/HeN mice treated in this fashion did undergo IgA anti-LPS PFC responses, although the magnitude of this response was always significantly less than seen in identically treated C3H/HeJ mice. This pattern of response in C3H/HeJ mice is similar to our past studies involving induction of oral tolerance to the thymic-dependent antigen SRBC (13-15). In those studies, we consistently noted that GI of SRBC daily for two wk or longer resulted in systemic unresponsiveness in C3H/HeN mice, but a complete lack of oral tolerance was observed in C3H/HeJ mice (14, 15). In this regard, Challacombe and Tomasi (33) first observed that GI with either bacterial or soluble antigen induced IgA responses in external secretions. A possible explanation for this concurrent induction of systemic unresponsiveness and IgA responses in external secretions was provided by Richman and coworkers (34), who reported that GI of mice with large doses of ovalbumin induced helper T cells for IgA responses and suppressor T cells for IgG responses in Peyer's patches and mesenteric lymph nodes. Our result suggest that systemic unresponsiveness is limited to the IgM and IgG classes in C3H/HeN mice GI with Salmonella, but that splenic IgA anti-LPS PFC and serum and salivary IgA antibodies occur in these mice.

In conclusion, these data taken together suggest that in the course of the anti-LPS response, antibodies are directed to the three major regions of the endotoxic molecule. Furthermore, C3H/HeJ mice do not form significant anti-LPS responses to S-LPS or R-LPS, or to whole Salmonella, while oral immunization results in significant anti-LPS responses to all three major regions in this mouse strain.

References

1. Morrison, D.C., Ryan, J.L.: Adv. Immunol. 28, 293-450 (1979)
2. Von Eschen, K.B., Rudbach, J.A.: J. Immunol. 116, 8-11 (1976)
3. Rudbach, J.A.: In The role of Immunological Factors in Infections, Allergic and Autoimmune Processes. Beers, R.E. Jr., Bassett, E.G. Eds. Raven Press NY, P. 29-41 (1976)
4. Rudbach, J.A., Reed, N.D.: Infect. Immun. 16, 513-517 (1977)
5. McGhee, J.R., Michalek, S.M., Moore, R.N., Merghenhagen, S.E., Rosenstreich, D.L.: J. Immunol. 122, 2052-2058 (1979)
6. Watson, J., Riblet, R.: J. Exp. Med. 140, 1147-1161 (1974)
7. Galanos, C., Lüderitz, O., Westphal, O.: Eur. J. Biochem. 24, 116-122 (1971)
8. Mattsby-Baltzer, I., Kaijser, B.: Infect. Immun. 23, 758-763 (1979)
9. Rietschel, E. Th., Galanos, C.: Infect. Immun. 15, 34-49 (1977)
10. Rioux-Darrieulat, F., Parant,M., Chedid, L.: J. Infect. Dis. 137, 7-13 (1978)
11. Johns, M., Skehill, A., McCabe, W.R.: J. Infect. Dis. 147, 57-67 (1983)
12. Ziegler, E.J., McCutchan, J.A., Fierer, J., Glauser, M. P., Sadoff, J.C., Douglas, H., Braude, A.I.: N. Engl. J. Med. 307, 1255-1230 (1982)
13. Michalek, S.M., Kiyono, H., Babb, J.L., McGhee, J.R.: J. Immunol. 125, 2220-2224 (1980)
14. Kiyono, H., McGhee, J.R., Wannemuehler, M.J., Michalek, S.M.: J. Exp. Med. 155, 605-610 (1982)
15. Michalek, S.M., Kiyono, H., Wannemuehler, M.J., Mosteller, L.M., McGhee, J.R.: J. Immunol. 128, 1992-1998 (1982)

16. Kiyono, H., Michalek, S.M., Mosteller, L.M., Torii, M. Hamada, S., McGhee, J.R.: Scand. J. Immunol. 16, 455-463 (1982)

17. McIntire, F.C., Sievert, H.W., Barlow, G.H., Finley, R.A., Lee, A.Y.: Biochemistry 6, 2363-2372 (1967)

18. Galanos, C., Lüderitz O., Westphal, O.: Eur. J. Biochem. 9, 245-249 (1969)

19. Galanos, C., Rietschel, E. Th., Lüderitz, O., Westphal, O.: Eur. J. Biochem. 19, 143-152 (1971)

20. Fink, P.C., Galanos, C.: Immunobiol. 158, 380-390 (1981)

21. Rittenberg, M.B., Pratt, K.L.: Proc. Soc. Exp. Biol. 132, 575-581 (1969)

22. Kiyono, H., McGhee, J.R., Michalek, S.M.: J. Immunol. 124, 36-41 (1980)

23. Cunningham, A.J., Szenberg, A.: Immunology 14, 599-601 (1968)

24. Kiyono, H., Babb, J.L., Michalek, S.M., McGhee, J.R.: J. Immunol. 125, 732-737 (1980)

25. Engvall, E., Perlmann, P.: J. Immunol. 109, 129-135 (1972)

26. Voller, A., Bidwell, D., Bartlett, A.: Manual of clinical Immunology. Rose, N.R., Friedman, H. Eds. American Society for Microbiology, Washington D.C., P. 359-371 (1980)

27. Mita, A., Ohta, H., Mita, T.: Immunology, 45, 333-340 (1982)

28. Westenfelder, M., Galanos, C., Madsen, P.O., Marget, W. : Microbiology-1977. Schlessinger D. Ed. American Society for Microbiology, Washington D.C., P. 277-279 (1977)

29. Jirillo, E., Mastrolitti, G., Schena, F.P., Munno, I., Lattanzio, A., Fumarola, D.: In Bacterial Endotoxin and Host Response. Agarwal M.K. Ed.Elsevier-North-Holland, P. 95-104 (1980)

30. Ruttkowski, E., Nixdorff, K.: J. Immunol. 124, 2548-2556 (1980)

31. Karch, H., Gmeiner, J., Nixdorff, K.: Infect Immun. 40, 157-165 (1983)

32. Karch, H., Nixdorff, K.: J. Immunol. 131, 6-12 (1983)

33. Challacombe, S.J., Tomasi, T. B. Jr: J. Exp. Med. 152, 1459-1472 (1980)

34. Richman, L.K., Graeff, A.S., Yarchoan, R., Strober, W.: J. Immunol. 126, 2079-2083 (1981)

Work in part supported by M.P.I. (Ric. Scien. 1982, 40%).

LPS-INDUCED NON-SPECIFIC RESISTANCE TO IMMUNODEFECTIVE CBA/N MICE AGAINST SALMONELLA-INFECTION

Masayasu Nakano, Kazuyasu Onozuka, Tatsuo Saito-Taki

Department of Microbiology, Jichi Medical School, Tochigi-ken 329-04, Japan

Introduction

Since bacterial lipopolysaccharide (LPS) is a potent immunomodulator (1), it modifies various functions of host defense systems and stimulates the resistance of animals to microbial infection (2,3). Its stimulation of macrophage activities such as phagocytic ability and bactericidal capacity may result in the enhancement of host defense mechanisms against exogenous and harmful agents.

The CBA/N subline shows a number of immunologic defects (4) that are apparently related to an X-chromosome-linked recessive abnormality in the functional responses of B lymphocytes to some kinds of antigens, especially the response to thymic independent type 2 antigens. These mice are also considered to show abnormalities in mitogenic and polyclonal antibody responses against LPS (5). Moreover, CBA/N mice are known to be susceptible to Salmonella typhimurium, which causes a disease in mice that appears to be similar in its pathogenesis to typhoid fever in man, and the expression of this X-linked defect is also associated with increased susceptibility of the mice to the organisms (6).

In the present paper, we deal with enhancements of bactericidal and phagocytic capacities by LPS in the peculiar CBA/N-defective mice against virulent Salmonella enteritidis which

Immunopharmacology of Endotoxicosis

is one of the causative agents in murine salmonellosis as well as S. *typhimurium*.

Materials and Methods

Animals

Both male and female (CBA/N female X BALB/c male)F1 mice, 8 to 10 weeks of age, were used throughout the experiments. Since the CBA/N-defect is *X*-linked (4), CBA/N females mated with normal males (in this case, BALB/c) produced F1 males which expressed some defects in B lymphocyte functions and female litter-mates which, although heterozygous for the defective *X*-chromosome, were phenotypically normal.

Bacterial lipopolysaccharide (LPS) and other endotoxins

LPS was extracted from *Salmonella typhimurium* LT2 by the hot-phenol-water procedure as described by Westphal (6) and then purified by repeated centrifugation at 100,000 x *g* for 60 min. Lipid A was prepared from the LPS by hydrolysis with 1% acetic acid at 100°C for 4 hr, and then the lipid A was complexed (8) with bovine serum albumin (BSA) by the method of Galanos *et al*. (9). Alkali-detoxified LPS was prepared from the LPS by hydrolysis with 0.25N NaOH at 56°C for 1 hr according to the method of Neter *et al*. (10). These materials were dissolved in physiological saline and injected intraperitoneally (ip) into mice (0.2 ml/mouse).

Bacterial strain

Virulent *Salmonella enteritidis* No.11 strain was used. The microorganisms were cultivated at 37°C overnight on nutrient agar and then suspended in physiological saline.

Infection

Mice were challenged ip or intravenously (iv) with the bacterial suspension (0.2 ml/mouse). Viable numbers of the infecting organisms were determined by quantitative cultivation. Mortality of mice and mean survival days were estimated on the twenty-first day after the challenge.

Quantitative cultivation of inoculated organisms

For the determination of bactericidal activity in mice, viable numbers of the infecting organisms in the peritoneal cavity and spleen were counted 24 hr after the infection. The peritoneal cavity was repeatedly washed out with 5 ml aliqots of saline. The spleen was stirred with 5 ml of saline in a homogenizer. The washes and homogenates were serially diluted, and the dilutions were mixed with melted nutrient agar for pour plate cultivation of the organisms. Colony counts were made after 48 hr incubation.

Test of bacterial clearance from blood

Two tenths ml of the suspension of bacteria was injected into the tail vein of the mice. Immediately after the injection (within 1 min) and at indicated times thereafter, 0.025 ml of blood was collected from the retroorbital vein of the mice using a heparinized capillary tube. For the determination of viable organisms, each blood sample was placed in 2 ml of saline, mixed with melted nutrient agar and a pour culture made in a petri dish. If appropriate, the samples were further diluted prior to making the pour culture. After 48 hr incubation, the colonies were counted. The number of viable bacteria in the sample at each time was plotted against time, and the rate of the clearance of bacteria from the blood was assessed from the line drawn between the data points.

Results

Susceptibility of CBA/N-defective mice against *Salmonella enteritidis* No.11-infection and protective effect of LPS

To estimate the susceptibility of CBA/N-defective mice, F1 male and female mice were challenged with virulent No.11 organisms of *S. enteritidis*. As shown in Table 1, F1 male mice seemed to be more susceptible than F1 female mice. When mice were infected with 3×10^4 organisms, there was an obvious difference in the mortality rates between males and females. When a smaller number of organisms was used for infection, all the males were killed, but the females survived. However, if mice were infected with larger doses of organisms (3×10^5 organisms or more), all mice, regardless of whether they were males or females, were killed within several days after the infection. These data indicate an increased susceptibility of CBA/N-defective mice to *S. enteritidis* which was similar to that observed with *S. typhimurium*.

CBA/N-deficiency conferred susceptibility to No.11 organisms; only 10 organisms of this virulent strain are enough to kill the majority of mice. However, if these defective mice were injected with LPS 24 hr before the infection, they were able to survive the infection (Table 2). The results clearly demonstrate that the CBA/N-defective strain, which is known to be refractory to LPS on the B cell level, can respond to LPS and acquire non-specific resistance to virulent *S. enteritidis*.

Inhibitory effect of LPS on multiplication of infecting organisms in the peritoneal cavity and spleen

The susceptibility of CBA/N-defective mice to salmonella may be due to some deficiency in their bactericidal ability, and previous administration of LPS may improve this property.

Table 1. Mortality of (CBA/N female X BALB/c male)F1 mice after infection with *Salmonella enteritidis* No.11 strain

Number of infecting organisms	CBA/N-defect	No. of deaths / No. of mice	Mortality rate (%)	Mean survival days ± S. D.
3×10^5 ip	yes (F1 male)	21/21	100	5.9 ± 1.3
	no (F1 female)	18/18	100	6.9 ± 2.5
3×10^4 ip	yes	25/27	92	7.0 ± 1.7
	no	12/27	44	9.6 ± 0.9
3×10^3 ip	yes	10/10	100	7.2 ± 0.9
	no	0/10	0	
3×10^1 ip	yes	7/8	88	11.4 ± 1.5
	no	0/8	0	

Table 2. Protective effect of LPS on CBA/N-defective mice infected with *Salmonella enteritidis* No.11 strain

No. of infecting organisms	Treatment	No. of deaths / No. of mice	Mortality rate (%)	Mean survival days ± S. D.
1×10^1 ip	LPS*	3/13	23	12.0 ± 2.7
	none	12/13	92	10.9 ± 1.4

* 20μg ip, 24 hr before the infection

To examine this, F1 male and female mice were ip injected with LPS. Twenty-four hr later, these mice and other untreated control groups of mice were challenged with No.11 organisms, and 24 hr after the challenge, all mice were killed. The numbers of viable organisms in their peritoneal cavities and spleens are shown in Figure 1. Male mice had obviously less bactericidal activity since the recovery of organisms in their peritoneal cavities and spleens was greater than in female

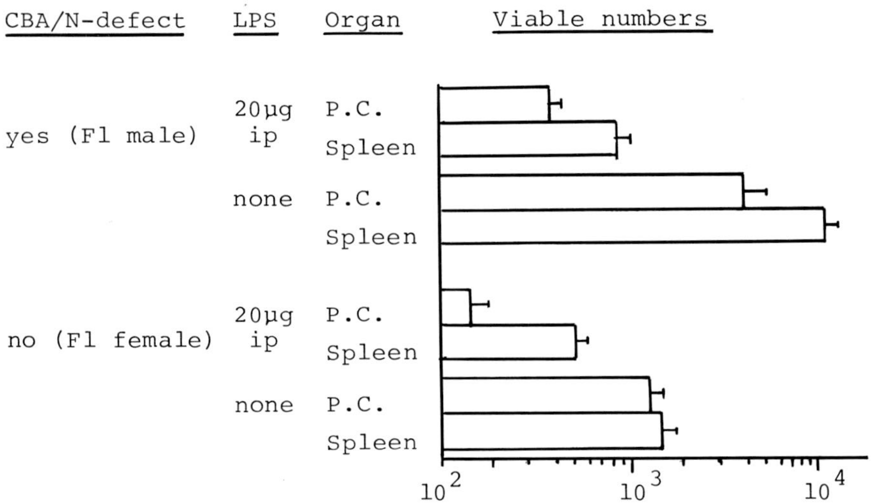

Figure 1. Effect of LPS on the infection of *Salmonella enteritidis* No.11 strain in (CBA/N female x BALB/c male)F1 mice. LPS (20µg) was ip injected 24 hr before the infection, and mice were ip challenged with 1 x 10^4 organisms of the No.11 strain. Viable numbers of the organisms in the peritoneal cavity (P.C.) and spleen were determined by quantitative cultivation at 24 hr after the infection. Results are expressed as the arthmetic mean of 3 mice ± the standard deviation.

mice. When male as well as female mice were treated with LPS before the infection, these mice showed an augmentation of bactericidal ability. The numbers of organisms in LPS-treated mice were clearly less than those in untreated control mice. The bactericidal effect induced by LPS was prominent in their peritoneal cavities rather than in the spleens. These results suggest that LPS has the ability to enhance bactericidal activity in CBA/N-defective mice, the same as in phenotypically normal female mice.

Dose-effect of LPS on bactericidal activity in the peritoneal cavity and spleen

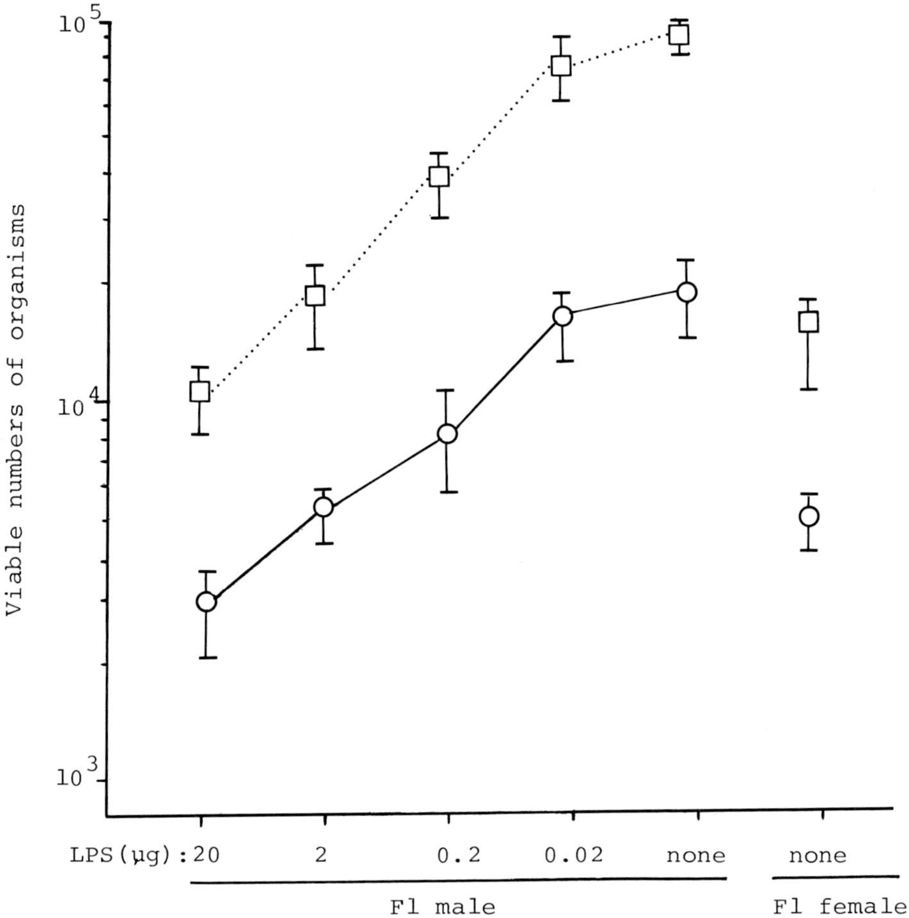

Figure 2. Anti-bacterial effect of LPS in the peritoneal cavity and spleen of CBA/N-defective mice. LPS was ip injected 24 hr before infection with S. enteritidis No.11 strain (4 x 10^5 organisms, ip). The viable numbers of infecting organisms in the peritoneal cavity (○) and the spleen (□) were determined by quantitative cultivation 24 hr after the infection. Each point represents the mean ± standard deviation.

Various doses (0.02 to 20 μg/mouse) of LPS were injected into CBA/N-defective mice, and 24 hr later, these mice and control female mice were challenged with No.11 organisms. All mice were killed 24 hr after the infection, and the numbers of viable organisms in their peritoneal cavities and spleens were determined. As shown in Figure 2, the effect of LPS-treatment seemed to be dose-dependent: increasing the doses resulted in a greater bactericidal effect.

Effect of multiple injections with LPS

Repeated injections of the LPS can increase its anti-bacterial effect in CBA/N-defective mice. Groups of F1 male mice were repeatedly ip injected with LPS (20μg each, 1 to 6 times, every other day) and then twenty-four hr after the last injection, these and untreated control mice were challenged with No.11 organisms. Twenty-four hr after the challenge, the viable numbers in the peritoneal cavities and spleens were examined. As shown in Figure 3, the mice receiving multiple injections of LPS showed greater effects than those injected only once; 5-injections seemed to induce the best effect.

Effect of time between LPS-injection and bacterial challenge on the bactericidal activity

The timing of the administration of LPS was a very important determinant for effective induction of the bactericidal action in these mice. Groups of CBA/N-defective F1 male mice were injected with 20 μg of LPS on day 7, 5, 3, 1 or 0 before the challenge with No.11 strain. On day 0, they were challenged, and 24 hr later, numbers of viable organisms in the peritoneal cavities and spleens were examined. As shown in Figure 4, the most obvious effect was seen in mice that had been injected with LPS at 24 hr before the challenge. Simultaneous injection of LPS and challenge organisms was not effective. The effect of LPS decreased when it had been injected several days before challenge.

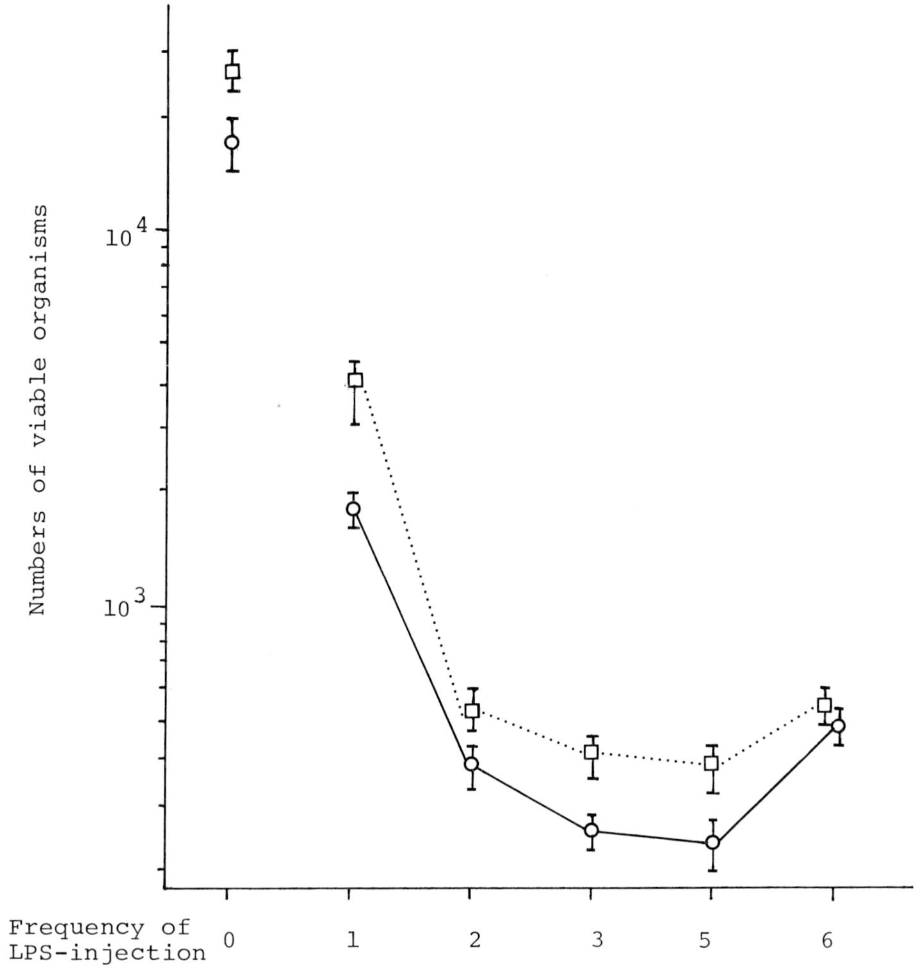

Figure 3. Effect of multiple injections with LPS on bactericidal killing in the peritoneal cavities and spleens of CBA/N-defective mice. Groups of F1 male mice were injected with LPS (20µg each, ip, every other day) at varying frequency as indicated in the abscissa. Twenty-four hr after the last injection, they were ip challenged with No.11 strain (4.5 x 10^5 organisms); and 24 hr later, viable numbers of infecting organisms in the peritoneal cavity (O) and spleen (□) were examined. Values represent the means ± standard deviation of 3 mice.

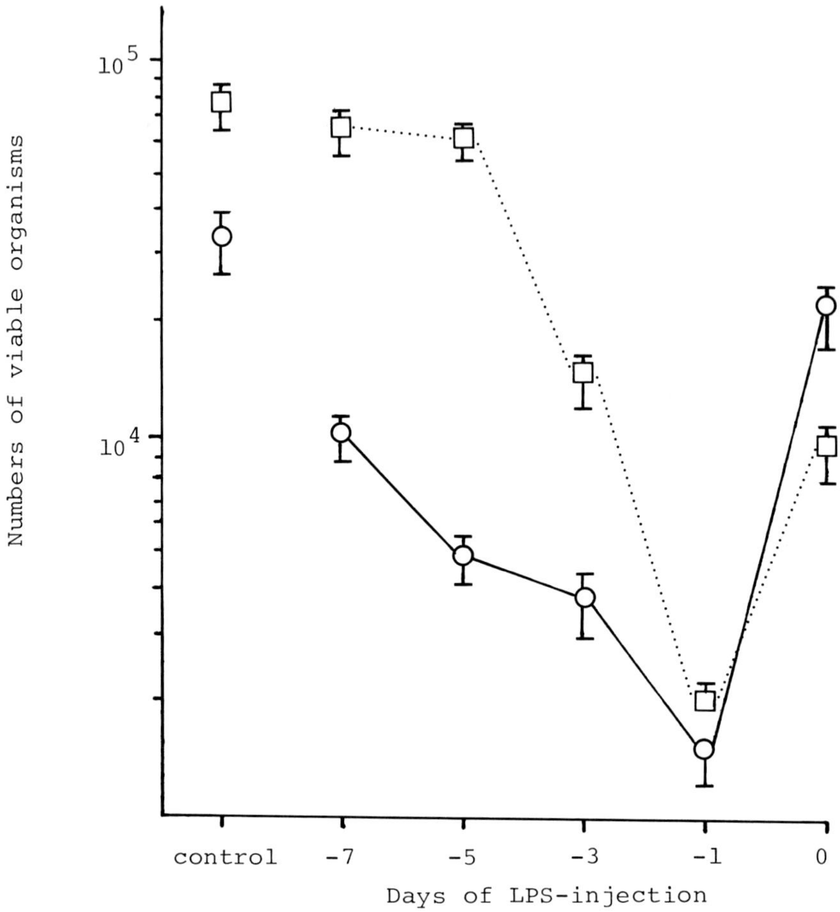

Figure 4. Effect of time between LPS-injection and bacterial challenge on the bactericidal activity of CBA/N-defective mice. F1 male mice were injected ip with 20 μg of LPS at varying times as indicated in the abscissa and challenged ip with S. enteritidis No.11 strain (4.0×10^5 organisms) on day 0. Twenty-four hr after the infection, viable numbers of infecting organisms in the peritoneal cavity (○) and spleen (□) were examined. Values represent the mean ± standard deviation of 3 mice.

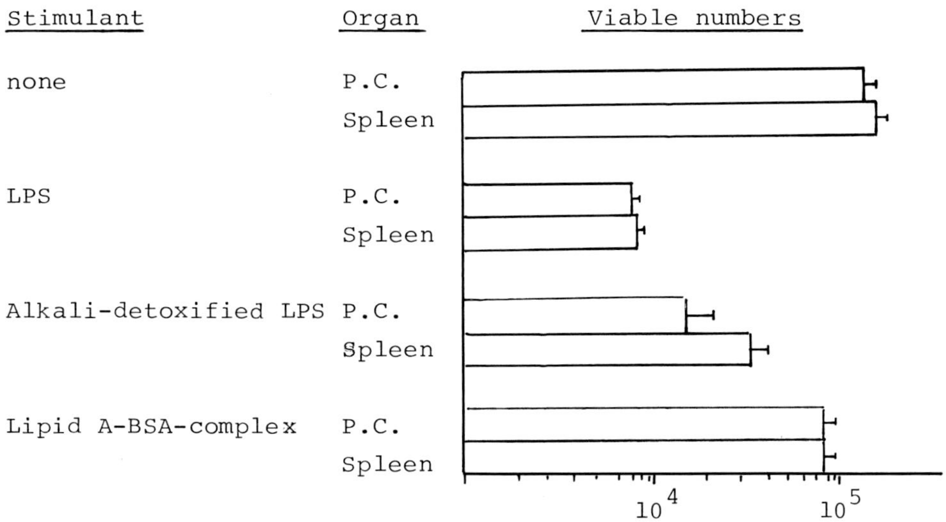

Figure 5. Effect of LPS, alkali-detoxified LPS or lipid A-BSA-complex on CBA/N-defective mice infected with S. enteritidis No.11 strain (8 x 10^5 viable organisms, ip). These stimulants were ip injected (20μg/mouse) 24 hr before the infection. Viable numbers of the organisms in the peritoneal cavity (P.C.) and spleen were determined by quantitative cultivation 24 hr after the infection. Values represent the means ± standard deviation of 3 mice.

Effect of lipid A-BSA-complex and alkali-detoxified LPS

To examine the active part of LPS which consists chemically of lipid A and polysaccharides, comparative studies were carried out using the lipid A-BSA-complex and alkali-detoxified LPS. The lipid A-BSA-complex is still biologically active as an endotoxin (9), while the alkali-detoxified LPS has lost the majority of the lipid molecules originally present in LPS and has quite a low toxicity. Induction of an anti-bacterial ability against strain No.11 in CBA/N-defective mice by these substances was examined. Groups of mice were injected with one of the substances 24 hr before challenge with No.11 organisms. Viable numbers of infecting organisms in the peritoneal cavities and the spleens of these mice were examined

24 hr after the challenge. As shown in Figure 5, the bactericidal activity induced by the detoxified LPS was stronger than that by lipid A-BSA-complex, but the original LPS showed the strongest effect among the substances examined.

Effect of LPS on bacterial clearance from the circulating blood

The bactericidal function in CBA/N-defective F1 male mice was apparently weak when compared with that in phenotypically normal F1 female mice, but their malfunction could be improved by previous administration of LPS. Bactericidal function is closely related to the phagocytic function of the reticuloendothelial phagocytic cells. Therefore, it is worthwhile to examine the phagocytic function of CBA/N-defective mice. Groups of F1 male and female mice were ip injected with LPS, and 24 hr later, these and other control groups were iv challenged with No.11 organisms. The phagocytic function of the reticuloendothelial system of these mice was estimated by the bacterial clearance test as described in "Materials and Methods". As shown in Figure 6, the bacterial clearance in defective males seemed to be slightly slower than that in normal females when it was estimated at 60 min after the challenge. Moreover, for the defective mice, LPS was apparently less effective on the acceleration of the phagocytic function in their reticuloendothelial system.

Discussion

O'Brien *et al.*(6) showed that CBA/N-defective F1 male mice were 1,000 times more susceptible to infection with *S. typhimurium* than were the normal F1 females; the *X*-linked gene that controlled B cell differentiation in the CBA/N-defective animals (*xid* gene) was probably also controlling the suscep-

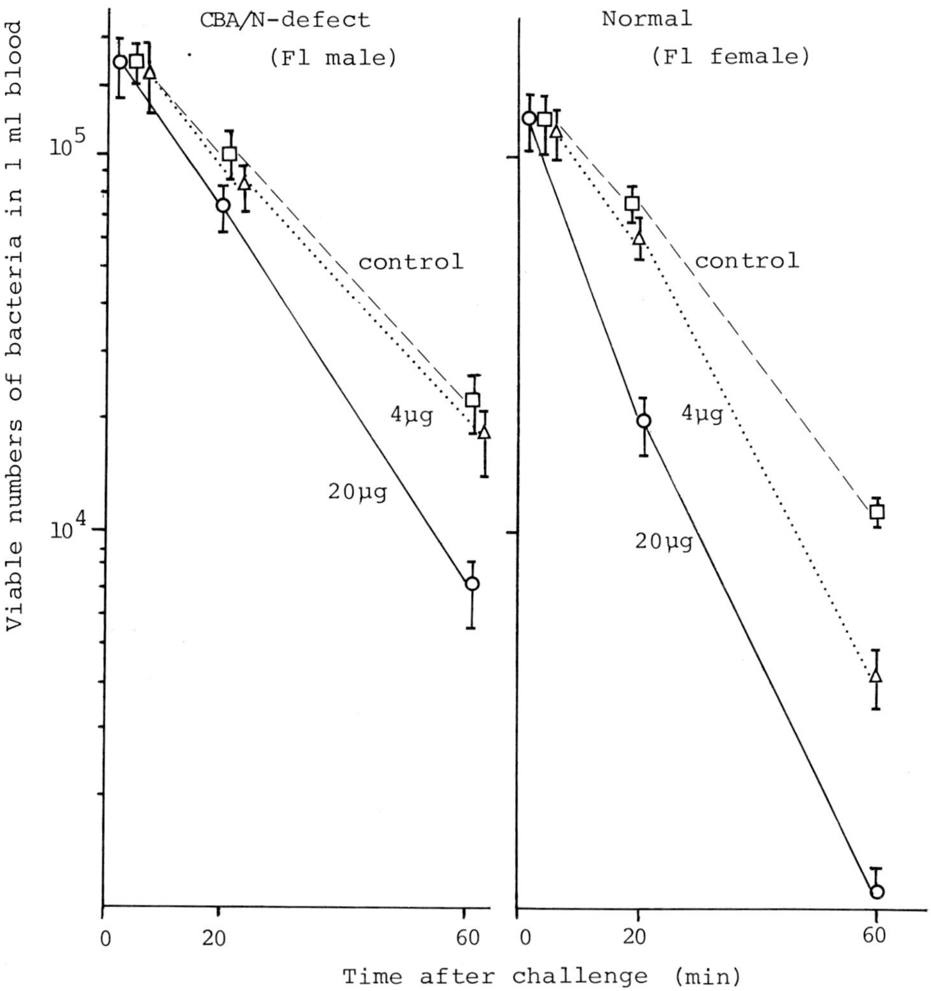

Figure 6. Effect of LPS on bacterial clearance from circulating blood in (CBA/N female x BALB/c male)Fl mice. Fl males (left panel) and Fl females (right panel) were ip injected with LPS, and 24 hr later, they were iv challenged with 1 x 10^6 No.11 organisms. The bacterial numbers in the circulating blood were determined after the challenge at the times indicated in the abscissa. □ , control. Δ , 4μg LPS. O , 20μg LPS. Values represent the mean ± standard deviation of 3 mice.

tibility of salmonella infection. They thought that the susceptibility of F1 male mice could not be attributed to the early failure of their reticuloendothelial cell organs to contain multiplication of the bacteria since in their experiments, the bacterial net growth in the spleens of F1 male and female mice was equivalent during the early stages of the infection (11). Furthermore, as the deficient males produced significantly less IgM and IgG antibodies against S. typhimurium than did the normal females, these investigators concluded that B-cells and humoral responsiveness play the primarily role in host resistance to S. typhimurium.

Salmonella enteritidis as well as S. typhimurium is a natural causative agent of murine typhoid (12,13). Mice infected with the virulent organisms develop a systemic illness that can be lethal (13). We demonstrated that the CBA/N-defective mice were highly susceptible to infection with S. enteritidis No.11 organisms (Table 1), and the bactericidal capacities in the peritoneal cavities and spleens of the defective mice were apparently affected in comparison with those of normal female mice (Fig. 1). This finding suggests that the xid-conferred susceptibility to salmonella may partially be attributable to malfunction of bactericidal activity in the early stage of infection, and in this point our experimental finding is inconsistent with that of O'Brien et al. (11).

CBA/N-defective mice were thought to be quite low responders to LPS since the mitogenic response and polyclonal antibody synthesis of splenic B lymphocytes were obviously poor after the LPS stimulation (5). To the contrary, previous administration of LPS bestowed non-specific resistance against virulent No.11 organisms of S. enteritidis (Table 2). These results indicate that the CBA/N-defective mice are not refractory to LPS with regard to the enhancement of non-specific resistance in hosts; in other words, phenotypic expression of xid including B cell dysfunction does not affect the LPS-induced non-specific resistance to salmonella. The defective CBA/N gene is thought to be expressed only in B lymphocytes and not in macrophages, as the macrophages respond normally to LPS and produce the monokine, lymphocyte-activating factor, and release prostaglandin (14). To the contrary, Kato et al. (15) demonstrated that the LPS-induced production of H_2O_2 in CBA/N-defective macrophages was weaker than in normal macrophages. Further information should be gathered about the functions of macrophages in this peculiar mouse strain. Anyway, the CBA/N macrophages activated by LPS probably acquire hightened ability in their bactericidal function, and these macrophages may play some role in the LPS-induced non-specific resistance in CBA/N-defective mice.

Bacterial LPS may enhance or inhibit the pathogenicity of infection, depending upon the infecting organisms, dose and route of injection of LPS, and the interval between administration of LPS and initiation of infection. A significant but transient increase in resistance to infection has been observed when LPS is given several hours before initiation of infection, and very low doses of LPS were found to be effective (2). In CBA/N-defective mice, LPS should be given before infection in order to get a certain bactericidal effect (Fig. 3). Simultaneous injections of LPS and challenge-organisms hardly increased the bactericidal activity in the peritoneal cavities and spleens. The most prominent effect was seen when LPS had been injected 24 hr before infection. The effect of LPS is dose-dependent (Fig. 2), and multiple injections every other day increased the effect (Fig. 4). The phagocytic function of the reticuloendothelial system in CBA/N-defective mice as well as normal mice was also activated by previous injection of LPS (Fig. 5). However, the capacity of bacterial clearance from the blood in normal F1 female mice seemed to be always greater than that of the defective F1 males, regardless of LPS-treatment or not. This may be due to a lack of opsonic antibody in the deficient male mice since they are already reported to have less antibodies against salmonella (11).

LPS consists of a specific polysaccharide covalently linked to lipid A. The lipid A is presumed to be an endotoxic principle but the lipid A can hardly show its activity as it is insoluble in water. When lipid A is solubilized by complexing with BSA, the solubilized complex can gain endotoxic potency (9). The major part of the lipid portion in LPS is removed by alkali treatment, and the lipid-deprived LPS becomes low in toxicity (10). The finding that the alkali-detoxified LPS is still effective (Fig. 6) suggests that the non-toxic carrier polysaccharide may be related to the potency for inducing bactericidal activity in these mice.

When endotoxin was readministered at appropriate intervals, the response to LPS decreased progressively, and endotoxin tolerance or acquired resistance to endotoxin was easily achieved (16). However, repeated injections of LPS in our experiments did not decrease but enhanced bactericidal activity in CBA/N-defective mice (Fig. 4). Therefore, it seems that the ability of LPS to enhance bactericidal activity may be different from its property to induce endotoxin tolerance.

Although the exact mechanism for the induction of non-specific resistance by LPS is presently unclear, we demonstrate here that LPS is capable of inducing non-specific resistance and can enhance bactericidal activity in genetically X-chromosome-linked CBA/N-defective mice with immunologic abnormalities expressing low B cell response to LPS and high susceptibility to salmonella.

Summary

Bacterial lipopolysaccharide (LPS)-induced non-specific resistance in (CBA/N female x BALB/c male)F1 mice against virulent strain No.11 of *Salmonella enteritidis* was studied. As CBA/N (*xid*)-deficiency is an *X*-chromosome-linked recessive abnormality, the F1 female mice are phenotypically normal, while the F1 male mice are known to show some immunologic defects such as dysfunction of B cell response to some kinds of antigen or LPS and high susceptibility to *Salmonella typhimurium*. When the mice were infected with No.11 organisms, F1 males were highly susceptible in comparison with F1 females. The bacterial numbers in the peritoneal cavities and spleens of F1 males 24 hr after infection were greater than those of F1 females. Previous injection of LPS 24 hr before the infection could give these mice a resistance against infection by a small number of organisms. The viable numbers of organisms in the peritoneal cavities and spleens of LPS-treated F1 mice 24 hr after the infection were significantly less than those in the untreated control.
The LPS-induced bactericidal effect was dose-dependent, and multiple injections of LPS every other day could increase th effect. The effect was most prominent when mice were injected with LPS 24 hr before infection. Simultaneous injections of LPS and organisms hardly showed an effect. Injection of LPS several days before the infection was also less effective. Alkali-detoxified LPS prepared from original LPS was still effective. Though the phagocytic function of the reticuloendothelial system in CBA/N-defective mice seemed to be slightly impaired, previous injection of LPS could elevate this function.

References

1. Morrison, D.C., Ryan, J.L.: Adv. Immunol. 28, 283-450 (1979).

2. Cluff, L.E.: Microbial Toxins Vol. V (Kadis, S., Weinbaum, G., Ajl, S.J., eds), 399-413, Academic Press, New York (1971).

3. Parant, M.: Beneficial Effects of Endotoxins (Nowotny, A., ed.), 179-196, Plenum Press, New York (1983).

4. Scher, I.: Adv. Immunol. 33, 1-71 (1982).

5. Scher, I., Zaldivar, N.M., Moiser, D.E.: Microbiology-1977 (Schlessinger, D., ed.), 310-320, American Society for Microbiology, Washington, D.C. (1977).

6. O'Brien, A.D., Scher, I., Champbell, G.H., MacDermott, R. P., Formal, S.B.: J. Immunol. 123, 720-724 (1979).

7. Westphal, O., Lüderitz, O., Bister, F.: Naturforsch. 76, 148-155 (1952).

8. Nakano, M., Saito, T., Asou, H.:Jpn. J. Microbiol. 19, 403-406 (1975).

9. Galanos, C., Rietschel, E.T., Lüderitz, O., Westphal, O., Kim, Y.B., Watson, D.W.: Eur. J. Biochem. 31, 230-233 (1972).

10. Neter, E., Westphal, O., Lüderitz, O., Gorzynski, E.A., Eichenberger, E.: J. Immunol. 76, 377-385 (1956).

11. O'Brien, A.D., Scher, I., Metcalf, E.S.: J. Immunol. 126, 1368-1372 (1981).

12. Collins, F.M.: Bacteriol. Rev. 20, 371-402 (1974).

13. Ushiba, D., Saito, K., Akiyama, T., Nakano, M., Sugiyama, T., Shirono, S.: Jpn. J. Microbiol. 3, 231-242 (1959).

14. Rosenstreich, D.L., Vogel, S.N., Jacques, A., Wahl, L.M., Scher, I., Mergenhagen, S.E.: J. Immunol. 121, 685-690 (1978).

15. Kato, H., Yaoita, H., Saito-Taki, T., Nakano, M.: see the preceding paper in this book.

16. Greisman, S.E.: Beneficial Effects of Endotoxins (Nowotny, A., ed.), 149-178, Plenum Press, New York (1983).

ANTITUMOR ACTION OF ENDOTOXIN IN THE MOUSE

Nanne Bloksma, Frans M.A. Hofhuis, Jan M.N. Willers
Laboratory of Microbiology, State University of Utrecht, Catharijnesingel 59, 3511 GG Utrecht, The Netherlands

Introduction

In 1931 Gratia and Linz (1) reported hemorrhagic necrosis and regression of a transplanted liposarcoma of the guinea pig after injection of culture filtrate of *Escherichia coli*. Shortly after similar effects were described in murine sarcomas and carcinomas (2-4) and many other reports followed (reviewed in 5). Endotoxin the lipopolysaccharide of gram-negative bacteria was recognized as the common incitant in 1942 (6). Since the early observations on the antitumor action of endotoxins, study of its effects has continued but up till now the mode of action is still a matter of discussion (7-13). It is generally agreed, however, that effects are mediated by the tumor-bearing host as already suggested by Shapiro in 1940 (14) because the necrotizing carbohydrate was not toxic for tumor cells in culture. This was elegantly confirmed otherwise by Männel et al. (15) using tumor-bearing C3H/HeJ mice which are hyporesponsive to many effects of endotoxin. Necrosis and regression could not be induced by this agent unless bone marrow cells of endotoxin-responsive histocompatible C3H/HeN mice were transferred.

This short review compares our recent data on the antitumor effects of endotoxin with some theories proposed in literature. Our studies concerned the effect of endotoxin on solid well established subcutaneously transplanted tumors. Endotoxin or its derivatives were shown to be also effective in ascitic tumor models and (semi)-prophylactic tumor models (7, 16-18). Different properties of endotoxin are probably involved in these tumor models. Butler et al. (19, 20) distinguished two different models namely the induction of tumor necrosis and regression and the induction of tumor resistance. These would be mediated by two distinct factors induced by endotoxin namely tumor necrosis factor (TNF) and tumor resistance enhancing

Immunopharmacology of Endotoxicosis

factor. Further the lipid A-KDO part of endotoxins was shown to be the minimal structure able to induce tumor necrosis (21). Regrettably the effect on subsequent tumor growth was not reported. The polysaccharide part, however, seemed to be essential for induction of enhanced tumor resistance (20). Therefore to avoid confusion (18, 22, 23) the tumor model studied or discussed when antitumor activities of endotoxin are concerned, must be clearly stated or reckoned with.

Some more or less generally accepted features of the tumor model we studied, are briefly summarized here. Hemorrhagic necrosis and regression of solid subcutaneously transplanted tumors can be induced upon a single intravenous injection of endotoxin. We have used a methylcholanthrene-induced Meth A fibrosarcoma and an IgA plasmacytoma, MOPC315, both syngeneic to BALB/c mice. Others found that primary and spontaneous tumors were also susceptible (5, review), as well as intramuscularly implanted tumors (24). Intraperitoneal and intralesional routes of injection of endotoxin were found to be effective too, but the oral and subcutaneous routes were usually ineffective (5, review). Only tumors above a certain size are susceptible to necrosis. Subsequent regression is generally only observed when tumors are somewhat larger but again not too large (7, 8, 13, 25, 54). An intact T cell immune system seems not to be required for induction of necrosis but is needed for induction of regression (7-9, 13).

Proposed Models of Endotoxin Action

Shwartzman reaction model

The induction of hemorrhagic necrosis of tumors by endotoxin has frequently been related to or identified with the local Shwartzman reaction (1, 7, 13, 26). Indeed both reactions are very alike. Whereas elicitation of a local Shwartzman reaction requires local priming with endotoxin followed by systemic elicitation with the same agent, tumor necrosis can be provoked by a single injection with endotoxin. Gratia and Linz (1) suggested therefore that tumor tissue would behave as though continually prepared for the Shwartzman reaction. Recently indeed a very weak Shwartzman active substance could be isolated from tumor cells (13). The histology of the events preceding hemorrhagic necrosis in the local Shwartzman reaction is featured by severe local inflammation and transient occlusion of small vessels by aggregation of platelets and leukocytes (27). Thomas (28), however, showed that grossly identical hemorrhagic necrosis can be induced in rabbit skins in the virtual absence of leukocyte infiltration and vascular thrombi. He used adrenaline as priming and endotoxin as eliciting agent. Induction of a local inflammatory infiltrate during priming was even inhibitory. Similar reactions could be induced with endotoxin and adrenaline as priming and eliciting agents respectively (29). Further, unlike the Shwartzman reaction the latter reaction was not inhibited or even potentiated by heparin and nitrogen musterd, while on the contrary glucocorticoids were inhibitory (28). These data suggest that endotoxin may induce two kinds of local hemorrhagic necrosis: the Shwartzman type needing cellular infiltrates, the other type (preferably) not. Some data indicate that the latter reaction fits tumor necrosis induced by endotoxin, because tumors

Table 1. Effect of anticoagulants on endotoxin-induced necrosis and regression of Meth A sarcoma

Treatment[a]	Necrosis[b]		Complete cure
	Incidence	Extent	
Heparin/endotoxin	8/11	58 ± 8	7/11
Heparin/saline	0/ 5		0/ 5
Saline/endotoxin	7/ 9	58 ± 3	7/ 9
Saline/saline	0/10		0/10
Warfarin/endotoxin	10/10	76 ± 3	6/10
Warfarin/saline	2/10	49 ± 13	0/10
Vehicle/endotoxin	10/11	75 ± 5	5/11
Vehicle/saline	0/10		0/10

a BALB/c mice received 10^6 viable Meth A cells s.c. at day 0 and 25 µg endotoxin (LPSw from *E. coli* 0111:B4) i.v. at day 9. Heparin (100 U) was injected i.p. 15 min before endotoxin. Sodium warfarin was solved in phosphate buffered saline (pH 7.2). It was injected i.p. at day 7 (0.5 mg) and added to the drinking water at days 9 and 10 (Mean uptake: 0.35 mg/mouse/day).

b Incidence and extent of necrosis were measured 2 days after injection of endotoxin. Extent of necrosis was expressed as mean ± SEM of the ratio of mean diameters of necrotic area and tumor x 100.

accidently contaminated with bacteria appeared much more resistant to induction of hemorrhagic necrosis than sterile tumors (reviewed in 5). In addition we found that both heparin and warfarin did not inhibit the induction of tumor necrosis in our Meth A system (Table 1). This is partly conflicting with earlier data of Parr et al. (7), who found that heparin reduced tumor necrosis in the murine F36 fibrosarcoma. In the L 5178Y lymphoma, however, only a slower appearance of tumor necrosis was observed. Histological examination of subcutaneous Meth A tumors 4 h and 24 h after intravenous injection of endotoxin has shown that hemorrhagic necrosis was induced in the absence of clear inflammatory infiltrates and fibrin deposition (30). Also North (31) stated that the number of infiltrating leukocytes in the tumors of endotoxin-treated mice was relatively small in comparison to the number of tumor cells that eventually must be destroyed. Our preliminary results with enzymatically disaggregated Meth A tumors confirm these data (32). Kodama et al. (13), however, recently described a marked infiltration of murine fibrosarcoma transplants 4 h after injection of endotoxin, but the route of endotoxin administration was not clearly stated. More study is needed to elucidate this further.

Vascular reactivity theories

The above mentioned Shwartzman model assumes induction of necrosis due to tissue hypoxia as a consequence of vascular occlusion by intravascular thrombi. Another theory assumes that necrosis is due to stasis of blood supply of the tumor as a consequence of the hypotensive action of endotoxin. This theory is based on the observation that intraperitoneal administration of a lethal dose of endotoxin caused a progressive slowing down of the capillary blood flow in an intramuscularly transplanted murine sarcoma which reached a minimum by 4 h (24). In addition a temporary mechanically induced vascular occlusion appeared to cause immediate hemorrhage in tumors but not in normal control tissue (33). Pradhan et al. (34), however, tested 37 pharmacological drugs having either hypertensive or hypotensive action (carotid blood pressure) in mice on their capacity to induce hemorrhagic necrosis of murine Sarcoma 37. Agents belonging to both groups were active but the majority of the agents inducing hypotension did not cause tumor damage. A clear relation between systemic blood pressure and tumor damage seems therefore absent. A more local action of the dam-

Table 2. Antitumor effects of tumor necrotizing agents[a]

Agent (dose per mouse)	MACROSCOPY				MICROSCOPY				
	Necrosis		Growth	Complete	Hyperemia	Mitotic arrest		HN	Non-HN
	Incidence	Extent	stop	cure	at 4 h	at 4 h/	at 24 h	at 24 h	at 24 h
Endotoxin (25 µg)	9/10	++++	8/10	3/10	+++	+++	+++	+	+++
Poly I:C (200 µg)	10/10	++++	10/10	3/10	+++	+++	+++	+	+++
Concanavalin A (200 µg)	6/10	++++	2/10	1/10	+++	+++	+++	-	+++
Serotonin (200 µg)	7/10	++	1/10	0/10	++	+++	+	+	++
Histamine (1 mg)	4/10	+	1/10	0/10					
Adrenaline (10 µg)	10/10	++	3/10	0/10	+++	++	+	+	+
Isoproterenol (200 µg)	4/10	++	3/10	0/10	++	+	+	+	+
TNS (0.5 ml)	9/10	++++	8/10	3/10	+++	+++	++	+	++
NMS (0.5 ml)	1/10	+	0/10	0/10	+	++	++	+/-	+
Interferon (5.6 x 10^4 U)	2/10	++	2/10	1/10					
Saline	0/20	-	0/20	0/20	-	-	-	-	+

a BALB/c mice received 10^6 viable Meth A cells s.c. at day 0 and a necrotizing agent i.v. at day 9. Adrenaline was injected into the tumor. Endotoxin was from *E. coli* 0111:B4 (Westphal). Sera were from Swiss mice and were heated at 56°C for 10 min prior to assay. TNS was obtained after treatment with *C. parvum* and endotoxin. Subsequent macroscopical necrosis, growth stop (no increase of tumor size during 2 days), complete cure and microscopical hyperemia (number of transsected vessels, ø >20 µm, per tumor section), mitotic arrest, hemorrhagic necrosis (HN) and non-HN were scored. Symbols ++++: very strong; +/-: very moderate; -: absent. For further details see 30, 35, 36 and 37.

aging vasoactive agents, however, may be involved (see below).

In order to get a better insight in the mechanisms involved in the tumor damage, effects of endotoxin and a number of other agents including vasoactive agents were compared (35-37). The results are compiled in simplified form in Table 2. On gross observation endotoxin, poly I:C, concanavalin A and TNS were superior to the remaining agents in the induction of tumor necrosis. Also effects on subsequent tumor growth were more pronounced. Histology showed that all agents as far as studied induced hyperemia by 4 h, which had disappeared for the major part by 24 h. Again the agents mentioned above were more active in this respect and appeared also to be the best inducers of mitotic arrest and non-hemorrhagic necrosis. Hemorrhagic necrosis was induced by all agents but concanavalin A and was only apparent at the ventral outside of the tumors. Otherwise hyperemia was the most prominent at the same site of the tumor. In none of the tumor sections studied significant inflammatory infiltrates could be observed. Moreover we did not detect clear fibrin deposition using anti-fibrin antibodies and a peroxidase stain. All data together permit ample discussion, which will be given here and in the following section. The degree of tumor necrosis seems to be related to the degree of hyperemia and mitotic arrest preceding this phenomenon and to the subsequent effects on tumor growth. This may be partly questioned by results obtained with the noncompetitive α-adrenoceptor blocking agent, phenoxybenzamine. This agent is not only an antagonist of adrenaline but also of serotonin and histamine (38). We found that it inhibited necrosis and subsequent regression of Meth A induced by endotoxin (32, 39). Histology showed, that it completely abolished the induction of hemorrhagic necrosis while it did not affect the induction of mitotic arrest and non-hemorrhagic necrosis. When compared to endotoxin alone, however, hyperemia was slightly reduced by 4 h, but much more pronounced by 24 h. Therefore the role of hyperemia in tumor destruction is not yet completely clear. We have used the number of dilated vessels per tumor section as a measure for hyperemia. It is interesting that the "tumor necrosis type" of hemorrhagic necrosis in rabbit skins, already discussed above, was also preceded by vasodilation, ultimately resulting in a complete standstill of bloodflow (28). Therefore our observation that all tumor necrotizing agents induced vasodilation may indicate that a similar mechanism underlies necrosis in this tissue.

Some data suggest that necrosis induced by endotoxin may be mediated by release of one or more vasoactive agents. Endotoxin is able to release vasoamines such as adrenaline, serotonin and histamine (40, 41), all found to induce tumor necrosis (34, Table 2). Also phenoxybenzamine, which antagonizes the action of these vasoamines (38) inhibited the induction of tumor necrosis by endotoxin (39). Nevertheless the mode of action of vasoamines in the vasodilation of the "tumor necrosis type" of necrosis remains enigmatic. A (partial) explanation may be the following: Concerning the observations in the rabbit skin (28), the priming injection with adrenaline may induce a local state of tachyphylaxis. A subsequent injection of endotoxin releases a variety of vasoactive agents among which adrenaline in the circulation. This agent causes strong vasoconstriction in the periphery (42). Blood vessels in the primed skin, however, would be unable to constrict, leading to passive accumulation of blood manifest as vasodilation and an ultimate standstill of blood flow with subsequent tissue hypoxia and necrosis. The reaction may be amplified by the capacity of endotoxin to sensitize to the action of adrenaline and histamine (43, 44). Why tumor tissue does not need to be primed may be due to an "innate" state of tachyphylaxis. Tumor vasculature is characterized by a lack of smooth musculature and adrenergic nerves within the vessel walls (45, 46). An injection of endotoxin, would therefore lead to the same reactions as observed in rabbit skins prepared with adrenaline. This reasoning may also explain why reactions can occur in the absence of cellular infiltrates. A difference with the local Shwartzman reaction may be that in this instance the priming injection with endotoxin would induce local procoagulant activity caused by infiltrated cells (47). This may facilitate vascular occlusion by intravascular thrombi when the intrinsic coagulation cascade is activated by i.v. injection of endotoxin.

Cytotoxic factor models

Endotoxin is known to induce various cytotoxic factors (48, review). One of these, tumor necrosis factor (TNF), has been described first by Carswell et al. in 1975 (49). Various reviews on TNF have been published since (10, 12, 48, 50, 51). Therefore we will only discuss data relating to our recent findings. TNF is a factor which mimics the induction of tumor necrosis and regression by endotoxin (49). Unlike endotoxin, this factor has cytostatic or cytolytic action against a variety of tumor cell lines *in vitro* (50). TNF is abundantly present in serum of mice 90 to 120 min after injection of endotoxin if they are treated one to two weeks earlier with a macrophage-activating agent such as BCG, zymosan or *Corynebacterium parvum*. Serum obtained after either treatment is virtually devoid of TNF (49, 52). Therefore TNF was considered to be a factor released by endotoxin from activated macrophages. In literature various cytotoxic factors released from macrophages are called TNF without the verification that the factor has necrotizing activity *in vivo*. Recent data of Kull and Cuatrecasas (53) indicate that this may be incorrect. Therefore in this paper the term TNF is only used in its original meaning i.e. a factor having necrotizing activity *in vivo*. TNF was suggested to be the mediator directly responsible for the tumor damage observed after injection of endotoxin (10, 12). This implies both that TNF is actually produced in endotoxin-treated tumor-bearing mice and that it acts directly on the tumor cells. The following data make this doubtful. We have been unable to detect TNF in the serum of endotoxin-injected tumor-bearing mice using 0.5 ml serum while 0.1 ml tumor necrosis serum (TNS) from mice treated with *C. parvum* and endotoxin was still active in our assay (35, 54). As far as we know mediation by TNF has not been demonstrated by others too. It seems unlikely that absence is due to a direct interaction with the tumor cells as repeated absorption of TNS with Meth A cells at 4^{0}C (35, 55) or repeated incubation with Meth A cells at 37^{0}C for 1 h did not substantially remove the necrotizing and regressing activity (54). Furthermore nude (nu/nu) mice were found to be unable to produce TNF after treatment with *C. parvum* and endotoxin (52), but injection of endotoxin into tumor-bearing nude mice resulted in tumor necrosis (13). More indirect data also make mediation by TNF less likely: a) Of all agents able to induce tumor necrosis (Table 2) only endotoxin could induce TNF in *C. parvum*-treated mice (Table 3). Poly I:C has been designated earlier as an inducer of TNF (49), but we found that the necrotizing activity of post-poly I:C sera is sensitive to RNase, while endotoxin-induced TNF is stable to this enzyme (36, Table 3). Hence activity of post-poly I:C sera is most probably due to residual poly I:C. b) Although induction of TNF by endotoxin requires a considerably activated mononuclear phagocytic system (MPS) (52, 55, 56) we found that tumor necrosis could be induced by endotoxin in mice with normal MPS function as measured by carbon clearance and

Table 3. Antitumor factors in sera of *C. parvum*-treated mice elicited with a tumor necrotizing agent[a]

Treatment of serum donor	Interferon	Cytostatic factors	TNF
Cp/endotoxin	+++	+++++	+++[b]
Cp/poly I:C	+++	+++	-[b]
Cp/concanavalin A	+++	+++++	-
Cp/serotonin	+/-	+++	-
Cp/histamine			-
Cp/adrenaline	-	++	-
Cp/isoproterenol	+/-	+++	-
Cp/saline	-	++	-
- /saline	-	+	-

a Swiss random mice received 1 mg *C. parvum* (Cp) and 14 days later a necrotizing agent i.v. (for doses see Table 1). Sera were prepared from blood collected 90 min after the last injection. Interferon assay: plaque reduction test. Cytostatic factor assay: ^{3}H-thymidine incorporation into Meth A cells after 40 h of culture with a heated (56°C, 10 min) serum sample. TNF assay: incidence and extent of necrosis upon injection of 0.5 ml heated serum into Meth A-bearing mice. For further details see references (35-37).

b Necrotizing and regressing activity of post-Cp/endotoxin and post-Cp/poly I:C sera were stable and labile to RNase respectively (36).

bactericidal capacity (54, 55).

The observation that TNF is only present in the circulation for a brief period after injection of endotoxin in *C. parvum*-treated mice (52, Table 4) and the data from the absorption experiments mentioned above already indicate that TNF, like endotoxin, does not act predominantly by a direct interaction with the tumor cells *in vivo*. Other data support this: i. Activity of murine TNS against tumor cells *in vitro* is only apparent after a latent period of 16 h or more (49, 57), while marked hyperemia and mitotic arrest could already be observed within 4 h after injection of TNS; ii. Susceptibility of tumors was shown to be dependent on the age of the tumors (49, 54); iii. The α-adrenergic receptor antagonists phenoxybenzamine and phentolamine inhibited the induction of necrosis and regression by TNS

Table 4. *In vitro* and *in vivo* antitumor activity against Meth A of sera obtained at different intervals after injection of endotoxin[a]

Treatment of serum donor	Blood withdrawal	Cytostatic activity	Necrosis Incidence	Necrosis Extent	Complete cure
Cp/endotoxin	60 min	28.0 ± 0.7	7/10	60 ± 5	5/10
- /endotoxin	"	144.7 ± 3.3	1/10	57	0/10
Cp/endotoxin	90 min	20.1 ± 3.8	10/10	62 ± 4	1/10
- /endotoxin	"	85.9 ± 2.3	1/10	61	1/10
Cp/endotoxin	180 min	3.0 ± 0.1	1/ 5	24	0/ 5
- /endotoxin	"	101.6 ± 2.1	0/ 5		0/ 5
Cp/endotoxin	270 min	7.8 ± 0.3	1/ 5	23	0/ 5
- /endotoxin	"	97.6 ± 2.4	0/ 5		0/ 5
- /saline (NMS)		207.2 ± 3.7	0/10		0/10

a Swiss mice received either 1 mg *C. parvum* or nothing i.v. at day 0 and 25 µg endotoxin (LPSw from *E. coli* 0111:B4) i.v. at day 14. Blood was obtained at various times after endotoxin. Cytostatic activity of the sera was determined as described earlier (57) and is represented as mean DPM ± SEM x 10^{-3} of sera of 5 individual mice, each measured in quadruplicate. Necrosis and regression was determined in Meth A-bearing mice after injection of 0.2 ml heated serum.

(39); iv. The strong hyperemia induced by TNS suggests an indirect interaction with blood vessels rather than with the tumor cells (Table 2).

With respect to the relation of the *in vitro* and *in vivo* activity of TNS, Carswell et al. (49) have stated "there has been no discrepancy between the TNF activity of sera against Meth A *in vivo* and their toxicity for L cells *in vitro*". We have compared the activity of various sera against Meth A *in vivo* and *in vitro*. The *in vitro* assay was based on ^{3}H-thymidine incorporation of Meth A cells after incubation with a serum sample (57). We have found various discrepancies between *in vivo* and *in vitro* activity of TNS: i. Whereas TNF is stable after heating at 56°C for 1 h (58) heating of TNS at 56°C up to 40 min gradually decreased its cytostatic activity (59). Compared to normal mouse serum the cytostatic activity of which was already maximally decreased after 5 min at 56°C, heated TNS remained far more cytostatic. This confirms earlier data (60) suggesting that endotoxin induces not only heat-stable but also heat-labile cytotoxic factors. Also, Kull and Cuatrecasas (53) found several distinct heat-stable cytotoxins in TNS. Only one had necrotizing activity; ii. Maximal TNF activity was found 60-90 min after injection of endotoxin in *C. parvum*-treated mice, while maximal cytostatic activity was found after 180 min (Table 4); iii. We found that C3H/HeJ mice, in contrast to C3H/HeN mice, were unable to produce TNF after treatment with *C. parvum* and endotoxin (LPSw from *E. coli* 0111:B4) (61), which confirms an earlier statement (62). The cytostatic activity of these C3H/HeJ sera, however, was as high as sera of similarly treated C3H/HeN mice[a] (61); iv. Detoxified endotoxin from *Salmonella typhimurium* Re mutant (Ribi ImmunoChem Research Inc.) hardly induced any TNF in *C. parvum*-treated Swiss random mice, but elicited a cytostatic activity comparable to other endotoxins like LPSw from *E. coli* 0111:B4 (64).

Although our *in vitro* assay differs in several respects from the L cell assay used by Carswell et al. (49, 57) our data suggest that *in vitro* activity of TNS against Meth A is not predictive for its *in vivo* activity against this tumor. This is stressed by data of Männel et al. who found high cytolytic activity against L929 cells in sera of nude mice treated with BCG and endotoxin (63), although these mice do not produce TNF (52).

Endotoxin induces, besides TNF, various other factors with established antitumor activity (48). We have obtained some data on the role of interferon in the induction of necrosis and regression by endotoxin. The tumor damage observed after injection of 5.6 x 10^4 U L cell interferon (36, Table 1) suggests that part of the action of endotoxin is due to induction of this factor. However, several data indicate that it would play at most a moderate role. The amount of interferon injected corresponded to that found in 0.5 ml fresh TNS, but in sera of endotoxin-injected normal and Meth A-bearing mice about 6 to 7 times lower amounts were found (35). This amount of interferon had no detectable effect against Meth A (36). Interferon present in TNS appeared to be inactivated after heating at 56°C for 10 min (60, 65); yet the activity of heated and fresh TNS against Meth A *in vivo* was comparable (35, 52). The same holds for the antitumor activity of endotoxin and heated TNS (Table 1) although heated TNS in contrast to

a In a cytolytic L cell assay, sera of C3H/HeJ mice treated with BCG and endotoxin were found to be devoid of activity (63).

endotoxin did not induce detectable levels of interferon in normal mice (65). The *C. parvum* strain NCTC 10387 was also shown not to induce interferon (66) but it caused similar effects as endotoxin in our Meth A system on i.v. administration (55). Further observations showed that the induction of interferon by endotoxin was almost completely abolished by prior administration of the α-adrenoceptor antagonist phentolamine while this agent hardly influenced the effects of endotoxin against Meth A (65). A minor role for interferon is also suggested by data of Tanamoto et al. (17) which show that different regions of endotoxin were essential for induction of interferon and antitumor effects, but the latter effect was measured against ascitic tumors. Recently, however, we found that detoxified endotoxin from *S. typhimurium* Re mutant, which is capable to induce interferon (64, 67), was a very weak inducer of tumor necrosis and regression (64). The only observation that may plea for a role of interferon or a closely related mechanism, is the observation that, with exception of heated TNS, solely agents which were capable of induction of high levels of interferon were the most effective in our Meth A system (35, 36, Tables 2 and 3).

Toxic shock theory

Endotoxin is one of the most potent antitumor agents known but clinical application is hampered by its toxicity. Many studies suggest that the toxicity of endotoxin is implicated in its antitumor action (5, review). Also various non-endotoxin inducers of tumor necrosis (and regression) were found to be only active when toxic doses were applied (5, 13, 54). Gardner et al. (68) already stated in 1939 that shock would not be the consequence of but the cause of damage to the tumor. Although the cause of the toxicity of endotoxin is still a matter of controversy, the adrenals may play an important role. The adrenergic system has been related to both harmful and beneficial effects of endotoxin. Involvement in its antitumor action is suggested by the observation that a portion of the antitumor effect of endotoxin could be mimicked with adrenaline (34, Table 2). In addition the α-adrenergic blocking agent phenoxybenzamine inhibited endotoxin-induced tumor necrosis (39, 69), regression (39) and release of TNF, interferon and cytostatic factors (35, 65). Furthermore the release of tumor resistance enhancing factor, interferon and TNF, was shown to be inhibited by prior administration of adrenaline, probably by a tachyphylactic mechanism (19, 65, 70). Because β-adrenoceptor blockade with propranolol slightly potentiated tumor necrosis and regression and induction of interferon by endotoxin (39, 65) and because endotoxin itself has β-adrenergic blocking properties (71), α-agonistic effects seem to be involved in the effects of endotoxin mentioned. Unfortunately this receptor seems also to be involved in the toxic action of endotoxin because α- and β-adrenoceptor blocking agents were shown to decrease and to increase the toxicity of endotoxin respectively (39, 41, 72, 73)[b]. Also the incapability of non-toxic endotoxin of *S. typhimurium* Re mutant to induce TNF and tumor necrosis and regression of Meth A sarcoma (64) points at the involvement of toxic mechanisms in these effects. A non-toxic polysaccharide-rich endotoxin preparation, however, was found to induce a tumor-hemorrhaging factor in sera of normal BCG-treated mice (20). Subsequent necrosis and regression were not scored. For this reason and because *in vivo* elicitation of TNF was shown to require substantial prior macrophage activation (52, 55, 56), it is questionable whether the hemorrhaging factor is identical to TNF as was suggested. The observations that non-toxic endotoxins and derivatives have overt antitumor activity in other animal tumor models (18, 20, 75-77) especially on admixture with several adjuvants, may stress that antitumor action of endotoxin is due to various distinct activities of this agent and that the properties of endotoxin required to exert antitumor activity are defined by the host tumor model.

Concluding Remarks and Summary

We have reviewed and discussed our studies of the antitumor effect of endotoxins on solid well established immunogenic tumors of the mouse. In our opinion data from literature and our data suggest that induction of tumor necrosis and regression is the result of a cascade of concurrent and sequential, possibly synergistic effects, not attributable to a single factor or mechanism. Endotoxin-induced vasoamines are likely to be involved in

b Protection by α-blockade was not confirmed in an other study (74).

induction of hyperemia and hemorrhagic necrosis, which are probably due to abnormal vascular reactivity of the tumor which may be related to its age. Induction of these effects on their own, however, appeared not sufficient for subsequent regression to occur. This suggests additional mediation systems. The strong cytostasis of the tumor cells observed shortly after injection of endotoxin, concanavalin A and poly I:C may be related to hypoxia due to the induced extreme vasodilation, but also induced cytostatic factors may play a role. A causative role of TNF in necrosis and other effects, however, could not be proven. Interferon has probably at most a moderate effect. Nevertheless mediation of the initial events by humoral factors seems likely because substantial cellular infiltrates could not be detected. Because also fibrin deposition was absent and because anticoagulants were not inhibitory, the Shwartzman phenomenon is probably not involved. The role of the immunogenicity of the tumor in the effects observed was not studied, but the need for an intact T cell system in endotoxin-induced regression has been well established (7-9, 13). Even a state of concomitant immunity was shown to be essential, also explaining the temporal effectivity of endotoxin. These data suggest that some kind of immuno-adjuvanticity of endotoxin is implicated in addition, although this is apparently too weak to abolish a state of T-suppressor cell activity (31). Additional adjuvant treatment may improve this, as already indicated in other studies (75-77).

Several direct and indirect data suggested that a toxic mechanism underlies induction of necrosis and regression and TNF. The latter factor, however, mimicked these effects without overt toxicity. Like endotoxin its *in vivo* effect is probably not, at least not predominantly, due to a direct interaction with the tumor cells.

Although production of TNF may be dependent on a toxic event, the lack of both toxicity and species-specificity of TNF (49, 51) may make clinical application of this factor in therapy of tumors feasible.

References

1. Gratia, A., Linz, R.: C.r. Séanc. Soc. Biol. Ses. Fil. 108, 427-428 (1931).
2. Shwartzman, G., Michailovsky, N.: Proc. Soc. Exp. Biol. Med. 29, 737-741 (1932).
3. Apitz, K.: Z. Krebsforsch. 40, 50-70 (1933).
4. Duran-Reynals, F.: Proc. Soc. Exp. Biol. Med. 32, 1517-1521 (1935).
5. Shear, M.J., Perrault, A.: J. Nat. Cancer Inst. 4, 461-476 (1943).
6. Zahl, P.A., Hutner, S.H., Spitz, S., Sugiura, K., Cooper, F.S.: Am. J. Hyg. 36, 224-242 (1942).
7. Parr, I., Wheeler, E., Alexander, P.: Br. J. Cancer 27, 370-389 (1973).
8. Berendt, M.J., North, R.J., Kirstein, D.P.: J. Exp. Med. 148, 1550-1559 (1978).
9. Berendt, M.J., North, R.J., Kirstein, D.P.: J. Exp. Med. 148, 1560-1569 (1978).
10. Hoffmann, M.K., Oettgen, H.F., Old, L.J., Mittler, R.S., Hammerling, U.: J. Reticuloendothel. Soc. 23, 307-319 (1978).
11. Jones, G.R.N.: J. Cancer Res. Clin. Oncol. 93, 245-254 (1979).
12. Oettgen, H.F., Hoffmann, M.K., Clarkson, B.D., Old, L.J.: Pontif. Acad. Scient. Scripta Var. 43, 121-144 (1979).
13. Kodama, M., Mizukuro, T., Yamaguchi, N., Yoshida, S., Katayama, Y.: In Bacteria and Cancer. Eds. Jeljaszewicz, J., Pulverer, G., Roszkowski, W. Acad. Press, London pp. 167-179 1982.
14. Shapiro, C.J.: Am. J. Hyg. 31 (B), 114-126 (1940).
15. Männel, D.N., Rosenstreich, D.L., Mergenhagen, S.E.: Inf. Immun. 24, 573-576 (1979).
16. Yang, C., Nowotny, A.: Infect. Immun. 9, 95-100 (1974).
17. Tanamoto, K., Abe, C., Homma, J.Y., Kojima, Y.: Eur. J. Biochem. 97, 623-629 (1979).
18. Kiso, M., Nishiguchi, H., Hasegawa, A., Okumura, H., Azuma, I.: Agric. Biol. Chem. 45, 1523-1525 (1981).
19. Butler, R.C., Abdelnoor, A.M., Nowotny, A.: Proc. Natl. Acad. Sci. USA 75, 2893-2896 (1978).
20. Butler, R.C., Nowotny, A., Friedman, H.: In Immunomodulation by Bacteria and their Products. Eds. Friedman, H., Klein, T.W., Szentivanyi, A. Plenum Press, New York pp. 181-197, 1981
21. Wasilauskas, B.L., Cameron, J.A.: Cancer 27, 217-222 (1971).
22. Tanamoto, K., Homma, J.Y.: J. Biochem. 91, 741-746 (1982).

23. Rietschel, E.Th., Galanos, C., Lüderitz, O., Westphal, O.: In Immunopharmocology and the Regulation of Leukocyte Function. Ed. Webb, D.R. Marcel Dekker Inc. New York pp. 183-229, 1982

24. Algire, G.H., Legallais, F.Y., Park, H.D.: J. Nat. Cancer Inst. 8, 53-62 (1947).

25. Andervont, H.B.: Am. J. Cancer 27, 77-83 (1936).

26. Thomas, L.: Natl. Cancer Inst. Monogr. 44, 137-139 (1976).

27. Stetson, C.A.: J. Exp. Med. 93, 489-504 (1951).

28. Thomas, L.: J. Exp. Med. 104, 865-880 (1956).

29. Rall, D.P., Kelly, M.G.: J. Exp. Med. 105, 643-652 (1957).

30. Kuper, C.F., Bloksma, N., Hofhuis, F.M., Bruyntjes, J.P., Willers, J.M.: Int. J. Immunopharmac. 4, 49-55 (1982).

31. North. R.J.: Progress in Cancer Research and Therapy 16, 113-124 (1981).

32. Kuper, C.F., Bloksma, N., Bruyntjes, J.P.: (unpublished observations).

33. Youngner, J.S., Algire, G.H.: J. Natl. Cancer Inst. 10, 565-580 (1949).

34. Pradhan, S.N., Achinstein, B., Shear, M.J.: Arch. Exp. Path. Pharmak. 230, 228-235 (1957).

35. Bloksma, N.: PhD. Thesis, State University of Utrecht, The Netherlands 1982

36. Bloksma, N., Kuper, C.F., Hofhuis, F.M.A., Benaissa-Trouw, B., Willers, J.M.N.: Cancer Immunol. Immunother. (in press)

37. Bloksma, N., Kuper, C.F., Hofhuis, F.M.A., Willers, J.M.N.: (submitted)

38. Weiner, N.: In The Pharmacological Basis of Therapeutics. Eds. Goodman Gilman, A., Goodman, L.S., Gilman, A., Macmillan Publishing Co., Inc. New York pp. 176-210, 1980

39. Bloksma, N., Hofhuis, F.M., Willers, J.M.N.: Immunopharmacol. 4, 163-171 (1982).

40. Rosenberg, J.C., Lillehei, R.C., Moran, W.H., Zimmermann, B.: Proc. Soc. Exp. Biol. Med. 102, 335-337 (1959).

41. Hinshaw, L.B.: In Bacterial Endotoxins. Eds. Kadis, S., Weinbaum, G., Ajl, S.J. Academic Press, New York pp. 209-275, 1971

42. Weiner, N.: In The Pharmacological Basis of Therapeutics. Eds. Goodman Gilman, A., Goodman, L.S., Gilman, A., Macmillan Publishing Co., Inc. New York pp. 138-175 1980

43. Kuratsuka, K., Shimazaki, Y., Funasaka, I., Watanabe, Y.: Experientia 34, 1483-1484 (1978).

44. Kuratsuka, K., Homma, R., Shimazaki, Y., Funasaka, I.: Experientia 31, 206-208 (1975).

45. Krylova, N.V.: Bibl. Anat. 10, 301-303 (1969).

46. Mattsson, J., Appelgren, L., Hamberger, B., Petersson, H.I.: Cancer Lett. 3, 347-351 (1977).

47. Morrison, D.C., Ulevitch, R.J.: Am. J. Pathol. 93, 526-618 (1978).

48. Ruff, M.R., Gifford, G.E.: J. Immunol. 125, 1671-1677 (1980).

49. Carswell, E.A., Old, L.J., Kassel, R.L., Green, S., Fiore, N., Williamson, B.: Proc. Natl. Acad. Sci. USA 72, 3666-3670 (1975).

50. Old, L.J.: Clinical Bulletin 6, 118-120 (1976).

51. Oettgen, H.F., Carswell, E.A., Kassel, R.L., Fiore, N., Williamson, B., Hoffmann, M.K., Haranaka, K., Old, L.J.: Rec. Res. Cancer Res. 75, 207-212 (1980).

52. Green, S., Dobrjansky, A., Chiasson, M.A., Carswell, E., Schwartz, M. K., Old, L.J.: J. Natl. Cancer Inst. 59, 1519-1522 (1977).

53. Kull, F.C., Cuatrecasas, P.: J. Immunol. 126, 1279-1283 (1981).

54. Bloksma, N., Hofhuis, F.M.A., Willers, J.M.N.: Eur. J. Cancer Clin. Oncol. (in press)

55. Bloksma, N., Hofhuis, F.M.A., Willers, J.M.N.: Ann. Immunol. Hung. 23 (in press)

56. Bloksma, N., De Leeuw, R., Hofhuis, F.M., Westerdaal, N.A., Benaissa-Trouw, B., Willers, J.M.N.: Ann. Immunol. (Inst.Pasteur) 134C, 159-164 (1983).

57. Bloksma, N., Schetters, T.P., Figdor, C., Van Dijk, H., Willers, J.M.: Int. J. Immunopharm. 2, 95-100 (1980).

58. Green, S., Dobrjansky, A., Carswell, E.A., Kassel, R.L., Old, L.J., Fiore, N., Schwartz, M.K.: Proc. Natl. Acad. Sci. USA 73, 381-386 (1976).

59. Bloksma, N., Hofhuis, F.M.A., Schetters, T.P., Willers, J.M.N.: INSERM 97, 223-230 (1980).

60. Kato, N., Nakashima, I., Ohta, M., Naito, S., Kojima, T.: Microbiol. Immunol. 23, 383-394 (1979).

61. Bloksma, N., Hofhuis, F.M.A., Benaissa-Trouw, B., Willers, J.M.N. (manuscript in preparation)

62. Koenig, S., Hoffmann, M.K., Thomas, L.: J. Immunol. 118, 1910-1911 (1977).

63. Männel, D.N., Meltzer, M.S., Mergenhagen, S.E.: Infect. Immun. 28, 204-211 (1980).

64. Bloksma, N., Hofhuis, F.M.A., Benaissa-Trouw, B., Willers, J.M.N.: (manuscript in preparation)

65. Bloksma, N., Hofhuis, F., Benaissa-Trouw, B., Willers, J.: Cancer Immunol. Immunother. 14, 41-45 (1982).

66. Neuman, C., Sorg, C.: Eur. J. Immunol. 8, 582-589 (1978).

67. Ribi ImmunoChem Research, Inc. (Hamilton, Montana, USA), Product information

68. Gardner, R.E., Bailey, G.H., Hyde, R.R.: Am. J. Hyg. (Sect. B) 29, 1-14 (1939).

69. Pradhan, S.N., Achinstein, B., Shear, M.J.: Cancer Res. 16, 1062-1068 (1956).

70. Jensen, M.M.: Proc. Soc. Exp. Biol. Med. 130, 34-39 (1969).

71. Schreurs, A.J.M., Verhoef, J., Nijkamp, F.P.: Eur. J. Pharmacol. 87, 127-132 (1983).

72. Gourzis, J.T., Hollenberg, M.W., Nickerson, M.: J. Exp. Med. 114, 593-604 (1961).

73. Filkins, J.P.: Circ. Shock 6, 99-107 (1979).

74. McLean, R.A., Berry, L.J.: Proc. Soc. Exp. Biol. Med. 105, 91-95 (1960).

75. Ribi, E., Parker, R., Strain, S.M., Mizuno, Y., Nowotny, A., Von Eschen, K.B., Cantrell, J.L., McLaughlin, C.A., Hwang, K.M., Goren, M.B.: Cancer Immunol. Immunother. 7, 43-58 (1979).

76. Takayama, K., Ribi, E., Cantrell, J.L.: Cancer Res. 41, 2654-2657 (1981).

77. Amano, K., Ribi, E., Cantrell, J.L.: Biochem. Biophys. Res. Commun. 106, 677-682 (1982).

Acknowledgements

This work was supported by the Stichting Koningin Wilhelmina Fonds, Netherlands Cancer Foundation. We are indebted to Mr. Pieter Hiemstra for performing some experiments.

THE ENDOTOXIN SKIN REACTION IN HUMAN CANCER PATIENTS

Yutaka Katayama

Department of Microbiology, Kyoto Prefectural University of Medicine
Kyoto, Japan

Nozomi Yamaguchi

Department of Preventive Medicine, Kyoto Prefectural University of Medicine
Kyoto, Japan

Masaaki Kanou

Department of Gynecology, Kyoto Prefectural University of Medicine
Kyoto, Japan

Hideo Hayashi, Yoichi Fujita, Masami Oshima

Department of Medicine, Kyoto Second Red Cross Hospital
Kyoto, Japan

Kenji Ogino

Department of Medicine, Ika Hospital
Kinomoto, Shiga, Japan

Masashi Kodama

Department of Surgery, Shiga University of Medical Sciences
Ohtsu, Shiga, Japan

Introduction

Bacterial endotoxins (LPS) have been known for their wide varieties of biological activities in the experimental animals, such as pyrogenicity, increased non-specific defense, adjuvanticity, Shwartzman activities, interferon induction, activation of macrophage, neutrophile, B lymphocyte and alternative com-

Immunopharmacology of Endotoxicosis

plement pathway, coagulation, tumour hemorrhage and necrosis, abortion and so on.

For a human being, they were also expected to have strong as well as wide activities from the clinical observations in patients of infectious diseases by LPS-containing bacteria (1) and several experiments were carried on by human beings concerning their toxicity and pyrogenicity (2-5), mobilization of bone marrow granulocyte reserve (6-9) and increased nonspecific defense including beneficial effects on rheumatoid arthritis (10-11).

Besides these activities, endotoxins have the capacity to cause the erythema on the abdominal skin of rabbits (12), and on the human skin (13-15). Applying this method, we showed that the endotoxins can induce erythematous reaction on the human skin by injecting about 10 ng amount of Escherichia coli LPS which is 100 times less than that needed for the rabbit skin reaction. This method may consist of one of the simplest and most basic forms of the endotoxin activities on human beings and we tested the applicability of this reaction for human cancer patients.

The phenomenon of hemorrhage and necrosis in solid malignant tumour tisssues following the i.v. or i.p. injection of LPS was first reported by Gratia and Linz in 1931 (16) and since been established as a consistent finding in malignant tumour tissues by many investigators (17-25). Although, specific tumour systems express variation in susceptibility to the LPS and for unexplained reasons, the response is not observed during the first several days after the inoculation of tumour cells (20,23,25), the phenomenon is specific and common to all malignant tumour tissues, which suggests the existence of a specific and common substance in all malignant tumour cells. The similarity of this phenomenon to the local Shwartzman phenomenon led Antopol to propose that Shwartzman active substance (s) may exist in malignant tumour tissues of man and mice (26).

Shear and his colleagues demonstrated that tryptic digestion of mlaignant tumour tissues resulted in a product having a strong Shwartzman activity as compared to normal tissues in which only a weak reaction was demonstrated after tryptic digestion (27). Shear later reported that the weak reaction in normal tissues after tryptic digestion was due to the contamination of the LPS-containing bacteria (28). Recently our own group has demonstrated the existence of weak Shwartzman active substance(s) in malignant tumour cells of rats and of human cancer patients (29,30).

If the supposition that the Shwartzman active substances are contained in all malignant tumour cells is correct, it then follows that the malignant tumour-bearing animals or human patients may be sensitized by the Shwartzman active substance liberated from the occasional death of tumour cells and may thus "prepare" the host in the Shwartzman sense. From analogy with mice chronically infected with the endotoxin containing bacteria whose skin is more sensitive to the intracutaneous inoculation of bacterial LPS (31), we assumed that the cancer patients are more sensitive to the intradermal injection of bacterial LPS than are normal persons (32). This report will presents some supporting data of this hypothesis.

Method

Endotoxin : The LPS of Escherichia coli O55 was obtained by the method of Boivin (33). The stock solution of 20 μg per ml was made by physiological saline and sterilized by heating at 80 C for 20 min or 100 C for 5 min. The LPS solutions of various concentrations were made by diluting LPS stock solution with sterile physilogical saline and sterilized by heating at 80 C for 20 min.

Results

1) Toxicity of the endotoxin

The toxicity of E. coli LPS was tested by injecting 500 ng of LPS subcutaneously thrice a week for three months to the authors themselves. No harmful effects were observed except for local pain, erythema and induration.

2) Trials on human skin injection

By i.c. injections of E. coli LPS, the erythematous reactions were observed one or two hours after injections and their colours were gradually intensified reaching their peak by about four hours, but the diameter of the erythema still increased very slowly thereafter until 8 to 10 hours. Injecting various doses ranging from 10 ng to 500 ng of the LPS, the larger the amounts of the LPS injected, the stronger was the erythematous reactions, and about 40 ng of the LPS caused the moderate erythema of about 15 mm of the diameter 4 hours after the injection, so 40 ng LPS disssolved in 0,2 ml of physiological saline was determined as the standard method in this trial. As for the place of injection, the center line of the flexor surface of the upper arm was chosen as the standard method. To obtain highly reproducible results, it was essential to inject exactly the same amount of LPS reagent intracutaneously. If injected subcutaneously, the reaction was usually very weak by the fast diffusion of the reagent. Therefore 3 or 4 injections were carried out for one person at a time and only the results of larger diameter were read. For the reading of the results, the major axis and the minor axis of the erythema were measured.

3) Reproducibility of the LPS-skin reaction

Fourty ng of E. coli LPS dissolved in 0.2 ml of physiological saline was injected to the same person repeatedly one week apart and the diameter of the erythema 4 hours after the LPS injection was measured. As shown in the Table 1, the reproducibility of the method was very high.

4) Instability of the LPS reagent

In the course of the skin reactions repeatedly carried out in

Table 1.
The human skin reactivity for the bacterial endotoxin.

Experiments	Diameter of the erythema (mm)		
1	15,	15,	14
2	15,	15,	14
3	15,	14,	14
4	15,	14,	14
5	14,	14,	13

Forty ng of E. coli LPS (Boivin) dissolved in physiological saline was injected intracutaneously, and the diameter of the erythema was measured 4 h after the injection. The tests were carried out for the same person 1 w apart.

our laboratories, it was found that the activity of the diluted LPS solution was not stable during storage in the refrigerator after several months, that the activity was rapidly deteriorated after freezing the reagents, and that the LPS solutions made by the double distilled purest water appeared to be more unstable than the reagents made by running tap water.

5) The effect of Mg ion on LPS activity

From these facts and the knowledge that Mg or Ca ions are important to keep the macromolecular form of the LPS (34) and to exhibit its anticomplementary activity (35), the effect of Mg and Ca ions on the LPS skin test was tested. The LPS solution of 200 ng per ml containing 0, 4, 50, 500 μM of $MgCl_2$ were tested for their human skin reactivities. This Mg containing LPS reagent induced much clearer red erythema than Mg free reagent and also more intense red erythema than Ca containing LPS solutions. From these observations, the effect of Mg ions was tested further.

If Mg ion was added, the LPS solution was found to be stable for over 6 months in a refrigerator as shown in the Table 2. To increase the stability of the LPS, the concentrations of 5

Table 2.
The stability of the endotoxin solution containing $50\mu M$ of Mg ion.

Test at	Diameter of the erythema (mm)		
0 month	16,	16,	15
1	16,	16,	15
2	15,	15,	15
3	15,	15,	14
4	15,	15,	14
5	15,	15,	14
6	15,	15,	14

The LPS skin tests were carried out for the same person one months apart for the same person using the LPS reagents containing 50 μM of Mg ion. Details are in the text.

to 500 μM of Mg ion seemed to be satisfactory and few differences were observed between the LPS solution containing 5μM, 50μM and 500μM of Mg ion. As for the stock solution of 20μg per ml of LPS containing 50μM of $MgCl_2$, it was found to be stable over 2 years in refrigerator at 4 to 18 C.

6) Dose-response of LPS skin reaction.

Solutions containing 10, 20, 40, 60, 80 and 100 ng of LPS in 0.2 ml of saline and 50μ M of $MgCl_2$ were prepared, sterilized by heating at 80 C for 20 min and injected intracutaneously for the same person. As observed in Table 3, the skin reactivity increased slowly by increasing the amount of the LPS injected.

7) Personal differences in LPS skin reaction

From the above observations, the standard method of the LPS skin reaction in our laboratory was defined as follows : Forty ng of E. coli O55 LPS (Boivin) dissolved in the physiologicsal saline of 0.2 ml containing 50μM of $MgCl_2$ and 0.5 % of phenol as the preservative and sterilized by heating at 80 C for 20 min, was injected intracutaneously to the center line of the flexor surface of the upper arm and the diameters of major axis and minor axis was measured 4 hours after the LPS injection.

Table 3.
Endotoxin skin reaction : dose-response.

Dose	Diameter of the erythema (mm)	
10ng	9×8 ,	8×6
20	12×11,	12×11
40	14×13,	13×12
60	18×16,	18×15
80	18×17,	18×18
100	19×17,	18×18

The LPS skin tests are carried out for the sampe person 1 w apart changing the amount of the E. coli LPS. The diameters of majaor axis and minor axis are measured 4 h after the i.c. injection of the reagents.

The major axis and the minor axis were measured but the results were usually expressed simply by the value of major axis. Usually 2 or 3 injections were made for one person at a time and the stronger reactions were taken as the result, not the mean of the readings.

This reaction was carried out on 20 healthy persons including the authors. The personal deviations ranging from 13 mm to 20 mm were observed as shown in the control in Fig.1. The reaction was carried out for another 32 healthy volunteers using the standard method. The results with 52 healthy volunteers ranged from 13 to 21 mm in diameter with the mean value of 16.06 and standard deviation (SD) of 3.85. Therefore the mean + 3 SD was 21.8 and the mean - 3 SD was 10.3.

8) Reactivity of malignant tumor patients

The LPS skin tests were performed at several insitutions independently. The patients were all hospitalized and most were receiving chemotherapy or radiotherapy as treatment for inoperable tumours or disseminated metastasis after surgical operation. The patients who were tested before surgical operation and before receiving any chemotherapy usually showed increased

Table 4. The results of the endotoxin-skin test of human malignant tumor patients.

	Hyperreactive	Hyporeactive	Sum
Healthy Control	1 (1.9%)	51 (98.1%)	52
Neoplasms of			
Digestive organ	12 (30.0%)	28 (70.0%)	40
Lung	11 (44.0%)	14 (56.0%)	25
Gynecological organ	14 (56.0%)	11 (44.0%)	25
Leukemia	6 (24.0%)	19 (76.0%)	25
Others	5 (50.0%)	5 (50.0%)	10
Sum	48 (38.4%)	77 (61.6%)	125

The LPS skin tests were carried out for normal healthy volunteers and for cancer patients. In this table, the results were expressed as Hyperreactive : erythema of larger than 22 mm and as Hyporeactive : less than 10 mm.

endotoxin skin sensitivity over 22 mm (the mean of healthy control + 3 SD) in diameter. In contrast, those with disseminated metastasis or receiving heavy doses of chemotherapy exhibited strongly depressed LPS skin sensitivity giving values of less than 10 mm in diameter. This suppression was especially manifest in many of the leukemia patients. A number of patients were tested several times with very similar results, but the results of the first test are cited in this study, and for the patients in the Department of Surgery or Gynecology, only the preoperational test values are given in order to eliminate the effects of immediate post-surgical trauma. Patients with metastatic foci were classified by the primary tumour site. The results are shown in Table 4 and Fig 1. Of the 125 patients included in this report, 47 demonstrated increased skin sensitivity having greater than 22 mm (the mean value of healthy controls + 3 SD) in diameter (22 - 35 mm). The 23 patients with advanced cancer receiving large doses of chemotherapy or radiotherapy showed a marked decrease in skin reactivity with the test of less than 10 mm (the mean value of healthy control - 3 SD) in diameter. The remaining 55 cancer

Fig.1. The results of the endotoxin-skin test of human malignant tumor patients.

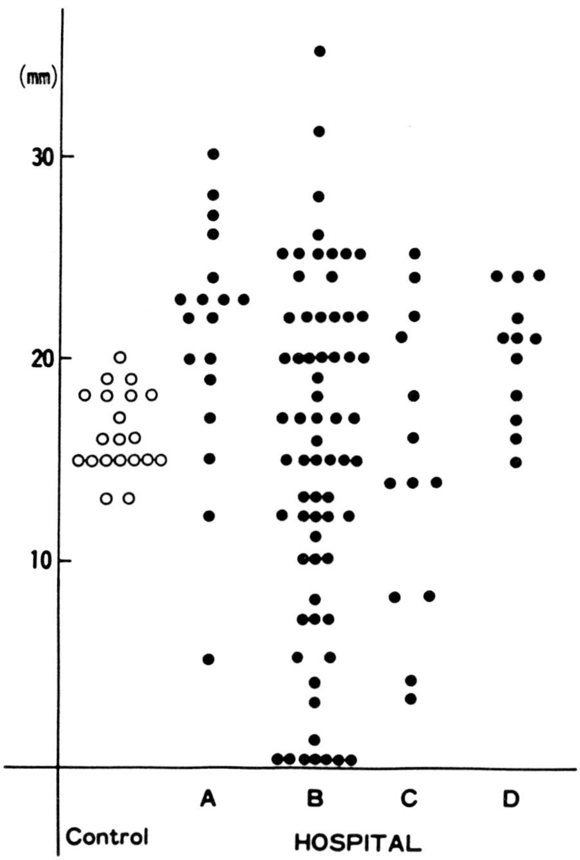

The LPS skin tests were carried out for healthy volunteers and for the patients with malignant tumours at four institutions independently. Details are described in the text.

patients responded with skin reactions that were not different from those observed in the healthy control population (11 - 21 mm : mean - 3 SD   mean + 3 SD).

9) Sensitivity to phenol solution

To test the effects that may have been induced by the presence of 0.5 % phenol, a control injection of 0.5 % phenol alone was injected into each patient on the opposite arm. No marked differences were observed in the erythema between the cancer patients and normal volunteers, all of whom showed a skin reactivity to the phenol solution of less than 5 mm in diameter from one hour to several hours post injection. These findings may also suggest that a nonspecific skin irritant does not cause a stronger skin reaction in cancer patients as compared to normal volunteers.

Discussion

Bacterial endotoxins are known to have many interesting biological activities. As one of them , they have been known to cause the erythematous reaction of the abdominal skin of the rabbits (15). Applying this method to the human beings, we showed that human skin was sensitive enough to show the reaction with 10 ng of E. coli LPS which is 100 times less than that needed for the rabbit skin reaction. This method was simple, easy to handle and to read the results, and without harmful effects for man except local pain, erythema and induration for 1 or 2 days. One of the advantage of this method was that several reactions could be carried out for one person at a time. This test may be used to express and compare the endotoxin action on human beings.

The LPS human skin test was found to be highly reproducible as shown in Table 1, but the LPS solution was not stable enough to keep over 6 months in regrigerator. The LPS solutions containing 5 to 500 μM of Mg ion were found to have the activity on human skin reactions of more intense and warmer red colour and were stable for keeping over 6 months as shown in Table 2.

The strength of the reaction was easily expressed by the diameter of the erythema 4 hours after intracutaneous injection of the LPS, and it was dose-dependent as shown in Table 3. From these observations, this human skin reaction may be used for the expression of the endotoxin action on human beings.

Bacterial endotoxins have been injected for human beings to test their pyrogenicity, side effects (2-5), mobilization of bone marrow granulocyte reserve (6-9), increased nonspecific defense including beneficial effects on rheumatoid arthritis (10-11) and skin reactivity (13-15). For these purposes, the amount of LPS ranging from 0.05 to 10 ng per Kg body weights or 100 to 200 ng per person of Salmonella abortus equi or E. coli LPS has been injected intravenously. Compared with the i.v. injection of 100 to 200 ng of the endotoxins, i.c. injection of 10 to 100 ng amount showed minute side effects.

Histological observations have clarified that the host responses to the i.c. injection of 10 ng or more of the LPS mainly consisted of neutrophil infiltrations, especially at the early stage of the responses (13-15,36-38). Therefore, the strength of the LPS skin test may be interpreted as the expression of the reactivity of hosts' neutrophils against the LPS.

One question is the difference from the Arthus reaction. The Arthus reaction in rabbits is shown to require the existence of the specific IgG antibody of about 500 μg/ml and the strength of the reaction is dependent upon the serum concentrations of the specific antibody (39). The precipitating antibody for E. coli 055 LPS was not detected in the authors' serum by the ring test, which can detect about 20μg/ml of the antibody (40). Furthermore, as shown in the Table 3, the strength of the skin test are dependent upon the amount of the LPS injected and not on the serum antibody titers. From these observations, the nature of the LPS skin test are essentially different from that of the Arthus reaction.

The application of this test may be to compare the individual difference of the human sensitivitiy to the LPS.

The increased susceptibility to the lethal action of the LPS was observed in experimental animals inoculated with living

Increased Susceptibility to Endotoxin in YOSHIDA Sarcoma-Bearing Rats

Experiment	Rat	Endotoxin injected (μg)	Saline (ml)	Number of Rats	Deaths	Death Rate (%)
I	Solid tumor rat	800		10	10	100
		400		10	10	100
		40		10	10	100
			0.5	5	0	0
	Normal rat	800		5	0	0
II	Ascitic tumor rat	800		10	0	0
		400		10	0	0
			0.5	5	0	0
	Normal rat	800		5	0	0

Table 5 The increased susceptibility for the endotoxin of the rats bearing Yoshida sarcoma cells. Donryu strain rats of 6 weeks of age were inoculated with half million of Yoshida sarcoma cells i.m. or i.p., and 20 days (solid tumour rats) or 7 days (ascitic tumour rats) later, rats were injected i.p. with 800, 400, or 40μg of E. coli O55 endotoxin (Westphal) and observed 48 h later.

or dead LPS-containing microorganisms such as Vibrio (41), Escherichia (42), Hemophilus (43) or Brucells (44). Similar phenomena were observed in animals infected with B.C.G. (45), mouse hepatitis virus (46), lymphocytic choriomeningitis virus (47), or Coxiella burnetti (48).

Aged animals are known to be more sensitive (49) and this phenomenon is especially manifest in nude mice (50).

Thymectomized (51) or adrenalectomized animals (52) or animals deficient in the components of complement system (53,54), animals in pregnancy (55) or in malignant tumor bearing (20,56) are also shown to be more susceptible. We found a markedly increased susceptibility to the lethal action of the E. coli LPS in rats bearing Yoshida sarcoma cells as shown in Table 5. From these observations and analysis, we assumed an increased skin sensitivity of human cancer patients for the bacterial LPS.

When 40 ng of E. coli endotoxin was injected intracutaneously into healthy volunteers and cancer patients, and the diameters of induced erythematous reactions were measured at 4 hours, a marked difference between the two populations was observed. The response in 52 healthy volunteers ranged from 13 - 21 mm in diameter and only one person had a reactivity of 21 mm in diameter. In contrast to this, skin testing of cancer patients demonstrated that 47 patients among 125 demonstrated a heightened skin responsiveness yielding the erythema greater than 22 mm (the mean + 3 SD) in diameter, whereas 23 of these patients demonstrated hyporesponsiveness to the skin test giving lesions of less than 10 mm (the mean - 3 SD) in diameter, and 55 patients showed the normal range of reactivity of 10 to 21 mm diameter. Thus, 70 of the cancer patients (56 %) had abnormal skin responsiveness to intracutaneous injection of the LPS.

Host response to the i.c. injection of 10 ng or more of the LPS were shown to consist mainly of a neutrophil leucocyte (13-15,36-38). Therefore, the strength of the LPS skin test may be the expression of the biological activities of neutrophils, and the results obtained in this study may demonstrate that the activities of neutrophills of many cancer patients increased as compared with those of normal healthy persons.
It has been reported that neutrophils of human cancer patients are increased over that of normal healthy adults when measured by the NBT test or by phagocytosis or by the bacterial killing test (57).
Recently Klostergaard et al. has shown that the serum of patients with solid tumours has highly significant stimulatory effects on the phagocytic activity of neutrophils in contrast with the depressed phagocytic activity of granlocytes associated with various forms of leukemia and lymphoma (58).
These results are in contrast to a general tendency of decreased biological activity of leucocytes especially of T-lymphocytes (59) or of macrophages (60) in cancer patients.
This general decrease in leucocyte activity is often enhanced

following chemotherapy or radiotherapay resulting in the increased susceptibility to infections (61-63). It is interesting to note that for some patients in our trial who were tested by both the NBT test and the LPS skin test, the results of these tests were not always parallel (data not shown).

The erythematous reaction by the LPS may be induced by the generation of anaphylatoxin, which can induce the accumulation of neutrophils (64). The problem as to whether increased LPS skin test is partly contributed by the increased productivity of the anphylatoxins in cancer patients remains to be studied.

An increase in susceptibility to the lethal action of bacterial LPS has been observed in tumour-bearing animals (20,56). The data obtained in this report clearly demonstrate the hyperreactivity of the skin of some cancer patients following i.c. injections of E. coli LPS. The mechanisms underlying both phnomena may not be the same and remain to be clarified.

Finally , we choose to interpret our data as supporting the view that all malignant tumour cells contain Shwartzman gene which produces the Shwartzman active substance and is closely linked to the malignancy-determining gene and that both of these genes must be strongly co-repressed in normal cells (65).

Summary

Human skin sensitivity to bacterial endotoxins was measured by injecting 40 ng of Escherichia coli 055 endotoxin dissolved in 0.2 ml of physiological saline containing 50 μM of $MgCl_2$ and 0.5 % phenol intracutaneously to the center line of the flexor surface of the upper arm. The diameter of the erythematous reactions were measured 4 hours after endotoxin injection. The LPS skin test of 52 healthy volunteers gave reactions of erythema that ranged from 13 mm to 21 mm in diameter and only one person showed erythema of 21 mm in diameter.

Patients hospitalized for malignant tumours were similarly

skin tested and these studies were performed independently at four different institutions. Among 125 patients so tested, 47 showed increased skin sensitivity with greater than 22 mm diameter erythema (the mean of healthy control + 3 SD). However, patients with advanced cancer receiving large doses of chemotherapy or radiotherapy showed markedly decreased reactivity to the challenge with endotxin and responded with erythematous areas that were less than 10 mm in diameter (the mean of healthy control - 3 SD).

The authors hope to express cordial appreciation to Professor Sidney Harshman of Vanderbilt Univeristy School of Medicine, for his helpful discussions and reading this manuscript.

References

1. Hjort,P.F., Rapaport,S.I.: Ann.Rev.Med. 16.135-168 (1965).

2. Favorite,G.O., Morgan, H.R.: J.Clin.Inv.21. 589-599(1942).

3. Morgan, H.R. : J. Clin. Inv. 27. 706-709 (1948).

4. Pickering, G.W. : Lancet i . 59-64 (1958).

5. Ollodart, R.M., Hawthorne, I., Attar, S. : Amer. J. Surg. 113. 599-607 (1967).

6. Keller,H.M., Heilmeyer,I.:Klin.Wschr.37.1003-1005(1959).

7. Cradock,C.G.Jr., Perry,S., Ventzke,L.E., Lawrence,J.S. : Blood 15. 840-855 (1960).

8. Athens, J.W., Haab, O.P., Raab, S.O., Mauer, A.M., Aschenbrucker, H., Cartwright, G.E., Wintrobe, M.W. : J. Clin. Inv. 40. 989-995 (1961).

9. Fink,M.E., Calabresi,P.:Ann.Intern.Med.57.732-742 (1962).

10. Woratz, G. : Muench. med. Wschr. 97. 1531-1534 (1955).

11. Berg, G., Brichzy, W., Braunhofer, J., Schricker, K.Th. : Deut. med. Wschr. 81. 1156-1160 (1956).

12. Larson, C.L., Ribi, E., Milner, K.C., Lieberman, J.E. : J. Exp. Med. 111. 1-20 (1960).

13 Sell, S., Braude, A.I.: J. Immunol. 87. 119-125 (1961).

14 Greisman, S.E., Wagner, H.N., Iio,M., Hornick, R.B. : J. Exp. Med. 119. 241-264 (1964).

15 Greisman, S.E., Hornick, R.B.: J. Immunol. 109. 1210-1222 (1972)

16. Gratia, A., Linz, R. : C.R.Sco.Biol. 108. 427-429 (1931).

17. Schwartzman, G., Michailovsky, N. : Proc. Soc. Exp. Biol. Med. 29. 737-741 (1932).

18. Apitz, K. : Zeitschr. f. Krebksforsch. 40. 50-70 (1933).

19. Zahl, P.A., Hutner, S.H., Spitz, S., Sugiura, K., Cooper, F.S. : Amer. J. Hyg. 36. 224-242 (1942).

20. Shear, M.J., Perrault, A. : J. Nation. Cancer Inst. 4. 461- 476 (1943).

21. Havas, H.F., Froesbeck, M.E., Donnelly,a. J. : Cancer Res. 18. 141-148 (1958).

22. Donnelly, A.J., Groesbeck, M.E., Donnelly, A.J. : Cancer Res. 18. 149-154 (1958)

23. Steinebring, W.R., Stevens, D. : Proc. Soc. Exp. Biol. Med. 156. 229-235 (1977).

24. Berendt, M.J., North, R.J., Kirstein, D.P. : J. Exp. Med. 148. 1550-1559 (1978).

25. Berendt, M.J., North, R.J., Kirstein, D.P. : J. Exp. Med. 148. 1560-1569 (1978).

26. Antopol, W. : J. Infect. Dis. 61. 334-337 (1937).

27. Landy, M., Shear, M.J. : J. Exp. Med. 106. 77-97 (1957).

28. Merler, E., Perrault, A., Trapani, R.-J., Landy., M., Shear, M.J.: Proc.Soc.Exp.Biol.Med. 105. 443-445 (1960).

29. Kodama, M., Yajima,H., Kubo,Y., Hashimoto, I., Katayama, Y. : Japan. J. Med. Sci. Biol. 26. 3-6 (1973).

30. Kodama, M., Yajima, H., Hashimoto, I., Katayama, Y. : Japan. J. Med. Sci. Biol. 30. 77-81 (1977).

31. Arndt,N.F., Schneider,H.A.: J.Exp.Med. 112. 167-185(1960).

32. Katayama, Y., Kodama, M. : Japan. J. Med. Sci. Biol. 33. 35-36 (1980).

33. Boivin, A., Messrobeanu, I., Mesrobeanu, L. : C. R. Soc. Biol. 113. 490-492 (1933).

34. Galanos,C., Luederitz,O.: Eur.J.Biochem.54.603-610(1975).

35. Galanos,C., Luederitz,O.: Eur.J.Biochem.65.403-408(1976).

36. Karsner, H.T., Moritz, A.R.: J.Exp.Med. 60. 37-48 (1934).

37. Gerber, I.E. : Arch. Path. 21. 331-350 (1936).

38. Stetson, C.A.Jr. : J. Exp. Med. 93. 489-504 (1951).

39. Kabat, E.A., Mayer, M.M. : Experimental Immunochemistry, C.C.Thomas Publ., U.S.A. 188 (1961).

40. Davis, B.D., Dulbecco, R., Eisen, H.N., Ginsberg, H.S., Wood,W.B.Jr. : Microbiology,Harper & Row,N.Y., 394 (1973).

41. Sanarelli, G. : Ann. Inst. Pasteur 38. 11-72 (1924).

42. Bennett, I.L. : J. Exp. Med. 88. 267-278 (1948).

43. Kind, L.S. : J. Immunol. 70. 411-420 (1953).

44. Abernathy, R.S., Bradley, G.M., Spink, W.W. : J. Immunol. 81. 271-275 (1958).

45. Suter, E., Ullman, G.E., Hoffman, R.G. : Proc. Soc. Exp. Biol. Med.: 99. 167-169 (1958).

46. Gledhill,A.W.:J.Gen.Microbiol.18.Proceedings xvii (1958).

47. Barlow, J.L. : Fed. Proc. 22. 628 (1963).

48. Pirsch, J.B., Mika, L.A., Van der Maaten, M.J. : Proc. Soc. Exp. Biol. Med. 96. 376-380 (1957).

49. Smith, R.T., Thomas, L. : Proc. Soc. Exp. Biol. Med. 86. 886-809. (1954).

50. Jutila, J.W. : J. Inf. Dis. 128. S.99-S.103 (1973).

51. Salvin, S.B. : J. Lab. Clin. Med. 61. 1004-1022 (1965).

52. Filkins, J.P. : Proc.Soc.Exp.Biol.Med. 140. 212-215(1971).

53. Johnson, K.J., Ward, P.A.: J.Immunol.106.1125-1127 (1971).

54. May,J.E.,Kane,M.A.,Frank,M.M.:J.Immunol.109.893-895(1972).

55. Wong, T-C. : Amer. J. Obstet. Gynecol. 84. 786-797 (1962).

56. Nowotny, A., Golub,S., Key, B.: Proc. Soc. Exp. Biol. Med. 136. 66-69 (1971).

57. Bruce,L.,Hancock,B.W.,Richman,J.: J.Physiol.269.48P(1976).

58. Klostergaard, J., Evans, J.T., Fonte, G.C.A. : J.Med. 11. 49-63 (1980).

59. Fahey, J.L., Bronsman, S., Dorey, F. : Cancer Res. 37. 2875-2878 (1977).

60. Hausman, M.S., Bronsman, S., Snyderman, R., Mickey, M.R., Fahey, J.L. : J.Nation. Cancer Inst. 55. 1047-1054 (1975).

61. Atkinson, K., Kay,H.E.M., McElwain, T.J. : Brit. Med. J. iii. 160-161 (1974).

62. Deinard, A.S., Fortuny, I.E., Theologides, A., Anderson, G.L., Boen,J., Kennedy,B.J. : Cancer 33. 1210-1218 (1974).

63. Rodriguez, V., Burgess, M., Dodey, G. P. : Cancer 32. 1007-1012 (1973).

64. Shin, H.S., Snyderman, R., Friedman, E., Mellors, A., Mayer, M.M. : Science 162. 361-363 (1968).

65. Kodama, M., Mizukuro, T., Yamaguchi, N., Yoshida, S., Katayama, Y. : Bacteria and Cancer. Pulverer, G., Jeliaszewicz, K., Roskowski, W. (Eds.)., Academic Press, London. 167-179 (1982).

EFFECTS OF IMMOBILIZED LIPOPOLYSACCHARIDE ON THE Vx2 TUMOR-BEARING RABBITS .

Tohru Tani, Totaro Oka, Kazuyoshi Hanasawa, Yoshihiro Nakane,
Masashi Kodama
First Department of Surgery, Shiga University of Medical Science School of Medicine, Seta Otsu-Shi SHIGA, 520-21 JAPAN

Introduction

The structural component of the bacilli with pyrogenic activity had been clinically suspected to have an anti-tumor effect(1). As a matter of fact, the results of the clinical trial by W.B.Coley(2)(3) indicated an anti-tumor effect of this component of the bacilli. Subsequently the active component of bacilli was found to be lipopolysaccharide(LPS) by many researchers. The mechanism of an anti-tumor effects was argued in many ways as: an action of the T and B lymphocytes, an effect of adjuvant(4), the inducing effects of Interferon(5) and Tumor Necrosis Factor(TNF)(6)(7). Furthermore, many different opinions about the active site of LPS on an anti-tumor activity have been reported. Homma reported that active site of LPS resides in the protein portion of that of Pseudomonas aeruginosa(8). On the other hand, E.Ribi reported that it resides in lipid A as well which had almost all other biological activities(9). Therefore, LPS was not introduced in clinical use, because it had the strong toxicities for the living body. The studies and trials to introduce the anti-tumor effects of LPS into the clinical use was only the following report by E.Ribi et al. He showed the anti-tumor effect of the fractions of Lipid A on the line-10 tumor of guinea pig by injecting the fraction on the tumor itself with trehalosedimycolates(TDM) or acetone precipitate of chloroform methanol(ACP), without toxicities(10). But in this use, LPS

Immunopharmacology of Endotoxicosis

was less effective than BCG used independently. The materials induced by LPS, the tumor necrosis factor(TNF), was reported by K. Haranaka who proceeded with its separation and purification(11). TNF is also of clinical without toxicity. We invented a novel fiber on which LPS was immobilized. If this fiber is used in place of LPS, the contact with LPS and many immunogenic cells or plasma can be controlled freely without dispersing LPS into the body. Here, the sterilization, the safety and a few biological activities of the LPS immobilized fiber are desclibed. After the effect of LPS on the Vx2 tumor was examined, an anti-tumor efficiency of the LPS immobilized fiber was investigated by the direct hemoperfusion in the Vx2 tumor-bearing rabbits.

Materials and Methods

A novel LPS immobilized fiber(LPS-F) was obtained by adding a condensing agent for peptide synthesis to a mixture of the insoluble carrier and an aqueous solution of the LPS. LPS was extracted from Escherichia Coli 055 :B5 with trichloroacetic acid (Difco Laboratories), which contained 10.7% protein. The concentration of the LPS in the immobilizing mother solution and the washing solution were determined by phenol/sulfuric acid method. The amount of immobilized LPS was estimated from that of LPS remaining in the mother solution. The density of LPS-F was 3.7mg/g. Bloodtreating column for DHP on the tumor-bearing rabbits was prepared by filling 0.8g of bundled LPS-F. Experiment I: a) 0.5meter of LPS-F was washed with 30ml saline and the concentration of LPS in this solution after washing was estimated by PYRODICK TEST. In two out of 19 samples LPS-F were treated with γ-ray(2.5 Mrad). b) 0.5meter LPS-F was incubated at 37°C for one hour with 9ml serum of rabbits. Then this activated serum(0.5ml) was injected into the vein of the ear of rabbits. Body temperature was monitored. c) To test toxicity of LPS-F 1ml of human serum treated in the following 3 different ways was injected intravenously to ICR mice and their mortality was examined at 48 hours : by incubation at 37°C for 1 hour with 0.5meter, or 0.25meter LPS-F, and the control group. d) To test the complement activation by LPS-F, the following parameter was measured in the human serum incubated with either carrier fiber only, LPS-F, or LPS-F treated with γ-ray. C3, C4, C5 and C3 activator were measured by the immunodiffusion method, and CH50 by Mayer's method.

Experiment II; 1) Tumor-bearing rabbit was obtained by intradermal inoculation of $4x10^5$ Vx2 tumor cells into the back of NZW white male rabbits. Tumor grew into a size of about 7 mm in diameter 7 days after inoculation, 12-30 mm in diameter by 14 days thereafter. LPS (E.Coli 0111 :B4, Difco Laboratories) was injected intravenously 108 - 200 mcg/kg on 7th or 14th day. The mortality of the rabbits and tumor growth curve were observed. Macroscopic and microscopic findings of tumor tissue were studied immediately after injection. 2) Direct hemoperfusion with the column containing 0.8g of LPS-F was conducted on Vx2 tumor-bearing rabbits(2200 - 3000g) obtained in the same way as 1). In the first group, eleven rabbits were treated with DHP 4 - 5 days after inoculation. In the second group, 10 rabbits were done 7 - 11 days after inoculation. DHP was performed via carotid artery and jugular vein cannulation under pentabarbiturate anesthesia (i.p). Flow diagram is displayed in Fig. 1. Priming volume was about 18 ml. After pre-perfusion with 200 ml saline containing 2000 U heparin, DHP was started at a flow rate of 15 ml/min for 90 minutes. Initial Heparinization was 500 U/kg because the carrier adsorbed 12000 U/g. Tumor size and body weight of the rabbits were monitored thereafter. Tumor size was calculated by the formula: the diameter of the tumor=$\sqrt{L1xL2}$, L1 and L2 means the longest diameter of the tumor and its rectangular one respectivly.

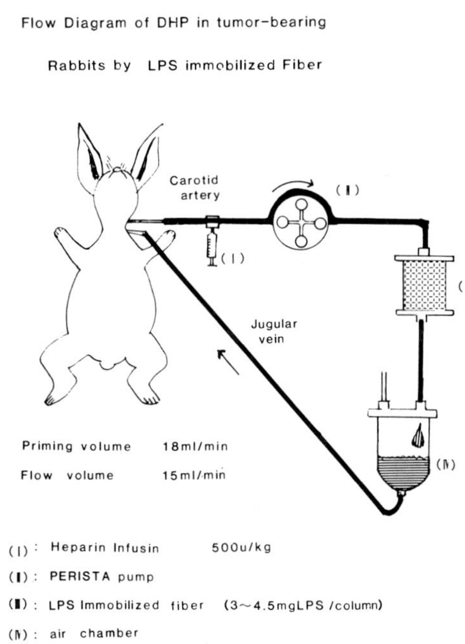

Fig 1: Blood flowdiagram.

Results

Fig 2:

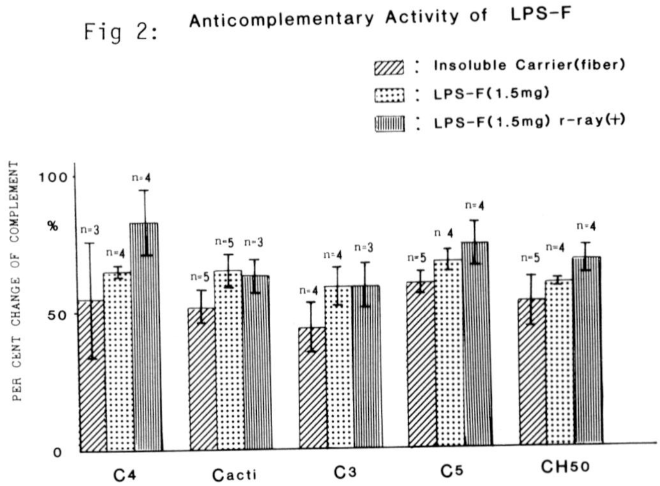

100% means the value of complements in control serum.

Experiment I: a) LPS was not detected in the saline solution after washing the γ-ray treated LPS-F. Less than 0.4ng/ml of LPS was found in saline after washing in the treated LPS-F. No increment of LPS concentration in the serum was observed after human serum was observed after human serum was incubated with LPS-F. Based on these results no significant dissolution of LPS from LPS-F was present. b) Rabbits did not show an elevation of body temperature. Immobilized LPS on the fiber did not have pyrogenicity. c) Mortality was 10 %(1/10) in the control group, 17 %(1/6) in 0.25 meter LPS-F and 40 %(2/5) in 0.5 meter LPS-F group. d) Concentration of C4, C3, C5 and C3 activator decreased after the incubation in all three different ways. General trend was that of decrement being largest in the fiber only, treated group followed by LPS-F, and LPS-F treated with γ-ray. The activity of CH50 showed the same results(Fig2). Experiment II: 1) When 100 mcg/kg of LPS was injected 7 days after inoculation, 4 out of 8 rabbits died within 24 hours in the 2 out of 4 remissions. When 108 - 200 mcg/kg of LPS was injected, 2 weeks after inoculation 6 out of 10 rabbits died and 2 out of 4 rabbits had complete remission. Upon

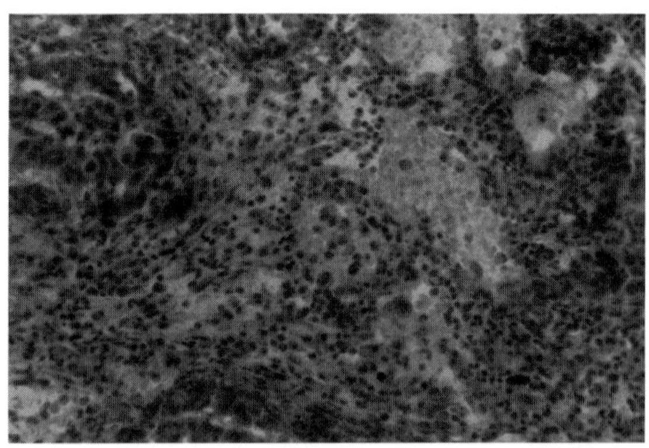

Fig 3: Focal hemorrhage & necrosis in the tumor after LPS injection.

microscopic examination of the tumor tissue of rabbits died from shock, the focal hemorrhage around the tumor was observed (Fig 3). In the first group which underwent DHP 4 - 5 days after inoculation, general tendency that DHP suppressed the growth of the tumor was found(Fig 4). In one examination(the pair of solid and open triangles), DHP were signifcant statistically(p<0.01). But in another examination(the pair of solid and open circles), since some rabbits failed the inoculation for, all cases of remission in DHP treated ones were rejected statistically. In the comparison of body weight of the rabbits, the variance of change in the body weight was great. No specific trend of the change was found, but body wéight of the DHP treated rabbits decreased temporarily for 10 days after treatment. In group 2, which underwent DHP at 7 - 11 days after inoculation, a similar trend to the 1st group was found(Fig 5). This group also had two experiments like the 1st group. In one experiment(the pair of solid and open circles), tumor growth curves were significant statistically(p<0.01). The

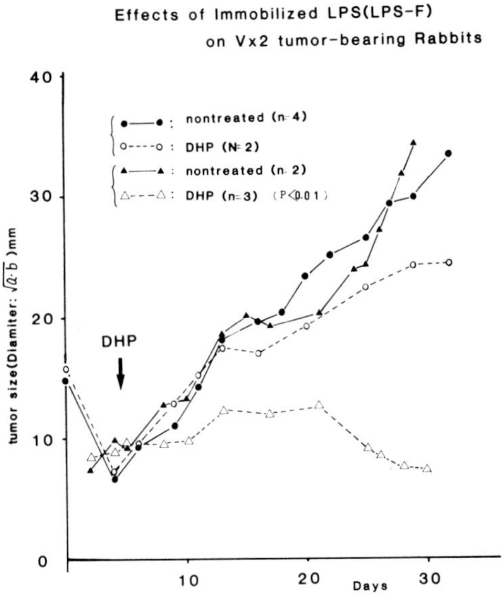

Fig 4: The results of 1st group mean tumor size = a.b
a=short diamiter, b=long one

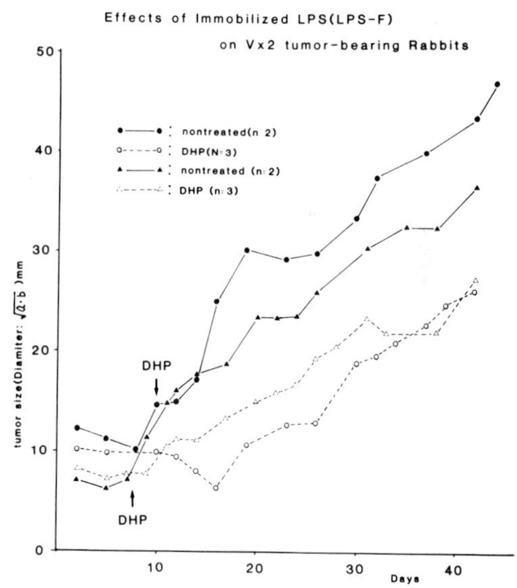

Fig 5: The results of 2nd group.

microscopic findings of the tumor tissue of rabbits treated by DHP revealed focal hemorrhage similar to the one finding in the 1st group(Fig 6).

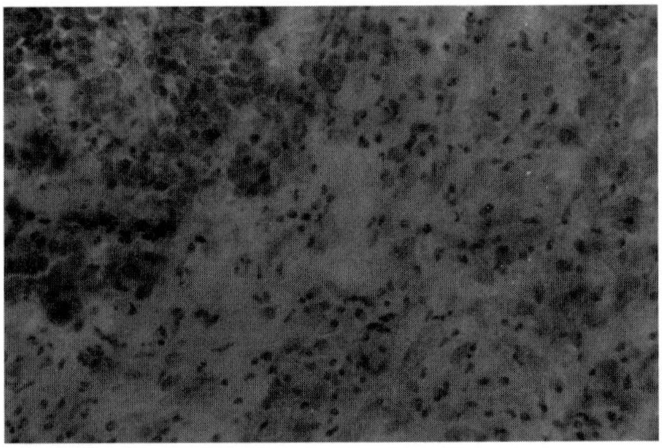

Fig 6: Focal hemorrhage around the tumor after DHP treatment.

In both groups, 4 rabbits underwent DHP which contained only insoluble carrier; 2 out of them were found to be more advanced in metastasis than nontreated ones. Finally, 25 rabbits were treated with DHP. Two rabbits died after DHP. Mortality directly related to DHP was absent. There were 2 technical trables with the DHP: the obstruction by blood clot and dislodgement of the cannulation. Both of them managed subsequent sequela. Coagula in the column were found sometimes.

Discussion

Up to date, the trials to introduce the LPS into clinical use without toxicities have followed the following three courses. First, it aimed to separate the toxic site and an anti-tumor site contained in LPS(9)(10). Next, it was aimed to induce an active material by means of LPS inoculation in to other animals(11). Finally, it was hoped to make an Endotoxoid (12). In all three courses, there was no idea to immobilize the LPS. There was only one report like ours that LPS ought to be applied to the living body directly(13). He immobilized LPS to Epoxy activated Sepharose 6B, and his material was more weak than that of free LPS in the immunogenic and immunoadsorptive activities, and could not be used practically. The novel LPS-F had slightly weaker anti-tumor and anticomplementary activities, but no pyrogenicity. Mouse lethal activity persisted a little. From our studies it was that the direct activity of LPS was reduced because of the immobilization. When LPS- F was treated with 2.5 Mred of

γ-ray, its anticonplementary activity decreased(14), because LPS from E. Coli was used in the LPS-F. DHP treatment is continuously able to produce a controllable quantity of activated materials in the body directly. In this method, therefore, even the activated material with significantly short half life can act effectively. But larger animals are necessary for DHP in the experiments. Moreover Mori et al published that LPS induced local Schwarzman reaction in the Vx2 tumor. The results of LPS injection on the tumor-bearing rabbits showed the same microscopic hemorrhage around the tumor as DHP treatment, with 50 % complete remission. Sensitivity to LPS of the tumor-bearing rabbits also showed significant increment(15 - 30 times of normal rabbits). The tumor-bearing rabbits underwent DHP perfusion safely without dying from endotoxin shock. Blood pressure and blood gas analysis did not show shock state. The counts of platelets, white blood cells and lymphcytes showed ordinary change at the extracorporeal circulation. DHP treated rabbits suppressed and/or remitted Vx2 tumor except the cases of failure of inoculation. The appropriate day of DHP application was not concluded. The mechanism of the anti-tumor effectiveness of the LPS-F could not be concluded clearly, but the microscopic findings of the tumor tissue after DHP were similar to those of free LPS injection. Perhaps its mechanism is the same as injection of free LPS. Hereafter, new design for an increased sensitivity to LPS-F and TNF induction by inoculation of BCG(15) ought to be planned. More dense immobilization of LPS, or immobilization of other kinds of LPS, will be examined in future.

Conclusions

1) LPS(E. Coli 055: B5) could be immobilized to insoluble carrier(fiber) which could be used in direct activation of the plasma and serum; its density of LPS was 3.7 mg/g (LPS-F).

1) Elution of LPS from LPS-F was not significant.

1) LPS-F had no pyrogenicity but weak anticomplementary activity.

1) LPS-F showed an anti-tumor activity in Vx2 tumor of rabbit when they were treated with DHP.

1) LPS-F was safe in DHP treatment.

1) Mechanism of anti-tumor activity of LPS-F was suspected to be the same as that of free LPS.

References

1. Bush, N. N.: Berlin Klin. Wochschr. 3, 245 (1866).
2. Nauts, H. C., Swift, W., Coley, B. L.: Cancer Res. 6, 205 (1946).
3. Nauts, H. C., Fowler, G. A. A and Bagatko, F. H.: Acta. Med. Scand. Suppl, 1-103 (1953).
4. Condie, R. M. et al.: Federation Proc. 14, 459 (1955).
5. Wagner, R. R. et al.: J. Immunol. 83, 87 (1959).
6. Carswell. E. A., Old, L. J. et al.: Proc. Natl. Acad. Sci. USA. 72, 3666-3670 (1975).
7. Green, S.: J. Natl. Cancer. Inst. 59, 1519-1522 (1977).
8. Ken-ichi, T., Chiyoji, A., Yuzuru, H.: J. Biochem. 83, 711-718 (1978).
9. Viiai, N. Nigam.: Cancer research. 35, 628-633 (1975).
10. Kuni, T., Edgar, R. et al.: Cancer research. 41, 2654-2657 (1981).
11. Katsumasa, H. et al.: Nippon Rinsho. 40, 186-193 (1979).
12. Nowotny, A.: Nature. 197, 721 (1963).
13. J. Fox., K. Hechemy.: Infection and Immunity. 20, 867-868 (1978).
14. Joseph, J. P., Y. Chang , El-Bisi, H. M.: J. Bacteriology. 93, 1607-1614 (1967).
15. Emanuel Suter. et al.: Proc. Scc. Exp. Biol. and Med. 96, 376 (1957).

THE ACTIVATION OF COMPLEMENT SYSTEM BY BACTERIAL ENDOTOXINS AND ITS EFFECT ON MALIGNANT TUMOR TISSUES OF MICE.

Yutaka Katayama

Department of Microbiology, Kyoto Prefectural University of Medicine
Kyoto, Japan

Nozomi Yamaguchi, Shun-ichi Yoshida

Department of Preventive Medicine, Kyoto Prefectural University of Medicine
Kyoto, Japan

Tomoyuki Mizukuro, Masahiro Horisawa, Masashi Kodama

Department of Surgery, Shiga University of Medical Sciences
Ohtsu, Shiga, Japan.

Introduction

The phenomenon of hemorrhage and necrosis of solid malignant tumour tissues by the i.p. injection of bacterial endotoxins (LPS) was found by Gratia and Linz (1). This phenomenon was confirmed as the constant finding for malignant tumours by many investigators (2 -18), especially by M.J.Shear's group, who had shown that the LPS injection regularly caused the selective hemorrhage and necrosis in chemically induced malignant tumours of 750 mice (7).

This phenomenon is highly selective in that it affects only malignant tumour cells, suggesting the existence of a tumour-specific and common substance(s) which will determine the susceptibility for the LPS action in all malignant tumour cells.

The study of this phenomenon may help clarify the nature

Immunopharmacology of Endotoxicosis

of the tumour cell and could lead to the finding of an effective therapy for cancer patients (19-22). In this report, the data will be shown which supports the assumption that the participation of the complement component(s) is essential for this phenomenon.

Antitumour Activity of the LPS

We have used two systems, MH134 hepatoma of C3H/He mice and MC -B1 fibrosarcoma of BALB/c mice which is established in our laboratory by the injection of methylcholanthrene. The C3H/He mice were inoculated with one million ascitic MH134 hepatoma cells intradermally in the abdomen or hind region, and the tumour reached about 7 mm in diameter seven days after inoculation,and all the mice died by the 40th day. When 50 or 100 μg of E. coli O127 LPS (Westphal or Boivin) was injected intraperitoneally at day 7, hemorrhage and necrosis were observed only at the tumour site without exception, and about 50 % of the mice were cured completely. All the control mice injected with saline died by the 40th day (Fig. 1). The cured mice showed specific immunity to inoculation of MH134 tumour cells of 5 million intradermally and this specific immunity was able to be adoptively transfered to the syngeneic mice by using splenic cell suspensions. When the LPS was injected from one to four days after the tumour implantation, no visible changes were observed in tumour tissues.

The similar finding was obtained also by MC-B1 fibrosarcoma - BALB/c mice system. The BALB/c mice were inoculated with half million MC-B1 cells intradermally which were maintained in tissue culture. The tumour reached about 9 mm in diameter 12 days after tumour inoculation and all the mice died by the 65th day. When 50 μg of E. coli LPS was injected intraperitoneally on day 12, the selective hemorrhage and necrosis of solid tumour tissues were observed in 100 % of the

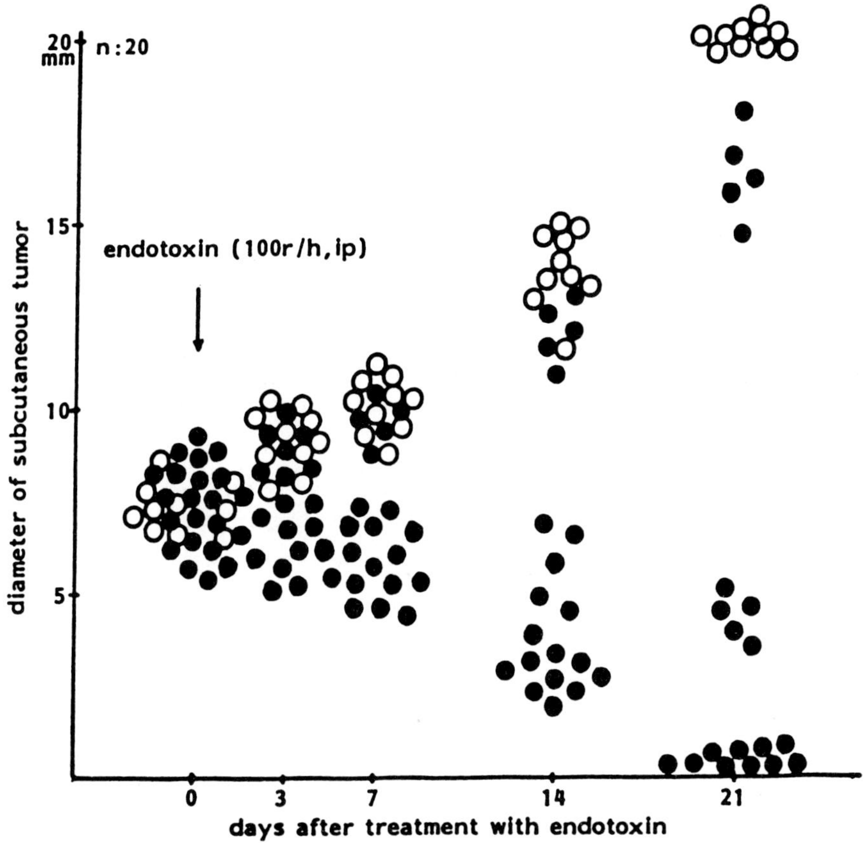

Fig. 1 The effect of LPS on MH 134 Hepatosarcoma of C3H/He mice. Mice were injected with 100 μg of E. coli LPS (●) or with saline (○) intraperitoneally.

mice and the complete cure of the tumour was obtained in 30 % of the mice. All control mice injected with saline died by the 65th day. When the LPS was injected several days after the tumour inoculation, the tumour hemorrhage was not observed

By the histological examinations, we observed platelet thrombus and hemorrhage at the periphery of the tumour mass at 1 h, hemorrhage and marked infiltration of neutrophils in the tumour tissues at 4 h, degeneration and necrosis of the tumour cells at 12 h, and infiltration of monoculear cells at 24 h.

These changes were completely similar with the necrotic

vasculitis in the local Shwartzman phenomenon (3, 5, 9, 23) and in the Arthus reaction, suggesting that these phenomena share some pathogenetic processes in common, and that the neutrophils are the major effector cells in these phennomena.

The Participation of the Complement System

A. Historical analysis

The hemorrhage and necrosis of tumour tissues after i.v. or i.p.injection of the LPS is very similar to the local Shwartzman phenomenon. Since the histological manifestation of both phenommena are the same, and both are provoked by the inoculation of the bacterial LPS. The tumour hemorrhage can also be induced by the injection of antigen-antibody complex, or of agar (3) or of glycogen (9), all of these substances are known to have the capacity to provoke the local Shwartzman phenomenon. These substances have the common capacity to activate complement and to form anaphylatoxin (25-29). These relationships are tabulated in the Table 1, which shows the year when these activities were reported.

These facts may suggest the participation of the complement component in tumour hemorrhage and necrosis after the LPS injection.

It is well established that the LPS does not accumulate in the tumour tissues after in vivo administration (30,31), and we confirmed this phenomenon using MH134 hepatosarcoma-C3H system with the peroxidase antibody method. This fact shows the existence of some mediator substance(s) of the antitumour activity of the LPS.

The existence of the antitumour activity in the post-LPS sera of mice was shown by Shear's group (32), and by Nowotny's group (33). The existence of an antitumour substance, tumour necrosis factor (TNF), was reported by Old's group in the sera of mice primed with B.C.G., Propionibacterium acnes or zymosan

Table 1

The ability to provoke the local Shwartzman phenomenon, the hemorrhage and necrosis in malignant tumour tissues, and to activate complement (Anaphylatoxin formation)

	Local Shw.rn	Tumor hem.	C activn
Endotoxin (LPS)	1928	1931	1968
Antigen-antibody complex	1931	1933	1911
Agar	1931	1933	1913
Starch	1935		1913
Glycogen	1943	1951	1913
Kaolin	1949		1911
Dextran	1952		1954
Protein A		1978	1972

The number shows the year of report.

10 to 20 days before and then injected with LPS (34-37). The TNF was shown to have not only a direct cytotoxic activity on some malignant tumour cells in vitro but also the selective antitumour activity in vivo (34-42). The priming injection of B.C.G., zymosan or P. acnes was reported to be essential for the production of the TNF (34-36), since without priming the TNF was not detected completely even by the injection of lethal amount of the LPS (42). As observed in the previous section, tumour hemorrhage and necrosis, and even complete cure, were always observed without the priming of mice with B.C.G., zymosan or P. acnes (20-22).

This fact shows that the TNF is not the major mediator, even if it has a strong antitumour activity, and that another mediator substance(s) is required for this phenomenon. From these, we have assumed that : (i) the complement is essential for the antitumour activity of the LPS; (ii) the LPS action is mediated by the complement component and; (iii) the activated complement fragment(s) is the major mediator of this phenomenon.

B. The effect of anticomplement substances.

To test the assumption that complement is essential for the antitumour activity of the LPS, cobra venom factor (CVF), zymosan or silica was administered intraperitoneally tumour-bearing mice 24 h before the LPS injection. CVF of Naja naja Kaouthia (Sigma Chem.Co., St.Louis,Mo,USA) was purified using DEAE-cellulose (43). BALB/c mice inoculated with MC-B1 fibrosarcoma cells 11 days previously were inoculated with 0.2 mg of CVF. With this dosage of CVF, no visual change in tumour tissues was observed in any of the mice, but CVF inoculation suppressed the tumour hemorrhage induced by the LPS injection 24 h later in 5 mice out of 7; the extent of hemorrhage in two mice was very weak and no complete cure of the tumour was obtained after the LPS injection.

The effect of zymosan (Sigma Chem.Co,St.Louis,Mo,USA) was investigated similarly. Intraperitoneal injection of zymosan (100 mg) one day before the LPS injection suppressed the tumour hemorrhage after the LPS injection in 6 mice out of 10, hemorrhage in 4 mice was very weak, and complete cure of tumours was not observed in any mouse tested.

The i.p. injection of silica (5 mg) 24 h before the LPS injection also suppressed tumour hemorrhage after LPS injection in 5 mice out of 10 , the degree of hemorrhage in positive mice was weaker, and no complete cure was obtained in any mouse tested.

The anticomplementary substance K76COONa, whose major effect is on the C5 step of the complement cascade (44), extracted from Stachybotrys complementi (supplied by Ohtsuka Chem.Co.,Tokushima,Japan) in 10 mg amounts was inoculated intraperitoneally for 3 consecutive days and 24 hours after the last injection, E.coli LPS (100 μg) was injected in tumour bearing mice. As observed in Fig.2, tumour hemorrhage and necrosis were suppressed when compared with the tumour mice injected with the LPS without pretreatment with K76COONa, as in Fig.1.

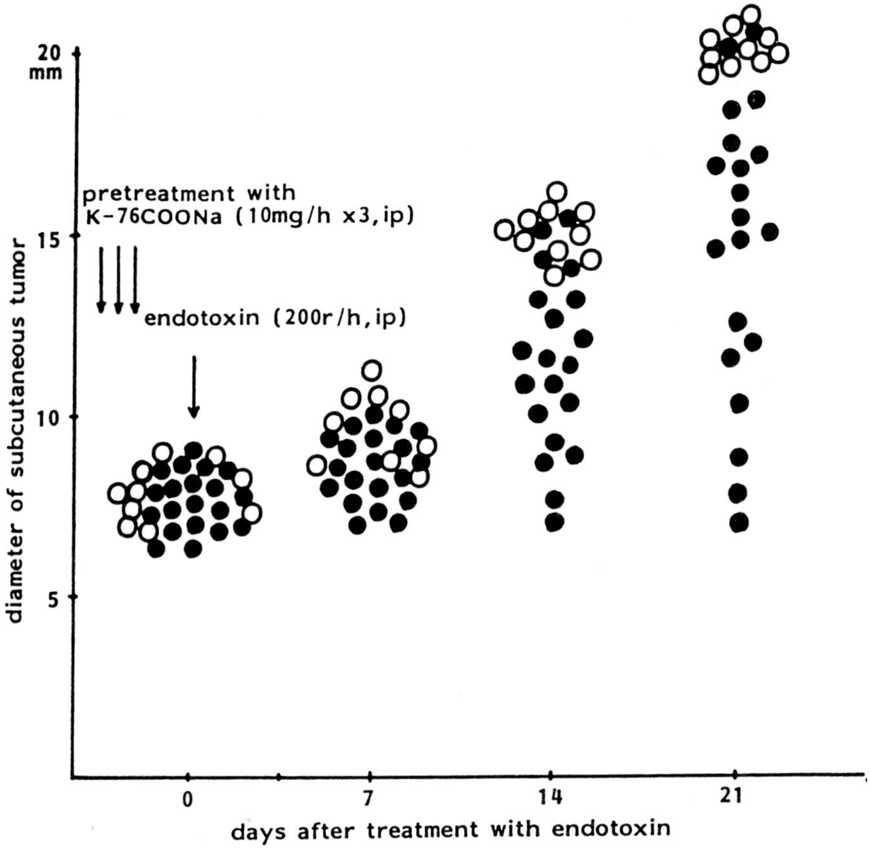

Fig. 2 The effect of K76COONa upon the antitumour activity of the LPS. MH134 Hepatosarcoma-inoculated C3H/He mice were injected i.p. with 10 mg of K76COONa for 3 consecutive days. Twenty-four h later, the mice were inoculated with 200 μg of E. coli LPS (●) or with saline (○).

From these observations, it may be concluded that complement is essential for the full expression of LPS activity on solid malignant tumour tissues.

To test the assumption that the antitumour activity of the LPS is mediated by complement, a large amount of CVF was administered intraperitoneally to the tumour-bearing mice. When 0.6 mg of CVF was administered to 9 tumour-bearing BALB/c mice one mouse died within 4 h, and the remaining 8 animals showed strong hemorrhage only at the tumour site, but 5 of them died within 24 h; 0.4 mg of CVF caused selective hemorrhage in the solid malignant tumour tissues of C3H/He mice in 4 out of 7 and no mice died by this dose. The histological observation of the tumour after the i.p.injection of CVF showed hemorrhage and neutrophil infiltration in the tumour tissues which are very similar to that observed after the LPS injection.

The hemorrhage in Sarcoma 37 in mice inoculated with the moccasin venom was observed by Shimkin and Zon in 1942 (45) but its meaning has not been discussed since then.

The CVF (27,28,46,47) and the LPS (48) have the ability to activate complement and this is the only biological activity shared by these two substances. Therefore, it suggests that activated complement can cause hemorrhage in solid malignant tumour tissues of the mice.

The antitumour activity of the activated complement component

The third assumption that the activated complement component (s) has the antitumour activity was tested.

Fresh human or mouse serum was mixed with one hundredth volume of yeast cells, or one twentieth volume of soft agar, or with solid endotoxin, made by coupling the LPS to the polymethylmetacrylate fiber (1.5 mg of E. coli LPS (Boivin) per one meter of the fiber), in the ratio of 0.5 mg LPS per 10 ml of the serum, and incubated at 37C for various length of time with or without the addition of epsilon aminocaproic acid (EACA) to the final concentration of 1 M. For controls, fresh mouse or humana serum was incubated at 56 C for 30 minutes and then treated in parallel with these activating agents.

As a preliminary test, the effect of human serum treated with baker's yeast at 37 C for 45 min without EACA was tested on MH134 tumour of about 7 mm in diameter in C3H/He mice by injecting 3 ml of the serum intraperitoneally on five consecutive

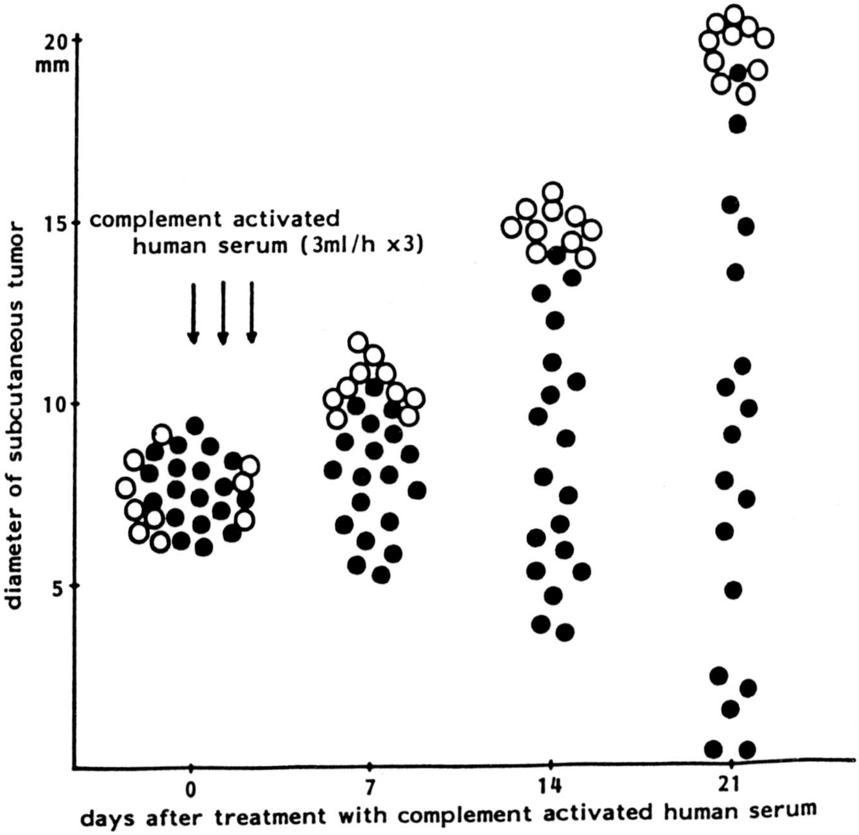

Fig. 3 The effect of the complement-activated human serum on mice tumour. The fresh human serum was mixed with 1/20 volume of 0.3 % soft agar and incubated at 37 C for 120 min and the agar was removed by centrifugation. The serum was diluted 5-fold with sterile saline. Three ml of the diluted activated serum was injected i.p. to the C3H/He mice carrying MH134 tumour for 3 consecutive days (●). The control mice (○) were injected with the human serum pretreated at 56 C for 30 min and then treated same as the test material.

days. Suppression of tumour growth was observed in all mice but there were no complete regressions. Similarly, treating the control human serum preheated at 56 C for 30 min and then with baker's yeast did not have any effect. The i.p.injection of activated mouse serum similarly treated, or of the activated human serum with the addition of EACA, did not show any better

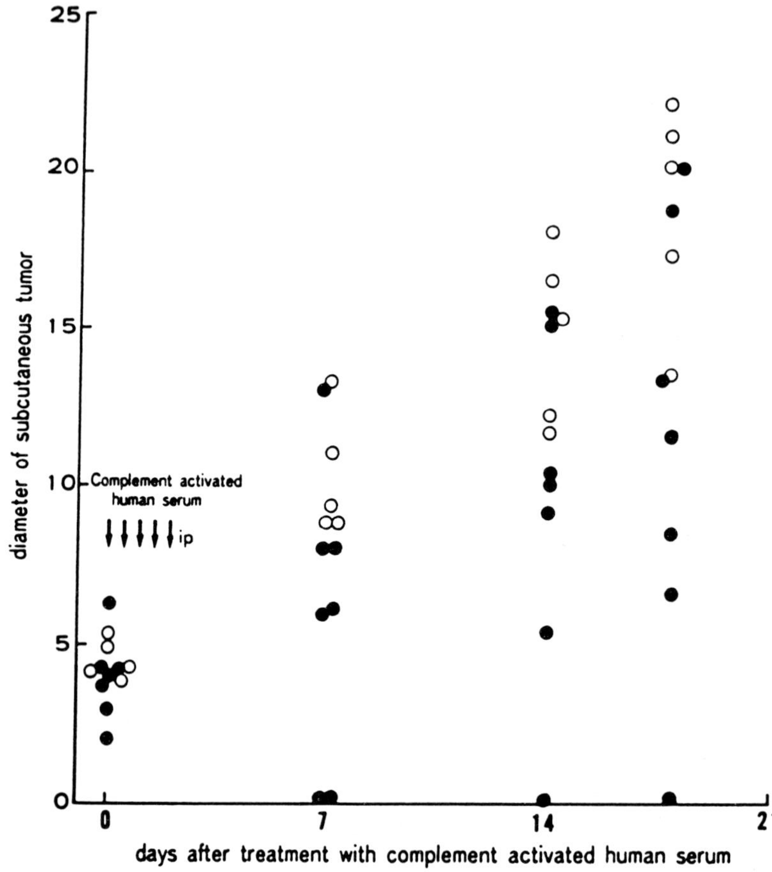

Fig. 4 The effect of the fresh human serum activated with the solid endotoxin on mice tumour. The fresh human serum was mixed with the solid LPS and incubated at 37 C for 120 min. The activated serum was diluted 6-fold with saline. Three ml of the diluted serum was injected i.p. to the BALB/c mice carrying MC-B1 tumour of about 7 mm diameter 5 times in one day (●). The control mice (○) were injected with the human serum pretreated at 56 C for 30 min and then treated same as the test material.

results than those without EACA.

The fresh pig serum was activated by soft agar or by solid endotoxin at 37 C for 180 min without EACA and tested on MC-B1 tumours in BALB/c mice of about 7 mm in diameter. With i.p. injection of activated pig serum of 3 ml for 3 consecutive days, ten in ten mice died, showing strong hemorrhage in the liver, kidney, spleen and pancreas. This experiment suggested that enough active material did not enter the blood stream but remained intraperitoneally, and therefore, further experiments were carried out by i.v. injection of the activated serum or by i.p. injection of the activated serum diluted 3 to 6 times by sterile saline.

The i.p. injection of 0.5 ml human serum activated with soft agar without EACA for 120 min and diluted to 3 ml with sterile saline for 3 consecutive days in MH134 tumour-bearing mice caused hemorrhage in the central portions of solid tumour tissues and the tumour growth was suppressed. Complete cure was obtained in two among 20 mice tested and all control mice administered the human serum pretreated at 56 C for 30 min and then incubated with soft agar, died by the 40th day (Fig.2).

The human serum (0.5 ml) activated with solid LPS at 37 C for 180 min without EACA and diluted to 3.0 ml with sterile saline was injected intraperitoneally five times the same day for mice bearing MC-B1 tumour of about 5 mm diameter; the hemorrhage was observed in solid tumour tissues and cure was obtained in two mice out of seven , but later the tumour reappeared in one mice as shown in the Fig.4. The administration of the control human serum pretreated at 56 C for 30 min and treated in parallel did not show any effect on tumour tissues of the mice and all control mice died eventually.

The i.p. injection of the activated and diluted human serum (3.0 ml) for 5 consecutive days showed suppressive effect on MC-B1 tumour tissues but the effect was weaker when compared with the tumour mice injected with the same amount of the activated and diluted serum (3.0 ml) five times in one days as described above.

The fresh hummman serum (about 3 ml) activated by the solid LPS at 37 C for 180 min was concentrated to about one fourth volumes by using polyvinyl alclohol and injected intravenously in MC-B1 tumour-bearing BALB/c mice. The hemorrhage was observed in the central portion of the solid tumour tissues in 3 mice out of four. Necrosis and scar formation of the tumour tissues was evident in these mice but the malignant tumour resumed from the peripheral portions and all the tumour mice died eventually. These results were confirmed repeatedly. The control serum pretreated at 56 C for 30 min and then treated in parallel with the solid LPS did not show any effect on the tumour tissues and the all control mice died by 40th day. Histological examination showoed hemorrhage in small vessels and massive infiltration of the neutrophil leucocyte at the peripheral portion of the necrotic tumour masses and these changes were essentially similar to that after the LPS or the CVF injections but the time needed for the appearance of these changes was much slower requiring two or three days.

From these data, it may be concluded that complement activated human serum can suppress the growth of the solid tumours of mice and even cause a complete cure. The control serum pretreated at 56 C for 30 min and treated with the activation procedures in parallel had no activity on the tumour mice and all the control mice died eventually.

The activation procedures are performed carefully to avoid bacteriaal contamination. The contamination may be excluded by the facts that the control serum preheated at 56 C for 30 min and treated in parallel with the test serum did not show any antitumour activity in any experiment. The i.p. injection of 50 μg of E. coli LPS on the 12th day after MC-B1 tumour implantation caused tumour hemorrhage in all mice but the injection on the 3rd day did not induce any effect on tumour tissues and inhibited the tumour hemorrhage and necrosis by the LPS injection on 12th day in all mice tested. This phenomenon of a kind of tolerance was observed in the case of the LPS injection but was not observed for the activated serum as described above. This fact may not only neglect the contamination of the LPS in the activated serum but may also suggest the clinical applicability of the activated serum.

Discussion

TNF problem.

Tumour necrosis factor (TNF) was reported to be produced by the LPS administration for the experimental animals which were primed with B.C.G., zymosan or Propionibacterium acnes one to three weeks befor LPS injection; the TNF was shown to have a direct cytotoxic activity on many malignant tumour cells in vitro, and also to cause the hemorrhage and necrosis of solid malignant tumour tissues of mice in vivo (34-42). It may be assumed that the tumour hemorrhage and necrosis of malignant tumour tissues by the LPS are caused by the action of the TNF, but there are several questions concerning this assumption.

a. For the production of the TNF, it is essential to prime the animals by B.C.G, zymosan or P. acnes (34-36), but the tumour hemorrhage and necrosis was always observed without these priming procedures in all experiments carried out by us (20-22) and by many other workers (1-18). It is also shown that no TNF activity is observed in the serum of the animals injected with the large amounts (100 μg) of the LPS without the priming by B.C.G., zymosan or P. acnes, but tumour necrosis was observed by the samll amount of the LPS (1 to 5 μg) without any priming (42).

b. In the case of nude mice, BALB/c nu/nu, it is reported that these animals can produce TNF when primed with B.C.G. but not with P. acnes (36). In our experiment, tumour hemorrhage and necrosis was observed in the nude mice without any priming by B.C.G.,zymosan or P.acnes., although complete cure of the nude mice was not obtained.

c. Closely similar phenomena as tumour hemorrhage by the LPS was also observed by the injection of the antigen-antibody complex, agar (3), glycogen (9), and CVF with the same histological manifestations.

d. TNF is reported to have a direct cytotoxic activity on malignant tumour cells (34-42), leading to an eventual necrosis in the tumour cells, but in the phenomenon of the tumour necrosis by the LPS , the major pathological change is the hemorrhage in small vessels and strong infiltration of the neutrophilic leucocytes, which is essentially identical with the necrotic vasculitis of the Arthus reaction, or the local Shwartzman phenomenon (3,5,9,23).

e. The direct cytotoxic agents like TNF might be assumed to be more effective on the free ascitic tumour cells than the solid tumour masses, but the effect of the LPS is much more marked on the solid tumour tissues than on the free ascitic cell tumours.

f. The cytotoxic agents might be much more effective on the minimum size tumour tissues in early stage after tumour implantation rather than on large tumour masses, but the LPS can not cause the tumour hemorrhage, necrosis and complete cure during the first several days after the tumour inoculation (7,16,19).

These several points are difficult to explain by the activities of the TNF produced after the LPS injection. We propose that the antitumour activity of the LPS is mainly mediated by the activated complemnt fragment(s), which is known to be mobilized by many drugs. There are some observations which support this assumption.

a. Many microorganisms which have no LPS are reported to have the antitumour activity similar to the LPS and cause the accumulation of neutrophils in malignant tumour tissues. *Saccharomyces cerevisiae*, *Candida*, *Staphylococcus*, *Streptococcus* and *P. acnes* are included in the group of materials (49-53), whose complement activating capacities are well established (28.54-57).

b. There are many macromolecular polysaccharides including zymosan which have been reported to have the antitumour activity (58-62). Most of these reagents are reported to have the complement activating capacity and vice versa (63,64).

c. Extracorporeal circulation of the tumour in dogs, or in patients with Staphylococcus aureus Cowan I, can cause selective hemorrhage and necrosis of malignant tumour tissues which is similar to the effect of the bacterial LPS macroscopically and microscopically (65-68). This procedure has been assumed to remove the inhibitory antibody for the specific cell-mediated immunity against malignant tumour cells. By this procedure, the complement titer decreased and C3 precipitated to the tumour tissues (66).

The complement activating capacity of the Protein A is well established (69), and in our preliminary experiment, the fresh mice serum treated with Sepharose coupled Protein A had antitumour activity on the MH134 tumour of C3H/He mice. The serum obtained from the tumor mice was active, but serum obtained from normal healthy mice treated by this procedure were also found to have antitumour activity (Tanaka et.al. unpublished). This may mean that the main point of this procedure is not to remove something from the tumour animals but to produce something new, and the activated complement fragment(s) may be one of the candidate for this active principles.

Nature of the malignant tumour cells

It is well established that the LPS injection can cause the selecive hemorrhage and necrosis of the malignant tumour tissues without the effect on normal tissues as described above. Shear's group has shown that this phenomenon occured regularly in 750 mice whose malignant tumours were induced with the chemical carcinogen (7).

To explain the selective occurrence of this phenomenon in malignant tumour tissues, we must assume the existence of the cancer specific substance(s) in all malignant tumour cells. Based on the similarity between the tumour hemorrhage and necrosis caused by the LPS and that of the local Shwartzman phenomenon, Antopol (70), Shear's group (71,72), and our own group (73,74) have all tried to demonstrate Shwartzman active substance(s) in malignant tumour tissues with varying degrees of success. If every cancer cell has the cancer specific substance in common, it can be explained only by assuming an intimate linkage between the gene determining cancer speific substance and the gene determining the malignant behaviour of the cancer cells. It is very difficult to assume that these genes are synthesized de novo by the action of many different carcinogenic substances; therefore, it may be natural to assume a most simple model of cancer cells where these genes are kept strongly repressed in all normal cells. This leads to the conclusion of the existence of a gene determining the malignant characters.

If the gene determining the malignant character exists, and the cancer specific and common substance exist in all malignant tumour cells, then it may be possible to assume the method to diagnose and to cure the cancer by using this tumour specific and common substance as a target. The method to use the bacterial LPS or the activated complement fragment(s) may be one of them.

Summary

Antitumour activity of the *E. coli* LPS was studied using MH134 hepatosarcoma cells of C3H/He mice and MC-B1 cells of BALB/c mice. The i.p. or i.v. injection of 50 to 100 μg of *E. coli* LPS caused selective hemorrhage and necrosis of the tumour tissue of 6 to 10 mm diameter and complete cure of the tumour in about 50 or 30 % of the mice respectively.

The antitumour activity of the bacterial endotoxin was shown to require the complement, it was mediated by the complement component, and fresh human serum containing activated complement fragment(s) was shown to have the antitumour activity in these systems. The effector cells of this phenomenon seem to be the neutrophil leucocytes.

References

1. Gratia, A., Linz, R.: Compt. Rend. Soc. Biol 108, 427-429 (1931).
2. Shwartzman, G., Michailovsky,N.: Proc. Soc. Exp. Biol. Med. 29.737-741 (1932).
3. Apitz, K.: Zeitschr. f. Krkebsforsch. 40. 50-70 (1933).
4. Shwartzman, G.: Arch.Path. 21, 284-297 (1936).
5. Gerber, I.E., Bernheim, A.I.: Arch. Path. 26. 971-983 (1938).
6. Zahl, P.A., Hutner, S.H., Spitz, S., Sugiura, K., Cooper, F.S.: Amer. J. Hyg. 36. 224-242 (1942).
7. Shear, M.J., Perrault. A.:J.Nation.Cancer Inst. 4. 461-476 (1943)
8. Zahl,P.A., Starr, M.P. Hutner. S.H.: Am. J. Hyg. 41-56. (1945).
9. Stetson, C.A.Jr. ; J. Exp. Med. 93. 489-504 (1951).
10. Creech, H.J., Hankwitz, R.F.Jr.: Cancer Res. 14. 824-829 (1954).
11. Havas, H. F., Groesbeck. M. E., Donnelly. A.J. : Cancer Res..18.141-148 (1958).
12. Donnelly, A.J., Havas, H. F., Groesbeck. M.E. : Cancer Res. 18. 149-154 (1958).
13. Nowotny,A.,Golub,S.,Key,B..: Proc.Sco.Exp.Biol.Med. 136. 66-69 (1971).
14. Wasilauskas, B.L., Cameron, J.A.: Cancer 27. 217-222 (1971).
15. Parr,I.,Wheeler,E.,Alexander,P. Brit.J.Cancer 27.370-389 (1973).
16. Steinebring,W.R., Stevens,D. Procs.Soc.Exp.Biol.Med.156. 229-235 (1977).
17. Berendt,M.J.,North,R.J.,Kirstein,D.P. J.Exp.Med. 148. 1550-1559 (1978).
18. Berendt, M.J.,North,R.J.,Kirstein,D.P. J.Exp.Med. 148. 1560-1569 (1978).

19. Katayama, Y., Kodama, M. : Japn. J. Med. Sci. Biol. 33. 35-36. (1980).

20. Kodama, M., Mizukuro, T., Yamaguchi, N., Yoshida,S., Kaktayama,Y. : in "Bacteria and Cancer" . Pulverer, G., Jeliaszewicz,K., Roszkowski,W. (Eds.)., Academic Press, London. 167-179 (1982).

21. Yoshida, S., Uozumi, G., Watanabe, Y., Yamaguchi, N., Katayama, Y. : Japan. J. Med. Sci. Biol. 35. 121 (1982).

22. Kodama, M., Mizukuro, T., Yamaguchi, N., Yoshida, S., Katayama,Y: Japan. J. Med. Sci. Biol. 36. 119-120 (1983)

23. Karsner,H.T., Moritz.A.R.: J.Exp.Med. 60, 37-48 (1934).

24. Gerber, I.E.: Arch. Pathol. 21. 3311-350 (1936).

25. Friedberger, E.: Z. Immunforsch. Exp. Ther.I. 4. 636- (1910).

26. Bordet, J.: Compt. Rend. Soc. Biol. 74. 225-227 (1913).

27. Friedberger,E., Mita,S., Kumagai,T.: Z. Immun. Forsch. 17. 506-538 (1913).

28. Coca,A.F. : Z. Immunforsch. 21. 604-622 (1914).

29. Nathan.,E.: Z.Immunforsch.Exp.Ther.I.23. 204-220 (1915).

30. Seligman,A.M., Shear,M.J., Leiter,J., Sweet,B.: J.Nation Cancer Inst. : 9. 13-18 (1948).

31. Nowotny,A., Golub,S., Key,B.: Proc.Soc.Exp.Biol.Med. 136 66-69 (1971).

32. O'Malley,W.E., Achinstein,B.,Shear, M.J. : J.Nation. Cancer Inst.: 29. 1169-1175 (1962).

33. Butler,C., Abdelnoor,A.M., Nowotny,A.: Proc.Nation.Acad. Sci.,U.S.A. 75. 2893-2896 (1976).

34. Carswell,E.A., Old,L.J., Cassel,R.L.,Green,S., Fiore,N., Williamson, B.: Proc.Nation.Acad.Sci.,U.S.A. 72. 3666-3670 (1975)

35. Helson,L., Green,S., Carswell,E., Old,L.J. : Nature 258. 731-732 (1975).

36. Old,L.J. : Clin. Bull. 6. 118-120 (1976).

37. Green,S., Dobryansky,A., Carswell,E.A., Cassel,R.L., Old, L.J., Fiore,N., Schwartz,M.K. : Proc. Nation. Acad. Sci. U.S.A. 73. 381-385 (1976).

38. Hoffmann,M.K., Oettgen,H.F., Old,L.J., Mittler,R.S., Hammerling,U. : J.Reticuloendoth.Soc. 23.307-318 (1978).

39. Ruff,M.R., Gifford,G.E.: J.Immunol.125. 1671-1677 (1980).

40. Matthews,N., Ryley,H.c., Neale,M.L.: Br.J.Cancer 42.416-422 (1980).

41. Ruff,M.R., Gifford,G.E. : in "Lymphokines" Vol.2. Pick, E., Academic Press , London. 235-272. (1981).

42. Kull,F.C., Cuatrecasas,P.:J.Immunol.126.1279-1283 (1981)

43. Cochrane,C.G., Mueller-Eberhard, H. J., Aikin,B.S. : J. Immunol. 105. 55-69 (1970).

44. Hong,K., Kinoshita,T., Miyazaki,W., Izawa,T., Inoue,K. : J. Immunol. 122. 2418-2423 (1979).

45. Shimkin, M.B., Zon,L.: J.Nation.Cancer Inst. 3. 379-382 (1942).

46. Shin, H.S., Gewurz, H., Snyderman. R.: Proc.Soc.Exp.Biol Med. 131. 203-207 (1969).

47. Cochrane, C. G., Mueller-Eberhard, H. J., Aikin, B. S.: J.Immunol. 105. 55-69 (1970).

48. Snyderman, R., Gewurz, H., Mergenhagen, S. E.: J.Exp.Med. 128. 259-275 (1968).

49. Mankowski,Z.T., Diller,I.C., Fisher,M.E. Cancer Res. 17. 382-386 (1957).

50 Mankowski,Z.T., Yamashita,M., Diller,I.C. : Proc.Soc.Exp. Biol. Med. 96. 79-80 (1957).

51. Minden,P., McClutchy,J.K., Brunda,M.J. : J.Nation.Cancer Inst. 61. 535-538 (1978).

52. Weinberg, J.B., Hibbs, J.B.Jr.: J.Nation.Cancer Inst.63, 1273-1278 (1979).

53. Brunda, M.J., Mathews, H.L., Fergusson, H.R., McClutchy, J.K., Minden, P. : Cancer Res. 40. 3211-3213 (1980).

54. Ray, T.L., Wuepper, K.D. : J.Invest.Derm. 67. 700-703 (1976).

55. Verbrunch, H.A., van Dijk, W.C., Peters, R. : Immunol. 37. 615-621 (1979).

56. Tauber, J.W., Polley, M.J., Zabriskie, J.B.: J. Exp. Med. 143. 1352 (1976).

57. McBride, W.H. : Clin. Exp. Immunol. 143. 143-147 (1975).

58. Bradner, W.T., Clarke, D.A., Stock, C.C.: Cancer Res. 18 347-351 (1958).

59. Bradner,W.T., Clarke,D.A.: Cancer Res. 19.673-678 (1959).

60. Diller,I.C., Mankowski,Z.T., Fisher,M.E. : Cancer Res. 23 201-208 (1963).

61. Chihara, G., Hamuro, J., Maeda, Y., Arai, Y., Fukuoka,F. Cancer Res. 30. 2776-2781 (1970).

62. Chihara, G., Hamuro, J., Maeda, Y.Y., Arai, Y., Fukuoka, F. : Nature 225. 943-944 (1970):

63. Hamuro, J., Hadding, U., Bitter-Suermann, D. : Immunol. 34. 695-705 (1978).

64. Kondo,M.,Ikezaki,M.,Imanishi,H.: Gann,66.675 (1975).

65. Bansal, S.C., Bansal, B.R., Thomas, H.L., Siegel, P.D., Rhoads, J.E.Jr., Cooper, D.R. Terman, D.S. Mark, R. : Cancer 42. 1-18 (1978).

66. Terman, D.S., Yamamoto, T., Mattioli, M., Cook, G., Tilllquist, R., Henry, J., Poser, R., Daskal,Y. : J.Immunol 124. 795-805 (1980).

67. Terman, D.S. : Fed. Proc. 40. 45-49 (1981).

68. Ray, P.K., Raychaudhuri, S., Allen, P. : Cancer Res. 42. 4970-4974 (1982).

69. Stalenhaim, G., Gotze, O., Cooper, N.R., Sjoquist, J., Mueller-Eberhard, H.J. : Immunochem. 10. 501-507 (1972).

70. Antopol, W. : J. Infect. Dis. 61. 334-337 (1937).

71. Landy, M., Shear, M.J.: J. Exp. Med. 106. 77-97 (1957).

72. Merler, E., Perrault, A., Trapani, R.-J., Landy., M., Shear, M.J.: Proc.Soc.Exp.Biol.Med. 105, 443-445 (1960).

73. Kodama, M., Yajimaaa, H., Kubo, Y., Hashimoto, I., Katayama, Y. : Japan. J. Med. Sci. Biol. 26, 3-6 (1973).

74. Kodama, M., Yajima, H., Hashimoto, I., Katayama, Y. : Japan. J. Med. Sci. Biol. 30, 77-81 (1977).

ELICITATION OF THE SHWARTZMAN REACTION BY A COMBINATION OF ENDOTOXIN AND AGENTS WHICH ACTIVATE THE COMPLEMENT SYSTEM: MICROVASCULAR EVENTS

Henry Z. Movat and Clement E. Burrowes

Division of Experimental Pathology, Department of Pathology, Medical Sciences Building, University of Toronto, Toronto, Canada M5S 1A8

Introduction

The Shwartzman reaction was described as a "phenomenon of local skin reactivity to B. typhosus culture filtrate" in 1928 by G. Shwartzman (1). It was induced by intradermal injections, followed 24 hours later by an intravenous (i.v.) injection of the culture filtrate and resulted in a thrombo-hemorrhagic necrosis of the intradermally (i.d.) "prepared" site. Confusion arose mainly from the introduction of the term "generalized Shwartzman reaction" by Apitz (2). The historical events are described in a recent brief review (3) and in more detail in a chapter in the Handbook of Physiology (4). Shwartzman summarized and described his findings and that of others in a monograph published in 1937 (5). Evidence indicated that the reaction could be induced only with 2 spaced injections of endotoxin, which, as we know today is a lipopolysaccharide, in which lipid A is the most important constituent (6).

Immunopharmacology of Endotoxicosis

At the symposium on endotoxin held in Paris 3 years ago we described the quantitation, kinetics and ultrastructural changes in the classical Shwartzman reaction elicited with 2 spaced injections on endotoxin (7). Since then we have made attempts to gain further insight into the mechanisms of this reaction. The main question we tried to answer was: can endotoxin be substituted by other substances, administered as a preparative of challenging injection? Since we know that the i.v. challenge with endotoxin is associated with thrombo-hemorrhagic changes (3, 4) we first attempted to substitute the challenging endotoxin with agents which could bring about microthrombosis and microhemorrhage.

Methods and Results

The methodology was essentially that described before (7). Briefly the developing skin lesions were quantitated by mainly 2 methods. Platelets labelled with ^{51}Cr (8) or ^{111}In (9) were used to assess the formation of microthrombi and ^{59}Fe-labelled erythrocytes (RBCs) were the markers for the amount of hemorrhage occurring in a lesion (10). ^{125}I-labelled fibrinogen, administered together with ^{131}I-serum albumin was also used for assessing formation of microthrombi (7). Of these methods the radiolabelled platelets are most reliable for quantitation of intravascular formation of microthrombi, whereas the ^{59}Fe-RBCs indicated both extravasated erythrocytes and at times stasis or thrombosis. The intra- and extravascular processes could be ascertained by histology on the sections of the affected dermis, using 1-2 μm thick methacrylated-embedded tissue.

The classical Shwartzman reaction was elicited by injecting 50-6.25 μg endotoxin (*E. coli* LPS, 055: B5, Difco Labs., Detroit, MI, USA; control No.682197) intradermally and 18 hours later 10 μg/kg i.v. into New Zealand white female rabbits. This procedure was modified by substituting the i.v. injection of endotoxin with antigen-antibody precipitates, zymosan or zymosan-activated rabbit serum. Two types of experiments were done with antigen-antibody (immune) complexes.

Either hyperimmunized rabbits (antigens: bovine serum albumin, BSA, or horse spleen ferritin) were given intradermal injections of endotoxin and 18 hours later they received antigen i.v. (10 mg/kg body weight). To visualize the BSA in the precipitates the antigen was fluorescein-labelled (11) with fluorescein isothiocyanate (Zeiss Photomicroscope II). The ferritin-antiferritin precipitates were assessed by electron microscopy, using a Phillips EM-400 transmission electron microscope. In other experiments preformed immune complexes were injected i.v. (precipitates containing 5 mg antibody protein per kg body weight). In another group of rabbits endotoxin was injected i.d. and 18 hours later zymosan (10 mg/kg body weight) was given i.v.. In a 4th group rabbit serum was activated with zymosan (ZAS), as described before (12) (5-50 mg/ml; 37°C; 60 min.), but serum was used instead of plasma, since heparin partially inhibits the Shwartzman reaction. These animals received (instead of the challening i.v. endotoxin) 10 ml of ZAS per kg body weight i.v. administered over a period of 30-40 minutes.

All rabbits were given i.v. radiolabelled RBCs and platelets and then the appropriate i.v. challenge 18 hours after the i.d. endotoxin. All were sacrificed 2 hours later. As described in the methodological publications (8, 9, 10) the thrombo-hemorrhagic lesions were punched out with a standard steel punch and counted in a γ-spectrometer. The ^{111}In- or ^{51}Cr-platelets are expressed as counts per minute (cpm) per lesion, whereas by taking a sample of blood at the end of an experiment, the ^{59}Fe-RBCs could be expressed as µl blood per lesion.

In animals of all groups tested, the preparative lesions showed ultrastructurally regressive (degenerative) changes in extravasated polymorphonuclear leukocytes (PMNs), but not in the monocytes, as described before (7). Similarly the inflammatory reaction at the i.d. injection sites had subsided by 6-8 hours after the i.d. injection of endotoxin, when measured for ^{51}Cr-leukocyte emigration, transient ^{111}In-platelet aggregation, increase in vascular permeability (^{125}I-albumin escape from leaky vessels) and ^{59}Fe-erythrocyte diapedesis (7). The experimental group consisting of actively hyper-

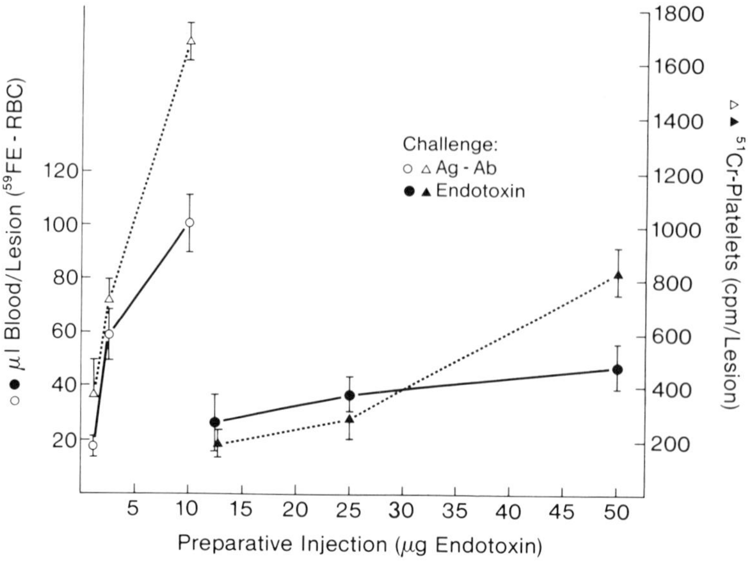

Fig. 1. Induction of the local Shwartzman reaction in rabbits with endotoxin i.d. (50-12.5 μg LPS) and challenged with i.v. endotoxin, or given endotoxin i.d. (10-0.156 μg LPS) and challenged i.v. with BSA in rabbits hyperimmunized with BSA.

immunized rabbits, which 18 hours after the i.d. injection of endotoxin were challenged i.v. with antigen (BSA or ferritin) consistently showed thrombo-hemorrhagic lesions, whereas only 60-70% of the animals challenged i.v. with endotoxin had grossly visible lesions. Furthermore hyperimmunized animals required 5-10 fold less i.d. endotoxin to obtain lesions comparable to those challenged with endotoxin (Fig. 1). When hyperimmunized rabbits were challenged with fluoresceinated BSA, numerous precipitates were seen in vessel lumina of the pulmonary microcirculation, as observed before (13). However, labelled antigen or immune precipitates did escape the pulmonary vascular bed and could be located in static vessels of the thrombo-hemorrhagic lesions of the skin (Fig. 2a and b) and in some of these vessels they were definitely in leukocytes (Fig. 3a and b).

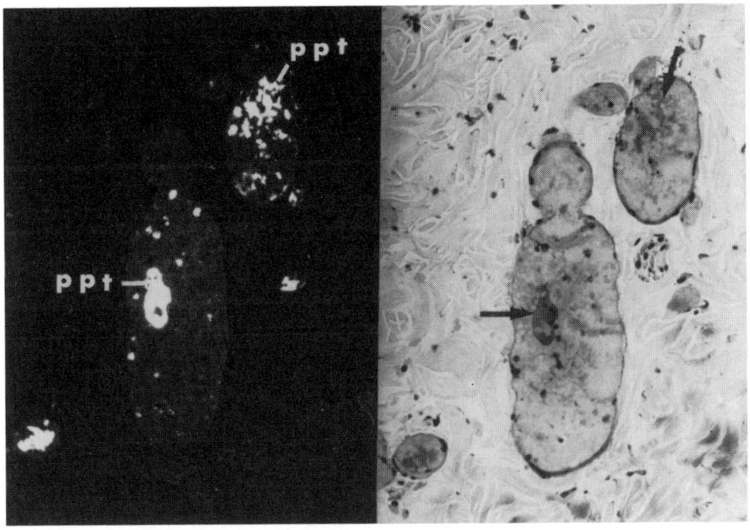

Figs. 2a and b. Immunofluorescent and corresponding light microscopic photomicrographs of a Shwartzman reaction elicited with fluorescein-labelled BSA in a hyperimmunized rabbit. The fluorescent precipitates (ppt) are mostly extracellular in static venule with the areas corresponding to the precipitates marked by arrows in the micrograph taken with ordinary light.

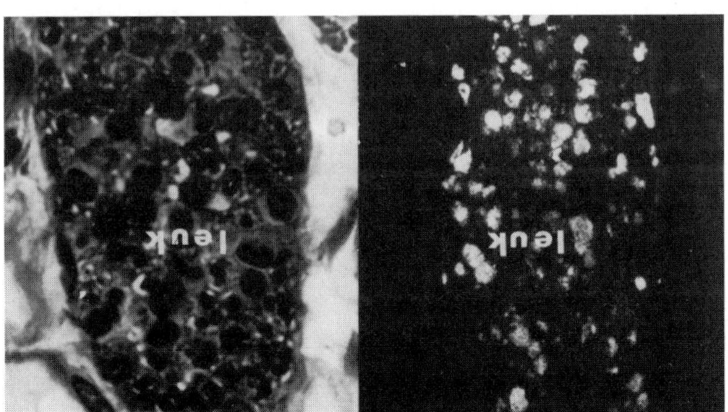

Figs. 3a and b. Illustrations similar to those in 2a and b, but showing the fluorescent immune precipitates exclusively in leukocytes.

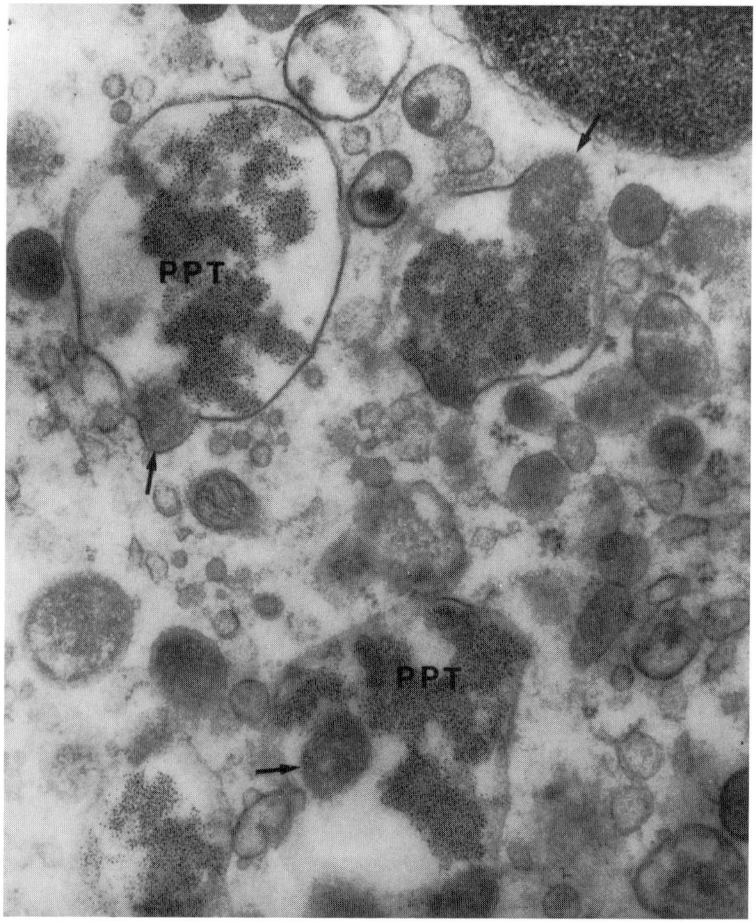

Fig. 4. Part of a PMN in a venule of the dermis of a rabbit hyper-immunized against ferritin. The rabbit was injected i.d. with endotoxin and given an i.v. injection of ferritin 18 hours later. The precipitates of ferritin-antiferritin (PPT) are in phagolysosomes and granules, devoid of a limiting membrane, are also present within the phagolysosome (arrows).

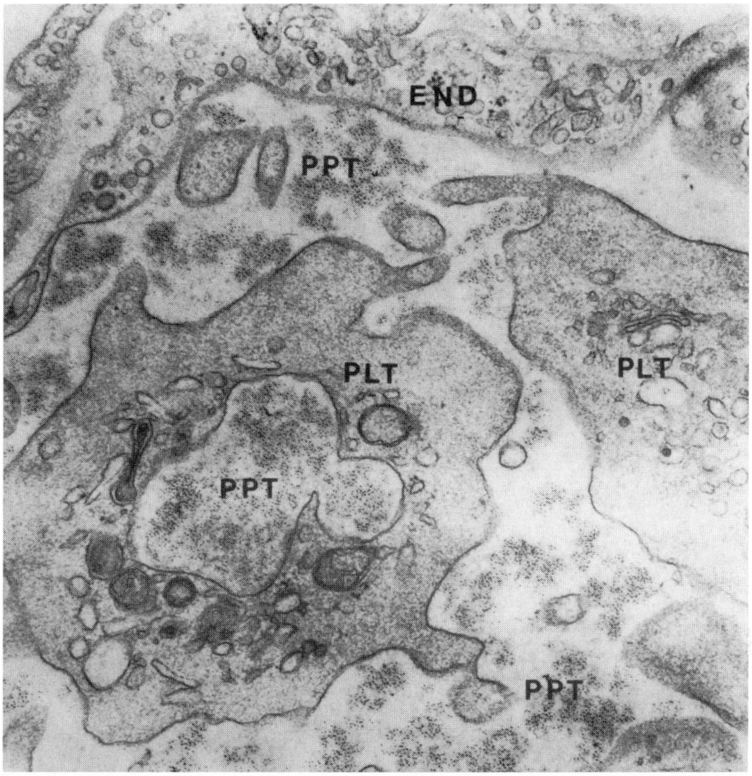

Fig. 5. Ferritin-antiferritin (PPT) is seen in a venule both free in the lumen and in one of two platelets (PLT). END = endothelium. Shwartzman reaction as in Fig. 4.

When ferritin was given i.v. to hyperimmunized rabbits, immune precipitates were observed in PMNs (Fig. 4), in monocytes and in platelets (Fig. 5). The latter were seen occasionally as a single corpuscle, but more frequently in aggregates of various sizes, the peripheries of which had undergone degranulation and swelling. When preformed immune precipitates were injected, no precipitates were seen beyond the lungs. However, like with the above hyperimmunized rabbits, in the dermal vessels there were leukocyte-platelet-fibrin-RBC

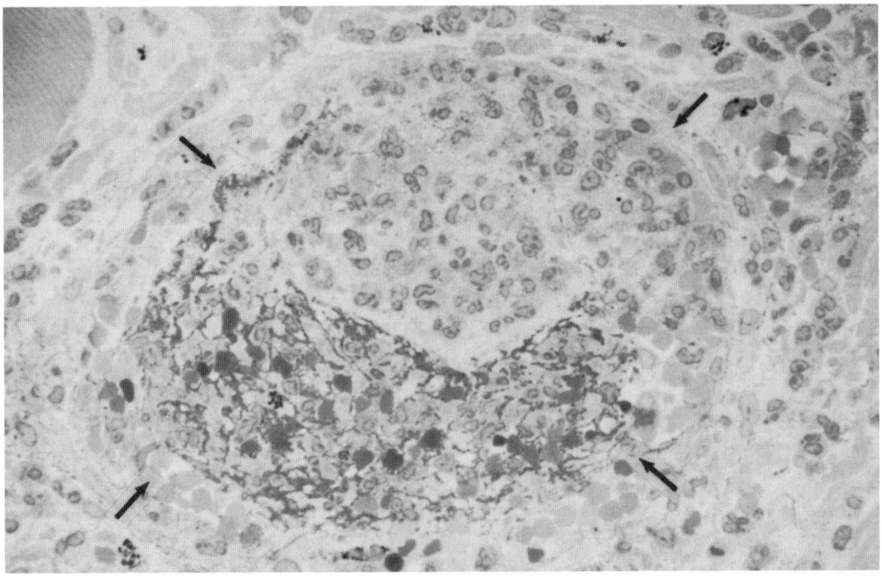

Fig. 6. The appearance of a microthrombus in a small vein of a dermal Shwartzman lesion elicited with preformed immune precipitates of BSA-anti BSA. The vessel is outlined by arrows.

thrombi (Fig. 6). By electron microscopy the leukocytes seen in Figure 6 showed degenerative changes, similar to the ones noted also with hyperimmunized animals (Fig. 7). Figure 8 illustrates all components of the microthrombi at the ultrastrucutral level.

Another way to challenge the rabbits prepared with endotoxin i.d. was to administer 18 hours after the local injections zymosan i.v. (Table 1, Fig. 9). However particulate matter, such as immune precipitates or zymosan were not essential for the i.v. challenge. It was possible to elicit reactions with ZAS administered i.v. over a period of 30-40 minutes (Fig. 10).

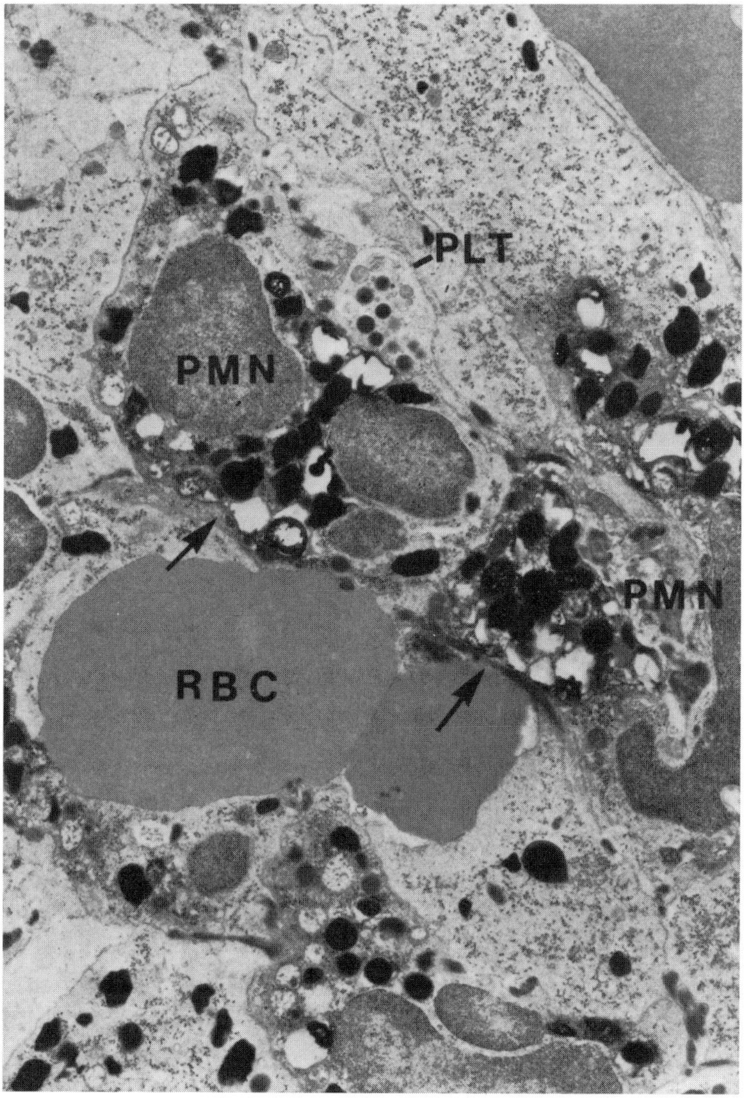

Fig. 7. Part of the microthrombus illustrated in Fig. 6. The arrows point to degenerating PMNs which form part of the microthrombus.

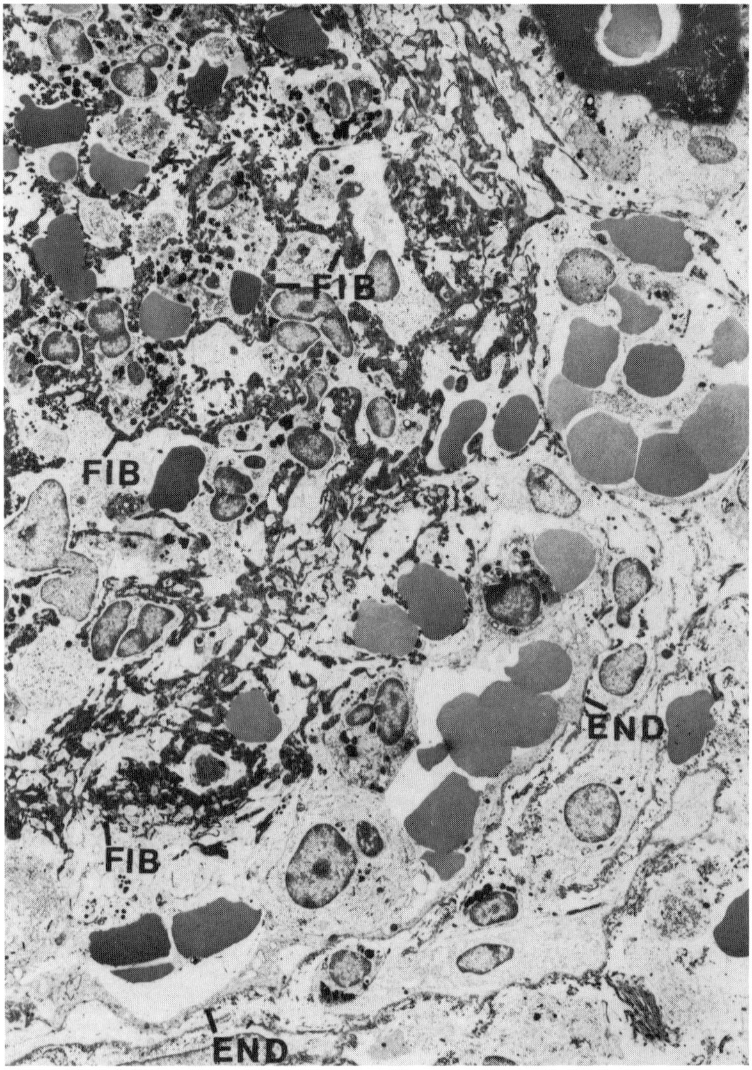

Fig. 8. Another area of the microthrombus illustrated in Fig. 6. Leukocytes, platelets and RBCs are surrounded by abundant fibrin (FIB). Areas marked END (endothelium) outline the wall of the small vein.

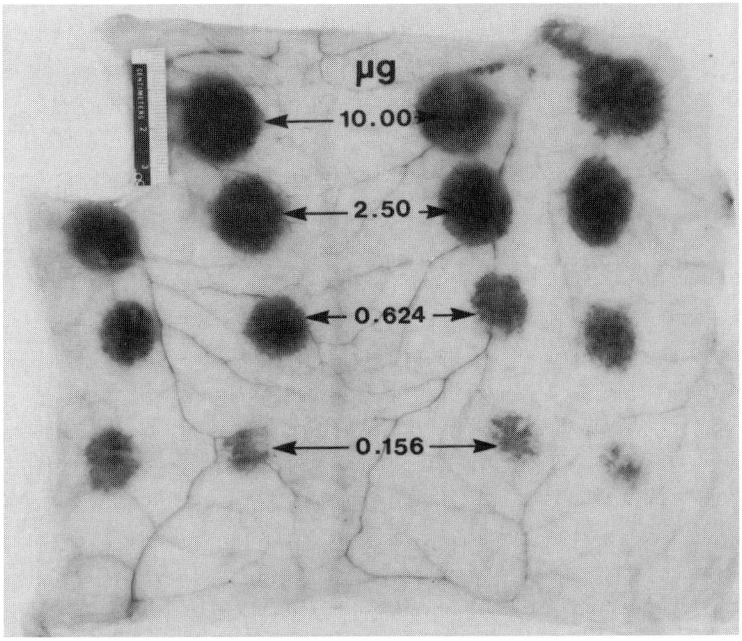

Fig. 9. Hemorrhage on the undersurface of the skin of a rabbit injected i.d. with 10-0.156 µg of endotoxin and challenged i.v. with zymosan.

Discussion

It is well established that i.v. injection into hyperimmunized animals produces aggregate anaphylaxis, which is distinct from that produced with homocytotropic, - i.e. IgE-antibody. We have previously reported that in aggregate anaphylaxis immune precipitates form, primarily in the lung and that this is associated phagocytosis of the precipitates by leukocytes and platelets and with aggregation of both formed elements of the blood (13). The observation that immune precipitates can substitute for i.v. endotoxin in a Shwartzman reaction is not a new one. Without the availability of probes available today, we had earlier reported on the qualitative changes in Shwartzman reactions elicited i.v. with antigen in hyperimmunized rabbits (14).

Table 1. Elicitation of a Shwartzman-type reaction with i.d. endotoxin and i.v. zymosan

Endotoxin(i.d.)	Thrombosis(^{111}In-platelets) cpm/site	Hemorrhage (^{59}Fe-RBC) µl blood/site
50.0	3489 ± 726	8.5 ± 1.5
25.0	1629 ± 96	5.4 ± 0.3
12.5	1044 ± 67	4.8 ± 0.2
NS	493 ± 32	2.6 ± 0.3

Challenge: 10 mg zymosan per kg i.v.; age of thrombo-hemorrhagic lesions: 2 hours.

However, particulate matter was not essential for the development of the dermal lesions. As reported in this publication skin lesion developed when immune precipitates could be detected only in the lungs and not in the prepared sites, when a relatively small amount of preformed immune complexes were injected i.v. This implies that the clotting cascade was set into motion and the microvasculature of the dermis was embolized by leukocytes, platelets, fibrin and erythrocytes. We could however, in a single experiment, deposit the same amount of immune complexes by bypassing the pulmonary circulation and administering the precipitates via a cannulated carotid artery, as in our studies with blood flow and intraarterially administered radiolabelled microspheres (15). Under these circumstances precipitates could be demonstrated morphologically in the skin lesions.

It would appear that aggregation of leukocytes and platelets and activation of the clotting cascade (both extrinsic and intrinsic pathways) represent a sine qua non in the modifications and modulations of the Shwartzman phenomenon. Other means to bring this about was

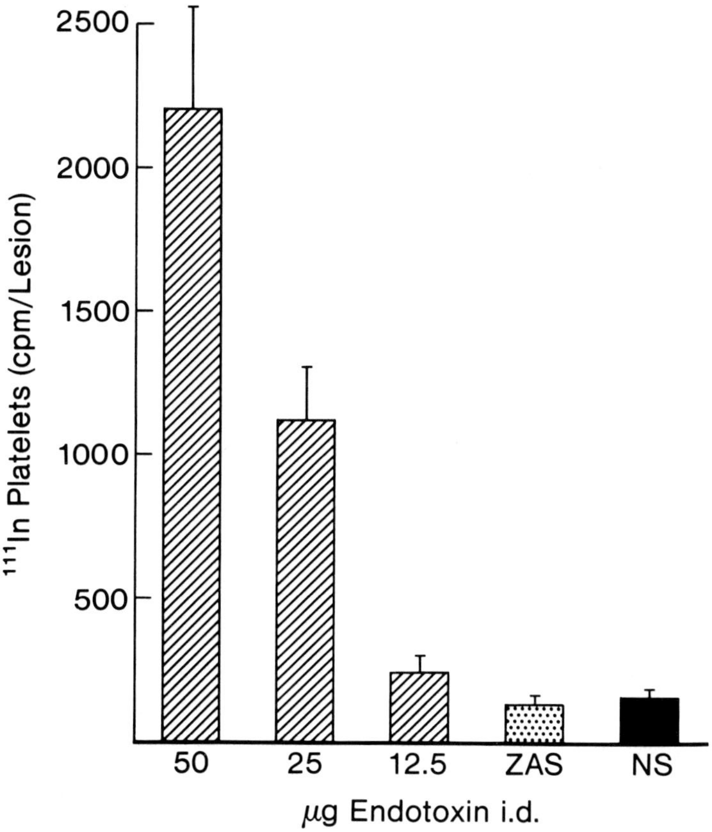

Fig.10. Accumulation of ^{111}In-labelled platelets, as part of microthrombi, in lesions injected with endotoxin i.d. and challenged with ZAS given i.v. over a 30-minute period. Sites injected, at the time when the endotoxin was injected intraterminally, with ZAS were no different from sites injected with normal saline (NS).

to inject zymosan or ZAS. With the former phagocytosed zymosan particles were demonstrable in leukocytes (not shown) and it is known that zymosan can activate the complement system and inter alia C5 is activated to C5a and this in turn is cleaved to $C5a_{des\ Arg}$ (16, 17). It was however, not necessary to inject particulate matter or induce its formation (Ag-Ab ppt.) in vivo. If preformed $C5a_{des\ Arg}$ was perfused in the form of ZAS it served as a challenging injection. Hence a soluble factor, $C5a_{des\ Arg}$, which is present in plasma, together with an anionic potentiator (18) and possibly $C3a_{des\ Arg}$ and other, as yet unidentified substances, can bring about leukocyte and platelet aggregation and indirectly conversion of fibrinogen to fibrin and hence the microthrombosis and microhemorrhage. We have previously reported that $C5a_{des\ Arg}$ can induce chemotaxis, increase in vascular permeability and hemorrhage when injected locally (12, 19). All the substances injected i.v. instead of endotoxin can induce chemotaxis in vitro and in vivo. It remains to be ascertained whether other synthetic leukocyte chemoattracting substances, - e.g. the synthetic oligopeptide N-formyl-methionyl-leucyl-phenylalanine (20), the semisynthetic platelet activating factor (PAF), known to be acetyl glyceryl ether phosphocholine (AGEPC) (21) and synthetic leukotriene B_4 (22) by themselves or in combination can substitute for the i.v. challenging dose in the Shwartzman reaction.

Summary and Conclusions

Instead of 2 spaced injections of i.d. and i.v. endotoxin, the latter can be substituted by immune precipitate (antigen into hyperimmunized rabbits or preformed antigen-antibody complexes), zymosan or zymosan-activated serum (ZAS). The fact that a soluble non-phagocytosable factor can be used is significant. ZAS contains the chemotactic substance $C5a_{des\ Arg}$ and the particulate substances (immune precipitates, zymosan) generate $C5a_{des\ Arg}$. Locally injected, these agents induce chemotaxis of leukocytes, but when administered i.v. they induce leukocyte and platelet aggregation, release of lysosomal contents and seemingly initiate the coagulation cascade, with formation of blood clots, which embolize and impact in the microcirculation with a predilection for the sites of locally injected endotoxin, ensuing in thrombo-hemorrhagic lesions, which are identical to lesions seen clinically in DIC (disseminated intravascular coagulation). It is possible that at the injected site the endotoxin has rendered the endothelial lining similar to that in marginating organs (lung, liver, spleen). However, it is also possible that the endothelial lining has been altered by the endotoxin and that we see not an embolization but a primary alteration of the endothelium followed by local microthrombosis.A better understanding of these phenomena will, hopefully, lead to the prevention or amelioration of DIC mediated by endotoxin or endotoxin and other parasite-derived and other host-derived substances.

References

1. Shwartzman, G.: Proc.Soc.Exp.Biol.Med. 25, 560-561 (1928).
2. Apitz, K.: J.Immun. 28, 255-266 (1935).
3. Movat, H.Z.: Surv.Synth.Path.Res. 1, 241-250 (1983).
4. Movat, H.Z.: In Handbook of Physiology, Sect. 2, The Cardiovascular System IV. Eds. Renkin, E.M., Michel, C.C., The Amer.Physiol.Soc., Washington, 1983 (in press).
5. Shwartzman, G.: Phenomenon of Local Tissue Reactivity and its Immunological, Pathological and Clinical Significance. P. Hoeber, New York, pp. 1-461, 1937.
6. Galanos, C., Lüderitz, O., Rietschel, E.T., Westphal, O.: In International Review of Biochemistry. Biochemistry of Lipids. Vol. 14. Ed. Goodwin, T.W., University Park Press, Baltimore, pp. 239-335, 1977.
7. Movat, H.Z., Jeynes, B.J., Wasi, S., Movat, K.W., Kopaniak, M.M.: In Bacterial Endotoxin and Host Response. Ed. Agarwal, M.K., Elsevier/North-Holland, Amsterdam, pp. 179-201, 1980.
8. Issekutz, A.C., Movat, H.Z.: Lab.Invest. 42, 310-317 (1980).
9. Jeynes, B.J., Issekutz, A.C., Issekutz, T.B., Movat, H.Z.: Proc.Soc.Exp.Biol.Med. 165, 445-452 (1980).
10. Kopaniak, M.M., Issekutz, A.C., Burrowes, C.E., Movat, H.Z.: Proc.Soc.Exp.Biol.Med. 163, 126-131 (1980).

11. Holborow, E.J., Johnson, G.D.: In Handbook of Experimental Immunology. Ed. Weir D.M., Blackwell Sci.Publ., Oxford and Edinburgh, pp. 571-598, 1967.

12. Issekutz, A.C., Movat, K.W., Movat, H.Z.: Clin.Exp.Immuno. 41, 505-511 (1980).

13. Movat, H.Z., Uriuhara, T., Taichman, N.S., Rowsell, H.C., Mustard, J.F.: Immun. 14, 637-648 (1968).

14. Taichman, N.S., Uriuhara, T., Movat, H.Z.: Lab.Invest. 14, 2160-2176 (1966).

15. Hay, J.B., Johnston, M.G., Hobbs, B.B., Movat, H.Z.: Proc.Soc.Exp.Biol.Med. 150, 641-644 (1975).

16. Fernandez, H.N., Henson, P.M., Otani, A., Hugli, T.E.: J.Immun. 120, 109-115 (1978).

17. Webster, R.O., Hong, S.R., Johnston, Jr. R.B., Henson, P.M.: Immunopharm. 2, 201-219 (1980).

18. Perez HD, Goldstein IM, Webster RO, Henson PM: Enhancement of human $C5a_{des\ Arg}$ by an anionic polypeptide (cochemotaxin) in normal human serum and plasma. J.Immunol. 126, 800-804 (1981).

19. Issekutz, A.C., Movat, K.W., Movat, H.Z.: Clin.Exp.Immun. 41, 512-520 (1980).

20. Schiffmann, E., Corcoran, B.A., Wahl, S.M.: Proc.Natl.Acad.Sci. US 72, 1059-1061 (1975).

21. Hanahan, D.J., Demopoulos, C.A., Liehr, J., Pinckard, R.N.: J.Biol.Chem. 255, 5514-5516 (1980).

22. Movat, H.Z., Burrowes, C.E., Johnston, M.G., Rettl, C.: In Chemical Mediators of Inflammation. Satellite Symposium of 5th Intern.Congress of Immun. Aug. 27-29, 1983. Ed. Takada, A., Hayashi, H., Academic Press, New York and London (in press).

Acknowledgements: The authors wish to express their gratitude to Ms. Marica Michael and Mrs. Otti Freitag for expert secretarial and technical assistance. The authors are indebted to Mr. Paul L. Schwartz for his assistance with photography. The work described in this publication was supported by the Ontario Heart Foundation (T-1-15). Dr. Movat is a Career Investigator of the Medical Research Council of Canada.

COMPLEMENT IN CLINICAL AND EXPERIMENTAL DISSEMINATED INTRAVASCULAR COAGULATION

Yasumasa Furukawa, Toshikazu Yoshikawa, Masashi Murakami, Motoharu Kondo

The First Department of Medicine, Kyoto Prefectural University of Medicine

Kawaramachi-Hirokoji, Kamikyo-ku, Kyoto, 602, Japan

Introduction

Total hemolytic complement activity and the third component of complement (C3) measured by immunodiffusion have been shown to be reduced in a variety of clinical conditions including disseminated intravascular coagulation (DIC) (1). The importance of the complement in initiating intravascular coagulation has been demonstrated by Zimmerman and Müller-Eberhard (2). They observed a complement-dependent activation of the coagulation system by bacterial endotoxin and proposed the concept that this initiation of blood coagulation is mediated by the alternative pathway of complement activation (3). We report here the changes in complement levels in patients with DIC, and in rats with experimental DIC induced by endotoxin.

Materials and Methods

1) DIC patients

   Serial blood samples were collected from 13 hospitalized patients with the initial clinical diagnosis of DIC; 8 associated with liver disease, 4 with bacterial infection and 1 with chronic myelogenous leukemia.

2) Animals

Immunopharmacology of Endotoxicosis

Wistar female rats weighing 200 to 220 g were used.

3) Endotoxin

Endotoxin (Escherichia coli 055: B5 lipopolysaccharide B) (Difco Lab., Detroit, Mich., USA) was dissolved in physiologic saline before every experiment.

4) Experimental Model of DIC

Experimental DIC was induced by a 4 hr sustaind infusion of endotoxin at a dose of 100 mg/kg in rats (4).

5) Collection of Blood Samples

Human blood samples were venously obtained serially. Blood samples of rats were collected from the abdominal aorta. Serum was stored at -70°C until used. The samples were withdrawn using plastic syringes and anti-coagulated with 0.01 M sodium ethylenediaminetetraacetic acid (EDTA) for platelet counts, and with 0.1 M sodium citrate for prothrombin time (PT), partial thromboplastin time (PTT) and fibrinogen determination.

6) Detection of Hemolytic Complement Activity

CH50 was determined by a standard assay method employing sensitized sheep erythrocytes (EA) and veronal buffer containing 0.1 % gelatin, 0.15 mM $CaCl_2$ and 0.5 mM $MgCl_2$ (GVB^{++}) (5).

7) Protein Concentration of the Complement Components

The protein concentration of the complement components was measured by single radial immunodiffusion using specific antisera against human C4, C3, C3A (Behring-Werke Co., Marburg, FRG), and rat C3 (Cappel Lab., Cochranville, Pa., USA).

8) Estimation of the Severity of DIC

The severity of DIC was determined with reference to various parameters, such as fibrinogen and fibrin degradation products (FDP), platelet counts, fibrinogen, PT, PTT, and the number of renal glomeruli having fibrin thrombi. FDP was determined by the latex agglutination test (FDPL test of Teikoku Hormone Mfg. Co., Tokyo, Japan) and expressed as μg/ml of equivalent to human fibrinogen base (6).

Platelets were counted by phase-contrast microscopy. Fibrinogen was determined according to the method of Ratnoff et al. (7). PT and PTT were measured according to the methods of Quick et al. (8), and Langdell et al. (9). Kidney tissue specimens were examined histologically by hematoxylin-eosin (HE) and Weigert's fibrin stain (10). Of 100 glomeruli examined those having fibrin thrombi were counted; the ratio was reported in percentages as %GFD (% glomerular fibrin deposits).

Results

1) <u>The Changes in CH50, C4 and C3 levels in Patients with DIC</u>

The CH50 levels of 13 patients rapidly decreased after the onset of DIC. However, the patients who responded to heparin showed a gradual increase in CH50 levels about 10 days after the onset of DIC; those who failed to respond continued to show low CH50 levels (Fig. 1). Similar results were seen in C4 and C3 levels (Fig. 2).

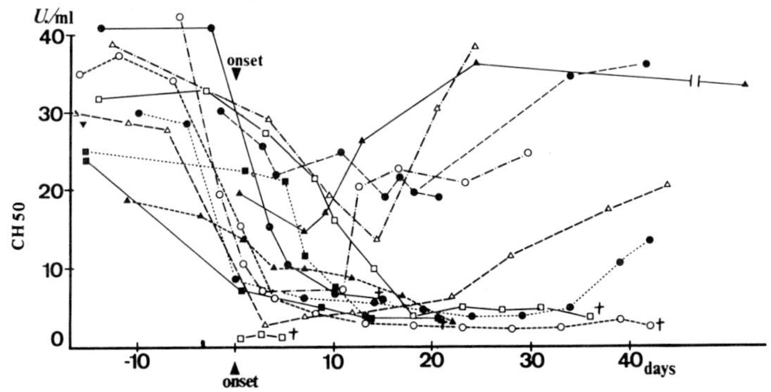

Fig. 1 CH50 levels in 13 patients with DIC.

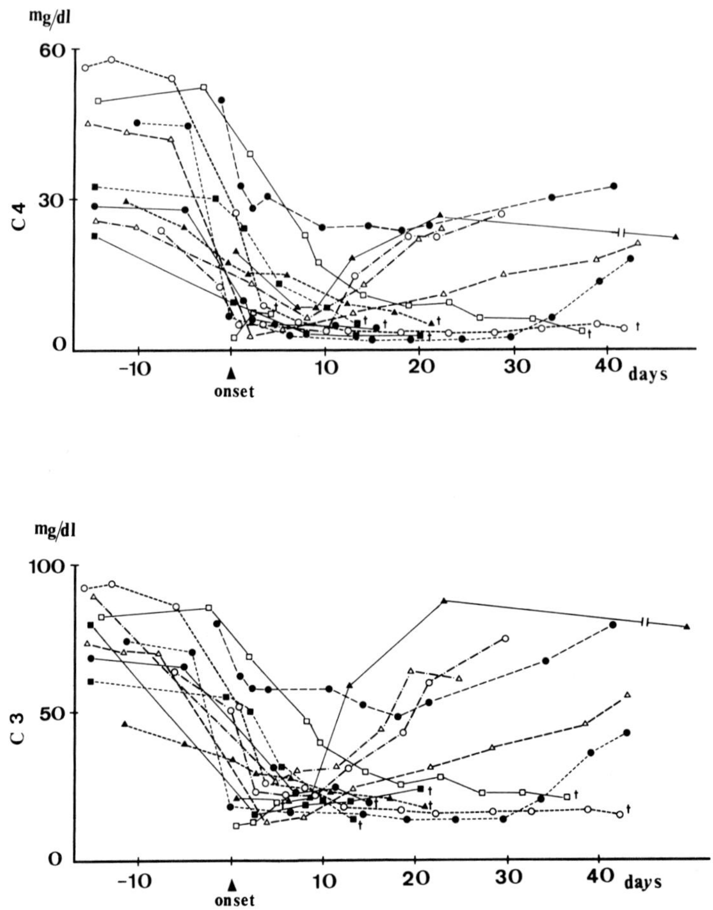

Fig. 2 C4 and C3 levels in 13 patients with DIC.

2) <u>Experimental DIC in rats induced by continuous infusion of endotoxin</u>

After the infusion of endotoxin, FDP levels and %GFD were increased, platelet counts and fibrinogen levels were decreased, and PT and PTT were prolonged.

3) The Changes in CH50 and C3 levels in rats induced by endotoxin

After the infusion of endotoxin, the levels of CH50 and C3 were decreased significantly (Fig. 3).

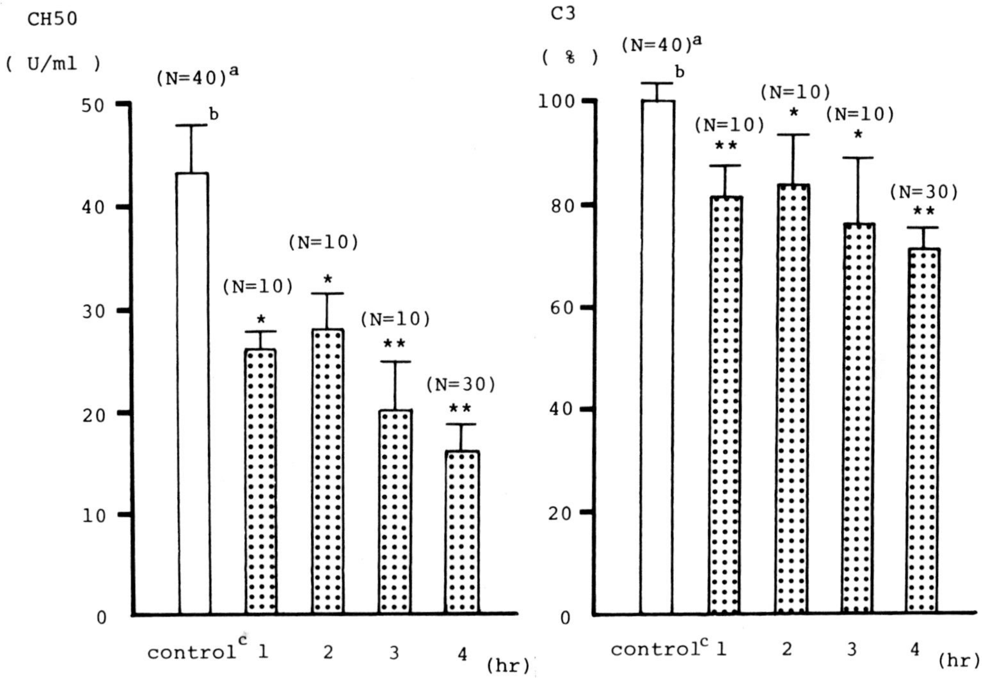

Fig. 3 The changes in CH50 and C3 levels in rats infused by endotoxin (100 mg/kg/4 hr).

a. Number of rats.
b. Data represent mean ± SEM.
c. Control rats were infused with saline.
* p<0.01, ** p<0.001 for difference to controls by Student's t-test.

Discussion

The results of this study demonstrated that complement levels decreased immediately after the onset of DIC, not only clinically but also experimentally. In addition, the reduction in complement levels were gradually restored in the patients who responded, suggesting that the examination of complement may be of clinical value for estimating the prognosis.

The demonstration of prolonged clotting time and retarded prothrombin consumption in rabbits deficient in the sixth component of complement (6) and the correction of these abnormalities with highly purified C6 suggests that this protein is a normal participant in the blood coagulation mechanism (11). Bacterial endotoxins and antigen-antibody aggregates activate complement, accelerate blood coagulation, and are implicated in intravascular blood clotting (1, 12, 13). Zimmerman et al. found that complement-activating materials initiate blood coagulation through a complement-mediated pathway (2).

The precise mechanism of interaction between the complement and blood coagulation system is uncertain. However, these data suggest an important role of the complement in DIC. The question of the importance of the activation of complement in initiating blood coagulation in the DIC state, is still open and can be resolved only by large studies.

References

1. Tomar, R.H., Kolchins, A.: Thromb. Diath. Haemorrh. 27, 389-395 (1972).
2. Zimmerman, T.S., Müller-Eberhard, H.J.: J. Exp. Med. 134, 1601-1607 (1971).
3. Gewurz, H., Shin, H.S., Mergenhagen, S.E.: J. Exp. Med. 128, 1049-1057 (1968).
4. Yoshikawa, T., Furukawa, Y., Murakami, M., Takemura, S.,

Kondo, M.: Res. Exp. Med. 179, 223-228 (1981).

5. Mayer, M.M.: Mechanism of hemolysis by complement. Ciba Foundation Symposium on Complement. Churchill, London, 4-32, 1965.

6. Ferreira, H.C., Murat, L.G.: Br. J. Haematol. 9, 299-310 (1963).

7. Ratnoff, O.D., Menzie, C.: J. Lab. Clin. Med. 37, 316-320 (1951).

8. Quick, A.J., Stanly, B.M., Bancroft, F.W.: Am. J. Med. Sci. 190, 501-511 (1935).

9. Langdell, R.D., Wagner, R.H., Brinkhaus, K.M.: J. Lab. Clin. Med. 47, 637-647 (1953).

10. Weigert, C.: Fortschr. Med. 5, 228-232 (1887).

11. Zimmerman, T.S., Arroyave, C.M., Müller-Eberhard, H.J.: J. Exp. Med. 134, 1591-1600 (1971).

12. Ishizaka, T., Ishizaka, K., Salmon, S., Fudenberg, H.: J. Immunol. 99, 82-91 (1967).

13. Mckay, D.G., Shapio, S.S., Shanberg, J.N.: J. Exp. Med. 107, 369-376 (1958).

EFFECTS OF STEROID, ANTI-THROMBOTIC AGENTS AND DEFIBRINOGENATION ON ENDOTOXIN-INDUCED DISSEMINATED INTRAVASCULAR COAGULATION IN RATS

Toshikazu Yoshikawa, Yasumasa Furukawa, Masashi Murakami, Motoharu Kondo

First Department of Medicine, Kyoto Prefectural University of Medicine, Kamikyo-ku, Kyoto, 602, Japan

Introduction

Disseminated intravascular coagulation (DIC) is a pathological syndrome where formation of fibrin thrombi, depletion of specific plasma proteins, and loss of platelets, and activation of the fibrinolytic system suggest the presence of thrombin in the systemic circulation. Clinically these effects are manifested as diffuse hemorrhage and thrombosis (1). Treatment for DIC is to prevent formation and development of fine thrombi caused by stimulated coagulation, and dissolve such thrombi formed.

Recently, the antithrombotic effectiveness of glucocorticoids was further evidenced by prevention of the generalized Shwartzman reaction (2), characterized by glomerular capillary thrombosis. The pathogenesis of thrombosis involves many factors, among which platelet aggregation and adhesion play a great role in the initial stages of thrombus formation. The specific removal of plasma fibrinogen by some purified thrombin-like snake venom enzymes proved to be a unique approach for anticoagulation and reduction of blood viscosity in prophylaxis and treatment of thrombotic diseases (3).

In this study, the experimental model of DIC in rats was used to investigate the preventive effects of methylprednisolone, antiplatelet agents, such as ticlopidine, dipyridamole, vitamin E, and aspirin, and defibrinogenation by batroxobin against DIC.

Immunopharmacology of Endotoxicosis

Materials and Methods

Female Wistar rats (200 g) were used. Endotoxin (Escherichia coli 0.55:B5 lipopolysaccharide B); (Difco Lab., Detroit, MI, USA) was dissolved in pyrogen-free physiological saline before each experiment. Experimental DIC was induced by a sustained infusion of endotoxin into a femoral vein cannulated under light Nembutal (5 mg/100 g i.p.) anesthesia.

The severity of DIC was determined by changes in fibrinogen and fibrin degradation products (FDP), fibrinogen, prothrombin time (PT), partial thromboplastin time (PTT), platelet counts, and % glomerular fibrin deposits (%GFD); 100 glomeruli were counted, and those having fibrin thrombi were expressed in percentage. Blood was withdrawn from the abdominal aorta into plastic syringes. The samples used for blood counts, platelet counts, and hematocrit were anticoagulated with 0.01 M sodium ethylenediamine tetraacetic acid, and those for PT (4), PTT (5) and fibrinogen (6) determinations were diluted (1:9 vol.) with 0.1 M sodium citrate.

FDP was determined by the latex agglutination test (Teikoku Hormone Mfg. Co., Tokyo, Japan) and expressed in μg/ml of equivalent human fibrinogen base (7). Platelets were counted by phase-contrast microscopy (8), and red and white blood cells in an automatic blood cell counter (S-plus-II, Coulter Electronics Inc., Hialeah, FL, USA). Hematocrit was determined in duplicate in microhematocrit tubes (Propper Mfg. Co., Inc. Long Island, NY, USA) centrifuged at 10,000 g for 6 min. Kidney tissue specimens were examined histo logically with hematoxylin-eosin and Weigert's fibrin stain (9).

Methylprednisolone sodium succinate (Japan Upjohn Ltd., Tokyo, Japan); ticlopidine (Parcor, Toulouse, France); dipyridamole (Nippon C.H. Boehringer Sohn Co., Ltd., Osaka, Japan); DL-α-tocopheryl acetate and HCO-60 (polyethylene 60-hydrogenated castor oil, Eisai Co., Tokyo, Japan); aspirin (Sigma Chemical Co., St. Louis, MO, USA) and batroxobin (Pentapharm, Ltd., Basel, Switzerland) were used. All other chemicals were of reagent grade.

Results

1) Experimental model of DIC

Table 1. Dose-response to infused endotoxin. Rats were infused with 0, 25, 50, 100, or 200 mg/kg endotoxin for 4hr. Control rats (0 mg/kg endotoxin) were infused with 11.4 ml saline solution for 4hr.

Endotoxin (mg/kg)	FDP (μg/ml)	Fibrinogen (g/l)	PT (s)	PTT (s)	Platelet count (x10^9/l)	GFD[1] (%)	N[2]
0(Control)	2.4± 0.3	1.39±0.40	11.9±0.7	60.2±11.3	1024±246	0	n=10
25	16.2± 3.2*	<0.2*	21.7±3.7*	88.0±13.1*	166± 59*	23.3± 5.5*	n=10
50	23.8± 5.3*	<0.2*	22.0±4.3*	94.3±16.3*	276± 53*	70.0±12.0*	n=10
100	58.7±14.3*	<0.2*	25.6±5.5*	117.0±15.4*	229± 66*	78.1±14.3*	n=20
200	41.6±11.1*	<0.2*	16.4±2.9*	65.0±10.1	148± 58*	35.2± 5.3*	n=10

Results are expressed as mean value±SD
1) Glomerular fibrin deposits
2) Number of subjects
* P<0.001 for difference to controls by Student's t-test

Endotoxin dosage. Endotoxin(11.4ml), dissolved in saline solution at various concentrations, was infused for 4 hr,after which all rats were killed and severity of DIC was evaluated on the basis of the above parameters. The dose of 100 mg/kg appeared to be optimal for induction of experimental DIC in rats (Table 1).

Table 2. Time-dependent response to infused endotoxin (100mg/kg). Rats were infused 100mg/kg endotoxin for 1, 2, 3, or 4hr. Control rats (0hr) were not infused endotoxin.

Hour	FDP (μg/ml)	Fibrinogen (g/l)	PT (s)	PTT (s)	Platelet count (x10^9/l)	GFD[1] (%)	N[2]
0(Control)	2.1± 0.2	1.41±0.34	12.2±0.6	60.5±10.4	1157±228	0	n=10
1	3.3± 1.9	1.38±0.22	12.1±0.9	67.3± 3.5	829±224*	7.5± 2.3**	n=10
2	4.0± 1.1**	1.33±0.43	13.7±0.7**	68.8± 8.2**	492±108**	10.0± 4.2**	n=10
3	59.0±18.4**	0.55±0.10**	18.2±3.6**	93.7±14.3**	228± 88**	28.8±16.2**	n=10
4	58.7±14.3**	<0.2**	25.6±5.5**	117.0±15.4**	229± 66**	78.1±14.3**	n=20

Results are expressed as mean value±SD
1) Glomerular fibrin deposits
2) Number of subjects
* <0.01, ** P<0.001 for difference to controls by Student's t-test

Infusion time. The optimal dose of endotoxin was administered for 1, 2, 3, or 4 hr at a constant rate of 2.85 ml/hr. When endotoxin was infused for periods of less than 3 hr, no significant changes were observed in the parameters. Moreover, as the severity of DIC showed no change even at infusion periods greater than 4 hr, this infusion period was selected for the subsequent experiments (Table 2).

<u>Time-dependent changes during 4-hr infusion of endotoxin.</u> At 1 or 2 hr after the infusion, no marked changes were seen in any of the parameters, but at 3 hr FDP and %GFD increased (Fig. 1), PT and PTT were prolonged (Fig. 2), and fibrinogen and platelet count were reduced (Fig. 3). At 4 hr coagulation was further advanced and glomerular fibrin deposits were present in all rats (Fig. 4a, 4b).

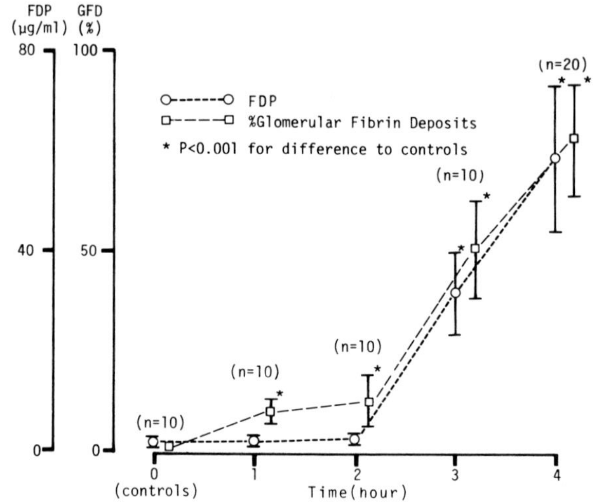

Fig. 1 FDP and %GFD in Experimental DIC IN Rats

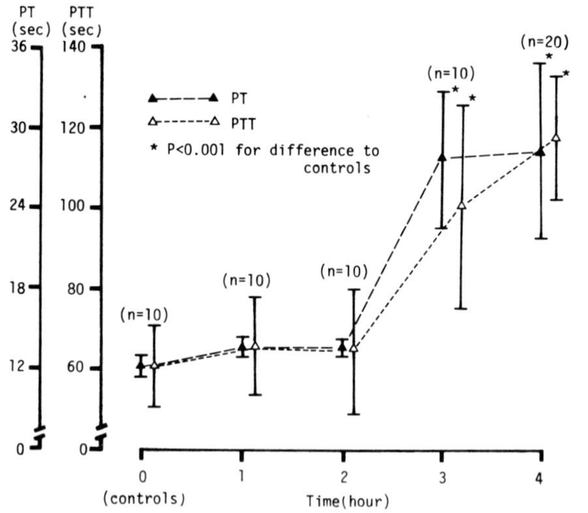

Fig. 2 PT and PTT in Experimental DIC in Rats

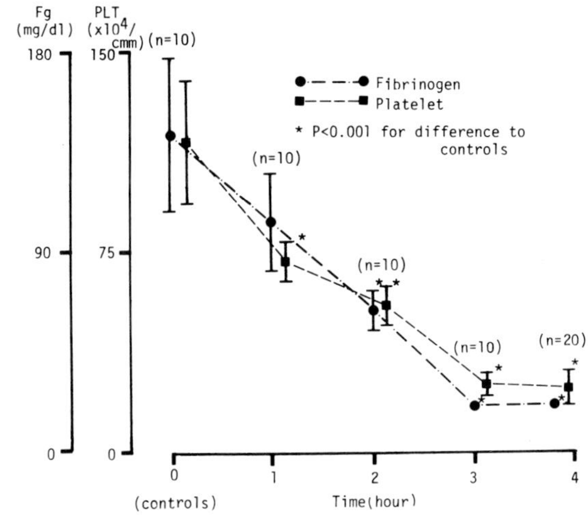

Fig. 3 Fibrinogen and Platelet counts in Experimental DIC in Rats

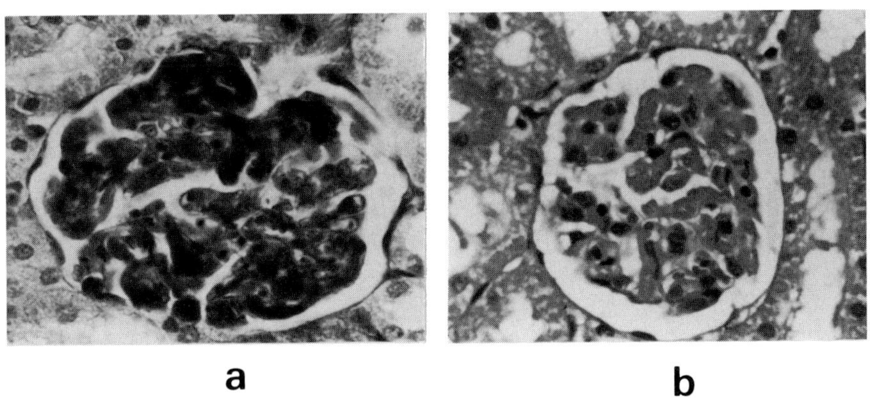

Fig. 4 Fibrin thrombi in glomerular capillaries.
a.(Weigert's fibrin stain, x400) b.(H and E x400)

2) Effect of various agents on experimental DIC

Methylprednisolone. The rats were injected with methylprednisolone at 0.1, 1, 10, 30 mg/kg, and immediately after the injection,they were infused with 100 mg/kg of endotoxin for 4 hr. A significant prevention was noted in all parameters, in rats pre treated with 1.0, 10.0 or 30.0 mg/kg of methylprednisolone (Table 3).

Table 3. Effects of methylprednisolone on experimental DIC in rats infused with 100 mg/kg of endotoxin for 4hr.

	FDP (μg/ml)	Fibrinogen (g/1)	PT (s)	PTT (s)	Platelet count (x10^9/1)	GFD[1](%)	N[2]
Control[3]	56.0±8.6	<0.2	30.8±8.2	146.6±15.4	79± 5	85.3±12.6	8
Methylprednisolone							
0.1 mg/kg	43.8±3.6*	0.38±0.04**	20.4±2.2*	65.8± 5.2**	285± 21**	83.1±14.0	8
1.0	8.8±0.6**	0.34±0.07**	15.3±1.8**	58.5± 7.2**	342± 59**	13.7± 4.6**	8
10.0	5.3±0.4**	0.39±0.06**	13.1±0.7**	61.9±10.4**	381± 69**	5.1± 1.3**	8
30.0	4.0±0.6**	0.49±0.09**	12.5±0.6**	51.6± 6.4**	656±124**	2.0± 0.5**	8
Normal rats[4]	2.0±0.1**	1.08±0.30**	12.2±0.6**	51.9± 3.9**	863±101**	0**	8

Results are expressed as mean value±SD
1) Glomerular fibrin deposits
2) Number of subjects
3) Saline was injected once intraperitoneally before the infusion of endotoxin
4) Normal rat group was treated with neither methylprednisolone nor endotoxin. The rats were infused continuously with 11.4 ml of physiological saline for 4hr.
* <0.01, ** <0.001 for difference to the values of rats pretreated with saline as controls by Student's t-test

Ticlopidine. The rats were intraperitoneally injected with 2.0, 20.0, 50.0, 100.0 or 200.0 mg/kg of ticlopicine, followed by the infusion of endotoxin. A preventive effect against DIC was noted in all the parameters in rats treated with 20.0, 50.0, 100.0 or 200.0 mg/kg of ticlopidine (Table 4).

Table 4. Effects of ticlopidine on experimental DIC in rats infused with 100 mg/kg of endotoxin for 4hr.

	FDP (μg/ml)	Fibrinogen (g/1)	PT (s)	PTT (s)	Platelet count (x10^9/1)	GFD[1](%)	N[2]
Control[3]	54.2±7.2	<0.2[4]	30.2±6.5	148.3±16.1	78± 5	84.9±12.1	15
Ticlopidine[5]							
2.0 mg/kg	48.8±6.9	<0.2	22.3±2.5*	109.4±21.7**	163±11**	67.5±16.5*	8
20.0	18.0±3.5**	0.32±0.07**	19.4±2.7**	83.5±10.1**	305±20**	40.1± 6.6**	8
50.0	12.3±2.6**	0.32±0.06**	14.8±1.7**	59.8±12.0**	343±41**	24.5± 5.0**	8
100.0	8.0±0.9**	0.40±0.11**	14.3±0.6**	56.4± 8.8**	359±66**	7.0± 1.9**	8
200.0	7.8±0.6**	0.47±0.08**	14.6±1.7**	54.4± 9.7**	515±87**	5.3± 2.3**	8
Normal rats[6]	1.9±0.2**	0.98±0.21**	12.6±0.5**	52.2± 4.0**	832±98**	0**	15

Results are expressed as mean value±SD
1) Glomerular fibrin deposits
2) Number of animals
3) Saline (1.0 ml) was injected once intraperitoneally before the 4hr. infusion of endotoxin (100 mg/kg)
4) Fibrinogen level was less than 0.2 g/l
5) Ticlopidine was dissolved in 1.0 ml of saline alone
6) Normal rats were infused with 11.4 ml/4hr of saline alone
* p<0.01, ** p<0.001 for difference to the values of control rats by Student's t-test

Table 5. Effect of dipyridamole on experimental DIC in rats infused with 100 mg/kg of endotoxin for 4hr.

	FDP (μg/ml)	Fibrinogen (g/1)	PT (s)	PTT (s)	Platelet count ($x10^9/1$)	GFD[1] (%)	N[2]
Control[3]	56.2±7.4	<0.2[4]	32.4±8.0	140.2±22.1	72± 6	85.1±12.9	8
Dipyridamole							
0.5 mg/kg	22.5±5.1**	0.24±0.04	24.9±9.6	108.6±32.3*	214± 58**	19.5±10.7**	9
5.0	20.5±4.9**	0.26±0.06*	22.7±5.3*	81.7±21.1**	255±100**	11.7± 7.3**	8
50.0	6.4±2.2**	0.44±0.10**	16.7±2.6**	52.1± 8.5**	307±129**	10.0± 5.9**	11
Normal rats[5]	3.1±1.2**	0.69±0.06**	11.4±0.4**	46.5± 9.8**	693±109**	0**	8

Results are expressed as mean value±SD
1) Glomerular fibrin deposits
2) Number of subjects
3) Saline was injected once intraperitoneally before the infusion of endotoxin
4) Fibrinogen level was less than 0.2 g/1
5) Normal rats group was infused 11.4 ml/4hr of saline and treated with neither dipyridamole nor endotoxin
* p<0.05, ** p<0.001 for difference to the values of control rats by Student's t-test

Dipyridamole. Before the infusion of endotoxin 0.5, 5.0 or 50.0 mg/kg of dipyridamole was injected intraperitoneally. A preventive effect against DIC was noted in all parameters in rats treated with 5.0 or 50.0 mg/kg of dipyridamole (Table 5).

Table 6. Effect of vitamin E (α-tocopheryl acetate) on experimental DIC in rats infused with 100 mg/kg of endotoxin for 4hr.

	FDP (μg/ml)	Fibrinogen (g/1)	PT (s)	PTT (s)	Platelet count ($x10^9/1$)	GFD[1] (%)	N[2]
HCO-60[3];							
0.2 ml/kg (control)	43.1±5.4	<0.2	22.9±3.8	121.6±12.4	235± 63	75.2± 9.7	12
α-tocopheryl acetate;							
0.01 mg/kg	40.7±5.9	0.21±0.04	21.9±3.7	102.1±16.4**	238± 66	61.1±12.5**	10
0.1	38.6±5.4	0.25±0.05**	21.3±3.4	92.3±20.4***	301± 61*	30.6± 8.1***	10
1.0	5.5±0.5***	0.32±0.06***	17.6±3.7**	67.8±12.0***	559±101***	4.9± 1.4***	10
10.0	5.1±1.0***	0.37±0.10***	15.1±2.3***	66.6±12.4***	662±111***	4.2± 1.5***	10
Normal rats[4]	2.3±0.5***	1.21±0.30***	11.9±0.8***	59.8± 5.2***	1076±281***	0***	14

Results are expressed as mean value±SD
1) Glomerular fibrin deposits
2) Number of subjects
3) HCO-60 was injected intraperitoneally for 4 successive days before the infusion of endotoxin
4) Normal rats group was infused 11.4 ml/4hr of saline and treated with neither α-tocopheryl acetate nor endotoxin
* P<0.05, ** P<0.01, *** P<0.001 for difference to the values of rats pretreated with HCO-60 as controls by Student's t-test

Vitamin E. DL-α-tocopheryl acetate, dissolved in HCO-60 at a final concentration of 50 mg/ml, was infused (0.01, 0.1, 1.0 or 10.0 mg/kg/day) intraperitoneally for 4 successive days; 10 min after the final injection, endotoxin was given. The preventive effect against DIC was noted in the rats treated with 1.0 or 10.0 mg/kg (Table 6).

Aspirin. The rats were intraperitoneally injected with aspirin at 0.03, 0.3,3.0 or 30.0 mg/kg, followed by the infusion endotoxin. A preventive effect was noted in FDP, PTT, platelet counts and the number of glomeruli with thrombi in rats treated with 0.03 or 0.3 mg/kg/ of aspirin, but not in fibrinogen level on PT (Table 7). Larger doses of aspirin showed less preventive effect against DIC.

Table 7. Effects of aspirin on experimeantal DIC in rats infused with 100 mg/kg of endotoxin for 4hr.

	FDP (μg/ml)	Fibrinogen (g/1)	PT (s)	PTT (s)	Platelet count ($x10^9/1$)	GFD[1] (%)	N[2]
Control[3]	54.2±7.2	<0.2[4]	30.2±6.5	148.3±16.1	78± 5	84.9±12.1	15
Aspirin[5]							
0.03 mg/kg	12.7±3.0**	<0.2	19.0±2.8**	79.4± 6.4**	164±26**	28.7± 5.8**	8
0.3	19.5±6.6**	<0.2	28.8±8.1	84.1±10.6**	122±27**	31.2± 6.6**	8
3.0	25.4±9.7**	<0.2	27.8±7.2	124.8±14.5*	75±13	35.0± 5.1**	
30.0	23.5±6.7**	<0.2	29.0±9.4	140.1±18.2	84±18	60.5±14.5**	8
Normal rats[6]	1.9±0.2**	0.98±0.21**	12.6±0.5**	52.2± 4.0**	832±98**	0**	15

Results are expressed as mean value±SD
1) Glomerular fibrin deposits
2) Number of animals
3) Saline (1.0 ml) was injected once intraperitoneally before the 4hr infusion of endotoxin (100 mg/kg)
4) Fibrinogen level was less than 0.2 g/l
5) Aspirin was dissolved in 1.0 ml of saline
6) Normal rats were infused with 11.4 ml/4hr of saline alone
* $p<0.01$, ** $p<0.001$ for difference to the values of control rats by Student's t-test

Changes in the clotting system after the injection of batroxobin. The fibrinogen level dropped rapidly in the first hour, reached the

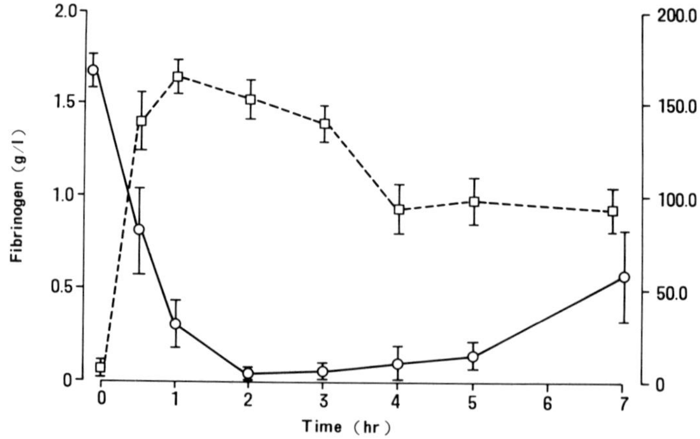

Fig. 5 Fibrinogen and FDP levels following the initiation of batroxobin injection at time 0. (○); fibrinogen, (□); FDP.

nadir at 2 hr and began to gradually increase at 5 hr (Fig. 5). Conversely, the FDP level showed a rapid increase in the first 30 min reached the maximal level at 1 hr and then gradually decreased.

PT and PTT were remarkedly prolonged 2 hr after the batroxobin injection (Fig. 6), whereas platelets, blood cell counts, and hematocrit level decreased only slightly or not at all.

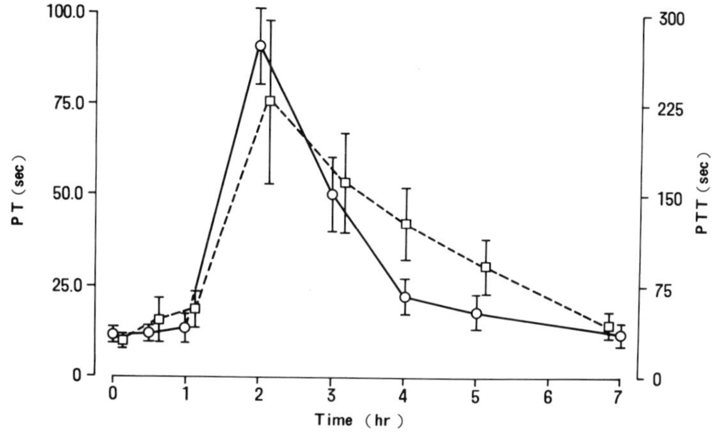

Fig. 6 PT and PTT following the initiation of batroxobin injection at time 0. (○); PT, (□); PTT

Batroxobin. One hour before the infusion of endotoxin, batroxobin (200 BU/kg) was injected intraperitoneally. The preventive effect against DIC was noted by the partial inhibition of the fall in the platelet counts and reduced number of renal glomeruli with fibrin thrombi (Table 8).

Table 8. Effect of defibrinogenation with batroxobin on experimental DIC in rats infused with 100 mg/kg/4hr of endotoxin.

	FDP (μg/ml)	Fibrinogen (g/l)	PT (s)	PTT (s)	Platelet count ($x10^9/l$)	GFD[1] (%)	N[2]
Control[3]	57.4± 8.8	0.04±0.02	30.5± 6.0	153.2±18.4	66± 6	87.6±13.3	16
Defibrinogenated[4]	137.3±18.7*	0.08±0.03	37.2±11.7	140.3±34.2	253±52*	11.6± 5.8*	8
Normal rats[5]	4.1± 0.6*	1.16±0.11*	12.5± 0.5*	59.4± 4.2*	625±96*	0*	8

Results are expressed as mean value±SD
1) Glomerular fibrin deposits
2) Number of rats
3) Saline (1.0 ml) was injected once intraperitoneally 1hr before the infusion of endotoxin
4) Batroxobin (200 BU/kg), diluted in 1.0 ml of saline, was injected intraperitoneally 1hr before the infusion of endotoxin
5) Normal rat group was infused with 11.4 ml/4hr of saline only
* P<0.001 for difference to the values of control rats by Student's t-test

Discussion

As DIC occurs by a very complicated mechanism, its study is difficult in many respects. Further, human DIC differs from the animal counterpart in various ways so that it is unclear whether data obtained in animals can be directly transferred to man. In the present investigation, however, experimental DIC induced in rats showed a close resemblance to human DIC as judged from such changes as an elevation in FDP and %GFD, prolongation of PT and PTT, and depression in fibrinogen and platelet counts. Using this experimental model of DIC, we investigated the effects of steroid, antiplatelet agents and defibrinogenation against DIC. Significant preventive effect was noted in all the above parameters in rats treated with methylprednisolone, ticlopidine, dipyridamole, DL-α-tocopheryl acetate, aspirin and batroxobin.

Acute massive doses of glucocorticoids protect against shock (10), and prolonged inschemia (11). On the other hand, they were shown to prevent thrombosis in hyperlipemic rats (12) as well as DIC induced by surgery (13), and cardiac arest (14). Little is known about the mechanisms of the antithrombotic action of glucocorti coids. In vitro, these steroids were shown to interfere with Hageman factor activation by leukocyte granules (15).

Besides blood coagulation, the response of microcirculation to endotoxin is also known to be an essential component of the generalized Shwartzman reaction (16). Glucocorticoids modify this vasomotor response; production of vasodilation and opening of arterio-venous shunts have been proposed to explain the action of glucocorticoids against shock(11). Massive doses of glucocorticoids such as methylprednisolone, 30 mg/kg, given to animals or man in shock can apparently prevent the aggravation. (17). The dose must be massive to cause vasodilatation. In our experiment also larger doses of methylpredrisolone prevented experimental DIC induced by endotoxin more effectively. Thus, glucocorticoids, appear to prevent DIC. Clinically, we found that this agent was effective for the treatment of DIC (18).

It is well established that platelets play an important role in the development of thrombosis. Ticlopidine is thought to inhibit platelet aggregation and prostaglandin synthesis from endogenous substrate by activating basal and prostaglandin (PG)E_1-stimulated activity of the adenylate cyclase, preventing PGE_2- induced depression of the adenylate cyclase activity and increasing the platelet adenosine 3',5'-cyclic monophosphate (c-AMP) level (19). Dipyridamole was reported to inhibit platelet aggregation (20). The anti-aggregatory effect of this agent was related in vitro to inhibition of platelet phosphodiesterase (21) and to inhibition of adenosine uptake by platelets leading to changes in the intra-cellular adenine nucleotide pool (22). Both effects may alter platelet c-AMP levels and platelet aggregability. Vitamin E inhibits platelet aggregation (23,24), and can prevent thromboembolism due to its antithrombin properties (25). In 1967 Weiss and Aledort reported that platelet aggregation and the release reaction were inhibited after ingestion of aspirin in man, and suggested that the drug might be a useful antithrombotic agent (26). There is now considerable evidence that aspirin inhibits platelet aggregation and secretion (27) by blocking the synthesis of prostaglandins (28) and thromboxane A_2 (29). This action is achieved by inhibition of the cyclo-oxygenase enzyme that converts arachidonic acid to prostaglandin G_2, the precursor of thromboxane A_2 (28,29).

The present study demonstrated the preventive effects of these anti-platelet agents against the experimental DIC. These results show that platelet aggregation may play a significant role in the onset of DIC. Aspirin was less effective in the prevention of experimental DIC, whereas ticlopidine almost completely prevented it. These differences in the effectiveness of these agents may be due to the potency of the anti-thrombotic action.

Specific fibrinogen removal, discontinuation of fibrin accretion in an established thrombus, a potential moderate thrombolytic action as well as reduction of blood viscosity by batroxobin have been applied in numerous animal trials to prevent and/or to treat experimental and arterial thrombosis and to investigate the role of fibrinogen in thrombogenesis (3). The anti-thrombotic effect of batroxobin is believed to be produced by both the decrease in fibrinogen and inhibition of platelet function (30). Large amounts of FDP in the plasma inhibit platelet aggregation (31). These findings might shed light on the possible mechanism of this agent in inhibiting the aggravation of endotoxin-induced experimental DIC. The decrease in fibrinogen and the increase in FDP in the DIC state might contribute in some way to the defence mechanisms against DIC.

In conclusion , the fact that these agents prevent the endotoxin-induced DIC in rats suggests that generalized Shwartzman reaction, platelet aggregation and fibrinogen metabolism play important roles in the DIC state.

References

1. Minna, J.D., Robby, S.I., Colman, R.W.: Disseminated intravascular coagulation in man. Charles C. Thomas, Springfield 1974.
2. Latour, J.G.. McKay, D.G., Nasu, K.: Am. J. Obstet. Gynec. 112, 863-867 (1972).
3. Stocker, K.: Fibrinolytics and Antifibrinolytics F. Markwardt (ed., Spronger, Berlin, Heidelberg, New York 1977, pp. 451-484.
4. Quick, A.J., Stanly, B,SG.M., Bancroft, F.W.: Am. J. Med. Sci. 190, 501-511 (1935).
5. Langdell, R.D., Wagner, R.H., Brinkhouse, K.M.: J. Lab. Clin. Med. 47, 637-647 (1953).
6. Ratnoff, O.D., Menzie, C.:J. Lab. Clin. Med. 37, 316-320 (1951).
7. Ferreira, H.C., Murat, L.g.: Br. J. Haemat. 9, 299-310 (1963).
8. Brecker, G., Cronkite, E.P.: J. Appl. Physiol. 3, 365-377 (1950).
9. Weigert, C.: Fortschr. Med. 5, 228-232 (1887).
10. Fine, J., Palmerio, C., Rutenberg, S.: Archs Surg. 96, 163-174 (1968).
11. Lillehei, R.C.: Archs Surg. 78, 464-471 (1959).
12. Renaud, S., Allard, C., Latour, J.G.: Am. Heart J. 72, 797-805 (1966).
13. Latour, J.G., Renaud, S.: Can. Med. Ass. J. 100, 20-26 (1969).

14. Latour, J.D., McKay, D.G., Parrish, M.H.* Thromb. Diath. Haemorrh. 27: 543-553 1972.

15. Cline, M.J., Melmon, K.K.: Science, 153, 1135-1138 (1966).

16. McKay, D.G., Latour, J.D., Lopez, A.M.: Thromb. Death. Haemorrh. 26, 71-76 (1971).

17. Mostsay, G.G., Alho, A., Jaeger, T., Dietzman, R.H., Lillehei, R.C.: Fed. Proc. 29, 1861-1873 (1970).

18. Yoshikawa, T., Murakami, M., Furukawa, Y., Onodera, H., Ueda, M., Kato, H., Takemura, S., Kondo. M.: Jpn. J. Anesthesiol. 32, 653-661, 1983.

19. Ashida, S., Abiko, Y.: Thromb. Haemostas. 41, 436-339 (1979).

20. Weis, H.J.: N. Engl. J. Med. 298, 1344-1347 (1978).

21. Moncada, S., Korbut, R.: Lancet 1, 1286-1289 (1978).

22. Pearson, J.D., Carleton, J.S., Hutchings, A., Gordon, J.L.: Biochem. J. 170, 265-271 (1978).

23. Fong, J.S.: Experientia 32, 639-641 (1976).

24. Steiner, M.: Tocopherol, Oxigen and Biomembrances, de Duve C, Hayashi, O (Eds), Elsevier/North-Holland Biomedical Ress, Amesterdam 1978, pp. 143-163.

25. Ochsner, A.: N. Engl. J. Med. 271, 211 (1964).

26. Weiss, H.J., Aledort, L.M.: Lancet 2, 495-497 (1967).

27. Packham, M.A., Mustaed, J.F.: Blood 50: 555-573 (1977).

28. Smith, J.B., Willis, A.L.: Nature (new Biol.) 231,235-237 (1971).

29. Samuelsson, B.: Prostaglandins in hematology, Silver, M.J., Smith, J.B., Kocsis, J.J. (Eds.), Spectrum, New York 1977, pp. 1-10.

30. Markwardt, F., Nowak, G.: Thromb, Res. 17: 103-111 (1980).

31. Kowalski, E., Budzynski, A.Z., Kopec, M., Lattallo, Z.S., Lipinski, B., Wegrzynowicz, Z.: Thrombos. Diathes. Haemorrh. 12, 69-86 (1964).

SEROLOGICAL AND IMMUNOHISTOCHEMICAL ANALYSIS ON THE ROLE OF COMPLEMENT IN ENDOTOXIN INITIATED LETHALITY IN MICE

Shigenobu Matsuo, Morio Totsuka, Hiroshi Hayasaka
First Department of Surgery, Sapporo Medical College
Sapporo, Japan

Kokichi Kikuchi
Department of Pathology, Sapporo Medical College
Sapporo, Japan

Introduction

The endotoxins activate the complement system mainly via the alternative pathway. Although the complement system is supposed to play an important role in the host defence mechanism, some of activated components are injurious to the host. Namely, C3a and C5a are known generally to have the anaphylatoxic activity, the C5a promotes the inflammatory response as a chemotactic factor, and $C\overline{142}$ has a kinin-like activity. Furthermore, $C\overline{567}$, $C\overline{5678}$ and $C\overline{56789}$ complexes bound to the cell membrane results in the lysis of cells. Moreover, the complement components cooperate with other humoral or cellur agents, and provoke several harmful effects in the host (1). It appeared that the complement system can trigger the endotoxic shock and exacerbate its progression (2). However, there have been several reports concluding that the complement is of little importance in endotoxin-induced lethality. Curry and Morrison, using the endotoxin resistant C3H/HeJ mouse, demonstrated that neither complement activation, nor the lack of it, is necessary for killing the mouse (3). To analyse the complement system of the endotoxin resistant C3H/HeJ mouse with endotoxemia seems to be important for understanding the mechanism

Immunopharmacology of Endotoxicosis

of endotoxin-induced lethality. In the present study, we have undertaken to study the complement activation following endotoxin injection in C3H/HeJ mouse.

Materials and Methods

1) Mice and bacterial endotoxin. 6 - to 8- week-old male C3H/HeJ and C3H/HeN mice were used. C3H/HeJ mice are known to be endotoxin resistant, and C3H/HeN mice to be endotoxin sensitive. Mice were bled from the orbital venous plexus into plastic tubes. After the blood had been allowed to clot for 15 min at room temperature, it was centrifuged at 2,000 rpm for 20 min at 4°C, and fresh serum was collected and used immediately. Fresh EDTA plasma was prepared as follows : 20 μl 0.1M EDTA-Na3 solution, pH 7.5 was previously suctioned into plastic pipet, and 160 μl blood was further collected from the orbital venous plexus into the same plastic pipet. It was immediately replaced into plastic centrifuge tube, and after centrifugation at 2,000 rpm for 10 min at 4°C, fresh EDTA plasma was collected and used immediately. A lipopolysaccharide type W derived from *E. coli* 0127: B8(Lot; 6784687, Difco Laboratories, Detroit, Mich.) was used as endotoxin. Appropriate quantities of endotoxin, suspended in 0.5 ml isotonic sterile saline, were injected i.p. to mice.

2) Crossed immunoelectrophoresis. It was performed according to the method of Laurel (4). 1.1 % agarose solution prepared by barbital buffer, pH 8.6, μ = 0.025 was used, and contained 0.005M EDTA to avoid conversion of β1C globulin during electrophoresis. The antigen of fresh EDTA plasma was put into the well, and antimouse C3 serum was mixed with agarose solution to obtain 8 % final concentration of antiserum. First and second electrophoresis were carried out at 4°C for 2 hr at 3 mA/cm. Rabbit antimouse C3 serum was prepared by the method of Tamura and Nakayama

(5, 6), and the specificity of this antiserum was confirmed by Ouchterloney's method, immunoelectrophoresis (7) and crossed immunoelectrophoresis.

3) Assay for C3 consumption. Groups, of six C3H/HeJ and C3H/HeN mice were bled at 30 min and 1, 4, 10, 24, 48 hr after injection of 5.0 mg/Kg endotoxin or saline. Levels of circulating C3 were determined by single radial immunodiffusion described by Mancini et al (8). Preliminary experiments showed that a 1 : 25 dilution of the antiserum provided optimal C3 precipitation using normal mouse serum at dilutions of 1 : 1, 1 : 2 and 1 : 4. Precipitation rings were analyzed in duplicate after reaction for 5 days at 4°C. Fresh serum was used as a sample, and the results were expressed as a percentage of the saline control.

4) Determination of the activity of alternative complement pathway. Groups of thirty C3H/HeJ and C3H/HeN each mice were injected with 5.0 mg endotoxin. Five mice were bled at 30 min, 1, 4, 10, 24 and 48 hr after injection, and the activity of alternative complement pathway(AP 50) was determined, using rabbit erythrocyte, by the method described by Dijk et al (9). Fresh serum was assayed as a sample.

5) Enzyme-labeled antibody method. PAP-Avidine-Biotin method described by Kamiya et al (10) was employed to investigate the histological localization and the degrees of C3 accumulation in the liver. Rabbit anitmouse C3 serum diluted 1 : 150 by phosphate buffer saline (PBS) was used as a first antibody. The liver of mouse injected with 5.0 mg/Kg endotoxin was perfused with 4 % paraformaldehyde supplemented with 0.1 M phosphate buffer, pH 7.4 through the portal vein. Fixed liver was cut into 2 - 3 mm slices in thickness, and these were embedded into paraffin after washing with PBS overnight. 4 µm thick paraffin sections were deparaffinized by xylene, hydrated with gradually decreasing concentrations of alcohol, and washed in PBS.

Table 1 Lethality following injection of LPS

Dose of LPS (mg/Kg)	C3H/HeJ		C3H/HeN	
	24hr	48hr	24hr	48hr
2.5	0/6	0/6	0/6	0/6
5.0	0/6	0/6	0/6	3/6
10	0/6	0/6	0/6	6/6
25	0/6	0/6	3/6	6/6
100	0/4	0/4		

Results

1) Mortality of mice. As shown in Table 1, none of C3H/HeJ mice died even 48 hr after injection with 100 mg/Kg of the endotoxin, whereas three of six C3H/HeN mice died with small dosage as 5.0 mg/Kg of the endotoxin.

2) Changes of serum C3 levels as AP50 values (Fig. 1). There were no significant differences in serum C3 levels between C3H/HeJ and C3H/HeN mice except for the C3 levels of mice 4 hr after injection and the AP50 values 1 hr after injection.

3) Cenversion pattern of C3. As shown in Fig. 2, C3H/HeJ and C3H/HeN mice did not show any differences with respect to the conversion pattern of C3. In the plasma of both mice, β1c, β1D and small amount of β1G globulin were present at 30 min. At 1 hr, larger amount of β1G globulin was detected in addition to β1C globulin. At 4, 10, 24 and 48hr, there appeared no differences in the precipitation curves between preadministrative and postadministrative serum of these mice.

4) C3 accumulation in the liver (Fig. 3). In both C3H/HeJ and C3H/HeJ and C3H/HeN mice, it was demonstrated that C3 was accumulated in Kupffer cells at 30 min, 1, 4 and 10hr. However, it appeared negative at 24 and 48 hr. C3 accumu-

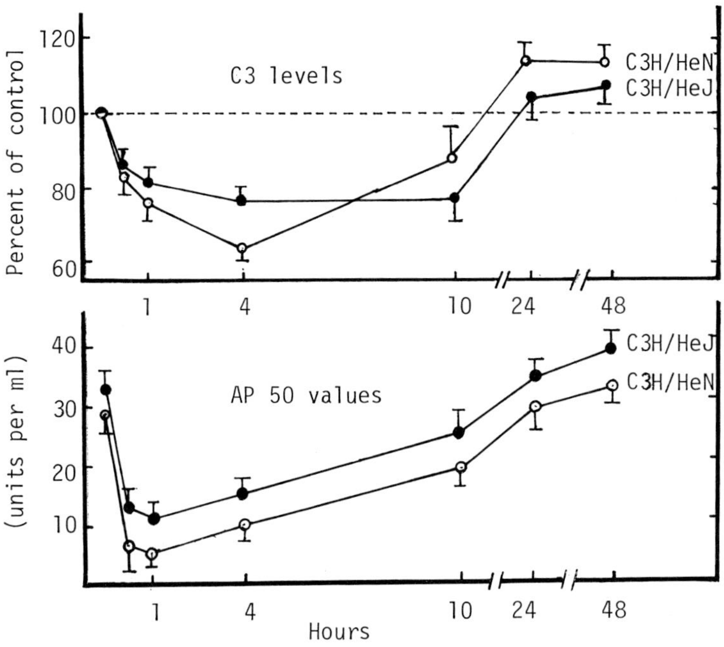

Fig. 1 The changes of C3 levels and AP 50 values in C3H/HeJ (●) and C3H/HeN (○) mice after i.p. injection with 5.0 mg/Kg endotoxin.

lation in the Kupffer cells of C3H/HeJ mice at 30 min and 1 hr seemed to be slightly decreased, when compared with those of C3H/HeN mice at the same time. Necrotic foci in the liver were observed in just one mouse of five C3H/HeN mice at 4 and 24 hr, but none of C3H/HeJ mice.

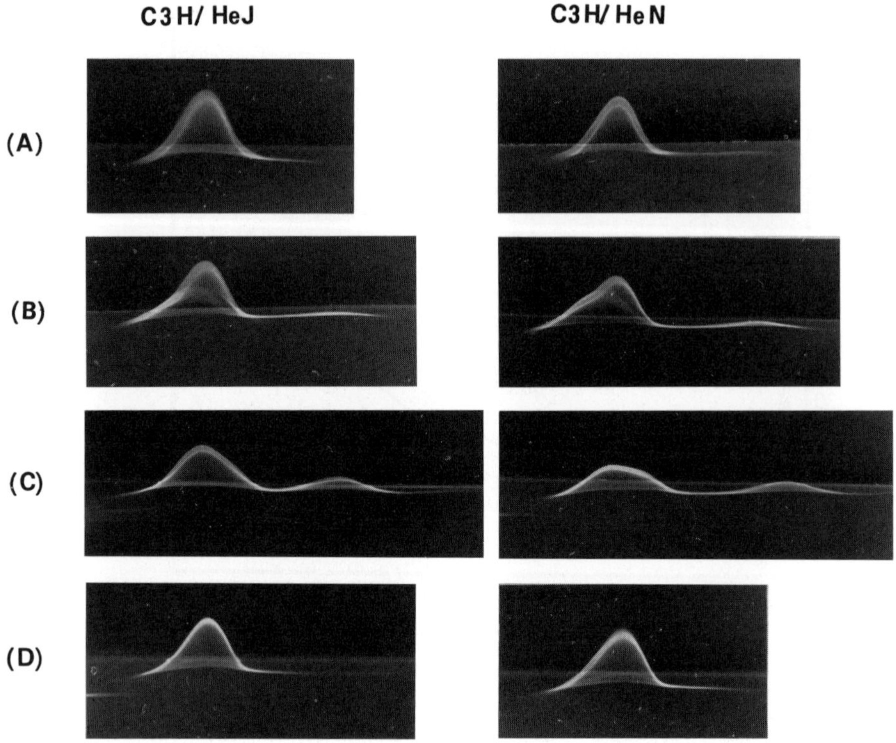

Fig. 2 Electrophoretic patterns of C3 at various time after i.p. injection with 5.0 mg/Kg endotoxin. C3H/HeJ mice are shown in the left, and C3H/HeN mice in the right. (A): control (B): 30 min (C): 1 hr (D): 4 hr.

Discussion

Our data indicate that serum complement activation and C3 accumulation in Kupffer cell in C3H/HeJ mouse showed little difference as compared with those in C3H/HeN mouse. Many investigators have previously studied the role of the comple-

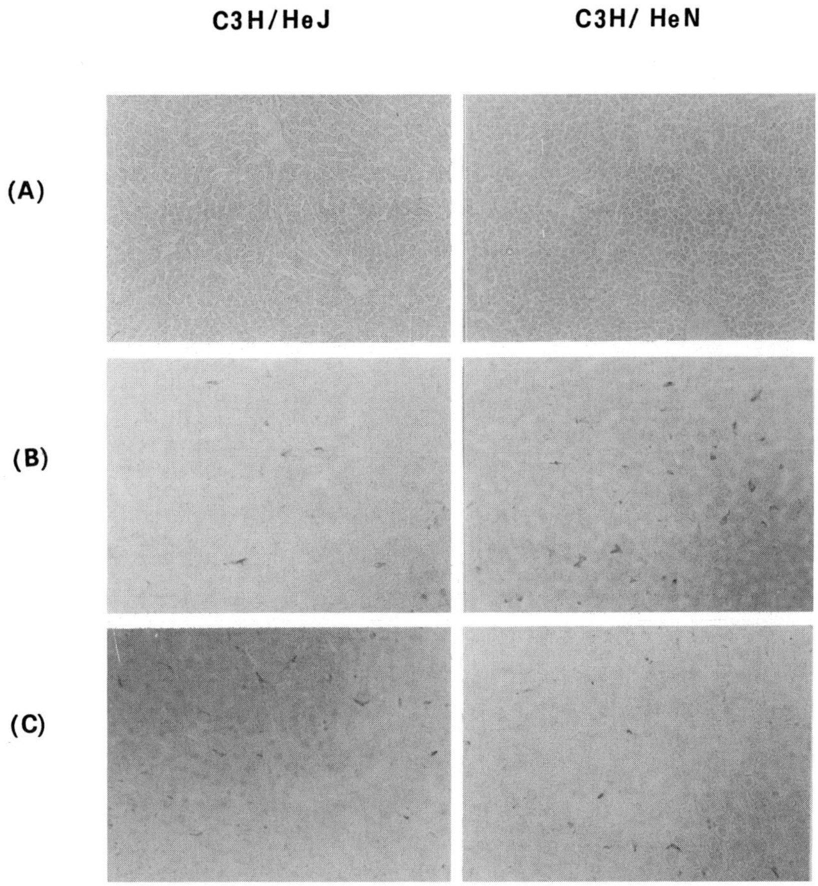

Fig. 3 The C3 accumulation in the liver of mice by immunohistochemical analysis with PAP-Avidine-Biotin method. The staining pattern of the liver of C3H/HeJ mice is shown in the left, and that of C3H/HeN mice is shown in the right. The liver was fixed by perfusion at 30min (B) and 10 hr (C) after injection with 5.0 mg/Kg endotoxin. (A): control mice injected with normal saline

ment system in the various pathophysiological effects of endotoxins. Most of these studies used the complement depleted or deficient animals. Brown and Lachman reported that all of sixteen normal rabbits died following the injection of 5 mg endotoxin, whereas no C3-9 depleted rabbits, treated with cobra venom factor, died. However, more recent experiment of Ulevitch and Cochrane (11), in which same endotoxin preparation and animal were employed, demonstrated completely different results.

C3H/HeJ mice have been recently utilized by many investigators to study the mechanism of the endotoxin initiated lethality. Hoffman reported that the serum of C3H/HeJ mouse did not show the fixation of C3 on sensitized erythrocytes (12). On the other hand, Curry and Morrison could not confirmed this observation, and they clarified that circulating C3 levels in C3H/HeJ mouse injected with endotoxin decreased similarly in endotoxin sensitive C3H/St mouse (3). Furthermore, administration of a weekly anticomplementary LPS preparation did not reduce the mortality in C3H/St mouse. Therefore, they considered that neither complement activation, nor the lack of it, is responsible for lethality in mice. The results obtained in the present study supported the concept that serum complement activation and C3 accumulation in the Kupffer cells may not be, by themselves, sufficient to induce the liver necrosis and the lethal response in mice. However it did not indicate that complement activation is not necessary for the induction of endotoxin initiated lethality. Most bacterial endotoxins seem to have a strong anticomplementary effect, and in case of clinical endotoxin shock, circulating C3 levels and haemolytic complement titers decreased in most patients (13). Therefore activated complement systems are considered to play some role in the initiation and progression of endotoxin shock. The study by Wahl et al. indicated that macrophages of C3H/HeJ stimulated with endotoxin are defective in prostaglandin synthesis (14). Recently, attention was given to arachidonic acid metabolites such as prostaglandins or leukotrienes having strong inflammatory effects (15). These investigators indicated that there is correlation between prostaglandin production by the macrophages and the lethal effects of the endotoxin preparations. They surmised that the released prostaglandins, in turn, induce hypotension, hypothermia and dehydration, all of which may contribute to the process that ultimately results in death. Recent study by Rutherford and Schenkein revealed that C3 cleavage products stimulate the release of prostaglandins from human mononuclear phagocytes in vitro (16). These two reports make us presume that complement activation in endotoxemia performs an important role in endotoxin initiated lethality by stimulating macrophages to promote the arachidonic acid metabolism. The results obtained from our study indicate that the C3H/HeJ mouse has no disorders in the serum complement activation and the accumulation of C3 in the macrophages in endotoxemia. The changes of the electrophoretic pattern of C3 following endotoxin injection observed in our study were different from the finding demonstrated by Mardiney and Muller-Eberhard (6). Our data showed that β1D globulin was present in the plasma 30 min after endotoxin injection, and β1G globulin appeared after 30 min with a peak at 1 hr. However, in contrast to our data, they showed that β1G globulin was present in the EDTA plasma at 10 min, and β1D globulin appeared at 2 hr following incubation with EAC 142a cells. Although the cause of these differences is not yet clear, it is possible that these differences may result from the experimental systems. In our experiments, the reaction of endotoxin and complement system was performed in vivo, whereas in their experiments, EDTA plasma was incubated with EAC 142a cells.

References

1. Osler, A.G.: Complement. Mechanism and Function, Prentice-hall, Inc., Englewood Cliffs, New Jersey (1976).

2. Eiseman, B., Sloan, R., Hansbrough, J., McIntosh, R.: Am. Surg. 46, 14-19 (1980).

3. Curry, B.J., Morrison, D.C.: Immunopharm. 1, 125-135(1979).

4. Laurell, C.B.: Anal. Biochem. 10, 358-361 (1965).

5. Tamura, N., Nakayama, H.: "Methods in Immunological Experiment VI" edited by Japanese Society for Immunology, (Jap) p.1949 Kanazawa (1977).

6. Mardiney, M.R., Jr., Muller-Eberhard, H.J.: J. Immunol. 94, 877-882 (1965).

7. Graber, P., Williams, C.A.: Biophys. Acta. 10, 193-194 (1953).

8. Mancini, G., Carbonara, A.O., Heremans, J.F.: Immunochem. 2, 235-254 (1965).

9. Dijk, H. Van, P.M. Rademaky, J.M.N. Williers: J. Immunol. Methods. 36, 29-39 (1980).

10. Kamiya, H., Masakatsu, I.: Clin. Immunol. (Jap) 14, 191 (1982).

11. Ulevitch, R.J., Cochrane, C.G.: Infect. Immun. 19, 204-211 (1978).

12. Hoffman, M.K.: J. Immunol. 121, 619-621 (1978).

13. Fearon, D.T.: Ruddy, S., Schur, P.H., McCabe, W.R.: N. Engl. J. Med. 292, 937-940 (1975).

14. Wahl, L.M., Rosenstreich, D.L., Glode, L.M., Sandberg, A. L., Mergehagen, S.E.: Infect. Immun. 23, 8-13 (1982).

15. Dahlén, S.-E., Björk, J., Hedoquist, P., Arfors, K.-E., Hammarstorüm. S., Lindgrem, J.-A., Samuelsson, B.: Proc. Natl. Acad. Sci. USA 78, 3887-3891 (1981).

16. Rutherford, B., Schenkein, H.A., J.: J. Immunol. 130, 874-877 (1983).

STUDIES ON ENDOTOXIN-LIKE PROPERTIES OF SEVERAL SIMPLE POLYSACCHARIDES

Takeshi Mikami, Toshihiko Nagase, Shigeo Suzuki*, and Masuko Suzuki
Department of Microbiology and *The Second Department of Hygienic Chemistry, Tohoku College of Pharmacy
Sendai 983 Miyagi, Japan

Introduction

It has been well-known that medical injections polluted with microbes cause severe febrile symptoms in the recipients. In order to avoid such accidents, many studies have been conducted for detecting and eliminating pyrogens, especially endotoxin of Gram-negative bacteria. However, it is now evident that a variety of materials besides LPS such as fungi, Gram-positive bacteria, mixovirus, poly I:poly C, colchichine, muramyl dipeptide (MDP), phorbol myristate acetate also elicite pyrogenic activity (1).

In the present pepar, we report that several simple polysaccharides, mannans and dextrans from *Saccharomyces cerevisiae* and *Leuconostoc mesenteroides*, respectively, are able to react with *Limulus* amoebocyte lysate (LAL), and to evoke pyrogenicity in rabbits. In addition to the above findings, the results of analytical study on correlation between chemical structures and these biological effects have also be presented.

Materials and methods

Rabbits. Male albino rabbits (Funabashi Farm, Chiba, Japan) weighing 1.8 to 2.7 kg, which did not show decrease of the

Immunopharmacology of Endotoxicosis

body weights during the feeding period, were used.

Polysaccharides. 1) Mannans. The neutral mannans of *Saccharomyces cerevisiae* wild type strain (bakers' yeast) and *S. cerevisiae* X2180-1A-5 mutant strain, abbreviated as WNM and MNM, respectively, were prepared in accordance with the previous description by Matsumoto et al. (2). Synthetic mannans, poly-α-(1→6')-anhydro-D-mannopyranose and poly-α-(1→2')-anhydro-D-mannopyranose, were the generous gifts of Dr. C. Schuerch. 2) Dextrans. The dextrans from *Leuconostoc mesenteroides* IAM 1046, NRRL B-1416, and NRRL B-1375 were prepared from the culture broths of those strains according to the method described by Niinobe and Kobayashi (3). Dextran T 2000 High-molecular dextran, both of which had molecular weights of 2,000,000, were purchased from Pharmacia Fine Chemicals, Uppsala, Sweden, and Meito Sangyo Co., Ltd., Nagoya, Japan, respectively. Dextran T 40, with a molecular weight of 40,000, was also purchased from Pharmacia Fine Chemicals. The soluble starch was a commercially available guaranteed class reagent supplied by Wako Chemical Industries, Ltd., Osaka, Japan.

Lipopolysaccharide (LPS). LPS of *Salmonella typhosa* 0901 strain was a commercial product of Difco Laboratories, Detroit, U.S.A.

Preparation of partially acid-hydrolysed mannan. This was conducted according to the description by Matsumoto et al. (2) by means of 0.33N H_2SO_4. The acid-modified mannan was designated as WNM-a.

Treatment of mannans and LPS with aqueous alkali solution. This was conducted in accordance with the previous description (4). To an aqueous solution of 4 mg of each sample per ml was added an equal volume of 200 mM NaOH. After incubation at 30°C, the mixture was neutralized with 100 mM CH_3COOH to pH 7, then dialyzed against water, in a 1,000-ml quantity, three times, for 24 hr, and the sample was recovered by lyophylization.

Limulus amoebocyte lysate. This was a commercial product prepared from amoebocytes of the Japanese horseshoe crab (*Tachy-*

pleus tridentatus) and was kindly supplied as a lyophilized powder in an ampoule, by Teikoku Hormone Mfg., Co., Tokyo, Japan.

Synthetic chromogenic substrate (SCS) for amidase activation assay of LPS. A reagent kit, PYRODIC®, was purchased from Teikoku Hormone Mgf., Co.

Reaction of polysaccharides with Limulus amoebocyte lysate. 1) Gelatoin assay. This assay was carried out according to the method reported previously (4). Briefly, the lyophilized Limulus amoebocyte lysate was dissolved in 100 μl of water in the ampoule. To the solution was added an equal volume of an aqueous solution of a known quantity of the test compound, the mouth of the ampoule was the covered with a piece of parafilm, and the mixture was incubated at 37°C for 1 hr. Negative and positive controls consisting of 100 μl of water, with or without endotoxin, 1 μg/ml, were run simultaneously. The presence of a solid gel (++) and marked increase in viscosity and turbidity (+) after incubation were scored as positive reactions. A weakly positive reaction (±) was recorded separately from the positive and negative (-) reactions. 2) Amidase activation assay. This assay was performed using PYRODIC® according to the manufacturer's mannual. The amount of liberated p-nitroaniline was quantitiated by reading the absorption at 405 nm.

Assay for pyrogenicity in rabbits. Rabbits were fed for one week at 22 ± 2°C and 50 ± 5% relative humidity. 1) Preparation and administration of injections. In order to avoid contamination from other pyrogens in the injection of samples, all buffers and solutions were tested for the absence of pyrogen before use. All glasswares and other utensils were sterilized for 30 min at 250°C in a hot-air sterilizer to abolish the effect of pyrogen. Each sample was dissolved in saline, but the sample for administration to the third cerebroventricle was dissolved in Dulbecco's phosphate-buffered saline (PBS). 2) Cannula implantation. In order to conduct intracerebroventricular (i.c.v.) administration of test materials, cannula was implanted on the cranium of each rabbit according the method of Kohno et al. (5). After anesthetization by s.c. administration of pentobarbital sodium salt (45 mg/kg), a guide cannula (stainless steel hypodermic needle (23 ga) cut to 25 mm lenght) was stereotaxically implanted into the third cerebroventricle from a small hole of cranium, according to an atlas by Sawyer et al. (6). The cannula was then fixed to the cranium with dental cement. The rabbits were kept feeding for 10 days in order to recover the wound of operation. 3) Mesurement of body temperature. This was conducted under the same conditions as the feeding environment. Each rabbit was fixed on the pyrogen test plate, and each sample solution was injected through ear vein. For administration to the third cerebroventricle, 50 μl of each solution was injected by means of an infusion cannula (stainless steel pipe, 0.6 mm diameter, connect tightly to mycrosyringe by polyethylene tube) through the guide cannula fixed to the cranium. Rectal temperature was recorded by means of THERMO-FINER® (Terumo Co., Ltd., Tokyo, Japan). Representation of the magnitude of generated fever was essentially the same as the description by Keen et al.(7), but the pyrogenicity of the samples was expressed as the fever index 6 hr (FI-6) which corresponds to an area under the fever curve when 1 hr and 0.5°C were plotted as 10 mm. The magnitude of fever was expressed as average value of FI-6 of four rabbits. 4) Treatment of LPS and mannans with sodium deoxycholate(SDOC). This was conducted by admixing equal volume of 0.5% SDOC saline solution to each sample solution. After leaving at room temperature for 20 min, the solution was administered to ear vein of rabbit (8).

Results

1) Chemical analysis of the polysaccharides. Analytical data for the polysaccharides are given in Table I. Nitrogen content of these polysaccharides was less than 0.3%, and phosphorus was not detected in any polysaccharide specimen.

Table I. Chemical Composition and Properties of Polysaccharides

Sample	Total carbohydrate (%)	Total phosphorus (%)	Total nitrogen (%)	Average length of branches	S_{20}
WNM	96.2	0.0	<0.3	2.4	3.2
WNM-a	95.2	0.0	<0.3	2.2	2.1
MNM	93.0	0.0	<0.3	1.1	2.2
Dextran IAM 1046	98.0	0.0	<0.3	-	-
Dextran NRRL B-1416	97.7	0.0	<0.3	-	-
Dextran NRRl B-1375	97.0	0.0	<0.3	-	-

2) Reaction of simple polysaccharides with LAL. 1. Gelation activity. Table II shows the results of the determination of minimum concentrations of several polysaccharides required for a positive LAL gelation. The neutral mannans, WNM and MNM, were positive in the reaction. WNM, a mannan containing very dense branches, gelated LAL at concentrations higher than 1 μg/ml. MNM possessing few branches and two synthetic mannans without branches, poly-α-(1→2')-anhydro-D-mannopyranose and poly-α-(1→6')-anhydro-D-mannopyranose, also gelated LAL at concentrations of 1, 10, and 500 μg/ml, respectively. Dextrans from *L. mesenteroides* IAM 1046, NRRL B-1416, and NRRL B-1375 strains containing many branches composed of 1→3-linked D-glucopyranosyl residues give a positive LAL reaction at a concentration of 100 μg/ml or more, but dextran T 2000 and High-molecular dextran, having no branches, did not show this reaction. Soluble starch, possessing very dense branches consisting of α-1→4-linked D-glucopyranosyl residues, also showed no reaction in the LAL gelation (1 mg/ml).

Table III shows the change of LAL gelation activity of LPS, WNM, and dextran IAM 1046 after mild alkali treatment.

Table II. Reaction of *Limulus* Amoebocyte Lysate with Various Polysaccharides

Sample	Minimum concentration required for gelation (μg/ml)
WNM	1
MNM	1
Poly-α-(1→6')-anhydro-D-mannopyranose	500
Poly-α-(1→2')-anhydro-D-mannopyranose	10
Dextran IAM 1046	100
Dextran NRRL B-1416	100
Dextran NRRL B-1375	100
High-molecular weight dextran	–[a]
Dextran T 2000	–[a]
Dextran T 40	–[a]
Soluble starch	–[a]

[a] Negative at a concentration of 1 mg/ml

Table III. Effect of Alkali Treatment of Some Polysaccharides and LPS on the *Limulus* Amoebocyte Lysate Gelation Activity

Sample	Minimum concentration required for gelation (μg/ml)
WNM	1
WNM-b	10
Dextran IAM 1046	100
Dextran IAM 1046-b	100
LPS	10^{-9}
LPS-b	10

The activities of WNM and dextran IAM 1046 did not decrease upon treatment with 100 mM NaOH at 30°C for 48 hr. The minimum concentrations (end point) for positive LAL gelation of the alkali-treated WNM and dextran IAM 1046 were 10 and 100 μg/ml, respectively, which were only slightly different from the untreated polysaccharides. On the other hand, the gelation activity of alkali-treated LPS was significantly reduced to one-ten billionth of the value of untreated LPS.

2. Amidase activity. Table IV shows the results of SCS test for LPS which seem to be consistent with those provided by the LAL gelation test. Treatment of WNM with alkali caused a decrease in the amidase-activating effect, while WNM-a did not lose this activity. Treatment of LPS with alkali also resulted in decrease of this effect, similar to that observed in the LAL gelation assay.

3) Pyrogenicity of the polysaccharides. Fig. 1 shows the results of WNM pyrogenicity assay. It can be seen immediately that each fever response curve exhibits two summits at 1 and 3 hr, respectively, when dimensions under each curve were expressed as FI-6. It is evident that a linear proportionality exists between logarithmus of dose and FI-6 values in WNM as shown in Fig. 2.

Table IV. Synthetic Chromogenic Substrate Assay of Mannans and LPS

Sample	Relative activity
WNM	100
WNM-b	31.2
WNM-a	97.5
MNM	28.3
LPS	4.76×10^7
LPS-b	104

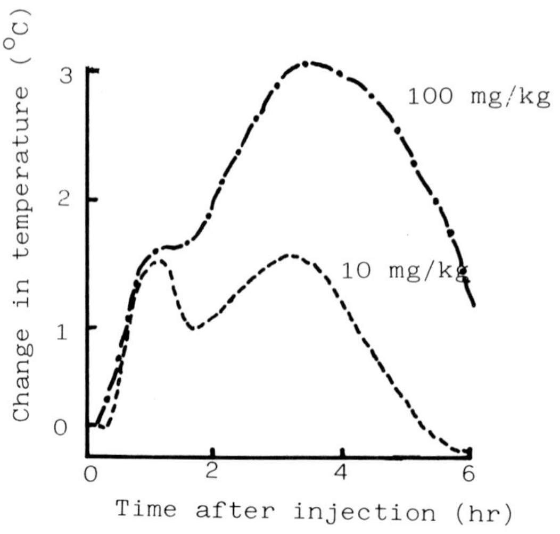

Fig. 1. Fever Response Curves of WNM in Rabbits

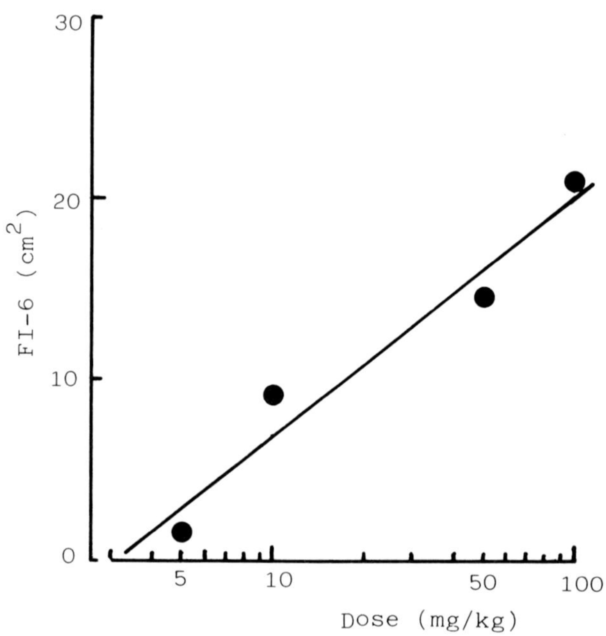

Fig. 2. Proportionality between Doses of WNM and FI-6 Values by i.v. Administration of WNM in Rabbits

Change of pyrogenicity by treatment with alkali. In order to compare the difference in mechanisms of pyrogenicity of WNM and LPS, both materials were treated with 100 mM NaOH at 30°C for 48 hr. The results of pyrogenicity assay indicated that LPS-b did not display the activity in the dose of 100 μg/kg, and no pyrogenicity was exhibited on WNM-b even in the dose of 300 mg/kg (data was not shown).

Effect of sodium deoxycholate (SDOC) treatment on pyrogenicity of LPS and WNM. Because SDOC was found to decrease pyrogenicity of LPS in less drastic manner than alkali-treatment, WNM was subjected to this treatment. As summarized in Table V, this treatment abolished pyrogenicity of LPS. Under similar conditions, however, SDOC-treated WNM showed pyrogenicity almost identical to that of the untreated mannan, demonstrating that lipoidal material was not involved in the pyrogenicity of WNM. As shown in Table I, molecular weight of WNM-a (S_{20} ; 2.1) was two-third of

Table V. Effect of Sodium Deoxycholate (SDOC) Treatment on the Pyrogenicity of WNM and LPS

Sample	Dose (mg/kg)	FI-6 (cm^2)	
		untreated	SDOC treated
WNM	50	14.6	8.8
LPS	0.001	21.9	0.6

Table VI. Pyrogenicity of Mannans and Dextrans in Rabbits

Sample	Dose (mg/kg)	FI-6 (cm^2)
WNM	10	9.7
WNM-a	10	9.9
MNM	10	0.0
	100	1.5
Dextran NRRL B-1416	100	7.0
Dextran NRRL B-1375	100	12.1

WNM (S_{20} ; 3.2), although a structural similarity, in terms of density of branching moieties, was observed in both mannans. On the other hand, MNM contained only small number of branching moieties (average length 1.1) as compared to those of WNM and WNM-a (average length 2.4 and 2.2, respectively). Although both mannans have similar α-(1→6') structures in the main chain, the results of assay of pyrogenicity of WNM-a and MNM revealed that the former mannan was able to show nearly identical pyrogenicity (in the dose of 10 mg/kg, FI-6 = 9.9) to that of WNM, while the latter mannan exhibited less activity (FI-6 = 1.5 in the dose of 100 mg/kg) as shown in Table VI. These fact indicate that density of the branching moieties of WNM is more important as an active site than length of the core in the expression of pyrogenicity by WNM.

Direct administration of WNM to hypothalamus temperature-regulatory center of rabbits. In order to obtain evidence for the presence of the target site(s), this mannan was administered through a cannula to the third cerebroventricle. After 2 hr, rabbits receiving administration of WNM in the

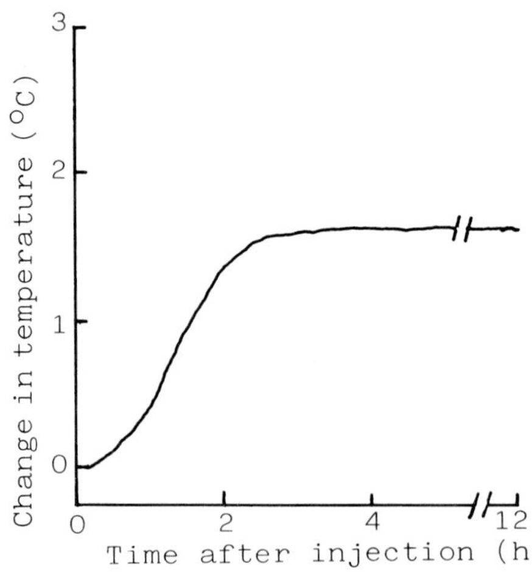

Fig. 3. Fever Response Curve of WNM obtained by Direct Administration into the Third Cerebroventricle of Rabbits in a Dose of 10 μg/kg

dose of 1 μg/kg showed a continuous body temperature elevation, of about 1.5°C up to of 12 hr (Fig. 3); WNM-b did not show this effect in the doses of 10 and 500 μg/kg by same administration route (data was not shown).

Discussion

In 1966, Radvany et al. (9) reported that pyrogenicity, Schwartzmann reactivity, and lethal activity of Serratia marcescens endotoxin in mice could be neutralized with homologous anti-S. marcescens antibody, and that the sugar residues corresponding to O-specific antigen participated in part in exhibiting these biological effects. One of the important findings obtained in the present study was also concerned with polysaccharides displaying these biological effects; WNM, WNM-a, and a dextran possessing dense branching moieties were able to exhibit positive pyrogenicity in rabbits by the i.v. route, while MNM and another dextrans containing few branches did not show this effect. The above finding seems to be different from that reported recently by Yamamoto and Iwata (10), where the chemical entity of a toxin of Candida albicans corresponding to a mannan-protein complex (canditoxin) was able to show the pyrogenicity in rabbits, while the mannans of C. albicans and S. cerevisiae, used as the standard materials, did not show this activity. The doses of the mannans used in their study, 50 μg/kg in maximum, appear to be low because WNM showed pyrogenicity in doses larger than 5 mg/kg in rabbits, (see Figs. 1 and 2). As reported by us previously (4), the two highly branched polysaccharides, WNM and dextrans IAM 1046, NRRL B-1416, and NRRL B-1375, were reactable with LAL, and this effect was not abolished by treatment with alkali. It was also revealed by us that the gel-forming effect of these polysaccharides with LAL was not due to the contamination of bacterial endotoxin. The present study also demonstrated that the pyrogenicity of WNM did not change by treatment with SDOC, while

LPS lost this activity by the same treatment (Table I). These findings eliminate the posibility for the participation of lipoidal material in the pyrogenicity of WNM, and demonstrate that the mannan itself is able to behave as a pyrogen. The fact that WNM-b was not pyrogenic in rabbits is another important finding obtained in the present study. The first peak in each fever response curve is attributable to the function of the third cerebroventricle, because direct administration of WNM to this organ evoked an elevation of body temperature within 2 hr. On the other hand, the second larger peak can be cosidered as the effect on the peripheral blood cells (11). Because WNM-b did not show pyrogenicity even by direct administration to the third cerebroventricle, it is highly presumable that the alkali-labile structure in WNM participate in the exhibition of pyrogenicity. This finding suggests that the two biological effects of WNM, reactivity with LAL and pyrogenicity in rabbits, are evoked by recognition of different structures in this mannan molecule. Treatment of WNM with alkali is assumed to involve an oxidative degradation of a part of sugar residues, because addition of a reducing agent, $NaBH_4$, in the reaction mixture of the alkali treatment prevents loss of reactivity (unpublished data). The above assumption seem to be substantiated by an additional finding on MNM where this mannan is able to exhibit only weak pyrogenicity although it possesses almost identical molecular weight as that of WNM. Therefore, density of branching moieties is a factor more important than the molecular size for exhibition of the pyrogenicity of these mannans. On the fate of yeast mannan administered to mice, Hojo et al. (12) and Kawasaki et al.(13) published consistent finding that ^{3}H-labeled mannan given intravenously disappeared from the blood within several minutes, and was then distributed in the liver and the lymphoid tissues. The results of the present study indicate that a mannan-binding receptor site also exists in third cerebroventricle, or in hypothalamus, and could be serve as the initiator of the pyrogenic response in the host animal. Recently, leucocytic pyrogen (LP) released from mononuclear and polymorphonuclear cells by the action of several pyrogens has been regarded to be important as the initiator of host-defence reaction and the effectors such as lymphocyte-activating factor (LAF) activity, prostaglandine E-producing activity, and augmentation of natural killer cell activity. Previously, we found that WNM displayed various immunomodulating effects such as tumor-growth inhibition (14) and protection against Staphylococcus aureus infection in mice (15). It was also reported by Fukazawa et al. (16) and Bennich et al. (17) that high-mannose type sugar chains, found in the carbohydrate moieties of macrophage-activating factor (MAF) and human IgE, participated in expressing their activities. Therefore, WNM and its degradation products might be utilized as a simple model and an efficient tool for analyzing the mechanism of LP production.

Summary

The present paper accounted for correlation between chemical structure and LPS-like biological effects of several simple polysaccharides including yeast mannans and bacterial dextrans by determining reactivity with Limulus amoebocyte lysate (LAL) and pyrogenicity in rabbits. The neutral mannan from Saccharomyces cerevisiae wild type strain (WNM) and dextrans from Leuconostoc mesenteroides IAM 1046, NRRL B-1416, and NRRL B-1375 strains possessing dense branches showed low reactivity with LAL as compared with LPS. The neutral mannan from S. cerevisiae X2180-1A-5 mutant strain (MNM) possessing linear structure exhibited even weaker reactivity than WNM. These polysaccharides retained LAL-gelation activities after treatment with 100 mM NaOH at 30°C for 48 hr in nearly identical extents to those of the parent mannans, whereas alkali-treated LPS decreased in activity to one-ten billionth of the untreated LPS.

WNM was found to be pyrogenic in rabbits, while MNM showed

a lower effect than WNM. The dextran from L. mesenteroides was also pyrogenic. However, alkali-treatment of these polysaccharides abolished pyrogenicity. The above results suggest that pyrogenicity of the polysaccharides is not related to the reactivity with LAL, and is derived from unknown alkali-labile structure(s) of the polysaccharides.

Acknowledgment

This investigation was supported by research grants from the Ministry of Education, Science and Culture, Japan.

References

1. Bernheim, H. A., Block, L. H., Atkins, E.: Ann. Int. Med. 91, 261-270 (1974).
2. Matsumoto, T., Takanohashi, M., Okubo, Y., Suzuki, M., Suzuki, S.: Carbohydr. Res. 83, 363-370 (1980).
3. Niinobe, M., Kobayashi, T.: Nippon Nogei Kagakukaishi 46,81-88 (in Japanese) (1972).
4. Mikami, T., Nagase, T., Matsumoto, T., Suzuki, S., Suzuki, M.: Microbiol. Immunol. 26, 403-409 (1982).
5. Kohno, H., Sakurada, T., Suzuki, T., Kisara, K., Satoh, H. : Japan. J. Pharmacol. 31, 863-873 (1981).
6. Sawyer, C. H., Everett, W. J., Green, J. D.: J. Comp. Neurol. 101, 801-824 (1954).
7. Keene, W. R., Silbermann, H. R., Landy, M.: J. Clin. Invest. 40, 295-301 (1961).
8. Ribi, E., Anacker, R. L., Brown, R., Haskins, W. T., Malmgren, B., Milner, C. K., Rudbach, A. J.: J. Bacteriol. 92, 1493-1509 (1966).
9. Radvany, R., Neale, N. L., Nowotny, A.: Ann. N. Y. Acad. Sci. 133, 763-786 (1966).
10. Yamamoto, Y., Iwata, K.: Jpn. J. Mycol. 21, 274-285 (in Japanese) (1980).
11. Dinarello, C. A., Goldin, N. P., Wolff, S. M.: J. Exp. Med. 139, 1369-1381 (1974).
12. Hojo, H., Uchiyama, M.: Yakugaku Zasshi 92, 1486-1489 (in

Japanese) (1972).

13. Kawasaki, T., Mizuno, Y., Masuda, T., Yamashita, I.: J. Biochem. 88, 1891-1894 (1980).

14. Suzuki, S., Suzuki, M., Hatsukaiwa, H., Sunayama, H., Suzuki, T., Uchiyama, M., Fukuoka, F., Nakanishi, M., Akiya, S.: Gann 60, 273-277 (1969).

15. Okawa, Y., Okura, Y., Hashimoto, K., Matsumoto, T., Suzuki, S., Suzuki, M.: Carbohydr. Res. 108, 328-334 (1982).

16. Fukazawa, Y., Kagaya, K., Miura, H.: Microbiol. Immunol. 25, 1163-1172 (1981).

17. Bennich, H., Johansson, S. G. O.: Adv. Immunol. 13, 1-55 (1971).

BONE RESORBING POTENTIAL OF ENDOTOXIN AND ITS IMMUNOPHARMACOLOGICAL MODULATIONS

A. Nowotny, F. Sanavi, A.M. Nowotny, E. Kovats, J. Rothmann, D. Siegler, K. Sallay, P.H. Pham

Center for Oral Health Research, University of Pennsylvania School of Dental Medicine, Philadelphia, PA 19104 U.S.A.

Introduction

The great majority of investigators agree that periodontitis is an infectious disease. As shown by Loe et al. (1,2) and other workers (3-5), a direct association exists between periodontal disease and the presence of bacterial plaque and calculus on the teeth. Further, removal of these materials by professional and personal care results in the resolution of gingival inflammation and slowing in the rate of loss of tooth support. The causative agents of periodontal disease are certain pathogenic bacteria accumulating in the subgingival plaque. Several gram negative strains could be isolated from this site. Jordan and Keyes (6) initiated in vivo studies which demonstrated that animals, when infected with these bacterial isolates, showed loss of alveolar bone, much like human periodontal disease. Human volunteers were used to study the effect of bacterial accumulation on the development of gingivitis. Several animal models using dogs, monkeys, hamsters, and conventional or gnotobiotic rats have been developed in an attempt to study periodontal disease. According to Page and Schroeder (7) it is clear however, that none of the above models is a true analogue of human periodontitis in spite of the fact that considerable similarities in some events are unequivocally manifested. What remains uncontested is that bone resorption is the cardinal sign of the destructive phase of periodontitis.

Immunopharmacology of Endotoxicosis

Bacterial products in periodontal disease

It is generally agreed that bacterial products are more important than the bacterial themselves in producing gingival inflammation (8). Lack of microorganisms in the gingival tissues tissues of patients with periodontal disease indicates that bacteria cannot invade the gingival tissue but does not exclude exclude the possibility that bacterial products can penetrate through the epithelial layer. The bacterial product with the most convincing, but still indirect evidence for a role in periodontitis is endotoxin. Endotoxin content in the saliva, and gingival sulcus has been correlated with the severity of gingival inflammation (9-11). The penetration of endotoxin through the gingival epithelium was investigated by Schwartz and associates (12). Their study demonstrated the passage of tritiated endotoxin through intact crevicular epithelium in dogs. On the other hand, Rizzo (13,14) found in rabbits that damage to the crevicular epithelium was necessary for passage of endotoxin. Differences between these two studies may be due to the fact that rabbits may have more keratinized epithelium than dogs. One of the critical points about periodontal disease is that bone is resorbed and the teeth are ultimately lost. Hausmann and associates (15) found that endotoxin stimulates bone resorption in tissue culture. Horton and associates reported (16) that a new soluble mediator was found in the supernatant fluid from cultures of human peripheral blood leukocytes stimulated with antigenic sonicates of human dental plaque deposits. This soluble factor produced bone resportion in organ cultures of fetal rat bones as measured by the same procedure used by Hausmann. As these studies suggested the endotoxin has the potential to induce bone destruction either directly by stimulating the bone resorption mechanism or indirectly via mediators.

Results of In Vivo Resportion (BR) Experiments Using Germ Free Rats.

1. <u>Immunosuppression can facilitate bone resorption.</u> As an immunosuppressive treatment of germ-free rats we used irradiation. The immunosuppressed status of rats was determined during these experiments using heat-killed <u>Eikenella corrodens</u> (<u>Ec</u>) bacteria as O-antigens and measuring the humoral anti-O-antibody response by passive hemagglutination. The Lymphoblasts response of the sacrificed rats was also measured to establish their cellular response capacity to the same antigen. It was established that 300 r irradiation one day before monoinfection with live Ec bacteria and repeated every three weeks afterwards kept the rats immunosuppressed to the bacterial O-antigen.

   Irradiation rapidly enhanced the alveolar bone resorption in the monoinfected rats. Table 1 shows that as early as the third week after monoinfection severe bone resorption was visible while the non-irradiated monoinfected controls did not show any signs of lesion during the same period.

Table 1

Effect of radiation-caused immunosuppression on the alveolar bone resorption

Weeks	Bone loss (a)	
After Irradiation	Irradiated	Control
3	1.50	0.25
4	1.75	0.37
6	2.00	0.50
8	2.00	0.50
10	1.87	1.00
12	2.00	1.75
14	1.87	1.90

(a) 2 rats were sacrificed at time indicated.

2. Immunization with Ec whole cell sonicate reduced alveolar bone loss in monoinfected rats. A crude vaccine preparation was prepared from Ec bacteria by sonicating the cells and adjusting the dry weight content of the sonicate to 1 mg/ml. A series of injections was given intramuscularly, starting with complete Freund's adjuvant mixed with cell sonicates. Another series of immunization was given into the general area of the salivary glands which has been reported to be an effective method for the induction of salivary antibodies. Three more injections of the antigens were given on weeks 2, 4 and 6. Animals were sacrificed on week 10, and their bone loss was determined. The serum antibody titers of the animals was determined by passive hemagglutination and their lymphoblasts reactivity to Ec sonicate was also measured. The most important points of the results are shown in Figure 1 and 2, which indicate that the best immunization schedule was by the intramuscular route. This gave the highest antibody titer and the lowest level of bone resorption. The nonimmunized infected controls showed the lowest level of antibodies to Ec and the greatest level of bone loss. According to these results there is an inverse relationship between antibody levels to the monoinfecting agent and the bone resorption measured in vivo (17).

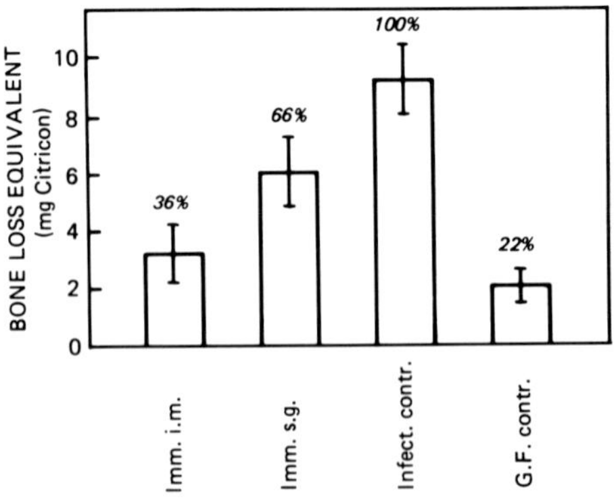

Figure 1

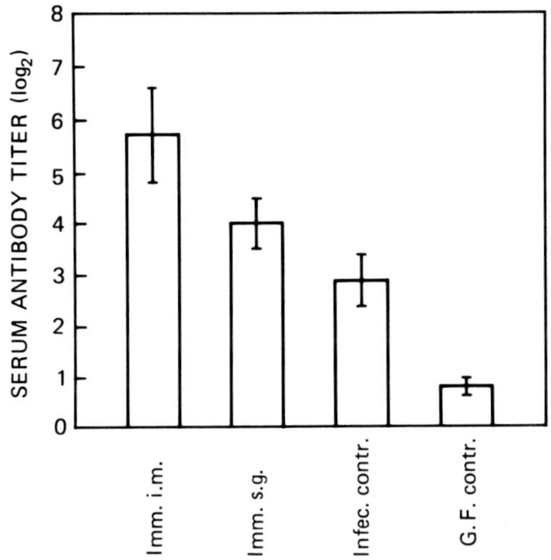

Figure 2

Results of In Vivo Bone Resorption Experiments Using Conventional Rats.

Due to the considerable limitations of the germ-free rat model which we originally used for the in vivo experiments, we adopted the use of the ligatured rat model of Rovin et al (18). Sterilized black braided 000 silk ligature was placed around the cervix of the secondary maxillary molar or rats anaesthetized with Innovar-Vet injection. When the bilateral ligatures were placed, one side was used for macroscopic and the other for microscopic evaluation. The ligature was knoted on the palatinal side of the tooth resulting in supragingival position palatinally and partially in subgingival position buccally. Bacteria rapidly accumulated on and around the silk ligature. This treatment caused inflammation in the gingival tissues of the animals and if the animals were sacrificed fourteen days after placing the ligature, concave bone loss could be seen with a nadir at the second molar. The surface of the buccal alveolar bone around the same area showed roughness. Horizontal bone destruction could be seen on the palatinal side. The ligature also caused leukocytosis in the peripheral blood. The number of mononuclear cells rose singificantly while the number of polymorphonuclear cells did not change. We used this ligature rat model in the following experiments.

1. Immunosuppression by cyclophosphamide (CY) caused rapid and extensive bone resorption. 75 mg/kg body weight CY was injected i.v. 24 hrs after ligature. Four days later CY injection was repeated. Mortality was recorded and all animals were sacrificed nine days after ligation. The results showed that in the group of rats treated with CY very extensive bone destruction could be seen in nine days.

   Figure 3 shows the defleshed maxilla of one of the experimental rats treated with ligature and CY. As comparison, Figure 4 shows the defleshed maxilla of a rat fourteen days after placing the ligature but without CY injections (19,20).

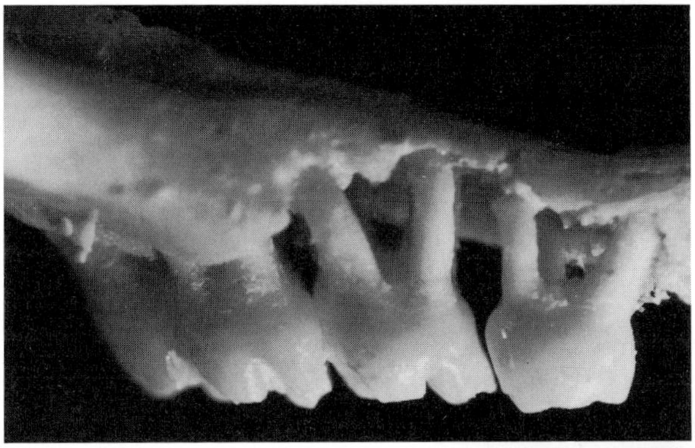

Fig. 3 Maxilla of rat received CY and ligature.

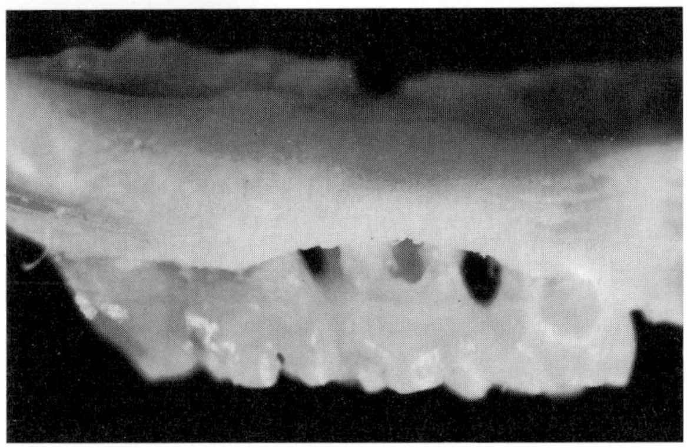

Fig. 4 Maxilla of rat received only ligature.

In the group of rats which received two CY treatments after placing the ligature, eight out of twelve animals developed septicemia and six out of them died on the seventh and eighth days. According to some elementary bacteriological studies, mixed culture of Pseudomonas aeruginosa, E. coli and Klebsiella pneumoniae were isolated from the animals. Histology of the ligature treated animals showed severe osteoclastic osteolysis without any sign of bone formation. The neck of the ligated tooth showed an accumulation of mass of bacteria, mainly gram-positive on the surface and gram-negatives underneath. Tissue necrosis was also visible which was densely colonized by gram-negative bacteria (19,20).

2. Antibiotic treatments prevented the bone loss in the ligated and immunosuppressed animals. Gentamycin sulfate, Clindamycin, Tetracycline and Chlorhexidine were used to eliminate the microorganisms. The results showed that antibiotic treatment prevented alveolar bone loss not only in the ligated rats but also in those which received ligature and two injections of CY. None of these animals developed septicemia. These results indicate clearly that in this model, bacteria are causing the observed tissue damage. Neither ligature or CY treatments could induce bone resorption in the absence of bacterial accumulation (20).

3. Induction of tolerance (non-specific immunity) to endotoxins prevented bone loss in ligature treated rats. Tolerance to the various effects of endotoxins can be induced by their repeated injection. It can be manifested by unresponsiveness to various endotoxic reactions such as pyrogenicity or lethality. The tolerance is immunologically non-specific since endotoxin from any gram-negative bacterium can induce tolerance to other endotoxins which are serologically unrelated to the endotoxin used for tolerization. We assumed that if the bone resorption is caused by endotoxin induced activation of osteoclasts or other mechanims, tolerance to endotoxin may render the animals resistant to the bone resorbing effect of bacterial products accumulating on and around the ligature.
   Accordingly, we induce tolerance with an endotoxin preparation obtained from Serratia marcescens, a non-oral microorganisms. The induction of tolerance has been tested by measuring hypothermia and lethality of the rats. It was found that injections of 8 μg Serratia marcescens endotoxin given daily for three days is sufficient to render the rats tolerant to unrelated endotoxins such as

Salmonella minnesota LPS. We induced therefore tolerance and applied ligature after three days of tolerizing injections. Animals were sacrificed 8 days later and their bone resorption was measured quantitatively by direct microscope measurement of the distance between the alveolar crest and the cemento-enamel junction. In addition, radiographic examinations were also conducted to record the interproximal bone loss.

The results showed that almost complete prevention of the alveolar bone loss could be achieved by 3 x 8 μg or by 1 x 50 μg injections of Serratia marcescens endotoxin. At the same time no detectable level of Serratia marcescens antibodies could be found. The least effective was a single 8 μg injection which was also the least effective in inducing tolerance. Figure 5 shows normal rat's teeth. In Figure 6 the effect of tolerization on the ligature-caused bone resorption can be seen which can be compared to the effect of ligature on the alveolar bone of nontolerized rats shown in Figure 7 (21,22).

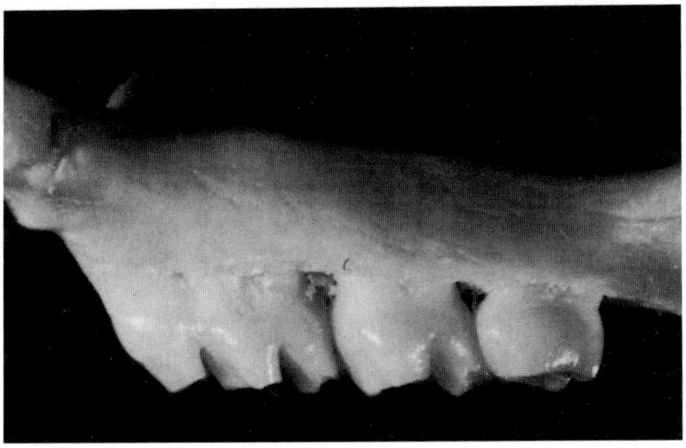

Fig. 5 Normal rat maxilla, no ligature

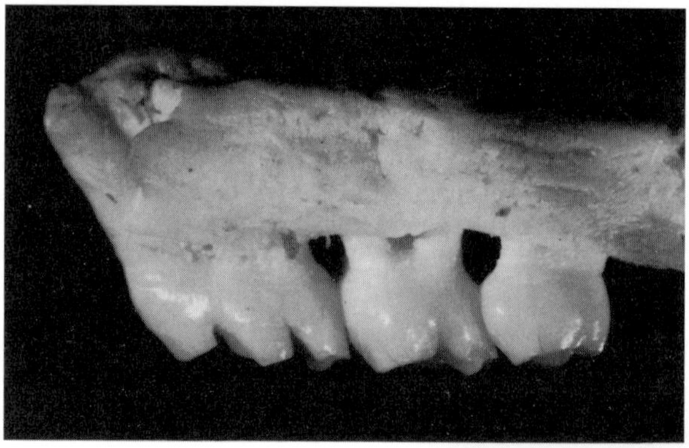

Fig. 6 Tolerized rat maxilla, with ligature.

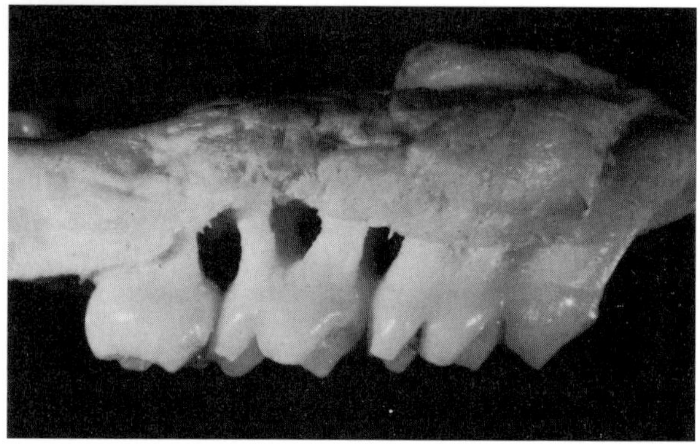

Fig. 7 Normal rat maxilla with ligature

Results of In Vitro Bone Resorption (BR) Studies

None of the above in vivo models are useful to test and compare killed bacteria or bacterial products for their bone resorbing capacity. We realized therefore the need to establish an assay which can be used quantitatively for above purposes and by which several bacterial preparations can be tested and compared for bone resorbing activity simultaneously. Our choice was the in vitro bone resorption system developed by Raisz and Niemann (23). In this system 18 day pregnant Sprague-Dawley rats are injected with 360 μCi of 45 Ca. The next day, the rats are sacrificed and fetal radii and ulnae are removed. The calcified portion of the bones are dissected and cultured in BGJ medium. The treatment group receives various concentrations of test substances dissolved in the medium. Bones are kept in culture for 5 days in 5% CO2 atmosphere at 37oC with changes of the medium on days 2 and 4. Activity of ^{45}Ca in the bone and in the medium is measured by liquid scintillation. Bone resorption is measured as the percent of ^{45}Ca released from bone into the medium during the culture period.So far the following experiments were carried out using this model.

1. Sonicates of whole bacterial cells are active in in vitro BR. Sterile whole cell sonicates of the following Lyophilized bacteria were tested: Eikenella corrodens (Ec), Actinobacillus actinomycetemcomitans (Aa) Y4 and Aa 627, Serratia marcescens 08, Salmonella minnesota 1114, Salmonella minnesota R5 and R7, Strep. sanguis and Bacterionema matruchottii. 50 μg of the suspension dispersed in BGJ medium was added to the bone cultures. The results shown in Table 3 indicate that the assay system can be used for screening killed bacteria for BR activity. It is important to note that the gram-negative oral isolates were the most active and the two gram-positive were inert.

Table 3
In vitro bone resorption by various killed whole bacterial cell sonicates

Strains	% 45 Ca release	T/C
Eikenella corrodens	75.8	3.37
A. actinomycetemcomitans Y4	67.6 ± 7.7	3.00
A. actinomycetemcomitans 627	54.8 ± 3.4	2.43
S. minnesota R5	45.3 ± 2.2	2.01
S. minnesota R7	33.4 ± 3.9	1.48
S. minnesota 1114	51.4 ± 8.9	2.28
Serr. marcesc. 08	38.4 ± 5.1	1.70
Strep. sanguis	20.1 ± 2.0	0.89
Bact. matruchottii	23.9 ± 1.9	1.06
Control	22.5	

T/C = treated/control ratio

2. Non-toxic derivatives of endotoxins can induce in vitro BR. Extensive studies were carried out in the past by us as well as by numerous other laboratories attempting to reveal the active sites in the endotoxin molecule (24). Almost all results agree that the lipid moiety (most frequently called lipid A) of the endotoxic LPS structure is responsible for the toxic manifestations. We have shown that non-toxic products, where the lipid moiety is either absent or partially destroyed (detoxified), could elicit several reactions attributed to the lipid moiety. A number of other laboratories confirmed our findings. We wished to investigate and identify the structural subunit of the endotoxic molecule which is responsible for bone resorption in vitro.

We used a) polysaccharide-rich, lipid-free and nontoxic fragments of endotoxin, b) chemically detoxified preparations obtained by CH_3OK treatment and c)sodium hydroxide detoxified endotoxin samples. Some of the results are summarized in Table 4. We demonstrated that not only toxic endotoxin preparations can induce bone resorption but completely non-toxic derivatives of it are almost equally able to induce in vitro BR.

Chemically detoxified endotoxins (CH_3OK treatment) were almost as active as the fully toxic starting materials. Detoxification by alkaline hydrolysis (0.1 N NaOH at 37°C temperature for various lengths of time) saponified the lipid moiety by the destruction of this part of the structure did not lead to the inactivation of the BR component. 60' hydrolysis under above conditions completely abolished the toxic properties of the preparation but fully maintained or even increased the BR activity of the samples (25).

Table 4

Bone resorbing effect of various non-toxic endotoxin derivatives

Preparation after	% 45 Ca release	T/C*	LD50**
Hydrolysis with 0.2 N acetic acid, 60 min 100 C	63.9 ± 16.2	3.09	0.50
Hydrolysis with 0.2 N acetic acid, 120 min 100 C	23.1 ± 2.0	1.11	>1.0
CH_3OK detoxification, 0.1N, 61C, 60 min	72.6 ± 13.8	3.51	>1.0
NaOH detoxification, 0.1N, 6 hrs, 37C.	51.3 ± 8.3	2.48	>1.0

* T/C is the ratio of treated to control.

** Chick embryo lethality. Samples were given i.v. LD50 of the toxic starting endotoxin sample was 0.022 μg/embryo.

We also studied the kinetics of the hydrolysis leading to the liberation of water soluble lipid free BR active components during acidic hydrolysis. Figure 8 shows that 60 min. hydrolysis with 0.02 N acetic acid is adequate to liberate the BR active component into the supernatant. Acidic hydrolysis longer than 2 hrs at 100°C water bath destroyed the BR activity in both the supernatant as well as in the precipitate (25).

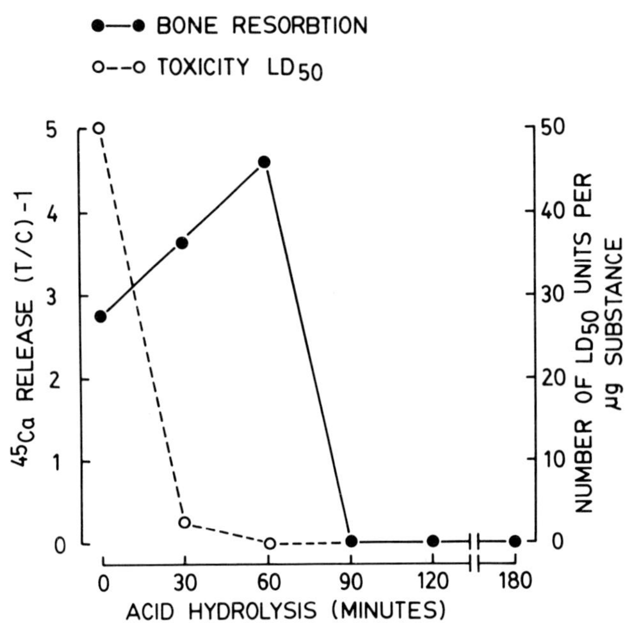

Fig. 8

3. Bacterial components unrelated to endotoxin can also induce in vitro BR B. White used 0.2 normal acetic acid hydrolysis of whole bacteria to isolate the sero-reactive components from the bacterial cell wall surface. We refer to this preparation as White-type polysaccharides (WPS). Recently we isolated such White-type polysaccharids (WPS) preparations from smooth as well as rough mutants of Salmonella minnesota, from Serratia marcescens and from the oral isolates Aa, Y4 and 627 and two gram-positive bacteria. While all smooth WPS preparations were active in bone resorption (25), as Table 5 indicates, some of the rough mutants' WPS preparations did not show such activity. The gram-positive WPS samples were also negative in BR.

Table 5

Bone resorbing effect on White-type polysaccharide (WPS)

WPS from	% ^{45}Ca release	T/C
Serratia marcescens	38.2 ± 3.1	1.99
S. minnesota 1114	47.3 ± 3.5	2.46
S. minnesota R5	41.4 ± 3.0	2.16
S. minnesota R7	25.7 ± 2.7	1.34
Eikenella corrodens	39.8 ± 1.8	2.07
A. actinomycetem. Y4	45.1 ± 3.5	2.35
Bacterionema matruchottii	18.1 ± 1.1	0.94
Strep. sanguis	21.10 ± 1.2	1.09
Control	19.2 ± 1.4	—

T/C is the ratio of treated to control.

4. Microvesicles released by Actinobacillus actinomycetemcomitans (Aa Y4) strain contained endotoxin and endotoxin unrelated toxic and bone resorption active cell products.

Shedding of gram-negative bacterial cell wall components into the surrounding medium occurs not only during disintegration of the bacterial cell body but some strains release these components during their regular growth (26). It has been described by Hammond and associates that Aa. Y4 strain releases microvesicles into the surrounding medium in large quantities. These microvesicles were isolated and we measured first the endotoxin content of these microvesicles using characteristic reactions of endotoxicity such as local Shwartzman test, pyrogenicity and Limulus lysate clotting assays. According to our findings the endotoxin content in the vesicles is approximately one-tenth to one-fifth of their dry weight (27,28). Endotoxic Y4 LPS was also isolated by the phenol-water method. In vitro bone resorbing activities of Y4 LPS and Y4 microvesicles were compared and they were found to be approximately equal on a weight basis. This was in spite of the fact that the vesicles contained only approximately 10 to 20% endotoxins. We compared the heat resistance of the bone resorbing activity of Y4 endotoxin and microvesicles. Y4 endotoxin, as expected was not affected by heat. The bone resorbing activity of microvesicles showed only a slight reduction. Proteolysis using a fungal proteinase K preparation, did not alter the bone resorbing activity of the Y4 LPS which observation is in agreement with the known resistance of such preparations to proteolytic enzymes. The vesicle preparation on the other hand, lost approximately 70-90% of its bone resorptive potency as Figure 9 indicates. This finding showed the presence of proteolysis sensitive but heat-resistant bone-resorbing component in the vesicles.

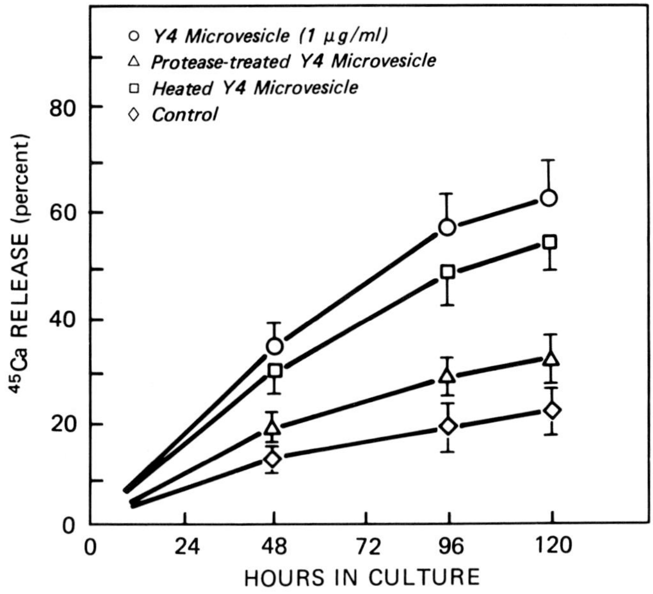

Fig. 9

The unusually high toxicity of the microvesicle preparations was measured by the chick embryo lethality test. The vesicles were slightly more toxic on a weight basis than the isolated endotoxin from the same strain. According to these results a highly toxic substance is associated with the microvesicles. The effect of heat and proteolysis on the toxicity of the microvesicles was also studied and the results showed that while the toxicity was relatively heat stabile, it could be greatly reduced by proteolysis (Table 6). According to these findings, the Y4 microvesicles contained a proteolysis-sensitive but heat resistant toxin also (27,28). Independently from our results, Hammond and associates, and Taichman and co-workers found that Y4 microvesicles also contained a toxin which destroys neutrophils. They showed that this leukotoxic substance is sensitive to both proteolysis and heat treatment; therefore, is it not identical to any of the active components of the microvesicles we discovered.

Table 6

AaY4 microvesicles and AaY4 endotoxic LPS in various biological assays

Preparation		Toxicity (a)	Shwartzman (b)
Microvesicles,	normal	0.046	0.19
"	heated	0.021	0.20
"	proteolyzed	1.0	0.30
Endotoxin,	normal	0.069	1.21
"	heated	0.032	1.41
"	proteolyzed	0.032	1.42

a) LD50 determined by i.v. injection of 11 day old chick embryos. Values are expressed in μg.

b) Local skin lesion in rabbits. Values are areas of hemorrhage expressed in square cm.

Accordingly, there may be at least three biologically highly active components in the microvesicles in addition to endotoxin. One of them is a toxin which was more toxic to the chick embryos than endotoxin. There is a component in the microvesicles which is non-endotoxic but a bone resorption inducer, and similar to the above mentioned toxin in its sensitivity to proteolysis and heat treatment. The third component present in these preparations is the leukotoxin. Two of the above preparations can induce bone resorption. Considering the very significant biological activities concentrated in these small vesicles which are released in large numbers by the Y4 bacteria into the crevicular environment, it is most likely that such microvesicles have a high pathogenic potential.

References

1. Loe, H., Theilade, E. and Jensen, B.: J. Periodont. 36, 177 (1965).
2. Theilade, E., Weight, W. H., Jensen, B. and Loe, H.: J. Periodont. Res. 1, 1 (1966).
3. Greene, J. C.: Am. J. Public Health 53, 913 (1963).
4. Ramfjord, S. P., Emslie, R. D., Greene, J. C., Held, A.-J. and Waerhaug, J.: Am. J. Public Health 58, 1713 (1968).
5. Friedman, L. A. and Klinkhamer, J. M.: J. Periodont. 42, 702, (1971).
6. Jordan, H. V. and Keyes, P. H.: Arch. Oral. Biol. 9, 401 (1964).
7. Page, R. C. and Schroeder, H. E.: Periodontitis in man and other animals, A comparative review, eds. Page and Schroeder, Karger, Basel, Switzerland (1982).
8. Glickman, I.: New Eng. J. Med. 284, 1071 (1971).
9. Simons, B. I., Goldman, H. M., Ruben, M. P. and Baker, E.: J. Periodontol. 41, 81 (1970).
10. Shapiro, I., Lodato, F. M., Courant, P. R. and Stallard, R. E.: J. Periodontol. 43, 591 (1972).
11. Evans, R. T. and Mergenhagen, S. E.: Proc. Soc. Exptl. Biol. Med. 119, 815 (1965).
12. Schwartz, J., Stinson, F. L. and Parker, R. B.: J. Periodontol. 43, 270 (1972).
13. Rizzo, A. A.: J. Periodontol. 41, 201 (1970).
14. Rizzo, A. A.: J. Periodontics 6, 65 (1968).
15. Hausmann, E., Raisz, L. G. and Miller, W. A.: Science 168, 862 (1970).
16. Horton, J. E., Raisz, L. G., Simmons, H. A., Oppenheim, J. J. and Mergenhagen, S. E.: Science 177, 793 (1972).
17. Behling, U. H., Sallay, K., Sanavi, F., Pham, P. H. and Nowotny, A.: Inf. & Immunity 33, 801 (1981).
18. Rovin, S., Costich, E. R. and Gordon, H.: J. Perio. Res. 1, 193 (1966).

19. Sallay, K., Sanavi, F., Ring, I., Pham, P., Behling, U. H. and Nowotny, A.: IADR Abstract No. 707, Chicago (1981).

20. Sallay, K., Sanavi, F., Ring, I., Pham, P., Behling, U. H. and Nowotny, A.: J. Periodontol Res. 17, 263 (1982).

21. Nowotny, A., Behling, U. H., Nejman, G., Hayatghaib, H., Beideman, R. W. and Sanavi, F.: IADR Abstract No. 1441, New Orleans (1982).

22. Nowotny, A. and Sanavi, F.: Inf. & Immunity 39, 873 (1983).

23. Raisz, L. G. and Niemann, I.: Endocrinology 85, 446 (1969).

24. Nowotny, A.: Bact. Rev. 33, 72 (1969).

25. Sanavi, F., Behling, U. H., Nejman, G., Kovats, E. and Nowotny, A.: IADR Abstract No. 1440, New Orleans (1982).

26. Nowotny, A.: Shedding bacteria, in "Pathological Membranes" ed. A. Nowotny. Plenum Publishing Corp. New York 1983.

27. Nowotny, A., Behling, U. H., Hammond, B., Lai, C.-H., Listgarten, M., Pham, P. H. and Sanavi, F.: J. Dent. Res. 60 (Special Issue A): 552 (1981).

28. Nowotny, A., Behling, U. H., Hammond, B., Lai, C.-H., Listgarten, M., Pham, P. H. and Sanavi, F.: Inf. & Immunity 37:151 (1982).

CATECHOLAMINE-HYPERRESPONSE IN ENDOTOXEMIA OF MICE

Kazuo Kuratsuka* and Reiko Homma
National Institute of Health. Tokyo 141, Japan

During the course of study (1) we found by chance that the mice previously given endotoxin through intravenouse route died hyperreactive death on a sublethal adrenaline dose. We named this phenomenon the lethal adrenaline-hyperreaction induced by endotoxin (2, 3). The hyperreaction was studied from a few points of view (4, 5, 6). Recently we found also similar phenomenon due to noradrenaline. Now we include these two kinds of hyperreaction as a lethal catecholomine-hyperresponse. In this paper we will present the characteristics of this hyperresponse, comparing the two kinds of hyperreaction from several points of view.

Material and Method.

Mice : Strain of ddy/S female 11-week-old mice were mainly used.
Endotoxin : *E. coli* LPS preparation (purified by Westphal's method) was principally used as endotoxin. Both Westphal and Boivin type of the LPS preparations from *S. typhosa* and *Sh. flexnerii* were randomly selected and tested. These LPS

---

*Present address : Dpt. Microbiology, Tokai University, School of Medicine, Isehara, 259-11 Japan

Immunopharmacology of Endotoxicosis

preparations were purchased from Difco Co., USA. The preparations were solubilized into saline and autoclaved for 20 min at 120 C. As sublethal dose, adequate amount of LPS preparation in a volume of 0.2ml was given iv. per mouse to put mice into state of endotoxemia. As the administration time of LPS solution in the day affects intensely the response to be tested, in later studies, the time at about 14:00 was adopted as the principal administration of LPS solution.
Cathecolamine : Adrenaline solution (1mg/ml) or dl-noradrenaline (1mg/ml) from Sankyo Co. diluted with sterilized saline was given ip. to mice in sublethal doses such as 50 or 65ug per mouse in case of adrenaline and 450 or more than 450ug per mouse in case of noradrenaline. The sublethal dose of both chathecolamine solutions were given mainly 4hr after LPS administration.
Catecholamine hyperresponse : The catecholamine-hyperresponse induced by endotoxin was observed 1hr or 20hrs after the challenge of adrenaline or noradrenaline as death rate of mice pretreated with LPS administration.
Transsectomy of spinal cord in mice : Under anaesthesia of nembutal thoracal spinal cord was transsected blindly at the 4-5 level of the thoracic vertebra with an injection needle. On the following day the survived mouse was selected for experimental use, of which both hind legs were paralysed. The rate of the operational success was about 80% at most.

Results

1) Time-response relation of lethal adrenaline-hyperreaction. Table 1 shows that incidence of lethal adrenaline-hyperreaction varies depending on the periods (h) between LPS-and adrenaline administrations. Under these columns of the adrenaline lethality, denominators are omitted and numerators only are shown. Omitted denominators are

Table 1

Incidence of leathal adrenaline-hyperreaction due to LPS administration

Exp. No.	LPS control Preparation	Dose µg/mouse	Death rate	Adrenaline lethality after and before LPS in following periods (h) -2^h	-1	1	2	3	4	5	6	8	18	24^h	Adrenaline control Death rate
1	E.coli (W)	100	0/14			0	1	3	4*		4*				0/14
2	〃	100	0/14			2			7***		3	3	1	0	0/14
3	〃	20	0/20					7**	9***	0					0/20
4	〃	10	0/20	0	0				5*						0/20
5	S.typhosa (W)	100	0/20				6**		15****						0/20
	S.typhosa (B)	100	0/20				7**		11****						
	Sh.flexneri (B)	100	0/20				3		8**						

Female mice were iv injected with LPS solution. (W) means Westphal's extraction method, (B) Boivin's method.

Challenge dose of adrenaline: 50 µg/mouse/ip

Numerator represents the number of dead mice; denominator the total number of mice tested.

Significance of difference in lethality between the LPS adrenaline groups and the adrenaline control groups.

Significance level, * 5 %, ** 2.5 %, *** 1 % and **** 0 1 %.

the same as those of LPS control and adrenaline control groups. Death rates in both control groups were zero%, respectively. Death rates were recorded in general 20hrs after the challenge of adrenaline.

Exp. 1 shows that 100ug of Westphal's type of LPS preparation was given to a mouse of each group of 14 mice. Each group of mice was challenged with 50ug of adrenaline in every indicated hours after LPS administration. Significant enhancement of adrenaline lethality was observed in the groups of 4 and 6hrs after LPS administration. In Exp. 2, such an enhancement was also observed 4hrs thereafter, but the larger the interval between LPS and adrenaline, the weaker the enhancement. Lastly, no enhencement was observed 24hrs after LPS administration. The time-response indicated the gradual appearance of the enhancement and the peak of response between 4 and 6hrs after LPS administration. Then the response gradually decreases and finally disappears 24hrs thereafter.

Exp. 4 shows that the enhancement was not observed in mice with adrenaline 1 or 2hrs before LPS administration, but only after LPS administration. The finding suggests that the enhancement of adrenaline lethality may be in relation to the state of endotoxemia.

Exp. 5 shows that similar enhancement was observed in several randomly selected LPS preparations other than *E. coli* preparation, regardless of the LPS extraction methods and the origin of bacterial cells.

2) Mouse sex factor in lethal adrenaline hyperreaction. As for sex difference of mice in lethal adrenaline-hyperreaction, female mice are high responder but male mice low responder. Table is omitted. Then female mice were used throught the experements.

Table 2. Mouse age and lethal adrenaline-hyperreaction

Age of mouse	LPS dose ug/mouse/iv				Adrenaline control
	3.125	12.5	50.0	Total	50ug/mouse ip
5 week	0/20	1/20	1/20	2/60	0/20
11 week	3/20	7/20**	6/20**	16/60***	0/20
20 week	1/20	2/20	6/20	9/60	0/20

Female mice were challenged with adrenaline 4 hrs after the iv administration of LPS solution.

Numerator represents the number of dead mice 20 hrs after the adrenaline challenge: denominator, the total number of mice used for the adrenaline challenge.

3) Mouse age factor in lethal adrenaline-hyperreaction. As

for age factor of female mice, three age groups of mice viz : 5-, 11-, and 20-week-old, were tested (Table 2). 11 week-old mice were the least sensitive among them. Then female 11-week-old mice were used through almost all experiments.

4) Dose-response relation (A) : Constant dose of endotoxin with varying dose of adrenaline.

Six groups of mice given in advance 40ug dose of LPS were challenged with varying doses of adrenaline, that is, in serial 1.3-fold dilutions, varying from 22.8 to 84.5ug. Deaths were tabulated 20hrs after the challenge of adrenaline.

Fig. 1 shows that a linear dose-response relation was realized between log dose of adrenaline and probit value of death rate in percentage. In addition, the dose-response line is realized in very narrow range of adrenaline dose. Therefore, the slope of the line is very steep. The characteristics will be discussed later.

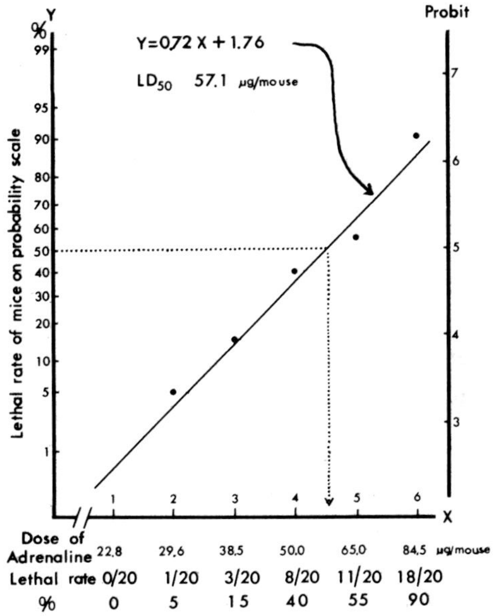

Fig. 1 Dose-response line (A)

5) Dose-response relation (B) : Constant dose of adrenaline with varying doses of endotoxin.

Fig. 2 shows two linear dose-response relations between log dose of endotoxin and probit value of death rate in percentage. From the same experiment death rates at 1hr and 20hrs after adrenaline challenge were tabulated at the bottom of the figure. The dose-response line of 20hrs-tabulation was found to be linear over wide range of LPS doses, but the line of 1hr-tabulation was realized over the first three endotoxin doses exclusive of 2000ug dose. It is evident that the lethal response of 2000ug dose in the 1hr-tabulation period was intensely suppressed. Such a paradoxical suppression of the response caused by large dose of endotoxin disappeared 20hrs after the challenge, giving the linear relation over the four doses inclusive of 2000ug dose. The paradoxical suppression will be discussed later.

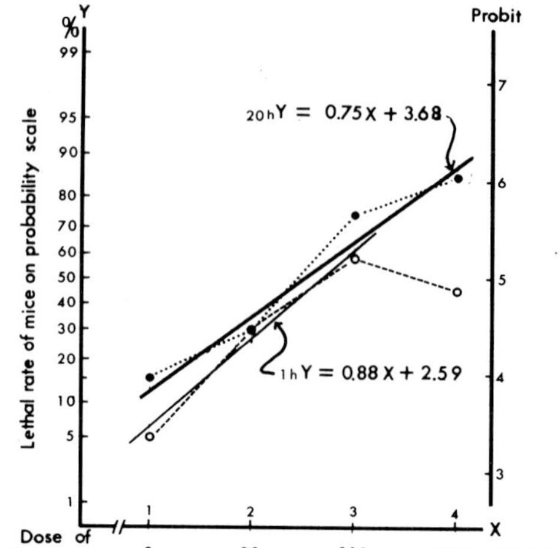

Dose of Endotoxine		2	20	200	2000 μg/mouse.
Lethal rate	20h*	3/20	6/20	14/19	17/20
	%	15	30	74	85
	1h**	1/20	6/20	11/19	9/20
	%	5	30	58	45
	# Difference	2/20	0/20	3/19	8/20

Fig. 2 Dose-response line (B)

6) Effect of hydrocortisone, histamine and serotonin on the lethal adrenaline-hyperreaction.

It is known that substances such as hydrocortisone, histamine, and serotonin are physiologically released *in vivo* after endotoxin administration (15-17). A preliminary study on the effect of administration of these three substances on the lethal adrenaline hyperreaction was conducted. Table 3 shows that suppressive effect on adrenaline hyperreaction was observed for each substance

Table 3. Effects of some biologically active substances on the lethal adrenaline-hyperreaction

Substances tested	dose/Ms	Treatment-period before and after LPS -2h	+2h	Without treatment Control	Adrenaline Control
#1 Hydrocortisone	1.25 mg	3/40** a)	8/40	11/40	0/40
#2 Histamine	6 mg	8/20	5/20	14/20	0/20
	10 mg	1/20**** b)	____		
Serotonin	6 mg	0/20****	0/20****		

#: Experimental number.

LPS: 40ug dose of *E. coli* LPS (W) was intravenously given to mice as a constant dose.

Adrenaline: The mice of exp. #1 were challenged ip. with 50ug dose of adrenaline contained 0.1ml of saline 4hrs after LPS administration. But the mice of exp. #2 were challenged ip. with 65ug dose of adrenaline.

a) Numerator represents number of dead mice challenged with adrenaline; denominator, total number of mice tested.

b) Significant difference between the lethal rates of LPS-adrenaline group and of adrenaline control group; Significance level: ** 2.5%, *** 1.0%, **** 0.1%

within the experimental conditions. These findings will be discussed later in relation to the paradoxical suppression of the hyperreaction caused by large dose of endotoxin.

7) Effect of hydrocortisone on the adrenaline-hyperreaction.

In the previous section it was shown that hydrocortisone had suppressive effect on the adrenaline hyperreaction. Furthermore, it is known that hydrocortisone is a potent therapeutic agent for endotoxin shock. Therefore, the

Fig. 3

The effect of hydrocortison-pretreatment on the lethal adrenaline hyperreaction

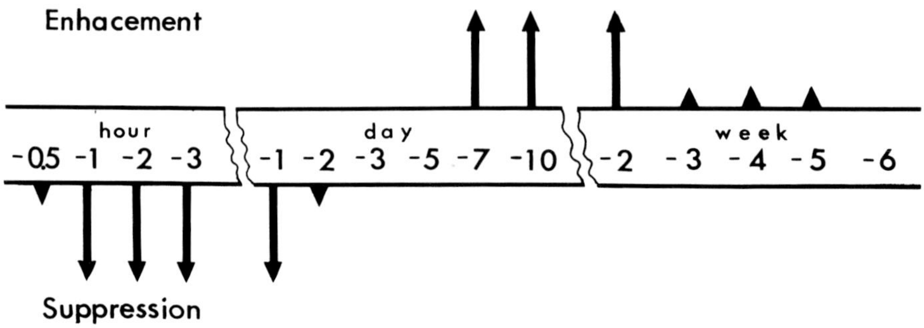

effect of hydrocortisone on the adrenaline-hyperreaction was studied in detail (5). The effects are not simple and they are shown schematically in Fig. 3.

Hydrocortisone acetate (2500ug per mouse) was given to mice in various periods before LPS administration. Then 4hrs later, the mice were challenged with 65ug of adrenaline. The effects of a single dose of hydrocortisone are quite different, depending on the pretreatment periods of hydrocortisone. Suppressive effect was observed in the case of short pretreatment periods such as less than 2 days. On the contrary, enhancing effect was noted when adrenaline was given 1 to 2 weeks in advance. No significant effect was found between 3 and 5 days and in much longer periods such as 6 weeks.

8) Circadian rythms of the lethal adrenaline-hyperreaction.

It was noticed that there was some difference in the intensity of responses between the experiment performed in the morning and that in the afternoon. Then the intensity of the lethal adrenaline-hyperreaction was tested using the same LPS solution administered at different time of the day such as at 11.30, 16.00, 24.00 and 8.00.

Fig. 4 shows that the response intensity in the midnight and in the morning is weak but the intensity near noon and in the afternoon is strong. This figure demonstrates a circadian rhythm of responses. Therefore the adrenaline-hyperreaction is of interest from the chronopharmacological points of view.

9) Time-response relation of the lethal noradrenaline-hyperreaction.

Recently we found that mice in a state of endotoxemia died hyperreactive death by the administration of sublethal dose of noradrenaline instead of adrenaline. The new hyperreaction was named the lethal noradrenaline-hyperreaction which is similar to the lethal adrenaline-hyperreaction mentioned above.

Fig. 4

Circadian response of lethal adrenalinehyperreaction

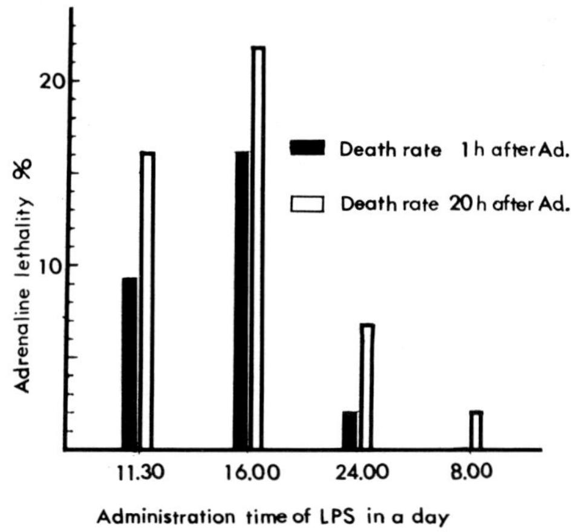

LPS: 40µg/0.2 ml iv/mouse
Ad: 50µg/0.1 ml ip/mouse, 4 h after LPS.

Table 4 shows the time-response relation of the noradrenaline-hyperreaction, which appears intensely as early as 1hr after endotoxin administration and persists until 4 to 5hrs and then decreases in intensity or disappears 24hrs thereafter. The time-response relation of the noradrenaline-hyperreaction is quite

Table 4 Incidence of lethal noradrenaline-hyperreaction by LPS

LPS control		Noradrenaline lethality after LPS in following periods (h).b)							Noradrenaline control		
ug iv		1	2	3	4	5	6	24 h	450	600	900 ug
40	0/20[a)]	8[b)]	9*	14****	9*	13***			3/20[a)]		
40	0/20	11***	12***		10***	11***	2	0		2/20	
40	0/20	11***	12***		9*	2	4	2			3/20

a) Numerator represents the number of dead mice; denominator the total number of mice tested.

b) Under the columns of noradrenaline lethality, denominators are omitted and numerators only are shown. The omitted denominators are the same as those of LPS control and noradrenaline control-groups.

*Significance levels; see the legend of Table 1.

different from that of the adrenaline-hyperreaction in that response appears rapidly after noradrenaline challenge. In addition, as shown in the columns of noradrenaline controls, the hyperreaction required large amounts of noradrenaline such as 450 to 900ug per mouse. On the contrary, the adrenaline-hyperreaction required much smaller amount of adrenaline such as 40 to 65ug per mouse. The facts suggest that noradrenaline may be less toxic than adrenaline as a therapeutic drug of endotoxin shock.

10) Dose-response relation (A) and (B) and circadian rythms of response in case of the lethal noradrenaline-hyperreaction.

Table 5 shows that a linear dose-response relation (A)

Table 5 Dose-response relation. Varying dose of NA. LPS constant dose

LPS control ug iv		Dose of Noradrenaline (ug) 56.2	112.5	225	450	NA control 450^{ug}
40	0/15	0/15	3/15	9/15****	13/15****	0/15
0.4	0/15	0/15	2/15	9/15****	10/15****	

Challenge of NA : 4h after LPS administration

is realized with varying doses of noradrenaline in serial 2-fold dilutions and certain constant doses of endotoxin such as 40 and 0.4ug per mouse. The slope of the dose-response line is not as steep as that after adrenaline. Furthermore, in spite of a 100 times difference between 40ug and 0.4ug of endotoxin, the dose-response relations of the two doses of endotoxin are similar. This fact suggests that it may be difficult to obtain a linear dose-response relation (B) with varying doses of endotoxin and a constant dose of nonadrenaline. This estimation was confirmed by actual experiments.

Also, the noradrenaline-hyperreaction did not show any significant difference between the intensity of response in the early morning and that in the afternoon. The fact suggests that the noradrenaline-hyperreaction would not have circadian rhythms of response unlike the adrenaline-hyperreaction.

In summary several different points between the adrenaline and the nonadrenaline-hyperreactions were described.

11) Effect of spinal transsectomy on lethal adrenaline-hyperreaction.

Egdahl demonstrated that secretion of adrenaline was increased from the adrenaline medulla of dogs given endotoxin but the secretion was lost from the medulla of dogs whose spinal cord had been transsected at the level of

Table 6 Effect of spinal transsectomy on lethal adrenaline-hyperreaction

Operation			-	+	+	-
LPS	40^{ug}		+	+	-	-
Adrenaline	65^{ug}		+	+	+	+
Death rate after Ad. Challenge						
1 h		%	69	6	6	0
20 h		%	75	59	6	0

the 7th cervical vertebra (13). The result suggested that secretion of adrenaline caused by endotoxin administration was brought about by nervous stimulation of the target brain cells by endotoxin.

Then the effect of spinal transsectomy on the lethal adrenaline-hyperreaction was studied in mice. In Table 6 the first column is LPS control column and the next one shows the column of the experimental purpose. In 1hr after the adrenaline challenge, the death rate(6%) of the mice operated surgically was suppressed significantly, as compared with that (69%) of the LPS control. In 20hrs after the adrenaline challenge, however, the suppressed death rate(6%) of the operated group of mice increased to 59% again. Such reincrease of death rate in the 20hrs tabulation was observed only in this group of mice previously given endotoxin,but not in the operation control group of mice that lack endotoxin administration,shown in the third column.

Mechanisms of the results obtained are still insoluble but they will be later discussed briefly.

In addition, the spinal transsectomy brought about also similar effects on the lethal noradrenaline-hyperreaction.

Discussion

The mechanisms of the lethal adrenaline-hyperreaction is not understood. Some workers reported the release of adrenaline *in vivo* by endotoxin administration (7 - 13). Such a release of adrenaline may be involved also in endotoxemia. The exogeneous addition of a sublethal dose of adrenaline to the possibly increased adrenaline contents in blood in a state of endotoxemia may result in the shock-like death of mice because of an untolerable amount of adrenaline. In the lethal adrenaline-hyperreaction the death rate of mice within 1hr after adrenaline challenge must correspond to the shock-like death of mice. As shown in Fig. 1, the dose-response line is realized in narrow range of adrenaline such as in serial 1.3-fold dilutions. Consequently the slope of the line is steep. The fact suggests that the change of challenge dose of adrenaline could be a critical factor in this hyperreaction and that fine quantitative analysis of the adrenaline-hyperreactivity induced by endotoxin may be possible with such a dose-response relation. In case of noradrenaline-hyperreaction, similar dose-response line was observed in narrow range of noradrenaline doses such as in serial 2-fold dilutions. Consequently the slope of the line is relatively steep but not as steep as in case of adrenaline-hyperreaction. It may safely be said that the change of noradrenaline could also be a critical factor in the noradrenaline-hyperreaction. On the other hand, the dose-response line with a constant dose of adrenaline and varying doses of endotoxin, as shown in Fig. 2, is realized in wide range of endotoxin doses in serial 10-fold dilutions such as 2 to 2000ug per mouse. The fact suggests that the adrenaline-hyperreacting activity of endotoxin preparations could be estimated singly or by parallel line assay method using such a dose-response line.

Furthermore, Fig. 2 shows that a paradoxical suppression of the hyperreaction due to large sublethal dose of endotoxin such as 2000ug per mouse was observed in the 1hr-tabulation after adrenaline challenge but such a suppression disappeared in the 20hr-tabulation. The fact suggests that there are some differences between the mechanisms of the early lethality such as death of mice within 1hr and the mechanisms of the later one tabulated 20hrs after adrenaline challenge. Large sublethal dose of endotoxin such as 2000ug per mouse may release large amount of physiologically active substances such as hydrocortisone, histamine and serotonin which have in fact suppressive effect on the lethal adrenaline-hyperreaction, as shown in Table 3. In consequence the large dose of endotoxin could induce such a paradoxical suppression. In the lethal noradrenaline-hyperreaction, however, linear dose-response relation could not be observed in wide range of endotoxin doses in serial 10-fold dilutions, as suggested from Table 5. The paradoxical suppression observed in adrenaline-hyperreaction was not seen with of noradrenaline-hyperreaction. These facts and the different time-response, as shown in Table 4, suggest that there are differences in the mechanisms concerned between two kinds of the hyperreaction but the real reason remains to be solved. Especially, the noradrenaline-hyperreaction was recently noticed and is still under investigation.

As shown in Fig. 4, the lethal adrenaline-hyperreaction demonstrates a circadian rhythm whose response intensity is weak in the midnight and in the morning, and strong near noon and in the afternoon. Such a circadian rhythm may depend largely on the function of adrenal cortex of nocturnal aminals like mice. We suppose that glucoccorticoids play a role in this circadian rhythm. Without knowing the exsistence of the circadian rhythms, it would have been impossible to obtain reproducible results for the hyperreaction experiments.

As for the effect of hydrocortisone on the adrenaline-hyperreaction, the effects are quite different depending on the pretreatment periods of a single large dose of hydrocortisone acetate such as 2500ug per mouse (see Fig. 3). Suppressive effect in case of short pretreatment periods of less than 2 days is reasonable judging from the preliminary results as shown in Table 3. However, the enhancing effect in case of 1 to 2 weeks pretreatment periods would seem to be paradoxical. The reason could be that large dose of hydrocortisone brought about hypofunction of the adrenal cortex and resulted in lack of secretion of glucocorticoids, in other words, in an "exhausted state" of cortex function. Consequently the hyperreaction might be enhanced by the lack of suppressive substances like hydrocortisone. This enhancement of the hyperreaction would be restored to the original state due to the normalization of cortex function by 6 weeks after the hydrocortisone administration.

As for the effect of spinal transsectomy on the adrenaline-hyperreaction (Table 4), the results in 1hr-tabulation are consistent with Egdahl's study (14), that endotoxin did not directly act on adrenal medulla to release adrenaline but via central nervous system. However, the suppressed death rate in 1hr-tabulation increased again in the 20hr-tabulation. The reason is not clear at present.

As shown in Table 1 and 4 and in Fig. 1, the fact that noradrenaline is less toxic than adrenaline for mice in a state of endotoxemia would be worthy of notice in clinical use of adrenaline or noradrenaline in endotoxin shock. The results in Fig. 3 suggest that administration of glucocorticoid preparations may give beneficial effect to the untoward response of adrenaline in endotoxin shock.

Summary

Catecholomine-hyperresponse in mice induced by endotoxin included the lethal adrenaline and noradrenaline-hyperreactions of which characteristics were compared in the following points:the time-response relation, the dose-response relation with a constant challenge dose of the catecholamines and varying doses of endotoxin and vice versa, the paradoxical suppression of response due to large sublethal dose of endotoxin, the circadian rhythms of response and the effect of spinal transsectomy.

Acknowledgements

We are grateful to Mr.Y.Shimazaki for his helpful assistance and Miss H.Kohama for her preparing and typewriting of the manuscript.

References

1. Kuratsuka,K.,Homma,R.,Shimazaki,Y.,Funasaka,I.:*in* Animal, Plant,and Microbial Toxins, ed.by Osaka,A.,Hayashi,K., Sawai,Y. vol. 1, 521-534, Plenum Press, New York, London (1976)
2. Kuratsuka,K.,Shimazaki,Y.,Funasaka,I.: Japan. J. Bact.31: 110 (1976)
3. Kuratsuka,K.,Shimazaki,Y.,Funasaka,I.,Watabe,Y.: Experientia 34: 1483-1484 (1978)
4. Shimazaki,Y.,Funasaka,I.,Kuratsuka,K.: Japan. J. Bact.35: 224 (1980)

5. Kuratsuka,K.,Shimazaki,Y.,Funasaka,I.,Homma,R.: Japan. J. Bact. 35:225 (1980)
6. Kuratsuka,K.,Shimazaki,Y.,Funasaka,I.,Homma,R.: Japan. J. Bact. 36:249 (1981)
7. Boquet,B.,Izard,Y.: Proc.Soc.exp.Biol.Med. 75:254 (1950)
8 Thomas,L.:Ann.Rev.Physiol. 16:467 (1954)
9. Rosenberg,J.C.,Lillehei,R.C.,Moran,W.H.,Zimmerman,B.: Proc.Soc.exp.Biol.Med.: 102:335 (1959)
10. Heiffer,M.H.,Mundy,R.L.,Mahlmare,B.: Am. J. Phsiol. 198: 1307 (1960)
11. Alican,F.: Am. J. med. Sci. 244 (1962)
12. Schayer,R.M.: *in* Bacterial Endotoxins, ed. by Landy,M., Braun,W., 182-186, Rutgers Univ. Press, New Brunswick, (1964)
13. Jacobson,E.D.,Mehlman,B.,Kalas,J.P.: J. clin. Invest. 43: 1000 (1964)
14. Egdahl,R.H.: J. clin. Invest. 38:1120-1125 (1959)
15. Hinshaw,L.B.,Jordan,M.M.,Vick,J.A.: Am. J. Physiol. 200: 987 (1961)
16. Hinshaw,L.B.,Jordan,M.M.,Vick,J.A.: J. clin. Invest. 40: 1631 (1961)
17. Kobold,E.,Katz,W.,Thal,A.P.: Fedn.Proc. 22:430 (1963)

MODULATION OF ENDOTOXICOSIS BY STEROIDS AND DIABETOGENIC AGENTS IN RESPONDER AND REFRACTORY MICE STRAINS.

M. K. Agarwal and G. Lazar[1]
Centre National de la Recherche Scientifique and Laboratory of Physio-Hormono-Réceptérologie, Université Pierre et Marie Curie, 15 rue de l'Ecole de Médecine, 75270 Paris Cedex 06, France.

[1]Institute of Pathophysiology, Medical School of Szeged, Szeged, Hungary.

Introduction

Two distinct sets of host parameters have classically been related to endotoxin lethality in experimental animals. The activity of the reticuloendothelial system (RES) was believed to play the primordial role in this respect nearly four decades ago (1) but this view underwent drastic revisions in subsequent studies (2-3).
Negative carbohydrate balance and hypoglycaemia are consistently associated with endotoxicosis and some six decades ago it was believed that they may be causal to endotoxin lethality (4). However, glucocorticoid hormones can protect against endotoxin lethality without restoring these parameters to the control level in normal mice (5), and elevated glycaemia, per se, does not lead to protection (6-8).
Ever since the discovery, the non-responder C3H/HeJ mouse has become a potent tool to probe the site and nature of endotoxin action (9). Data reported here explore the relationship between the two aforementioned sets of host parameters in normal and refractory mice strains whose sensitivity to bacterial endotoxins had been modified by streptozotocin (SZN), a diabetogenic agent, alone or in combination with RES-active agents such as glucan, and steroid hormones.

Immunopharmacology of Endotoxicosis

Materials and Methods

Male, C3H/HeJ and C3H/eb mice (CNRS, Orléans, France) and OF_1 animals (IFFA-CREDO, France) were housed with free access to pellet food and water ad libitum. An automatic switch assured 12 hours of darkness and 12 hours of light in a temperature and humidity controlled environment.
Streptozotocin (SZN: 2-deoxy-2(3-methyl-3-nitrosoureido)-D-glucopyranose) was obtained from Upjohn, Kalamazoo, Michigan; this antibiotic was dissolved in citrated saline, pH 4.5, immediately prior to intraperitoneal (ip) injection in a 0.5 ml volume containing the desired amount of product.
Particulate glucan (Standard Brands, New York, lot IF 5500) was suspended in saline prior to intravenous (iv) injection in 0.2 ml volumes containing 1 mg of the product. Glucan was injected on days 7 and 5 (total of 2 mg) prior to ip challenge on Day 0 with different doses of Salmonella typhimurium W LPS (Difco, lots 644809 and 654477) in 0.5 ml saline. The survivors were recorded 24 and 48 h later.
Triamcinolone acetonide (TA) was dissolved in saline and injected ip in 1 mg amounts in a volume of 0.5 ml in all cases. This steroid was obtained from Sigma in the non-radioactive form whereas the tritiated product was purchased from New England Nuclear.
The functional activity of the RES was assessed by the carbon clearance test deatiled previously (10). Briefly, 20 mg/ml of carbon (Gunther Wagner, Hanover, Germany, C 11/1431 a) in physiological saline containing 10 U/ml heparin was given iv in 0.2 ml volumes. Blood samples of 50 µl from the retro-orbital plexus were withdrawn 2 and 7 min later and lysed in 4 ml of 0.1% sodium carbonate. The quantity of carbon was measured at 675 nm and compared against a standard curve. Body and organ (liver and spleen) weights were determined on an electronic balance.

Blood glucose levels were quantitated on heparinized samples from the retroorbital plexus centrifuged to obtain clear plasma. Samples of 20 µl were mixed with 2.5 ml of the Gluci-Biotrol reagent (Biotrol, Paris) containing phosphate buffer, pH 7.5, *Aspergillus niger* glucose oxidase, horseradish peroxidase, amino-4-antipyrine, and hydroxybenzoate. In this procedure, D-glucose is transformed into gluconic acid and peroxide, in presence of glucose oxidase, which is thereafter converted to a coloured chromogen in presence of peroxidase. The intensity of the stable pink colour is measured at 510 nm and compared with a standard D-glucose curve calculated as grams per liter.

Liver glycogen was determined by hydrolysis into glucose and subsequent transformation into hydroxymethylfurfural quantitated with a standard glucose curve at 520 nm. Briefly, 50-100 mg liver tissue was homogenized in 5 ml of 5% TCA-0.1% silver sulphate, heated in a boiling water bath for 15 min, and centrifuged at 5000 g for 15 min (4°C). Aliquots of 1 ml were mixed with 6 ml of 36 N H_2SO_4, heated in a boiling water bath for 6 min, cooled in an ice bath, and quantitated in a spectrophotometer. The glycogen levels are expressed as milligram percent.

Hormone receptor assays were performed on 0.5 ml aliquots of the 105,000 g liver supernate in 0.01 M Tris-HCl, pH 7.4, incubated (60 min, 4°C) with the radioactive steroid. The free radioactivity was removed by the addition of 0.2 ml of a 50 mg/ml suspension of activated charcoal (Sigma), incubation (10 min, 4°C), and centrifugation (3000 g). 0.2 ml samples were counted in 10 ml Unisolve (Kochlight). Protein was determined by the Biuret method. All results are expressed as CPM/mg protein, as previously described (11-13).

The standard deviation, the Student *t* test, and the Chi^2 test, were used to analyze results. Further details are provided in appropriate legends.

Results

Data in table 1 show that OF_1 male mice are sensitized to endotoxin lethality by SZN in a dose dependent manner. The 5 mg dose of this antibiotic permitted 40% survival after the toxin which, in control animals, produced 78% survival. This 5 mg dose did not stress mice in any apparent way and was chosen for all subsequent studies.

Table 1. Dose dependent sensitization to endotoxin lethality by Streptozotocin in OF_1 male mice.

Streptozotocin	Living/ Total 48 h	% Survival	Statistics
1. None	47/60	78	
2. 1 mg	14/20	70	2 vs 1: N.S.†
3. 3 mg	11/20	55	3 vs 1: N.S.
4. 5 mg	8/20	40	4 vs 1: $p < 0.005$
5. 10 mg	2/20	10	5 vs 1: $p < 0.001$

†N.S. = Not significant

The indicated dose of SZN was given concurrently with 500 µg endotoxin,ip, in all cases and the survivors were recorded for 48 h thereafter. For further details see (7).

Data in Fig.1a show that blood glucose level was significantly elevated as early as 24 h after a single ip injection of 5 mg SZN ($p < 0.001$). After a maximal level within the first 48 h, glycaemia remained elevated for at least 15 days after SZN. Endotoxin lethality increased significantly when SZN was given concurrently with 250 µg of the toxin ($p < 0.02$) and did not change further 2, 4, 11, or 15 days after SZN pre-treatment. The lower survival after 15 days of SZN is probably related to the secondary consequences of diabetes. It seems therefore that glycaemia is not related to endotoxin lethality since mice were sensitized to a very comparable extent at all times after SZN. Furthermore, this antibiotic appears to have an extra-pancreatic site of action. Thus, glycaemia is elevated only 24 h after SZN but sensitization to endotoxin seems immediate. This must be borne in mind when SZN is used for the treatment of various endocrinian tumours (14). The nature of this extra-pancreatic site remains unknown.

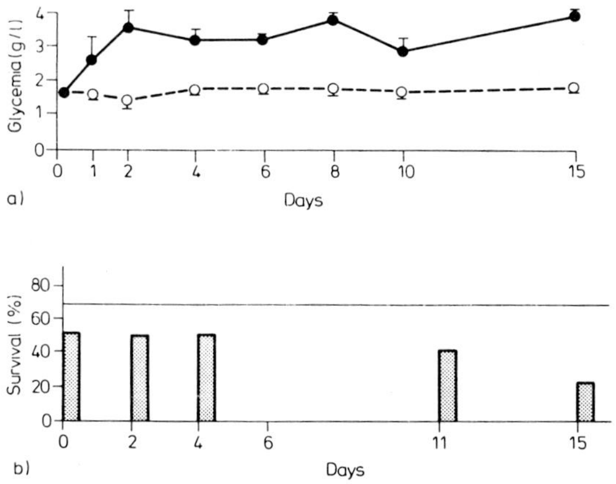

Fig. 1. Effect of Streptozotocin on glycaemia and endotoxin lethality in OF_1 mice.

Animals were given 5 mg ip SZN (●) or saline (O). All values are expressed as g/l $\pm$ the standard deviation for Fig. 1a. For Fig. 1b, mice were challenged, all on the same day, with 5 mg SZN ip and 250 µg endotoxin. Control (——); SZN (▨). For further details see (15).

Data in table 2 show that either SZN or glucan, alone, significantly aggravated endotoxin lethality and that the combined effect of these two materials was additive. Only 15% of the animals given SZN and glucan survived endotoxin challenge compared to the 50% survival in the group given SZN alone, and 45% survival in the glucan treated group. Triamcinolone acetonide (TA) significantly protected all groups of mice challenged with endotoxin although the presence of glucan interfered with TA protection.

Table 2. Survival of OF_1 mice challenged with endotoxin in presence of various combinations of Glucan, Streptozotocin, and Triamcinolone acetonide.

Treatment	Living/Total	Survival %	Statistics
1. E	53/75	70	
2. E + TA	30/30	100	2 vs 1 $p < .001$
3. E + SZN	35/70	50	3 vs 1 $p < .02$
4. E + SZN + TA	30/30	100	4 vs 1 $p < .001$
5. G + E	9/20	45	5 vs 1 $p < .05$
6. G + E + TA	15/20	75	6 vs 5 $p = .05$
7. G + E + SZN	3/20	15	7 vs 5 $p < .05$
8. G + E + SZN + TA	12/20	60	8 vs 7 $p < .005$

E = Endotoxin 250 µg ip; TA = triamcinolone acetonide 1 mg ip with E; SZN = streptozotocin 5 mg with E; G = glucan 1 mg iv on days 7 and 5 each prior to challenge with E on day 0. For further details see (15).

Does the RES contribute to endotoxin lethality under these conditions? Data in table 3 show that glucan significantly increased the K value, the liver and the spleen weights. SZN, within the first 4 h, did not influence any parameter of the RES in either glucan pretreated (table 3) or control (not shown) mice.

Within 4 h, triamcinolone acetonide significantly lowered the K value in normal (not shown), the glucan treated mice, or in animals given both glucan and SZN (table 3). TA was without effect on other parameters of the RES under these conditions. Thus, the steroid was able to protect against endotoxin lethality (table 2) at a time when it actually depressed the rates of carbon clearance (table 3). In other words, activation of the RES was found with sensitization to, and protection against, endotoxin lethality in OF_1 male mice. Furthermore, the RES does not appear to be the extra-pancreatic site of SZN action mentioned above.

Table 3. Influence of Glucan, Streptozotocin, and Triamcinolone acetonide on various parameters of the RES in OF_1 mice.

Treatment	K Value	Spleen wt (mg)	Liver wt (mg)
1. Saline	0.032 ± 0.02 (10)	90 ± 18	1755 ± 155
2. Glucan	0.203 ± 0.06 (12)	176 ± 47	1928 ± 145
3. G + SZN	0.188 ± 0.03 (5)	166 ± 34	1936 ± 193
4. G + TA	0.091 ± 0.02 (5)	165 ± 33	2052 ± 110
5. G + SZN + TA	0.120 ± 0.04 (5)	159 ± 20	2022 ± 116
Statistics			
2 vs 1	$p < 0.001$	$p < .001$	$p < 0.02$
3 vs 2	N.S.	N.S.	N.S.
4 vs 2	$p < 0.001$	N.S.	N.S.
5 vs 2	$p < 0.005$	N.S.	N.S.
5 vs 3	$p < 0.02$	N.S.	N.S.
5 vs 4	N.S.	N.S.	N.S.

All assays 4 h post treatment shown as mean ± S.D. for the number in parentheses. N.S. = not significant. Body weights were 26-28 g in all cases. Further details in (15).

Attention was next directed to blood glucose levels under these conditions.

Table 4. Interaction between SZN, endotoxin, and TA on blood glucose levels in glucan pretreated OF_1 mice.

Treatment	Glycaemia (g/l)	Statistics
1. Saline	1.80 ± 0.16 (12)	
2. Glucan	1.69 ± 0.27 (12)	2 vs 1; N. S.
3. G + SZN	1.71 ± 0.19 (7)	3 vs 2; N. S.
4. G + TA	1.70 ± 0.17 (7)	4 vs 2; N. S.
5. G + SZN + TA	1.84 ± 0.32 (7)	5 vs 2,3,4; N. S.
6. G + E	0.75 ± 0.40 (8)	6 vs 2; $p < 0.001$
7. G + E +SZN	0.92 ± 0.35 (7)	7 vs 2; $p < 0.001$;vs6=NS
8. G + E + TA	0.95 ± 0.65 (6)	8 vs 2; $p < 0.001$;vs6=NS
9. G+E+SZN+TA	1.02 ± 0.59 (5)	9 vs 2; $p < 0.001$;vs6-8=N.S.

All assays 4 h post treatment in glucan pretreated mice shown as mean ± S.D. for the number in parentheses. N.S. = not significant. For further details see (15).

Data in table 4 show that glucan pretreatment did not alter glycaemia. SZN or hormone, alone or concurrently, did not change blood glucose levels within the first 4 h in either normal (not shown) or glucan pretreated (table 4) mice. Endotoxin significantly decreased blood glucose within 4 h in glucan pretreated mice and this was not further influenced by either SZN or TA, or SZN + TA (table 4). Thus, the additive effect of SZN plus glucan on endotoxin lethality (table 2) is not manifest on glycaemia (table 4). Furthermore, TA protection against endotoxin death, too, does not appear to be related to blood glucose levels, at least within the first 4 h which are known to be critical to the outcome of endotoxicosis (2,3,8,16).

Table 5. Effect of SZN, Endotoxin and Triamcinolone acetonide on liver glycogen levels in glucan pretreated OF_1 mice.

Treatment	Liver glycogen (mg %)	Statistics
1. Saline	12.63 ± 3. 18 (11)	
2. Glucan	12.94 ± 2. 83 (10)	2 vs 1; N.S.§
3. G + SZN	15.70 ± 5. 75 (7)	3 vs 2; N.S.
4. G + TA	28.57 ± 6. 58 (7)	4 vs 2; $p < 0.02$
5. G + SZN + TA	22.51 ± 4. 00 (7)	5 vs 2; $p < 0.01$
6. G + E	2.04 ± 0. 90 (7)	6 vs 2-5; $p < 0.001$
7. G + E + SZN	2.11 ± 1. 38 (5)	7 vs 2; $p < 0.001$
8. G + E + TA	3.25 ± 2. 25 (6)	8 vs 2; $p < 0.001$
9. G+E+SZN+TA	3.88 ± 2. 61 (5)	9 vs 2; $p < 0.05$

§N.S. = Not significant. All assays 4 h post treatment. All values are mean ± S.D for the number shown in parentheses.

Data in table 5 show that glucan pretreatment does not affect liver glycogen in OF_1 mice and that this is unaltered by SZN within the first 4 h. TA alone provoked very significant glyconeogenesis which remained unaltered by SZN. Endotoxin alone depleted liver glycogen levels within 4 h and this was

not further changed by either TA, SZN, or TA + SZN. Thus, neither TA protection against, nor the additive effect of SZN + glucan on, endotoxin lethality (table 2) is manifest at the level of liver glycogen in normal (not shown) or glucan pretreated (table 5) mice of the OF_1 strain.
Data in table 6 show that five, daily, ip, injections of increasing doses of endotoxin induced a state of tolerance to an eventual lethal challenge with the toxin in control mice. The establishment of tolerance to endotoxin was less pronounced in mice that had been pretreated with SZN than in saline injected counterparts. This sort of impaired induction of tolerance was also evident when tolerance inducing doses of endotoxin were given concurrently with those of SZN (not shown). In other words, the combination of SZN + endotoxin challenge was more lethal both in acute lethality (table 2)of, and tolerance to, endotoxin.
Attention was next directed to the possible mechanisms by which SZN could conceivably interfere with the establishment of endotoxin tolerance.

Table 6. Impaired tolerance to endotoxin in OF_1 mice as a result of streptozotocin administration.

Endotoxin	Living/ Total	% Survival	Statistics
Buffer + E	20/20	100	
1. 0.5 mg	15/15	100	
2. 1.0 mg	16/20	80	
3. 2.0 mg			
SZN + E			
4. 0.5 mg	20/20	100	4 vs 1 = N. S.[+]
5. 1.0 mg	20/20	100	5 vs 2 = N. S.
6. 2.0 mg	6/20	30	6 vs 3 $p < 0.005$

[+]N. S. = Not significant; see (7) for further details.

Table 7. Activity of the Reticuloendothelial System during impaired tolerance to endotoxin following treatment with Streptozotocin.

Treatment	K Value	Spleen wt	Liver wt(mg)
1. Buffer	0.046 ± .017 (6)	91 ± 11	1526 ± 175 (8)
2. Endotoxin	0.079 ± .026 (7)	206 ± 22	1733 ± 120 (8)
3. Streptozotocin	0.059 ± .009 (6)	96 ± 19	1499 ± 107 (8)
4. SZN + LPS	0.078 ± .022 (8)	204 ± 76	1602 ± 209 (8)
Statistics			
2 vs 1	$p < 0.02$	$p < 0.001$	$p < 0.02$
3 vs 1	N. S.	N. S.	N. S.
4 vs 1	N. S.	$p < 0.001$	N. S.
4 vs 3	$p = 0.01$	$p < 0.001$	N. S.

For further details see text and (7).

Data in table 7 show that the rate of carbon clearance (the K value), the total body weight, the spleen and liver weights, were nearly identical in both the buffer and the SZN pretreated groups. Endotoxin increased all parameters of the RES under these conditions in control OF_1, Swiss albino mice. The toxin significantly increased spleen weights and lowered body weights in SZN pretreated mice (table 7). Thus, SZN impaired the development of tolerance to endotoxin (table 6) but the activation of the RES accompanying tolerance remained unaffected. These further undermine the role of the RES during endotoxicosis.

The nonresponder C3H/HeJ strain of mice has been very helpful in revealing various mechanisms of endotoxin mediated host dysfunction. Hitherto, these studies have suggested that this strain can not be rendered fully responsive to endotoxins due primarily to the so called lps gene (reviewed in 9). It was therefore of more than passing interest to investigate whether the mechanisms of action of SZN and glucan are also expressed in the C3H series. In addition, the lps gene hypothesis could furthermore be analyzed in this context.

Data in table 8 show that 1 mg ip endotoxin did not kill the nonresponder mice at all. However, animals were sensitized to LPS if they had been pretreated with either 5 mg ip SZN (40% survival) or 2 mg iv glucan (10% survival); 1 mg TA protected animals in both groups very significantly. SZN + glucan produced an additive effect on endotoxin lethality (0% survival);TA was less effective than in groups sensitized to LPS by only one of the two agents.

Table 8. Additive effect of Streptozotocin and Glucan on Endotoxin lethality in the non-responder strain.

Treatment	Living/ Total (48 h)	% Survival	Statistics
1. E	10/10	100	
2. SZN+E	4/10	40	2 vs 1 $p < 0.005$
3. SZN+E+TA	9/10	90	3 vs 2 $p < 0.05$
4. G+E	1/10	10	4 vs 1 $p < 0.001$
5. G+E+TA	10/10	100	5 vs 4 $p < 0.001$
6. G+SZN+E	0/10	0	6 vs 1 $p < 0.001$
7. G+SZN+E+TA	6/10	60	7 vs 6 $p < 0.005$

E = 1 mg endotoxin; SZN = streptozotocin 5 mg ip with LPS;
G = glucan 1 mg iv on days 7 and 5 each prior to LPS challenge
TA = 1 mg ip with endotoxin challenge.

These data show that glucan and SZN are just as effective in C3H/HeJ nonresponder mice as in OF_1 strain.The influence of the two materials was additive rendering the non-responder totally responsive to endotoxin lethality. In addition, TA effectively protected C3H/HeJ nonresponder against LPS, just as with OF_1 mice.

In an effort to understand the mechanisms of sensitization to endotoxin lethality, various parameters of the RES were analyzed in the C3H/HeJ strain. The K value, spleen weight (SW), and liver weight (LW), but not the body weight (BW), increased significantly after glucan pretreatment in the nonresponder strain. The K value and the spleen size continued to increase 7 days after glucan in the nonresponder (table 9).

TA alone profoundly decreased the K value with smaller but statistically significant reduction in BW and SW as compared to the normal control. Thus, as in the OF_1 animals, TA protection is associated with a decrease in most parameters of the RES whereas glucan sensitization increases RES activity and SZN sensitization to endotoxin has no influence on RES at all (not shown).

Table 9. Analysis of the Reticuloendothelial Function in the Nonresponder strain of mice.

Treatment	K Value	Spleen wt	Liver wt (mg)
1. Saline	0.026 ± 0.006 (12)	62 ± 7	1165 ± 133
2. Glucan	0.070 ± 0.016 (8)	171 ± 16	1437 ± 133
3. Glucan	0.104 ± 0.020 (6)	255 ± 36	1406 ± 82
4. TA	0.008 ± 0.003 (10)	36 ± 6	1210 ± 108
Statistics			
2 vs 1	$p < .001$	$p < .001$	$p < .001$
3 vs 1	$p < .001$	$p < .001$	$p < .005$
3 vs 2	$p < .01$	$p < .001$	p = N. S.
4 vs 1	$p < .001$	$p < .001$	p = N. S.

2 x 1 mg iv glucan on days 7 and5; assays on day 2 (line2) or day 0 (line 3); TA 48 prior to assay (line4). Details in 10

Data in table 10 confirm that SZN sensitization to LPS in the nonresponder is not related to the diabetogenic effect of the antibiotic. Glucan, too, did not lower blood sugar which was reduced significantly within 4 h after endotoxin. TA alone, as with OF_1 animals, did not influence glycaemia within the first 6 h that are critical to the outcome of endotoxicosis (not shown). Thus, glycaemia can not account for either resistance or sensitization to endotoxin lethality.

In still other studies, TA induced liver glycogen and LPS decreased it in C3H/HeJ animals. This was similar to the behaviour noted earlier in the responder OF_1 strain. In other words, the carbohydrate balance did not seem to account for the state of resistance to endotoxin lethality that is characterstic of the C3H/HeJ strain (not shown).

Table 10. Influence of sensitization to endotoxin on blood sugar levels in the nonresponder strain.

Treatment	Glycaemia	Statistics
1. Saline	1.68 ± 0.18 (10)	
2. Glucan	1.59 ± 0.18 (11)	2 vs 1, N.S.
3. Endotoxin	1.06 + 0.23 (5)	3 vs 1 $p < 0.001$
4. Streptozo-tocin	1.71 ± 0.18 (10)	4 vs 1, N.S.

E = endotoxin and SZN = streptoztocin, both 5 mg each ip 4 h prior to assay; G = glucan 1 mg iv each on days 7 and 5 prior to assay on Day 0. All values are averages ± standard deviation for the number of samples in parantheses. See also (10).

In an effort to understand the mechanism of TA protection against LPS lethality, the liver glucocorticoid receptor was quantitated. Data in table 11 show that, at all concentrations, TA association to the liver glucocorticoid receptor in the sensitive C3H/eb parent was similar to that in the resistant C3H/HeJ mouse. Similar results were obtained when OF_1 mice were used (not shown). Thus, liver glucocorticoid receptor levels do not seem to account for LPS resistance in the non-responder C3H/HeJ strain.

Table 11. Liver Glucocorticoid Receptor levels in the C3H series of mice.

^{3}H-TA	CPM/mg protein	
(M)	C3H/eb	C3H/HeJ
10^{-9}	0	0
5×10^{-9}	123	138
10^{-8}	497	452
5×10^{-8}	858	665
10^{-7}	1233	1045

For further details see (10).

Discussion

The results described in this chapter challenge numerous views currently in vogue to explain endotoxicosis. The activity of the RES, blood sugar levels, the overall carbohydrate balance, and glucocorticoid receptor dysfunction, could not be related to various phases of endotoxin lethality in a causal manner. Although endotoxin lowers most of these parameters, sensitization to the lethal effects of the toxin was possible by streptozotocin which does not influence either of them, at least in the first 24 h that are critical for the eventual outcome of endotoxicosis. Elevation of glycaemia, as a result of the diabetogenic action of the antibiotic, ensues only 24 h after SZN and does not reach maximum at least until 48 h. In still other studies, C3H/HeJ mice required several weeks of antibiotic pretreatment to become diabetic yet SZN sensitized them to endotoxin lethality just like the Swiss, albino OF_1 strain where diabetogenic action is manifest within 48 h. These dissociate the pancreatic effects of streptozotocin from site(s) that are responsive to endotoxin.

Glucan was just as effective in OF_1 animals as in the C3H/HeJ mice. In fact, both SZN and glucan appeared to exert greater sensitization to endotoxin lethality in the C3H/HeJ strain than in the OF_1, though statistical analysis was not done to fully understand the exact degree of difference. Nonetheless, the additive effect of these two very different materials, on endotoxin lethality, was clearly evident in the responder and the nonresponder strains of mice. This appears to be the first report where total reversal of refractoriness could be observed when glucan plus streptozotocin were used to study endotoxin lethality in the C3H/HeJ strain of mice. Again, TA was able to effectively protect against endotoxin lethality in all strains without influencing sugar balance and with some diminution in the activity of the RES. Thus, TA sensitive processes were equally operative in both strains.

These data support the possibility that glucan may act via activation of the RES but other sites of glucan action can not be ruled out. Similarly, TA protection against endotoxin lethality in glucan pretreated mice was actually associated with RES depression. Thus, RES activity seems to be a related effect of these materials and may not be causal to the outcome of endotoxin lethality.

The hyporesponsiveness of the C3H/HeJ strain is said to be controlled by an lps gene (for discussion see 8). It seems, however, inconceivable that glucan, streptozotocin, and a steroid hormone, known to influence different sets of host parameters, should exert their influence via the same lps gene, especially since all materials have comparable qualitative effects in the responder and the nonresponder.

An inescapabale conclusion of the lps gene hypothesis would be the genesis of a peptide in response to endotoxin. This endogenous mediator would then have to act on a wide variety of cell types. One would further have to assume that small quantities of this mediator are elicited even in the non-responder C3H/HeJ strain. Materials that sensitize to endotoxin can be envisioned to lower the level at which various cellular functions begin to respond to even minute quantities of this mediator, more effectively so in the refractory than in the sensitive strain. All this seems pure conjecture and it is perhaps wise to revise the classical lps gene viewpoint. Rather, endotoxin may act at several targets independently (16). The site(s) of endotoxin action thus remain elusive.

Acknowledgments

Professor G. Lazar was supported by the International Fellowship programme of the I.N.S.E.R.M., France. Some of the data reported here were aided by grants from the C.N.R.S. and the Faculty of Medicine Broussais, Hotel Dieu, Université Pierre et Marie Curie.

References

(1) Biozzi, G., Benacerraf, B., Halpern, B.N., Br. J. Exptl. Pathol., 34: 441 (1953)

(2) Agarwal, M. K. Die Naturwissenschaften, 62: 167 (1975).

(3) Agarwal, M. K. and Lazar, G., Microbios, 20: 183 (1977).

(4) Delafield, M.E., J. Pathol. Bacteriol., 35: 53 (1932).

(5) Agarwal, M. K., Biochem. Pharmacol., 23: 2577 (1974).

(6) Agarwal, M. K. and Berry, L. J., Biochem. Med., 2: 274 (1969).

(7) Agarwal, M. K., Paillard, J., Philippe, M. and Blondel, D. Biochem. Pharmacol., 30: 925 (1981).

(8) Agarwal, M. K. (ed) Bacterial Endotoxins and Host Response Elsevier/North Holland, 1980.

(9) Agarwal, M. K. and Lazar, G., Microbios (in press).

(10) Lazar, G. and Agarwal, M. K., Biochem. Med., 28: 310 (1982).

(11) Agarwal, M. K. (ed) Principles of Recepterology, Walter de Gruyter, 1983.

(12) Agarwal, M. K. (ed) Hormone Antagonists, Walter de Gruyter, 1982.

(13) Agarwal, M. K. (ed) Antihormones, Elsevier/North Holland, 1979.

(14) Agarwal, M. K. (ed) Streptozotocin: Fundamentals and Therapy, Elsevier/North Holland, 1980.

(15) Lazar, G. and Agarwal, M. K., Med. Microbiol. Immunol., 171: 179 (1982).

(16) Agarwal, M. K., FEBS Letters, 122: 9 (1980).

PATHOGENESIS IN THE AGGRAVATION OF ACUTE CHOLANGITIS : ITS RELATION TO THE INTERPLAY OF ENDOTOXEMIA AND HOST RESISTANCE

Hiroshi Shimada,* Gizo Nakagawara,* Fumihiko Kito, Tetsuo Abe
Mamoru Kobayashi, Shuzi Tsuchiya

*First Dep. of Surgery, Fukui Medical School, Fukui, Japan
Second Dep. of Surgery, School of Medicine, Yokohama City University, Yokohama, Japan

Simoaizuki 23-aza, Matsuokacho, Yoshidagun, Fukui 910-11
Japan

Intrduction

Acute cholangitis whose chief complaints are Charcot's triad (1), ie., abdominal pain, fever and jaundice, is sometimes accompained by shock or disseminated intravascular coagulation. Recently it is reported that endotoxemia and gram-negative bacteremia trigger the aggravation of acute cholangitis (2). But from our experience not all of the cases of acute cholangitis which are accompanied by endotoxemia and gram-negative bacteremia become severe. Their clinical course is affected by various factors.
In order to clearly examine pathogenesis in the aggravation of acute cholangitis, patients who suffered from acute cholangitis were divided into two groups. The first group consisted of 42 patients suffering from "severe cholangitis" accompanied by shock with a systolic blood pressure of less than 80 mmHg and the second group consisted of 49 patients suffering from "mild cholangitis". Both groups were compared according to the incidence of endotoxemia, status of host defense and clinical profile.
An experiment was also performed to clarify the relation between the level of serum CH50 and the clearing of endotoxin from the blood in order to give us a profile of the severity

Immunopharmacology of Endotoxicosis

of the endotoxemia, especialy the disturbance in coagulation and in the fibrinolytic system.

Material and Method

1) Clinical study: 91 patients with acute cholangitis admitted to the Second Dep. of Surgery, Yokohama City Univ. Hospital and its related hospitals during a recent ten year period were studied.

Tests of endotoxin in the blood, blood cultures, and other blood laboratory tests were carried out preoperatively with the mild cholangitis patients having a high fever and with the severe cholangitis patients in a state of shock. Bile cultures were done at the time of laparotomy, PTC(percutaneous transhepatic cholangiography) or PTC-D (PTC-drainage). Endotoxin in the blood was determined by the method of Levin, et al (3). Bacteria in the blood were detected by the incubation of arterial or venous blood vacuum drawn into the culture bottle containing a thyoglycolate medium. Bacteria in the bile were detected by incubation of bile in the blood and brain-heart infusion agar and by the isolation of the respective strains. Plasma fibronectin was measured by a fibronectin kit (Boehring Manheim, Germany) using a kinetic turbidimetric method based on an antigen-antibody reaction. Haptoglobins, transferrin, C3, C4, C5, IgG, IgM and IgA were measured by the immunological single radial diffusion method using commercially available kits. Phagocytic index, K, was determined by injecting chondroitin sodium iron sulfate, collecting blood samples at certain intervals, measuring serum iron in the blood samples, and making calculations in accordance with Halpern, et al (4).

2) Experimental study: 23 adult male dogs weighing 10kg to 18kg were used and each animal served as its own control. They were divided into three groups each having a different level

of serum CH50. The first group had a high level of serum CH50, n=9. They had been administrated an injection of 0.1 ml of E.coli with a concentration of 10^9 cell/ml into the common bile duct on the seventh day before the injection of endotoxin. The second group had a level of serum CH50, n=6. They had been treated by intravenous injection of endotoxin (E.coli 0111 B4) at a does of 0.5 mg/kg. They received the same intravenous dosage of endotoxin 24 hours later. The third group consisted of healthy dogs (n=8) who were observed to have a normal level of serum CH50 over seven days before they received an injection of endotoxin.

All dogs in the three groups were intravenously injected with endotoxin (E.coli 0111 B4) at a does of 0.5 mg/kg. The determination of endotoxin in the blood was performed at intervals using Levin's method, CH50 by Mayer's method, FDP (fibrinogen-fibrin degradation product) by FDPL kit (Teikoku-zoki Company, Japan).

Results

1) Clinical study: The 42 patients with severe cholangitis consisted of 26 patients with benign diseases and 16 patients with malignant diseases (Table 1). There is no difference in primary diseases between severe and mild cholangitis.

The 78.6% frequency of endotoxemia in severe cholangitis was significantly higher than the 32.6% in mild cholangitis (Table 2). Of 9 cases of negative endotoxin in severe cholangitis, 7 were severely icteric cases with serum total bilirubin of more than 10 mg/dl. The 52.9% frequency of gram-negative bacteremia in severe cholangitis was significantly higher than the 25.0% in mild cholangitis. The severe cholangitis patients (including 1 case of candidiasis and all other 18 cases of gram-negative bacteremia) did not include any cases of only gram-negative bacteremia. Most strains of gram-negative bacteremia in severe cholangitis cholangitis are

Table 1 Primary disease of acute cholangitis

Primary disease	Severe cholangitis	Mild cholangitis
Hepatic abscess	2	0
Intrahepatic stone	5	5
Common duct stone	10	17
Residual or recurrent stone	6	5
Gallbladder stone	1	3
Bile duct stricture	2	1
Bile duct carcinoma	8	8
Pancreatic head carcinoma	5	6
Carcinoma of Vater's papilla	1	0
Gallbladder carcinoma	1	2
Metastastic obstructive jaundice	1	2
Total	42	49

Table 2 Incidence of endotoxemia and gramnegative bacteremia

	Severe cholangitis	Mild cholangitis
Incidence of endotoxemia	78.6%(33/42)	32.6%(15/46)
Incidence of gramnegative bacteremia	52.9%(18/34)	25.0%(11/44)

E.coli or Klebsiella. In the strains of bacteria detected in the bile there was no difference between severe and mild cholangitis. In all cases the number of same species of bacteria greater than the one species detected in the blood could be also detected in the bile (Table 3).

From these clinical data and pathological study of the liver in severe cholangitis, we proposed three main routes to explain the occurrence of endotoxemia and bacteremia (Fig. 1). The most common route is cholangio-venous reflux of contaminated bile due to the elevation of bile pressure in number 1 of Fig. 1.

The mean WBC of severe cholangitis was 21,480, significantly greater than that in mild cholangitis (Table 4). The mean platelet count in severe cholangitis was 7.4×10^4,

Table 3 Blood and bile culture

Organism	Severe cholangitis Blood	(Bile)	Mild cholangitis Blood	(Bile)
Klebsiella	9	(15)	3	(15)
E,coli	7	(11)	4	(16)
Pseudomonas	1	(11)	2	(19)
Citrobacter	1	(3)	1	(4)
Enterobacter	1	(3)	2	(2)
Proteus	1	(2)	0	(2)
Serratia	1	(0)	0	(1)
H.influenzae	0	(0)	1	(0)
Entercoccus	0	(3)	1	(2)
Staphylococcus	0	(0)	1	(1)
Streptococcus	0	(0)	2	(0)
Cl.perfrigens	1*	(1)	0	(0)
Candida	3**	(2)	0	(0)

* Mixed infection with Cl.perfrigens, Klebsiella, E.coli and Citrobacter

** 1 case had mixed infection with Candida and Klebsiella. 1 case had mixed infection with Candida and Pseudomonas.

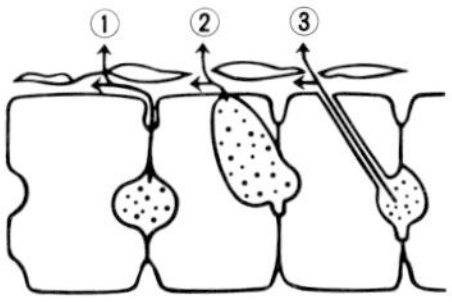

1) Cholangio-venous reflux of contaminated bile due to the elevation of the bile duct pressure.
2) Direct inflow of hepatic abscess into the blood vessels
3) Cholangio-venous reflux of contaminated bile due to the formation of traumatic shunt between bile duct and blood vessels induced by PTC or PTC-D

Fig. 1 Mechanism of occurrence of endotoxemia and bacteremia

significantly less than that in mild cholangitis.
PT (prothrombin time) and APTT (activated partial thromboplastin time) were prolonged in severe cholangitis. Plasma

fibrinogen was less than that in mild cholangitis. FDP, s-total bilirubin, s-GOT, s-GPT, and alkalinephosphatase showed no difference in both groups. Serum creatinine in severe cholangitis was significantly greater (Table 4). According to Matsuda's DIC (disseminated intravascular coagulation) criteria (5) determined by platelet count, plasma fibrinogen level and FDP level, 32 cases (76.2%) of severe cholangitis were complicated with DIC, (Table 5). As shown in the figures in the parentheses, the mortality rate is directly due to the complication of DIC.

Table 4 Laboratory findings 1

Determination		Severe cholangitis (mean)	Mild cholangitis (mean)	Significant difference
WBC	cells/mm	21,480	13,190	p<0.001
Platelets	cells/mm	7.4 10	25.6 10	p<0.001
PT	sec	15.0	11.3	N.S.
APTT	sec	44.2	32.5	N.S.
Fibrinogen	mg/dl	370.7	540.6	N.S.
FDP	g/dl	39	20.2	N.S.
Total bilirubin	mg/dl	10.3	7.9	N.S.
s-GOT	IU	368.9	153.0	N.S.
s-GPT	IU	214.4	130.4	N.S.
Al-phosphatase	Bessey.L.U	409.1	397.6	N.S.
Creatinine	mg/dl	2.9	0.9	p<0.001

Table 5 DIC as complication

Grade of DIC	no. of severe cholangitis	no. of mild cholangitis
DIC diagnosed definitively	24 (20)*	0 (0)
DIC suspected	8 (3)	0 (0)
DIC not diagnosable	5 (1)	0 (0)
DIC absent	5 (0)	49 (0)

* Figure in the parentheses represents the number of deaths

Table 6 Laboratory findings 2

Determination		Severe cholangitis (mean)	Mild cholangitis (mean)	Significant difference
s-Albumin	g/dl	2.6	3.6	N.S.
CH50	U/ml	23.4	48.4	$p<0.001$
C3	mg/dl	50.3	122.9	$p<0.001$
C4	mg/dl	22.5	41.4	$p<0.001$
C5	mg/dl	9.5	15.4	$p<0.001$
p-Fibronectin	μg/dl	145.0	930	$p<0.01$
Haptoglobin	mg/dl	320.0	143.1	$p<0.01$
Transferrin	mg/dl	285.0	182.5	N.S.
IgA	mg/dl	296.1	439.5	N.S.
IgM	mg/dl	117.1	135.4	N.S.
IgG	mg/dl	1204.5	1599.1	N.S.
Phagocytic index(K)		0.011	0.048	$p<0.01$
Reduction rate by NBT test*	%	36.5	30.5	N.S.

* NBT test means nitrosoblue tetrazolium test of neutrophil

Factors considered to be related to individual host resistance were investigated (Table 6). Serum albumin levels, CH50, C3, C4, C5, plasma fibronectin and haptoglobin were lower in severe cholangitis than that in mild cholangitis. Phagocytic index (K value) which is thought to represent the clearance of endotoxin was significantly lower in severe cholangitis. But the reduction rate shown by the NBT test (6) which shows the reduction activity of nitrosoblue tetrazolium by neutrophils did not show a difference in either group.

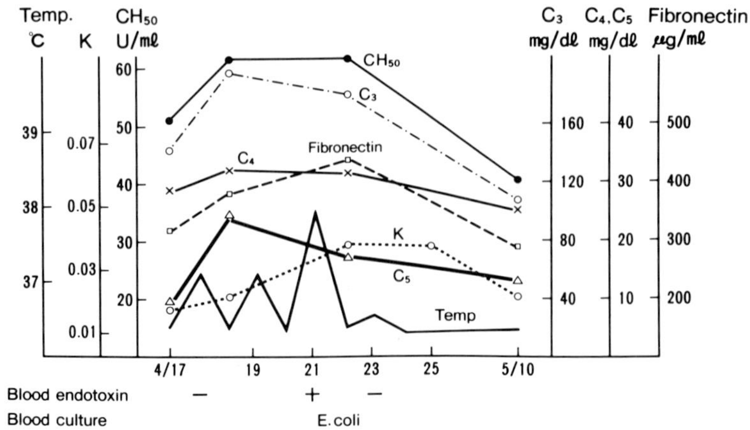

Fig. 2-1 Case 1 K. N. Choledochal stone with liver chirrhosis

Case report. Case 1: A 67-year-old male patient with choledocholithiasis complicated by chirrhosis of the liver (Fig. 2-1). On admission, when he had slight fever, CH50 and C3 levels were elevated. 4 days after admission, a high fever developed, and endotoxin and E.coli were detected in the blood. However, high CH50 and C3 levels persisted, and fibronection and the K value increased. As a result, endotoxin in the blood became negative, and the patient recovered. In this patient, the host's defense afainst infection, such as phagocytic function affected by complements was enhanced; and probably for this reason he was able to eliminate the endotoxin from the blood, improving his general condition.

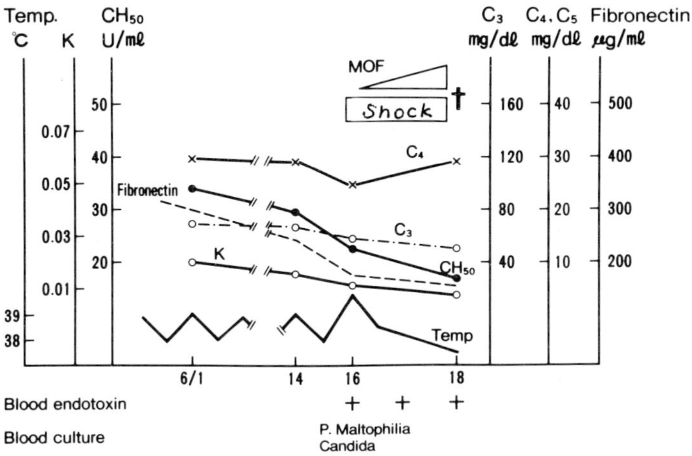

Fig. 2-2 Case 2. M. E. Gall bladder carcinoma with biliary fistula

Case 2: A 77-year-old female, a patient with gallbladder carcinoma and an external biliary fistula (Fig. 2-2). As a remmitent fever of 39°c persisted, C3, CH50 and fibronectin decreased and the K value tended to decline. The patient was becoming the so called "consumptive opsoninopathy". During this period, a high fever of 40.5°c was developed, and endotoxin, P.Maltophilia and Candida were detected in the blood. CH50, C3, fibronectin and K value tended to decline further. Endotoxemia persisted and multiple organ failure occured causing death. In this patient the consumption of blood opsonic factors may have caused deterioration of her host defense against infection. The depressed host's defense system may have failed in overcoming opportunistic infection.

These cases show that factors related to host defense against infection, such as blood opsonic factors, may have affected the clearance of endotoxin in the blood and the general clinical profile.
There was a significant correlation between the phagocytic

index represented by the K value and CH50 (Fig. 3). It could be assumed that when CH50 was high, phagocytosis was enhanced, and that when CH50 was low, phagocytosis was depressed. That is, the level of CH50 affects the phagocytic activity.

The level of CH50, C3 and fibronectin correlated significantly with platelet counts (Fig. 4). Thus decreased CH50 and C3 or fibronectin suggested a decrease of platelet count, and a possibility for DIC.

2) Experimental study (Fig. 5) In the first group of dogs the mean level of CH50 was 32.6± 6.2 U/ml (Mean S.E), 8.8± 0.6U/ml in the second group, 18.6± 3.8 U/ml in the third group. In the first group after the injection of endotoxin, the clearance rate of endotoxin in the blood was highest, and the decrease of serum CH50 level and platelet count and increase of FDP level were least of those in three groups. Thus the group with the highest level of serum CH50 was most tolerant to the intravenous injection of endotoxin.

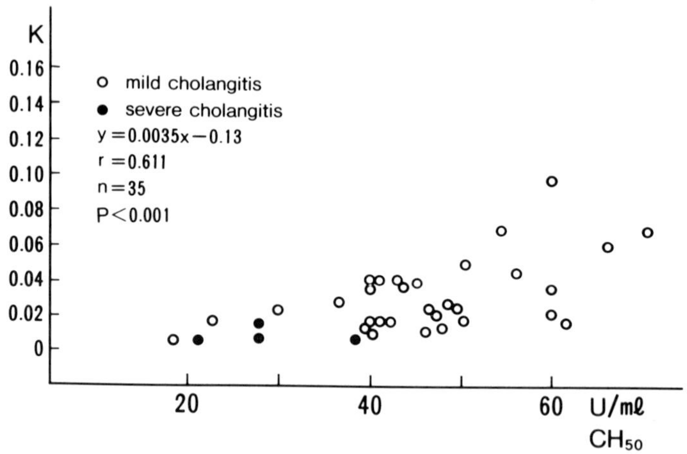

Fig. 3 Correlation between K value and CH50

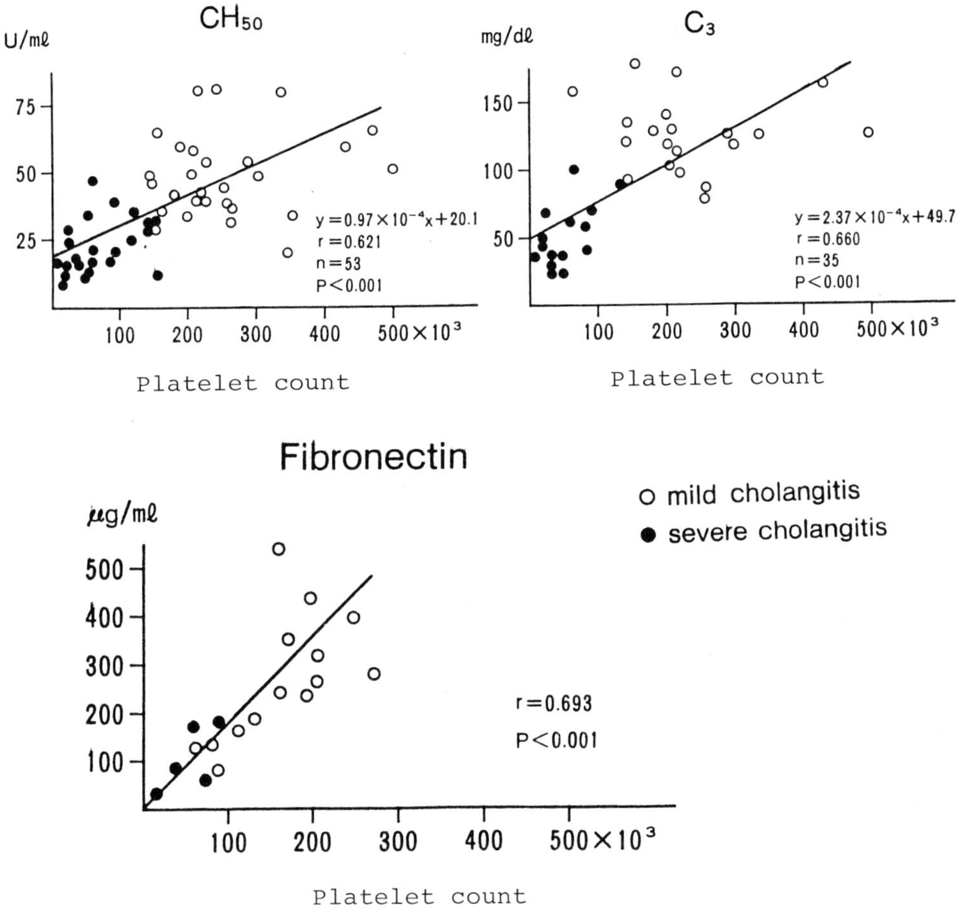

Fig. 4 Correlation complements and fibronectin, and platelet count

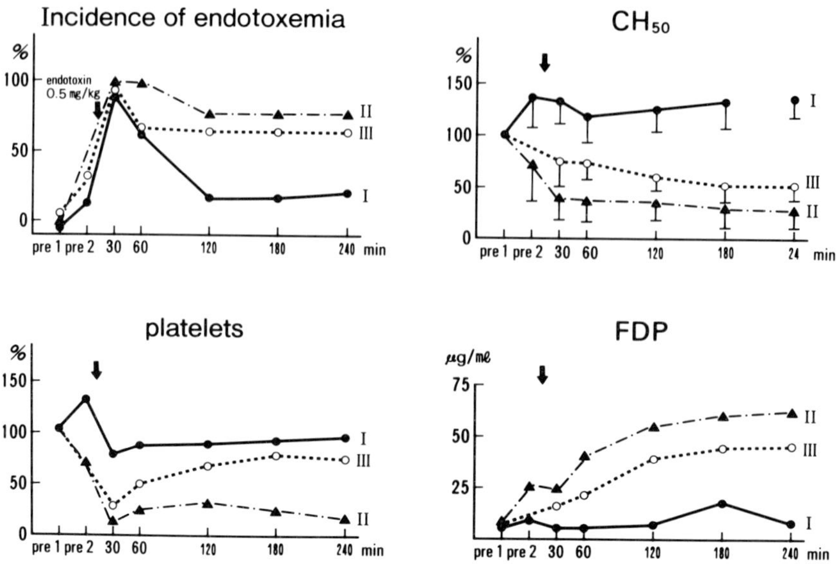

Fig. 5 Experiment

I: Group with high level of serum CH50 (n=9)
II: Group with low level of serum CH50 (n=6)
III: Group with normal level of serum CH50 (n=8)

Discussion

Since Levin(1970) (3) demonstrated endotoxin in the blood of patients having acute cholangitis for the first time, endotoxin has been noticed as the direct aggravating factor of acute cholangitis as reported in our paper (21). Considering the fact that blood bile acid inhibits the limulus test, a higher positive rate of endotoxemia in severe cholangitis than that shown in our paper might be expected. Gram-negative bacteremia was considered to be caused by the mechanism of cholangio-venous reflux because the same bacteria in the blood was detected in the bile. Because endotoxin was a part of the cell wall of gram-negative bacteria, endotoxemia

was thought to be caused by the same mechanism.
Shock or DIC has been considered to occur through the following mechanism (7.8); There is clinical evidence that among cases of gram-negative bacteremia, only those showing a decrease in C3 or alternative pathway factors (9) had shock as a complication and thus had a poor prognosis. This clinical report suggests that shock was induced by the C3a and C5a, the anaphylatoxin formed during the activation of an alternative pathway by blood endotoxin.
On the other hand endotoxin is thought to induce DIC because it enhances blood coagulation, breaks down platelets and damages the endothelium. The present study also reveals a significant correlation between platelet count and CH50 or C3 as shown in Fig. 4. This result shows that the consumption of complements by blood endotoxin is closely associated with the decrease of platelet count, and accordingly the occurrence of DIC.
However, in the present study of the 48 patients with endotoxemia, 29.2% had mild cholangitis cases, 33.3% survived after shock, and 37% died. This finding suggests the clinical profiles of acute cholangitis patients to be affected by host defense as well as the occurrence of endotoxemia or gram-negative bacteremia.
It is not easy to accurately judge the degree of host defense against infection in individual patients. In the present study, the following factors decreased in severe cholangitis patients: C3, C4, C5 and CH50, transferrin, fibronectin, and haptoglobin which has a property of combining with hemoglobin and is considered to act as an opsonin. This finding suggests depressed host defense against bacterial infection in severe cholangitis patients. For such a state, Alexander (10) proposed the concept " consumptive opsoninopathy ".
Endotoxin is generally considered to be ingested by reticuloendothelial cells of the liver, lung and spleen (11,12, 13), particularly Kupffer cells. Neutrofils also take in endotoxin when ingesting gram-negative bacteria. Recently,

platelets have attracted attention as taking up small particles such as endotoxin by pinocytosis and transporting them to phagocytes of the reticuloendothelial system. That is, platelets combine with endotoxin to such a high degree that they are regarded as having receptors for endotoxin.
Macrofages such as Kupffer cells have receptors for Fc and C3b. Their ingestion of endotoxin has been considered to be more efficient in the presence of opsonization. So far C3 (14,15), IgG (16), IgM (16), Factor eight (16), transferrin (17) and certain α-globulin fractions decomposing into subunits (18) have been reported as opsonic factors. Kaplan (19) recently stated that fibronectin should be regarded not as an opsonin for bacteria but as an opsonin for fibrin, collagen and platelet-bound endotoxin. His opinion may be supported by our results demonstrating good correlation between fibronectin and platelet count.
The presence of phagocytic activity was determined by iron uptake based on the finding that Kupffer cells take up iron (20). The results showed the highest correlation between the phagocytic activity of Kupffer cells and CH50. Also a good correlation was seen between the course of blood endotoxin and CH50. That is, the prognosis was known from changes in CH50.

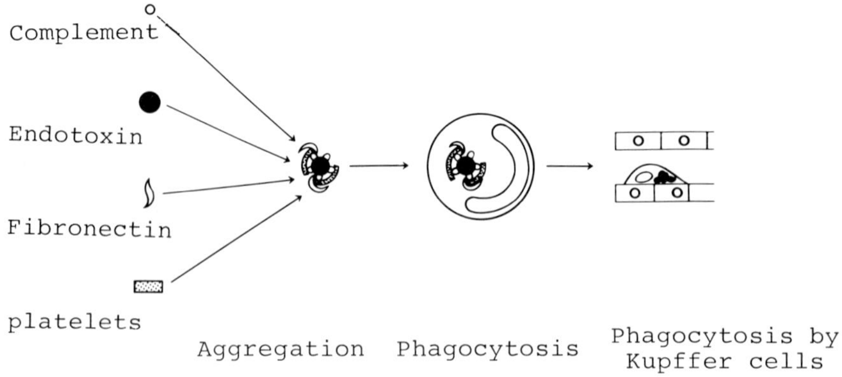

Fig. 6 Conception of endotoxin phagocytosis

Our experimental study also showed that the condition of a high level of serum complement was very effective in the clearance of endotoxin. These findings indicated that components of complements affecting CH50, such as C3, ie., nonspecific humoral factors, are the most important for processing endotoxin.

From the viewpoint mentioned above, the conception of endotoxin phagocytosis can be suggested (Fig. 6). Platelets bind to the endotoxin through the medium of complement and platelets. Serum fibronectin helps in fastening the complex formation of endotoxin, complement and platelets.

Summary

The pathogenesis in the aggravation of acute cholangitis was investigated based on the course of endotoxemia and the host defense. Acute cholangitis starts to become severe when endotoxin or gram-negative bacteria appear in the blood. A clinical profile of severe cholangitis shows DIC combined with a decrease of serum opsonins. However the general condition of the patient improves and the patient recovers, if CH50, C3 and fibronectin increase and phagocytic activity is elevated. In cases where complement or fibronection is consumed or fails to be produced, endotoxemia persists, causing cytolysis, cytotoxicity, and accelerated coagulation, which will lead to DIC or MOF (multiple organ failures).

References

1. Charcot, JM.: Leçons sur Les Maladies du Foie des Voies Filaries et des reins Paris; Faculte de Medecine de Paris 1877:1-102.
2. Longmire, WP.: Suppurative cholangitis in critical surgery illness edited by Hardy, JD. W.B.Saunders,Co 1971 p380-408.

3. Levin, J., Tomasulo, PA., Oser, RS.: J Lab Clin Med 75,903 -911 (1970).

4. Halpern, B and Biozzy, BB.: Brit J Exp Path 34,426-435 (1953).

5. Matsuda, Y.: Rinsho Byori (The Japanese Clinical Patholopy) 32,85-100 (1978), in Japanese.

6. Tomonaga, M.: Rinsho Ketsueki 17,1266-1275(1976), in Japanese.

7. Corrigan, JJ.: J Pediat 91,695-700(1970)

8. McKay, DG.: Diathes Haemorrh 29,11-25 (1973).

9. Fearon, DJ.: N Eng J Med 292,937-940(1975).

10. Alexander, JW., Meclellan, MA., Ogle, C. et al.: Ann Surg 184,672-678(1976).

11. Nolan, JP et al.: Gastroenterol 75,445-449(1978).

12. Cremer, N et al.: Proc Soc Exp Biol Med 95,510-513(1959).

13. Munford, RS.: Gastroenterol 75,532-535(1978).

14. May, JE., et al.: J Immunol 109,893-895(1972).

15. Golub, S. et al.: J Reticuloendothel Soc 5,324-339(1968).

16. Seeling, HP., Liehr, H., Seeliry, R., et al.: Immunofluorescenese studies of Kupffer cells in various liver in man. The RES and the pathogenesis of liver disease pp317-323 Liehr, H and Grün, M. Elsevier, North Holland, New York, Oxford 1980.

17. Weinberg, ED.: J Infect Dis 124,401-410(1971).

18. Von Eschen, KB and Rudbach, JA.: J Infect Dis 129,21-29 (1974).

19. Kaplan, Je.: Plasma fibronectin and resistance to thrombosis during sepsis. Advance in Shock research. Rechard, SM and Fletcher, JR editors Alan, R New York, 7,159-172(1982).

20. Dullman, J., Wulphekel, U and Hausman n, K.: Difference between Kupffer cells and endothelial cells of human liver in iron storage and release under normal condition after i.v. iron administration. Kupffer cell and other liver sinusoidal cells. pp 233-241, Wisse, E and Knod, 01 eds, Elsevier/North Holand, Amsterdam, 1977.

21. Shimada, H.: Nihon Geka Gakkai Zashi (Japanese J of Surgery) 83,1321-1330(1982), in Japanese.

ENDOTOXEMIA IN PREGNANCY DUE TO CHLORAMPHENICOL ADMINISTRATION

Hiroshi Irie, and Wataru Mori
Department of Patholgy, Faculty of Medicine, University of Tokyo,
3-1, Hongo 7-chome, Bunkyo-ku, Tokyo Japan.

Introduction

Acute fatty liver in pregnancy has been frequently reported since first described by Sheehan (1, 2, 3, 4). Many reports indicated a possibility of DIC (disseminated intravascular coagulation) to have a pathophysiological significance in this disease (5, 6, 7, 8, 9). DIC is induced by endotoxemia. Close correlation has been recognized between systemic endotoxemia and intravascular coagulation (10, 11, 12, 13, 14). On the other hand, liver injuries induced by antibiotics such as tetracycline, aureomycin, chloramphenicol, etc. have been reported (15, 16, 17). Especially a fulminant form of liver injury has been found to appear in pregnant women (15). In 1979, we achieved production of an experimental model of acute massive hepatic necrosis by injection of endotoxin preparation into the bile duct, i.e.,Shwartzman mechanism (18). This hepatic necrosis occurred more frequently and severely in pregnant rabbits than non-pregnant rabbits. Pregnant rabbits also ocassionally developed severe fatty liver under the same conditions. It is well known that DIC in human corresponds roughly to a generalized Shwartzman reaction in rabbits. The present study was designed to clarify the possible inter-relationship among fatty liver in pregnancy, endotoxemia, DIC and antibiotics -induced liver injury. The purpose of experiment performed in this study included; (1) whether or not a massive dose of chloramphenicol induced endotoxemia in rabbits and (2) if so, what kind of pathohistological picture appears in systemic organs including the liver in rabbits? The finding obtained suggested the pathogenesis of fatty liver in humans. The results are outlined and the readers are solicited for kind and frank comments.

Immunopharmacology of Endotoxicosis

Materials and methods

Twenty-four pregnant rabbits (gestation period: approximately 20 days, weighing 3500 to 5000 g) and 12 non-pregnant rabbits (weighing 2500 to 3000 g) were used in the experiment. These rabbits were divided into four groups, A, B, C and D. Group A consisted of 18 pregnant rabbits. Group B consisted of 6 pregnant rabbits. Groups C and D consisted of 6 non-pregnant rabbits each.

Experimental procedure. Ten milliliters of whole blood were collected from all the rabbits prior to experiment. Groups A and C were given a large dose of chloramphenicol (300 mg/kg, Sankyo Seiyaku: Chloromycetin) dissolved in sterile water by intramuscular injection. Groups B and D were used as controls by injection of an equal volume of saline. Twelve hours after the injection, 10 ml of whole blood were collected again, and intramuscular injection of chloramphenicol (300 mg/kg) to groups A and C and saline groups B and D was administered. Twenty-four hours after the reinjection, final 10 ml of whole blood were collected. Each blood sample was divided into 2 portions. All the rabbits were killed 40 hours after the beginning of the experiment under anesthesia with somnopentyl.

Liver function test. Using 5 ml of venous whole blood from each rabbits, serum was separated for measurement of serum glutamic-oxaloacetic transaminase (GOT) and glutamic-pyruvic transaminase (GPT).

Endotoxin assay. Endotoxin was assayed basically by the method of Levin et al (19). Five milliliters of heparinized venous blood from each rabbit were placed into a silicon-coated pyrogen free tube and centrifuged at 2000 rpm for 30 seconds at 4°C. After centrifugation, the platelet-rich plasma layer was collected. Zero-point-two milliliter of chloroform was added to 0.8 ml of platelet-rich plasma. This emulsion was vigorously mixed for 60 min with a vibrator at room tempertature and then separated by centrifugation at 2000 rpm for 14 min at 4°C into following 3 layers; top clear aqueous layer, middle cloudy layer and bottom layer that contained particles materials. Zero-point-one milliliter of middle layer was mixed with amebocyte and used for detection of endotoxin. assay of endotoxin was carried out in an incubator at 37°C. The components of the assay were mixed and then evaulated at 4 hours and, after a long interval, at 24 hours. The development of marked viscosity or gelation was defined as definitely positive for the presence of endotoxin. A slight increase in viscosity and turbidity was difined as weakly positive. The mixture that underwent no change after 24-hour incubation was defined as negative. One microgram of E. coli endotoxin (0111-B4) dissolved in distilled water was a positive control. Zero-point-one milliliter of saline added to the lysate was used as a negative control. All the glasswares and instruments had been sterilized by autoclaving.

Histological examination. The liver, kidney, lung, adrenal, heart and spleen were examined carefully for macroscopical changes. Furthermore, histological examination was done for the liver, kidney, lung, adrenal, heart and spleen. After fixation, in 10% formalin, all the tissue were embedded in paraffin and stained with hematoxylin and eosin. Fatty changes of the liver were classified into four grades as follows. Grade 1: No evidence of fatty change either macroscopically or microscopically. Grade 2: No fatty change was observed macroscopically but a few droplets were shown microscopically. Grade 3: Macroscopical observation showed a yellowish brown colored surface of the liver and microscopical examination revealed a marked fatty change. Grade 4: The severity of the fatty change largely exceeded that of Grade 3.

Furthermore, in some of the experimental animals from groups A and B, bacterial culture of the stool obtained on necropsy was performed.

Results

None of the rabbits of groups A, B, C and D died during the experiment. The frequent presence of loose brown stool in the alimentary tract in groups A and C was a noteworthy finding at the time of necropsy. Table 1 lists the frequency of endotoxemia in each group. In group A, 5 (27.8 %) of 18 rabbits were definitely positive and 3 (16.7%) of 18 rabbits

Table 1

Frequency of endotoxemia in groups A, B, C and D

	Cases	Limulus test Positive	weakly Positive	Negative
Group A	18	5 + 1*	3	6
Group B	6	0	0	6
Group C	6	0	0	6
Group D	6	1*	0	5
Total	36	7	3	26

* Endotoxin positive case unrelated to chloramphenicol injection

were weakly positive after chloramphenicol injection. However, one (Rabbit No. 32) was positive in the limulus amebocyte lysate test of the blood collected prior to chloramphenicol injection. On the other hand, none of the animals in groups B and C were endotoxin positive. However, endotoxin in blood was detected in one (Rabbit No. 5) of group D. An asterisk (*) in table 1 indicates an endotoxin positive case unrelated to

Table 2

Histological severity of fatty liver in groups A, B, C and D

	Cases	Histological grade of fatty liver 1	2	3	4
Group A	18	1	6	7	4
Group B	6	2	1	2	1
Group C	6	2	4	0	0
Group D	6	5	1	0	0
Total	36	10	12	9	0

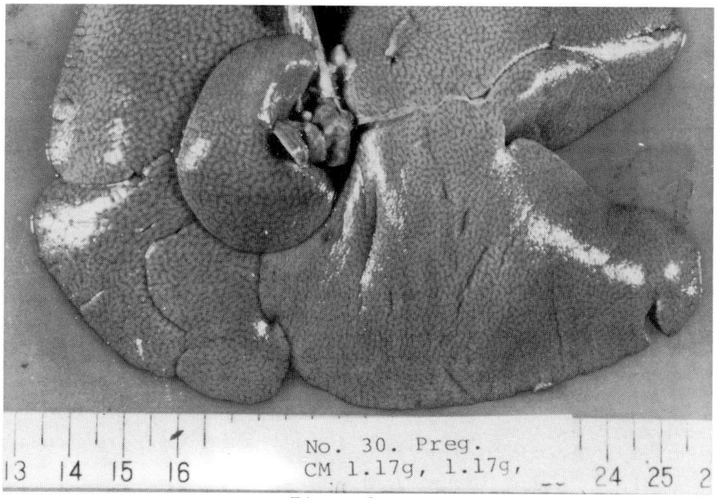

Fig. 1

Marked fatty liver of pregnant rabbit shown after chloramphenicol injection. Endotoxin in blood was positive in this case (Rabbit No. 30)

chloramphenicol injection. Table 2 shows histologically classified severity of fatty liver. As indicated in the table, histological fatty change of the liver was relatively more severe in groups A and B than in groups

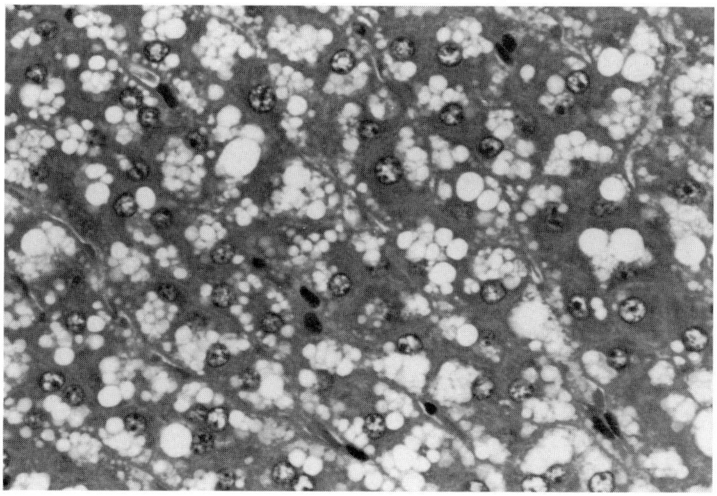

Fig. 2

Histological picture of fatty liver of the same case in Fig. 1.

C and D. Macroscopically the liver of pregnant rabbits, whether treated with chloramphenicol or not, generally showed a yellowish brown colored surface. Moderate to marked fatty changes were the main pathohistologic change seen in groups A and B. Especially, marked fatty liver was frequent in group A. Figure 1 shows the marked fatty liver from rabbit No. 30 of group A. On comparing between groups C and D, mild to moderate fatty changes were more frequent in group C. It was noted that the general histological pattern of fatty liver was multivacuolated fatty droplet deposition in the hepatocytes. Figure 2 shows the histological picture of fatty liver from the pregnant rabbit No. 30. On the other hand, dispersed focal necrosis of the liver tissue was observed in 5 animals of group A and one of group B. However, there was no remarkable Kupffer cell proliferation in the sinusoidal space. Pericholangial inflammatory cell infiltration observed in many rabbits of all the groups may not be evaluated as significant. There was no microscopical finding of fibrin and fresh blood thrombi in the liver, kidney and adrenal in any of the rabbits. Table 3 shows the incidence of multiple petechial pulmonary bleeding in each group. Petechial pulmonary bleeding was noted in 5 animals of group A and occurred more frequently (50%) in the endotoxin positive

Table 3

Incidence of multiple petechial pulmonary bleeding in each group

	Multiple petechial pulmonary bleeding	
	+	−
Group A	5 (ETX positive 4, ETX negative 1)	13
Group B	0	6
Group C	0	6
Group D	0	6
Total	5	31

ETX : endotoxin

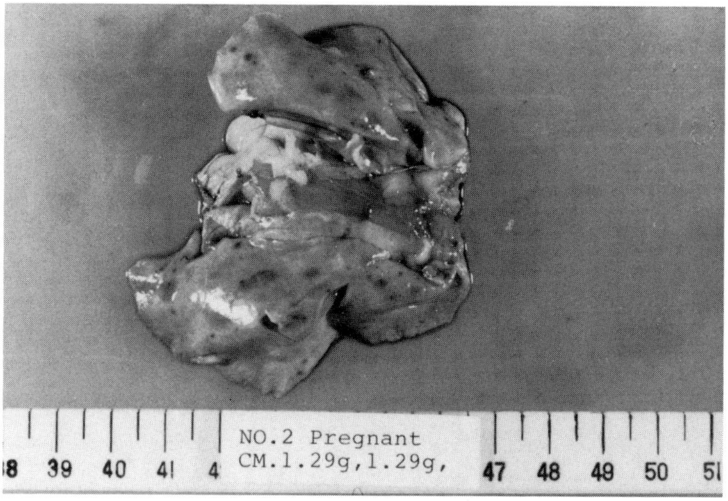

Fig. 3

Multiple petechial pulmonary bleeding in endotoxin strongly positive pregnant rabbit (Rabbit No. 2) after chloramphenicol injection

cases including weakly positive cases, as compared with the endotoxin negative cases. Figure 3 and 4 illustrate the pulmonary lesion in the pregnant rabbits (group A) after chloramphenicol injection. Endotoxemia was found in the blood of both the cases. Table 4 lists the levels of

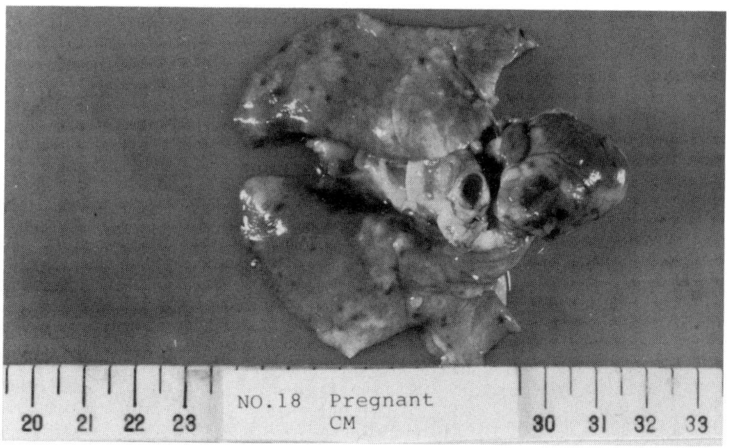

Fig. 4

Multiple petechial pulmonary bleeding in endotoxin positive pregnant rabbits (Rabbit No. 18) after chloramphenicol injection

Table 4

Levels of s-GOT and s-GPT

s-GOT (unit/100 ml)

	0 hrs.	12 hrs.	36 hrs.
Group A (18)	45.4 ± 13.2	36.6 ± 12.8	41.5 ± 15.7
Group B (6)	48.5 ± 22.5	60.5 ± 40.2	33.0 ± 11.5
Group C (6)	19.0 ± 16.0	38.0 ± 29.5	42.0 ± 34.0
Group D (6)	29.5 ± 15.0	36.2 ± 15.2	42.0 ± 22.0

s-GPT (unit/100 ml)

	0 hrs.	12 hrs.	36 hrs.
Group A (18)	32.1 ± 10.5	37.8 ± 14.2	31.8 ± 11.4
Group B (6)	26.0 ± 10.5	30.2 ± 14.2	17.5 ± 6.8
Group C (6)	28.0 ± 18.2	35.0 ± 15.0	48.3 ± 32.0
Group D (6)	36.2 ± 15.8	58.5 ± 30.4	81.0 ± 52.0

( ) numbers of cases

Table 5

Stool culture at necropsy

	Group A			Group B
	Rabbit No. 28	No. 29	No. 30	Rabbit No. 27
Endotoxin in blood	-	-	+	-
Stool culture	Negative	Streptococcus faecium	Pseudomonas maltophilia	Streptococcus faecium

s-GOT and s-GPT. Extremely high levels of s-GOT and s-GPT were noted in none of the groups. However, the s-GOT level was relatively higher in groups A and B than groups C and D. The s-GPT level showed no significant differences in groups A, B, C and D. Table 5 shows the results of bacterial culture of the stool in some animals of groups A and B. Pseudomonas was identified in one endotoxin positive case and no organism was identified in one of group A.

Discussion

Chloramphenicol is a broad spectrum antibiotic known to inhibit protein synthesis in the bacterial ribosome. On the other hand, endotoxin is a constituent of the cell wall of gram negative bacilli. The flora of the gastrointestinal tract is the only major source of endotoxin in the absence of systemic gram negative infection. Chloramphenicol acts on gram negative organisms in the flora of colon to release endotoxin. The release of some other enterogenic toxic substances might also cause loose brown stool and diarrhea as frequently observed in groups A and C. It is generally postulated that endotoxin from organisms in the alimentary tract passes continuously into the portal system and is removed by the reticuloendothelial system of the liver. Therefore, endotoxin does not appear in the systemic circulation under the normal circumstances. However, endotoxemia occurs in the systemic circulation under some pathological circumstances. Endotoxemia has been reported in fulminant hepatitis (20) and liver cirrhosis (21, 22). Endotoxemia appears in these diseases, because the reticuloendothelial function is overwhelmed or suppressed. Pregnancy is also estimated to be the state of markedly systemic reticuloendothelial function including the liver (23, 24). This suggests that endotoxin is detected in the blood of pregnant rabbits after chloramphenicol injection. In this study it was considered that chloramphenicol administration conspicuously increases endotoxin in the circulation as a result of destruction of alimentary organisms. Endotoxemia was actually seen in some cases (27.8%) of pregnant rabbits after chloramphenicol injection in this study. It was considered that in rabbits which lost reticuloendothelial function on pregnancy, endotoxemia in systemic circulation resulted from the lack of ability to detoxify massive endotoxin produced by destruction of gram negative organisms by chloramphenicol injection. There were two endotoxin positive rabbits without relation to chloramphenicol administration. These may be suspected to be associated with cholangitis seen in many rabbits. Multiple petechail pulmonary bleeding was frequently observed on necropsy in endotoxin positive rabbits (50%). This finding strongly suggested frequent occurrences of the systemic circulatory disturbance such as shock in endotoxin positive rabbits. Moreover, the possibility of mild generalized Shwartzman reaction was also suggested. However, an essential lesion such as renal cortical necrosis or alimentary tract hemorrhage existed in no rabbit. In other words, the lesion in the rabbits experimented could not be identified as a pattern corresponding to clinical human DIC. There were also no distinct microthrombi in the capillaries of the liver, kidney, adrenal or lung. In this study, the degree of fatty liver was higher in severity as well as frequency in groups A and B than in groups C and D. Moreover, the results indicated obviously that chloramphenicol injection produced increased occurrences and intestified the fatty change. In tetracycline-induced fulminant hepatic failure in pregnant women, a high dose of tetracycline induced marked fatty liver. And yet in this disease s-GOT and s-GPT do not exhibit high titers (15, 25), because of mild necrosis of hepatocytes. A marked fatty change may be related to inhibition of protein synthesis by tetracycline in the hepatocytes. A similar mechanism is also considered to be involved in cases of chloramphenicol administration. Moreover, a decrease in VLDL (very low density lipoprotein) synthesis would also play a role in this acute fatty liver. On the other hand, the pathophysiological change of the host response in pregnancy cannot be overlooked as another most important underlying factor for the occurrence of endotoxemia and fatty liver. Furthermore, an evidence of circulatory endotoxin positive has been reported in children with Reye's syndrome (26). Fulminant hepatic failure by an advanced fatty change was proposed to be a major factor for systemic endotoxemia (26). A definite correlation between severity of fatty change and endotoxin was not established in this study. Nevertheless, the results are highly suggestive of pathogenesis of acute fatty liver in pregnancy, especially when viewed in combination with our previous experimental model of massive hepatic necrosis or acute fatty liver produced by a univisceral Shwartzman mechanism with endotoxin prepared injection into the bile duct (18). So far as we know, the studies on endotoxemia in pregnant acute fatty liver are non-existent. We believe that the present study is the first to have confirmed the possibility of pregnant rabbit fatty liver to develop endotoxemia after chloramphenicol injection. It may be reasonable to interprete that elevation of the blood endotoxin level by chloramphenicol injection is closely associated with the development or progression of fatty change in the liver. There still remains, for future studies, more detailed investigation on the differences between endotoxin positive rabbits and negative pregnant rabbits and the role of enterogenic toxic substance, apart from endotoxin, in relation to changes in bacterial flora after chloramphenicol administration.

Conclusion

Intramuscular injection of a large dose of chloramphenicol induced endotoxemia in some pregnant rabbits. In endotoxin positive cases, a fatty change of the liver tended to be more severe and frequent than in controls. Furthermore, multiple petechial pulmonary bleeding was frequent in endotoxin positive pregnant rabbits. A possibility was suggested that systemic circulatory disturbance such as shock and mild Shwartzman reaction developed from endotoxemia. These results explained some pathogenic aspects of acute fatty liver in pregnancy.

Acknowledgment

We are thankful to Miss Seiko Suzuki, Laboratory of Mitsui Memorial Hospital, for her competent technical support.
We are grateful to Dr. Hiroshi Ōami, Division of Pathology, Clinical laboratory of National Medical Center Hospital, for his comment.

References

1. Sheehan, H.L.: Amer.J.Obst.Gynec. 81, 472-440 (1961).

2. Joske, D.J., McCully, D.J., Mastoglia, E.L.: Gut 9, 489-493 (1968).
3. Iwata, K., Kakazawa, J., Endoh, T., Kohgame, S.: Acta.Pathol.Jap. 18, 473-483 (1968).
4. Stander, H.L., Cadden, J.F.: Amer.J.Obst.Gynec. 28, 61-69 (1934).
5. Cano, R.I., Delmon, M.P., Pitchumoni, C.S., Lev. R., Rosenthal, W.S.: J.A.M.A. 231, 159-191 (1975).
6. Hatfield, A.K., Stein, J.H., Greenberger, N.J., Abernetty, R.W., Ferris, T.F.: Amer.J.Diq.Dis. 17, 167-178 (1972).
7. Zwierina, W.D., Kurz, F., Judmaier, G., Mikurz, F., Gattringer, K.: Dtsch.med.Wschr. 108, 703-707 (1983).
8. Yoshimura, R., Fujiyama, S., Nakazawa, H., Harihara, S., Kadona, J., Yamamoto, S., Yamaguchi, K., Nakamori, H., Junkawa, K.: Jap.J.Hepatol. 17, 691-697 (1976).
9. Holzbach, R.T.: Obstet.Gynecol. 43, 740-744 (1974).
10. Kojima, M., Shimamura, K., Kanashiro, T.: Igaku no Aumi. (supple) ed. by Aoki, N.,Matsuda, T., Mori, W., Tamakuma, E., pp 66-72, Ishi-yaku shuppan, Tokyo, Japan, 1979.
11. Nolan, J.P.: Gastroenterol. 69, 1346-1256 (1975).
12. Kondo, G., Takemura, S.: Jap.J.Gastroenterol. 73, 1373-1379 (1977).
13. Corrigan, J.J., Bell, B.M.: J.Lab.Clin.Med. 77, 802-810 (1971).
14. Matsuda, T.: Nihon Rinshō 32, 89-95 (1974).
15. Kunelis, C.T., Peters, J.L., Edmondson, H.A.: Amer.J.Med. 38, 359-377, (1965).
16. Lepper, M.H., Zimmerman, H.J., Caldwell, E.R., Carroll, G., Spies, H. W., Wolfe, C.K., Dawling, H.F.: A.M.A.Archiv.Intern.Med. 88, 284-295, (1951).
17. Dowing, H.F., Lepper,M.H.: J.A.M.A. 188, 307-309 (1964).
18. Mori, W., Shiga, J., Kato, A.: Virchows Archiv.A.Path.Anat.and Histol. 382, 179-190 (1979).
19. Levin, J., Tomasulo,P.A., Oser, R.S.: J.Lab.Clin.Med. 75, 903-911 (1970).
20. Wilkinson, S.P., Arreyo, V., Gazzard, B.G., Modie, H., Williams,R.: Lancet 1, 551-524 (1974).
21. Prytz, H., Holst-Christensen, J., Korner, B., Lieber, H.: Scandinav. J.Gastroenterol. 11, 875-863 (1974).

22. Tarao, K., Endo, O., Ikeuchi, T., Soo, K., Mori, T., Fukushima, K.: Jap.J.Gastroenterol. 73, 1366-1372 (1976).

23. Mckay, D.G., Merrill, S.J., Weider, A.E., Hertig, A.T., Reid, D.E.: Amer.J.Obst.Gynec. 66, 507-539 (1953).

24. Burrow,G.N., Ferris, T.F.: Medical complications during pregnancy. pp 42-44, Saunders Comp. Philadelphia, 1975.

25. Mori, K., Yoshikawa, K., Akashi, Y., Tanaka, R., Ono, K.: Jap.J. Hepatol. 15, 529-535 (1974).

26. Cooperstook, M.S., Tucker, R.P., Baublis, J.V.: Lancet. 1, 1272-1273 (1975).

GENERAL DISCUSSION

The general discussion was divided into four themes, each led by a discussion leader for approximately 30 min each, as follows:

Theme 1.	Dr. M. K. Agarwal
Theme 2.	Dr. S. Niwa
Theme 3.	Dr. M. Yoshida
Theme 4.	Dr. N. Bloksma

The discussion published here was transcribed by the editors' from that recorded on the tape, in the order of proceedings. The discussion leaders, discussants, and other contributors did not have the opportunity to verify their remarks prior to publication. Errors of interpretation, thus, remain the sole responsibility of the editors'.

The questions raised at the end of individual papers have been grouped together under the topic most appropriate to their content. They may sometimes appear out of place due to heterogeneous nature of remarks. Indulgence is asked of the reader. It is nevertheless felt that inclusion of this section would make the volume that much more valuable in provoking further thought.

GENERAL DISCUSSION

A. The Contribution of the lps Gene in Endotoxicosis.

Agarwal. The observations leading to a so called LPS "low responder" or "non responder" mouse strain go back some two decades. In our own laboratory, the C3H/HeJ low responder can be upto 50 times more resistant than the responder C3H/eb control when LPS lethality is titrated. What I thought one could do today was to discuss whether all responses provoked by endotoxins can be mediated by the modulation of the so called the one and the same lps gene.

The accompanying table lists the complexity and the diversity of LPS mediated responses. Endotoxin can interact with two major cell types viz. the mesodermal and the endodermal which then differentiate into more specialized tissues in different organs. The initial, direct interaction between these cell types and the LPS gives rise to a whole array of mediators that interact with a second target that is more or less well defined. Finally, a whole series of effects can be detected.

Can we explain all this diversity of reactions just by assuming that endotoxin mediates all its effects by a single lps gene. Genetically, the lps gene product, a protein, would have to act on all these cell types but there is no evidence for that. We may assume that the lps gene product acts only at the level of the LPS receptor but, again, no evidence for such a receptor exists at the level of all the cell types known to respond to bacterial endotoxins. Furthermore, there is no direct, causal evidence that LPS-receptor interaction is required for endotoxin activity in all cases. Thus, I am not very convinced that the lps gene hypothesis can explain all the mysteries of endotoxin poisoning.

Origin	FIRST TARGET	MEDIATOR	SECOND TARGET	EFFECT
MESODERM → RES IN THE HOST / B-cells	White blood cells	Leucocyte endogenous mediator	Various cells	Fever, Hypergammaglobulinemia, Leucocytosis
	Monocytes	Leucocytic pyrogen	Liver	Serum amyloid A synthesis, Secondary amyloidosis
	Peritoneal exudates	Endogenous pyrogen	Fever centres	Fever
	Bone marrow	Procoagulants	Complement	DIVC
	Liver RES	Monokines, GAF	Hepatocytes	Enzymatic dysfunction
	Spleen lymphocytes	Lymphokines	Macrophage Fc receptor	LPS nonresponder → Responder
	Granulocytes, Macrophages	Histamine & Serotonin.	Various cells	Vasoactive imbalance
		Procoagulants	Complement	Thrombosis
		MIRA, MILA	Liver, pancreas adipocyte etc.	Glucose, glycogen dysregulation
		Lysosomes	Various cells	Kinin release, degradation products
	Neutrophils	Prostaglandins	Various cells	Hypertension, ischemia, hypoxia
	Platelets	Thromboxane	Vaisseaux	Altered permeability
MESODERM → T-cells	Spleen	Lymphokines	Suppressor T-cells	Adjuvanticity, Immunogenicity, Mitogenicity
	Thymocytes	? ? ?	ibid	
ENDODERM →	HEPATOCYTES	cAMP	none	Bile production

FIG. 1. ENDOTOXIN MEDIATED REACTIONS IN THE HOST.

Jirillo. The C3H/HeJ mice are a potent tool to assay the purity of various endotoxin preparations. When an LPS preparation produces a mitogenic response in this strain, that endotoxin should be discarded since proteins associated with lipid A or the polysaccharide will also stimulate a similar response.

Endotoxin from Bacteriodes fragilis used by Michalek et al (see this volume) does not contain KDO and also differs in the sequence of fatty acids. This lipid A is less toxic than other, conventional endotoxin preparations but it stimulates the C3H/HeJ mice. When spleen cells are isolated, in order to remove macrophages and leave only B cells, this Bacteriodes LPS is unable to stimulate mitogenicity of polyclonal activation in the C3H/HeJ refractory strain. Thus, carbohydrate moiety is extremely important in stimulating macrophages which in turn induce mitogenic effects on lymphocytes. The F_1 hybrids of the C3H/HeJ X C3H/HeN backcross gave an intermediate response to this endotoxin confirming a single lps gene for antibody production to LPS.

Agarwal. Since you are studying only one or two responses that are very closely related, a single gene may control them. However, the entries listed in the table I above, are all functioning at a very low level in the C3H/HeJ strain; so how would you envision a single lps gene protein controlling all of them.

Jirillo. In C3H/HeJ mice there is very little expression of the Fc receptor on the macrophage and this can be reversed by interferon. Not all aspects have been explored in this strain. May be there are different genetic alterations for each tissue and organ and it is very difficult to answer your question at this point.

Nakano. I think that the lps gene controls many responses of endotoxins but we presented evidence for another gene as well. The CBA mice have some defect in the response to endotoxin which may be duc to X-linked recessive genes and the whole problem should be analyzed more extensively.

Bloksma. The induction of interferon by endotoxins is completely abolished by prior administration of alpha-adrenergic agents. The adrenals of C3H/HeJ mice do not appear to synthesize glucocorticoids normally. In addition, the adrenergic or the sympathetic nervous system is also involved.

Agarwal. The glucocorticoid receptor levels are influenced by endotoxins and there is just no evidence of any linkage between the lps gene and the glucocorticoid receptor gene. So I feel convinced that we have to entertain the involvement of more than one gene or foci in overall LPS action.

Matsuo. The unresponsiveness of the C3H/HeJ mouse is a B cell defect, primarily concerned with LPS mitogenicity, but its relation to endotoxin lethality is not clear. Several defects become evident after injection of endotoxin in C3H/HeJ mice, such as prostaglandin synthesis inhibition, and the event(s) causal to LPS lethality remain unknown.

Hirata. I would like to talk about four other parameters induced by endotoxin. First, endotoxin induced cytotoxic action in bone marrow is mainly mediated by granulocytes and monocytes; second, endotoxin induced increase in procoagulant activity and in tissue thromboplastin, in bone marrow cells, is generated from monocytes; third, endotoxin induced decrease in nucleated cells in the marrow is due to the migration of the former into the circulation; fourth, endotoxin induced the production of FDP after activation of the fibrinolytic system which is an important parameter for the genesis of DIC.

None of these reactions could be induced in the HeJ mice. In fact, the HeJ mice were 100 times less responsiveness than conventional mice in the elicitation of the procoagulant activity. As to the FDP production, different endotoxin preparations, some containing 10. 4% protein (Boivin type), even in very high doses (500 µg) did not produce an increase in FDP. Thus, I think that neither endotoxin nor lipid A associated protein (LAP) has any effect on blood coagulation in HeJ mice. It is therefore likely that in our experiments all these responses are under the influence of the lps gene. Of course I do not know whether all endotoxin responses are regulated by the one lps gene or several.

Jirillo. How about the butanol extracted LPS which also contains protein.

Hirata. We only used Boivin type of endotoxin.

Jirillo. Which strain did you use.

Hirata. We used *E. coli* 0111:B4, for LPS extraction.

Yoshida. The defect in the C3H/HeJ mice is cellular since humoral factors such as the complement system are normal. To my knowledge, the coagulation system, Hageman factor and so on have not been analyzed. Dr. Agarwal mentioned the receptor systems. Such examples indicate that the whole question is very complicated and at the present time we do not have precise mechanisms of action.

Agarwal. The moral of the story is that we have good hard work to do back home and come with fresh data in three years from now to Toronto for a more meaningful discussion of the uniqueness or multiplicity of genes involved in LPS action.

B. The Relative Importance of Lipid A and other Constituents in Endotoxin Toxicity.

Agarwal. At one time lipid A was said to be causal to the toxicity provoked by endotoxins. In recent years the carbohydrate moiety has been gaining ground, like the Shinkansen. So we could have some discussion on the relative importance of these two constituents of the LPS in endotoxicosis.

Niwa. In so many interesting presentations today different endotoxin preparations were employed by different investigators. So it is important to elucidate the relationship between chemical structure and biological activities of endotoxins. Last November, during an international symposium at Susono City, Dr. Westphal et. al. stressed that lipid A was the most important component of endotoxins. Almost all the endotoxic activities could be attributed to lipid A. However, many problems remained unanswered. For example, some synthetic analogues of lipid A were synthesized by Dr. Shiba of Osaka University which elicited adjuvant activity but, as far as I know, no pyrogenicity or limulus positivity. Some biological activity was found in the polysachharide portion such as bone resorption (Nowotny), colony stimulation and so on. Today, Dr. Suzuki presented endotoxin like activities of mannan which is very weak; poly I:C and muramyl dipeptide also have weak endotoxin like activity. Since polysachharides other than endotoxins may contribute to deeper understanding of endotoxin poisoning I would call upon Dr. Suzuki to tell us some more about the biological activity of mannans.

Suzuki. Our present study is limited to the biological effect of simple polysachharides and does not include lipid A or other lipoidal material. The neutral mannan fraction (WNM) is highly toxic in mice and induces anaphylaxis like syndrome. The LD_{50} of WNM in ddy strain of mice was 12 mg/kg (iv). It consists largely of α-1,6 linkage and very small side chains.

Synthetic mannan consisting of α-1,2 linkage shows very high LD_{50} values. The 1025 fraction is the same as WNM but also contains phosphate, glucosamine and protein. This acidic yeast mannan has an LD_{50} of more than 200 mg/kg. The artificially phosphorylated WNM, containing 1.2% phosphate, also shows small LD_{50} values. Dextran T-2000 and other dextrans from Leuconostoc mesenteroides also show high LD_{50} values. So it occurred to us that the mannopyranosyl residue is an important factor in the lethal activity; dense branches consisting of α-1,2 and α-1,3 linked residues are also important. Non reducing terminals with phosphate groups decreased LD_{50} values.

D-mannose and N-acetyl-D-glucosamine was effective in preventing death. The toxic effect of this mannan is attributable to blood coagulation caused by hyperactivation of arachidonic acid metabolism since aspirin decreased lethality. These findings constitute examples of the biological effects of polysachharides which does not contain lipid A.

Jirillo. The induction of antibody response requires polysachharides as a first factor and lipid A as the second factor. Free lipid A, intraperitoneally, is able to induce anti-Lipid A response in C3H/HeJ mice.

Suzuki. In the responder C3H/HeSL strain, LD_{50} of WNM was 9.5 mg/kg whereas in the C3H/HeJ nonresponder it was 75 mg/kg.

Yoshida. Is there some hydrophobic part in your mannans? Are they chemically heterogeneous?

Suzuki. Yes, generally, polysachharides consist of heterogeneous species. We could not detect lipoidal material during gas chromatography or alkali treatment.

Yoshida. Activity of lipid A might depend upon heterogeneity.

Niwa. The next problem is the importance of protein moiety of endotoxins.

Nakano. We used Boivin antigen which contains lipid A but which is very potent in eliciting hydrogen peroxide. Lipid A and lipid A - BSA complexes were used in some of our experiments; the former was not effective due to solubility problems. Lipid A - BSA complexes are not very potent in eliciting nonspecific resistance in mice. In contrast, alkali detoxified LPS has very strong protective activity. We do not have enough data about the protein moiety but I think that the carbohydrate portion of LPS is very important for nonspecific protection against Salmonellosis.

Agarwal. Can you get hydrogen peroxide production by either the protein or the carbohydrate preparations?

Nakano. We have not checked that yet.

Yoshida. I would like to mention two points. We have to take care of the physical status of lipid A for solubility. It is very difficult to control the solubility of lipid A uniformly, even in the LPS of R-forms. Some samples are easily soluble while others are not. Some say that polysachharides have lipid A activity but this could just be contamination. It is possible to assay for picogram amounts of LPS by colorimetric procedures or the limulus lysate test.

Niwa. Dr. Homma of Kitasato Institute found interferon induction, and sometimes tumour regression, by the protein fraction of LPS.

Jirillo. Oral administration of Salmonella renders the C3H/HeJ mouse responsive to LPS. This may be a novel finding that could be exploited in future to solve many dogmas.

Suzuki. We treated WNM with 100 mM NaOH at 7° C for 3 h. Alkaline treatment of WNM resulted in the loss of gelation activity but lethal activity was retained completely.

McGhee. Dr. Suzuki, what tests have you done to eliminate endotoxin contamination in your carbohydrate preparation.

Suzuki. We used gas chromatography, alkali treatment, and limulus lysate tests. Our polysachharide was reactive to limulus lysis.

Nowotny. In view of tremendous heterogeneity of endotoxin and lipid A preparations, it is absolutely essential to know how did you obtain lipid A and how did you purify it.

Nakano. I just followed the method of Galanos et.al.

Agarwal. Dr. Jirillo, you mentioned three regions of LPS; are they recognized by the same lps gene? Did you use carbohydrate preparations?

Jirillo. All three regions of LPS are recognized by the same lps gene. We did not use carbohydrates but crossover studies sometimes clarify the specificity of the lethal response.

Niwa. I am wondering what part of endotoxin is responsible for bone resorption. Have you tested Re-glycolipid, Dr. Nowotny.

Nowotny. Re-glycolipid is also active but less so than the smooth type endotoxin and this is not due to solubility problems.

Niwa. Your polysachharide may contain some degraded lipid A.

Nowotny. No fatty acids were present though we detected phosphorous and glucosamine.

Niwa. Do you think that any polysachharide may be active or just the core polysachharide.

Nowotny. Polysachharide rich fraction, which is a horrible mixture of all partially degraded hydrolytic products in the supernatant after you have precipitated all lipid, is active.

McGhee. Can you distinguish the toxic effects of lipid A from bone resorption.

Nowotny. Yes, that is quite clear.

McGhee. Have you done any other biological tests for mitogenicity and so forth for Eikenella.

Nowotny. Mitogenic response due to Eikenella LPS was normal.

McGhee. What about free lipid A.

Nowotny. We do not stick out our neck very far for the omnipotence of free lipid A which certainly seems responsible for the toxic effects of LPS; other components from Gram negative bacteria that contaminate LPS may be responsible for the beneficial effects.

Agarwal. How do you explain the beneficial effect of antibiotics on LPS activity.

Nowotny. Very good question. Bacterial penetration into gingival tissue seems to be of prime importance in our model. When we apply a ligature and two injections of cytoxan, the bacteria form an artifical pocket around the ligature. As long as they remain there nothing happens but damage is produced if they move downwards and this latter step is prevented by the antibiotic.

McGhee. The LD_{50} of your Serratia LPS was only 100 μg for a rat of 200 g or so which may be a good model for LPS action.

Agarwal. Have you tried Gram positive infections since a lot of native flora in the mouth is really Gram positive.

Nowotny. Do you mean the monoinfected rat?

Agarwal. Yes.

Nowotny. Other authors have tried that but bone resorption is not restricted to Gram negative organisms in the germ free model.

C. Humoral Factors in Endotoxicosis.

Agarwal. Bacterial endotoxins elicit a very large variety of mediators from different cell types all of which can find their way into the blood stream. Thus the problem of humoral factors is really very complex. Since Pr. Yoshida has taken so much trouble in the organization of this workshop perhaps he can take some more and lead the discussion in this area.

Yoshida. I would like to divide this theme into two sections. First we could discuss the role of complement and second that of neurotransmitters, histamine and so on. I now ask Dr. Katayama for his viewpoint.

Katayama. Although endotoxin is toxic to the host, it also has many beneficial effects. My major concern is the antitumour activity of endotoxins. In fact, LPS is the strongest antitumour agent in nature and can cause destruction of the tumour tissue without affecting normal cells. In other words, endotoxins recognize tumours and the latter recognize LPS or its derivative. Therefore, we can use endotoxins for analyzing the nature of the neoplastic cell. Since endotoxin does not accumulate in the tumour, its action is derived by mediated substances such as interferon, DNAse, interleukins etc. all of which are stimulated by LPS. Our conclusion is that the major effectis mediated by the complement. If we treat normal mouse serum, or serum from a tumour bearing mouse, with protein A coupled to Sepharose, both sera have antitumour activity. The influence of protein A is not to remove something but to produce something new. It has not been proven at this time but complement may be a candidate for this effect of protein A.

Bloksma. Tumour bearing C3H/HeJ Mice injected with endotoxin do not show tumour necrosis although they have a normal complement system. These are in conflict with the general role assigned to complement and its activation in LPS induced tumour necrosis. I think that a high amount of activated complement is so vasoactive that it can induce tumour damage although it may not mediate factors in tumour bearing mice injected with endotoxin.

Yoshida. Some feel that complement is very important in endotoxicity whereas others doubt that.

Bloksma. It may be an additional factor but it is not the major factor.

Furukawa. I emphasize that complement plays an important role in pathogenesis of endotoxin induced DIC. In decomplemented rats treated with CVF, DIC was inhibited.

Yoshida. The studies of Drs. Furukawa and Yoshikawa emphasize the significance of complement in DIC.

Yoshikawa. Complement is not the major factor in the anti-tumour activity . Many immunopotentiators such as levan, BCG, PSK, OK432, activate the complement system. Complement is not the major mediator since it alone can not kill the tumour. The AKR mouse, defective in C5, easily die of leukemia; tumour growth is inhibited when C5 is administered into the AKR mice. So complement is necessary for anti-tumour activity of LPS.

Yoshida. In bone marrow reactions, there is decrease in nucleated cells due to leukocyte migration into the circulation; red blood cells in the marrow are increased due to hemorrhage; marrow cells are cytotoxic. All these three reactions occurred in the AKR or A mice defective in complement components.

Matsuo. In our studies, the complement system in itself is sufficient for the lethal effects of endotoxin.

Yoshida. Can you provide some evidence for the role of complement in endotoxin lethality.

Matsuo. In our recent studies, FUT 175,which is a protease inhibitor, reduced endotoxin lethality. Administration of CVF prior to endotoxin challenge reduced LPS lethality in the rat.

Yoshikawa. In our experiments, DIC was also significantly decreased in decomplemented rats. Although FUT 175 is anti-complementary, it also has other effects such as antiplasmin, antithrombotic, and so on. Thus, the protective role of FUT 175 is not necessarily due to its anti-complementary activity. If one could find an agent capable of inhibiting complement pathways, a definitive answer may become feasible but at the moment such a material does not exist.

Matsuo. FUT 175 inhibits not only complement but also other proteases such as trypsin, plasmin. FUT 175 has stronger inhibitory effect on complement activation than other protease inhibitors, although it may not be specific. CVF is an inhibitor of the alternate pathway but has an effect of weakening animals; therefore CVF is not very good in studying endotoxin lethality.

Hirata. I would like to emphasize that many kinds of anti-complement drugs are being used by different authors. Are they specific on the complement system? In endotoxin induced DIC, cell derived thromboplastin has an important role. So one should check the efffect of drugs on leucocytes or other cell types.

Niwa. The antitumour activity of endotoxins is very impressive but can we expect this in clinical use in the near future. I am rather pessimistic about it.

Katayama. There is a long history of using mixed bacterial vaccines, consisting of Streptococcus and Serratia strains, which was said to have cured over 200 patients suffering from inoperative cancer. However, purified LPS is not as good as crude vaccine.

Agarwal. Can anyone tell me whether the CBA animals are cmplement deficient. Dr. Nakano reported that CBA mice are 10 fold more sensitive to endotoxin, as compared to normal mice, and if they are defective in complement the latter may not be important in endotoxin lethality.

Yoshida. The interaction of complement and endotoxin is very confusing in vivo. The interaction between these two materials in vitro is very important to analyze the molecular mechanisms of endotoxin action.

Agarwal. Would anyone like to comment upon the endotoxin detoxifying component which was very fashionable at one time, and the role of complement in that system.

Yoshikawa. Complement levels decreased after either LPS or the cobra venom factor (CVF); we can protect against LPS induced DIC by CVF. Have you tried to administer CVF before endotoxin injection.

Matsuo. We did not use CVF; there are conflicting reports as to the protective role of CVF in endotoxin lethality.

Yoshikawa. Do the catalases or scavangers of free radical protect mice against endotoxin toxicity.

Sakaguchi. 50 mg/kg trypan blue had a very good effect on the relationship between free radical formation and the scavanger system.

Yoshikawa. AKR mouse is complement deficient. Did you try this strain.

Matsuo. No.

Niwa. Is there any difference in thrombocytopenia after LPS in the responder versus the nonresponder mouse.

Natsuo. We do'nt know.

McGhee. Would you speculate as to how the macrophages become activated in CBA mice.

Nakano. Some macrophages are directly activated by the LPS. In some cases low serum antibody titer may affect phagocytic ability and intracellular killing but we think that direct activation of macrophages is more important for protection against Salmonella infection.

Niwa. Besides complement, blood coagulation factor is very important in LPS lethality.

Matsuo. Ulevitch and Cochran have reported that C3H/HeN mice exhibit low complement titer after LPS but DIC, thrombocytopenia, or thrombosis was not observed in C3H/HeJ animals.

McGhee. Is it possible to induce resistance in the CBA animal and can that be transferred by either cells or polyclonal antibodies into normal recipients.

Nakano. We did not assay for antibody and did not attempt passive cell transfer. O'Brien has already showed that these animals have low antibody levels.

Movat. The studies described by Dr. Yoshida bring up an important point with respect to the procoagulant activity in the blood. As you know, in man it has been established that monocytes, other than polymorphonuclear leucocytes, are responsible for the procoagulant activity and this has been confirmed in rabbits. Semeraro, not long ago, showed that unlike the rabbit, in rats no procoagulant activity could be elicited with endotoxin. Can you comment upon the differences between rodents.

Yoshida. We have no experience with other experimental animals; I think that in human DIC the cells may be different from those in laboratory animals.

Movat. Did you perform two stage clotting assays to see if your procoagulant activity is a tissue factor like material.

Hirata. We only performed one stage assays and showed that procoagulant activity is tissue thromboplastin.

McGhee. You mentioned that your antibody may be acting in possibly two different ways. Since both IgG and IgM work, it does not appear to be a complement mediated antibody effect.

Yoshida. We have no experience concerning the complement. Both antibodies are however important.

McGhee. Have you ever looked at immunocomplexes of endotoxin and the anti-Re antibody injected into the animals for their effect on the bone marrow.

Yoshida. No.

Agarwal. Have you tried protease inhibitors like leupeptin

and ε-aminocaproic acid etc... to see if they would have some protective effect.

Hirata. We have used one protease inhibitor, trans AMCHA, which has antifibrinolytic activity. It also inhibited the bone marrow reaction, especially the bone marrow reaction and decrease in nucleated cell counts in the marrow. The inhibition was however partial, not complete.

Agarwal. Dr. Movat, you mentioned platelet accumulation. Do they have an eventual role in Shwartzman.

Movat. The overall reaction is very complex. Although the monocytes, macrophages, etc... produce the procoagulant activity, the hemorrhage seems to me to be primarily polymorphonuclear or neutrophil mediated.

Yoshida. Next we should like to discuss the neurotransmitter. I would like to have the comments of Dr. Bloksma on the presentation of Dr. Kuratsuka.

Bloksma. I found his data very interesting and they fit very well with my hypothesis. The toxicity of adrenaline seems to be mediated by the α-adrenergic receptor. Endotoxins are capable of β-adrenergic blockade and facilitate the action of α-adrenergic receptor. Concomitant administration of β-adrenergic receptor blockers together with adrenaline increases the toxicity of the latter. So, it may be that endotoxin increases the toxicity of adrenaline by its β-blocking action thereby facilitating the α-adrenergic effect which is related to the toxicity of adrenaline.

<u>Kuratsuka</u>. I wish to ask Dr. Bloksma about the histamine receptor. More than 10 years ago I tested the histamine sensitizing activity of antitumour drugs isolated from Japanese fungi. The product, called Lentinan, is now commercially available and has potent histamine sensitizing action in mice. Can you comment upon the histamine sensitizing activity, or histamine receptor, in the antitumour activity of drugs or of endotoxin.

<u>Bloksma</u>. Very difficult question. We found that the H1 receptor blocking agents, given prior to endotoxin, diminish the antitumour action slightly. Administration of H2 receptor blocking agents somewhat potentiates the antitumour effect of endotoxin. In my opinion, histamine itself has weak antitumour activity against certain sarcomas. I think that the effects mediated by the H1 receptor are mainly vasoactive effects whereas those mediated by the H2 receptor are the activation of H2 suppressor cells. The slight increase in the antitumour activity of endotoxin, when it is combined with H2 receptor blockers, is due to enhanced H1 receptor activity mediated by the vasoactive component of endotoxin induced tumour necrosis along with the blockade of activation of T-suppressor cells which are switched on by H2 agonists. I believe that the administration of histamine has a facilitating effect and an inhibitory effect because it switches on H2 receptor positive T-suppressor cells.

<u>Hirata</u>. I am very encouraged by Dr. Bloksma's data because not many reports emphasize the role of histamine in endotoxicosis. About 10 years ago we reported that, in the bone marrow reaction, the cytotoxicity, decreased nuclear cells, and marrow hemorrhage, could all be inhibited with antihistamines and an H1 antagonist - diphenhydramine. Histamine alone could also lower nucleated cells and provoke hemorrhage. The activity of histamine alone does not persist for a long time though it was very similar to endotoxin activity.

Therefore, I think that not only the histamine released after endotoxin, but also nascent histamine, or intrinsic histamine formed from histidine by the action of histidine decarboxylase, is also involved in endotoxin action. What do you think about the role of intrinsic histamine.

Bloksma. As far as I know, endotoxin stimulates the decarboxylase so the endogenous histamine is very important. May be the release of histamine from mast cells could add to the antitumour effect of endotoxin. I think that the capacity of endotoxin to potentiate histamine action, together with the increase in histamine production, are all involved in the antitumour action of LPS.

Movat. Dr. Nowotny do you think that endotoxin acts directly on osteoclasts or via a host derived mediator.

Nowotny. Osteoclast activating factor is released from the macrophage after the latter interact with endotoxin. We do not think that this is the exclusive mechanism. In recent experiments, chick bone osteoclasts, which are only 90% pure, exposed either directly to LPS or indirectly through an intermediate membrane, do a beautiful job of absorption of calcium labelled bones.

Bloksma. Do you have any evidence for Interleukin 1 in bone resorption.

Nowotny. Not yet.

Agarwal. Have you tried materials that modify host response to LPS to assess their influence on bone resorption.

Nowotny. We have only tried indomethacin which had no effect on bone resorption in vitro.

D. The Reticuloendothelial System and Endotoxin Poisoning.

Agarwal. The history of the involvement of the RES during endotoxicosis dates back to at least 1947. Contrary to the initial premiss, it now appears that enhanced RE function may actually be harmful to the evolution of endotoxin poisoning. Dr. Bloksma can perhaps help us decide as to the true contribution of the RES in host response to endotoxins.

Bloksma. It is indeed a dilemma that RE-activation as well as RE-inhibition increase susceptibility to bacterial endotoxins. I wanted to ask Dr. Sakaguchi whether RE-activation could influence HDL production since RE-inhibition by trypan blue and lead acetate sensitizes to LPS and decreases HDL levels.

Sakaguchi. This relationship is not yet very clear. I think that apo serum amylase protein increase may be responsible for the detoxification of endotoxin.

Shimada. A good crrelation appears to exist between DIC and RES. CH50 levels correlate well with K values and platelet counts. High complement levels may be necessary for the removal of endotoxin from the circulation.

Bloksma. Is the removal of endotoxin from the blood, as a result of elevated K values, enough to ensure against LPS mediated damage.

Agarwal. I think that release of mediators from RE cells has to be considered. Glucocorticoids may be inhibiting this latter parameter and thereby protecting against LPS mediated damage.

Yoshikawa. The relationship between complement, DIC and endotoxin shock is not clear.

Jirillo. The role of complement in clinical situations is especially difficult to determine.

Matsuo. Does endotoxin poisoning involve an actual decease in the synthesis of complement or an accelerated consumption.

Shimada. I think that low CH50 levels are an expression both of increased utilization and diminished production.

Matsuo. Within 30 min of endotoxin administration, complement and C3 concentrations dropped greatly but recovered to normal within 1 h with an increase above control by 24-48 h. Thus, small doses of LPS may increase, and larger doses decrease, the levels of complement in vivo.

Bloksma. Do different mice strains vary in this regard.

Nakano. The three strains of mice have phenotypic similarities but different underlying mechanisms.

Bloksma. What are these mechanisms.

Nakano. I do not know but some of the differences are due to different experimental systems.

Bloksma. Is there any correlation between the resistance of C3H/HeJ macrophages to LPS, their susceptibility to Salmonella infection, and their capacity to produce H_2O_2.

Nakano. H_2O_2 production, alone, can not explain all this.

Bloksma. Since prior endotoxin administration stimulates H_2O_2 production, can other macrophage activating agents do the same.

Nakano. C3H/HeJ mice could respond to muramyl dipeptide but CBA/N mice and beige mice could not.

McGhee. What might be the relationship between the lack of macrophage H_2O_2 production and the ability of the animal to handle an intracellular infection with Salmonella.

Kato. The injection of LPS into immunologically defective mice improved bacterial clearance and antibacterial activity.

Nakano. CBA mice are really susceptible to Salmonella infection but H_2O_2 production alone can not explain susceptibility.

McGhee. Would you speculate that the macrophage defect is more important in the susceptibility of these mice to infection.

Nakano. I think that there is partial relationship.

Movat. Do you know if the ability to produce H_2O_2 in the resistant mouse can be restored by BCG, lymphokines, or interferon.

Nakano. We have never checked that so far but we would like to try it.

Yoshida. Did you examine the activity of macrophages in vitro after streptozotocin, Dr. Agarwal.

Agarwal. No, we only studied the carbon clearance, the spleen and the liver weights.

Yoshida. Is it possible that Streptozotocin may change the character of macrophages such as BCG does.

Agarwal. I do'nt know. However, to my knowledge BCG does not totally reverse the resistance to LPS lethality in the C3H/HeJ strain as was possible by a combination of zymosan and streptozotocin. I am wondering whether Dr. Movat used soluble or particulate zymosan for his studies on the Shwartzman reaction.

Movat. Particulate zymosan was used in our studies.

Agarwal. Do you know if soluble zymosan is effective.

Movat. I do'nt know.

Katayama. Zymosan was claimed to have no activity to provoke Shwartzman but you showed the contrary when zymosan was given intravenously and the route is important. Although zymosan has complement activating property, even very large amounts, given intraperitoneally, had no effect on mouse tumour. Thus, either it should be injected intravenously or the activated serum should be diluted for intraperitoneal injection.

Movat. Thank you. Zymosan activated plasma was injected, or infused, over a 30 min period to a total of 30 ml of zymosan activated serum per rabbit.

McGhee. Do'nt you think 30 ml/kg was a large dose.

Movat. Not over a 30 min period. Animals were normal.

Agarwal. Did you try leupeptins in your system.

Movat. No we did not.

Yoshida. The number of Kupffer cells is very flexible and this should be kept in mind when we talk about RE functional modifications.

Agarwal. I am afraid that time is running out and I pass on the floor to Pr. Yoshida.

Yoshida. I would like to thank everyone for participating and I declare this workshop closed.

AUTHOR INDEX

Subject Index